DYNAMIC RESEARCH, INC.

DRI: 95003864

ELECTRONIC BRAKING, TRACTION, AND STABILITY CONTROLS

VOLUME 2

Other SAE books in this series:

Multiplexing and Networking, Volume 2
(Product Code: PT-128)

Automotive Software
(Product Code: PT-127)

Electronic Engine Control Technologies, Second Edition
(Product Code: PT-110)

Sensors and Transducers, Second Edition
(Product Code: PT-105)

Electric and Hybrid Electric Vehicles
(Product Code: PT-85)

On- and Off-Board Diagnostics
(Product Code: PT-81)

Electronic Transmission Controls
(Product Code: PT-79)

Multiplexing and Networking
(Product Code: PT-78)

Electronic Steering and Suspension Systems
(Product Code: PT-77)

Electronic Braking, Traction, and Stability Controls
(Product Code: PT-76)

Navigation and Intelligent Transportation Systems
(Product Code: PT-72)

For information on these or other related books, contact SAE by phone at (724)776-4970,
fax (724)776-7090, e-mail: publications@sae.org, or the SAE website at www.sae.org.

Electronic Braking, Traction, and Stability Controls
Volume 2

PT-129

Edited by

Ronald K. Jurgen

Published by
Society of Automotive Engineers, Inc.
400 Commonwealth Drive
Warrendale, PA 15096-0001
U.S.A.
Phone: (724) 776-4841
Fax: (724) 776-5760
www.sae.org
April 2006

For permission and licensing requests contact:

SAE Permissions
400 Commonwealth Drive
Warrendale, PA 15096-0001-USA
Email: permissions@sae.org
Tel: 724-772-4028
Fax: 724-776-3036

Global Mobility Database®

All SAE papers, standards, and selected books are abstracted and indexed in the Global Mobility Database.

For multiple print copies contact:

SAE Customer Service
Tel: 877-606-7323 (inside USA and Canada)
Tel: 724-776-4970 (outside USA)
Fax: 724-776-0790
Email: CustomerService@sae.org

ISBN-10 0-7680-1786-6
ISBN-13 978-0-7680-1786-1
Library of Congress Catalog Card Number: 2006922021
Copyright © 2006 SAE International

Positions and opinions advanced in this publication are those of the author(s) and not necessarily those of SAE. The author is solely responsible for the content of the book.

SAE Order No. PT-129

Printed in USA

INTRODUCTION

Interrelationships are Key

The previous edition of this book, PT-76, contained 62 papers covering progress in the technologies of electronic braking systems, including antilock breaking systems (ABS) and brake-by-wire; traction control systems (TCS or ASR); and stability control systems (VDC). This new edition features 81 new papers on those subjects, none of which appeared in PT-76

All of these systems are interrelated. For example, electronic stability controls extend the capabilities of antilock braking and traction control systems (see 2004-01-2090), and many, if not most, of the papers herein reflect such interrelationships. This makes categorization of any one paper somewhat arbitrary. The approach taken here was to classify a paper according to its main subject matter, e.g. ABS, even though that paper also contains information on other systems.

It should be noted that information on the use of electronics to detect objects, provide collision warnings, and to initiate actions to avoid collisions are beyond the purview of this book. The interested reader is referred to PT-70 in this automotive electronic series, "Object Detection, Collision Warning, and Avoidance Systems."

Time triggered protocols (TTPs) are mentioned in several papers (2005-01-1538, for example) but extensive coverage of TTPs is also beyond the scope of this book. However, the interested reader is referred to PT-78 and PT-128, both of which contain TTP papers.

In the final section of this book, information on future trends is presented. This information is taken from papers in this book as well as from *Automotive Engineering International*.

*　　*　　*　　*　　*　　*　　*　　*　　*　　*　　*

This book and the entire automotive electronics series is dedicated to my friend Larry Givens, a former editor of SAE's monthly publication, *Automotive Engineering International*.

Ronald K. Jurgen, Editor

Table of Contents

Introduction

Brake-by-Wire

Antilock Braking Systems (ABS)

Traction Control Systems (TCS, ASR)

Stability Control Systems (VDC)

What the Future Holds

BRAKE-BY-WIRE

Robust Wheel-Slip Control for Brake-by-wire Systems

Kunsoo Huh and Daegun Hong
HANYANG University

Paljoo Yoon, Hyung-Jin Kang and Inyong Hwang
MANDO Co.

ABSTRACT

Wheel-slip control systems are able to control the braking force more accurately and can be adapted to different vehicles more easily than conventional ABS systems. But, in order to achieve the superior braking performance, real-time information about the vehicle status variables such as wheel slip ratio, tire force, etc is required. In this paper, a wheel slip controller based on the estimated braking force is developed for brake-by-wire systems. The proposed wheel slip control system is composed of the braking force monitor and robust slip controller. In the brake force monitor, the tire braking forces as well as the brake disk-pad friction coefficient are estimated. The robust wheel slip controller using the estimated tire braking force is designed based on the sliding mode control technique. This system determines the braking pressure as the control input and maintains the wheel slip at any given target slip. The performance of the proposed wheel-slip control system is evaluated in various simulations.

INTRODUCTION

The wheel slip control system using the brake-by-wire actuator has been studied widely in the vehicle brake research. The wheel slip control system controls the wheel slip ratio that determines the braking force depending on the road type. With the wheel slip control system, the braking force can be accurately adjusted to various vehicles, operating conditions and road conditions. By maintaining a specified wheel slip for each tire, the braking force can be maximized for the shortest stopping distance or can be distributed to each tire so that the steering stability can be improved from a higher-level controller such as ESP (Electronic Stability Program).

In order to control the wheel slip accurately in the maneuvering vehicles, real-time information about wheel slip ratio, tire force at each wheel, etc is required. In particular, the tire braking force can be calculated by the wheel dynamics equation if the braking torque can be measured or known. The brake-by-wire system can be classified into two types: EHB (Electro-Hydraulic Brake) which uses hydraulic power and EMB (Electro-Mechanical Brake) which uses electric power. In the case of EMB, the braking torque can be measured based on the motor current. In the case of EHB, the braking torque cannot be measured directly, but can be approximated based on the characteristics of the brake disk-pad friction. However, the friction characteristics can change significantly depending on aging of the brake, moisture on the contact area, heat, etc. Müller et al. [1] demonstrates that the brake disk-pad friction can change more than 50 % under various braking conditions. Regarding the accurate wheel slip calculation, there are some reported techniques to estimate vehicle velocity using GPS (Global Positioning System) [2], Kalman filter [3], Fuzzy rule [4], etc.

Various control methods have been reported for the wheel slip control. In the case of EMB, the braking force is calculated from the measured braking torque and the wheel slip controller is designed based on the feedback linearization method [5]. Sliding mode controllers have been designed to control the wheel slip assuming that the braking force is known [6] or the upper bound of the force is known [7]. Several control methods were developed for the wheel slip control system [8],[9],[10] to adjust the brake torque as the control input. In the case of EHB, it is difficult to know the braking force directly due to the brake disk-pad friction variation. Besides, the control input should be formulated in terms of braking pressure, not braking torque. An adaptive control system for EHB has been proposed [11] to control the wheel slip with the estimating road friction coefficient and the brake gain from a tire model.

In this paper, a wheel slip control system is developed for EHB brake-by-wire systems. The dynamics of the wheel slip ratio is utilized to represent the dynamic relation between the braking pressure and the slip ratio. The dynamic relation includes not only the longitudinal tire force, but also the brake disk-pad friction coefficient. Because the tire force and the friction coefficient cannot be easily measured, they are estimated utilizing the simplified vehicle model and the Kalman filter technique.

A sliding mode controller is designed based on the dynamic relation so that the wheel slip is controlled to follow the target wheel slip ratio. The controller gain is determined for the robust control performance by considering the bound of the estimation error. The performance of the proposed wheel-slip control system is evaluated in various simulations. A full vehicle model [12] is used as a real vehicle in simulations.

WHEEL SLIP CONTROL SYSTEM

Figure 1 shows the wheel slip control system proposed in this paper. The proposed system consists of the braking force monitor, wheel slip controller and EHB actuator. As shown in Fig. 1, the monitor block estimates the tire braking forces as well as the brake disk-pad friction coefficient from the measured brake pressure, wheel speed, longitudinal/lateral acceleration and steer angle. The wheel slip controller calculates the pressure variation as the control input to compensate the error between the target slip and the current slip ratio. The calculation of the control input requires the on-line information of the tire braking forces and disk-pad friction coefficient. The EHB actuator is a pressure servo type actuator and applies the control input brake pressure to a brake caliper, which is determined by the wheel slip controller.

TIRE BRAKING FORCE MONITOR

In order to design the tire braking force monitoring system, the dynamic motion of the vehicle is considered as shown in Fig.2 and modeled into a three degree-of-freedom longitudinal, lateral and yawing dynamics.

$$m(\dot{v}_x - r v_y) = \sum_{i=1}^{2}(F_{xi}\cos\delta_f) + \sum_{i=3}^{4} F_{xi} - 2F_{yf}\sin\delta_f$$

$$m(\dot{v}_y + r v_x) = \sum_{i=1}^{2}(F_{xi}\sin\delta_f) + 2F_{yf}\cos\delta_f + 2F_{yr}$$

$$I_z \dot{r} = l_f \sum_{i=1}^{2}(F_{xi}\sin\delta_f) + l_f \cdot 2F_{yf}\cos\delta_f - l_r \cdot 2F_{yr}$$

$$+ t_f \sum_{i=1}^{2}(-1)^i (F_{xi}\cos\delta_f) + t_r \sum_{i=3}^{4}(-1)^i F_{xi}$$

(1)

where m is vehicle mass, I_z is moment of inertia in yaw-direction and v_x, v_y, and r denote the longitudinal velocity, lateral velocity and yaw rate, respectively. F_{xi} stands for the longitudinal tire force at each wheel. F_{yf} and F_{yr} are the mean values of the front and rear lateral tire forces, respectively. δ_f is the steering angle of the front wheel.

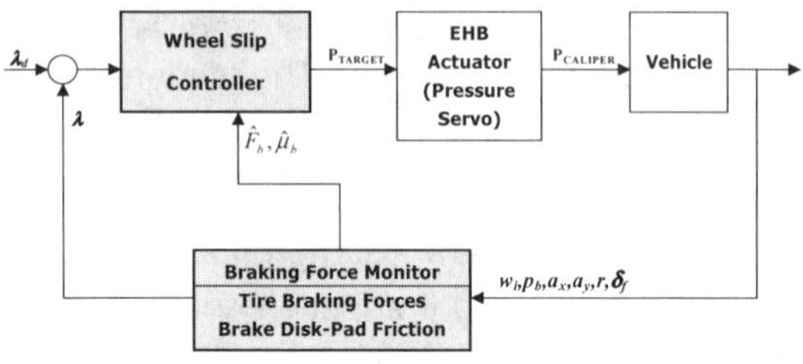

Fig. 1. Block diagram of the wheel slip control system

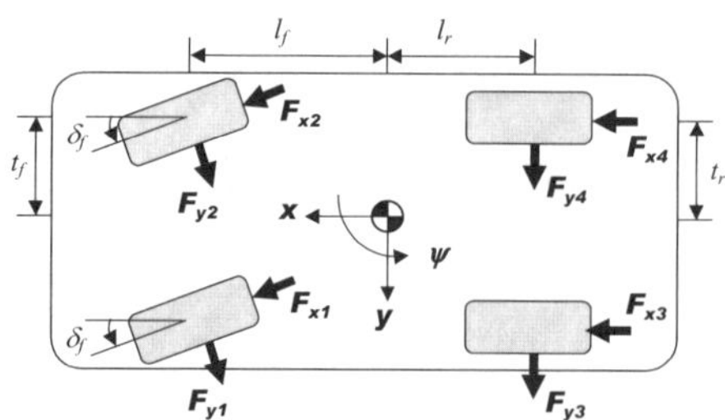

Fig. 2. Force diagram of the vehicle model

During braking, the wheel torque is expressed in terms of braking pressure, brake disk-pad friction coefficient, etc.

$$T_{wi} = -2 \cdot P_{bi} A_{pi} R_{bi} \mu_b, \quad (i = 1 \sim 4)$$

(2)

The wheel dynamics equation in braking can be expressed as follows by substituting the above relation.

$$J_{wi}\dot{\omega}_i = T_{wi} - r_w F_{xi} = -2 \cdot P_{bi} A_{pi} R_{bi} \mu_b - r_w F_{xi}$$

(3)

where T_w stands for wheel torque, P_b is hydraulic braking pressure, A_p is the piston area, R_b is the effective radius between the center of disk rotor and pad, and μ_b represents the brake disk-pad friction coefficients. ω denotes the rotational wheel velocity, J_w is the moment of inertia of wheel, and r_w is the wheel radius, respectively.

In order to estimate the tire braking force, a monitoring model is constructed considering the variation of the brake disk-pad friction and is utilized for the Extended Kalman Filter design. In the nonlinear vehicle model as expressed in Eq. (1) and Eq. (3), the longitudinal tire forces (F_{xi}), lateral tire forces (F_{yf}, F_{yr}) and brake disk-pad friction coefficient (μ_b) are unknown variables and cannot be easily measured using physical sensors. For estimating these unknown variables, they are augmented as state variables in the state equation and the time derivatives of the augmented variables are assumed to be zero. Thus, the nonlinear state equations of Eq. (1) and Eq. (3) are formulated into the following monitoring model.

$$\dot{v}_x = r v_y + \frac{1}{m}\cdot\left(\sum_{i=1}^{2}(F_{xi}\cos\delta_f) + \sum_{i=3}^{4}F_{xi} - 2F_{yf}\sin\delta_f\right)$$

$$\dot{v}_y = -r v_x + \frac{1}{m}\cdot\left(\sum_{i=1}^{2}(F_{xi}\sin\delta_f) + 2F_{yf}\cos\delta_i + 2F_{yr}\right)$$

$$\dot{r} = \frac{1}{I_z}\cdot\left(l_f \sum_{i=1}^{2}(F_{xi}\sin\delta_f) + l_f \cdot 2F_{yf}\cos\delta_f - l_r \cdot 2F_{yr}\right.$$

$$\left. + t_f \sum_{i=1}^{2}(-1)^i(F_{xi}\cos\delta_f) + t_r \sum_{i=3}^{4}(-1)^i F_{xi}\right)$$

(4)

$$\dot{\omega}_i = (-2 \cdot P_{bi} A_{pi} R_{bi} \mu_b - r_w F_{xi})/J_{wi}, \quad (i = 1 \sim 4)$$

$$\dot{\mu}_b = 0$$

$$\dot{F}_{xi} = 0, \quad (i = 1 \sim 4)$$

$$\dot{F}_{yf} = 0$$

$$\dot{F}_{yr} = 0$$

In the monitoring model, the output variables are the longitudinal acceleration, lateral acceleration, yaw rate and wheel velocities of the four wheels.

$$y = \begin{bmatrix} a_x & a_y & r & \omega_i \end{bmatrix}^T, \quad (i = 1 \sim 4) \tag{5}$$

where a_x and a_y represent the longitudinal and the lateral accelerations, respectively.

Based on the monitoring model in Eq. (4) and the output equation in Eq. (5), the extended Kalman filter [13] is designed for estimating the tire braking force at each wheel and the brake disk-pad friction coefficient simultaneously [14].

ROBUST WHEEL SLIP CONTROLLER

SLIP DYNAMICS

When a brake is engaged in the wheel, the tire braking force is generated between a tire and road surface due to friction as illustrated in Fig. 3. In the case of the hydraulic disk brake as shown in Fig. 4, the braking torque is determined from the brake parameters and brake pressure.

$$T_b = 2 \cdot P_b A_p R_b \mu_b \tag{6}$$

Based on Fig. 3 and Eq. (6), the wheel dynamic equation is formulated.

$$J_w \dot{\omega} = -2 \cdot P_b A_p R_b \mu_b + r_w F_b \tag{7}$$

where T_b and F_b are the braking torque and braking force expressed as positive values, respectively.

The wheel slip generated between a tire and road surface during braking is defined as the relative difference between the vehicle longitudinal velocity and the wheel velocity.

$$\lambda = \frac{v_x - r_w \omega}{v_x} \tag{8}$$

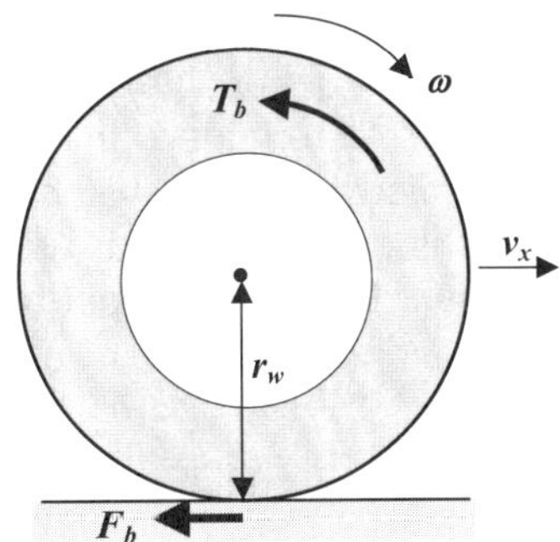

Fig. 3 Force diagram of the wheel during braking

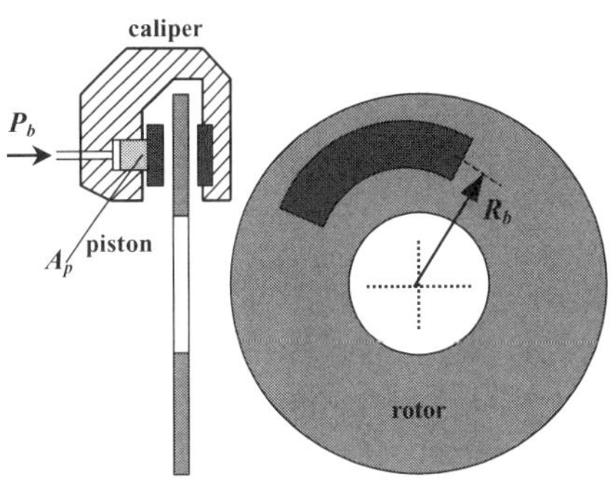

Fig. 4 Hydraulic disk brake

By differentiating Eq. (8) and substituting Eq. (7), the slip dynamics equation can be obtained.

$$\dot{\lambda} = -\frac{\dot{v}_x}{v_x}(\lambda - 1) - \frac{r_w}{J_w v_x}\left(r_w F_b - 2 A_p R_b \mu_b P_b\right) \tag{9}$$

SLIDING MODE CONTROLLER DESIGN

In this study, the sliding mode control method is applied in designing the wheel slip controller because of the nonlinear characteristics of the slip dynamics. The control objective is to drive the wheel slip, λ, to the target wheel slip, λ_d, and the control input is applied as the pressure value in the EHB actuator. The control input brake pressure based on the sliding mode control can be formulated as follows.

$$P_{b,c} = P_{eq,c} - k \, \text{sgn}(s) \tag{10}$$

where $P_{b,c}$ is the control input brake pressure, $P_{eq,c}$ is the equivalent control brake pressure, s is the sliding surface and k is the control gain. The sliding surface, s, can be defined as the error between the current slip and target slip.

$$s = \lambda - \lambda_d \tag{11}$$

The equivalent control brake pressure, $P_{eq,c}$, can be obtained from solving the equation that the time derivative of s becomes zero.

$$\dot{s} = 0 \tag{12}$$

Assuming that the target slip is constant, $\dot{\lambda}_d = 0$, the equivalent control brake pressure can be obtained by substituting Eq. (9) into Eq. (12).

$$P_{eq,c} = \frac{1}{2 A_p R_b \mu_b}\left(r_w F_b + \frac{J_w \dot{v}_x}{r_w}(\lambda - 1)\right) \tag{13}$$

Because the tire braking force and the brake disk-pad friction coefficient cannot be easily measured in passenger cars, the estimation method explained in the previous section is utilized to estimate the tire braking force and the disk-pad friction coefficient. Consequently, the control input brake pressure can be determined from Eq. (10) and Eq. (13).

$$P_{b,c} = \frac{1}{2 A_p R_b \hat{\mu}_b}\left(r_w \hat{F}_b + \frac{J_w \dot{v}_x}{r_w}(\lambda-1) \right) - k\,\mathrm{sgn}(s) \qquad (14)$$

where $\hat{F}_b$ and $\hat{\mu}_b$ indicate the estimated braking force and the estimated brake disk-pad friction coefficient, respectively.

In Eq. (14), the control gain, k, should be designed such that the control performance is robust to the estimation errors in braking force and brake disk-pad friction coefficient. The control gain is determined from the following sliding condition [15].

$$\frac{1}{2}\frac{d}{dt}s^2 \leq -\eta|s| \qquad (15)$$

where η is a design parameter and a strictly positive constant. Substituting Eq. (9) into Eq. (15) gives the following condition.

$$s\dot{\lambda} \leq -\eta|s|$$
$$\Rightarrow\ s\left[-\frac{\dot{v}_x}{v_x}(\lambda-1) - \frac{r_w}{J_w v_x}\left(r_w F_b - 2 A_p R_b \mu_b P_b \right) \right] \leq -\eta|s| \qquad (16)$$

Using the control input braking pressure in Eq. (14), the above equation becomes the following condition.

$$-s\frac{\dot{v}_x}{v_x}(\lambda-1)\left(1-\frac{\mu_b}{\hat{\mu}_b}\right) - s\frac{r_w^2}{J_w v_x}\left(F_b - \frac{\mu_b}{\hat{\mu}_b}\hat{F}_b \right)$$
$$-s\frac{r_w}{J_w v_x}2 A_p R_b \mu_b k\,\mathrm{sgn}(s) \leq -\eta|s| \qquad (17)$$

If the sliding condition in Eq. (15) is changed into the following form,

$$\frac{1}{2}\frac{d}{dt}s^2 \leq -\eta'|s| \quad where,\ \eta' = 2 A_p R_b \mu_b \eta \qquad (18)$$

it is confirmed that η' is also a strictly positive parameter and the sliding condition is still valid. By defining the control gain, k, as follows,

$$k = \frac{J_w v_x}{r_w}(N+\eta) \qquad (19)$$

and utilizing Eq. (18) as a new sliding condition, Eq. (17) can be modified into the following form.

$$-s\frac{\dot{v}_x}{v_x}(\lambda-1)\left(1-\frac{\mu_b}{\hat{\mu}_b}\right) - s\frac{r_w^2}{J_w v_x}\left(F_b - \frac{\mu_b}{\hat{\mu}_b}\hat{F}_b \right)$$
$$-2 A_p R_b \mu_b (N+\eta)s\,\mathrm{sgn}(s) \leq -\eta'|s| \qquad (20)$$

$$-s\frac{\dot{v}_x}{v_x}(\lambda-1)\left(1-\frac{\mu_b}{\hat{\mu}_b}\right) - s\frac{r_w^2}{J_w v_x}\left(F_b - \frac{\mu_b}{\hat{\mu}_b}\hat{F}_b \right)$$
$$-2 A_p R_b \mu_b (N+\eta)|s| \leq -2 A_p R_b \mu_b \eta|s| \qquad (21)$$

By canceling out the same term,

$$s\left[-\frac{\dot{v}_x}{v_x}(1-\lambda)\left(\frac{\mu_b}{\hat{\mu}_b}-1\right) + \frac{r_w^2}{J_w v_x}\left(\frac{\mu_b}{\hat{\mu}_b}\hat{F}_b - F_b\right) \right] \leq 2 A_p R_b \mu_b N|s| \qquad (22)$$

The parameter, N, has to be designed to ensure the sliding condition specified in Eq. (18). Because the following factors have positive values in Eq. (22),

$$-\frac{\dot{v}_x}{v_x} \geq 0,\ (1-\lambda) \geq 0,\ \frac{r_w^2}{J_w v_x} \geq 0 \qquad (23)$$

Choosing the parameter N as the following form ensures the sliding condition specified in Eq. (18).

$$N \geq \frac{1}{2 A_p R_b \mu_b}\left(\frac{-\dot{v}_x(1-\lambda)}{v_x}\left|\frac{\mu_b}{\hat{\mu}_b}-1\right| + \frac{r_w^2}{J_w v_x}\left|\frac{\mu_b}{\hat{\mu}_b}\hat{F}_b - F_b\right| \right) \qquad (24)$$

In the above equation, three uncertainties can be considered with respect to the estimated tire braking force, estimated disk-pad friction coefficient and real disk-pad friction coefficient. Assuming that these uncertainties are bounded.

$$\left|\frac{\mu_b}{\hat{\mu}_b}-1\right| < B_1,\ \left|\frac{\mu_b}{\hat{\mu}_b}\hat{F}_b - F_b\right| < B_2,\ \mu_b > B_3 \qquad (25)$$

where B_1, B_2 and B_3 indicate the bound values of the uncertainties, the parameter N and the control gain k can be determined for the robust control performance.

$$N = \frac{1}{2 A_p R_b B_3}\left(\frac{-\dot{v}_x(1-\lambda)}{v_x}B_2 + \frac{r_w^2}{J_w v_x}B_1 \right) \qquad (26)$$

$$k = \frac{J_w v_x}{r_w}\left[\frac{1}{2 A_p R_b B_3}\left(\frac{-\dot{v}_x(1-\lambda)}{v_x}B_2 + \frac{r_w^2}{J_w v_x}B_1 \right) + \eta \right] \qquad (27)$$

By substituting Eq. (27) into Eq. (14), the control input brake pressure can be determined. In order to reduce chattering due to the discontinuous switching, sgn($\bullet$), the discontinuous switching is replaced with continuous switching, sat($\bullet$). Then, the control input brake pressure can be formulated as follows.

$$P_{b,c} = \frac{1}{2 A_p R_b \hat{\mu}_b}\left(r_w \hat{F}_b + \frac{J_w \dot{v}_x}{r_w}(\lambda-1) \right)$$
$$-\frac{J_w v_x}{r_w}\left[\frac{1}{2 A_p R_b B_3}\left(\frac{-\dot{v}_x(1-\lambda)}{v_x}B_2 + \frac{r_w^2}{J_w v_x}B_1 \right) + \eta \right]\mathrm{sat}\left(\frac{s}{\Phi}\right) \qquad (28)$$

where Φ is a design parameter representing the boundary layer around the $s=0$ sliding surface and is a small positive parameter.

SIMULATIONS

In order to evaluate the proposed wheel slip control system, the simulation is conducted on the various road conditions. The CarSim™ [12] software, a commercial vehicle simulation tool, is used to represent the real vehicle model. Table 1 shows the vehicle parameters used in simulation. The EHB actuator adjusts the brake pressure and has the response speed of 50 MPa/sec. In the simulation, the EHB actuator is modeled into a simple 1st order linear transfer function with 0.005 sec time constant.

$$G_{EHB}(s) = \frac{1}{0.005s + 1} \qquad (29)$$

The tire braking force monitor, robust wheel slip controller and the simple EHB actuator model are integrated as shown in Fig. 1 and constructed using the CarSim™ [12] and Simulink®. The simulations are conducted in real-time using the RT-LAB program [16], QNX real-time OS and CarSim™ real-time module. The target slip values should be selected from the specific objectives such as maximizing the braking force or maintaining the vehicle stability.

Fig. 5 ~ Fig. 7 show the simulation results on normal road, where the target slips for front and rear wheels are 0.2 and 0.1, respectively. The estimation results of the brake disk-pad friction coefficient and tire braking force are illustrated in Fig. 5 where their estimation accuracy is good enough to be used for the wheel slip control. The maximum steady-state errors for the disk-pad friction coefficient and tire braking force are 5.7% and 3.2%, respectively. The vehicle velocities and wheel slip control results are illustrated in Fig. 6. As shown in Fig. 6, the wheel slips of the front and rear tires are controlled well to follow the target slip values. Fig. 7 shows the control inputs in the braking pressure for front and rear tires. It is confirmed that the size of the control input is within the pressure limit of the EHB actuator.

Fig. 8 ~ Fig. 10 show the simulation results on μ-change road, where the road friction coefficient is changed from the normal condition to the slippery condition at 1.75 sec simulation time. The estimation results of the brake disk-pad friction coefficient and tire braking force are illustrated in Fig. 8. The maximum steady-state error in estimating the disk-pad friction coefficient is 5.7% in the normal road condition and 5.2% in the slippery road condition. In the braking force estimation, the maximum error is 3.2% in the normal road condition and almost zero in the slippery road condition. Fig. 9 shows the vehicle velocities and wheel slip control results on the μ-change road. As shown in Fig. 9, the wheel slips of the front and rear tires follow the target slip values well even with the estimation errors and road disturbance. The

control inputs in the braking pressure for front and rear tires are illustrated in Fig. 10.

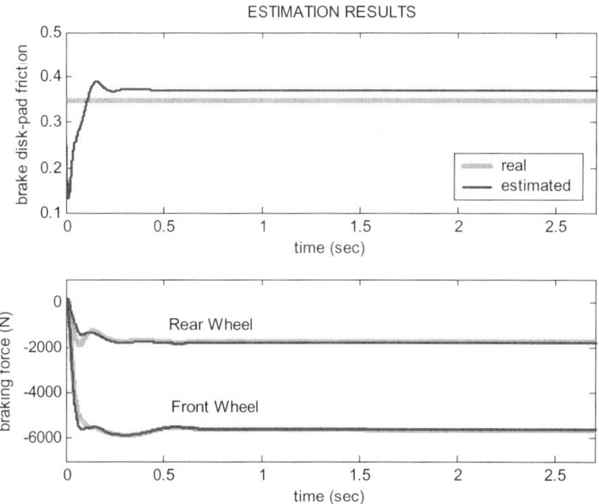

Fig. 5 Estimation results on normal road

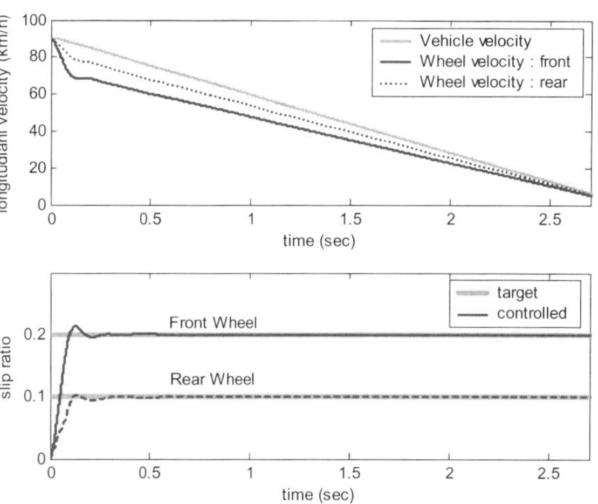

Fig. 6 Control results on normal road

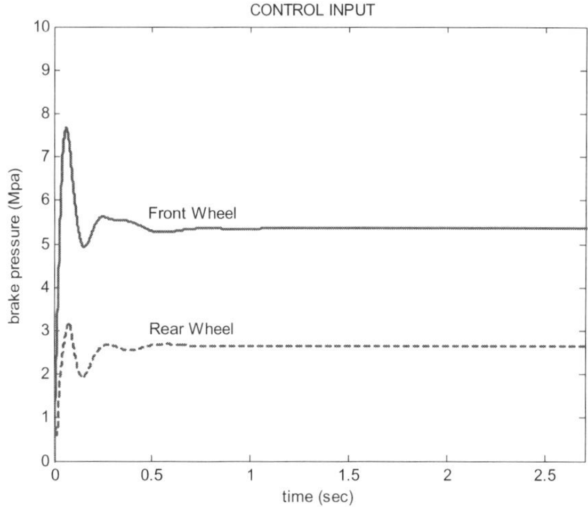

Fig. 7 Control input on normal road

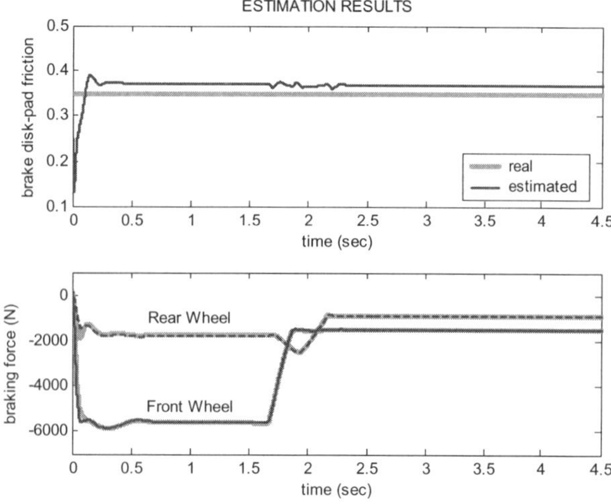

Fig. 8 Estimation results on μ-change road

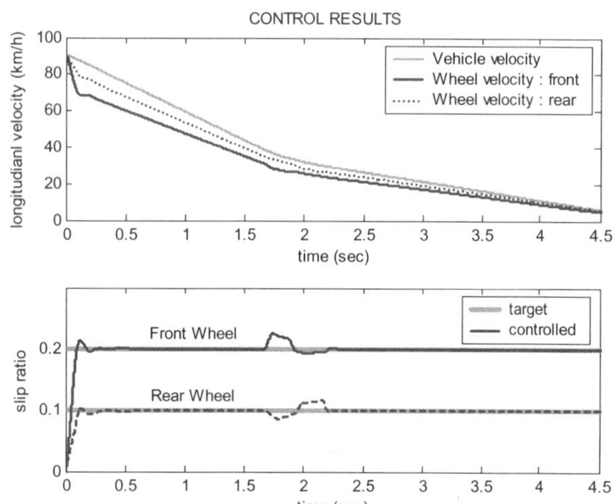

Fig. 9 Control results on μ-change road

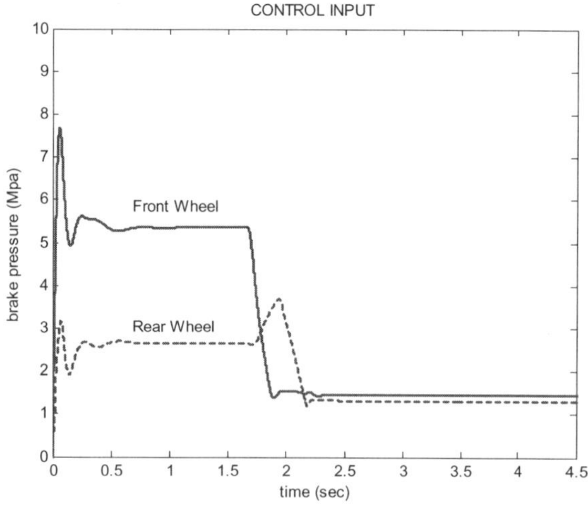

Fig. 10 Control input on μ-change road

Table 1. Vehicle parameters

parameter	value	parameter	value
m	1707 kg	J_w	0.9 kgm^2
I_z	2741.9 kgm^2	r_w	0.301 m
l_f	1.014 m	$A_{p1,2}$	3931.848 mm^2
l_r	1.6760 m	$A_{p3,4}$	2670.227 mm^2
t_f	0.770 m	$R_{b1,2}$	0.109 m
t_r	0.765 m	$R_{b3,4}$	0.107 m

Table 2. Uncertainties and bound values

Uncertainty	Bound value	Simulation (normal road)
$\left\|\dfrac{\mu_b}{\hat{\mu}_b}-1\right\|$	B_1 : 0.06 (upper bound)	0.057
$\left\|\dfrac{\mu_b}{\hat{\mu}_b}\hat{F}_b - F_b\right\|$	B_2 : 500 (upper bound)	49.2
μ_b	B_3 : 0.3 (lower bound)	0.35

The above estimation results demonstrate that the estimation error is larger in the normal road condition than the slippery road condition. Table 2 shows the bound values of the uncertainties chosen and the estimation errors from the simulation results. As shown in Table 2, it is confirmed that the uncertainties from the estimation error are within the bound values chosen in designing the controller.

CONCLUSION

A robust wheel slip control system is developed for the EHB brake-by-wire system. The proposed wheel slip control system is composed of the braking force monitor and robust wheel slip controller. The tire braking force at each wheel and the brake disk-pad friction coefficient are estimated using the vehicle model, measurable motion variables and extended Kalman filter. Based on the estimated tire braking forces and brake disk-pad friction coefficient, the wheel slip controller is designed based on the sliding mode control method. The controller determines the size of the brake pressure as the control input and is designed to be robust to the estimation errors. The performance of the proposed wheel slip control system is evaluated in simulations using the full vehicle model and the simulation results demonstrate the effectiveness of the proposed system.

ACKNOWLEDGMENTS

The support of the Ministry of Commerce, Industry and Energy in Korea (No. 10005253) is gratefully acknowledged.

REFERENCES

1. S. Müller, M. Uchanski and K. Hedrick, "Slip-Based Tire-Road Friction Estimation During Braking", Proceedings of IMECE'01, 2001.
2. S. L. Miller, B. Yonugberg, A. Millie, P. Schweizer, J. C. Gerdes, "Calculating Longitudinal Wheel Slip and Tire Parameters Using GPS Velocity," Proceedings of the American Control Conference, pp. 1800~1805, 2001.
3. A. Daiß, U. Kiencke, "Estimation of Vehicle Speed Fuzzy-Estimation in Comparison with Kalman-Filtering," Proceedings of the 4th IEEE Conference on Control Applications, pp. 281~284, 1995.
4. S. Semmler, D. Fischer, R. Isermann, R. Schwarz, P. Rieth, "Estimation Of Vehicle Velocity Using Brake-By-Wire Actuators," IFAC 15th Triennial World Congress, 2002.
5. S. Semmler, R. Isermann, R. Schwarz and P. Rieth, "Wheel Slip Control for Antilock Braking Systems using Brake-by-Wire Actuators", SAE World Congress, No. 2003-01-0325, 2003.
6. K. R. Buckholtz, "Reference Input Wheel Slip Tracking Using Sliding Mode Control", SAE World Congress, No. 2002-01-0301, 2002.
7. T. Kawabe, M. Nakazawa, I. Notsu and Y. Watanabe, "A Sliding Mode Controller for Wheel Slip Ratio Control System", International Symposium on Advanced Vehicle Control, pp. 797~804, 1996.
8. I. Petersen, T. A. Johansen, J. Kalkkuhl, J. Lüdemann, "Wheel Silp Control in ABS Brakes Using Gain Scheduled Constrained LQR," European Control Conference, 2001.
9. M. Schinkel, K. Hunt, "Anti-Lock braking Control using a Sliding Mode like Approach," Proceedings of the American Control Conference, pp. 2386~2391, 2002.
10. C. Ünsal, P. Kachroo, "Sliding Mode Measurement Feedback Control for Antilock Braking Systems," IEEE Transaction on Control Systems Technology, Vol. 7, No. 2, pp. 271~281, 1999.
11. J. Yi, L. Alvarez, R. Horowitz, "Adaptive Emergency Braking Control With Underestimation of Friction Coefficient," IEEE Transaction on Control Systems Technology, Vol. 10, No. 3, 2002.
12. CarSim™, Version 5.16b, Mechanical Simulation Corporation, 2004.
13. M.S. Grewal, A. P. Andrews, Kalman Filtering Theory and Practice, Prentice Hall; 1993.
14. P. Yoon, H. Kang, I. Hwang, K. Huh, D. Hong, "Braking Status Monitoring for Brake-By-wire Systems," SAE World Congress, No. 2004-01-0259, 2004.
15. J. J. E. Slotine, W. Li, Applied Nonlinear Control, Prentice Hall, 1991.
16. RT-Lab, Version 6.2, Opal-RT Technologies Inc., 2002.

CONTACT

Kunsoo Huh (Associate Professor)
School of Mechanical Engineering
Hanyang University
17 Haengdang-Dong, Sungdong-Ku,
Seoul 133-791, Korea
Tel : 82-2-2290-0437 Fax : 82-2-2295-4584
E-mail : khuh2@hanyang.ac.kr

Performance Evaluation of Vehicle Network Protocols for X-By-Wire Using Simulation

Jung Hwan Choi
Hyundai Mobis

Kyung Chang Lee, Man Ho Kim and Suk Lee
Pusan National University

ABSTRACT

Automakers have been pushing harder for the development of a safety-critical bus for the past years since there are many advantages of x-by-wire systems including steer-by-wire, brake-by-wire and throttle-by-wire. Such systems allow automakers to eliminate heavy hydraulic actuators and they offer the possibility of creating smarter and more efficient components that can be connected to a network bus. However, automakers and vendors know that a reliable, fault-tolerant bus is needed for such applications. Controller Area Network (CAN) buses, which are commonly used for powertrain and other automotive controls, are not considered reliable enough for drive-by-wire. The problem with CAN is that it is only event-based, so there is always a possibility that a message won't get through. For important applications, time-triggered architectures are needed because as time goes on, a slot for important messages is always assigned. In this paper, the performance evaluations of CAN, TTP and Byteflight are presented to explain which aspects are different and how the system should be designed for these protocols for x-by-wire systems. For evaluations, the simulation models were developed by using a discrete event simulation language, SIMAN. Based on these models numerous experiments were conducted for the comparable factors such as the probability of transmission failure, the average transmission delay and the maximum transmission delay. From the performance evaluation results, we can conclude that TTP exhibits a very stable performance over an extended traffic load while it lacks the flexibility for system changes. In addition, CAN and Byteflight show very good performance under low traffic while the performance deteriorates very fast as traffic exceeds a certain threshold.

INTRODUCTION

The latest research and development topic in the automotive industry is X-By-Wire. X-By-Wire replaces throttle, brake, and front and rear steering system in a conventional vehicle with an electronic system. In the current vehicle system having mechanical linkages, there are many issues involving design space allowed, weight, and design steps of a vehicle. If X-By-Wire system replaces the conventional parts mentioned, gas mileage will improve due to decrease in weight. Not only will there be advantages of effectiveness and increase in functionality, but also design and module production are promoted such that more effective manufacturing process can be expected.

In the area of communication systems, it is necessary to have a fault-tolerant function that stops fault becoming an error in case of fault situation and a reliable network protocol that can predict data transfer time between the modules connected by X-By-Wire. Since it was decided that CAN, the current real time communication protocol, does not satisfy these requirements, a new X-By-Wire system protocol is in the process of being developed. Thus, Time-triggered protocol method, instead of Event-triggered protocol method, is being developed.

TTP and FlexRay are the representative Time-Triggered Protocol. In this research, from CAN, TTP, and Byteflight simulations, characteristics of these protocols can be found and network performance can be evaluated. The reason for including Byteflight for simulation instead of FlexRay is that Byteflight is the basic protocol for FlexRay, the next generation X-By-Wire system protocol.

This paper consists of five sections including this introduction. In the following section, CAN, TTP, and Byteflight are explained in detail in the media access method point of view. Next, simulation model is explained by using flow chart. Then, simulation performance results are analyzed by the occurrence of non-periodic and periodic messages. Finally, conclusions are made from this research.

1. CAN, TTP AND BYTEFLIGHT

1.1 CAN

CAN (Controller Area Network) is a representative serial communication protocol that connects sensors and actuators inside the vehicle. It was developed by BOSCH in 1980 and became a representative communication system between ECU's (Electronic Control Units) due to high reliability by using a multi-master broadcast method.

Depending on the content of message, CAN allows ID (identifier) and determines message's priority. For example, when stations are starting to transmit messages instantaneously, stations compare IDs and check whether the station's own ID is lower than the others (priority is higher). A station with the lowest message ID can send and other stations receive data.

1.2 TTP

TTP (Time Triggered Protocol), made by TTTech, is a time-triggered protocol and it's also an appropriate protocol for hard real-time communication that belongs to Class C of SAE. When TTP transmits a message, jitter does not occur most of the time. TTP has a fault-tolerant function where one node's break down does not interfere with the rest of the nodes' communication. Also, it has a through error examination and provides recovery function. Lastly, it provides a membership service such that each node always knows the other nodes' conditions.

TTP uses TDMA (Time Division Multiple Access) of the reserved slot method. As time passes by, the right to approach a media is assigned by a slot to each node. The length of TDMA slot is determined by taking the transmitter and receiver's execution time into consideration. The relationship between a slot and transmitted data and transmitted node's order is determined before the system configuration. This series of planning is stored in MEDL (Message Descriptor List)

1.3 BYTEFLIGHT

Byteflight is a high performance data bus developed by BMW. The purpose of developing Byteflight was to achieve a communication protocol that integrates synchronous and asynchronous transmitting advantages. The highest transmission speed of 10Mbps and 250 µs data update are possible. Also, flexible integration is a plus.

Byteflight's MAC (media access control) sends a message in a prior order in between synchronization pulses. The slot count's ID of each node increases by 1 when the transmission is complete or after an idle time. Within one cycle, each node compares its own ID with a slot count ID and if they match, the node sends a message.

2. SIMULATION MODEL

For simulation, SIMAN (discrete event simulation language) is used for changing system condition as time goes by and for recording important information. By doing so, CAN, TTP, and Byteflight's simulation models are developed and quantitative evaluation and comparison are done.

2.1 CAN

CAN simulation model is assumed to have a normal condition without an error at the time of message transmission. Also if a created message is inserted into que, protocol's transmission buffer of size 1, newly created messages are ruled to be deleted before a former transmission is not complete.

The subject of the model is CAN 2.0A in which arbitration field is expressed in 11 bits and can have 2^{11} ID values. In a real controller, ID is compared with bit by bit, but in this simulation model bits were converted into an overall value for comparison. Topology is assumed to be a bus and changes in the number of stations and message's data size is possible.

CAN's simulation model flow chart is shown in figure 1.

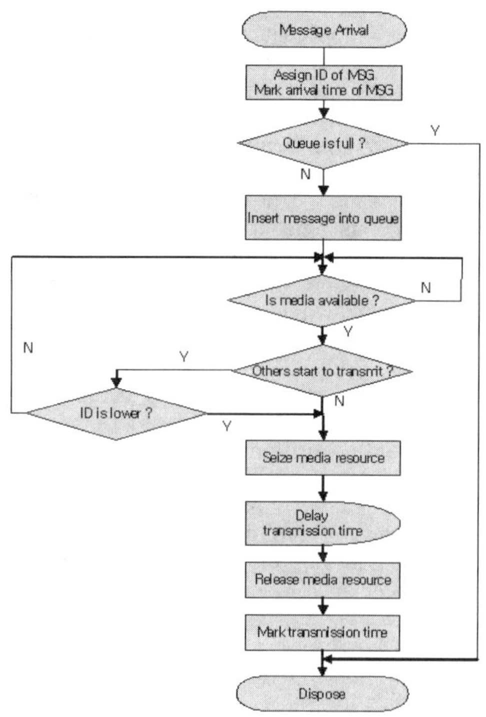

Figure 1. Flow chart of CAN

2.2 TTP SIMULATION MODEL

In case of TTP simulation model, TDMA (Time Division Media Access) was used as MAC. Like CAN, error does not occur at time of transmission and buffer's size was 1 when modeling.

In a normal network condition, TTP with an overhead of 4 bits sends a message composed of 2~16 bytes of data and 2 bytes of CRC. The minimum time interval between the messages is 5 μs and one transmission cycle can be defined at the system level of design.

Therefore, in TTP simulation model, it is assumed to have the same speed (1 Mbps) as CAN. The time interval between the messages are defined to be 5 μs. A cycle length is a time needed for transmitting overhead, data and CRC values. With the assumption of all the stations transmitting once, one cycle interval was defined by the time that requires all stations to transmit once. The flowchart is shown in Figure 2.

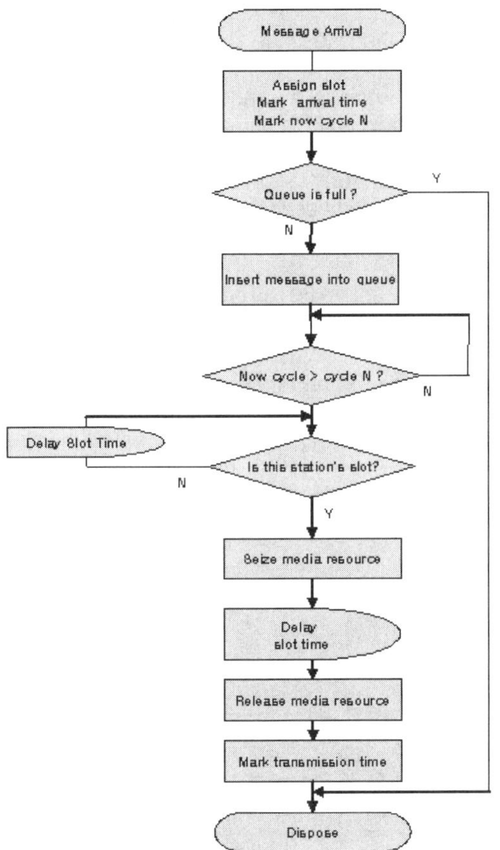

Figure 2. Flow chart of TTP

2.3 BYTEFLIGHT SIMULATION MODEL

Byteflight's data transmission speed is set to be 10Mbps by using an optical fiber. This is very different from CAN and TTP's physical properties. Compared to the CAN's maximum speed of 1 Mbps and TTP's maximum speed of 1~2 Mbps, this is the cause of difficulty in comparison. Considering this difference, instead of comparing the quantitative measurements between the Byteflight and

CAN and TTP, characteristics caused by traffic can be observed.

A Byteflight station is composed of 16 buffers (buffer's size 1). Messages with same ID cannot be stored in a buffer. In other words, messages with 16 different ID can be in buffer. This condition is same as CAN and TTP. Transmission failure condition is characterized by a condition of messages existing in a buffer and messages with same ID being produced. Network is assumed to have no occurrence of error.

Byteflight's flow chart is shown in Figure 3.

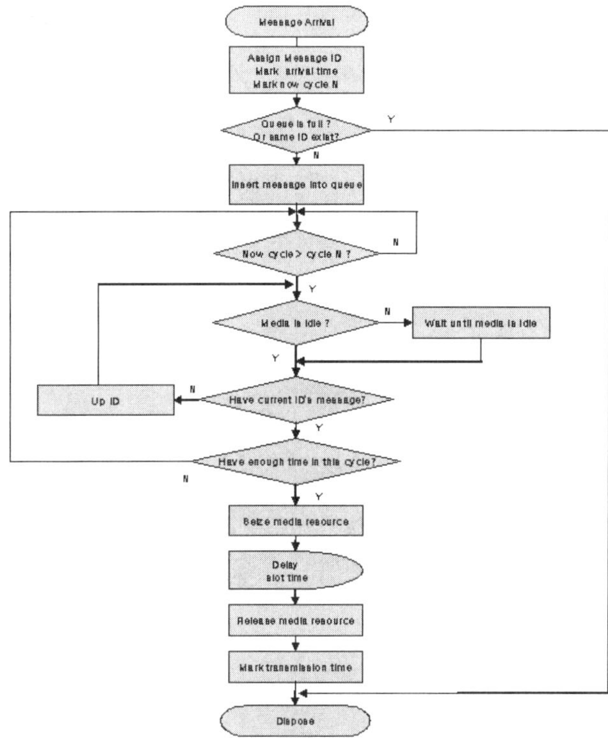

Figure 3. Flow chart of Byteflight

3. SIMULATION RESULTS

Equally distributed non-periodic messages and periodic message will be generated. Simulation performance will be observed as the network's traffic varies by changing data length and number of stations. Transmission failure probability also will be observed.

3.1 NON-PERIODIC MESSAGE

Messages are generated with uniformly distributed average interval of 1024 μs and simulation is performed until 1500 message are transmitted properly. In case of Byteflight, a message generation with average interval of 102 μs is used considering Byteflight's speed of 10Mbps.

Transmission failure probability increases as data length and number of stations increase, as shown in figure 4 and figure 5. The reason is that as data length increases a time needed to transmit a message increases. As the number of stations increase, the messages are being produced more and more. Thus, the number of transmission trials increases. As a result, trying transmission at the same time occurs for CAN and transmission failure probability increases because messages with low priority cannot be transmitted. For TTP, more than one transmission trials occur in one cycle and as a result, transmission failure probability increases. Also, as shown in Figure 4 and 5, TTP has a lower transmission failure probability than the one in CAN. Considering the relationship between the occurrence and transmission period, simulation's condition has a positive effect on TTP's transmission failure probability.

Figure 6 shows Byteflight's transmission failure probability. In Figure 6, transmission failure probability suddenly increases with Byteflight. The transmission failure cause in Byteflight is due to an existence of messages with same ID in the buffer. While an ID with low priority could not be transmitted in the previous cycle, messages with same ID reappear and are thrown away.

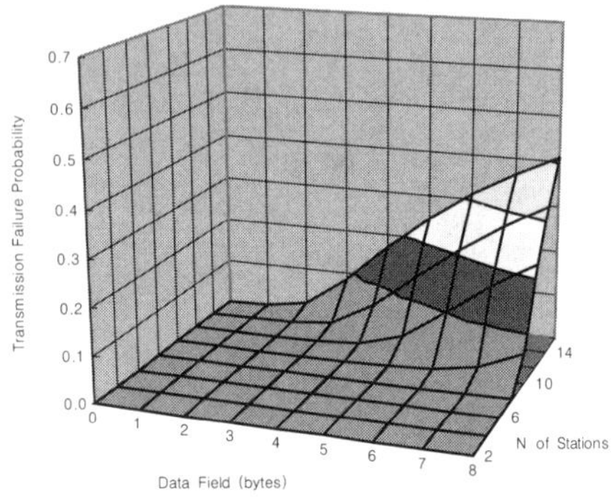

Figure 4. Transmission failure of non-periodic messages with CAN

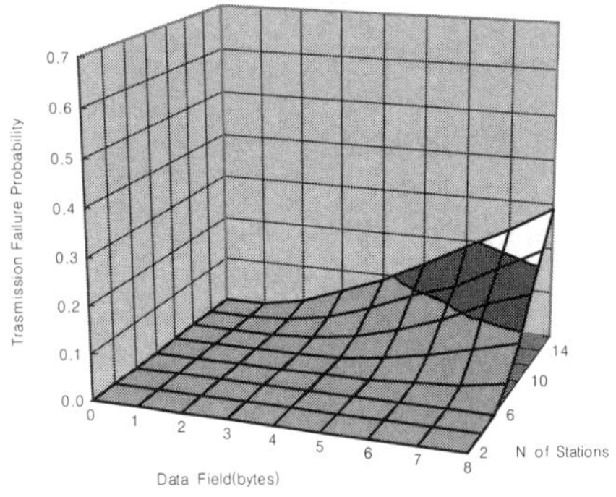

Figure 5 Transmission failure of non-periodic messages with TTP

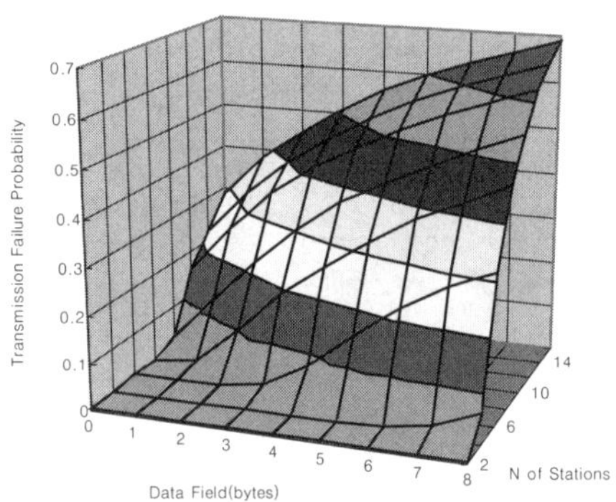

Figure 6 Transmission failure of normalized non-periodic messages with Byteflight

Figure 7 and 8 show CAN and TTP's average transmission delay, respectively. Delay is defined by the time starting from inserting a message produced by the station for transmission up to completing the transmission of that message's last bit.

In case of two protocols' average delay, CAN has a lower delay than the one in TTP since CAN tries to transmit at the time of message occurrence. However, TTP needs to prepare a slot for transmitting that message in the next cycle after the message occurs. Thus, the average delay for TTP is higher than the one in CAN.

In figure 9, the peak value of Byteflight's average delay is about 2200 μs compared to the 820 μs for CAN and 1700 μs for TTP. When the average delays are compared among the protocols, CAN shows the smallest delay. This is due to the advantage of event-triggered method over the time-triggered method when there is low traffic, since event-triggered method tries to transmit right after the request is sent.

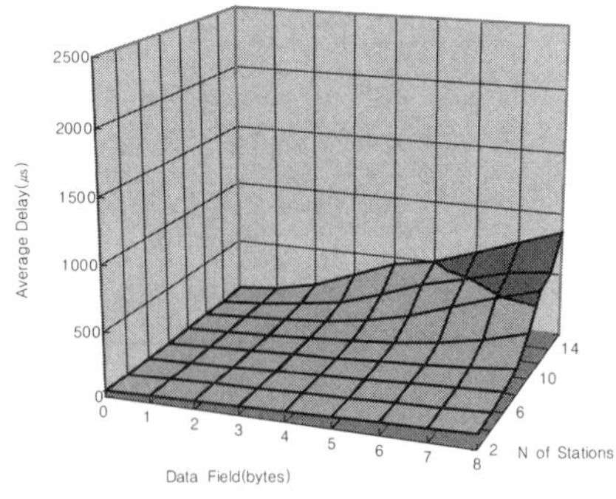

Figure 7. Average delay of non-periodic messages with CAN

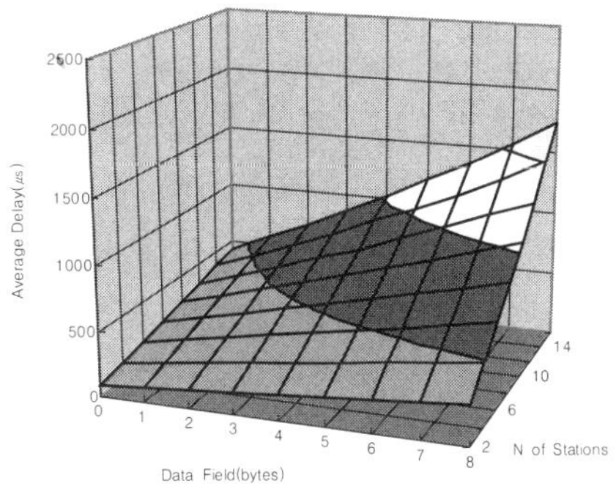

Figure 8. Average delay of non-periodic messages with TTP

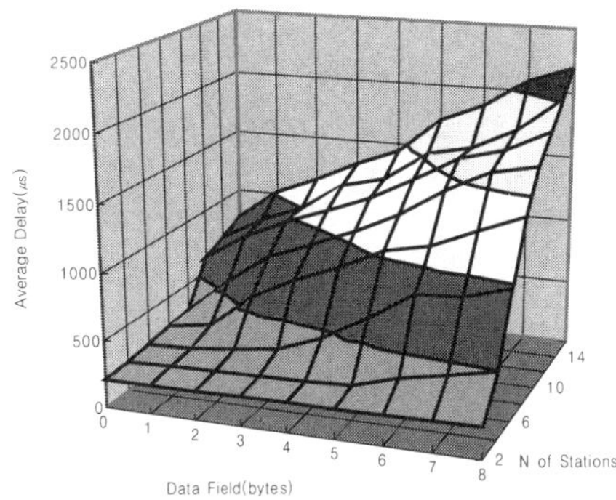

Figure 9. Average delay of normalized non-periodic messages with Byteflight

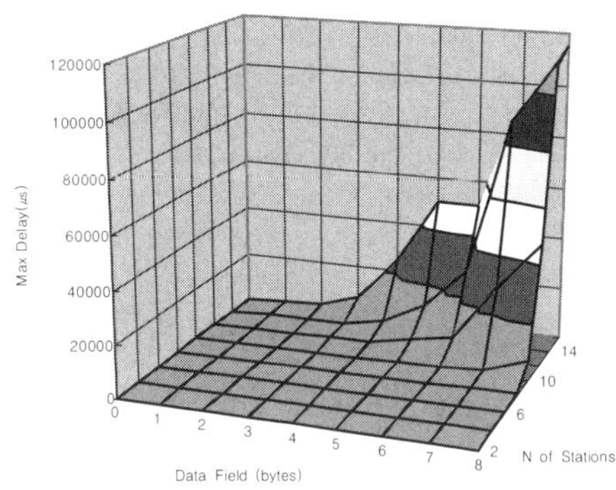

Figure 10 Max delay of non-periodic messages with CAN

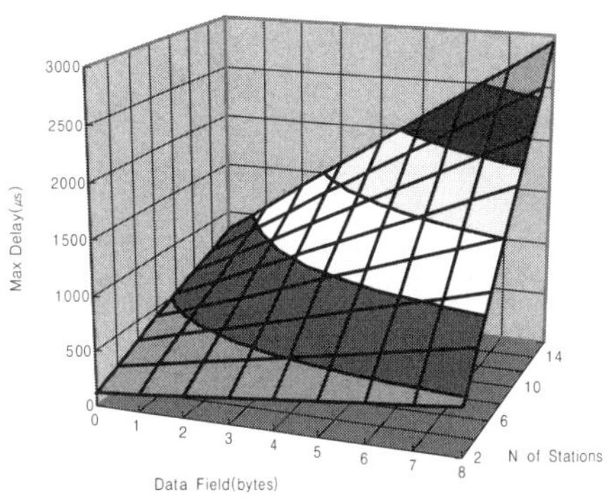

Figure 11 Max delay of non-periodic messages with TTP

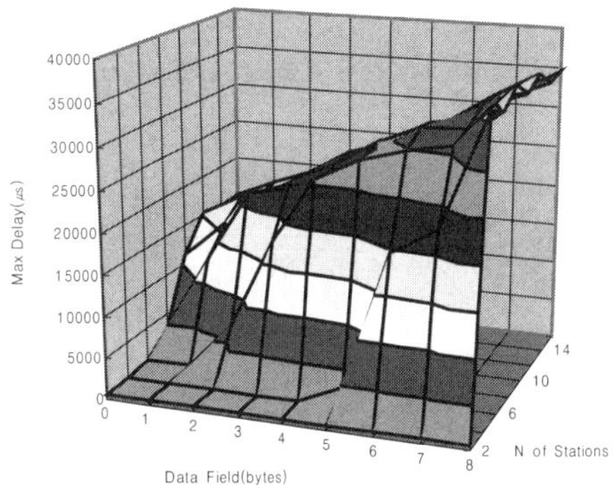

Figure 12 Max delay of normalized non-periodic messages with Byteflight

Figure 10 and 11 show the maximum delay in CAN and TTP, respectively. From these figures, the difference between these two communication protocols, CAN and TTP, can be found. As the traffic increases, message with a low priority has a good chance of waiting at the buffer for CAN.

In the worst case, in order for the message with lowest priority to be transmitted, it should wait at the buffer until the message with a higher priority is transmitted.

Tendency shown in Figure 7 is certainly different from the one in Figure 10. As traffic increases, maximum delay sharply increases in CAN. However, in case of TTP, a message must be transmitted in the next cycle even though there's a high traffic. This shows that maximum delay does not increase sharply. These characteristics guarantee a predictable transmission necessary for X-By-Wire system. The completion of transmission time can be assumed to be predicted considering delay time caused by waiting time at the buffer due to time difference between occurrence of message and transmission. On the other hand, with Byteflight, modified TDMA method is being adopted. Media access method causes maximum delay increase as the traffic increases due to ID's priority. Like CAN, the message with a low priority waits at the buffer until the message with a higher priority is transmitted. This shows a result that lowers delay performance while obtaining flexibility.

3.2 PERIODIC MESSAGE

In order to evaluate the performance of periodic message occurrence, message period occurred at a fixed period of 1024 μs instead of equally distributed 1024 μs. But, with Byteflight, in order to make same traffic conditions, messages with a period of 102 μs were issued.

CAN and TTP's transmission failure probability is shown in Figure 13 and 14, respectively, and the case for Byteflight is shown in Figure 15. In case of occurrence of periodic message in CAN, a similar trend with the non-periodic message generation is shown. Compared to CAN, transmission failure probability drops in TTP. In case of Byteflight, the similar characteristics for transmission failure probability are shown compared to the non-periodic message.

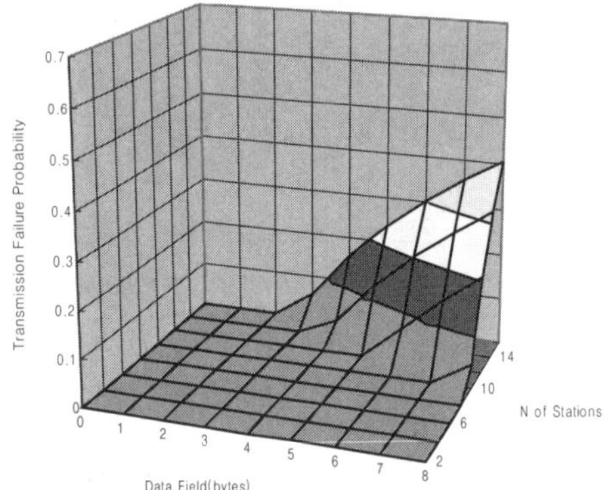

Figure 13 Transmission failure of periodic message with CAN

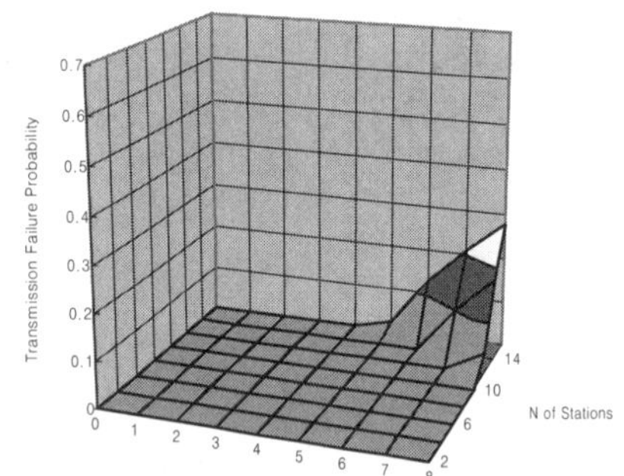

Figure 14 Transmission failure of periodic message with TTP

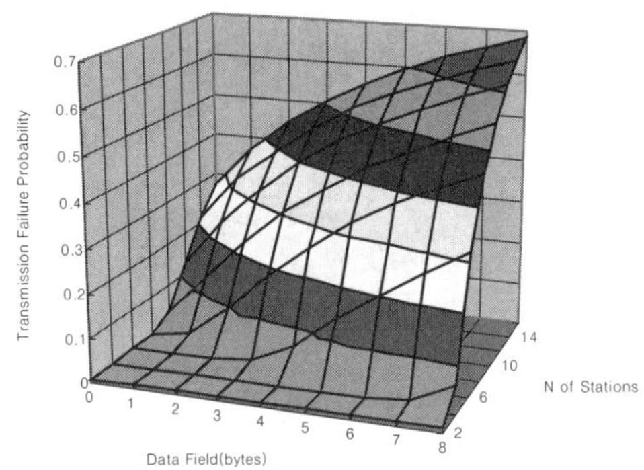

Figure 15 Transmission failure of normalized periodic message with Byteflight

Average delay of periodic message for CAN and TTP are shown in Figure 16 and 17, respectively. As traffic increases, the requests for sending messages increase. The waiting chance in buffer increases, so average delay takes longer in both CAN and TTP. In Figure 16, it is shown that the result of TTP have nonlinear characteristics in high traffic compared to periodic message generation. This is due to the coincidence between message occurrence and transmission cycle.

Namely, as traffic increases, a transmission cycle is longer and the number of transmission request in a cycle increase. Considering coincidence will make quick transmission possible.

Average delay of Byteflight shown in Figure 18 is 200 ~ 2300 μs and similar tendency with non-periodic message is observed.

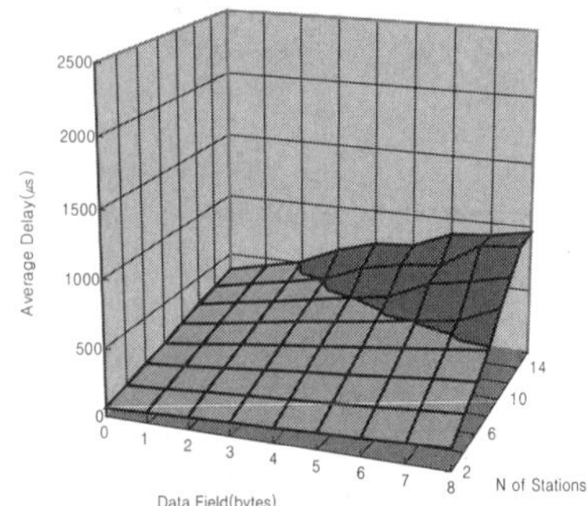

Figure 16 Average delay of periodic message with CAN

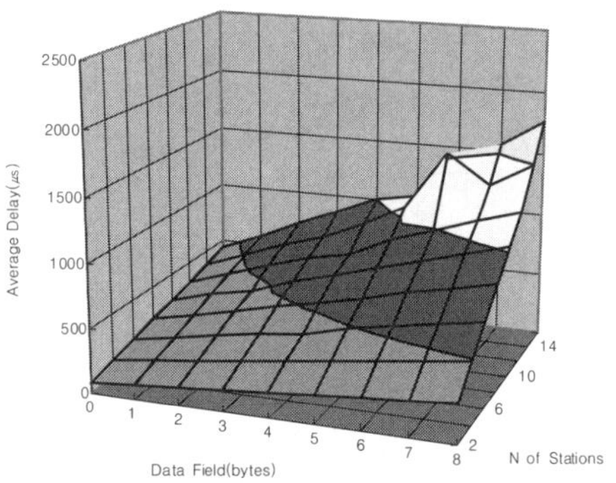

Figure 17 Average delay of periodic message with TTP

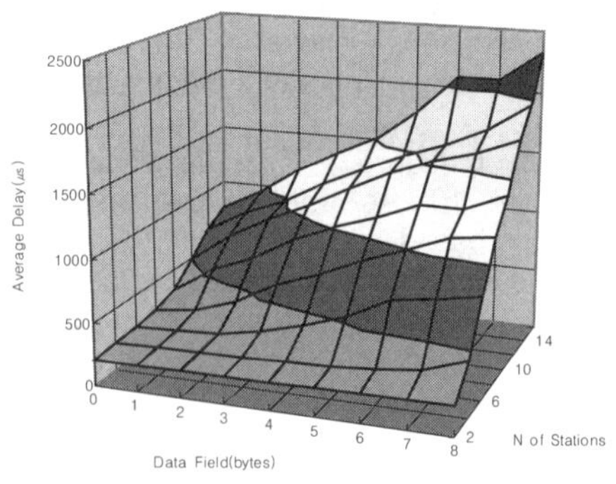

Figure 18 Average delay of normalized periodic messages with Byteflight

Maximum delay of CAN and TTP, shown in Figure 19 and 20, also have similar tendency with non-periodic messages. In CAN, messages are transmitted based on the ID that gives priority. In case of high traffic, CAN attempts to transmit a message with a low ID and it cannot predict how long it will wait in the transmission buffer until other messages with higher IDs are sent. But TTP allows a node to always have its own slot in which the message is sent in a cycle.

According to simulation experiments, maximum of 130,000 μs waiting time in buffer is observed in CAN. But in TTP, even in the worst case, there is no longer delay than two cycle time(2848 μs).

Figure 21 shows the maximum delay of normalized periodic messages with Byteflight. Byteflight also have transmission priority based on the ID. Even though a maximum delay of Byteflight is not recorded longer than that of CAN, considering average delay of 2500 μs and 34600 μs is notably high. Since transmission order is decided by message priority, it seems that message with low priority has a high maximum delay.

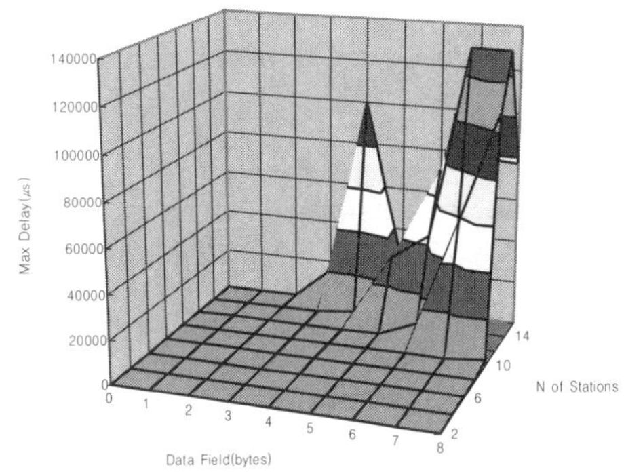

Figure 19 Max delay of periodic messages with CAN

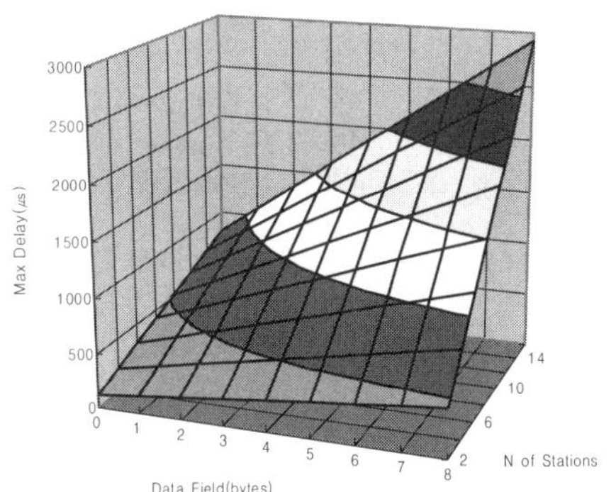

Figure 20 MAX delay of periodic messages with TTP

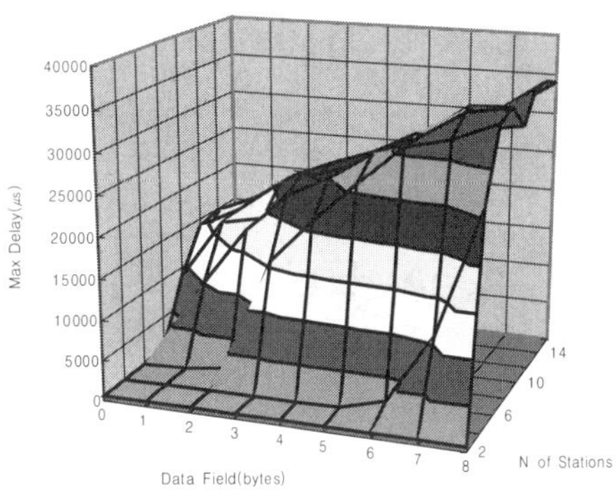

Figure 21 MAX delay of normalized periodic messages with Byteflight

The preceding results show that performance of TTP is affected by synchronization with cycle. So in this simulation, message is generated synchronized with TTP Cycle. TDMA cycle length varies with data length and the number of stations. The result is shown in Figure 22 and 23.

The transmission failure probability of TTP, which is zero, is shown in Figure 22. It means that all the messages are processed properly. This is possible because messages were generated at the start of TDMA cycle. This shows that network design of TTP is important to fulfill the system demand. The predictable network systems are accomplished by considering TTP protocol property.

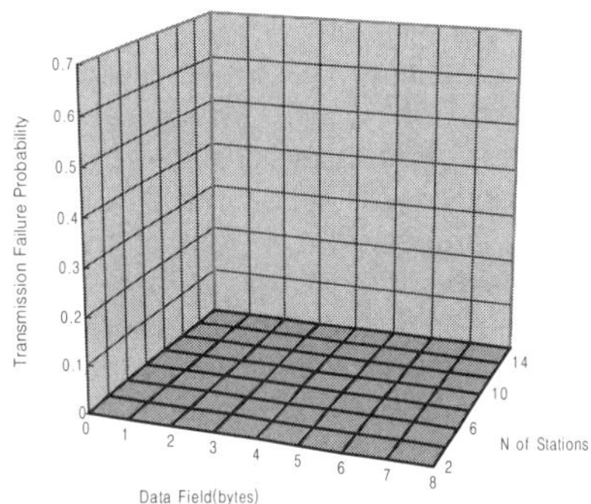

Figure 22 Transmission failure of TDMA cycle messages with TTP

Figure 23 shows the maximum transmission delay with TTP. As traffic becomes heavier, TDMA cycle gets longer; therefore, transmission delay takes longer. It is observed that waiting time in the buffer for transmission increases proportionally. TDMA cycle with data length of 8 bytes and 16 stations is 1424 μs, and does not exceed twice its value, which is 2848 μs.

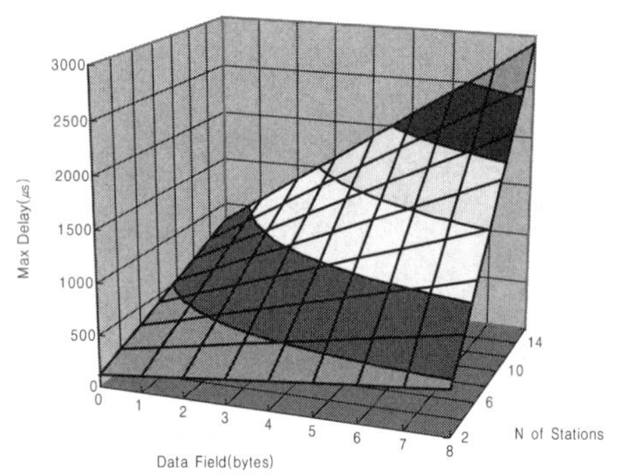

Figure 23 Max delay of TDMA cycle messages with TTP

3.3 PERFORMANCE EVALUATION

Up to now, messages were non-periodic and uniformly distributed and periodic(1024 μs) messages were generated. Network performance experiment were carried out and quantitative analysis (transmission failure probability, average transmit delay and max transmit delay) were shown.

In case of CAN, transmission failure probability and average delay of non-periodic message is better than other protocols. CAN also shows a good performance in maximum delay in low traffic. But, in high traffic condition, message transmission time cannot be estimated due to low priority message.

In case of TTP, transmission delay is longer than CAN under low traffic situation. But transmission delay is in the range of twice TDMA cycle while traffic increases. Transmission failure probability is related to the message generation frequency cycle developed in TTP simulation model. Based on the assumption that one message can be sent in a cycle by a node, more than one message generation causes message transmission failure. TTP protocol shows a superior performance in TDMA cycle message generation condition compared to non-periodic message generation condition. This means that if length of TDMA cycle is determined based on message generation frequency in network system design step, it ensures a reliable transmission under a wide range of traffic.

Byteflight has 10Mbps transmission speed although CAN and TTP are assumed to have 1Mbps speed. Thus, comparison of CAN and TTP is somewhat difficult and modification for setting same conditions is unreasonable. Even though quantitative performance is not directly accepted, simulation shows that Byteflight has a defect of sending low priority message in high traffic network.

CONCLUSION

In this paper, simulation models for CAN(Controller Area Network), TTP(Time Triggered Protocol) and Byteflight were developed by using SIMAN. Protocol performances were evaluated on message generation condition by using the simulation models. As a result of the simulation, conclusions on these protocols are listed blow.

(1) In case of a low traffic, CAN shows a high performance but in case of a high traffic CAN cannot guarantee processing time for all the message.

(2) TTP ensures message transmission to be carried out over a wide range of traffic.

(3) In TTP, transmission failure probability is affected by message generation time.

(4) In high traffic, CAN is not suitable for X-By-Wire protocol due to a wide rage of delay. TTP has restrictive expansion ability after system design.

(5) Byteflight protocol uses modified TDMA method and has system expansion ability.

(6) Byteflight protocol has a wide range of delay under high traffic.

CAN, today's representative real time vehicle network, does not have a transmission boundary limit so that traffic should be kept under a certain value in designing the system. A careful selection of message ID for a real-time or non real-time is needed. TTP has a difficulty in expanding the system, so assigning a slot to allow expansion is needed afterward. It is also pointed out that Byteflight has the same defect as the one in CAN and real-time ID needs to be carefully selected.

REFERENCES

1. Jung-A Yun, etal. (1998). Performance Evaluation of Multiplexing Protocols, SAE paper, 981105.
2. Pegden, C., R. Shannon and Sadowski. (1990). Introduction to Simulation Using SIMAN, McGraw-Hill.
3. CAN specification version 2.0, Robert Bosch GmbH, 1991
4. Time-Triggered Protocol TTP/C High-Level Specification Document version 1.0.0, TTTech, 2002
5. Byteflight specification version 0.5, BMW, 1999.
6. B. Hedenetz, R.Belschner, "Brake-by-Wire Without Mechanical Backup by using a TTP-Communicaion Network", SAE paper, 1998

DEFINITIONS, ACRONYMS, ABBREVIATIONS

Probability of transmission failure: the number of transmission failure divided by the number of transmission attempts over a given simulation period.

Delay: the time interval from the instant when a message is generated to the instant when the message's last bit is transmitted.

Traffic: Ratio of network capability to data field length

Design of Two Families of Electric Park Brake and Their Optional Suite of Functionality & Operational Features

Dennis Plunkett and Nui Wang
PBR Australia PTY. LTD.

ABSTRACT

Cost reduction, performance and feature trade offs are amongst the list of challenges that automotive engineers face in the attempt to develop new products. Knowing your product, its domain and the systems with which it interacts allows suppliers to accurately cost development and manufacture, assess warranty and reliability issues. These key trade offs have the greatest impact when the product is a new implementation of a highly developed and cost optimised vehicle function.

This paper presents the challenges posed by hidden system interactions and the solutions that PBR created while developing its future product range of By-wire Electric Park Brake Systems (ePARK™).

KEY WORDS

By-wire, Electric Park Brake, Electric Actuator, Park Brake Lever, Park Brake Force, Passive Safety, Kidney Spool.

INTRODUCTION

Electric motors are becoming more common place in the modern automobile. This can be attributed to the development of low cost electronic controls and the improved efficiency and performance of modern electric motors [1]. The addition of an electric park brake will significantly change the appearance and use of the park brake in more and more vehicles in the future: the park brake lever will disappear and be replaced by a switch. Engaging a push button or a knob will apply the required park brake force [2]. This simple step has the potential to increase the design flexibility for vehicle architects and improve passive safety.

The concept of electric park brakes is not new. Pulling brake cables with electric actuators is over 20 years old [3]. It would be easy to assume that cost has been the prohibitive factor in the past. By-wire systems and system integration have long been seen as an opportunity to increase both vehicle flexibility and improve performance whilst serving the needs of cost reduction, improved features and functions. Recent advancements in controls including software, tools and cost effective silicon has facilitated the use of new technology in an old field. However it should not be expected that an electric park brake has a direct cost advantage [3], or is there?

Knowledge management experts argue that we should radically alter the practices and routines around the processes: always questioning 'why' to locate any other hidden advantages [4][5].

If there is no apparent component cost advantage, then what are the hidden system advantages in this case? Why should it be done? Investigation of the National Highway Transport Safety Authority (NHTSA) recall data since 1980 [6], unveiled amongst other findings over 100 vehicle safety recalls for the park brake lever, affecting in excess of 22 million vehicles in the USA alone! Replacing the lever with a simpler device will dramatically reduce the number of these recalls, as well as their associated cost.

PROBLEM DESCRIPTION

Existing mechanical park brakes are based on the principle of a lever pulling a cable that acts on a park brake (see Figure 1). Variants include:-

Leading-trailing service drum brake
Duo-servo drum-in-hat park brake
Integral caliper park brake

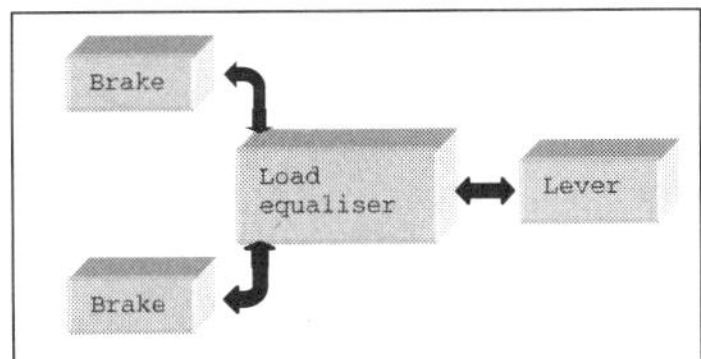

Figure 1. Basic lever operated park brake

This current design approach was used as the starting point to develop the in-depth product scoping analysis. In the conversion from mechanical to electrical systems there are three key stake holders:-

> The end user
> The Original Equipment Manufacturer (OEM)
> The supplier

Driving the end user needs are reliability, ease of use, convenience and advanced safety features. In addition to the previous features the OEM is focused on reliability, reduced weight, cost, performance, manufacturing flexibility and design freedom. Motivating the supplier is a desire to develop the next generation of product that adheres to Federal standards, reduces manufacturing cost and captures future business.

The understanding of the design approach is covered in the following sections:

Analysing Mechanical Park Brakes introduces the problems at the solution level and highlights areas that would impinge the solution.

Design of Two Variants provides a description of the electric park brake solutions, including both the electric Park Brake cable puller (ePB) and electric Park Brake Hydraulic Caliper (ePBHC) actuators and the vehicle electrical interface.

Discussion covers changes that can be made to both the way the park brake is used, and the Federal regulations that affect design constraints.

Conclusion wraps up the entire electric park brake product development.

ANALYSING MECHANICAL PARK BRAKES

The driving force behind the approach was to develop an electric park brake that emulates the current mechanical device, identification of features that add value and integrate these into an evaluation platform, all whilst being conscious of the key stakeholder needs.

MECHANICALS

The relentless approach for cost / performance improvement is apparent in the design of the park brake mechanicals: they are nominally validated to perform 40,000 to 60,000 cycles holding a fully laden vehicle on a grade of 30% and to withstand an operator applied force distribution covering the 95% percentile (see Figure 2).

The ePB as designed fulfilled this requirement at a cable retention force of 1300N. OEMs are indicating that product acceptance is dependant on many factors, including functionality enhancements like automatic apply / release and hill start. This significantly increases the number of operations to 100,000 cycles or more.

The interaction with other systems and environment needs to be taken into account to determine the overall safety [7]. This interaction methodology was applied to evaluate the automatic apply / release and hill start function dependencies on other system components.

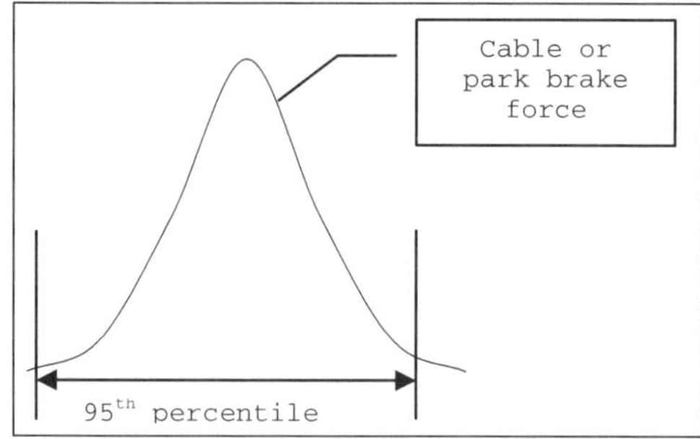

Figure 2. Typical cable load

To interface an electric park brake with the number of operation cycles higher than current vehicle design requires careful consideration to mitigate redesign and validation factors (see Figure 3). These constraints include:-

> Tolerance stack
> Vehicle mass
> Inclination
> Current park brake design
> Bowden cable load

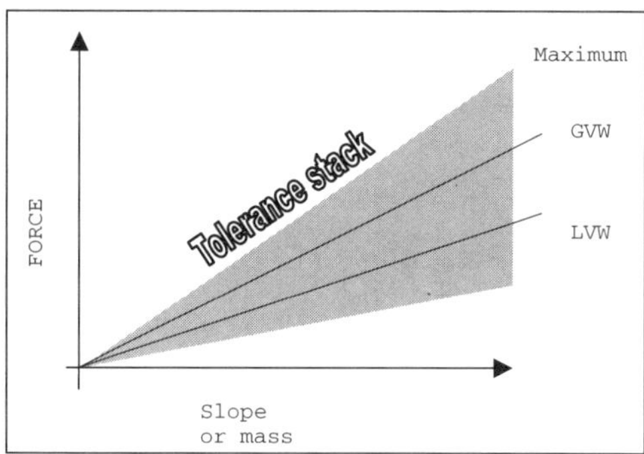

Figure 3. Cable force VS slope mass and tolerance

Ignoring these constraints or making a worst case allowance could lead to over-applying the cable force (an issue identified as an undesired function). Other simpler solutions are to redesign the park brake mechanicals to accommodate the increased force and/or apply frequency.

In a "modulated manual apply" of an electric park brake, the upper level of the 95th percentile apply force limit is an issue with respect to the number of applies. Operators will find it increasingly easy to inadvertently assert full force applies. It is possible to conceive that the operator's cognitive reaction is just the same as current mechanical systems. For example: the first time a vehicle moved after a partial park brake apply, the operator would invariably perform a full apply.

Emulating the current mechanical system was the first solution approach, however is there another way? The second solution concept entered into was designing a system (including the mechanicals) for the park brake from scratch, as this may emerge as an optimum 'vehicle system' result, whilst providing the most flexible solution. This approach has a number of inherent benefits when the actuator interfaces to an integrated caliper: electric Park Brake Hydraulic Caliper. Envisaged vehicle cost savings are: the automatic gearbox with the removal of the parking pawl, the removal of Bowden cables, ease of assembly, no need for an inclination sensor and reduced component count. Furthermore, deployment on electric vehicles and achieving the required Federal regulations of 8% hill hold with one failure is practicable.

<u>Manual Apply of a Park Brake</u>

The lever or pedal Human Machine Interface (HMI) presents its own individual challenges. Operators are used to being able to pull or push on a lever or pedal to set the park brake. The HMI provides the functions of:-

> Allowing the operator to activate and deactivate the park brake
> Providing the driver with a feedback to the amount of load that is being applied
> Indicating that the system needs adjustment
> Indicating the current state of the park brake (ON/OFF)

These functions are provided in two vehicle operational states: static apply (vehicle stationary) and dynamic apply (vehicle in motion). The governing factors were mapped into a relationship diagram (see Figure 4).

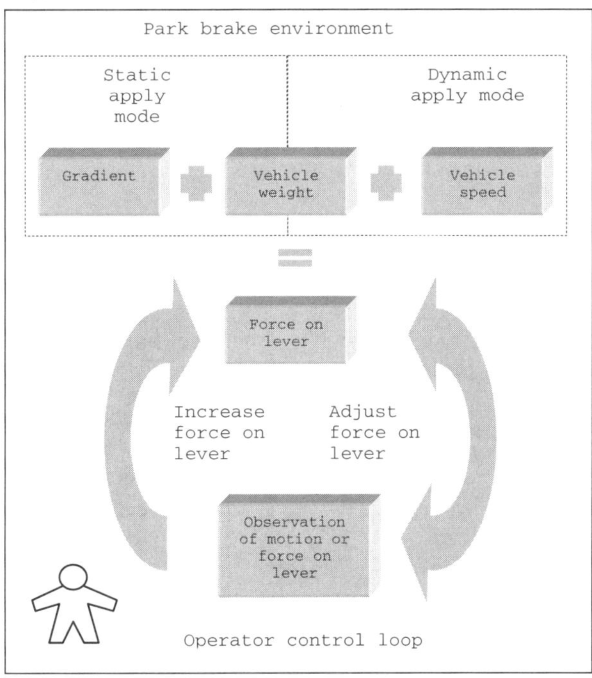

Figure 4. Manual apply HMI to vehicle relationship

With a low entry point system which provides a simple ON/OFF park brake function (no modulation), control of the park brake may not be easy. The operator interacts with the HMI and feedback is provided by an observation of motion, while the operator performs successive ON/OFF switching. If there is no lever to adjust the apply force, would the system be cognitively acceptable in both static and dynamic apply?

Applying the 'why' question, why does the Economic Commission for Europe (ECE) [8] dynamic apply test exist? Is it because of service brake failure? Service brakes are designed to have a 'no fail mission profile' (split system, front-rear or left-right) and single point failure [8] safety critical devices designed 'not to fail'. It is safe to assume that a dynamic apply is required from any vehicle speed (as conventional park brake systems provide this). Altering the perception of a dynamic stop i.e. remove it because service brakes provide this; can aid in reducing system cost by removing a modulated HMI and limiting the number of sensors required.

Variances in the environment, tolerances and vehicle weight affect dynamic stopping. Consequently a dynamic stop can require more or less force than a static park force (see Figure 3). The use of an ON/OFF HMI controlled by the operator may require force or positional control of the actuator and an input to indicate speed. This has potential safety impacts. For example: the loss of the speed sensor could allow an electric park brake to perform a full static apply in a dynamic situation; conversely a faulty speed sensor could also allow a dynamic apply when a static apply is required. A second assurance value is required to resolve this conflict.

<u>Electrical Interfaces</u>

The electrical interfaces (see Figure 5) required include:-

> HMI
> Red Status Lamp [8]
> Yellow Status Lamp [8]
> Ignition Signal
> Energy Supply

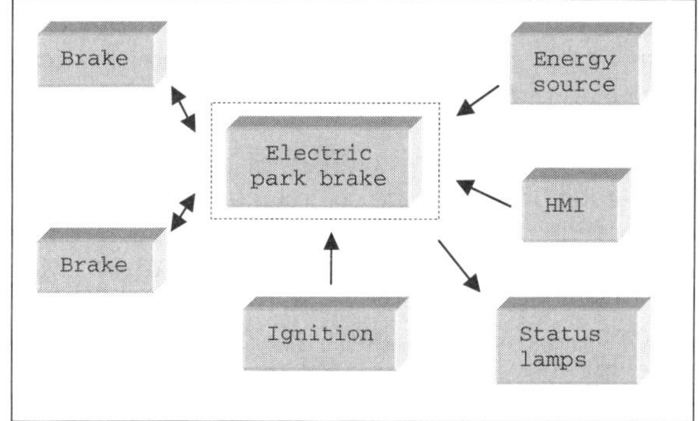

Figure 5. System components and system boundary

The addition of added-value functions such as automatic apply / release and hill start require transmission gear state and conceivably an additional inclination sensor. If the vehicle is a manual then perhaps clutch position. Other signals that may be used to aid the control include drive train torque, vehicle speed and engine RPM.

A functional hierarchy diagram was produced to describe the electrical interface needs (see Figure 6). As the number of interface wires increases so too does the system cost, error condition / matrix and the number of potential sneak paths [9][10]. A wire interface also reduces the possibility to expand the system by not taking cost effective advantage of existing vehicle sensors and functions.

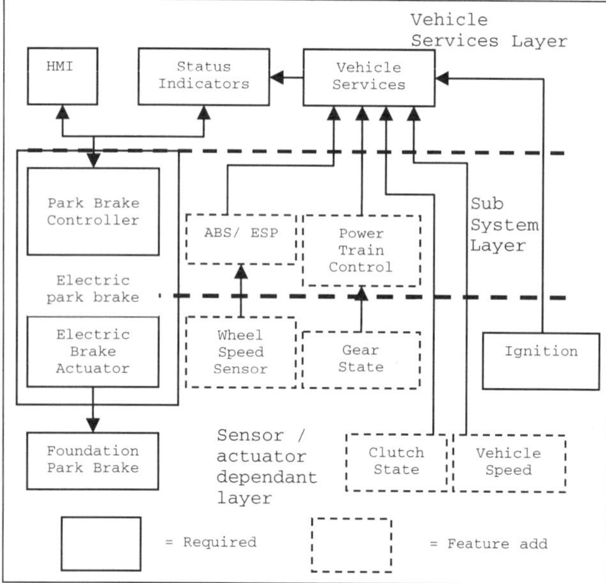

Figure 6. Functional analysis as a hierarchy

An open interface definition based on a communications bus would enable data to be passed from the vehicle services to the electric park brake irrespective of the supplier [11]. Autosar [12] is currently investigating such collaboration. Using a common format would enable suppliers to save development effort across different OEM platforms, and at the same time allow the suppliers to concentrate on the functionality aspects that differentiate one vehicle to another.

ECE requirements include the use of a yellow lamp to indicate to the driver that the control mechanism is faulty [8], and a red lamp to indicate failure. Application of the knowledge management 'why' ideal, highlights that current mechanical park brakes are not required to have these lamps, and that the operator is not made aware of a failure until they attempt to apply the park brake.

DESIGN OF TWO VARIANTS

FUNCTIONAL ANALYSIS

To provide the optimum solution choice for the market place it was seen fit to develop two types of electric park brake each specifically targeted towards the markets needs: a cable puller and integrated calliper.

Using the concept of Figure 1 and Federal regulations, a system component diagram was constructed and a system boundary was placed around the electric park brake (Park Brake Controller and Brake Actuator) (see Figure 5 and Figure 6). The ePB can be mounted on the vehicle axle or on the chassis, and pulls cables connected to the park brake whereas the ePBHC actuator is directly fitted to hydraulic calipers. The HMI is located in the cabin along with the status lamps. Ignition is required to prevent inadvert release of the park brake, and the energy to run the system is provided by the vehicle.

For the functional architecture a three layer hierarchy was developed (see Figure 6). At the lowest layer are the actuators and sensors. The middle layer provides the actuator controller (Park Brake Controller) fusion elements and other potential support functions. The top layer is the vehicle support system that provides communications and high level operator interfaces and integration patterns.

The layered architecture makes it possible to demonstrate the interactions and control strategies without focusing on the actual 'hows'. This illustrates that the Electric Brake Actuator and Park Brake Controller are common elements irrespective of which ever electric park brake product is being deployed, and that the interfaces can be minimised and made common with other By-wire interfaces (see Figure 6). The hierarchy also shows that many of the inputs that an electric park brake requires for additional functions are already present on the vehicle infrastructure.

Electric Park Brake (ePB) Cable Puller Mechanicals [13]

A kidney shaped rotary spool was devised in order to save packaging space for a given cable stroke and to assist vehicle plant assembly. The spool has a groove on its entire circumference and a groove dividing it into

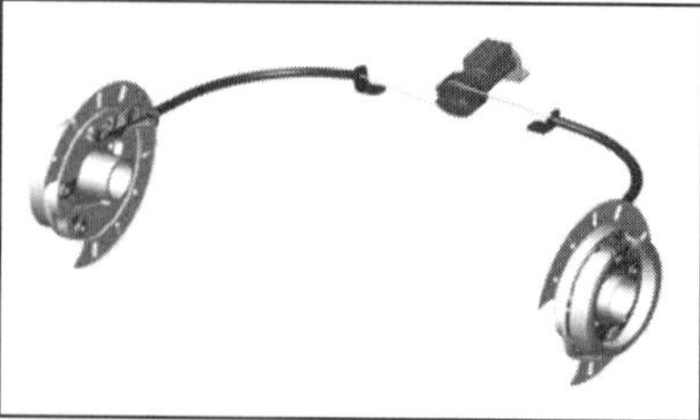

two even lobes. A continuous cable runs over these smoothly blended grooves in a letter 'S' configuration, with each end of the cable connected to a park brake, either directly or via

Figure 7. Example installation

Bowden cables (see Figure 7).

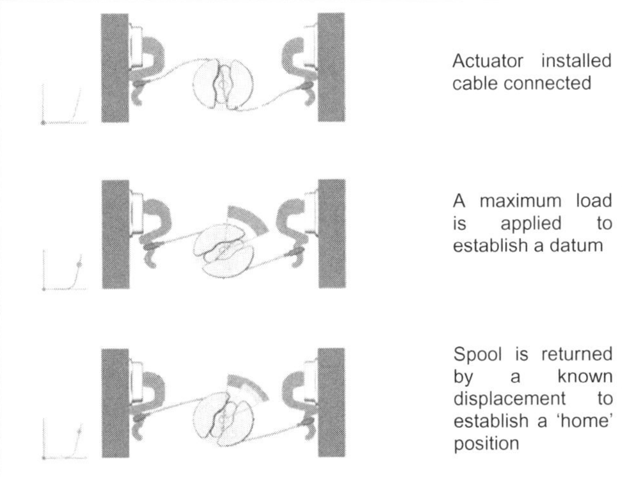

Figure 8. Automatic cable adjust

Figure 9. ePB

Rotating the spool in one direction tightens the cable, whilst rotating in the opposite direction releases the cable (see Figure 8). Equalisation occurs in two stages, slippage occurs in the cable during the initial slack take-up stage. Once the tension is on, the friction at the cable/groove will prevent further slippage. The second stage of equalisation is facilitated by a pair of parallel steel blades doubling up as support brackets, attached between the actuator and the vehicle (see Figure 9 and Figure 12).

The spool is powered by a Permanent Magnet (PM) motor through a worm and wheel gear at the high speed stage and a reduction helical gear train at the low speed end. These gears are designed to be non-back drivable for cable load retention. This arrangement has resulted in an actuator that is light weight, compact and exceptionally quiet (see Figure 10).

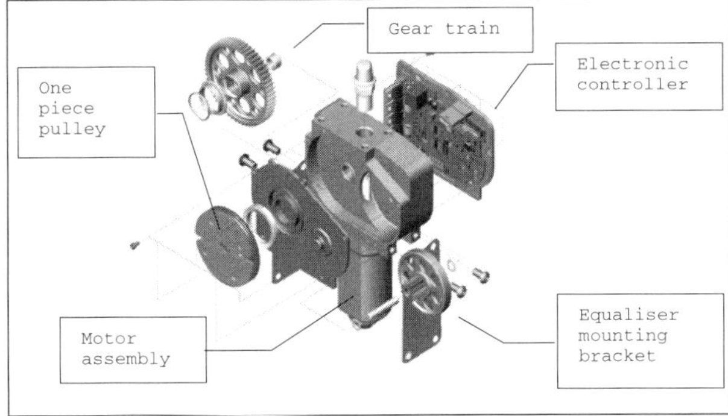

Figure 10. Assembly diagram

For some high cable load applications, a unique patented load intensifier [14] can be used. When this intensifier is used, the actuator can be rigidly mounted as the intensifier itself provides the necessary equalisation (see Figure 11).

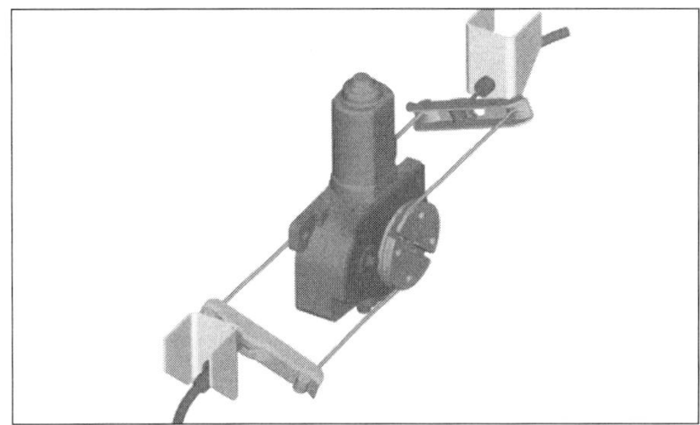

Figure 11. Load intensifier

Figure 12. Installation with Bowden cables

<u>Electric Park Brake Hydraulic Caliper (ePBHC) Mechanicals [15]</u>

To cater for the rear disc brake market an electric park brake integrated into a 'conventional' hydraulic caliper was developed. This caliper comprises a power screw operated by a PM motor and gear box, applying clamping to the service brake friction material.

Unlike many conventional integral park brake arrangements, the power screw is arranged separate from the hydraulic piston bore. The gearbox shares a similar layout to the ePB actuator – worm and wheel then a helical reduction unit. This gear arrangement like its cousin excels in low noise operation and load retention is provided by the power screw.

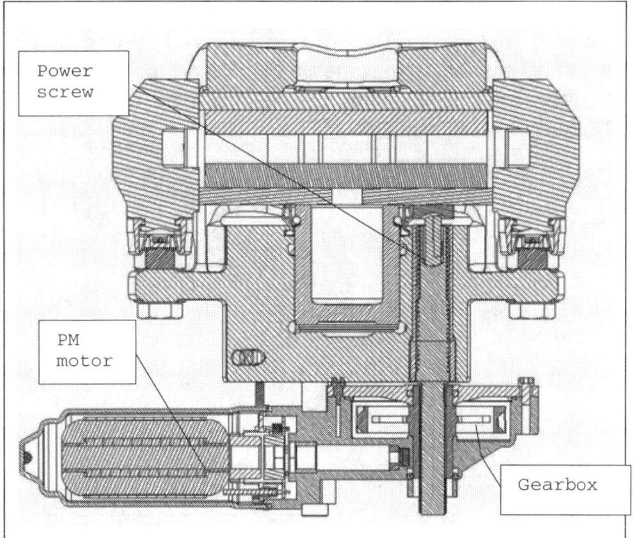

Figure 13. Cross section of an ePBHC caliper

The power screw method has several advantages including decoupled failure modes and decoupled potential interactions between the hydraulic system and electromechanical counterpart (see Figure 13). Additionally, the hydraulic seal does not experience piston displacement when parked, which eliminates deformation of the seal on the piston over extended parking periods which often results in poor pedal feel. The product can be installed on either a single or twin piston caliper (see Figures 14, 16 and Figure 15). In either configuration the actuator can be installed on the caliper body in a wide angular range. This provides maximum flexibility for creating clearance to suspension components and operator access.

 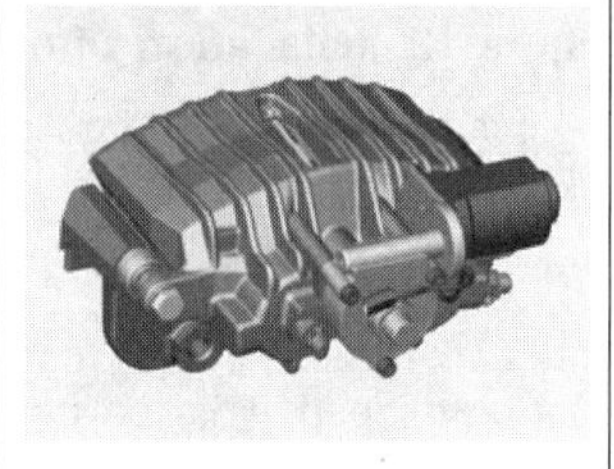

Figure 14. Single piston caliper arrangement　　**Figure 15. Twin piston caliper arrangement**

Compared to a hydraulic piston, the power screw can be placed closer to the caliper bridge, increasing the effective clamp radius. At the same time it reduces the bending moment in the bridge arm which partially offsets the increased stress in the bridge caused by eccentric loading. The bridge section is strengthened accordingly to accommodate the increased stress. Despite an increase in caliper stiffness in hydraulic mode, there is a reduction in stiffness in park brake mode that is advantageous against clamp relaxation when a hot disc and pads cool down.

Offset applied parking clamp force on the pads has been extensively analysed, tested and supported by previous data from similar cable operated brake product [16].

Figure 16. Installation on a caliper

ELECTRIC BRAKE ACTUATOR

Both the electric park brakes use an actuator connected to a controller (see Figure 6). The ePBHC system differs in that it has two actuators, but the functions of both solutions are the same:-

Move actuator to position X or
Apply force Y

This simple concept enabled the development of a single actuator controller interface including the electronics and control software. Items such as motor cogging torque [17], Pulse Width Modulated (PWM) induced Electro Magnetic Interference (EMI) [18] and motor position / force control were addressed as a single design. The issues of fail silent and fail current state were captured and modelled as a simple machine. This design philosophy provides the potential to migrate the design to an Application Specific Integrated Circuit (ASIC).

<u>Software and Control Features</u>

Developing a single actuator control enabled the development of a layered hierarchy and flexible control interface that sits between the actuator and the vehicle services layer (Park Brake Controller, see Figure 6 and Figure 17). This ensures that both products can cater for either wire or bus interface communications and fulfil End Of Line (EOL) and service diagnostics.

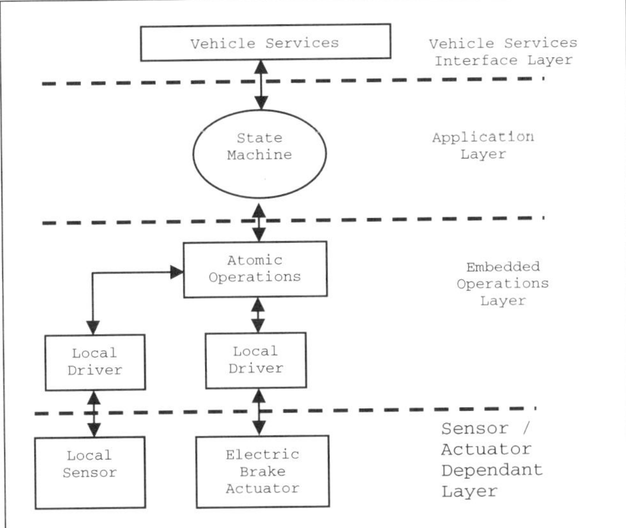

Figure 17. Software decomposition as a hierarchy

A layered hierarchy provides flexibility by abstracting away from the product (see Figure 17). Virtual drivers are used to allow communication between the core of the product and both the vehicle services and embedded operations layers. The core application layer is implemented by 'a programmable state machine framework' (Application layer), which provides the product functionality and interaction to other services. Using this layered hierarchy facilitates the simple addition and subtraction of features; two examples being automatic apply on key out and wake up and re-clamp of the park brake (as the brakes cool down the clamp load in some applications relaxes).

For each application only the physical interfaces to the vehicle needs to be described: the timing elements and dependencies, and the sequence of events through the state machine. This methodology assisted in the development of executable requirements and a model of the interactions between the vehicle and the electric park brake. This method adds flexibility, allowing the OEM to tailor systems to suit each vehicle platform's individual needs with little impact on the development cost and time.

In manual dynamic apply the number of steps (resolution) is modelled on current mechanical systems. Stability under dynamic braking is equivalent to today's Mechanical Park Brake systems; if access to items such as vehicle wheel speed is provided, stability can be significantly improved in a dynamic stop. A feature matrix is provided in Appendix A.

<u>Safety Concepts</u>

Developing a safety based system that meets Federal regulations can be a daunting task. The ECE require that a failure does not cause the park brake to operate [8]. This type of operation is known as *'fail silent'*. The status indicators are designed to show if the park brake is operational and must fail ON, known as *'fail assert'*.

The park brake actuator and its mechanicals can fail in either of two states, ON or OFF (as in park brake applied or park brake not applied), this is *'fail current state'* and the park brake must be releasable with tools provided on the vehicle [8]. Accompanying these fail states are the input fail states such as speed input, ignition and gear states. The enormity of this task is greatly simplified with the adaptation of the functional analysis as shown in Figure 6, placing boundaries around each subsystem component 'safety barriers' and assigning the desired fail state attribute (see Figure 18). This method can be likened to the Henry Ford principle of "Nothing is particularly hard if you divide it into small jobs" [19].

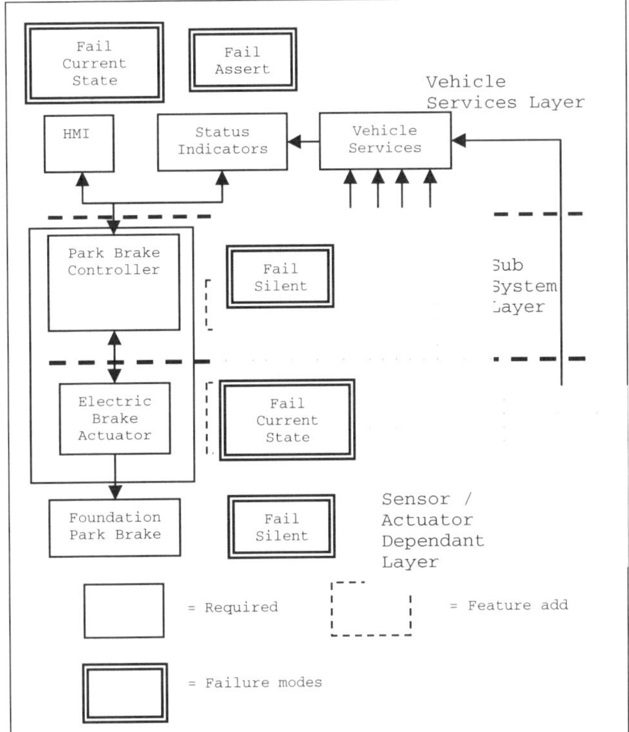

Figure 18. Safety boundary

The safety strategy consists of ensuring the integrity through ensuring the operations and integrity of the sub system components (Park Brake Controller, Electric Brake Actuator and ancillary items such as mechanical interfaces and connectors); the safety barriers aid in identifying the interfaces and the controls for each sub system or components safe state. Integrity is assured though the use of this and close coupling of hardware and software. Standard integrity tests are used (sensor plausibility and data checks). Vehicle based assurance values are only used if they are derived from independent sources. Continued service is provided by recovery / dependability values which where derived using Markov models [20].

ELECTRIC PARK BRAKE ADVANTAGES

Integrating new systems into vehicles can be an arduous task. Often for many subsystems the data or sensor values required are already present on the vehicle. For example: the ignition provides information to systems, the outside air temperature sensor provides information to Heating Ventilation and Air Conditioning (HVAC) and engine management, the gearbox provides vehicle speed to odometer and trip computer. The described electric park brake solutions are able to obtain information from many sources, enabling the supply of products from simple emulations of the current mechanical park brake to systems with numerous integrated features and functions.

Removing of the park brake lever will reduce associated safety recalls and increase passive safety. No lever; will allow vehicle architects greater flexibility in cabin design. Operator lever apply force will no longer limit park brake design. Incidents of vehicle run away from incorrect operator apply of the park brake can become a distant memory [2]. Deploying a single solution across an OEM platform range will reduce the parts inventory and development costs. OEM 'DNA' can be preserved whilst allowing the OEM to differentiate between vehicle platforms though tailoring of the programmable state machine frame-work (Application layer).

DISCUSSION

Changing ECE requirements for dynamic stop and status lamps has the potential to simplify the products and reduce the cost of a non-integrated solution. Status lamps improve the safety of an electric park brake, by indicating to the driver that a fault has occurred. The use of the status lamp when compared with a mechanical system is a cost increase. Removing or altering the function of the status lamp, or using another existing warning device, would aid in bringing electric park brakes in line with a conventional mechanical counterpart.

Changing the use expectation and Federal regulations of dynamic stop would aid in assisting the integration of electric park brake systems on current vehicles. With the correct sensors or changes in the park brake design it is possible to remove modulation of the park brake from the operator. Re-visiting knowledge management questions is essential to ensure that products are not over engineered or that past performance limitations are re-designed into new product.

Common architectures for communications and interfaces reduce time to market and invariably lower costs, allowing developers to concentrate on the features and functions of the product, whilst allowing OEMs to tailor individuality into the vehicle platform.

CONCLUSION

PBR has designed two types of ePARK^{TM} solutions (a cable puller (ePB) and an integrated caliper (ePBHC)). An aim of the project was to design common system with 'reusable' components, the heart of all these product variations is a trouble-free, robust, small and exceptionally quiet actuator which achieved these aims whilst demonstrating flexibility of mounting arrangements coupled with flexible software and vehicle interfaces.

This solution approach enables an OEM to integrate an electric Park Brake into a vehicle platform while providing the OEM with design flexibility, improved performance and increased features, starting from a base product that mirrors current mechanical park brakes, to products that include numerous automated functions and improved passive safety. Whilst enhancing flexibility that will enable OEMs to mix and match depending on the vehicle platform demands, further enabling implementation of alternative control strategies.

The virtual drivers and programmable state machine frame-work allow a myriad of interfaces and functions to be deployed, with the limitations only being restricted by imagination and cost. Other advantages demonstrated included simplification of corner or axle manufacture, reduction in component count, improved passive safety, flexibility in cabin layout and improved integration with existing vehicle infrastructure.

Applying radical changes in thinking using the 'why' question revealed many elements that have always been practiced that way, with no apparent use in a modern reliable environment. Integration and sharing of resources, high data rate buses and the relentless pursuit of cost savings is not a matter of 'if' but more a 'must'. The electric park brake solutions provide an excellent platform on which to gain knowledge on integration issues, including EOL test, field service and supplier support.

The design of what is a simple vehicle feature highlights the interaction between several sub systems, and that in order for a supplier to introduce a product, they must understand the impact on other proprietary systems.

ACKNOWLEDGEMENTS

ePARK^{TM} is a trade mark of PBR.

Authors appreciate the cooperation of the Research Centre for Advanced By-wire Technologies (RABiT) during the development of the electric Park Brakes.

REFERENCES

1. C. Peter Cho, "Permanent Magnet Motors/ Generators for Automotive Applications", ETechno-Group Carmel IN. www.arnoldmagnetics.com/mtc/ tech_presentations.htm
2. www.siemenvdo.com/com/pressarticle2003.asp?Atri cleID=17030e accessed September 2004.
3. Leiter Ralf, "Design and control of an Electric Park Brake", 2002-0102583, SAE 2002.
4. Harmer M, "Reengineering work Don't Automate, Obliterate", Harvard Business Review p68 104-112.

5. Davenport T, "Process innovation: Reengineering through Innovation Technology", Harvard Business School Press 1993.
6. www.nhtsa.org accessed September 2004.
7. Plunkett D, "Qualitative Methods Used in the Design of a Concept Automotive Seat Controller", Available from Royal Melbourne Institute of Technology library 2002.
8. ECE R13 Annex 4.
9. Price C.J. et al., "Automated Sneak Identification", Department of Computer Science, University of Wales. www.aber.ac.uk/compsci/Research/mbsg/publications/aieng96.pdf
10. European Corporation for Space Standardisation, "Sneak Analysis – Part 1 Method and Procedure", ISSN 28-396X, 1997.
11. Coelingh E. et al., "Open-Interface Definitions for Automotive Systems", 2002-01-0267, SAE 2002.
12. www.autosar.org Autosar web content V21.0 accessed September 2004.
13. Cable puller Australian provisional patent PR6360.
14. Cable load intensifier Australian provisional patent PR 2004902840.
15. ePBHC Australian provisional patent PR6888
16. Wang N, "Lock Actuator Parkbrake – A New Concept In Disc Brake Parking Mechanisms", 870257, SAE 1987.
17. Studer C, Keyhani A, "Study of Cogging Torque in Permanent Magnet Machines", IEEE IAS annual meeting New Orleans, Louisiana October 1997.
18. Athalye G. et al., "Design of Pulse Width Modulation Based Smart Switch Controller for Automotive Applications", Ohio State University, Columbus Ohio. www.ece.osu.edu/ems/OSUpdf/FilesPdf/dasc_geeta.pdf
19. Henry Ford US automobile industrialist (1863 - 1947)
20. Bondavalli A, et al., "An Adaptive Approach to Achieving Fault Tolerance in a Distributed Computing Environment", Journal On Systems and Architectures. Vol 47 No 9 pp 763-781 2002.

CONTACT

Dennis Plunkett
PBR. 264 East Boundary Road, East Bentleigh Victoria 3165 Australia.
Telephone: +613 9575 2000
Email: dennis_plunkett@PBR.com.au

Nui Wang
PBR. 264 East Boundary Road, East Bentleigh Victoria 3165 Australia.
Telephone: +613 9575 2000
Email: nui_wang@PBR.com.au

DEFINITIONS, ACRONYMS, ABBREVIATIONS

ASIC	Application Specific Integrated Circuit
ECE	Economic Commission for Europe
EMI	Electro Magnetic Interference
EOL	End Of Line
ePARK™	Electric Park Brake Systems
ePB	Electric Park Brake
ePBHC	electric Park Brake Hydraulic Caliper
GVW	Gross Vehicle Weight
HMI	Human Machine Interface
HVAC	Heating Ventilation and Air Conditioning
LVW	Lightly loaded Vehicle Weight
NHTSA	National Highway Transport Safety Authority
OEM	Original Equipment Manufacturer
PM	Permanent Magnet
PWM	Pulse Width Modulation
RPM	Revolutions Per Minute

APPENDIX A

Electric Park Brake Attributes	
Manufacture	Substantially reduced plant installation cost- "plug-and-play", auto pad clearance setting
	Eliminate variants of Bowden cable configuration for multiple version platforms – reduced inventory cost
	Automatic pad wear take up. Remove the need for on-line cable setting
Maintenance and service features	Brake wear adjustment warning for drum-in-hat brakes
	Pad replacement warning
	Automatic clamp/cable retraction for pad replacement – no special tools required
	Automatic initiation of system on first use
	Manual apply/release in case of system/power failure.
Occupant safety	Removal of lever or pedal minimises risk of leg and foot injuries in case of accident
	Auto apply on key out
	Auto apply on door open
	Automatically regulated emergency stop– controlled deceleration.

Feature list
- Automatic apply and release
- Constant deceleration
- Key out apply
- Hill start Assist
- Driver modulation of brake effort
- Lining wear indication
- Servicing mode (Cable service or pad replacement)
- Re-apply
- Fail silent
- Thermal protection
- Automatic burnishing
- Automatic cleaning

Digital By-Wire Replaces Mechanical Systems in Cars

Ronald W. Stence
Freescale Semiconductor

ABSTRACT

Currently, a large number of hydraulic control systems in vehicles are used to transform human inputs through mechanical means to control braking systems, engine torque or vehicle speed, steering and other mechanical systems throughout the vehicle. Replacing these systems with digital electro-mechanical control systems can reduce the system cost, increase reliability and flexibility, when compared to the traditional solutions. With the advent of digital by-wire communication systems and advanced semiconductor solutions to enable a new wave of subsystems such as: electric power assisted steering, hybrid technology, ABS, traction and stability control systems.

INTRODUCTION

The requirements and selection criteria can be very complex for any embedded design; however, the embedded processor selection for an automobile can be especially difficult. Many embedded processors within the modern vehicle are moving beyond the 16-bit microcontroller to very complex 32-bit microcontrollers. The 32-bit microcontrollers of the next generation will require an increasing amount of S/W written in high level languages. The H/W design is relying on integration and very complex S/W to solve the new problems in engine management. The communications between each of these systems will increase the demand for data to be shared between ECUs around the car.

When a critical system is controlled through a communications channel, it is referred to as "by-wire." By-wire denotes a control system that replaces traditional hydraulic or mechanical linkages with electronic connections between control units that drive electromechanical actuators. By-wire technology was originally used in the aerospace field and has migrating into ground vehicles. The early application of by-wire technology in ground transportation can be found in many heavy truck applications. The throttle-plate is "wide open" and the diesel engine is controlled by the ECU receiving data of the driver's foot position on the accelerator peddle to adjust the amount of fuel going into the cylinders. Automotive by-wire includes three broad application areas: throttle by-wire, brake by-wire, and steer by-wire.

The throttle by-wire system replaces mechanical cables connecting the gas pedal and throttle body plate with an electrical connection. The throttle is electronically controlled to match the driver's request for more or less torque, balanced with S/W routines to provide a more efficient operation than the mechanical cable can provide. In addition, electronic throttle units can enhance safety systems such as a traction control system (TCS) or electronic stability program (ESP).

A brake-by wire system uses electrical connections to connect the four braking "corners" to the pedal and to each other. This system provides better control of pedal stiffness, traction control, vehicle stability, and brake force distribution.

A steer by-wire system replaces the steering column with control units linked by a fault-tolerant network. The driver's steering controller is connected through the network to motors that are connected to the steering rack or individual corners of the vehicle. A steer by-wire system can enhance safety, increase fuel economy, provide varying levels of "road feel", and allow car designers more flexibility under the hood of the vehicle.

Multiple systems will be implemented to change the dynamics of by-wire systems to improve communications, reliability and debugging while adding a growing number of features.

ELECTRONIC THROTTLE CONTROL

The linkage between the driver's foot-peddle and the throttle body in the standard gasoline engine can be replaced with a small motor and some wiring. There are several advantages to this which can result in lower manufacturing cost and improved fuel efficiency while maintaining exceptional reliability.

Electronic throttle control has several advantages over the conventional throttle cable:

- The vehicle's on board electronic systems are able to control all of the engine's operation with the exception of the amount of incoming air.

- The use of throttle actuation ensures that the engine receives the proper amount of throttle opening for a given powertrain control situation.
- The optimization of the air supply reduces exhaust emissions while marinating derivability.
- Electronically linking the throttle, adaptive cruise control, traction control, idle speed control and vehicle stability control systems can improve vehicle drivability and safety.
- Simplifies the mechanical placement of cables and linkages for variations in engine placement and left-hand and right-hand drive options.
- Reducing the mechanical components with electronics, reduces the number of moving parts (and associated wear) and therefore requires minimum adjustment and maintenance.
- Greater accuracy of data improves the drivability of the vehicle, which in turn provides better response and economy.

To insure integrity of the system, electronic throttle control systems use a primary embedded processor and a watchdog processor to ensure proper operation. The primary processor will be the engine control processor in many systems with an 8-bit watchdog processor that frequently checks the throttle petal position against the current throttle plate position. The alternative implementation will deploy a more powerful processor to make all of the calculations and motor positioning adjustments with another processor acting as the watchdog processor.

In many systems, the foot peddle position sensor and the throttle body position sensors are redundant to reduce possible catastrophic failures. The communications protocol between the control system and the throttle body has to be fault tolerant.

ELECTROMECHANICAL BRAKING SYSTEMS

Electromechanical braking systems (EMB), or brake by-wire, replace conventional hydraulic braking systems with electrical components that are free of hydraulics hoses from a central point in the vehicle, typically under the hood. This occurs by replacing conventional actuators with electric motor driven units. This move to electronic control system eliminates many of the manufacturing, maintenance, and environmental concerns associated with hydraulic systems.

The electrical system must be fault tolerant once the hydraulic system is removed to maintain proper safety levels within the vehicle. The implementation of EMB requires features such as a dependable power supply, fault-tolerant communication protocols (i.e., TTCAN and FlexRay™), and some level of hardware redundancy.

Electrohydraulic braking (EHB), EMB is designed to improve connectivity with other vehicle electronic systems to enable less complex integration of higher-level functions. Examples of higher-level functionality that can be integrated into an EHB or EMB system are traction control and vehicle stability control. This integration may vary from embedding the function within the EMB system, as with ABS, to interfacing multiple systems using communication links.

EHB and EMB systems can eliminate the large vacuum booster found in conventional linkage based braking systems. Replacing the full mechanical system with an electronics module allows for flexibility in the tight spaces of the engine bay and greatly reduces the problems surrounding right- and left-hand drive vehicle variants. An increase in flexibility for the placement of components is also provided by EMB systems, compared to those of EHB, with the total elimination of the hydraulic system.

Diagram 1 shows one configuration for integrating stabilization, EPAS, ABS and brake-by-wire for a vehicle. These systems must support fail-safe modes and deterministic communications between the nodes within the system and ECU modules across the vehicle. The communication between the modules must be fault tolerant, high speed and deterministic with clearly defined specification. The processing performance required for each of these functions can vary between an 8-bit MCU to a high performance 32-bit RISC processor. The foot peddle position ECU in many applications is a 32-bit MCU with another 8-bit processor that checks the results of the main processor.

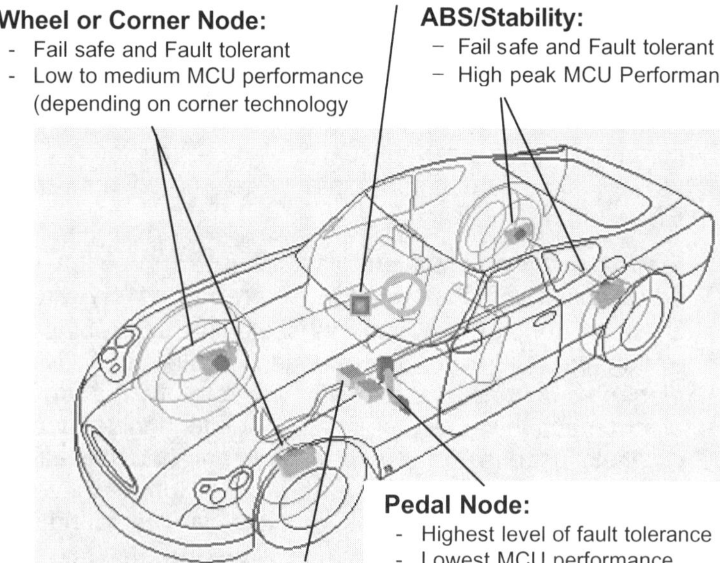

Diagram 1: X-By-Wire Control System

Other ECU's within this Brake-by-Wire system MCUs will require different levels of MCU performance.

The use of electric brake actuators can take advantage of a higher powered 42 volt electrical system as it becomes common in the next generation of the vehicle. The requirement of high temperature with unprecedented quality and reliability will be placed on the electronics of these new systems.

In addition to supporting existing communications standards such as CAN and K-line, EMB systems require the implementation of deterministic, time-triggered communications, such as those available with FlexRay, to assist in providing the required system fault tolerance. The EMB nodes may not need to be individually fault tolerant, but they help to provide fail-safe operation and rely on a high level of fault detection by the electronic components.

ELECTRONIC STABILITY PROGRAM

The ability to link and extend the anti-lock braking system (ABS) and traction control system (TCS) to an electronic stability program (ESP) will greatly improve the safety and drivability of the vehicle. The ESP is designed to detect a difference between the driver's control inputs and the actual response of the vehicle. When differences are detected, the system intervenes by providing braking forces to the appropriate wheels to correct the path of the vehicle. This automatic reaction is engineered for improved vehicle stability by reducing over-steering and under-steering skidding. The TCS will dramatically improve the safety during severe cornering and on slick or low-friction road surfaces.

To implement ESP functionality in an ABS system, additional sensors must be included along with the appropriate H/W and S/W. In systems were the ABS is a 16-bit MCU or a low-end 32-bit MCU, additional compute power will be required to handle the additional complexity. A steering wheel angle sensor is used to detect driver input with a yaw rate sensor and a low-G sensor that measure the vehicle response to changing direction and other driving conditions. Some ESP systems include a connection to the powertrain controller for the vehicle to enable reductions in engine torque when required.

ESP systems often emerge from existing ABS systems through additional enhancements. The inclusion of ESP and analog electrohydraulic valves requires closed-loop, current-controlled pulse width modulation (PWM) outputs from the electronic control unit (ECU). The yaw and low-G sensors are typically located in a separate cluster near the vehicle's center of gravity. Additional communication interfaces are required to enable data transfer between the powertrain ECU, ABS module and sensor cluster.

Safety is a major concern for ESP and ABS systems. The need to communicate reliably with other ECU modules is critical to improving the total safety of the vehicle. The ability to communicate reliably is even more important when the mission is to keep the driver within the limits of the vehicle under changing road conditions. ESP systems must operate as fail-safe device during the event of a fault. The system relies on the electronic components to provide a high level of operational fault coverage, to provide early fault detection and to correctly initiate the fail-safe condition.

ACTIVE AND SEMI-ACTIVE SUSPENSION - Semi-active suspensions are often placed in higher end production sports cars and feature damping control on all four corners of the vehicle. The control objective of semi-active suspension is to counteract for heave, roll and pitch. This control is achieved by pairing a microprocessor with the four shock absorbers that have a continuously variable (and controllable) damping coefficient. On occasion, these dampers are paired with pneumatic springs to provide ride height and leveling control.

Potential driver benefits of the semi-active suspension system include:

- An adjustable ride, optimized for comfort or handling performance

  o Suspension will automatically adjust according to road conditions

  o User has the option to select the firmness of the suspension

- There is no change in size from traditional suspension systems.

A semi-active suspension system will drive new mechanical hardware and sensor input requirements to properly control the system. The performance of the system relies in part on the response time of the shock absorbers and their damping coefficient range.

Mechanical Hardware Requirements - Variable damping coefficient shock absorbers, with a response time of less than 10 milliseconds, are required for the most basic semi-active suspension systems. These dampers can be hydraulic or pneumatic and are usually controlled with electromagnetic switches or pumps.

Sensor Input Requirements - The possibilities of a semi-active suspension system are virtually limitless in the number of input variables which they are able to counteract. This allows the system designer to determine the amount of control a system offers under various conditions. A semi-active suspension system can monitor for:

- Vehicle speed

- Steering angle velocity

- Steering angle position

- Brake condition

- Vertical acceleration

- Lateral acceleration

- Vehicle level position

ELECTRIC POWER ASSISTED STEERING

Electric power assisted steering (EPAS) removes the mechanical linkage between the driver's steering wheel and the hydraulics that control the front wheels. The EPAS system uses an electric motor to control to the direction driver is choosing for the vehicle. An EPAS system can use a variable assist, which allows for more assistance as the speed of a vehicle decreases and less assistance from the system during high-speed situations. The control function is a precise balance of power and control that is being made available through new innovations in communications between multiple ECUs.

EPAS systems do not require mechanical or hydraulic pumps from the engine to operate. Thus, a vehicle equipped with an EPAS system may achieve two to five percent greater fuel economy than the same vehicle with conventional hydraulic power steering. The improvement comes from not having to maintain hydraulic pressure, even if the vehicle is at high way speeds, where power assisted steering is at a minimum. As an added benefit, more of the engine's power is transmitted to its intended location—the wheels. Often the EPAS system is powered by a 42-volt DC source to provide the high power necessary to move the front wheels at a dead stop.

The EPAS system brings an additional benefit from removing the mechanical linkage from the front of the vehicle straight to the within inches of the driver. In a head-on collision, the linkage is designed to break-away from entering the passenger compartment with fairly good success. However, the EPAS system will totally eliminate the possibility of the steering column being driven into the passenger space during severe head-on crashes.

Power steering applications require that the assist device mimic the driver's inputs at the steering wheel precisely with the wheels on the ground. The inputs are typically precise course corrections followed by periods of inactivity, which is an interesting motor design challenge. The motors must operate for extended periods in a harsh under-hood environment that can sometimes reach temperatures of up to 150 degrees Celsius with little or no maintenance. The EPAS system can run self diagnostics and report problems earlier than traditional hydraulic systems.

A multiple-poled "brushless" DC motor can be designed to handle the high power at low speeds and precise movement at high speed, followed by inactivity. Their brushless design moves the electrical windings to the stator which eliminates the need for motor brushes and can help to improve overall efficiency. This combination enables the motors to be used in the harshest of conditions with a very long service life.

NEW SOLUTIONS

The new X-by-wire systems are driving improvements in systems integration, communications and improved debug tools. The system integration will bring new levels of miniaturization, improved reliability, lower power consumption and lower system cost.

System integration can be seen with the integration of timers, SRAM and flash technology with an embedded processor. Technology nodes beyond 180nm will enable higher integration and smarter peripherals. FlexRay is an example of the coming integration of a very sophisticated, yet reliable communications protocol.

The ability to debug system elements in real time is critical for many X-by-wire applications. The IEEE-ISTO Nexus 5001 Standard addresses the requirement for a real time debug and calibration interface.

FLEXYRAY COMMUNCICATION

CAN has served as the mainstay of ECU module communications for more than a decade. However, CAN is limited is several major areas that would not provide solutions for mission critical applications. A new high speed automotive networking protocol has been architected specifically for mission critical communications, called FlexRay. The major features of FlexRay are:

- **Synchronous and asynchronous data transmission**

- **High net data rate of 5 Mbit/sec; gross data rate approximately 10Mbit/sec**

- **Deterministic data transmission, guaranteed message latency and message jitter**

- **Support of redundant transmission channels**

- **Fault tolerant and time triggered services implemented in hardware**

- **Fast error detection and signalling**

- **Support of a fault tolerant synchronised global time base**

- **Error containment on the physical layer through an independent "Bus Guardian"**

- **Arbitration free transmission**

- **Support of optical and electrical physical layer**

- **Support for bus, star and multiple star topologies**

FlexRay is a network communication system targeted specifically at the next generation "by-wire" automotive applications. The by-wire applications demand high-speed bus systems that are deterministic, fault-tolerant and capable of supporting distributed control systems. Major automotive OEM, Tier 1, independent S/W design firms and semiconductor suppliers are working together to develop and establish FlexRay as the standard for next generation applications.

The FlexRay communication system encompasses more than just being a communications protocol. FlexRay includes specifically designed high-speed transceiver and the definition of hardware and software interfaces between various components of a FlexRay 'node'. The FlexRay defines the protocol format and function of the communication process within an automotive networked system.

FlexRay is designed to meet key automotive requirements like dependability, availability, flexibility, and a high data rate to complement the major in-vehicle networking standards - CAN, LIN and MOST.

New applications requiring improved communication and higher data transfer rates between the vehicle's ECUs will drive by-wire applications to the forefront. FlexRay is initially targeted for a data rate of approximately 10Mbit/sec; however, the design of the protocol allows much higher data rates in the future.

FlexRay provides for a scalable communication system with both synchronous and asynchronous data transmission, or a mixture of both. The synchronous data transmission enables time triggered communication to meet the requirement of dependable systems. The asynchronous transmission technique has its fundamentals based on work from the Byteflight™ protocol which allows each node to use the full bandwidth for event driven communications.

When FlexRay is operated in a synchronous data transmission mode, the messages are deterministic with a message latency and message jitter guaranteed as built-in feature of the standard. The time base is controlled by a redundant, fault-tolerant synchronized clock. The FlexRay clock is used as a global time base to keep the schedule for all the network nodes within a tight, predefined precision window.

The physical layer acts as an independent bus guardian by insuring that no data collisions can occur even in the event of a fault in the communication controller. It is critical that by-wire applications can have the information in a distributed system linked by a communications protocol which guarantees that messages will transfer in a known timeframe.

FlexRay has been designed to support both optical and electrical physical layers, to provide for flexibility in the wiring system choices now and into the future. A FlexRay system supports networking from a single channel bus to dual channel multi-star topology with full redundancy.

For additional information on FlexRay visit the web site:

www.flexray.com

NEXUS DEBUG STANDARD

Real time debug is critical when debugging and calibrating electromechanical systems which often cannot tolerate the embedded processor performance being altered or halted for interrogation by a development tool. Today's systems engineer can take advantage of improvements of development tools that were not possible a few years ago.

The IEEE-ISTO Nexus 5001 Consortium, or Nexus Forum was created to solve the problem of real time data and instruction trace across multiple processor core types. The Nexus Forum released the original specification in 1999 and an update in 2003.

The Nexus 5001 specification includes standard features for setting breakpoints and watchpoints on data and instructions using non-intrusive debug techniques. The specification will deploy several unique features for tracking down the toughest S/W and H/W bugs. Some of these new features include ownership trace messaging, data trace, memory substitution, port replacement, program trace, overrun and error messaging. While many of these features have been deployed on microprocessors over the years, none have implemented all of these and a real time debug interface.

The past decade of calibration and debugging methodology has deployed a "must see every cycle" mindset for debug and calibration powertrain systems. The Nexus 5001 in methodology makes four assumptions about the debugging scenario that frees it from the "must see every cycle" mindset.

1. The source code and object code are available to the debugging tools. This allows the host based tools to track or make calculations as to the program flow without any direct address or data bus visibility.
2. The only change of flow instructions required from the target system to the host based debugging tool. When the host calibration/debugging tool has access to the object code, proper synchronization can be maintained between the embedded processor and the host tool with just change of flow instruction addresses being transmitted over the debugging interface. The Nexus 5001 specification will send out a synchronization message if a change of flow has not forced a synchronization address within 255 instructions. This would be a rare event since most C compiled programs

generate a change of flow once every 5 to 10 instructions for virtually all embedded processors.

3. Only a limited number of the data locations have to be displayed in real time, while most can be examined during a halted condition or updated due to a special event. The ability of the Nexus 5001 interface to trace on data values will be new for many engineers. Often this is accomplished with a powerful logic analyzer tracing the address bus and triggering on the data bus writes to a specific memory location. This is a very cumbersome task that is made nearly impossible with the advent of large data caches and system SRAM on-chip.

4. Fourth, if or when an error occurs, the user must be informed by the debugging environment. The Nexus 5001 specification provides for a variable sized FIFO buffer in the transmitter section. If the FIFO overflows, the interface will transmit an error condition. The user has the option to specify that when an overflow error condition occurs, to enforce a stale of the embedded processor or to continue with a new synchronization message.

Chip designs that might not need the full capabilities of the Nexus Specification, can chose from a variety of features grouped together in four Classes. The Class 3 provides for real-time instruction and data trace over a single port. The Nexus 5001 port can be configured based on the amount of information that is being captured by the development or calibration tool. The Nexus 5001 specification has been deployed on multiple CPU architectures by several IC venders for a variety of applications from cell phones to automotive to hard disk drive controllers to video processors.

CONCLUSION

A luxury vehicle today can have 3 miles of wiring. A proper networked system can reduce and simplify much of the additional complexity, weight and cost from the point-to-point communications found in existing cars. FlexRay is being adopted rapidly around the world as the networking solution of choice for the next generation of applications.

Graph 1 shows the growth rate of the silicon content in the vehicle over a decade. The dramatic growth of 16 : 1 is a result of the increased need for higher feature and functional content in the car. Also, the advances that electronics can bring in a cleaner vehicle that is safer and much more efficient.

The functionality that a high-speed, fault tolerant networked protocol can provide car manufactures will be a major enabling technology for new and safer vehicles in the next couple of years.

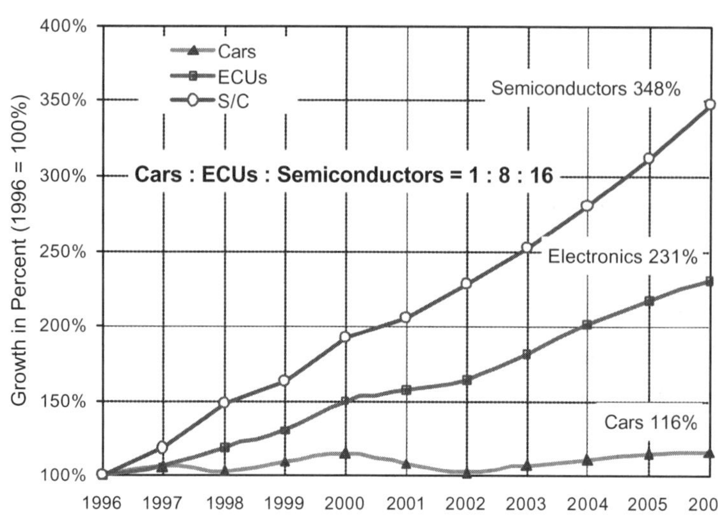

Source: ZVEI -Trendanalysis

Graph 1: Growth of vehicles : electronics : semiconductors in cars

The road to a vehicle that is absolutely safe is still years off. There will be several near term advancements that can bring significant improvements with TCS, EPAS and adaptive cruise control culminating with an automated vehicle and highway system. All of these near term and future systems will require fault tolerant, deterministic and high speed networking communications protocol.

REFERENCES

1. John G. Kassakian (Massachusetts Institute of Technology), "Automotive Electrical Systems - The Power Electronics Market of the Future"
2. Strategy Analytics, "Brake By Wire - A Mass Market Solution?" June 2004 - Viewpoint
3. Strategy Analytics, "Automotive Multiplexing Trends December 2002 - Viewpoint
4. Gabriel Leen (PEI Technologies), Donal Heffernan (University of Limerick), "In - Vehicle Networks: Expanding Automotive Electronic Systems", January 2002 IEEE 0018-9162/02
5. F. Bogenberger, A. Krüger (Motorola), "FlexRay– The Communication System for Advanced Automotive System for Advanced Automotive Control Systems Control Systems" SAE 2001
6. Daniel McKay, Gary Nichols and Bart Schreurs (Delphi), "Delphi Electronic Throttle Control Systems for Model Year 2000; Driver Features, System Security, and OEM Benefits. ETC for the Mass Market" SAE 2000-01-0556
7. Ronald Stence (Motorola), "Real Time Calibration and Debug Techniques of Embedded Processors with the Nexus 5001 Interface" SAE Paper 03AE-144

CONTACT

The author has worked in the electronics industry for 21 years. Most of this time he spent working in the semiconductor industry with Motorola and Freescale Semiconductor. Prior to working for Freescale, Mr. Stence worked for Advanced Micro Devices (AMD) and Tellabs.

Mr. Stence has made significant contributions toward establishing the IEEE-ISTO Nexus 5001 Debug Interface as an industry wide standard and currently serves on the Steering Committee and as the Chairman of the Nexus 5001 Consortium. He has received 4 Motorola Corporation awards for his contribution to the Nexus 5001 standard.

Mr. Stence currently holds 7 U.S. Patents in the area of embedded processors and graduated from Texas A&M University in 1986 with a BS in Computer Science.

Email: ron.stence@freescale.com

DEFINITIONS, ACRONYMS, ABBREVIATIONS

ABS – ANTI-LOCK BRAKE SYSTEM

An electrical system that detects when the wheel have locked. Improves drivability and helps control a vehicle during extreme braking by reducing the potential for skidding.

ACC – ADAPTIVE CRUISE CONTROL

The ACC system scans the area ahead of a car for objects - and applies the brakes automatically to slow or stop if a collision is likely to occur.

CAN – CONTROLLER AREA NETWORK

A communications protocol deployed in most vehicles. Provides basic communications between different ECUs. Maximum transfer rate of 1Mbit/sec.

ECU – ELECTRONICS CONTROL UNIT

Electronics module that controls electromechanical systems.

EMB – ELECTROMECHANICAL BRAKING SYSTEMS

Also referred to as Brake-by-Wire system. Replaces conventional hydraulic and mechanical components with an eclectically controller braking system.

ESP – ELECTRONIC STABLITY PROGRAM

Often tightly coupled with an ABS system or added on as an extra feature to the ABS system. The system uses G-sensors to detect difference between the driver's intended course and the vehicles response with the intent of minimizing skidding.

EPAS – ELECTRONIC POWER ASSISTED STEERING

A system that uses an electric motor to move the front wheels for steering.

ETC – ELECTRONIC THROTTLE CONTROL

A system that uses an electric motor to open/close the throttle plate based on input from an acceleration peddle.

FLEXRAY

An emerging communications standard and consortium. The major advantages are the fault tolerate, time triggered communications and higher data rate over other protocols.

www.flexray.com

MECHATRONICS

The introduction of electronics into mechanical components. An example of Mechatronics is the integration of power control semiconductors into a connector at the point of use.

NEXUS 5001 STANDARD

A real time debug, rapid prototyping and calibration interface standard based on the IEEE-ISTO Nexus 5001 Specification.

www.ieee-isto.org

JTAG – JOINT TEST ACCESSS PORT

Originally defined as a boundary scan port under the IEEE 1149.1 for board assembly level testing. Extended use as a slow speed run-control debug port by the Nexus 5001 Specification.

PWM – PULSE WIDTH MODULATION

A wave form that is modulated by H/W or S/W. The base frequency and the duty cycle can be varied by S/W to produce a wave form.

RISC – REDUCED INSTRUCTION SET COMPUTER

A processor Architecture that has many attributes that defines it as a different type of processor core (i.e. DSP, CISC). The major attributes of a RISC processor are: 1) fixed length instructions, 2) load/store architecture, 3) single cycle execution of instructions 4) hard coded instructions, 5) high-level language support.

TCS – TRACTION CONTROL SYSTEM

A technique where the vehicle suspension is modified to improve the drivability of a vehicle. The suspension can be stiffened or relaxed to modify the ride or compensate for changing road conditions, changes in the weight of the vehicle and compensate for skidding.

SOC – SYSTEM-ON- A-CHIP

A highly integrated embedded processor that is focused at a single market. Often contains the part or all of the system memory.

Control of an Electromechanical Brake for Automotive Brake-By-Wire Systems with an Adapted Motion Control Architecture

Chris Line, Chris Manzie and Malcolm Good
The University of Melbourne

ABSTRACT

A disk brake clamp force controller for electromechanical brakes (EMB) in automotive brake-by-wire systems may be obtained from a standard motion control architecture with cascaded position, speed and current control loops by replacing the outer position control loop with a force control loop. When implemented with proportional, integral and differential (PID) controllers this architecture generally performs well for standard motion control problems, but the EMB control problem is differentiated by a large operating range in which non-linear load disturbances such as friction become significant at high clamp forces of up to 30kN. This paper investigates the feasibility of a cascaded PI control architecture for an EMB with the intention of establishing a baseline standard against which the performance of future control schemes may be compared. Simulation results are presented based on an accepted EMB model.

INTRODUCTION

Borrowed from the idea of fly-by-wire and applied within the automotive industry, the concept of drive-by-wire is to replace all mechanical linkages between the driver and vehicle with electronic systems. The benefits of drive-by-wire include component reduction, weight reduction, potential for improved vehicle performance, increased cabin space, removal of the steering column, ergonomic and crash compatible mounting of controls, complete access to vehicle dynamics control, 'plug, play and bolt' modularity, software upgrading, and potential for better fuel economy. To realize the full benefits of drive-by-wire requires the electronic integration of the three primary vehicle control systems, throttle, steering and braking. Of these it is the latter, brake-by-wire, that is the subject of this paper.

Currently there are two implementations of brake-by-wire with disc brake calipers and these are differentiated by their mode of actuation. While one system uses electromechanical brakes the other retains current hydraulic brakes in an adapted electrohydraulic system. Because they only require some modification of an existing system, electrohydraulic brakes are considered the first approach to brake-by-wire [1]. With proportional valves, electrohydraulic brake systems have greater resolution in brake pressure levels than standard hydraulic systems. The other characteristic that determines brake performance is the rate at which an actuator can generate and dissipate brake pressure at the disc brake caliper. Typically hydraulic systems perform well for high load applications, but an alternative is offered by electromechanical brakes that have high resolution in caliper clamp force levels as well as the potential for high brake force apply and release rates.

A number of patents exist for different electromechanical brake designs. From a survey of EMB patents held by the United States Patent and Trademark Office two categories were identified, these being electromechanical drum brakes and electromechanical disk brakes. While drum brakes can exhibit the phenomenon of self-energisation, this introduces a non-linear brake characteristic. Further, drum brakes have less heat resistance to thermal loads [2]. It is partly for these reasons that electromechanical disk brakes are being pursued more vigorously.

From patent designs of the EMB disk brake calipers [3-15] four features were identified as being common. Most EMBs have an electric motor drive, some reduction gearing that converts rotational to translational motion, a floating disc brake caliper with brake pads and some physical and communications interface with the vehicle. Running on a 42 volt direct current (DC) power supply [16] and using a safety critical time triggered communication protocol, the EMB receives brake instructions from the central vehicle dynamics controller via an in-vehicle network. These instructions command either brake torque, brake clamp force, or a particular mode of operation such as standby, off, or anti-lock braking. For the purposes of this paper it is assumed that a desired brake clamp force has been requested by the vehicle dynamics controller and the focus is on delivery of this command from the local EMB controller. It is here that the control problem for electromechanical brake actuators is introduced. A controller is required to drive the EMB and respond as best as possible to brake commands from the vehicle dynamics controller.

Control of a servo motor is a standard problem for linear motion systems. A common solution is to implement a motion control architecture with cascaded position, speed and current control loops. This architecture may be adapted to control the EMB brake clamp force by replacing the outer position control loop with a force control loop as shown in Figure 1.

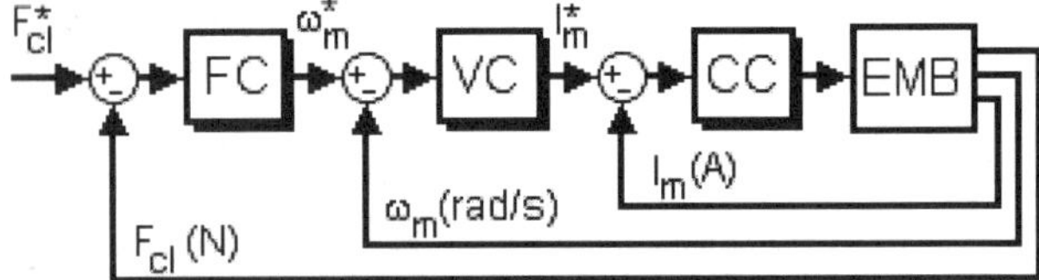

Figure 1: EMB brake clamp force control architecture with cascaded force (FC), motor velocity (VC) and motor current (CC) control loops

Although brake torque control is preferred for management of vehicle dynamics, brake clamp force is instead chosen as the control variable because robust measurements are available from a force sensor embedded in the EMB. For control of vehicle dynamics it is then possible to use the brake clamp force to estimate the brake torque.

The cascaded control architecture shown in Figure 1 requires feedback measurements of motor current, motor speed and caliper clamp force. Typically, these are available from Hall current sensors on the motor, a motor position resolver or encoder, and a load cell or strain gauge configuration on the EMB to measure the caliper clamp force. While Schwarz *et al.* proposed a clamp force estimation scheme that used the remaining current and position sensors [17], the issue of sensor reduction is not investigated in this paper.

Although cascaded control architectures generally perform well for standard motion control problems, the EMB control problem is differentiated by a large operating range of between 0 and 30 kN for a full brake apply. At higher clamp forces non-linearity is introduced by a load dependent component of the Coulomb friction. Further non-linearities exist due to the non-linear caliper stiffness and saturations of the power supply voltage, safe motor current limits and the maximum motor speed. Given these considerations it is not immediately obvious how well a controller with cascaded force, velocity and current loops will perform. In an attempt to address this potential shortcoming, an investigation was undertaken to assess the feasibility of a cascaded PI control architecture for an EMB.

Having an architecture with brake clamp force control rather than position control is acceptable during brake apply. During standby however, position control is required to manage the air gap between the brake pads and brake rotor. Both modes of operation are achieved by switching between outer position and force control loops across the contact point between the brake pads and disc brake rotor.

This paper is structured beginning with nomenclature and a description of the EMB simulation model that was developed and validated for the purposes of simulation and controller design. This is followed with a definition of the control problem and statement of typical performance requirements for the EMB. The design of the controller is then detailed and results presented to indicate its performance. This leads to discussion of the results and concluding remarks.

NOMENCLATURE

The following symbols are used throughout this paper.

α = load dependency of Coulomb friction (Nm/N)
F_{cl} = caliper clamp force (N)
F_{cl}^* = caliper clamp force set-point command (N)
I = controller integral gain
i_m = motor current (A)
i_m^* = motor current set-point command (A)
J = system inertia lumped at motor axis (kgm^2)
K_b = back EMF coefficient (Vs/rad)
K_c = lumped stiffness of system (N/m)
K_t = motor torque constant (Nm/A)
L_m = motor inductance (H)
N = gear ratio (m/rad)
P = controller proportional gain
φ_m = motor position (rad)
R_m = motor resistance (Ω)
T_C = torque due to Coulomb friction (Nm)
T_E = applied external torque (Nm)
T_F = torque due to friction (Nm)
T_m = motor torque (Nm)
T_S = torque due to stiction (Nm)
T_V = coefficient of viscous friction (Nms)
T_0 = constant Coulomb friction offset (Nm)
ω_m = motor velocity (rad/s)
ω_m^* = motor velocity set-point command (rad/s)
ω_s = Stribeck velocity (rad/s)
x_{sp} = caliper spindle position (m)

MODELLING AND SIMULATION

In the first instance an EMB model was developed and validated against test data for the purposes of simulation and controller development. Previous models from Maron *et al.* and Schwarz *et al.* were used as a basis for developing the EMB model [17-19]. Using their approach the model was simplified by lumping the system inertias, stiffnesses, and damping. Their simplified brake model was modified to include the motor dynamics with back electromotive force (EMF) and voltage saturation. Also, the friction model was changed to include stiction plus Coulomb, viscous and Stribeck friction in accordance with the classical friction model stipulated by Olsson *et al.* [20]. A diagram outlining the shape of the friction model is shown in Figure 2.

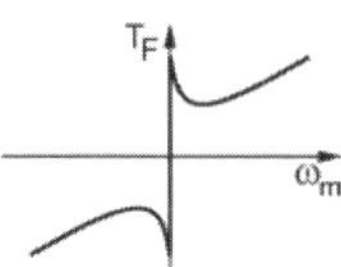

Figure 2: Friction model as a function of motor velocity with stiction plus Coulomb, viscous and Stribeck friction

Using the function given by Olsson *et al.* [20] the classical friction model is,

$$T_F = \begin{cases} T(\omega_m) & \text{if } \omega_m \neq 0 \\ T_E & \text{if } \omega_m = 0 \text{ and } |T_E| < T_S \\ T_S sign(T_E) & \text{otherwise} \end{cases} \qquad (1)$$

where T_F is the frictional torque, ω_m the motor velocity, $T(\omega_m)$ an arbitrary function, T_S the torque due to stiction and T_E is any external torque that is applied about the motor axis.

The function $T(\omega_m)$ is given by,

$$T(\omega_m) = T_C + (T_S - T_C)e^{-|\omega_m/\omega_s|^{\delta_s}} + T_V \omega_m \qquad (2)$$

where T_C is the torque due to Coulomb friction, T_V the coefficient of viscous friction and ω_S is the Stribeck velocity.

The Coulomb friction is a constant offset, T_0, with some load dependency, α, and is given by,

$$T_C = T_0 + \alpha F_{cl} \qquad (3)$$

where F_{cl} is the EMB clamp force.

The Karnopp remedy [20] for zero velocity detection was employed in the friction model using a zero velocity interval $\pm \omega_0$.

With the friction model, motor dynamics, and back EMF included, a simulation of the EMB model was implemented in Simulink. A block diagram of the model is shown in Figure 3.

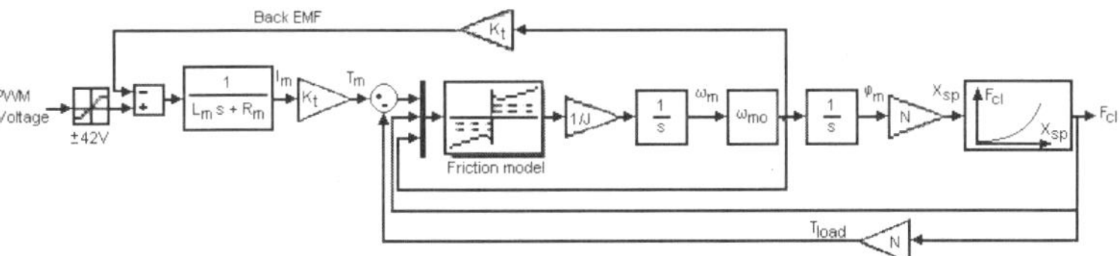

Figure 3: Block diagram of EMB model

The input to the model is the pulse width modulation voltage across the motor and this is saturated at the maximum power supply limits of ±42V. Briefly stepping through the block diagram, the back EMF is assumed proportional to the motor velocity and is subtracted from the supply voltage. The motor current is then determined by a first order lag that models the motor impedance. The motor torque is calculated from a proportional relationship between the current and the motor torque constant. A torque balance is then considered about the motor axis, with the effects of friction included. The net torque divided by the equivalent system inertia referred to the motor axis gives the motor acceleration. The acceleration is integrated to give the motor velocity and then zero velocity detection is employed to match that performed in the friction model. The motor velocity is integrated to give the motor position. The motor position multiplies the gear ratio to give the spindle position, that in turn determines the caliper clamp force via a non-linear stiffness characteristic. Finally, the EMB clamp force is the model output.

CONTROLLER DESIGN

DEFINITION OF THE CONTROL PROBLEM

With a cascaded control architecture the EMB control problem is subdivided into three lesser problems relating to each of the control loops. The clamp force control problem is to maintain the EMB clamp force, F_{cl} (N), at the set-point value, F_{cl}^* (N), by adjusting the set-point motor velocity demand, ω_m^* (rad/s). In turn, the motor velocity control problem is to maintain the motor angular velocity, ω_m (rad/s), at the set-point value, ω_m^* (rad/s), by adjusting the set-point current demand, i^* (A). Finally, the current control problem is to maintain the motor current, i (A), at the set-point value, i^* (A), by adjusting the supplied motor voltage, u (V).

DESIGN APROACH

Normally controllers with cascaded control loops are designed beginning with the inner most control loop and then working outwards. Consequently, the motor current controller is addressed first, then the motor velocity controller, and lastly the caliper clamp force controller. Classical control theory is applied at each stage with the general approach being to linearise the system, design a controller for the linear system, and then allow feedback to handle non-linear disturbances when the controller is implemented with the actual system. To improve the controller response, integral anti-windup is incorporated as well as feed forward compensation of the back EMF and Coulomb friction.

PERFORMANCE REQUIREMENTS

The following performance specifications for each of the control loops were chosen to give an overall EMB performance that is approaching the hydraulic brake standard.

<u>Motor current controller</u> - The current controller is required to follow a step command between the maximum safe motor currents of ±25 A with less than 1% overshoot and a ±2% settling time under 1 millisecond.

Motor velocity controller - Establishing a performance requirement for the velocity controller is more difficult because this depends on both the caliper clamp load, which increases during an apply, and also the direction of motion. Consequently, the performance requirement has been selected as reaching 95% of the maximum motor speed of 420 rad/s ($0.95 \times 420 \approx 400$ rad/s) in less than 0.02 s for an apply with zero initial air gap.

EMB clamp force controller - During an apply with no initial air gap the EMB must obtain the maximum caliper clamp force of 30 kN with less than 5% overshoot and a $\pm 2\%$ settling time of 0.15 s. The same requirements should also be met for force commands of less than 30 kN.

CONTROLLER DESIGN

The following section describes the design of each control loop subject to the required specifications listed previously.

Motor current controller - The current controller is a feedback controller with proportional and integral (PI) gains, integral anti-windup and feed forward compensation of the back EMF and Coulomb friction. Back EMF may be considered as a disturbance and this is shown in Figure 4.

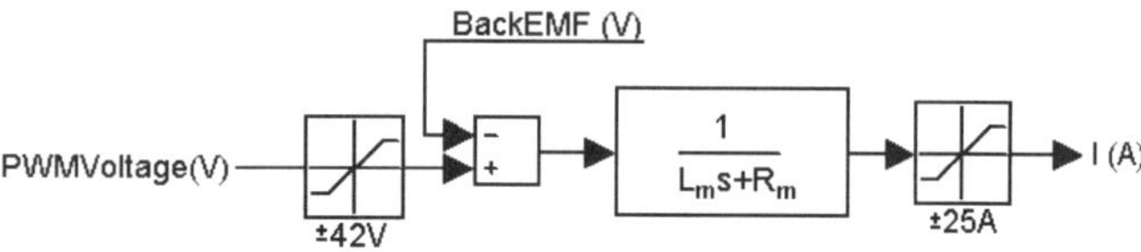

Figure 4: System showing back EMF as a disturbance

The back EMF is proportional to the motor velocity, but this is affected by the non-linear system characteristics. A feed-forward compensation term $(+K_t \omega_m)$ in the controller is used to reject the back EMF $(-K_t \omega_m)$, and this relies on knowledge of the motor speed and torque constant.

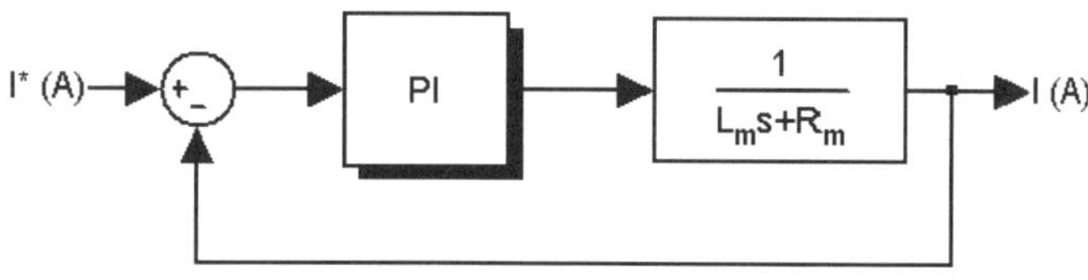

Figure 5: Current control loop

The closed loop transfer function can be shown to be,

$$\frac{I(s)}{I^*(s)} = \frac{P_I s + I_I}{L_m s^2 + (P_I + R_m)s + I_I} \qquad (4)$$

where P_i and I_i are the proportional and integral gains for the current controller.

The proportional and integral gains were tuned for the linear system by first specifying a proportional gain and the desired overshoot and then using a search algorithm to find the corresponding integral gain. The search algorithm used a step response in the time domain that was derived via an inverse Laplace transformation.

For this system it was found that the gains could be increased to give arbitrarily better rise times. However, the system performance is limited by the ± 42 V maximum power supply voltage and the current protection limits of ± 25 A that are included to avoid overheating the motor. These are shown in Figure 6.

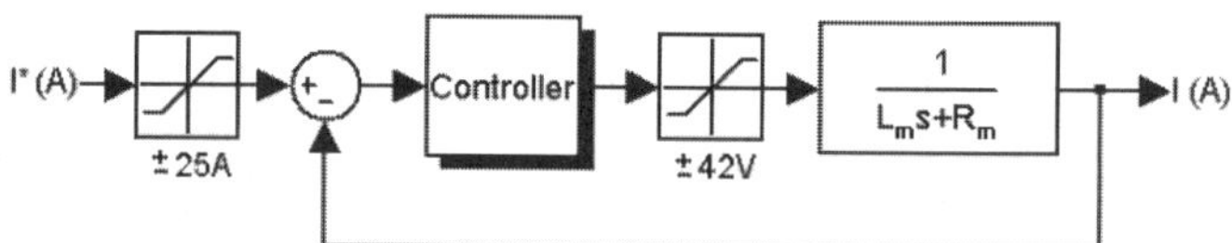

Figure 6: Current control loop with saturation of the power supply voltage and safe motor current limits

To avoid voltage saturation for a 25 A step command when the integral term in the controller is zero, the proportional gain must be 42/25 = 1.68. For a 1% overshoot requirement the corresponding integral gain was found to be 3221. Actually, the most demanding step command occurs when the system is running at –25 A and a current of +25 A is demanded, but avoiding the voltage saturation in this case was found to give an overly conservative controller. To ensure the controller demand did not exceed the ± 42 V constraint a saturation function was applied to the control output.

Motor velocity controller - The motor velocity controller is a PI controller with integral anti-windup. The location of the velocity controller within the cascaded architecture is shown in Figure 7.

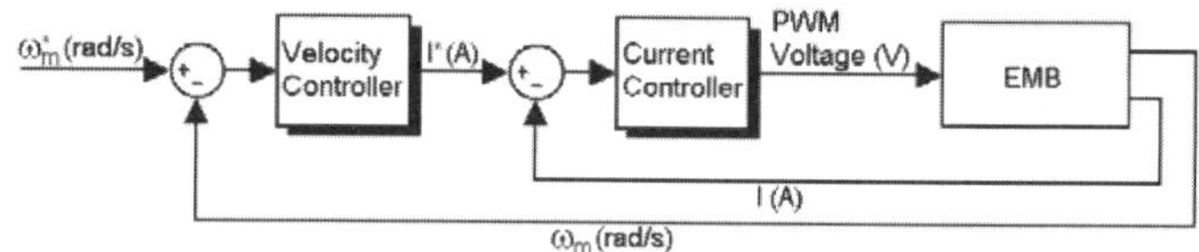

Figure 7: Control architecture with cascaded velocity and current loops

To arrive at a transfer function for the velocity loop the system must first be made linear. To do this the non-linear stiffness is approximated with a linear function and the technique of feed forward compensation is used to reject the back EMF and Coulomb friction disturbances. These disturbances are shown in the system diagram in Figure 8.

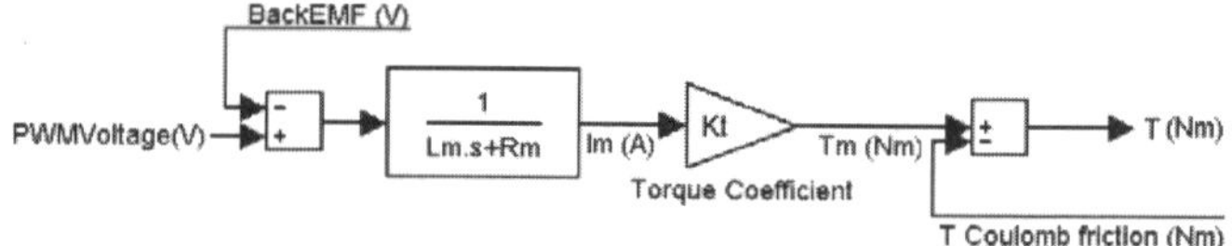

Figure 8: System diagram showing back EMF and Coulomb Friction as system disturbances

Accurate calculation of the Coulomb friction using a classical friction model assumes knowledge of the Coulomb friction offset and the coefficient of load dependency. These parameters must be measured prior to operation. The Coulomb friction calculation also requires real-time sensor measurements of the EMB clamp force and motor velocity.

Once a value for the Coulomb friction is available then feed forward compensation may be implemented. Given that the current loop has a high bandwidth and unity DC gain, the computed Coulomb friction may be introduced as an addition to the set point current command, scaled by the inverse of the motor torque constant K_t.

For the linearised system the closed loop transfer can be shown to be,

$$\frac{\omega_m(s)}{\omega_m^*(s)} = \frac{K_t(P_v s + I_v)}{Js^2 + (T_v + K_t P_v)s + K_c N^2 + K_t I_v} \qquad (5)$$

where P_v and I_v are the proportional and integral gains for the velocity controller. The DC gain of this transfer function is less than unity:

$$\frac{\omega_m(0)}{\omega_m^*(0)} = \frac{K_t I_v}{K_c N^2 + K_t I_v} < 1 \qquad (6)$$

The resulting steady-state error may be reduced by setting the integral gain large. Specifying a steady-state error requirement of less than 1%, the integral gain is calculated to be $I_v = 8.457$.

The proportional gain may now be selected to give the desired rise time and overshoot. However, the performance of the EMB is limited by the ±42 V voltage saturation. During a typical apply the motor can reach a maximum velocity of approximately 410 rad/s. Depending on the set-point motor velocity the motor speed error can instantaneously reach values in the order of 100-200 rad/s. To avoid the ±42 V saturation the proportional gain must be P_v = 42/200 = 0.21.

<u>EMB clamp force controller</u> – The EMB clamp force controller is a PI controller with integral anti-windup. This controller determines the commanded motor velocity, but since the EMB response is limited by the maximum motor speed the controller should avoid excessive demands. The limitation of the maximum motor velocity is used to determine the proportional gain.

Neglecting external forces, the maximum theoretical motor speed is reached when the back EMF fully cancels the applied voltage across the motor. Using $K_t = 0.1$ as the back EMF coefficient, and setting the back EMF equal to the maximum applied voltage of 42 V we find $\omega_{max} = 420$ rad/s.

During simulations of a brake apply a maximum motor speed of 410 rad/s was observed. This is close to the theoretical limit, but reduced somewhat due to external forces such as friction and load. The maximum speed is similar to a saturation, but instead of having a sharp cut-off value the acceleration is reduced as the maximum speed approaches. The PI force controller should avoid the effective speed saturation and not demand unattainable motor speeds. For a more conservative controller the maximum set-point motor velocity command is limited to 85% of the observed maximum speed $\omega^* = 410 \times 0.85 = \pm348.5$ rad/s. To avoid demanding a motor speed of more than 348.5 rad/s with a force error signal of 30 kN at the beginning of a full apply, the proportional gain must be 348.5/30000 = 0.0116. A small integral gain of 10^{-3} is chosen to help eliminate the steady state error.

Having now designed a controller for each of the control loops, the details are summarized in Table 1.

Control loop	Controller	Gains	Variations
EMB motor current (A)	PI	P = 1.68 I = 3221	Anti-windup Back EMF compensation
EMB motor velocity(rad/s)	PI	P = 0.21 I = 8.457	Anti-windup Coulomb friction compensation
EMB clamp force (N)	PI	P = 0.1 I = 0.001	Anti-windup

Table 1: Specification of the EMB clamp force controller with cascaded force, velocity and current control loops

It was found that using the gains shown in Table 1, the performance requirements for each control loop were satisfied.

RESULTS AND DISCUSSION

The performance of an EMB is in part determined by how quickly the unit can respond to a brake apply and how well it can track fine modulations about a particular caliper clamp load. These characteristics are important to provide both a rapid brake response during a critical brake maneuver and also allow implementation of anti-lock braking. Hence, a series of tests were simulated to determine the EMB step and frequency responses. Before presenting the results however, the validation of the model is first addressed.

In an attempt to validate the model, the simulation output for a 30 kN full brake apply is compared with experimental data from an actual EMB provided by an industry partner. The results are shown in Figure 9.

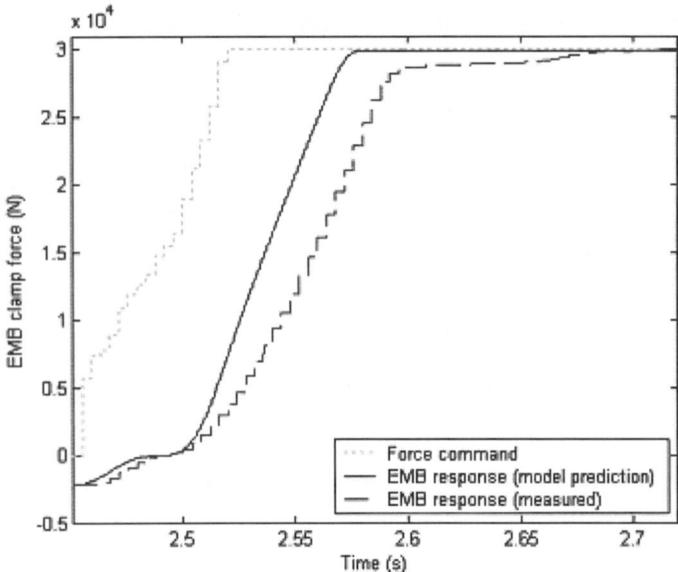

Figure 9: EMB response to full brake apply simulated for the newly designed controller and measured with a former gain scheduled controller

In Figure 9 it may be observed that the response predicted by the present EMB model is somewhat faster than the experimental result. In part this may be attributed to the model containing ideal sensors and a controller with continuous rather than discrete operation.

Importantly, the predicted response has a maximum slope that agrees with the measured result. This indicates that the simulation is accurately modelling the physical limitations of the actuator. Similar observations were made for other comparisons between simulation and actual EMB responses.

Management of the separation between the EMB brake pads and brake rotor is incorporated within the controller by switching the outer control loop between position and clamp force control across the contact point between the pads and rotor. Continuous operation between the regions of position and force control may be observed in Figure 9 where the clearance is represented on the force plot using an imaginary negative force.

In Figure 9 it can be seen that the EMB and controller give satisfactory performance for a full brake apply of 30 kN. However, the same performance requirements should be met for all step commands within the range of operation. To test this a series of step responses were conducted. The results are shown in Figure 10.

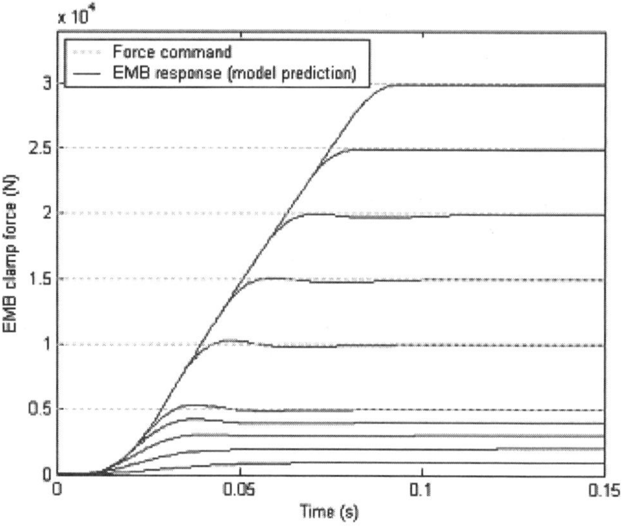

Figure 10: EMB response to force step commands of 1, 2, 3, 4, 5, 10, 15, 20, 25 and 30kN

In Figure 10 it may be observed that the EMB controller satisfied the performance requirements for a series of step commands. However, upon closer examination it is seen that the controller responds slower to clamp force commands less than 5 kN. The reason for this is that the controller gains were conservatively chosen to avoid the system saturations for the maximum demand. However, for small force commands the reduced error signal results in less effort being commanded via the proportional control. This can be a problem when responding to small clamp force commands or even more so when tracking modulations about a large clamp force. In these cases the error signal and proportional gain may not be sufficiently high to overcome the Coulomb friction. Instead, we find the response is delayed until the integral control ramps up the command and friction is overcome. A solution to this problem may be offered either by friction compensation or gain scheduling based on the clamp force error signal. With this scheduling approach higher gains may be used to obtain a better response for small error signals in the clamp force command.

As well as responding to step commands it was mentioned earlier that the EMB should be able to track modulations about a particular clamp force. To assess this performance objective a series of tests were conducted to observe the response to sinusoidal modulations about a given clamp force. The results from these tests are presented in Figures 11 and 12.

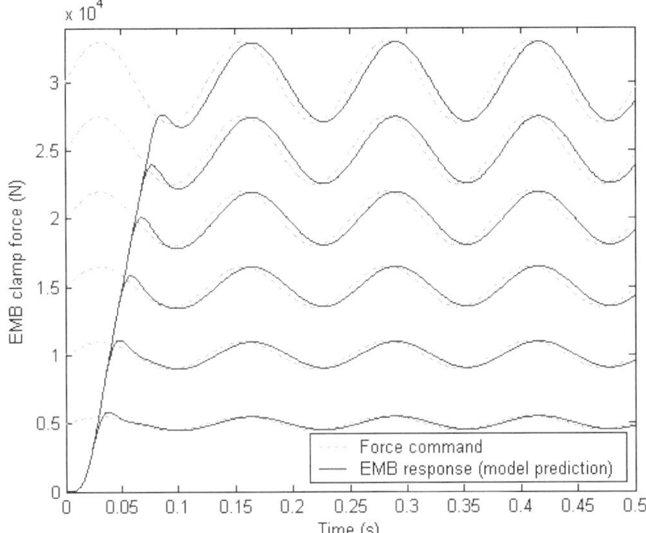

Figure 11: EMB response to force commands of 5, 10, 15, 20, 25 and 30 kN with 50 rad/s (7.95 Hz) sine wave, amplitude 10% of the step value

In Figure 11 it may observed how the controller tracks a sinusoidal modulation about a desired clamp force. In this case the modulation was arbitrarily selected as a 50 rad/s sine wave with an amplitude that is 10% of the step value. While this gives the response to a single input frequency, it is useful to know the full frequency response. This is determined by repeating the tracking test across a range of frequencies. The result of this is shown in Figure 12.

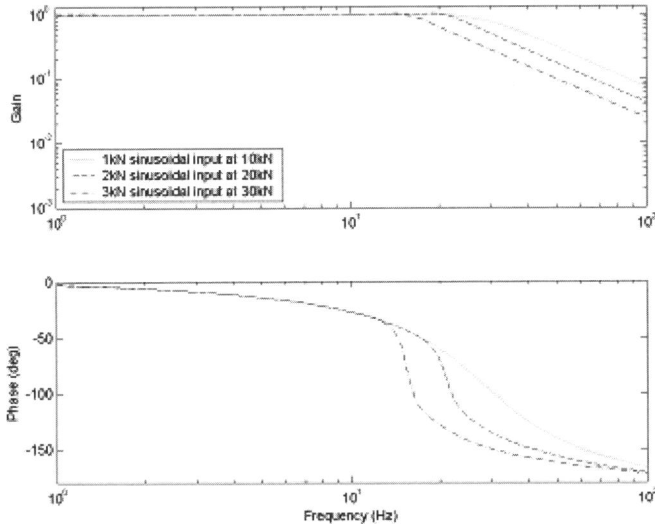

Figure 12: EMB frequency response to a 10% sinusoidal input at the given clamp forces

In Figure 12 it may be observed that for a 10% sinusoidal modulation about a given EMB clamp force the frequency response drops off at around 20 Hz. Non-linearity of the system may also be observed with the bandwidth decreasing at higher clamp loads.

It should be noted that the frequency response in Figure 12 was achieved using both the back EMF and friction compensation detailed in the controller design. Initially there was no compensation included in the controller, but

this was later added for improved disturbance rejection. To assess its effectiveness, the EMB response was compared with and without compensation. The results are shown in Figures 13, 14 and 15.

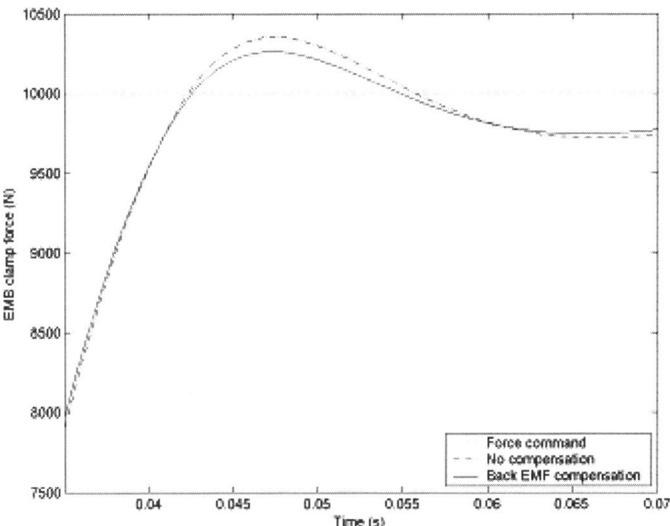

Figure 13: EMB response to a step force command of 10kN with and without back EMF compensation.

Figure 13 compares the EMB response to a 10kN step force command with and without back EMF compensation. It may be observed that the response with compensation marginally leads and suffers slightly less overshoot than the case without. However, the improvement with back EMF compensation appears minor and has been included in the controller mainly to demonstrate the best achievable results with feed forward compensation of the system nonlinearity.

In order to test the impact of Coulomb friction compensation on the frequency response, a slowly varying frequency chirp signal was input into the model about a steady state clamp force. The fundamental harmonic of the model output was then used to estimate a linear transfer function of the EMB system. The accuracy of the linear approximation was determined using the standard technique of coherence evaluation. In this approach, power spectral densities of the input and output, P_{XX} and P_{YY} respectively, are used along with the cross spectral density, P_{XY}, to calculate the coherence, C_{XY}, according to the following relationship

$$C_{XY} = \frac{|P_{XY}|^2}{P_{XX} P_{YY}} \in [0,1] \qquad (7)$$

Coherence values close to 1 indicate that there is good linear correlation between the input and output, and hence the transfer function estimate is a good approximation of the system behavior. The coherence is plotted in Figure 14 for each of the corresponding bode plots.

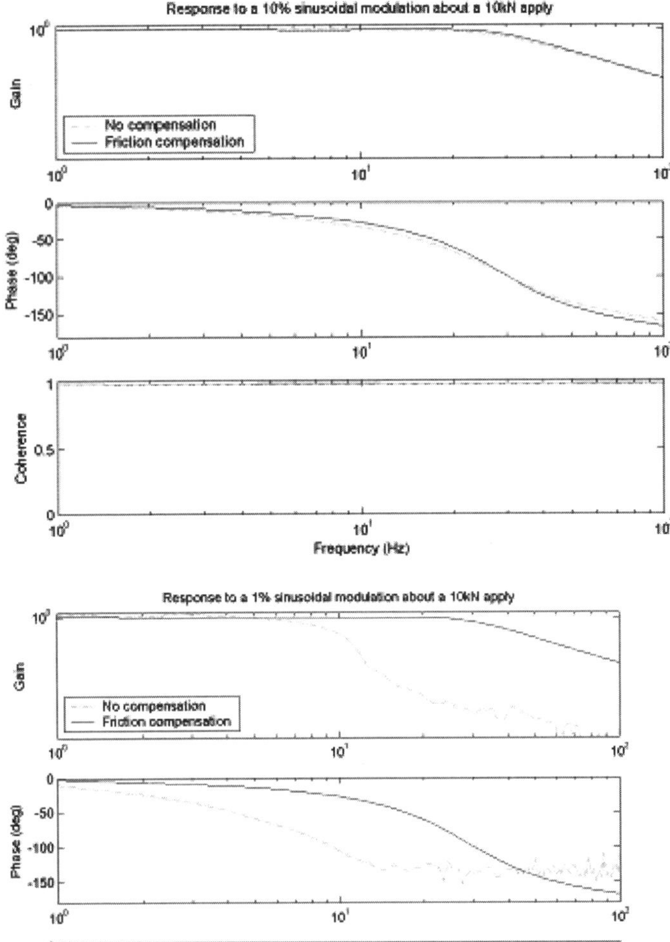

Figure 14: EMB frequency response to a sinusoidal input of amplitudes 10% and 1% about a 10kN brake apply with and without friction compensation

In Figure 14 it may be observed that the EMB response without friction compensation is attenuated for small modulations at high frequency and this is attributed to the effect of stiction. It is also seen that the friction compensation is more beneficial for finer clamp force modulations, and although there is little improvement in the response to a sinusoidal force modulation with amplitude of 10% about a 10kN apply, a significant improvement is observed for the smaller 1% modulation. This occurs because at finer levels of force modulation the friction becomes more significant when compared with the magnitude of the force error that drives the controller response.

The coherence plots indicate this is a reasonable representation of the overall system frequency response, especially for frequencies under 10Hz. This is due to the controller being able to overcome 'slow' disturbances such as stiction through integral action. The drop in coherence for the 1% modulation scenario at approximately 10 Hz indicates this is the fastest rate at which the existing integral action can be successfully applied in this context.

Figure 15 demonstrates the resulting output with and without friction compensation at 8 Hz, which is close to the corner frequency observed in the gain plots of Figure 14.

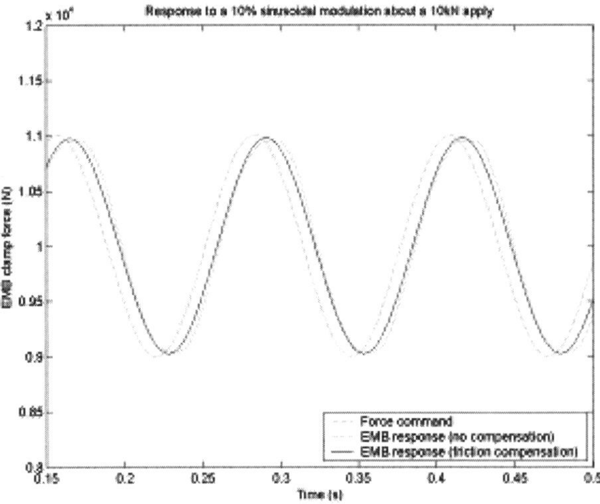

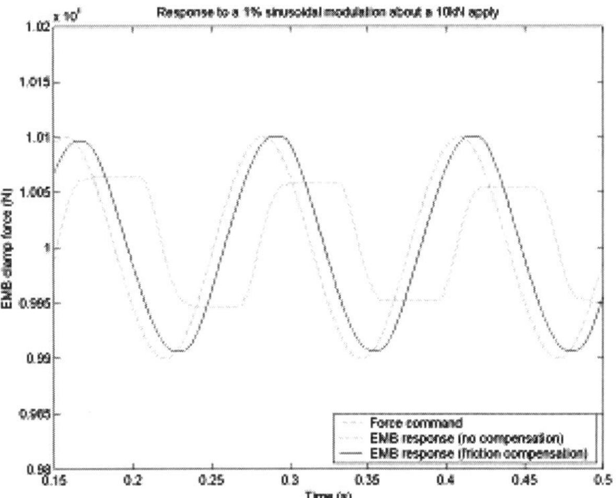

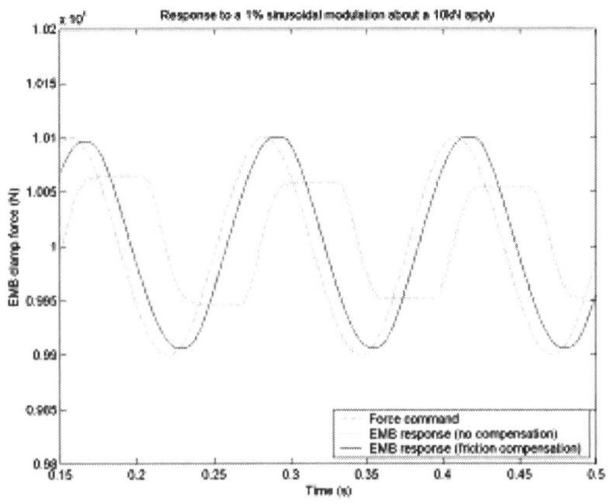

Figure 15: EMB response to a sinusoidal input at 50 rad/s (7.95 Hz) about a 10kN brake apply with and without friction compensation

Figure 15 compares the EMB response to a sinusoidal force command with and without friction compensation, and as before, the friction compensation appears to be more beneficial for finer modulations. In particular, it is seen that the uncompensated system suffers stiction to a greater extent for smaller amplitudes of modulation. Importantly, this disturbance is partly overcome when friction compensation is included in the controller to command more effort from the EMB actuator.

Satisfactory disturbance rejection and control robustness are important issues for an EMB in a safety-critical brake system. It is necessary for the controller to handle parameter uncertainties such as variations in friction levels with wear and also characteristic stiffness with temperature and wear of the brake pads. This is a topic for future work on robust control.

As a secondary outcome of modeling an EMB and designing a controller it was realized that EMB design is a difficult optimization problem. The mechanical unit must provide high and reliable performance over many operating cycles during its lifetime. Further, the unit must endure high loads, high temperatures up to $800°C$ at the brake pads, and an environment that can be both dirty and rough. All this must be achieved with an actuator that is compact and constrained in size by the limited available space. To assist with the design task it would certainly be helpful to understand what parameters were most influential in limiting the performance of an electromechanical brake. A high performance motor and appropriate gear ratio are important. It is also desirable for the motor and gearing to be compact and efficient with low friction. Ultimately however, it was observed that the EMBs performance is limited by the maximum power supply voltage, maximum motor speed and the limits on the safe motor current.

CONCLUSION

Automobile brakes are a critical vehicle system and consequently the performance demands are high. Prior to this work it was unclear how well a cascaded EMB clamp force controller would handle the large operating range in which non-linear load disturbances such as friction become more significant at high clamp forces of up to 30kN. In this paper the design of a cascaded EMB controller is presented and its capacity to supply and track brake caliper clamp force commands is observed from its response to step and sinusoidal inputs. It was found that provided the controller gains were selected to avoid saturations of the maximum motor speed, maximum power supply voltage, and limits on the safe motor current then the conventional motion control architecture with an outer force loop provided a satisfactory response to clamp force commands over a wide range of operation. However, it is suggested that for operation about low clamp forces, where the selected controller gains were more conservative than necessary, the performance may be further improved with the inclusion of gain scheduling. Additionally it was found that the ability to perform fine modulations about a steady clamp force level is significantly improved if feed forward compensation of the load dependent friction force is implemented based on an accurate friction model.

The important results of this paper include:

- the design of an EMB controller.
- the validation of an EMB model simulation.
- characterization of the step and frequency responses for the EMB.
- an assessment of the effectiveness of back EMF and Coulomb friction compensation.

Future research directions include implementing the controller on an EMB and updating the model to include uncertainties before addressing the problem of robust control.

ACKNOWLEDGMENTS

The authors would like to acknowledge and thank their industry partner, Pacifica Group Technologies for providing both data and assistance. We would also like to acknowledge that this work was supported by the Research Centre for Advanced By-Wire Technologies (RABiT).

REFERENCES

[1]   W.-D. Jonner, H. Winner, L. Dreilich, and E. Schunck, "Electrohydraulic brake system - The first approach to brake-by-wire technology," *SAE Technical Paper*, vol. 960991, 1996.

[2]   E. Gohring and E.-C. v. Glasner, "Performance comparison of drum and disc brakes for heavy duty commerical vehicles," *SAE Technical Paper*, vol. 902206, 1990.

[3]   A. G. Taig, "Electrically actuated disc brake," in *United States Patent 4850457*. United States of America: Allied-Signal Inc., 1989.

[4]   Y. Fujita, T. Aral, and M. Ogura, "Motor disc brake system," in *United States Patent 5107967*. Japan: Honda Giken Kogyo Kabushiki, 1992.

[5]   S. S. Shaw and D. E. Schenk, "Electric disc brake," in *United States Patent 5219048*. United States of America: General Motors Corporation, 1993.

[6]   G. Halasy-Wimmer, K. Bill, J. Balz, L. Kunze, and S. Schmitt, "Electromechanically actuated disc brake system," in *United States Patent 5829557*. United States of America: ITT Automotive Europe GmbH, 1998.

[7]   A. W. Kingston, R. L. Ferger, T. Weigert, S. Oliveri, L. Tribe, and H. L. Linkner, "Electric actuation mechanism suitable for a disc brake assembly," in *United States Patent 5931268*. United States of America: Kelsey-Hayes Company, 1999.

[8]   K. Bill, J. Balz, and V. Dusil, "Electromechanical disc brake," in *United States Patent 6158558*. United States of America: Continental Teaves AG & Co. OHG, 2000.

[9]   K. L. Holding, "Electrically-operated dics brake assemblies for vehicles," in *United States Patent 6145634*. United Kingdom: Lucas Industries public limited company, 2000.

[10]   J. Dietrich, B. Gombert, and M. Grebenstein, "Electromechanical brake with self-energization," in *United States Patent 6318513*. United States of America: Deutsches Zentrum fur Luft- und Raumfahrt e.V., 2001.

[11]   K. Takahashi and K. Kawase, "Electric braking device," in *United States Patent US 6279691 B1*. Japan: Akebono Brake Industry Co., Ltd., 2001.

[12]   T. Usui and Y. Ohtani, "Electric disc brake," in *United States Patent US 6491140 B2*. Japan: Tokico, 2002.

[13]   F. Keller, "Electromagnetic wheel brake device," in *United States Patent US 6536561 B1*. United States of America: Robert Bosch GmbH, 2003.

[14]   A. H. E. A. Olschewski, H. J. Kapaan, C. Druet, T. W. Fucks, M. Antensteiner, A. C. Rinsma, J. Gurka, and A. J. C. D. Vries, "Actuator having a central support, and brake calliper comprising such actuator," in *United States Patent US 6554109 B1*. Netherlands: SKF Engineering and Research Centre B.V., 2003.

[15]   D. B. Drennen, E. R. Siler, G. C. Fulks, and D. E. Poole, "Caliper with internal motor," in *United States Patent US 6626270 B2*. United States of America: Delphi Technologies, Inc., 2003.

[16]   M. Smith, "42 Volts - enabling a technological revolution," in *Auto Briefing*, Special Eddition May 2001 ed: Knibb, Gormezano and Partners, 2001.

[17]   R. Schwarz, R. Isermann, J. Bohm, J. Nell, and P. Rieth, "Clamping force estimation for a brake-by-wire actuator," *SAE Technical Paper*, vol. 1999-01-0482, 1999.

[18]   C. Maron, T. Dieckmann, S. Hauck, and H. Prinzler, "Electromechanical brake system: Actuator control development system," *SAE Technical Paper*, vol. 970814, 1997.

[19]   R. Schwarz, R. Isermann, J. Bohm, J. Nell, and P. Rieth, "Modeling and control of an electormechanical disk brake," *SAE Technical Paper*, vol. 980600, 1998.

[20]   H. Olsson, K. J. Åström, C. C. d. Wit, M. Gäfvert, and P. Lischinsky, "Friction models and friction compensation," *European Journal of Control*, 1998.

Braking Status Monitoring For Brake-By-Wire Systems

Paljoo Yoon, Hyung-Jin Kang and Inyong Hwang
MANDO Co.

Kunsoo Huh and Daegun Hong
HANYANG University

ABSTRACT

Recently, wheel slip controllers with controlling the wheel slip directly have been studied using the brake-by-wire actuator. The wheel slip controller is able to control the braking force more accurately and can be adapted to various different vehicles more easily than the conventional ABS systems. The wheel slip controller requires the information about the tire braking force and road condition in order to achieve the braking and stability control performance. In this paper, the tire braking forces are estimated considering the variation of the friction between brake pad and disk due to aging of the brake, moisture on the contact area or heating. In addition, the road friction coefficient is estimated without using tire models. The estimated performance of tire braking forces and the road friction coefficient is evaluated in simulations.

INTRODUCTION

Conventional ABS (Anti-lock Brake System) retains steerability by preventing the wheel lock while braking. However, it has some limitations as follows: the slip oscillates around the peak friction point, many experiments are required for tuning the ABS system in trial and error manner, etc. Recently, in order to overcome these limitations of the conventional ABS, the wheel slip controller that controls the tire slip directly have been studied widely in vehicle brake industry [1-3]. The wheel slip controller is able to control the tire braking force more accurately through maintaining a specified tire slip for each wheel and can be migrated to different vehicles more easily than the conventional ABS. In addition, the steering stability can be improved by deciding the slip set point through a higher-level controller such as ESP (Electronic Stability Program), etc. Wheel slip control can be realized using brake-by-wire actuators such as EHB (Electro-Hydraulic Brake) and EMB (Electro-Mechanical Brake).

In the wheel slip controller, the information on the tire braking force and road condition is essential in order to achieve the good control performance. The tire braking force can be calculated by wheel dynamics equation if it is possible to measure or know the braking torque. In case of EMB, the braking torque can be measured by measuring the motor current. However, in EHB systems it is difficult to calculate the braking torque from the measured braking pressure because the friction between brake disk and pad can change due to ageing of the brakes, moisture on the contact area or heating during extended braking maneuvers. Müller et al. [4] demonstrates that the brake disk-pad friction can change more than 50 % under various braking conditions and calculates the total braking force by estimating the brake torque gain.

Regarding the road condition monitoring during vehicle driving, many research papers have been reported. The research trend is divided into two groups as follows: One group is direct sensing approach using a physical sensor such as optical sensors, acoustic sensors or strain sensors. The other group is indirect estimation approach such as slip-based estimation, model-matching estimation, and so on. In the slip-based estimation, the slip-slope is estimated using Kalman Filter, RLS (Recursive Least Square), etc, and the road condition is identified using the estimated slip-slope (Gustafsson [5], Müller et al. [4]). In the model-matching method, the μ-slip curve over a whole slip range is identified from the estimated parameters of the Magic Formula tire model, the Burckhardt friction model, etc. (Kageyama et al. [6], Kiencke et al. [7]). In other estimation methods, the road friction coefficient is estimated using the probability method [8] and an adaptive observer method [9]. Pasterkamp and Pacejka [10] applies the neural network to estimate the road friction coefficient. However, these methods require some specific decision rule or tire models to be applied.

In this paper, a monitoring system is developed to estimate the tire braking force, friction between brake disk and pad, and the road friction coefficient. The vehicle and brake systems are modeled as a seven degree-of-freedom dynamics model and Extended Kalman Filter is designed based on this model for estimating the tire braking force of the four wheels individually. In addition, for identifying the road condition, the common characteristics of the tire-road friction are derived regardless of road condition and the road friction coefficient is estimated based on these characteristics without using a tire model. The estimated performance of tire braking forces and the road friction coefficient is evaluated in various simulations.

The rest of this paper is organized as follows: Section 2 derives the vehicle model and estimation algorithm for estimating the tire braking force and the brake disk-pad friction. Section 3 describes the road friction estimation algorithm and Section 4 verifies the estimation performance by simulations.

TIRE BRAKING FORCE ESTIMATION

VEHICLE MODELING

Based on the force diagram of the sprung mass as illustrated in Fig. 1, the motion of vehicle can be modeled into three degree-of-freedom system of the longitudinal, lateral and yawing motion as expressed in Eq. (1).

$$m(\dot{v}_x - r v_y) = \sum_{i=1}^{2}(F_{xi}\cos\delta_f) + \sum_{i=3}^{4} F_{xi} - 2F_{yf}\sin\delta_f$$

$$m(\dot{v}_y + r v_x) = \sum_{i=1}^{2}(F_{xi}\sin\delta_f) + 2F_{yf}\cos\delta_f + 2F_{yr}$$

$$I_z\dot{r} = l_f\sum_{i=1}^{2}(F_{xi}\sin\delta_f) + l_f\cdot 2F_{yf}\cos\delta_f - l_r\cdot 2F_{yr}$$

$$+ t_f\sum_{i=1}^{2}(-1)^i(F_{xi}\cos\delta_f) + t_r\sum_{i=3}^{4}(-1)^i F_{xi} \tag{1}$$

where m is vehicle mass, I_z is moment of inertia in yaw-direction and v_x, v_y, and r denote the longitudinal velocity, lateral velocity and yaw rate, respectively. F_{xi} stands for the tire braking force at each wheel. F_{yf} and F_{yr} are the mean values of front/rear lateral tire force as expressed in Eq. (2). δ_f is the steering angle of the front wheel.

$$F_{yf} = (F_{y1} + F_{y2})/2$$
$$F_{yr} = (F_{y3} + F_{y4})/2 \tag{2}$$

When a brake is engaged, the tire braking force is generated between tire and road surface due to friction as illustrated in Fig.2. In case of the hydraulic disk brake as shown in Fig.2, the brake torque is calculated as expressed in Eq.(3) and the wheel dynamic equation is determined as expressed in Eq.(4).

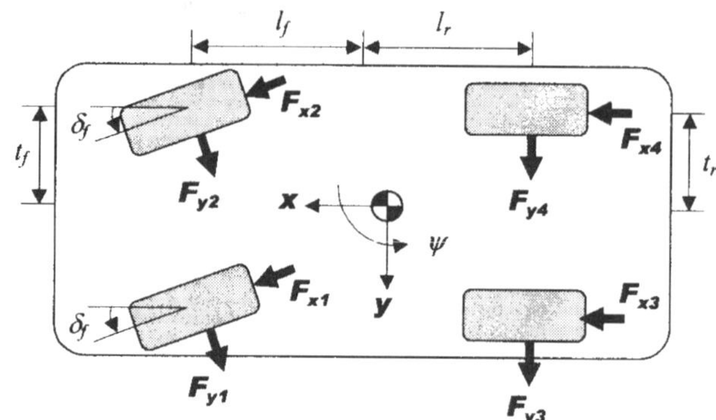

Fig.1 Force diagram of the sprung mass

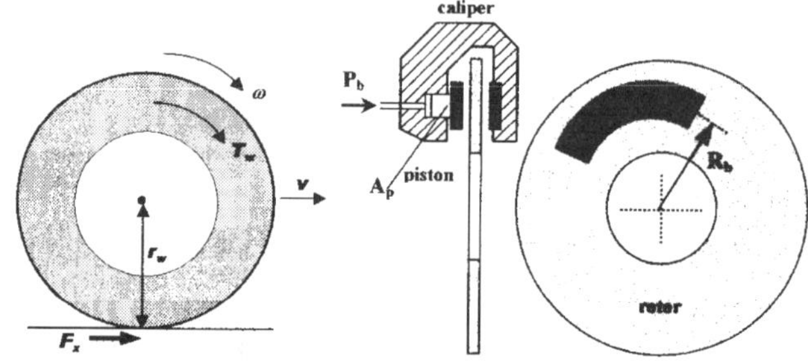

Fig. 2 Wheel dynamics and hydraulic disk brake

$$T_{wi} = -2\cdot P_{bi}A_{pi}R_{bi}\mu_b, \quad (i = 1\sim 4) \tag{3}$$

$$J_{wi}\dot{\omega}_i = T_{wi} - r_w F_{xi} = -2\cdot P_{bi}A_{pi}R_{bi}\mu_b - r_w F_{xi} \tag{4}$$

where T_w stands for the braking torque, P_b is the hydraulic braking pressure, A_p is the area of piston, R_b is effective radius between the center of disk rotor and pad, and μ_b represents the brake disk-pad friction. ω denotes the rotational velocity of wheel, J_w is the moment of inertia of wheel, and r_w is the wheel radius.

EXTENDED KALMAN FILTER DESIGN

In order to estimate the tire braking force, a new monitoring model is proposed considering the variation of the brake disk-pad friction and is utilized for Extended Kalman Filter design. In the nonlinear vehicle model derived at the previous chapter, the braking force (F_{xi}), lateral tire force (F_{yf}, F_{yr}) and brake disk-pad friction (μ_b) are unknown factors but cannot be easily measured using physical sensors. For estimating these unknown factors, they are augmented as state variables in the state equation and assumed to be slowly time varying compared to the motion variables of a vehicle. Therefore, the time derivatives of these augmented state variables are assumed to be zero and the whole nonlinear state equation of the monitoring model is expressed in Eq (5).

$$\dot{v}_x = r\,v_y + \frac{1}{m}\left(\sum_{i=1}^{2}(F_{xi}\cos\delta_f) + \sum_{i=3}^{4}F_{xi} - 2F_{yf}\sin\delta_f \right)$$

$$\dot{v}_y = -r\,v_x + \frac{1}{m}\left(\sum_{i=1}^{2}(F_{xi}\sin\delta_f) + 2F_{yf}\cos\delta_i + 2F_{yr} \right)$$

$$\dot{r} = \frac{1}{I_z}\left(l_f \sum_{i=1}^{2}(F_{xi}\sin\delta_f) + l_f 2F_{yf}\cos\delta_f - l_r 2F_{yr} \right.$$

$$\left. + t_f \sum_{i=1}^{2}(-1)^i (F_{xi}\cos\delta_f) + t_r \sum_{i=3}^{4}(-1)^i F_{xi} \right)$$

$$\dot{\omega}_i = (-2\cdot P_{bi} A_{pi} R_{bi}\mu_b - r_w F_{xi})/J_{wi}, \quad (i=1\sim4) \qquad (5)$$

$$\dot{\mu}_b = 0$$

$$\dot{F}_{xi} = 0, \quad (i=1\sim4)$$

$$\dot{F}_{yf} = 0$$

$$\dot{F}_{yr} = 0$$

In this study, the measured variables are the longitudinal acceleration, lateral acceleration, yaw rate and wheel velocities of the four wheels, and expressed as the output equation as follows:

$$y = \begin{bmatrix} a_x \\ a_y \\ r \\ w_1 \\ w_2 \\ w_3 \\ w_4 \end{bmatrix} = \begin{bmatrix} \dfrac{1}{m}\left(\sum_{i=1}^{2}(F_{xi}\cos\delta_f) + \sum_{i=3}^{4}F_{xi} - 2F_{yf}\sin\delta_f \right) \\ \dfrac{1}{m}\left(\sum_{i=1}^{2}(F_{xi}\sin\delta_f + F_{yi}\cos\delta_i) + 2F_{yr} \right) \\ r \\ w_1 \\ w_2 \\ w_3 \\ w_4 \end{bmatrix} \qquad (6)$$

where, a_x and a_y represent the longitudinal acceleration and the lateral acceleration, respectively.

Based on the monitoring model composed of nonlinear state equation and output equation as shown in Eq. (5) and (6), Extended Kalman Filter which estimates the state variables of the nonlinear state equation is designed for estimating the tire braking force at each wheel and the brake disk-pad friction simultaneously. The design procedure is expressed in Eq. (7).

$$x_k = f(x_{k-1}, k-1) + w_{k-1}$$

$$y_k = h(x_k, k) + v_k$$

$$P_k(-) = F_{k-1}\,P_{k-1}(+)F_{k-1}^T + Q_{k-1}$$

$$K_k = P_k(-)H_k^T\left[H_k P_k(-)H_k^T + R_k \right]^{-1} \qquad (7)$$

$$P_k(+) = \left[I - K_k H_k \right]P_k(-)$$

$$\text{where, } F_{k-1} = \left.\frac{\partial f(x,k-1)}{\partial x}\right|_{x=\hat{x}_{k-1}(+)}, \; H_k = \left.\frac{\partial h(x,k-1)}{\partial x}\right|_{x=\hat{x}_k(-)}$$

$$\hat{x}_k(+) = \hat{x}_k(-) + K_k\left(y_k - \hat{y}_k\right)$$

$$\hat{x}_k(-) = f(\hat{x}_{k-1}(+), k-1)$$

$$\hat{y}_k = h(\hat{x}_k(-), k)$$

where, $f(\cdot)$ and $h(\cdot)$ represent the state equation and the output equation, respectively. $w_k{\sim}N(0,Q_k)$, $v_k{\sim}N(0,R_k)$ represent the plant model uncertainty and the sensor noise, respectively. P is the error covariance matrix and K is the filter gain matrix.

The measurement noise covariance, R, is selected to satisfy the Lyapunov stability proposed by Boutayeb et al. [11] as shown in Eq. (8).

$$R_k = \xi_1 H_k P_k(-)H_k^T + \xi_2 I \qquad (8)$$

ROAD FRICTION COEFFICIENT ESTIMATION

TIRE-ROAD FRICTION CHARACTERISTICS

During vehicle braking, the *operating friction*, μ, is defined as the ratio of tire braking force (F_x) and normal force (F_z) on one braked wheel as expressed in Eq. (9). Also μ can be computed from the estimated tire braking force and the calculated normal force. The tire braking force can be obtained by the estimation algorithm described in Section 2. The normal force can be determined by the load transfer calculation [12], but its robustness is evaluated in experiments.

$$\mu = \frac{F_x}{F_z} \qquad (9)$$

The μ varies according to the slip ratio of tire. The slip ratio, λ, is defined as the relative difference of a braked wheel's angular velocity and vehicle absolute velocity as shown in Eq. (10).

$$\lambda = \frac{v_x - \omega \cdot r_w}{v_x} \qquad (10)$$

The μ variation according to the slip ratio is shown in Fig. 3, which is plotted using the Magic Formula tire model (Bakker et al. [13]) and is called μ-slip curve. Fig 3 also shows the μ-slip curve variation according to road conditions such as road friction coefficient (μ_H). The μ has the maximum value, μ_{max}, around λ=0.15 on normal road (μ_H=1.1) and the μ_{max} is almost the same with the road friction coefficient ($\mu_H = \mu_{max}$). The μ_{max} decreases according to decreasing μ_H and the λ corresponding to μ_{max} also decreases. In particular, if one connects all the μ_{max} points at μ-slip curve of each road condition, a linear line can be drawn and assumed as the *peak-line*. Then, the *peak-line* is the 1st linear equation of the slip ratio.

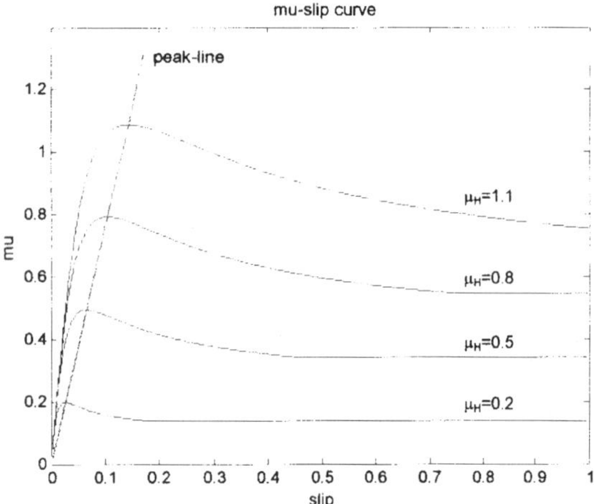

Fig. 3 Tire-road friction characteristics (MF model)

ESTIMATION ALOGORITHM

The road friction coefficient can be determined through estimating the maximum operating friction (μ_{max}) at the μ-slip curve. However, it is difficult to know the μ-slip curve over the whole slip ratio range even if μ can be calculated at any operating point during braking. In this study, the road friction coefficient is estimated utilizing the difference between the calculated μ and the value on the *peak-line* at the operating slip ratio. Fig. 4 shows that this difference is getting smaller as the road friction coefficient decreases. It is shown that this difference δ can be calculated as shown in Eq. (11). The δ with respect to slip ratio is plotted in Fig. 5 on various road conditions.

$$\delta = \mu - \mu_{peak\,line}(\lambda) = \mu - C\lambda = \frac{F_x}{F_z} - C\lambda \qquad (11)$$

where, $\mu_{peak-line}$ stands for the value on the *peak-line* and C is the slope of *peak-line*, so it is the coefficient of the 1st polynomial.

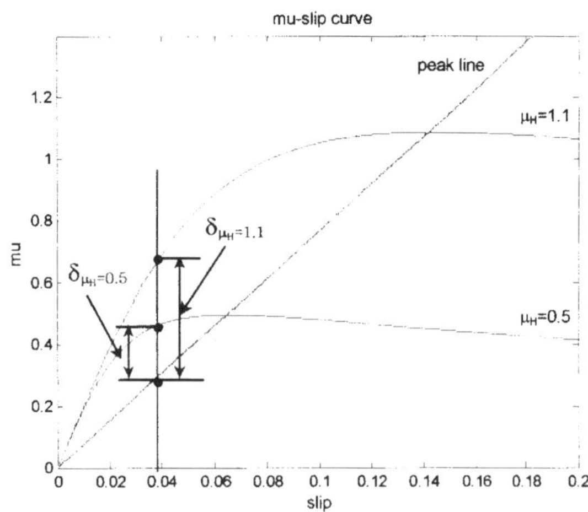

Fig. 4 Difference between operating friction and peak line

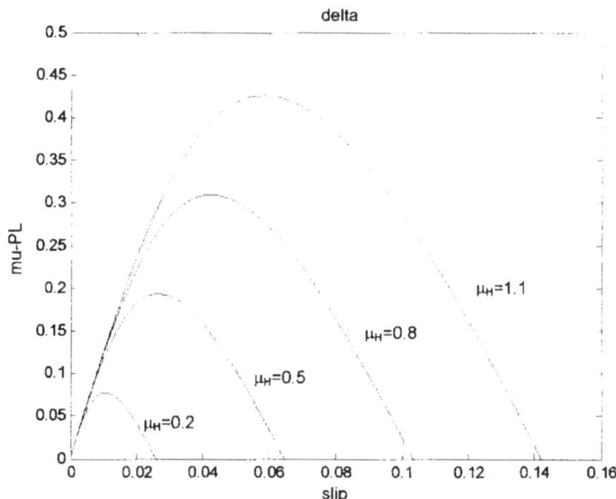

Fig. 5 δ with respect to slip

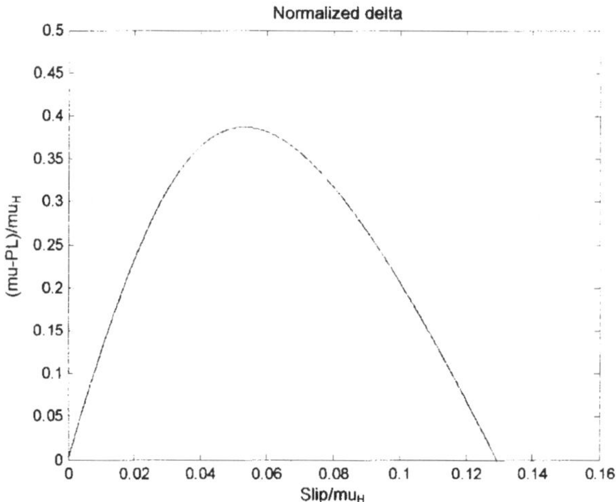

Fig. 6 Normalized δ according to normalized slip

In order to normalize the δ curve as shown in Fig. 5, the slip ratio (x-axis in Fig. 5) and δ (y-axis in Fig 5) is divided by road friction coefficient (μ_H). As a result, the normalized δ curve can be obtained as shown in Fig. 6, and it is confirmed that the normalized δ curve is almost same on various road condition. Consequently, the normalized δ curve has no relation with road condition.

In this paper, we propose the estimation algorithm of road friction coefficient using this normalized δ curve. The normalized δ curve can be expressed with 3rd polynomial of λ/μ_H as shown Eq. (12).

$$\frac{\delta}{\mu_H} = f\left(\frac{\lambda}{\mu_H}\right) = a_3\left(\frac{\lambda}{\mu_H}\right)^3 + a_2\left(\frac{\lambda}{\mu_H}\right)^2 + a_1\left(\frac{\lambda}{\mu_H}\right)^1 \qquad (12)$$

Then, Eq. (12) can be expanded as the road friction coefficient variable, and thus, the 2nd polynomial about the road friction coefficient is obtained as expressed in Eq.(13). Finally, the road friction coefficient can be estimated easily by solving this 2nd polynomial. Therefore, the final numerical formula of road friction coefficient estimation can be expressed as Eq. (14). It is a simple algorithm for estimating the road friction coefficient compared with other algorithms.

$$(a_1\lambda - \delta) \cdot \mu_H^2 + (a_2\lambda^2) \cdot \mu_H + (a_3\lambda^3) = 0 \qquad (13)$$

$$\mu_H = \frac{-(a_2\lambda^2) + \sqrt{(a_2\lambda^2)^2 - 4(a_1\lambda - \delta)(a_3\lambda^3)}}{2(a_1\lambda - \delta)} \qquad (14)$$

PRE-PROCESSING

In order to use the algorithm described in the previous chapter, the slope of peak-line (C) and the coefficients of 3rd equation in Eq. (13) should be acquired beforehand through pre-processing. In this paper, in order to acquire the necessary values, simulation is conducted with full braking and ABS operation on normal road. Fig. 7 shows the μ-slip curve and Fig. 8 shows the normalized δ. Utilizing these data, the necessary values are acquired as follows: $C = 7.4$, $a_1 = 18.897$, $a_2 = -233.99$, $a_3 = 693.06$.

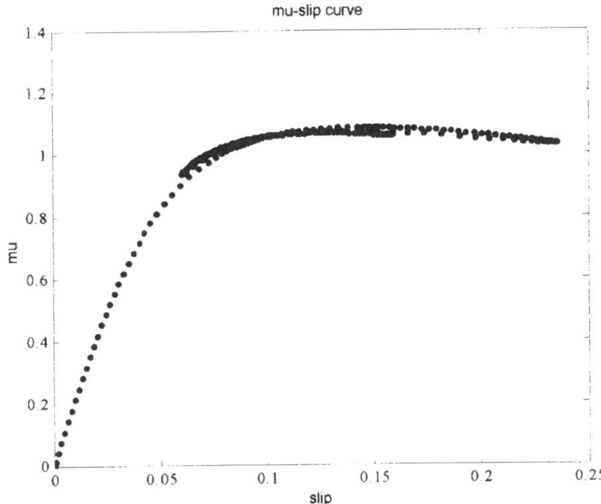

Fig. 7 μ-slip curve during full braking on normal road

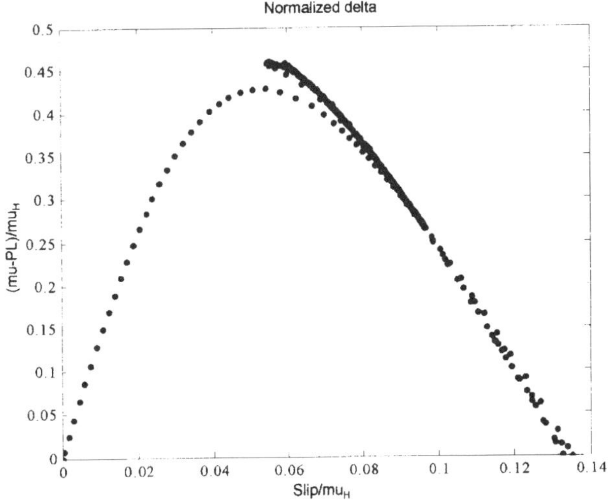

Fig. 8 Normalized δ during full braking on normal road

SIMULATION

SIMULATION OF THE TIRE BRAKING FORCE ESTIMATION

In order to evaluate the performance of tire braking force estimation, simulation is conducted with operating ABS and varying the driving and road condition. The fourteen degree-of-freedom full vehicle model with the Magic Formula tire model (Bakker et al. [13]) is used as a real plant. The calculated tire braking force and brake disk-pad friction from this full vehicle model is used as true value and compared with the estimated tire braking force and brake disk-pad friction.

Estimation results for the brake disk-pad friction and the left-front/left-rear tire braking force are shown in Fig. 9 ~ Fig. 12.

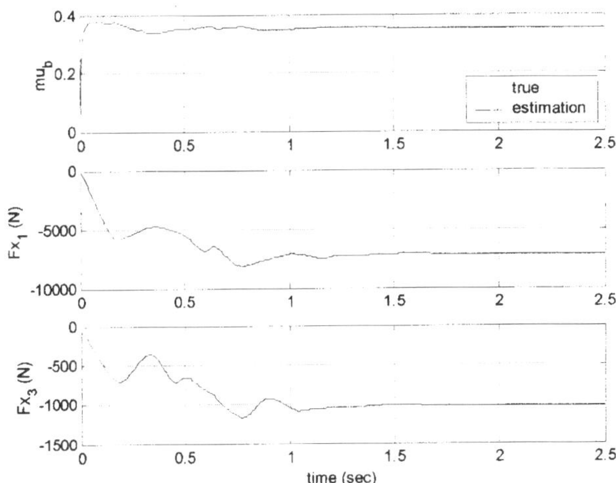

Fig. 9 Simulation results in braking on normal road

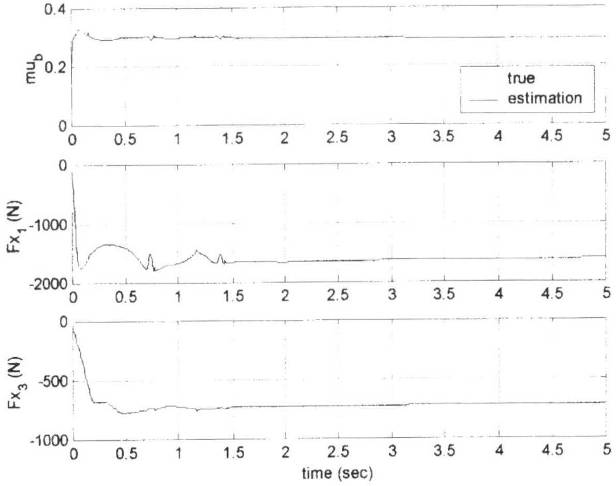

Fig. 10 Simulation results in braking on slippery road

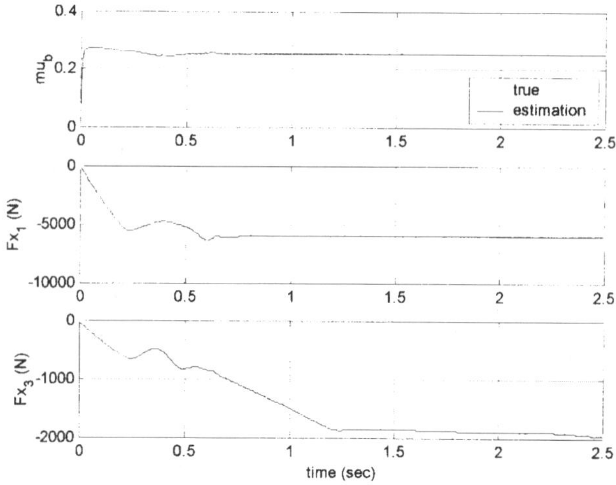

Fig. 11 Simulation results in turning-braking

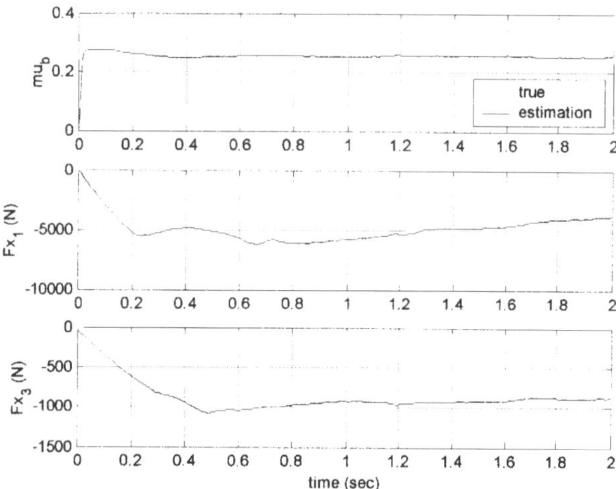

Fig. 12 Simulation results in braking on μ-split road

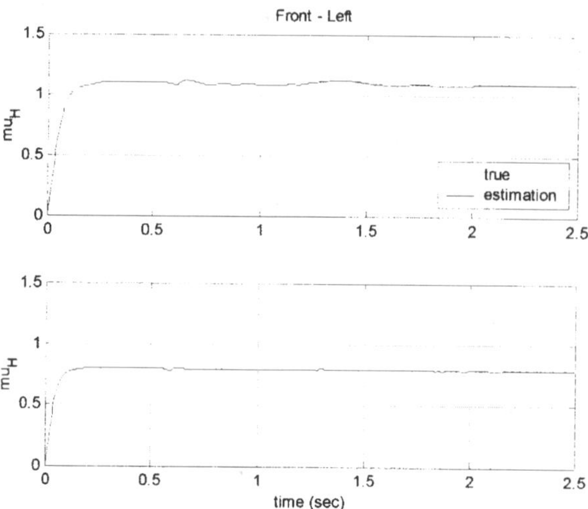

Fig. 13 Simulation result on μ_H =1.1 & μ_H =0.8 road

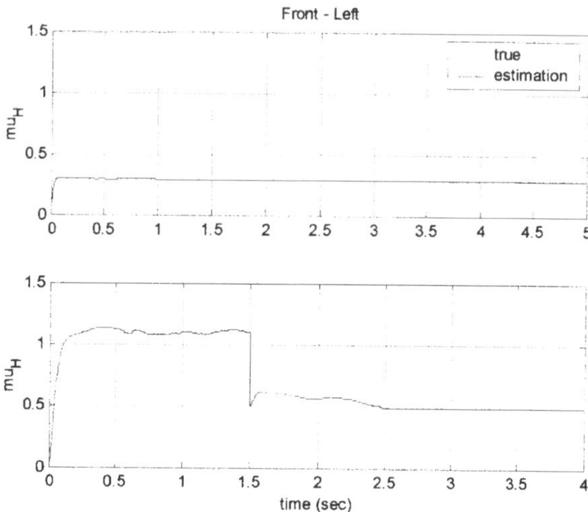

Fig. 14 Simulation result on μ_H =0.3 & μ_H =1.1→0.5 road

Fig. 9 shows the results in case of braking on normal road (μ_H =1.1) and the brake disk-pad friction is 0.35 and Fig. 10 shows the results in case of braking on slippery road (μ_H =0.3) and the brake disk-pad friction is 0.3. In turning-braking case on normal road, the estimation results are shown in Fig. 11 and estimation results for braking on μ-split road (left: μ_H =1.1, right: μ_H =0.3) are shown in Fig. 12. The estimated brake disk-pad friction and tire braking force at each wheel coincides well with the true values at the transient as well as at the steady state.

SIMULATION OF THE ROAD FRICTION COEFFICIENT ESTIMATION

The road friction coefficient estimation results based on the proposed method are verified with simulation on four types of road condition as follows: normal road (μ_H =1.1 & μ_H =0.8), slippery road (μ_H =0.3) and μ-change road (μ_H =1.1 → μ_H =0.5). The verification results are shown in Fig. 13 and Fig.14 and the estimated road friction coefficient coincides well with the true value.

CONCLUSION

The tire braking force and road friction coefficient estimation algorithm is proposed in this paper in order to achieve the good performance of the wheel slip control system in case of EHB. The tire braking force acting at each tire is estimated considering the variation of the brake disk-pad friction due to various circumstances. The estimation algorithm is designed utilizing seven degree-of-freedom vehicle model and Extended Kalman Filter. The road friction coefficient is estimated without a tire model. The common characteristics of the tire-road friction without relating road condition are derived and a simple algorithm using these characteristics is developed. The estimated performance of tire braking forces and the road friction coefficient is evaluated in various simulation and the results demonstrate that their estimation performance coincide pretty well with the true values regardless of various road conditions. It is believed that the proposed tire braking force and road friction coefficient estimation methods are promising technique for realizing the wheel slip control system replacing the conventional ABS.

ACKNOWLEDGMENTS

The support of the Ministry of Commerce, Industry and Energy in Korea (No. 10005253) is gratefully acknowledged.

REFERENCES

1. I. Petersen, T. A. Johansen, J. Kalkkuhl, J. Ludemann, "Wheel Silp Control in ABS Brakes Using Gain Scheduled Constrained LQR," European Control Conference, 2001.

2. J. Yi, L. Alvarez, R. Horowitz, "Adaptive Emergency Braking Control With Underestimation of Friction Coefficient," IEEE Trans. on Control Systems Technology, Vol. 10, No. 3, 2002.

3. Kenneth R. Buckholtz, "Reference Input Wheel Slip Tracking Using Sliding Mode Control," SAE 2002-01-0301, 2002.

4. S. Müller, M. Uchanski, K. Hedrick, "Slip-Based Tire-Road Friction Estimation During Braking," Proceedings of IMECE'01, 2001.

5. F. Gustafsson, "Slip-based Tire-Road Friction Estimation," Automatica, Vol. 33, No. 6, pp. 1087-1099, 1997.

6. I. Kageyama, T. Katayama, "A Study of Anti-Lock Braking Systems with μ Estimation," Proceedings of AVEC'96, pp. 695-703, 1996.

7. U. Kiencke, L. Nielsen, *Automotive Control Systems*, pp. 283-289, Springer, 2000.

8. L. R. Ray, "Nonlinear Tire Force Estimation and Road Friction Identification," Automatica, Vol. 33, No. 10, pp. 1819-1833, 1997.

9. H. Nishira, T. Kawabe, S. Shin, "Road Friction Estimation Using Adaptive Observer with Periodical σ-modification," IEEE Intl. Conf. on Control Applications, pp. 662-667, 1999.

10. W. R. Pasterkamp, H. B. Pajeka, "The Tyre As A Sensor To Estimate Friction," Proceedings of AVEC'96, pp. 839-854, 1996.

11. M. Boutayeb, H. Rafaralahy, M. Darouach, "Convergence Analysis of the Extended Kalman Filter Used as an Observer for Nonlinear Deterministic Discrete-Time Systems," IEEE Trans. Autom. Control, Vol. 42, No. 4, pp. 581-586, 1997.

12. D.E., Smith and J.M., Starkey, "Effects of Model Complexity on the Performance of Automated Vehicle Steering Controllers: Model Development, Validation and Comparison," *Vehicle System Dynamics*, Vol. 24, pp. 163~181, 1995.

13. E. Bakker, H. B. Pacejka, L. Lidner, "A New Tire Model with an Application in Vehicle Dynamics Studies," SAE 890087, pp. 83-95, 1989.

An Original Method to Predict Brake Feel: A Combination of Design of Experiments and Sensory Science

Victoire Dairou and Alain Priez
RENAULT, Research Division

Jean-Marc Sieffermann and Marc Danzart
Ecole Nationale Supérieure des Industries Alimentaires (ENSIA)

ABSTRACT

A brake by wire research vehicle was developed and equipped with an electrical active pedal feel emulator. With this device, the relationships between brake pedal force, pedal travel and vehicle deceleration are fully adjustable. We can tune 11 technical parameters (7 for the force/travel pedal relationship and 4 for the deceleration/travel relationship). A Plackett-Burman design of experiments was used to generate 12 unique braking laws. These laws were evaluated by a sensory descriptive analysis panel on seven criteria. Partial Least Square (PLS) analysis was used for data treatment. This analysis sorted out the most influential technical parameters on brake feel. As a result, a software application able to predict brake feel using technical parameters was implemented.

INTRODUCTION

In a conventional braking system, the wheel brake force is directly provided by the transmission of the driver's muscular energy through the pedal. The shape of the pedal is constrained by its function: by leverage, it amplifies the force of the driver and transmits this force to the braking device.

In a Brake by Wire (BBW) system, the driver's braking intention is no longer transmitted to the brake system mechanically but electronically. Therefore, the braking energy is not directly provided by the driver [1]

In a BBW system, the pedal is decoupled from the rest of the brake system. This creates new opportunities in the definition of driver/brake pedal interface [2]

In order to improve braking comfort, we need to know what kind of sensations and feedback are important for the drivers to brake in optimum conditions of comfort and safety.

A lot of information is available about brake feel. Newcomb described precisely the driver behavior during braking [3]. Harries published what he thought the best

range of pedal travel / effort / deceleration relationships was, to obtain an acceptable pedal feel [4]. Ebert *et al.* presented their work about the establishment of a brake feel index (BFI). This BFI is used as a predictor of subjective brake feel [5].

Markus published an article in 1999 called "where does good brake feel come from?" [6].

These documents relate to conventional hydraulic braking. In our case, we take into account the opportunities given by BBW.

In a BBW device, a braking law can be divided in 2 independent relationships: 1) the pedal characteristic which describes the relationship between pedal force and travel. 2) the braking command law which describes the relationship between the driver request and the vehicle deceleration.

We look for a driver/brake pedal interface which allows fast and precise deceleration control of the vehicle. We study on one hand the feedback to be emulated to the driver (force level, stroke length,...) and on the other hand the shape of the law of deceleration.

This investigation implies two different phases: 1) Modelization of brake feel in order to understand the relationships between brake design parameters and brake feel. 2) Identification of customer's preferences about brake feel in order to tune brake systems and to meet customer's expectations.

This paper describes the first phase of this investigation: Modelization of brake feel.

In order to improve our knowledge of brake feel, we begin a systematic investigation of the relationships between different braking laws and the resulting brake feel.

To describe and characterize the brake feel, the methods of sensory analysis were utilized.

Sensory science tools are widely used in food and cosmetic industries. The sensory techniques strive to allow objective measurement of sensations (ISO Norm, [7]). It uses human subjects as instruments. It induces the need to minimize the variability of data provided by humans and control the bias by making full use of the best existing techniques in psychology and psychophysics. A sensory study generates a huge amount of data, their treatment rely on statistical multivariate analysis [8].

We established a connection between our purpose and a development of new products: our purpose consists in the optimization of braking laws.

The first question to solve is: what are the factors to optimize? The factors are the technical parameters identified to define a braking law.

The second question is: which braking laws are relevant to test? A design of experiments is used to generate the fewest number of braking laws to estimate the impact of each technical parameter on brake feel.

The third question is: Is it possible to predict brake feel from technical parameters?

To answer this question we've developed an original method of optimization based on the use of Design of Experiments and sensory profiling.

MATERIALS AND METHODS

VEHICLE

To investigate the brake pedal feel, we need to collect driver sensations in a decelerating vehicle. We think that a static or dynamic test using a driving simulator is not relevant because it is too far from reality.

We could have chosen to study brake feel in different vehicles but this would have presented problems due to brand image of vehicles and the large dispersion of geometry, positions of pedals, seats, vehicle behavior, etc. In order to eliminate as many secondary influences as possible, a unique vehicle was developed. This prototype is equipped with Electro-Hydraulical Brake (EHB) and an electrical active pedal feel emulator (Figure 1).

With this device, we are able to modify by software the brake pedal characteristics and the associated deceleration while driving. The technical possibilities of this vehicle allow us to study a wide range of braking configurations in one common environment (seat, pedal geometry, etc.).

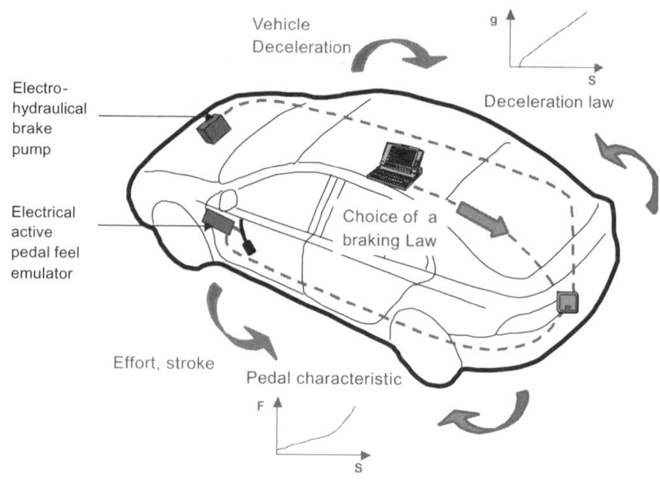

Figure 1: functional drawing of the research vehicle

The dynamic parameters of vehicle deceleration are another source of variability; such parameters are the variation of friction brake pad as they relate to temperature, patterns of pad wear etc. In order to eliminate these factors of variability, the braking system of the vehicle is controlled by deceleration instead of pressure.

BRAKING CONFIGURATIONS

In order to answer our first question (what are the parameters to optimize?), a braking law was modelized.

Braking law

A braking law is a set of 2 curves: 1) the relationship between brake pedal force and pedal travel (brake pedal characteristic). 2) the relationship between the driver request and the deceleration of the vehicle (command law).

Brake pedal characteristics can be described with pedal force, pedal travel and pedal applying speed. Seven parameters are identified to describe the curve: threshold of effort (F0), stiffness 1 (S1) and stiffness 2 (S2), travel at stiffness change (T0), damping (D), hysteresis (H), maximum stroke (T1) (Figure 2a).

The hysteresis represents the quantity of possible release of effort without any pedal displacement. The dry friction coefficient represents the value of this parameter. The slope of the curve Effort = f(applied pedal speed) represents the value of damping.

The command law can be described with pedal stroke and vehicle deceleration. Four parameters are used to describe this curve: dead stroke (DS), slope 1 (Sl1), stroke at slope change (Sc) and slope 2 (Sl2) (Figure 2b).

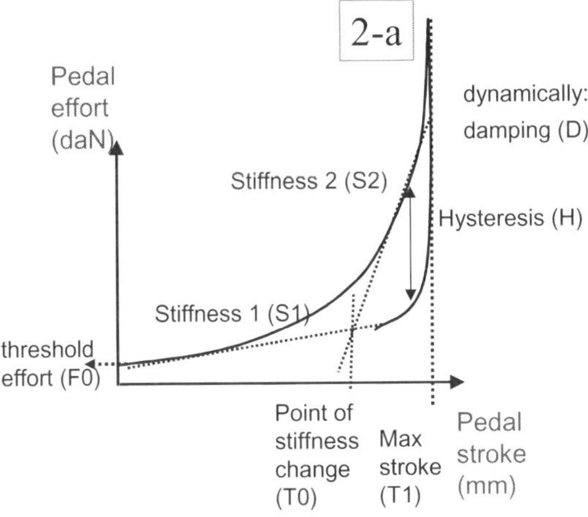

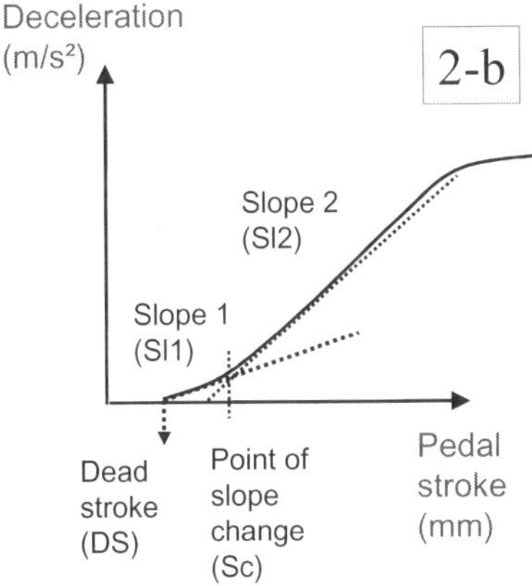

Figure 2 : Braking Law 2-a: Pedal Characteristic and 2-b Command Law

A braking law consists of these two curves and the 11 parameters identified to represent it. For the following sections, we name the 7 pedal characteristics parameters: pedal1 to pedal7 and the 4 command law parameters: decel1 to decel4.

Experimental designs

We identified 11 parameters to be optimized. To test all possible combinations of the 11 parameters on two levels, 2^{11} = 2048 experiments would need to be conducted.
To reduce the number of combinations, we built a Plackett-Burman design of experiments with all the technical parameters [9]. We studied a maximum level (max) and a minimum level (min) of every technical parameter. The design consisted of 12 braking laws presented in table 1.
To improve the system understanding, the Design of Experiments tool is used to discover the effects of the technical parameters on the brake feel.
Then, the impact of each parameter on the brake feel criteria will be estimated.

Table 1: Construction of 12 braking laws according to a Plackett-Burmann design of experiments.

	Pedal 1	Pedal 2	Pedal 3	Pedal 4	Pedal 5	Pedal 6	Pedal 7	Decel 1	Decel 2	Decel 3	Decel 4
L1	max	max	min	max	max	max	min	min	min	max	min
L2	min	max	max	min	max	max	max	min	min	min	max
L3	max	min	max	max	min	max	max	max	min	min	min
L4	min	max	min	max	max	min	max	max	max	min	min
L5	min	min	max	min	max	max	min	max	max	max	min
L6	min	min	min	max	min	max	max	min	max	max	max
L7	max	min	min	min	max	min	max	max	min	max	max
L8	max	max	min	min	min	max	min	max	max	min	max
L9	max	max	max	min	min	min	max	min	max	max	min
L10	min	max	max	max	min	min	min	max	min	max	max
L11	max	min	max	max	max	min	min	min	max	min	max
L12	min	min	min	min	min	min	min	min	min	min	min

EXPERIMENTAL CONDITIONS

Driving experiments

We studied braking in comfortable and normal braking situations ie deceleration inferior to 5 m/s². This deceleration range represents more than 99.9 % of braking situations encountered by the customers [10].

Emergency braking was not used for these experiments.

The experiments were carried out on the RENAULT test tracks, in Normandy (France).

SENSORY EVALUATION

In order to evaluate braking laws, we searched for a specific lexicon, i.e. sensory descriptive criteria. The best way to elaborate a lexicon was to set up a sensory profile.

Flash profile to identify sensory criteria [11]

We chose Flash profile methodology for its rapidity [12]. We set up a sensory profile panel composed of 5 subjects, 2 technical and 3 non-technical drivers. Each subject chose his/her own words to describe his/her sensations in different braking situations. The subjects evaluated 5 cars, from the market, equipped with conventional braking devices. To complete the study, each subject spent between 2 and 8 hours in the cars. The subjects who spent 8 hours for the experimentation had triplicated their evaluations. We obtained a grid of 7 sensory attributes to characterize the driver's sensations when braking: **travel, idle travel, force level, responsiveness, deceleration perceived, ease of balance** (or ease of modulation) **and graduality of the braking** e.g. the controllability of braking" and the ability to brake smoothly.

<u>Sensory profiling of braking laws</u>

In order to describe the sensory characteristics of the 12 braking laws of the design of experiments (Table 1), we set up a sensory profile.

Subjects

The 8 assessors, 6 men and 2 women, aged from 24 to 52 were recruited among the personnel of Renault. 5 panelists were technical drivers and 3 were non-technical drivers. These 3 subjects were trained to evaluate their sensations during braking.

The assessment procedure

We adapted the QDA® procedure described by Stone *et al.* [13] and by ISO Norm 11035 [14] to our very situation.

In an introductory session, the experimenter explained the objectives of the study, the time commitment, gave the security recommendations and presented the vehicle.

Sessions 1 to 2: Panelists were individually presented with a range of braking laws. Descriptive terms issued from Flash Profile study were presented and discussed. Terms definitions were clarified and different evaluation procedures were tested.

Sessions 3 to 4: Subjects developed an agreed-on grid of terms. For each term, the grid was constituted of:

1. A definition,
2. An evaluation procedure describing the action to do to evaluate the term
3. The condition of test (vehicle speed, pedal applying speed, level of deceleration, …)
4. Examples of braking laws represented maximum and minimum references for this term
5. Synonymous terms.

Sessions 4 to 8: Training: Subjects were familiarized with the scale and its use for developing a unified approach to scaling. References were used for calibrating intensity scores. Repeated trials were made on the 12 braking laws by the 3 non-technical drivers for each attribute: means and standard deviations were used to monitor assessor's reliability. Analysis of Variance (ANOVA) was also carried out to monitor panel performance.

Sessions 9 to 11: Evaluations of the 12 braking laws (triplicated for the 3 non-technical drivers) were conducted. Subjects evaluated the 12 braking laws term by term. The braking laws were presented in a sequential monadic way according to a balanced latin square design to carry over the effect of presentation. A 15-cm unstructured line scale, anchored with the words "low" and "high" from each end, was used to rate intensity of the attributes.

The mean time spent by the subjects to complete the study was about 8 hours.

DATA ANALYSIS

<u>Sensory evaluation</u>

Data quality

Individual data treatments were performed to assess quality of the data. Analysis of variance (ANOVA) was applied to the descriptive ratings. ANOVA is the statistical method used to compare more than two samples in a single study. The source of variation was the braking laws. One way-ANOVA measures the reliability of a subject for one variable (an attribute) and allows the experimenter to decide if a subject is consistent or not for that variable.

Results presentation

Principal Component Analysis (PCA) provides a way to summarize data collected on a large number of variables in fewer dimensions [15]. PCA of the matrix of mean attribute ratings across samples was carried out to sort out relationships among the sensory attributes and differences among the twelve braking laws. PCA is the statistical technique used to identify the smallest number of latent variables, called 'principal components', that explain the greatest amount of observed variability. PCA analyses the correlation structure of a group of multivariate observations and identifies the axis along which the maximum variability occurs. This axis is called the first principal component. The second principal component is the axis along which the greatest amount of remaining variability lays subject to the constraint that the axes must be perpendicular to the previous one, etc.

PCA generates graphical outputs which clearly illustrate the relationships both among the variables and between the observations.

The technical parameters of every law were added as supplementary variables in the PCA analysis. They are represented in the form of vectors on the plot of attributes. Xlstat®, a set of Microsoft Excel® macros, was used to perform PCA.

<u>Design of Experiments</u>

Partial Least Squares Regression (PLS)

The PLS regression is a multivariate technique [16] PLS derives factors, i-e linear combinations of the predictor variables (x-variables), that explain large portions of the variability in the x-variables and simultaneously correlate, to as great a degree as possible, with the dependent predictor variable y. PLS ensures that each factor identified has maximal predictive power on y.

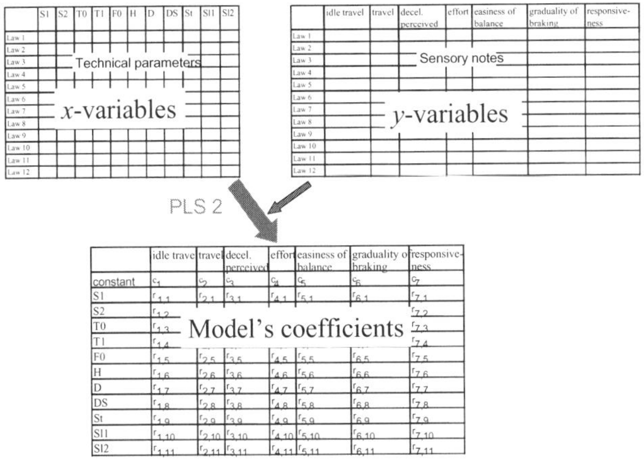

Example for construction of the model
Effort predicted = c4 + S4 (r4,1) + S2 (r4,2) + T0(r1,3) + T1(r4,4) + F0(r4,5) + H(r4,6) + D (r4,7) + DS (r4,8) + St (r4,9) + SI1(r4,10) + SI2(r4,11)

Figure 3: Principle on use of PLS regression to construct a model

PLS readily extends to predicting more than one dependent variable simultaneously. When a single dependent variable is predicted the analysis is called PLS1. When several variables are predicted the analysis is called PLS2. We used the reference computer programs that perform PLS: SIMCA® [17].

PLS2 was performed with the experimental design shown in Table 1 as *x*-variables and with sensory ratings as *y*- variables (Figure 3).

RESULTS

SENSORY DESCRIPTION OF BRAKING

Data quality

Three of the 8 subjects triplicated their evaluations. We could only analyze their data with ANOVA. The 5 technical drivers did not triplicate their evaluations, so we could not analyze their performance. However, because they were trained experts in subjective descriptions of braking we had good reasons to be confident in their data.

Table 2: Significance levels from ANOVAs carried out on each attribute

	Subject A	Subject B	Subject C
idle travel	< 0,001	< 0,001	< 0,001
Travel	< 0,001	< 0,001	< 0,001
Force level	< 0,001	< 0,001	< 0,001
Responsive	< 0,001	< 0,001	0,0504
Ease of balance	< 0,001	< 0,001	< 0,001
Decel. perceived	< 0,001	< 0,001	< 0,001
Graduality of braking	0,024	0,015	< 0,001

Results of ANOVA are presented in table 2. Every term is significant at the level of 5% or less. So each subject uses every term to differentiate the 12 braking laws.

These 3 subjects appeared to be consistent and reliable. We had good quality data.

Sensory evaluation

PCA is used to depict the relationships of 12 braking laws evaluated on seven attributes. In our analysis the first two principal components explained 77 % of the original variability. PCA is used to display, in two dimensions, the relative "locations" of the samples (Figure 4).

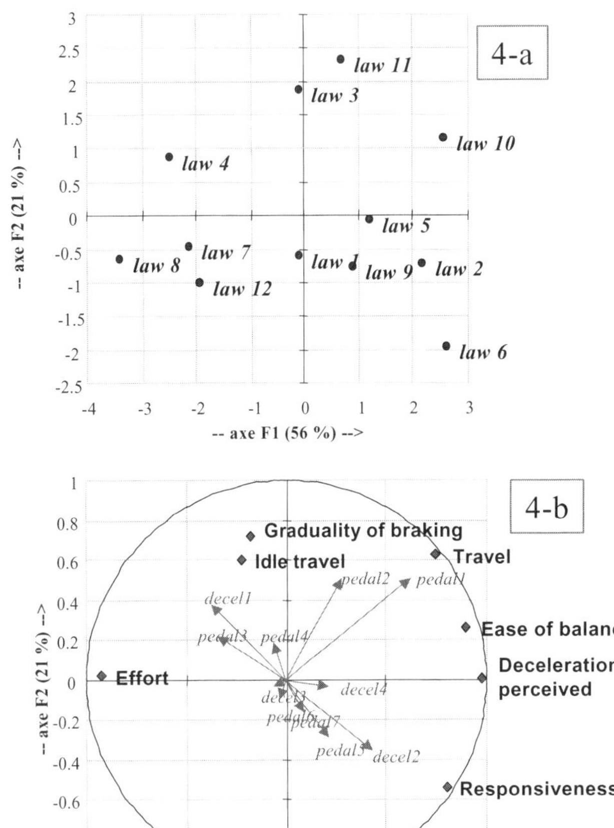

Figure 4: Sensory maps of the 12 braking laws obtained by PCA (first 2 principal components) 4-a: mean samples plot; 4-b: variables plot with technical parameters in supplementary variables

We notice that the laws are evenly distributed in the sensory maps. The Plackett-Burman design created laws which cover all the sensations of braking.

The braking laws are mainly opposed on the characteristics of effort level and deceleration perceived. The laws **4**, **8**, **7** and **12** are characterized by short **travel**, and a high level of **effort**. They are perceived as low **efficiency** and not **easy to balance**. The very **responsive** laws, like law **6**, are difficult to measure. A law with a long **idle travel** is perceived as **non-responsive**.

Information brought by the supplementary variables:
- The longer the vector, the more influence the parameter has on the brake feel. In our study the most influential parameters are: *pedal1*, *decel2*, *pedal2*, *decel1*, *decel4* and *pedal5*. The closer the parameters are to the center the less influence they have on the brake feel.
- the direction of the vector indicates which direction this parameter influences the sensation; for example an increase of the parameter *pedal1* is going to increase the sensations of **travel**, **ease of balance** and to decrease the sensation of **effort**. Inversely, if we want more **responsiveness** of the braking, we will increase the level of *decel2* and decrease *decel1*.
This plot displays the most influential parameters on brake feel and allows us to understand in what way these parameters influence the sensations.

MODELIZATION OF BRAKE FEEL

PCA on figure 4 gives us qualitative indications on the influence of the technical parameters on brake feel, but to quantify and modelize these influences, we must perform Partial Least Square regression. PLS2 is performed with the technical parameters shown in Table 1 as **x**-variables and with sensory ratings as **y**- variables (Figure 3). The result of PLS2 is a matrix of coefficients. A linear combination of these coefficients allows to predict for a braking law, the sensory notes for the seven criteria from technical description of this braking law (11 technical parameters). This research lead to the creation of a numerical model to simulate the brake feel. As a result, we implemented a software application able to predict brake feel. The inputs of the model are the braking laws technical parameters and the outputs are sensory characterization of these laws, which consist of a score in the seven sensory criteria. This model is called model-1.

Validation of the model:

We constructed 20 braking laws, different from those located in-table 1 which served for constructing the model. The technical parameters of these 20 laws were entered into the software application to obtain the predicted sensory scores.
At the same time, these 20 braking laws were implemented in the research vehicle and evaluated by the subjects of the descriptive panel.
A comparison of the predicted scores and the observed scores was made by calculation of the coefficients of correlation between every predicted and observed attribute.
In figure 5, histograms show that the correlations go from 0.98 for the attribute **travel** to 0.56 for the **graduality of braking**. We consider that a percentage of correlation superior to 0.7 allows to obtain a reliable qualitative indication of the braking feel.

The predictive model of brake feel allows the user to obtain a predicted sensory profile of a braking law from its technical parameters. The model is reliable concerning the sensations of **effort, responsiveness, idle travel, stroke** and **deceleration perceived**. The least predicted attributes are **ease of balance** and **graduality of braking**.

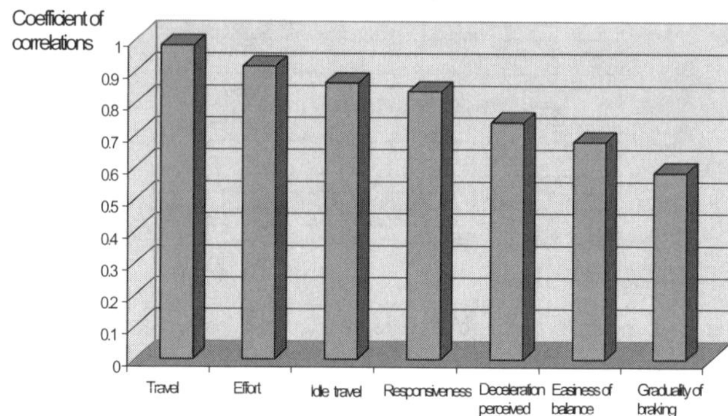

Figure 5: Histograms of coefficients of correlation between predicted and observed sensory notes for the seven descriptive terms.

In the same way that we model brake feel from technical parameters, we built another model to predict the technical parameters of a braking law from expected brake feel (model-2). To build this model-2, PLS2 is performed with sensory ratings as **x**-variables and with the technical parameters shown in Table 1 as **y**-variables. Thanks to this model-2, we could easily tune a brake by wire vehicle according to customer brake feel expectations. But the predictive power of this model-2 is not as good as the predictive power of the model-1 based on technical parameters. One of the reasons could be that the model-2 of braking laws is constructed from subjective data with variance (sensory notes) whereas the brake feel model is constructed from exact data (level of technical parameters).

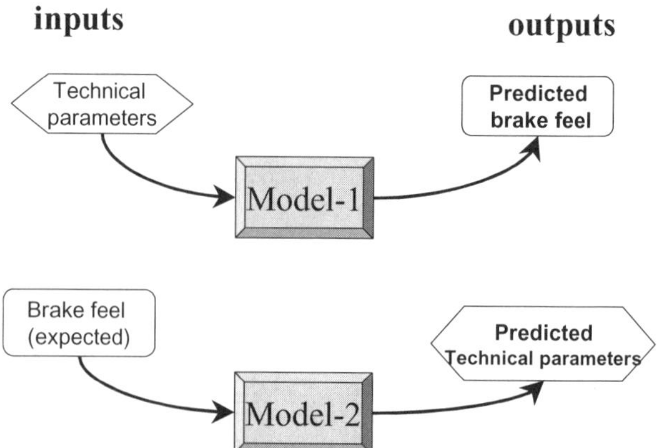

Figure 6: Schematic representation of the inputs and outputs of model-1 and model-2

CONCLUSION

The main results obtained during the study are:
1. Development of a sensory descriptive vocabulary for braking and construction of a tool to teach the sensory braking attributes.
2. Determination of how technical parameters influence brake feel.
3. Development of a predictive model of braking sensations which will allow the user to qualitatively predict the sensory characterization of a braking law from the shape of the laws.

The model simplifies the development of new software functionalities and enables the optimization of future Brake by wire systems.
Even if, sensory profiling is time-consuming, its use warrants robust and reliable results on driver's perceptions. Combining Sensory Science and Design of Experiments is a relevant and efficient tool to optimize motorist's brake feel.
The next step of this work is the identification of customer's expectations about brake feel. We will be inspired by Kowalski and Ebert methodology published in 1993 [18] and by preference mapping for consumer's data treatment.

ACKNOWLEDGMENTS

Authors wish to thank N. Herbeth, S. Baerd and J. Gonidec for their help and support in conducting the study and A. Richard and R. Pothin for their help in developing the vehicle.

REFERENCES

1. Hedenetz, B. and R. Belschner, *Brake by Wire without mechanical backup by using a TTP-Communication Network*. SAE Technical Paper Series, 1998. **981109**.
2. Bill, K., M. Semsch, and B. Breuer, *A new approach to investigate the vehicle interface driver / brake pedal under real road conditions in view of oncoming brake-by-wire systems*. SAE Technical Paper Series, 1999. **1999-01-2949**.
3. Newcomb, T.P., *Driver behaviour during braking*. SAE Technical Paper Series, 1981. **SAE 810832**.
4. Harries, D.A., *Pedal feel with power braking systems*. Lucas Engineering Review, 1978. **7**(3): p. 65-69.
5. Ebert, D.G. and R.A. Kaatz, *Objective characterization of vehicle brake feel*. SAE Technical Paper Series, 1994. **940331**.
6. Markus, F., *Where does "good brake feel" come from?* Car & Driver, 1999. **45**(2): p. 97-101.
7. ISO., *ISO Norm 8586-1. Sensory analysis - General guidance for the selection, training and monitoring of assessors - part I: selected assessors.*, I.O.f. Standardization, Editor. 1993, American National Standards Institute, 11 west 42nd street, NY 10036.: NY.
8. Meilgaard, M., G.V. Civille, and B.T. Carr, *Sensory Evaluation Techniques*. Third ed. 1999, Boca Raton: CRC Press. 375 p.
9. Montgomery, D.C., *Design and Analysis of Experiments*. 5th ed. 2001, Hardcover: John Wiley & Sons. 696.
10. Mortimer, R.G. and P.L. Olson, *Some factors limiting driver-vehicle performance*. SAE Technical Paper Series, 1973. **730017**.
11. Dairou, V., et al. *Sensory Evaluation of Car Brake Systems -The Use of Flash Profile as a Preliminary Study Before a Conventional Profile*. in *4th Pangborn Sensory Science Symposium*. 2001. Dijon, France: INRA.
12. Dairou, V. and J.M. Sieffermann, *A comparison of Fourteen Jams Characterized by Conventional Profile and a Quick Original Method, the Flash Profile*. Journal of Food Science, 2002. **67**(2): p. 826-834.
13. Stone, H. and J.L. Sidel, *Sensory Evaluation Practices*. 2nd ed. Food Science and Technology, ed. S. Taylor. 1993, San Diego: Academic Press, Inc. 337 p.
14. ISO., *ISO Norm 11035. Sensory analysis - Identification and selection of descriptors for establishing a sensory profile by a multidimensional approach.*, I.O.f. Standardization, Editor. 1994, American National Standards Institute, 11 west 42nd street, NY 10036: NY.
15. Pigott, J.R. and K. Sharman, *Methods to Aid Interpretation of Multivariate Data*, in *Statistical Procedures in Food Research*, E.S. Publ, Editor. 1986, Pigott, J.R: Essex.
16. Tenenhaus, M., *La régression PLS - Théorie et pratique*. 1998, Paris: Editions Technip. 256 p.
17. Umetrics, A., *SIMCA-P 9 - A New Standard in Multivariate Data Analysis. User guide and tutorial*. 2001: Umetrics AB. 333 p.
18. Kowalski, M.F. and D.G. Ebert, *Establishing Brake Design Parameters for Customer Satisfaction*. SAE Technical Paper Series, 1993. **930799**.

CONTACT

Victoire DAIROU
RENAULT, Research Division, Ergonomics Department
TCR RUC T55
1, avenue du golf,
78288 Guyancourt Cedex
France
Tel: +33 (0) 1 30 03 28 37
Fax: + 33 (0)1 34 95 90 80
e-mail: victoire.dairou@renault.com

DEFINITIONS AND SYNONYMS

Synonyms terms:

1) descriptive term = descriptor, attribute, sensory criteria.

2) panelist = assessor, subject

Wheel Slip Control for Antilock Braking Systems Using Brake-by-Wire Actuators

Sascha Semmler and Rolf Isermann
Darmstadt University of Technology

Ralf Schwarz and Peter Rieth
Continental Teves AG & Co. oHG

ABSTRACT

This paper describes an approach of how to control the wheel slip of a vehicle using brake-by-wire actuators. The advantage of brake-by-wire actuators - such as the electro-hydraulic (EHB) and the electro-mechanical brake (EMB) - is that the caliper pressure or the clamping force, respectively, are known. It will be shown by measurement results that the wheels of a research vehicle equipped with an EHB system and the new control approach can be kept at any desired wheel slip on different surfaces, i.e. ice, snow, and dry asphalt.

INTRODUCTION

The objective of Antilock Braking Systems (ABS) is to increase the safety of a vehicle in emergency braking situations. This is accomplished by maintaining steerability and stability of the vehicle as well as reducing braking distance by keeping the wheels at an optimum brake slip.

An ideal Antilock Braking System (ABS) would be able to keep the slip of the wheels at a point of maximum vehicle deceleration which would lead to a minimum braking distance. The function which shows the relation between wheel slip λ and the road friction coefficient μ is called μ-λ curve. The shape of this curve is defined by many factors, e.g. tire, surface, load, temperature etc. The curve itself can be divided into stable and unstable areas in respect to the wheel slip control.

Fig. 1 illustrates the nonlinear behavior of the tire traction in relationship with the wheel slip. At the optimum wheel slip $\lambda=\lambda_{opt}$, the gradient $d\mu/d\lambda$ is equal to zero. Conventional ABS usually cycle around this optimal point.

Their control algorithms are based on a combination of wheel slip and wheel acceleration control in order to prevent wheels from locking. The input signals are the wheel speeds. Additionally, four-wheel driven vehicles sometimes have a longitudinal acceleration sensor. Output signals are brake pulses which increase or decrease the caliper pressure.

The ABS braking cycle of conventional ABS systems available on the market is divided into different phases, hereby trying to keep the slip of the wheels at the optimum wheel slip with maximum tire traction as long as possible. Due to limited information about the road and the tire conditions as well as the caliper pressure, the efficiency of ABS is limited. Good systems reach an efficiency of $\eta_{ABS}\approx0.95$ [1].

However, an ideal ABS would be able to keep the wheel slip at a point of maximum tire traction and reduce the braking distance to a minimum.

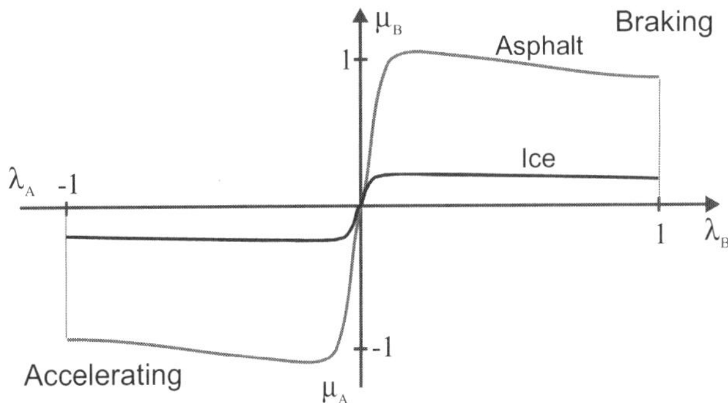

Figure 1.   Two typical μ-λ-curves

Neglecting the air resistance and assuming a constant normal force and road friction coefficient μ, Eqn. 1 shows the resulting braking distance depending on the initial vehicle velocity v_0 with g as the gravity constant.

$$s_{min} \approx \frac{1}{2} \cdot \frac{v_0^2}{\mu \cdot g} \tag{1}$$

It shows that the braking distance is linear to the reciprocal value of the tire traction achieved during braking. Therefore, the value of μ during ABS braking is the value to be optimized to reduce the braking distance.

Continental Teves and the Institute of Automatic Control are currently developing a new generation of ABS for brake-by-wire actuators. This ABS is capable of keeping the wheel at any desired slip with high accuracy - whether in stable or unstable areas of the μ-λ curve. In the following, a wheel slip controller is introduced using brake-by-wire technology. The advantage is that additional sensors such as pressure sensors can be used for data evaluation and control of the wheel slip. Also, the desired pressures are given directly to the electronic control unit of the pressure controller. The ABS does not have to use brake pulses any more.

DESCRIPTION OF THE BRAKE-BY-WIRE SYSTEM

Brake-by-wire systems increase the range of functionalities and improve driving safety. The brake-by-wire system used for this research is the electro-hydraulic brake system developed by Continental Teves AG & Co. oHG. A detailed description of the system can be found in [2].

The hydraulic connection between brake pedal and brake is separated in EHB systems. An electronic pedal module with pedal feel simulator and electronic sensor pickup replaces the classic brake actuation. The driver's demand is transmitted "by wire" to a hydraulic unit with an integrated electronic control unit (ECU). The wheel brakes are still conventional. The EHB is quiet, comfortable, and compact. [3]

Figure 2.   Research vehicle VW Golf IV equipped with the EHB system

A Volkswagen Golf IV is used as a research vehicle. It is equipped with the mentioned EHB. The following sensors are accessible to the new ABS approach:

- four wheel speed sensors
- four caliper pressure sensors
- longitudinal and lateral acceleration sensors
- yaw rate sensor

For the new control approach, only the wheel speed, the caliper pressure, and the longitudinal acceleration sensors are used.

MODELING AND CONTROLLER DESIGN

The new ABS described in this paper is based on the control of the wheel slip. This is reasonable since ABS as well as the Electronic Stability Program (ESP) can use the wheel slip controller for their purposes. The objective of the ABS is to keep the wheel at a slip that provides maximum deceleration whereas the ESP aims at increasing the slip only on selected wheels in order to stabilize the vehicle.

The advantage of brake-by-wire systems is that the clamping force or the caliper pressure, respectively, are known. In this case, the caliper pressure will be used by the wheel slip controller which makes it possible to hold the wheel slip at any desired slip.

To calculate the longitudinal force at the wheel which is useful for many submodules of the new ABS and especially the wheel slip controller, the torque provided by the engine and the brakes as well as the inertia of the wheel have to be taken into account. Eqn. 2 gives the mathematical expression [1], [4].

$$J \cdot \dot{\omega} = r_{dyn} \cdot F_x - e \cdot F_N - M_F + M_A - M_B \tag{2}$$

J is the inertia of the wheel, $\dot{\omega}$ stands for the derivative of the rotational velocity of the wheel, r_{dyn} is the dynamic radius of the wheel, F_x is the longitudinal force at the wheel, F_N denominates the normal force, and M_F, M_A and M_B are the torques due to bearing friction, engine and brakes. Fig. 3 gives a graphical interpretation.

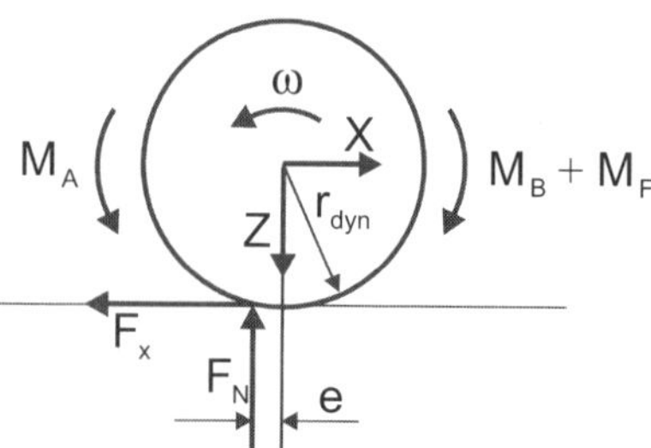

Figure 3.   Forces and torque acting on the wheel

Since the term $e \cdot F_N + M_F$ is treated as disturbance, it is replaced by

$$M_* = e \cdot F_N + M_F \tag{3}$$

Eqn. 2 can therefore be rewritten as

$$J \cdot \dot{\omega} = r_{dyn} \cdot F_x - M_* + M_A - M_B \tag{4}$$

The braking torque is calculated using the caliper pressure p_B in combination with three parameters, the effective piston area A, the brake factor c^*, and the effective friction radius r_B.

$$M_B = A \cdot c^* \cdot r_B \cdot p_B$$
$$= k_B \cdot p_B \tag{5}$$

Describing the wheel slip, it has to be considered whether the wheel is accelerated or braked. In our case, only the wheel slip during braking is relevant which is defined by Eqn. 6.

$$\lambda = \frac{v - \omega \cdot r_{dyn}}{v} \tag{6}$$

v stands for the velocity of the wheel in longitudinal direction of the wheel's inertial system.

The forces which can be applied by the wheels in connection with the surface are limited. The force at each wheel depends on a variety of parameters as illustrated below. It can be changed by varying the wheel slip λ.

$$\mu_x = \frac{F_x}{F_N} = f(\lambda, v, F_N, \ldots) \tag{7}$$

To receive a mathematical model of the wheel dynamics based on the above equations, Eqn. 6 is derived by time and the derivative of the angular velocity is replaced by Eqn. 2.

$$\dot{\lambda} = -\frac{\dot{v}}{v} \cdot (\lambda - 1) - \frac{r_{dyn}}{J \cdot v} \cdot \left(r_{dyn} \cdot F_x + M_A - M_B - M_* \right) \tag{8}$$

The structure of Eqn. 8 is given by

$$\dot{x} = f(x) + G(x) \cdot u$$
$$y = h(x) \tag{9}$$

with a state feedback control

$$u = \alpha(x) + \beta(x) \cdot v \tag{10}$$

and a change of variables represented by

$$z = T(x) \tag{11}$$

The objective is to find a suitable controller. Using the exact feedback linearization [5], a control law which is capable of keeping the wheel at the desired slip is

$$M_B = M_A - M_* + r_{dyn} \cdot F_x + \frac{J \cdot \dot{v}}{r_{dyn}}(\lambda - 1) + \frac{J \cdot v}{r_{dyn}} \cdot \xi \tag{12}$$

where $p_B = M_B / k_B$ is the desired pressure to be built up at the brakes. The difference can be compensated for by

$$\xi = k_P \cdot (\lambda_d - \lambda) + k_D \cdot (\dot{\lambda}_d - \dot{\lambda}) \tag{13}$$

λ stands for the current wheel slip and λ_d for the desired wheel slip with k_P and k_D as amplification factors.

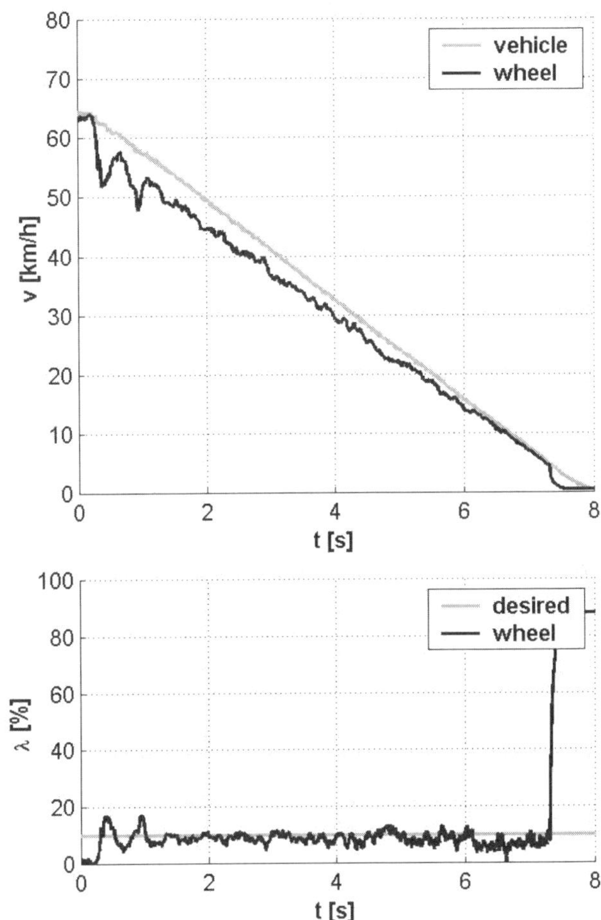

Figure 4. Vehicle and wheel velocity, wheel slip (λ_d=10%, snow)

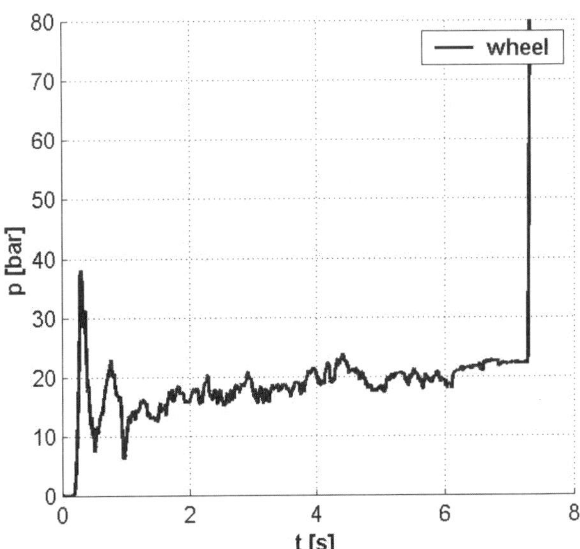

Figure 5. Brake caliper pressure (λ_d=10%, snow)

The PD controller with exact feedback linearization satisfies the stability conditions since the overall transfer function of the wheel dynamics is finally given by

$$G(s) = \frac{\lambda_{\mathrm{d}}(s)}{\lambda(s)} = \frac{1 + s \cdot \left(\dfrac{1 + k_{\mathrm{D}}}{k_{\mathrm{P}}} \right)}{1 + s \cdot \dfrac{k_{\mathrm{D}}}{k_{\mathrm{P}}}} \qquad (14)$$

This is a PDT_1 transfer function which is stable for $k_P > 0$ and $k_D > -0.5$. This controller is able to stabilize the wheel at any desired slip.

MEASUREMENT RESULTS

The measurements shown in the following are carried out on the mentioned VW Golf IV research vehicle shown in Fig. 1. The vehicle is equipped with the brake-by-wire system EHB. The new ABS program code is running on a Power-PC. The front axis is driven in ABS mode. Only the rear wheels are running free in order to provide the reference velocity of the vehicle. Also, the stability of the vehicle at high slip values (i.e. λ_d=70% as can be seen below) can be assured by preventing the rear axis from braking.

BRAKING MANEUVERS ON SNOW - Fig. 4 shows a braking maneuver on snow conducted at the Continental Teves test center in Arvidsjaur, Sweden. Usually, the disturbances on snow are very high due to the uneven ground. But even on this surface, the wheel slip controller is able to keep the wheel at the desired wheel slip of 10%. Also, it can be seen from Fig. 5 that the caliper pressure changes slightly due to the disturbances by the road.

In Fig. 6 and 7, the same maneuver was conducted, but now at a desired wheel slip of 44% which is in the unstable area of the μ-λ-curve. It has to be pointed out that the wheel slip at the beginning of the braking maneuver can be caught at the desired slip and does not overshoot. Even disturbances by the ground can be sufficiently be compensated for.

BRAKING MANEUVER ON ICE - In Fig. 8, the vehicle is stopped on ice at a desired wheel slip of 70% which in reality is only reasonable at selected wheels for a stability control system such as the ESP. The wheel slip is very smooth whereas the pressure is changing quickly in order to stay at the desired wheel slip. Fig. 9 shows the caliper pressure.

BRAKING MANEUVERS ON DRY ASPHALT - The maneuvers on dry asphalt were carried out at the Continental Teves test track in Frankfurt, Germany. Fig. 10 and 11 represent a measurement at a desired wheel slip of 7% which is close to the optimal braking slip.

But even at a large wheel slip of 31%, the wheels can be stabilized as it can be seen from Fig. 12 and 13.

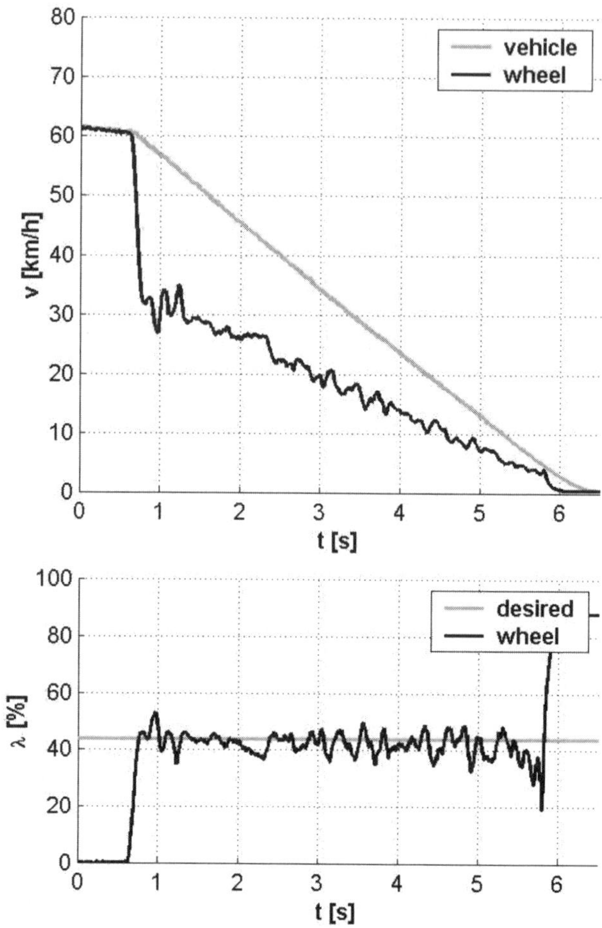

Figure 6.   Vehicle and wheel velocity, wheel slip (λ_d=44%, snow)

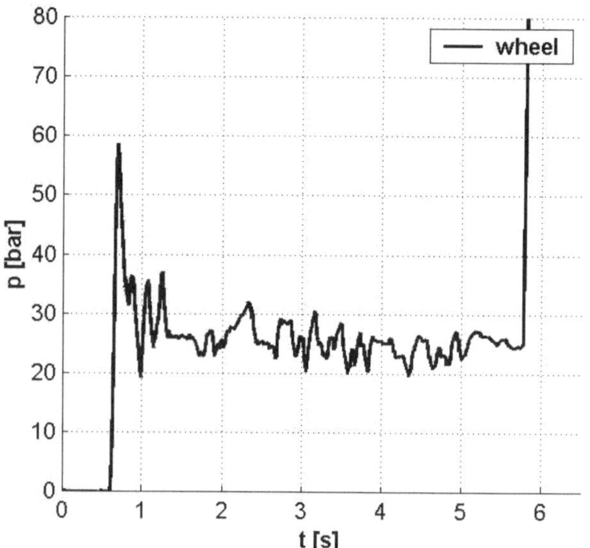

Figure 7.   Brake caliper pressure (λ_d=44%, snow)

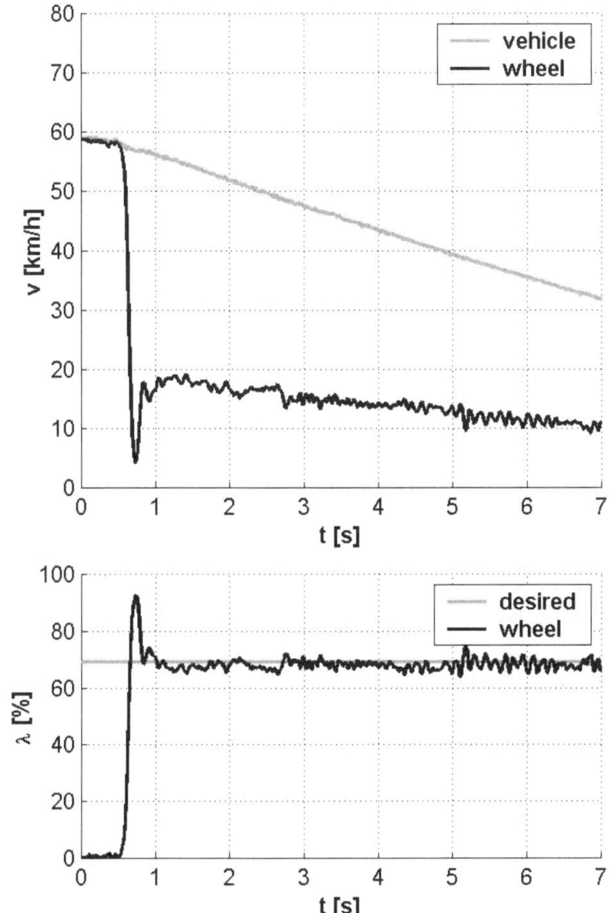

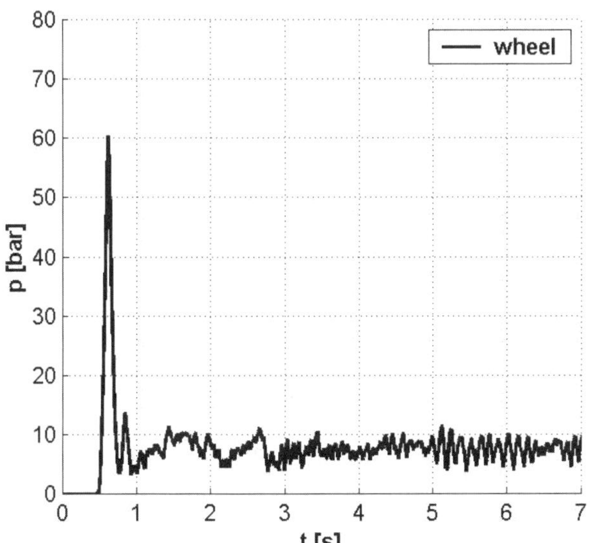

Figure 8. Vehicle and wheel velocity, wheel slip (λ_d=70%, ice)

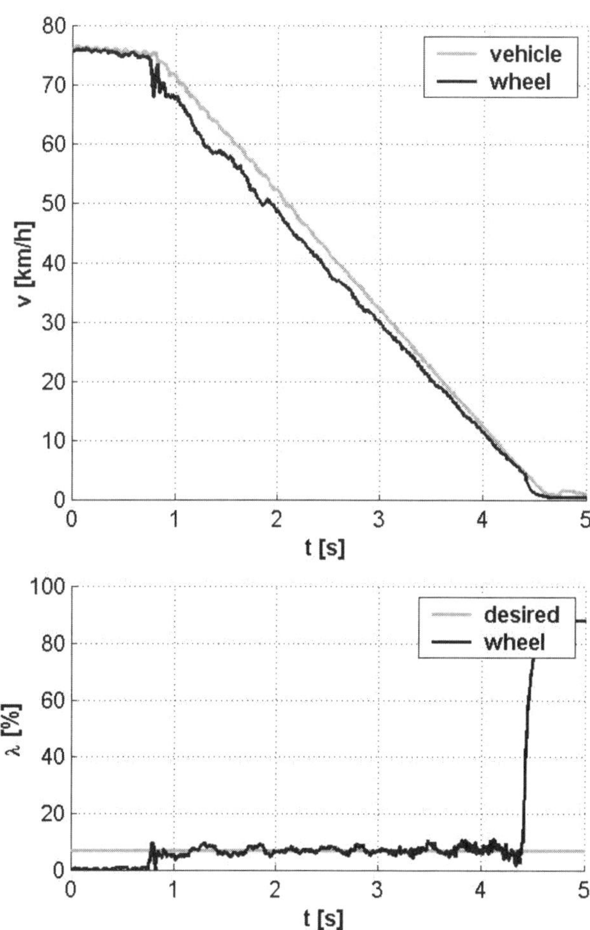

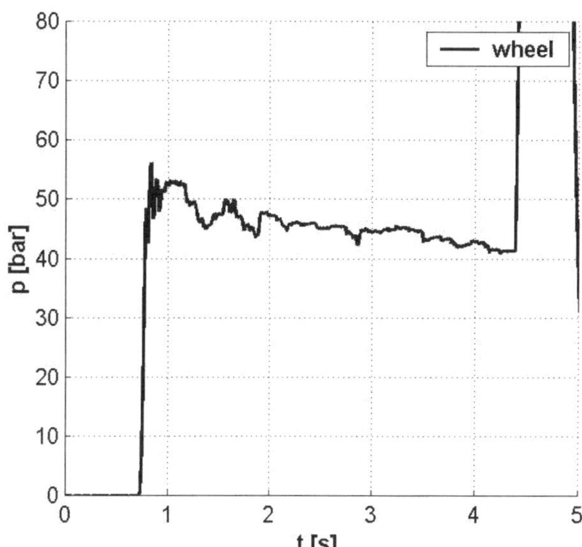

Figure 10. Vehicle and wheel velocity, wheel slip (λ_d=7%, dry asphalt)

Figure 9. Brake caliper pressure (λ_d=70%, ice)

Figure 11. Brake caliper pressure (λ_d=7%, dry asphalt)

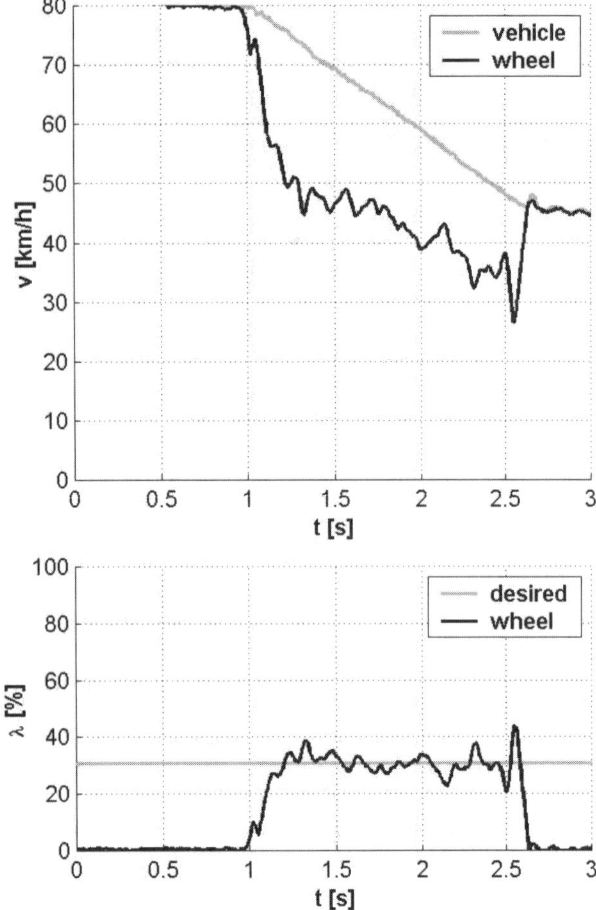

Figure 12. Vehicle and wheel velocity, wheel slip
(λ_d=31%, dry asphalt)

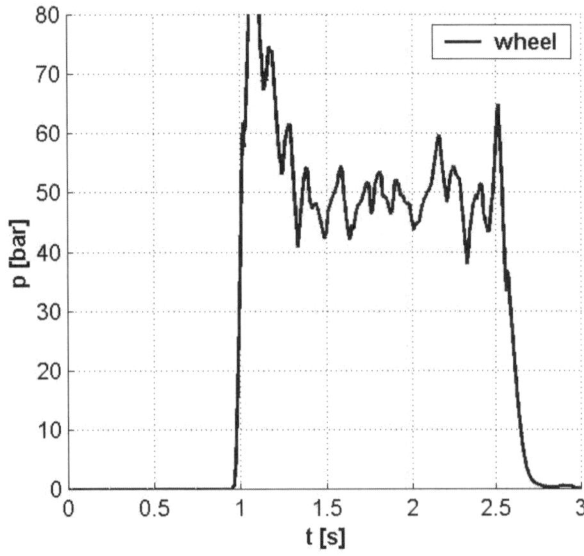

Figure 13. Brake caliper pressure
(λ_d=31%, dry asphalt)

CONCLUSION

This paper described an approach of how to control the wheel slip of a vehicle both in stable and unstable areas of the μ-λ curve and at an optimum braking slip using brake-by-wire actuators. The advantage of brake-by-wire actuators - such as the electro-hydraulic (EHB) and the electro-mechanical brake (EMB) - is that the caliper pressure or the clamping force, respectively, are known. It was shown by simulation and measurement results that the wheels of a research vehicle equipped with EHB and the new ABS algorithm can be kept both in stable and unstable areas of the μ-λ curve at any slip on different surfaces, i.e. ice, snow, and dry asphalt.

ACKNOWLEDGEMENTS

The authors wish to thank all colleagues who supported the research collaboration between Continental Teves and Darmstadt University of Technology.

REFERENCES

[1] Burckhardt, M.   (1993).   *Fahrwerktechnik: Radschlupf-Regelsysteme*. Vogel Verlag.

[2] Stoelzl, S. et al. (2000). *Das elektrohydraulische Bremssystem von Continental Teves*. VDI Berichte Nr. 1547.

[3] Continental Teves AG & Co. oHG (2000). *Facts & Figures 2000*.

[4] Mitschke, M. (1995). Dynamik der Kraftfahrzeuge Band A. Springer, Berlin Heidelberg New York.

[5] Khalil, H. K. (1996). Nonlinear Systems. 2nd Edition. Prentice Hall, New Jersey.

Virtual Design of a 42V Brake-by-Wire System

Joachim Langenwalter and Bryan Kelly
Synopsys Inc.

ABSTRACT

X-by-Wire implementations can lower manufacturing costs by reducing packaging problems and assembly costs. It also offers weight reductions combined with new safety features as well as an opportunity and challenge to couple electronic control and mechanical subsystems via mechatronics. Specialized software tools can expedite the design of X-by-Wire systems and enhance system reliability to steer around expensive recalls or liability problems. A brake-by-wire system will serve as an example for this advanced virtual design process using the iQBus™ design environment and the Saber® Simulator.

INTRODUCTION

Today's transportation systems consist of different technologies that move people and goods quickly and safely around the world. The frames of such vehicles operate for many years in harsh and aggressive environments. Functions inside the vehicles, such as multimedia entertainment systems, must have the latest developments to remain competitive. This means systems have to change during the lifetime of the product line. The frames hold more than 30,000 combinations of electronic, mechanical, hydraulic, pneumatic, software, and magnetic components, which must meet variations for the environment, customers, and legislative mandates. Because of the huge number of variants, companies cannot test or predict the behavior of the final product with real hardware prototypes in the short time that exists for development. Companies must use an integrated tool package to design, simulate, and document systems used in automobiles, trucks, busses, trains, and other vehicles. Such tool packages must support the design of system architectures, mechatronic systems, X-by-Wire systems, hybrid systems, bus systems (CAN, TTP, Flexray, MOST etc.), power systems, cabling, and other design applications. Designers need to manage the design, prototype, and tests of their latest systems in a virtual environment to make sure that possible variants comply with specifications. All data must go into a design editor, which allows an engineer to create a schematic that defines a vehicle's functionality. The tool also should offer links to 3D-mechanical CAD software that allow placement of functions into frames and chassis.

BRAKE-BY-WIRE DESCRIPTION

A brake-by-wire Electro Mechanical Brake (EMB) system replaces the traditional hydraulic linkage between the brake pedal and the wheels with an electronic controller interface. It is composed of mechanically decoupled sets of actuators and controllers connected through multiplexed in-vehicle computer networks, which are typically TTP or FlexRay bus systems.

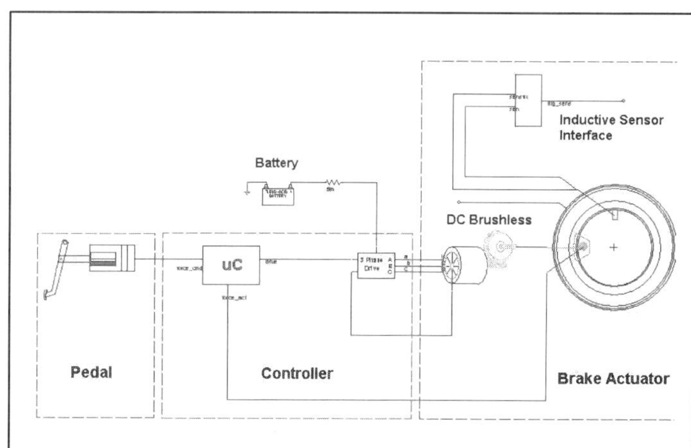

Figure 1: Three major parts of the EMB system

To achieve the required level of fault tolerance, a brake-by-wire must be designed as a redundant, distributed system composed of a number of fault tolerant units connected by a reliable real time communication system.

It does not require hydraulic fluids to store or load at the assembly plant and permits more modular assembly thus reducing the number of parts to be handled during production. Figure 1 shows the three major parts of such a system: the pedal with its force emulator, the controller, and the wheel together with the brake actuator. The pedal contains sensors to provide information about the requested brake force. Depending on the current state of the vehicle and wheel, the controller determines the actual brake force at the wheel. A dc-brushless motor regulated by the controller is driven by a 3-phase drive. The motor shaft is then connected via a worm gear/rack and pinion combination or a lead screw drive to the brake piston and caliper assembly. Piezo force and inductive sensors respectively provide feedback of the actual brake force and wheel speed to the controller. Often referred to as an Electro Mechanical Brake (EMB) system, it enables many new driver interfaces and performance enhancements. Overall it offers a wide flexibility in the tuning of vehicles for safety and performance. While it promises many benefits, they must be carefully designed, analyzed, and verified for their functionality.

DESIGN PROCESS

A new design starts by investigating system architectures and defining major modules, their positions in the system, and their connectivity. The topology analysis leads to requirements for the subsystem design. Here, the partitioning of the functionality into software and hardware implementation occurs. The components are then defined and connected by wires (2D/3D cable drawings). This process is tightly integrated with the in-house product data management (PDM) system. A virtual verification using simulation follows every revision to check the functionality against specifications and to apply a modification or redesign if the system does not match specifications. Real subsystems then replace virtual subsystems step by step. This requires that a virtual subsystem runs in real time to evaluate the system with real hardware in the loop. Finally, the designer must document the system and send it to manufacturing.

BRAKE-BY-WIRE ARCHITECTURE

While design variations affect performance attributes, costs, etc., typical architectures of brake-by-wire systems have been proposed for either single or dual voltage systems. Figure 2 illustrates a brake-by-wire architecture with redundant batteries for the EMB. The dual voltage bus shows both 42-Volt and 14-Volt bus implementation. The electrical housekeeping on the busses includes among other things various heaters and fans, lights, an audio system, navigation, EMB, and
 speed dependent loads such as EMVT (Electromechanical Valve Train) and ignition. Note in this particular application, the battery, fuses, wires, and generators are modeled with dynamic thermal effects.

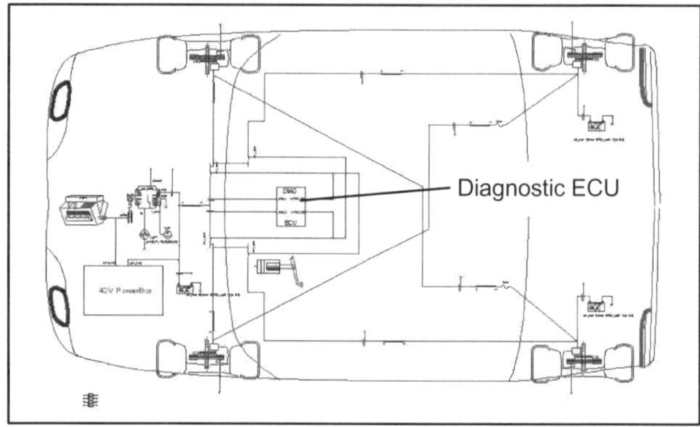

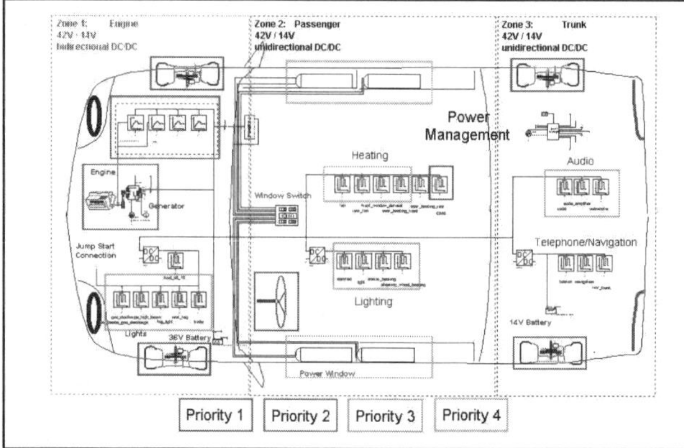

Figure 2: EMB architecture and redundant power distribution system with load priorities

For optimal load management, it is important to test different power management algorithms to ensure that the essential loads get sufficient power in the event of a low state of charge of the batteries. To prevent this condition, the loads are grouped into different priorities as illustrated in Figure 2. Depending on their priority, the loads get reduced or switched off completely by the power management to maintain essential functionality such as braking the vehicle. All the systems were tested with a standard highway drive cycle (see figure 3) with the following main loads activated:

- Electrical Valve Train   (2.5 kW peak)
- Brake-by-Wire   (2 kW peak)
- Steer-by-Wire   (1.8 kW peak)
- Rear/Front Window Defrost   (1.5 kW peak)
- PTC Heating   (1 kW peak)
  (Positive Temperature Coefficient)
- Electrical Fan   (0.8 kW peak)
- Seat Heating Front   (0.7 kW peak)
- Active Suspension   (0.6 kW peak)
- Lights   (0.5 kW peak)
- Air Pump   (0.4 kW peak)
- Power Window   (0.3 kW peak)
- Wipers   (0.3 kW peak)

The System includes the following source and storage devices:

- Integrated Starter Generator (ISG)   (6kW/4.5kW)
- 36V Battery   (50 Ah)
- 2 Backup 36V Batteries   (5 Ah)
- 12V Battery   (50 Ah)

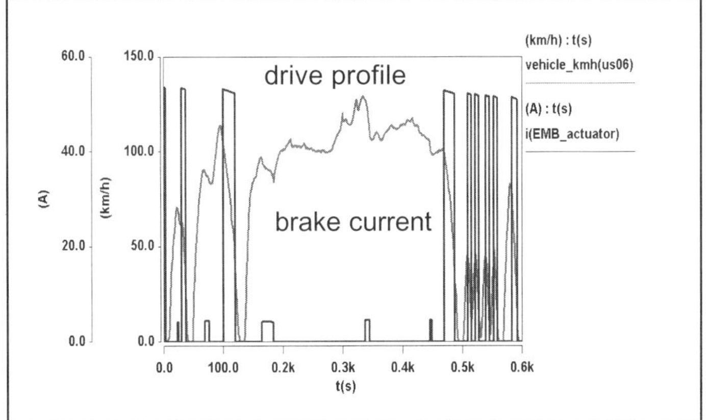

Figure 3: Vehicle test drive and brake actuator current of one wheel

By adding the peak values together (12.3 kW) it is obvious that the generator is not capable to deliver the peak power installed (see figure 4). Therefore, it is quite important to apply accurate user profiles. They are unique for every car manufacturer and are defined by monitoring against real drive cycles, e.g. one for Asia, Europe, and the USA. The user profile defines which load is switched on and off at a certain time.

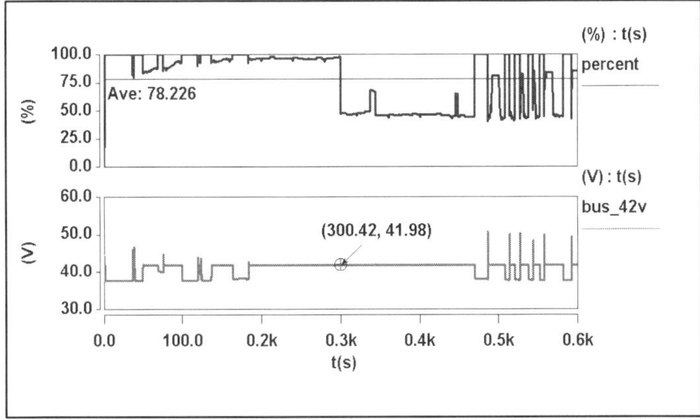

Figure 4: 42V bus voltage and percent generator usage

THE RIGHT MODEL ABSTRACTION

Depending on the accuracy requirements, the right abstraction of the model has to be used because the more detailed models consume longer simulation times. Figure 5 shows the simulation results where the brake actuator was implemented on the component level (see figure 1). Here the three phase voltage signals, the mechanical forces, vehicle performance etc. can be examined. In this actuator, a worm-gear is attached to the shaft of the brushless DC motor. The worm-gear has a modulation and self-locking mode, which respectively enables or disables the rotation of its output drive shaft. The self-locking mode helps to preserve the applied brake force almost without the need for continued application and consumption of battery energy. This level of actuator is used for sizing the individual components of a subsystem, but can be later replaced with a higher level representation after this process is complete to allow more efficiency when carrying out performance studies of the entire system. When sizing the subsystems in the whole architecture, it is not recommended to include all component effects because the simulation time will be too long to obtain results in reasonable time. The architectures are typically tested with standard drive profiles such as driving down a mountain pass for 1/2 hour, and various city and highway drive profiles. The results in figure 3 and 4 illustrate a 10 min drive cycle with a higher level representation of the brake actuator, which simulated in 1 min. Hence the simulation time is 10 times faster than real time. However with the brake actuator implemented on the component level, the simulation time is about 10 times slower then real time. A Pentium 3/750 MHz processor was used for this simulation.

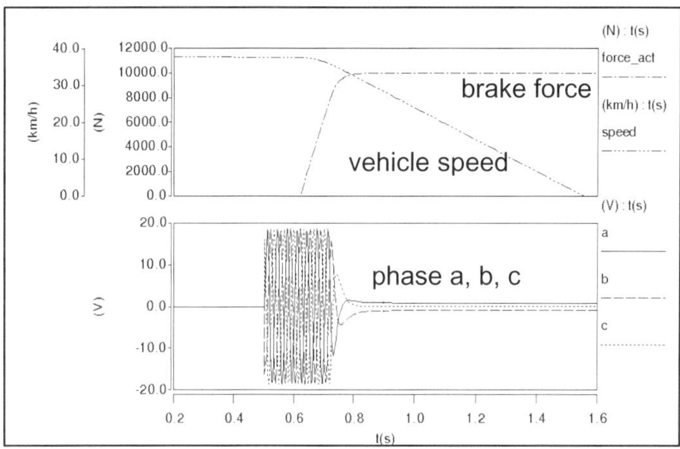

Figure 5: Three Phase motor drive signal (a,b,c / [V]), vehicle speed (speed/[km/h]) and brake force (force_act/[N])

CONCURRENT ENGINEERING

Designers split larger designs into multiple schematic sheets which allow separate teams to work independently. Most sheets share common data such as power supply and ground. Other sheets may share components such as electronic control units (ECU) and communication systems. Having different engineers work on separate sheets can jeopardize data integrity. Therefore, a check-in/check-out mechanism and a conflict resolution feature prohibit multiple users from concurrently manipulating shared/split objects. Advanced development processes also need an integrated flow from conception through manufacturing. A design package on top of a single database serves as a intuitive tool for placing and connecting components, allowing system engineers and harness designers to become familiar with the data. The database eliminates data translation errors downstream in the design. Data integrity remains intact even when the user exchanges data between the schematic editor and the 3D-CAD tools.

MODEL AVAILABILITY

Building a virtual prototype of the design requires "models" to enable the analysis; however, finding all the required models serves as the biggest challenge for assembling a virtual prototype. In a typical brake-by-wire system the following types of parts and technologies are used: batteries (lead acid or lithium Ion), ISG (Integrated Starter Generator), MOSFETs, sensors, fuel cells, DC/DC converters, TTP/FlexRay transceiver, receivers and so on. Ideally, the models all exist in the libraries of the tool used, but this is not always the case. When additional models are required, one avenue is to contact the manufacturer or vendor of the component, who often offers simulation models for their latest products. iQBus^{TM} offers options to help address this roadblock. This product offers a suite of characterization tools that allow users to quickly and conveniently create many types of models from any available data. Input for model characterization can come from a combination of datasheet curves, measurement data, or simulation coupled with optimization. A built-in optimizer automatically matches the model performance data to measured values. As an example, Figure 6 illustrates a snapshot of the waveform window of a lead-acid battery model characterization tool. In this case, 12 V lead-acid battery dc and transient (shown) measurement data was loaded into a battery characterization tool and a built-in optimizer was used to quickly characterize and validate the required battery model. The 36V battery consists of three times the number of cells of the 12 V pack.

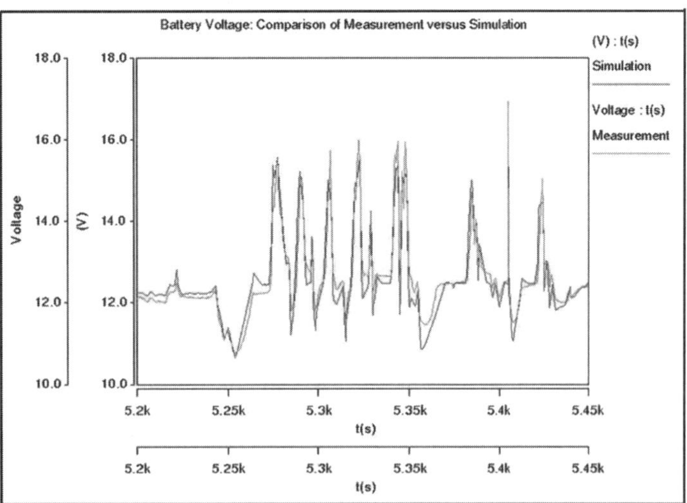

Figure 6: 12V battery voltage: Measurement versus Simulation

The deviations between transient measurement data and actual model performance are small. This is due to the higher-level abstraction of the battery model. The tradeoff is this level of model simulates fast and provides sufficient accuracy for performing load management analyses of the entire system.

iQBus^{TM} also includes the comprehensive Saber® library, which contains 30,000 parts (models) spanning different engineering disciplines (e.g. electronic, mechanical, hydraulic, pneumatic, magnetic, control, thermal, optical parts). Parts range from high level control blocks to physical component implementations. Multiple model abstractions in the library give designers flexibility to select the right model to help determine an optimum balance between simulation speed and accuracy as described earlier. Designers can implement a microprocessor and its software in several ways. One way includes using a control algorithm from the control library as a behavior model. Implementation also can occur by importing a foreign-function, C-Code call from a design tool such as Mathworks' Simulink/Stateflow, or using an instruction set microprocessor model written in SystemC/C++ with the target code running on it. Also the bus functional model of the communication bus (e.g. FlexRay) is typically implemented the same way.

FAILURE MODE EFFECT ANALYSIS

There are project management and technical objectives in simulation. From a project management perspective, one learns "more" earlier. This offers more information up front to permit better decisions on how to increase product effectiveness and return on investment, as well as how to reduce the time to market, and increase the bottom-line profitability. From a technical perspective, the design engineer can quickly and inexpensively experiment with new or existing designs. He can also verify that the chosen design approach will work before building a real system, efficiently debug or troubleshoot existing designs, and test the system for quality and manufacturability. For example, tools such as the Experiment and FMEA Editor enable the design engineers to determine the limits of robustness for the current design configuration to ensure driver safety. FMEA permits engineers to perform "what-if" studies on brake-by-wire systems like virtual testing of battery wire disconnection or shorts, bus communication faults, mechanical failures and other design variations. Figure 7 illustrates a typical output from a fault analysis performed on a brake-by-wire system. The Experiment Editor allows engineers to put together simulation and analyses tests virtually on the fly. At the end of the day, the goal of simulation should increase the reliability of the final product.

Faults		Test	Test	Test	Test	Test	Test	Test	Test	Test
		Brake Cycle	fl	fr	fr	rl	fl_label	fr_label	rr_label	rl_label
/batt_cable.	short	37.7	3805.0	6.4	2.9	4117.1	danger	failure	failure	okay
/wirep_f.1	open	37.7	3812.0	3977.6	4124.4	4116.4	danger	danger	okay	okay
/wirep_f.1	short	37.7	34.3	3791.5	4114.4	4115.8	danger	danger	okay	okay
/wirep_f.2	open	37.7	3943.9	3948.9	4125.4	4119.1	danger	danger	okay	okay
/wirep_f.2	short	37.7	3790.2	3731.6	4046.9	4115.1	danger	danger	okay	okay
/gen_cable.	short	37.7	3941.8	3941.8	4118.6	4118.6	danger	danger	okay	okay
/gen_cable.	open	37.7	3664.5	3664.5	4074.9	4074.9	danger	danger	okay	okay
/wirep_f.4	open	37.7	3942.4	3824.8	4127.1	4118.2	danger	danger	okay	okay
/wirep_f.4	short	37.7	3883.3	0.0	4123.9	4014.1	danger	failure	okay	okay
/wirep_f.5	open	37.7	3977.6	3812.0	4116.4	4124.4	danger	danger	okay	okay
/wirep_f.5	short	37.7	3887.2	34.8	4115.8	4052.7	danger	failure	okay	okay

Figure 7: Example partial FMEA report with wire faults

The above report shows that some faults in the first column are not detected by the diagnostic algorithm implemented in the ECU (see figure 2), which results in a 'danger' or 'failure' message. For example the first short circuit (short) at a power distribution point of the main 42-Volt battery connection cable (/batt_cable) produces a failure message at the front right (fr/fr_label) and rear right wheels (rr/rr_label). In this test case these wheels do not have enough electrical power to produce the brake energy necessary to stop the vehicle in time. After every design change the previously defined failure modes are automatically applied to the system and displayed in the same report format. The failures can be analyzed further and the causes removed by interpreting the diagnostic messages in the systems. At the end, no possible fault must lead to a complete system failure.

HARDWARE IN THE LOOP

It is not recommended to replace all 50 virtual subsystems in one step with real devices because in case of an error, it is almost impossible to diagnose the cause. Therefore, an alternative is to replace them step by step with real parts. In order to do that the virtual system has to run in real-time. This can be achieved by choosing the right abstraction level as described earlier. The real model is connected through appropriate interface cards (e.g. CAN, ADC) to the HIL platform (e.g. dSPACE or Applied Dynamics (ADI)) and a PC which is running the virtual system in SaberRT. Figure 8 illustrates such a configuration with the 1kW PTC-heating connected to the virtual vehicle systems running in real-time.

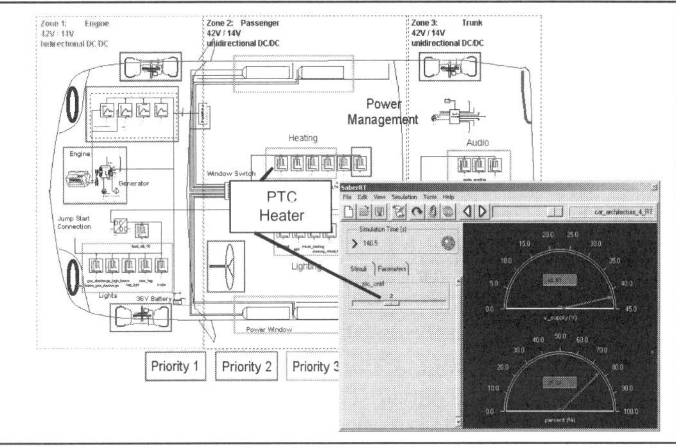

Figure 8: Hardware in the loop configuration of PTC-heater with generator load and voltage monitored online

The PTC heating system was selected because it is a high power load (1 kW peak) and its effects on the brake system supply could be significant. This is especially true under cold starting conditions and a low SOC of the battery. If during PTC operation a fault occurs, it could compromise the reliability of the EMB system.

FINAL STEPS TO MANUFACTURING

The database stores all the information about components, connectivity, and wires. The CAD designer can position the components and route the wires in the 3D model. iQBus[TM] can import added geometrical data, which allow wires and components to have all needed information for manufacturing. Many tools require different formats to exchange data. A table manager lets the designer quickly extract data and generate tables in any format. A designer can then pass the data to a manufacturer.

CONCLUSION

With appropriate software tools it is straightforward to apply and test various cost-effective strategies to improve brake-by-wire system performance. It is possible to shave off time/costs of the hardware design/test prototyping cycle. Many "what-if" studies and virtual prototype iterations can take place and eliminate a life-threatening circumstance like a brake failure. A design engineer may make a simple incorrect assumption for example, that the battery (and/or SuperCap – if implemented) has sufficient energy for stopping the vehicle. In fact, that may not be the case, particularly if a fault occurs during this time. Furthermore, once preliminary issues of a high-level design are worked out, more detailed virtual prototyping can be introduced for various sub-system components. This makes the overall simulation study even more realistic and at the same time the design more robust and competitive.

ACRONYMS

ADC	Analog Digital Converter
CAN	Controller Area Network
EHB	Electro Hydraulic Brake
EMB	Electro Mechanical Brake
EMVT	Electromechanical Valve Train
FMEA	Failure Mode Effect Analysis
FL	Front Left Wheel
FR	Front Right Wheel
HIL	Hardware in the Loop
ISG	Integrated Starter Generator
MOST	Media Oriented Systems Transport
PTC	Positive Temperature Coefficient
RL	Rear Left Wheel
RR	Rear Right Wheel
SOC	State of Charge
TTP	Time Triggered Protocol

CONTACT

Joachim Langenwalter, joachiml@synopsys.com
Bryan Kelly, bkelly@synopsys.com

REFERENCES

1. " A system-safety process for "by-wire" automotive systems", Sanke Amberkar, Barbara J. Czerny, Joseph D'Ambrosio, Brian Murray, Joseph Wysocki, Automotive Engineeering International, 2000 Technical Achievements, pp. 8-12, September 2000

2. iQBus^{TM} user manual, Synopsys., September 2002 http://www.synopsys.com/products/avmrg/iqbus_ds.html

3. "The X-by-Wire Concept: Time-Triggered Information Exchange and Fail Silence Support by New System Services", Elamar Dilger, Thomas Führer, Bernd Müller, Stefan Poledna, Electronic Steering and Suspension System, Progress in Technology, Automotive Electronics Series, pp. 239-248, SAE 980555, 1999

4. Joachim Langenwalter, "Avant!'s Saber Product Line Enables the 42-volt PowerNets in Cars", Electronic Journal, pp. 35-39, Mar/Apr 2001,

5. Verification of the Control Algorithm for a High Power Load by the Use of the Tool Chain Matlab/Simulink – Targetlink – Saber, G. Florissen, R. Grosse, R. Schoenen FKA Aachen, C. Amsel IKA Aachen University of Technology; European ASSURE Meeting, Munich, 14th November 2001

eBrake® – The Mechatronic Wedge Brake

Henry Hartmann, Martin Schautt, Antonio Pascucci and Bernd Gombert
eStop® - innovative brake technology - GmbH

ABSTRACT

eBrake® (1, 2) - a new "brake-by-wire" technology, was developed at the German Aerospace Centre, DLR e.V.. It is based on an electric powered controlled friction brake with high self-reinforcement capability. To avoid jamming the brake a special control technology was developed. Thus, by intelligently controlling a brake wedge, the kinetic energy of a vehicle is transformed into braking power. Furthermore an advanced design was found to deal with a broad variation of the friction coefficient. The physical effects involved lead to a significant reduction of energy consumption of the brake actuator compared to "conventional" brake-by-wire systems.

INTRODUCTION

The development work done on the "brake-by-wire" technology by the DLR based on the principle of a purely electromechanical braking systems corresponds fully with the general trend to replace, where and whenever possible, hydraulic or pneumatic systems with a clean and intelligent controllable electromechanical or mechatronic system. Mechatronic systems will most definitely become far more common and will penetrate all areas of the future industrial world and service industry and will represent a majority of the most innovative products in the future. Their importance for the economy in general will increase steadily during the next few years. The fields which will mainly profit from this development are the automobile and aircraft industry, fields which in the future will be much more dependent on export and international competitiveness (current examples for competition: ICE-TGV, Airbus-Boeing).

This integration of mechanical and electronic elements, as well as information technology (i.e. computer capacity) in order to create intelligent, controllable systems and machines will be replacing the classical standards of mechanical engineering in ever increasing fields and will eventually lead to a renaissance hereof with a new character. This integration often has to be achieved within a very small space – often within the scope of micro systems engineering. The optimal functionality of mechatronic systems can only be achieved by optimally combining electronics, information technology and mechanic elements into one system which works as a whole.

Already today there exist quite a few examples of mechatronic systems: CD players, ink jet-printers, tool machines as well as robots belong to this group just as "fly-by-wire" planes do, or furthermore, ABS-, TCS- and ESP systems in cars, airbags which recognize the weight on respective seats, etc. The scope of applications will increase significantly in the future.

PROBLEM DESCRIPTION

The increasing requirements a modern braking system has to meet today – i.e. anti-lock braking systems, driving stability control systems, traction control systems – make wheel-selective braking necessary. To date, it has been possible to meet these requirements with conventional braking systems, which have been continuously upgraded by adding hydraulic pumps or magnetic valves. All of these solutions have however resulted in vibration, resonance and damping problems within the hydraulic pipes and difficulties in addressing the hydraulic module accumulator or in other words, the magnetic valves. Due to the characteristics of these magnetic valves, which are highly non-linear two-step controls, the possibility of achieving high control quality regarding the braking pressure is also rather limited.

All attempts so far to develop an electrical driven brake face the seemingly insuperable obstacle of extremely high actuator forces and the resulting high energy requirement of the actuator. The energy source, usually electric motors which have to supply quite significant torques and power, are large, heavy and accordingly expensive. For this reason a successful and profitable "all electric" brake-by-wire concept has not yet been developed.

NEW SOLUTION APPROACHES

The new DLR "brake-by-wire" technology is based on an electric powered controlled friction brake with high self-reinforcement capability (eBrake ®).
Whenever engineers deploy existing forces elegantly, usually the simplest concept is the most convincing. By intelligently controlling a brake wedge, the kinetic energy (momentum) of a vehicle is transformed into braking power.

MECHANICAL MODEL

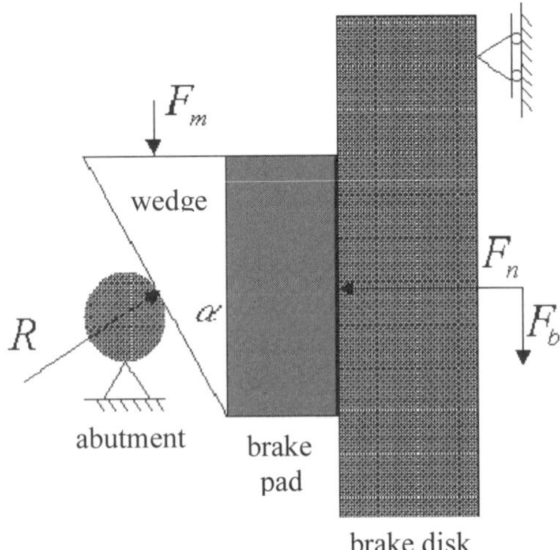

Fig.1: Mechanical Model

The brake lining is equipped with a wedge on its backside which is rested on an abutment, e.g. a bolt (Fig. 2). The actuator presses the brake lining in between the abutment and the brake disc with the motor force F_m. The braking force F_b resulting from the contact between the brake disc and the brake lining acts in the same direction as the motor force which results in the anticipated self-reinforcement.

From the force balance can be derived

$$F_m = F_b \frac{\tan\alpha - \mu}{\mu},$$

for the characteristic brake factor C* then applies:

$$C^* = \frac{F_b}{F_m} = \frac{2\mu}{\tan\alpha - \mu}$$

CONTROL SYSTEM AND FIRST TESTING

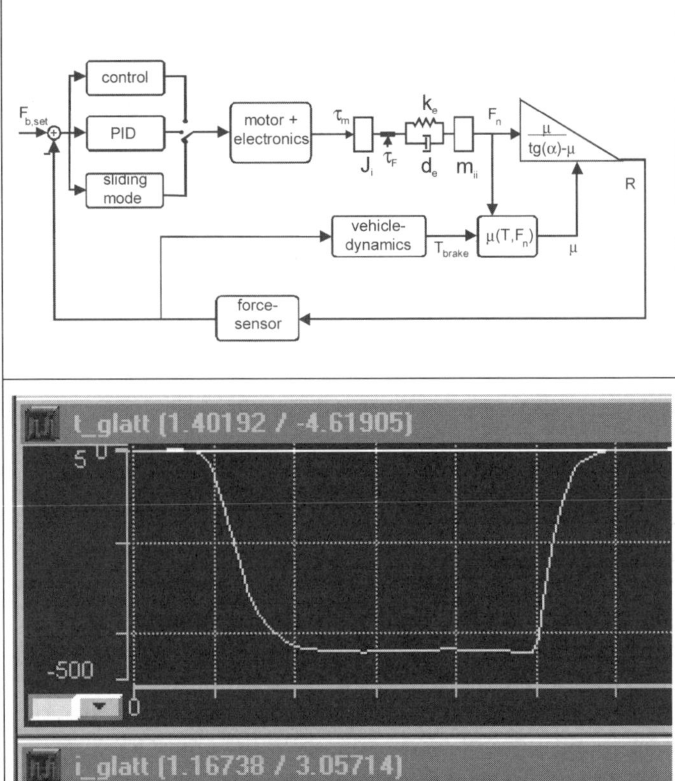

Fig.2: Control structure and braking test proving stability

The main problem with this simple but very efficient method of braking was to find a way to avoid the jamming of the brake or better said, to "control" this jamming advantageously. DLR was successful in solving this problem. A special control technology developed under Matlab/Simulink and dSPACE stops the wedge from getting stuck.
In order to prove the general concept of a controlled wedge brake a prototype of DLR eBrake® and a test bench was build (fig.3). In this first approach a stable ring construction was realized avoiding elasticity problems within the mechanical structure. Fig.2 shows the control structure and a braking test at 440 Nm braking moment proving stability of the system (see also 3). At this point it has to be said that due to restricted testing facilities the maximum braking torque of the set-up was limited to 500 Nm.

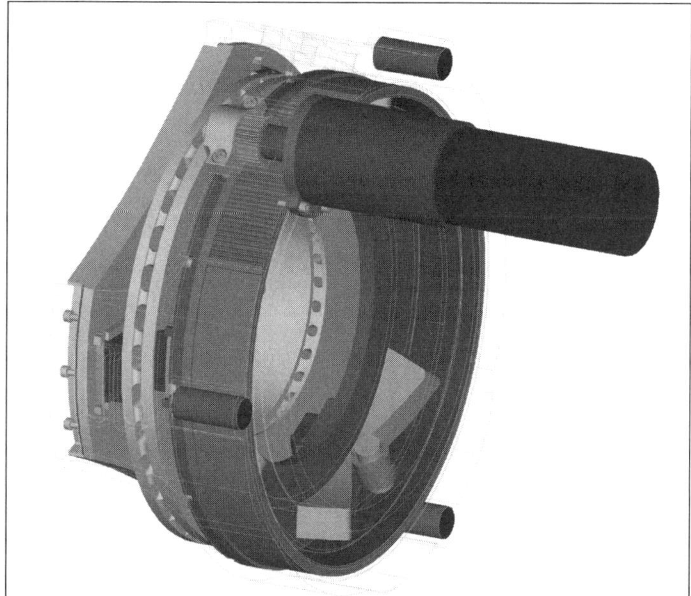

Fig.2: First Prototype of DLR eBrake ® and test bench

THE "PUSH-WEDGE" PRINCIPLE

Let's have a look at the so called "push-wedge" principle (Fig. 4), meaning $\tan \alpha > \mu$.

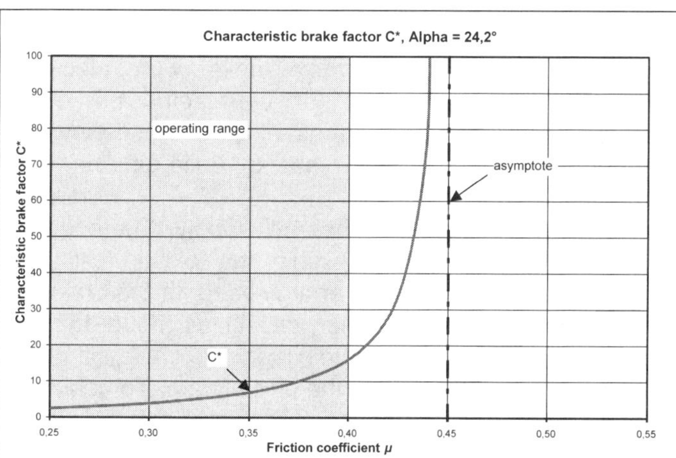

Fig. 4: Characteristic brake factor C* in relationship with the friction coefficient μ for a "push-wedge" principle.

In order to avoid jamming of the wedge the operating range has to stay well fond of the asymptote which is characterized by the condition $\mu = \tan \alpha$.

This requires a design with big wedge angles corresponding with low over all self-reinforcement capability. In this first approach to avoid the jamming problem it was thought, that reaching the critical point, i.e. the asymptote, during brake operation will cause the worst case, i.e. a complete blockage or even destruction of the brake. In reality this problem does not occur – but why?

THE "PUSH-PULL-WEDGE" PRINCIPLE

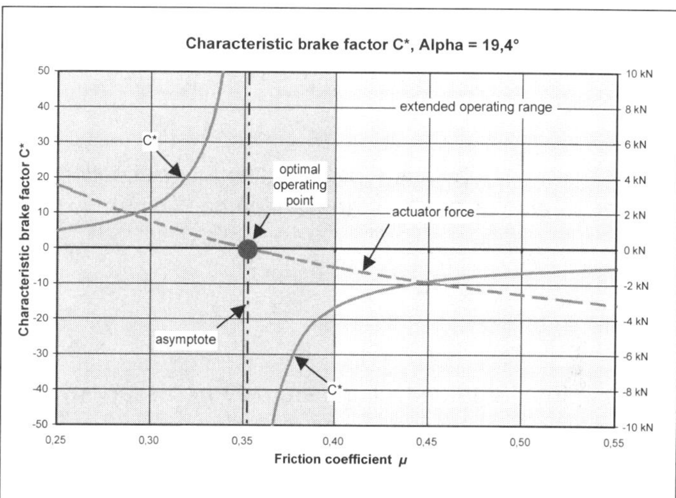

Fig.5: Characteristic brake factor C* in relationship with the friction coefficient μ for a "push-pull-wedge" principle.

In a real set-up the actuator, e.g. a motor+ spindle unit, is designed to deliver thrust and pull forces anyway. This is essential for a stable and fast control of the system, as there are bi-directional dynamic loads due to inertia within the power train itself. Looking at an increasing friction factor, e.g. during brake operation with rising temperature of the friction material, we reach a said critical point. But here the actuator force, i.e. the motor force, doesn't become infinite – it becomes <u>zero</u> (Fig.5)! Once the friction coefficient rises further, the force direction within the power train will change from pushing to pulling. This means the actuator has to hold resp. pull the wedge out. Looking at the graph of the actuator force in fig.5 it becomes obvious that this said critical point is the most effective operating point of all: <u>the self-reinforcement reaches infinity!</u>

FURTHER WEDGE SHAPE OPTIMIZATION

Let's for a moment suppose that the coefficient of friction would be constant and $\alpha = \arctan(\mu)$. In that case only a position control of the wedge would be necessary.

The position of the wedge corresponds with a defined widening of the elastic caliper leading to

$$F_n = c \cdot \tan(\alpha) \cdot x,$$

with the caliper stiffness c and the wedge position x. This means that the normal force is a function of the wedge position.
Following this idea we come to the first important insight how to reduce necessary maximum actuator forces in an overall approach:

<u>The wedge position corresponds to a defined normal force!</u>

Looking at the demanded maximum value for the braking force and taking into account the variation of the friction coefficient:

$$F_{b,max} = \mu \cdot F_n = const.$$

then it becomes obvious that the maximum deflection of the wedge also depends on the friction coefficient:

$$x = \frac{F_{b,max}}{\mu \cdot c \cdot \tan(\alpha) \cdot}$$

meaning that large deflections are only necessary when the coefficient of friction is low, e.g. due to fading. We conclude:

<u>The wedge angle for extreme wedge positions can be independently optimized for each value of the friction coefficient!</u>

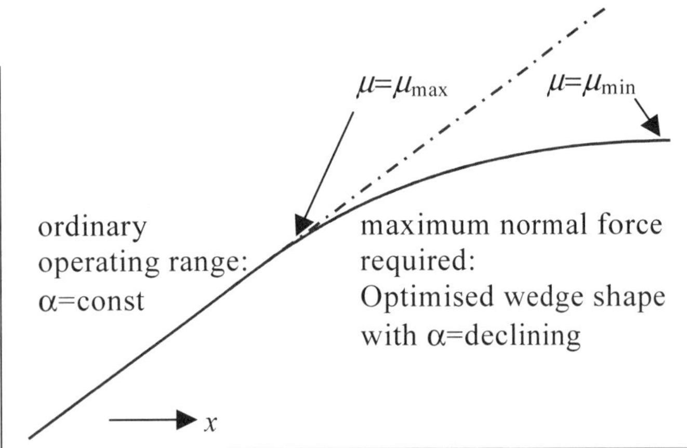

Fig.6: Optimized wedge shape meeting maximum normal force requirements with minimized actuator force.

This second insight leads to future potential for wedge shape optimization with declining wedge angle over increasing wedge deflection (Fig.6). Therefore the power train design can be further downsized.

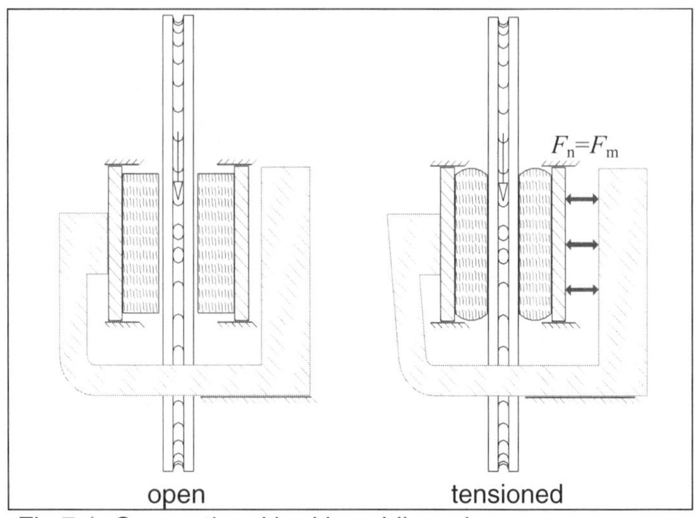

Fig.7a): Conventional braking philosophy

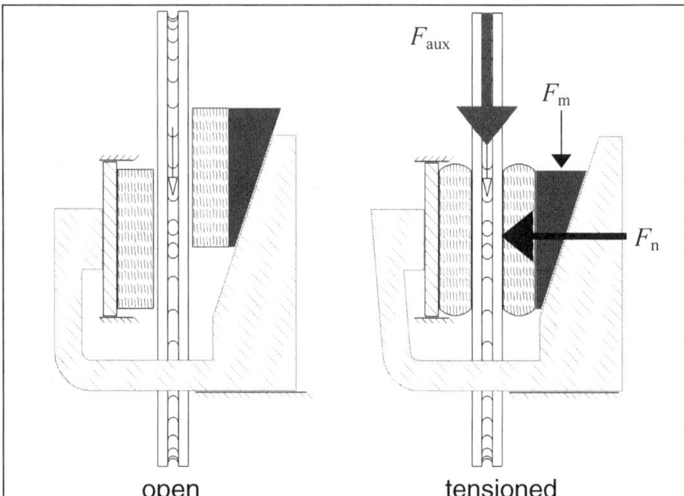

Fig.7b): eBrake® – electric powered controlled friction brake with mechanical self-reinforcement (4)

The conventional braking philosophy (fig.7a) assumes that the normal force has to be build up directly, actively and in full height. New concepts merely try to replace hydraulic or pneumatic brake units with electro-hydraulic, electro-pneumatic or all electric powered solutions. Talking about an all electric brake-by-wire approach the electric motor has to build up the full normal force.
Furthermore the actuator has to supply the energy absorbed by resp. stored within the elastic caliper. Therefore an extremely high energy need for the braking actuator arises. Even further capacities have to be supplied in order to guarantee high system dynamics looking at high inertia within the power train. To achieve these goals large, heavy and expensive electric actuators, spindle units or gears have to be used leading to increased space requirements and increased primary damped masses within the chassis.
Instead the DLR eBrake® uses the vehicle's momentum to support the electric actuator (fig.7b). An auxiliary force derived from the self-reinforcement effect is used to

build up the normal force. Therefore the braking actuator only has to supply a small portion of the required normal force. Furthermore the energy needed to widen the caliper is also taken from the vehicle's kinetic energy.

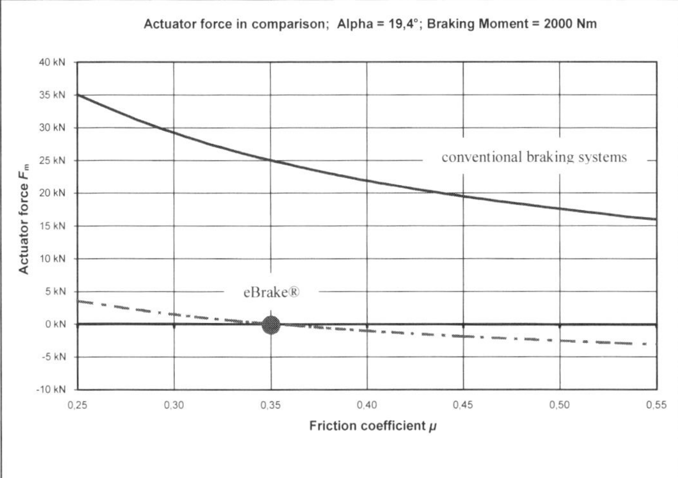

Fig.8: Braking power in comparison

The physical effects involved lead to the following advantages as opposed to "conventional" braking system in general and brake-by-wire systems in particular:

1. The average energy consumption of the actuator can be significantly reduced, because
   - the required actuator forces drop dramatically – down to <u>zero</u> in the optimal operating point (fig.8)
   - the energy needed to widen the caliper has not be supplied by the electrical supply system – it is taken from the kinetic energy of the vehicle
2. The actuator can be enormously downsized, thus
   - the requirements for the installation space of the actuator can be reduced
   - the weight of primary damped masses within the chassis can be reduced
   - the costs of the braking unit can be reduced
3. No conversion of the board system to 42V is needed
4. Increased dynamics, controllability and stability

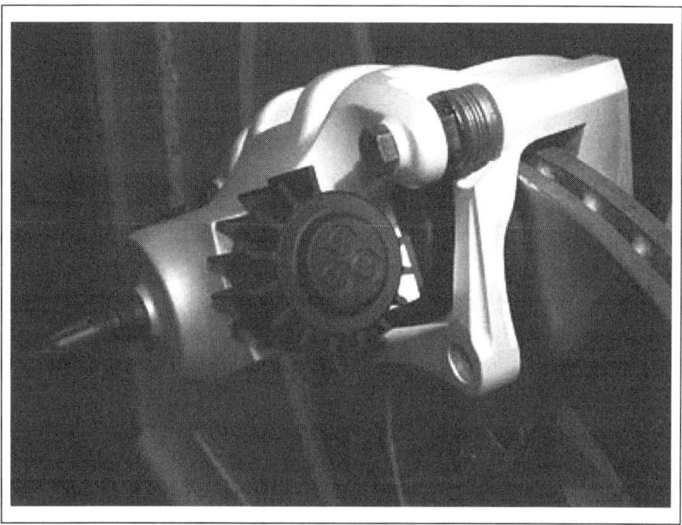

Fig.9: Second Prototype

THE NEXT GENERATION

A follow-up prototype (fig.9) already includes fail-safe concepts, e.g. a second motor unit delivering half the work load under normal conditions but capable of emergency braking as well as solutions for an electric park brake using defined slack within the power train. Most important feature of the power train concept is the active compensation of said slack while operating within the zero-force-control-mode (optimal operating point).

CONCLUSION AND OUTLOOK

eBrake® - the new and unique "brake-by-wire" technology, developed at the German Aerospace Centre, DLR e.V. is described. It is based on an electric powered controlled friction brake with high self-reinforcement capability. By intelligently controlling a brake wedge, the kinetic energy (momentum) of a vehicle is smartly transformed into braking power.

A control technology is outlined which avoids the jamming of the brake. The mechanical model is described and the basic formula for the systems behavior is deduced. The progressions of the characteristic brake factors of two different eBrake®-solutions are discussed and further optimization potentials are shown.

A comparison with other braking systems, following the "conventional" braking philosophy, in particular brake-by-wire systems, is given and the following advantages of the eBrake® concept are pointed out:

- the average energy consumption of the actuator can significantly reduced
- the actuator can be enormously downsized (installation space, costs, weight)
- no conversion of the board system to 42V is needed
- system dynamics, controllability and stability are increased

Most notably increased dynamics will lead to improvements in reaction time, shortening of ABS-cycles and thus reduced braking distances. This is an important most drivers will be braking with a system, which our ancestors were already able to rely on totally.

ACKNOWLEDGMENTS

We would like to thank the staff members of the Institute of Robotics and Mechatronics at the German Aerospace Center, DLR e.V., in Oberpfaffenhofen near Munich under the leadship of Gerd Hirzinger, where the basic research work was done.
Furthermore we acknowledge the support of BBA friction who supplied different friction materials for testing.

REFERENCES

1. US patent 6,318,513 - elektromechanical brake with self-energization
2. German patent 19819564 – Elektromechanische Bremse mit Selbstverstärkung
3. Martin Semsch, „Neuartige Mechatronische Teilbelag-scheibenbremse", XIX. Internationales $\tilde{A}$ Symposium, 29./30. Oktober 1999, Bad Neuenahr, ISBN 3-18-340512-1, pp 53-74
4. http://www.estop.de/eStop-flyer-Englisch.pdf

CONTACT

Henry Hartmann
eStop GmbH
An der Hartmühle 10
D-82229 Seefeld
Tel.: +49(0)8152/9936-15
Fax: +49(0)8152/9936-11
henry.hartmann@estop.de

DEFINITIONS, ACRONYMS, ABBREVIATIONS

F_m: motor force
F_n: normal force
F_{aux}:   auxiliary force
F_b: braking force
$F_{b,max}$:   maximum braking force (nominal value)
R: reaction force
α: wedge angle
c: caliper stiffness
x: wedge position/ deflection

APPENDIX

eStop® und eBrake® are registered Trademarks of eStop® - innovative brake technology - GmbH

Development of Vehicle Dynamics Management System for Hybrid Vehicles - ECB System for Improved Environmental and Vehicle Dynamic Performance -

Masayuki Soga, Michihito Shimada, Jyun-ichi Sakamoto and Akihiro Otomo
Toyota Motor Corporation

ABSTRACT

In anticipation of the increased needs to further reduce exhaust gas emissions and improve fuel consumption, a new brake-by-wire system called an "Electronically Controlled Brake" system (hereafter referred to as "ECB") has been developed. With this brake system, which is able to smoothly control the hydraulic pressure that is applied to each of the four wheel cylinders on an individual basis, functional enhancements can be added by appropriately modifying its software. This paper discusses the necessity of the ECB, the system configuration, and the results of its application on hybrid vehicles.

Keywords: brake, vehicle dynamics, integrated control by-wire, hybrid vehicle

Foreword

The development of hybrid vehicles is being promoted out of concerns for environmental protection. In this regard, the Estima Hybrid, which was announced in June 2001, has adopted the world's first brake-by-wire system called ECB[*1], in order to achieve the three requirements at high levels: environmental technology, safety, and driving enjoyment. Through the use of the Vehicle Dynamic Management, which was developed under a new concept to comprehensively control the braking and driving functions of the hybrid system, a high level of dynamic performance has been realized. This paper gives an outline of the ECB system, which manages comprehensive control, as well as the improvements realized in the vehicle dynamics and environmental performance through the adoption of the comprehensive Vehicle Dynamics Management system.

[*1]ECB: Electronically Controlled Brake System

Outline of the ECB System

1. Functional Requirements of a Brake System

This hybrid vehicle uses a regenerative brake system that recovers the braking energy in order to improve fuel consumption.

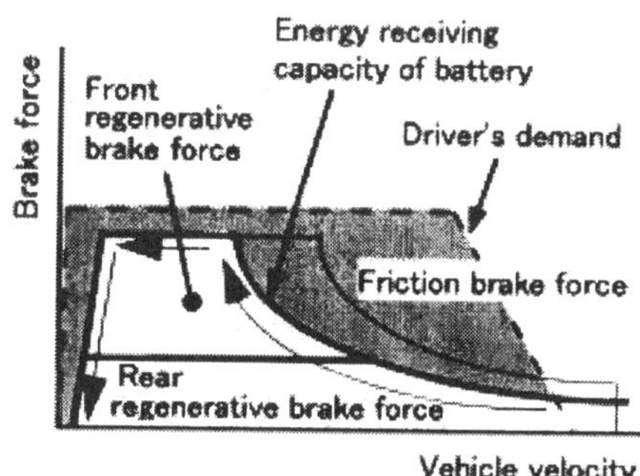

Figure 1: Example of coordination functions of the regenerative and friction brakes

As Figure 1 shows, this vehicle responds to the driver's brake force requirement by utilizing the maximum possible amount of regenerative brake force and resorting to the friction brakes for the amount of brake force that is lacking.

To make this possible, the system must:

Be able to linearly control the hydraulic brake force in the normal operating range.

Be able to generate the required brake force , without allowing the driver to feel the coordination functions of the regenerative and friction brakes.

Furthermore, out of concerns for active safety, the system is required to independently control the brake force of each wheel in a highly responsive and precisely .

2. ECB System Configuration

To satisfy the aforementioned requirements, the ECB is configured as shown in Figure 2. It consists of mainly two parts , that electricallty detects the pedal operation of the driver, and that controls the hydraulic pressure to the wheel cylinders, in order to achieve a so-called by-wire system that electrically controls the hydraulic pressure that is applied to the wheel cylinders. Furthermore, the ECB uses linear valves that are arranged in pairs, to control the hydraulic pressure of the brakes in all ranges, for normal braking or vehicle dynamic control, such as ABS , the Vehicle Stability Control (VSC).

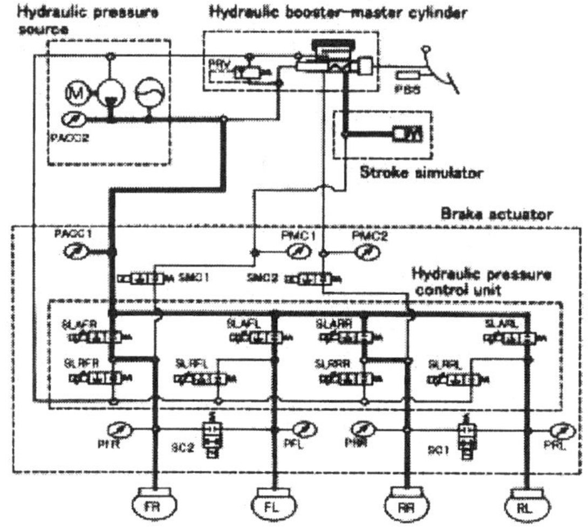

Figure 2: Configuration of ECB hydraulic circuits

3. Linear Hydraulic Pressure Control

Figure 3 shows the basic configuration of the linear hydraulic pressure control of the ECB.

The linear hydraulic pressure control consists of the difference between the actual hydraulic pressure and the target hydraulic pressure (which is added to the feed-back term) and the valve-opening current that varies with the difference in pressure upstream and downstream of the valve (which is added to the feed-forward term).

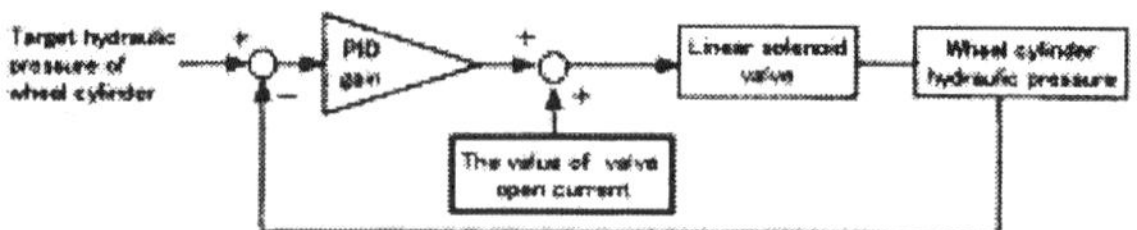
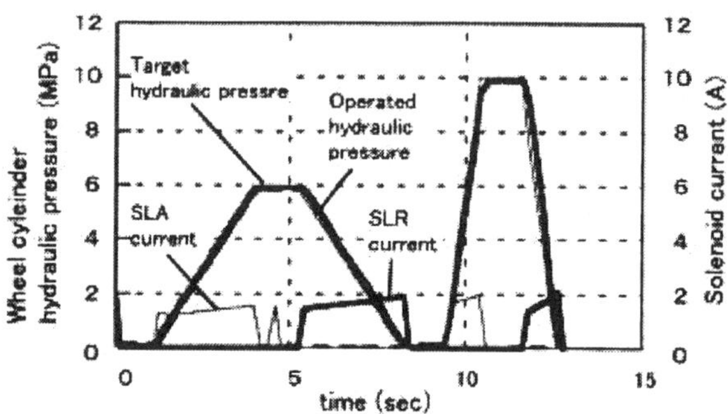

Figure 3: Configuration of hydraulic pressure control

Figure 4 shows a control example of the linear hydraulic pressure control. It controls the responsiveness and the controllability of the actual hydraulic pressure against

Figure 4: Example of linear hydraulic pressure control

the target hydraulic pressure within a range that satisfies the requirements of the various types of control applications, which will be discussed later.

4. Outline of the Failsafe Function of the ECB System

For the purpose of sustaining the friction brake function in case of various types of system failures that could occur, the failsafe function of the ECB system provides ample brake force by detecting a failure, and ultimately applying the hydraulic pressure generated by the hydraulic booster-master cylinder , to the four wheel cylinders.

If the regenerative function that is unique to the hybrid vehicle fails, the ECB control continues to apply the hydraulic pressure brake in order to promptly compensate for the amount of regenerative braking.

Outline of the Vehicle Dynamics Management System

The configuration of the entire system, which comprehensively controls the braking and driving functions through the combination of the ECB system and the hybrid system, and an outline of the control processes, are discussed in the following section.

1. Hardware Configuration of the System

Figure 5 shows the hardware configuration. In contrast to the aforementioned ECB system, the hybrid system is the drive system that contains front and rear drive motors to drive the respective axles. In the front unit, the engine and motor are laid out parallel to each other, and the drive force is transmitted via the CVT. The rear unit consists of

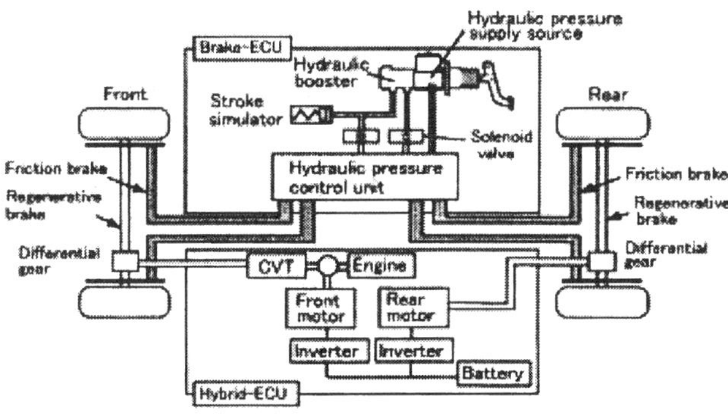

Figure 5: Hardware configuration of the system

only a motor and a reduction unit. Thus, it is an independent four-wheel drive system that does not require a transfer unit or a propeller shaft.

2. Configuration of the Comprehensive Control Process

Figure 6 shows the control process of this system. The system detects the requirements of the driver through a sensor that detects the pedal input and a steering angle sensor, and computes the vehicle's dynamic targets. At the same time, the system detects the driving conditions of the vehicle through a wheel speed sensor, yaw rate sensor, and acceleration sensor. In accordance with the conditions of these sensors, the system selects and executes the respective control modules in the comprehensive braking and driving control logic. The commands for the brake force and the driving force, which are required by the wheels, are directed to and executed by the ECB hydraulic pressure control module and the hybrid computer, respectively.

The advantages of this configuration are described below. A new concept of comprehensive control that expands the control range from the critical limit of Vehicle Stability Control (VSC) to the normal operating range, this configuration improves the driver's comfort and environmental performance of the vehicle. This system is hereafter referred to as "Vehicle Dynamics Management".

On one hand, this configuration enables the system to compute the required brake force and driving force from the normal operating range represented by the regenerative coordination control, in order to maximize the regeneration of energy. On the other hand, this configuration facilitates the execution of continuous control by regulating the brake hydraulic pressure and the driving torque near the limit range.

The condition assessment modules, which were previously computed and executed on a module-by-module basis by the ABS and VSC brake control modules, have been integrated and organized. By implementing the control targets through the wheel cylinder hydraulic pressure, the actuator driving modules could be separated from the respective software applications in order to effect comprehensive control. As a result, the quality of the software has been improved and the application time has been shortened.

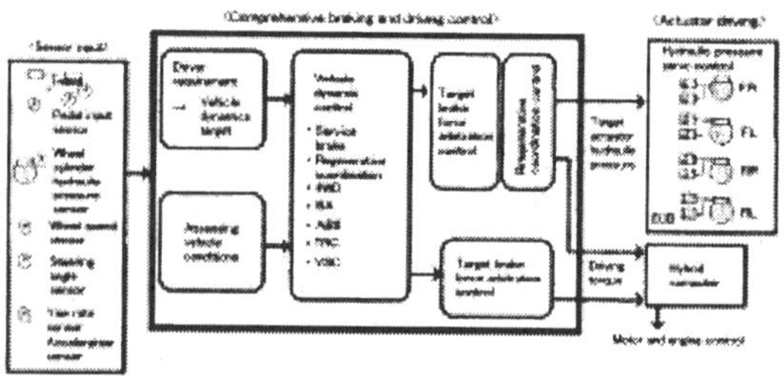

Figure 6: Vehicle Dynamics Management System Processing

Vehicle Dynamics Control through Vehicle Dynamics Management

This section gives specific examples of the improvements in the vehicle dynamic performance and environmental performance that have been realized through the aforementioned system and control.

1. Method for Conforming the Brake Hydraulic Pressure

As previously described, the ECB system has linear control valves and wheel cylinder pressure sensors attached to the four wheels. In contrast to the previous system that regulated the duration of the solenoid valves by turning them ON and OFF, the ECB system can express the control targets of the applications through the use of a physical unit, namely the wheel cylinder hydraulic pressure. Thus, it can seamlessly link the controls and eliminate their gaps.

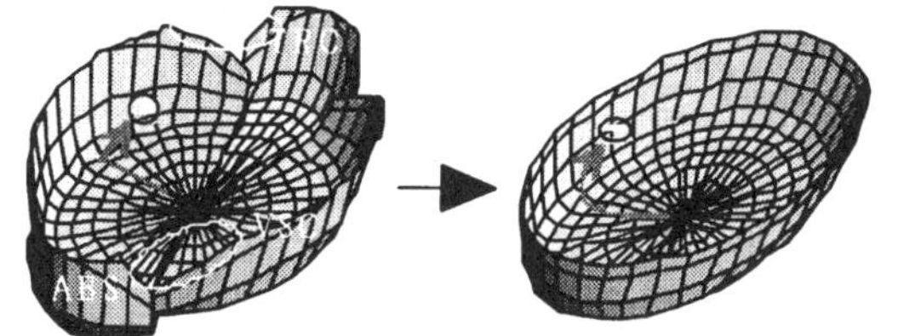

Figure 7: Conceptual image of control links (ball in a bowl)

2. Service Brake Control and Regenerative Coordination Control

The service brake, which controls the longitudinal deceleration of the vehicle, detects and computes the driver's brake application. Then, it distributes the target deceleration between the hydraulic pressure

brake and the regenerative brake. For the purpose of maintaining the distribution of the front-rear brake force at a constant level, both the front and rear motors of this vehicle prioritize fuel efficiency while effecting the regeneration of energy. Figure 8 shows an example of the actual regenerative and hydraulic pressure coordination control of the front brakes, and Figure 9 shows the rear brakes. The coordination control alone improves the vehicle's fuel efficiency by approximately 20 percent, as verified by the 10-15 mode fuel consumption measurement.

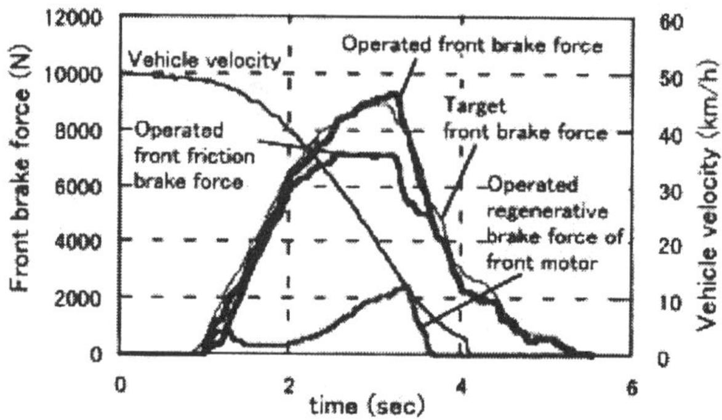

Figure 8: Example of hydraulic-regenerative brake control of the front wheels

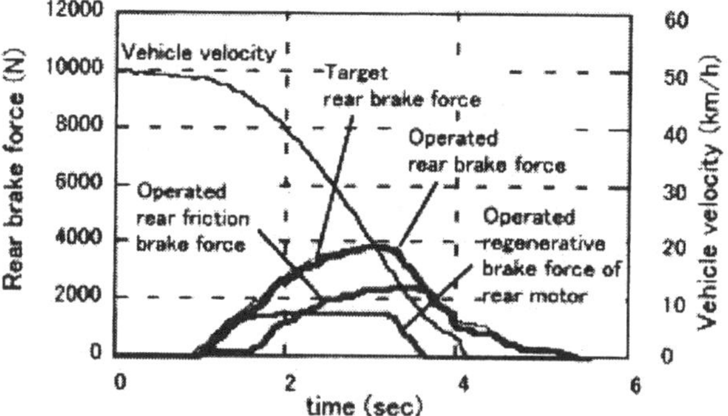

Figure 9: Example of hydraulic-regenerative brake control of the rear wheels

3. ABS Control

The ABS control is also effected through hydraulic pressure control with feedback of wheel cylinder hydraulic pressure values, just as with normal brake control. Various means of improving the braking efficiency are incorporated in the control: accelerating the target hydraulic pressure during pressurization by referring to the hydraulic pressure value information at the beginning of the depressurization; and making it easier to hold the hydraulic pressure in the vicinity of the μ peak, than in the previous system (see Figure

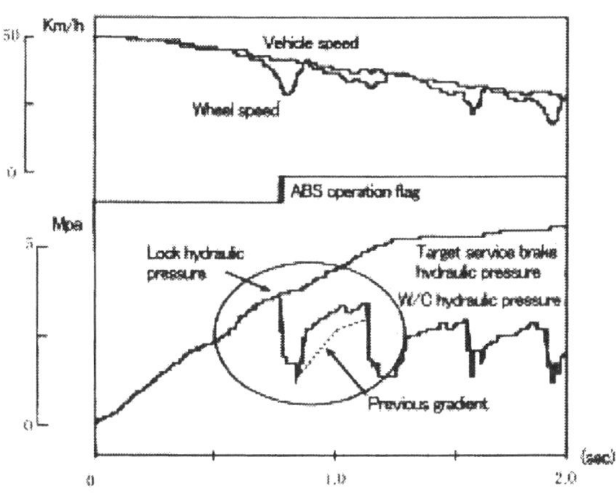

10).

Figure 10: Example of braking efficiency improvement during ABS control

4. Cornering and Startoff Performance

As part of dynamic performance improvement technology that extends from the normal to the critical limit, the driving performance during cornering and the startoff performance while driving on low-μ surfaces are described below.

4.1. Cornering Performance

Figure 11 shows a comparison between the Vehicle Dynamics Management (VDM) and the previous Vehicle Stability Control (VSC), with respect to the vehicle behavior and the driver's steering input, while the driving locus is constrained to the following conditions: μ = 0.35, R = 150m, and initial speed = 19.4 m/s.

In the previous VSC control (Figure 11-a), the hydraulic pressure control, when control steps in near the limit and when control is operating, is effected by an ON/OFF valve through duty cycle control. In contrast, the control that has recently been realized by the VDM gradually applies hydraulic pressure to the wheel cylinders from before the limit. This improves the follow-up of the hydraulics to the target yaw rate and reduces the amount of steering correction made by the driver (Figure 11-b). Furthermore, because the amount of control itself is minimized and the control converges quickly (Figure 11-c, -d), smooth driving is realized, while the deceleration of the vehicle that occurs when the control steps in is kept small.

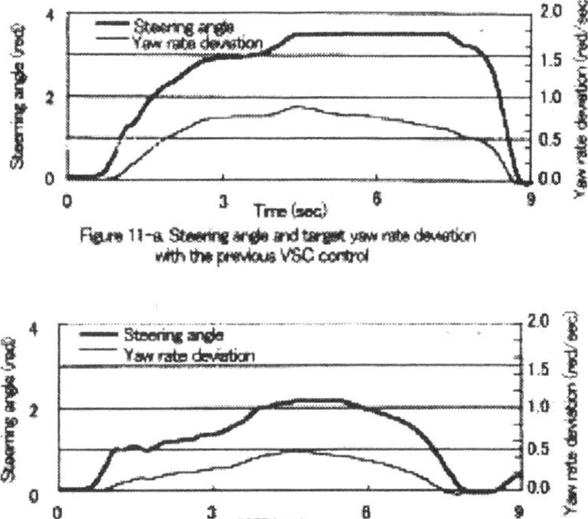

Figure 11-a: Steering angle and target yaw rate deviation with the previous VSC control

Figure 11-b: Steering angle and target yaw rate deviation with the Vehicle Dynamics Management

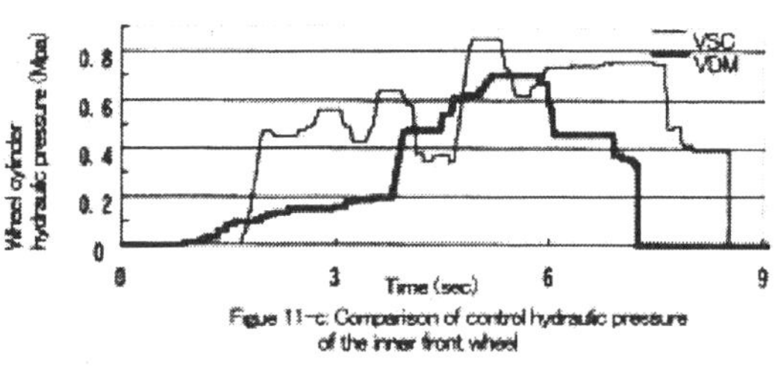

Figure 11-c: Comparison of control hydraulic pressure of the inner front wheel

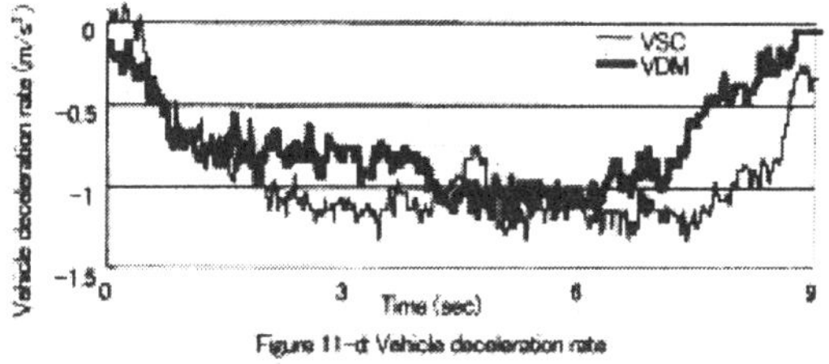

Figure 11-d: Vehicle deceleration rate

4.2. Cornering Startoff Performance on Extremely Low-μ Surfaces

When a vehicle starts off at an extremely low-μ surface intersection to turn right or left, its startoff performance is improved if it is a four-wheel-drive vehicle. However, if the front wheels slip, the drive torque applies to the rear wheels, causing the occurrence of the so-called "push-under" phenomenon, which pushes the rear of the vehicle out.

With respect to the vehicle behavior simulating a left turn at an intersection with a μ=0.35 surface, Figure 12 shows the following:

[1] The actual vehicle data using the new system and control (Figure 12-a)

[2] The responsiveness equivalent to gasoline engine, with the control logic remaining as is (150 ms lag, Figure 12-b)

[3] Only the VSC control effected, without the drive force control (Figure 12-c)

The results are shown in Figure 12-d, according to the amount of deviation of the actual yaw rate in comparison to the target yaw rate per the aforementioned three pieces of data. Although the previous system can effect drive torque control in accordance with the deviation between the target yaw rate required by the driver and the actual yaw rate that is generated by the vehicle, the new system realizes a smoother cornering startoff performance by controlling the motor (whose torque can be controlled more responsively and linearly than in vehicles with ordinary engines) starting at the normal range.

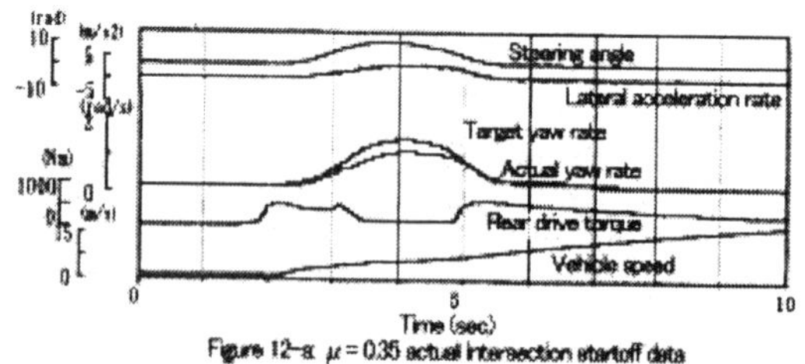

Figure 12-a: $\mu = 0.35$ actual intersection startoff data

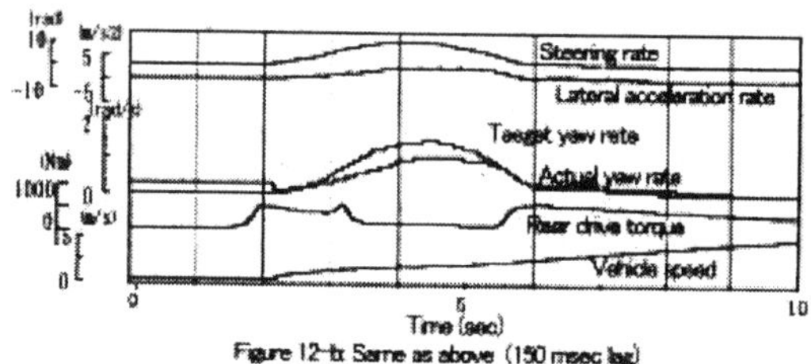

Figure 12-b: Same as above (150 msec lag)

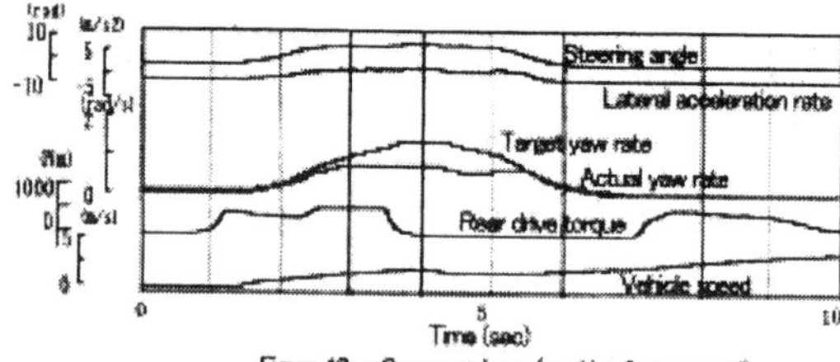

Figure 12-c: Same as above (no drive force control)

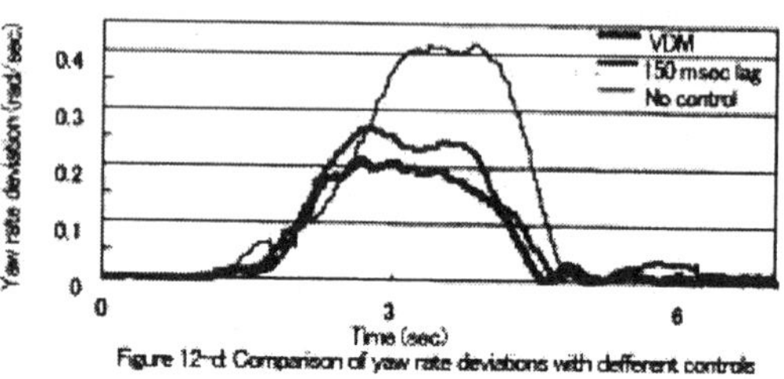

Figure 12-d: Comparison of yaw rate deviations with defferent controls

Conclusion

For the purpose of improving the environmental performance (through the coordination with the regenerative brakes at the front and rear wheels), preventive safety, and vehicle dynamic performance, a brake-by-wire system called an Electronically Controlled Brake (ECB) system (which linearly controls the hydraulic pressure that is applied independently to the four wheels) has been developed, and the Vehicle Dynamics Management (which comprehensively controls the braking and driving forces) has been incorporated in this system. As a result:

Excellent brake feel has been realized through regenerative coordination control, while improving the fuel consumption rate by approximately 20%.

The driver comfort has been improved with respect to the vehicle behavior from the normal range to the critical limit.

Because this technology can also be applied to ordinary vehicles in addition to hybrid vehicles, it is believed that this technology can contribute to enhancing the dynamics of the vehicles.

REFERENCE

[1]W.D.Jonner et al,"Electrohydraulic Brake System-The First Approach to Brake-by-wire Technology",SAE960991

[2]A.Sakai,et al,"Toyota Braking System for Hybrid Vehicle with Regenerative System",EVS13

[3]A.T.Van Zanten,et al,"VDC Systems Development and Perspective",SAE980235

Development of Electronically Controlled Brake System for Hybrid Vehicle

Eiji Nakamura, Masayuki Soga, Akira Sakai, Akihiro Otomo and Toshikazu Kobayashi
Toyota Motor Corporation

ABSTRACT

We expect to reduce exhaust gas emissions further and improve fuel consumption, by developing a new brake system (called brake-by-wire system) to control the friction brake force and the regenerative brake force of the two motors, one each at front and rear axle. Within this new system we developed the new technology listed below.
1. To compensate the changes of the regenerative brake force of front and rear motors, the friction brake force is controlled by adjusting the wheel cylinder hydraulic pressures.
2. The pressure of each wheel cylinder is controlled by linear solenoid valves. So the hydraulic pressure of wheel cylinders is controlled individually and smoothly.

This brake system also operates ABS, VSC, TRC functions. The vehicle stability performance is improved by controlling the braking and driving torque of two motors and also controlling the friction brake torque cooperatively. By using this new brake system, it is possible to make maximum use of the regenerative brake, which reduces exhaust gas and improves fuel consumption with the same brake pedal operation as in conventional vehicles. This brake system, that is able to control the hydraulic pressure of four wheel cylinders individually and smoothly, has high potential to add functional enhancements with proper modifications of software. So that this brake system is expected to become the core technology for further improvement of the vehicle stability and convenience.

INTRODUCTION

In response to growing social needs for conserving the global environment and saving energy, the hybrid car has been studied actively.

It is required for the brake system of a hybrid passenger car to make maximum use of regenerative brake for further reduction of emissions, improvement of fuel consumption and also to ensure higher vehicle stability equivalent to that of ordinary passenger cars. In order to meet such requirements, we developed a new four-wheel drive (4WD) hybrid passenger car with two driving motors, one each at front and rear axle, and we have developed a brake system named "Electronically Controlled Brake

System" (hereafter referred to as "ECB"). The ECB integrates brake control functions such as the cooperative control between the regenerative brakes of the front/rear motors and the hydraulic brakes, ABS (Antilock Brake System), VSC (Vehicle Stability Control), TRC (TRaction Control), etc. The ECB is a brake-by-wire system that controls the wheel cylinder hydraulic pressure electronically. Requirements of the brake system in a hybrid vehicle, ECB composition, results of its application to vehicle, failsafe performance and the effect on fuel consumption will be discussed in this paper.

REQUIREMENTS OF BRAKE SYSTEM OF HYBRID VEHICLE

4WD HYBRID SYSTEM - The 4WD-hybrid system developed anew is equipped with two driving motors, one each at front and rear axle (Figure 1). These front and rear motors are used as generators during braking, and the electrical energy recovered into the battery is re-used when starting and running the vehicle. Owing to the above, further reduction of exhaust gas emissions and improvement of fuel consumption are attained. Friction brake done by driver's brake pedal operation is replaced partially or entirely by regenerative brake. The regenerative brake also provides the deceleration done by engine brake when the accelerator pedal is released in this system.

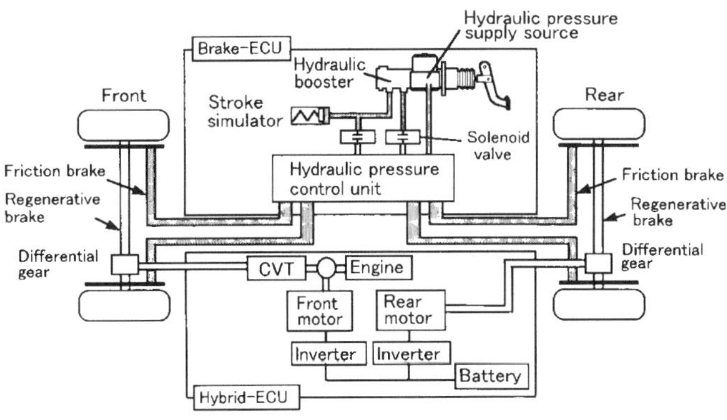

fig. 1 Hybrid System and Brake System

CONTROL OF BRAKE FORCE - As shown in Figure 2, brake forces are divided into the brake force corresponding to the engine brake and the brake force of service brake generated by operating the brake pedal. The engine brake force is determined by the vehicle speed and accelerator pedal operation. The service brake, on the other hand, is activated by operating the brake pedal in the same way as in ordinary vehicles. At this time the regenerative brake force and the friction brake force are so distributed that the energy can be recovered at the highest rate during service brake.

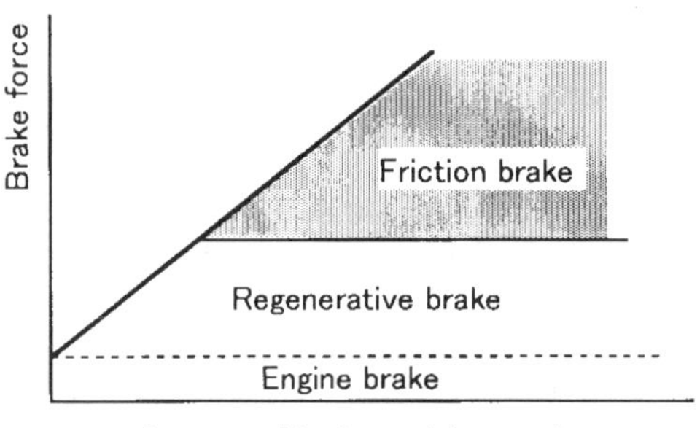

fig. 2 Allotment of Brake Force

A specific example of the above is shown in Figure 3. The regenerative brake force is determined by the power of battery (the energy receiving ability of battery per hour). As the power of battery is constant, the brake force that can be generated becomes smaller as the vehicle speed becomes higher because of the physical properties. Since the vehicle speed is high and the regenerated brake force is small in the initial phase of braking, the brake force required by the driver cannot be covered by the regenerative brake force alone. Therefore, the friction brake force works frequently in the initial phase of braking. Then the regenerative brake force increases as the vehicle speed decreases, and the friction brake force decreases accordingly.

When the vehicle is stopped, no regenerative brake force is generated. Therefore the regenerative brake force decreases gradually from a certain vehicle speed while the friction brake force increases gradually to ensure an adequate brake force required by the driver. In order to increase the energy recovery rate with the regenerative brake at that time, a quick switch-over between the friction brake force and the regenerative brake force within a short period of time is important. For that reason, it is necessary to adjust the friction brake force keeping higher response characteristic.

It is also necessary to adjust the friction brake force for front and rear wheels independently so that the brake distribution force between the front and rear wheels remains the same even if the regenerative brake force is changed. A specific example of friction brake force adjustment is shown in Figure 4. For this adjustment, it is necessary to control the wheel cylinder hydraulic pressures to provide high response performance individually for the front and rear sides during a normal operation of service brake, as the friction brake force is controlled by the adjustment of the wheel cylinder hydraulic pressure.

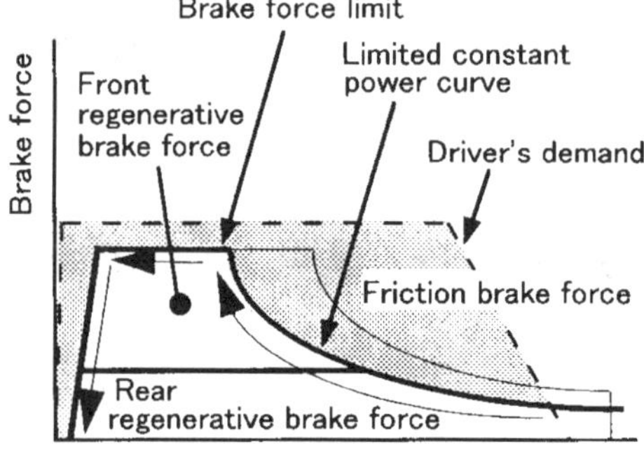

fig. 3 Change in Regenerative Brake force

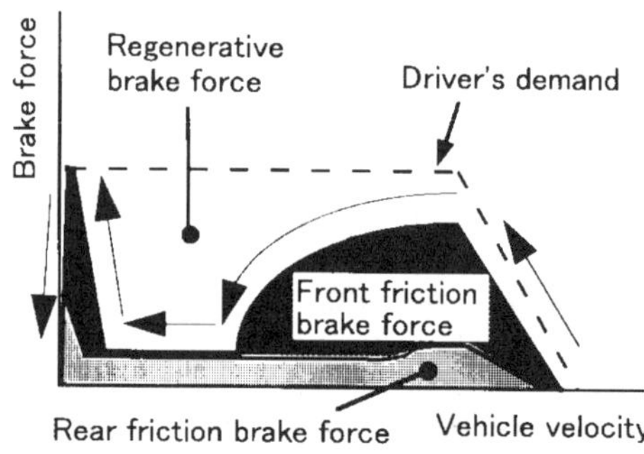

fig. 4 Change in Friction Brake force

INTEGRATION OF ABS, VSC AND TRC FUNCTION - The brake system of the hybrid passenger car is required of not only the reduction of exhaust gas emissions and the improvement of fuel consumption, but also ensuring equivalent vehicle stability to that of ordinary vehicles, by means of ABS, VSC, TRC, etc.

In order to attain the regenerative brake function described in the foregoing based on conventional brake systems, however, it is necessary to add the hydraulic pressure adjustment valves to control the hydraulic pressure of front and rear wheel cylinders individually, and also to add driving electric circuits for these valves. Application of such components would make concerns such as larger size, higher cost and heavier weight of the system. When constructing the ECB, it is important to integrate the regenerative brake, ABS, VSC, TRC, etc. more reasonably.

OTHER REQUIREMENTS - Items required for ECB are as follows.

1. To be free from influence of vacuum pressure variation on the brake forces that may be caused by stopping of engine, as the hybrid vehicle often runs with the motors alone.
2. Capability to control wheel cylinder hydraulic pressures properly without generating unpleasant noise and vibration, as the wheel cylinder hydraulic pressures are always controlled in braking operations
3. Capability to provide equivalent brake pedal feeling as ordinary passenger car by preventing the brake pedal vibration caused by the adjustment of wheel cylinder hydraulic pressure, etc.

ECB SYSTEM COMPOSITION

The ECB system consists of the following items as shown in Figure 5 in order to meet the requirements described in the foregoing.

1. Input unit to detect brake pedal force that is provided by the driver and to create proper feeling in brake pedal operation
2. Hydraulic pressure control unit to adjust wheel cylinder hydraulic pressures individually for each one of four wheels.
3. Hydraulic pressure supply source as the energy source for friction brake force

With the brake-by-wire system which separates the input unit to be operated by the driver from the output unit of the wheel cylinder hydraulic pressure, it has become possible to control the wheel cylinder hydraulic pressures freely without any influence on brake pedal operation.

Moreover, the wheel cylinder hydraulic pressures can be created regardless of vacuum pressure variation caused by stopping of engine or brake pedal operation, which is caused by using the high-pressure accumulator as the energy source of brake force.

Furthermore, the hydraulic pressure control unit is provided with the function to adjust the pressure smoothly and individually for each one of four wheels. So the hydraulic pressure adjustment of wheel cylinders, which is necessary for kinds of braking devices such as the regenerative brake, ABS, VSC and TRC, can be integrated.

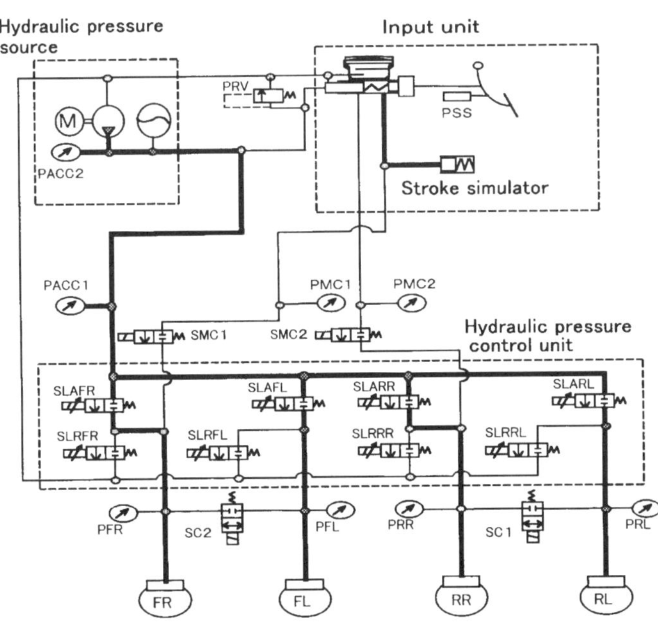

Fig. 5 ECB Composition

ACTIVATED MOTIONS IN SYSTEM CONTROL MODE - Upon detection of driver's intention for braking by means of the brake pedal stroke sensor (PSS in Figure 5) and the master cylinder hydraulic pressure sensors (PMC1 and PMC2 in Figure 5), the master cylinder cut valves No. 1 and No. 2 (SMC1 and SMC2 in Figure 5) are closed. The master cylinder hydraulic pressure is cut off from the each

wheel cylinder, in order to prevent the adverse effect of hydraulic pressure adjustment on the feeling in brake pedal operation. At the same time, the master cylinder hydraulic pressure is led to the stroke simulator in order to generate the pedal reaction force and stroke. The stroke simulator consists of two types of coil springs and rubber units to create a natural feeling in brake pedal operation.

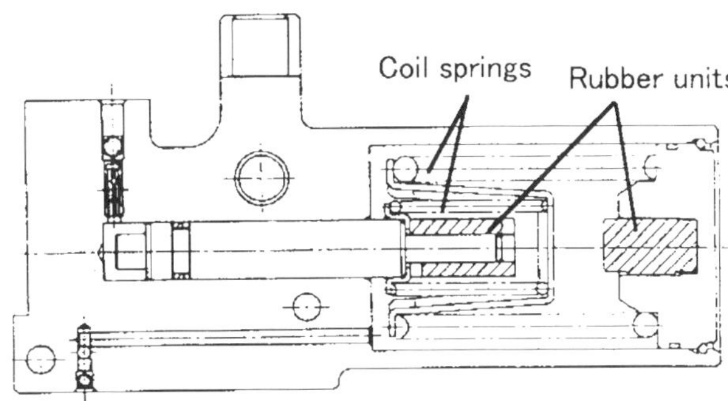

fig. 6 Stroke Simulator

Figure 7 shows a diagram of the force applied to brake pedal and the brake pedal stroke, while Figure 8 shows a diagram of the force applied to brake pedal and the deceleration. Equivalent feeling in braking to that of ordinary vehicle can be obtained.

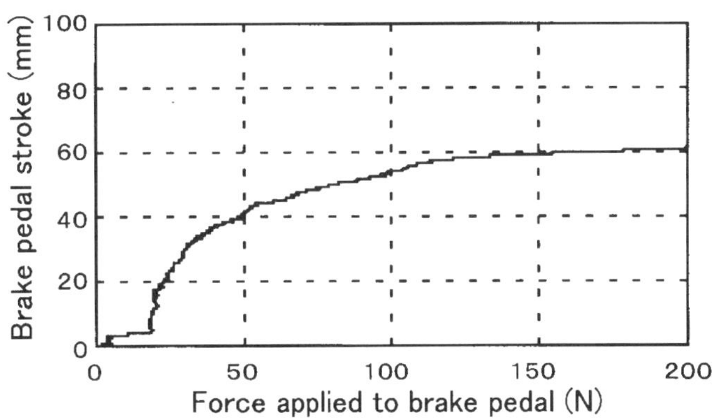

fig. 7 Force Applied to Brake Pedal and the Brake Pedal Stroke

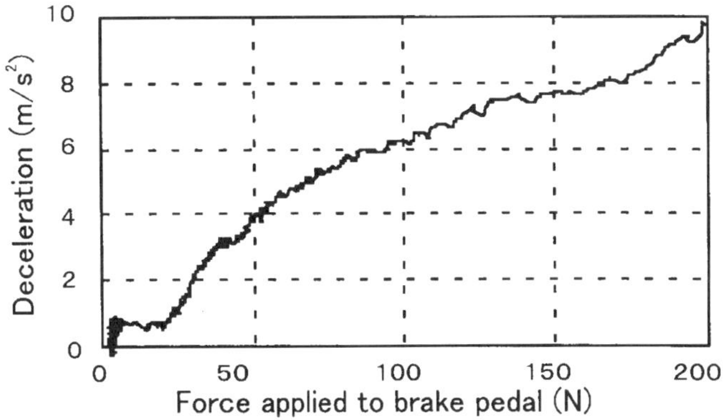

fig. 8 Force Applied to Brake Pedal and the Deceleration

The wheel cylinder pressure is supplied from the accumulator kept under high pressure by the hydraulic pump. Also the right-left separation valves (SC1 and SC2 in Figure 5) are closed to separate the right and left wheels, and hydraulic pressure paths to each wheel is formed independently.

The hydraulic pressure control unit consists of linear solenoid valves for pressure increase (SLA** in Figure 5) and linear solenoid valves for pressure decrease (SLR** in Figure 5), in order to adjust the hydraulic pressure for each wheel smoothly. The stroke rates of the valves SLA** and SLR are controlled by adjusting the electric current, and then the brake fluid flow rate is adjusted to control of wheel cylinder hydraulic pressures quietly and smoothly. The wheel cylinder hydraulic pressures are adjusted likewise not only for the service brake but also for control the ABS, VSC and TRC by the linear solenoid valves.

RESULTS OF APPLICATION TO VEHICLE

Figure 9 shows the structure of control system. Brake-ECU integrates brake control functions such as the service brake (the regenerative brake), ABS, VSC, TRC.

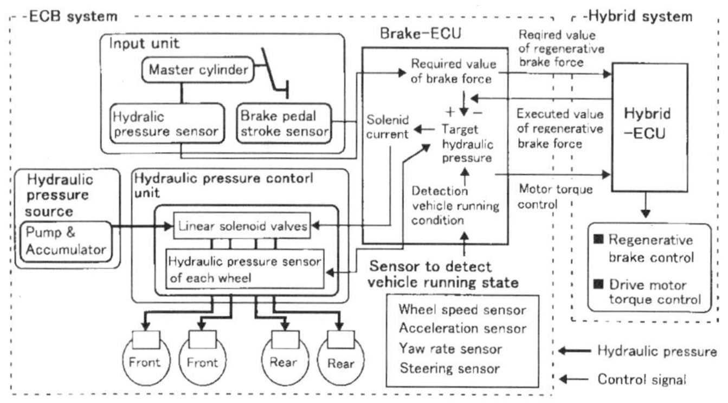

fig. 9 ECB Control Composition

SERVICE BRAKE OPERATION - The service brake system consists of the regenerative brakes and the friction brakes (hydraulic brakes). The regenerative-frictional blending control is done by the cooperative operation between the hybrid ECU to control the motors and the brake ECU to control the friction brakes.

As shown in Figure 9, the brake force required by the driver can be calculated with the brake ECU according to the value of hydraulic pressure generated in the master cylinder and the value of brake pedal stroke. Then the distribution of brake force between front and rear wheels is calculated according to the ideal brake force distribution, and this brake force is transmitted to the hybrid ECU as the required value of the front wheel regenerative brake force, and that of rear wheels. (The upper limit is set to the required value of the regenerative brake force for each front and rear wheels because of fail-safe.) The hybrid ECU calculates the executable regenerative brake force according to the vehicle speed and the energy receiving ability, and also calculates the executed values of individual regenerative brake forces of front and rear

wheels. In case of this hybrid vehicle, the energy recovery rate on the rear wheel is higher than the front, so that the regenerative brake on the rear wheel is executed much more than that of front. And the hybrid ECU transmits the executed values of individual brake forces of front and rear wheels to the brake ECU. Then the brake ECU compares the required values of regenerative brake force calculated and the values actually executed. It also calculates the necessary friction brake forces - i.e., target values of wheel cylinder hydraulic pressures of individual wheels. The brake ECU actuates the linear solenoid valves to control the hydraulic pressure of each wheel individually. Owing to such functions of the system, blending control between the regenerative braking forces generated from the front and rear motors and the friction braking forces is ensured.

<u>Performance of Regenerative - Frictional Blending Control</u> - Examples of the brake force with the execution of blending control for the friction brakes and the regenerative brakes by means of front and rear motors are shown below. Figure 10 shows front side; Figure 11 shows rear side. To compensate the changes of regenerative brake force with front and rear motors, the friction brake forces is controlled accurately. So that the brake force required by driver is executed precisely. Further more these figures show that, as the energy recovery rate on rear wheel is higher than the front, so that the regenerative brake force is distributed on the rear wheel much more than the front at high speed.

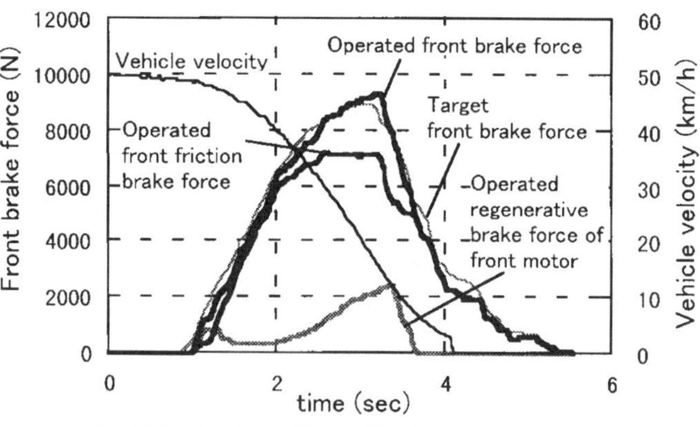

fig. 10 Brake Force (Front Side)

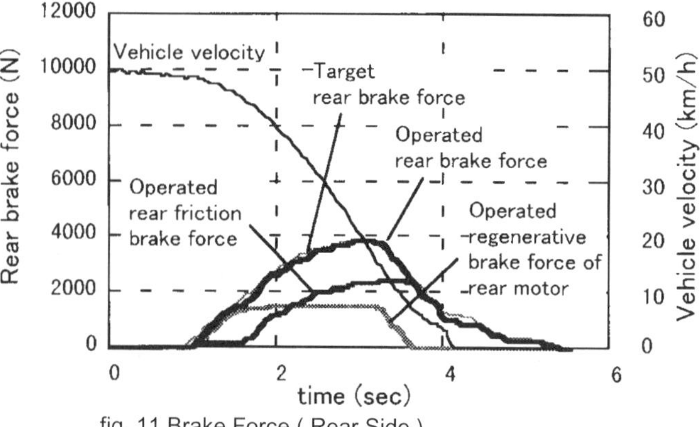

fig. 11 Brake Force (Rear Side)

ABS, VSC AND TRC OPERATION - ABS, VSC and TRC are controlled with the cooperative control between the brake ECU and hybrid ECU. As shown in Figure 9, the brake ECU detects the vehicle running conditions according to the signals sent from the sensors for wheel speed, acceleration, yaw rate and steering. Then the brake ECU actuates the linear solenoid valves to control the hydraulic pressure of each wheel, and executes the control of ABS, VSC and TRC.

At the same time, it determines the optimum driving motor output torque individually for front and rear motors, and transmits the value to the hybrid ECU. The hybrid ECU executes the control on engine and motor driving forces. The skid, the deterioration of steering performance, the driving wheel slip, etc. can be suppressed with this cooperative control of the braking and driving forces of the front/rear motors, and also the friction brake force. So the vehicle stability is improved furthermore.

ABS Performance - The ABS performance is equivalent to that of ordinary vehicles in terms of both stopping distances in braking and vehicle behavior. As the ABS is controlled in the state that the brake pedal and the wheel cylinders are separated from each other, the vibration of brake pedal that found in conventional ABS does not occur. Since the linear solenoid valves are used for the adjustment of wheel cylinder hydraulic pressures, the valve activation noise is also reduced.

Vehicle Stability Performance - An example, that the vehicle stability is improved by controlling the braking and driving torque of the front and rear motors and also controlling the friction brake torque cooperatively, is described below.

In case that the slip of front wheel is detected when starting or accelerating the vehicle, appropriate starting/acceleration performance is ensured by reducing the front wheel driving force, and activating the rear driving motor. When starting or accelerating the vehicle during a turning operation, however, the steering performance may drop due to this generation of rear driving force. As countermeasures against such a problem, the ECB executes the TRC and the optimum control of the driving force generated by the rear motor according to the extent of vehicle turning. So that the deterioration of steering performance is prevented and an adequate vehicle starting performance is ensured at the same time.

FAIL-SAFE PERFORMANCE

On construction of fail-safe performance, the following measures have been taken based on the concept "to ensure an adequate brake force with the friction brakes" against every kind of system failure. Since it is a brake-by-wire system that controls the wheel cylinder hydraulic pressure electronically, (1) precise detection of the brake force provided by the driver, and (2) measures against electric failures of the control system and failures of the hydraulic pressure system are quite important.

DETECTION OF BRAKE FORCE PROVIDED BY DRIVER - The system to detect the brake force provided by the driver consists of three sensors - two master cylinder hydraulic pressure sensors (PMC1 and PMC2) and one pedal stroke sensor (PSS)(Figure 12). Owing to this sensor composition, abnormal sensor signals can be identified and the brake force provided by the driver can be detected without fail. Conversely, the sensors are capable of judging positively that the driver is not engaged in brake pedal operation, so the generation of any unintended brake force is prevented.

MEASURES AGAINST ELECTRIC FAILURES OF CONTROL SYSTEM AND FAILURES OF HYDRAULIC SYSTEM - The regenerative brake is stopped on occurrence of any abnormality in communications between the hybrid ECU and the brake ECU or driving motor system, but the brake control function continues to generate the brake force required by the driver, with the friction brakes alone. Failures associated with the regenerative brakes are detected according to the communications between the hybrid ECU and the brake ECU.

Upon detection of an electric failure such as abnormal signal from the hydraulic sensor, disconnection of wiring system, short-circuit, abnormality in brake ECU, etc. or a failure of hydraulic system such as the failure of linear solenoid valves, the drop of pressure in hydraulic pressure source caused by the failure of pump or motor, the brake control function is stopped and the cut valves SMC1/SMC2 and the separation valves SC1 and SC2 are opened to form an ordinary two - circuit hydraulic pressure brake system. Then the hydraulic pressure generated in the hydraulic booster is led into the wheel cylinders of four wheels to ensure an adequate brake force.

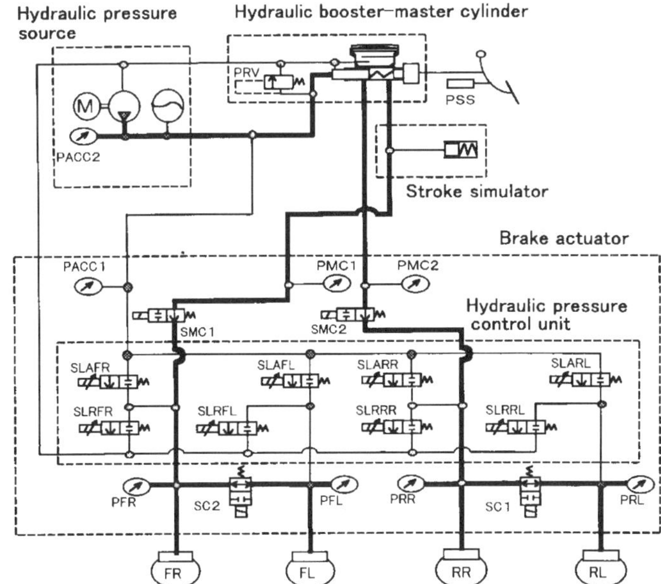

fig. 12 Hydraulic Circuit Diagram (System Non-control Mode)

IMPROVEMENT FUEL CONSUMPTION

We verified the fuel consumption effect of this system by carrying out the Japanese 10-15 mode fuel efficiency test. In the 10-15 mode test which simulates street driving conditions in Japan, the vehicle's speed is comparatively low and the ratio of engine brake in proportion to the brake force is large. As figure 13 shows, we can verify the effect of improving fuel consumption with this brake system.

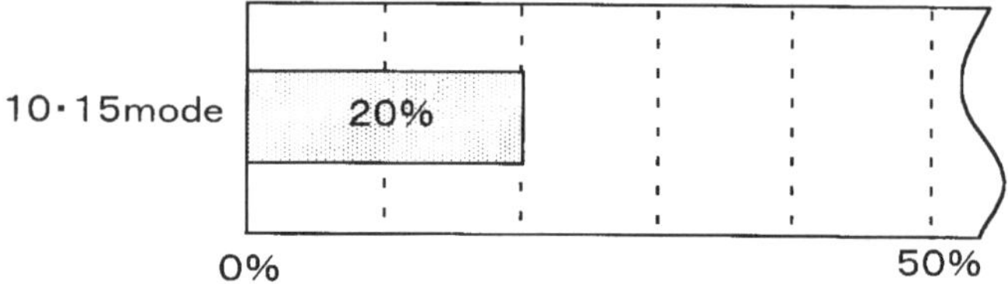

fig. 13 Amount of improvement in fuel consumption due to the regenerative brake

CONCLUSIONS

New technologies have been established for smooth control of wheel cylinder hydraulic pressure independently for each one of four wheels by means of linear solenoid valves. A brake-by-wire system named ECB has been thus developed, which is capable of faithful reflection of driver's braking and the creation of natural feeling in braking.

With this new brake system installed in the hybrid passenger car, the energy recovery rate of regenerative brake has been maximized, and the reduction of exhaust gas emissions and the improvement of fuel consumption have been attained at the same time.

Furthermore the vehicle stability has been improved by controlling the braking and driving torque of the front and rear motors and also controlling ABS, VSC, and TRC functions cooperatively.

This brake system, that is able to control the hydraulic pressure of four wheel cylinders individually and smoothly, has high potential to add functional enhancements with proper modifications of software. So that this brake system is expected to become the core technology for further improving the vehicle stability and convenience.

REFERENCES

1. Masamoto Ando, N.Enomoto: Advanced Braking System for Electric Vehicles, EVS13.
2. Akira Sakai: Toyota Braking System for Hybrid Vehicle with Regenerative System, EVS14.
3. Akihiro Otomo: Development of Regenerative Braking System for Hybrid Vehicle, IPC10.

Advances and Challenges in BBW AWB Dispulsion for Future Automotive Vehicles

Bogdan Fijalkowski
Krakow University of Technology

ABSTRACT

Current implementation of novel advanced electro-mechanically (E-M) activated brake-by-wire (BBW) all-wheel-driveable (AWB) dispulsion mechatronic control hyposystems may be comparatively costly. Though, as the cost of macroelectronic commutator (macrocommutator) and microelectronic processor (microprocessor) based controllers becomes increasingly cheap, fixed with bettered short-stroke linear brake-force-actuator motors, novel advanced E-M-activated BBW AWB dispulsion mechatronic control hyposystems become a potential practical choice in the not-too-distant future. The size and mass of these short-stroke tubular linear DC-AC macrocommutator brake-force-actuator E-M motors may be comparatively small and light, respectively, in the nearest future with bettered interior permanent magnet (IPM) materials. Specific advantages of the novel advanced E-M-activated BBW AWB dispulsion mechatronic control hyposystem are cleanliness, lack of fluid and flexibility.

This paper presents a very advanced anti-lock and/or anti-spin BBW AWB dispulsion mechatronic control hyposystem with E-M-activated friction disc, ring and/or drum brakes for controlling not only the braking forces during normal riding and cornering but also the braking forces between the inner and outer motorized and/or generatorized wheels (M&GW) in a hard turn independently.

INTRODUCTION

Electro-mechanical (E-M) fly-by-wire (FBW) aviation or aerospace mechatronic control systems have been used successfully for over twenty years in aerial vehicles, that is, aircrafts and helicopters. In these mechatronic control systems, the control commands for the aircraft or helicopter are not transmitted in a fluido-mechanical (F-M) or pneumo-mechanical (P-M) mode but through E-M one, that is, electrical wires (by-wire). In this subject, the advantages against conven-tional F-M- or P-M-activated FBW mechatronic control systems have established to be so significant that the technique is used in other subjects as well. Since the reliability of hardware and software in this class of system plays a dominant role, mechatronic control systems in aircrafts and helicopters were designed with multiple redundancies.

Automotive scientists and engineers also predict the integration of the E-M ride-by-wire (RBW) or x-by-wire (XBW) automotive mechatronic control system's steering (diversion), driving (propulsion), braking (dispulsion) and absorbing (suspension) technologies into integrated safety systems (ISS). In certain circum-stances, this may presume responsibility for either avoiding a collision or at least minimizing the effect of a collision. There is considerable concern, only now being realized by automotive manufacturers, about the significance of the invention of high-speed intelligent data bases (IDB), to collect information and commu-nicate their estimations and references between varieties of electronic control units (ECU) of an extremely complicated RBW automotive mechatronic system. Conventional F-M- or P-M-activated brake-by-wire (BBW) all-wheel-brake-able (AWB) dispulsion mechatronic control hyposystems as they are used in auto-motive vehicles today are based on F-M- or P-M-activated hyposystems, where safety is guaranteed by using two independent F-M or P-M brake loops. In the past few years, safety and comfort (S&C) of conventional BBW AWB dispulsion mechatronic control hyposystems have been improved by the applica-tion of ECUs, like the anti-blocking system (ABS), anti-slip regulation (ASR), electronic stability program (ESP) or the brake assistant (BA). All these ECUs, however, can be shut down in case of a failure while the conven-tional FM- or PM-activated BBW AWB dispulsion mechatronic control hyposystem is still accessible as a backup one.

A very advanced BBW AWB dispulsion, a promising automotive technology modified from aviation and aerospace advances, promises safety and comfort (S&C) advantages for motorists and high development for its providers.

Future very advanced BBW AWB dispulsion mechatronic control hyposystems for automotive vehicles have potential to be so scientifically complicated that they will in effect intervene human driver input. By explaining the intelligence of the human driver through a mechatronic inter-face and a large number of fused sensors, it 'intelligently' disconnects – and takes advantage of mechatronic control to handle the most rigorous braking dispulsion condition. Determined by the circumstances, this intercession indicates a significant supremacy on motorists' driving and further on the S&C of braking dispulsion. The actual advantages that accumulate stand for a considerable jump forward in automotive very advanced technology.

The author predicts, that in the 2010s motorists will observe an extra generation of very advanced BBW AWB dispulsion mechatronic control hyposystems that will utilize short-stroke tubular linear DC-AC macrocommutator brake-actuator E-M motors at each wheel or axle, that is, on all wheels or axles, thus removing the fluidics or pneumatics and related hardware. These BBW AWB dispulsion mechatronic control hyposystems will insert additional improvements to act perfectly with very advanced vehicle dynamic mechatronic control, adaptive cruise mechatronic control, and ultimately, intelligent transportation systems (ITS).

Significant betterments have been seen in automotive vehicle riding and cornering performance in recent years as a result of advances achieved, especially in wheel-tire and vehicle-suspension technology. Due to these betterments, automotive vehicle handling characteristics during braking have taken an added importance. Recently every automotive brake maker is working on full-time anti-lock BBW AWB dispulsion mechatronic control hyposystems [1-14]. One disadvantage of using fluidic (hydraulic) or pneumatic pressure - the standard for automotive vehicle's brakes for 60 years - is that modern automotive vehicles have difficulty fitting the parts under the bonnet or hood as the Americans say. The BBW AWB dispulsion will come in a hybrid form first, retaining some fluidics (hydraulics) or pneumatics [2,5-7]. Now it may be mechatronic. In the nearest future, automotive vehicle makers will be compelled to manufacture for the global market ultra low-emission vehicles (ULEV) and virtual zero-emission vehicles (VZEV), for instance, as the *Poly-Supercar* (advanced ultralight hybrid) that offer emission-free and moderate speed in certain areas [1,3,4,8,9]. An ultralight (*600 kg*), *4* to *5* passenger supercar with series hybrid-electric drive-by-wire (DBW) all-wheel-driveable (AWD) propulsion, advanced anti-lock BBW AWB dispulsion, predictive and adaptive absorb-by-wire (ABW) all-wheel-absorbable (AWA) suspension and dual-mode hybrid steer-by-wire (SBW) all-wheel-steerable (AWS) diversion mechatronic control hypo-systems and simplified design can achieve *1.5 - 4.5 l/100 km*. Automotive E-M components for full-time DBW AWD propulsion, BBW AWB dispulsion, ABW AWA suspension and SBW AWS diversion mecha-tronic control hyposystems of the *Poly-Supercar*, which is a VZEV, will be well-suited to help reduce local air pollution emissions. The electrical energy required to dispel such VZEV is generated by the clean-burning, hydrogen fueled automotive gas turbo-generator (AG-TG) with the brushless AC-DC macrocommutator fly-wheel-disc generator or the Fijalkowski engine-generator/motor (FE-G/M) which directly converts into electrical energy the mechanical energy supplied to the pistons by the combustion of gaseous hydrogen installed on the board of a VZEV. At the VZEV user end, the electrical energy may be stored in DC chemo-electrical (C-E) storage batteries, superconductive magnetical energy storage (SMES) ultra-inductors, SEES super-capacitance electrical energy storage (SEES) ultra-capacitors and inertial mechanical energy storage (IMES) ultra-flywheels.

For instance, the core of the very advanced anti-lock and/or anti-spin BBW AWB dispulsion mechatronic control hyposystems can be not only four E-M friction disc, ring and drum brakes with the novel short-stroke linear tubular brushless DC-AC macrocommutator IPM) brake-force-actuator E-M motors, single smart reciprocating E-M brake-pedal actuator, driven by the short-stroke linear tubular brushless DC-AC macrocommutator brake-pedal-actuator E-M motor for the left-foot brake pedal (normal braking), but also four in-dependently-suspended front and rear electromechanical/mechanoelectrical (E-M/M-E) motorized and/or generatorized wheels (M&GW) with the brushless DC-AC/AC-DC macrocommutator IPM wheel-hub E-M motors/M-E generators (regenerative braking).

E-M-activated friction disc, ring and drum brakes for M&GWs may provide bettered default modes compared to conventional FM- or PM-activated friction disc, ring and drum brakes for conventional all-mechanically internal combustion engine (ICE) driven wheels plus significantly bettered serviceability. Increased diagnostic capabilities may be available through the use of mechatronics.

Automotive developers expect to use a spring-damper to give *'pedal feel'* to a brake pedal, which will push against a mechatronic movement-sensing device. The sensor will signal an electronic control unit (ECU). For instance, for hybrid anti-lock BBW AWB dispulsion mechatronic control hypospystems (for example, now developed by *Delphi*), electrical energy for rear M&GWs will be generated by an automotive brushed AC-DC electronic-commutator M-E generator and stored in a DC chemoelectrical (C-E) storage battery and, fluidic pressure for front M&GWs will be generated by a mechanofluidical (M-F) pump and stored in a fluidic accumulator. The ECU will direct it to apply EM-activated friction drum brakes for rear wheel drive (RWD) and F-M-activated friction disc brakes for front wheel drive (FWD) as required. If the electronics fail, pushing harder on the brake pedal will work as back-up emergency F-M braking hyposystem [2,7].

A very advanced anti-lock and/or anti-spin BBW AWB dispulsion mechatronic control hyposystem eliminates fluidics (hydraulics) or pneumatics. Electrical energy will work, for instance, the brake calipers. Less brake bulk in front helps better crash deceleration for lower occupant injury. Other advantages are reduced mass, better brake control in anti-lock and/or anti-spin braking accident situations, more accurate brake force distribution, and potentially better reliability as well as better pedal feel.

Rather than attempt to adjust the proportion-ing directly, very advanced anti-lock and/or anti-spin BBW AWB dispulsion mechatronic control hyposystems sense the M&GW lockup occurs, release the E-M-activated friction disc, ring or drum brakes momentarily on locked M&GWs, and reapply them when the M&GW spins up again.

At *Krakow University of Technology's Automotive Mechatronics Institution*, novel very advanced E-M-activated BBW AWB dispulsion mechatronic control hyposystems and their practicability were considered.

The objective is to replace the conventional F-M- or P-M-activated BBW AWD mechatronic control hyposystem with an advanced E-M-activated BBW AWB dispulsion mechatronic control hyposystem without a pneumatic or fluidic backup one.

The features of such a very advanced E-M-activated BBW AWB dispulsion mechatronic control hyposystem are: easier adaptation of assistance system (for instance, ABS, ASR, ESP, BA) as they could be realized only by soft-ware; cost reduction by easier construction and packaging; and necessary functionality without complicated F-M or P-M gears.

42 V_{DC} - AUTOMOTIVE ELECTRICAL ENERGY AND SIGNAL DISTRIBUTION (E^2&SD) SYSTEM

Increased electrical energy demands on automotive electrical energy and signal distribution (E^2&SD) systems coupled with the automotive industry's need to improve specific fuel consumption (SFC) and exhaust emissions (EE) are making current single-voltage (14 or 28 V_{DC}) E^2&SD systems inadequate.

Automotive manufacturers are closing the electrical energy gap with single-voltage (42 V_{DC}) and dual-voltage (14/42 V_{DC}) E^2&SD systems to in-crease electrical energy and efficiency as well to enable additional loads and new ride-by-wire (RBW) or x-by-wire (XBW) technologies.

The transition to a single-voltage (42 V_{DC}) E^2&SD system will affect a broad range of automotive vehicles. The new voltage standard could also improve BBW AWB dispulsion mechatronic control hyposystems, enabling the use of more powerful rotary and/or linear DC-AC macrocommutator brake-force-actuator motors. As automotive vehicles complete the transition from a dual-voltage (14/42 V_{DC}) E^2&SD system, its architecture will be reconfigured, resulting in reduced wire gauge, smaller wire bundles, smaller wire connectors, and decreased size and mass as well as cost. In addition, installation and routing of the wiring system will be simplified, creating new design opportunities.

The single-voltage (42 V_{DC}) E^2&SD system will also enhance BBW AWB dispulsion technology, allowing for smaller mechatronic brake-actuators that feature lower mass and size and improved performance. It will facilitate the development of E-M-activated friction disc-, ring or drum brakes at each wheel or axle, that is, on all wheels or axles, as well as advanced RBW or XBW automotive mechatronic control systems, including BBW AWB dispulsion mechatronic control hyposystems.

E-M-ACTIVATED FRICTION DISC, RING AND DRUM BRAKES

Considering manually operated (muscular-energy) friction brakes, the brake pedal or lever may be connected to the actual brake either all-mechanically, by means of rods or wires, or electromechanically, by means of an electric current in a wire. Before considering these connections, however, one must deal with the E-M-activated friction brake them. The lower brake factor (that is a mechanical advantage that can be utilized in brakes to minimize the actuation effort required) of E-M-activated friction disc and ring brakes need higher actuation effort, and development of integral parking brake features has been required before friction disc and ring brakes could be used for all wheel positions. Automotive F-M-, P-M- and/or E-M-activated friction drum brakes have seen common usage in the world [5-7], because of their high brake factor and the easy incorporation of parking brake features. On the negative side, drum brakes may not be as consistent in torque performance as disc brakes

The E-M-activated friction disc, ring or drum brake is operated by direct current from a maintenance-free application specific integrated matrixer (ASIM) DC-DC or DC-AC macrocommutator in the rotating brushed DC-AC mechanocommutator IPM brake-force-actuator E-M motor or short-stroke linear tubular brushless DC-AC macrocommutator IPM brake-force-actuator E-M motor's terminal box, respectively. DC and/or AC pulse-width modulation (PWM) operation provides smooth braking, a low closing brake application armature current and the opportunity to vary the actuating time of the brake. Since braking takes place with runner displacement, the type of transmission can be selected at will, so that gearless runner, for example, can be used, and then can be set up in an arbitrary position without affecting the braking. For instance, easily replaceable long-life brake lining of sintered material permit a high operating intensity, limited only by the performance of the rotating brushed DC-AC mechanocommutator IPM brake-force-actuator E-M motor [2,5-7,9] or short-stroke linear tubular brushless DC-AC macrocommutator IPM brake-force-actuator E-M motor [3,4,8,11-14]. The braking torque is adjustable, practically constant and independent of wear. The braking effect is not diminished by water or dust. The E-M-activated friction disc, ring or drum brake is free from oil and the operating brake-force-actuator motor's armature winding is not affected by moisture or vibration. A major advantage could also be in reducing the mass of the E-M-activated friction brakes: discs, rings and drums can be made of silicon-carbide-reinforced aluminum instead of cast iron. Disengagement of the E-M-activated friction disc, ring or drum brakes can be easily arranged electrically by connecting the brake's ASIM DC-AC macrocommutators to the automotive vehicle's onboard mains in front of the brake-force-actuator E-M motors.

E-M-ACTIVATED FRICTION DISC AND RING BRAKES - With the E-M-activated friction disc or ring brake, a silicon-carbide-reinforced aluminum disc instead of a cast iron disc rotating with the M&GW is braked by brake application voltage indirect acting on a pair of self-adjusting friction pads lined with a friction material.

Figure 1 shows a layout of E-M-activated friction disc (Left) and ring (Right) brakes with the novel short-stroke linear tubular brushless DC-AC macrocommutator IPM brake-force-actuator M-E motors. In this type of

the E-M-activated disc and ring brakes, force is applied equally to both sides of disc or ring rotor and braking action is achieved through the frictional action of inboard and outboard brake friction pads against the disc or ring rotor. The pads are contained within a caliper, shown in Figure 1, as is the M&GW brake actuator. Although not a high-gain type of E-M-activated friction brake, disc and/or ring brakes have the advantage of providing relatively linear braking with lower susceptibility to fading than E-M-activated friction drum brakes.

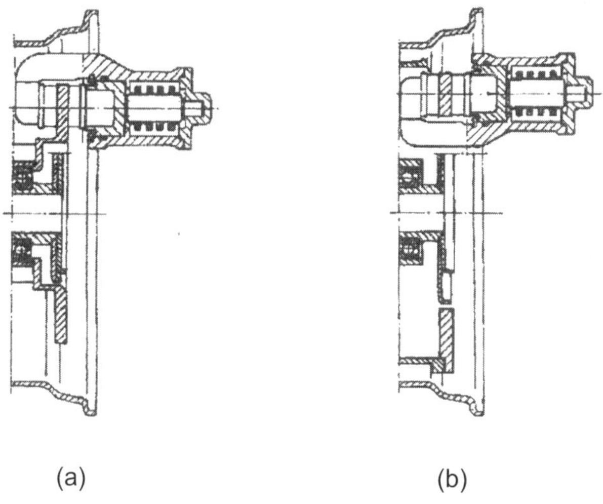

(a)   (b)

Figure. 1.   E-M-activated friction disc (a) and ring (b) brakes with the short-stroke linear tubular brushless DC-AC macrocommutator IPM brake-force-actuator E-M motors.

The static brake torque T can be calculated using the following equation:

$$(1) \qquad T = k\,i\,E\,R$$

where: k - constant coefficient, depending on a construction of the short-stroke linear tubular brushless DC-AC macrocommutator IPM brake-force-actuator E-M motor;

i - brake application armature current of the short-stroke linear tubular brushless DC-AC macrocommutator IPM brake-force-actuator E-M motor;

E - effectiveness factor (ratio of the disc or ring rubbing surface to the input force on the shoes);

R - brake radius. Brake force applied to the disc rotor by the pads is a function of brake application armature current in the BBW AWB dispulsion mechatronic control hyposystem and the constant coefficient of the M&GW brake-force-actuator E-M motor.

The static brake force F can be calculated with the following relationship:

$$(2) \qquad F = T/r,$$

where r - M&GW tire rolling radius.

E-M-ACTIVATED FRICTION DRUM BRAKES - With the E-M-activated friction drum brake; a brake drum rotating with the M&GW is braked by brake application voltage indirect acting on a pair of curved shoes lined with a friction material. Pull-off springs retract the shoes and the E-M brake-force-actuator IPM plunger (runner) when brake-pedal application voltage is released, and shoe adjusters can be fitted to compensate for lining wear.

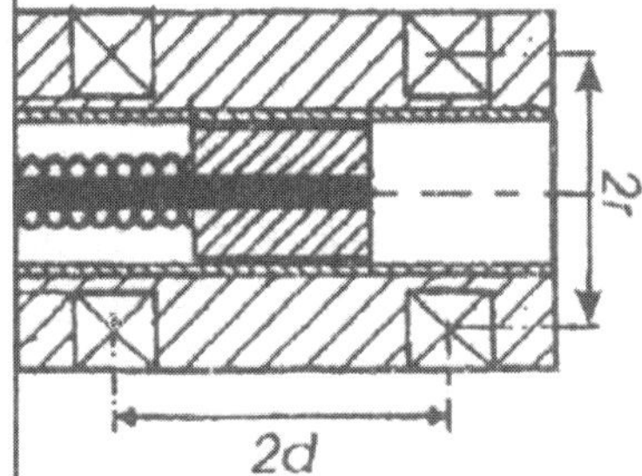

Figure 2. An E-M-activated friction drum brake with the novel short-stroke linear tubular brushless DC-AC macrocommutator IPM brake-force-actuator E-M motor.

Figure 2 depicts a layout of an E-M-activated friction drum brake with the novel short-stroke linear tubular brushless DC-AC macrocommutator IPM brake-force-actuator E-M motor. I
In an E-M-activated friction drum brake; force is applied to a pair of brake shoes in a variety of configurations, including leading trailing shoe (simplex), duo-duplex, and duo-servo. The short-stroke linear tubular brushless DC-AC macrocommutator IPM brake-force-actuator E-M motor can be applied to the traversing of a drum brake apply lever, for instance, replacing the conventional brushed DC-AC mechanocommutator brake-force-actuator E-M motor, gearing driveshaft and control equipment. Maintenance is simplified and, as the short-stroke linear motor is unaffected by atmospheric conditions, the like-hood of breakdown is reduced.

An E-M-activated friction drum brake features high gains relative to E-M-activated friction disc or ring brakes, but some configurations tend to be more nonlinear and sensitive to fading and other brake lining coefficient-of-friction changes.

The static brake torque equation previously presented for E-M-activated friction disc or ring brakes, Eq. (1), is equally applicable to E-M-activated friction drum brakes with design-specific changes for brake radius and effectiveness factor. By design, the brake radius for an E-M-activated friction drum brake is one-half the drum diameter. The effectiveness factor represents the major functional difference between E-M-activated disc and ring brakes; the geometry of friction drum brakes may allow a static brake torque to be produced by the friction force on the shoe in such a manner as to rotate it against the drum and increase the friction force developed. The action can yield a mechanic advantage that significantly increases the gain of the brake and the effectiveness factor as compared with E-M-activated friction disc or ring brakes. The dynamic brake force calculation for E-M-activated friction drum brakes is more complex since the brake lining coefficient of friction is a function of temperature; as the lining heats during a braking maneuver, the effective coefficient of friction increases and less brake application voltage/current is needed to maintain a constant brake torque. High-gain E-M-activated friction drum brakes maximize the torque capability and minimizes electric energy requirements. High gain is achieved through utilization of mechanical self-energizing. Dynamic stability of the gain is achieved by closed-loop control technology. For instance, at an operating torque of $0.5\ kNm$ each E-M-activated friction drum brake can respond in a closed-loop control mode at rates of up to $5\ kNm/s$ to continuously adjust dynamic brake output [4]. Integral to the drum brake is a rotating brushed DC-AC mechanocommutator IPM brake-force-actuator E-M motor, gear train, and ball screw/nut mechanism [2,5] or a short-stroke linear tubular brushless DC-AC macrocommuta-tor IPM brake-force-actuator E-M motor only [1,3,8,11-14]. In the first case, the ball-screw converts rotary-motion torque to linear-motion force. This in turn actuates a conven-tional friction surface drum brake mechanism through a system of apply and stationary levers.

The E-M-activated friction drum brake, as shown in Figure 2 must have the ability to mechanically reduce braking to an acceptable level upon electrical energy removal; it must not remain energized in cases of electrical energy interruption during braking. This requires a highly efficient system with a return spring mechanism to 'astern-drive' the E-M-activated friction drum brakes without the assistance of electric energy. The 'astern-drive' capability needs a separate parking-brake latch mechanism. Thus, the ability to automatically release braking torque during an electric energy interruption demands the use of a highly efficient brake-apply-lever actuator and se-parate parking-brake holding latch. No electric energy is required to remain applied.

NOVEL PROOF-OF-CONCEPT E-M-ACTIVATED BBW AWB DISPULSION MECHATRONIC CONTROL

In a novel proof-of-concept advanced E-M-activated BBW AWB dispulsion mechatronic control hyposystem, the F-M or P-M brake-force actuators and associated assemblies are replaced by a novel short-stroke tubular linear DC-AC macrocommutator brake-force-actuator E-M motor that directly controls a brake-stem interior permanent magnet (IPM) plunger position.

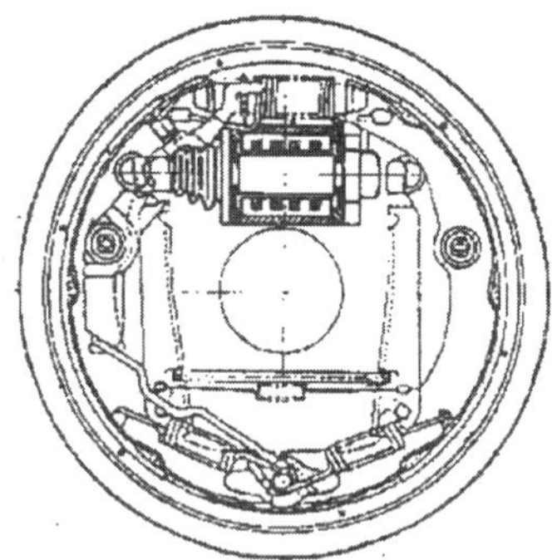

Figure. 3.　An E-M-activated friction drum brake with the short-stroke linear tubular brushless DC-AC macrocommutator IPM brake-force-actuator E-M motor.

Conceptually, the author can consider such a mechatronic M-E-activated BBW AWB dispulsion mechatronic control hyposystem to be simply a novel short-stroke tubular linear DC-AC macrocmmutator brake-force-actuator E-M motor's IPM brake-stem position mechatronic control hyposystem with its armature comprised of the ã-titanium-aluminide (TiAl) or sintered-silicon-nitride (SNN) ceramic brake-stem IPM plunger, as shown in Figure 3.

DESIGN PHILOSOPHY OF THE DC-AC MACRO COMMUTATOR BRAKE-FORCE-ACTUATOR E-M MOTOR - The short-stroke tubular linear DC-AC macrocommutator brake-force-actuator E-M motor consists of an IPM brake-stem plunger located in the electromagnetic field of two-or even multi-phase armature-winding stator with reversibly connected phase inductors and fed by DC electrical energy supply through a DC-AC macro-electronic ASIM application specific integrated matrixer (ASIM) or customer speci-fied integrated matrixer (CSIM) converter commutator, that is, so-called a DC-AC macrocommutator [9].

The brake-stem IPM plunger is submitted to an E-M force F_e along the horizontal axis of the two- or multi-phase armature-winding stator that is given by

(3) $\qquad F_e = M\ dH/dx$

where:　M - magnetic moment of a brake-stem IPM plunger [$Nm/A/m$];

　H　electromagnetic field strength of a stator's armature winding [A/m];

　x - displacement of a brake-stem IPM plunger [m].

The E-M force supplied by the brake-force-actuator E-M motor

(4)　$\quad F_e = (\ L_{af}/\ x)\ i_f\ i_a = L_{af} * I_{f0}\ i_a = c\ i_a$

where:　$L_{af} = L_a\ L_{af}$ -mutual-inductance between the stator armature winding and plunger IPM exciter of the brake-force-actuator E-M motor [H];

　L_a　-self-inductance of the stator armature winding of the brake-force-actuator E-M motor H];

　L_f　- virtual self-inductance of the plunger's IPM exciter of the brake-force-actuator E-M motor [H];

　$i_f = I_{f0}$　- virtual constant IPM-exciter current of the brake-force-actuator E-M motor [A];

　i_a　- armature current of the brake force-actuator E-M motor [A];

　$c = L_{af} * I_{f0}$　- constant E-M interlinkage coefficient [N/A].

The E-M force F_e is proportional to the armature current i_a and depends on ratios d/r and x/r where $2d$ is the distance between the phase coils and r is the phase-coil radius. With ratio d/r between $3/2$ and 1 a best possible compromise on a large displacement of a brake-stem IPM plunger x (approximately $x = d$) and an E-M force F_e which is nearly constant with a rather high value may be get.

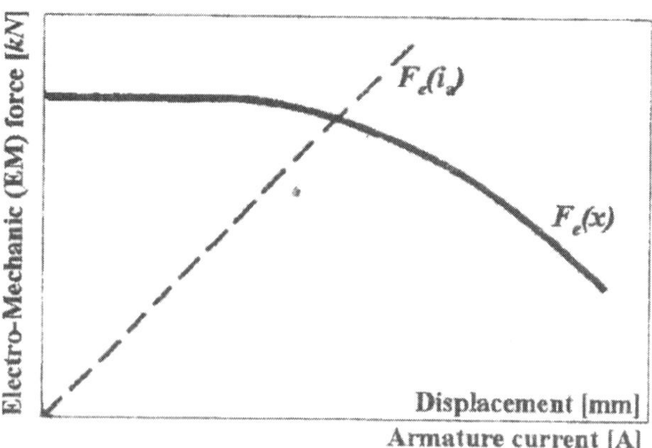

Figure 4.　E-M force changes with armature current and displacement

The short-stroke tubular linear DC-AC macrcommutator brake-force-actuator E-M motor is very simple; the aim is to get linear characteristics. There is no external magnetic circuit and phase-coil section is rather small. The length of the brake-stem IPM plunger (with a magnetization $J = 1.2\ T$) is also very important with regard to the characteristics linearity.

Figure 4 shows how the E-M force F_e [kN] varies with the displacement of a brake-stem IPM plunger x [mm] from the center of the brake-force-actuator E-M motor for the nominal value of an armature current I_{aN} [A] and also with the armature current i_a [A] in the linear area; this latter characteristic is a straight line with the constant E-M interlinkage coefficient c [N/A].

The mathematical model for a DC-AC macrocommutator brake-force-actuator E-M motor driving an inertial mass load, that is, a TiAl or SSN ceramic E-M-activated brake-stem IPM plunger is

(5) $\quad M\ d^2x/dt^2 = c\ i_a -- b\ dx/dt - \ddot{A}F_c\ sign\ dx/dt$

(6) $\qquad\qquad F_e = c\ i_a$

(7) $\quad L_a\ di_a/dt = u_a - R_a\ i_a - \ddot{A}u_a\ sign\ i_a - c\ dx/dt$

(8) $\qquad\qquad e_a = c\ dx/dt$

where: m - mass of the brake-stem IPM plunger [kg];

     x -displacement of the brake-stem IPM plunger [m];

     c -constant E-M interlinkage coefficient [N/A];

     i_a - armature current of the brake-force-actuator E-M motor [A];

     b - viscous friction coefficient of the brake-stem IPM plunger [kg/s];

     $\ddot{A}F_c$ - M force drop due to dry (Coulomb) friction of the brake-stem IPM plunger [N];

     F_s - M force supplied by the brake-force-actuator closing spring [N];

     F_e - E-M force supplied by the brake-force-actuator E-M motor [N];

     L_a - self-inductance of the armature winding of the brake-force-actuator E-M motor [H];

     u_a - driving armature voltage of the brake-force-actuator E-M motor [V];

     R_a - resistance of the armature of the brake-force-actuator E-M motor [$\grave{U}$];

     $\ddot{A}u_a$ - armature voltage drop of the brake-force-actuator E-M motor [V];

     e_e - induced armature voltage of the brake-force-actuator E-M motor [V].

The E-M-activated BBW AWB dispulsion control problem is to move the brake-stem IPM plunger through a target (reference) value of displacement x^* using readings of the displacement and the rate-of-change of displacement as control variables. The con-trol output is a driving armature voltage to a short-stroke tubular linear DC-AC macrocommutator brake-force-actuator E-M motor.

E-M FORCE CONTROL OF THE BRAKE-FORCE-ACTUATOR E-M MOTOR - An E-M force control for a brake-force-actuator E-M motor that must take a brake-stem IPM plunger without destroying it is needed. That is why, in certain cases, it is better not to have any overshoot during the transient states. The E-M force control in order to get a first order response to a reference E-M force step will be realized. The armature current in the phase coils of an armature winding, since the E-M force is proportional to it, is controlled. The actual value of an armature current is getting by measuring the voltage drop across a shunt. The 25 kHz harmonics, due to the DC-AC macrocom-mutator, is eliminated by means of a low-pass filter. The useful displacement of the brake-stem IPM plunger is about the phase-coil radius, which is quite small. However, using several phase coils of an armature winding with the proper DC-AC or DC-DC macrocommutator would allow quite a larger displacement. The DC-AC macro-commutator brake-force-actuator E-M motor is used for controlling the braking E-M force or the position of an E-M-activated brake gear as well.

Brake-Force-Actuator E-M Motor with Smart Displacement Sensor - In the early E-M-activated BBW AWB dispulsion mechatronic control hypo-system, the brake-force-actuator E-M motor acceleration/deceleration usually consists of a toothed lining attached to the brake stem, and a pick-up placed near to the periphery of the brake stem. The pick-up is essentially a horse-shoe magnet with a winding, and the projections of the brake stem act as a succession of keepers which bridge the poles of the magnet thus momentarily causing an increase in the magnetic flux through the winding and setting up a electric current in it. The frequency of this electric current depends on the linear velocity of the brake-stem IPM plunger and the rate at which that frequency changes is proportional to the acceleration/deceleration of the brake stem. This rate is measured relatively simply electronically and then is used to supply a signal for the control of the E-M-activated brake-actuators. A sensor is provided for each brake-force-actuator E-M motor to be controlled. For fixed parameters, this control problem can be treated using a standard proportional-integral-differ-ential (PID) microcontroller, but the simplicity of the example allows the principles of neural network (NN) and fuzzy logic (FL), that is, neuro-fuzzy (NF) control to be introduced without the distraction of control system complexity. The construction of a NF microcontroller consists of two main parts. The first of the data of input/output parameters and the second of the rule base, which are specific for a broad class of problems, and is the internal processing of that data. The output from the NF microcontroller is a driving armature voltage of the brake-force-actuator E-M motor and the input to the NF microcontroller is the displacement control-error value $\ddot{A}x = x^* - x$, that is, the difference between the target (refer-ence) value x^* and actual (measured) value x of the brake-stem IPM plunger displacement measured by a smart or linear potentiometer displacement sensor.

BBW AWB DISPULSION MECHATRONIC CONTROL HYPOSYSTEM ARCHITECTURE – is a fault-tolerant one. The very advanced anti-lock and/or anti-spin BBW AWB dispulsion mechatronic control hyposystem architecture consists of a set of three redundant microprocessor-based electronic control units (ECU) for both the BBW AWB telerobotic-driver (TD) and single ECUs, one for each E-M-activated friction disc, ring or drum brake. The microprocessor-based ECUs are interconnected by three replicated buses. The ECUs for the E-M-activated friction disc-, ring- or drum brakes are not designed for redundant action. The failure of a single M&GW brake is not considered to be serious as the other M&GW brakes will still remain operational under the condition that the brake-force actuator opens in case of failure.

An automotive communication system (ACS) based on the time-triggered protocol (TTP) ought to be used. It can provide deterministic data transmission and supports a fault-tolerant architecture. There exist following paradigms for the design of real-time (RT) distributed ACSs:

- event-triggered (ET) ACS; and
- time-triggered (TT) ACS.

In an ET ACS, all activities – activation of tasks, transmission of data and so on – are triggered by events. The advantage of an ET ACS is flexibility; the disadvantage, however, is non-determinism. For safety-critical (SC) ACSs, this is not acceptable.

In a time-triggered ACS , all activities – activation of tasks, transmission of data and so on – driven by the progression of a global time base (GTB). All task and communication actions are periodic. For an RT distributed ACS, this requires all clocks to be synchronized with a known precision.

Time-triggered ACSs rely on stronger assumption about the regularity of the processes to control in their environment and are therefore less flexible but far easier to analyze and test.

The TTP is an integrated communication protocol for time-trigerred ACS. It has been designed to fulfill the specific requirements of SC RT automotive applications and therefore provides services needed to build fault-tolerant RT distributed ACSs:

- integrated network management;
- distributed redundancy management;
- GTB (clock synchronization service);
- deterministic data transmission; and
- error detection with short latency.

A node in an RT TT ACS consists of a host microprocessor and the communication microprocessor (TTP controller). This entity is called a fail silent unit (FSU). Three FSUs may be composed to a fault-tolerant unit (FTU) that is intended to tolerate a single-node failure.

The communication network topology is a broadcast bus where bus access is controlled by a static time division multiple access (TDMA) scheme. The TTP has been designed to tolerate any single physical fault in any one of the constituent parts (nodes, bus).

BBW AWB DISPULSION MECHATRONIC CONTROL HYPOSYSTEM OPERATION - At this instant the very advanced anti-lock and/or anti-spin BBW AWB dispulsion mechatronic control hypo-system intervenes and releases the E-M-activated anti-lock and/or anti-spin BBW AWB dispulsion mechatronic control hyposystems are capable of releasing the E-M-activated friction disc, ring or drum brakes before the M&GW goes to lockup, and modulating the level brake application voltage on reapplication to just hold the M&GW near peak slip conditions.

A very advanced anti-lock and/or anti-spin BBW AWB dispulsion mechatronic control hyposystem consists of an ECU, a rotary brushed DC-AC mechanocommutator IPM brake-force-actuator E-M motors or short-stroke linear tubular brushless DC-AC macrocommutator IPM brake-force-actuator E-M motors for releasing and reapplying brake application voltages to disc, ring or drum brake-force actuators, and a M&GW angular-speed sensors. Short-stroke linear tubular brushless DC-AC macro-commutator IPM brake-force-actuator E-M motors have translational instead of rotary motion. They can be applied to the drive of a disc- or ring-brake calipers or of a drum-brake applies levers. The ECU normally monitors automotive vehicle velocity through the M&GW angular-speed sensors, and upon E-M-activated friction disc, ring or drum brake application begins to compute an estimate of the diminishing automotive vehicle velocity. Actual (measured) values of M&GW angular speeds can be compared against the computed reference values of them to determine whether an M&GW is slipping excessively, or the deceleration rate of an M&GW can be monitored to determine when the M& GW is advancing toward lock-up.

Different anti-lock and/or anti-spin BBW AWB dispulsion mechatronic control hyposystem designs use different combination of these physical variables to determine when locking or spinning is imminent and E-M-activated friction disc, ring or drum brake release is warranted.

At that point a command reference signal is sent to the rotary, brushed DC-AC mechanocommuta-tor IPM brake-force-actuator E-M motor or short-stroke linear tubular brushless DC-AC macrocommutator IPM brake-force-actuator E-M motor to release the brake application voltage, allowing the M&GW to spin back up. Once the M&GW regains its angular speed, the brake application voltage is increased again. Depending on the refinement of the control algorithms, the brake application voltage rise rate and the final value of a brake application voltage may be controlled to minimize cycling of the E-M-activated friction disc, ring or drum brakes.

During the stop of a automotive vehicle with the very advanced anti-lock and/or anti-spin BBW AWB dispulsion mechatronic control hyposystem when the E-M-activated friction disc, ring or drum brakes are first applied, actual values of M&GW angular speeds diminish more or less in accordance with the actual value of a automotive vehicle velocity. If the E-M-activated friction disc, ring or drum brakes are applied to a high adhesion coefficient M&GW level, or the road surface is slippery, the actual values of angular speeds of one or more M&GWs begins to drop rapidly, indicating that the M&GW's tire has gone through the peak value of the friction-slip curve and is heading toward lockup. At this instant the very advanced anti-lock and/or anti-slip BBW AWB dispulsion mechatronic control hypo-system intervenes and releases the E-M-activated friction disc, ring or drum brakes on those M&GWs before lockup occurs. Once the actual value of an M&GW angular speed picks up again the brakes are reapplied.

The objective of the very advanced anti-lock and/or anti-spin BBW AWB dispulsion mechatronic control hyposystem is to keep each M&GW's tire on the automotive vehicle operating near the peak value of the friction-slip curve for that M&GW's tire. In the latest automotive vehicles, braking is achieved with the assistance of the M&GWs with the brushless DC-AC/AC-DC or AC-AC macrocommutator IPM wheel hub E-M motors/M-E generators. The human driver (HD) selects a level of braking effort and this is transmitted to the very advanced anti-lock and/or anti-spin BBW AWB dispulsion mechatronic control hyposystems.

Initially, the E-M-activated friction disc, ring or drum brakes are operated, whilst the brushless DC-AC/AC-DC or AC-AC macrocommutator IPM wheel-hub E-M motors/M-E generators are configured into brushless AC-DC or AC-AC macrocommutator IPM wheel-hub M-E generators, thus developing braking forces. Once developed, regenerative braking is normally sufficient, the E-M-activated friction disc, ring or drum brakes are not required and are *'blended'* out. In theory, a very advanced DBW AWB propulsion mechatronic control hyposystem can brake right down to zero speed, though the very advanced anti-lock and/or anti-spin BBW AWB dispulsion mechatronic control hyposystem as a *'brake'* is not fail safe. Losing the DBW AWB propulsion mechatronic control hypo-system can mean losing all braking.

The very advanced anti-lock and/or anti-spin BBW AWB dispulsion mechatronic control hyposystem can quickly intervene and is capable of braking the automotive vehicle without the DBW AWD propulsion mechatronic control hyposystem. Nevertheless, for complete safety at slow speed, the DBW AWD propulsion mechatronic control hyposystem is faded out (not to be confused, for instance, with pad/disc fade) and the very advanced anti-lock and/or anti-spin BBW AWB dispulsion mechatronic control hyposystem brings the automotive vehicle to a halt. This *'blended BBW AWB'* approach, for instance, has the advantage of reducing pad wear significantly and, in many cases, the regenerated electrical energy from the brushless AC-DC or AC-AC macrocommutator IPM wheel-hub M-E generators can be put back into the DC chemoelectric al (C-E) storage battery and/or IMES ultra-flywheel for use during fast-acceleration starting, hill climbing and high-speed passing, thus saving energy. When this is not possible, the electric energy may be dissipated across a braking resistor located on the automotive vehicle for heating in the winter (known as rheostatic braking).

A trend that will impact very advanced anti-lock and/or anti-spin BBW AWB dispulsion mechatronic control hyposystems is the automotive industry's desire to reduce automotive vehicle wiring through the use of multiplexing techniques. As increasing numbers of emerging automotive vehicles are outlined with anti-lock and/or anti-spin, this trend is expected to result in an increased number of conventional anti-lock BBW AWB dispulsion mechatronic control hyposystems communicating with other automotive-vehicle mechatronic control hyposystems (for example, DBW AWD propulsion, ABW AWA suspension and SBW AWS diversion mechatronic control hyposystems) through a multiplex link. In addition to the wheel angular speed/ vehicle velocity data available from the anti-lock and/or anti-spin BBW AWB dispulsion mechatronic control hyposystem, the anti-lock ECU could benefit from this technology by being able to receive ICE, transmission, M&GW, steering angle, and other mechatronic control hyposystems data.

Another trend in very advanced anti-lock and/or anti-spin BBW AWB dispulsion mechatronic control hyposystems is automotive vehicle dynamics control during propelling (non-braking maneuvers) as well as during dispelling (braking). This is accomplished through use of the DBW AWD propulsion mechatronic control ECU normally inte-grated in enhanced anti-lock and/or anti-spin electric modulators, that are ASIM DC-AC macro-commutators, the addition of sensors to more accurate-ly determine the dynamic state of the automotive ve-hicle, and communication links with the DBW AWD propulsion mechatronic control.

Automotive vehicle dynamics control holds the promise of safer its operation through bettered stability in all maneuvers. Automotive vehicle behavior during braking in a hard turn includes spin and drift-out phenomena. The former can be effectively controlled by distribution the braking forces to the left and right rear M&GWs independently; the latter can be con-trolled by adopting a method for preventing M&GW lockup. This paper presents a novel very advanced anti-lock/or and anti-spin BBW AWB dispulsion mecha-tronic control hyposystem for controlling as well the braking forces between the inner and outer M&GWs in a hard turn independently.

Significant betterments have been seen in automotive vehicle cornering performance in recent years as a re-sult of advances achieved in wheel tire and ABW AWA suspension technology. Due to these betterments, automotive vehicle handling characteristics during braking have taken an added importance. The analytic-al results show that decreasing the yaw moment before M&GW locking or spinning occurs is effective in achiev-ing stable handling. An effective approach to decreas-ing the yaw moment is to control the braking forces between the inner and outer M&GWs independently. An independent braking force mechatronic control hyposystem for regulating the distribution of brake ap-plication voltages to the left and right rear M&GWs include examples that combine a linkage load-sensing input and a lateral acceleration-sensitive input. Current-ly *Robert Bosch GmbH* is developing a mechatronic yaw rate sensor for use in very advanced anti-lock and/or anti-spin BBW AWB dispulsion mechatronic control hyposystems. Production was starting in mid-1998. It is three times lighter than existing yaw rate sensors (only *70 g* against *210 g*).

The very advanced BBW AWB dispulsion mechatronic control hyposystem used in the *Poly-Supercar* [1] incorporates two linkage load-sensing inputs installed at the back of the rear suspension member. These two inputs provide independent control over the brake application voltage distribution to the left and right rear M&GWs. The reason for configuration the BBW AWB dispulsion mechatronic control hyposystem in this way are ex-plained below: it facilitates easy adjustment of the brake application voltage applied to the left and right rear M&GWs to maintain good automotive vehicle stability when braking in a hard turn; the brake application voltage characteristic is tuned to match the load at each rear M&GW; by combining this mecha-tronic control hyposystem with the very advanced anti-lock and/or anti-spin BBW AWB dispulsion mechatronic control hyposystem at all four M&GWs, it is possible to lower the yaw acceleration forces causing both the drift-out and spin moments. The operation performed in braking (decelerating) is the reverse of that carried out in driving (accelerating). In the latter the thermal energy of the fuel is converted into the mechanic energy and, the mechanic energy of the prime mover is converted into kinetic mechanic energy of the *Poly-Supercar*, whereas in the former the kinetic mechanic energy of the automotive vehicle is converted into thermal energy. Again, just when driving the automotive vehicle the torques of M&GWs produce tractive efforts at the peripheries of the driving M&GWs, so, when. the brakes are applied the braking torques introduced at the brake discs, rings or drums produce negative tractive efforts or retarding efforts at the peri-pheries of the braking M&GWs. As the acceleration possible is limited by the adhesion available between the driving M&GWs and the ground, so the decelera-tion possible is also limited. Even, so, when braking from high automotive vehicle velocity to a halt, the rate of retardation is considerably greater than that of full-ahead booster-induction-adjuster ('induction throttle') acceleration.

Consequently, the friction mechanic energy dissipated by the brakes, and therefore the thermal energy gen-erated, is correspondingly large. When E-M-activated friction brakes are applied to M&GWs or an automotive vehicle, forces are immediately introduced between the M&GWs and the road, tending to make the M&GWs keep in turning. The decelerations are proportional to the braking forces, the limiting values of which depend on the normal forces between the M&GWs and the road, and on the coefficients of frictions, or of adhesion, as they are called, corres-pondingly. Since the braking forces do not act along a line of action passing through the barycenter (the center of gravity) of the automotive vehicle, there is a tendency for an automotive vehicle to turn so that its rear M&GWs rise into the air. The inertia of the automotive vehicle introduces internal inertia force acting at the barycenter in the opposite sense of direction to the braking forces.

The magnitude of the inertia force is equal to that of the braking force. The two forces constitute a couple tending to make the rear M&GWs rise as stated. Since actually the rear M&GWs remain on the ground, an equal and opposite couple must act on the automotive vehicle somewhere so as to balance the overturning couple. A need to measure, in some way, a quantity that is poorly understood (adhesion or creep) and the use that measurement to control an enhanced anti-lock and/or anti-spin BBW dispulsion sphere that has some uncertainty associated with it may be identified. It appears that neural networks (NN) may offer some advantages in mapping the adhesion/creep character-istic and for generating an estimate of the current adhesion level. Given this esti-mate - which prones to some uncertainty in its own right - and the uncertainty in the very advanced anti-lock and/or anti-spin BBW AWB dispulsion mechatronic control hyposystem, then the ASIC NF microcontroller can be used to control M&GW slip and slide.

The DBW AWD propulsion and BBW AWB dispulsion mechatronic control hyposystems are still mechatronic-ally-controlled by their respective mechatronic control systems, but the human driver-demand signal passes through the slip/slide ASIC NF microcontroller before reaching the formers. In this way, the slip/slide ASIC NF microcontroller acts to optimize, where possible, the performance of the automotive vehicle. The effects of the NN adhesion estimator and FL slip/slide ASIC NF microcontroller when compared with a conventional ABS (anti-lock brake system), for the case of DBW AWD propulsion mechatronic control are interesting. For reasonable adhesion conditions there is little difference, as expect-ed, but for oil on the road surface the new control system performs better with the effects of overspeed much reduced and a corresponding bettering in acceleration.

In a very advanced anti-lock and/or anti-spin BBW AWB dispulsion mechatronic control hyposystem, an in-vehicle sensor (IVS) may to be applied to sense not only the actual (measured) values of vehicle acceler-ation but also its jerk (the time rate of change of ac-celeration of a vehicle body), and emulate responses of well-skilled and -experienced human driver.

CONCLUSION

Further R&D work is required in several areas before these techniques can be applied to real DBW AWD propulsion and BBW AWB dispulsion as well as ABW AWA suspension and SBW AWS diversion mecha-tronic control hyposystems, although the R&D work to date demonstrates some of the benefits that can be had by applying these methods to automotive vehicle chassis-motion control.

Fundamentally, for instance, in the case of the *Poly-Supercar* [1, 3,4,12], the front M&GWs must lock earlier than the rear ones. Otherwise, the vehicle stability is dangerously lost and the vehicle would skid, because the locked M&GW tires cannot contribute to side stability. Thus, yaw torques cannot be balanced. Consequently, in automotive vehicles without a very advanced anti-lock and/or anti-spin BBW AWB dispul-sion mechatronic control hyposystem, the braking force between front and rear M&GWs is distributed in a relation of approximately *0.7* to *0.3*.

The very advanced anti-lock and/or anti-spin BBW dis-pulsion mechatronic control hyposystem is continuous-ly monitored for faults such as open- and short-circuits. Precise knowledge of such faults allows instant re-configuration of the very advanced anti-lock and/or anti-spin BBW AWB dispulsion mechatronic control hyposystem for best managed stability, that is not only three-wheel diagonal or front/rear wheel split braking but also inner/outer wheel split braking independently.

The very advanced anti-lock and/or anti-spin BBW AWB dispulsion mechatronic control delivers better-ment in performance, human driver convenience, auto-motive vehicle design and assembly flexibility, as well as mass reduction and default mode operation. E-M-activated friction disc, ring and drum brakes provides bettered default modes compared to F-M and/or P-M ones plus significantly increased diagnostic capabil-ities, proactive default mode management, and bet-tered serviceability. Increase diagnostic capabilities are available through the use of mechatronics. Many hu-man drivers of the very advanced anti-lock and/or anti-spin BBW AWB dispulsion mechatronic control hypo-system-equipped automotive vehicles, faced with an emergency stop, do not sustain the initial panic-strength brake-pedal voltage through-out the stop, so that maximum anti-lock and/or anti-spin braking ceases when it is still very much needed. The simplest to be fitted as standard to automotive vehicles, is the elec-tronic actuation system (EAS). The EAS recognizes the fact that the driver is panic-braking by sensing the rate at which the brake pedal is pushed, and that the driver then subconsciously eases off. By harnessing the help of the enhanced anti-lock and/or anti-spin BBW AWB dispulsion mechatronic control system's booster induc-tion adjuster and master ASIC NF microcontroller, it automatically keeps the brake application voltage at anti-lock and/or anti-spin braking operating level until the automotive vehicle stops. Electronic stability sys-tem (ESS) is another handling product that uses the very advanced anti-lock and/or anti-spin BBW AWB dispulsion mechatronic control hyposystem. To help correct an understeering or oversteering skid, this hyposystem momentarily applies E-M-activated friction disc, ring or drum brakes to an individual M&GW.

Robert Bosch GmbH for a conventional ABS first developed ESS, and *Mercedes-Benz* was the first automotive vehicle maker to adopt it. It is currently found on the *S-* and *E-*class *Mercedes* [6].

The *Krakow University of Technology's Automotive Mechatronics R&D Team* has already started on a version of automotive vehicle disabling system (AVDS) capable of immobilizing automotive vehicles. The AVDS only works if the automotive vehicle is fitted with novel very advanced anti-lock and/or anti-spin BBW AWB dispulsion mechatronic control hyposystems. AVDS works by releasing the brake application voltages in an automotive vehicle's E-M-activated friction disc, ring or drum brakes, thus immobilizing them. The automotive developers explain that most construction work automotive vehicle's theft involves moving the automotive vehicle onto a transporter. But, *'if they cannot get the E-M-activated friction disc, ring or drum brakes off, they cannot easy move it'* - automotive developers note.

Beyond the above rather obvious benefits of E-M-activated brake-force actuators, which have been previously covered in the literature [1-14], there are a number of an intriguing very advanced anti-lock and/or anti-spin BBW AWB dispulsion mechatronic control hyposystem consequences of their use that have not been considered.

REFERENCES

1. B, Fijalkowski, *Modele matematyczne wybranych lotniczych i motoryzacyjnych mechano-elektro-termicznych dyskretnych nadsystemów dynamicznych* [Mathematical Models of Selected Aviation and Automotive Discrete Dynamic Hypersystems]. Monografia 53, Politechnika Krakowska im. Tadeusza Koœciuszki, Kraków: 1987 (In Polish).
2. R. Wells, and J. Miller, "Electric brake system for passenger vehicles - Ready for production", *Proc. ISATA 93*, Aachen, Germany 13th - 17th September 1993, Paper 93ME115, pp. 349-356.
3. B. Fijalkowski, "The concept of a high performance all-round energy efficient mechatronically-controlled tri-mode supercar", Special Issue on Automotive Electronics: Part 2 (Guest Editor: B. T.. Fijalkowski), *Journal of Circuits, Systems and Computers*, Vol. 5, No. 1 (March 1995), pp. 93-107.
4. B. Fijalkowski, and J. Krosnicki, "Concepts of electronically-controlled electromechanical/mechano-electrical steer-, autodrive- and autoabsorbable wheels for environmentally-friendly tri-mode supercars", Special Issue on Automotive Electronics: Part 1 (Guest Editor: B.T. Fijalkowski), *Journal of Circuits, Systems and Computers*, Vol. 4, No. 4 (December 1994), pp. 501-516.
5. J. Balz, K. Bill, J. Böhm, P. Scheerer und M. Semsch: "Konzept für eine elektromechanische Fahrzeugbremse". *ATZ Automobiltechnische Zeitschrift*, 98 Jahrgang, No. 6, Juni 1996, pp. 328-333 (In German).
6. E. Chew, "Bosch deal, Lucas talks put focus on brakes", *Automotive News Europe*, No. 9, (May 1996), p. 13.
7. M. Scarlett, "New wave of brake technology is coming", *Automotive News Europe*, No. 9, (May 1996), p. 13.
8. B, Fijalkowski, "Emerging and future enhanced anti-lock & anti-spin brake-by-wire dispulsion spheres", *Proc. Dedicated Conference on Mechatronics – Efficient Computer Support for Engineering, Manufacturing, Testing & Reliability, including Fuzzy Systems/Soft Computing in conjunction with the 30th ISATA: International Symposium on Automotive Technology & Automation*. Florence, Italy 16th – 19th June 1997, pp. 557-564
9. Automotive Division, WMRC, "Braking - Automotive Brakes in the Electronic Era", pp. 145-149, in Business Briefing: *Global Automotive Manufacturing & Technology – An analysis of the automotive manufacturing & technology industry and perspectives on the future. Includes exclusive CD-ROM, Published by World Markets Research Centre, London: 2000, p. 240
10. Robert Bosch GmbH, *Driving-safety systems*, 1999
11. B. Fijalkowski, "DBW 4WD propulsion & BBW 4WB dispulsion control system for intelligent automotive vehicles", *Proc. Telematics Automotive 2000 – Conference and Exhibition*, Birmingham, England, 11, 12 & 13 April 2000
12. B. Fijalkowski, "Design and Electronics - Advanced Chassis Engineering", pp. 109-116, in Business Briefing: *Global Automotive Manufactu-ring & Technology – An analysis of the automotive manufacturing & technology industry and perspect-ives on the future. Includes exclusive CD-ROM, Published by World Markets Research Centre, London: 2000, p. 240
13. B. Fijalkowski, "Mechatronic active RBW automotive control system", *Proc. Programme Track on Chassis Engineering in conjunction with the ISATA 2000: Automotive & Transportation Technology*, Dublin, Ireland, September 25 – 29, 2000, No 00CE001, pp. 277-284
14. B. Fijalkowski, Unmanned vehicles (UV) for terrestrial/water military operations and noiseless law-enforcement special operations", *Proc. Fall 2000 Panel Business Week and Symposium on UNMANNED VEHICLES (UV) FOR AERIAL, GROUND AND NAVAL MILITARY OPERATIONS*, NATO, Research & Technology Agency, Ankara, Turkey, 9 – 13 October 2000, pp. 1-12

The Kinetic Brake Booster

Valentin Ivanov
Belarusian State Polytechnic Academy

Joseph Lepeshko and Vladimir Boutylin
Scientific Center for Machine Mechanic Problems, National Academy of Sciences of Belarus

ABSTRACT

Main defect of traditional structure of the brake boosters is the necessity of an external energy source. The analysis of redistribution of power streams occurring during at braking of the automobile shows that it is possible to use the force component of the driving automobile kinetic energy for the drive of a booster (so-called kinetic booster). The power consumed by a booster is taken from power developed during brake action thereby it promotes vehicle slowing down.

In the paper for the booster work the schemas of power take-off from an engine, onboard electric system, transmission, single wheel are considered. Especially for brake-by-wire systems the project of pilot management is probed. It allows applying the serial x-by-wire components both on small automobiles andon vehicle with the great load-carrying capacity and trailers.

INTRODUCTION

The automotive braking properties in many respects depend on the capabilities of the brake booster. It is especially important for cars and vehicles with a hydraulic brake drive. The problem consists here in finding of the tradeoff between reliability, response speed and braking efficiency. The brake-by-wire (BBW) technology allows to considerably increase the speed of operation, and yet BBW has some problems with the reserve and parking brake system and with the stability of the electrical part characteristics, what is not significant for the commonly accepted hydraulic or pneumatic systems. But, on the other hand, in such constructions the additional external energy source is required for work of the booster, for example, the vacuum or hydraulic pump or the hydraulic-accumulator system. In the aggregate with the power take-off devices for the operation of anti-lock braking systems (ABS) that will perceptibly influence on the operating and the cost parameters of the vehicle.

Irrespective of an embodiment it is possible to offer a following functional diagram for all types of braking

drives with boosters as automatic follow-up control systems with parallel action, Fig. 1. With its help it is possible to eliminate the common imperfections for such systems.

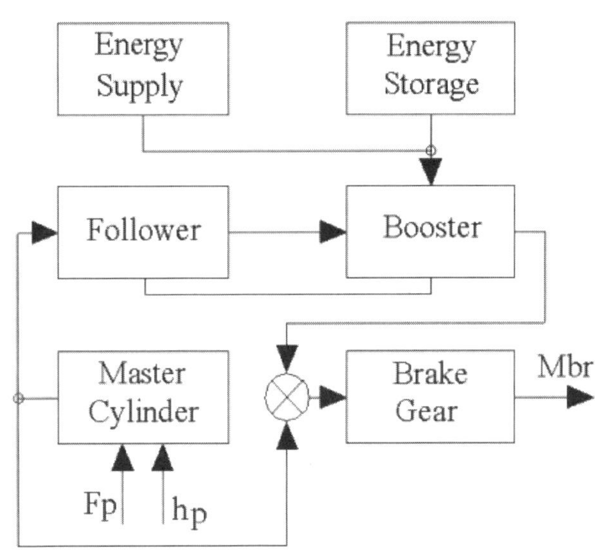

Fig. 1: The Functional Diagram of Brake Drive with the Booster

1) The brake drive needs at external energy source, therefore in its absence or by booster failure the uncontrolled braking is taking place, because the driver can not physically ensure a vehicle deceleration required.

2) The availability of power take-off gear and control system by the power accumulator unit makes a structure as a whole more complex.

3) The certain consumption of energy is necessary for the drive operation and maintenance of reserve in accumulators what has an influence on the vehicle fuel efficiency.

ALTERNATIVE OPERATION PRINCIPLE FOR BRAKE BOOSTER

The brake drive with the kinetic booster allows overcoming above-mentioned imperfections [1, 2]. This system represents the direct action drive for which the required effort for creation of brake torque is formed at the expense of braking forces or inertial forces of the vehicle interaction with base surface [3], Fig. 2.

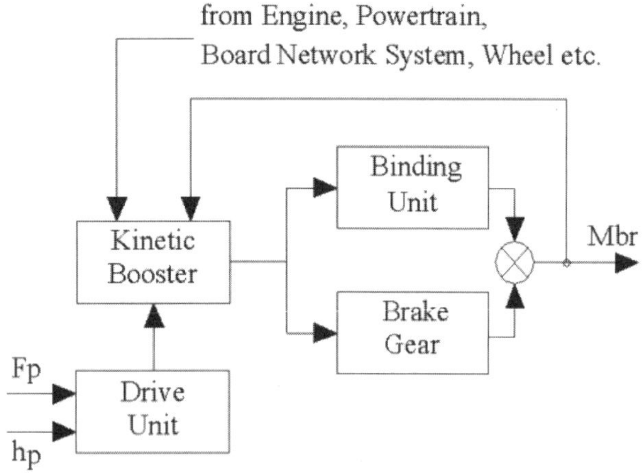

Fig. 2: The Functional Diagram of Brake Drive with the Kinetic Booster

It should be particularly emphasized the booster does not require the additional energy source (air-free or hydraulic pumps, compressed air etc.) for own activity. It consumes a mechanical energy from a source (engine, transmission, wheel etc.) for transformation of its force component F_{bs} for the brake drive. The power N consumed by the booster from power developed during braking is taken promoting thereby the vehicle deceleration. By this means the energy consumption for booster activity are formed out of supplied energy for the brake drive performance A_{br} and dissipation energy Aq, Fig. 3, which is defined as

$$Aq = \int_0^t F_{br}(t)\left(V - \frac{dX_{br}}{dt}\right)dt \,. \tag{1}$$

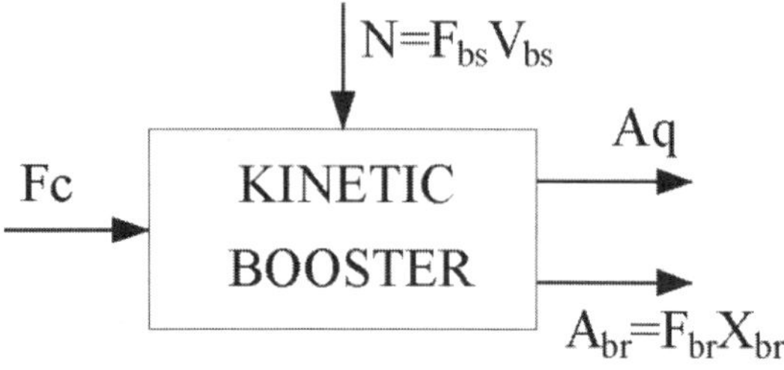

Fig. 3: The Interchange of Energy for Kinetic Booster

By this means is that the energy consumption of the kinetic booster take place only during braking. At a motion mode without of braking, the kinetic booster works on an idling without the energy consumption, as distinct from other types of boosters.

CONSTRUCTIVE MODIFICATIONS OF KINETIC BOOSTERS

In Fig. 4-7 are shown the different design concepts for kinetic boosters.

At a power take-off from the vehicle engine for booster work, fig. 4, the auxiliary brake caliper (booster) 3 is connected to the master cylinder 5 by mechanical binding. Under the running engine 7 and at pressing of the pedal 1 on primary cylinder 2 will be accomplished a forcing down of brake shoes of the auxiliary brake 3 to a brake drum or disk. As this takes place, the auxiliary brake will activate the master cylinder 5; the liquid will come in cylinders of wheel brake gears 6, transferring them into the hindered condition.

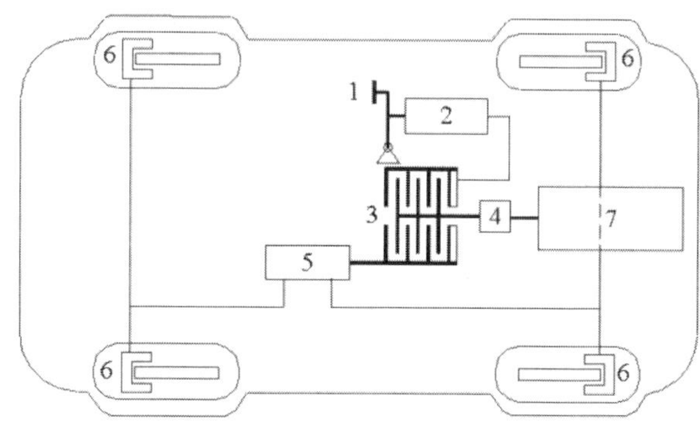

Fig. 4: The Kinetic Booster with Power Take-off from Engine

At unbrake phase the auxiliary brake (booster) terminates a power take-off and the braking drive is returning to its original position under the action of an accumulated potential energy and elastic deformation of the drive.

Merits:

- Sufficient response speed of drive due to a narrow effective range of the engine;

- The reverse of brake torque of the auxiliary brake is not required.

Demerits:

- The stationary binding with the vehicle wheels is needed when engine doesn't operative;

- It is necessary to provide a power take-off during the idling.

At a power take-off for booster work from the board electric mains of vehicle, fig. 5, auxiliary brake (booster) 3 is connected through the matching reducer 4 to the electric motor 9 which is supplied through a management system 8 from accumulator batteries 7. The auxiliary brake caliper as well as in the previous case is connected to the master cylinder 5. At pressing a pedal 1 under acting electric motor 9 the brake-booster works from cylinder 2 and it influences wheel brake gears 6 through the master cylinder 5.

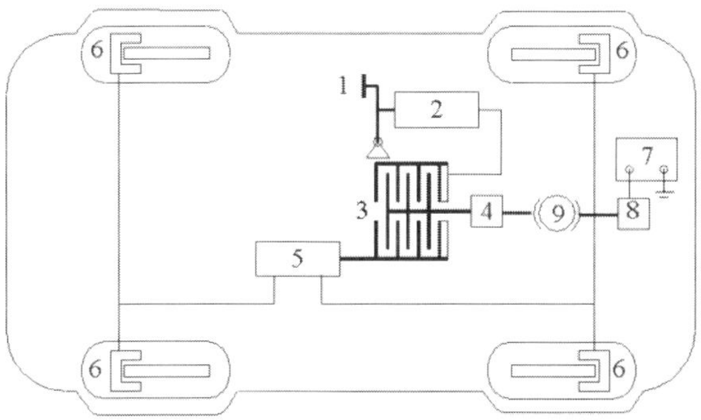

Fig. 5: The Kinetic Booster with Power Take-off from Board Network System

At unbrake phase the pedal effort is released and power take-off from the electric motor ceases. The brake drive returns into initial state and after full pedal release the electric motor is disconnected with the board network system takes place.

Merits: The long-lived workability of the braking drive under idling engine (up to exhaustion of the accumulator batteries).

Demerits: The drive operation depends on capacity of accumulator batteries.

In the following scheme the power take-off for booster work is achieved by vehicle transmission, fig. 6. The auxiliary brake gear (booster) 3 is controlled from pedal 1 and cylinder 2. The booster through the matching reducer 7 is connected to the transmission part 8 which is bounded with driving wheels. The brake-booster caliper is coupled through the reverse device 4 to the master cylinder 5.

After pedal is pressed the brake blocks of the auxiliary brake press the brake drum or disk. It will swivel because of coupling with transmission shaft, and its caliper will activate the master cylinder 5 and wheel brake gears 6 through the reverse device.

Merits:

- Comparatively high response speed owing to a simplicity of kinematics connections;

- There are no limitations on parameters of a master cylinder.

Demerits: The device for reverse of a booster brake torque is necessary.

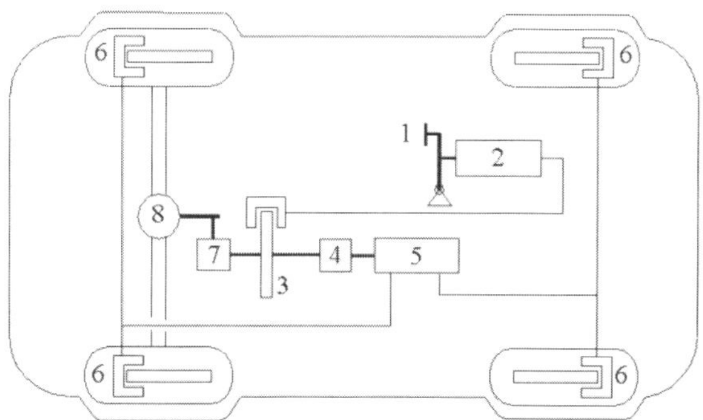

Fig. 6: The Kinetic Booster with Power Take-off from Transmission

The simplest scheme can be implemented by the power take-off for booster work from a single vehicle wheel, fig. 7. In this case the cylinder 2 operates by a brake cylinder of only one wheel and its brake gear 3 fulfils the function of a booster. The booster brake caliper is connected by the reverse device 4 (as well as in the previous case) to the master cylinder 5, actuating braking gears of remaining wheels 6.

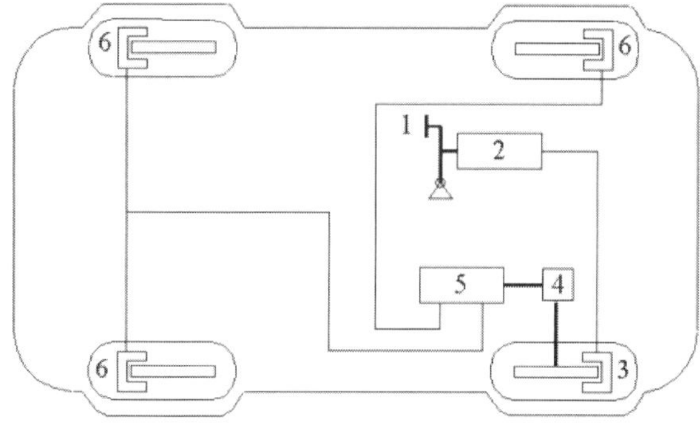

Fig. 7: The Kinetic Booster with Power Take-off from Wheel

Pressing a pedal the brake-booster 3 begins to work. At the vehicle motion its wheel with brake shoes pressed to a drum or disk will be rotated up to an angle limited by gaps in remaining wheel brake cylinders and by fluid rate required for the master cylinder 5. Through the reverse device and master cylinder all wheels will have braked.

The brake-booster will be in equilibrium owing to braking forces in a wheel-road contact and of effort on the reverse device depending on wheel brake cylinders pressure of remaining gears.

Merits:

- Simple design;

- The scheme matches up not only for single vehicles, but also for trailers;

- High response speed proportionates to an initial braking speed.

Demerits: The reverse device for a brake torque booster is necessary.

Thus it is possible to emphasize the next characteristic features of kinetic boosters.

1) The brake systems with kinetic boosters can promote the application of the hydraulic and electric-hydraulic drive to heavy automobiles. This is reflected in the fact that the driver will need only an effort for brake-booster control.

2) At vehicle motion the kinetic booster does not consume energy.

3) The kinetic booster produces the controlled brake torque.

4) The brake torque for a parking brake system is provided with gravitational forces (at a power take-off from a transmission or single wheel).

More detailed investigation [2, 4] has shown the schemes with a power take-off from a transmission and from a single wheel for booster work should be preferred to. For a 16t-truck the experimental designs with a hydraulic drive were calculated. They have brought the following results.

The kinetic booster with a power take-off from a single wheel:

- Brake torque on each wheel, kNm........................ 15;

- Drive pressure, bar... 150;

- Wheel brake cylinder diameter, mm..................... 52;

- Pedal cylinder diameter, mm............................. 13;

- Master cylinder diameter, mm............................ 10;

- Maximum pedal effort, N................................. 420;

- Drive response time (by initial speed of braking 60 kph), ms ....0,6.

The kinetic booster with a power take-off from an vehicle transmission:

- Brake torque on each wheel... ....15 kNm;

- Drive pressure, bar 150;

- Wheel brakes cylinder diameter, mm ... .52;

- Pedal cylinder diameter, mm 16;

- Master cylinder diameter, mm........................... 31;

- Booster brake torque, kN................................... 1;

- Brake-booster cylinder, mm............................... 36;

- Maximum pedal effort, N................................420;

- Drive response time (by initial speed of braking 50 kph), ms...23.

The presented data characterize the advantages of kinetic boosters.

KINETIC BOOSTER AS BRAKE-BY-WIRE COMPONENT

A great gain in braking efficiency can be obtained for systems brake-by-wire by application of the kinetic booster. It is well-known that now the BBW expansion is limited to the passenger-class cars, as electrical brakes of trucks, buses and sporting cars need the large power for feeding part and control system. We consider the embedding of the kinetic brake booster will allow, for example, for 5...10 t trucks equipped with standard brake gears to apply the electric drive from a brake system of a passenger car.

As perspective designs of electric brakes for BBW are proposed both the disk gears (for example, designed by Continental Teves [5]) and the drum gears (for example, designed by Delphi Automotive System [6]). However the drum gears are now closest to full-scale production and therefore we shall consider the design of kinetic booster as applied to them, fig. 8. In this case in one brake drum 1 two brake gears are located. The brake gear 2 is connected to an electric line and operates the brake gear 3. In such a manner the role of the kinetic booster executes the brake 2, which interacts with a brake drum and creates active brake torque perceived by the automobile. The reactive torque controls the brake 3 through an expansion-type device. On a brake drum the brake torques of both gears are summed.

In the offered scheme a rigid connection between the caliper of brake-booster and the expansion-type device of the main brake is necessary.

The work of drive is carried out as follows. At action onto the control pedal the electric expansion-type device of the brake-booster 2 and its shoes are pressed to the drum 1. At rotation of a brake drum the brake-booster is rotated in the same direction. Its caliper rotates the expansion-type device of the main brake 3 and drives it to action. As the brake shoes of the main brake and brake-booster are pressed to the brake drum the actual brake torque is equal the sum of their brake torques.

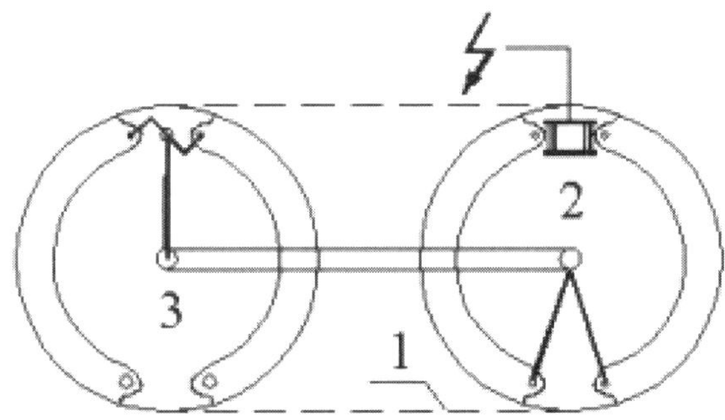

Fig. 8: The Kinetic Booster for Brake-by-Wire

Hence the active brake torque of the brake-booster creates additional brake torque owing to a power take-off and the reactive torque will be used for the main-brake control.

CONCLUSION

The modern technologies in automotive industry require the alternate solutions at designing of components and aggregates of the vehicle. For the brake system can be open here new scopes by application of the kinetic boosters. It is necessary to underline especially the results which can be obtained at their embedding.

- The kinetic boosters do not require additional energy sources for work. The power take-off is made through energy of the driving vehicle or the gravitation.

- Onto trucks, buses and the truck-tractor-trailers hydraulic and combined electric-hydraulic drives can be extended.

- By integration of the kinetic booster with brake-by-wire the electric power drain is not considerably increased under the high brake torques.

REFERENCES

1. Lepeshko J. and Sucheil M. (1994): Mechanical booster in brake system, Proc. of 4th conference 'Brakes of Road Vehicles', Lodz, Poland: 114-119 pp. (in Russian)
2. Sucheil M. (1997): Development of a technique of calculation for the braking drive with the mechanical booster. Ph. D. thesis Belarusian State Polytechnical Academy, Minsk, Belarus. (in Russian)
3. Patent BY No. 186: The booster for hydraulic brake system: B 60 T 13/08: Belarus.
4. Ivanov V., Boutylin V. and Lepeshko J. (2000): Intelligente Fahrsicherheitssysteme für Nutzfahrzeuge und Busse: Proc. of the Polish-German Symposium on 'Science, Research, Education 2000', Zielona Gora, Poland: 329-334 pp.
5. Schwarz R. (1999): Bremskraft-Rekonstruktion für elektromechanische Fahrzeugbremsen: Automobiltechnische Zeitschrift: Bd. 101, Nr. 6: Wiesbaden: Verlag Vieweg: 402-412 pp.
6. Electric Brake System: Delphi Automotive System: Information DC-98 WC-029 12/98: 2 pp.

CONTACT

Dr. Valentin Ivanov, Dept. of Automobiles, Belarusian State Polytechnical Academy. F. Skaryny 65, 220 027 Minsk Belarus. E-mail: vivanov@altavista.com.

DEFINITIONS

A_{br}: Brake drive work

A_q: Energy Dissipation

F_{br}: Booster brake force

F_{bs}: Force component of power N

F_c: Control effort

F_p: Pedal effort

h_p: Pedal stroke

M_{br}: Brake torque

N: Power consumed by the booster

V_{bs}: Speed component of power N

X_{br}: Displacement component of work A_{br}

ANTILOCK BRAKING SYSTEMS (ABS)

Influence of Hydraulic ABS Parameters on Solenoid Valve Dynamic Response and Braking Effect

Xuele Qi, Jian Song and Huiyi Wang
State Key Laboratory of Automotive Safety and Energy, Tsinghua University

ABSTRACT

The performance of Hydraulic Anti-lock Braking System (ABS) is important to vehicle brake safety. In this paper, finite element models and simulation models for studying dynamic response of solenoid valve, a critical part of ABS, are established. Based on the calculation results, which are proved in experiment, several effective methods for improving the performance of the solenoid valve are presented. Furthermore, an ABS-vehicle model is developed to study the influence of hydraulic ABS parameters on braking effect. The simulation results have good agreement with the data of in-vehicle driving experiments.

INTRODUCTION

Hydraulic Anti-lock Braking System (ABS) is of significant importance to vehicle brake safety. During emergency braking, 4-channel ABS can control the wheel slip by independently regulating the actual brake pressure of each wheel. In this way, the stability and steerability of vehicles can be maintained on various road surfaces. The hydraulic ABS typically contains four wheel speed sensors, an electronic control unit (ECU) and a hydraulic control unit (HCU) which is the primary object studied in this paper. The HCU mainly contains following components: hydraulic pump, low pressure accumulator, restrictor, isolation solenoid valves and release solenoid valves. Further study and experiment provide that several key parameters of the hydraulic units play an important part in the dynamic response of the HCU, which influences the whole ABS performance.

The high frequency dynamic response of HCU is primary determined by the high performance solenoid valves, and the dynamic behavior of such solenoid valve is result from the integrated influence of electromagnetism, mechanics and hydrokinetics[2].

Based on ANSYS®, finite element analysis is used to develop the ABS solenoid valve model. In this model, nonlinearity of magnetic flux density to magnetic field intensity is taken into account. As functions of the coil current and plunger position, the magnetic pulling force and coil inductance are calculated. Another finite element model of the incompressible turbulent fluid flow fields in solenoid valve is established to study the relations between hydrokinetic force and the plunger position. Based on MATLAB/SIMULINK®, a simulation model is developed to simulate the dynamic response of the solenoid valve by using the finite element results. Compared with the results of experiment, the validity of simulation solutions and the results of finite element models are verified. Furthermore, several effective methods to improve the performance of the solenoid valve can be provided according to the results of simulation.

Finally, introduce another form of model to study the dynamic response of solenoid valve is established by using a software named AMESim®, which contains many general magnetic and hydraulic components suitable for simulating ideal dynamic behavior based on component performance parameters. For further study, a model contains the whole hydraulic ABS and a vehicle model described by seven degrees of freedom is developed to research the influence of hydraulic ABS parameters on braking effect.

FINITE ELEMENT MODEL OF MAGNETIC FIELD

SOLENOID VALVE MAGNETIC MODEL – The structure of solenoid valve is shown as Fig. 1.

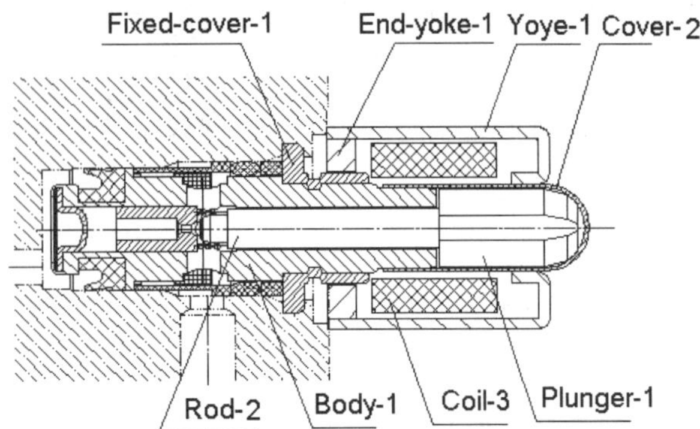

1-Magnetizable Material   2-Weak Magnetizable Material
3-Non-Magnetizable Material
Other non-marked parts are all the Non-Magnetizable Material

Fig. 1 The structure of solenoid valve

Base on ANSYS®, the finite element model of solenoid valve is built and shown as Fig.2. In order to improve the efficiency of calculation and abbreviate the process of modeling, several idealizations about geometric complexity have been employed. The idealizations, which cannot decrease the accuracy out of acceptable deviation range, mainly include: (1) Neglecting the grooves on the plunger so as to assume the solenoid valve as an axisymmetric model; (2) Since the relative magnetic permeability μ_r of weak magnetizable material parts, such as the cover and the rod, is approximate to the relative magnetic permeability value of air, these parts have been omitted in the finite element model. (3) Some geometric details have been smoothed[2].

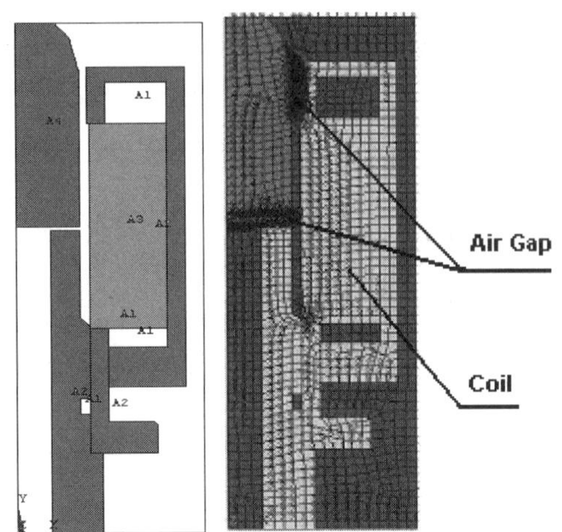

Fig. 2 Finite element model of solenoid valve

Figure 2 also presents the finite element mesh. Each element is two-dimensional (planar and axisymmetric) magnetic fields and defined by eight nodes and has up to four degrees of freedom per node: magnetic vector potential, time-integrated electric scalar potential, electric current, and electromotive force. The elements of air gaps have been detailed meshed to improve the accuracy of the solution, since the magnetic field strength gradient is great in these places. Moreover, the electric current of every coil element should be coupled together. The elements of magnetizable material have nonlinear magnetic capability for modeling B-H curve[3], which is, in the solenoid valve, determined as Fig.3.

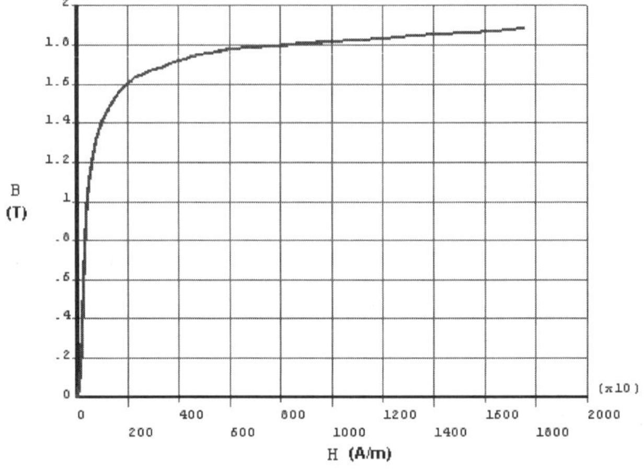

Fig. 3 B-H curve of magnetizable material

FORMULATION OF MAGNETIC FIELD – The basic formulation of magnetic field is Maxwell Equations:

$$
\begin{cases}
\nabla \cdot \vec{D} = \rho \\[4pt]
\nabla \cdot \vec{B} = 0 \\[4pt]
\nabla \times \vec{E} = -\dfrac{\partial \vec{B}}{\partial t} \\[4pt]
\nabla \times \vec{H} = \vec{J} + \dfrac{\partial \vec{D}}{\partial t}
\end{cases}
\tag{1}
$$

where $\vec{D}$ denotes the electric displacement, ρ is the charge density, $\vec{B}$ is the magnetic flux density, $\vec{E}$ is the electric field intensity, $\vec{H}$ is the magnetic field intensity, and $\vec{J}$ is the current density vector.

In order to abbreviate the solution of Maxwell Equations, it is necessary to introduce two variables: the vector magnetic potentials $\vec{A}$ and the scalar electric potentials Φ, which satisfy the following equations:

$$\vec{B} = \nabla \times \vec{A} \tag{2}$$

$$\nabla \cdot \vec{A} = -\mu\varepsilon \frac{\partial \phi}{\partial t} \tag{3}$$

take (2) and (3) into (1), the variables of magnetic field and electric field can be separated to calculate, the equations are expressed as

$$\begin{cases} \nabla^2 \vec{A} - \mu\varepsilon \dfrac{\partial^2 \vec{A}}{\partial t^2} = -\mu\vec{J} \\[2mm] \nabla^2 \phi - \mu\varepsilon \dfrac{\partial^2 \phi}{\partial t^2} = -\dfrac{\rho}{\varepsilon} \end{cases} \tag{4}$$

where μ is magnetic permeability, ε is the dielectric constant and ∇^2 is the Laplace's arithmetic operators.

Hence, vector magnetic potentials $\vec{A}$ and the scalar electric potentials Φ can be calculated in this way firstly, later the other variables, such as the magnetic flux density $\vec{B}$ and the magnetic pulling force $\vec{F}m$ which is calculated by the Maxwell stress method as follow:

$$\vec{F}m = \frac{1}{\mu_0} \oiint_S [(\vec{B}\cdot\vec{n}_e)\vec{B} - \frac{1}{2}B^2\vec{n}_e]dS \tag{5}$$

where μ_0 is the permeability of vacuum, S is the area of curved surface around the air gap, $\vec{n}_e$ is the unit normal vector.

Since the coil inductance L is also important to dynamic simulation, the equation for computing L is as follow:

$$L = \frac{N\int_C \vec{A}dC}{i} \tag{6}$$

where N is the winding turn, i is the coil current, and C is the coil perimeter.

RESULTS OF MAGNETIC FIELD – Load 12V step voltage on the elements of coil, the magnetic flux distributions is shown as Fig.4.

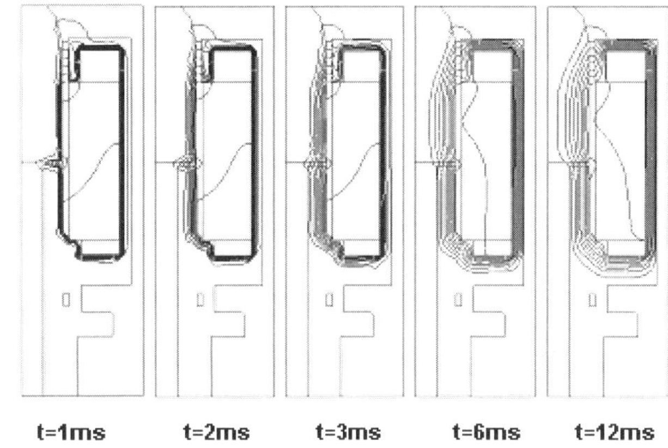

Fig. 4 Magnetic flux distributions when voltage is loaded (valve on)

According to the result, it is not difficult to find that the magnetic flux mainly distributes in the magnetizable material parts, such as plunger, body, fixed-cover, end-yoke, and yoke. Only a small amount of flux leak into the air. The result also shows that the magnetic flux mostly distributes on surface of the plunger at the beginning, and only a little of them occur in the interior of the plunger when the magnetic field has established. Hence, it is an effective way to improve the dynamic response by decreasing a few of material in the interior of the plunger.

To simulate the dynamic response of solenoid valve, it is necessary to obtain the force-current-position table and the inductance-current-position table, which are all the two dimensional tables[1]. Both of the magnetic pulling force $\vec{F}m(x,i)$ and coil inductance $L(x,i)$ are the functions of plunger position x and coil current i. The stroke length of plunger is 0.2mm, and is divided into 10 steps. It means that the increment of plunger position is 0.02mm when establishing the solenoid valve model. On every fixed plunger position, the curves of function $\vec{F}m_x(i)$ and $L_x(i)$ can be obtained. Hence, the two tables of $\vec{F}m(x,i)$ and $L(x,i)$ can be calculated by increasing the plunger position. Figure 5 plots the results.

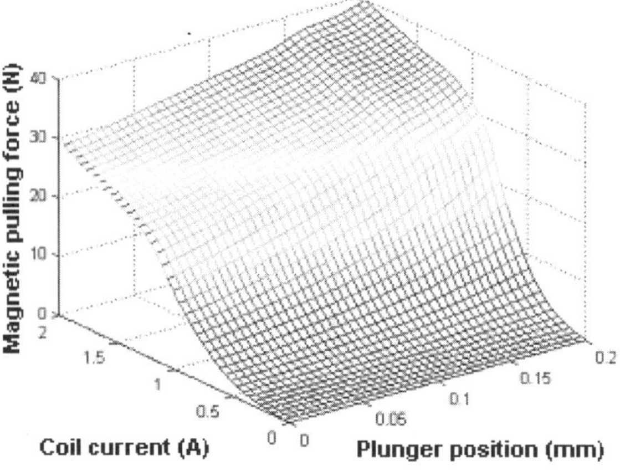

Fig. 5a Force-current-position table

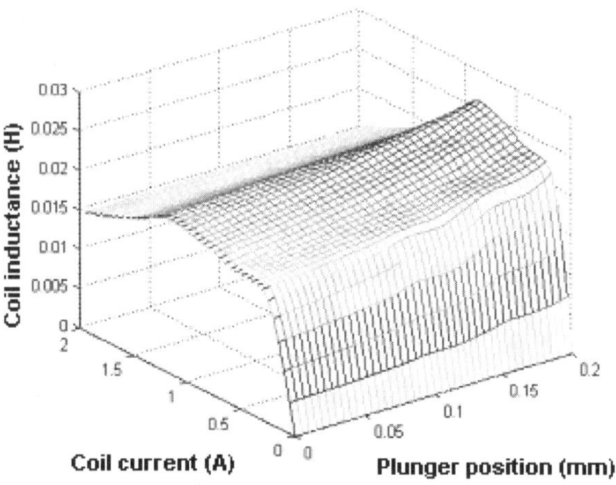

Fig. 5b Inductance-current-position table

Figure 5a shows the magnetic pulling force as a function of coil current and plunger position. For a fixed plunger position, the force is rising with coil current up to 1A, this is the linear region of magnetic material. When coil current is beyond 1A, the rising rapidity is slow down, for magnetic saturation occurs. Figure 5b shows the coil inductance as a function of coil current and plunger position. Obviously, the coil inductance decreases when the coil current beyond 1A. Therefore, the magnetic saturation phenomenon can reduce the value of coil inductance. For a fixed current, neither the force nor the inductance has significant increase with the plunger position. These two tables will be used in simulation model.

FINITE ELEMENT MODEL OF FLUID FIELD

SOLENOID VALVE FLUID MODEL – The hydrokinetic force is of great importance to the dynamic response of solenoid valve, and it is calculated by using the finite element fluid model, which is established according to the channel for brake oil. The model is shown as Fig.6.

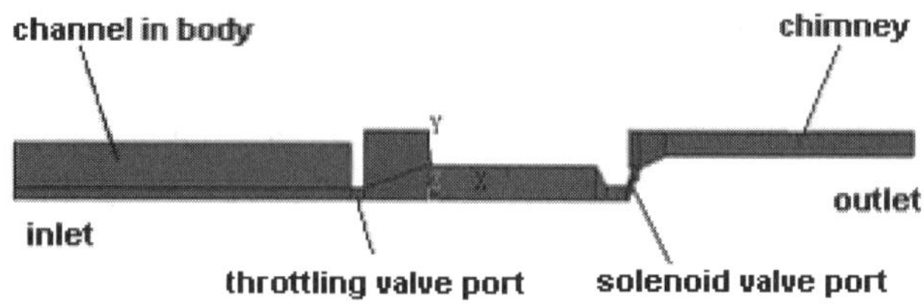

Fig. 6 Finite element model of fluid field

In this model, solenoid valve port, the gap between rod and rod seat, is changeable because of the movement of plunger. In other words, when the plunger position changes to 0.2mm, the solenoid valve port is completely closed. However, the size of throttling valve port is fixed for a certain solenoid valve. The chimney is an extended region of the actual outlet, the reason for modeling the chimney is as follow: If the flow has not fully developed at the outlet, ANSYS® is forced to adjust it across the last row of elements to satisfy the boundary condition, which may cause an unexpected error sometimes[2]. Hence, in order to prevent this from occurring, it is necessary to build an additional region to ensure that the flow can develop completely.

Figure 7 shows the partial finite element mesh near the solenoid valve port. Before meshing the model, it is important to make assumptions about where the gradients are expected to be the highest, and to adjust the mesh accordingly. In this model, the higher gradients

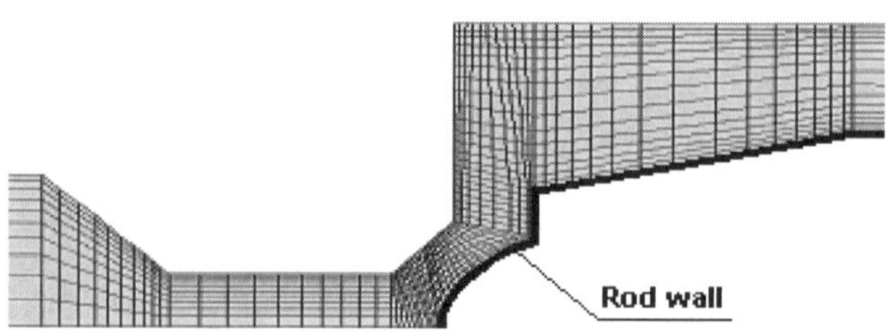

Fig. 7 Finite element mesh near the solenoid valve port

appear near the regions of the two valve ports (solenoid valve port and throttling valve port) and the wall, as a result, elements of these regions may small in size than others. Since this is the turbulence model, the foregoing regions have a much denser mesh than would be needed for a laminar problem. If it is too coarse, the original mesh may not capture significant effects brought about through steep gradients in the solution.

RESULTS OF FLUID FIELD –To simulate the dynamic response of solenoid valve, it is necessary to obtain the hydrokinetic force $F_p(x)$ on rod as a function of plunger position x. For a fixed plunger position, the hydrokinetic force is the surface integral of the force on every element node adjacent to the rod wall (bold black line shown as Fig.7). Hence, it is not difficult to work out the hydrokinetic force $F_p(x)$ by calculating at each position step, the result is shown as Fig.8.

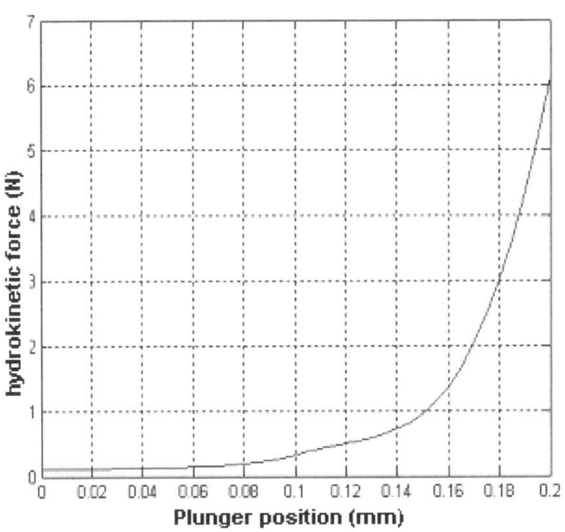

Fig. 8 Hydrokinetic force –plunger position curve

According to the result, it is easy to find that when the solenoid valve is completely closed (plunger position is 0.2mm), the hydrokinetic force on rod is the highest, then reduces rapidly as the valve is opening. When the plunger position is lower than 0.1mm, the hydrokinetic force is almost near 0N. This curve will be used In simulation model.

SIMULATION MODEL OF SOLENOID VALVE

FORMULATION OF SOLENOID VALVE MODEL – The dynamic response of solenoid valve can be formulated by equations for coil current and Newton's law for the translational motion of the plunger. The coil current equations are as follow:

$$\begin{cases} U = R \cdot i + \dfrac{d\psi}{dt} \\ L(x,i) = \dfrac{\psi}{i} \end{cases} \quad (7)$$

where U is the voltage on the coil, R is the coil resistance, and ψ is the magnetic flux linkage. The second equation of (7) can be expressed as

$$\frac{d\psi}{dt} = \frac{d(Li)}{dt} = \frac{dL(x,i)}{dt}i + L\frac{di}{dt} = \frac{\partial L}{\partial x}vi + (\frac{\partial L}{\partial i}i + L)\frac{di}{dt} \quad (8)$$

Take (8) into the first equation of (7), the coil current equation can be expressed as

$$\frac{di}{dt} = \frac{1}{L + i \cdot \dfrac{\partial L}{\partial i}} \cdot (U - R \cdot i - \frac{\partial L}{\partial x} \cdot i \cdot v) \quad (9)$$

The equations for Newton's law for the translational motion of the plunger are as follow:

$$\begin{cases} \dfrac{dv}{dt} = \dfrac{1}{m} \cdot [F_m(x,i) - K(x+G_0) - F_p(x) - bv - F_f] \\ \dfrac{dx}{dt} = v \end{cases} \quad (10)$$

where v is the plunger velocity, m is the plunger mass, K is the restoring spring constant, G_0 is the spring precompression, b is the viscous damping constant, and Ff is the frictional force.

Hence, (9) and (10) are the formulation of solenoid valve dynamic response. The magnetic pulling force $\vec{F}m(x,i)$, coil inductance $L(x,i)$ and the hydrokinetic force $F_p(x)$ are all the results which have been calculated by the finite element models.

SIMULATION MODEL OF SOLENOID VALVE – Based on MATLAB/SIMULINK, the simulation model of solenoid valve dynamic response is established according to the equations (9) and (10), shown as Fig.9.

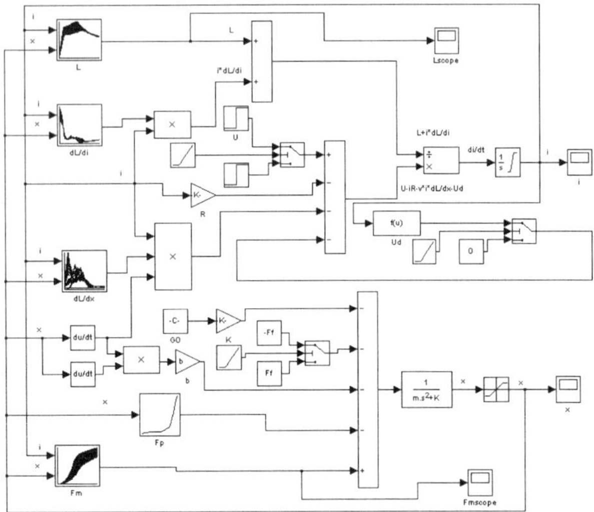

Fig. 9 Simulation model of solenoid valve dynamic response

In this model, the simulation block "Look-up Table" is widely used in order to look up the force-current-position table, inductance-current-position table, hydrokinetic force-position table and so forth. Load 12V step voltage on the coil on 0s and unload it on 0.08s, the simulated result of coil current dynamic response is shown as Fig.10a.

According to the result, a few of expected phenomena can be obtained: (1) At the very beginning, coil current rises up quickly, so as the magnetic pulling force; (2) When the magnetic pulling force rise up to a certain value, the spring force cannot hold the plunger on its original position, and the plunger begins to move; (3) This motion in magnetic field can build a counter electromotive force to the coil, so current descends for a little while; (4) When the rod touches the rod seat, the plunger also stops moving, therefore, the current turns to rise again up to the ultimate value. On the contrary, it is an opposite process for the coil current after the voltage eliminates on 0.08s.

The solenoid valve Dynamic Response Time of Loading voltage (LDRT), defined as the period from the time that loading step voltage on coil to the time that plunger stops moving, is presented as *ton* in Fig.10a. In the same way, the Dynamic Response Time of Unloading voltage (UDRT), defined as the period from the time that the voltage removes from coil to the time that plunger stops moving, is presented as *toff* in Fig.10a. According to the results of simulation, the LDRT is 2.9ms, and the UDRT is 3.1ms.

Figure 10b shows the coil current curve measured in experiment. Compared with the foregoing result, the experiment curve is not changed so observably with the movement of plunger, but it is not difficult to get the response time data by magnifying the figure on screen of test-bed. Hence, the LDRT of test solenoid valve in experiment is measured as 2.7ms, and the UDRT is 2.8ms. The percentage difference between calculated and measured dynamic response time is no more than 10%.

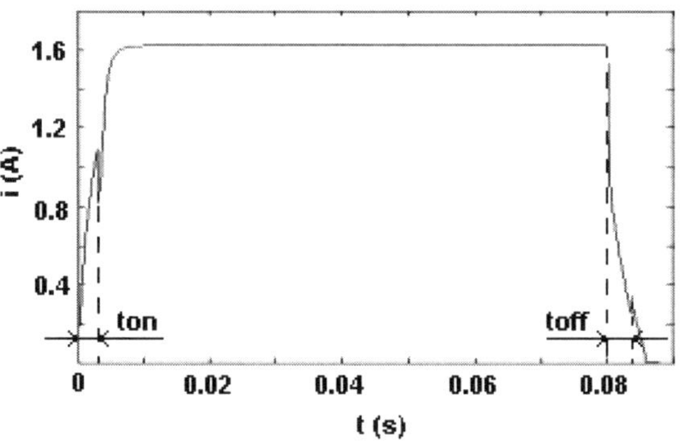

Fig. 10a Dynamic response of coil current (Simulation)

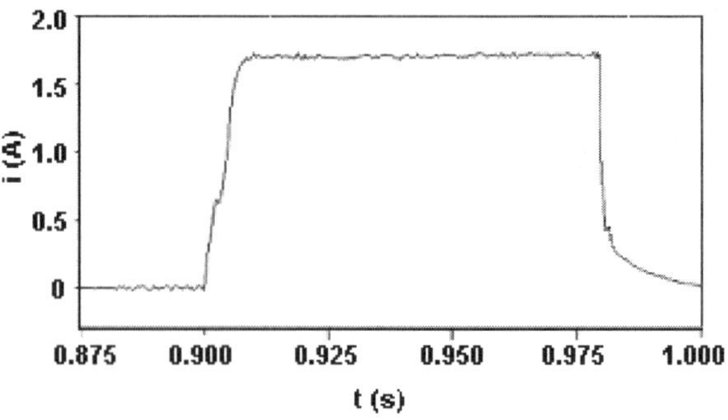

Fig. 10b Dynamic response of coil current (Experiment)

Compared with the results of experiment, the validity of simulation solutions and the results of finite element models are verified.

ADVICES FOR IMPROVING SOLENOID VALVE DYNAMIC RESPONSE

Designers can modify the hydraulic ABS parameters, such as plunger mass, restoring spring constant, coil resistance, etc, in the simulation and finite element models. It is not difficult to study the influence of these parameters on the dynamic response in this way. Some effective methods are advised to improve the performance of solenoid valve.

The plunger mass is one of the main factors about dynamic response time. It is an effectual way to improve the motion frequency by reducing the plunger mass. However, the reduction must be proper, because it usually goes with the decrease of plunger size, and may results in the reduction of magnetic pulling force.

Increase the restoring spring constant can enhance the original force on rod. Hence, the UDRT can be reduced in this way, but on the opposite aspect, the LDRT will be delayed. So if the UDRT of a solenoid valve is not perfect enough, designers can increase the restoring spring constant while the LDRT must be taken into account.

The higher rate of current change is partly depends on the lower coil resistance, which is a function of wire diameter and winding turn for certain wire material. Obviously, reduce the winding turn is a method to diminish the coil resistance, but this may tremendously decline magnetic pulling force which is unacceptable for designing. Increase the wire diameter is another way, but the serious disadvantage is the augment of coil size and cost. Therefore, it is necessary to reduce the coil resistance reasonably without carrying negative consequences.

The air gap length (0.2mm in the model) is also an important parameter about the dynamic response time. If we reduce the length, for example, 0.15mm in some kinds of solenoid valve, the magneto resistance can be cut down, so the magnetic pulling force will lift up to improve the valve performance. However, the air gap length directly influences the size of solenoid valve port, which can affect the rate of pressure change. Hence, the length is always determined by requirement of the whole brake system.

The magnetizable material has significant impacts on dynamic response. For the purpose of keeping high level of magnetic pulling force, it is unacceptable to choose the material with low saturation.

Besides above parameters, there are several others. Though their actions are not all the same, they more or less influence the performance of solenoid valve.

In sum, while hydraulic ABS parameters improve the solenoid valve dynamic response, they usually bring some such-and-such disadvantages. Therefore, designers should consider various factors to carry their points. In other words, it is an integrative process to optimize parameters for a high performance solenoid valve.

SOLENOID VALVE MODEL ON AMESIM

The finite element model is an effective method for studying dynamic response of solenoid valve. However, the process of finite element modeling is not so simple for designers to finish in a few minutes. When some physical structures of the valve are changed, some of the modeling phases must be reworked consequently to calculate a new result. Therefore, another form of model to study the dynamic response of solenoid valve is established by using the software named AMESim®, which contains many general magnetic and hydraulic components suitable for simulating ideal dynamic behavior based on component performance parameters. The model is shown as Fig.11.

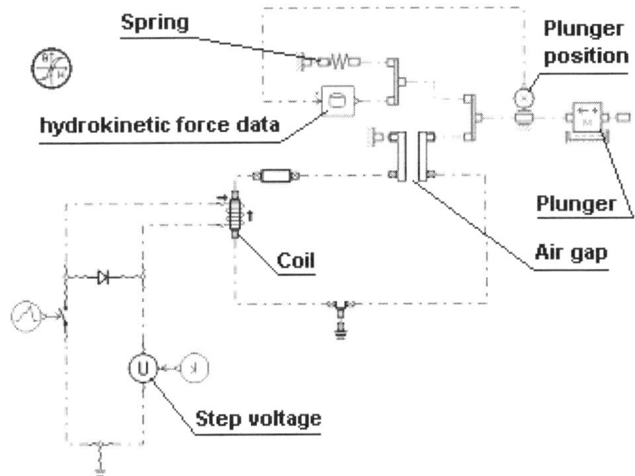

Fig. 11 Solenoid valve model on AMESim

In this model, since every component contains a few of variables which can be easily obtained from the result, it is convenient for designers to observe history of every interested variable. For instance, the histories of magnetic flux density (shown as Fig.12) and magnetic field intensity, difficult to present in the SIMULINK® model, can be plotted automatically in this calculation result.

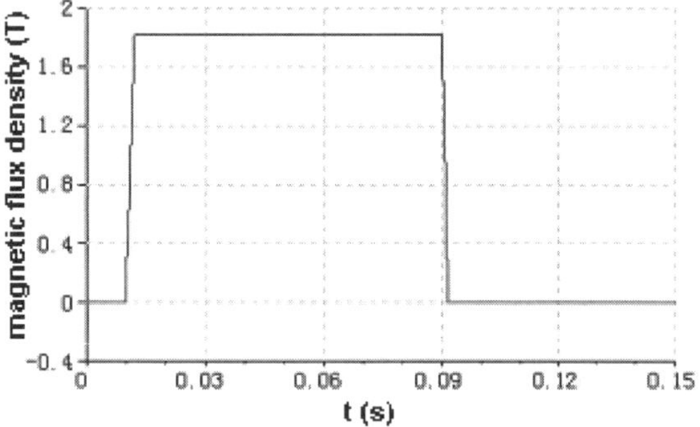

Fig. 12 History of magnetic flux density (AMESim)

In this model, the B-H characteristic of the magnetizable material is presented by two lines which may bring some differences to the simulating results. Therefore, the percentages difference between calculated and measured dynamic response time, no more than 14% for LDRT and 19% for UDRT, are not so accurate as the ones of SIMULINK® model, but it is convenient to research the influence of parameters on solenoid valve dynamic response, because that whether the value of a parameter could lead to high performance is obviously by reading the result figures. Models built on AMESim® have several advantages as follow: (1) The process of establishing model is greatly abbreviated. (2) Many variables' histories can be observed easily. (3) Since relative physical components are connected directly in the model, it is intuitional to study the whole system by logistic thinking.

INFLUENCE OF HYDRULIC ABS PARAMETERS ON BRAKING EFFECT

Recently, attention of the hydraulic ABS has been focused on the in-vehicle driving tests and the hardware-in-loop experiments. In this way, a designer can research the influence of such parameters on braking effect. Consequentially, these works have taken the designers a mass of times to match the HCU for different types of the vehicles. In order to solve this complicated problem, we build an ABS-vehicle model on AMESim® which contains three components—that are vehicle model described by seven degrees of freedom, HCU model and ECU model, shown as Fig.13. The vehicle model and ECU model are both established by using AMESim Control Library, but the HCU model based on AMESim Hydraulic Library.

By modifying the values of HCU parameters in the model, various dynamic response curves of the brake pressure, flow rate, wheel speed, etc can be presented. Moreover, designers can also obtain the vehicle's overall stopping distance and time.

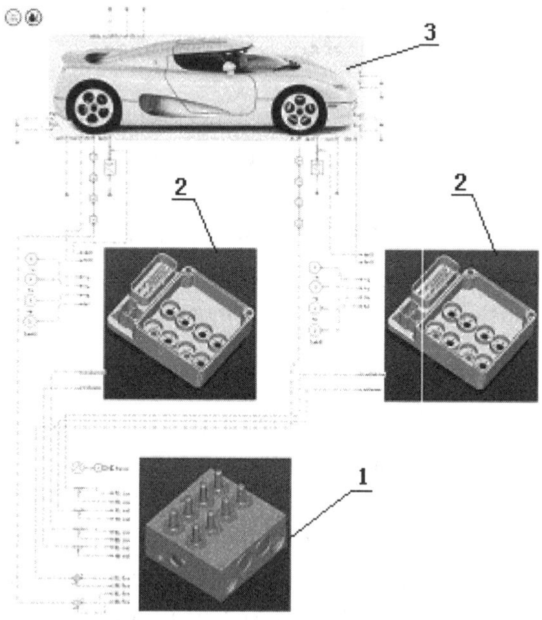

Fig. 13 ABS-vehicle AMESim model

The process of matching proper ABS parameters for different vehicles in this model is not difficult. Take a certain type of vehicle for example: (1) According to the parameters of actual vehicle, input each value to globe parameters table in the model. In this way, it is easy to abstract an actual vehicle to the seven degrees of freedom vehicle model. (2) Set all the values of HCU parameters and ECU thresholds to default or the same as the ones which have been matched for the similar type of vehicles. (3) Run the ABS-vehicle model and check the results of brake pressure, wheel speed, etc. (4)

If the wheel speed reduces to zero or other little values, which means that the wheel tends to be locked, it is necessary to check the ability of releasing pressure. What can be done is to increase the size of release solenoid valve port, the capability of low pressure accumulator and the hydraulic pump speed. (5) If the brake pressure can not increases rapidly, it is necessary to enlarge the size of isolate solenoid valve port or throttling valve port. (6) If the performance of ABS can not still meet the requirement, it is an effective way to adjust the values of ECU threshold. (7) Change adhesion coefficient of the road and repeat the steps (3) to (6), ensure that these parameters can adapt various road surfaces. (8) Finally, do fine adjustment to the values of HCU parameters and ECU thresholds so that the stopping distance and time of vehicle model are close to the minimum.

According to the HCU parameters' values chosen in the simulation, an actual solenoid valve with the same values has been fixed in the test vehicle. The in-vehicle driving experiment is on high adhesion coefficient road, and the initial speed of test vehicle is 80km/h.

The equipment in vehicle for gathering the experimental data mainly contains a sampling system made by B+S company in Germany, circuits for filtering waves and serial data transfer, wheel speed sensors and pressure sensors. The software used for observing test results is developed on Labview. In this way, the experimental data of the test vehicle can be obtained.

Figure 14 shows the dynamic response of brake pressure calculated in simulation and sampled in experiment. Figure 15 shows the dynamic response of wheel speed in the same way. Table 1 presents the comparison between simulation and experiment in stopping distance and time.

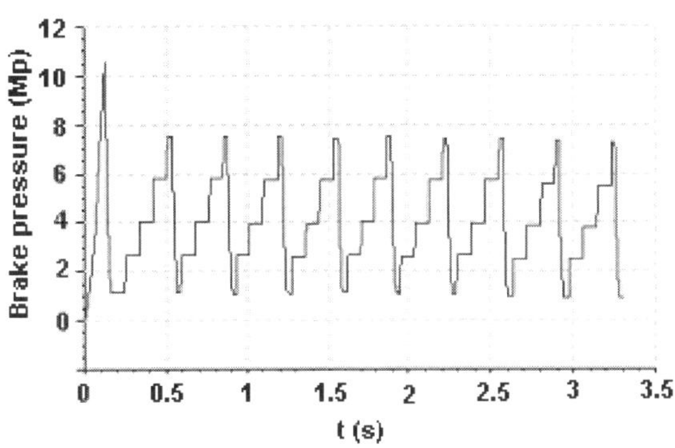

Fig. 14a Dynamic response of brake pressure (Simulation, 80km/h)

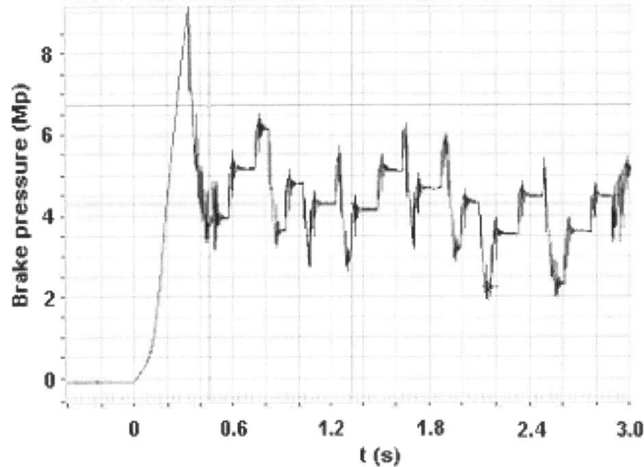

Fig. 14b Dynamic response of brake pressure (Experiment, 80km/h)

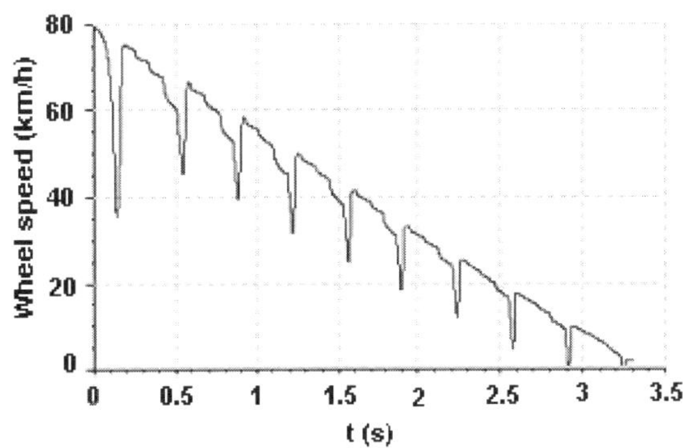

Fig. 15a Dynamic response of wheel speed (Simulation, 80km/h)

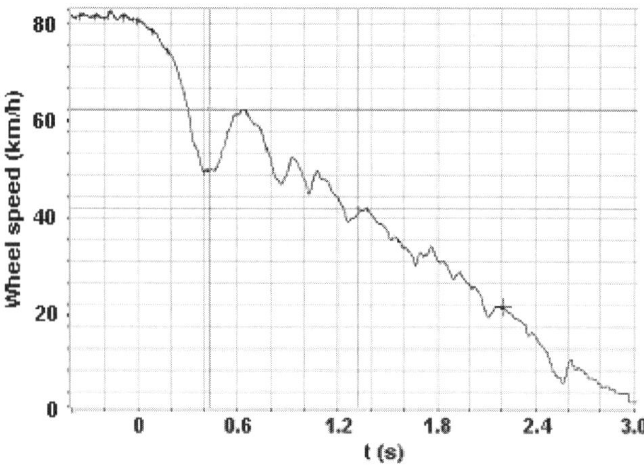

Fig. 15b Dynamic response of wheel speed (Experiment, 80km/h)

Tab.1 Comparison between simulation and experiment

	Simulation	Experiment
Stopping distance	37.9m	34.1m
Stopping time	3.3s	3.0s

Calculated results, brake pressure, wheel speed, stopping distance as well as the stopping time, have good agreement with the experiment data. Therefore, the ABS-vehicle model is an effective tool for studying the influence of hydraulic ABS parameters on braking effect. Furthermore, it is not necessary for designers to do such a lot of in-vehicle driving tests and the hardware-in-loop experiments.

CONCLUSION

In this paper, finite element analysis is used to develop the ABS solenoid valve model. Based on ANSYS®, the magnetic field model and fluid field model are built to calculate the magnetic pulling force, coil inductance and hydrokinetic force. A simulation model with the finite element results is established on MATLAB/SIMULINK® to compute the dynamic response of solenoid valve. Compared with the data of experiment, the results of simulation solutions and finite element models are proved. Furthermore, several effective methods for improving the performance of the solenoid valve are presented.

To abbreviate the process of modeling, another form of model to study the dynamic response of solenoid valve is established on AMESim®, an effective simulation tool. For further study, an ABS-vehicle model for studying the influence of hydraulic ABS parameters on braking effect is developed on the same software flat. The simulation results have good agreement with the in-vehicle driving experimental data.

REFERENCE

1. Song-min Wang, et al, "Electromagnetic Field Analysis and Dynamic Simulation of a Two-Valve Solenoid Actuator". IEEE TRANSACTIONS ON MAGNETICS, VOL.29, NO.2, MARCH 1993
2. Yonggang Xu, "Design and Simulation Study of ABS Solenoid Valves". Master thesis. Tsinghua Univ. 2000
3. Fred W.McMann. "Automotive magnets. Materials engineering (Cheveland)", V108, n2, Feb 1991.

Driver Crash Avoidance Behavior: Analysis of Experimental Data Collected in NHTSA's Vehicle Antilock Brake System (ABS) Research Program

Graeme F. Fowler and Robert E. Larson
Exponent, Failure Analysis Associates

Laura A. Wojcik
Packer Engineering

ABSTRACT

As part of the National Highway Traffic Safety Administration's (NHTSA) Light Vehicle Antilock Brake System (ABS) Research Program a study was conducted to examine driver crash avoidance behavior and the effects of ABS on drivers' ability to avoid a collision in a crash-imminent situation. The test track study, described in detail in the SAE paper "Driver Crash Avoidance Behavior with ABS in an Intersection Incursion Scenario on Dry Versus Wet Pavement" [1], was designed to examine the effects of ABS versus conventional brakes, ABS brake pedal feedback level, and ABS instruction on driver behavior and crash avoidance performance.

Exponent has obtained the electronic data collected by NHTSA in the dry pavement study and analyzed the steering inputs to better understand how drivers respond to emergency avoidance situations. The results of this study can also be used to put the steering magnitudes, rates, and patterns used by test drivers performing idealized avoidance-type maneuvers on the skid-pad into the proper perspective by comparing them with what truly surprised drivers apply when confronted with an emergency situation requiring a steering response.

INTRODUCTION

As part of the National Highway Traffic Safety Administration's (NHTSA) Light Vehicle Antilock Brake System (ABS) Research Program a study was conducted into driver crash avoidance behavior in a simulated crash-imminent event. The protocol for the study, along with a summary of the results, is provided elsewhere [1 and 2]. According to NHTSA, the study was designed to examine the effects of ABS versus conventional brakes, ABS brake pedal feedback level, and ABS instruction on driver behavior and crash avoidance performance. The protocol for their investigation involved naïve subjects driving on a test track roadway under the guise that they were participating in a study of how average drivers steer and maintain speed in typical driving conditions. The driver traveled through an intersection three times where two vehicles were stationary at the stop lines of the crossing lane. Before the driver approached the intersection for a fourth time, the stationary vehicles were replaced with realistic looking stryofoam mockups. As the driver approached intersection, the styrofoam vehicle on the right intruded 6 feet into their lane of travel. The resulting driver steering and braking inputs as well as the vehicles' response were recorded with on-board instrumentation and video cameras. The protocol targeted an approach speed of 45 mph.

The experiment was designed to elicit emergency steering response under true surprise conditions. It was not intended to be representative of all crash avoidance-type scenarios or pre-crash on-road maneuvers. However, as pointed out by the NHTSA [1], the study is "useful in determining not only the extent to which drivers are able to maneuver the vehicle, but also drivers' physical capacity to supply control inputs to the vehicle. Insight into drivers' ability to maintain control during a panic maneuver and ability to avoid a collision can also be gained from this research." An understanding of typical drivers' response to an emergency avoidance is of interest to researchers in the areas of vehicular crash reconstruction as well as vehicle handling, controllability and rollover resistance. In fact, over the years numerous avoidance type maneuvers that purportedly model this situation have been proposed to investigate vehicles' limit handling and stability. The most notable being the International Standards Organization 3888 Part 2 Double Lane Change (ISO-DLC) and Consumer's Union Short Course (CU-SC) maneuvers. Both are discussed extensively in [3].

The NHTSA driver performance study represents the most extensive investigation into driver behavior in a crash-imminent situation using actual vehicles since the 1970's [4, 5, 6]. In the dry pavement experiment, 192 subjects were involved. Also of significance is the fact that the study utilized modern vehicles - a 1995 Chevrolet Lumina and a 1996 Ford Taurus - whose steering response is representative of passenger vehicles in the driving fleet today.

Exponent has obtained the electronic data collected by NHTSA in the dry pavement study for analysis of the drivers' behavior. In this paper we summarize the drivers' control responses obtained in the experimental study with a focus on their steering magnitudes and rates. Since data from this study are likely to be used by researchers as a reference for the capability of lay drivers to apply steering inputs in an emergency situation, the control inputs of those subjects who recorded the highest steering amplitudes are discussed in some detail. Next, the data from subjects who responded to the mock intersection collision by steering in both directions (reverse steered) were singled out for examination. Reverse steer inputs are intentionally generated in emergency lane change maneuvers using the ISO-DLC and CU-SC protocols with the intent of stressing vehicles' limit handling characteristics. It is of interest then, to compare the steering responses of typical drivers in this study to those of test drivers when performing the ISO-DLC and CU-SC maneuvers as a way of understanding the severity of these tests and their relevance to real-world crash situations. This comparison is provided in the final section.

ANALYSIS OF DRIVER BEHAVIOR

NHTSA's experimental design included multiple variables, such as different vehicles with unique ABS properties, response time allowed prior to the impending collision (time to intersection, TTI), conventional braking versus ABS, driver gender, and ABS driver instruction. With the exception of conventional versus ABS brake systems' impact on the steering and braking response, we have not addressed the impact of these additional factors. (These are discussed in detail in [1 and 2].)

An examination of the in-vehicle logs documenting the notes of the onboard observer indicated that 16 of the 192 subjects anticipated the vehicle crossing into the intersection. Of those 16 subjects, the in-vehicle logs also reported that only one (6%) failed to avoid impacting the incursion vehicle. In contrast, 41% of the subjects that did not anticipate the event impacted the styrofoam vehicle. Since the drivers who anticipated the incursion may have reacted differently to those who were truly surprised, they were not considered in our evaluation of driver behavior.

DEFINITION OF "EVENT"

A review of the electronic data files for the 176 subjects who did not anticipate the vehicle incursion led us to define an "event" window based on either time or vehicle speed during which the driver and vehicle data were to be analyzed. For the analysis presented here, data were analyzed for 5 seconds starting from the movement of the foam incursion vehicle or until the subject's vehicle speed dropped below 10 mph, if that occurred prior to the end of the 5 second period. It should be noted that movement of the incursion vehicle into the subject vehicle's lane began at either 2.5 or 3.0 seconds before the subject vehicle would have reached the intersection, based on the subject's actual approach speed. Five seconds was sufficient for the subject vehicle to either pass the incursion vehicle or, if heavy braking was applied, complete the primary avoidance maneuver. Steering inputs after the vehicle's speed had dropped below 10 mph were assumed to not be associated with the primary avoidance maneuver. Initial speeds ranged from 39 to 50 mph. It was noted that, within the defined event window, some subjects had begun to accelerate again and apply steering inputs which were probably associated with re-aligning the vehicle on the roadway after the avoidance maneuver. With respect to these variances in driver behavior, the extent of the event window was considered to be conservative.

SUMMARY OF DRIVER RESPONSES

Given the above definition of the "event," the data were processed to analyze the drivers' response. Initial review of the steering and braking data indicated that – even with this relatively simplistic crash-imminent scenario – a diverse range of outcomes was observed. This variability is clearly demonstrated in **Figure 1**, where the maximum absolute steer angle for each subject is plotted against the change in speed during the event. The change in speed was used as a surrogate for braking effectiveness since subjects implemented a wide range of brake forces and application times. Steering amplitudes ranged from 0 to 271 degrees and speed changes ranged from a small increase to a 37 mph reduction.

In order to broadly classify the subjects' braking and steering response, we assumed that a speed change of 6 mph or less corresponded to zero or minimal braking effort. (For these subjects, peak braking pedal forces were at or below 35 lbs). Peak steering magnitudes of 25 degrees or less were assumed to be indicative of minimal steering response. With these thresholds in mind, the majority of subjects (61%) steered and braked to avoid colliding with the obstacle (i.e., their speed change exceeded 6 mph and they steered in excess of 25 degrees). The next largest group – representing 28% of the subjects – principally braked in response to the emergency, without significant steering. Only 7% of the subjects applied steering only, which notably, is the control response modeled in both the ISO-DLC and CU-SC dynamic maneuvers. A number of subjects (3.4%) appeared to not respond in any meaningful way to the emergency situation. In the experiment, approximately 40% of the subjects "crashed" into the encroaching vehicle, which provides an indication of the severity of the emergency scenario simulated.

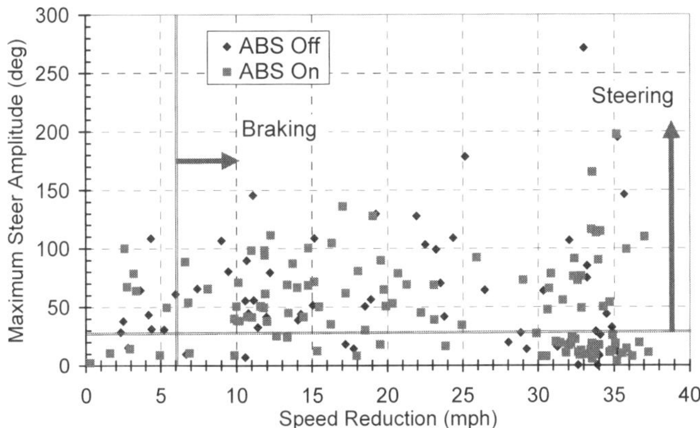

Figure 1: Maximum Absolute Steer Angle v. Change in Speed During Event, No Anticipation

In **Figure 1**, vehicles with ABS active and ABS deactivated (simulating vehicles with conventional braking systems) are delineated. It can be observed from the Figure that the higher peak steering angles (for example, 150 deg and greater) were associated with heavier braking, regardless of the braking system. In their paper [1], Mazzae et al. reported the average avoidance steer magnitudes for subjects with ABS and those with conventional brake systems as 49 and 61 degrees, respectively. Although the difference in steer angle was not statistically significant, the authors did note that "drivers appeared to alter their steering behavior based on the degree to which they felt the steering inputs were affecting the motion of the vehicle in the desired direction. Subjects with ABS made smaller steering inputs and used lower steering input application rates than subjects with conventional brakes. The reason for this difference is believed to be that subjects made increasingly large steering with conventional brakes since, with locked wheels, their steering inputs were not effective in directing the vehicle's motion." This observation was confirmed when we looked specifically at the 60 subjects in the conventionally braked vehicles. Seven of these subjects applied the brakes with sufficient magnitude to lock the front wheels (based on the wheel speed data). The average peak steering amplitude for the 7 subjects was 122 degrees compared to 54 degrees with no lockup. Because of the potential confounding influence of ABS as apposed to conventional brakes, in the analysis that follows we will display the data with the different braking systems delineated.

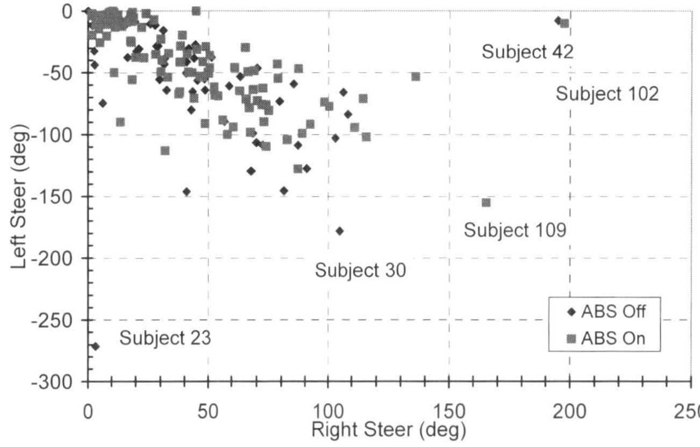

Figure 2: Peak Left and Right Steer Angles

Figure 2 displays the peak left and peak right steer angles recorded for the 176 subjects during the event. The data reveal that, in the experimental crash-imminent scenario, 95% of the subjects input steering amplitudes of less than 130 degrees in an attempt to avoid the encroaching vehicle. Only 3% (5 subjects) exceeded 150 degrees of steering input. Three of these subjects were in conventionally braked vehicles (subjects 23, 42 and 30, representing 5% of that group) while two subjects (numbered 102 and 109) operated vehicles with the ABS activated (less than 2% of that sample).

TOP 5 STEERING EVENTS

The steering and braking responses of the five "large-steer" subjects, based upon the electronic data recorded during the event and information provided in the in-vehicle log, will be discussed below. Relevant steering, braking and vehicle response data for each subject are shown in **Appendix A, Figures A1 - A3 (conventional brakes) and A4 – A5 (ABS brakes).**

Conventional Brakes

Subject number 23 recorded the largest peak steer amplitude of 271 degrees. In response to the incursion vehicle, this subject first braked, which caused the front wheels to lock up. Approximately 0.75 seconds after braking – when the vehicle speed had decreased from approximately 43 mph to 27 mph – the subject steered to the left 271 degrees at a peak rate of 570 degrees/second.[1] At about the time of the maximum steer angle, the vehicle had almost come to a stop. During the steer, the subject continued to brake and, as a result of front wheel lockup, the vehicle did not respond to the steer input as evidenced by the negligible lateral acceleration generated (A_y). According to the in-vehicle log, the vehicle remained within the original lane of travel (right lane) in spite of the large left steer and did not collide with the obstacle.

---

[1] The steering rate was calculated by differentiating the data after implementing a 2 Hz low-pass filter.

To avoid a collision with the obstacle, subject 42 applied a right steer of 195 degrees at a peak rate of 356 degrees/second. Braking was initiated approximately at the same time as the steering input. Due to the braking, lateral accelerations of only 0.3 g's were developed during the "event." The vehicle migrated to the right to just straddle the road edge according to the in-vehicle log. This subject did impact the incursion vehicle.

Subject 30 braked hard and then steered to the left 178 degrees during his or her avoidance maneuver. Due to front wheel lock up at the beginning of the maneuver, the vehicle did not initially respond to the steering input. After the vehicle had decelerated to approximately 20 mph, the driver reduced braking and the vehicle responded with approximately 0.3 g lateral acceleration. The driver then steered to the right 105 degrees and began to accelerate. According to the in-vehicle logs, this subject did successfully avoid the obstacle by steering into the left lane.

ABS Brakes

Subject 102 braked heavily and steered right 197 degrees in response to the obstacle in his/her path. The peak steering rate was 320 degrees/sec and the vehicle left the roadway to the right. Although the speed approaching the intersection was 45 mph, the speed at the time of the peak steer input was 28 mph. This subject impacted the incursion vehicle. The peak lateral acceleration was 0.56 g's.

Subject 109 also initially steered right to avoid the incursion vehicle. The 165 degree right steer was followed by a 155 degree reverse steer, resulting in the largest peak-to-peak steering input of all the subjects (320 degrees). The peak steering rate was 870 degrees/sec. At the initiation of the event, the subject was traveling 44 mph. At the point of the maximum right steer the vehicle had slowed to 33 mph, at maximum left steer it was at 26 mph. As a result of the control inputs, the vehicle partially encroached the right shoulder and did impact the obstacle vehicle (according to the in-vehicle logs). The peak lateral acceleration generated in the maneuver was only 0.5 g's due to the heavy braking.

As can be established from **Figures A1 – A5** and the descriptions of the "large-steer" events above, all five of the subjects with steer amplitudes exceeding 150 degrees also braked heavily in response to the incursion vehicle. As a consequence of the braking, the maximum lateral accelerations developed during the avoidance maneuvers performed by the five "large steer" subjects was only 0.56 g's - substantially below the maximum lateral acceleration capability of the test vehicles. Three of the five subjects steered primarily in <u>one direction</u> during the course of the "event." In fact, the largest steer amplitudes - approximately 200 to 270 degrees - were from the subjects who steered in a single direction. The remaining 2 subjects steered in both directions to avoid the encroaching vehicle.

DRIVER AVOIDANCE INVOLVING REVERSE STEER INPUTS

We now restrict our attention to those subjects who steered in both directions during the "event" and whose peak steering amplitude in each direction is relatively comparable. It is assumed that included in this subset of drivers are those that made an attempt to steer around the incursion vehicle and then steered in the opposite direction to recover their original lane – the type of maneuver modeled in the ISO-DLC and CU-SC procedures. To this end, subjects were included if their peak-to-peak steer angle exceeded 50 degrees and the ratio of peak left and right steer amplitudes fell within the range of 0.33 to 3.0. This selection criterion is somewhat arbitrary, but removes from consideration those subjects who steered only a minimal amount or steered principally in just one direction.

Figure 3 documents the peak left and right steer angles for the 104 subjects meeting this requirement. Within this group, the average maximum steer angle input was 72 degrees, with more than 98% of the subjects utilizing peak steer amplitudes of less than 150 degrees.

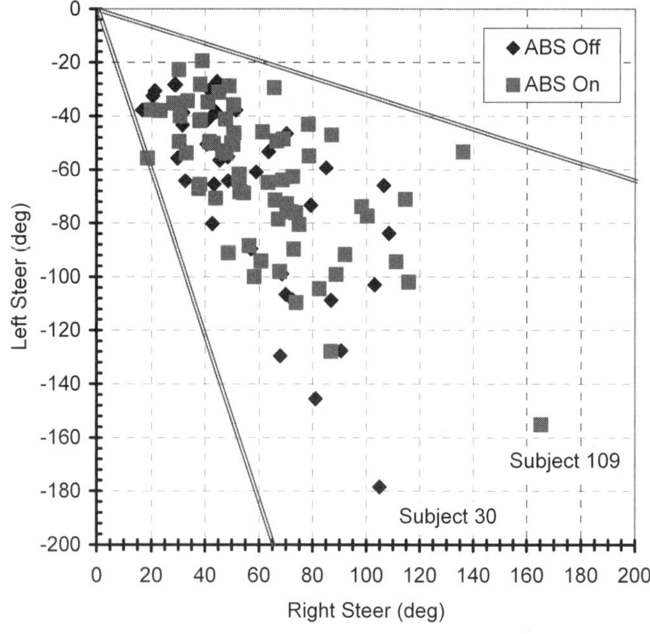

Figure 3: Peak Left and Right Steer Angles - Reverse Steer Subjects

In **Figure 4** the distribution of the peak-to-peak steer angles for the subjects who input reverse steers is provided. We define the peak-to-peak steer angle as the sum of the peak left and right steer amplitudes recorded for each subject during the "event." The average peak-to-peak steer angle for these subjects is 125 degrees. Only two subjects (109 and 30) exceeded peak-to-peak angles of 230 degrees, using 320 and 283 degrees, respectively. Both of these subjects braked heavily during the avoidance maneuver, with one slowing to a stop in three seconds, and the other slowing to 11 mph in two seconds and then accelerating.

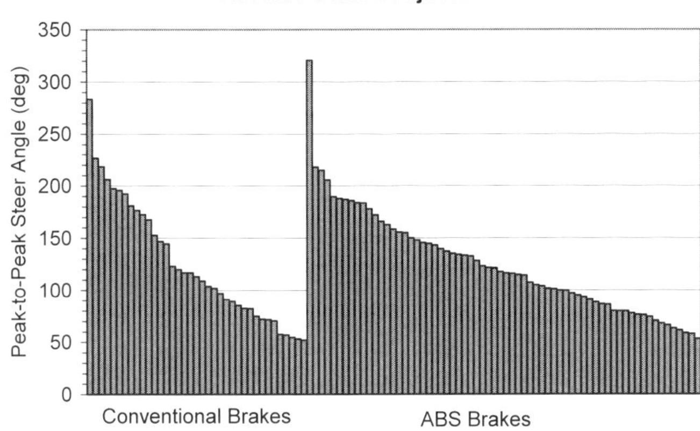

Figure 4: Peak-to-Peak Steer Angle
Reverse Steer Subjects

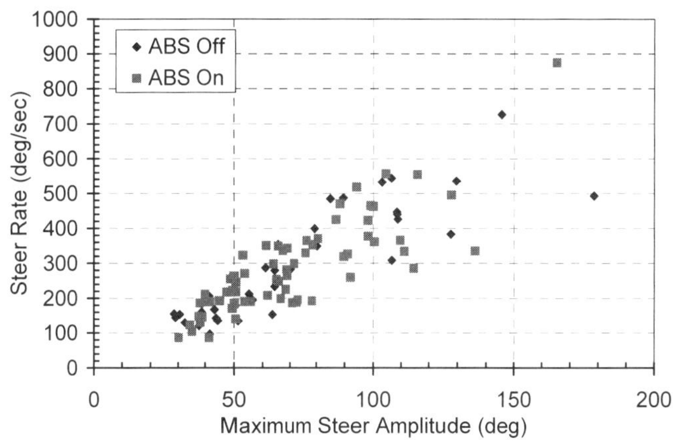

Figure 6: Maximum Steer Rate v Steer Amplitude
Reverse Steer Subjects

Figure 5 is a plot of the peak steer amplitude for each subject versus their speed approaching the intersection for the reverse steer subjects. For approach speeds above 45 mph, the maximum steer angles recorded did not exceed 120 degrees. The largest steer angle amplitude for these subjects was 179 degrees, associated with an initial speed of approximately 40 mph (subject 109).

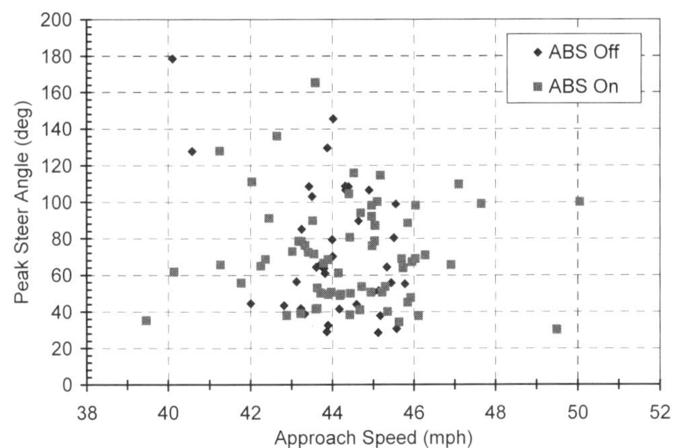

Figure 5: Peak Steer Angle v Approach Speed
Reverse Steer Subjects

Given the nature of the experiment, it is not surprising that an essentially linear relationship existed between the maximum steer amplitude and the steering rate for those subjects who input comparable left and right steering angles during their avoidance attempt. As shown in **Figure 6**, the average steering rate recorded for these subjects was approximately 290 deg/sec, with 98% of the subjects below 600 deg/sec. The maximum steer rate was 875 deg/sec.

The peak absolute value of lateral acceleration recorded during the "event" as a function of the maximum steer amplitude is plotted in **Figure 7**. Ninety four percent (94%) of the subjects generated lateral acceleration magnitudes of 0.6 g's or less. The highest lateral acceleration recorded for those subjects who input a reverse steer in the experiment was 0.73 g's[2]. Interestingly, the subjects experiencing the highest lateral accelerations did not apply the largest steering inputs due to the confounding influence of braking. Peak absolute values of lateral accelerations exceeding 0.6 g's were generated with steering amplitudes in the range of 90 to 110 degrees. Furthermore, the two subjects who developed the highest lateral accelerations (0.7 and 0.73 g's) <u>both exited the roadway to the left</u> (according to the in-vehicle logs). This loss of lane position suggests that lay drivers who steer aggressively in an avoidance maneuver and generate relatively high lateral accelerations are likely have difficulty keeping the vehicle within the driving lanes of a two-lane roadway.

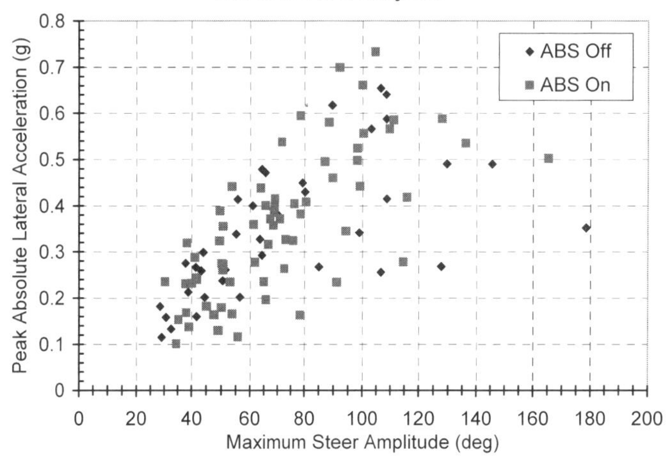

Figure 7: Peak Lateral Acceleration v Steer Amplitude
Reverse Steer Subjects

---

[2] The peak lateral acceleration of 0.73 g's was actually the highest recorded among all the subjects who did not anticipate the incursion vehicle.

COMPARISON WITH ISO-DLC AND CU-SC MANEUVERS

In **Figure 8** the peak left and right steering inputs recorded in the NHTSA driver study for those subjects responding to the simulated emergency by applying a reverse steer are compared to the steering amplitudes used by test drivers when performing the ISO-DLC and CU-SC maneuvers. The test driver data were obtained from testing performed by the NHTSA as part of their Phase IV experimental evaluation of maneuvers that may induce untripped rollovers [3]. The data points for the ISO-DLC and the CU-SC maneuvers represent the peak left and right steer angles at the maximum maneuver entrance speed for each of the four test vehicles in their nominal loading conditions[3].

The maximum peak-to-peak steering angles reported in the Phase IV tests were 454 degrees for the ISO-DLC and 713 degrees for the CU-SC. As discussed in the previous section, in the driver response study the maximum peak-to-peak steer angle recorded was 320 degrees, with 98% of the drivers applying steer reversals of 230 degrees or less. The steering magnitudes used by the test drivers to successfully negotiate the ISO-DLC and CU-SC courses cleanly at the maximum speeds well exceeded that which lay drivers were found to use in the simulated collision scenario.

We next looked at the steering inputs utilized by lay drivers who maneuvered the vehicle in a fashion that more closely followed the emergency lane change scenarios modeled by the ISO and CU maneuvers. In this case, we only considered subjects who avoided impacting the incursion vehicle by steering into the left lane according to the in-vehicle logs. As the ISO and CU lane change maneuvers do not involve braking, we further restricted the data set to include subjects whose speed change during the "event" was less than 15 mph - representing a relatively small amount of braking at most. Using these filters, the maximum steering amplitudes recorded by test subjects was 111 degrees and the maximum steering rate was 544 degrees/second.

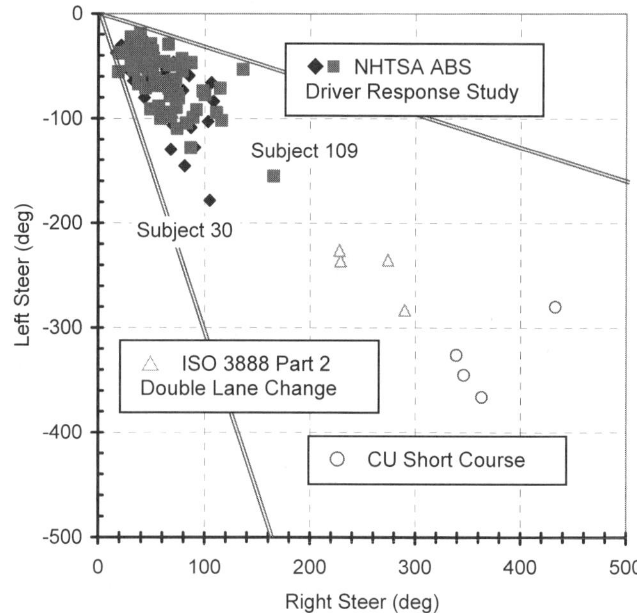

Figure 8: Comparison of Steering Inputs Recorded in NHTSA's Driver Response Study and Phase IV Research Program

In **Figure 9** the steering input, lateral acceleration, and vehicle speed for a Chevrolet Blazer driven by the three NHSTA's test drivers during their maximum speed clean CU-SC runs is compared to the two most severe steering reversals found in NHTSA's driver response study. As can be seen from the figure, the steering inputs used by the naïve test subjects (operating either a Chevrolet Lumina or Ford Taurus) were significantly smaller than those used by the test drivers in the CU-SC. The initial speeds in the driver response study were as high or higher than the CU-SC maximum speed clean runs, but in the driver response study the drivers who input the largest steering angles also utilized braking and their speed dropped off more rapidly. The lateral accelerations observed in the driver response study were also considerably lower than found from the CU-SC. This difference is due to a combination of factors, including the lower steering magnitudes, the reduced speed due to braking, and, in the case of Subject 30, a reduced lateral acceleration capability of the vehicle due to front wheel brake lock up.

CONCLUSIONS

This NHTSA study produced a diverse range of driver responses to a relatively simple accident avoidance situation. Most subjects steered and braked in response to the simulated emergency. The next largest group used their brakes as the principal method of avoidance. A relatively small number of subjects (7%) used only steering to avoid the incursion vehicle.

---

[3] 2001 Chevrolet Blazer, 2001 Toyota 4Runner, 1999 Mercedes ML 320, 2001 Ford Escape. Stability control enabled on vehicles with that that feature.

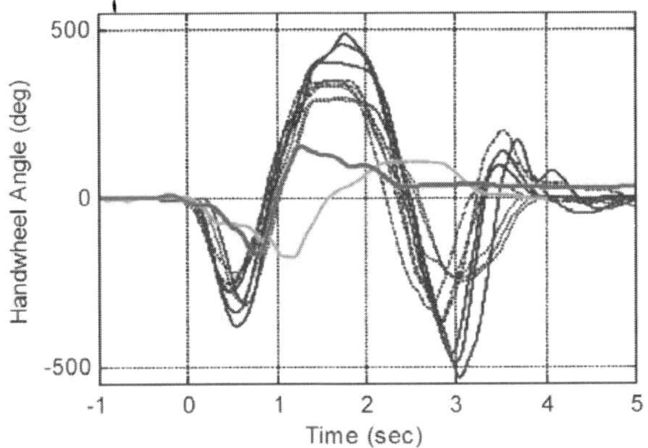

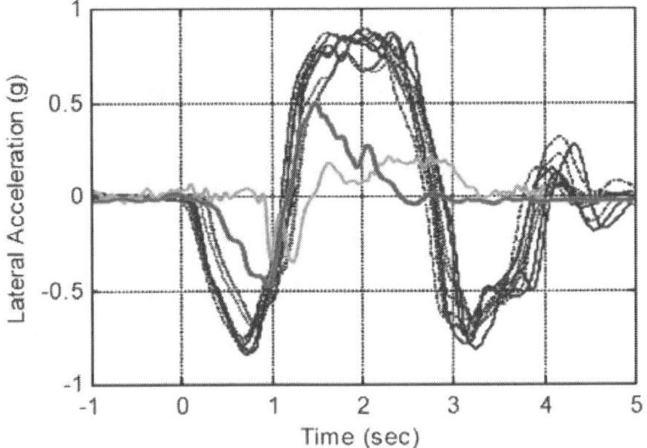

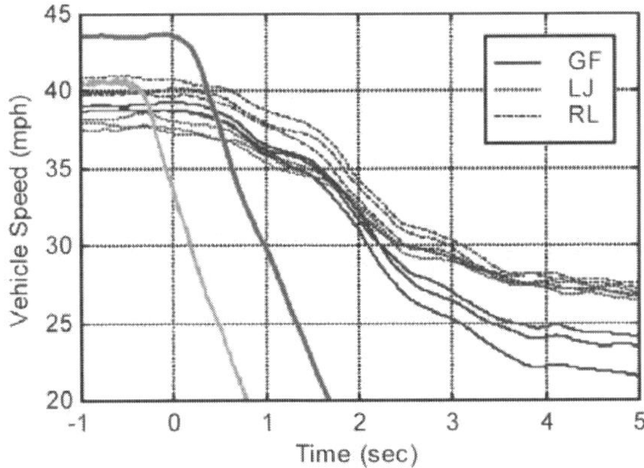

— CU Short Course Testing,
NHTSA Figure 12.7 [3]
— Driver Response Subject 109,
with ABS
— Driver Response Subject 30,
with conventional brakes

Drivers of conventionally-braked vehicles that experienced front wheel lockup in the avoidance maneuver recorded, on average, higher maximum steering amplitudes (122 degrees) compared to those who did not lock up their front wheels (54 degrees). The average peak steering amplitude for the ABS vehicles was 49 degrees – almost identical to that of the subjects in the conventionally-braked vehicles that did not lock up the front wheels in the avoidance maneuver. Consequently, drivers appeared to alter their steering input based on the degree to which they felt that their steering inputs were affecting the motion of the vehicle.

The largest steering amplitudes were recorded by subjects who steered primarily in one direction – often associated with front wheel lockup with the conventional brake system.

When we focused our attention on subjects who responded to the unexpected incursion with a reverse steer input, the average peak steer amplitude was 72 degrees, with the great majority of subjects (98%) inputting maximum steer amplitudes of less than 150 degrees. The average peak-to-peak steer angle was 125 degrees, with 98% of the subjects not exceeding 230 degrees. Only two subjects applied steering rates in excess of 600 degrees/second. The maximum steering rate measured in the experiment was 875 degrees/second.

For subjects who maneuvered the vehicle in a fashion that most closely represents that of the emergency lane change scenario modeled by the CU and ISO maneuvers, maximum steering amplitudes were less than 111 degrees.

Most subjects (94%) experienced lateral accelerations below 0.6 g's in response to their control inputs. The two subjects who recorded the highest lateral accelerations (0.7 and 0.73 g's) were not able to keep the vehicle within the driving lanes.

The ISO-DLC and CU-SC tests represent very severe vehicle handling maneuvers requiring steering angles and rates well in excess of what typical drivers implemented in the experimental collision avoidance scenario.

The peak steering amplitudes and rates applied by test drivers during skid-pad obstacle avoidance-type maneuvers should not be considered representative of what lay drivers might use when faced with a real-world collision scenario or when attempting to regain control on the highway.

ACKNOWLEDGMENTS

The research presented in this paper was sponsored by American Suzuki Motor Corporation. The authors gratefully acknowledge this support.

REFERENCES

1. "Driver Crash Avoidance Behavior with ABS in an Intersection Incursion Scenario on Dry Versus Wet Pavement," Mazzae, Barickman, Baldwin and Forkenbrock, SAE 1999-01-1288.

2. "NHTSA Light Vehicle AntiLock Brake System Research Program Task 5.2/5.3: Test Track Examination of Drivers' Collision Avoidance Behavior Using Conventional and Antilock Brakes," DOT HS 809 561, March 2003.

3. "A Comprehensive Experimental Evaluation of Test Maneuvers That May Induce On-Road, Untripped, Light Vehicle Rollover, Phase IV of NHTSA's Light Vehicle Rollover Research Program," Forkenbrock, Garrott, Heitz and O'Harra, DOT HS 809, October 2002.

4. "An Experimental Study of Automobile Driver Characteristics and Capabilities," Rice and Dell'Amico, Calspan Report No. ZS-5208-K-1, prepared for General Motors Corporation, March 1974.

5. "Automobile Controllability – Driver/Vehicle Response for Steering Control," Systems Technology Inc., DOT HS 801 407, February 1975.

6. "Performance of Driver-Vehicle System in Emergency Avoidance," Maeda, Irie, Hidaka, and Nishimura, SAE Paper 770130, 1977.

Appendix A: Response Data for the Five "Large-Steer" Subjects.

Figure A1: Subject 23 (Conventional Brakes)

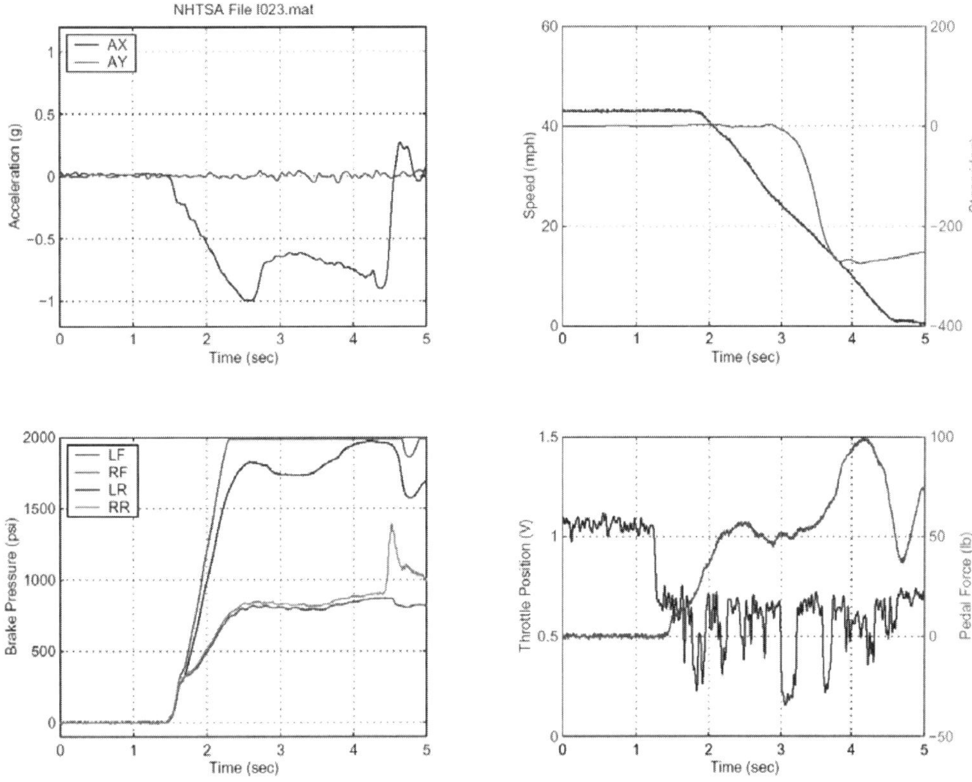

Figure A2: Subject 42 (Conventional Brakes)

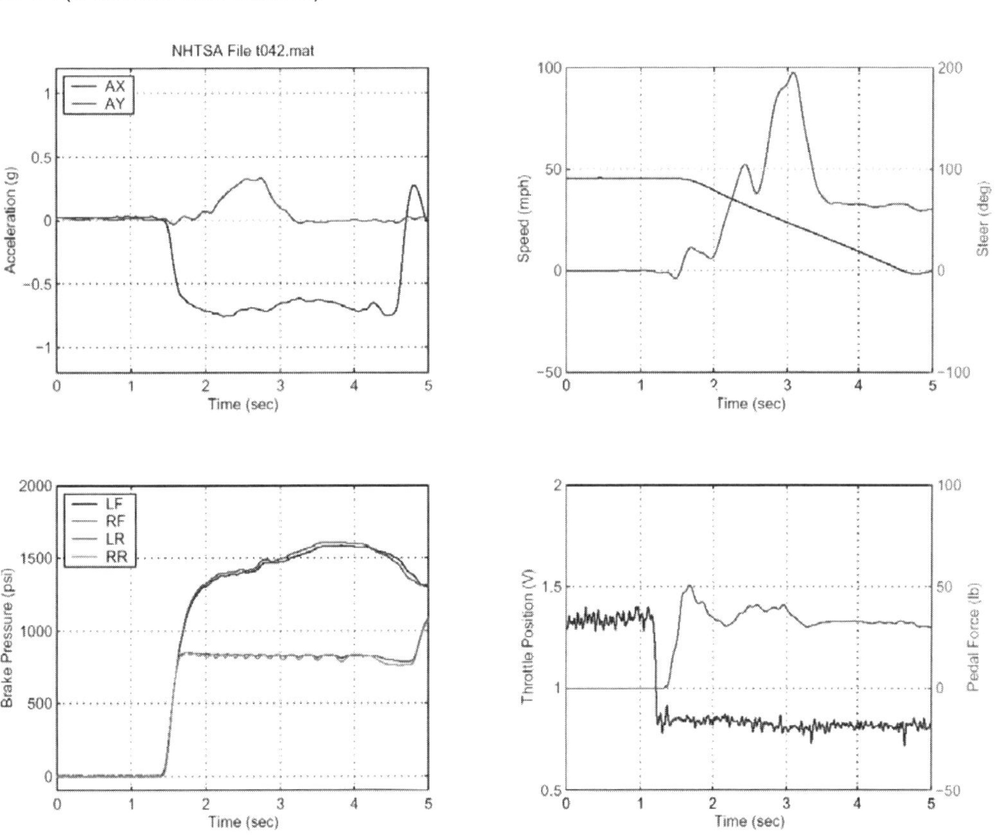

Figure A3: Subject 30 (Conventional Brakes)

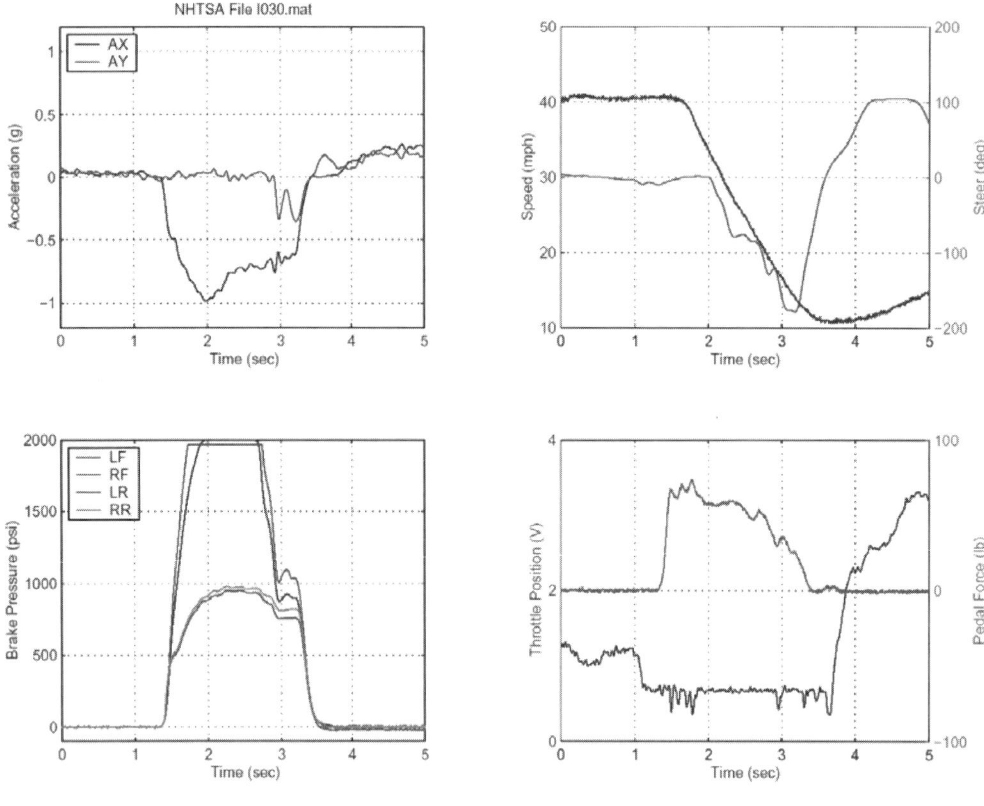

Figure A4: Subject 102 (ABS Brakes)

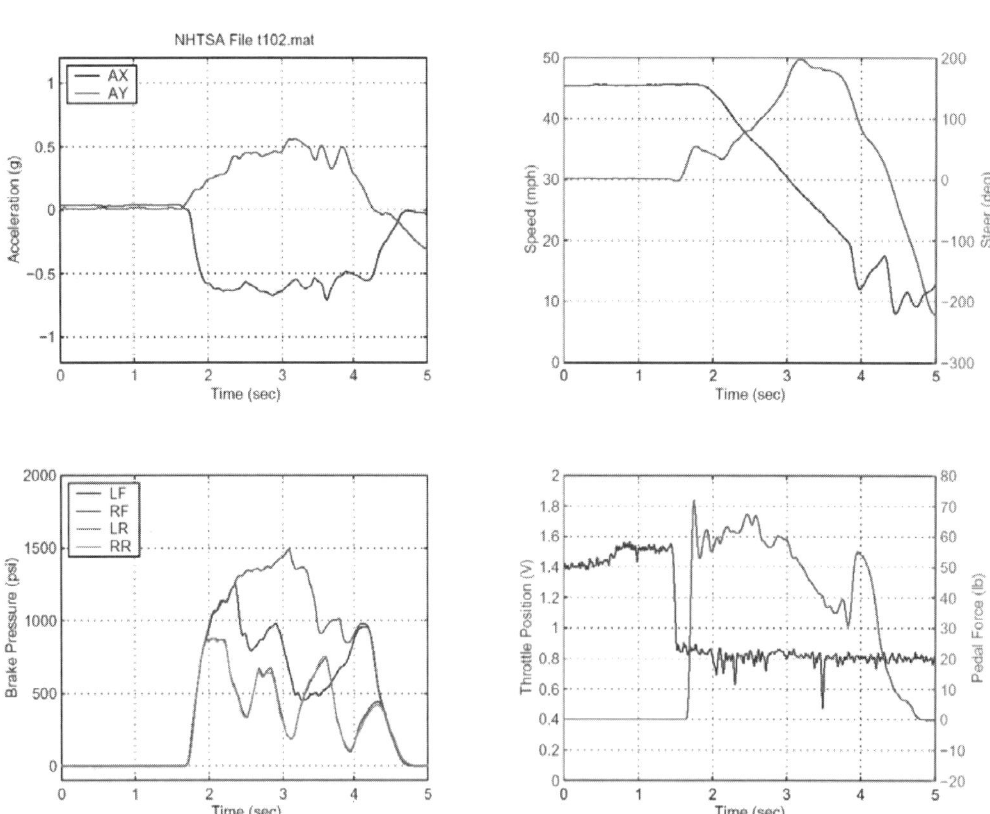

Figure A5: Subject 109 (ABS Brakes)

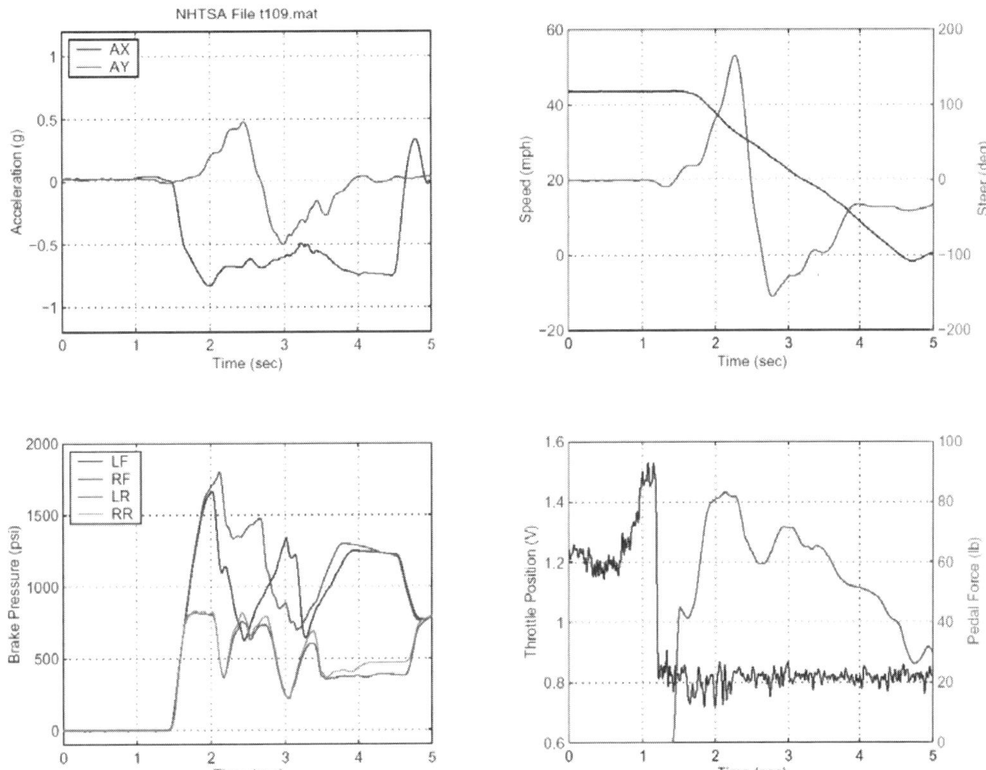

Semiconductor Solutions for Braking Systems: New Partitioning and New Safety Concepts Increase Safety and Reduce System Cost

Andrea Diermeier and Wilhard von Wendorff
Infineon Technologies

ABSTRACT

Braking systems require a high system safety level: New safety concepts need to be implemented by reducing the system complexity. Microcontrollers with special safety functions are available with implemented features, self detecting and compensating different types of faults.

Today usually two microcontrollers are used to check each other. Power devices provide microcontroller supplies and drive motors and valves; internally the functions are supervised to avoid incorrect system behaviour due to wrong voltages, currents, missing loads or other malfunctions. Bus interfaces, signal conditioning and interfaces for high voltage signals are integrated into the power system ICs.

Latest BIPOLAR-CMOS-DMOS power technologies enable the power semiconductors to integrate logic functions. This capability drives new partitioning of power ICs and microcontroller in ABS and ESP systems: Calculation tasks and microcontroller supervision functions migrate from the microcontroller into the power devices - Power devices get intelligence.

Today, the microcontroller is loaded with current measurement and the generation of duty cycles to control the target pressure for each wheel. Intelligent power devices take over the current regulation control loop.

The increasing data transfer between components and modules requires advanced interfaces as well as enhanced data coding and supervision.

New balancing between the pure CMOS microcontroller and BCD mixed-signal power ICs is established. Individual component cost may increase, but the overall system cost is reduced with a higher safety level.

INTRODUCTION

Tailoring of safety features to the specific needs of an application is economically and from the safety point of view no longer possible by microcontroller only. Hardware features are interlinked with system architecture and application software to achieve a cost effective and safe system. The goal of tomorrow's safety critical system design is no longer the individual consideration of components but the interaction between semiconductor products and vendors. New partitioning and concepts combine safety architecture, safety prepared products, software, and mechanics.

When dealing with the safety of an electronic controlled system, a clean evaluation of the expected types of faults has to be performed. This evaluation has to take into account every system layer. The system vendor together with the vehicle integrator has to quantify the severity of faults from the system point of view resulting from the identified fault scenario. Counter actions have to be implemented. This paper will show examples of different implementations and new partitioning to meet the requirements cost effective.

Latest smart power technologies enable the power semiconductors to integrate complex logic functions. This capability drives a new partitioning. Safety functions, supervision features, and direct valve control are relocated into the smart power devices. The power devices get intelligence.

TODAY´S SYSTEM

A typical ABS or ESP braking electronic ECU uses a microcontroller to calculate the extended regulation algorithm. A fail safe controller supervises this controller. The fail safe controller may be a separate microcontroller or both the controller and the fail safe controller are combined in dual core architecture. The supply for the microcontroller and sensors, signal conditioning, physical interfaces etc. are combined in a power system IC. The valves are controlled by smart low side switches. In addition pump control and safety switch for the valve branch are required.

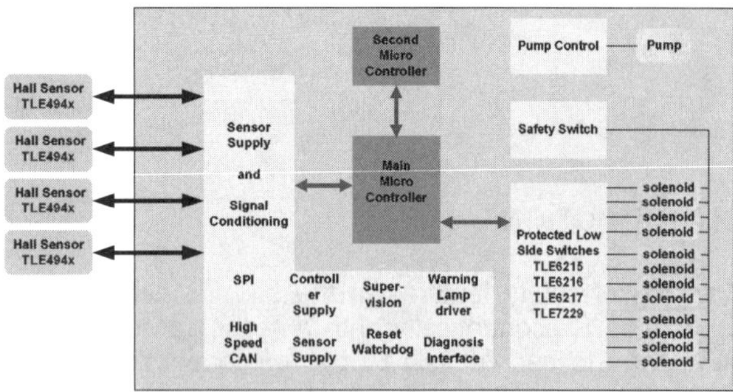

Figure 1: Today's typical partitioning of an ESP System

The most expensive parts in this system are the two microcontrollers. To ensure high safety level and meet the legal aspects, the second controller can not simply be removed. This paper shows possibilities to relocate functions from the two microcontrollers / cores to the power ICs and using new safety features within the controller. The overall system cost is reduced, as the safety controller is no longer required. By implementing new features the safety is increased. For electrical park-brake, electro hydraulic brake and brake by wire the increased safety aspect is even more important as the hydraulic and mechanic fall backs are reduced or finally eliminated.

SPT5 TECHNOLOGY CAPABILITIES

BCD technologies (Bipolar CMOS DMOS) enable current handling up to several Amperes as well as driving those switches, regulators etc. with its bipolar and analogue CMOS devices. Functions as precise references, bus-interfaces – e.g. CAN physical interface, or supervision functions are designed into one integrated circuit. Beside the analogue and DMOS functions also digital CMOS is available. Latest technologies as the SPT5 technology not only increased the DMOS capability and the analogue CMOS function, but add extended logic capabilities. The logic density was multiplied by more than ten. Not only simple logic functions could be realized for a reasonable price, but more complex functions as compound state-machines can be implemented.

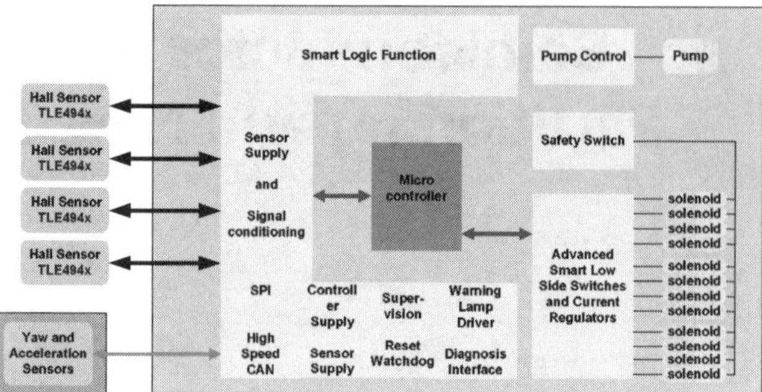

Figure 2: New Partitioning

The presupposition for the integration of safety functions into highly integrated system ICs and advanced smart switches is granted. Figure 2 shows how the partitioning may look like in the future. Detailed descriptions of functionality and features are shown below.

WHEEL SPEED SENSOR DATA

To compute the control of the hydraulic pressure at each wheel, wheel speed information is required. The wheel information is detected by hall sensors and communicated by a current interface. Simple hall sensors send the pure wheel tooth modulated currents. In principle the information can easily be detected with comparators in the system IC. However many disturbances are coupled into the cables between the sensors, located at each wheel, and the detection circuit in the system IC. So the signals require filtering to eliminate disturbances.

Latest advanced hall sensors as the TLE4942 not only give the wheel speed information, but supply also air gap information, direction information and detect the standing wheel. The data is transmitted by PWM modulating the wheel speed signal bit length. Other sensors use a 3 level current interface and add Manchester coded information. In addition to the detection and sensor specific filtering the decoding of the data can be integrated easily into the mixed signal system IC.

The wheel speed information needs to be provided to the microcontroller in real time and therefore one output for each wheel is necessary. The extended sensor information is transmitted by a serial data interface.

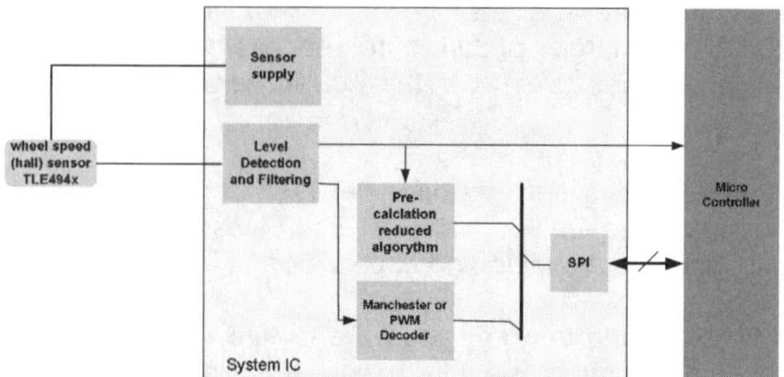

Figure 3: Signal Conditioning and wheel speed data processing in the power system IC

WHELL SPEED DATA PROCESSING IN THE SMART POWER DEVICE

The wheel speed raw data have to be computed in the Microcontroller due to the extended algorithm required. If implementing functional redundancy a simplified algorithm is processed by the second microcontroller. However if the same microcontroller type is used, common mode errors may be present. Processing tasks can be performed with the power device. This enables a complete independent state machine based implementation of the simplified calculation (spatial and functional redundancy). The microcontroller performs the main calculations and reads the state machine result by SPI interface to judge its own calculations.

CAN COMMUNICATION

Dependent on the system complexity additional information is required by the ECU. For example ESP systems require acceleration-, gear-, and yaw-rate information as well as steering angle information. Those data are detected locally at the electrical power steering control unit and central gear / yaw rate modules. The data is shared via a private high speed CAN bus.

As this data is an essential input to the stability calculation algorithm, it is necessary to look into the safety of the CAN communication. Safe data communication especially requires deterministic data communication and data consistency. Any data errors have to be detected by the microcontroller and errors of the physical layer have to be detected in the transceiver.

To support temporal redundant software architecture, a deterministic behaviour can be achieved for CAN nodes by a time-triggered CAN extension TTCAN. The time-triggered CAN protocol is fully compliant to the existing CAN protocol and is according to the confirmed standardization proposal for the ISO 11898-4. The time-triggered functionality can be added as higher layer extension (session layer) to the CAN protocol enabling temporal redundant software. The new features allow a deterministic behaviour of a CAN network and the synchronization of networks.

Special redundancies are built into the AUDO (automotive unified processor) serial CAN. The MultiCAN node contains of 4 independent CAN nodes, representing the communication interfaces for spatial redundancy.

The 4 Full-CAN MultiCAN nodes operate independently or exchange data and remote frames via a gateway function. All CAN nodes share /a common set of message objects, allocating each message object to the respective CAN node. Besides serving as a storage container for incoming and outgoing frames, message objects may be combined to build gateways between the CAN nodes or to setup a FIFO buffer (e.g. to allow temporal redundant message communication).

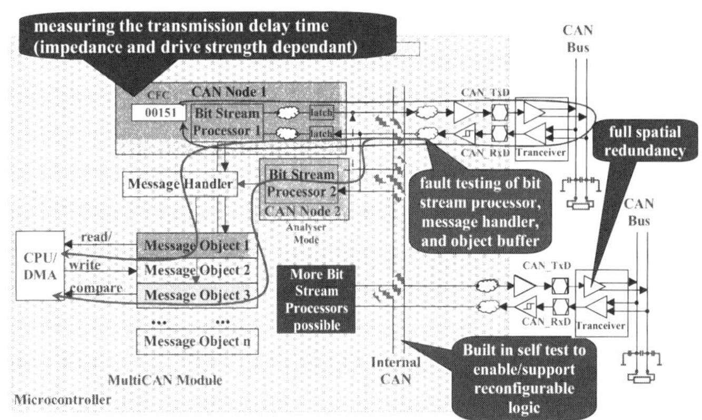

Figure 4: MultiCAN Implementation in AUDO 32 bit microcontrollers.

By storing this data in parallel to the master CAN node, software is enabled to detect receipt and transmit faults. To react on discovered faults and implement fault tolerance, all CAN nodes can communicate between each other by using an internal CAN network instead of an external network. Therefore erroneous CAN nodes are identified by checking all nodes with the spatial redundant nodes. Furthermore the CPU can check on this internal bus and also identify erroneous nodes. These erroneous nodes can than be disabled and replaced by fault-free nodes.

CAN PHYSICAL INTRFACE

Even when data processing and protocol handling is correct, still the physical layer and the connection between controller and the transceiver require fault checking. This is mainly implemented within the transceiver. Of cause device faults or physical bus faults as short to battery or GND of the CANH or CANL line are detected to protect the component itself; however fault-information is given by status pins or SPI bits to the microcontroller enabling respective reactions.
Latest CAN physical interfaces according to the ISO11898-5 proposal also include additional features to crosscheck data consistencies. The High Speed CAN interface detects a permanent dominant state at the bus. If the transmit data from the microcontroller are low for more than typically 0,6 ms - a time not possible by the CAN protocol - indicating that something is wrong with the CAN module, the driver section is automatically switched off.

SUPERVISION AND WATCHDOG FUNCTIONALITY

A major task of the fail safe controller is the supervision of the main controller. Today simple watchdog functionality is implemented in system ICs.

To ensure proper function of microcontroller watchdog supervision is a common function of power system ICs.

Within a special time-window a watchdog function needs to be retriggered. The retriggering is usually either an edge at a dedicated input or a SPI command needs to be written. If this retriggering is not recognized, the system switches into a failsafe state; the microcontroller is reset and will be reconfigured. To increase the safety level of this function, it is extended further.

Beside the microcontroller usually at least 2 smart power ICs for example the system IC and intelligent low side switches or current regulators are available.

The watchdog functionality can be implemented in any of these components. The trigger information can be sent between any of these devices. So a circular supervision is possible.

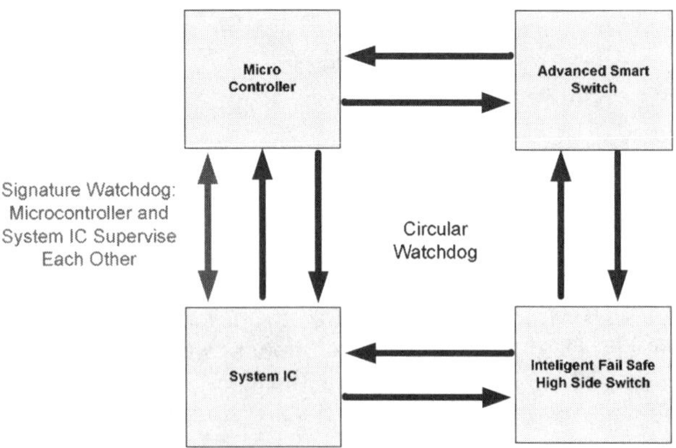

Figure 5: Circular watchdog

SMART WATCHDOG

To grant a properly working software and hardware not only a simple timing has to be supervised. The different "members" in the watchdog chain have to perform processing tasks to check both software and hardware. One member of the chain generates pseudo random request data and transmits the data via SPI or Microsecond bus to the other members. Any partner must calculate an answer using a respective algorithm and answers within a defined timely window. If the response is as expected in the time and data domain, the watchdog is triggered. As this method can be initiated by any member in the chain, all devices can supervise each other for high data retention.

Watchdog functions are also implemented within the microcontroller.

SIGNATURE WATCHDOG (TESTED MODULE TECHNIQUE)

The AUDO products extend the standard window watchdog timer functionality by a so called signature feature. This enhancement allows the implementation of a simple "Tested Core Technique". The core executes a deterministic test programme, testing all the critical paths within the CPU achieving a given test coverage. This is a feature widely implemented. The critical issue is the compare operation of the final and intermediate results of this test routine to the fault free values. This is often executed by the core under test and therefore opens risks for bad test coverage scenarios.

The Watchdog Timer (WDT) of the AUDO architecture therefore provides a highly reliable and secure way to detect and recover from software or hardware failure. The WDT helps to abort an accidental malfunction of the AUDO products in a user-specified time period. When enabled, the WDT will reset the system if not serviced within a user programmable time interval (standard window watchdog functionality). Hence, routine service of the WDT confirms that the system is functioning properly.

Because servicing the Watchdog are critical functions that must be prevented in case of a system malfunction, a sophisticated scheme is implemented which requires a password and guard bits during accesses to the WDT control register (signature enhancement). Any write access that does not deliver the correct password or the correct value for the guard bits is regarded as a malfunction of the system, and a Watchdog reset is triggered.

The basic idea is not to provide this password to the core, but calculate the password by the respective test routine. Therefore the core can only retrigger the watchdog, if the test routine has been correctly executed and generated the correct result retriggering the signature watchdog. In addition, even after a valid access has been performed the Watchdog imposes a time-limit for this access window. If the WDT has not been properly set again before this limit expires, the system is assumed to malfunction, and a Watchdog reset is triggered.

A further enhancement of WDT is its reset pre-warning operation ("soft fault technique"). Instead of immediately resetting the device on the detection of an error, as known from standard Watchdogs, the WDT first issues a non-maskable interrupt (NMI) to the CPU before finally resetting the device at a specified time period later. This gives the CPU a chance to detect evolutionary increasing of the watchdog service time and implement respective counteraction, e.g. switching to limb home mode.

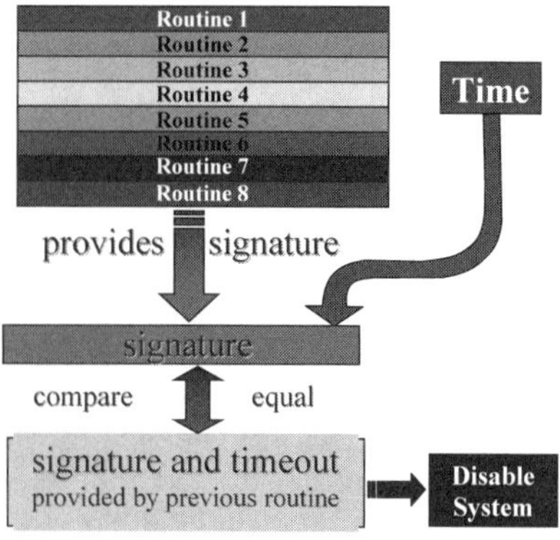

Figure 6: Functionality of Signature watchdog

The signature watchdog can also be used to monitor the correct flow of software (functional redundancy). The password is an 8-bit field that can be set by software to any arbitrary value during a Modify Access. Settings of this field have no effect on the operation of the WDT, other than access and retrigger condition. The purpose of this field is to support further enhancements to the password protection scheme. For the following description, it is assumed that software does at least not fully compute the value for the Password Access, but uses a predefined constant, embedded in the instruction stream, for the password (this is at least necessary for the user-definable password field). For example, software can modify this field each time it executes a Watchdog service sequence. The next service sequence needs to take this new value into account for its password access. And it again changes the value during its Modify Access. Up to 256 different password values can be used. In this way, each service sequence is unique. If a malfunction occurs, that, for instance, would result in the omission of one or more of these service sequences, the next service sequence would most probably not write the correct password. This service sequence would rely on the password value programmed during the normally preceding service sequence. However, if this one was skipped, the password value required by the contents of the Watchdog registers is the one programmed at the last service sequence executed before the malfunction had occurred. A watchdog error condition would be detected in this case. In the same manner, the WDT would detect the malfunction if a service sequence would be executed twice due to a falsely performed jump.

Coming back to the partitioning of a braking ECU: The main task of an ABS or ESP system is the manipulation of the pressure in the hydraulic system. This is done by changing the current in the related valves. The following chapters will look into the realization of the valve control.

FAIL SAFE HIGH SIDE CONTROL

Usually the valves are controlled in low side configuration. For safety reason a main valve switch is required. This high side switch will disable all valves in case of any malfunction of the controller or the system. For example the valves need to be off, when the microcontroller software is in a dead look. This switch is realized as semiconductor switch. One possibility is a high side driver with external MOSFET. Instead of a simple FET a full protected high side switch could be used. The Infineon PROFET® family ICs are protected against malfunctions as overload, over temperature and are equipped with under-voltage shut down and over-voltage clamping. A control output feeds back the status of the switch and a sense current proportional to the load current.

This safety switch can be directly controlled from the microcontroller or to increase the safety by the smart power IC. In case its extended system supervision

detects any malfunction the PROFET® can be switched off.

The integrated current feedback can be used in the control loop to judge additionally the current consistency in the valve branch.

By combining the advantages of latest semiconductor technology with the automotive proven chip on chip packing technology, new advanced parts can be defined: In the base chip the high current switching function is realized with the over-temperature protection. In a top chip advanced control functions and supervision functions are realized. Safety functions and intelligence can be added as shown above.

Figure 7: Chip on chip technology

VALVE CONTROL

By means of the valves the pressure in the hydraulic system of each wheel can be increased or decreased in order to regulate the stability of the car under severe environmental condition. Dependent on the class of system ABS, ESP or traction control between 8 and 12 valves are controlled. Some valves simply open, or close the hydraulic flux while others regulated the hydraulic pressure with high accuracy in the related branch or wheel.

Valves are low side controlled. Smart low side switches supervise themselves. Malfunctions as open, shorts, over temperature, over current or any signal loss are monitored. Dependent on the severances of the detected fault the switch is disabled and / or the controlling microcontroller is informed of the malfunction.

For a precise regulation of the pressure at the wheels, the current through the related valves has to be controlled very precisely. PWM modulated switching defines the mean current through the valve. Usually the low side switches are controlled by the microcontroller with PWM digital signals. As the valve current depends on many factors as the supply voltage or the inductance, the actual current flow has to be measured. AD channels are occupied; the software is loaded with the calculation of the required duty cycle to close the loop. With the extended logic performance of the smart power technology SPT5 the control function can be implemented within the advanced smart switch. Target current and timing calculations can be removed from the microcontroller. Current is measured and the calculation of required duty cycle to reach the target current value is done within the intelligent power current regulator. In addition – as described above - advanced watchdog functions from the system watchdog concept could be implemented.

However low side switches can be integrated in the system IC. Different advantages and disadvantages are asserted to this partitioning:

+ Less packages
+ Reduced PCB space
- Less flexibility for various system complexities
- increased power dissipation requires advanced thermal concept
- no additional watchdog member

Which and how many low side switches are implemented into the system IC depends on the overall system and safety concept. All the points need to be considered and balanced to find the best solution.

SAFE COMMUNICATION BETWEEN COMPONENTS

Between the microcontroller and the smart system IC or smart switches data is exchanged. As safety functions are relocated from the microcontroller to the smart power devices, more and more data need to be exchanged. Data is very sensitive. Usually the serial synchronous peripheral interface SPI is used for bidirectional point to point communication. For unidirectional high speed data transfer the microsecond bus is also available to increase the transfer rate.

Independent of SPI or the microsecond bus usage, new methodologies are implemented as part of the safety concept: Data is accepted only if the correct number of clock edges is counted within the chip select (CS) period as well as a defined timeframe. With the new partitioning the transmitted data consist of address identification, the data packages and CRC check sums. The CRC is chosen according to the bit stream and the required hamming distance.

At the receiver all data are stored until the whole data stream or all data frames are received without violating the CS and clock conditions and the CRC check sum is verified. Afterwards the data is interpreted and computed (firewall).

Requested data or status information are extended with address information and CRC checksums and attached to the next data transfer.

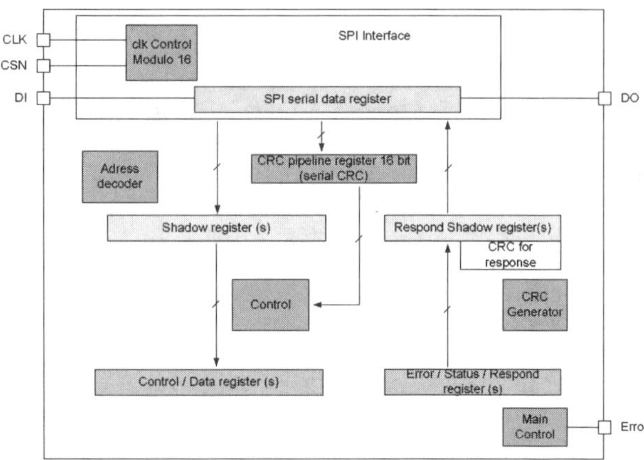

Figure 8: Secure SPI interface with CRC coding and decoding

CONCLUSION

Today's common braking systems with ABS and ESP functionality already require a high rate of safety. However, with new systems as electro-hydraulic brake, electric brake with fewer hydraulic or mechanic backup rate and finally brake by wire require even more extended safety considerations.

Many different concepts ensure the high safety within a braking system. There is no one feature that by its own can meet all the requirements of the future applications. However, by combining different functions in a chipset of microcontrollers with extended safety features, smart power system IC and advanced smart power switches / current regulators combine the individual advantages to a cost efficient system. Presupposition for an optimized solution is the availability of microcontrollers, sensors and a variety of semiconductor technologies and packaging options. The combination in a system and the partitioning has to be defined in close cooperation between the semiconductor vendor and the braking system supplier.

CONTACT

Andrea Diermeier
Infineon Technologies AG
Senior Manager Application and Product Engineering
Power System ICs
Mail Address:
P.O.Box 800949, D-81609 Munich, Germany
Office Address:
Thomasiusplatz, D-81541 Munich, Germany
Email: andrea.diermeier@infineon.com
Telephone: +49 (+89) 234-21362
Facsimile: +49 (+89) 234-9550412

Visit www.infineon.com/auto

Electro Hydraulic Braking System Modelling and Simulation

Luigi Petruccelli
Fiat Auto

Mauro Velardocchia and Aldo Sorniotti
Politecnico di Torino

ABSTRACT

The first step toward a braking system 'by wire' is Electro-Hydraulic Braking System (EHB). The paper describes a method to evaluate through virtual experimentation the actual improvement in vehicle behaviour, from the point of view of both handling and comfort, including also pedal feeling, due to EHB. The first step consisted in modelling the hydraulic unit, comprehensive of sensors. Then it was conceived a control logic devoted to medium-low intensity braking manoeuvres, without ABS intervention, to determine an optimal braking force distribution and pedal feeling depending on the manoeuvre. A failsafe strategy, complete of on board diagnosis, to prevent dangerous system behaviour in the eventuality of a component failure was carried out and tested. Finally, EHB wheel pressure sensors were used to improve both ABS performance, increasing the adherence estimation, and Vehicle Dynamics Control (VDC) performance, through a more precise actuation. New ABS/VDC control strategies integrated with EHB were conceived and some of the main results obtained are presented.

INTRODUCTION

The trend towards the development of chassis control systems to increase vehicle safety and performance involves significantly brake systems. Biggest improvement were when Antilock Braking Systems were put on the market in the second half of the 70s, later Anti Schlupf Regulierung (ASR) and especially Vehicle Dynamics Control (VDC) applied braking system to active vehicle safety increasing. There has always been a physically direct correlation between the effort on the brake pedal (input) and calliper braking torques (output). Traditionally, braking systems have a mechanical link between the pedal and the calipers, which is good from the point of vehicle safety, but not always is good from the point of view of the pedal feeling or braking system performance. For example, it is quite usual to have a disturbing change in pedal travel when pads temperatures change after some braking manoeuvres, due to pad stiffness variation. However, if the driver can understand that the braking system is in a critical

condition, the vehicle safety should be increased. Generally, the biggest freedom in a brake manoeuvre can be obtained separating the pedal from the callipers actuation. Such behaviour can be achieved through brake-by-wire systems, which have a lot of problems from the point of view of reliability. A simpler way to determine the same result consists in Electro-Hydraulic Braking (EHB) system, maintaining a failsafe hydraulic connection between brake pedal and the callipers if any fault occurs. The present paper main aim is to verify the actual effectiveness of EHB through simulations, by using both commercial, like Matlab/Simulink and AMESim, and especially conceived mathematical models.

EHB HYDRAULIC MODEL

A realistic EHB hydraulic model was carried out using a commercial software (AMESim). The control logic and the vehicle model were implemented in Matlab/Simulink. Figure 1 represents an EHB general layout. The scheme add to electromechanical components and sensors typical of a VDC, a pressure sensor for each calliper.

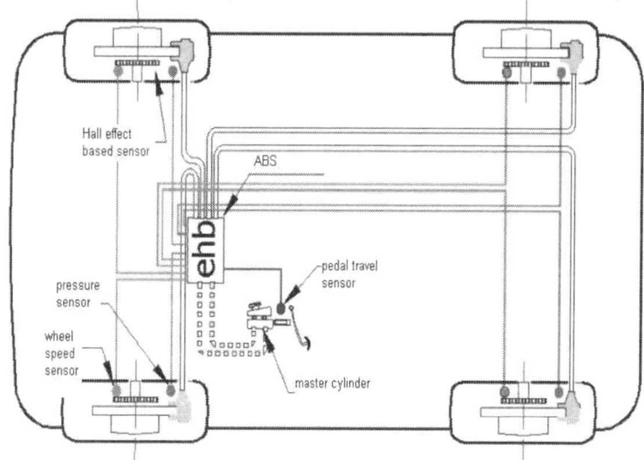

Figure 1 – Electro Hydraulic Brake (EHB) general layout

The pedal effort simulator modelled, consists of two independent hydraulic chambers, similar to those of a tandem master cylinder. The difference is due to the fact that during normal braking actions there is no flow rate to the calipers, since the failsafe valves, connecting the simulator to the calipers, are normally closed. The pedal effort simulator is characterized by an elastic system whose deformation permits pedal travel. All the remaining part of the EHB hydraulic unit consists of eight valves (two for each wheel), four failsafe valves, a pump and an accumulator. All the valves are normally closed if there is no electric input on them.

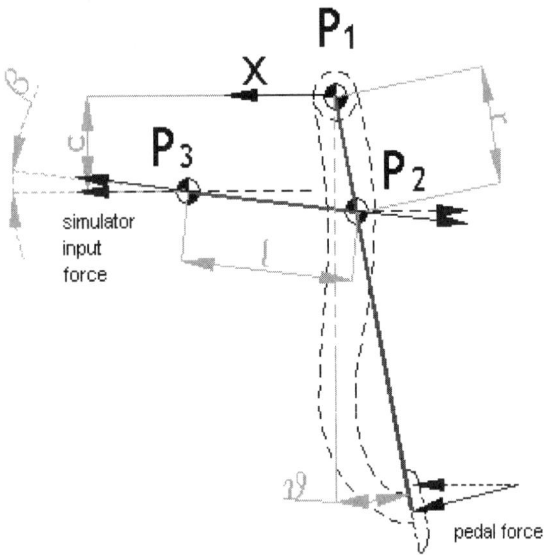

Figure 2 - Brake pedal layout

Figure 2 is a scheme of the brake pedal. The model includes the lever ratio variation depending on the travel (Figure 3).

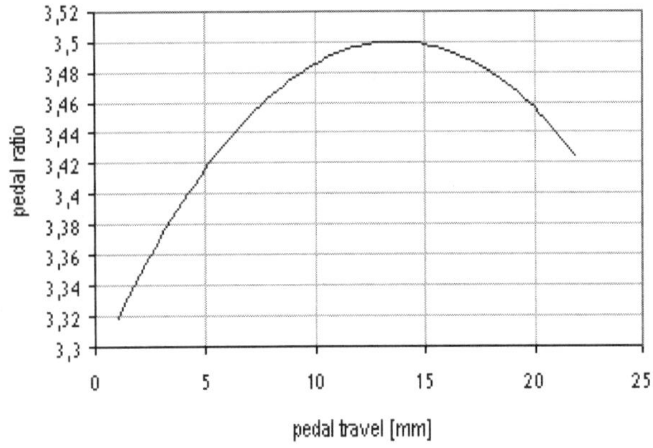

Figure 3 - Pedal ratio depending on pedal travel

The pedal effort simulator stiffness was modeled. As a starting basis, it was chosen the displacement - force characteristic (Figure 4) of an actual hydraulic braking system, experimentally measured at Politecnico di Torino braking systems test bench. It was decided to reproduce a similar characteristic, a little stiffer for high values of displacement, according to subjective evaluations [2]. It was necessary to design the pedal simulator as a tandem master cylinder with closed holes (corresponding to the failsafe valves) and a third piston, actuating the force feedback for the driver. A first

solution for pedal travel simulator is presented in [4]: the elastic element is a rubber disc having well defined mechanical properties. The solution here presented consisted in putting two conical springs one after the other to have the desired behaviour (Figure 5). The system conceived derived from a common production tandem master cylinder.

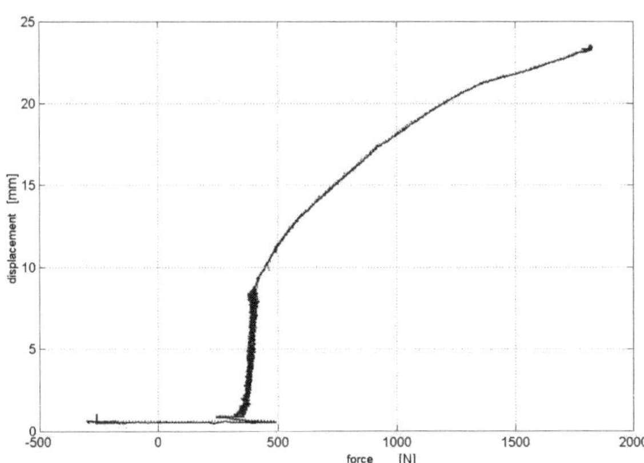

Figure 4 - Force-displacement characteristic of a standard braking system measured on Politecnico di Torino braking systems test bench

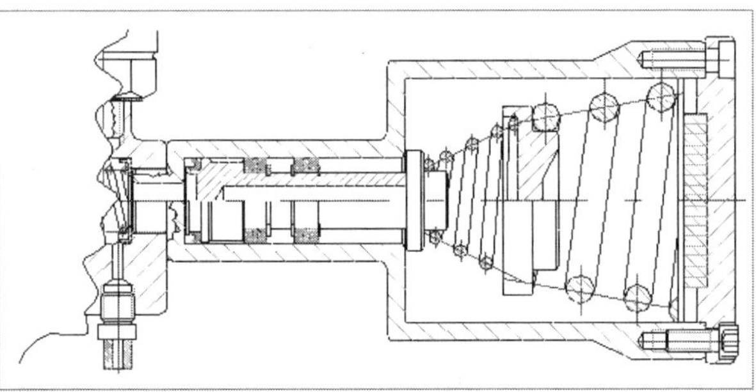

Figure 5 - Pedal simulator conical springs

Figure 6 is the sketch of the pedal effort simulator first approximation model: it is possible to see the two hydraulic circuits (active only in case of fault) and spring 7, with a force-displacement characteristic corresponding to that of the two springs of Figure 5. In EHB, the pressure source is a pump (similar to ABS pump) feeding an accumulator. Usually the accumulator permits the pump to work almost independently on the braking system requested performance. The control logic, on the base of an accumulator pressure sensor measurement, evaluates when activate or deactivate the pump, typically at well defined pressure levels (for example, activation at 160 bars and deactivation at 200 bars), without an excessive power request. Figure 7 shows 'com' and 'sca' valves: the first put in communication each calliper with the accumulator, the second each calliper with the tank. The control strategy decides when opening each of those valves, dependent on driver action on the brake pedal.

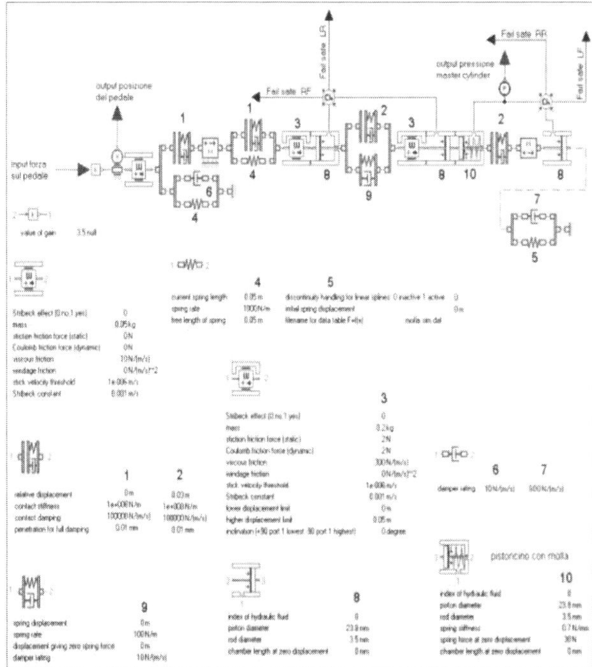

Figure 6 - Pedal effort simulator AMESim model

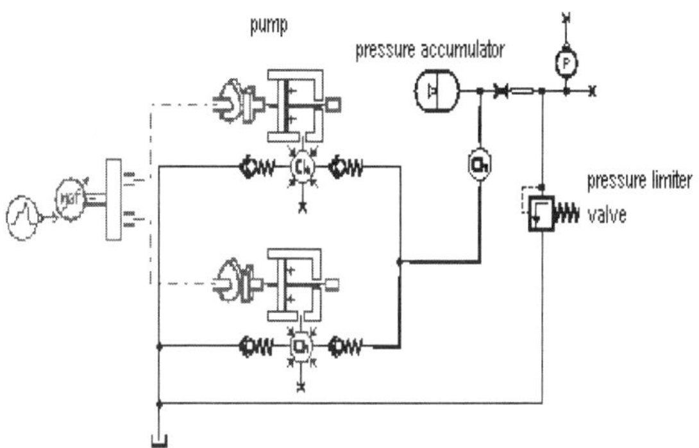

Figure 6 - Pump model example

to the Electronic Control Unit (ECU), where is a look-up-table having pedal travel as input and pressure reference for each wheel as output. Then a PID controller, according to each calliper pressure level with that due to the pedal simulator travel sensor comparison, transmits the proper signals to the valves.

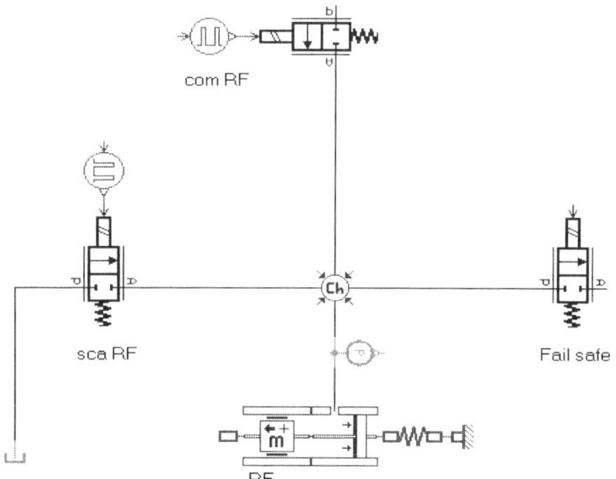

Figure 7 - Quarter EHB simple model

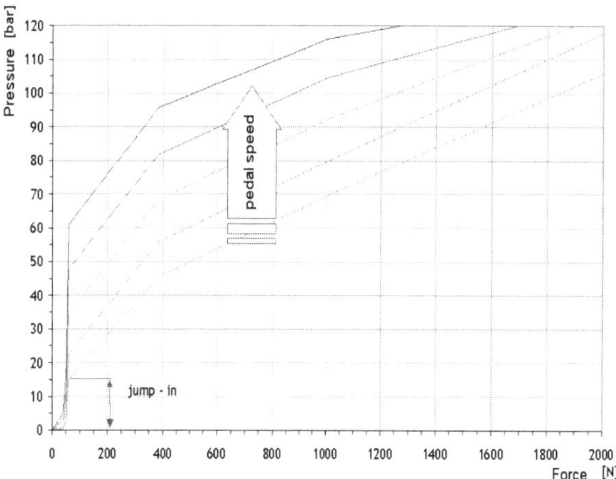

Figure 8 - Front callipers pressure versus input rod force for different values of pedal travel rate

If the control unit recognizes a fault, it deactivates the valves 'com' and 'sca' and opens the fail safe valves. The 'com' and 'sca' valves work differently from the fail safe valves: the first are piloted through Pulse Width Modulation (PWM) (or have a particular current control, [2]), the second are on/off valves with no modulation. It was supposed that valves have the characteristics typical of those of a common ABS hydraulic unit. When the failsafe valves get open, it is important the correct design of the pedal travel simulator, which has to guarantee the same performance of a passive braking system from the point of view of pedal travel, pedal effort and wheel pressures, with a failure at the booster.

EHB has a pressure sensor for each wheel, then the number of the additional sensors requested by such a system in comparison to a traditional one is equal to seven (four wheel pressures, pedal travel, simulator/master cylinder pressure, accumulator pressure).

In the following is described how the EHB modelled works during a standard braking action. When the driver brakes, the pedal travel sensor sends such information

EHB control logic can be naturally much more complex than that shortly described formerly. For example, subjective analyses ([5], [6]) on drivers demonstrated that it is better to have a bigger jump-in in panic brake manoeuvres, characterized by high values of pedal speed, and the opposite during parking manoeuvres. As a consequence, it is possible to make EHB work with different pressure references according to pedal speed (Figure 8) and also, for example, vehicle speed.

Figures 9 and 10 are a comparison of traditional braking system (results were experimentally obtained through Politecnico di Torino braking systems test bench) and EHB systems (results were obtained through simulation) during respectively a semistationary and a very fast brake apply. Figure 10 shows that the response time is much smaller for EHB. For example, at point 'a' the pressure in the EHB system is at least 20 bar higher than that of the passive system.

143

By simulation, it was tried to develop also a 'soft stop' control strategy, in order to reduce the vehicle acceleration felt by the vehicle passengers at the end of the brake manoeuvre. It consisted in a proper wheel pressures reduction during the vehicle stop. The strategy obviously uses also wheel speed sensors measurements.

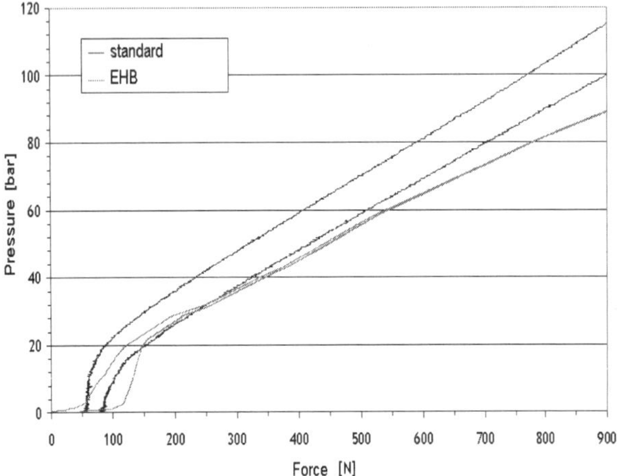

Figure 9 - Front callipers pressure versus input rod force in a semistationary test (blue - standard system, red - EHB)

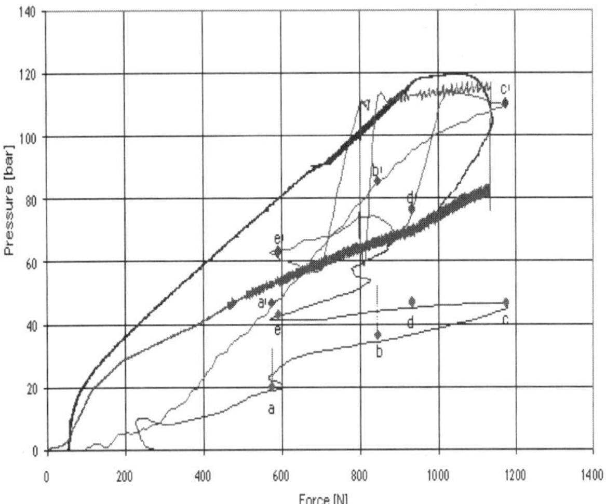

Figure 10 - Front callipers pressure versus input rod force in a dynamic test (blue - standard system, red - EHB)

To resume the main simulation activity results, it is possible to highlight the following EHB main advantages over traditional systems:

- calliper pressure level and pedal travel are always the same, due to PID controller, although eventual components characteristic variation (brake pad stiffness, disc consumption, etc.);

- the pressure gradients in a panic bake manoeuvre are bigger than those obtainable with a conventional system, since the accumulator can be set at a high pressure level (160-200 bar), so determining a very fast pressure increase at the beginning of the manoeuvre;

- EHB permits a different brake pressure rise depending on the brake manoeuvre; for example, as a result of an analysis of drivers' subjective impressions ([6]), it was stated that it is better to have a bigger jump-in for high pedal travel rates;

- EHB has a very big flexibility from the point of view of pressure distribution at the four wheels, very useful for a good brake distribution also in extreme conditions, like μ-split conditions;

- there is no dependence of braking torques on engine vacuum level.

EHB FAILSAFE STRATEGY DEFINITION

The second step was the failsafe strategy definition. It is particularly important to define a failsafe strategy for those electronic elements, like the sensors, which are necessary for EHB but absent in a standard braking system.

The activity was articulated in the following steps:

- recovery definition in case of a sensor failure

- definition of a strategy to understand which failure has occurred. Such provision is useful not only to minimize the inconveniences due to the fault, but also to have a diagnosis of the fault itself

The first task was carried on by producing a large number of simulations of brake manoeuvres. EHB hydraulic model was interfaced with a Simulink vehicle model [8], with different kinds of failures and different kinds of recovery strategies, by considering EHB firstly as a stand alone system and secondly integrated with vehicle dynamics. For example, in brake manoeuvres in straight ahead travel, the main tasks were to reduce at the maximum level stopping distance and secondly not to have too big yaw acceleration values [7] due to asymmetrical faults or recovery strategies. The results obtained in this part of the activity are summarized in the following lines:

- pedal travel sensor is the fundamental device to make the system work in the proper way: if it has a fault, it is necessary to open all the fail safe valves;

- wheel pressure sensors are important but their fault determines the opening of the failsafe valve only for the wheel affected by the fault and not for all the other wheels. This strategy reduces stopping distances but can cause some problems from the point of view of body yaw rate, especially if the failure regards the front axle;

- pedal travel simulator pressure sensor failure does not determine important effects from the point of view of vehicle performance. From a theoretical point of view, this sensor is not useful. Instead, this sensor is fundamental from the point of view of the failsafe strategy: it would be possible with a fault at this sensor to make the braking system work in a full way, but there would not be the chance of determining the occurrence of other new faults. As a consequence, it was necessary to put in fault a wheel (it was decided a rear wheel) to maintain a sufficient diagnosis capability in correspondence of a failure of this sensor;

- it was possible to distinguish, only by having the signals from these seven sensors, accumulator pressure sensors faults and pump faults. When pressure sensor has a failure, the system can work more or less in the same way: of course there is a change in the pump actuation strategy. When the pump itself has a failure, the only chance is to open all the failsafe valves and make the system work as a passive one.

EHB DIAGNOSTIC STRATEGY EXAMPLES

Figure 11 is an example of a portion of the diagnostic strategy flow chart, comprehensive also of multiple faults. The basic principle of this failsafe strategy is the comparison between different signals and the evaluation of their compatibility [3].

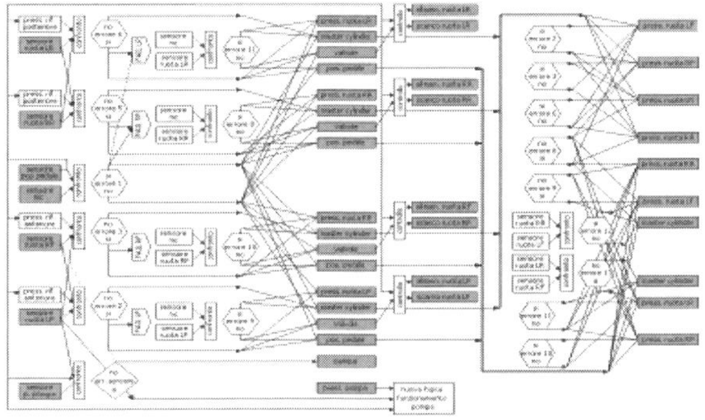

Figure 11 - Scheme of the fail safe control strategy

As an example, a brief description of a particular case will be done in the following, in order to describe the diagnostic algorithm basis. Let us suppose that during a brake maneuver there is a fault at the pressure sensor of the right front wheel. Firstly the fail safe algorithm compares the wheel pressure levels and reference pressure levels. Such 'step1' it is performed continously, step by step. In the specific case, there is a consistent difference between p_{RF} and the reference pressure, whereas everything will work in the proper way for the other wheels.The control logic can now close 'com' and 'sca' valves and open the failsafe valve of the right front wheel (step 2). It is not yet possible to understand if the fail is due to the wheel pressure sensor or the pedal travel sensor, even if the first chance is much more probable since the other wheels seem to work in the proper way. The second comparison (comparison 2) is between pedal effort simulator pressure sensor value and the simulator estimated pressure level that can be computed through the pedal travel sensor. Last comparison (comparison 3) is between pedal simulator pressure level and wheel pressure level (now the fail safe valve is open). If comparison 2 does not signal problems whereas comparison 3 does, the diagnostic algorithm states that the fault regards the right front wheel pressure sensor, otherwise the fault is attributed to the pedal travel sensor. This strategy can appear too complex, but all the other tempted strategies gave origin to consistent mistakes, especially in the case of multiple failures. The most difficult task was to make this complex diagnostic strategy efficient in all the possible maneuvers: it had to promptly recognize the failures but at the same time it had not to signal non-existing faults. By simulation we tested the control algorithm also in unusual conditions, for example with pads stiffnesses different from the standard and asymmetrical on the wheels of the same axle, as it can happen by changing the pads or when exist very high temperature levels at the callipers. Through simulation extensive campaigns of fault injection were performed considering the various EHB components in a number of manoeuvres. Figure 12 plots, as example, the times required to recognize a pedal travel sensor fault occurred in different situations (1 to 14).

ABS AND VDC IMPROVEMENT DUE TO EHB

The additional sensors introduced by EHB, especially those devoted to wheel pressure measurement, can be used to increase ABS and VDC performance, so improving the vehicle safety. As far as VDC, the control strategy can verify the desired pressure, so obtaining the desired targets more faster and better than without wheel pressure sensors. For ABS, the problem is more complex: the target was the through an improved modulation, due to the presence of EHB pressure sensors. ABS standard systems usually present pressure oscillations with an amplitude of about 20 bar; a reduction of these oscillations e.g. of 10 bar, is positive for braking smoothness and stopping distances reduction.

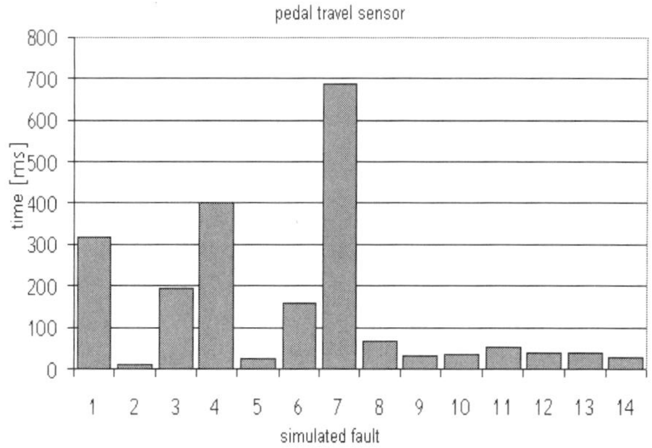

Figure 12 – Time required to recognize pedal travel sensor fault in different situations

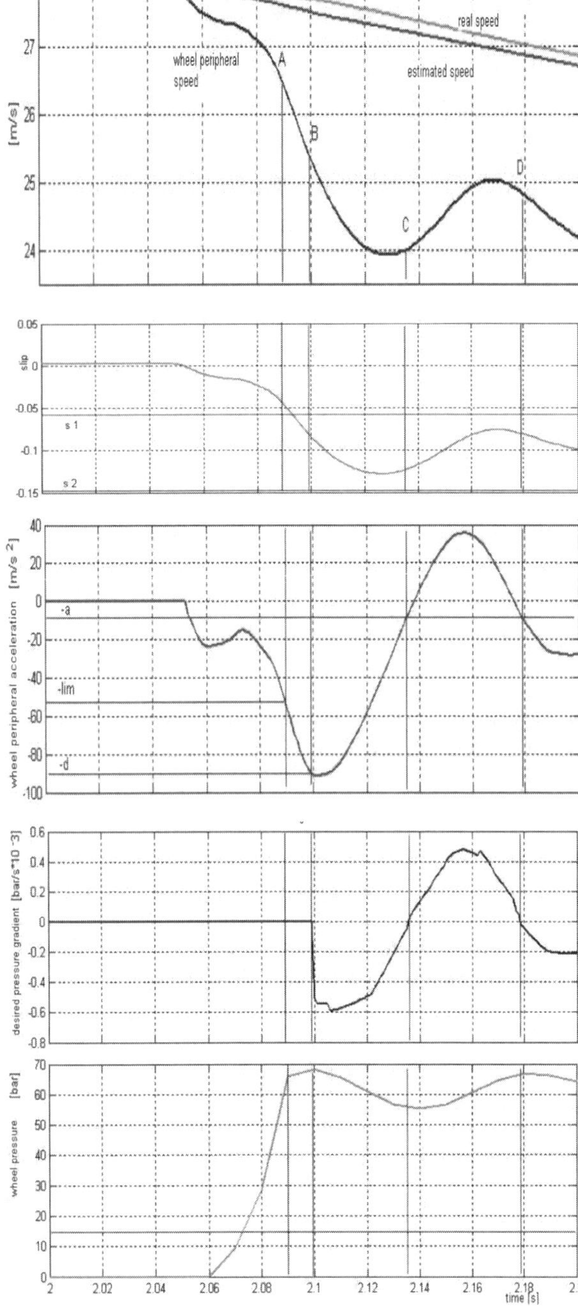

Figure 13 - ABS intervention in a panic brake manoeuvre

This result was obtained through a control logic which introduces a variable pressure gradients. It is possible to use wheel pressure sensors also to improve ABS adherence estimation, since wheels tend to lock at different pressures according to adherence conditions. A quite consistent reduction of stopping distances was obtained in comparison with a traditional ABS system. To perform the comparison, it was used on the same vehicle model both an ABS tier's black box and an ABS control strategy conceived by the authors.

Figure 13 presents the actual and estimated vehicle speed, peripheral wheel speed, tyre longitudinal speed, tyre peripheral acceleration and wheel pressure versus time, during ABS first control cycle. Until point 'A', ABS does not work. When the value of wheel deceleration '-lim' is overreached, pressure increase rate is limited. In the meanwhile, wheel acceleration continues increasing, until it reaches the '-d' value, when ABS starts reducing calliper pressure. There is also a tyre longitudinal slip control, getting active in the case the slip overreaches the value s_2, even if wheel acceleration is smaller than '-d'. Pressure gradients are proportional to wheel acceleration. Calliper pressure decays until wheel acceleration reaches a '-a' value: then pressure can increase again. '-a' depends on estimated adherence and wheel inertial properties, depending on the gear used. 's_1' and 's_2' limit too big longitudinal slip increase/decay; they are function of the estimated adherence.

Figure 14 shows that, during a panic brake in straight ahead travel, wheels longitudinal slip has a particularly regular behaviour, unknown for a traditional system.

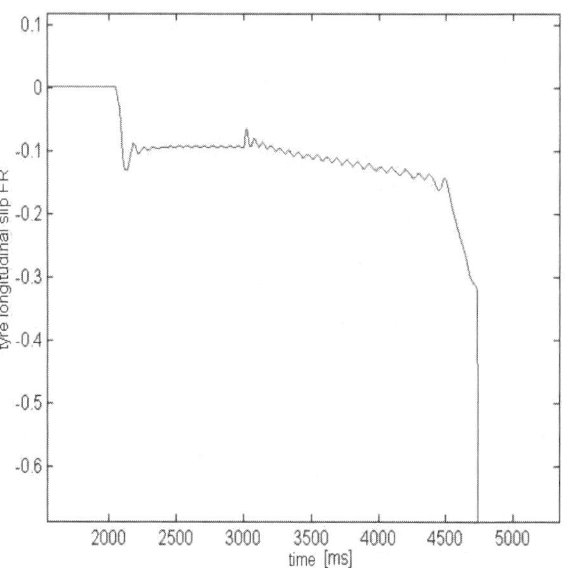

Figure 14 - Tyre longitudinal slip versus time in a panic brake manoeuvre

ABS interventions in an EHB equipped vehicle do not provoke the usual brake pedal vibrations due to pump interventions.

Figure 15 summarizes the simulation results of EHB adapted ABS control logic, in terms of stopping distances for different adherence conditions between tyres and ground and for different values of wheels inertia, depending on the gear ratio imposed by the driveline. No particular criticalities are evident, much less than with a traditional ABS with which, e.g., it is quite usual to have a performance variation depending on wheels inertia with gear.

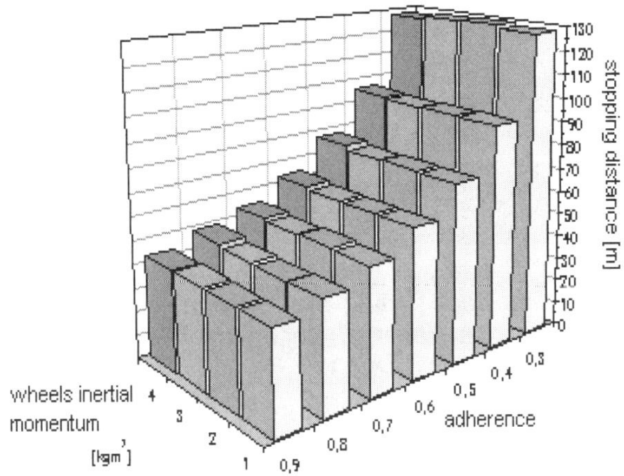

Figure 15 - Stopping distances considering different adherence and different wheels inertia values

The ABS/EHB control strategy conceived with wheel pressure measurement availability was virtually tested also in panic braking while cornering manoeuvres. Figure 16 presents an example of trajectories with and without ABS, demonstrating that it improves not only stopping distances but first of all the cornering behaviour of the vehicle. Figure 17 plots, in the same manoeuvre, vehicle actual and estimated longitudinal speed and wheels peripheral speeds. All the characteristics are very regular, much better than what can be observed in an usual ABS. Body lateral acceleration, as Figure 18 shows, is bigger for the active vehicle, because of the interaction between longitudinal and lateral tyres forces. Figure 19 plots wheels pressures during the same manoeuvre: it is evident that the wheels external to the bend have, as it is right, the biggest values. It is quite easy to obtain good results by integrating EHB with a pre-existing VDC [8], because in a standard VDC one of the most difficult tasks consists in estimating the desired values of calliper pressures without having any sensors. Figure 20 shows a comparison between the behaviour of a passive and a VDC-equipped car, in an extreme step steer manoeuvre, in low adherence conditions.

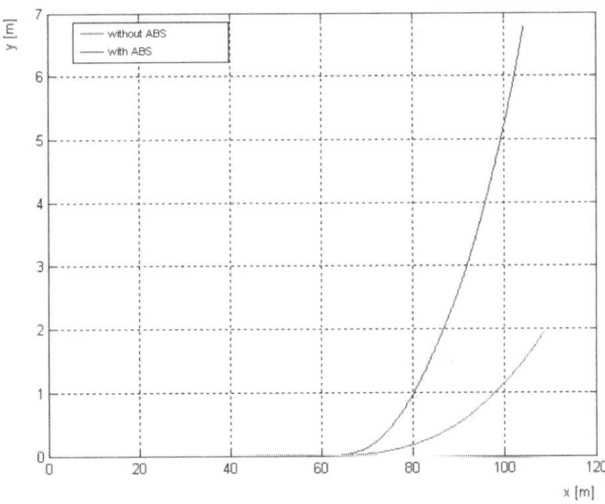

Figure 16 - Trajectory during a panic brake manoeuvre (blue - active vehicle, red - passive vehicle)

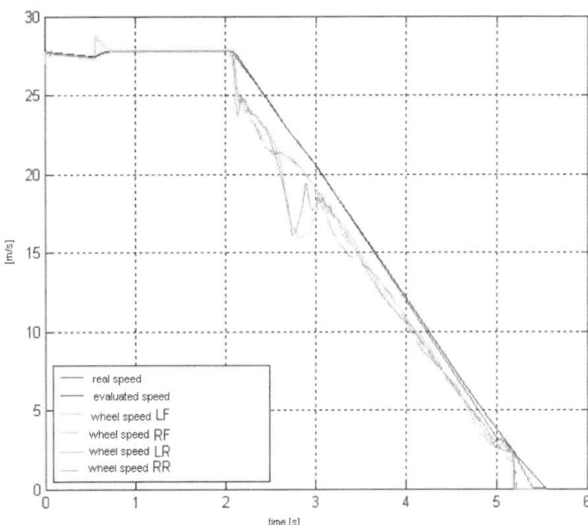

Figure 17 - Vehicle and wheels speed versus time

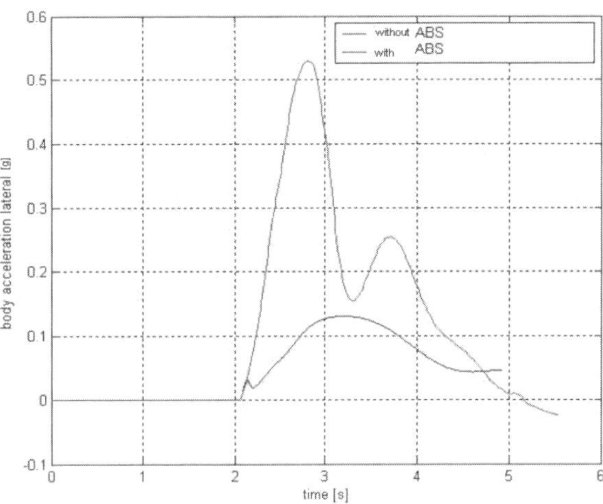

Figure 18 - Body lateral acceleration versus time (blue - active vehicle, red - passive vehicle)

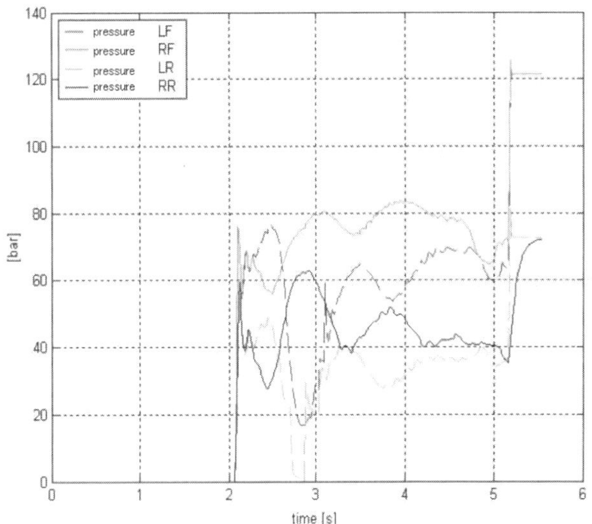

Figure 19 - Wheels pressures versus time

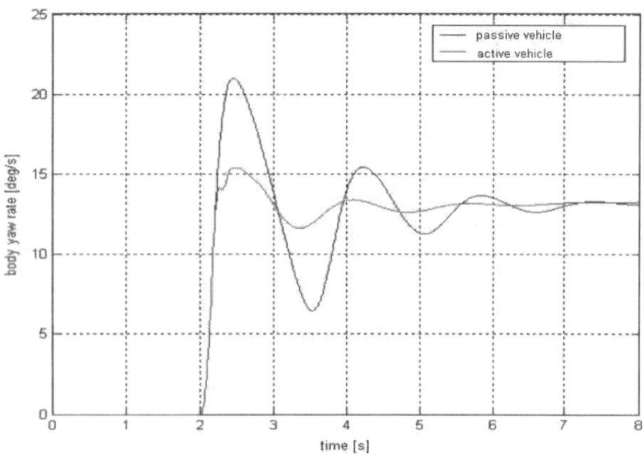

Figure 20 - Body yaw rate versus time (blue - active vehicle, red - passive vehicle)

Figure 21 - Politecnico di Torino braking systems test bench

CONCLUSIONS

This paper demonstrates that through virtual experimentation it is possible to investigate about the characteristics and perform a first approximation design of an active system like EHB to efficiently prepare the road tests on the system. EHB can determine consistent advantages, if compared to traditional braking systems, from the point of view of pressure gradients in panic brake manoeuvres, pedal feeling and ABS and VDC performance improvement. Finally, failsafe and diagnosis strategy were positively investigated through virtual experimentation.

Acknowledgments

The authors belonging to Politecnico di Torino Research Team on Vehicle Dynamics wish to thank FIAT Auto to have sustained the presented work both through grants and with technical suggestions.

BIBLIOGRAPHY

1. Leffler, 'Electronic Brake Management EBM - Prospects of an Integration of Brake System and Driving Stability Control', SAE 960954
2. Jonner, Winner, Dreilick, Schunck, 'Electrohydraulic Brake System-The First Approach to Brake by Wire Technology', SAE 960991
3. Fennel, Ding, 'A Model-Based Failsafe System for the Continental TEVES Electronic-Stability-Program (ESP)', SAE 2000-1635
4. Zehnder, Kanetkar, Osterday, 'Variable Rate Pedal Feel Emulator Designs for a Brake-by-Wire system', SAE 1999-01-0481
5. Ebert, Kaatz, 'Objective Characterization of Vehicle Brake Feel', SAE 940331
6. Bill, Semsch, Breuer, 'A New Approach to Investigate the Vehicle Interface Driver/Brake Pedal Under Real Road Conditions in View of Oncoming Brake-by-wire Systems', SAE 1999-01-2949
7. R. Limpert, 'Brake Design and Safety', ed. SAE, 2000
8. Morgando, Sorniotti, Velardocchia, 'Vehicle Dynamics Control based on Yaw Rate and Sideslip Angle', EAEC Conference 2003, Paris

AUTHORS

Prof. Mauro Velardocchia
Ph. Dr. St. Aldo Sorniotti
Department of Mechanics Politecnico di Torino
C.so Duca degli Abruzzi, 24 - Torino -Italy
Tel: 0039 011 564 6931/6947
Fax: 0039 011 564 6999
Email: mauro.velardocchia@polito.it
        aldo.sorniotti@polito.it

Eng. Luigi Petruccelli
Fiat Auto - P.&P.E. -A.C.V.
Vehicle Dynamics – Braking Systems
C.so Settembrini, 40 –Torino - Italy
tel. 0039 011 0034894
fax 0039 011 003 5379
Email: luigi.petruccelli01@fiat.com

Development of Hardware-in-the-Loop Simulator and Vehicle Dynamic Model for Testing ABS

Soo-Jin Lee and Young-Jun Kim
Hyundai Mobis

Kihong Park
Kookmin University

Dong-Goo Kim
Hyundai Mobis

ABSTRACT

In-vehicle driving tests for evaluating performance of vehicle control devices are often time-consuming, expensive, and not reproducible. Using hardware-in-the-loop simulation scheme, actual control devices can be easily tested in real time in a closed loop with a virtual vehicle. This advantage has made HILS systems popular as testbench lately in automotive industries. This paper describes a PC-based HILS system for ABS that has been developed in Matlab environment with real-time rapid prototyping tools. Also presented in this paper is a semi-empirical vehicle dynamic model that has been designed to account for kinematic and compliant characteristics of the suspension system from rig tests.

INTRODUCTION

As the vehicle electronic control technology grows fast and becomes complicated, a more efficient way than the traditional in-vehicle driving test is demanded for extensive testing of ECUs. For this purpose, hardware-in-the-loop simulation(HILS) scheme is very promising since it can replace significant portions of actual driving test procedures. It incorporates hardware components of prime concern in numerical simulation environment, yielding results with better credibility than pure numerical simulations. HILS runs in real time and it provides time- and cost-effectiveness over actual driving test. It also makes possible test procedures that are difficult or even impossible in actual driving tests. In this study, a hardware-in-the-loop simulation system has been developed for ABS. The system consists of hardware part which includes the ABS module, brake system, and interface devices, and software part that virtually implements vehicle dynamics and that performs visualization.

Despite significant advantages over numerical simulation and actual driving tests, the ABS HILS systems in the previous research often involved high cost to construct [1]. In this research, an ABS HILS system has been developed with objectives of achieving economy and easiness to use as well as reliability of results close to that from actual driving tests. To this end, Matlab by the MathWorks, Inc. has been adopted as the programming environment. The MathWorks has recently released products for real-time rapid prototyping in PC environments. Using these and other Matlab design tools, the sequence of system modeling, simulation, and hardware implementation could be made continuously. The Matlab products used in this research include Simulink, Real-Time Workshop, xPC Target, and Virtual Reality Toolbox.

The validity of HIL simulation is largely contingent upon accuracy of the vehicle models. To achieve high validity, efforts were made in this study to properly incorporate the experimental data in the mathematical models of the vehicle and tire. For the vehicle dynamic model, kinematic linkage and compliant characteristics of the suspension were first modeled as a functional form for unconstraint body, and then the suspension characteristics obtained from kinematics and compliance (K&C) test [1] were used to determine the model parameters. For the tire model, theoretical equations were derived by physics and then the coefficients were obtained through experiments with further corrections when necessary.

The rest of this paper is organized as follows. The following section introduces the real-time simulation and

---

[1] K&C test measures the kinematics and compliance characteristics of the suspension when known displacement and force inputs are applied to the wheel center.

visualization tools used in development of the HILS system. The structure of the HILS system is also described here. The third section explains the vehicle dynamic model and the tire model. The fourth section presents results of the experiments performed to validate the HILS system. In the last section conclusions are drawn.

ABS HILS SYSTEM

The ABS HILS system consists of hardware part that includes the brake system, and software part that includes the dynamic and visualization models of the vehicle. This section describes the programming environment for developing these models, and describes structure of the hardware part.

REAL-TIME SIMULATION AND VISUALIZATION TOOLS

MathWorks has recently released a product called xPC Target. This is a tool that enables real-time simulation environment on a PC platform without requiring a high cost DSP device. Using this solution, the ABS HILS system in this research could be constructed at a fractional cost of the ones built in the past. Also since it seamlessly integrates with other Matlab design tools used in this research, the current system could be constructed in a significantly short period.

The following describes the sequence ranging from modeling to real-time simulation in construction of the ABS HILS system of this research. First the vehicle dynamic model was developed using Simulink. Using Real-Time workshop, a C code for real-time simulation was generated. The C code was compiled and downloaded to the real-time kernel that xPC Target provides. Then, the model was executed in real time. The xPC Target provides means for interlinking the hardware and software to form a closed loop. This is done by first selecting the input/output blocks provided by Simulink at modeling stage, and then by setting parameters for the selected blocks.

For visualization of the vehicle response, the Matlab toolbox, Virtual Reality Toolbox, was employed in this research. In the ABS HILS system, visualization is not just for displaying test results. It must run in real time so that the operator can respond interactively while the HIL simulation is being performed. Virtual Reality Toolbox provides an economical solution for real-time visualization on a PC platform. The graphic model is programmed using VRML format, and then it is connected to the simulation model by selecting the virtual reality block at Simulink modeling environment. To prevent failure of real-time simulation due to excessive computation for visualization, two computers are used. The visualization model runs in the host computer while the simulation model runs in the target computer where the real-time kernel is loaded. The target computer is under control of the host computer.

STRUCTURE OF HILS HARDWARE PART

The hardware part of the HILS system is composed of the brake system, steering wheel and pedals, visualization equipments, computers for simulation and visualization, and interface devices. Each component is explained below.

Brake system [7]. Commercial ABS modules are manufactured with ECU(Electronic Control Unit) and HCU(Hydraulic Control Unit) integrated together. Due to the HCU, the HILS system for testing commercial ABS modules requires a brake system similar to the one in actual vehicles. Basic components of the brake system include the brake pedal, booster, master cylinder, hydraulic tube, and calipers and brake disks for each wheel. In addition, a vacuum pump is needed in the HILS system to provide vacuum to the booster in the absence of the engine. A pressure sensor is needed at each wheel to measure the brake pressure. The pressure measurements are fed to the PC that performs real-time simulation, then they enter the vehicle dynamic model as system inputs after being converted into brake torques. This forms interface from hardware to software in the closed loop of the ABS HILS system. The pressure of the master cylinder was also measured in this study for validating purpose.

Computers. Since commercial ABS modules require the speeds of four wheels, the vehicle dynamic model need be simulated in real time so that the wheel speeds are updated at every sampling period of the ABS module. As will be shown in the next section, the vehicle dynamic model of this research has 23 degrees of freedom, and this imposes high numerical load on the CPU. To achieve fast implementation, a Pentium IV PC was used solely for doing numerical simulation of the vehicle dynamics. The real-time simulation environment was provided by using xPC Target that provides real-time kernel on the target PC. Real-time visualization was done by using Virtual Reality Toolbox in a separate host PC. With real-time visualization, the operator can monitor the vehicle behavior, and take steering action interactively, while the HIL simulation is being carried out.

Steering wheel and pedals. The HILS system of this research is equipped with the steering wheel, acceleration pedal, and brake pedal. A control set for computer driving simulation game was used for the steering wheel and acceleration pedal, and the brake pedal for an actual vehicle was used.

Interface. Now described are the devices that were used for interface of aforementioned hardware and software models. First, interface is needed to provide linkage between the actual brake system and the virtual vehicle dynamic model. This was done by measuring the wheel cylinder pressures at four wheels and feeding them to the vehicle model. To achieve this, used are the pressure sensors, amplifiers and A/D board. In actual vehicles, the wheel speeds are measured by the wheel speed sensors and they enter the ABS ECU as input. In the ABS HILS system, however, since the wheels do not rotate, electric signals similar to the wheel speed sensor outputs must be generated from the current wheel speeds computed from the vehicle dynamic model. For this, four function generators were used to produce four sinusoidal signals. These machines allow frequency modulation by external voltage. In this study, the external voltage is provided through D/A conversion from the vehicle dynamic model. The output signals from the steering wheel and acceleration pedal are continuous voltages, and they are transferred to the vehicle model through A/D board. The simulation PC and the visualization PC are connected via Ethernet communication.

Fig. 1 shows the ABS HILS system built in this research.

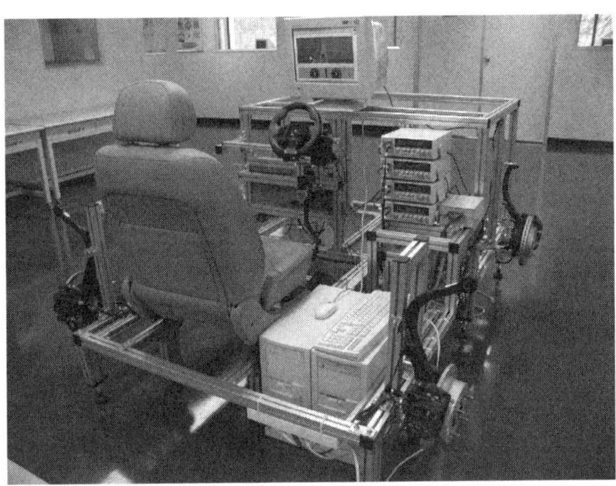

Fig. 1 ABS HILS system

VEHICLE DYNAMIC MODEL

A vehicle dynamic model has been developed in this research. For its application in the HILS environment, following aspects were considered. The vehicle dynamic model must run in synchronization with the ABS ECU that operates in real time. The model must allow steering maneuvers so that the situation of avoiding obstacles in braking can be tested. The model must support simulating braking maneuvers in various road conditions such as split or low friction road.

The vehicle dynamic model in this study has 23 degrees of freedom, which consist of 3 rotations of the vehicle about the body axes, 3 translations of the center of gravity of the vehicle, 3 wheel-suspension deflections for each wheel, 4 wheel rotations about the spindle axes, and 1 connecting-rod displacement at steering column.

EQUATIONS OF MOTION

The vehicle model in this study is composed of 5 bodies, that is, 4 wheels and one sprung mass as shown in Fig. 2. Each wheel is modeled as a spring that allows relative translational motion in 3 directions of the sprung mass. The kinematic and compliant characteristics of the suspension are included as functions of relative wheel

displacements. These characteristics are obtained through K&C rig test.

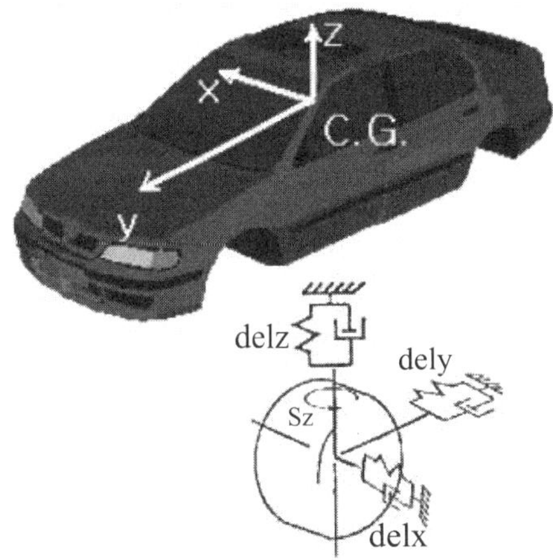

Fig. 2 Schematic diagram of vehicle model

WHEEL ORIENTATIONS

This section shows how to calculate the individual wheel orientations with respect to the sprung mass. These orientations are determined by the camber, caster and toe angles of each wheel. The variable delz in Fig. 2 represents the total deflection of the wheel center relative to the sprung mass. Fig. 3 shows that the camber angle of the front wheel is a function of delz and steering wheel angle.

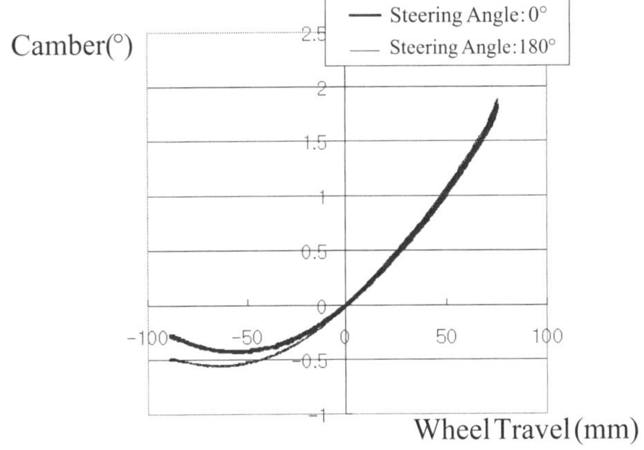

Fig. 3 Camber angle versus wheel travel

Based upon Fig. 3, the camber angle was modeled as a nonlinear function of two variables. The characteristic curves of the toe and caster angles were obtained from the K&C rig test. Let {wna1, wna2, wna3} and {wnn1, wnn2, wnn3} be the normal vectors of the wheel center with respect to the sprung mass and the global coordinate, respectively. The wheel-ground intersection unit vector in the global coordinate,{wgi1, wgi2, 0}, is defined in Eq. 1.

$$\varphi : \text{toe angle (rad)} = f(\text{delz, steer angle})$$
$$\phi : \text{camber angle (rad)} = f(\text{delz, steer angle})$$
$$\theta : \text{caster angle (rad)} = f(\text{delz})$$
$$wna1 = -\sin\varphi \times \cos\theta \qquad (1)$$
$$wna2 = -\sin\varphi \times \sin\theta \times \sin\phi + \cos\varphi \times \cos\phi$$
$$wna3 = \sin\varphi \times \sin\theta \times \cos\phi + \cos\varphi \times \sin\phi$$
$$[wnn] = [A][wna]$$
$$wgi1 = \frac{wnn2}{\sqrt{wnn1^2 + wnn2^2}}$$
$$wgi2 = \frac{-wnn1}{\sqrt{wnn1^2 + wnn2^2}}$$

When driving force is applied to a wheel, moving direction of the wheel is {wgi1, wgi2, 0} with respect to the global coordinate.

SUSPENSION FORCES

Suspension forces act between the sprung mass and each wheel. In this study, they are defined by 3 directional components of the body-fixed coordinates shown in Fig. 2. The vertical force is determined by the suspension deflection and its rate that are obtained through K&C rig test. The longitudinal and lateral forces are determined in similar ways. It is assumed that the suspension forces show linear characteristics due to small variations of deflections.

TIRE MODEL

The tire force characteristics are commonly described by the relationship between longitudinal friction coefficient and slip ratio and by the relationship between lateral friction coefficient and slip angle [4-6]. Fig. 4 shows typical longitudinal friction coefficient characteristics for braking with no lateral slip. The solid line represents dry pavement condition, and the dotted line et pavement condition.

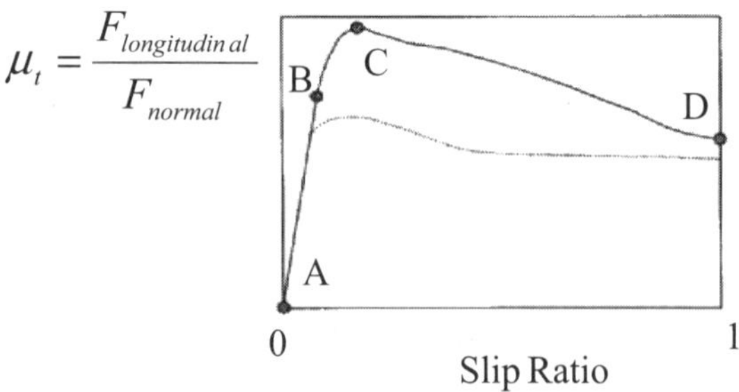

Fig. 4 Longitudinal tire force characteristics [4]

In the longitudinal tire force characteristics, the initial linear section displays nearly the same slope regardless of road conditions. This means that, at low slip ratio where the force is below the peak force that can be supported by adhesion limit of the surface, the tire behavior is dominated by the tire's intrinsic elastic properties. At high slip ratio, however, the tire force is highly dependent on friction of the surface in contact.

Fig. 5 shows typical lateral friction coefficient characteristics for a pure lateral slip. The tire is free-rolling as longitudinal loads are zero.

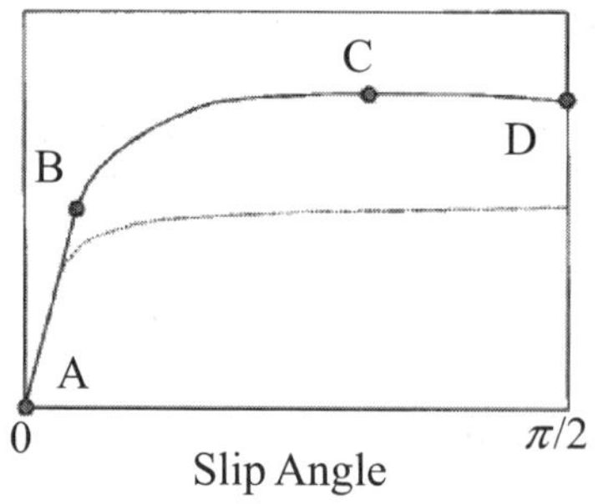

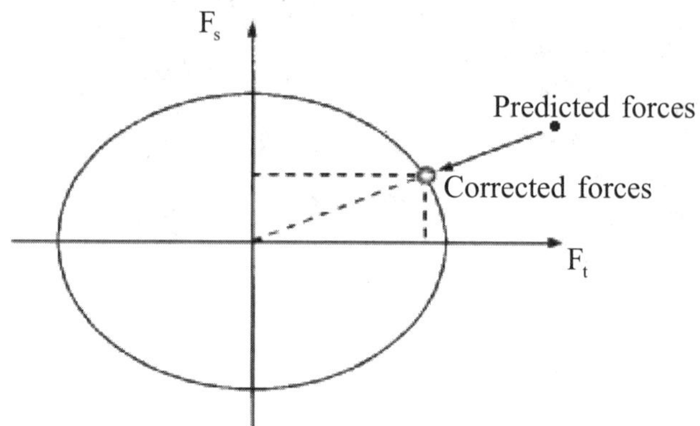

Fig. 5 Lateral tire force characteristics [4]

The case of combined tire forces with both longitudinal and lateral components can be solved using the concept of the friction ellipse [4,5] as shown in Fig. 6.

Fig. 6 Friction ellipse – correcting force for sliding [4]

The equation for the friction ellipse is as shown in Eq. 2.

$$\left[\frac{F_t}{\mu_t}\right]^2 + \left[\frac{F_s}{\mu_s}\right]^2 \leq F_n^{\,2} \qquad (2)$$

where F_t represents the longitudinal tire force, μ_t the longitudinal friction coefficient, F_s the lateral tire force, μ_s the lateral friction coefficient, and F_n the normal force.

The combined tire characteristics change further if the tire slides, and correction must be made accordingly in computation of the tire forces. This is done by the following procedure. First, it is checked whether Eq. 2 is satisfied using longitudinal/lateral friction force and coefficient obtained from Figs. 4 and 5 and the normal force obtained from tire deflection. If Eq. 2 is not satisfied, the tire is in a sliding mode. Then correction is applied based on the force limits that friction can sustain.

Both the lateral and longitudinal forces are corrected by the factor in Eq. 3.

$$\sqrt{\frac{RHS}{LHS}} \qquad (3)$$

In Eq. 3, RHS and LHS are the values of the right-hand side and the left-hand side of Eq. 2, respectively. Fig. 6 shows how this correction is done.

DRIVING AND BRAKING TORQUES

The driving torque is a function of the wheel angular velocity, acceleration pedal position, and difference between slip ratios of driving wheels. The following shows the procedure of its computation. First, the maximum torque is determined by the engine tractive effort map and average angular velocity of the driving wheels. Then, the average torque is determined by the current acceleration pedal position. Finally, each wheel's driving torque is determined according to the difference between slip ratios of driving wheels. Braking torque is a function of the brake cylinder pressure and friction coefficient of brake pad.

When a front driving wheel starts to rotate, suspension force is induced by the deflection between the sprung mass and a wheel. In the current vehicle model, the suspension force can either drive or brake the vehicle.

EXPERIMENTS

The performance of the ABS HILS system in this research was tested under the following driving conditions. The road has split friction coefficients: the left friction coefficient is 0.8 and the right friction coefficient is 0.3. The initial vehicle velocity until the brake is applied is 100 km/h. During braking, the steering wheel can be manually operated if necessary to keep the vehicle on intended track. The test was also conducted without operation of the ABS ECU for performance comparison.

Fig. 7 shows the vehicle response before and after putting on brakes in the ABS HILS test. Fig. 8 shows the vehicle trajectory in the plane during braking. The results in Figs. 7 and 8 demonstrate that the vehicle with the ABS can stay on its lane in a μ-split condition while the vehicle without the ABS deviates and rotates significantly from the lane to the side of high frictional road. Although not shown, the operator is providing appropriate steering wheel motion in both cases based upon real-time visualization of the vehicle response.

(a) Before putting on the brakes

(b) After braking without ABS

(c) After braking with ABS

Fig. 7 HIL Simulation results I in panic braking

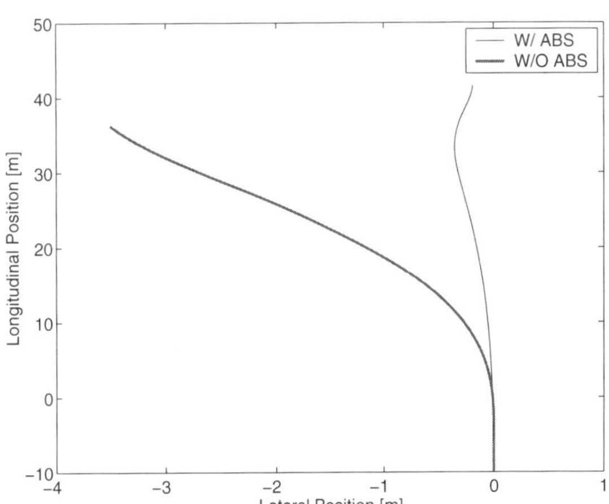

Fig. 8 HIL Simulation results II in panic braking

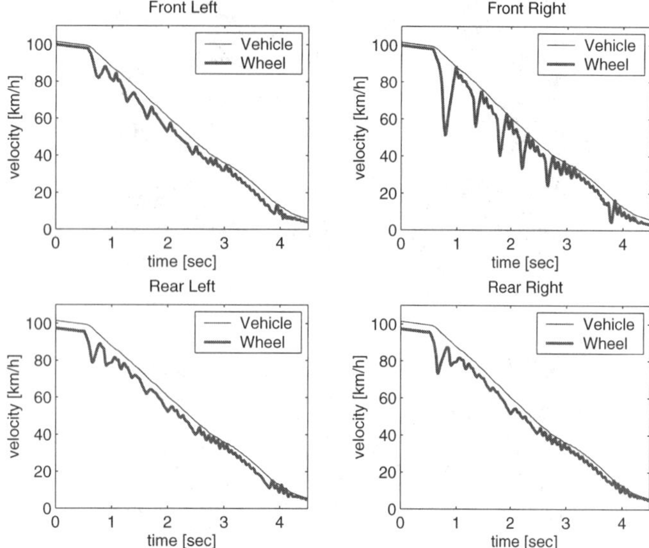

Fig. 9 HIL Simulation results III in panic braking

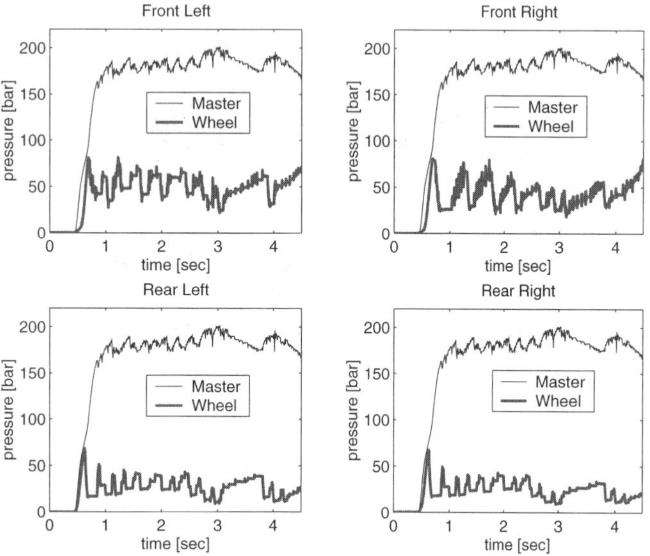

Fig. 10 HIL Simulation results III in panic braking

Fig. 9 shows the velocities of each wheel and vehicle in the ABS HILS test under the same driving condition as above. Fig. 10 shows the pressures of the wheel cylinders and the master cylinder. From these figures, it can be seen that the ABS HILS system developed in this research is performing well providing a testbench for extensive tests of the ABS module.

CONCLUSION

A PC-based hardware-in-the-loop simulation system has been developed for ABS in this research. The system consists of hardware part which includes the ABS module, brake system, and interface devices, and software part that virtually implements vehicle dynamics and that performs visualization. Aspects that were emphasized in design of the system include low cost, easiness to use, and credibility of the test results. The first two aspects were achieved by employing for the programming environment the following Matlab products; Simulink, Real-Time Workshop, xPC Target, and Virtual Reality Toolbox. They provide a total solution for real-time simulation and visualization in cost- and time-effective ways. For high credibility of the test results, a semi-empirical vehicle model has been constructed that can account for kinematic and compliant characteristics of the suspension that are obtained from rig tests. Through experiments the performance of the ABS HILS system was examined. The results of the experiments indicated that the system could serve as a testbench for extensive tests of the ABS module.

REFERENCES

1. Suh, M.W., Seok, C.S., Kim, Y.J., Chung, J.H., and Kim, S.M., "Hardware-In-The-Loop Simulation for ABS", SAE Paper 980244, 1998.
2. Kiffmeier, U., Otterbach, R., Schutte, H., "Real-Time Simulation of 3-D Vehicle Dynamics Model on the DEC Alpha Processor", 5th IMACS World Congress, 1997.
3. Allen, R.W., "A Low Cost PC Based Driving Simulator for Prototyping and Hardware-In-The-Loop Applications", SAE Paper 980222, 1998.
4. Noronha, P.F., "A robust lumped-parameter tire model developed for real-time simulation", SAE paper 980243, 1998.
5. Wong, J.Y., "Theory of ground vehicles", ISBN 0-471-52496-4, 1993, pp.3-58.
6. Milliken, W.F., "Race car vehicle dynamics", ISBN 1-56091-526-9, 1995, pp.13-82.
7. Park, K., Oh, S., and Kang, S.M., "Performance Validation of ABS Electronic Control Unit using Hardware-in-the-loop Simulation", Jounal of Engineering Technology, Kookmin University, No. 24, 2001, pp.225-232.
8. Nikravesh, P.E., Computer-Aided Analysis of Mechanical Systems, ISBN 0-13-164220-2-0 025, Prentice-Hall, Inc., 1988, pp.347-352.

Development of Fuzzy Logic Anti-Lock Braking System for Light Bus

Sun Jun

Department of Information and Control Engineering, Tongji University

Abstract

This paper presents a fuzzy ABS controller designed for 6700 series light bus. First, the fuzzy control theory is simply introduced. The performances and demands of ABS and the control algorithm using in ABS at present are described also. Second, The design of the fuzzy ABS controller for 6700 series light bus is discussed. Finally its test results are illustrated, then, some conclusions and further works are given .

1. Introduction

The central concept of L.A. Zadeh's fuzzy set theory is the membership function which represents numerically the degree to which an element belongs to a set. The membership function is assessed subjectively in any instance, lower values representing a lower degree of membership and higher values representing a higher degree of membership[1].

Based on fuzzy set theory, fuzzy controllers differ from conventional controllers, which are mathematically described by differential equations, transfer functions or state equations, in that they use a set of fuzzy IF-THEN decision rules to describe the mapping relationship between input and output. This relationship denoted by fuzzy IF-THEN decision rules is call fuzzy model or linguistic model.

Fuzzy controllers is composed mainly of four parts. First, fuzzification: which transform the precision value of input into a fuzzy value. Second, knowledge database : which is composed of database and fuzzy control rule sets . It includes also the knowledge from special application area and the demanded control aim. Third, fuzzy reasoning: which is based on implication relationships and reasoning rules of fuzzy logic. It is the core of the fuzzy controller. The last, defuzzification: which convert the fuzzy control value from fuzzy reasoning into control value for special application area[2].

2. Purpose of ABS Control

Antilock braking systems (ABS) are closed-loop devices within the braking system that prevent wheel lock-up during heavy braking or on slippery surfaces. The system allows a driver to apply the brakes fully while maintaining steering control to avoid a hazard or lessen a collision.

The most effective braking can occur only when the retarding force on the wheel brake matches the grip imposed between the tire and the road surface. Road grip is dependent upon the vertical weight imposed on the wheel and the frictional resistance generated between the tire tread and road surface. Should the retarding force be too great, the weight on the wheel will be insufficient to keep the wheel rolling. It will thus cause the wheel to lock and skid. If the retarding force is below the maximum required to bring the wheel to the point of lock, then the retardation of the wheel will be much lower than its possible maximum. It therefore adds lo the minimum possible stopping distance of the vehicle[3].

It is difficult for the driver to judge the optimum amount of brake pedal force to bring a wheel to rest in the minimum time on a good road surface, and it is practically impossible on a wet, muddy or icy surface, as the minimum wheel grip on each wheel will be continuously varying. To prevent the individual wheels from locking and consequently skidding when bulking, the driver should rapidly apply and release the brake pedal repeatedly until the vehicle speed has been reduced to a safe driving speed relative to road conditions. This difficult technique, known as cadence braking, is usually only performed by professional racing drivers, and even when executed the braking of individual wheels will not necessarily match the changing road surface conditions.

An anti-lock braking system consists of an electronic control unit(ECU), sensors (inputs) and brake pressure control(actuator). The control unit compares signals from each wheel sensor and measures the acceleration or deceleration of an individual wheel. From these data and preprogrammed look-up tables, brake pressure to one or more of the wheels is regulated by four operating scenarios: normal braking, pressure release, pressure holding and pressure build-up.
The control effect of ABS lies on the control algorithm using in ABS mainly.

At present, the ABS control algorithm can be divided into two kinds by control parameters. One is the continuous control system by controlling wheel slip. It is a best algorithm from the view of theory. It can get a smooth, continuous, chatter-free braking process, but is difficult to practice because it need to measure the vehicle speed or acceleration. The other system uses logic control algorithm to control the wheel angular acceleration or deceleration. This algorithm preset some thresholds on control parameters. During braking, ECU compares the calculated real-time parameters and thresholds corresponding to determine the movement status of wheels, and regulate the brake pressure of wheels. This algorithm is widely used in today's vehicle, because it avoids complex theoretical analysis and only need to mensurate the wheel speed. But it is not a best algorithm. The thresholds under varied road situations are experiential data from repeating tests, so the developing periods of ABS is long and the control quality is difficult to assure[4].

Fuzzy logic is a useful tool for developing robust controllers for non-linear, time-varying, higher-order systems. In this paper, we intend to develop a fuzzy controller for 6800 series light bus.

3.   The Fuzzy ABS Controller for 6800 series light bus

The test bus, HFC6801A, has a dual circuit air brake system formerly. Now we rebuild it with a 4 sensors, 4 channels ABS, which has a sensor and a air brake valve in each wheel.

The key approach of developing a logic control ABS is to determine the control states during the anti-lock braking process. Fig.1 is a representative anti-lock braking process which has four control states. State 0 is the initial phase. In this phase, the brake torque is increase, and the deceleration is increase also. When the deceleration is over some preset thresholds, i.e. in state 1, the brake torque and the deceleration are decrease until the deceleration is close 0. Now it is in phase 2. The brake torque is maintained. The deceleration is from decrease to increase until the deceleration is close 0 again. The phase change into phase 3., which is same as phase 0, but the brake pressure is build-up in small gradient. When the deceleration is over some preset thresholds, the system is into phase 0 again[5].

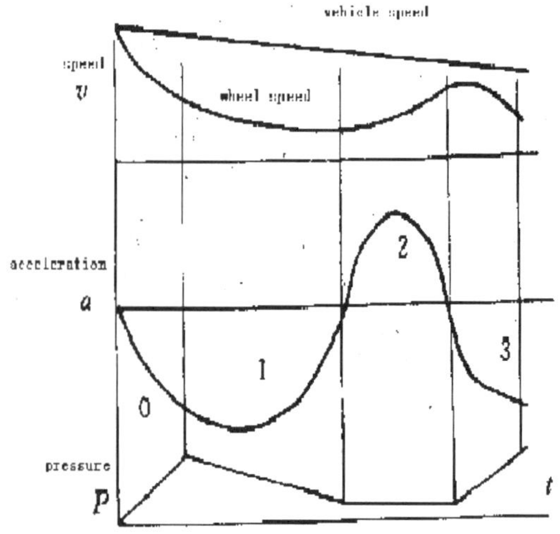

Fig.1   the sketch of anti-lock braking process

The fuzzy controller for 6800 series light bus has three input variables. Two continuous variables are difference between wheel deceleration and its expectation SE and differential of the difference SEC. They are fuzzified with piecewise linear membership functions and described using five linguistic variables "NL", "NM", "NS", "ZE", "PS", "PM", and "PL". Another input variable is the previous control state. It is a precise variable and defined in terms of four crisp variable "normal braking", "pressure release", "pressure holding" and "pressure build-up". The output variable is the current control state. It is also a precise variable.

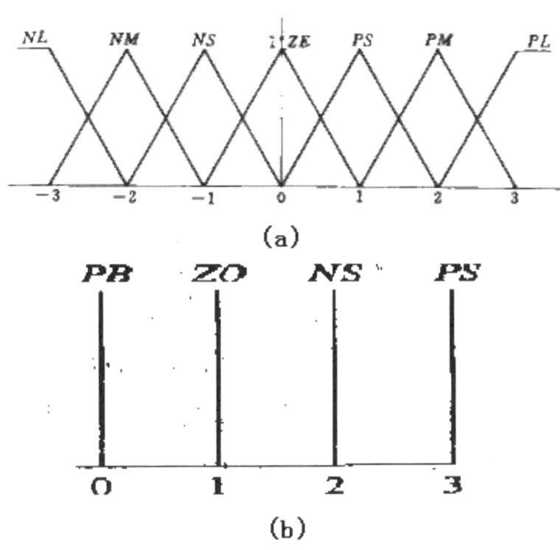

Fig.2 Membership function of (a). the difference between wheel deceleration and its expectation SE and differential of the difference SEC (b). the previous control state

The controller has a set of nine logic rules. Four rules govern braking on dry roads, three on ice, two on wet. For real-time control, the reasoning results of all logic are calculated off-line and stored in ECU in the form of control data table. This allows the ABS controller to respond immediately once a new situation has been identified.

4. Test result and conclusions

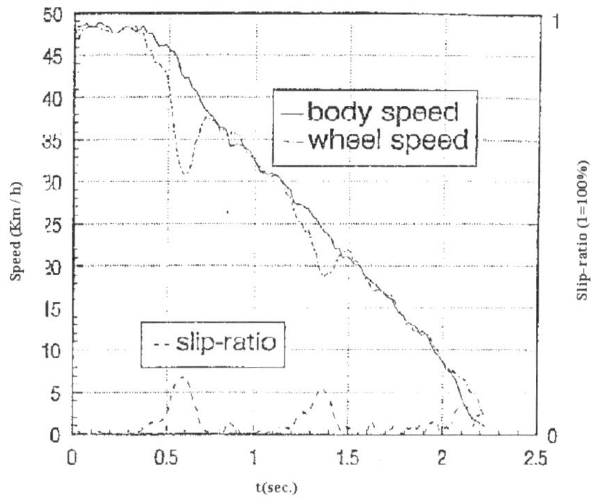

Fig.3 The test result on dry road

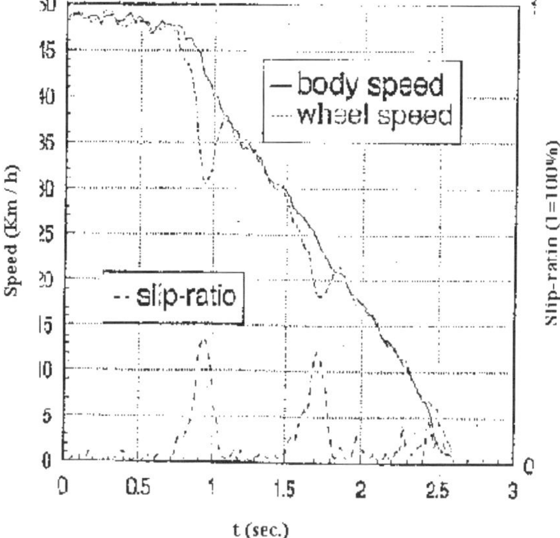

Fig. 4 The test result on wet road

Fig.3 and fig.4 are the test result of HFC6801A light bus braking on dry (about $\mu = 0.7$) and wet road (about $\mu = 0.3$).

The test showed that a fuzzy controller in combination with decision logics for estimation of the road condition is a rapid and effective means to provide brake pressure control over operating condition ranging from dry road to icy road. The controller is found to be generally quite robust. Move over, the elaboration of the control strategy is simplified by its formulation with linguistic rules which easily dialog with experts of the system to control.

References

[1]   L.A. Zadeh. Fuzzy algorithm. Information and control, Vol.12,1968

[2]   M.Sugeno and M.Nishida. Fuzzy control of model car. Fuzzy Sets Syst., Vol.16,1985

[3]   L. Austin and D.Morrey. Recent advances in antilock braking systems and traction control systems. Proc Instn Mech Engrs, Vol.214,Part D,2000

[4]   Guy Kokes,Tarunraj Singh. Adaptive Fuzzy Logic Control of an anti-lock braking system. Proceedings of the 1999 IEEE. International Conference on Control Applications.

[5]   Cheng Jun. A study on anti-lock brake system based on fuzzy control. Automotive Engineering (in Chinese), Vol.19,No.4,1997

The Evaluation of ABS Performance

Yongping Hou and Zhuoping Yu
Clean Energy Automotive Engineering Center, Tongji University

ABSTRACT

On the basis of existing methods, a new method of evaluating ABS performance is proposed in the paper. This method can evaluate lateral stability of vehicle during vehicle braking process. It is verified by simulation. The most advantage of this method is that it can evaluate the performance of ABS during design phase. It also can short the developing cycle of ABS, and save the developing expenditure of ABS. This method can be applied to the design and simulation study of ABS.

1 INTRODUCTION

Currently simulation computation has become the developing trend of manufacture in place of experiments, which can not only shorten the design cycle of product, but also extremely reduce the manufacture cost. Many auto plants have achieved great advancement on the aspect.

With respect to ABS, how to comprehensively evaluate its performance involves evaluation index for ABS performance. So far, there is no effective index for evaluating ABS performance, which results from complicated braking that relates to automotive maneuverability and other factors. It is difficult to fully take into account. ECER13 regulation presents that an adhesion coefficient utility ratio ε can be an index for evaluating ABS performance. ABS is used to secure the brake effectiveness and maneuverability during braking. The adhesion coefficient utility ratio ε pays great attention to whether ABS improves automotive brake effectiveness, but does not take maneuverability into proper consideration. For instance, suppose that adhesion coefficient and slip ratio accord with the curve shown in Figure 1, if two ABS systems, ABS_C and ABS_D, are separately mounted to a vehicle, under the same working condition, ABS_C and ABS_D can keep each wheel on point C and D. It can be shown in Figure 1 that the lateral adhesion coefficient on point C is equal to that on Point D, which means $\mu_{bC} = \mu_{bD} = \mu_{b1}$. If the two ABS systems are evaluated by adhesion coefficient utility ratio, they will be considered to be identical. In fact, from Figure 1, it can be seen that their lateral adhesion coefficients are much different although the longitudinal adhesion coefficient on point C is the same as that on point D, and the lateral adhesion coefficient μ_{lC} on point C is much larger than that μ_{lD} on point D, which indicates that the lateral stability of a vehicle with ABS_C is much better than that with ABS_D. It is obvious that the performance of ABS_C is better than that of ABS_D. From this it can be found that adhesion coefficient utility ratio emphasizes on evaluating the brake effectiveness during automotive braking, but does not pay enough attention to maneuverability, which is extremely important. In addition, the adhesion coefficient utility ratio is an index for test evaluation and in the design phase of ABS, its performance cannot be evaluated. [1][2] Based on the above two points, this paper presents an index for evaluation, which take both lateral and longitudinal adhesion into account. Besides, the index can also be conveniently used into ABS design and during design phase the performance of ABS can be analyzed, then further improved. To do so can shorten the design cycle of ABS and greatly reduce its manufacture cost. This is also current developing trend for automobile manufacture.

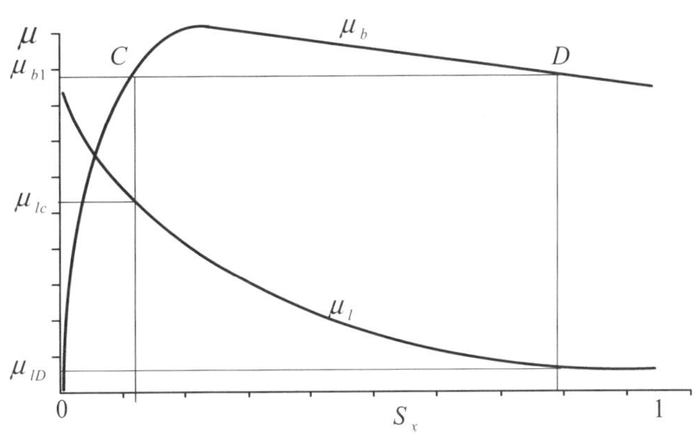

Figure 1 Slip Ration-Coefficient Curve

2 E EXPONENT TIRE MODEL OF STEADY STATE

In reference $[3]$, steady state lateral force F_y, steady state longitudinal force F_x, aligning torque M_z are all expressed as follows:

$$\begin{cases} \overline{F} = 1 - \exp(-\phi - E_1\phi^2 - (\frac{1}{12} + E_1^2)\phi^3) \\ F_x = \mu_x F_z \cdot \overline{F} \cdot \phi_x/\phi \\ F_y = \mu_y F_z \cdot \overline{F} \cdot \phi_y/\phi \\ D_x = (D_{x0} + D_e)\exp(-D_1\phi - D_2\phi^2) - D_e \\ D_y = F_y/K_{cy} \\ M_z = -F_y \cdot D_x + F_x \cdot D_y \end{cases} \quad (1)$$

Where:

$$\phi_x = \frac{K_x S_x}{\mu_x F_z} \quad \phi_y = \frac{K_y S_y}{\mu_y F_z} \quad \phi = \sqrt{\phi_x^2 + \phi_y^2} \quad (2)$$

$$S_x = \frac{V_{sx}}{V_r} = \frac{V_x - \omega R}{|\omega R|} \quad S_y = \frac{V_{sy}}{V_r} = (1 + S_x) \cdot \tan\alpha \quad (3)$$

$V_r = |\omega R|$ =Rolling Speed $\quad \omega$ —Rotational Speed

R —Rolling Radius $\quad V_x$ —Wheel Speed

μ_x —Longitudinal Friction Coefficient

μ_y —Lateral Friction Coefficient

K_{cy} —Lateral Stiffness $\quad K_y$ —Cornering Stiffness

V_{sy} —Lateral Slip Speed $\quad \alpha$ —Cornering Angle

$\overline{F}$ —Non-dimensional Force

3 ESTABLISHING EVALUATION INDEX AND SIMULATION VERIFICATION

Figure 2 is a diagrammatic sketch of longitudinal and lateral forces during braking. Suppose that under the control of ABS, slip ratio S_x retains on point S_A during braking, then the curve OAC shows the variation of longitudinal force, and EBD indicates that of lateral force. During the course of braking, from the optimal slip ratio to wheel skidding, the variation of lateral force is larger than that of longitudinal force, correspondingly, the area variation under the lateral force curve is larger and that under the longitudinal force curve is relatively small.

Additionally, during braking, if the vehicle needs to keep on comfortable lateral stability, which requires larger lateral force, the area under the lateral force curve will be much greater. According to this point, define the following index:

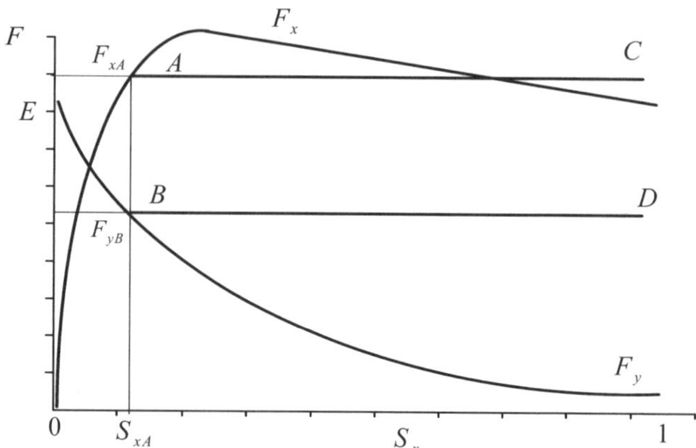

Figure 2 the Schematic Plan of longitudinal Force and Lateral Force during Braking Process

$$G = \frac{A_{F_y}}{A_{F_x}} \quad (4)$$

Where:

A_{F_y} —The area under the lateral force curve

A_{F_x} —The area under the longitudinal force curve

If vehicle's lateral stability is quite good, the value A_{F_y} is greater, and G is also greater. Then the greater the value G is, the better vehicle's lateral stability is. In addition, in Eq.(4), the index G includes not only the lateral force, but also the longitudinal force. In order to conveniently utilize G, Eq.(4) can be transformed. Figure 2 shows that:

$$A_{F_y} = \sum F_y \cdot \Delta S_x \quad (5)$$

$$A_{F_x} = \sum F_x \cdot \Delta S_x \quad (6)$$

Substitute Eq.(5) and Eq.(6) into Eq.(4):

$$G = \frac{\sum F_y \cdot \Delta S_x}{\sum F_x \cdot \Delta S_x} = \sum \frac{F_y \cdot \Delta S_x}{F_x \cdot \Delta S_x} = \sum \frac{F_y}{F_x} \quad (7)$$

And $G = \sum \dfrac{F_y}{F_x} \quad (8)$

Obtained by Eq.(1):

$$\frac{F_y}{F_x} = \frac{\phi_y}{\phi_x} \tag{9}$$

Under non-steady state condition, lateral effective slip ratio and longitudinal effective slip ratio are calculated as follows respectively[4],

$$\begin{cases} \phi_y = \dfrac{K_{y0} \cdot S_y}{\mu \cdot F_z} \\[2mm] \phi_x = \dfrac{K_{x0} \cdot S_x}{\mu \cdot F_z} \end{cases} \tag{10}$$

Where

K_{y0} —steady state cornering stiffness,

K_{x0} —steady state longitudinal Stiffness

By Eq.(10):

$$\frac{\phi_y}{\phi_x} = \frac{K_{y0}}{K_{x0}} \cdot \frac{S_y}{S_x} \tag{11}$$

Substitute Eq.(3) into Eq.(11):

$$\frac{\phi_y}{\phi_x} = \frac{K_{y0}}{K_{x0}} \cdot \frac{(1+S_x) \cdot \tan \alpha}{S_x} \tag{12}$$

Substitute Eq.(12) into Eq.(9):

$$\frac{F_y}{F_x} = \frac{K_{y0}}{K_{x0}} \cdot \frac{(1+S_x) \cdot \tan \alpha}{S_x} \tag{13}$$

Because K_{y0} , K_{x0} are all constant, then

$$K = \frac{K_{y0}}{K_{x0}} \tag{14}$$

So,

$$\frac{F_y}{F_x} = K \cdot \frac{(1+S_x) \cdot \tan \alpha}{S_x} \tag{15}$$

F_y / F_x are function of longitudinal slip ratio S_x in Eq.(15).

Substitute Eq.(15) into Eq.(8), then

$$G = \sum K \cdot \frac{(1+S_x) \cdot \tan \alpha}{S_x} \tag{16}$$

Because Eq.(16) is sum expression, more amount of data, more bigger the value of Eq.(16), Obviously, this value could not be as evaluation index, modify Eq.(16) as follows,

$$G(S_x) = \frac{1}{n} \sum K \cdot \frac{(1+S_x) \cdot \tan \alpha}{S_x} \tag{17}$$

n —Total amount of data

After the mathematical treatment, $G(S_x)$ is a quantity independent of total amount of data points. The greater the value $G(S_x)$ is, the better the lateral stability of vehicle during braking will be. On the contrary, the lateral stability of vehicle will be worse. The evaluation index $G(S_x)$ mainly emphasizes on evaluating the lateral stability of vehicle when braking.

This paper adopts three methods, fuzzy control, slide mode control and fuzzy slide mode control, to create ABS simulation model [5] and conducts simulation computation of vehicle's brake process, by which the index $G(S_x)$ can be obtained. Table 1 shows some related vehicle parameters [5] used in simulation, and Table 2 is the value $G(S_x)$ calculated by simulation. Figure 3 and Figure 4 are the curves of simulation computation.

On the basis of the above analysis results, it can be concluded from Table 2 ($G(S_x)$) that the lateral stability with fuzzy control is the best, fuzzy slide mode control the second, and slide mode the worst. This conclusion can be explained by the simulation curve (Figure 3) of slip ratio and F_y / F_x simulation curve (Figure 4).

Name and Symbol	Value
Mass M/kg	1850
Normal Force F/N	18125
Wheel Moment of Inertia J/kg·m^2	2
Braking Pressure Range p/Kpa	0~700
Braking Factor C/Nm/Kpa	21
Braking Initial Speed V/m/s	30
Gravitational Acceleration g/m/s^2	9.8
Wheel Rolling Radius R/m	0.52
Coefficient parameters Peak Value μ_h Slip Value μ_g	0.8 0.7
Peak Slip Value S_0	0.2

Table 1 Vehicle Parameters of Single Wheel Model [5]

In Figure 3 it can be found out that during the course of simulation, the slip ratio with fuzzy control is controlled to be 0.13 or so, slide mode control 0.22 or so, and fuzzy slide mode control 0.18 or so. In Figure 2, it can be shown that within a certain range, the lateral force F_y will decrease along with the increasing of slip ratio. Therefore, the lateral force by fuzzy control is the greatest with the best lateral stability, which is indicated by the greatest value F_y/F_x (Figure 4), the lateral force by slide mode control is the least and its lateral stability is the worst, which can be indicated by the least value F_y/F_x (Figure 4).

	T_max(s)	G
Slide Control	3.530	0.2071
Fuzzy Control	5.500	0.359
Fuzzy Slide Control	3.702	0.249

Table 2 Simulation Result

From the above analysis, it can be concluded that the lateral stability index $G(S_x)$ indicates the lateral stability of vehicle during braking. It can be an index for evaluating the performance of ABS. The index can be used to evaluate the performance of ABS in the design phase, which is a great advantage. One side this can shorten the design period of ABS, on the other side can reduce the expense of design & development.

Many evaluation indexes must be combined to fully evaluate the performance of ABS and any evaluation index cannot comprehensively indicate the performance of ABS.

4 CONCLUSION

From the above discussion, we can concluded:

1. Based on the existing evaluation index, the paper presents an index emphasizing on evaluating the lateral stability during vehicle braking. Through simulation computation and verification, it can be an index for evaluating the performance of ABS.
2. The most attractive advantage of the index $G(S_x)$ is to evaluate the performance of ABS in the design phase, which can not only shorten the development period of ABS, but also save the development cost.
3. Many evaluation indexes must be combined to fully evaluate the performance of ABS and any evaluation index cannot comprehensively indicate the performance of ABS.

REFERENCES

1. Haitao Ding, Preliminary study on the fuzzy control of auto ABS, MS thesis, Jilin University of Technology, !999.
2. Yixuan Li, Simulation, performance evaluation and parameters selection of ABS, MS thesis, Jilin University of Technology, !999.
3. Konghui Guo, Lei Ren, Yongping Hou. A Non-Steady Tire Model for Vehicle Dynamicsimulation and Control. 4th International Symposium on Advanced Vehicle Control, Sept 1998, Nagoya Japan, Paper 060.
4. Yongping Hou. "A Study of Non-Linear Non-steady State Side Slip Property of Tire and Its Semi-Empirical Model", Ph.D thesis, Jilin University of Technology, 1999.
5. Chen Jun, Theory and practice of automobile ABS, Publishing House in Beijing Institute of Technology, Sep. 1999.

CONTACT

Yongping Hou, Ph.D, Associate Professor, Major: Vehicle Engineering, mail address: yp215@263.net.

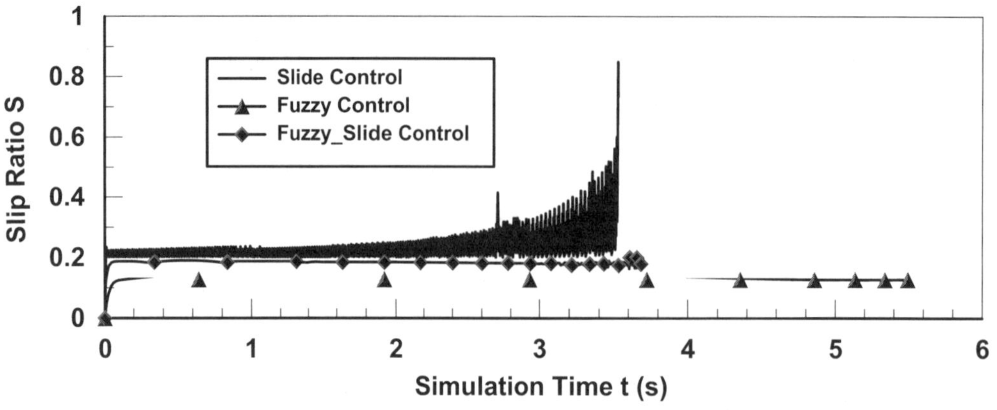

Figure 3 Simulation Curve of Slip Ration

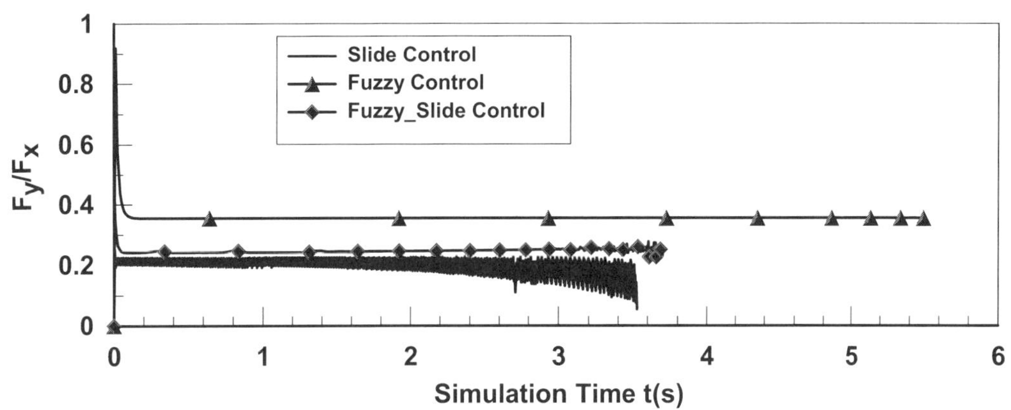

Figure 4 F_y/F_x Simulation Curve

New Generation ABS Using Linear Flow Control and Motor Speed Control

Seongho Choi, Jinkoo Lee and Inyong Hwang
Central R&D Center, MANDO Corporation

ABSTRACT

Recent trends of research & development for vehicular ABS are focused on the NVH as well as braking performance. For the conventional systems, we used On-Off control strategy, which is composed of pulse-up pressure rise, pressure dump and pressure hold mode. The main problems associated with the discrete (On-Off) nature of this control are severe noise and vibration.

Especially, among the three pressure modes of conventional control, the pressure pulse-up is the major source of noise and vibration. To reduce pulse-up noise & vibration of ABS, we introduce Linear Flow Control (LFC) concept, which allows continuous pressure rise mode instead of pressure pulse-up. Continuous pressure rise can be obtained by controlling the magnetic force of solenoid valve, which needs current control of the solenoid valve. In this regard, we proposed using pulse width modulation (PWM) method to control the current and to compensate for the discrete nature of actuator dynamics by duty control.

Another source of noise and pedal kickback during ABS action is excessive pumping of brake fluid caused by unnecessarily high motor speed. Because the level of noise and vibration caused by the motor is directly related with the motor speed, we introduced a new Motor Speed Control (MSC) method that could compromise pumping performance and noise.

Our new generation ABS also achieves better performance in stopping distance compared with conventional control method by reducing average pressure dump and maintaining optimal slip band.

In this paper, we will present recent results of LFC & MSC for our new generation ABS in a view of performance and NVH.

INTRODUCTION

A vehicular ABS system is to regulate the wheel slip ratio around the optimal slip ratio range so that the road adhesion coefficient is maximized and directional stability is achieved.

For the conventional systems, we used On-Off control strategy, which is composed of pulse-up pressure rise, pressure dump and pressure hold mode. The main problems associated with the discrete (On-Off) nature of this control are severe noise and vibration. Under many driving conditions, ABS allow the driver to brake a vehicle more rapidly while maintaining steering control even during situations of extreme, panic braking. However, normal drivers sometimes cannot properly control the vehicle because of harsh noise and pedal kickback caused by ABS operation. This is the reason why recent trends of research & development for vehicular ABS are focused on the NVH as well as braking performance. Various methods were tried to mitigate the noise and vibration due to ABS operation. Many brake experts gradually recognized that continuous flow/pressure control, which eliminates abrupt pressure change, is the most powerful countermeasure for above problems. Continuous flow rate can be obtained by controlling the magnetic force of solenoid valve, which needs current control of the solenoid valve.

Especially, among the three pressure modes of conventional On-Off control, the pressure pulse-up is the major source of noise and vibration. One objective of this paper is to introduce our Linear Flow Control (LFC) concept, which enables continuous pressure rise to eliminate pulse-up noise & vibration of ABS. For the control of magnetic force of solenoid valve, we proposed using pulse width modulation (PWM) method to control the current and to compensate for the discrete nature of actuator dynamics by duty control.

During the cyclic operation of ABS, the pumping of dumped flow from accumulator to hydraulic circuit is another source of noise and pedal kickback. To reduce such noise and pedal kickback, excessive pumping of brake fluid caused by unnecessarily high motor speed must be controlled by adequate method. To do this, we introduced a new Motor Speed Control (MSC) method that could compromise pumping performance and noise.

In this paper, we will present recent results of LFC & MSC for our new generation ABS in a view of performance and NVH.

LINEAR FLOW CONTROL

For ABS hydraulic actuator, we have used so-called solenoid-solenoid type actuator. This type of actuator is originally designed for On-Off (discrete) control and, for continuous flow or pressure control, the proportional control valve is generally used. However, this type is much more expensive and bigger than conventional type On-Off valve. So, research engineers in this area were trying to find new methods, which enables continuous flow control with conventional On-Off valve. Our recent research shows that current control, using conventional On-Off valve with minor modifications, enables continuous flow control. We developed new control method, named LFC, which uses PWM duty modulation as a current control method and adopted this technique to our ABS. As mentioned above, among the three pressure modes of conventional On-Off control, the pressure pulse-up is the major source of noise and vibration. Figure 1 shows the pressure profile of our LFC compared with conventional one. Our LFC allows continuous pressure rise mode instead of pressure pulse-up by controlling magnetic force of solenoid valve using PWM method to control the current and to compensate for the discrete nature of actuator dynamics by duty control.

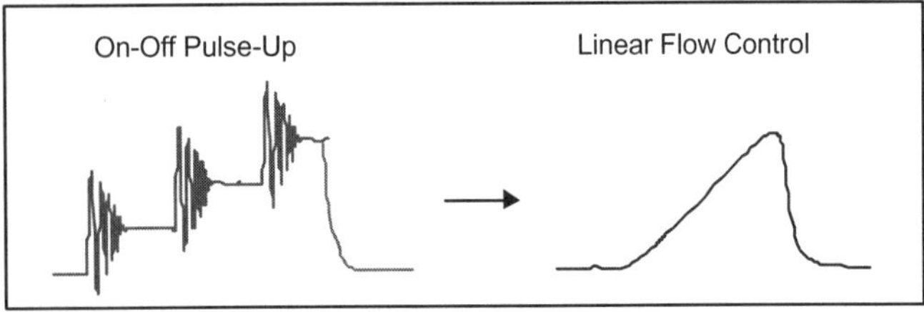

Figure 1: Comparison of pressure pulse-up pattern

The realization of stable and repetitive continuous flow control needs sophisticated skills considering road state, pedal efforts and electrical environments. To develop such skills, we tried to model the electric and mechanical behavior of solenoid valve and verified it with In-door test.

ACTUATOR MODELLING

The brake actuator (hydraulic actuator) converts the control commands for pressure modulation in the wheel brakes by use of the solenoid valves. It acts as the hydraulic link between the master cylinder, or pressure accumulator, and the wheel cylinders of individual wheel brakes. Using PWM as a method of current control, the characteristic of solenoid valve must be modeled in terms of PWM parameters. To do this, we studied the static & dynamic characteristics of electrical & electro-hydraulic behavior of solenoid valve theoretically and experimentally.

Modeling of Solenoid Valve

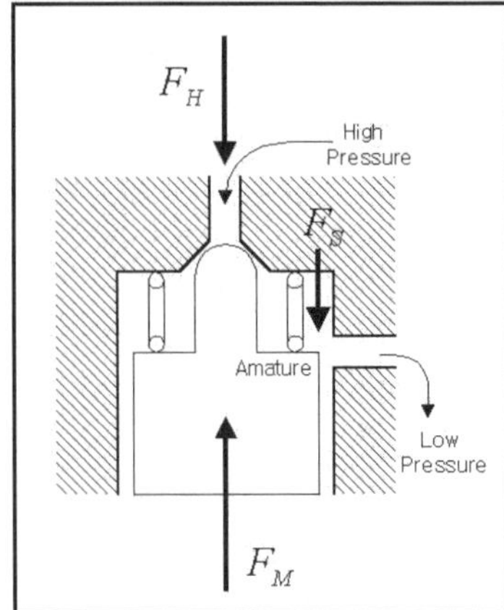

Figure 2: Force balance on solenoid valve

The force relationship in the normally open solenoid valve depicted in Figure 2 is as follows;

$$F_M = F_H + F_S$$

where F_M : magnetic force, F_H : hydraulic force, F_S spring force of solenoid valve.

The continuous pressure rise of LFC is possible when the magnetic force is balanced with the sum of hydraulic and spring force near the closed position. This force balance allows small amount of flow passage and results continuous pressure rise of wheel cylinder. The relationship between the current of solenoid coil and the magnetic force can be represented as following;

$$F_M \propto i^{\acute{a}}$$

where i : current of solenoid coil, $\acute{a}$: exponential factor(about 1~2).
By controlling the current of solenoid coil, we can control the magnetic force of solenoid valve and achieve the force balance needed for our LFC.

Current of Solenoid Coil under PWM

To achieve the function needed for our LFC, exact control of coil current is essential. There are two methods for current control of solenoid coil, i.e. direct current control and PWM duty control. For the following reasons, we prefer PWM duty control to direct current control.

- PWM is cost effective and compact.
- PWM is easy to handle.
- PWM is good in heat characteristic and robust in electrical environments.

On the contrary, use of PWM as a current control method needs such kinds of additional consideration as solenoid coil states and source voltage, e.t.c.
Consider general PWM controlled system subject to one PWM input;

$$\dot{x} = f(x) + \overline{u}g(x)$$

where $\overline{u}$ is PWM control input defined as;

$$\overline{u} = \begin{cases} 1 & for \quad t_k < t \le t_k + \hat{o}\,(x(t))T \\ 0 & for \quad t_k + \hat{o}\,(x(t))T < t \le t_k + T \end{cases}$$

where T is period and $\hat{o}(x)$ is duty ratio.

From the characteristics of the PWM control, it generally follows that

$$x(t+T) = x(t) + \int_t^{t+\hat{o}T} [f(x(t')) + g(x(t'))]dt' + \int_{t+\hat{o}T}^{t+T} f(x(t'))dt'$$

$$= x(t) + \int_t^{t+T} f(x(t'))dt' + \int_t^{t+\hat{o}T} g(x(t'))dt'$$

The ideal average model of the PWM controlled system is obtained by allowing the duty cycle frequency to tend to infinity, that is, the period T approaches zero. In the limit, the above relationship yields

$$\lim_{T\to 0} \frac{x(t+T)-x(t)}{T} = \lim_{T\to 0} \frac{1}{T}\left[\int_t^{t+T} f(x(t'))dt' + \int_t^{t+\hat{o}T} g(x(t'))dt' \right]$$

i.e.,

$$\dot{x} = f(x) + \hat{o}\,(x)g(x)$$

For the solenoid coil under PWM control, the governing equation can be represented as below;

$$Ri + L\frac{di}{dt} = \overline{u}V$$

where R : resistance of solenoid coil, L : inductance of solenoid coil, V : source voltage.

Using the above result, the solution of ideal average model of this PWM system can be easily obtained as following;

$$i(t) = \hat{o}\cdot I_s - \hat{o}\cdot I_s\, e^{-\frac{t}{\hat{o}}} + I_0\, e^{-\frac{t}{\hat{o}}}$$

where $I_s = V/R$, I_0 : initial current of solenoid.

So, the average current of solenoid coil converges to the value, which is proportional to duty ratio.

$$I_{average}(\hat{o}) = I_s \cdot \hat{o}$$

This equation means that the change of PWM duty ratio is directly proportional to the change of current of solenoid coil.

Experimental Study of Solenoid Valve Behavior

As mentioned before, the characteristic of solenoid valve must be modeled in terms of PWM parameters to use PWM as a method of current control. Therefore, we studied the static & dynamic characteristics of electrical & electro-hydraulic behavior of solenoid valve experimentally in our test bench shown in Figure 3.

As the first step for the development of our LFC, we checked the opening condition of solenoid valve in various situations and tried to find the relationship among opening duty of valve, coil resistance and pressure difference between wheel cylinder and master cylinder (Δp). Figure 4 shows the results, which represents the relationship between opening duty of valve and Δp.

From this figure, we can find some information how to determine the opening duty of valve. However, we need more accurate information about this relationship to concretely determine the opening duty in LFC. To get such relationship, we additionally consider the source voltage and coil resistance and normalize the opening duty of valve. The new relationship between normalized opening duty and Δp is depicted in Figure 5.

Figure 3: LFC Test Bench

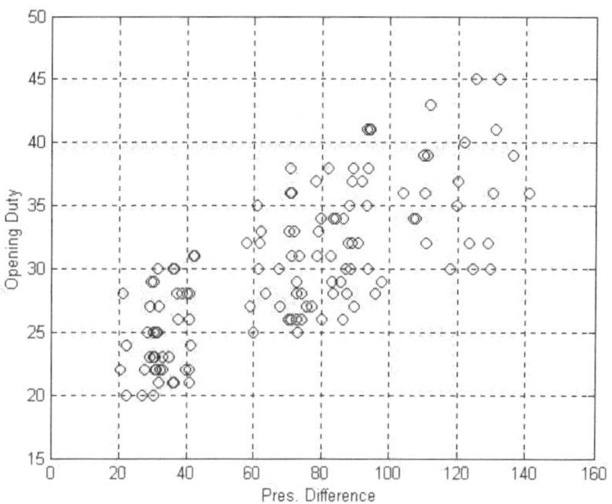

Figure 4: The relationship between opening duty of valve and Δp

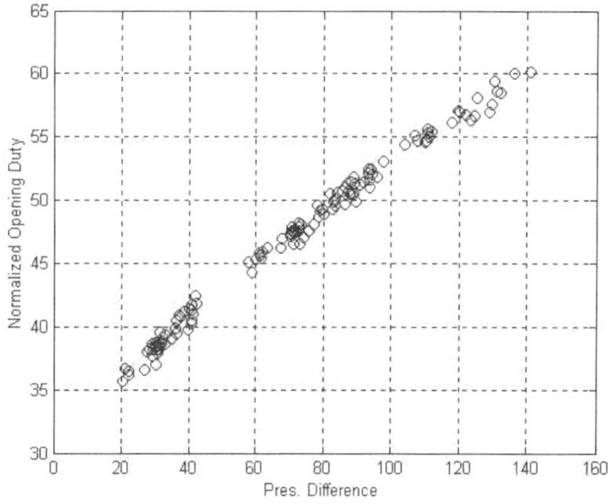

Figure 5: The relationship between normalized opening duty of valve and Δp

From the strongly linear relationship shown in Figure 5, we convinced that Δp, source voltage and coil resistance could be used to decide the opening duty of valve. Once we knew the above relationship, the test data could be transferred to any operating point. Figure 6 shows the data transferred to various Δp, preserving the basic relationship.

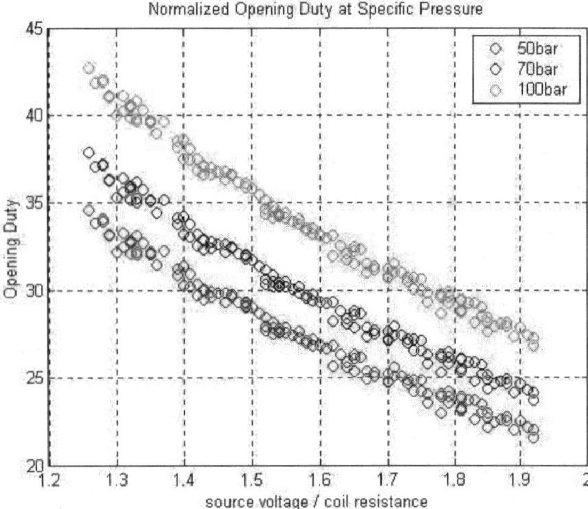

Figure 6: The relationship between normalized opening duty of valve and Δp

DUTY COMPENSATION

The duty compensation is the core function of LFC. As shown in Figure 7, LFC logic consists of two major parts, that is, duty pattern generation and duty compensation. The duty pattern generation module decides the duty pattern from the desired pressure rise rate that is required by ABS logic. This duty pattern must be compensated considering road state, driving situation and electrical environments. In our LFC logic, three categories of duty compensation are carried out as followings;

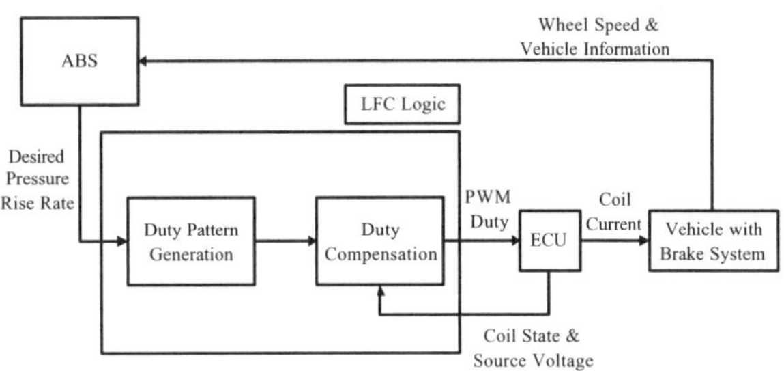

Figure 7: The block diagram of ABS with LFC Logic

Cyclic Compensation

At every control cycle of ABS, LFC observes the road state, driving situation and determine whether prior duty pattern is proper or not. The results are reflected in the next control cycle.

Special Road or Situation Compensation

Special road such as μ_jump should be considered in PWM duty, for it means abrupt change in Δp. Such situation as rough road and double braking also must be considered.

Coil Resistance & Source Voltage Compensation

In Figure 5, we recognized that Δp, source voltage and coil resistance could be used to decide the opening duty of valve that is core parameter of duty pattern of LFC. We can measure source voltage directly and also can obtain coil resistance indirectly by observing source voltage and coil current. These values are used for duty compensation.

MOTOR SPEED CONTROL

Another source of noise and pedal kickback during ABS action is excessive pumping of brake fluid caused by unnecessarily high motor speed. To find out the dominative factor to the motor noise, we conducted repetitive noise tests changing such parameters as motor PWM period, PWM duty and tried noise test with specially designed drive patterns. From the various tests, we concluded that the level of noise and vibration caused by the motor is directly related with the motor speed and the other factors are indirectly related via motor speed.

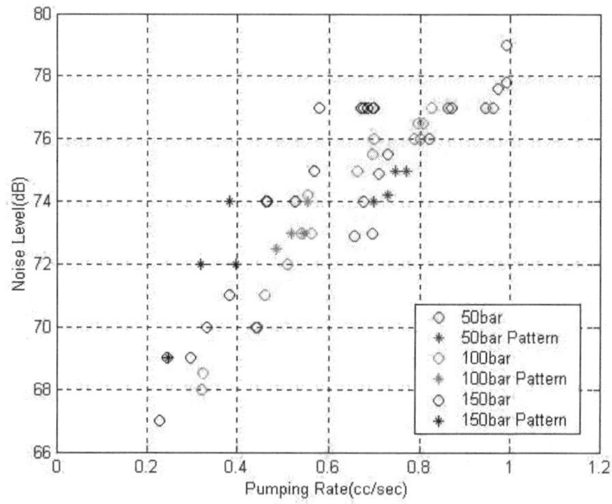

Figure 8: The relationship between motor noise and motor speed

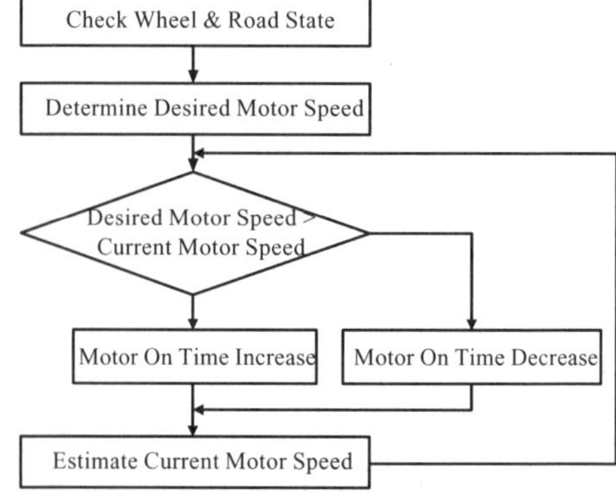

Figure 9: Flowchart for motor speed control

Figure 8 supports this observation, which represents the relationship between motor noise and motor speed in indoor test. Because the level of noise and vibration caused by the motor is directly related with the motor speed, we introduced a new Motor Speed Control (MSC) method that could compromise pumping performance and noise. The brief flow chart for MSC is depicted in Figure 9.

EVALUATION RESULTS IN FIELD TEST

An evaluation of the above LFC & MSC concepts was made in actual vehicle in comparison with conventional On-Off strategy in braking performance and NVH. Our new generation ABS with LFC & MSC achieves better performance in stopping distance compared with conventional control method by reducing average pressure dump and maintaining optimal slip band. We got drastically reduced ABS operation noise by introducing continuous pressure rise instead of pressure pulse-up of conventional On-Off control. Also, our MSC added positive effects for NVH by controlling the motor speed to adequate level. No pressure pulse-up of LFC and gentle pumping of MSC mitigate the harsh pedal feeling.

BRAKING PERFORMANCE

The LFC showed improved braking performance compared with conventional On-Off control because of optimized slip band. As shown in the figure 10, the LFC has continuous and linear pressure rise rate which prevent abrupt wheel locking behavior. As a result, the detection point of pressure dump is much more accurate than conventional On-Off control, resulting in lowered pressure dump.

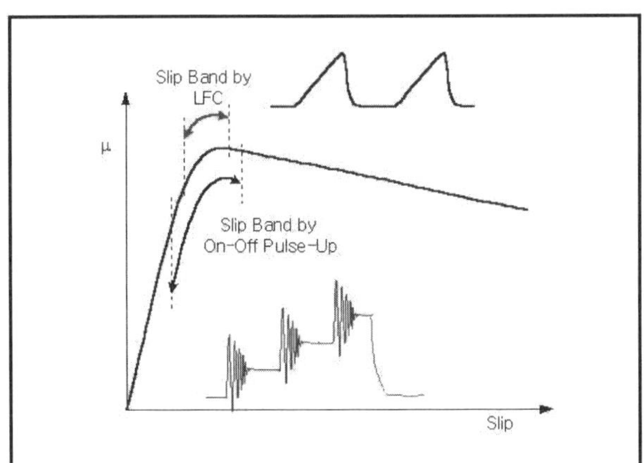

Figure 10: Control loop of LFC and conventional On-Off ABS on μ_slip curve

Figure 11 and Figure 12 show typical results of LFC and conventional On-Off ABS trace in dry asphalt. Above result can be seen in Figure 13, which shows the relationship between wheel pressure and estimated slip ratio of aforementioned trace for LFC and conventional On-Off ABS. The narrow band of slip ratio trajectories of LFC shows that it is optimally controlled compared with conventional On-Off ABS. As a result, our LFC shows improved braking performance compared with conventional On-Off control.

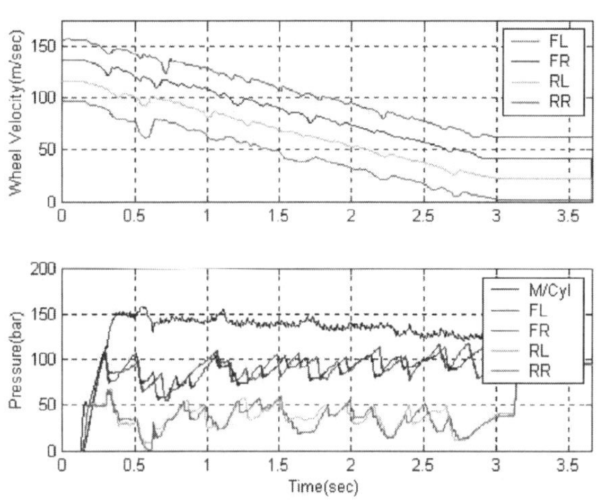

Figure11: Example traces on dry asphalt : LFC (100kph)

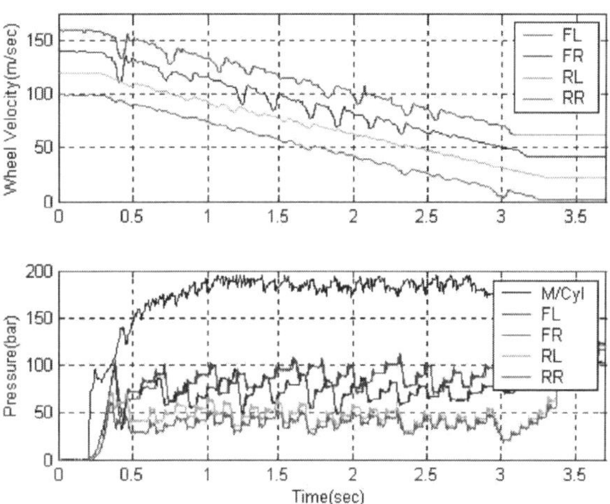

Figure 12: Example traces on dry asphalt : On-Off (100kph)

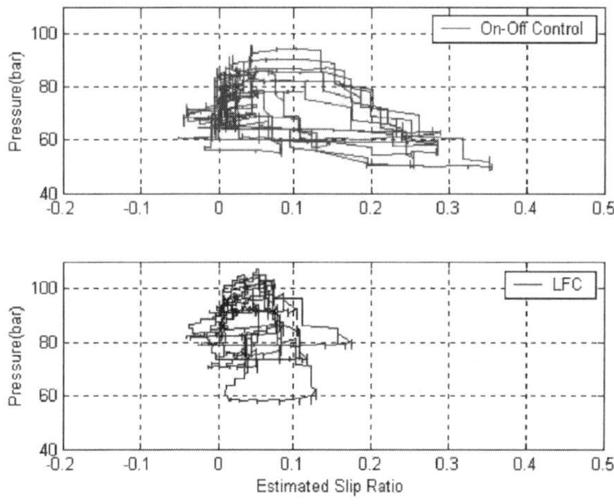

Figure13: Control loop of LFC and conventional On-Off ABS on wheel pressure and slip curve (front wheels)

NOISE REDUCTION

The noise generation during ABS operation mainly caused by the pulse-up pressure rise and pulse-down dump of conventional On-Off control. The main objective of the development of LFC is to reduce the pulse-up noise by introducing continuous pressure rise. This was successfully accomplished and we got drastically reduced ABS operation noise. In addition, MSC also added positive effects for NVH by controlling the motor speed to adequate level. Our noise test results for various road conditions are summarized in Table 1.

Conditions	High-μ	Medium-μ	Low-μ
On-Off	79 ~ 80dB	77 ~ 79dB	75 ~ 76dB
LFC	76 ~ 77dB	73 ~ 75dB	67 ~ 68dB
Noise Improve	About 3dB	About 4dB	About 8dB

Table 1: Control loop of LFC and conventional On-Off ABS on μ_slip curve

PEDAL FEELING

Pedal kickback is the force and stroke variation due to the pressure and volume change in the master cylinder.

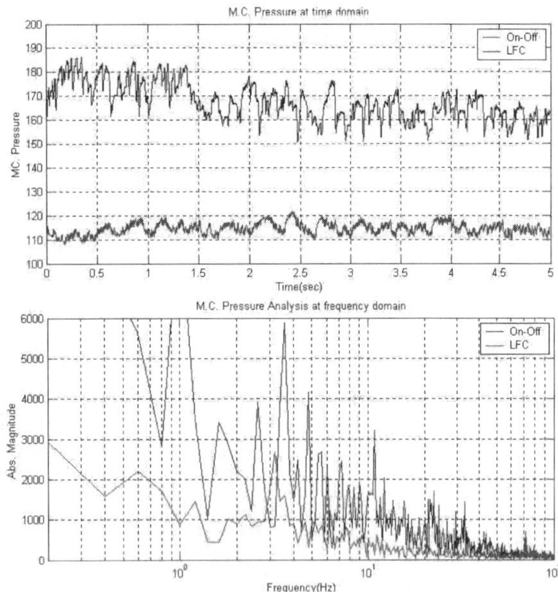

Figure 14: Spectral density of pressure profile of LFC & MSC compared with conventional On-Off ABS

This variation in pressure and volume is caused by pressure pulse-up and pumping of motor. Pressure pulse-down in wheel cylinder does not make such variation directly but makes indirect influence via pumping of motor. Introducing LFC & MSC both takes good effects on pedal feeling directly and indirectly. No pressure pulse-up of LFC and gentle pumping of MSC mitigate the harsh pedal feeling. To figure out the effects of LFC & MSC on pedal feeling, the pressure profiles of master cylinder under conventional On-Off control and LFC & MSC were analyzed in frequency domain. Figure 14 shows the spectral density of pressure profiles. As we can see in this Figure, the positive effects of LFC & MSC on pedal feeling are very clear.

CONCLUSION

In this paper, we reviewed recent results of LFC & MSC for our new generation ABS in the view of performance and NVH. We evaluated our LFC & MSC concepts in actual vehicle in comparison with conventional On-Off strategy. Our conclusions are as follows;

- Our new generation ABS with LFC & MSC achieves better performance in stopping distance compared with conventional control method by reducing average pressure dump and maintaining optimal slip band.
- We got drastically reduced ABS operation noise by introducing continuous pressure rise instead of pressure pulse-up of conventional On-Off control. Also, our MSC added positive effects for NVH by controlling the motor speed to adequate level.
- No pressure pulse-up of LFC and gentle pumping of MSC mitigate the harsh pedal feeling.

REFERENCES

1. Seongho Choi, Dong-woo Cho, "Design of Nonlinear Sliding Mode Controller with Pulse Width Modulation for Vehicular Slip Ratio Control", International Journal of Vehicle Mechanics and Mobility, Vol.36 No.1, July 2001, pp. 57-72.
2. R. Kazemi, B. Hamedi and B. Javadi, "A New Sliding Mode Controller for Four Wheel Anti-Lock Braking System (ABS)", SAE paper No. 2000-01-1639.
3. Huiyi Wang, Bo Gao and Jian Song, "Approach of Debugging Control Laws of ABS Combined with Hardware-in-the-Loop Simulation and Road Experiment", SAE paper No. 2001-01-2729.
4. Li Jun, Zhang Jianwu and Yu Fan, "An Investigation into Fuzzy Control for Anti-Lock Baking System Based on Road Autonomous Identification", SAE paper No. 2002-01-0599.

A Model Based Design Analysis of Hydraulic Braking System

Ijin Yang, Woogab Lee and Inyong Hwang
Central R&D Center, MANDO Corporation

ABSTRACT

Several approach have been used to design Mando MGH ABS hydraulic system. Each component is analyzed with parameter study and finally made to a customized model that reflect the real components. Parameters selected are applied to design of series production ABS(MGH-20). This process has been used to the components of solenoid valves, pump, motor, inflow conduit, hydraulic chamber and fluid property throughout system operating range. This allows the function of complex hydraulic system to be estimated at limiting conditions such as low temperature, low voltage and mixed conditions as well as at normal conditions. This paper presents a model-based component design analysis by comparing with test data and a parameter study of hydraulic braking system for optimal pressure generation. New generation ABS, ESP, EHB hydraulic braking system of MANDO are investigated more accurately and easily with combination of proven component model early in the design engineering process. This approach is of great use to take account of systematic analysis of hydraulic braking system at all operating conditions, to pre-estimate the system in combination of components with different characteristics and to focus on the way of improvements in required function. In addition, the design analysis can be readily performed just connecting the accumulated component model and product reliability can be secured from the beginning

INTRODUCTION

In recent days brake slip control systems are required to meet sophisticated customer's demands of specific functions and endurance of the product. The technologies for the advanced braking system of Active Safety Control System such as Electronic Stability Program(ESP) and Brake-By-Wire(EHB, EMB) are needed in the market to secure the driver's safety also. Highly contested market urges manufacturer to develop a new product in a short period and in addition to that at a lowest price.

Developing with only experience spends too much time and lack of confidence makes long time debugging cycling. Finally this leads to a tightly regulated tolerance of component dimensions that is at the high price.

The one way to avoid that kind of troublesome development process is to use mathematical modeling techniques. This method reduces prototype building time cycles, helps to find the critical design parameters and secure robust design through out predetermined boundaries of system guarantee from the beginning.
Brake slip control system is composed of many sub-systems that have peculiar characteristics such as mechanical, electrical, electromagnetic and combined ones.
The analysis of the hydraulic braking system is performed with AMESim® for the hydraulics and MAXWELL® for the elecromagnetic field. Each sub-system is made into a customized model by doing parameter study for the every component. Essentially advanced hydraulic braking system itself is the combination of the sub-systems such as solenoid valves, motor, pump, hydraulic chamber, pipe and working fluid. In that respect, defining the intrinsic characteristics of these sub-system is makes it much easier to analyzed the complex system just by organizing the proven sub-system simulation models. Sensitivity analysis is conducted with linked sub-system with worst case conditions like low voltage, high pressure and extreme temperature for the brake slip control system as well as each component. This gives a robust hydraulic braking system design.

The next step is co-simulation and interface with another analytical tools of MATLAB® for the control logic, vehicle dynamics and system integration based on the hydraulic component models determined here.

This paper describes the modeling process of each component of an Anti-lock Braking System. In order to validate the model, a hydraulic circuit simulation of an ABS brake event using the model was compared to an actual vehicle ABS braking test.

SUBMODELS FOR BRAKE SLIP CONTROL SYSTEM

- **BRAKE FLUID**

Starting with numerical analysis of each component of Anti-lock Braking System, one must define the property of working fluid, brake oil, to reflect the dynamic behavior of pressure more precisely according to temperature and pressure variation. To consider real conditions, fractional air content by volume, saturation pressure are introduced and property data of the brake fluid are used [1][2].

$$\frac{V_{air}}{V_{liq} + V_{air}} \tag{1}$$

Density can be defined by Taylor's approximation and have the relation with bulk modulus

$$\rho = \rho_o + (\frac{\partial \rho}{\partial P})_T (P - P_o) + (\frac{\partial \rho}{\partial T})_P (T - T_o) \tag{2}$$

$$\beta = \rho \frac{dP}{d\rho} \tag{3}$$

where ρ is the density of the fluid air mixture, P its pressure and β is the bulk modulus.

Figure 1 shows the calculated bulk modus with applied brake pressure at different temperature. This result is reflected consistently for hydraulic system simulation with variation of temperature.

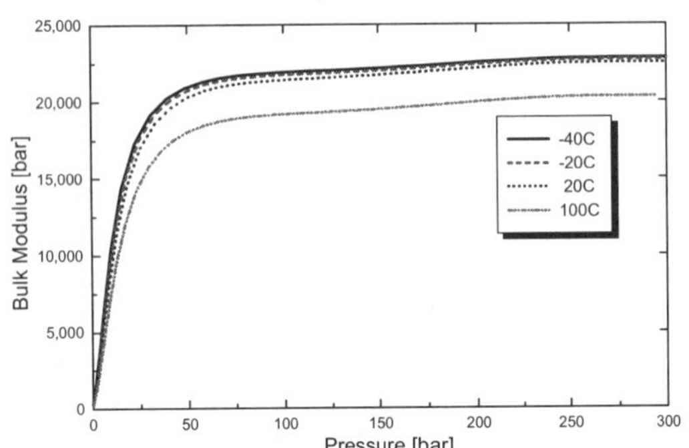

Figure 1. Calculated bulk modulus of brake fluid with variation of temperature

● **Brake Wheel Cylinder Modeling**

Actual brake cylinder modeling is cumbersome because it should reflect material property and mechanical data. For convenience, pressure and volume displacement curve is modeled by the equation

$$V_{cyl} = aP_{cyl}^{b} \tag{4}$$

where V_{cyl}, P_{cyl} are the cylinder volume and pressure respectively. a and b are constants.

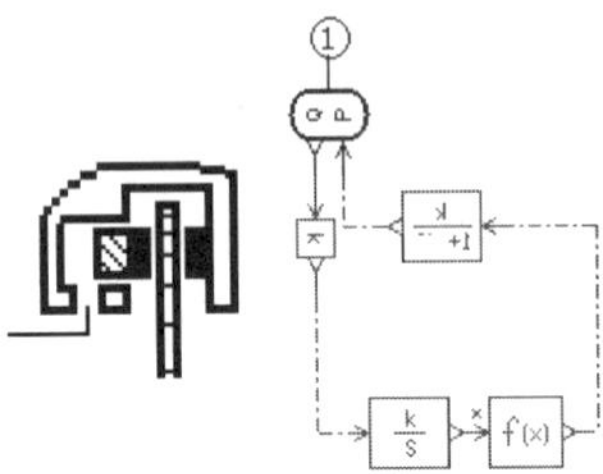

Figure 2. Pressure and volume characteristics model of the wheel cylinder

● **Solenoid Valve Modeling**

Analysis of solenoid valve functional characteristics can be investigated with several methods. Here we discuss the time response parameters concerned switching time of moving parts, electromagnetic field and hydraulic behaviors of orifice.

As soon as the voltage is applied to the coil's terminals, the current begins to rise and magnetomotive force grows and finally overcomes mechanical and hydraulic force. The time elapsed from the applying voltage to the beginning of the armature is defined waiting time. Considering air gap, while remains, causes much greater magnetomotive force drop than all ferrous parts, electric field can be described

$$U = iR + \frac{d\psi}{dt} = iR + L_o \frac{di}{dt} \tag{5}$$

where

U : applied voltage

R : resistance of coil

i : current

L_o : initial inductance with maximum air gap

ψ : Magnetic flux

The solution of equation (5), waiting time is

$$t_w = \frac{L_o}{R} \ln \frac{I_y}{I_y - i_w} = \frac{L_o}{R} \ln \frac{1}{1 - \frac{i_w}{I_y}} \tag{6}$$

where

I_y : U/R(maximum current)

i_w : Current that armature starts to move

The magnetic force can be computed

$$F_m = \frac{1}{2} i^2 \frac{dL}{d\delta} \tag{7}$$

where δ is air-gap. The solution of this equation

$$i_w = \sqrt{\frac{2 F_{mech} \delta_0}{L_0}} \tag{8}$$

where $F_{mech} = F_{hyd} + F_{spring}$

Finally the waiting time is described

$$t_w = \frac{L_o}{R} \ln(\frac{1}{1 - \frac{R}{U} \sqrt{\frac{2 F_{mech} \delta_o}{L_o}}}) \tag{9}$$

Rising time is defined the time elapsed from the beginning of the armature movement to the its full stroke. The armature, striving to close the air gap, increases thereby the resulting inductance of the solenoid, decreasing simultaneously the total resistance of its magnetic circuit. Therefore, a drop in current is observed. With some numerical simplification by defining current and magnetic flux as a 3rd order polynomial similar to exponential increase. The rising time can be expressed

$$t_r = \sqrt[3]{\frac{3 \delta_o m}{U \sqrt{\frac{F_{mech}}{2 \delta_o L_o}} - R \frac{F_{mech}}{L_o}}} \tag{10}$$

With the parameters of equations above, geometric boundaries and relations with exciting coils are pre-described in terms of time response.

Geometrical arrangement between ferrous materials and the coil, in addition, material selection should be investigated in the process of solenoid valve design. FEM analysis with Maxwell[®] EMSS package is carried out with respect to magnetic characteristic values of all solenoid valves of brake slip control system. This analysis assures reliable magnetic force against worst conditions like low voltage, high temperature and big pressure. Test results with special fixture at these conditions show a good agreement with the simulated ones. Finally developing time is shortened with the numerical analysis compared to trial-and-error method with the engineer's intuition.

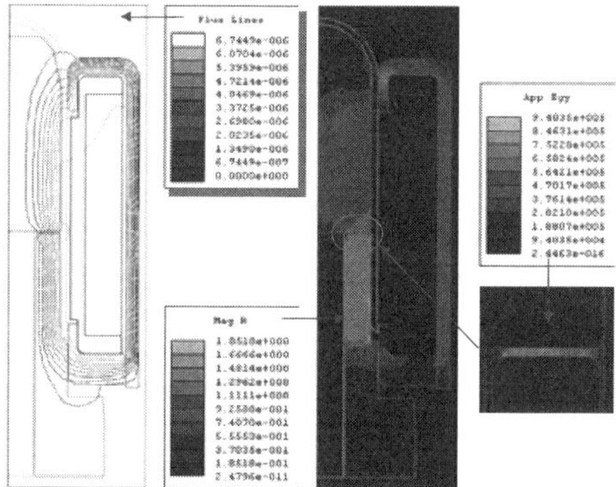

Figure 3. Magnetic flux lines and flux density (B) of the Inlet solenoid valve

Hydraulic behaviors of the solenoid valve at static and dynamic conditions are simulated regarding pressure rise and dump rate with variation of orifice size and seat geometry. Inlet solenoid orifice should have the orifice size that drivers don't feel different pedal feel compared to conventional braking system also should be easy to modulate pressure during slip control. The rise rate of inlet solenoid valve should be grouped to cope with car variations due to volume vs. pressure characteristics of brake wheel cylinder.

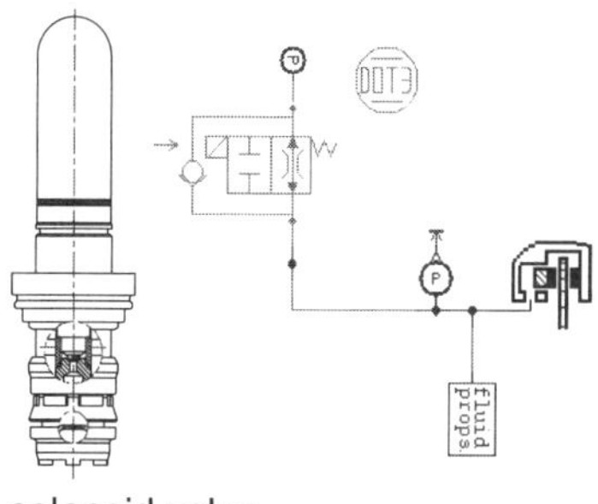

Figure 4. Schematics of model based inlet solenoid valve

Parameter study and sensitivity analysis of orifice size are executed by comparing test data of constant pressure apply from the inlet port. The next step is pressure pulse-up modulation with frequency of 2 Hz operation with open command of duration 6ms.

Continuity equation of fluid is described by equation (11)(12)

$$\Sigma Q_{in} - \Sigma Q_{out} = \frac{V}{\beta}\frac{dP}{dt} + \frac{dV}{dt} \tag{11}$$

$$Q = C_d A \sqrt{\frac{2\Delta P}{\rho}} \tag{12}$$

where β is bulk modulus of elasticity, Q is volume flow rate, P and V is the pressure and volume at the brake chamber respectively, C_d is the discharge coefficient of the orifice, A is the orifice section area.

Damped pressure rise with static condition and dynamic switching pressure pulsation are observed at the pipe line by operating inlet valves in Figure 5. Orifice sizes are selected to guarantee adequate pressure rise by several levels. Comparing to the test results of bench tester at room temperature, the errors remain within 5% for all orifice size. With parameters selected, change of the pressure pattern is easily estimated by the effect of low temperature and pipe length. Hydraulic pipe sub-models are referred to the lumped parameter pipe model with compressibility of oil expansion of line walls, air release and cavitation, gravitational effects, pipe friction, fluid inertia[1].

Figure 6 shows pressure rise curves in the Conventional Braking System(CBS) and in ABS system at maximum orifice size selected. One can find out the considerable decease of rise rate at low temperature and with long pipe length due to increase of dead volume, pipe surface friction and compressibility effects.

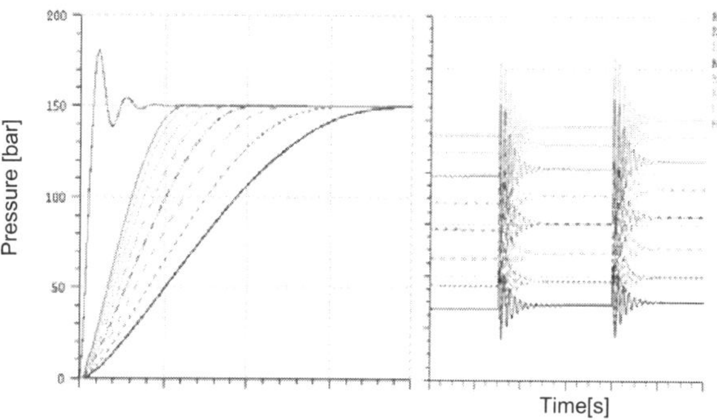

Figure 5. Inlet valve pressure rise pattern with different orifice size

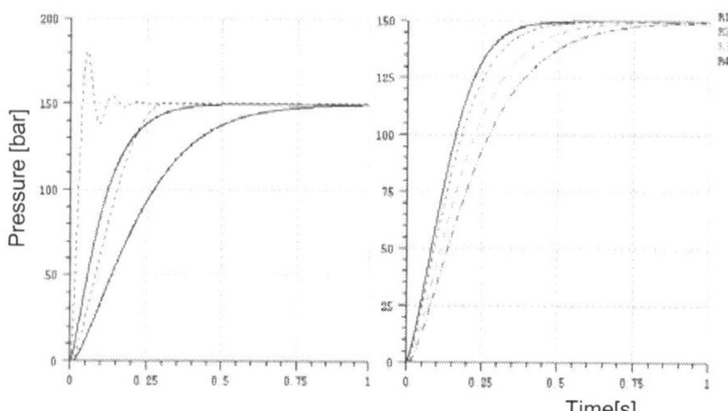

Figure 6 Inlet valve pressure rise pattern with variation of temperature(20℃/-20℃) and pipe length(10mm, 1m, 3m, 5m)

Outlet solenoid valve has a function of pressure dump from the wheel cylinder to prevent wheel locking. Usually orifice size of outlet solenoid valve is selected to get a same or faster response of pressure than that of inlet solenoid valve. Parameters are searched to get an appropriate orifice size and sensitivity analysis is accomplished for the geometrical tolerance of each component such as seat diameter, angle and orifice size tolerance. With same flow coefficient for the orifice flow, pressure dump rate remains within 3% tolerance of test data.

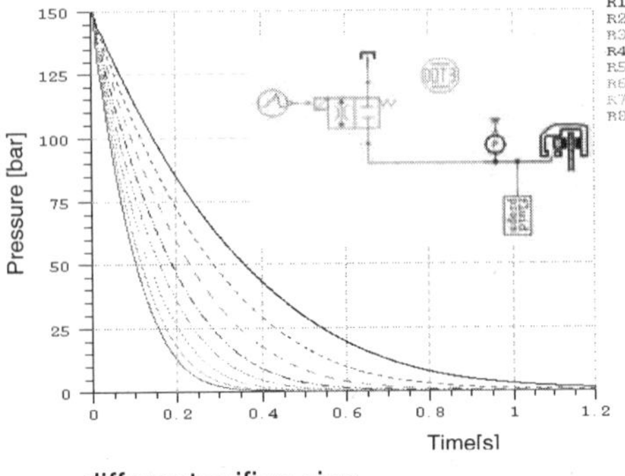

Figure 7. Outlet valve pressure rise pattern with different orifice size

In the same manner high viscosity of operating fluid at low temperature delays pressure dump rate as can be in Figure 8

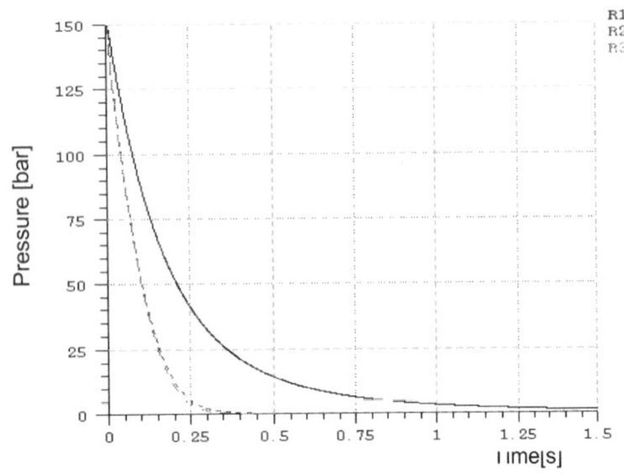

Figure 8. Outlet valve pressure dump pattern with variation of temperature(-20℃/20℃/100℃)

- **DC Motor and Reciprocating Hydraulic Pump Modeling**

DC Motor

Designing DC motor itself is a hard job. Considering the objective of this paper motor performance is simplified by the equation (13)(14).

$$N = N_O - kT \tag{13}$$

$$T = AP_{cyl}\, e \sin(\theta) \tag{14}$$

where

N_O : rotation speed without external load

k : coefficient of decrease rate of rotation speed

T : torque exerted by hydraulic piston

A : piston section area

θ : rotational angle of the motor shaft

e : eccentricity of the motor shaft

For convenience and short simulation time, the assumption of rotating speed of motor is widely used as a constant value. This is only suitable for the pressurized piston motion with inlet pressure, but for active braking system such as Traction Control System (TCS) or Electronic Stability Program(ESP), motor speed variation with time should be considered from the beginning of command signal to the motor.

Simulation model for motor is generated with this concept and is shown in Figure 9.

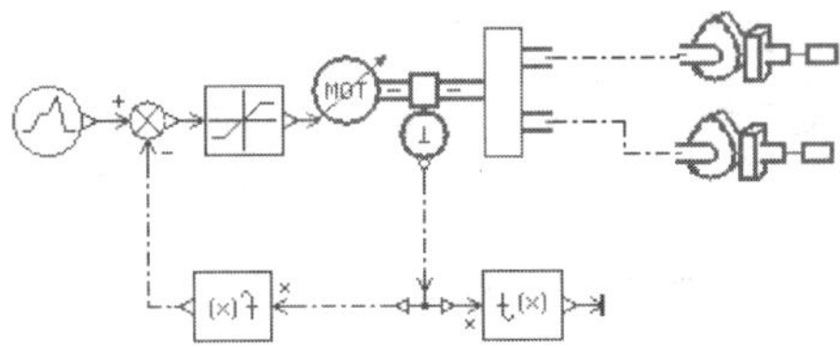

Figure 9 Model based DC motor model

The advantage of this modeling is to monitor the rotational speed and torque changes with time domain at different load pressures exerted by reciprocating piston pump. This makes it possible to find out the duty cycle appropriate to the control concept with Pulse Width Modulation(PWM) of an Anti-lock Braking System. In addition, required motor performance curves can be decided easily for the new generation motor.

Reciprocating Hydraulic Piston Pump

Reciprocating hydraulic pump plays an important role in brake slip control system as a pressure generator. It should have a capability to discharge the fluid rapidly from the Low Pressure Accumulator (LPA) to prevent wheel lock and to generate pressure to the wheel cylinder by making fluid inflow from the reservoir.
Pump is comprised of low pressure and high pressure check valves. First one is attached to the moving piston the other is stationary. This type of piston is easy to install to the valve block and has a small number of components.

To develop brake slip control system with coverage from the small passenger cars to the heavy-duty trucks, all design parameters of pump with combination of motor are investigated before making functional samples. Parameters investigated are outlet valve spring characteristics and its closing pressure, outlet valve seat diameter at constant valve closing pressure, master cylinder counter pressure, piston diameter, pump inlet pressure and eccentricity of motor shaft with combination of all operation temperature and applied voltage to the motor. Finally robust design is achieved with respect to stability of hydraulic system.
Figure 10 shows the pumping flow rate comparison with parameter selected simulation model and test data at the same inflow pressure and master cylinder counter pressure

Jet force value is used to describe check valve motions[3]. When a valve is opened, a net force is produced in a direction that always closes the valve. Its value is

$$F_{jet} = \rho \frac{Q^2}{S(x)} \cos\theta \tag{15}$$

where $S(x)$ is open area and θ is jet force angle.

Figure 11 shows inlet check valve lift and compression chamber pressure at temperature 20℃ and −20℃ with counter pressure 150 bar. Higher viscosity at low temperature lead to bigger inlet valve opening lift and increased pumping flow rate.

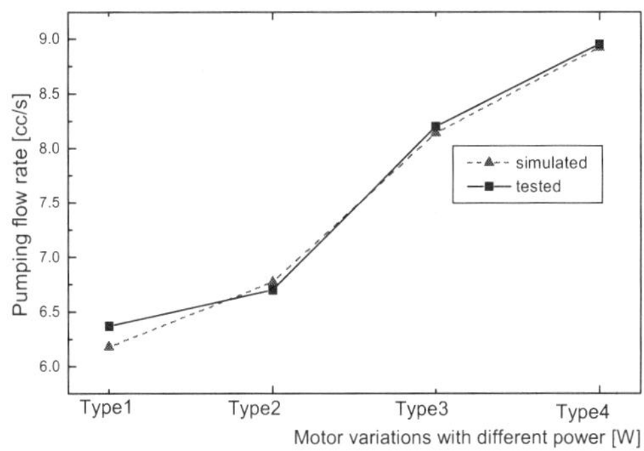

Figure 10 Volumetric pumping flow rate with different motor power

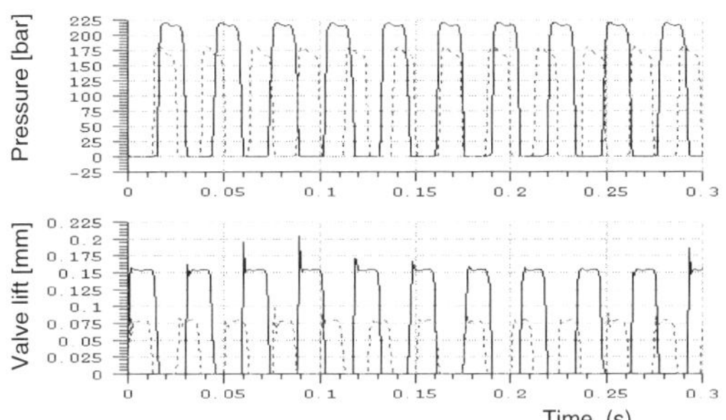

Figure 11 Inlet check valve lift and compression chamber pressure at temperature 20℃ and −20℃

The pump performance is not influenced significantly by the closing pressure determined by the outlet valve spring force. A more significant influence on the pump performance is produced by the piston diameter and motor eccentricity. The pump performance decreases continuously for a decrease in piston diameter even at higher master cylinder counter pressure. With a same motor, decreased piston diameter Φ 7 compared to Φ 8 shows 17% drop of pumping rate, and increased eccentricity e=1.0 compared to e=0.8 with Φ 8 and Φ 7 shows 10% and 15% increase of pumping rate respectively. The main reason for higher rise of pumping flow rate at small diameter piston is the increase of rotation speed caused by reduced torque of motor. Therefore, if the piston diameter is varied, the motor performance has to be adjusted to achieve an adequate pump performance.

For the active braking systems, pump induces inflow by making pressure drop at the compression chamber during suction stroke. In that case the pressure drop from the reservoir to the pump inlet should be as lower as possible. Actually reducing flow resistance is the way of improving pump efficiency. Flow resistance can be reduced by shorter installation distance from the reservoir with bigger inner diameter pipe and by optimum hydraulic circuit with lower hydraulic resistance in the valve block. Figure 12 shows the pressure generation in the specific wheel cylinder according to the pressure drop between reservoir and pump inlet during same operating time of motor. (-) signed pressure means under gauge pressure. Considering the deterioration of pressure rise, hydraulic unit installation conditions should be carefully examined to a vehicle.

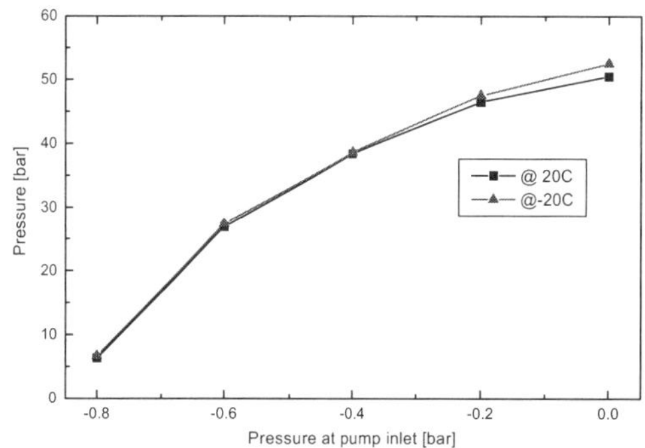

Figure 12 Pressure generation at the brake cylinder with variation of pressure drop

VALIDATION OF MODEL-BASED COMPONENT WITH VEHICLE TEST

The sub-assembly models of anti-lock braking system are integrated together into a hydraulic system just connecting the input and output port easily. Vehicle test of straight stop at the speed of 50kph on the dry asphalt is simulated and compared with vehicle test data of project car. Operating command to the solenoid valve and motor are realized together with data file. Figure 13 shows a schematic diagram of hydraulic circuit of Mando ABS system(MGH-20) established by the model-based component.

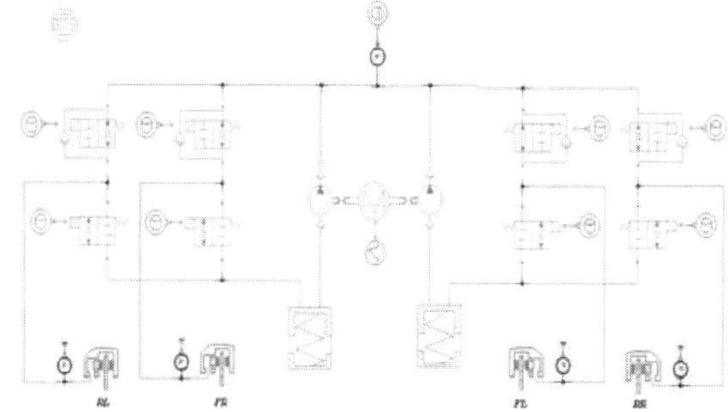

Figure 13 Schematic circuit diagram of ABS system

The simulation results compared with vehicle test are shown in Figures 14,15

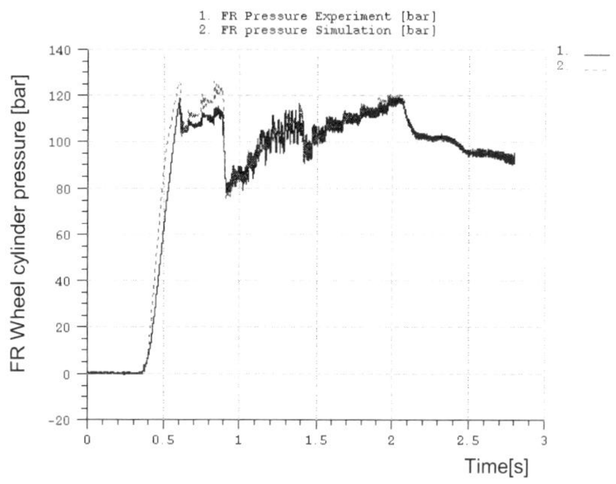

Figure 14 Comparison of brake pressure FR on the dry asphalt condition, 50kph

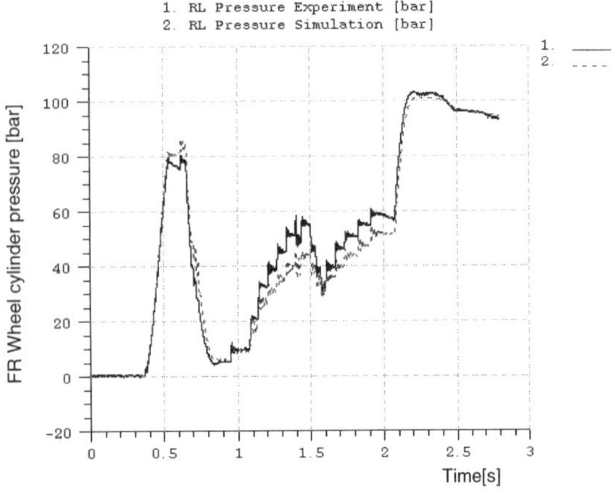

Figure 15 Comparison of brake pressure RL on the dry asphalt condition, 50kph

The differences in the pressure pulse-up modulation period are caused by
(i) Simplified valve model without time delay and dynamic motion to reduce the simulation time
(ii) Simplified caliper modeling without considering the dynamics of caliper or drum brake

The results show a considerably good estimation of brake pressure with the vehicle test data.

CONCLUSIONS

Development process for the hydraulic system of brake slip control system is established with numerical analysis. All the hydraulic and electronic components of an Anti-lock Braking System are made into customized models using numerical analysis together with simulation program of AMESim® and Maxwell® EMSS package.

By the way of parameter study, each sub-component of an Anti-lock Braking System is customized by verifying the data of test bench. The parameters selected as a typical value in a component is used same in a combined hydraulic system except for the properties of an operating fluid.

The whole hydraulic circuit of braking system is established just by adding pre-described sub-component and linking together with hydraulic lines and control commands. And it is simulated reliably with changing global parameters concerning brake fluid properties at normal and extreme conditions. And performance data of electromechanical sub-component of solenoid valves and DC motor are used at each corresponding situation. Advantage of this approach is an easy analytical access to complicated systems such as Electronic Stability Program(ESP) and Electro-Hydraulic Brake System(EHB).

Finally estimation of the pressure behaviors at vehicle level is possible easily. With the aid of numerical analysis, developing time for a new model is shortened and the robust design against the worst operating conditions is acquired from the beginning of design process. This helps engineers to understand hydraulic, electromagnetic characteristics and mixed situations visually by changing design parameters of dimensions and tolerance of every component. Sensitivity analysis is carried out for the critical dimensions simultaneously for the brake slip control system.

REFERENCES

1. "Hydraulic pipe/hose submodels, technical bulletin no. 105", IMAGINE S.A, 2002
2. Dipl.-Ing.Robert Mutscheler, "Untersuchung zur Konzeption einer elektrohydraulischen Energieversorgung für zukünftige Pkw-Bremssysteme unter besonderer Berücksichtigung der Pulsationsminderung," 1997
3. "AMESim Hydraulic Component Design Library User Manual", IMAGINE S.A., 2002
4. Qingyuan Li, Keith W. Beyer and Quan Zheng, "A Model-Based Brake Pressure Estimation Strategy for Traction Control System", Delphi Automotive Systems, 2001
5. Aravind S. Bharadwaj, "Multi-Domain Modeling and Simulation of Mechatronic Systems", GM Powertrain Group, 1997

ABS System Validation: Integrating Tone Rings and Wheel Speed Sensors in HIL Simulation

Mark Bennett and Michael Tober
Bendix Commercial Vehicle Systems, LLC

ABSTRACT

The utility of Hardware-In-The-Loop (HIL) simulation for Antilock Braking System (ABS) development and validation is well known. A continuing challenge is the simulation of wheel speed signals normally produced on a real vehicle by magnetic pickup wheel speed sensors and toothed exciter wheels (tone rings).

This paper presents motivation for, and implementation details of, a state-of-the-art HIL implementation using actual wheel speed sensors and tone rings, velocity-controlled electric motors, and real-time vehicle models connected to air brake system hardware. This implementation differs from prior HIL systems by allowing ABS system validation using actual wheel speed sensors and tone rings.

The inclusion of tone rings in the HIL environment allows simulation of proving-grounds-type testing in a controlled laboratory environment, while preserving the realism of actual sensors and tone rings. Sensors and tone rings of interest could be new or modified designs, damaged units returned from the field, or test units purposely damaged to represent real or imagined worst case scenarios. The basics of HIL simulation are covered only in summary, but details can be found in the other SAE papers referenced herein. This paper assumes the reader has some familiarity with HIL systems.

INTRODUCTION

The state-of-the-art in ABS control system development requires stand-alone desktop simulation capability, HIL simulation capability, and real vehicle testing. In the case of the HIL environment, the engineer is challenged with choosing what system components to model and simulate, and what components to connect in their actual form via HIL hardware. Historically this has resulted in the system supplier's prototype or production components being included in their actual form, while the remainder of the system is modeled mathematically. For example, the typical HIL system for a brake system might include some or all of the production brake components in real form, yet the vehicle body, drivetrain and suspension, road profile, and driver themselves are modeled in some mathematical fashion. Conversely, the implementation could be exactly the opposite, whereby a prototype control system is implemented on an HIL platform and is then connected to a real vehicle for testing and development. The majority of this paper will focus only on the case in which the vehicle itself is modeled and the brake control system is connected, via the HIL platform, as real components.

Implementing an HIL system presents the issue of separating real components from modeled components. Deciding what components to model and what components to connect in their real form introduces issues of cost, complexity, and risk of inaccuracy. In contrast, a full vehicle test at the proving grounds introduces issues of repeatability, speed, and environmental noise, among others.

In a typical HIL system, the bridge from reality to virtual reality is the desktop PC and specialized hardware to allow the modeled components to interact in real time with the actual production or prototype components. Specific to the topic of this paper, it is typical to simulate the wheel speed sensor and tone ring combinations using a variety of electronic means to produce sine wave signals similar to those produced by the actual wheel speed sensor. Practically speaking, this often means off-the-shelf sine wave generators, digital I/O boards, and simple circuits to accomplish synthesis of wheel speed signals. In contrast, this paper details an HIL system that includes actual wheel speed sensors and tone rings in the realized side of the HIL environment. The critical element of such an implementation is an electric motor control system capable of physically replicating the rotational wheel dynamics of a real vehicle, yet easily coupled to commercially available HIL hardware.

MOTIVATION

Clearly, the addition of electric motors to an HIL system entails additional expense, complexity, and space requirements as compared to typical systems. To gauge the necessity of such a system, a further understanding of the wheel speed sensor system itself must be obtained, along with an understanding of common ways to simulate and synthesize that system.

THE WHEEL SPEED SENSING SUBSYSTEM

As an oversimplification, the wheel speed sensor provides a signal to the Electronic Control Unit (ECU) that is a sine wave with frequency proportional to the rotational speed of the wheel. The ECU simply calculates the wheel speed from the frequency. To simulate wheel speed signals to this level of realism, a voltage-controlled sine wave generator is sufficient.

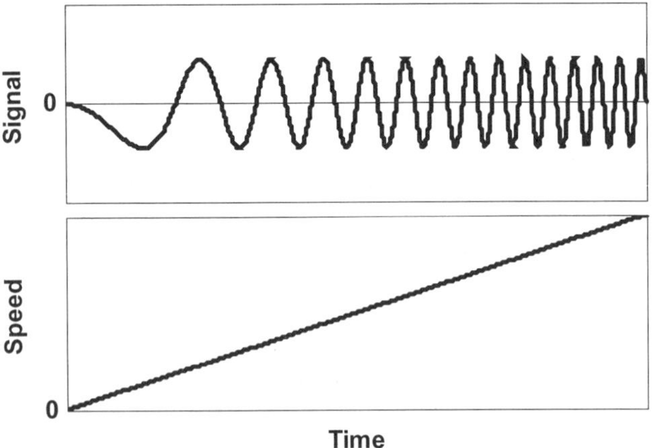

Figure 1: Simplified Wheel Speed Sensor Sine Wave and Corresponding Wheel Speed

Adding one degree of realism, the wheel speed signal *amplitude* is also proportional to speed. The ECU calculates the wheel speed from the frequency, but can only measure frequencies of sufficient amplitude. Perhaps more importantly, the signal's amplitude is used for fault detection, troubleshooting, and failsafe ABS modes of operation. To simulate wheel speeds to this level of realism, the aforementioned sine wave generator is needed as well as some simple circuits to scale the sine wave amplitude as a function of speed.

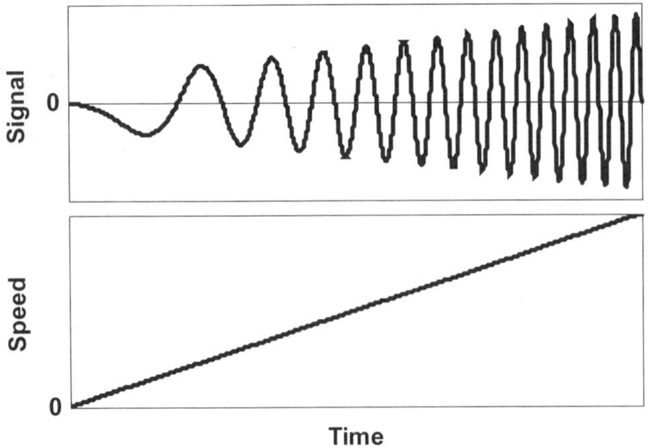

Figure 2: Amplitude Modulated Wheel Speed Sensor Sine Wave and Corresponding Wheel Speed

Adding yet another degree of reality, the wheel speed signal is not always a perfect sinusoidally-shaped waveform. Magnetic and physical effects of the spinning tone ring can induce a forward slant to the wheelspeed signal, as shown in Figure 3.

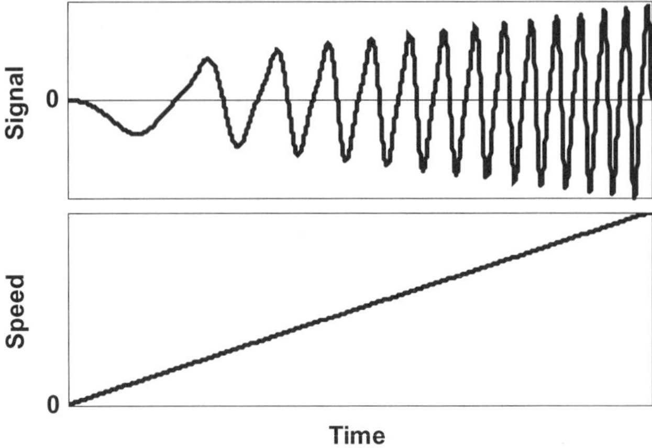

Figure 3: Practical Wheel Speed Sensor Sine Wave and Corresponding Wheel Speed

The final and perhaps most pervasive degree of realism is that of wheel speed *noise*. In general, noise can come in many forms and have many effects that are detrimental to performance. Subsequently, it will be claimed that these attributes of wheel speed signals are extremely time-consuming and somewhat difficult to model – meanwhile the accuracy of such models is suspect.

In the case of heavy truck ABS systems, the noise source can be mechanical or electrical in nature, and can be part of the wheel speed sensing system itself or some external source. Specifically, some examples of wheel speed noise sources can be:

- Tone ring axial and/or radial runout
- Non-uniform tone ring tooth pattern
- Chipped or missing tone ring teeth
- Hammer dents or other damage done to tone ring during service
- Sensor loose and vibrating within socket
- Tone ring not firmly affixed to hub
- Tone ring teeth ground flush by loose axle bolt
- Sensor wire chafed by rotating drivetrain component
- Brake squeal, screech, howl, or judder
- Torsional drivetrain vibration and/or slack
- Suspension vibration
- Sensor core intermittently grounding to tone ring
- Radio Frequency Interference

The heavy truck industry is unique in that the OEM often has design responsibility for wheel-end components, including tone rings and sensor mounting and cabling. Additionally, more than one supplier's brake system can be fitted to a given OEM platform.

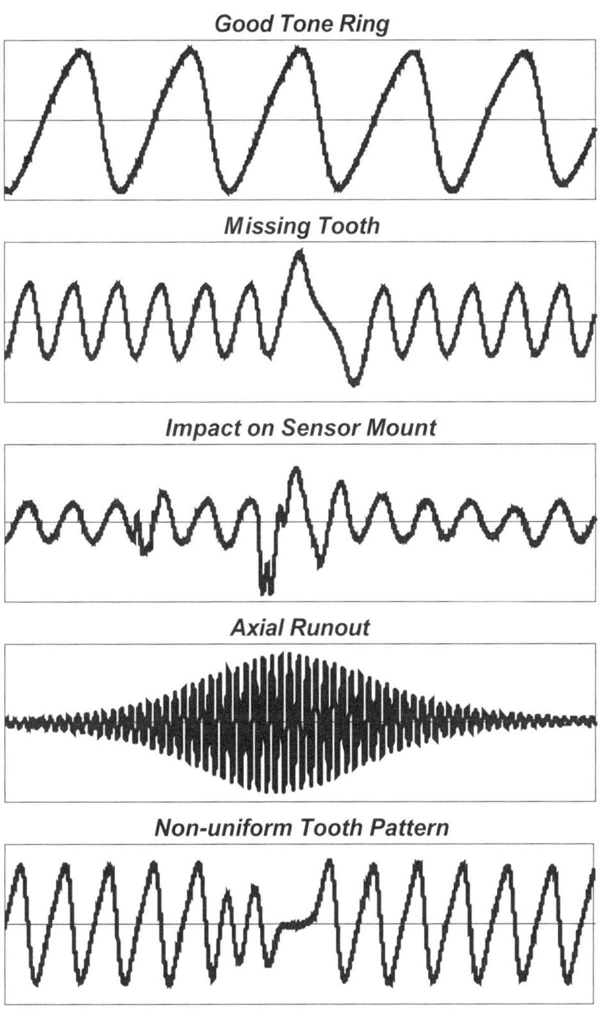

Figure 4: Actual Wheel Speed Sensor Signal Oscilloscope Traces

realism mentioned in the previous section is very challenging.

Consider an example of a tone ring with one or more chipped teeth. To model this component, the engineer must first measure the signal from such a tone ring. He then must find a way to express this tone ring mathematically. Even if he does this successfully, he must then account for different sensor gaps, different tone ring configurations, different sensor types, and any other relevant variances that the OEM may include on their platform. His model will likely be cumbersome and hard to maintain, and require much empirical tuning, experience, and intuition to use effectively. The combination of high effort and questionable results provides motivation to find a better solution.

The general solution proposed is to include the tone ring and wheel speed sensor in the HIL system as real, rather than modeled, components. The particular solution discussed is to use velocity-controlled motors in closed-loop with a vehicle model. In this manner, the motors must be able to *accelerate* the tone rings as quickly as a truck could accelerate its own wheels (a full throttle launch on ice, for example). Similarly, the motors must be able to *decelerate* the tone rings as quickly as a truck's brakes would decelerate its own wheels (a panic stop on ice, for example).

It should be emphasized that such motors do not simply sit on a benchtop and spin up and down randomly or based on user input, such as turning a knob on a power supply. Instead, the HIL system is designed to emulate the proving grounds environment, and operates similar to a tightly controlled driving simulator. That is, the wheel speeds do not simply follow preprogrammed patterns or user input – rather, they respond according to the vehicle dynamics, which in turn are calculated responses to the operator's driving actions. Again, the motors must carry out acceleration and braking demands with the same effectiveness as the real vehicle's drivetrain and braking system would.

THE BASIC HIL SYSTEM (WHEEL SPEEDS SIMULATED)

A basic HIL system is shown in Figure 5. As depicted, the bulk of the vehicle dynamics is modeled in the PC. The portions that are not modeled are the entire pneumatic (tanks, valves, tubing and hose), electro-pneumatic (ABS modulators), and electrical and electronic portions (ECU and wiring harness) of the braking system. As mentioned previously, the wheel speed signals themselves are produced by sine wave generators. The vehicle model uses pressure sensors in the brake chambers to determine how much torque to apply to the modeled brakes and wheels of the modeled vehicle. The resultant calculated behavior of the vehicle dynamics model produces wheel speeds in physical units of distance/time. It is these wheel speeds that are input to the ABS system as sine wave frequencies.

In summary, wheel speed noise arises from a variety of sources, many of which are difficult to diagnose, are unrepeatable, and/or whose source may not even be known in the present day. The challenge for the OEM and brake system supplier is to provide systems that are robust to any and all forms of noise, some of which may be unknown. For these reasons, it is imperative that the OEM and brake system supplier comprehensively include wheel speed noise immunity in design requirements and system validation.

ELECTRIC MOTORS AS AN HIL COMPONENT

It has been previously alluded that, in an HIL system, a logical choice is to model those things that are easily modeled, and connect these models via PC HIL hardware to real components that are either intended for production or are not easily modeled. In the case at hand, the idealized wheel speed is quite easy to model using sine wave generators and simple circuits. However, modeling wheel speed signals to the levels of

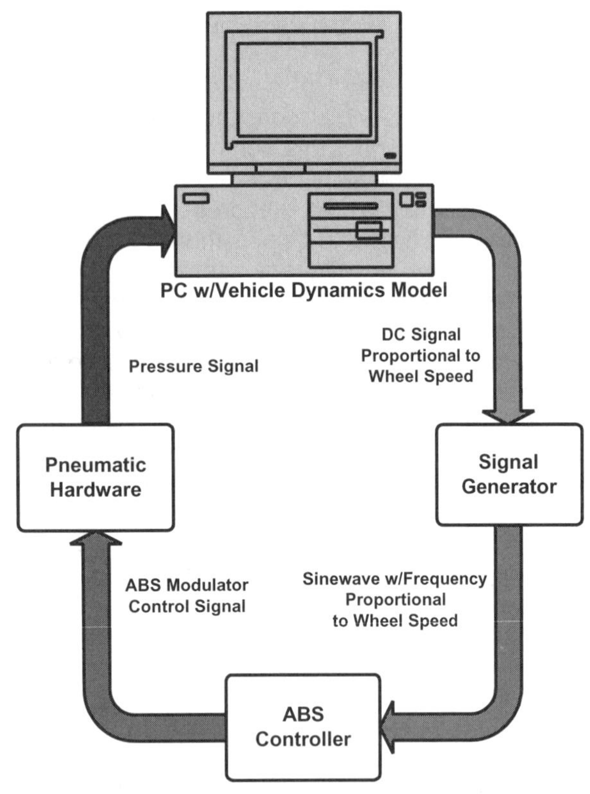

Figure 5: Basic HIL System

THE HIL SYSTEM USING REAL TONE RINGS AND WHEEL SPEED SENSORS

The inclusion of real wheel speed sensors and tone rings into the HIL environment prompted the following requirements for the resultant HIL system:

- Commercially available motors and controllers

- Motors must be able to follow reference with < 1% error on worst case experiments such as fast and deep cycling ABS behavior

- Compatibility with all versions of all current ABS manufacturer system offerings for both tractor and trailer

- Compatibility with all current tone ring types except for those that are cast into the wheel hubs themselves

- Compatibility with any HIL software and/or hardware provider (Matlab, MATRIXx, dSPACE, ETAS, *et al*)

- Precision adjustability for sensor gap, sensor offset, tone ring axial runout, and tone ring radial runout

- Expandable and configurable for future requirements

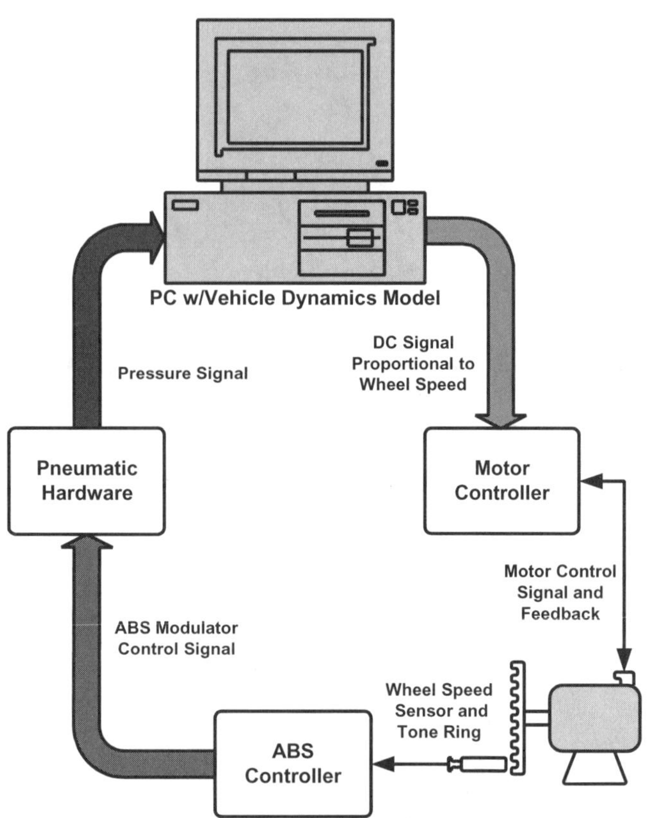

Figure 6: HIL System with Integrated Velocity Controlled Motors

Compared to the Basic HIL System presented, the changes made are strictly to the connected HIL hardware – that is, no modeling or software changes are necessary. Comparing Figure 5 and Figure 6, it is clear that the electric motors are substituted for the sine wave generation circuit. Real wheel speed sensors and tone rings are used in the same manner as on a real vehicle. The sensors and tone rings tested can be from the following sources:

- Field and warranty returns

- Units that represent potential or actual manufacturing errors

- Units that represent damage during service or assembly

- New types of sensors/tone rings not used before

Figure 8: Detail Picture of a Single Fixture

CONCLUSION

Existing HIL systems were found lacking in their capability to accurately produce wheel speed signals like those found on real vehicles. Most forms of wheel speed noise are difficult to model and simulate, and difficult to test with statistical significance on a real vehicle. The proposed solution of integrating electric motors in the HIL environment allows suppliers and OEMs to further ensure that production brake systems and vehicles are robust to sensor and tone ring anomalies that can reduce braking performance.

ADDITIONAL SOURCES

The following sources provide further detail and information on basic HIL systems:

1. Bennett, Mark A., and William P. Amato. "Improved Heavy Vehicle Dynamics Model using MATLAB/SIMULINK", SAE 1999-01-3707
2. Amato, William P. and Mark A. Bennett. "Verification of Heavy Truck EBS and ABS Using MATRIXx Hardware In The Loop Tools", SAE 1999-01-3713
3. Hatipoglu, Cem and Amer Malik. "Simulation Based ABS Algorithm Development", SAE 1999-01-3714

Figure 7: Photo of Cabinet and Four Motor/Tone ring Fixures

Magneto-Optic Tachometers for Automotive Vehicles' Chassis

Wojciech Weglarski
Krakow University of Technology

ABSTRACT

Anti-lock braking systems (ABS) are an accident eva-sion system (AES) that by incorporating relevant sen-sory systems (SS), for instance, such as magneto-optic (M-O) tachometers, accelerometers etc., can avoid wheel locking during hard braking in an emergency, especially when the road surface is slippery. Auto-motive vehicles (AV) lose steering when the front wheels (FW) lock. An ABS uses sensors at each wheel to monitor deceleration when the fluidomechanic (F-M) or electromechanic (E-M) drum, ring or disc brakes are applied. If any of the wheels begin to lock, the ABS will modulate the brake pressure or voltage, thus *pumping* the fluidomechanic (F-M) or electromechanic (E-M) drum, ring or disc brakes, respectively, at the rate faster than the average human driver (HD) could. This will allow the wheel to continue rotating so avoiding locking, and the wheel will remain to react to the steering wheel (SW).

INTRODUCTION

In the 2010s, there is an increasing necessity for reli-able angular position, velocity and acceleration/decel-eration sensors in automotive applications. Two main reasons for this are the increasing number of elec-trically-assisted ride-by-wire (RBW) or x-by-wire (XBW) automotive mechatronic control systems in automotive vehicles (AV) requiring feedforward and feedback con-trol and the high level of reliability requested by both automotive manufacturers and automotive customers.

Contactless angular position, velocity and acceleration/ deceleration sensor solutions exist today but their introduction in mass manufacture is still waiting for low cost, at least of the same price order than the former solutions. The emerging trend in automotive applica-tions is to use new types of sensors that do not require mechanical parts in contact.

There exist different transducing techniques based on different physical principles [1-10]:

- ➢ variable reluctance (VR);
- ➢ magnetic (magnetoresistive, Hall effect, LVDT);
- ➢ magneto-optic;
- ➢ optic (incremental, absolute encoder);
- ➢ capacitive.

As listed transducers need specific signal-conditioning electronics and so they are more sensitive to tempe-rature and electromagnetic interference (EMI) compar-ed to other devices.

Thin and thick film deposition techniques for magneto-resistive materials seem to be very promising for the development of linear position and angular sensors that can be of the absolute or incremental type.

Automotive application examples of these devices are sensors that use thin film deposition of ferromagnetic materials (MRE) that have interesting characteristic of stability, linearity and resistance in rough environmental condition [7].

The severe environmental conditions of vehicle sys-tems make the optic or capacitive solution less suitable for applications in such area even though there are some studies, for instance, on automotive capacitive position sensor [7].

The use of contactless tachometers remains in any case dependent on achievement of costs competitive in the automotive market. By means of the suitable Faraday effect magneto-optic (M-O) tachometers, an anti-block braking system (ABS) permits a human driver (HD) to steer an automotive vehicle (AV) during emergency braking. Traction control system (TCS) is another accident evasion system (AES) used to resist uncontrolled wheel spin and to keep AVs from sliding side-ways on slippery surfaces.

In most AVs with TCS, the system works only during low-velocity acceleration and is then disconnected. In more expensive system operates at all vehicle velocities. The same M-O tachometers that can detect wheel lock during braking with an ABS can be used to sense uncontrolled wheel spin. In operation, the TCS senses when one driving wheel (DW) spins faster than the other and applies pulses of brake pressure or voltage, respectively, to limit its slippage. This will then allow other wheels to receive and transmit internal combustion engine (ICE) power, thus improving steerability and traction.

The operation of the TCS can be linked to ICE control where, if the system is in operation for more than a given time (~ 2 s), the ICE microcontroller cuts liquid fuel delivery to the ICE cylinders or turns-off electric energy to the steered, motorized and/or generatorized wheels (SM&GW), respectively, on a selective basis to limit the wheel angular velocity. This helps to minimize stresses on the drum, ring or disc brakes, transmission and ICE prolonged TCS operation. A dashboard display can be provided to show when the TCS is in operation. As the advances in other types of the SS for automotive vehicles' chassis application continue, so also does the growth in the use of M-O tachometers. Some applications of M-O tachometers will be described.

TACHOMETER DESIGN PHILOSOPHY

Tachometers have been used in an increasing number of automotive applications. Primary usage has been in the field of angular position, velocity, acceleration/deceleration and sense of rotation sensing, with major application being in ICE and transmission management control and ABS wheel-hub angular velocity monitoring mechatronic control system. Current, legislative initiatives have introduced new requirement on the performance of such mechatronic control systems that have resulted in requirement for high performance gear tooth sensing (GTS). In response to these trends, contactless tachometer GTS technology has focused on the development of self-calibration techniques that dramatically increased usable airgaps, switching time accuracy, and overall reliability of the sensor system [1-10].

PASSIVE SENSORS - The contactless variable reluctance (VR) transducer used for sensing gear teeth of a moving target, for example, such as a ferromagnetic disc, tire-wheel hub or internal combustion engine (ICE) flywheel include VR tachometer functions (see Figure 1). The basis for this design is to integrate a VR transducer, an instrumentation amplifier, and a comparator [7].

The VR transducer is itself of the ferromagnetic coil with a constant current passing through it. When sensor is subjected to an orthogonal magnetic field, voltage pulses that are proportional to the applied magnetic field will appear across the sensor. This very low-level voltage signal is amplified and the result is a high-level digital output signal that is proportional to actual values of the angular velocity.

ACTIVE SENSORS - The contactless magneto-resistive (MR) transducer used for sensing gear teeth of a moving target, for example, such as a ferromagnetic disc, tire-wheel hub or internal combustion engine (ICE) flywheel include MR tachometer functions (see Figure 1). The basic principle used in the currently offered sensors is simple, that is, a magnetic field generator moves relatively to a magnetic flux concentrating system [3]. A magnetic field detector measures the magnetic field intensity or sense of direction variation induced by the movement. The MR sensors consist of alternating very thin layers of magnetic elements (*FeNi, Co, Ni*) and non-magnetic elements (*Cu, Au, Ag*). The layer thickness varies between a few angstrom [*Å*] to about *100 Å*. These layers were first deposited by molecular beam epitaxy (MBE). Sputtering, a standard process in application specified integrated circuits (ASIC) manufacturing, is now being used. Depending on the non-magnetic layer thickness, two adjacent magnetic layers have opposite magnetization (antiparallel coupling) or parallel magnetization (parallel coupling). In the case of MR one uses the effect that the resistance due to electronic band structure effects depends on the relative align-ment of the electron spin and the magnetization of the layers. The electrons having a spin up interact more with layers of one sense of magnetic direction and less with the layers having the opposite sense of direction (antiparallel layer). When the layers are antiparallel coupled, each electron family interacts by at least every second magnetic layer. This is due to the fact that the electrons mean free path is larger than the layer thicknesses. When an external magnetic field forces the magnetic layers to be all aligned in one sense of direction one of the electron families is no longer strongly scattered.

The Hall-effect transducer used for sensing gear teeth of a moving target, for example, such as a ferromagnetic disc, tire-wheel hub or ICE flywheel also include Hall effect tachometer functions (see also Figure 1). The basis for this design is to integrate a Hall cell, an instrumentation amplifier, and a comparator with base magnetic field cancellation [6]. The Hall cell is itself of the piece of silicon with a constant current (source of electrons) passing through it.

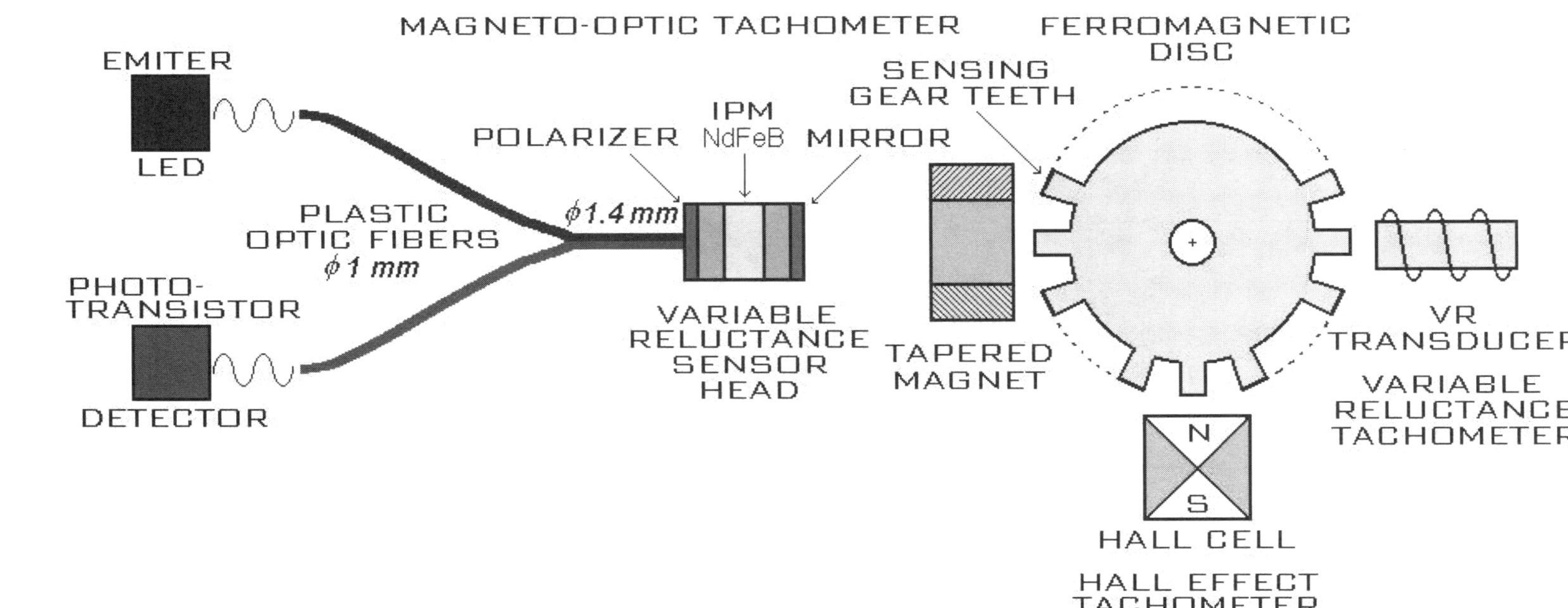

Figure 1: Principle layout of the Faraday effect magneto-optic (M-O) (reflexion) tachometer

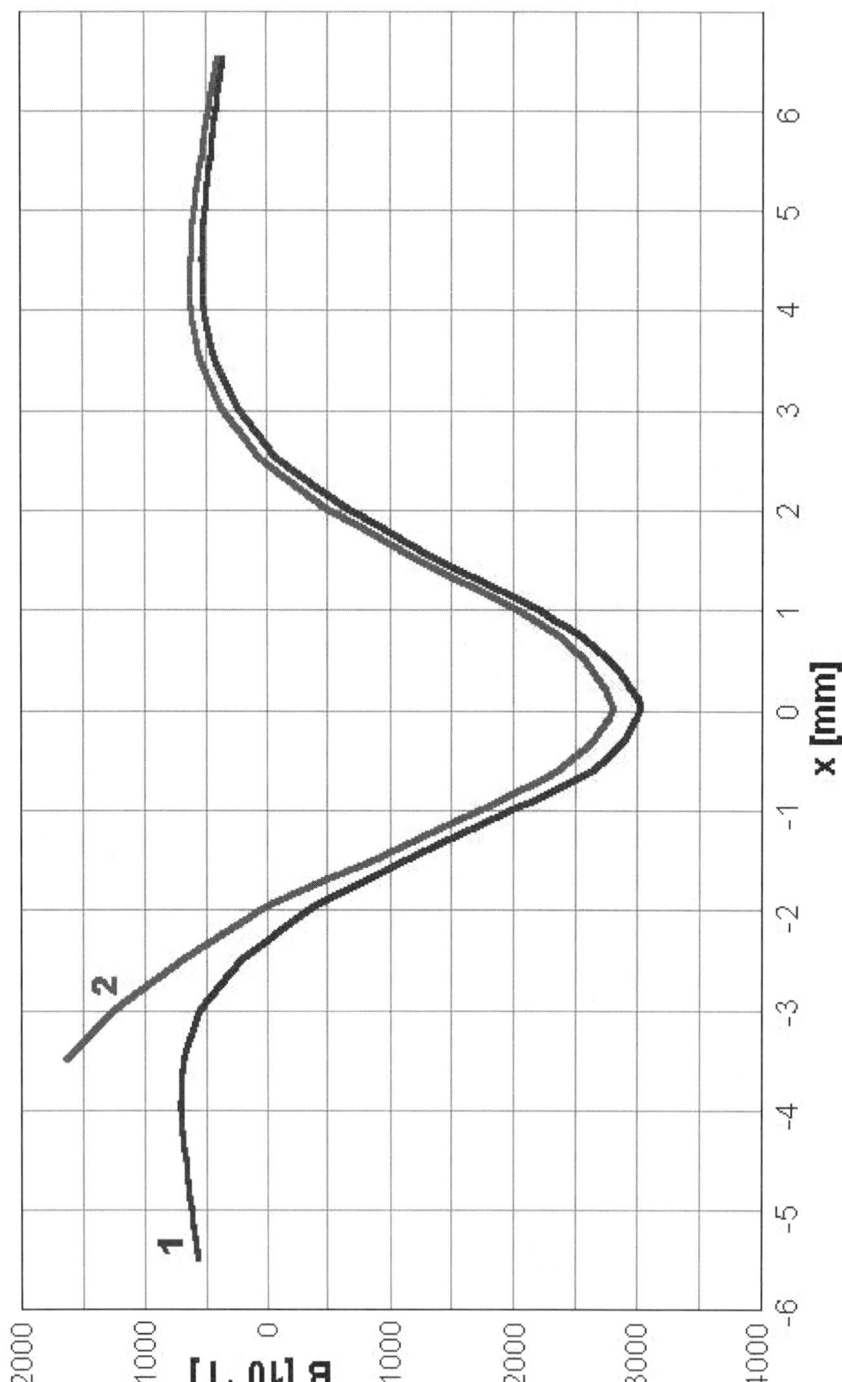

Figure 2: Dependence of the magnetic induction B [10^{-4} T] as a function of the axial Faraday rotator position x [mm]

TABLE 1

	Passive sensors	Active Sensors		
	Variable Reluctance	Magneto-resistive	Hall effect	Magneto-optic
Minimum value of speed	~ 0.56, 0.84 m/s	0 m/s	0 m/s	0 m/s
Maximum value of frequency	~100 Hz	~1 MHz	~5 MHz	>10 MHz
Linear response	NO	yes	yes	yes
Resistance to EMI	LOW	moderate	moderate	high
Number of conductors	2	2	2	1
Sensor price	LOW	medium	medium	high
ECU price	HIGH	high	medium	low

When sensor is subjected to an orthogonal magnetic field, voltage that is proportional to the applied magnetic field will appear across the sensor. This very low level signal is amplified and compared to the average magnetic field and the result is a high-level digital output signal. A discrete with external signal conditioning would be temperature and voltage unstable and would require user trim. By integrating the sensor with an amplifier and the temperature current source it is able to produce a sensory system (SS) with accuracy and noise immunity than can be matched using a discrete system. For instance, the Hall-effect transducer has a very sensitive threshold of $\pm\ 15\ 10^{-4}\ T$ and can operate with a wide airgap between the sensor and its target. An application would have the Hall effect transducer placed in between an interior permanent magnet (IPM) and a moving target such as a ferromagnetic disc or flywheel. The motion of a passing gear tooth would disturb the magnetic field and would be sensed by the electronic circuit. It would produce a high level digital output in response to the small change in magnetic field.

The Faraday effect M-O transducer used for sensing gear teeth of a moving target, for example, such as a ferromagnetic disc, tire-wheel hub or ICE flywheel include M-O tachometer functions as well (Figure 1). The basis for this design is to integrate a VR sensor head with a polarizer, an interior permanent magnet (IPM) and a mirror, as well as a tapered magnet [4]. A VR sensor head receives an input IR light signal by means of plastic optic fibers from a light-emitting diode (LED) emitter. An output IR light signal is sent to a phototransistor detector. The motion of a passing gear tooth would disturb the polarized IR light and would be sensed by the VR reluctance sensor head. It would produce a high-level digital output signal in response to the small change in IR light. The IR light is polarized based on the Faraday effect. M-O tachometers will be finding increasing applications in RBW automotive mechatronic control systems. In Figure 2 is shown a dependence of the magnetic induction $B\ [10^{-4}\ T]$ as a function of the axial Faraday rotator position $x\ [mm]$.

CONCLUSION

Integrated tachometers (sensors and signal conditioners) offer many benefits over their discrete and computer-controlled counterparts. The ability to integrate highly accurate sensors and signal conditioners on single chip and to trim out any errors at the manufacture enables the automotive customer to realize savings in accuracy, cost, mass and optic fiber wiring requirements. There are many solutions of tachometers (sensors and signal conditioners) that can be combining to manufacture application and ASICs with analog, digital or angular frequency output signals [1-10].

In Table 1 is given a comparison of the parameters of various passive and active tachometers (angular velocity) sensors, such as a passive sensor (variable reluctance) and active sensors (magnetoresistive, Hall effect and magneto-optic) [4].

The availability of low-cost and reliable sensors is probably the most critical factor in ECUs realization. For this reason integration can be a solution that allows the sensors to be shared amount different ECUs, with a reduction in the number of required sensors, and consequently cost reduction. In these devices the specific conditioning and self-diagnosis electronics will be integrated into the sensors (smart sensors) [4]. Undoubtedly, the automation of AVs has offered one of the greatest opportunities in use of automotive sensors.

REFERENCES

[1] Baxter, L.K.: Capacitive sensors, *IEEE Trans.*, 1996.

[2] Fiorentin, S.R.: Sensors in automobile applications. Chapter 2 in the book: *Automotive Sensory Systems*. (C.O. Nwagboso, Ed.), Chapman & Hall, 1993.

[3] Genot, B., and W. Clemens: Giant magnetoresistive (GMR) based contactless position sensors. *Proc. 30th International Symposium on Automotive Technology & Automation*, Mechatronics/Automotive Electronics, Pa-per 97AE018, 1997, pp. 967-974.

[4] Guerrero, H., E. Bernabeu, and F.J. Mustieles: Magneto-optical tachometers for anti-lock braking systems: a comparison with conventional sensors. *Proc. ROVA '97 INTERNATIONAL: 3rd International Conference on Road Vehicle Automation*, 1997

[5] Kato, S., and J. Nakaho: High precision angular precision sensor. *SAE Paper 940631*, 1994.

[6] Milano, S., and R. Vig: Digital powertrain speed measurement using self-calibrating Hall effect gear tooth sensing technology. *Proc. 30th International Symposium on Automotive Technology & Automation*, Mechatronics/Automotive Electronics, Paper 97AE038, 1997, pp. 1003-1011

[7] Nwagboso, C.O., Ed.: *Automotive Sensory Systems*. Chapman & Hall, 1993.

[8] Rizzoni, G., and W.B. Ribbens: Sensors and systems for crankshaft position measurements. Chapter 4 in the book: *Automotive Sensory Systems* (C.O. Nwagboso, Ed.), Chapman & Hall, 1993.

[9] Wells, R.F.: Non-contacting sensors for automotive applications, *SAE Paper 880407*, 1988.

[10] Wells, R.F.: Automotive steering sensors, SAE Paper 900493, 1990.

The Theoretical Concepts for Pre-Extreme ABS

Valentin Ivanov
Belarusian State Polytechnical Academy

Michael Vysotsky, Vladimir Boutylin and Joseph Lepeshko
National Academy of Sciences of Belarus

ABSTRACT

The analysis of existing systems of automotive active safety (SAS) shows that the antilock braking system (ABS) is a kernel for anyone of them. However circle of the tasks, which should be decided by SAS, is complicated, and the algorithms approaches are traditionally based on threshold control.

Basing on the researches of interactions within a chain 'automobile - wheel – road' a new so-called pre-extreme control philosophy for ABS may be offered. It is based on the discovered regularities between the tire grip and wheel slippage.

For the given philosophy the family of ABS algorithms is offered which allows realising both discrete and continuous control of wheel operation.

INTRODUCTION

The analysis of present-day state of the automotive engineering makes possible to speak about the systems of active safety as individual equipment class. *The system of active safety* is a complex of mechanical and electronic units integrated with the informating-controlling and energy channels for the purpose of crash avoidance at an appearance of critical road situation.

In a general case, SAS may be categorised according to their functionality, Fig. 1. As all represented classes in the special literature have been studied a number of conclusions for goals of presented paper can be made.

1) The brake system is a kernel of a SAS majority. In addition to deciding of conventional tasks it is used for supporting of vehicle stability and controllability by means of cohesion characteristics control for wheel-road-interaction. The brake system interacts with other vehicle components both in braking mode and in traction mode.

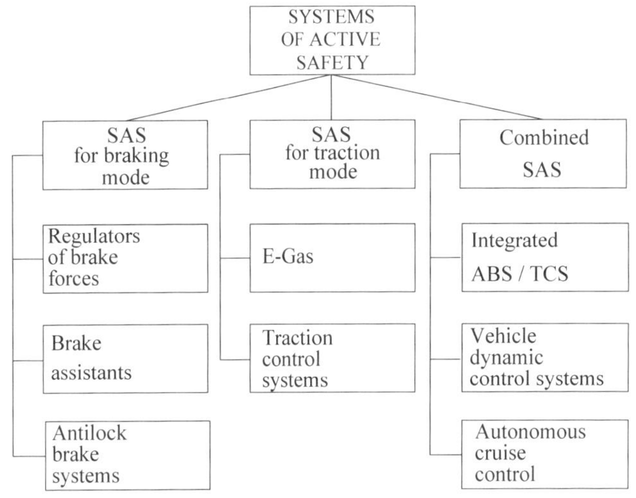

Figure 1. The Functional Classification for Systems of Active Safety

2) The clearly defined transition from unitary, single-purpose systems like ABS or TCS to the integrated systems which cover the various movement modes has taken place.

3) In modern SAS the traction and/or braking control effort is regulated with the tracing of most important characteristics of automotive safety: longitudinal and lateral wheel sliding, slip and course angle, distance between the vehicles and others.

4) The SAS functions to remove the vehicle from critical road situation are supplemented with the new functions which allows a prediction and forestalling for such situations.

5) At the regulation process a majority of SAS uses the brake gear control. This fact allows to develop an integrated system on the basis of ABS.

By this means the following hidden self-contradiction can be noted for modern SAS. The antilock brake system as the most important element of SAS gains new informat-ing-controlling linkages and new intelligent tasks but its control philosophy have not been practically varied. The threshold, discrete control methods were used despite on the fact that they have exhausted their possibilities for solution of the control prognostic tasks of automotive active safety. Because of this it makes sense to consider alternative ways for ABS control philosophy.

CLASSIFICATION OF CONTROL PHILOSOPHY FOR ABS

By a control philosophy is meant a frame bounded with the effect of control parameters on organisation and al-gorithms of work for automatic control system.

The most important characteristics for ABS classification are dependencies between the tire grip factor or cir-cumference force factor μ and the relative wheel slip **s**, Fig. 2. The main task of ABS control is, by definition, to keep the vehicle stability and handling by simultaneous ensuring of high level of braking efficiency. The bound-ary for fulfilment is located in extreme area of μ-**s**-curves, which is control area in ideal case. Hence the following directions for the ABS control can be evolved, Fig. 3.

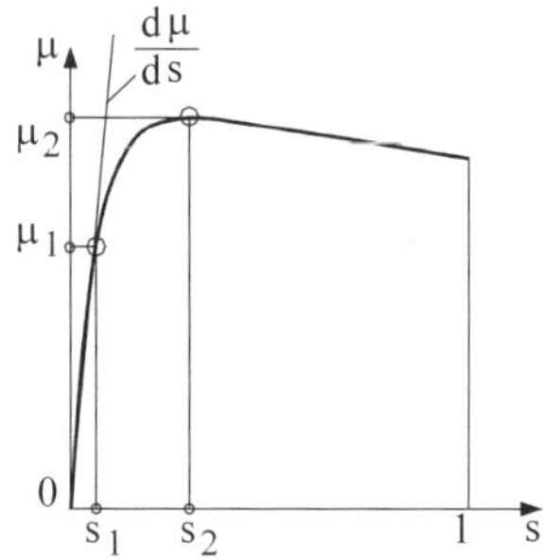

Figure 2. μ-**s**-dependence

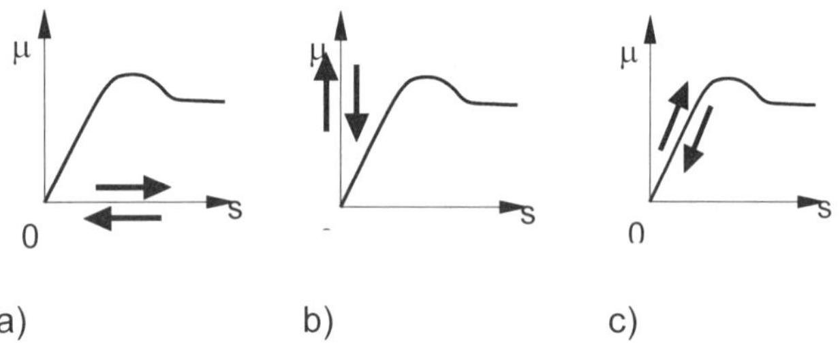

a)   b)   c)

Figure 3. The Directions of ABS Control

The control philosophy for ABS has the largest distribu-tion whereby the location of extreme area is traced in the line of changing of relative wheel slip, i.e., **s**-direction with reference to the μ-**s**-curves, so-called **s**-control, Fig. 3a.

Merits: well-developed theoretical footing; minimal sen-sor requirements.

Demerits: sensitivity to the initial conditions of braking; the relevant ways of information gaining for wheel sliding have high sensibility of disturbances; insufficient adapt-ability to the environmental effect.

The second kind of ABS organisation is an application of algorithms operated in μ-direction, Fig. 3b. Here the force in wheel-road-contact or the moment from this force put the information about magnitude of coefficient of the wheel's cohesion.

Merits: the direct information about wheel's cohesion is used; high braking efficiency can be achieved.

Demerits: sensor part of ABS requires the modification of vehicle construction; sensitivity to the road micropro-file.

The logical development of build-up concepts for ABS is the philosophy of pre-extreme control [1,2]. It is based on tracing of the optimal cohesion area of wheel opera-tion simultaneously both the μ-direction and the **s**-direction, Fig. 3c. For example, the derivative **dμ/ds** can be herein the control parameter.

PRE-EXTREME CONTROL PHILOSOPHY FOR ABS

LOGICAL GROUNDS

Maximal braking efficiency, dynamic stability and mini-mal energy losses in a wheel can be important factors in choosing of boundary conditions.

One of the boundary conditions is

$$d\mu / ds = 0 \ , \qquad\qquad (1)$$

This is in accord with the maximum of μ-**s**-curve. Beyond μ_{max} the wheel working is not desirable because of safety factors and also the tire wear. The braking power begins to redistribute in this situation more and more from the brake gear to the contact between wheel and road.

After transformation of the equation (1) it can be ob-tained that **dμ/dt = 0** and **ds/dt** is any value. In this way the tracing of μ_{max} ensures the automatic adaptation of the ABS operation to the drift of extremum along the **s**-axis.

HIERARCHY OF PRE-EXTREME CONTROL PRINCIPLES FOR ABS

Control principle with constant thresholds

The simplest variant of pre-extreme control principle for ABS comprises an assignment of constant thresholds for the exhaust, build-up and hold of brake pressure by analogy with usual systems. The release threshold is selected into the pre-extreme area, Fig. 2, point **s1**, for example, reasoning from the statistical manipulation of μ-**s**-curves. The threshold for immediate pressure boost is controlled by the boundary condition (1), i.e., in according to extremum, Fig. 2, point **s2**.

Conceivable problems and demerits for given control principle are in close agreement with **s**-control. The adaptation characteristics of such variant are slender. This can be illustrated by the averaged data for constant thresholds obtained through statistical manipulation of μ-**s**-curves for car's tires [2], Fig. 4, 5.

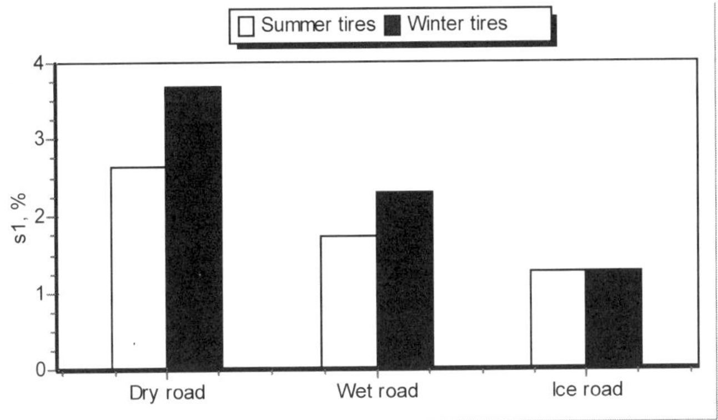

Figure 4. The Average Statistical Values for Point **s1**

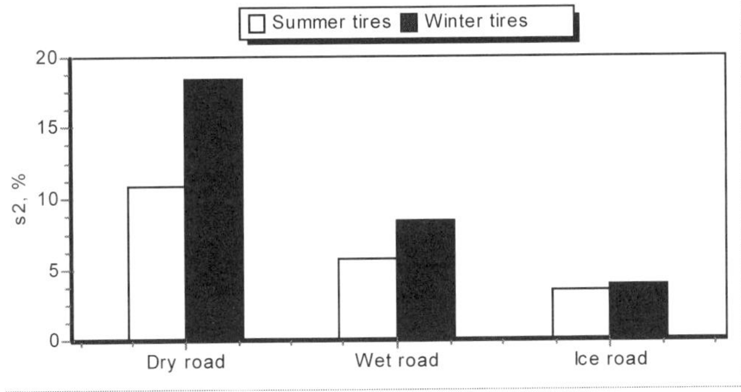

Figure 5. The Average Statistical Values for Point **s2**

It might be well to point out from these diagrams that the distinguishing points of μ-**s**-curves depend both on the road type and on the tyre type. This is just certain of the factors, which cause the curve drift.

Control principle with variable thresholds

The variable thresholds can raise adaptation characteristics for ABS. For the threshold correction methods to determine the well-known equation of the force balance for generalised wheel model is worth consideration:

$$m\frac{dV}{dt} = F_\mu + \left(1 - s\right)\left(\frac{M_{br}\left(t\right)}{r_d} - \frac{J}{r_d}\frac{d\omega}{dt} - F_\mu \right), \qquad (2)$$

Where **M**₍br₎ and **F**₍μ₎ are

$$M_{br} = k_m \cdot F_c, \qquad (3)$$

$$F_\mu = \mu \cdot R_z. \qquad (4)$$

If F_μ is chosen as control quality index and F_c as control response, so other components of eq. (2)-(4) can be subdivided into external disturbances and noise.

Generally the transformation coefficient of a brake gear k_m and angular deceleration of a wheel $d\omega/dt$ should be considered among noise, and the normal road reaction R_z and velocity of vehicle **V** among external disturbances.

The parameter k_m summarises an influence of such factors as the friction between brake-shoe lining and brake drum (disc) and as the brake gear geometry. The change of friction coefficient as well as the hysteresis may be compensated through the change of effort F_c. The quality index F_μ will be invariant at that.

The angular deceleration $d\omega/dt$ is the crucial factor for inertial moment loading a brake gear but not for brake force in the wheel-road-contact. For example, if $d\omega/dt$ value rises, so the control response F_c should be increased for compensation of $J \cdot d\omega/dt$ value.

With rise of initial velocity of vehicle **V** at braking the maximum of tire grip coefficient falls. This is because the physical processes operating into the contact area between a wheel and surface. On the other hand, by the increase of velocity the shift of brake force peak is observed to the lesser values of the sliding **s**, especially on the wet road and ice [3].

A change of normal road reaction R_z involves a change of value F_μ, with the loading of a wheel the maximal cohesion coefficient goes down. The value R_z influences the extremum location for μ(**s**)-dependencies relative to **s**-direction at the same time. This phenomenon is typical for all types of road surface.

On this basis the following scheme can be proposed for the pre-extreme control principle with variable thresholds, Fig. 6.

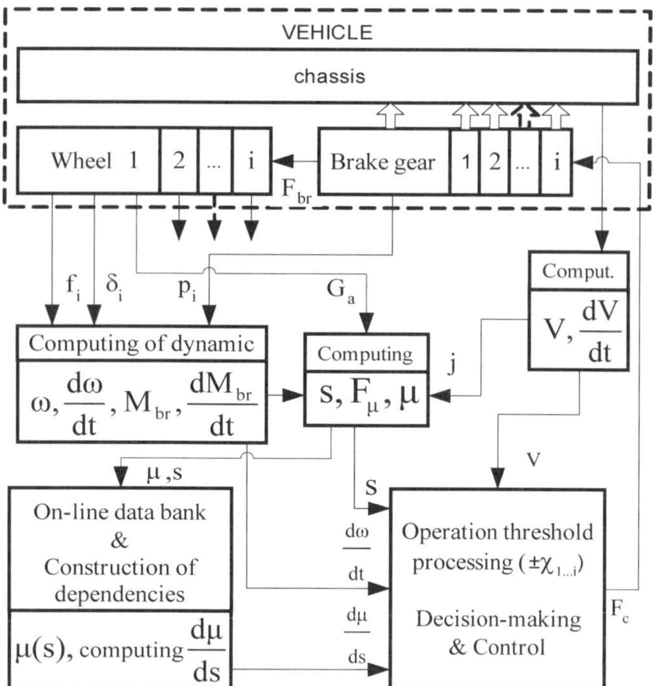

Fig. 6. Block Scheme of Pre-extreme ABS Controller

It can be noted that the problem of inserting of variable thresholds is connected with the performance of the preliminary experiments for investigations of cohesion characteristics for tyre and road. In addition, the availability of statistical database for μ-**s**-curves is appropriate.

Control principle with linearisation of pre-extreme area

By applying this control principle, the information about fluctuating of 1st derivative **dμ/ds** (or n-th derivatives when needed).

The researches disclose that in the pre-extreme field of μ-**s**-curves the initiating, practically linear upgrade of the characteristic takes place. It shows that transfer point from linear to non-linear section of μ-**s**-curve can be accepted as a first approximation for the operation threshold of ABS for the pressure exhaust.

Such pre-extreme algorithm can be analytically described as follows. ABS applies signal for the pressure release by realisation of criterions

$$dF_\mu / ds > 0 , \tag{5}$$

and

$$\left| \left(\frac{dF_\mu}{ds} \right)_i - \left(\frac{dF_\mu}{ds} \right)_{i-1} \right| \le \chi_1 , \tag{6}$$

where χ_1 is a control deviation for the pressure release. The attainment of value χ_1 is evidence for the completion of linear section of μ(**s**)-curve and for the tendency to the approach of an extremum.

The pressure release can begin both in the pre-extreme area and in the post-extreme area of μ(**s**)-curve depending on response time of a system. However the decrease of sliding **s** will take place in any case. The conditional-branching test for (5) and (6) not occurs in the process and the antilock brake system monitors the fulfilment of relations

$$dF_\mu / ds < 0 \tag{7}$$

and

$$\left| \left(\frac{dF_\mu}{ds} \right)_i - \left(\frac{dF_\mu}{ds} \right)_{i-1} \right| \le \chi_2 , \tag{8}$$

where χ_2 is a control deviation for the pressure build-up. This deviation indicates that the rating value for displacement from extremum takes place.

The described processes are further repeated until the critical situation comes to end. The ideal control cycles whereby the system become a signal for next pressure build-up in point of extremum are shown in Fig. 7 (taking the response lag of actuators to be 0).

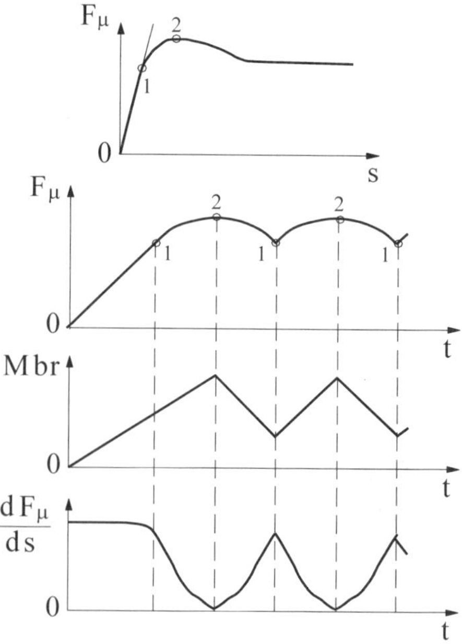

Fig. 7. The Ideal Control Cycles of Pre-extreme ABS

The information about **dFμ/ds** can be direct received with a sensor for the circumference force or moment from this force or can be controlled by hardware with usual sensors for brake pressure and angular velocity of a wheel. The analytical expression for dependence between the derivative **dFμ/ds**, the brake torque of brake mechanism **M_br** and the wheel's deceleration **dω/dt** can be obtained from equation (2):

$$\frac{dF_\mu}{ds} = m\frac{dj}{ds} + \frac{1}{r_d}\left(1-\frac{1}{s}\right)\left(\frac{dM_{br}}{ds} - J\frac{d\varepsilon}{ds}\right) - \\ -\frac{1}{r_d \cdot s^2}\left(M_{br} - J\cdot\varepsilon\right) \qquad (9)$$

In this manner the principles of not only discrete control, but also stepless control can be applied here contrary to the previous system.

<u>Control principle with prediction</u>

The control principle with prediction can be used for the intelligent systems of active safety. The foundation of this method is an assessment of potential and actual forces in the wheel-road-contact [4,5], Fig. 8.

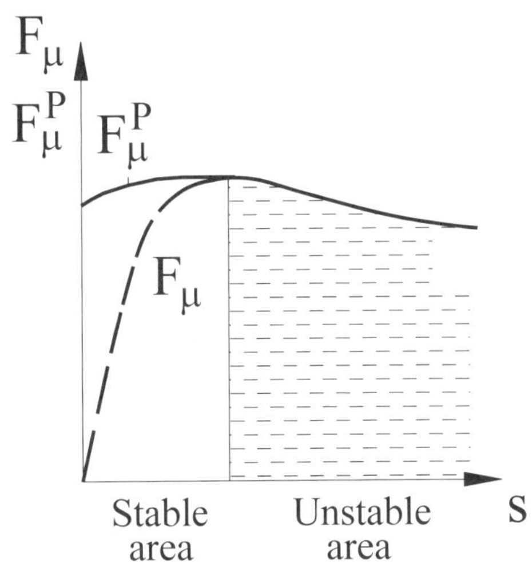

Fig. 8. The Potential and Actual Forces in the Wheel-Road-Contact

The potential force **Fμ^P** describes virtually the boundary, which defines the possibilities for the transference of brake forces from road to vehicle. The actual force **Fμ** determines the current brake force in the wheel-road-contact. When the brake torque to a wheel is applied the absolute magnitude of force **Fμ** aspires to **Fμ^P**.

The force **Fμ** amounts to its potential value in certain point corresponding to the extremun of μ-**s**-curve. By the ratio between the actual and potential forces within contact the reserve of wheel stability can be estimated in stable area, for example, through the some coefficient

$$k_\mu = 1 - \frac{F_\mu}{F_\mu^P} \qquad (10)$$

It is seen that in the absence of translational movement of a wheel **k_μ** is 1 and on reaching of potential boundary **k_μ** is zero.

Thus the pre-extreme ABS using control principle with prediction follows the reserve of wheel stability and forecasts simultaneously the approaching of potential boundary for tire grip. From the automatic control standpoint, the methods of quasicontinuous regulation would be appropriate for use in such an antilock brake system. Instead of the operation thresholds the behaviour of regulation curves is assessed here.

The following scheme can be proposed for the pre-extreme control principle with prediction, Fig. 9.

CONCLUSIONS

The above mentioned discourses and analytical computations allow to make a conclusion that the indicated control principles have good adaptation abilities due to capacity for forecasting of situation development and they can be used for creation of perspective intelligent systems of active safety.

REFERENCES

1. Ivanov V., Boutylin V., Liashchinski A. Structural Synthesizing of Intellectual Systems of Automobile Active Safety. - Detroit: SAE Technical paper series, 2000-01-1637.
2. Ivanov V. Die vorextremen Antiblockiersystemen, Brakes of road vehicles' 2001, Lodz, Poland, April 2001.
3. Wehner B. Einige Beobachtungen über der Gleitwiederstand auf winterglatten Strassenoberflächen // Automobiltechnische Zeitschrift.- 1961.- Bd. 63, Nr. 2.
4. Ivanov V., Lepeshko J., Boutylin V. Potential Forces of a Braking Mode and Their Realisation Within Wheel-Road Contact.- Bratislava, 8th European Automotive Congress, 2001.
5. Ivanov V., Lepeshko J., Boutylin V. About Interrelation between the Tire Grip Properties and the Wheel Sliding - Detroit: SAE Technical paper series, 2001-01-3338.

CONTACT

Dr. Valentin Ivanov, Dept. of Automobiles, Belarusian State Polytechnical Academy. F. Skaryny 65, 220 027 Minsk Belarus. E-mail: vivanov@tut.by

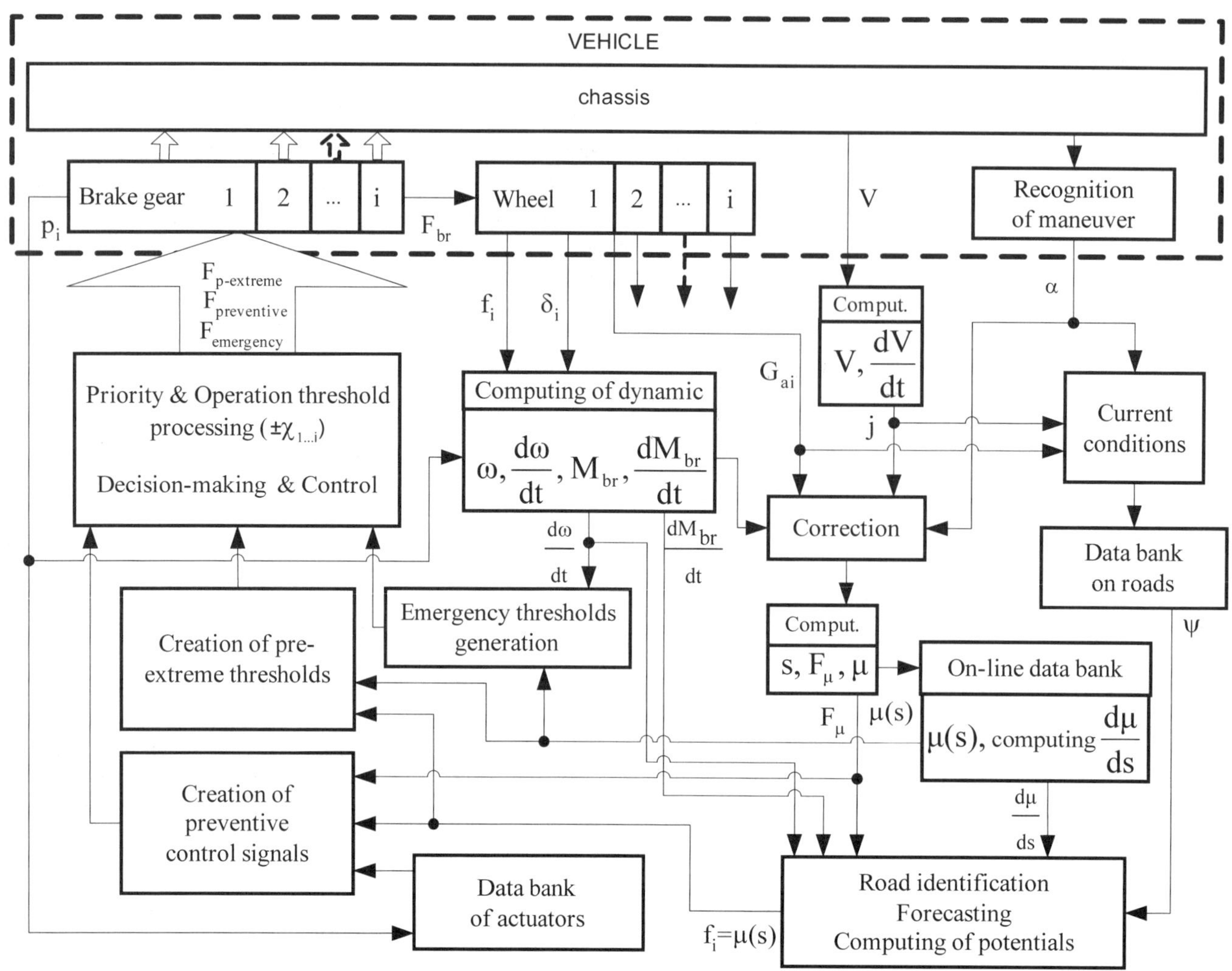

Fig. 9. Block Scheme of Forecast ABS Controller

DEFINITIONS, ACRONYMS, ABBREVIATIONS

F_c: Control effort on a brake gear

$F\mu$: Force within a wheel-road-contact

$F\mu^P$: Potential force within a wheel-road-contact

G_a: Wheel's loading

j: Vehicle acceleration

J: Moment inertia of a wheel

k_m: Transformation coefficient of a brake gear

k_μ: Safety factor

m: Weight reduced to a vehicle wheel

M_{br}: Brake torque of the brake mechanism

p: Brake pressure

r_d: Dynamic radius of a wheel

R_z: Normal road reaction

s: Wheel slip (sliding)

V: Velocity of vehicle

α: Steering angle

δ: Side-slip angle

ε: Angular acceleration of a wheel

μ: Specific force within a wheel-road contact, coefficient of the wheel's cohesion

ω: Angular velocity of a wheel

On Board Doppler Sensor for Absolute Speed Measurement in Automotive Applications

Thierry Ditchi and Stéphane Holé
Laboratoire des Instruments et Systèmes d'Île de France, Université Pierre et Marie Curie

Céline Corbrion and Jacques Lewiner
Laboratoire d'Électricité Générale, École Supérieure de Physique et de Chimie Industrielles

ABSTRACT

Many automotive applications, like anti-lock braking systems (ABS), anti collision radars, airbags, require an absolute speed accurate measurement, especially in high risk situations. On-board Doppler sensors may provide a better answer to such a problem than sensors which measure the speed of rotation of the wheels. Indeed the latter can for instance slide on wet or icy roads. However, conventional Doppler sensors cannot operate when the beam emitted by the radar does not encounter a scattering obstacle on the road surface.

In this paper, we present solutions, tested by simulations and by experiments, which drastically increase the probability of having reflecting obstacles in the antenna footprint of Doppler sensors, leading to accurate measurements.

INTRODUCTION

Many automotive applications, for instance anti-lock braking systems (ABS), anti collision radars and airbags, require a good knowledge of the absolute speed of the vehicle for any kind of ground surfaces. Measurements based on the rotation of the wheels may not reflect the reality, particularly in dangerous situations. This is the case, for instance, when the wheels are sliding on an icy or very wet road surface. Indeed in this case even though the wheels do not rotate, the vehicle may still have a significant speed. This can lead the safety systems of the vehicle to take erroneous decisions.

On-board Doppler sensors seem well suited to solve such problem since they give a measurement of the absolute speed of the vehicle, relative to ground. In such systems, and as is shown in Figure 1, an antenna emits towards the road surface a wave which is partially back scattered by the ground obstacles. The reflected wave frequency is shifted by an amount f_d proportional to the vehicle speed and to the cosine of the angle α under which the obstacle is seen by the antenna. For this reason and in order to make accurate measurements, conventional Doppler sensors use a narrow beam to well define the emission and reception angles. Unfortunately, the measurement can only be achieved if there is at least one scattering obstacle in the antenna footprint. However in the above-mentioned dangerous situations, the obstacle density may be so small that the measurement cannot be performed continuously.

In order to increase the probability that the wave emitted by the radar encounters a scattering obstacle, we have proposed to use a broad beam emission [1], but in this case, the angle α becomes an additional unknown variable. The use of a variable frequency allows the determination at the same time of the vehicle speed and of the angle α [2]. This makes accurate speed measurements possible in situations where narrow beam sensors would not operate, but it requires costly hardware and sophisticated data processing.

In the present paper, we propose a new approach using a broad beam antenna with a single frequency [3-5]. This leads to devices which can be easily mass-produced taking advantage of the development of Monolithic Microwave Integrated Circuit (MMIC) technology [6] and the acceptance of microwaves in automotive applications [7].

In the following section, the principle of this absolute speed sensor is introduced. In the next part, the tools developed in order to test this principle are presented. The signal analysis techniques are then explained and the validity and the accuracy of the solutions are discussed.

PRINCIPLE

The principle of the measurement is shown in Figure 1. A wave of frequency f_0 is emitted towards the ground during a time slot ΔT. This wave is reflected by the scattering obstacles on the road surface, present in the antenna footprint.

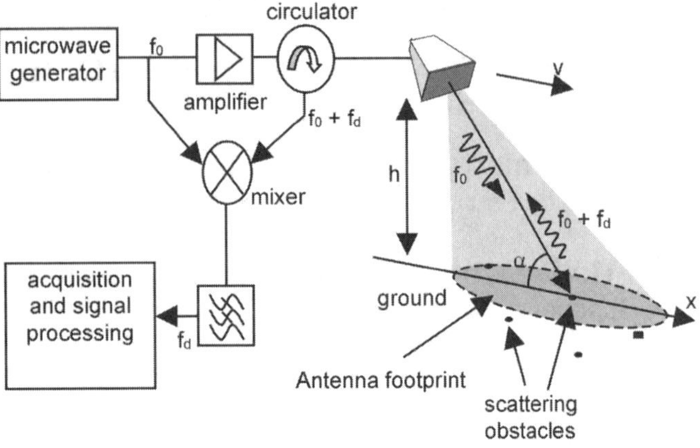

Figure 1: Principle of a Doppler effect speed sensor.

The Doppler frequency shift f_d of the reflected signal associated to a particular obstacle is given by:

$$f_d(t) = \frac{2 \, f_0 \, v \cos(\alpha)}{c} \, , \qquad (1)$$

where c is the propagation speed of the wave and v the absolute speed of the vehicle. Since the relative position between the scattering obstacle and the sensor changes during the time slot, due to the vehicle movement, the Doppler frequency associated with a ground obstacle also changes [8]. This variation depends on three p a-rameters: i) the vehicle speed v , ii) the obstacle position x at the beginning of the time slot relative to the proje c-tion on the road surface of the sensor, and iii) the height h of the sensor above the road. If the time slot is chosen to be short enough so that the speed of the vehicle can be considered constant during this time and since the frequency f_0 of the emitted wave is constant, then the frequency variation of the Doppler signal is only due to the vehicle movement. Thus, the time dependent Do p-pler frequency $f_d(t)$ can be expressed as:

$$f_d(t) = 2 \, f_0 \, v \left/ \left(c \sqrt{1 + h^2/(x-vt)^2} \right) \right. \qquad (2)$$

In this expression, the vehicle speed has been chosen positive if the vehicle approaches the obstacle and negative otherwise.

Examples of Doppler frequency variations as derived from equation (2) are plotted on Figure 2 for different values of the parameters v and x.The vehicle speed v can be determined by fitting this analytical function with the experimental Doppler frequencies associated to an obstacle during its displacement relative to the vehicle.

This fitting is carried out by adjusting the two parameters (v and x) of equation (2).

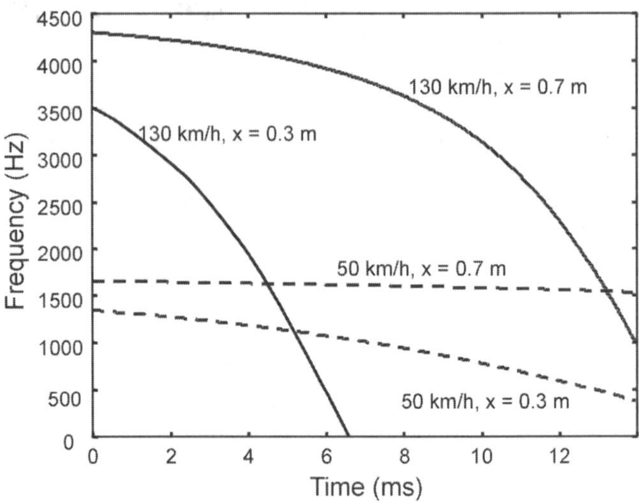

Figure 2: Time dependence of the Doppler frequency for different values of the parameters v and x.

An example is shown on Figure 3. The vehicle speed is equal to 70 km/h, and a unique obstacle is present in the antenna footprint. Its position at the beginning of the measurement is 35 cm. The dotted line represents the experimental frequency evolution and the solid line is obtained by the above described adjustment process. The best-fit parameters in this case are 35 cm for the position and 70.1 km/h for the velocity.

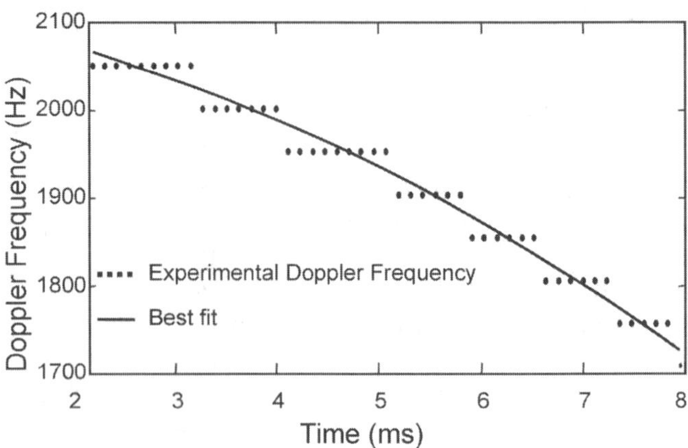

Figure 3: Comparison of the measured Doppler frequency with the ana-lytical one with the parameters obtained by the best fit procedure.

IMPLEMENTATION

The frequency evolution of the Doppler signal during the time slot ΔT is determined using time-frequency methods [9]. For each slot, we use a Fast Fourier transform (FFT) algorithm on moving windows of duration δT here after defined as analysis windows. These windows overlap in order to enable a good frequency tracking.

The accuracy of the speed measurement is directly dependent on the accuracy of the determination of the Doppler frequency. In order to reach a good precision, we need a high spectral resolution technique and as constant as possible a Doppler frequency. These two conditions may not be easy to fulfill at the same time. Indeed since the spectral resolution is inversely proportional to the duration δT, this would favor large δT. Unfortunately, a large δT leads to a significant vehicle displacement during the measurement and thus to a large variation of the Doppler frequency. A satisfactory compromise has been found for the moving window duration δT. It corresponds to a displacement of the vehicle within 5 to 10 cm. For similar reasons, the main direction of the antenna beam results from an other compromise. In order to have a small variation of α during the time slot, one would favor small angles. However, if the angle α is small, the amplitude of the reflected wave is also small. A satisfactory compromise has been found. It corresponds to $\alpha = 35°$.

The possible speed variations during the time slot ΔT, due to acceleration or braking, lead to uncertainties on the speed determination. They must thus be limited by using sufficiently short windows. When the speed of the vehicle is less than 100 km/h and considering a maximal acceleration of 14 m/s^{-2}, which corresponds to an hard braking reducing the speed from 100 km/h to 0 km/h in 2 s, we choose $\Delta T = 10$ ms which leads to an absolute error smaller than 0.5 km/h. For speeds above 100 km/h, we choose $\Delta T(ms) = v(km/h) / 10$. This leads to a relative error equal to 0.5%.

EXPERIMENTAL SETUP AND SIMULATIONS

In order to test the accuracy and the validity of the principles presented in this paper, we carried out simulations and made real measurements.

An experimental evaluation system has been developed to allow measurements with one or multiple obstacle configurations, with an optimal reproducibility. It is illustrated in Figure 4. A belt, which represents the road surface, is driven by a motor-pulley system, and its linear speed can be adjusted in the range 0-90 km/h. Scattering obstacles are fixed on the belt. Their number, shape and distribution can be set. Furthermore, this distribution can be discrete or continuous in order to simulate any kind of road surfaces. A device based on optical barriers is used in order to have an accurate speed reference. An antenna with main axis oriented at a 35° angle from the belt surface was used to emit a wave at a frequency of 19 GHz.

In addition to the experimental system, we have developed a simulation program. It makes it possible to precisely analyze the influence on the measurement of numerous parameters such as vehicle speed and acceleration, noise, road profile, antenna diagram, height above the road, duration of the time slot etc… The number, repartition and shape of the obstacles can be

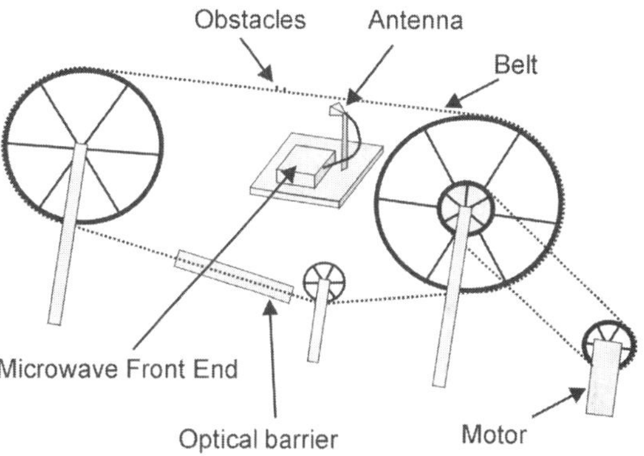

Figure 4: Laboratory set-up.

introduced in the program in order to allow the simulation for different road surfaces. These obstacles can be positioned anywhere on the road surface. The radiation pattern of the antenna and the influence of the antenna-obstacle distance on the signal amplitude can also be introduced. Colored noise of settable levels can be added to the signal in order to simulate noise from electronic components or mechanical vibrations.

SIGNAL ANALYSIS AND SPEED DETERMINATION

The Doppler signal depends on the scattering obstacles in the antenna footprint. If there is only one scattering obstacle in the antenna footprint, the Doppler frequency is non-ambiguous. If there are several widely distributed scattering obstacles, the associated frequencies are well separated. In both cases, a spectral analysis based on the Fourier transform with zero filling allows the determination of the frequencies associated to all the obstacles. An example is presented in Figure 5. One can see essentially two series of Doppler frequencies representing two obstacles.

If several obstacles are closely distributed in the antenna footprint, the associated frequencies are also close to each other and their contributions overlap in the global Doppler signal. Each obstacle produces an elementary spectrum which depends on the duration of the analysis windows δT and on the window function used to compute the FFT. For instance, using flat top windows leads to sinus-cardinal (sinc) functions as the elementary spectra. The decomposition of the calculated Doppler spectrum in elementary spectra allows the determination of the frequencies associated to each obstacle [2].

Once the frequencies associated to the different obst a-cles have been found for each moving window within the time slot, the procedure described above to fit equation (2) must be implemented. Two approaches have been investigated, either collecting in series the Doppler fre-quencies generated by each obstacle, or using a Radon-like Transform.

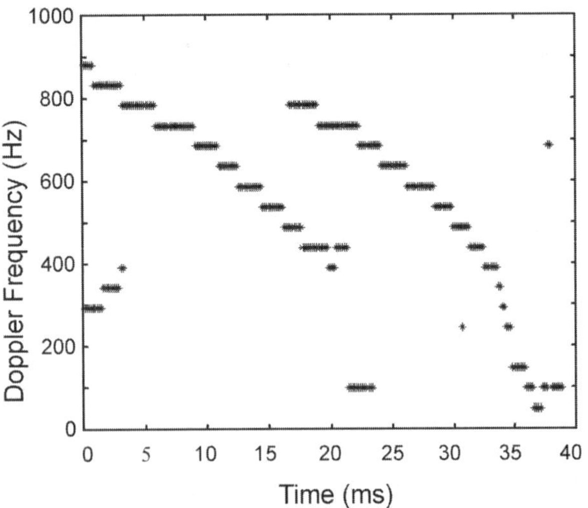

Figure 5: Example of Doppler frequency evolution with two scattering obstacles on the ground.

ASSOCIATION IN SERIES

The frequencies associated with the same scattering obstacle seen at several positions during the time slot are organized in order to form series. The determination of these series takes into account the speed measured at the previous time slot, the maximum possible values of the vehicle acceleration and also the fact that, during a measurement, the number of scattering obstacles in the antenna footprint may change. An example of such an association is shown on Figure 6 where two obst a-cles are identified.

Once the series have been determined, we fit the an a-lytical function (2) to each of them. If the vehicle has a height sensor, the fit is done by only varying two p a-rameters: the speed (v) and the obstacle position (x) in the analytical function. If no height sensor is available, there are three fit parameters: the speed, the position and the height above the road surface. The parameters for which there is the best fit between the frequency se-ries and the analytical function are chosen as the sol u-tions of the problem.

The optimization is achieved by minimizing an error function associated to the difference between the the o-retical and the experimental frequencies. In order to avoid weak local minima of the function, we combine a Monte-carlo method with a gradient method.

This approach leads to a speed of 99.8 km/h for the first obstacle and of 99.4 km/h for the second one for a real speed equal to 100 km/h, which leads to a relative error equal to 0.4%.

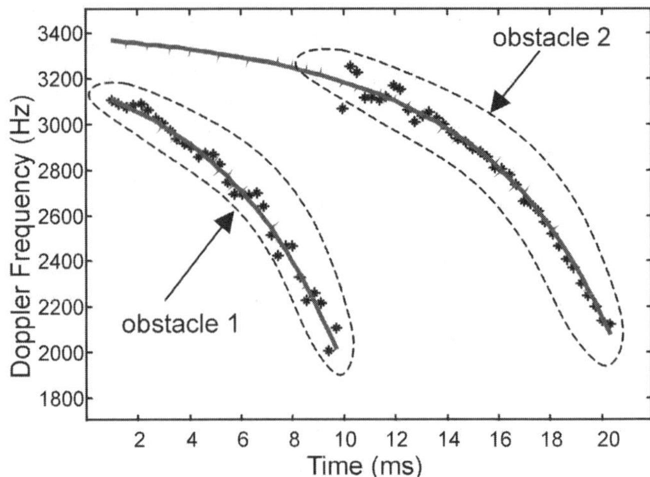

Figure 6: Frequency association by obstacle and fitting.

RADON-LIKE TRANSFORM.

The association in series and the fitting process can be performed by a modified Radon transform. The standard Radon transform [10] allows for the detection of linear feature in a noisy image. The modification we have d e-veloped allows for the detection of function (2) features in the image representing the Doppler frequencies ve r-sus time such as shown in Figure 6. The result of the transformation gives an other image. In this image the coordinate of one point is (x,v) and the color coded a m-plitude of each point is proportional to the probability that an obstacle at the initial position x has a relative speed v. Each point of the image of coordinate (x,v) is calc u-lated by determining the number of Doppler frequencies versus time points laying on the $f_d(t,x,v)$ function. In this case, the Radon-like transform R(x,v) can be expressed as

$$R(x,v) = \iint I(t,f_d)\,\delta(f_d - \frac{2\,f_0\,v}{c\sqrt{1+(h/(x-vt))^2}})\,dt\,df_d \quad (3)$$

where $I(t,f_d)$ is the image representing the variation of the experimental Doppler frequency with time and δ is the Dirac's function. As an example, applying this transform to the data of Figure 6 leads to Figure 7.

It can be seen that the speed of each obstacle can be derived directly. The darkest points give the higher probability to have an obstacle at position x with a velocity v. The Radon-like transform presented in Figure 6, leads to an estimated velocity of 101 km/h for the first obstacle and of 100.5 km/h for the second one for a real speed equal to 100 km/h, which corresponds to a relative error of 0.7%.

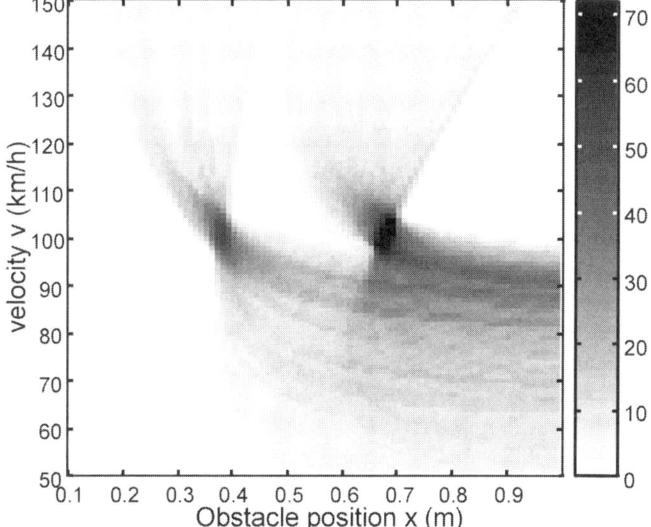

Figure 7: Radon-like Transform of the image shown in Figure 6.

If the third parameter h must be determined at the same time, the Radon-like transform can still be applied. A three dimensional image is then built where each point has three coordinates (x,v,h).

One major difficulty in the association in series is the choice of the convenient series. The Radon-like transform can help making this choice since the position of each obstacle is clearly identified. However as the resolution of Radon-like transform is increased to obtain the best accuracy in the estimation, the time it takes for the processes also increases. For real-time estimations, a coupling of the two methods is possible. First a low resolution Radon-like transform is performed to obtain a first estimation of each obstacle position x_i and velocity v_i. These informations are then used to associate the closest points to $f_d(t,x_i,v_i)$ in the ith series. Finally the parameters (x,v) of $f_d(t)$ are adjusted to obtain the best fit. The coordinates (x,v) give the accurate estimation of the velocity. In this case, the number of operation necessary to compute the velocity is mainly due to the time-domain frequency analysis. It requires roughly 10^6 operations to compute the 1024 point FFT of the 100 analysis win-dows. The low resolution Radon-like Transform and then the optimization need only less than 10^5 operations. Such an algorithm is directly implementable in real time (in less than 50 ms) with most of the actual DSP's.

RESULTS AND DISCUSSION

COMPARISON BETWEEN NARROW AND BROAD BEAM SENSORS

The broad beam sensor presented above is very acc u-rate when one only or a limited number of obstacles are present in the antenna footprint, situations associated with a high risk (snow, water, etc…). It is now possible to compare the performances of standard narrow beam Doppler sensors with the present broad beam sensor. Figure 8 shows the speed uncertainty for both types, as a function of angle of view of a unique scattering obstacle in the antenna footprint. The apertures of the narrow beam and of the broad beam antennas are respectively 10° and 34°. It can be seen that the use of a broad beam makes it possible to determine the speed for a much larger interval of obstacle positions than in the case of narrow beam sensors. In the illustrated case, for h=30 cm, the broad beam sensor leads to an accurac y of ± 0.5 % over a 50 cm range of obstacle position (from 14 to 64 cm), whereas the same accuracy of ± 0.5 % is obtained over a 7 cm only range for the narrow beam sensor (from 36 to 46 cm).

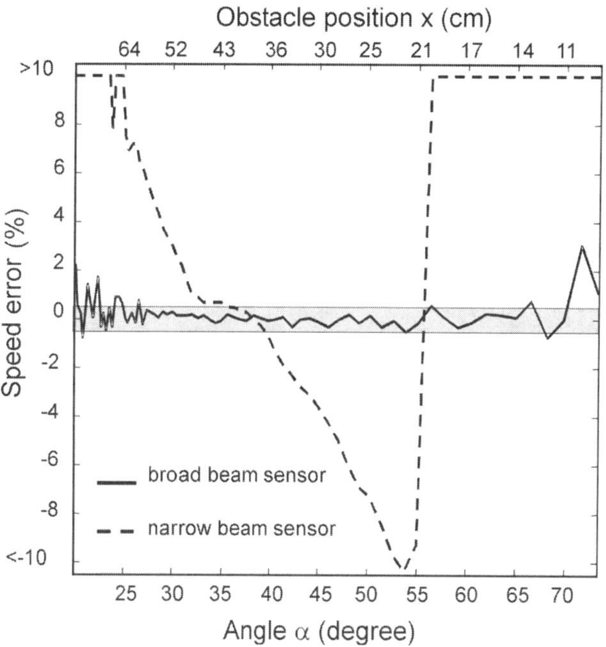

Figure 8: Comparison between broad and narrow beam sensors.

Thus when the obstacle density is larger than one o b-stacle every 10 cm, which does not correspond to a high-risk situation, then narrow beam Doppler sensors can be applicable with a good accuracy [5].

SENSITIVITY TO HEIGHT

The height h above the road surface is also a parameter of equation (2). Until now, it has been assumed to be known although its value can fluctuate due to the vertical movements or bad conditions of the road surface. Simulations have been performed to test the influence of such variations on the speed determination accuracy. The height variations around its average value can reach ± 5 cm in the case of an automobile with a 1 to 2 Hz fr equency oscillation. An error of ± 5 % on height h, i.e. 1 cm for a height of 20 cm, leads to a maximum speed error of ± 0.5 %. This variation of 1 cm can only be reached in a time larger than 25 ms. Thus these height variations can be neglected since the time slots have been chosen less than 20 ms.

Another factor must be taken into account: the variation of the average height, due for instance to variations of the vehicle load, the suspension aging, the chafing of the tires, etc… Such variations can typically reach 10 cm for a normal height of 20 cm. Using only the th eoretical height in the calculations leads to a maximal error of 6%. It is therefore necessary to determine this third param e ter along with the speed and the position of the obstacle. These three parameters can be determined by a best fit procedure between equation (2) and the experimental time dependent Doppler frequency data. This usually leads to speed errors less than 4%. In order to improve this result, the two and three parameter optimization procedures can be combined. Indeed, it is possible to use some heights determined at some previous time slots by three parameter fittings, in order to calculate an average height value. This value is then used to perform a two parameters fitting. After the two parameters opt i mization has been carried out, the error becomes smaller than 2.5%.

CONCLUSION

A Doppler sensor using a single fixed frequency and a broad beam of emission is particularly well suited for high-risk situations in automotive applications. We have shown that it combines the advantages of broad beam of emission Doppler sensors, i.e. a high probability of ha v ing a reflecting obstacle in the antenna footprint, and of single frequency emission devices, which allows easy implementation. It has been shown that the accuracy of this sensor is very high for low density obstacle distribu tions. It gives accurate results in a large range of obst a cle positions in comparison with narrow beam sensors. A modified Radon Transform allows a better automation of the process to separate frequency contributions due to several obstacles present in the antenna footprint. Finally, the same sensor can also give the value of the vehicle height above the road.

REFERENCES

1. J. LEWINER, E. CARREEL: "Method and apparatus for measuring the speed of a moving body ", US patent US 5 751 241, May 1998.

2. C. CORBRION, T. DITCHI, S. HOLÉ, E. CARREEL, J. LEWINER, "A broad beam Doppler effect sensor for the measurement of the absolute speed of a ve hicle", 8th European Automotive Congress, Brat i slava, Slovak republic, june 2001.

3. C. CORBRION, T. DITCHI, E. CARREEL and J. LEWINER, " Procédé et dispositif pour mesurer la vitesse d'un mobile ", French patent, FR 00 06494, 2000.

4. C. CORBRION, " Etude de nouveaux capteurs de vitesse absolue à effet Doppler et à grande ouve r ture angulaire ", PhD thesis, Université de Paris 6, December 2000.

5. C. CORBRION, T. DITCHI, S. HOLE, E. CARREEL and J. LEWINER, "A broad beam Doppler speed sensor for automotive applications", Sensor Review, Vol. 21, pp. 28-32, January 2001.

6. L.H. ERIKSSON, S. BRODEN: "High Performance Automotive Radar", Microwave Journal, pp.. 24-38, October 1996.

7. P. HEIDE: "Commercial Microwave Sensor Tech nology: An Emerging Business", Microwave Journal, pp 348-352, May 1999.

8. N. KEES, M. WEINBERGER, J. DETLEFSEN: "Doppler measurement of lateral and longitudinal velocity for automobiles at millimeterwaves", IEEE MTT-S digest, vol. 2, pp. 805-808, June 1993.

9. P. FLANDRIN, "Temps-Fréquence", Ed. Hermes, Paris, 1998.

10. A.C. COPELAND, G. RAVICHANDRAN, and M.M.TRIVEDI, ``Localized Radon Transform-Based Detection of Linear Features in Noisy Images,'' *Proc. IEEE Conf. on Computer Vision and Pattern Recog nition*, Seattle, WA, pp. 664-667, June 1994.

CONTACT

Thierry Ditchi, Assistant Professor, Université Pierre et Marie Curie, Laboratoire des Instruments et Systèmes d'Ile de France (LISIF)
Tel: (33) 1 40 79 45 71, thierry.ditchi@espci.fr

Stéphane Holé, Assistant Professor, Université Pierre et Marie Curie, Laboratoire des Instruments et Systèmes d'Ile de France (LISIF)
Tel: (33) 1 40 79 45 71, stephane.hole@espci.fr

Céline Corbrion-Filloy, Assistant Professor, Ecole Supé rieure de Physique et de Chimie Industrielles (ESPCI), Laboratoire d'Electricité Générale (LEG)
Tel: (33) 1 40 79 45 91, filloy@optique.espci.fr

Jacques Lewiner, Professor, Ecole Supérieure de Ph y sique et de Chimie Industrielles (ESPCI), Laboratoire d'Electricité Générale (LEG)
Tel: (33) 1 40 79 45 31, jacques.lewiner@espci.fr

The Rapid Development of Vehicle Electronic Control System by Hardware-in-the-Loop Simulation

Li Jun, Feng Jinzhi, Yu Fan and Zhang Jianwu
School of Mechanical Engineering, Shanghai Jiao Tong University

ABSTRACT

The paper describes the idea of constructing the rapid development system of vehicle electric control subsystem in details by using hardware-in-the-loop (HiL) simulation technology for cost reduction in today's competitive business environment. This rapid development system is probably most effectively used in parallel with a new vehicle electronic control product. Once a desired simulation results are achieved, it can be verified by setting up the same configurations in the new product and comparing results. The feasibility of using such rapid development system is demonstrated through vehicle ABS algorithm development. Significant reduction of developing cycle time and cost has been achieved with the aid of this powerful tool and promising effectiveness of ABS algorithm has been validated in vehicle field test. The ideal proposed in this paper can be applied to the development of more advanced control systems, such as traction control system, vehicle stability control system and so on.

INTRODUCTION

With more and more attention paid to vehicle performances of safety, emission and ride comfort, etc., the vehicle electronic control systems to improve vehicle overall performances have been under development by almost all major automotive companies for at least two decades. Because of fierce competition, the need to reduce cost and time to market is constantly pressuring the automotive industry to find better ways to develop its sophisticated vehicle electronic products.

The development of vehicle electronic control system involves mechanics, electronics, hydraulics, hardware and software design, field test and validation, etc. In early years, the product development for electronic control system heavily depended on a huge amount of vehicle filed tests to investigate the control law and a great deal of repeated modifications also had to be done. Moreover even in the most controlled enviroment, the conditions for vehicle field test cannot be repeatable. This leads to the increase of cost and the development process is also long. However, with the presence of powerful computer and software technology, the automotive industries have been investigating the application of advanced engineering tools to design and test increasing complex products and improve overall quality. A real-time hardware-in-the-loop (HiL) simulation has got a spotlight as an important tool for software and system engineers to design, analyze and test control systems, and more and more applications have been implemented in automotive industries.

The HiL simulation originated in the aerospace and defense industries where it was often impossible, impractical, or just too costly to test controllers on actual systems. The basic elements of HiL simulation system include computer, I/O resources, signals conditioners

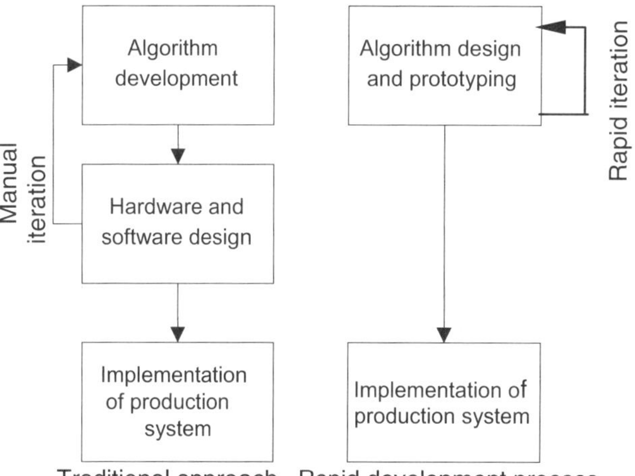

Fig.1 Comparison of traditional and rapid development process

and user interface. It can emulate the control system's sensors and actuators to recreate an operating environment, similar to the actual one, in a controlled setting. Recently, it has become an important tool and has been applied extensively in automotive industry to develop, validate, test and evaluate a vehicle control system.

The first generation HiL simulator for vehicle applications was designed in the mid-1980s, which was a predominant tool in use for system development and software verification. However, it was very complex and expensive due to its low-level electronic technology compared that of today and had not enough functionality to meet evolving sophistication of electronics controller. The second generation HiL was established in 1989 and upgraded in 1994. The heart of the system is Applied Dynamics Real-time Station (RTS), which executed analytical and empirical vehicle description, and performed extensive I/O operation in real-time. Although the HiL system has successfully demonstrated the effectiveness of the controller development process, there exists a need to transfer from a dedicated facility to engineering benches. Therefore, an affordable dynamics simulator needs to be identified and deployed for controller development [1~5].

The objective of this paper is to develop an economical and friendly interface real-time HiL simulator based on personal computers by utilizing a widely used simulation software package—Matlab/Simulink. It combines modeling, control algorithm, software and hardware design phases, thus eliminating potential bottlenecks. The process allows engineers to predict the results and rapidly iterate design process before expensive hardware is developed. Fig.1 shows the comparison of a traditional design method and a rapid development process.

This paper firstly introduces the whole process of developing a rapid prototyping system in Matlab/Simulink graphic environment, including system modeling, normal pure simulation, generic real-time simulation, real-time HiL simulation and vehicle field test validation. And then an example of ABS algorithm development is described using the proposed methods. Finally, the promising vehicle field tests are presented, which validated the effectiveness of the rapid development system for ABS algorithm.

THE RAPID DEVELOPMENT OF VEHICLE

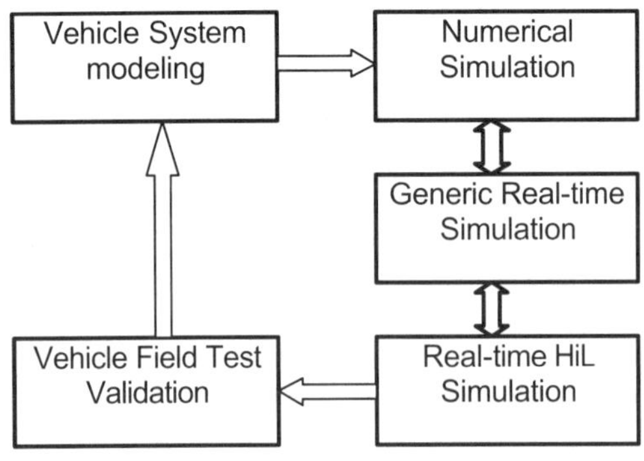

Fig.2 General ideal of rapid development system

ELECTRONIC CONTROL SYSTEM

The rapid development process for vehicle control system consists of five steps: vehicle system modeling, normal numerical simulation, generic real-time simulation, real-time HiL simulation and vehicle field-test validation. Fig.2 illustrates the general ideal of rapid development process.

VEHICLE SYSTEM DYNAMIC MODELING - The vehicle dynamic model is intended to have sufficient complexity to replicate the dynamic responses of a vehicle through the full range of conditions. However, in order to satisfy the real -time simulation, some simplifications on vehicle model are necessary to be done according to practical need.

NUMERICAL SIMULATION-The main program for vehicle dynamic control system simulation is illustrated as Fig.3 and it can be used for the simulation with variable time step. Hence, the control algorithm and model parameters can be optimized by simulations until the desired results are obtained.

GENERIC REAL-TIME SIMULATION-From simulation parameters dialog-box, the fixed time step is selected as 0.002s and the RK4 algorithm is used. The graphic program can be converted to optimized and stand-alone C code, and complied, linked automatically and downloaded directly to the local processor by activating the Real-Time Workshop (RTW)[6]. Then a generic real-time simulation can be running. By using Simulink external mode that established a communication channel between Simulink and the generated code by RTW, both simulation results can be monitored in the scope and model parameters can be tuned on the fly without stopping the simulation. The vehicle model and parameters of control scheme are refined until satisfactory results are obtained. Therefore, the vehicle model and control algorithm are verified by pure simulation.

REAL-TIME HiL SIMULATION-The real-time HiL simulation is based on previously developing block diagram plant models of sufficient fidelity for preliminary system design and simulation. Once the generic real-time simulation shows encouraging system performances, the controller block diagram is separated from the vehicle dynamic mode and I/O device drivers offered by I/O block library are attached as shown in Fig.4. With the aid of RTW and xPC_Target software, it is possible to replace any of these models with real hardware, bring their sensor information, generate the corresponding control signals and submit them to actual hardware via I/O block. After click building model command, the program shown as in Fig.4 is complied, linked, and automatically downloaded by the serial cable or network to the target computer or other target device such as dSPACE, which provides the real-time running environments and is equipped with physical connector (I/O device) to the real world such as sensors and actuators for the need of system implementation.

During real-time HiL simulation, the results can be tracing on the host computer and the parameters can be modified and automatically downloaded to the target computer on the fly by using Simulink external mode. Through analyzing results and tuning parameters, the actuator system is validated and the refined control algorithm is adapted better to the real world.

VEHICLE FIELD TEST VALIDATION- In the end, vehicle field test will be done to validate the same control algorithm as in real-time HiL simulation, and to estimate the correlativity between the real vehicle and vehicle model. The only changes of the program are that the real vehicle replaces vehicle model and the real sensors replace sensor model. Once more click building model command, a stand-alone program is generated and implemented on notebook computer with physical connector (I/O device) to the sensors and actuators equipped on the vehicle. Because of the difference between vehicle model and real vehicle, the data of vehicle dynamic variable are acquired and analyzed to tune the control parameters until the promising results are obtained, and also used to refine vehicle model. The lessons learned from the vehicle testing can also be used to modify the simulation environment to better correlate to the vehicle.

THE RAPID DEVELOPMENT VEHICLE ABS ALGORITHM

The schematic diagram of vehicle ABS rapid development system is given in Fig.5. It combines the vehicle dynamic model and ABS control algorithm with the hardware of brake system seamlessly and is proved to be an important tool for the rapid development of ABS and research of ABS overall performances. The numerical simulation, generic real-time simulation and real-time HiL simulation are all carried on this platform.

The ABS rapid development system mainly consists of software part, input/output (I/O) interface part and hardware part.

SOFTWARE PART- The software part is implemented on the computer, and made up of vehicle dynamic model, tire/road friction model, hydraulic model and ABS control algorithm.

Vehicle dynamic model-With the precondition of ensuring real-time simulation, the vehicle dynamics model is intended to have sufficient complexity to replicate braking and handling response of vehicle through the full range of non-crash conditions. In this paper the vehicle dynamic model is described by 8-DOF, e.g., 3 DOF for sprung mass, 4 DOF for wheels and 1 DOF for steering wheel [7].

Tire/road friction model-The tire model is very essential in vehicle dynamic simulation. Then many researchers had done a lot of work on it and a few typical of tire models have been published. This paper introduces the Pacejka's tire model with combined braking and cornering, which is popular in the vehicle dynamics simulation research. [8] This model has three inputs (tire normal force, wheel slip angle and slip ratio) and two outputs (longitudinal and lateral forces). Through the measurement of tire brake characteristic as related to wheel slip on different road surface (dry, wet or icy road)[9], a few group of corresponding parameters of Pacejka's tire model can be identified to describe high-μ, medium-μ and low-μ tire/road friction conditions.

Hydraulic model- According to the theory of hydrokinetics and lab test, the ABS hydraulic dynamic model is categorized as pressure increasing, pressure holding and pressure dumping model, which are described by following equation [10] and implemented in pure simulation.

$$\frac{dp_w}{dt} = \begin{cases} 35.7418(p_m - p_w)^{0.58} & \text{Increasing} \\ 0 & \text{Holding} \\ -36.3714(p_w - p_r)^{0.92} & \text{Dumping} \end{cases}$$

p_w -- Wheel cylinder pressure p_m -- Master cylinder pressure p_r --Remnant pressure

ABS control algorithm- With the fast development of control theory and computer technology, more and more advanced intelligent control theory such as fuzzy logic, neural networks and so on are applied into research on ABS control logic. Some fuzzy control ABS products were tested on vehicle and promising results have been obtained. Though as such that, the commercial ABS product now still adopts the threshold control logic. However, some new control theories had been added into this control algorithm. To compare with original product, the present paper has investigated adaptive thresh hold control algorithm.

I/O INTERFACE PART- I/O part is an important communication bridge between the software and hardware. The controller sends signals according to the received signals of vehicle dynamics states to drive solenoid valves so as to modulate the pressure of wheel cylinder through this important interface. This paper adopts PCI-6025E I/O cards and PCLD-786 relay to drive the solenoid valves and motor.

HARDWARE PART- Hardware part consists of the components of conventional brake system such as brake pedal, master cylinder and brake drum and disc and some auxiliary sensors such as pressure sensor and so on. Two PCs must be needed, one is high level and the other is normal one. Also a hub is included to support communication between the host computer and target computer.

PROCEDURE OF RAPID DEVELOPMENT OF ABS AlGORITHM- Firstly, the vehicle system dynamic model and design of ABS control algorithm are developed on host computer in Matlab/Simulink graphical environment. And then, the numerical simulation and self-target generic real-time simulation are running on host computer. The vehicle model and control algorithm are amended until promising performances are obtained according to the analysis of simulation results. Such as mentioned in above sections, the normal and generic real-time simulation validate the vehicle model and control algorithm. And then, the controller block diagram is separated from vehicle mode and the I/O device drivers offered by I/O block library are attached in line with the real system need, then activate the RTW and xPC_Target to transfer the program running on host computer into C language, which is complied, linked and downloaded automatically to target computer through network. So the real-time HiL simulation can be running on the target computer, the results will be uploaded by the TCP/IP and displayed on host computer. The digital value of vehicle dynamic model output (wheel speed) can be converted into analog electric value such as voltage and pulse throng the D/A conversion so as to simulate the function of sensors. The signals from the virtual sensors are transferred to memory of control algorithm through the A/D conversion. Then the controller sends control signals, based on calculated vehicle dynamic states (wheel speed and their corresponding derivative), to actuators such as solenoid valves and motor through the I/O port and power-amplified circuits. Then the signals from sensors of real actuators system again are transferred to vehicle dynamic model through the A/D device and a cycle of real-time HiL simulation is completed.

The vehicle model and control algorithm are revised again until encouraging performances come forth. The real-time HiL simulation validates the actuators of brake system and sensors. Finally, the real vehicle replaces the vehicle model and some necessary modifications have been done, and then, the control algorithm will be validated in vehicle field test. Following, the same work as before, that is, refining the vehicle model and optimizing the control algorithms.

FIELD TEST VALIDATION-To demonstrate the feasibility and effectiveness of the rapid developed system, the developed ABS control algorithm has been applied to vehicle field test. Three representative cases are presented here. Promising performances of vehicle braking have been obtained. The test vehicle and control system equipment are illustrated as shown in Fig.6 and Fig.7. Fig.8 is shown the vehicle velocity and wheel speed during braking process of filed test and real-time HiL simulation under the same conditions, such as same initial velocity (25m/s) and low friction road condition. The braking pressure of a wheel cylinder and control signals are also illustrated in Fig.8a Fig.9 is shown he similar case, but with different initial velocity (16m/s). Fig.10 is shown the third case that vehicle brakes on high friction road condition with initial velocity 26m/s. These results show that none wheel lock-up happens whether in high or low road conditions and the stopping distance is shorten apparently. Although, not exactly the same results (field test and real-time HiL simulation) have been achieved because of difference between real world and modeling, it is still proved that the effectiveness of ABS algorithm developed by proposed method

OTHER FUNCTIONS OF ABS RAPID DEVELOPMENT SYSTEM - The other functions of ABS rapid development system are described as following:

(1) Research on the hydraulic dynamic characters of ABS
(2) Design and development of ABS key components
(3) Validation of the whole functions of ABS products
(4) Research on system matching between the ABS product and vehicle
(5) Analysis the control logic of other commercial ABS product
(6) Extended functions, such as development of TCS and Vehicle Stability Control

CONCLUSIONS

The real-time HiL simulation is becoming increasingly an important tool for the development of much high-level quality vehicle electronic controllers with more and more complexity and sophistication. This paper provides a rapid and economical development method for vehicle electronic controllers and introduces the process of developing the vehicle electronic controller by using the present rapid development system. The development of ABS algorithm and field validation test is demonstrated as an example, encouraging performances of vehicle braking are obtained. This proves that it makes vehicle electronic control system development and parameter tuning fast and efficient when compared to in-vehicle development time and cost.

REFERENCES

1. Wagner, J., Brunts, R., and Kaster. A vision for automotive electronics hardware in the loop testing. Int. J. of Vehicle Design, 1999,Vol.22, Nos.1/2,pp.14-28
2. Cheng Jun. The integrated development system for vehicle control system. Automotive Engineering, 2000, Vol.22, No.2, pp.109~113
3. Suh, M.W., Chung, J.J., and Kim, S.M. Hardware-in-the-loop simulation for ABS on PC. Int. J. of Vehicle Design, 2000,vol.24, Nos.2/3,pp.157-170
4. Wanger, J.R. and Furry J.S. A real-time simulation environment for the verification of automotive electronics controller software. Int. J. of Vehicle Design, 1992,Vol.13, Nos.4, pp. 365– 377
5. William P.Amato and Mark A. Bennett. Verification of Heavy Truck EBS and ABS Using Matrix Hardware in

the Loop Tools. SAE Technical Paper Series, Paper No.1999-01-3713.

6. Mathworks Inc., xPC Target for use with Real Time Workshop, 1999

7. Li Jun, Zhang Jianwu and Yu Fan. An Investigation into Fuzzy Controller for Anti-lock Braking System Based on Road Autonomous Identification. SAE Technical Paper Series, SP-1576, Paper No.2001-01-0599.

8. Bakker, E., Nyborg, L. and Pacejka, H.B. Tyre Modeling for Use in Vehicle Dynamics Studies. SAE Technical Paper Series, Paper No. 870421, 1987

9. J.L. Harned, L.E.Johnston and G.Scharpf. Measurement of Tire Brake Force Characteristics as Related to Wheel Control System Design. SAE Technical Paper Series, Paper No.690214.

10. Liu Li. Construction of ABS test-bed and research on dynamic characteristics of hydraulic system. PhD dissertation, 2000, Jilin University of Technology, Changchun, China

CONTACT

LI JUN   Email: jli927@mail1.sjtu.edu.cn or lijun@uaes.com

INST. OF AUTOMOTIVE ENGINEERING

SCHOOL OF MECHANICAL ENGINEERING

SHANG HAI JIAO TONG UNIVERSITY, SHANGHAI 200030, P.R.China

APPENDIX

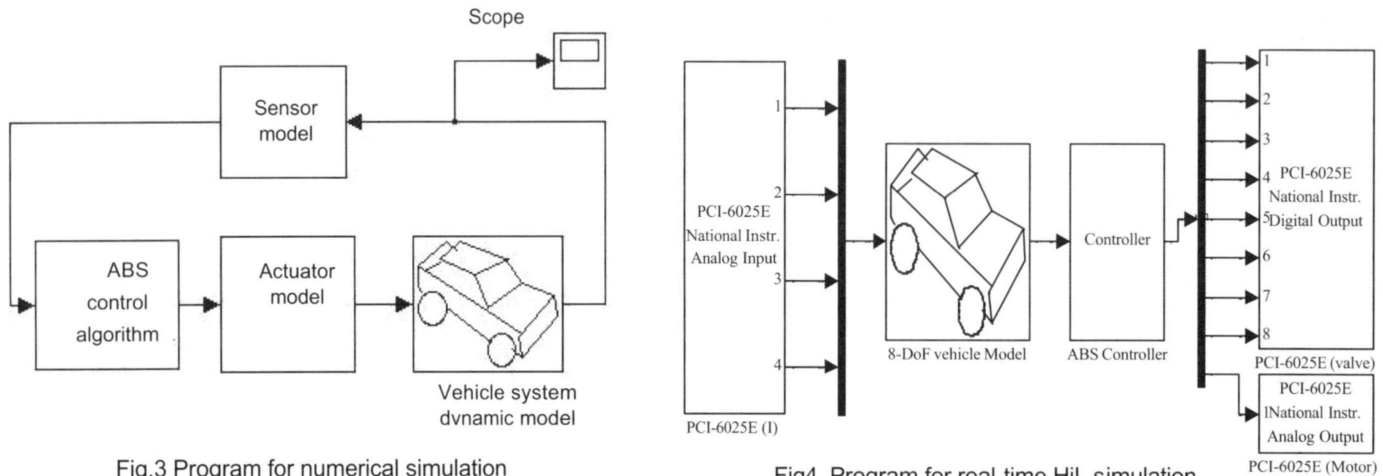

Fig.3 Program for numerical simulation

Fig4. Program for real-time HiL simulation

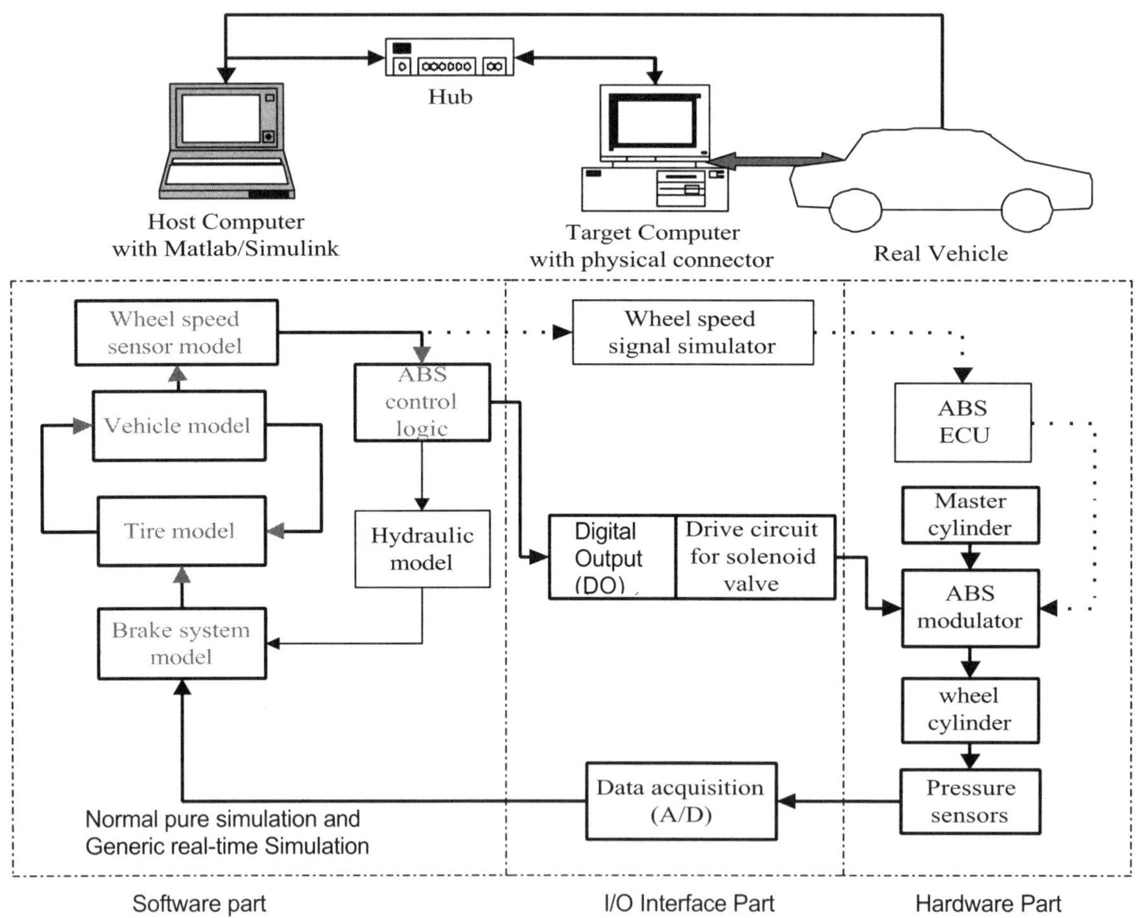

Fig.5 The rapid development system for ABS algorithm

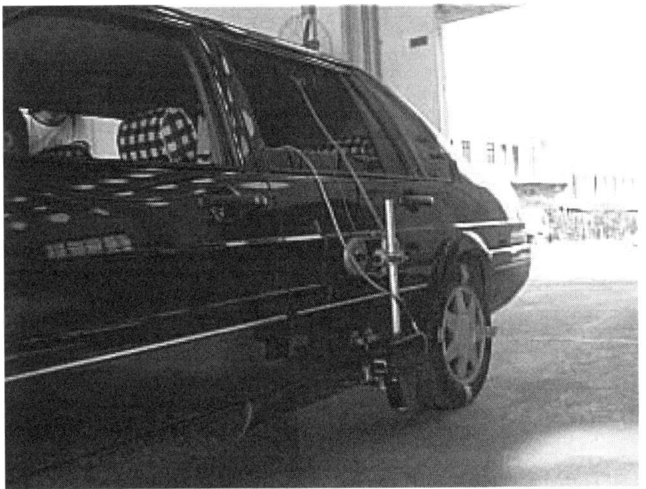

Fig.6 Vehicle for field test

Fig.7 Control system in-vehicle for field test

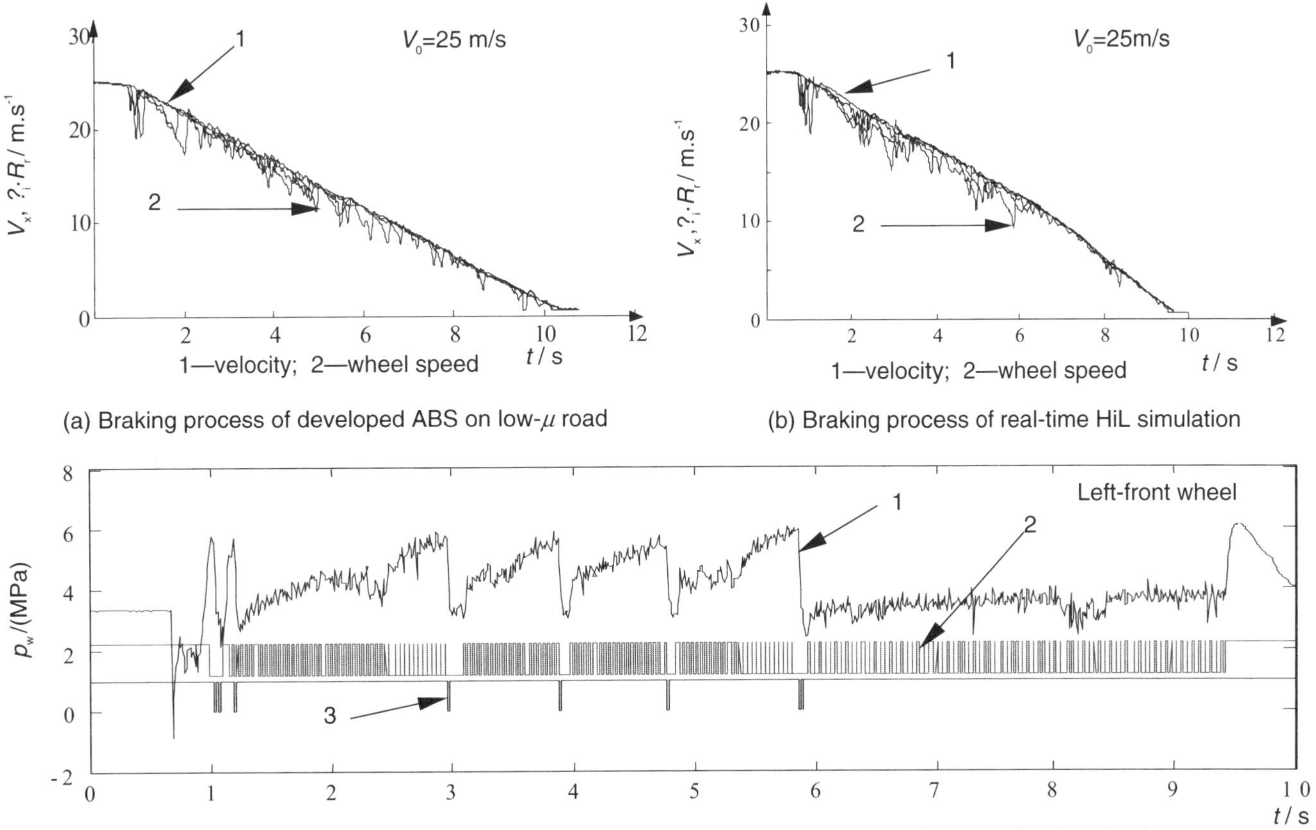

(a) Braking process of developed ABS on low-μ road

(b) Braking process of real-time HiL simulation

1—braking pressure; 2—control signal for normally open valve;3—control signal for normally closed valve

(c) Braking pressure and control signals during braking process of real-time HiL simulation (V_0=25m/s)

Fig.8 The validation of ABS algorithm on low-μ road and comparisons with real-time HiL simulation

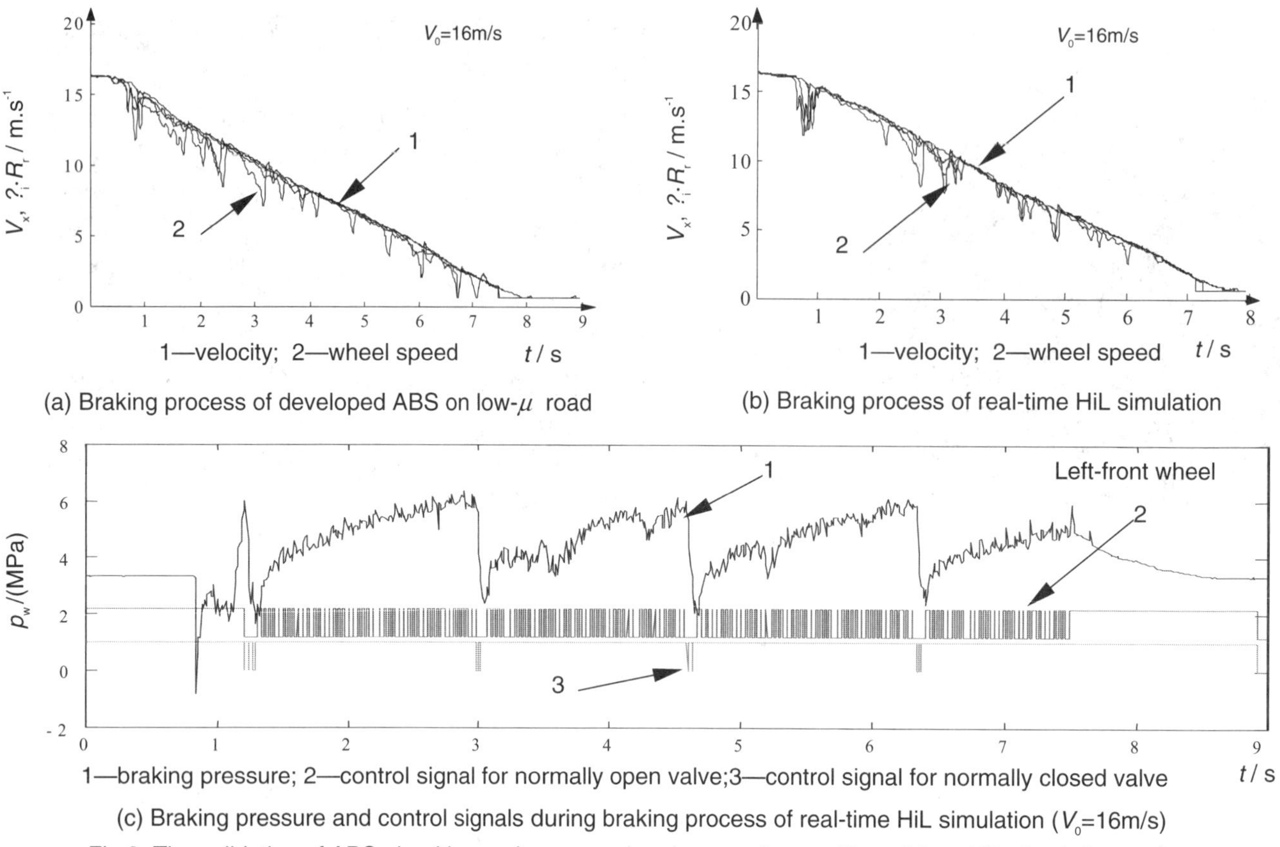

(c) Braking pressure and control signals during braking process of real-time HiL simulation (V_0=16m/s)

Fig.9 The validation of ABS algorithm on low-μ road and comparisons with real-time HiL simulation

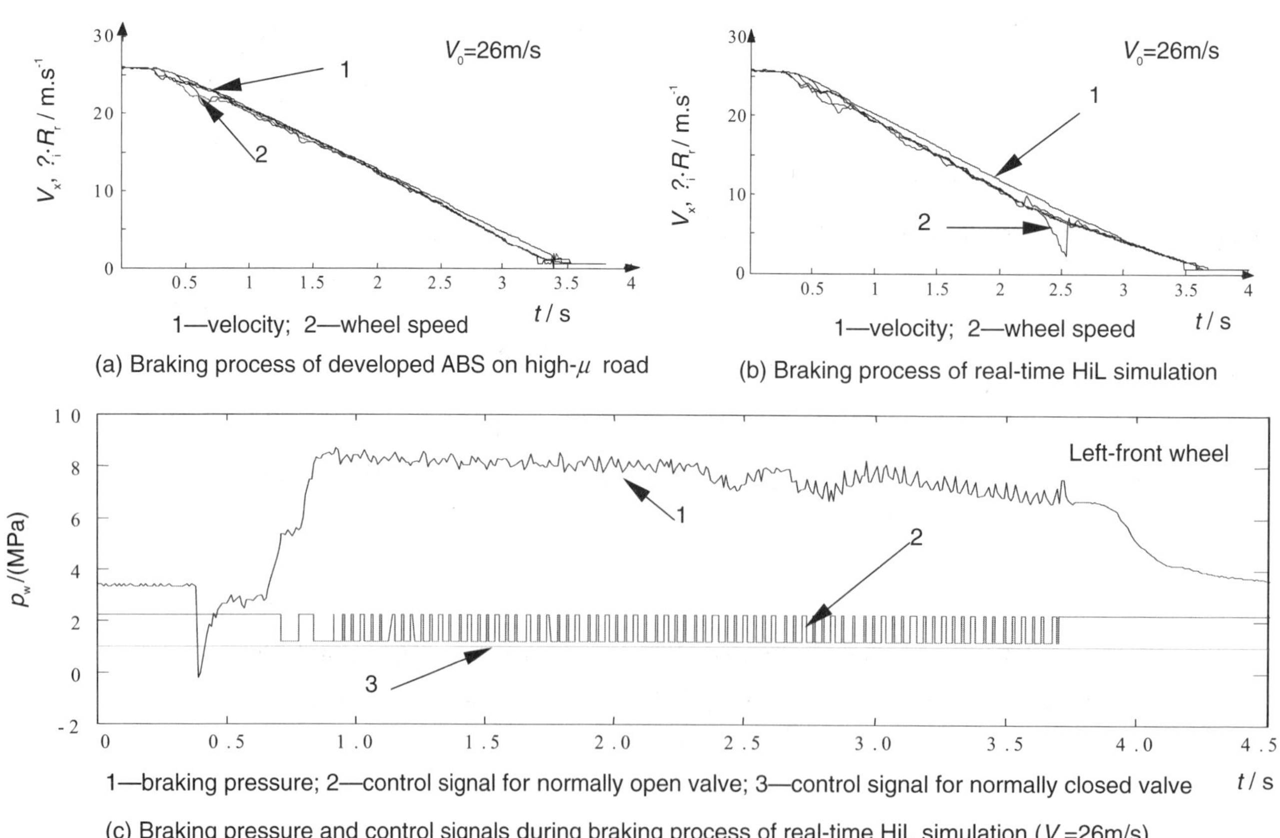

(c) Braking pressure and control signals during braking process of real-time HiL simulation (V_0=26m/s)

Fig.10 The validation of ABS algorithm on high-μ road and comparisons with real-time HiL simulation

Approach of Debugging Control Laws of ABS Combined with Hardware in the Loop Simulation and Road Experiment

Huiyi Wang, Bo Gao and Jian Song
State Key Laboratory of Automotive Safety and Energy, Tsinghua University

ABSTRACT

Anti-lock Braking System (ABS) is an important control system that can improve the automotive performance, reliability and safety obviously. Hardware-in-the-loop Simulation (HILS) testing is a promising method to assistant design automotive electronics system. In the State Key Laboratory of Automotive Safety and Energy, a HILS system was established to assistant design ABS. The HILS system is composed of a hydraulic control unit of ABS, a normal brake system, a commercial personal computer, a data acquisition card, and several signal-conditioning modules. Powerful software was programmed to perform managing input and output signals, solving 7-freedom and 15-freedom vehicle models, and acquiring response signals of the brake system.

Based on the simulations of HILS system, the road experiments with the same conditions of ABS are very necessary and important. In this method, the combination of HILS and road experiment rapidly reduce the cycle of design and debugging the control laws of ABS in automobile.

INTRODUCTION

With the development of the automotive electronic control technology, many testing methods based on micro-processor were devised, among which the most important one is the hardware-in-the-loop simulation technology. It is applied in the field of automobile more and more. This technology of hardware-in-the-loop simulation was first applied in the aviation and national defense industry, for in these fields the cost of test on the real controller is extremely expensive or even impossible. The hardware-in-the-loop simulation technology only needed a complicated giant computer system and just one suit of hardware of missile or flightier can fulfill any times experiments repeatedly and preciously. Nowadays, the cost of computer is low, the electric device is in high performance-price rate, the hardware-in-the-loop simulation method is possible for automotive design and testing.

Anti-lock braking system (ABS) can improve the vehicle braking stability and steering performance, through using electronic control devices to adjust braking pressure to prevent the wheel from locking. The testing of the ABS requires varied road condition and driving condition. But in fact, the expected road condition is not easy to get, and the testing is easy to be affected by the driver factor. In order to reduce the developing cost and duration, many big auto companies adopted the hardware-in-the-loop simulation technology. Through hardware-in-the-loop simulation the experiment can be repeated under the same condition conveniently, thus improve the credibility of the experiment.

Recently, the performance of the PC has improved greatly, so it is possible to carry out the hardware-in-the-loop simulation using PC, thus the system developing cost can be greatly reduced. In this paper the hydraulic ABS hardware-in-the-loop simulation system based on PC has been devised. And the research on a certain kind of Bosch ABS is also included. Finally, experiments of ABS were fulfilled on roads in the Proving Ground for Highway and Traffic, the Ministry of Communication.

HARDWARE IN THE LOOP SIMULATION SYSTEM

In order to investigate the performance of ABS in a car, a hardware-in-the-loop simulation system was established in the first.

SYSTEM STRUCTURE

Six sub-modules constructed hardware-in-the-loop simulation system mainly:

- Computer System. Using the common PC, carrying out the tasks of data processing, model calculating and wheel speed signal producing.

- I/O interface. Using high speed multi-function data sampling card, carrying out the task of Analog/Digital (A/D), Digital Input (DI), Digital Output (DO) and etc.
- Voltage/Frequency converter (V/F converter) circuit. In experiment rig, it is hard to get wheel speed signal from a rotary wheel, so pulse signals from those V/F converters to emulate the wheel speed signal preciously were adapted, depending on the vehicle wheel speed calculated by the PC computer.
- ABS electronic control unit (ECU). ECU is the core of ABS, main target of simulation. The aim of simulation is to study whether the ABS ECU can carry out anti-lock braking control properly.
- Actuators and Main Cylinder. The actuator of the simulation system will respond to the instruction of the ABS controller. They were the real actuators and real main cylinder of the vehicle being tested.
- Pressure sensors. Feedback components of the system, feeding back the pressure signals, which are the parameters of the simulation system to decide the states of the vehicle that time.

The system structure is shown in figure 1.

Figure 1 the hardware-in-the-loop simulation system structure

Table 1 shows the components used in the hardware-in-the-loop simulation system.

	Sub-system	Type	Configuration	Number
1	PC Computer	Pentium III	CPU: Intel Pentium III 500, RAM: 128 Mb	1
2	Data Acquisition System	Advantech PCI 1800L	Rate: 330kHz 16 ch A/D 2 ch D/A	1
3	V/F Converter	CE VZ02-T2MS1	Input: 0~5V Output: 0~2kHz	4
4	Actuators	IVECO	Disk Drum	2 2
5	Pressure Sensors	AK-4	Supply: 12V Range: 0~20Mpa Output: 0.5~5V	5

SIMULATION SOFTWARE DESIGN

In order to perform the simulation in the Hardware-in-the-Loop system well, software is very important. The software of the hardware-in-the-loop simulation system was developed with Visual C++, with the property of friendly interface and easy to use. The main function modules are listed as below:

1) Model selection: Different road condition (high-íroad, low-íroad, slip-í and jump-íroad) and different brake mode could be set and suitable model can be selected for different experiment object.

2) Parameters setting: The geometry parameters of the vehicle (wheelbase, centroid position and etc.) and physical parameters (suspension rigidness, absorber damp parameter and etc.) of the vehicle model can be modified, and different braking of different vehicle can be simulated. At the same time the road condition and the angel of steering braking can also be set.

3) Simulation control: Sending the wheel speed signals and braking start and end signals to the hardware of hardware-in-the-loop simulation system, the experiment will be performed.

4) Hardware state checking: Send out warning message when hardware fails.

5) Signal sampling: While the brake begins, the data acquisition card reads the brake pressure signals and the solenoid state signals every operating cycle.

6) Computing the vehicle Model: System will compute the mathematic model of vehicle during the signal sampling, then it will send the wheel speed signals of the operating result in the form of analog signal to the V/F converter circuits via D/A port of the data acquisition card.

7) Data saving: Save the operating results to the specified file every operating cycle.

8) Online display: Display the real-time values of the parameters, such as solenoid action curve, braking pressure curve and slip ratio curve, etc.

9) Offline display: After the simulation process, the former experiment results can be loaded to display for further analysis.

SYSTEM OPERATION

When the software running on the hardware-in-the-loop simulation system is successful, the simulation experiment can be progressed.
- Set initial condition: select model, set road condition and etc.
- Put down the start button, then the power supply that is 12V storage battery will be turned on. Normally, ABS will start self-diagnosing in 3 seconds. After passing the self-diagnosing, it steps to working mode.
- Send out initial wheel speed signal without brake pressure, let ABS pass the self-checking.
- Stepping the pedal as the real driver doing, start a braking operation.

- The pressure sensors measure the pressures of the cylinders, and then the pressures were fed back to the computer via A/D converter. And the computer calculates the wheel speed bases on the selected model.
- The wheel speeds calculated are sent to ABS ECU through D/A converter and V/F convert circuit. The simulation process continues until the vehicle speed is lower than the initial value (The default is 0.3km/h, far lower than 5km/h, which is the lowest limit of the real ABS).

SIMULATION EXPERIMENT AND ITS ANALYSIS

We carried out the simulation experiment on a certain type ABS with the hydraulic ABS hardware-in-the-loop simulation system. Normally, three types of road situation are chosen to have a hardware-in-the-loop system. There are high-ì road, medium-ìroad and low-ì road.

<u>High-ì road Simulation</u>

First, a high-ìroad was selected as the condition of vehicle running. The peak ìis equal to 0.8, initial brake speed V0 is equal to 20m/s, step t is equal to 2.5ms;

The experiment curves are shown in figure 2.

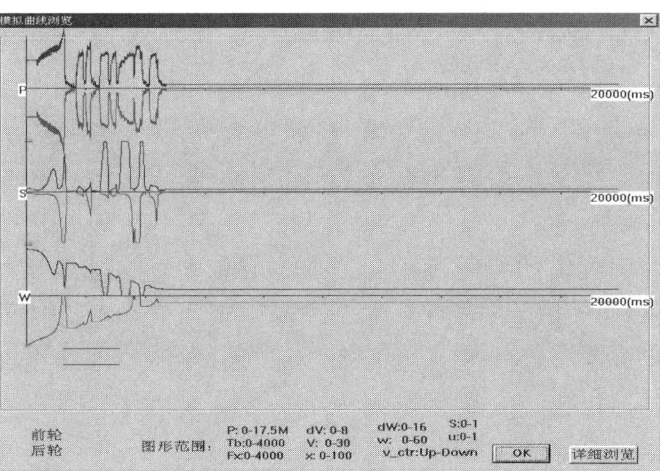

Figure 2 Simulation experiment of high-ìroad

We can get the result: brake duration is 4.7s; brake distance is 49.69m.

<u>Medium-ì Road</u>

The Experiment condition is: road adhesion coefficient ìis equal to 0.4, initial brake speed V0 is equal to 20m/s, step t is equal to 2.5ms.

The experiment curve is shown in figure 3.

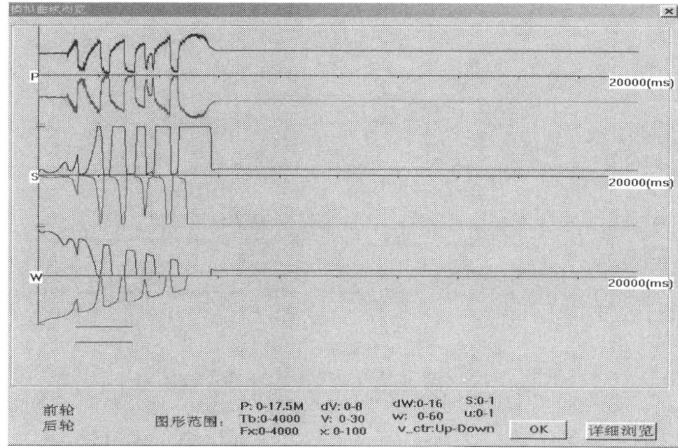

Figure 3 Simulation experiment of medium-ìroad

The simulation results are: brake duration T is equal to 5.893s, brake distance s is equal to 65.528m.

<u>Low-ì Road</u>

Experiment condition: peak ì is 0.15, initial brake speed V0 is20m/s, step time is 2.5ms.

The experiment curve is shown in figure 4.

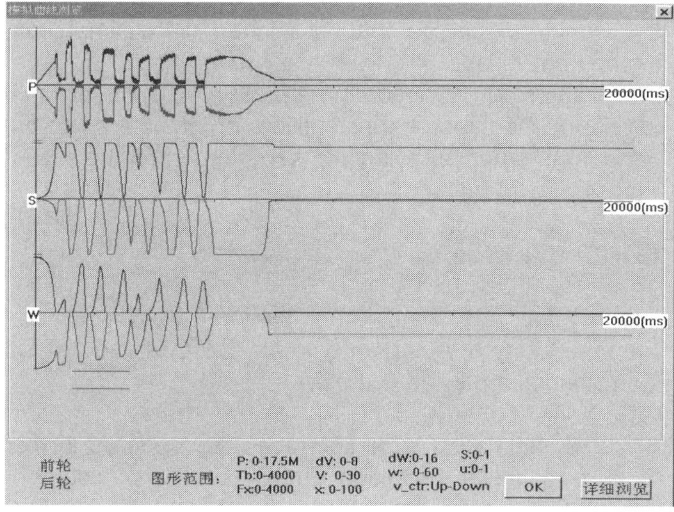

Figure 4 Simulation experiment of low-ìroad

We can get the result: brake duration is 7.96s; brake distance is 110.898m, both longer than the former.

ROAD EXPERIMENT

In order to compare with the simulation data, we carried out the road experiment with the ABS used in the simulation system. The experiment was carried out on high-ì road, the adhesive coefficient was about 0.7, initial brake speed V0 was 78.14km/h. The vehicle speed and wheel speed result are shown in figure 5.

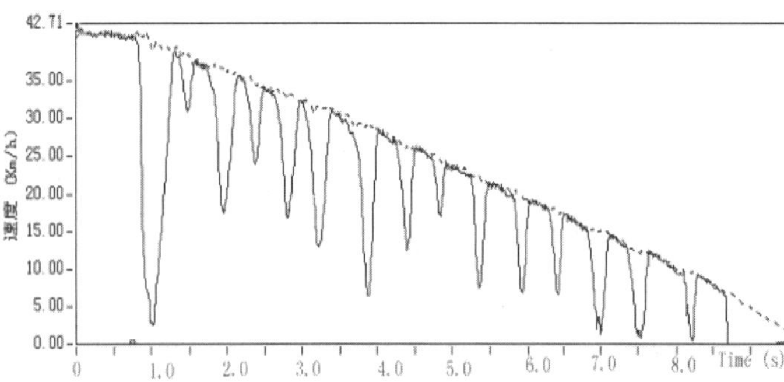

Figure 5 Speed during Road Testing

The curves above show that the simulation system can simulate the real operation of ABS, reappear the control logic of the controller. But there is still a certain disparity between the simulation result and the result of the road experiment. It is mainly because that it is hard to simulate the real road condition, and the experiment conditions are not the same, so it is difficult to compare the two in the same initial condition. What's more the model in the hardware-in-the-loop is simplified in some extend, and some of the parameters in the model are not exact, so there is disparity between the simulation and road experiment. On the other hand, this put forward higher requirement to the simulation system.

It is worthy to point out that we finished all the simulation experiment in one day, with just two persons. While in road experiment, it needs more person, and it is difficult to change the experiment condition. What more it is a bit dangerous when carrying out experiment on the low-μ road.

CONCLUSION

In this paper ABS hardware-in-the-loop was discussed, then the study of a certain type of ABS was concluded. Conclusions are listed as below:

- The simulation experiment can be repeated in the same initial condition (road condition, weather, vehicle state and etc.) without being affected by the difference of the initial condition.

- The simulation experiment can be repeated under different experiment condition (road condition, weather, vehicle state and etc.) and don't need so many person for the experiment, thus greatly reduce the cost and duration of the experiment.

- The simulation system (hardware/software) can be easily modified.

- The simulation experiment can simulate the extreme state without danger and destruct.

- It is closer to the real status compared with digital simulation, for the real hardware was brought in the system.

- The same idea can easily used in the more complicated system such as Traction Control System (TCS), Vehicle Dynamic control system (VDCS) and etc.

- There is a certain disparity between the simulation and the road experiment, which should not be replaced completely.

ACKNOWLEDGMENTS

Authors would like to express their appreciation to the Commit of "the ninth five-year" Project Programming.

CONTACT

Huiyi Wang, associate Professor in Tsinghua University. His email address is:hywang@Tsinghua.edu.cn.

REFERENCES

1. Si Lizeng. Anti-slip Control System. Beijing: National Transportation Publishing House, 1996
2. ABS Corporation. The Principal and Structure of Anti-lock Braking System (ABS). Beijing: National Mechanical Engineering Publishing House, 1995
3. M.W. Suh, C.S. Seok, etc. Hardware-in-the-loop Simulation for ABS. SAE Special Publications v1339, Feb 1998, SAE P91-97
4. Zhang Yadong. Simulation of Logic Threshold Method in Automotive Anti-lock Braking System (Master Thesis). Beijing: Tsinghua University, 1998
5. Chen Zaifeng. The Study of the Control Algorithm Based on the Maximum of Braking Decay Power of ABS and Its Realization (Ph. D Dissertation). Beijing: Tsinghua University, 2000

Roll Over Prevention Based on State of the Art ABS Systems

Gusztav Holler and Joseph Macnamara
Bendix CVS - Honeywell Inc.

ABSTRACT

The benefits of roll over prevention (ROP) or even roll over warning systems are obvious: besides potentially saving human lives, goods and vehicles they may save on insurance costs and help educate drivers and promote better driving habits. Various brake system manufacturers offer several ROP systems, but they are all common in one respect, namely that they are hosted by an Electronic Braking System (EBS) system. In general, based on sensory observation, the danger of a particular situation is judged by a central ECU and if necessary, brakes are actuated individually, independent of the driver's intention, to improve the vehicle's stability. Although the feature is very attractive, the market seems slow to accept new technologies, particularly in view of the additional price overhead of EBS compared to current ABS braking systems. A better short term low cost solution is to realize an ROP algorithm on common ABS systems for trailers.

INTRODUCTION

The main factors of rollover of a turning vehicle are the vehicle's speed, road curvature, center of gravity position and adhesion. Dynamic effects, such as center of gravity position change and nonlinear behavior due to moving liquid or livestock load are also significant. Avoidance may include actions like reducing speed, lessening lateral force components or changing the vehicle's parameters such as damper stiffness or air-bag inflation. We may classify these systems as passive (warning only) or active (autonomously intervening) systems. The key for a successful *warning* system is the reliable detection combined with non-misleading warning and a fast efficient way of counter-acting for a *preventive* (active) system.

The vehicle combination can be observed from the tractor or from the trailer. Tractor-only observers have the advantage of being compatible to virtually any trailer. The drawback, however, is that the semitrailer's state is difficult to observe from the tractor. A box style trailer with its rigid frame usually lifts the wheels on the tractor's driven axle while a flexible flatbed trailer tends to do the opposite: the trailer's inner wheels are the first

to leave the ground. A further problem is the compatible data connection between tractor and trailer, which allows the transfer of data between tractor and trailer. Currently Power Line Carrier for Trucks (PLC4Trucks) is the standard.

Leaders of the truck brake industry are continuously working on marketable ROP solutions. The major directions of algorithm development are:

- Tractor-based ROP: based on sensor located on the tractor, the lateral acceleration is estimated for the trailer. If this value exceeds a certain level, a so-called test pressure is applied to the trailer using the tractor's trailer brake control device (proportioning valve). If the low-level brake application is enough to lock the trailer's wheels, implying that the wheel(s) have little or no contact with the surface, the trailer's ABS system becomes active (due to locking wheels). The activation of wheel-end modulators on the trailer is observed by the tractor through the trailer's additional power consumption. The tractor then triggers the automatic brake application for the whole combination. Disadvantages of this approach are the relatively slow reaction time (especially pressure build on the trailer) and the need for an electric current sensor. EBS for the tractor and ABS for the trailer are a prerequisite for both warning and prevention. Also, this is not a satisfactory solution for trailers having a rigid frame, since the trailer's wheels will be lifted after the tractor wheels and as such is too late to prevent the rollover event.

- Trailer-based ROP: works very similarly to the tractor-based variant. The algorithm decides, if the current lateral acceleration can potentially cause a rollover. A low-level pressure is applied to the inner wheels (inside the curve) and ABS activity is monitored. In case of wheel locking - which is a sign of lifted wheels – a full brake application initiated by the trailer system prevents the rollover. If an appropriate data link exists, the driver may be warned, as well. An advantage of this approach is that it does not assume any interaction with the tractor. Poor performance with some rigid-body semi-trailers is a drawback. EBS on the trailer is required for both warning and prevention.

- Position monitoring: for a given trailer, the current roll angle can be monitored using roll rate sensor. This data combined with the speed and lateral acceleration tells if further increase in speed or lateral acceleration could lead to a rollover. This method involves previously known trailer specific structural/dynamic information. The knowledge of exact vertical position as a result of integration of the roll rate sensor is very crucial. Once the situation is judged dangerous, system initiated brake application on the trailer or other appropriate counter step(s) such as changing the suspension's characteristics may inhibit the rollover. The advantage for this algorithm is that it does not require any unintended drag force for the detection (no test pressure). EBS is required though for the brake application without the driver's intention.

All the efficient solutions for warning, face the problem of data communication and compatibility between tractor and trailer and all the prevention methods require EBS to be able to raise the braking pressure at a certain time to a specific level.

MARKET CONDITIONS

Although researchers and engineers are continuously addressing the problem of rollover around the globe, introducing any specific ROP algorithm is not easy. One reason is the different regulatory background of countries. A good example is the mandatory introduction of EBS on heavy-duty class tractors in Western Europe. Beyond the pre-existing requirement for compatible brake efficiency (the trailers and tractors are equipped with load sensing valves), the overall length of combination vehicles is strictly limited. Electrical connections must conform to well-defined standards. US regulations are much less strict, allowing doubles and triples on specific roads, not forgetting the Australian road trains. Another major problem is the profitability. It is hard to find someone in the US market who is willing to pay for an EBS system just because of improved safety and improved (braking) performance. Appreciation of safer systems by insurance companies, through lowered premiums is not enough today to compensate for the additional cost. As a result of these factors, no "global" solution for rollover prevention exists today.

POSSIBLE COST REDUTION

The need for cheaper but equally performing advanced systems led to the idea of the current investigation. The goal of this work is to

- address feasibility issues of a roll over protection algorithm implemented on a conventional trailer ABS system

- define the necessary additions to an ABS system to

allow for the implementation of roll over protection. As a core, the trailer-based solution was chosen as the algorithmic model. For performance tests, an in-house developed computer simulation program was used. The model had the following features:

- 4x2 European style tractor with no active brake control system (no EBS)
- driver model with a simple task of following the road curvature
- Two axle European style semi-trailer; detailed braking system model including wheel-end ABS modulators but no ABS logic.
- No trailer EBS features were assumed. The braking pressure was modulated with common ABS/traction valves.

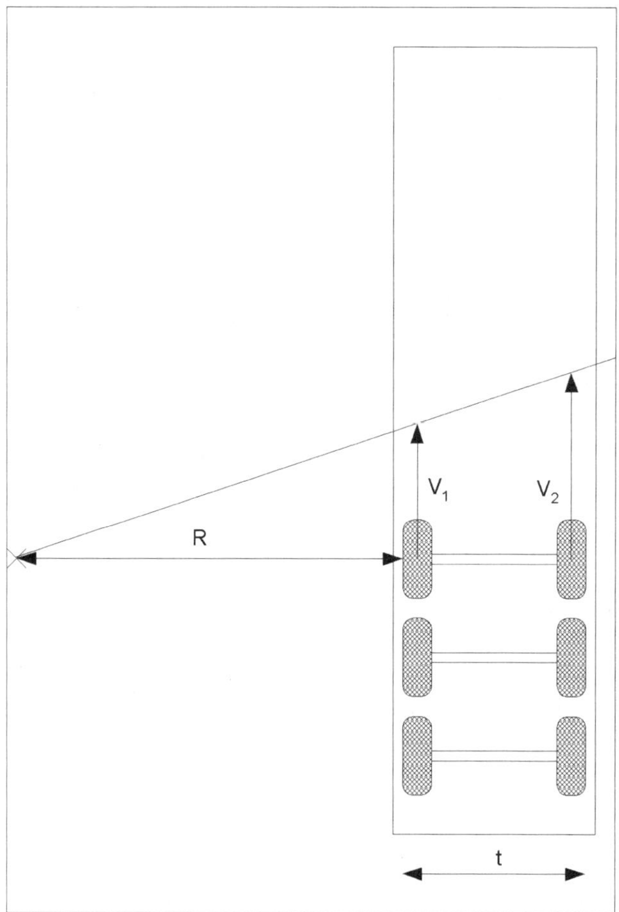

Figure 1: Data used for lateral acceleration estimation

For observing the vehicle's state, we could use the wheel speed data of the sensed axle. Using basic geometry the current turning radius for wheel (1) is defined by

$$\frac{V_1}{R} = \frac{V_2}{R+t}$$

The trailer's lateral acceleration at its center of gravity is then estimated with

$$a = \frac{\left(\dfrac{V_1 + V_2}{2}\right)^2}{\dfrac{t}{\dfrac{V_2}{V_1} - 1}}$$

Unfortunately, when the test pressure is present a drag force exists on the effected wheels and as such the slips are not predictable. This means that the ABS based ROP system must have a lateral acceleration sensor. The equation above is still useful for sensor offset determination.

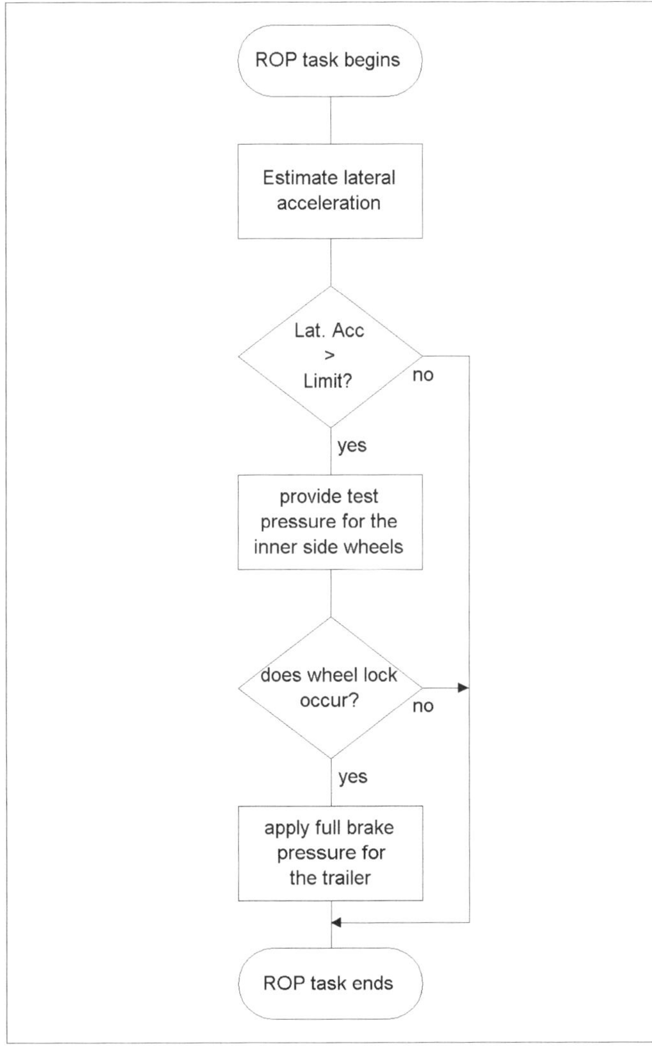

Figure 2: Flowchart of the rollover protection algorithm

While the lateral acceleration is monitored all the time, a so-called test pressure is only applied, when the lateral acceleration exceeds a level at which indicates that an impending rollover may occur. The application of brakes without the driver's intention is only possible, if we use a traction type valve. Lacking the EBS systems convenient proportioning valves, we must pulse the conventional ABS modulator(s) on the inner side of the curvature and place the outer ones in hold mode. This procedure fills the inner wheels' brake chambers to a level, where the drag force is not high enough to cause a wheel in contact with the surface to brake significantly, but to cause a wheel not in contact with the surface to approach lock up. This state of control remains in effect until the lateral acceleration drops below a given level. This state is referred to as "Alert mode". The exit level is lower than the entry level to provide a deadband. Once the system is in Alert mode, the inner wheels' speed is continuously checked for entry into a locking state. The idea behind this is that if the current lateral acceleration is enough to roll the vehicle over, the inner wheels will have less and less contact (normal force component) with the surface, as the roll angle advances. In this particular case, the modest braking torque on those wheels will be enough to lock them.

As soon as the wheels on the inner side begin to lock, indicating that the vehicle is possibly beginning to roll over, the ABS wheel-end modulators of the trailer on both sides are de-energized, allowing the tank pressure provided by the traction control valve to fill up the brake chambers. This will cause two things:

1. Reduction of the vehicle's speed resulting in a decrease in lateral acceleration.
2. Reduction of lateral tire forces that are primarily responsible for rollover.

The full brake application should last until the vehicle's speed is reduced enough to cause the lateral acceleration to drop below a given level where after control is relinquished and normal braking is reverted to.

SYSTEM CONFIGURATION

A conventional trailer ABS brake system usually consists of reservoir, ABS controller, wheel-end modulators (with integrated relay functionality), brake chambers, control and supply lines. In addition to these components, the rollover prevention function requires the following elements:

- Traction valve: supplies the full reservoir pressure for the ABS modulators, when activated, independent of the driver. This is a key element in avoiding the use of an EBS system.
- Check Valve: selects the higher control pressure between the driver's brake demand and the ROP traction control valve brake demand. Without it, the ROP algorithm could prevent the driver from braking the vehicle with a higher pressure than the test pressure level. The check valve allows the highest brake control pressure to reach the modulators.
- Pressure sensor for monitoring trailer control pressure: this is necessary for the ROP algorithm to recognize the driver's braking intention. It can be substituted with a cheaper pressure switch. Using this sensor/switch the algorithm recognizes the driver's braking command and places the ABS modulators into normal (load) mode to allow the control pressure to reach the brake chambers.

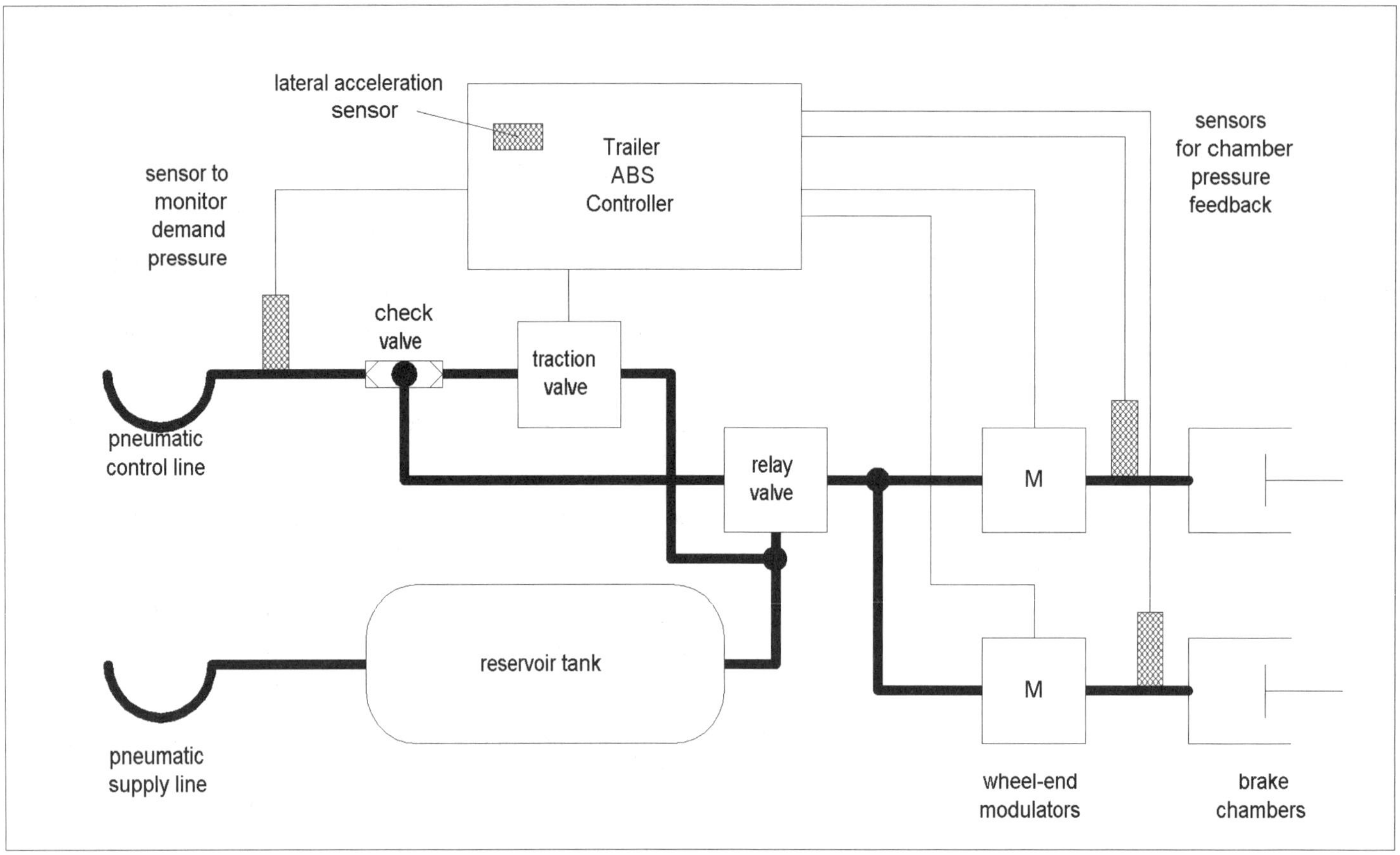

- Sensors for brake chamber pressure monitoring: since the normal open-loop control of the brake chamber can be inaccurate, it is better to use two feedback pressure sensors to be able to closely control the test pressure level.

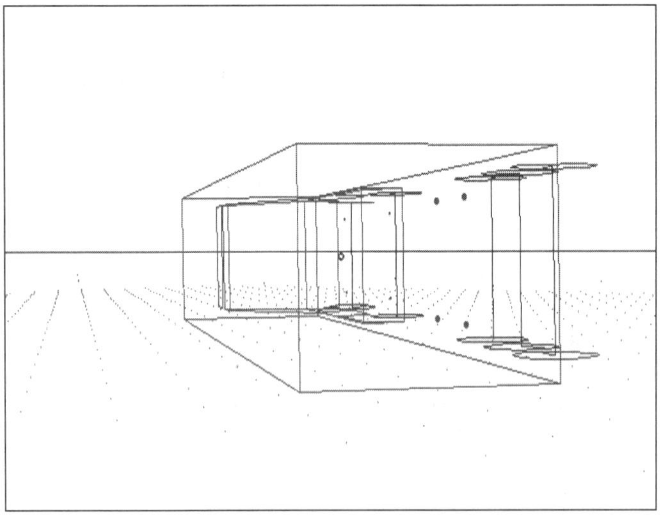

Figure 3: Overturned vehicle model (ROP was disabled)

VEHICLE SIMULATIONS

The objective of the simulations is to prove the feasibility of a rollover prevention algorithm implemented in a conventional trailer ABS environment. A single rollover situation was chosen and the vehicle configuration was. The driver model tried to follow the given path, which was a turn with about 100m radius. Initial speed was 85 km/h. Further blocks were added to our in-house developed model to simulate the behavior of the ABS modulators. Previously collected data from real experiments provided the basis for the parameter settings of the model. Another block served as the implementation of the algorithm. This was similar to that depicted on the flowchart of Figure 2.

The vehicle model we use is comparable in complexity to the tractor – semi-trailer model known from MSC's TruckSim™ program. Furthermore, it is capable of handling situations like rollover without "dropping" the overturned vehicle below the road surface, but this is only a visual aid. Figure 3 shows an overturned vehicle as a result of a turning maneuver without the ABS based rollover prevention logic.

RESULTS

After setting up the parameters with the proper values, the algorithm worked smoothly. It was however not fast enough in the case of very severe steering actions. In other words we were still able to roll it over intentionally, but in situations which are common in practice, it worked as expected.

The system acted very similarly to other EBS based ROP systems. It's preventive action was somewhat slower, due to the fact that it takes longer to fill up the brake chambers through the traction valve – relay valve – wheel-end modulator line than through the quick acting and accurate EBS proportional valves. On the other hand, it acted faster than the case of a tractor based ROP, where the tractor's trailer control valve command takes a certain amount of time to reach the back of the trailer. Figure 4 shows clearly the "alert" phase of the ROP system: brake chamber pressure p is regulated to about 1.3 bars (test pressure) to test for a rollover condition. This can be seen between 0.25 and 0.8 s on the time scale.

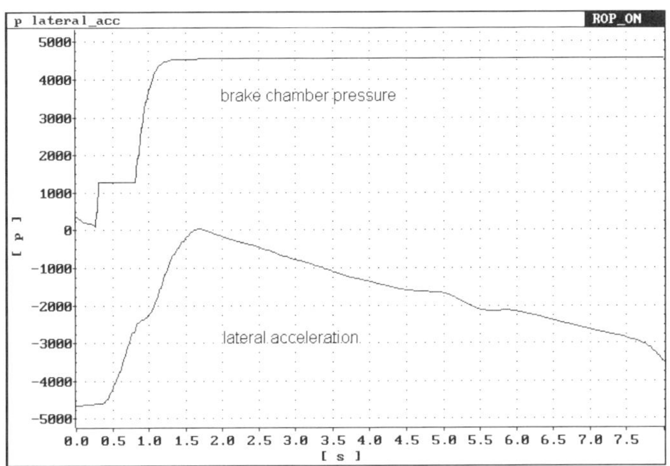

Figure 4: Brake chamber pressure of the inner-side wheels and estimated lateral acceleration

Since the driver model was not programmed to behave normally, it continues to keep to the original turning radius and speed in an attempt to follow the curve. This is why the high lateral acceleration for a longer period of time exists. A human driver would act differently than the driver model, once he recognized that the automatic intervention was taking place.

No data link was assumed between tractor and trailer. PLC4Trucks seems a viable solution for this, providing a communication channel with the tractor to send messages to warn the driver and instruct the truck's braking controller (if it is capable) to brake the tractor as well. This of course would require a PLC4Trucks capable brake ECU on both tractor and trailer.

CONCLUSION

The feasibility of an ABS-based ROP system has been verified by a computer simulation model. Although a special emphasis was made on modeling critical components, the method of verification has its limits. Several factors exist which will effect any implementation being considered to be put into practice:

- The trailer frame's flexibility was out of the scope of the current investigation, but it can be a crucial factor. Proper modification of the computer simulation model is needed to explore the algorithm's sensitivity to this parameter.
- It was assumed that the traction valve has enough capacity for feeding the relay valves. This must be verified in practice. Although the use of other electro-pneumatic valves could be considered if necessary. The conventional traction valve originally was chosen to save costs.
- Internal parameters of the algorithm must handle structural differences of various trailer types. Trailers carrying live stock or fluid or having many axles may behave more or less aggressively. Preprogrammed system parameters in the controller or proper identification methods could be used to handle special cases.
- The same trailer can be used in longer vehicle combinations such as doubles and triples. These cases must also be thoroughly examined and all possible issues must be addressed and solved before a possible practical implementation.

It is too early to answer all of these questions, but the possibility of having a ROP system at minimal cost overhead is promising.

Figure 5: Active ABS-based ROP system is capable of braking the trailer at the right moment to avoid rollover

DEFINITIONS, ACRONYMS, ABBREVIATIONS

Test pressure – a moderate level of autonomous brake application of the ROP system on the trailer. It allows the ROP logic to decide whether a wheel is barely touching the surface or not at all touching the surface.

Brake Onset Pressure – the minimal pressure level at which the brake linings has a contact with the drum or disk and braking torque is provided.

<u>EBS</u> – Electronic Braking System. Brake control is conducted by electronic devices, such as electro-pneumatic valves.

<u>ROP</u> – Roll Over Prevention. Electronic surveillance and prevention system on the top of a pneumatic braking ssytem to prevent vehicle rollover.

<u>PLC4Trucks</u> – Power Line Carrier for trucks. Power line carrier is a method of using power lines (power and ground) to transmit serial communications type data. In this way any devices that are connected to the same power lines can be capable of sending and receiving data.

An Investigation into Fuzzy Control for Anti-Lock Braking System Based on Road Autonomous Identification

Li Jun, Zhang Jianwu and Yu Fan
Institute of Automotive Engineering, Shanghai Jiao Tong Univ.

ABSTRACT

Automatic identification of road profiles ensures primarily right working condition of an anti-lock braking system (ABS). This paper presents a practical strategy to identify road conditions according to braking pressure, wheel slip ratio and angular deceleration. A fuzzy controller is designed to follow the target wheel slip ratio determined by road conditions. Simulation tests have been undertaken respectively on single and variable adhesion road surfaces by using a 7-DOF nonlinear vehicle model, accounting for nonlinearity of suspension and tire. It is shown that the ABS fuzzy controller can identify changes of road conditions precisely and therefore modulate the control scheme, so as to obtain the maximum road friction and good lateral stability. In comparison with a conventional PID controller, the present fuzzy controller possesses of more robustness and better performance.

1 INTRODUCTION

With an increase of the speed and density of road vehicles, the active safety performance is obviously desired and realized by anti-lock braking system developed by far and compulsorily equipped for a variety of vehicles in recent decades. A vehicle ABS system is to regulate the wheel slip ratio around the demanded slip ratio range so that the road adhesion coefficient is maximized and lateral stability is achieved. This in turn leads to the minimization of the vehicle stopping distance and the direction stability. While this desired slip ratio range is road surface dependent. An ABS system tries to regulate the wheel slip ratio to a value around that maximum. For instance, the optimal value of slip ratio for a dry asphalt road is different from the

optimal value for an ice road. Fig.1 presents typical plots of the road adhesion coefficient versus wheel slip for different road surfaces. Because of the different road characteristics of different types of road shown in Fig.1, the ABS controller must identify the prevailing road condition first, then be able to determine the

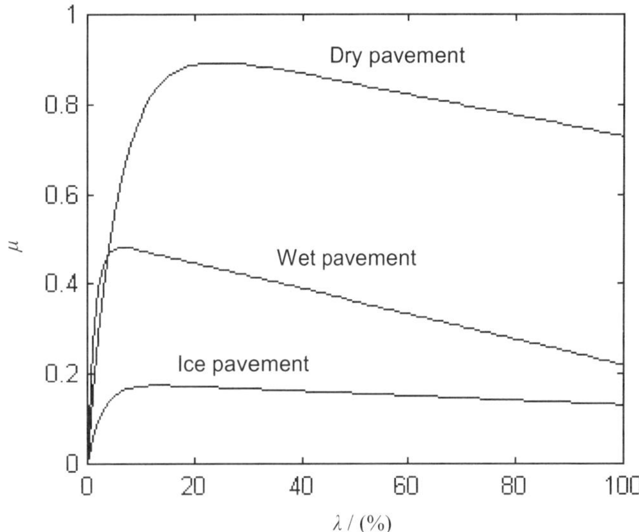

Fig.1 Typical plots of road adhesion coefficient versus slip

corresponding optimal wheel slip range on line without expensive road surface sensors would be much desired. Drakunov [1] proposed an ABS algorithm to calculate optimal slip ratio using a friction force observer. Mauer [3~4] presented a fuzzy logic surface identification scheme based on detected braking pressure, slip ratio and vehicle deceleration rate comparison. Lin *etal* [2] proposed an ABS controller using a linear feedback as well as sliding mode control.

Based on the previous studies, this paper proposes a practical strategy to identify road conditions according to braking pressure, wheel slip ratio and wheel angular deceleration. A fuzzy controller is designed along with to keep the target wheel slip ratio determined by road conditions. Simulation tests have been carried out based on a 7-DOF nonlinear vehicle model, which accounts for the non-linearity in suspension and tire. The feasibility and effectiveness of proposed controller are examined. The control system does not need excessive sensors and computation time is short because of the parallelism of fuzzy logic control algorithm. Compared with a conventional PID controller, the fuzzy controller can obtain more robust and better adaptive ability to the changes of road surfaces condition.

2 VEHICLE MODEL

The present research is interested in longitudinal vehicle braking performance and a single-track model, i.e., bicycle model, is thus used as shown in Fig.2. For motions in the vertical direction, the time is treated as a simple spring-damper system. Subscripts f and r denote

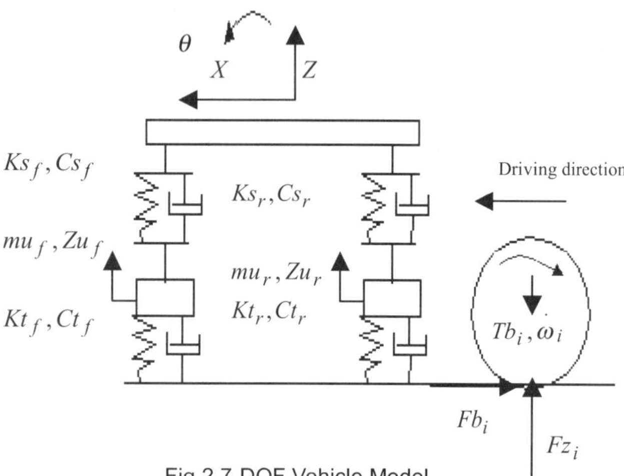

Fig.2 7-DOF Vehicle Model

front and rear wheel respectively. The motion equations of the bicycle vehicle model can be derived of the following form:

$$M \ddot{X} = -(Fx_r + Fx_f)$$

$$M \ddot{Z} = -(Fz_r + Fz_f)$$

$$I_y \ddot{\theta} = -Fz_r b + Fz_f a + (Fx_r + Fx_f)(h_0 + Z)$$

$$J_i \dot{\omega}_i = Tb_i - Fx_i (R_i + Zu_i) \qquad (i = r, f)$$

$$mu_i \ddot{Z}u_i = Fz_i - Ft_i \qquad (i = r, f)$$

in which

$$Fz_i = K_{si}\Delta_{si} + C_{si}\dot{\Delta}_{si} \qquad (i = r, f)$$

$$Ft_i = K_{ti}\Delta_{ti} + C_{ti}\dot{\Delta}_{ti} \qquad (i = r, f)$$

$$\Delta_{sr} = Z + b\theta - Zu_r; \qquad \Delta_{sf} = Z - a\theta - Zu_f$$

$$\Delta_{tr} = Zu_r; \qquad \Delta_{tf} = Zu_f$$

3 TIRE MODEL

The relationship between wheel slip and longitudinal friction force can be described by empirical curves. In this paper, Pacejka's Magic formula tire model [5] is used to present the relationships between wheel slip and normal tire force to longitudinal force. In Fig.1 are shown curves of road baking force *vs.* wheel slip ratio for a typical tire bearing a total vertical mass of 400kg, respectively, on dry, wet and ice road surfaces.

4 CONTROLLER DESIGN

In design of proposed control system, two main parts are presented and discussed, i.e., road identification and fuzzy controller respectively.

4.1 Identification of road condition

A correct identification for road surface condition is first important in an ABS control system design in order to ensure system working in right condition. Hence, a practical ABS controller needs to adjust its target slip ratio and control strategy to prevailing road condition. Otherwise, Further road adhesion provided by the interaction between tire and road may not be fully used, and wheel lock also would happen frequently.

In this paper a practical identification algorithm is proposed by the idea, which can be described by following equation:

$$T_b - F_z \mu(\lambda)R = J \dot{\omega}_c$$

in which F_z can be estimated by the following equation:

$$Fz = Mg/L \cdot (b + \mu(\lambda)h_0) + \Delta\Phi$$

The road identifier is designed as following description. Firstly, some typical plots for different road surfaces are pre-stored in the ABS controller for describing the relationships between road friction coefficients and wheel slip. The designed road identifier works along with a designed fuzzy logic ABS controller throughout whole braking procedure. When encountering road condition changes, the identifier specify the road type by comparing measured wheel angular deceleration with assumed wheel angular deceleration in different road surface conditions. In the comparison, the type of road with least absolute difference between calculated and measured wheel angular deceleration will be specified. Considering the disturbance and measurement noise, the calculation is made twice. Fig.3 shows the diagram of road automatic identification.

4.2 Fuzzy logic ABS controller design

The ABS fuzzy controller regulates wheel slip around an optimal slip range so that minimum stopping distance and directional stability can be achieved. Fig.4 illustrates the ABS fuzzy controller along with road identifier. The fuzzy rules can be adjusted according to road condition in theory, but it is not easy to realize in real control system. This paper tries to modify the gain schedule of the ABS controller to achieve the object of adjusting fuzzy rules on-line. For example, when vehicle brakes from dry pavement to wet one, if the gain value of the controller keeps constant, the wheel block will occur.

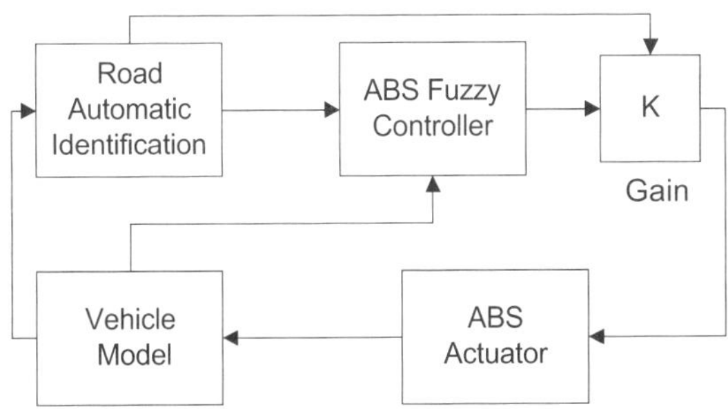

Fig.4 ABS fuzzy control with road Identification

The input variables of the fuzzy controller are the difference between actual slip value and target value, i.e., λ_{error}, and the changing rate $d\lambda_{error}/dt$, which are respectively described as,

$$\lambda_{error} = \lambda - \lambda_{t\,arg\,et}$$

$$\frac{d\lambda_{error}}{dt} \approx \frac{\Delta\lambda_{error}}{\Delta t}$$

The output is the expected difference of braking line pressure Δp. Hence the adapted pressure is,

$$p = p + \Delta p$$

The membership functions of the input language variables and the output language variable are respectively shown in Fig.5.

The control rules for the ABS fuzzy logic control system is shown in Table 1.

Table1. Fuzzy logic control rules

Δp		λ_{error}			
		PB	PS	ZO	NS
$\dfrac{d\lambda_{error}}{dt}$	PB	NB	NB	NB	NS
	PS	NB	NB	NS	PS
	ZO	NB	NS	ZO	PS
	NS	NS	NS	ZO	PB
	NB	NS	NS	PS	PB

In Table1, *PB* denotes Positive Big, *NS* for Negative Small and *ZO* for Zero, etc. In this study the Mandain method is used for fuzzy logic operation and the centroid method used for defuzzilization calculation.

5 SIMULATION RESULTS AND ANALYSIS

The main simulation program of ABS fuzzy control with road identification is shown in Fig.6. It consists of an ABS fuzzy controller, road identifier, a vehicle dynamic model, an ABS actuator and a tire model masked in vehicle model. After loading a data file, simulations in different road surface conditions can be carried out and the designed control system can be examined. The whole braking procedure can be presented by a designed monitor, i.e., scope. Hence, the working states of the control system can be seen and recorded conveniently.

The fuzzy logic ABS control system with the road identifier includes ABS fuzzy controller module, road identification module, ABS brake module and vehicle dynamic module. Tire module is inserted in the vehicle dynamic module.

Given a vehicle parameter data file, simulation tests can be carried out either on single friction road conditions or on variable friction coefficient road surfaces. By a designed monitor, i.e., the scope in the system, all operating states of ABS can be seen and recorded for each step throughout whole braking process.

Longitudinal dynamic behaviors of the model vehicle while braking on different road surfaces have been examined, and herein, only some typical braking tests on variable friction coefficient cases are presented. In Fig.7 and Fig.8 are shown results of velocity, slip ratio, braking torque, normal force of tire and braking distance, simulated by the ABS fuzzy controller, respectively on two different changeable friction coefficient conditions, i. e., one from high friction to low friction and the another from low friction to high friction. The figures present the braking performances, including output vehicle speed, wheel velocity, wheel slip, braking torque, normal force of front wheels and stopping distance. Fig.7 presents the simulation results in a three-section road surface, i.e., dry, wet and ice road. It is shown that the road identifier starts to work at the beginning of braking process. By comparing actual wheel angular velocity respectively with assumed wheel angular velocity in different road conditions, it makes a right judgment of being dry road with high friction coefficient. Hence a target wheel slip ratio, i.e., 22%, is selected correctly. Meanwhile, the fuzzy logic controller gain K is correspondingly adjusted. The actuator modulates braking pressure to keep wheel slip around the target value. At the time of 2 seconds when the road condition changes from dry road to wet road, the resultant slip is increased and road identifier make judgments of being wet road. So, the target slip ratio is re-selected being about 9%. At the time of 4 seconds when road condition is changed to be ice, the identifier along with the fuzzy controller works similar as described procedure above. The output wheel slip ratio is adjusted at a new target value and remained steady rapidly. For the whole braking process, the stopping distance is 61.67 meters with time of 5.79 seconds.

Fig.8 shows the simulation results in a changeable road condition, respectively including ice, wet and dry road sections. It can be seen that the road identifier works properly with very good adaptive ability to the changes of road surface condition. Similar good braking performances are obtained. The stopping distance is 113.6 meters long with time of 5.84 seconds.

Fig.9 and Fig.10 show the results of a conventional PID ABS controller in time domain. The figures describe variety of braking performances, such as vehicle velocity, wheel speed, wheel slip ratio, braking torque, normal force of the vehicle front wheel and stopping distance. Compared with the performances of the fuzzy controller shown in Fig.7 and Fig.8, the performances are worse, with significant fluctuation of controlled slip and poorer adaptive ability to the changes in road surfaces, especially when vehicle runs from dry to wet pavement, wheel lockage occurs. The resultant stopping distance is relatively long being 83.87 meters and braking time of 8.78 seconds under road condition verifying from high to low friction and from low to high friction and the stopping distance is 120.1 meters long and braking time of 6.18 seconds. The stopping distances are 22.20 and 6.5 meters longer than those of corresponding road conditions variation schedule. Wheel lockage occurring and actuator force frequently adjusting would damage the ABS system performance and also lead to lower life span of the ABS system.

6 CONCLUSIONS

In this paper is presented a practical strategy to identify road conditions according to the braking pressure, wheel slip ratio and wheel angular deceleration, and a fuzzy controller for ABS is designed to keep the target wheel slip ratio set by the road conditions. Simulation tests have been undertaken for the vehicle braking on the single and variable adhesion road by using a 7-DOFs nonlinear vehicle model, considering nonlinear influence of suspension and tire. It is shown that the ABS fuzzy controller can identify changes of road conditions precisely and therefore modulate the control scheme, so as to obtain the maximum road friction and good lateral stability. It is proved through further comparison with the conventional PID control that the fuzzy control is more robust and promising.

REFERENCES

1. Drakunov, S. *etal* (1995), ABS control using optimum search via sliding modes, IEEE Transactions on Control and Systems Technology, Vol.3, pp.79-85, March.
2. Lin. W.C., Dobner, D.J.(1993), Design and analysis of an anti-lock brake control system with electric brake actuator, Int J. of Vehicle Design, Vol.14, No.1,pp.13-43.
3. Mauer, G.F. (1995), A fuzzy logic controller for ABS braking system, IEEE Transactions on Fuzzy System, Vol.3, pp.381-388, November.
4. Georg F. Mauer, *etal,* Fuzzy logic continuous and quantizing control of an ABS braking system, SAE Paper 940830
5. Bakker, E., Nyborg, L. and Pacejka, H.B. Tyre Modeling for Use in Vehicle Dynamics Studies, SAE paper 870421, 1987

CONTACT

Mail stop:

Li Jun

Institute of Automotive Engineering

Shanghai Jiao Tong University No.A9906022#

No.1954, Huashan Road, Shanghai City, P.R.China

PostCode: 200030

Email: jli927@mail1.sjtu.edu.cn

Tel: +086-021-62933772   Fax: +086-021-62933093

APPENDIX

Nomenclature

M	Sprung mass	h_0	Height of center of gravity
mu_f, mu_r	Unsprung mass	K	Gain
θ	Pitch angle of sprung mass	$\dot{\omega}_c$	Theory wheel deceleration
ω_i	Speed of i th wheel $(i = r, f)$	$\dot{\omega}_r$	Real wheel angular deceleration
$\ddot{X}$	Longitudinal gravity acceleration of sprung mass	$\Delta\Phi$	Coefficient of suspension influence on normal force
$\ddot{Z}$	Vertical gravity acceleration of sprung mass	p	Braking pressure
$\ddot{Zu}_i$	Vertical gravity acceleration of i th unsprung mass	P_0	Pressure threshold
I_y	Moment of inertia for sprung mass about axis x	G_1	Effective coefficient of brake pads
J_i	Moment of inertia for i th unsprung mass	G_2	Area of wheel cylinder
F_{xi}	Longitudinal force acted on i th unsprung mass $(i = r, f)$	G_3	Average radius for bake pads
F_{zi}	Normal force acted on i th unsprung mass $(i = r, f)$	LF	Coefficient of heat fade
K_{si}, C_{si}	Stiffness and dump of suspension $(i = r, f)$	λ	Slip ratio
K_{ti}, C_{ti}	Stiffness and dump of tire $(i = r, f)$	μ	Adhesion coefficient
T_{b_i}	Braking torque $(i = r, f)$	b	Distance between center of gravity and rear axes
a	Distance between center of gravity and front axes	L	Wheel base

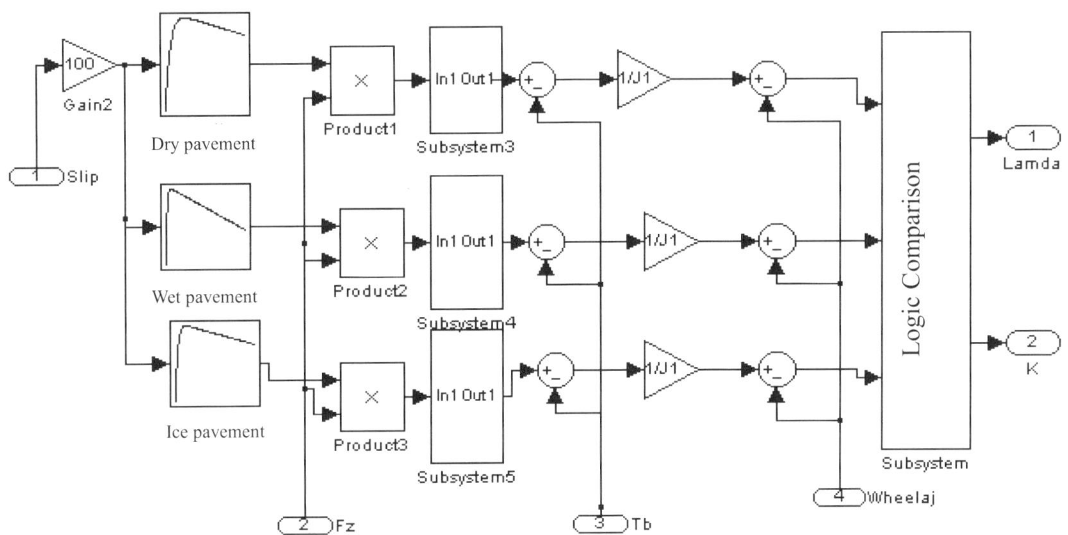

Fig.3 Main Program for road automatic identification

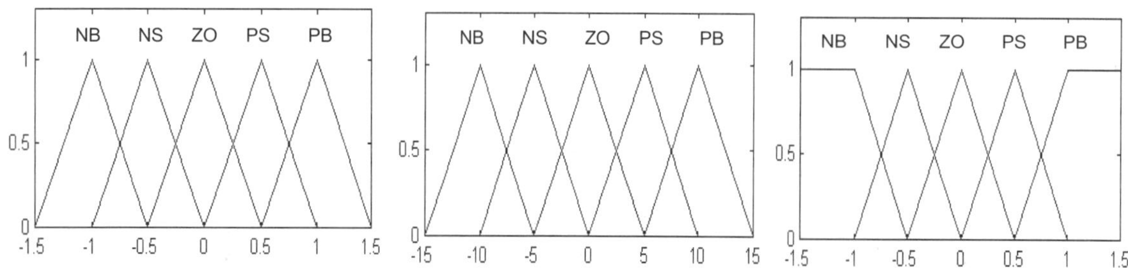

a. Membership function to λ_{error} b. Membership function to $d\lambda_{error}/dt$ c. Membership function to Δp

Fig. 5 Membership functions of input /output language variables

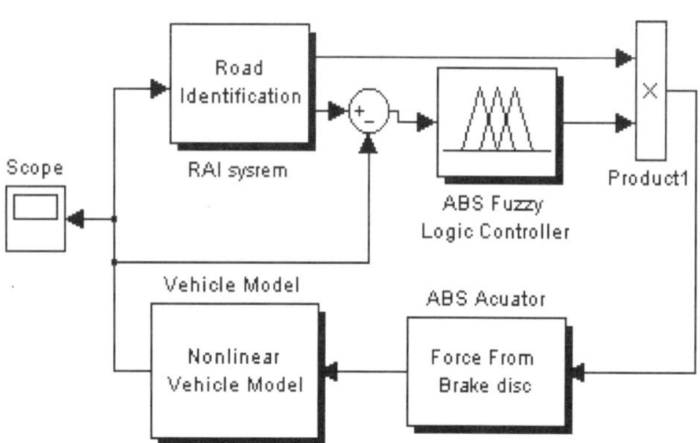

Fig.6 Main program for ABS fuzzy control

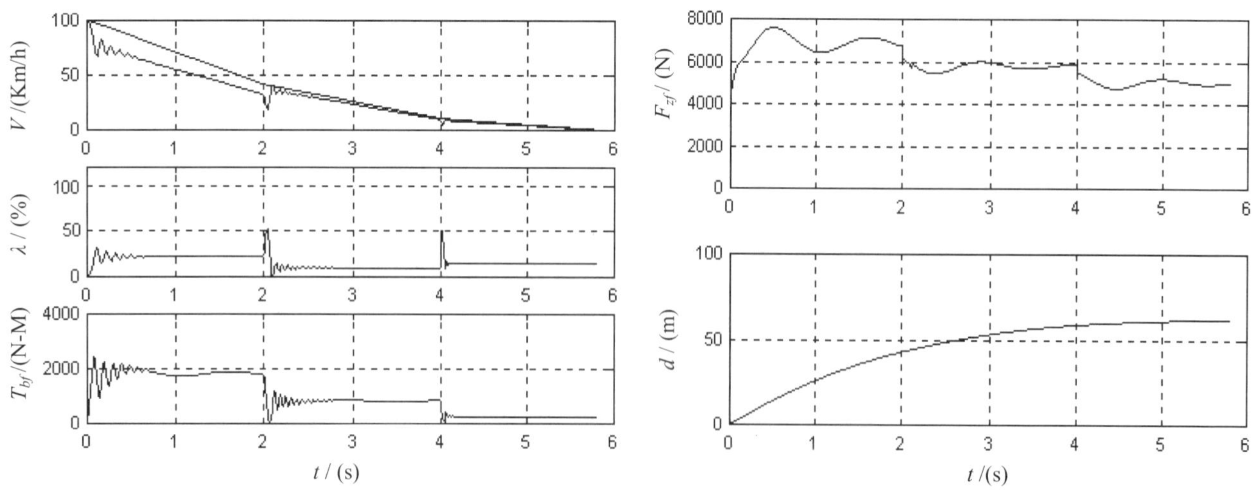

Fig.7 Results of the vehicle braking on a dry-wet-ice pavement by Fuzzy control

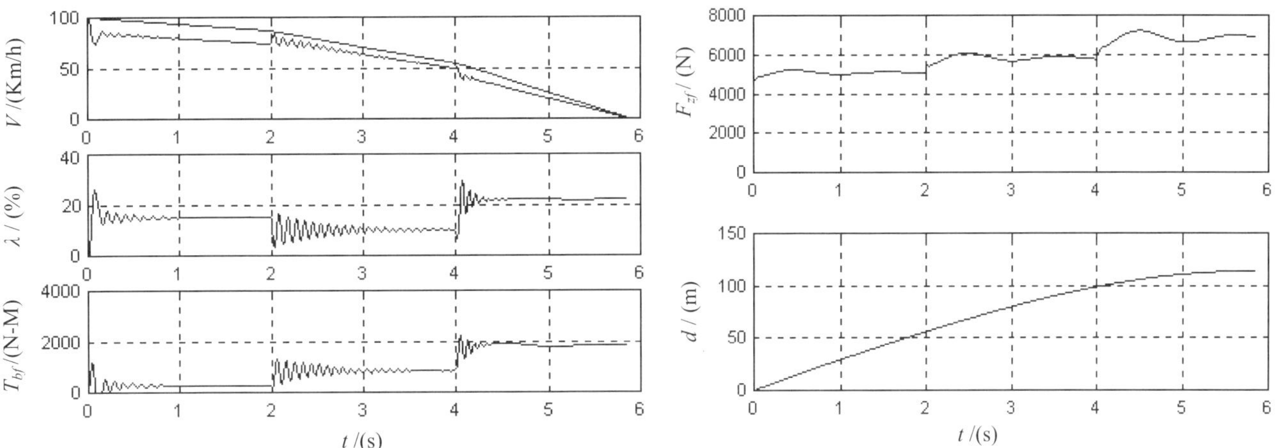

Fig.8 Results of the vehicle braking on an ice-wet-dry pavement by Fuzzy control

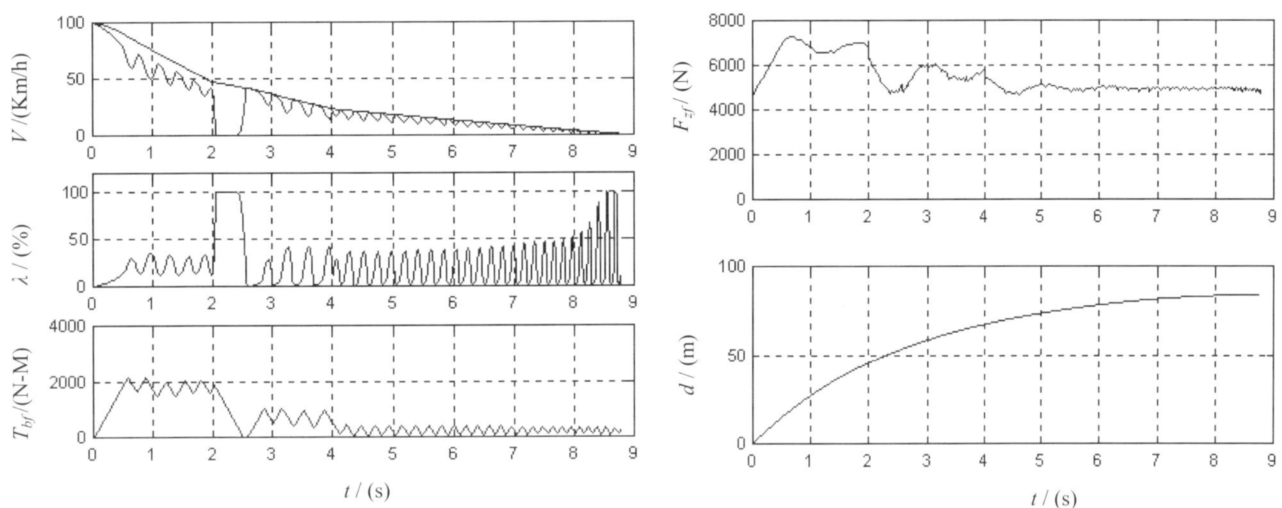

Fig.9 Results of the vehicle braking on a dry-wet-ice pavement by PID control

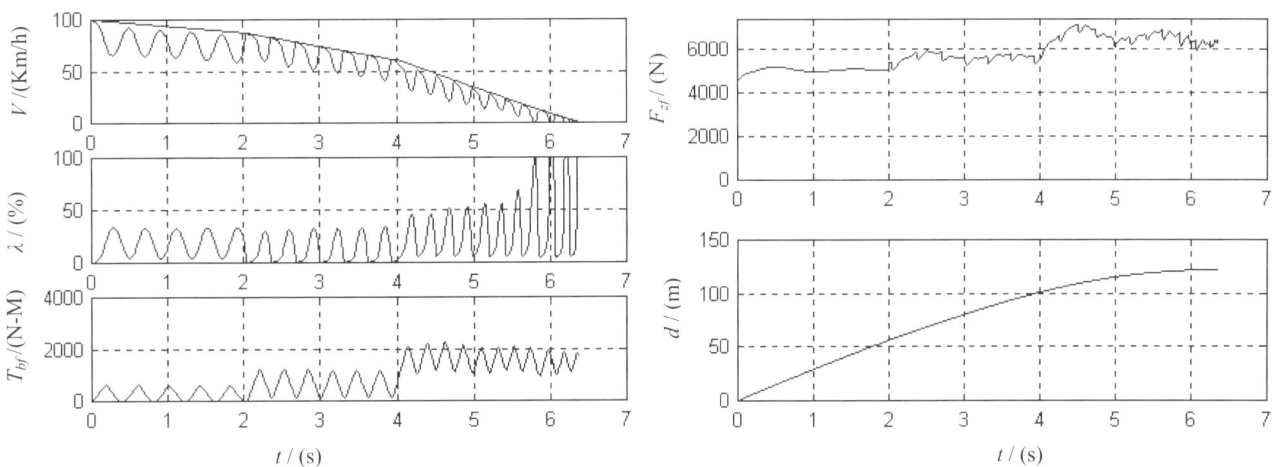

Fig.10 Results of the vehicle braking on an ice -wet- dry pavement by PID control

The Development and Testing of a Prototype Electromagnetic ABS for Drum Brakes

J. Andrew Ridnour and Frank H. Speckhart
The University of Tennessee

ABSTRACT

This paper describes a new design for ABS drum brakes. The design uses some components associated with electric trailer brakes.   Similar to electric trailer brakes, a cam is rotated by the friction force of an electromagnet acting on the brake drum.   The cam releases the brakes in proportion to the rotation of the cam.   The design has the advantage of providing proportional reduction in brake torque and fast response times. Expensive components such as pumps are not required and no pedal vibration is introduced.   Test data and testing procedure is included.

INTRODUCTION

Anti-lock brake systems (ABS) have become standard equipment on most vehicles sold in the United States and around the world.   Current ABS consists of five common components: wheel speed sensors, wheel brake cylinders, pressure modulator assembly, master cylinder, and an electronic control unit (ECU). The wheel brake cylinders and master cylinder are common to all hydraulic automotive braking systems. The wheel speed sensor is typically a pole-pin type sensing the teeth on a rotating sensor ring. The ECU is an on-board computer. The pressure modulator assembly consists of several components: up to four pairs of solenoid valves, two accumulators, two damper chambers, two orifices, and two return pumps.   There is one pair of solenoid valves for each wheel. Typically, there are one accumulator, one damper chamber, one orifice, and one return pump for each hydraulic circuit.

Figure 1 shows a typical ABS control loop and how the ABS components are related. The wheel sensor is continuously monitored by the ECU. When the wheel speed decreases at a rate that would cause wheel lockup (zero rotation rate), the ECU uses the pressure modulator to release and reapply the pressure to the wheel cylinder to prevent complete wheel lockup.

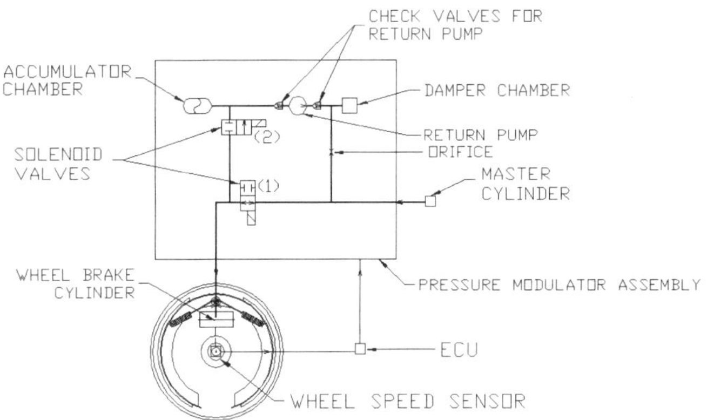

Figure 1: A Standard ABS Control Loop.

The pressure modulator operates at a rate of 4 to 10 cycles per second. The pressure modulator operates by first separating the wheel cylinder from the master cylinder by solenoid valve 1. Solenoid valve 2 opens and the fluid is transferred back to the master cylinder by the return pump. The fluid travels through an accumulator, damper, and an orifice on the way back to the master cylinder.   The accumulator absorbs the surge in fluid. The damping chamber suppresses pressure oscillations and the orifice acts as a flow restrictor.   When the deceleration decreases to an acceptable value, solenoid valve 1 between the master cylinder and wheel cylinder opens and solenoid valve 2 closes. The pressure is then reapplied from the master cylinder to the wheel cylinder. In theory, the pressure is modulated at a rate designed to keep the brake slip as close as possible to the peak of the braking coefficient curve, shown in Figure 2.

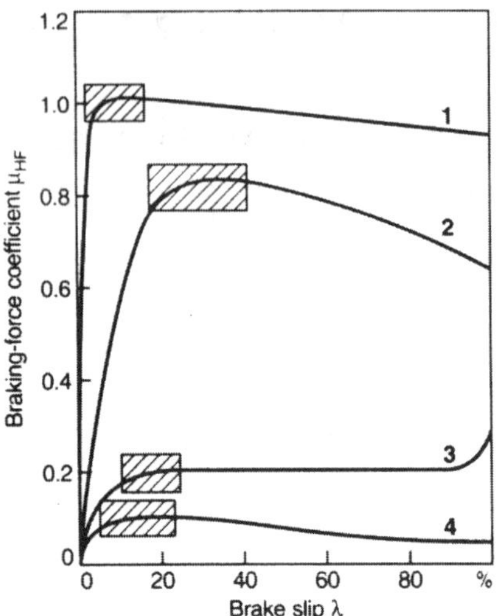

Figure 2: Braking-Force Coefficient μ_{HF} as A Function of Wheel Slip λ. Source: Automotive Brake Systems, 1st Ed., by Bosch Inc. Copyright 1995.

Brake slip is determined by the following equation:

$$\% \, \text{SLIP} = \frac{v - a * r}{v} * 100\%$$

where v is the vehicle speed (m/sec), r is the radius of the tire (m), and ω is the angular velocity (rad/sec). The wheel alternates between lockup, or near lockup, and rolling as the wheel cylinder pressure is modulated. The pressure modulation continues until the vehicle is stopped or the wheel is not longer sliding.

Wheel lockup decreases braking performance by providing a discontinuous braking torque. The frequency of modulation is limited by several factors, including the response time of the mechanical parts used to control the wheel cylinder pressure. The pulsation of the brake pedal due to the fluid being pumped back to the master cylinder and the noise from the pressure modulator may pose problems with some drivers who might decrease pressure on the brake pedal when the ABS is activated. This turns off the ABS, due to the decreased pressure from the master cylinder, and also reduces braking performance. The noise and pulsation of the brake pedal can be easily observed on some current production vehicles. The pedal pulsation and pressure modulator noise are noted in the owner's manuals of several manufacturers' vehicles.

This paper explores the possibility of replacing the current pressure modulated ABS with a simpler and less costly system that could improve anti-lock braking performance for drum brakes. This new system could completely replace rear wheel ABS and improve current four-wheel ABS. This paper covers the initial steps in developing a new ABS including the design and testing of the prototype electromagnetic ABS using a simple on-off control system.

DESCRIPTION OF PROTOTYPE SYSTEM-The proposed electromagnetic release system uses a variation of a braking system concept used in many electric trailer brakes. Electric trailer brakes have been used reliably for over 30 years. Electric trailer brakes have an electromagnet connected to a cam by an actuating lever. The cam is located between the ends of the brake shoes. The electromagnet is in close contact with the inside surface of the brake drum. When the electromagnet is energized, friction between the magnet and the drum surface causes the cam to rotate. Frictional force increases as the electrical current to the magnet increases. When the cam is rotated, the brake shoes are forced outward against the brake drum resulting in braking. Brake torque is a function of the rotation of the cam that depends upon the force of the electromagnet.

The proposed design uses a design similar to the electric brakes to release the brake instead of applying it. Similar to the current electric brake system, the energized magnet is attracted to the inside surface of the brake drum. The friction force between the electromagnet and rotating brake drum causes the cam and release-arm to rotate and consequently release the brake. The rotation of the release-arm is restrained by the torque of the return spring. When the magnet force is reduced, a return spring rotates the cam clockwise and reapplies the brake in proportion to the rotation of the release-arm. Consequently, the frictional force provided by the electromagnet controls the magnitude of cam rotation and the resulting brake torque reduction.

When the electromagnetic ABS is not activated, the brake functions normally. Because of the profile of the cam, in the normal operating position the force of the brake shoes on the cam does not tend to rotate the cam. If the electromagnetic ABS malfunctions, the brake operation operates as a non-ABS system.

There are several advantages to the proposed electromagnetic ABS. The system eliminates brake pedal vibration due to the pulsation of the hydraulic fluid for rear wheel ABS and reduces the vibration for four-wheel ABS. The noise from the pressure modulator is also eliminated for rear wheel ABS and reduced for four-wheel ABS. Since no pressure modulation is required, for the rear drum brakes, the components of this system should be inherently less expensive than those of current ABS. The electromagnetic ABS has the potential to operate at a rate faster than the current ABS.

The main advantage of the proposed design is the capability to proportionally control the reduction of brake torque. By using pulse-width modulation of electrical current, the force of the electromagnet can be continuously regulated which will result in a continuous reduction of brake torque. This would eliminate the lockup and roll of current ABS for drum brakes while keeping the wheel slip near the peak of the braking coefficient curve, thus improving braking performance.

DESCRIPTION OF PROTOTYPE DESIGN-The prototype design was intended to demonstrate initial feasibility of this concept. The proposed system is obviously not intended as a final design. Many of the problems associated with a production system have not been solved, optimized or addressed. For example, the mechanical design used readily available brake components that were adapted for the prototype. The technique used to measure wheel velocity was limited by the microprocessor that was used. Also, the rate at which the electromagnetic release system operates was limited by its control system. The system tested did not use pulse-width modulation. This means that on-off control was used for the electromagnetic.

Figure 3 shows the prototype design and a modified ABS control loop for the proposed system in the normal braking position. Figure 4 shows the prototype design in the anti-lock braking mode. The proposed system consists of six basic components: wheel speed sensors, wheel brake cylinders, electromagnetic release system, solenoid valve, master cylinder, and a microcontroller. Many components are the same as used in current ABS.

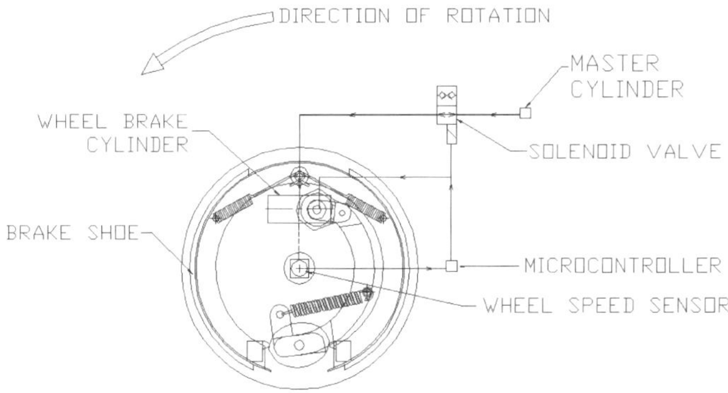

Figure 3: Modified ABS Control Loop in the Normal Braking Mode.

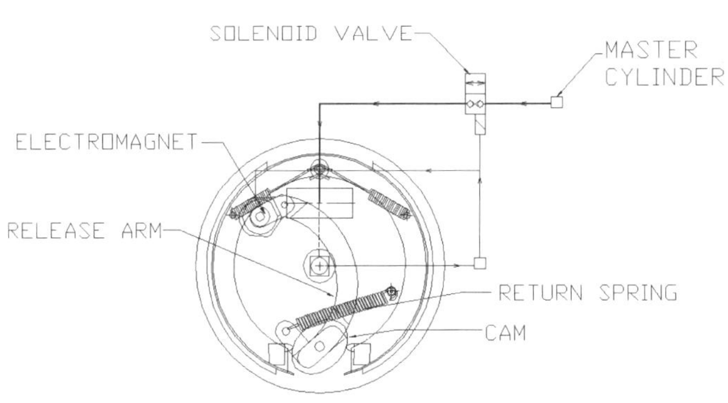

Figure 4: Modified ABS Control Loop in the Anti-lock Braking Mode.

BRAKE HARDWARE DESIGN

A drum brake from the rear end of a 1986 Ford F150 two wheel drive truck was selected for the prototype. The drum brake was a standard 279.4mm X 50.8mm brake. The parking brake and brake adjuster were removed due to the fact they were not used directly in the braking process.

Only 25.4 mm of space was available between the back of the back of the hub and the shoe attachment springs. For this reason the smallest possible electromagnet was selected for the design. The magnet was used on a 184mm X 38.1mm Fayette electric trailer brake. The magnet was 17.3mm thick after a small flange was removed. A 4.8-mm thick cold-rolled steel plate was connected to the wheel to provide an adequate thickness to guarantee a flat surface for the magnet to work properly

The cam was located as close to the bottom of the backing plate as possible and the magnet path was the largest radius from the cam pivot point as possible. The maximum radius in turn gave the maximum length lever arm and therefore the most torque on the cam. The selected magnet path is shown in Figure 5.

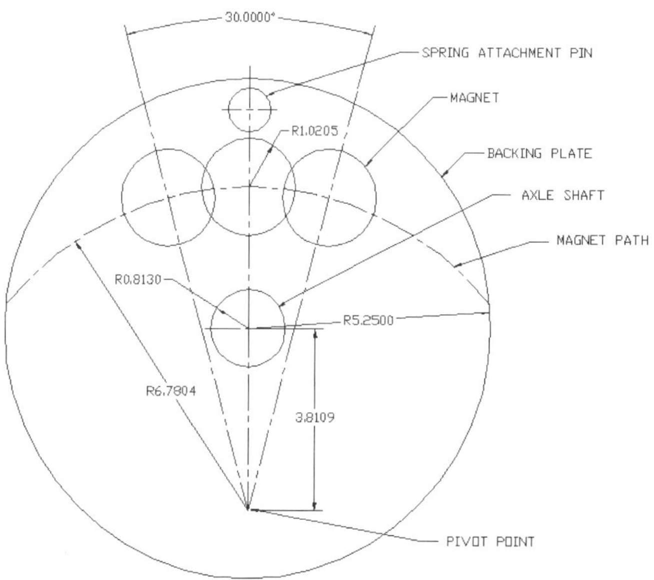

Figure 5: Selected Magnet Path (Units in inches).

As the magnet moves with the wheel the cam is rotated by the release-arm and the torque applied to the cam changes with rotation. The maximum torque is at the point where the magnet is directly above the centerline of the cam.

The following equations relate the rotation of the cam and the force of the magnet to the torque (T) applied to the cam, as shown in Figure 6:

$$T = F*R*\cos(\phi)$$

$$r2 = \sqrt{R^2 + r1^2 - [2*r1*R*\cos(\theta)]}$$

$$\frac{\sin(\phi)}{r1} = \frac{\sin(\theta)}{r2}$$

$$\Rightarrow \sin(\phi) = \frac{r1}{r2}*\sin(\theta)$$

$$\Rightarrow \phi = \sin^{-1}\left[\frac{r2*\sin(\theta)}{r1}\right]$$

R is the radius of the magnet path, r1 is the distance from the pivot point to the center of the axle, and r2 is the distance from the center of the axle to the center of the magnet. Also, F is the frictional force of the magnet acting on the wheel, θ is the rotation of the cam, ϕ is the angle shown in Figure 6. The distances R (172 mm) and r1 (96.8 mm) were previously determined from the selected magnet path.

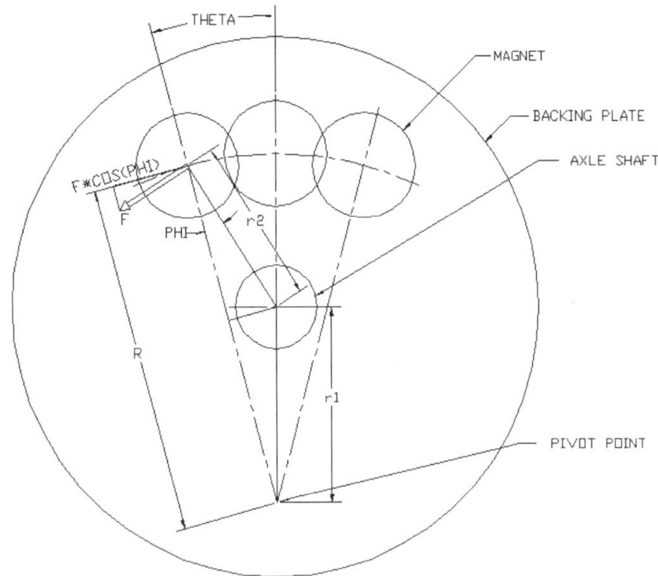

Figure 6: Diagram Showing the Relationship between the Rotation of the Cam, From Cam Centerline, and the Applied Torque.

As we can see from Figure 7, the percentage of magnet force translated to the applied torque decreases below 95.3% of the maximum possible generated torque after the cam is rotated 15 degrees past its centerline. Therefore, the system was designed to be operated over a 30 degree range, that is, 15 degrees to either side of the cam centerline.

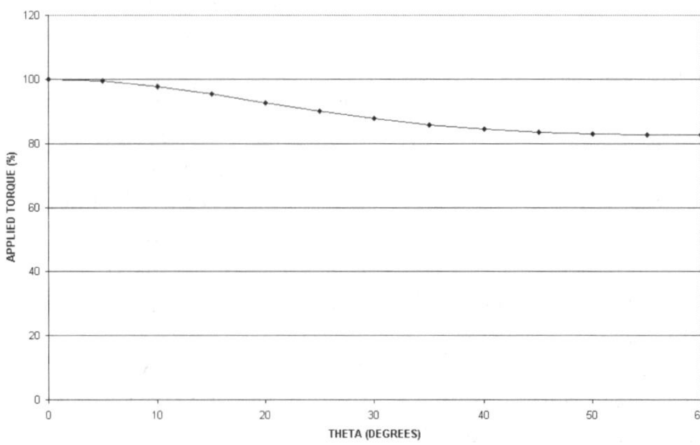

FIGURE 7: Relationship between the Rotation of the Cam, From Cam Centerline, and the Applied Torque.

The chosen cam profile had a constant velocity design of -8.5 mm/rad. This profile provided 4.44 mm of brake release movement for each brake shoe for 30 degrees of cam rotation. The cam base circle diameter 78.7 mm. Brass followers having a radius of 3.175 mm were used. This constant velocity cam design is not intended to represent an optimum design. Significant additional work would need to be done to determine the best cam design.

A return spring that is connected to the cam by a short-lever arm is used to reapply the brake. The length of the short-lever arm (spring arm) was designed to be as long as possible length such that a small return spring could be used.

The torque required of the return spring must be sufficient to reapply the brakes. The maximum return torque must be less that the torque provided by the magnet. For the prototype design, the maximum spring return torque was 5.07 N*m. For a maximum magnetic dynamics friction force of 33.7 N. the maximum torque was 5.81 N*m.

Using the calculated rotational mass-moment of inertia of the magnet and release-arm of 0.0071 kg*m^2, the brake could be released and reapplied in approximately 0.1 seconds

Figures 8 and 9 show the brake mechanism with the wheel hub removed. Figure 9 shows the electromagnetic anti-lock brake mechanism in the normal braking position, with 7.5 degrees of cam rotation. Figure 9 shows the electromagnetic anti-lock brake mechanism in the anti-lock braking position, with the full 30 degrees of cam rotation. Figure 10 shows a photograph of the electromagnetic anti-lock brake mechanism minus the wheel hub. Figure 11 shows a photograph of the magnet and plate contact.

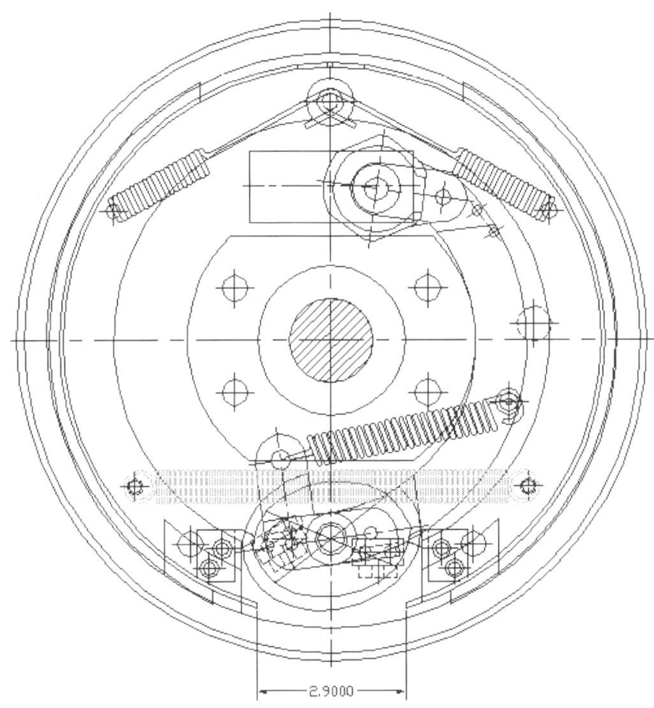

Figure 8: Brake Mechanism in the Normal Braking Position (Units in inches).

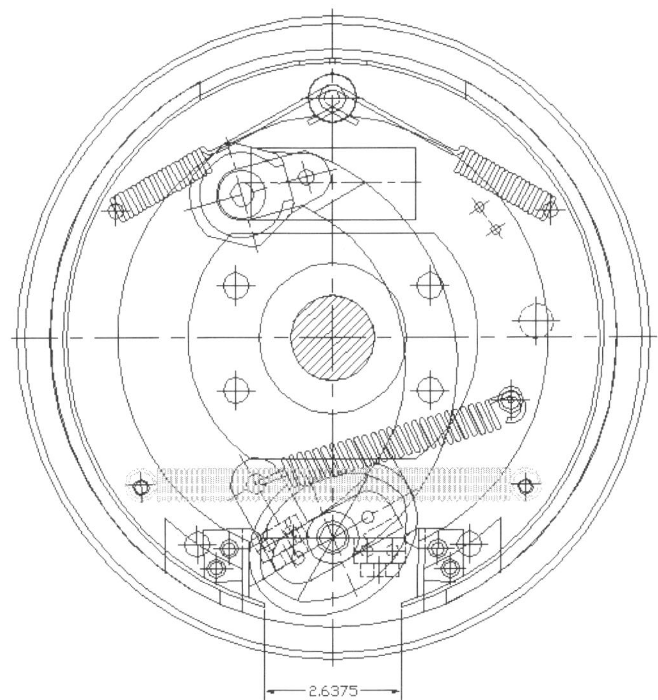

Figure 9: Brake Mechanism in the Anti-lock Braking Position (Units in inches).

Figure 10: Photograph of the Electromagnetic Anti-lock Brake Mechanism Minus Wheel Hub.

Figure 11: Photograph of the Magnet and Plate Contact.

A single piston master cylinder and lever arm were used to apply the brake, and a 12 VDC solenoid valve was used to limit pressure to the wheel cylinder when the ABS was energized.

The speed of the wheel was measured using a magnetic sensor that counted teeth on a gear mounted on the brake drum. The gear was a stock spur gear having 130 teeth and a pitch diameter of 330.2 mm. The inside of the gear was cut away to allow it to fit over the drum and was then held with set screws. Figures 12 and 13 show a photograph of the completed drum and ring gear.

Figure 12: Photograph of Front View of the Completed Electromagnetic Anti-lock Brake.

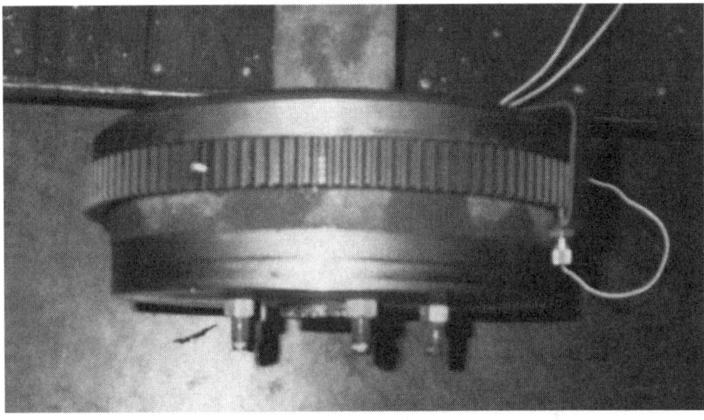

Figure 13: Photograph of Top View of the Completed Electromagnetic Anti-lock Brake.

CONTROL SYSTEM DESIGN

HARDWARE- The electromagnetic ABS was controlled by a Motorola 68HC11 microcontroller. The microcontroller had to be correctly interfaced with the brake system to achieve the desired performance and control. The system was installed on one wheel of a towed trailer for testing.

The input/output port A was used as the port for all the electronic connections. TTL (transistor transistor logic) was used for all the electronics. Figure 14 shows a schematic of the electronic circuitry and connections used.

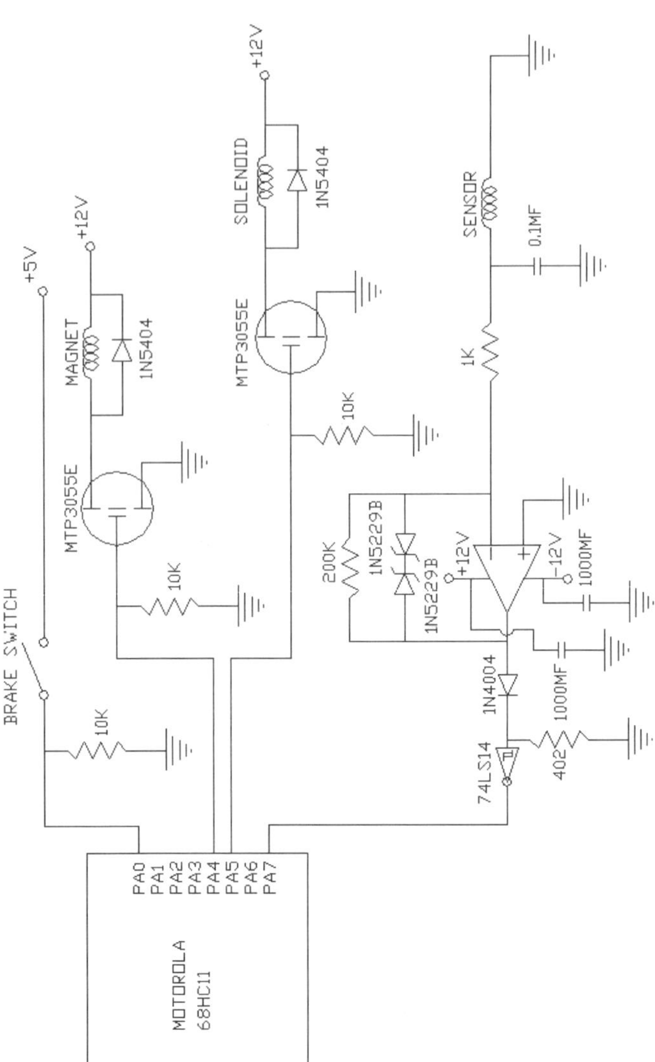

Figure 14: Schematic of the Electronic Circuitry and Connections Used.

There were two inputs to the microcontroller. The brake switch serves as one input to the microcontroller to determine if the brake is applied. The magnetic sensor serves as a second input to the microcontroller. The magnetic sensor sends pulses to the microcontroller for the purpose of determining wheel speed. The pulses from the magnetic sensor were counted and compared using the microcontroller's two accumulators (A and B).

The two outputs from the microcontroller were setup the same. MOSFET transistors were chosen to control the 12V supply voltage and currents required to operate both the magnet and solenoid valve.

Since the brake was to be tested on a trailer, the power required needed to be portable. Therefore, a single 12V lead-acid battery was used to power the system.

SOFTWARE- As shown in Figure 14, the electromagnetic ABS was controlled by a Motorola 68HC11 microcontroller. The software for the microcontroller was written in a text editor using assembly language and then assembled into machine language using a two-pass assembler.

The primary function of the braking program was to determine when to activate and then control the electromagnetic ABS. As previously stated, on/off control was used. Activation of the electromagnetic ABS is based on the angular deceleration for the wheel. The value for maximum angular deceleration was calculated by determining when wheel-slip first occurs. To insure the electromagnetic ABS would only be activated when the wheel would be slipping, a value of 2 for μ_f was used. By neglecting rolling and aerodynamic forces, the maximum value of angular deceleration can be determined from the following equation:

$$\alpha = m_t \cdot g \cdot \mu_f / (2 \cdot (m_t + m_{tk}) \cdot R)$$

Using the mass of the trailer (m_t = 258 kg), mass of the truck (m_{tk} = 1542 kg), the radius of the wheel (R = 348 mm), and the coefficient of friction of 2, the angular acceleration of the wheel for impending sliding is 4.05 rad/sec^2.

For purposes of ease of programming with the given microcontroller, the wheel speed was determined by counting the number of teeth over a 0.10 second period. Therefore, the wheel deceleration could be determined every 0.2 seconds.

Note that the error in pulse counts by the microcontroller can be +/− 1 tooth depending on where the magnetic wheel speed sensor is in relation to the gear tooth when counting is started and stopped. The +/−1 tooth error in actual wheel speed is +/−0.606 km/h. This error value is calculated by:

$$\frac{1\,\text{tooth}}{0.1\,\text{sec.}} * \frac{2186.56\,\text{mm}}{130\,\text{teeth}} * \frac{1\,\text{km}}{1E6\,\text{mm}} * \frac{3600\,\text{sec.}}{1\,\text{hr}} = 0.606\,\text{km/h}$$

where 2186.56mm is the circumference of a 225/70-R15 tire.

Since only one wheel of a small trailer was used to stop the vehicle, the angular deceleration of the wheel for impending wheel lockup is much smaller than for an automobile where all four wheels are braking.

Once it is determined that brake lockup is imminent, the electromagnetic release system is activated: the valve is turned off, and the magnet is turned on. The microcontroller once again counts pulses and compares the number of pulses to determine if the angular acceleration of the wheel has decreased to an acceptable level. If the angular acceleration is still too high the microcontroller continues to count with the brake released. If the angular acceleration has decreased below the lockup limit the microcontroller checks to determine if the brake is still on. If the brake is still on, the magnet is turned off with the valve still on and the brake is reapplied. The microcontroller resumes

counting and checks to determine if lockup is once again imminent. If the brake is released, the program is reset. The program controls the ABS system continuously at a frequency of 5 Hz.

To monitor and record wheel speed, the testing program outputs the value recorded in accumulator A every 0.2 seconds to a laptop computer by serial connection. This process can only be used for one accumulator due to the Motorola 68HC11 only having two accumulators, limiting the output resolution.

TESTING

The electromagnetic ABS was tested by mounting it to a trailer pulled by a pickup truck. Only one wheel was equipped with the anti-lock system. The brake from the other wheel was not used. The electromagnetic anti-lock brake and Ford rear end, described previously were mounted to the trailer using 225/70-R15 tires. The master cylinder was positioned in the bed of the pickup truck and is activated remotely. The total mass of the trailer was 258 kg.

A photograph of the trailer and complete testing setup are shown in Figures 15 and 16, respectively.

Figure 15: Photograph of the Electromagnetic ABS Testing Trailer.

Figure 16: Photograph of the Complete Testing Setup.

For the testing, the truck was accelerated to 32.19 km/h and 48.28 km/h, and then placed in neutral. This speed was limited by the length of the parking lot and for the safety of the personal conducting the test. The brake was applied with sufficient force to lock the wheel. The truck was then brought to a stop by using only the single anti-lock brake.

Approximately 60 test runs were made at different speeds and brake pressures. The brake pressures were monitored to ensure the brake pressure was appropriate for each test run and to determine if the solenoid valve was operating properly. The brake pressure ranged from approximately 4.1 MPa to 6.9 MPa. The indicator lights and braking wheel were also monitored during testing. During testing, the release and reapplication of the brake by the electromagnetic ABS could be seen by viewing the braking wheel. The release and reapplication of the brake was very evident during heavy braking due to high brake pressures and increased speeds. At low speeds, the indicator light was the only indication of the release and reapplication of the brake.

Discontinuous skid marks were apparent on the asphalt surface indicating wheel slip was occurring intermittently. After the trailer slowed to less two mph the wheel finally locked until the brake switch was turned off.

RESULTS

The trailer was connected to a Nissan pickup truck, as shown in Figure 16. The remote master cylinder was placed in the bed of the truck. The bundle of wires and hose were connected between the trailer and the truck. The testing was performed in an empty parking lot. The surface of the parking lot was asphalt and the test track was approximately 0.19 km in length.

The electromagnetic anti-lock brake performed as expected throughout the testing. Figure 17 shows the data obtained for the first test run in the form of speed vs. time, and a linearized vehicle deceleration. The linearized vehicle deceleration was calculated by simply dividing the difference between the initial and final wheel speeds by time. The linearized vehicle deceleration was found to be 0.716 m/sec^2. The assumption of constant deceleration was made only for comparison purposes, to estimate the average coefficient of friction, and to calculate wheel slip.

From Figure 17 it can be seen that the wheel initially decelerated rapidly, the electromagnetic ABS was activated and the wheel accelerated almost to the value of the assumed vehicle speed. The wheel once again decelerated rapidly and once again the electromagnetic ABS was activated. The electromagnetic ABS was activated a total of five times, in the same manner, as can be seen from the plot. Figure 17 shows, overall, a nearly linear wheel deceleration. This can be easily seen by comparing the wheel speed to the assumed vehicle

speed. This shows a controlled deceleration due to the electromagnetic ABS as opposed to the wheel lockup that would have occurred without the electromagnetic ABS. By assuming that the vehicle deceleration was linear, the wheel slip could be calculated.

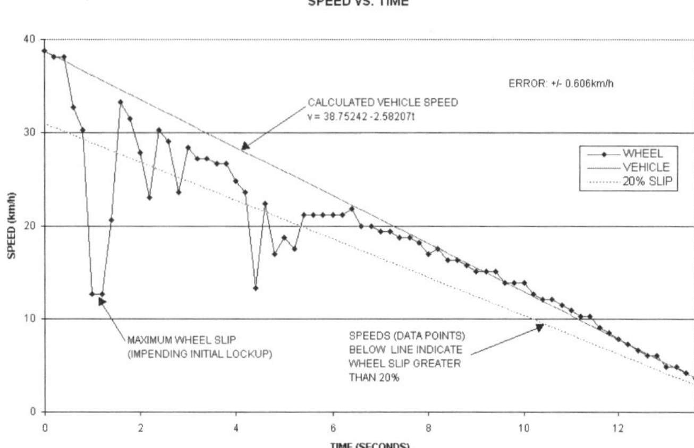

Figure 17: Speed vs. Time for Test Run 1.

Figure 17 shows the wheel slip vs. time for this same test run. Figure 17 shows that the rapid deceleration causes rapidly increased wheel slip. If the electromagnetic ABS had not been activated, the wheel slip would have quickly gone to 100 % and complete wheel lockup.

The wheel slip did not exceed over 65% in the test, showing lockup did not occur. When the wheel slip exceeded 20%, the electromagnetic ABS was activated. As wheel slip increases from 0 to 20 %, the coefficient of friction for braking also increases. This can be seen in Figure 18.

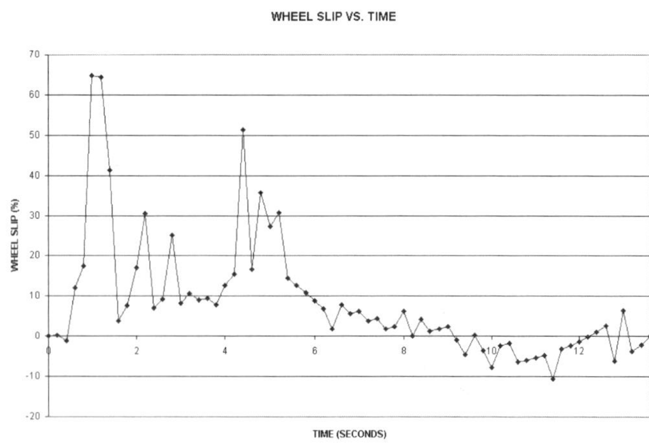

Figure 18: Wheel Slip vs. Time for Test Run 1.

Therefore, the wheel slip that occurred throughout the test when the electromagnetic ABS was not activated was desirable. The negative values of wheel slip can be explained by errors in the assumed vehicle deceleration and the actual wheel speed. The error in the actual wheel speed is +/−0.606 km/h. This error is from where the magnetic wheel speed sensor is in relation to the gear teeth when counting is started and stopped.

The linearized vehicle deceleration for the stop was found to be 0.716 m/sec^2. By neglecting all other drag forces, the total stopping force provided by the single wheel showed an effective coefficient of friction of 1.02.

$$\mu_f = \frac{a*m}{N} = \frac{0.716 \,\text{m/sec\textasciicircum}2 * 1796 \,\text{kg}}{129 \,\text{kg} * 9.81 \,\text{m/sec\textasciicircum}2}$$

(129 kg is the Weight on the Braking Wheel)

$\Rightarrow \mu_f = 1.02$

The results shown in Figures 17 and 18 proved to be typical of the results of the 60 test runs that were made. The test runs were made at different speeds, different brake pressures, and different lockup values of angular deceleration (comparison values). Increased speed only increased the time to stop. Increased brake pressure increased the initial wheel slip before the ABS was activated. This could easily be avoided by using a counting gear with more teeth to improve the resolution. The brake was designed to cycle at 10 Hz, with the current number of teeth the brake can only cycle at 5 Hz. The initial activation could be made significantly quicker by using a shorter counting time and a smaller lockup value. Using lockup values that only differed by one tooth had very little effect on the results due to the error being +/−1 tooth.

It should be noted that the indicator lights used during testing showed the electromagnetic ABS being activated more than can be seen on the charts after the vehicle was slowed. This electromagnetic ABS activation can probably not be seen due to the resolution of the output. The output indicated the wheel speed every 0.2 seconds using the first pulse counts. The electromagnetic ABS takes 0.1 seconds to completely release the brake and 0.05 seconds to reapply the brake, so the charts are missing one-half of the possible data. Quick electromagnetic ABS pulses could be easily missed. This could be solved by using a microprocessor with more accumulators to store and output data.

CONCLUSION

The electromagnetic ABS has many capabilities that were not tested due to the scope of the project. By using pulse width modulation, the electromagnetic ABS has the capability of proportionally reducing brake torque in response to wheel slip. This reduced brake torque would allow greater braking efficiency than current ABS. The control system used proved that impending wheel lockup can easily be detected, by simple electronics, and decrease brake torque well before wheel lockup could occur.

The electromagnetic ABS also eliminates any potential issue of drivers' tending to decrease the pressure being applied to the brake pedal due to pulsation of the hydraulic fluid and the noise from the pressure modulator for rear wheel ABS and reduce the pulsation and noise for four-wheel ABS. The components of the electromagnetic ABS are inherently less expensive than those used in current ABS. The magnet force used for this system has been used to sufficiently apply the brakes on heavy travel trailers, to the point of wheel lockup, for over 30 years.

The electromagnetic ABS demonstrated functional capability during repeated testing. This system proves to be an alternative for existing ABS in both hydraulic and pneumatic brake systems.

Comparison of Control Methods for Electric Vehicle Antilock Braking / Traction Control Systems

P. Khatun, C. M. Bingham and P. H. Mellor
University of Sheffield

ABSTRACT

The alleviation of environmental problems associated with personal, public and commercial transport in urban areas has become an important issue for both policy makers and the automotive industry. Future legislation in Europe and the USA is expected to introduce strict limits in vehicle emissions, and both electric and hybrid vehicles are considered to be strong contenders for meeting low / zero emissions targets. As a result, research into electrically driven powertrains, which have similar performance attributes as ICE (Internal Combustion Engine) vehicles, has led to the development of electrically actuated wheel technologies, with increasing attention being focused on research into novel antilock braking / traction control (ABS / TCS) strategies.

This paper describes a comparison of traction control schemes using real - time observer based estimates of μ-slip characteristics. In particular, a 'bang-bang' type strategy will be evaluated against fuzzy based control schemes to facilitate torque production from the powertrain to produce optimal traction.

Simulation studies and experimental trials on a laboratory test facility and simulation studies on a prototype electric vehicle will be used to evaluate and compare the response of the proposed techniques in real driving situations.

Furthermore, the paper will demonstrate that the application of observer techniques are appropriate for the on-line determination of the peak adhesion coefficient for different tyre-road surface conditions, and the fuzzy based techniques offer substantial potential for optimal control of wheel traction.

INTRODUCTION

Increasingly stringent regulations to reduce emissions from automotive vehicles has led to a resurgence of research activity to realise alternative electrically driven power-trains; the resulting localised control of wheel torque has allowed advanced, high-bandwidth anti-lock braking/traction control (ABS/TC) strategies to be realised.

A major obstacle to the development of robust ABS/TC has always been the real-time estimation of wheel-slip vs. adhesion-coefficient characteristics for different tyre types and road surface conditions [1]. Recent studies have investigated the application of observer/estimation schemes to obtain real-time data indirectly, the extended Kalman filter (EKF) receiving increasing attention [2]. However, published material to-date is primarily derived from simulation studies, with practical implementation and validation of the techniques being rare. Current commercial/passenger vehicles incorporating ABS/TC systems often employ look-up tables, which are based on experimental trials, and have been shown to provide adequate performance under many driving situations. This type of technique however, is limited by the fixed structure of the control strategy, and is often 'de-tuned' to accommodate worst-case scenarios, for example, traction control in icy conditions with old tyres. Consequently, a sub-optimal wheel-slip characteristic is imposed for most driving conditions.

To-date, a feature of many TC/ABS's for commercial IC-engine vehicles, is that the realisable performance is often limited by the mechanical bandwidth of the active actuation systems (typically 20Hz for ABS). However, with the advent of electrically powered vehicles into the marketplace, the means for controlling drive-torque at much greater bandwidths will be possible, and hence, some of the difficulties attributed to mechanical systems will be inherently addressed in this paper.

TRACTION CONTROL / ANTILOCK BRAKING SYSTEMS

The tractive force between a tyre and the road surface (Figure 1) is proportional to the normal load, F_z, the constant of proportionality being termed the adhesion coefficient, μ. The adhesion coefficient μ, is the ratio of tire brake force at the tire road interface and the normal load acting on the tire, i.e:

$$\mu = \frac{F_L}{F_Z} \qquad (1)$$

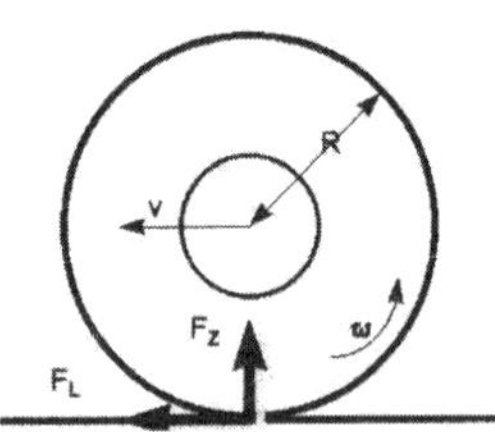

Figure 1. Wheel model [3].

The value of μ is highly dependent on tyre characteristics (compound, wear, ageing etc) and road surface conditions (dry, icy, gravel, tarmac etc), although it can be primarily regarded as a function of the relative slip, σ, between the two contacting surfaces. By definition slip is the normalised difference between the wheel speed and the contact point:

$$\sigma = \frac{v - R\omega}{v} \qquad (2)$$

where v is the vehicle velocity, R is the effective radius of the driven wheel, and ω is the angular velocity of the wheel. Typical μ-σ characteristics are shown in Figure 2.

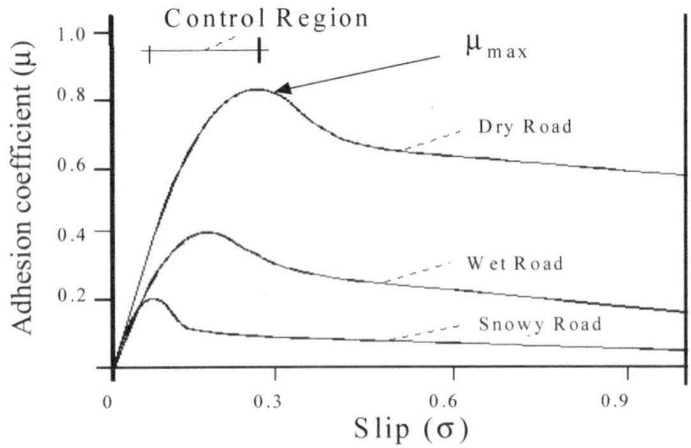

Figure 2. Example μ-slip characteristics for various road conditions [4].

It is clear from Figure 2 that increasing slip can increase the tractive force between the tyre and road surface by virtue of an increase in μ. However, once the peak of the characteristic is encountered, μ_{max}, any further increase

in slip will reduce traction, and consequently induce an unstable acceleration of the wheel until the drive torque is reduced.

The objective of an antilock braking system (ABS) is to manipulate the tractive force applied to the driven wheels in order to limit the slip, σ, between the road surface and the tyre, and consequently only operate within the stable region of the μ-σ characteristic. A major obstacle in the practical design of ABS control schemes is the determination in real-time of μ-slip characteristics, i.e. μ_{max} and σ. For example, Table 1 shows typical values of slip required to obtain the maximum adhesion coefficient, for various road conditions.

Road condition	Max. adhesion coefficient (μmax)	Optimum slip (σ)
Dry Road	0.85	0.35
Slippery / Wet Road	0.4	0.2
Snowy / Icy Road	0.2	0.1

Table 1. Max. values of μ for various road surfaces [4].

To overcome this problem, many manufacturers employ a slip-limiting control scheme to account for 'worst-case' conditions, typically icy roads. Although this constitutes the safest criteria to adopt for design purposes, it can impose conservative limits on the traction available under less severe conditions [5].

A second obstacle often encountered by designers of TC/ABS schemes, is the availability of appropriate test-tracks/skid-pans or other experimental rolling road facilities to evaluate proposed algorithms and provide repeatable conditions for comparative studies.

<u>Experimental Test Facility</u>

In order to address these issues, a low cost experimental test bench is used to facilitate the design and preliminary evaluation of TC/ABS schemes. The test facility illustrated in Figure 3, comprises of an electric traction drive connected to a standard three-phase induction machine, which is used to mimic the road load.

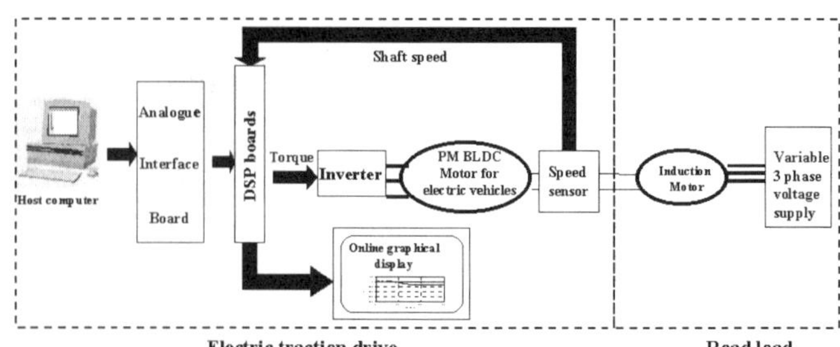

Figure 3. DSP based experimental test facilities.

The experimental electric traction drive (Figure 4) is a 6kW continuous, 12kW peak rated brushless permanent magnet motor supplied via a power controller from a 72V high performance Nickel Cadmium traction battery . The power electronic controller is constructed using three independent MOSFET / IGBT H-bridges [6] and is capable of full four-quadrant operation. The electrical machine and power converter has a combined efficiency of 92% at rated power. In the prototype electric vehicle, the traction motor is connected to the wheels via a step-down gearstage. However, for the proposed test-bench, the brushless permanent magnet motor is directly coupled to a 2.2kW, 50Hz, 3-phase, 6-pole, induction machine, the angular velocity of the connecting shaft being monitored via an optical shaft encoder interface. The induction motor is connected to the utility supply via a variac, which allows the supply voltage to be adjusted. To facilitate the implementation of the developed control schemes, a custom state-of-the-art, DSP hardware platform has been designed, based on the TMS320C31 floating-point processor, to integrate the system control algorithms with sensor interfaces, man-machine interface's, monitoring, and supervisory tasks. A foreground/background distribution of 'non-time-critical' and 'time-critical' functions has been employed to ensure the controller remains a high integrity process.

Figure 4 . All – Electric Racing Vehicle.

The key principle of operation of the proposed test facility for TC/ABS investigations is based on the correspondence between μ-σ characteristics of tyre/road interactions Figure 2, and the torque-slip characteristics of the induction machine when operated as a generator, Figure 5.

The brushless PM motor acts as the prime-mover, and drives the induction machine into the generator region of operation, thereby inducing a negative slip with respect to synchronous speed (1000rpm in this case). As the torque demand to the PM machine is raised further, the slip becomes increasingly negative, and the induction machine develops an opposing torque, eventually resulting in a stable steady-state condition.

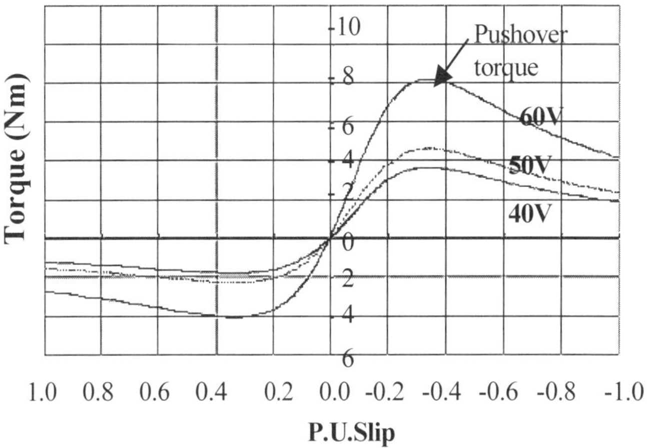

Figure 5. Induction machine torque-slip characteristics.

This procedure can continue up to the pushover torque limit of the induction machine. A subsequent increase in applied torque will then further decrease the slip and force the system into an unstable operating regime where the induction machine is not able to oppose the applied torque, and a runaway condition is encountered with the brushless motor accelerating uncontrollably. This is analogous to the classical μ-σ characteristics encountered in traction control systems when the slip at μ_{max} is exceeded. An algorithm which automatically controls the torque developed by the brushless PM motor to maintain the induction machine in the 'stable-slip' region of operation, is therefore also likely to be a good candidate to control the torque applied to the wheels of an electric vehicle to maintain tractive force in the stable region of the μ-σ characteristic. The torque-slip characteristic can be obtained from the standard equivalent circuit of an induction machine:

$$\tau_{ind} = \frac{3V^2 R_2 / s}{\omega_{sync}((R_1 + R_2 / s)^2 + (x_1 + x_2)^2)} \tag{3}$$

where: V is the applied voltage, R_1 is the stator resistance, R_2 is the referred rotor resistance, s is the slip, ω_{sync} is the synchronous speed, X_1 is the stator reactance, X_2 is the referred rotor reactance.

The magnitude and form of the torque-slip characteristics is dependent upon the supply voltage and impedance, and by adjusting these, a variety of tyre/road conditions can be readily simulated. It is noted that, due to the unsymmetrical nature of induction machine characteristics, different parameters for exciting the induction machine are required to simulate ABS's, compared with TC systems.

Simulation model of traction control test facility

To demonstrate the attributes of the proposed test facility for design and assessment of traction control algorithms for electric vehicles, a simulation model of the test facility as been employed, as illustrated in Figure 6.

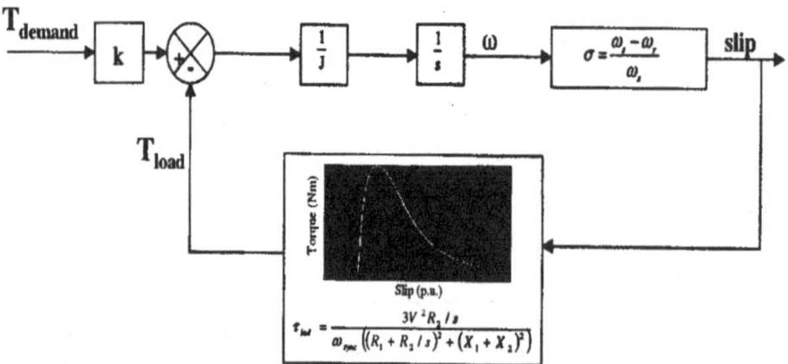

Figure 6. Simulation model of traction control test facility.

The load imposed on the brushless drive train is determined by the torque-slip characteristics of the induction machine, the parameters of the induction machine being varied to simulate alternative μ-σ road conditions. For example, to simulate a dry road condition, the induction motor has 60V excitation, providing a pushover torque capability of 9Nm. This simulates the adhesion characteristics of a dry road on the primary drivetrain, when referred back through the gearbox. The dynamic response of the system for a demanded drive train shaft torque of 12Nm is shown in Figure 7.

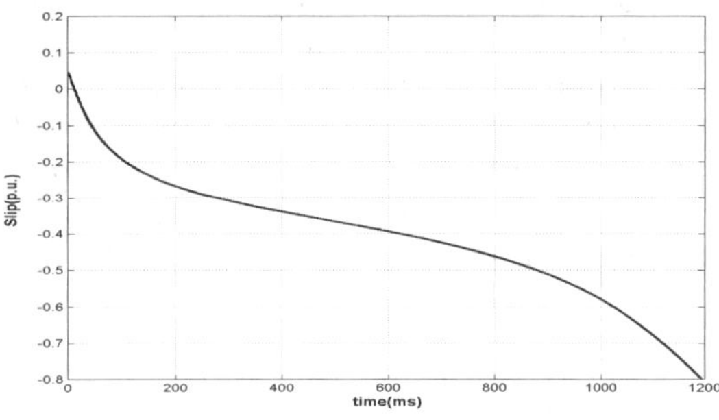

Figure 7.a. Slip

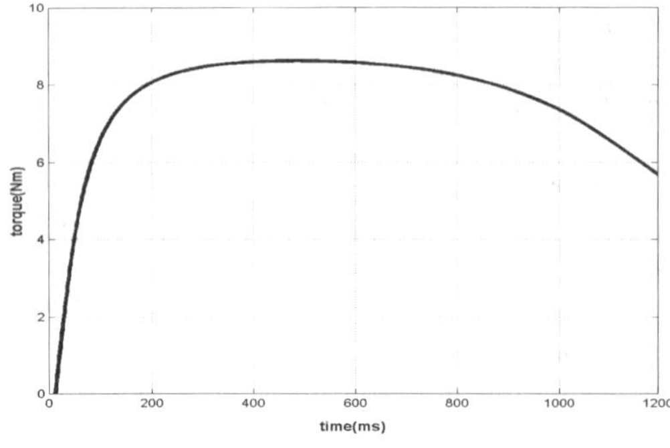

Figure 7.b. Torque on drivetrain

Figure 7.a.&b. Simulation results for road condition with no traction control.

By consulting Figure 7, it can be seen that the demanded torque is sufficient to force the induction machine past the pushover point and into the unstable torque-slip region, and hence, allowing the slip (and angular velocity) to increase uncontrollably. The results from the experimental test facility, Figure 8, confirm this dynamic behaviour.

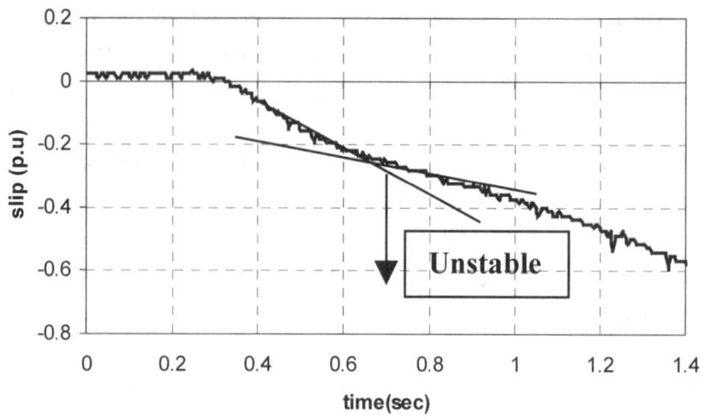

Figure 8.a. Slip

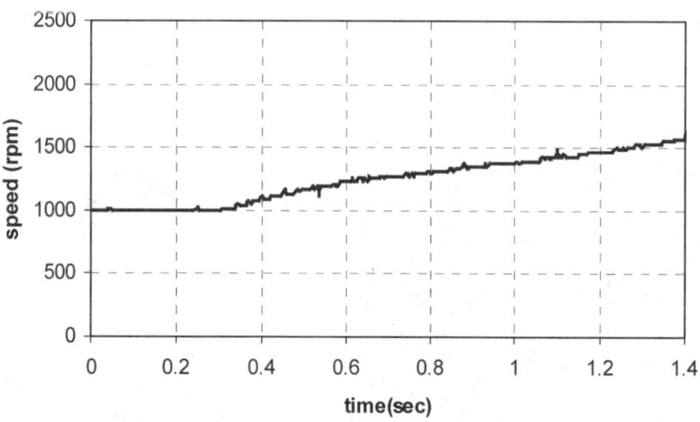

Figure 8.b. Speed of primary drive shaft

Figure 8. a & b. Experimental results from laboratory test facility without traction control.

SIMULATION AND EXPERIMENTAL RESULTS FROM TRACTION CONTROL TEST FACILITY

Initially, implementation of the "bang-bang" traction control scheme to limit the slip to -0.1 p.u.is carried out, resulting in a transient response which is shown in Figure 9a & b, where it is clear from the simulations that the induction motor is now operating in the stable region of the characteristic curve. Here, the controller is accommodating 'worst-case' (icy road) conditions.

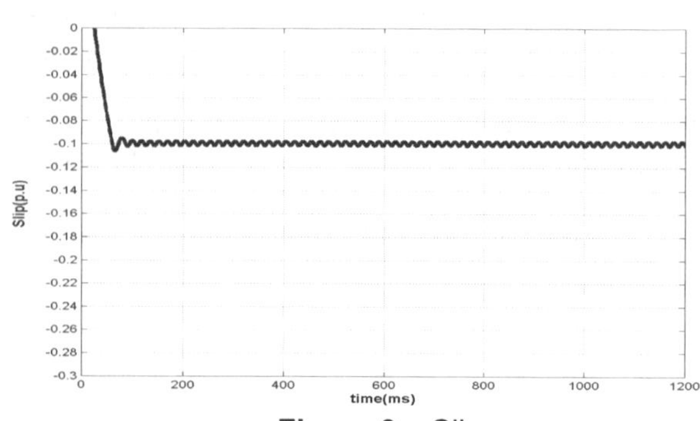

Figure 9.a. Slip

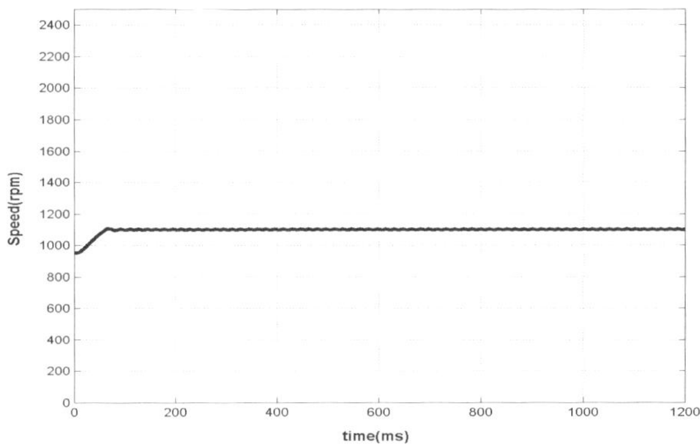

Figure 9.b. Speed of primary drive shaft

Figure 9 a & b. Simulated results with slip control scheme implemented.

Experimental results on the proposed test rig (Figure 10a & b) show the induction motor operating in the stable region, which is analogous to limiting the operation of a traction control scheme to the stable region of the μ-σ curve for icy road conditions, and confirm the simulation predictions.

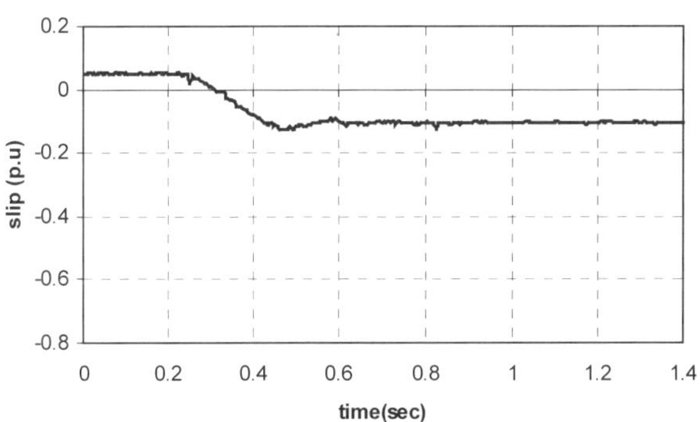

Figure 10.a. Slip

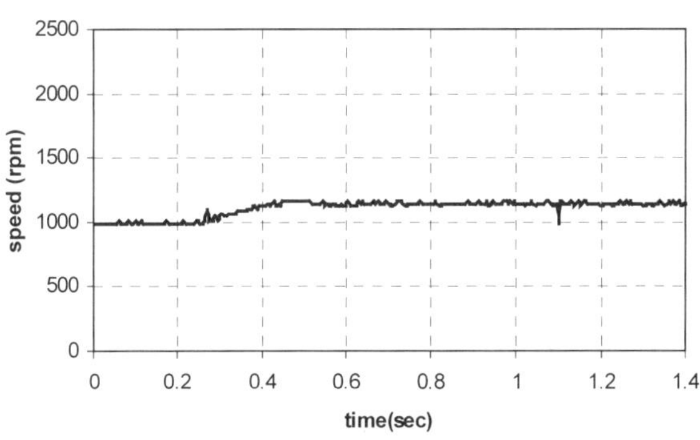

Figure 10.b. Speed of drivetrain

Figure 10.a & b. Experimental results with traction control scheme implemented.

Discrete-time traction control scheme

To obtain optimal traction between the tyre and road surface, a requirement is to possess prior knowledge of both the maximum value of adhesion coefficient, μ_{max}, and the slip at which it occurs, $\sigma_{\mu max}$. A traction control scheme would then prevent the slip between the road surface and the tyre from exceeding $\sigma_{\mu max}$. Clearly, both μ_{max} and $\sigma_{\mu max}$ are highly dependent on tyre and road surface conditions, and obtaining real-time data is a primary source of difficulty for practical controllers. With regard to the proposed experimental test facility, this problem is analogous to determining the pushover torque of the induction machine and the slip at which it occurs, without prior knowledge of the excitation voltage.

A state-observer technique to estimate the load/road-adhesion applied to the PM brushless-dc drive train, and, by knowledge of the slip, a straightforward method of obtaining optimal traction without prior knowledge of μ_{max} and $\sigma_{\mu max}$ is employed. The desired wheel torque is allowed to be transmitted to the wheel shaft until the traction controller detects that the slip has exceeded $\sigma_{\mu max}$, at which time, the demanded torque is set to zero. This has the effect of allowing the slip to reduce and re-enter the stable region of the μ-σ characteristic. Having re-entered the stable region, the desired torque is again allowed to be transmitted to the wheel shaft. This simple 'bang-bang' type of operation is particularly suited to electric traction control, since the bandwidth of electromagnetic torque production is relatively high compared to mechanical actuation, and the modulation of torque is 'smoothed' at the vehicle wheel due to inertia. The controller determines whether operation is taking place in the stable or unstable region by calculating the rate-of-change of load/adhesion with respect to slip i.e.

$$\mathrm{sgn}\left(\frac{d\mu/dt}{d\sigma/dt}\right) = \mathrm{sgn}\left(\frac{d\mu}{d\sigma}\right) = \begin{cases} +1 & operation\ in\ stable\ region \\ -1 & operation\ in\ unstable\ region \end{cases}$$

Practically, the time derivatives of μ and σ are required to be low-pass filtered to remove high frequency noise, however, since the inertia of the wheel and the maximum torque that can be applied to the wheel shaft is known a-priori, appropriate filtering does not unduly affect performance of the controller.

Under the assumption that the electromagnetic time-constant of torque production in the power electronic inverter is relatively small compared with the mechanical time-constant, the dynamics of the drive train can be modelled as a first order differential equation:

$$\dot{\omega} = \frac{k_t}{J}T_{dem} - \frac{1}{J}T_l \tag{4}$$

where ω is angular velocity, k_t is the motor torque constant, J is the inertia, T_{dem} is the demanded torque to the drive train, and T_l is the load/adhesion imposed on the drive motor. Discretizing, and defining $x_T = -T_l$, the resulting discrete-time dynamic equations of the power train become:

$$\begin{bmatrix} x(n+1) \\ x_T(n+1) \end{bmatrix} = \begin{bmatrix} 1 & -1/J \\ 0 & 1 \end{bmatrix} \begin{bmatrix} x(n) \\ x_T(n) \end{bmatrix} + \begin{bmatrix} k_t/J \\ 0 \end{bmatrix} T_{dem}(n)$$

$$\omega = \begin{bmatrix} 1 & 0 \end{bmatrix} \begin{bmatrix} x(n) \\ x_T(n) \end{bmatrix} \tag{5}$$

Whilst this does not constitute a controllable system, since there is no independent control of x_T, it is observable, and is appropriate for state-observation. Hence, real-time estimation of T_l or equivalently μ, is possible. For the experimental test facility, T_l represents the load imposed by the induction motor, whilst on the electric racing vehicle, it represents the adhesion between the road surface and the tyre, referred to the drive motor. The resulting traction control scheme is illustrated in Figure 11,

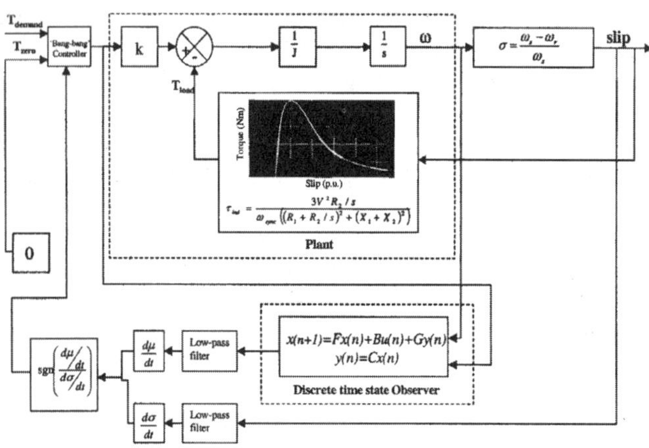

Figure 11. Discrete time traction control scheme

where F defines the dynamics of the observer, and G determines the poles ($z=0.951, z=0.949$) of the error convergence dynamics between the actual and observed states. The sampling period has been fixed at $T=1 \times 10^{-4}$ seconds.

Initial simulation studies of the discrete time based traction control scheme, show that the control of phase voltage of the induction machine only affects the magnitude of the torque-slip characteristic, with the value of slip at pushover remaining unchanged. Hence, by varying the phase voltage between 40V and 60V, it is possible to obtain different observed load torque characteristics, which are analogous to various adhesion coefficients for dry and icy road conditions as shown on Figure 12a and b. However, for both of these conditions, the peak slip remains the same.

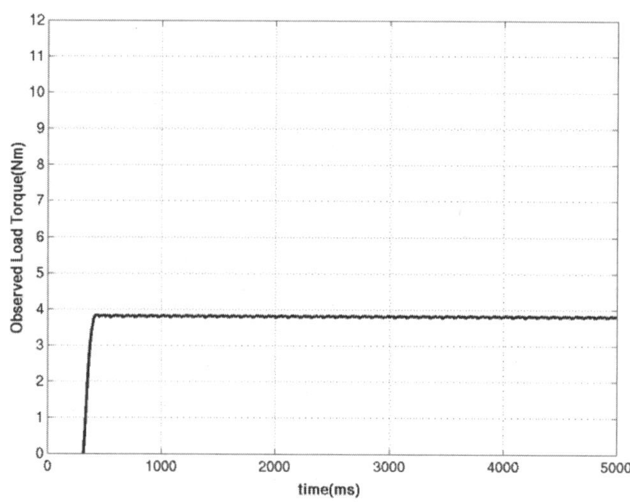

Figure 12.a. Simulation results for 40V.

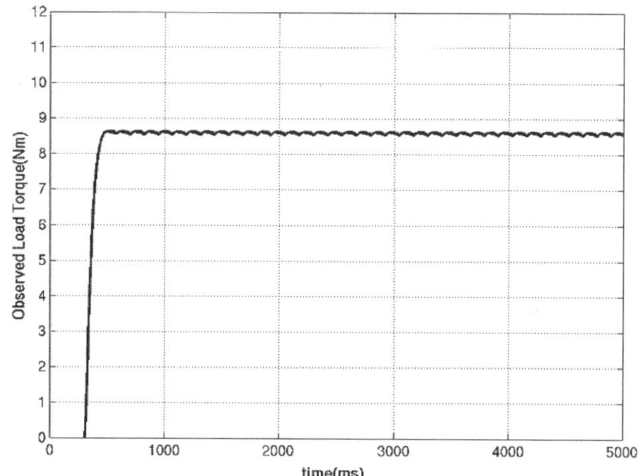

Figure 12.b. Simulation results at 60V.

Also, by implementing the proposed discrete-time observer based optimal traction control scheme, it is possible to allow operation upto, but not beyond, the stable region of the μ-slip curve (Figure 13). It can be seen that μ_{max} is -0.35 (p.u.) which represents a dry road surface on a classical μ-σ characteristic.

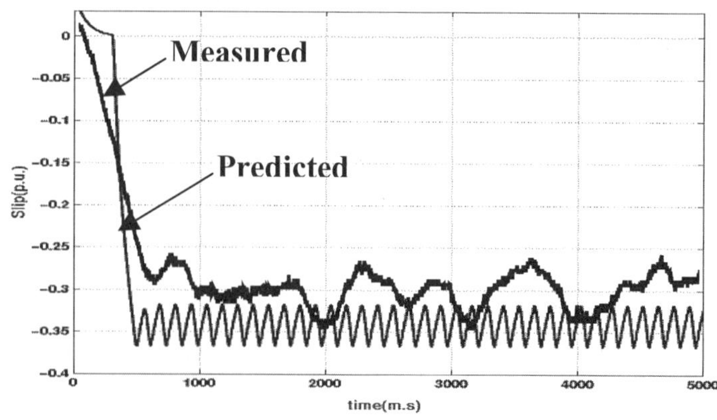

Figure 13. Discrete-time based 'optimal' bang-bang traction control scheme

To simulate the dynamic μ-slip characteristic over a wider range of road conditions, again the variation of phase voltage is employed. Hence, it is possible to observe the discretised load torque (or μ) over a time period of 12sec, whilst the phase voltage is varied between 55V and 48V (Figure 14). From the measurement it can be seen that the incorporation of discrete-time observation technique allows the dynamic μ-slip characteristics to be obtained without a-priori knowledge of $μ_{max}$.

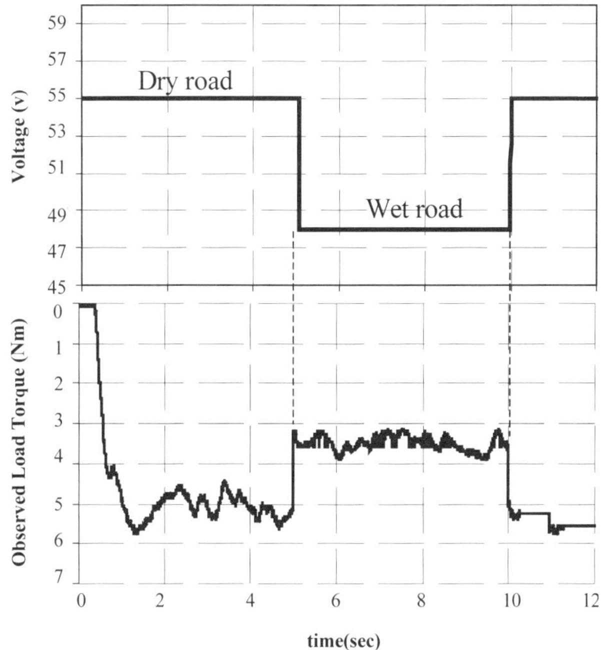

Figure 14. Measured results of observed load torque.

Fuzzy Logic based traction control scheme

In recent years fuzzy logic techniques have been applied to a wide range of systems. Many electronic control systems in the automotive industry such as automatic transmissions, engine control and Antilock Braking System (ABS) are currently being pursued in the United States. These electronically controlled automotive systems realise superior characteristics through the use of fuzzy logic based control rather than traditional control algorithms [7].

Fuzzy logic allows highly non-linear or mathematically complex systems to be modelled in a reliable and efficient manner. Moreover, fuzzy logic deals well with uncertain and noisy data. These characteristics offer fuzzy control as a potentially important tool for the development of robust traction control systems.

One of the strengths of fuzzy systems is their ability to express a confidence in reasoning results. Recent studies have shown that fuzzy systems are part of a class of universal approximators of continuous functions. The key advantage of fuzzy control is that there are many instances where TRUE and FALSE or ON and OFF fail to describe a given situation. These cases require a sliding scale where variables can be measured as PARTLY ON or MOSTLY TRUE and PARTLY FALSE [8].

Traditional set theory is based on bivalent logic where an object is either a member of a set or it is not. However, with fuzzy logic, an object can be a member of multiple sets with a different degree of membership in each set. A degree of membership in a set is based on a scale from 0 to 1; 1 being complete membership and 0 being no membership. Hence, by applying fuzzy logic control to this discrete time traction control system, an output (the torque demand) is calculated based on the amount of membership the input signals (observed load torque and slip) has in the configured fuzzy sets. Therefore, it is required that the system inputs go through three major transformations before becoming a system output. Figure 15 shows the overall fuzzy logic based traction control scheme, which includes the transformation process of fuzzification, fuzzy rule association and de-fuzzification [8].

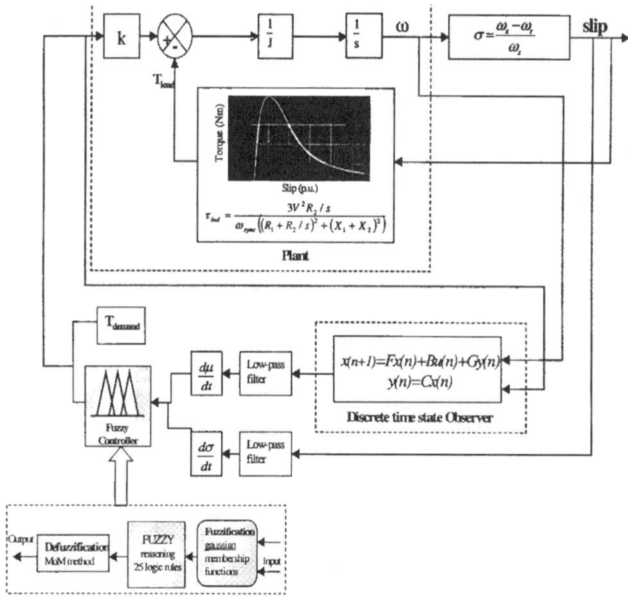

Figure 15. Fuzzy Logic based traction control scheme

In order to establish the two appropriate input fuzzy membership functions and the rule base, a thorough investigation of the ABS braking process has been undertaken. In this case the rate of change of observed load torque and the rate of change of slip were chosen. The two inputs are fuzzified into membership functions as shown on Figure 16a and 16b.

Although gaussian functions were initially used to investigate the ABS braking process, trials have shown that functions based on polysigmoid curves provide enhanced performance characteristics.

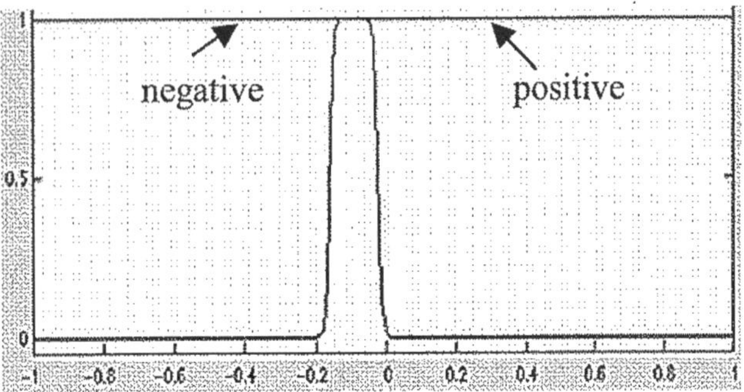

Input variable: Load torque

Figure 16.a. Membership functions for observed load torque.

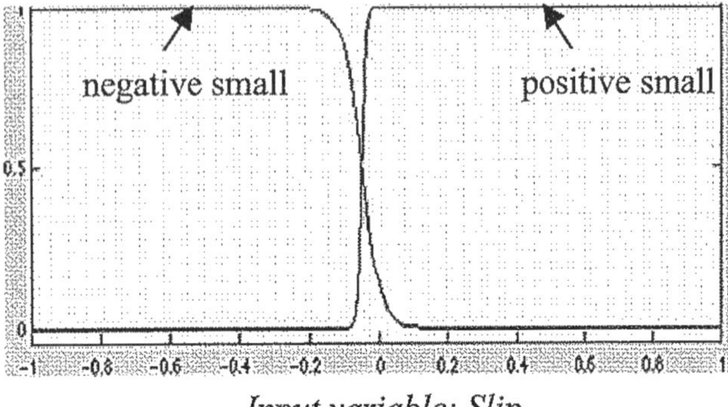

Input variable: Slip

Figure 16.b. Membership function for slip.

For the slip, the universe of discourse was chosen between 0 and 100%, since during the operation of an induction machine at 0 slip the system is stable and at 100% the system is unstable. This is analogous to the braking process of a vehicle, where a wheel can rotate freely with slip at 0 or be locked at slip equal to 1.

Both inputs, slip and observed load torque, are fuzzified into two membership functions. A smaller number of membership functions causes less sensitive control, but better computation speed, and vice versa. Therefore it is very important to understand and test the influence of the number of membership functions, as there is always a compromise between execution speed and sensivity of the fuzzy controller.

For the testing, the Fuzzy Logic based Traction Control scheme has two inputs (slip and load torque) to the controller and one output (torque demand), with 4 possible rules.

Rate of change of slip

	ps	ns
p	z	pb
n	pb	z

Rate of change of observ

Figure 17. Rule map for Fuzzy Traction controller.

All four rules are used for the controller and are determined from the fuzzy rule map shown in Figure 17. The 4 rules cover each state of input combinations during the braking process. The output of the fuzzy controller is the torque demand (or in the case of a vehicle the tractive force). The fuzzy membership functions for the output, torque demand, is shown on Figure 18.

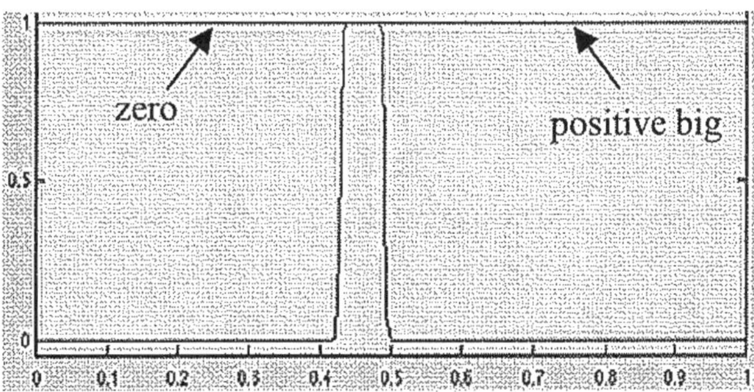

Output variable: Output Torque

Figure 18. Membership function for Induction machine torque demand.

The membership functions, shown in Figure 16, were created by experimentally testing the relationship between slip and torque of the induction machine. A fuzzy controller consists of fuzzification, rulebase and defuzzification procedures. Fuzzification, that is, creating fuzzy inputs from crisp inputs, is accomplished using the membership functions in Figure 16a and 16b. The fuzzy output is generated using the rulebase, shown in the rulemap in Figure 17. A crisp output is generated from the fuzzy output by defuzzification, using the membership functions shown on Figure 18. This is the basic fuzzy control algorithm for a traction control system. By "tuning" the membership functions the controller can be optimised.

Implementing the fuzzy control scheme, with an experimental setup analogous to a vehicle travelling on a dry road surface, Figure 19 shows the simulated performance of the system. In comparison with limited 'bang-bang' slip control, it can be seen that the fuzzy scheme automatically adapts the slip control algorithm, identifies the unstable region of the torque-slip curve and reduces the slip. Eventually the slip stabilises at –0.25 p.u. which is similar to the μ-slip characteristic of a dry road surface(where the peak slip is between –0.25 and –0.35 p.u.).

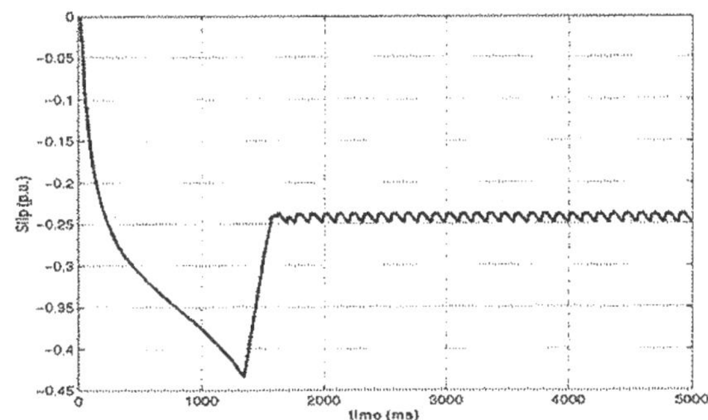

Figure 19. Simulation results for road conditions with Fuzzy Logic Control.

SIMULATION STUDIES OF ALL-ELECTRIC RACING VEHICLE TRACTION CONTROL SYSTEM

To fully explore the operation of the proposed traction control schemes, the all-electric racing vehicle was chosen as a suitable platform, since it could be easily reconfigured and safely tested.

Initially the vehicle performance was analysed using a Matlab system model (Figure 20).

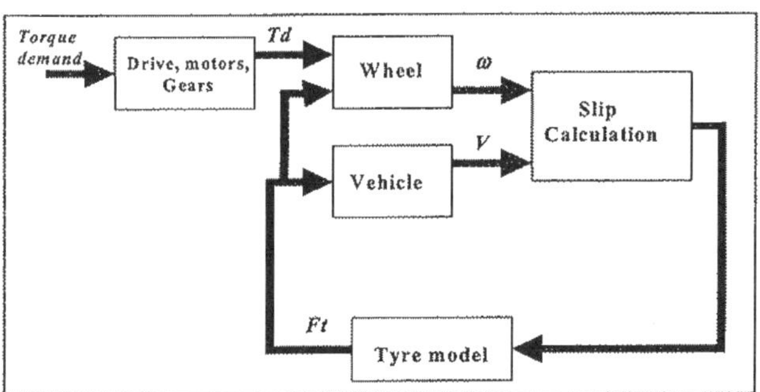

Figure 20. Matlab system model of electric vehicle.

Where wheel speed is defined as:

$$\omega = \frac{1}{J} \int_0^t T_d - F_t r \, dt \tag{6}$$

Vehicle speed:

$$V = \frac{1}{M} \int_0^t F_t - C V^2 \, dt \tag{7}$$

Tyre model is described as:

$$F_t = N \mu(\sigma) \tag{8}$$

Slip is defined as:

$$\sigma = \frac{V - r\omega}{r\omega} \tag{9}$$

where J is the inertia, T_d is the machine torque, F_t is the tractive force, r is the wheel radius, M is the mass, C is the drag coefficient, V is the vehicle speed, μ is the adhesion coefficient, ω is wheel speed and N is normal force ($\frac{mg}{4}$ m is mass, g is gravity).

From the simulation results (Figure 21), it can be seen that by limiting the slip at –0.1 p.u., the vehicle system model confirms the similarities in dynamic behaviour as the traction control test facility results shown previously.

The vehicle speed for the simulation model was measured using the tractive force acquired from the tyre model; the value of μ being obtained from a look-up table for μ-slip characteristics. However, in the case of measured data from the prototype electric vehicle, the front wheels will provide the measurements for wheel speed and the rotational speed of the rear axle will provide the vehicle speed; and any change between the front and rear wheels will indicate slip.

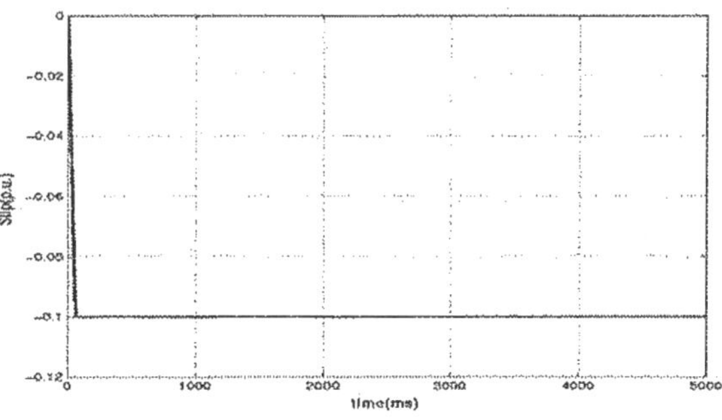

Figure 21. Matlab electric vehicle simulation result with slip control at –0.1 p.u.

CONCLUSION

An analysis of different traction control schemes is presented in this paper. Simulation studies and experimental trials on the laboratory test facility and simulation studies based on the prototype electric vehicle were used to evaluate the response of the proposed techniques in real driving situations. The paper confirmed that application of observer techniques are appropriate for on-line determination of peak adhesion coefficient for different tyre-road surface conditions, and fuzzy logic based control strategies offer substantial potential for optimal traction control of driven wheels.

FUTURE WORK

The following areas have been highlighted for future investigation:

- Implementation of novel intelligent control schemes to provide automatic adaption of slip control algorithms on the test facility.
- Implementation of control schemes on the prototype high-performance racing vehicle.

The results of these investigations will be published in due course.

ACKOWLEGEMENT

The authors would like to acknowledge the UK Engineering and Physical Sciences Research Council for providing a research studentship to Ms.Khatun.

CONTACT

Ms.Parvin Khatun can be reached through the Department of Electronic & Electrical Engineering, University of Sheffield, Mappin Street, Sheffield S1 3JD, U.K., at +44 114 2225195, or email: elp97pnk@sheffield.ac.uk.

Dr.C.M.Bingham can be reached at the University of Sheffield on +44 114 2225849, or e-mail: c.bingham@sheffield.ac.uk.

Prof.Phil Mellor can be reached through the Department of Electronic & Electrical Engineering, University of Bristol, Merchant Venturers Building, Woodland Road Bristol BS8 1UB, U.K., at +44 117 954 5259, or e-mail: P.H.Mellor@bristol.ac.uk.

APPENDIX

Vehicle parameters:

- Mass, m   160 kg
- Gravity, g $9.81 m/s^2$
- Machine Torque, T_d 12Nm
- Wheel radius, r   0.139 m
- Referred wheel Inertia, J_w $0.8 kgm^2$
- Machine Inertia, J_m $0.0.156 kgm^2$
- Drag coefficient, c   0.5
- Gear ratio,   5:1

REFERENCE

1. Harned, J.L., Johnston, L.E., Scharpf, G.,"Measurement of tire brake force characteristic as related to wheel slip(antilock)control system",*SAE Trans,Vol78, No.69214,1969.*
2. Ray,L,R.,"Non linear system and tire force estimation for advanced vehicle control",*IEEE Trans on Control Systems Technology,Vol3, No1, 1995.*
3. Klien,R., "Antilock Braking System and vehicle speed estimation using Fuzzy Logic",*FuzzyTECH Application paper,http//www.fuzzytech.com/,1999.*
4. Constantin Von Altrock., "Fuzzy Logic in Automotive Engineering", *Embedded Systems Conference, 1996.*
5. Robinson,B.J,Riley,B.S,"A study of various car antilock braking system",*Research report 340, Department of Transport and Road Research Laboratory, U.K.,1991.*
6. Mellor,P.H, Schofield,N., Hor,P.J.,"Brushless permanent magnet drives for all-electric racing karts",*Proc.29th Int. Symposium on Automotive Technology Automation,1996.*
7. Elting,D.,Fennich,M.,Kowalczyk,R.,Hellenthal,B.,"Fuzzy Antilock Brake System Solution", Application note, http://developer.intel.com/,1999.
8. De Koker,P.M., Gouwes,J., Pretorius,L., "Fuzzy Control Algorithm for Automotive Traction Control Systems", *IEEE Trans on Industrial Applications in Power Systems, Computer Science and Telecommunication*, vol.1, 1996.

Mixed H_∞ and Fuzzy Logic Controllers for the Automobile ABS

Farhad Assadian
Centre Technique PSA (Peugeot-Citroen)

ABSTRACT

A mixed of two different control strategies are proposed for stability and performance enhancement of the vehicle Antilock Braking Systems (ABS). The first is based on the H_∞ control theory in order to devise a robust control methodology for stabilizing the system under various road conditions. The second is based on the Fuzzy Logic concept for performance enhancement of the aforementioned system. The results of the mixed controllers are compared to the ideal condition for which access to the maximum tire-road coefficient of friction is available.

INTRODUCTION

Much has been said about the estimation of tire-road coefficient of friction and its use for automobile ABS. It is also understood that the shortest stopping distance by the automobile ABS can be accomplished when access to maximum tire-road coefficient of friction is available. There is continuous effort in developing tire-road friction estimation methodologies. However, due to the complexity of the tire-road friction estimation, none of the proposed methodologies has shown real promise for implementation on mass produced vehicles.

The development of a viable control strategy that can be used for continuous control of a properly selected proportional actuator for the automobile ABS, such as brake by wire method, is another area that much effort has been concentrated. However, again, the majority of the proposed methods are either too complex for implementation or do not result in an optimal solution with regard to the shortest stopping distance. Furthermore, all the aforementioned studies make no attempt in comparing their final results to an ideal solution in order to illustrate if their proposed solutions result in acceptable performances.

In this study, a reduced linear model of a nonlinear vehicle, including the braking system dynamics, is obtained in place of traditional linearization of the proposed nonlinear system and the linear H_∞ control theory is utilized to derive a linear controller for the aforementioned system. Then the simple linear controller is applied to the original nonlinear system, and it is illustrated that the controller guarantees system stability under all road conditions. The final results are compared to an ideal solution for which access to maximum tire-road friction coefficient is assumed. The results of the H_∞ controller do not generate optimal stopping distances due to large variations in tire-road coefficient of friction under different road conditions. Therefore, in order to accommodate for the uncertainties associated with different road conditions, fuzzy logic mapping methodology is used to enhance the performance of the H_∞ controller by varying the commanded longitudinal slip using vehicle acceleration as input. It is illustrated that the combination of the proposed simple linear H_∞ controller and the fuzzy logic rule based controller produce optimal results under all road conditions when compared to the ideal solution.

In this work, We utilize the one wheel vehicle model including a hydraulic brake system to derive our control strategies. This model is fully discussed in [11], however, for the sake of completeness, we briefly provide an overview of this dynamic model including the derivation of the state-space equations with the use of bond graphs [12].

DYNAMIC MODEL

The one wheel car model is illustrated in Figure 1. This model is attractive for control design development in a sense that the model is simple yet it includes some important dynamic characteristics of a full vehicle model. We use this model for deriving the state-space equations. Furthermore, the lateral and vertical dynamic capabilities of the model are not utilized in this study, and the model is used for straight-line motion only.

In this work, we make a comparison of our results to a so-called traditional ABS. Therefore, before proceeding further a brief description of a simplified version of this traditional ABS is provided.

A schematic of a simplified traditional ABS [1] is shown in Figure 2. With a traditional ABS, the controller measures the rotational speed of the wheels and it will estimate the linear speed of the vehicle. Based on this information, the controller computes the wheel slip, and if the wheel slip is greater than some optimal slip value then the controller modulates the supply brake pressure by opening and closing the solenoid valves as shown in Figure 2.

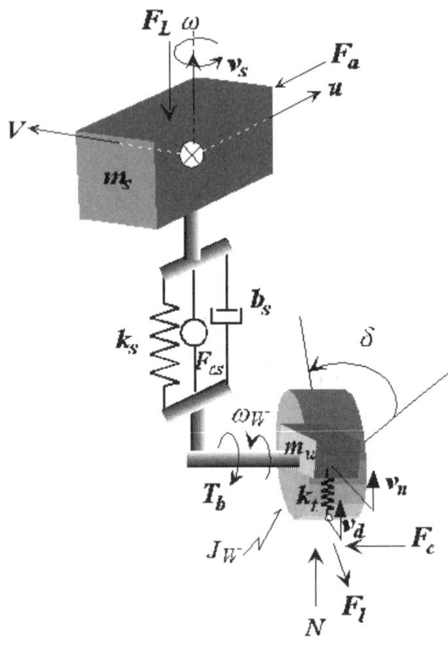

Figure 1. One wheel vehicle model.

The opening and closing the valves result in undesired control chattering. Some of the unwanted effects of control chattering are overshooting the optimal slip and excitation of higher dynamic modes of the tires. One

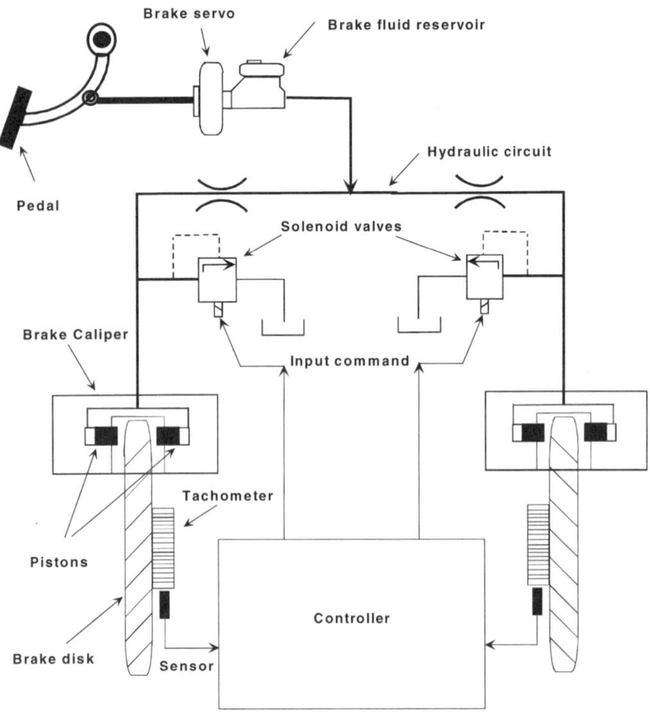

Figure 2. A simplified schematic of a traditional ABS.

way to alleviate this, is to discard the solenoid valves and to control the brake supply pressure with a new type actuator (e.g. an electric brake actuator) including a control strategy that is capable of proportionally control the supply pressure.

A simplified model of a brake system is depicted in Figure 3. Here, we assume that our new actuator generates the input force, F_{mcyl}, and for simplicity we ignore the dynamics of the actuator. Input force is transformed to the input pressure, p_{mcyl}, by a piston with the mass, m_{pmcyl}, and the surface area, S_{mcyl}. The input pressure, P_{mcyl}, modulates the brake supply pressure, P_e. The supply pressure generates the brake torque, T_b, by acting on the brake disk with the mass, m_{vpe}, the damping coefficient, C_{vpe}, and the nonlinear stiffness, k_{vpe}.

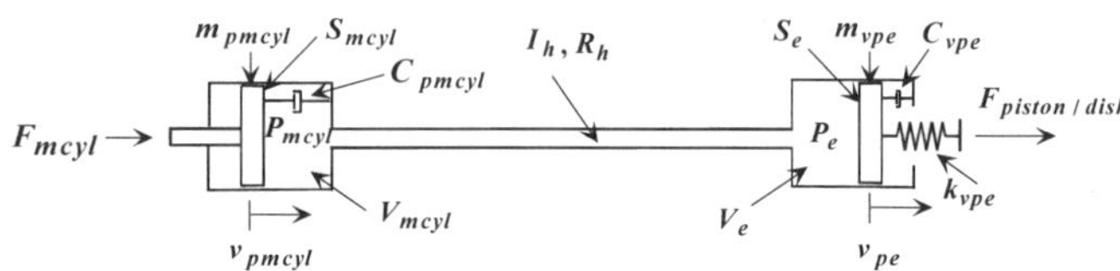

Figure 3. Model of a hydraulic brake system.

The bond graph model of the one wheel vehicle and the simplified hydraulic brake system is shown in Figure 4. We use this model for derivation of the state equations and for further model simplification for the purpose of control system analysis and design.

The state-space equations can be written as follows:

$$\dot{p}_u = \omega\, p_v + F_l \tag{1}$$

$$\dot{p}_v = -\omega\, p_u + F_c \tag{2}$$

$$\dot{p}_{J_w} = -T_b + R\,F_l \tag{3}$$

$$\dot{p}_{m_s} = k_s q_s + b_s\left(\frac{p_{m_u}}{m_u} - \frac{p_{m_s}}{m_s}\right) + F_{c_s} - F_d \tag{4}$$

$$\dot{p}_{m_u} = k_t q_t - \left[k_s q_s + b_s\left(\frac{p_{m_u}}{m_u} - \frac{p_{m_s}}{m_s}\right) + F_{c_s}\right] \tag{5}$$

$$\dot{q}_s = \frac{p_{m_u}}{m_u} - \frac{p_{m_s}}{m_s} \tag{6}$$

$$\dot{q}_t = v_d - \frac{p_{m_u}}{m_u} \tag{7}$$

$$\dot{p}_{vpmcyl} = F_{mcyl} - C_{pmcyl}\frac{p_{vpmcyl}}{m_{pmcyl}} - \frac{B\,S_{mcyl}}{V_{mcyl}}q_{mcyl} \quad (8)$$

$$\dot{q}_{vpmcyl} = v_{pmcyl}\,S_{mcyl} - \frac{p_h}{I_h} \quad (9)$$

$$\dot{p}_h = \frac{B}{V_{mcyl}}q_{mcyl} - \frac{B}{V_e}q_e - r_h\frac{p_h}{I_h} \quad (10)$$

$$\dot{q}_e = \frac{p_h}{I_h} - v_{pe}S_e \quad (11)$$

$$\dot{p}_{vpe} = \frac{B\,S_e}{V_e}q_e - C_{vpe}\frac{p_{vpe}}{m_{vpe}} - k_{vpe}(q_{vpe})\,q_{vpe} \quad (12)$$

$$\dot{q}_{vpe} = \frac{p_{vpe}}{m_{vpe}} \quad (13)$$

Where the horizontal motion dynamics consist of p_u and p_v, the forward and transverse momenta. The wheel dynamics include p_{J_W}, the angular momentum of the wheel. The vertical motion dynamics include p_{m_s} and p_{m_u}, the sprung and unsprung mass vertical momenta, q_s and q_t, which are the relative

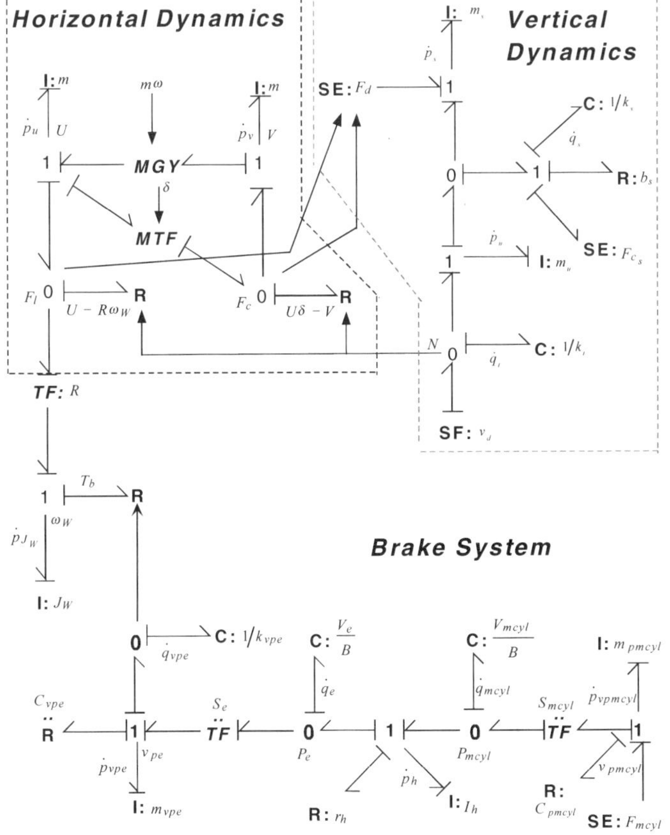

Figure 4. Bond graph of the one wheel vehicle model including the brake system.

displacements across the suspension and tire respectively. The break system dynamics consist of p_{vpmcyl}, the primary cylinder piston momenta, q_{vpmcyl}, volume displacement in the primary cylinder, p_h, hydraulic line pressure momentum, q_e, volume displacement of the secondary cylinder piston, p_{vpe}, the secondary cylinder piston momenta, and q_{vpe}, the relative displacement across the break disk.

The constitutive law for the modulated resistive element in the bond graph in Figure 4 for the generation of brake torque is given by,

$$\begin{cases} T_b = 0 & x_{vpe} < x_j \\ T_b = 2\,R_e\,\mu_e\,k(x_{vpe})(x_{vpe} - x_j) & x_{vpe} \ge x_j \end{cases} \quad (14)$$

where R_e is the effective radius of the brake caliper, and μ_e is the coefficient of friction between the piston and the brake disk.

H_∞ CONTROLLER

As stated in [1], the design of a linear H_∞ controller requires linearization of the prescribed nonlinear dynamic system. However, our approach is not to linearize the system, but to deal with system nonlinearities as perturbations. In this fashion, we can use the machinery of the linear robust control theory to devise a controller for the transformed nonlinear system as will be explained in the following paragraphs.

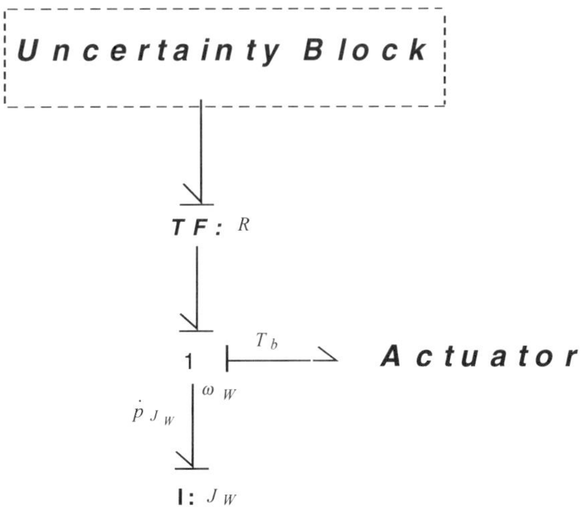

Figure 5. Equivalent bond graph of the one wheel vehicle model with the uncertainty block.

An equivalent bond graph for the one wheel vehicle model is illustrated in Figure 5. In this Figure, the uncertainty block includes the nonlinear dynamics portion which is responsible for generation of the longitudinal force F_l. Hence, the problem of the control design for a nonlinear system is transformed to a linear system control design problem. When a vehicle is derived on an asphalt road vs. a snowy road, there is significant change in the magnitude of the longitudinal force. The traditional controllers such as PID or PI that are designed around the nonlinear system can not compensate for this change in magnitude. This change in magnitude is equivalent to the gain variation of the control loop. Therefore, utilization of H_∞ control theory for devising a robust controller, which will compensate for this gain variation is sensible.

To reiterate the basics for the H_∞ control theory, the general block diagram of a standard SISO (Single Input Single Output) control problem of Figure 6 is provided.

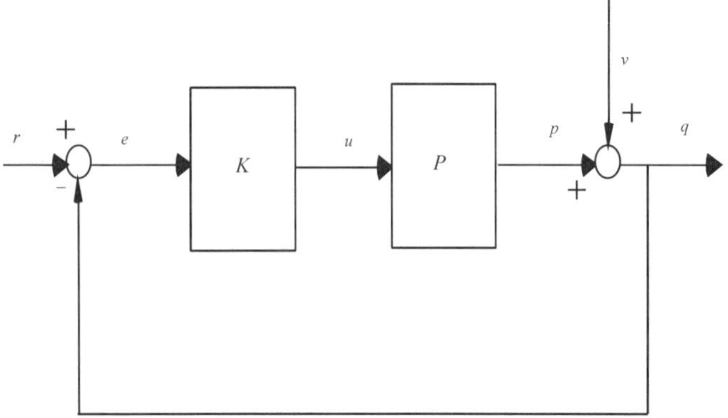

Figure 6. The standard SISO control block diagram.

The objective is to force the output $q(t)$ to follow $r(t)$ as closely as possible in the presence of the wide band disturbance v, via an output feedback $u = -Ke$. The following relationships are written in the frequency domain (using the same symbol for the signal and its Laplace transform):

$$e = Sr,$$

$$p = Tr$$

$$q = Tr,$$

where e is the error signal and

$$S = \frac{1}{1+PK}$$

is called the sensitivity transfer function. The transfer function

$$T = \frac{PK}{1+PK}$$

is called complementary sensitivity transfer function because of the identity

$$S + T = 1 \tag{16}$$

The performance of the control system is thus measured by the transfer function S, which should be kept small. In this case, the complementary sensitivity, T, will be close to unity, and the control loop will have good command following characteristic. Typically, one would want to minimize the scalar quantity $\|S\|_\infty$ (the maximum of the magnitude plot of S over all frequencies). However, it is also required to take into account the modeling uncertainties (plant perturbation) in P. Let the nominal plant have a transfer function P_0, and $P = (1 + \Delta P)P_0$. This is equivalent to the block diagram of Figure 7. Typically, one knows a frequency-dependent bound on ΔP, of the form $|\Delta P(j\omega)| < \delta(\omega)$.

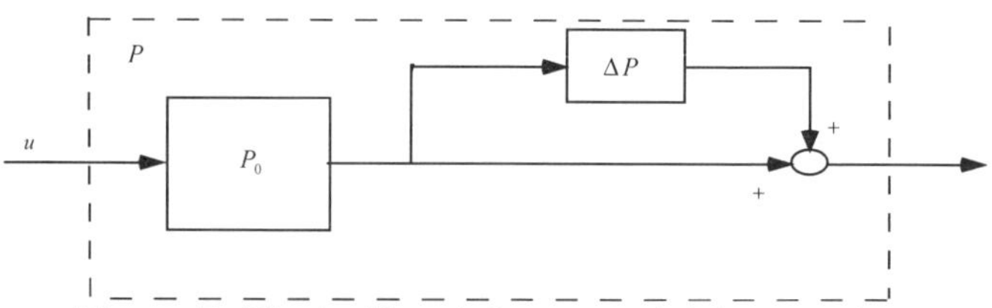

Figure 7. Plant with relative uncertainty operator.

Let S_0 and T_0 be the sensitivity and complementary sensitivity transfer functions constructed with P_0 instead of P. According to small gain theorem, robust stability requires that

$$|T_0(j\omega)|\delta(\omega) \le 1. \tag{17}$$

Because of the equation (16), it is not possible to simultaneously lower $|T_0(j\omega)|$ and $|S_0(j\omega)|$. In particular, (17) will result in $|S_0(j\omega)| \to 1$ as $\omega \to \infty$. Hence it is not possible to keep $|S_0(j\omega)|$ small at all frequencies.

One usually sets a performance objective in the form

$$|S_0(j\omega)|W_1(j\omega) \le 1 \tag{18}$$

where $W_1(j\omega)$ is a suitably chosen realizable transfer function. The frequency ω_0 at which $W_1(j\omega_0) = 1$ is called the cutoff frequency.

One also has to devise a rational function $W_2(s)$ such that $|W_2(j\omega)| > \delta(\omega)$ and furthermore such that $W_2 P_0$ is proper Considering the open-loop nominal system, without K and ΔP, as enclosed by the dashed box in Figure 8. Take u as the control input, v as the disturbance input, e as the measured output, and

$$z = \begin{pmatrix} W_1 e \\ W_2 p \end{pmatrix}$$

as the controlled output. The problem then is to devise a controller $u = Ke$ such that the transfer matrix T_{vz} from v to z satisfies $\|T_{vz}\|_\infty \leq 1$, ensuring both (17) robust stability and (18) performance.

It may be desirable to limit the control effort and thus add a third component in the controlled output, $z_3 = W_3 u$ as shown in Figure 8.

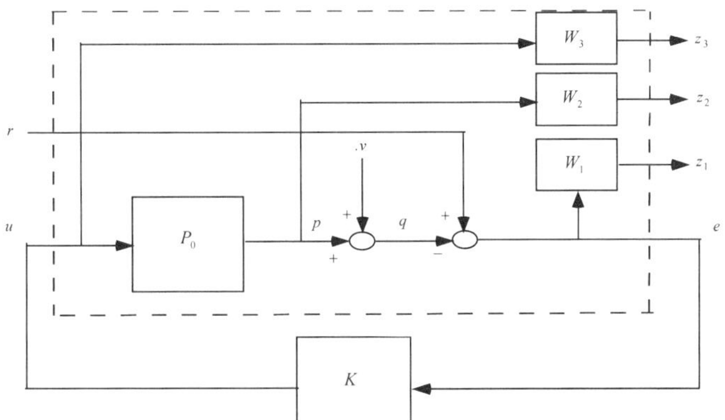

Figure 8. Sensitivity and complementary sensitivity reduction.

From the bond graph model of Figure 5 we can come up with an equivalent closed loop configuration of Figure 9.

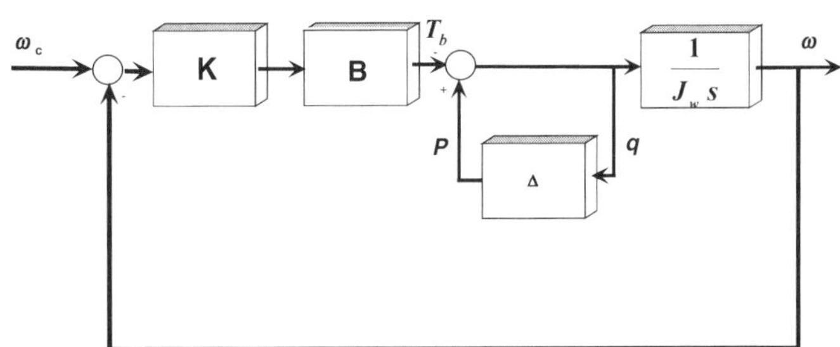

Figure 9. Closed loop system with uncertainty operator.

The block diagram in Figure 9 is setup to impose a constraint on the sensitivity transfer function S_0 for compensation of the plant gain variation due to the longitudinal force variation F_l. In figure 9, p and q are fictitious signals, where $H_{qp} = S_0$ and $H_{pq} = \Delta$, B is the transfer function of the actuator, and K is the controller. According to this block diagram, we can write,

$$-T_b\left(\frac{1}{J_w(j\omega)}\right)\left(\frac{1}{1-\Delta}\right) = \omega \tag{19}$$

If we select $\frac{1}{1-\Delta} = \alpha$, then α represents the loop gain variation. We select the loop gain variation to be

$$0.1 \leq \alpha \leq 1,$$

and based on the previous discussion on H_∞ control theory, $W_1 > \Delta$. Hence,

$$W_1 = \max\left\{\left\|1 - \frac{1}{\alpha_{min}}\right\|_\infty, \left\|1 - \frac{1}{\alpha_{max}}\right\|_\infty\right\} = $$
$$\max\left\{\frac{1}{\alpha_{min}} - 1 \,,\, 1 - \frac{1}{\alpha_{max}}\right\} \tag{20}$$

, and finally

$$\|S_0\|_\infty \leq \frac{1}{W_1} = \min\left\{\frac{\alpha_{min}}{1-\alpha_{min}} \,,\, \frac{\alpha_{max}}{\alpha_{max}-1}\right\} \tag{21}$$

Hence, based on the algebraic constraint of equation (16), $W_1(j\omega)$ can not be chosen to be a constant over all frequencies. We select $W_1(s)$ to be the following second order filter,

$$W_1(s) = \frac{\beta(\alpha s^2 + 2\varsigma_1\omega_{co}\sqrt{\alpha}s + \omega_{co}^2)}{\beta s^2 + 2\varsigma_2\omega_{co}\sqrt{\beta}s + \omega_{co}^2} \tag{22}$$

where α and β are the high frequency gain and the DC gain respectively, ω_{co} is the cross-over frequency, and ς_1 and ς_2 are damping ratios. Then, the filter parameters are selected to meet the constraint of (21) in the low frequency region. The problem of $j\omega$-axis plant pole can be solved via axis shifting technique. However, there is still a plant zero at infinity, which is also on the $j\omega$ axis. This can be taken care of by selecting,

$$W_2(s) = \frac{s}{\gamma} \tag{23}$$

where the differentiator makes the plant full rank at infinity, but also serves as the complementary sensitivity weighting function and limits the control of the system bandwidth to γ rad/sec.

It is important to note that the input command, tire angular velocity (ω_c), as shown in Figure 9, is computed using the following relationship,

$$\omega_c = \frac{U}{R_w}(1 - s_c) \tag{24}$$

where U is the vehicle velocity, R_w is the tire radius and s_c is the commanded slip, which is taken to be equal to a constant value of 0.2 for all road conditions.

In the next section we compare the performances of the H_∞ controller with an ideal controller for which the direct access to the generated tire force is possible.

SIMULATION RESULTS

In this section, we present the simulation results for the proposed control methodologies. As stated previously, the controllers are simulated on the one wheel vehicle model in rectilinear motion without including the effects of the road roughness on the braking process. The characteristics of the generated tire force can be described by specific curves which maps the slip, s, ($s = \dfrac{U_s}{U}$, where U_s is the difference between the vehicle and the wheel velocities) into the adhesion coefficient, μ. In this work, these curves are described by the parametric functions using the model of Bruckhardt [2],

$$\mu(s) = a_0(1 - e^{-b_0 s}) - c_0 s \qquad (24)$$

where a_0, b_0, and c_0 are constant coefficients. Figure 10 illustrates the aforementioned curves for several road surfaces for which the simulation results are obtained.

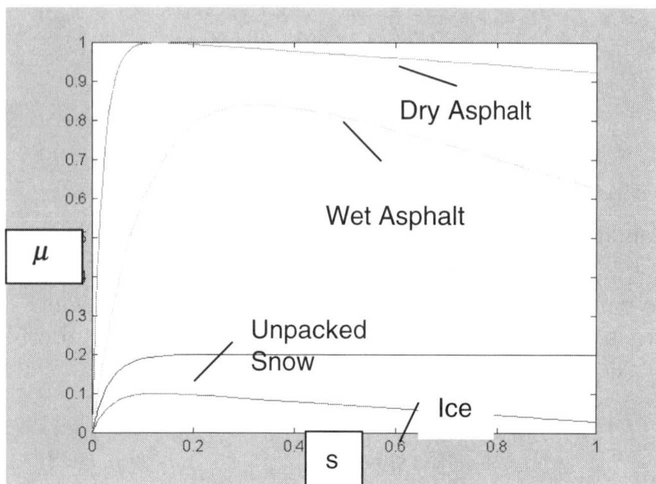

Figure 10. Coefficient of friction, μ, as function of slip, s, for several different road surfaces.

In order to observe the general performance of the proposed controllers vs. the traditional ABS system, Figures 11 through 16 are provided. These figures illustrate the slip vs. time and the vehicle/wheel velocities vs. time for the traditional ABS and the proposed controllers when the vehicle is derived over ice with the speed of 50 Km/hr.

As illustrated in Figures 11 and 12, the performance of the ABS using a traditional or on-off control strategy is drastically reduced when the vehicle is derived on ice. It can be observed that at lower speed, the traditional controller results in the wheel blockage. The results for the H_∞ control strategy are illustrated in Figures 13 and 14. As shown in these Figures, this control strategy drastically reduces the slip oscillation that is associated with the bang-bang or traditional ABS control. In addition, the controller is robust in a sense that the performance of

this control strategy will not deteriorate by the change of the road surface condition, e.g. ice to asphalt. Another important point is that the H_∞ controller is designed around a perturbed linear system, but by adjusting the aforementioned filter gains, W_1 and W_2, is successfully applied to a highly nonlinear system, including a time delay in the actuator dynamics. However, as illustrated, with this control methodology the slip is controlled at a nonoptimal constant value of 0.2. This value of the slip does not maximize the deceleration at most road surface conditions (based on the given curves of Figure 9, slip of 0.2 is near optimal in the case of snow). In this next section, we will alleviate the problem of the maximization of the vehicle deceleration, by mapping the vehicle acceleration to the longitudinal slip implementing the rules of fuzzy logic.

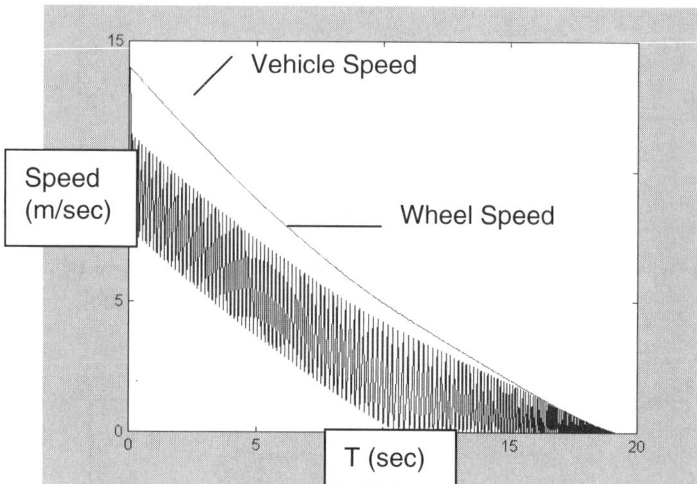

Figure 11. Vehicle/ wheel velocities vs. time for the traditional ABS controller while driving on ice.

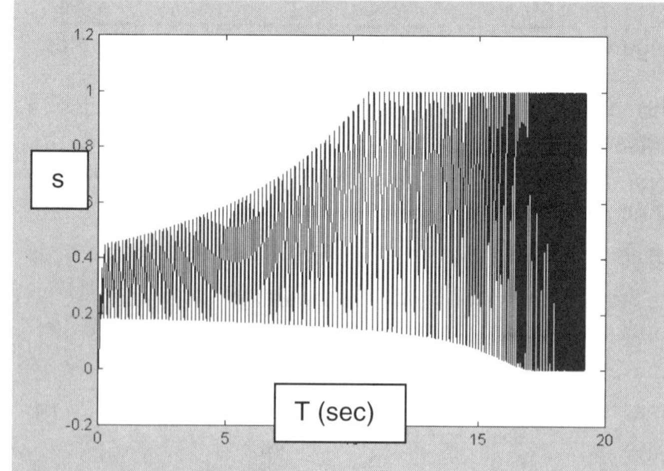

Figure 12. Slip vs. time for the traditional ABS controller while driving on ice.

The performance of the force controller is illustrated in Figures 15 and 16. Based on our ideal assumption, the force controller has access to the maximum tire-road coefficient of friction. Therefore, by feedforwarding this information and by the use of a simple controller, the force controller results in the maximum vehicle deceleration, therefore, minimum stopping distance. The difference in performance between the H_∞ and the force controller can not be easily extracted from the previous Figures. Therefore, in the next section, we propose a better way of comparing the results of the aforementioned controllers using the phase portrait, where the vehicle velocity is plotted vs. the distance traveled by the vehicle.

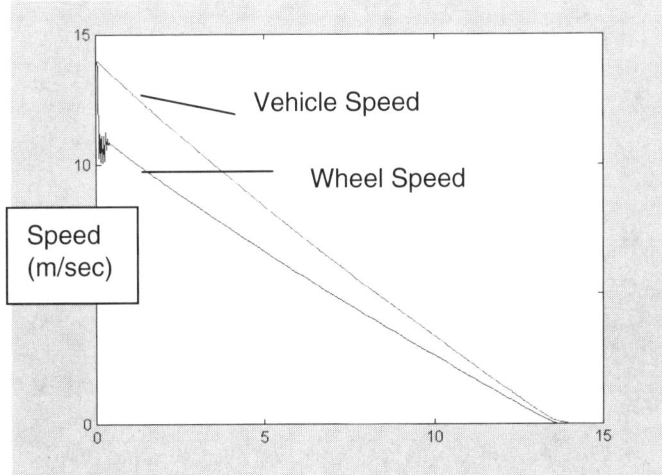

Figure 13. Vehicle/ wheel velocities vs. time for the H_∞ controller while driving on ice.

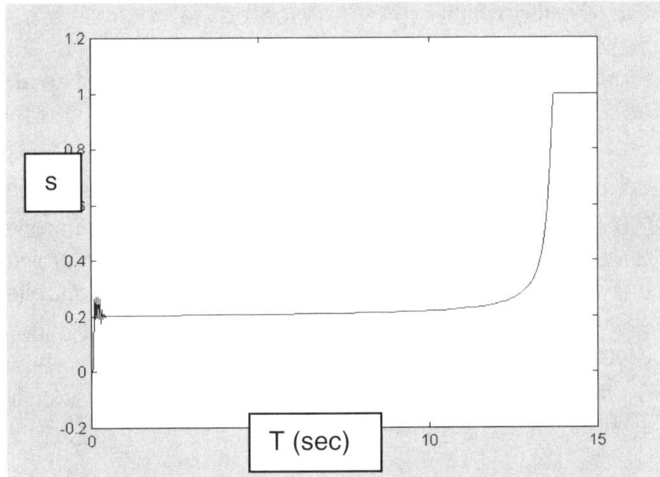

Figure 14. Slip vs. time for the H_∞ controller while driving on ice.

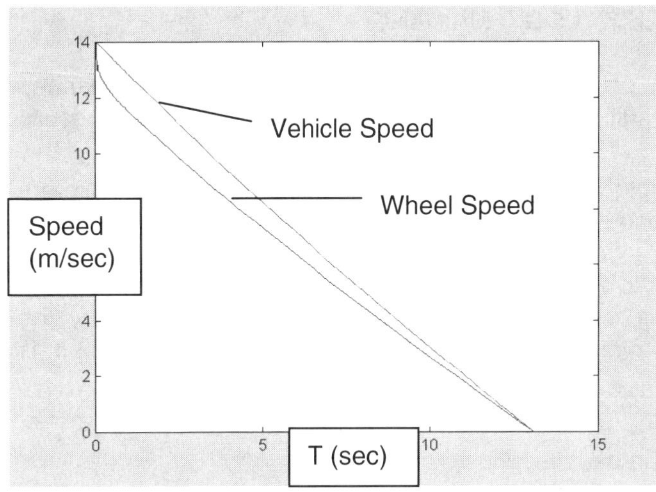

Figure 15. Vehicle/wheel velocities vs. time for the force controller while driving on ice.

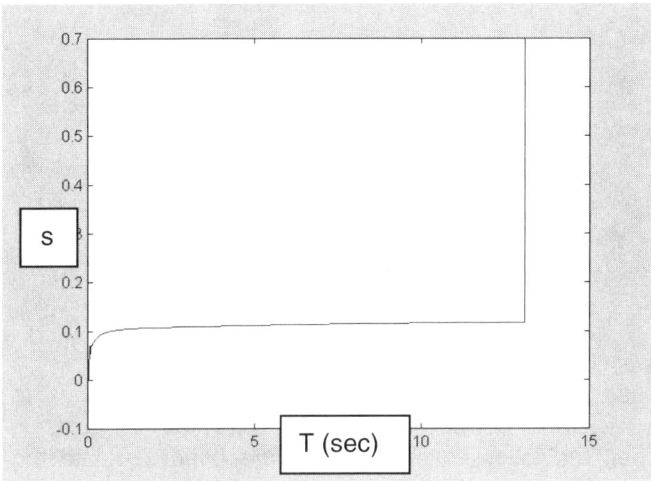

Figure 16. Slip vs. time for the force controller while driving on ice.

FUZZY LOGIC MAPPING

In the previous section, a constant value for commanded longitudinal slip was utilized in order to derive the results for the H_∞ controller. However, as illustrated, this method does not produce optimal results for different road conditions.

In this section, we utilize the rules of fuzzy logic to map the vehicle acceleration (deceleration) to the commanded slip for the H_∞ controller. Hence, it is illustrated that by mixing the fuzzy logic mapping with H_∞ controller, similar results to the ideal force controller can be produced.

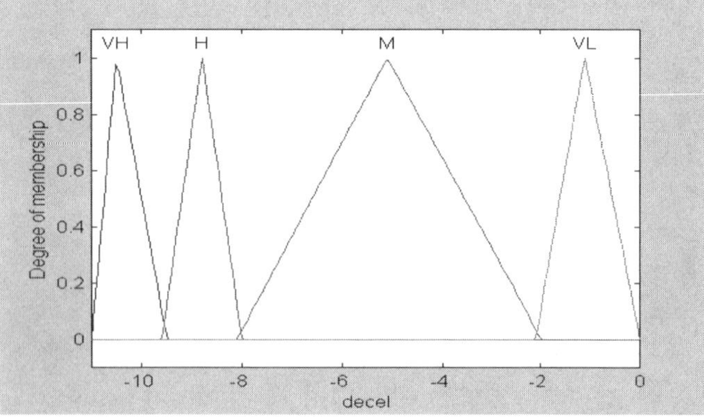

Figure 17. Fuzzification of the deceleration input.

Based on the fuzzy logic method, the first step is to fuzzify the input deceleration, as illustrated in Fiugre17. Then the following simple rules are used to relate the input deceleration to the commanded slip:

(1) If (deceleration is Very High) then (Slip is Very Low).

(2) If (deceleration is High) then (Slip is Very High).

(3) If (deceleration is Medium) then (Slip is High).

(4) If (deceleration is Low) then (Slip is Medium).

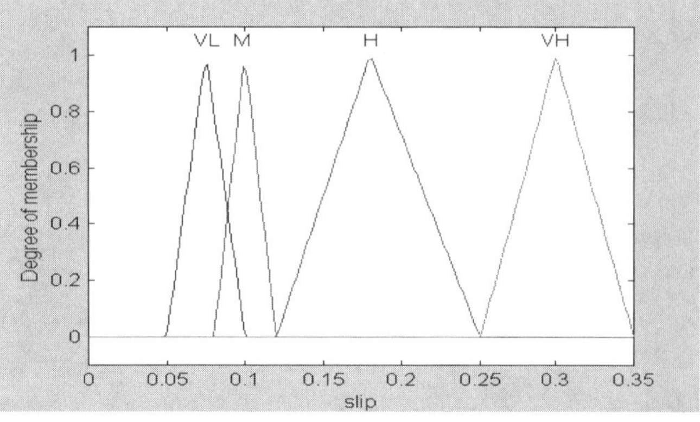

Figure 18. Defuzzification of the slip output.

Finally, we defuzzify the output in order to produce crisp values for the commanded slip, as illustrated in Figure 18. The commanded slip is used by the H_∞ controller in order to produce maximum vehicle deceleration under various road conditions.

As stated previously a better way of comparing the performance of the proposed controllers is the phase portrait, as illustrated in Figure 19.

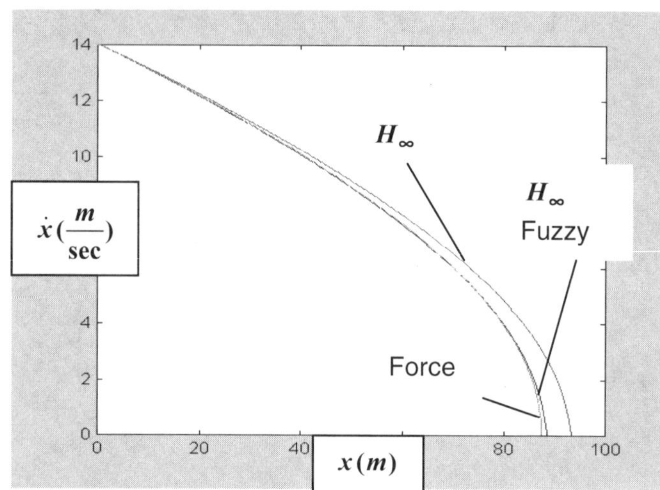

Figure 19. Phase portrait for driving on ice at 50 Km/hr of the proposed controllers.

Clearly, the relative performance of the controllers can be easily examined using this representation. In this Figure, $\dot{x}$ is the vehicle speed (previously defined as u), and x is the distance traveled by the vehicle during the braking. The braking stoppage distance gained over H_∞ controller by the mixed H_∞ and fuzzy controllers is near 7.0 meters, while the performance of the mixed controller is very near the performance of the ideal force controller. It is foreseen that better tuning of both the H_∞ controller and fuzzy maps will produce identical results to the ideal controller. It is also important to note that similar phase portrait for different road conditions and different vehicle speeds can be generated.

CONCLUSION

In this study, a nonlinear model of a one-wheel vehicle including a hydraulic brake system is reduced to a linear model with nonlinear perturbation. The machinery of the linear robust control theory is used to derive a controller for the vehicle ABS. The H_∞ controller is successful in compensating for different road surface conditions with a constant value of the commanded slip, however, maximization of the deceleration is not guaranteed when using this method. Furthermore, the fuzzy logic mapping methodology is used to vary the commanded slip based on the vehicle deceleration input. Combining these two strategies produces optimal results when compared to the ideal solution, where the tire-road friction coefficient is accessible.

REFERENCES

1. Assadian, F., "Modeling the Dynamics of an Automobile Brake System for the Purpose of Developing New Approaches for Control and Friction Estimation," Unpublished Internal PSA Technical Document.

2. Burckhardt, M., "ABS and ASR, Sicherheitsrelevante Radschlupf-Regelsysteme, Lecture Scriptum, University of Braunschweig, 1987.

3. Perronne, J. M., Renner, M., and Gissinger, G., "On Line Measurement of Braking Torque Using a Strain Sensor," SAE paper No. 940333.

4. Mauer, G. F., "A Fuzzy Logic Controller for an ABS Braking System," IEEE Transactions on Fuzzy Systems, Vol. 3, No. 4, 1995.

5. Lin, W. C., Dobner, D. J., and Fruechte R. D., "Design and analysis of an antilock brake control system with electric brake actuator," Int. J. of Vehicle Design, Vol. 14, No. 1, 1993.

6. Will, A. B., Hui, S., and Żak, S. H., "Sliding mode wheel slip controller for an antilock braking system," Int. J. of Vehicle Design, Vol. 19, No. 4, 1998.

7. Ünsal, C., and Kachroo, P., "Sliding Mode Measurement Feedback Control for Antilock Braking System," IEEE Transactions on Control Systems Technology, Vol. 7, No. 2, 1999.

8. Jacquet, C., "Modelisation et commande CRONE d'un system de freinage avec antiblocage (ABS) pour vehicule automobile," Diplome d'Etudes Approfondies, Universite Bordeaux I / ENSERB, 1996.

9. Margolis, D. L., "A one Wheel Vehicle Model for Dynamics and Control," Proceedings of 2nd IMACS International Multiconference, CESA 98, pp. 866-868, Nabeul-Hammamet, Tunisia, 1998.

10. Karnopp, D. C., Margolis, D. L., and Rosenberg, R. C., System Dynamics: A Unified Approach, John Wiley & Sons Inc. (1990).

11. Basar, T., and Bernhard, P., H^∞-Optimal Control and Related Minimax Design Problems: A Dynamic Game Approach, Birkhäuser (1995).

TRACTION CONTROL SYSTEMS (TCS, ASR)

The Effect of Tyre Dynamics on Wheel Slip Control Using Electromechanical Brakes

Okwuchi Emereole and Malcolm Good
The University of Melbourne

ABSTRACT

This paper investigates the response of the wheel slip to brake torque perturbations during a braking manoeuvre as the wheel slip state and the vehicle speed change. Its purpose is to better understand and model the wheel slip response characteristics during braking and to inform the development process of an antiskid wheel slip controller. The developed controller is implemented in the Simulink environment on a half-car vehicle model using a nonlinear electromechanical brake (EMB) model. A linearised analysis of the wheel slip response to brake torque perturbations about a nominal "trim" condition is first performed, based on a quarter-car vehicle model. It is shown that the inclusion of first-order tyre dynamics (characterised by the so-called "relaxation length") has a significant effect on the dynamics of the "plant" seen by the slip controller and these dynamics vary significantly with the trim value of the slip and the vehicle speed. The implications of these variations for controller design are discussed, taking into account the EMB actuator dynamics. A gain-scheduled PID wheel slip controller is developed and an analysis of the antilock performance of this controller is presented.

INTRODUCTION

An antilock braking system (ABS) aims to ensure that the directional stability of a vehicle is maintained at all times during a braking manoeuvre and especially during hard braking. Such systems have traditionally been hydraulic based. However, with increasing research interest in brake-by-wire and development of electromechanical brakes (EMB), suitable continuous antilock brake torque control techniques are being sought. These include fuzzy logic, sliding mode control, and gain-scheduling [1-4]. Some of the potential benefits and challenges facing EMB development are listed in [5, 6]. The dynamic relationship between the wheel slip and the coefficient of longitudinal friction on a given road surface is nonlinear and sensitive to tyre load and vehicle speed [7, 8]. In order to develop a model-based wheel slip controller, it is important to understand how the wheel slip dynamics are affected by parameters such as the vehicle speed and slip trim states, the brake

actuator dynamics and the effects of tyre relaxation lag. Tyre relaxation lag, if modeled, is normally represented in linear analyses as a fixed time lag. The tyre force lag effect is modeled here using a simple first order lag model characterised by the "relaxation length". In the following, a linearised analysis of the wheel slip response to brake torque perturbations about a nominal "trim" condition is developed, based on a quarter-car vehicle model. The effects of varying the trim values of the tyre slip and vehicle speed on the dynamics of the "plant" as seen by the slip controller are analysed with and without the first-order lag model representing the tyre lag dynamics. It is shown that the inclusion of the first-order dynamics has a significant effect on the plant dynamics, which vary significantly with the trim value of the slip, and the vehicle speed. The implications of these variations for controller design are discussed. A simple but robust gain-scheduled PID controller is developed and the simulated performance of the controller is demonstrated for straight-line braking on a number of road surfaces, including rapid transitions between high and low friction levels, using a half-car vehicle dynamics model (see Appendix). It is shown that scheduling the gains of the PID controller based on the vehicle speed results in robust closed-loop tracking of the target slip, even when this is in the "unstable" region of the mu-slip characteristic curve.

WHEEL SLIP CONTROLLER MODEL

A wheel slip control model is shown in Figure 1. The controller, actuator and plant dynamics blocks are indicated. The combined effect of the actuator and the plant dynamics are important in determining the overall dynamic response of the wheel slip, λ, to variations in the controlled variable, P_c (master cylinder pressure or clamp force for the EMB). The models of the actuator and the plant are investigated next.

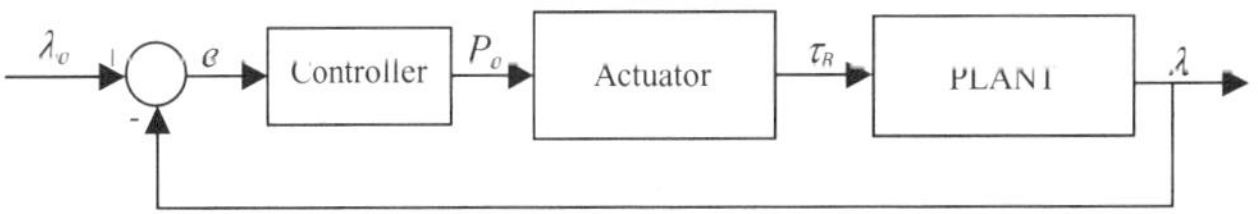

Figure 1: Wheel slip control model

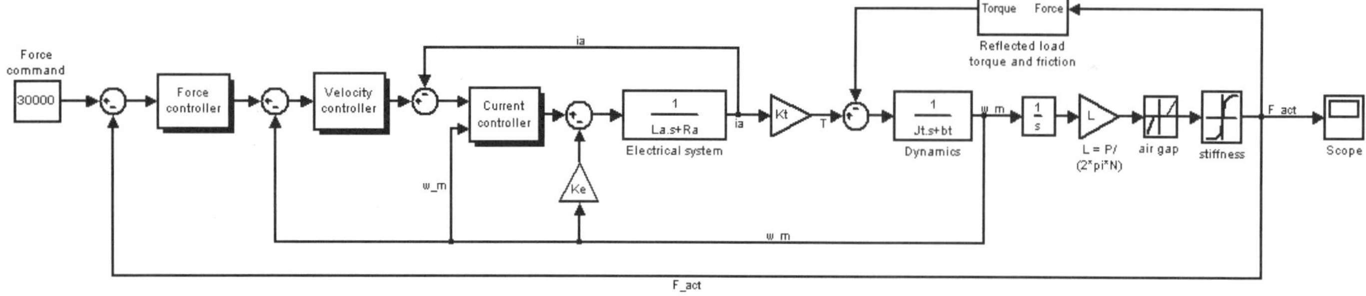

Figure 2: EMB simulation model showing control loops

ACTUATOR

The EMB model consists essentially of

1. An electric motor, which responds with an output torque to an input clamp force command. The force command is converted into equivalent velocity and current commands before generating the rotor torque. The brushless DC motor (BDCM) and the permanent-magnet synchronous motor (PMSM) are preferred servo-actuators mainly due to their ruggedness and high torque-to-weight ratios.
2. A gearing device to convert the servo's rotational motion to linear motion and to amplify and transmit clamping forces of an appropriate magnitude. Typically, an assembly that includes a planetary gear (PG) and a ball screw may be used for this purpose.
3. An appropriate contact between the linear actuator and the brake rotor through brake pads.
4. A control system (and associated power electronics) utilising sensed or estimated feedback from the contacting surfaces or rotating parts. The standard controller has a cascaded structure – current is controlled in the inner loop, then velocity and finally force control in the outermost loop.

Figure 2 shows the EMB model developed in Simulink. The current, velocity and force control loops implement a PI controller and integral anti-windup to handle excessive overshoots caused by saturation. The back emf "disturbance" to the current loop is rejected by feeding forward an estimated voltage based on the measured motor speed. This helps to linearise the system and improve its dynamic response. The EMB is assumed to have a peak supply (PWM) voltage of ± 42 V and a maximum safe operating current of ± 25 A. A bicycle model representation of the overall vehicle in Simulink is shown in Figure 24 (Appendix).

For control design purposes, the EMB actuator dynamics was represented using a 2nd order estimate of the EMB simulation model characteristics identified from modulations about the 30 kN clamp force level. The

30 kN level response was chosen since it provides a conservative (slower) estimate of the actuator dynamics, considering that the bandwidth of the EMB model response was observed to decrease with load. The model estimate is given as:

$$G_{EMB}(s) = \frac{135^2}{s^2 + 2(0.9)(135)s + 135^2} \qquad (1)$$

PLANT

By linearising the state along the tyre characteristic curve during a braking manoeuvre, a transfer function can be obtained to represent the responses of the wheel slip and the braking force to a brake torque input. From Figure 3, the slope of the force-slip curve at the point $(\lambda_0, \ F_0)$ is defined as the local tyre stiffness, $C_f = \partial F / \partial \lambda$. The influences of the trim or instantaneous vehicle speed and tyre slip (which determines C_f for a given tyre-road characteristic) on the slip and tyre force response characteristics can then be determined.

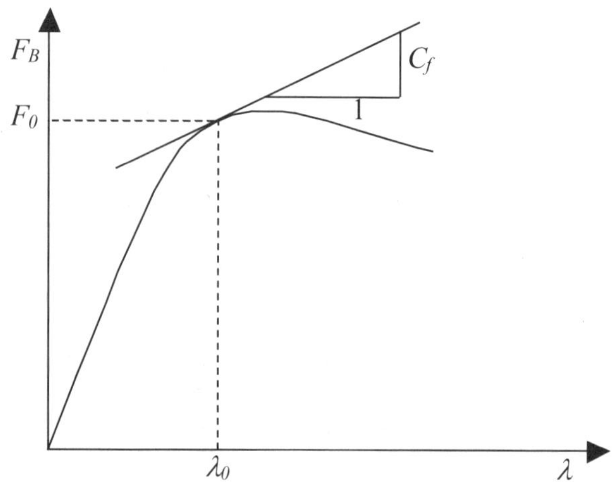

Figure 3: Linearising about a trim point on the mu-slip curve

Linearised plant equations

The dynamic equations of the vehicle during braking are developed next using a quarter-car model. Consider a vehicle of mass M and velocity V. A brake torque T_B is applied resulting in the following dynamic equations:

$$J\dot{\Omega} = r_l F_B - T_B \tag{2}$$

$$M\dot{V} = -F_B \tag{3}$$

$$F_B = f(\lambda) \tag{4}$$

where Ω is the wheel rotational speed and λ is the wheel slip. J, r_l and F_b are the wheel inertia, the tyre loaded radius and the brake force respectively. Consider the trim or average condition when $\lambda = \lambda_0$ (Figure 3). Linearising about this point (with initial vehicle velocity V_0 and local tyre slip stiffness C_f), we have for the vehicle and wheel dynamics:

$$\begin{aligned} \Omega &= \overline{\Omega} + \omega \\ \dot{\Omega} &= \dot{\overline{\Omega}} + \dot{\omega} \end{aligned} \tag{5}$$

and

$$\begin{aligned} V &= \overline{V} + \upsilon \\ \dot{V} &= \dot{\overline{V}} + \dot{\upsilon} \end{aligned} \tag{6}$$

where $\overline{\Omega}$ and $\overline{V}$ are average values at the trim point and ω and υ are small deviations from the trim. The trim force at $\lambda = \lambda_0$ is $\overline{F}_B = F_0$. For the wheel slip,

$$\begin{aligned} \lambda &= \lambda_0 + \tilde{\lambda} \\ \dot{\lambda} &= \dot{\tilde{\lambda}} \end{aligned} \tag{7}$$

where $\tilde{\lambda} = \Delta\lambda = \lambda - \lambda_0$ is a small slip change.

Without relaxation lag

Assuming no relaxation lag (i.e. lag in the development of tyre forces due to tyre relaxation concept), the average equations of motion are:

$$\overline{V}(t) = V_0 - \frac{F_0}{M}t \tag{8}$$

$$\lambda_0 = 1 - \frac{r_e \overline{\Omega}(t)}{\overline{V}(t)} \tag{9}$$

Hence,

$$\overline{\Omega}(t) = \frac{\overline{V}(t)(1-\lambda_0)}{r_e} = \frac{1-\lambda_0}{r_e}\left(V_0 - \frac{F_0}{M}t\right) \tag{10}$$

In the above equations, r_e is the effective tyre rolling radius. Differentiating (8) and (9), we obtain

$$\dot{\overline{V}} = -\frac{F_0}{M} \tag{11}$$

$$\dot{\overline{\Omega}} = -\frac{F_0}{r_e M}(1-\lambda_0) \tag{12}$$

Using the simplification that $r_l = r_e = r$, the trim brake torque can now be obtained as:

$$\overline{T}_B = r\overline{F}_B - J\dot{\overline{\Omega}} = rF_0\left[1 + \frac{J}{Mr^2}(1-\lambda_0)\right] \tag{13}$$

We now consider the effect of a small torque perturbation, τ_B. This will result in a small tyre reaction force, f_B. For the overall wheel angular dynamics, we have:

$$J\dot{\Omega} = J\dot{\overline{\Omega}} + J\dot{\omega} = r(F_0 + f_B) - (\overline{T}_B + \tau_B) \tag{14}$$

and

$$J\dot{\overline{\Omega}} = rF_0 - \overline{T}_B \tag{15}$$

Hence

$$J\dot{\omega} = rf_B - \tau_B \tag{16}$$

where

$$f_B = C_f \tilde{\lambda} \tag{17}$$

Similarly, we have for the vehicle longitudinal dynamics:

$$M\dot{\upsilon} = -f_B \tag{18}$$

Define the instantaneous wheel slip as:

$$\lambda = \lambda_0 + \tilde{\lambda} = \lambda_0 + \left.\frac{\partial \lambda}{\partial \Omega}\right|_0 \omega + \left.\frac{\partial \lambda}{\partial V}\right|_0 \upsilon \tag{19}$$

Differentiating the average slip equation (9), we have:

$$\frac{\partial \lambda}{\partial \Omega} = -\frac{r}{\overline{V}(t)} \quad \text{and} \quad \frac{\partial \lambda}{\partial V} = \frac{r\overline{\Omega}(t)}{\overline{V}^2(t)} \tag{20}$$

Substituting (20) into (19),

$$\tilde{\lambda} = -\frac{r}{\overline{V}(t)}\omega + \frac{r\overline{\Omega}(t)}{\overline{V}^2(t)}\upsilon \tag{21}$$

Differentiating and rearranging, we get:

$$\dot{\upsilon} = \frac{\overline{V}^2(t)}{r\overline{\Omega}(t)}\dot{\tilde{\lambda}} + \frac{\overline{V}(t)}{\overline{\Omega}(t)}\dot{\omega} = -\frac{f_B}{M} \tag{22}$$

Substituting $\dot{\upsilon}$ and $\dot{\omega}$ from (16) to (18) into (22), and rearranging, we obtain

$$\dot{\tilde{\lambda}} = -C_f\left[\frac{r^2}{\overline{V}J} + \frac{r\overline{\Omega}}{M\overline{V}^2}\right]\tilde{\lambda} + \frac{\tau_B r}{\overline{V}J} \tag{23}$$

Taking Laplace transforms and using the simplification in (9), the transfer function can now be determined:

$$\frac{\tilde{\lambda}(s)}{\tau_B(s)} = \frac{rM}{\overline{V}MJs + C_f[J(1-\lambda_0) + Mr^2]} \tag{24}$$

The transfer function in Equation 24 has a pole at $\frac{-C_f}{\overline{V}MJ}[J(1-\lambda_0) + Mr^2]$. The magnitude of this pole increases as C_f increases or as $\overline{V}$ reduces, leading to a faster slip response. Conversely, a decrease in C_f or an increase in $\overline{V}$ will slow the slip response. This can be confirmed using a root locus analysis in the stable region (C_f positive) with respect to $\overline{V}$ and λ_0 (which determines C_f) respectively. In the first case, as $\overline{V}$ is increased, the root approaches the zero at the origin from the negative real axis, indicating a slowing response. However, when λ_0 is varied (increased), the root begins at the origin and approaches negative infinity. In the unstable region, the roots in both cases are in the right half plane due to the change in sign of the slope C_f.

The above results indicate that in the response of the plant to brake torque control inputs becomes slower as the peak of the curve is approached, due to the decreasing value of C_f. this suggests that it becomes increasingly difficult to control the wheel slip as the optimum slip region is approached, and increasing controlling gains will be required. As the average vehicle speed decreases, the wheel slip response becomes faster. These characteristics are further analysed in the next section with the inclusion of a first order lag in to represent the tyre relaxation characteristics.

<u>With relaxation lag</u>

With the inclusion of the lag in tyre force development, the force f'_B resulting from a torque perturbation τ_B is not developed instantaneously but builds up to f_B according to a first order lag model:

$$\frac{\sigma}{\overline{V}}\dot{f}'_B + f'_B = f_B \tag{25}$$

with Laplace transform:

$$f'_B(s) = \frac{f_B(s)}{\left(\dfrac{\sigma}{\overline{V}}s + 1\right)} \tag{26}$$

where σ is the relaxation length and $\overline{V}$ is the average vehicle velocity. The wheel dynamics equations is then given by

$$J\dot{\omega} = rf'_B - \tau_B \tag{27}$$

For the wheel slip dynamics, we replace f_B with f'_B in (22) and substitute (27) to obtain:

$$\tilde{\dot{\lambda}} = -\left[\frac{r^2}{\overline{V}J} + \frac{r\overline{\Omega}}{M\overline{V}^2}\right]f'_B + \frac{\tau_B r}{\overline{V}J} \qquad (28)$$

Taking Laplace transforms and substituting (27) and (17) and rearranging, we obtain

$$\frac{\tilde{\lambda}(s)}{\tau_B(s)} = \frac{rM\left(\dfrac{\sigma}{\overline{V}}s + 1\right)}{\sigma JMs^2 + \overline{V}JMs + C_f[Mr^2 + J(1-\lambda_0)]} \qquad (29)$$

The numerator of (29) contains a relatively fast speed-dependent zero at $-\overline{V}/\sigma$, which approaches the origin as the vehicle velocity decreases. The effect is to increase the phase lead in the transfer function as the speed decreases, thus increasing the potential for oscillatory behaviour. This could present some challenges to slip controller design. The steady state value of this transfer function is the same as for the transfer function without relaxation lag (24). The poles of (29) are influenced in a complex manner by $\overline{V}$ and λ_0. This will be further analysed next. The characteristic equation of the transfer function in (29) is:

$$\sigma JMs^2 + \overline{V}JMs + C_f[Mr^2 + J(1-\lambda_0)] = 0 \qquad (30)$$

To determine the effect of varying $\overline{V}$ on the roots of (30), it is rearranged in the following form for root locus analysis:

$$1 + \overline{V}\frac{JMs}{\sigma JMs^2 + C_f[Mr^2 + J(1-\lambda_0)]} = 0 \qquad (31)$$

To determine the effect of varying λ_0 on the roots of (30), it is rearranged as follows for root locus analysis:

$$1 + K(\lambda_0)\frac{1}{\overline{V}JMs\left(\dfrac{\sigma}{\overline{V}}s + 1\right)} = 0 \qquad (32)$$

where

$$K(\lambda_0) = C_f[Mr^2 + J(1-\lambda_0)] \qquad (33)$$

In Figure 4, the root loci for constant velocities of 10 to 110 km/h (in steps of 10 km/h) and for constant slip trims of 0 to 12% (stable region), in steps of 2%, are shown for a typical high friction surface. We can deduce from the figure that if a particular wheel slip is to be maintained while the vehicle average speed is decreased from a high value, the slip response will initially be slow and non-oscillatory. At this stage, there is a fast and a slow root, both along the real axis, the slow root dominating the slip response characteristics. The roots approach each other as the average vehicle speed decreases. They eventually meet up and breakaway from the real axis. From this point onwards, the response is oscillatory. As the vehicle speed decreases further, the damping of the system reduces from 1 to zero, resulting in an increasingly oscillatory response at the natural frequency defined by the breakaway point. Due to the initial slow response, a large gain will be required to drive the system to a desired state in a given time. As the vehicle speed decreases and the roots approach each other, the overall response becomes faster and the required control gains will reduce. As the trim slip is increased, the breakaway point approaches the origin. That is, the slip response close to the peak slip (say, at 10% slip) will be much slower than the response at say 2% slip. In other words, as the peak slip/friction state is approached, it becomes increasingly difficult to drive the system to that state. It also suggests that the minimum required or necessary control gains will increase as the slip approaches the peak value. If a fixed gain, which is more practicable, were however selected for the entire slip region at a particular speed, its value should be the value required at the design slip, which will ideally be close to the peak slip. Figure 4 describes the linearised characteristics of the plant over the stated regions and significantly informs the design of a gain-scheduled controller.

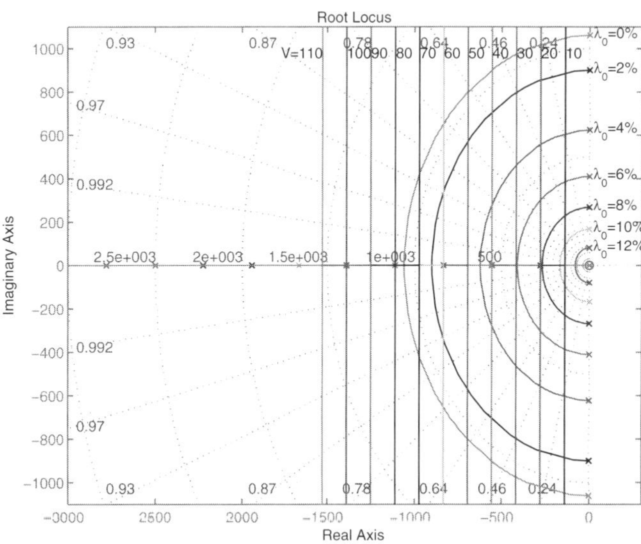

Figure 4: Root locus of plant transfer function for constant trim slip and constant speed in the stable region

In the unstable region of the λ - μ curve, there are right hand plane poles. The root loci for this region are shown in Figure 5 for constants slips of 14 to 20%. We see that the response in this region is non-oscillatory and initially quite slow. Hence, sufficiently large gains would be required both to stabilise the system and to improve its dynamic response whilst in the unstable region.

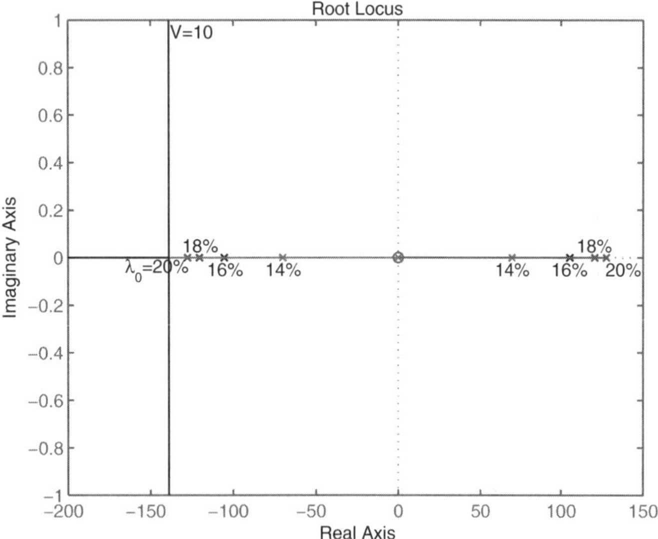

Figure 5: Root locus of plant transfer function for constant trim slip and constant speed in the unstable region

CONTROLLER DESIGN AND IMPLEMENTATION

The actuator and plant model transfer functions were combined and an appropriate controller was investigated using the MATLAB SISO Design Tool. The following parameters for the quarter-car vehicle model were used in this investigation: σ = 0.01, r = 0.3 m, J = 1 kg.m^2 and M = 435 kg. A PID (Proportional plus Integral plus Derivative) controller was found to be suitable at the different states and the controller gains were scheduled based on the vehicle velocity. For the stable region of the tyre friction curve, the controller was designed around the 10% trim slip point. The PID controller has a fixed pole at the origin and two zeros that can be adjusted for the control. By fixing one of these zeros, only the second zero and the gain needed to be adjusted to obtain satisfactory control. The controller was designed to achieve a closed loop bandwidth of 100 rad/s.

Figure 6 plots the parameters obtained for the PID controller. It can be seen that the P- and D- control gains increase with speed. For the integral gain, a fixed value was found to be sufficient in practice. The integral control contribution can be seen to be the most significant of the three. An anti-windup setup is used in the simulation model to prevent excessive integral errors. Another practical measure implemented was to limit the error signal into the controller to ±10% slip. This

helped to avoid large integral errors that occurred if the wheel locked up. The maximum EMB clamp force is limited in the simulation model to about 35 kN, which limits the maximum braking torque that can be developed through an appropriate clamp force to braking torque ratio.

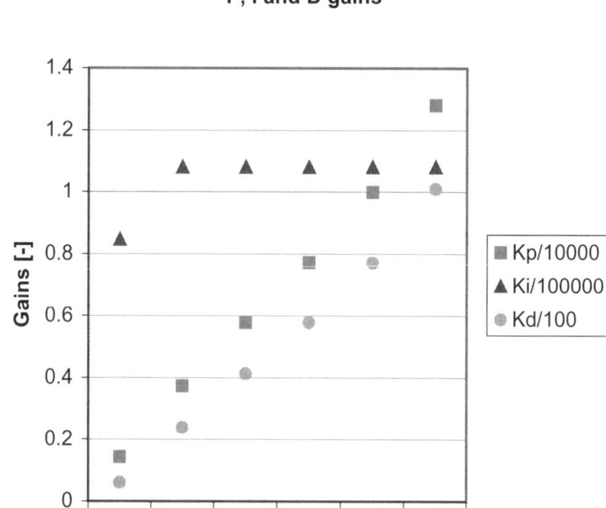

Figure 6: Gains for the PID ABS Controller, designed around a 10% trim slip

Evaluation of controller performance

The performance of the controller was initially investigated using a quarter-car vehicle simulation model. Once satisfactory control was achieved, the controller was then used in the half-car or bicycle vehicle model developed in [9]. The wheel slip controller performance for the half-car model is demonstrated in Figures 7 to 14 for a target slip of 10% and in Figures 15 to 22 for a target slip in the unstable region of 20%. Figures 7 and 8 show the control performance during a braking manoeuvre on a high friction surface. In Figure 7, the vehicle and wheel speeds, and the front and rear wheel slip are shown. An efficiency measure termed the Antilock Performance Index (API) is also shown in the latter two plots. The API is defined in [9] as:

$$API(t) = \frac{F_x(t)}{F_{x,\max}(t)} = \frac{\mu(t)}{\mu_p(t)} \tag{34}$$

from which the mean API value can be determined as:

$$\overline{API} = \frac{1}{\Delta t} \int_{t_{ABS}}^{t_1} \frac{\mu}{\mu_p} dt \tag{35}$$

where t_{ABS} is the ABS actuation time and t_1 is the ABS control stop time ($t_1 = t_{ABS} + \Delta t$). Figures 9 and 10 show the braking performance on a low friction surface. In Figures 11 and 12, there is a change in the road surface friction level from high to low friction after 2 seconds. In Figures 13 and 14, the surface change is from a low to a high friction surface and occurs after 3 seconds. The simulations are stopped at 5 km/h.

It can be seen that the wheel slip control maintains the target slip quite well especially in the stable region of the friction curve. After a surface change, the controller also recovers quickly in order to maintain the desired wheel slip. In the unstable region, the control response oscillates around the commanded wheel slip for the higher friction surfaces. The control performance on low friction surfaces in the unstable region is quite good. In Figure 23, the response of the system to a varying high-low wheel slip command sequence is demonstrated for a command slip targets ranging between 10% and 20%. The initial velocity if the vehicle is 100 km/h and there is a surface change after 2 sec from a high to a low friction surface. It can be seen that apart from the initial oscillatory response, the controller is able to perform reasonably well in the unstable region. There are no control difficulties in the stable region. This demonstrates that the EMB controller is quite robust.

CONCLUSION

It has been demonstrated that a systematic, linearised analysis can be used to develop a reliable wheel slip controller for an EMB. For the analyses, the dynamics of the vehicle and the electromechanical brake actuator, and the tyre relaxation characteristics were included in the plant models. A gain-scheduled PID wheel slip controller was found to be suitable, and it was demonstrated that the controller is quite robust and able to follow demanded wheel slips in the stable and unstable regions of the mu-slip curve quite well. It is important that the controller is supplied with appropriate wheel slip target commands. This is usually the function of a higher order controller such as a vehicle dynamics control system.

REFERENCES

[1]   M. Akey, 1995, "Development of fuzzy logic ABS control for commercial trucks," *SAE Technical Paper*, 952673.

[2]   G.F. Mauer, et al., 1994, "Fuzzy logic continuous and quantizing control of an ABS braking system," *SAE Technical Paper*, 940830.

[3]   K.R. Buckholtz, 2002, "Reference input wheel slip tracking using sliding mode control," *SAE Technical Paper*, 2002-01-0301.

[4]   S. Solyom and A. Rantzer, 2003, "ABS control - a design model and control structure," in *Nonlinear and hybrid systems in automotive control*, R. Johansson and A. Rantzer, Eds. Warrendale, PA.: SAE International.

[5]   R. Schwarz, et al., 1998, "Modeling and control of an electromechanical disk brake," *SAE Technical Paper*, 980600.

[6]   R.T. Bannatyne, 1998, "Advances and challenges in electronic braking control technology," *SAE Technical Paper*, 982244.

[7]   T.D. Gillespie, 1992, *Fundamentals of Vehicle Dynamics*. Warrendale, PA: Society of Automotive Engineers (SAE).

[8]   J. N. H. Sledge and K.M. Marshek, 1997, "Vehicle critical speed formula - Values for the coefficient of friction - a review," *SAE Technical Paper*, 971148.

[9]   O.C. Emereole, 2004, "Antilock Performance Comparison Between Hydraulic and Electromechanical Brake Systems," M.Eng.Sc. Thesis, Dept. of Mechanical and Manufacturing Engineering, University of Melbourne.

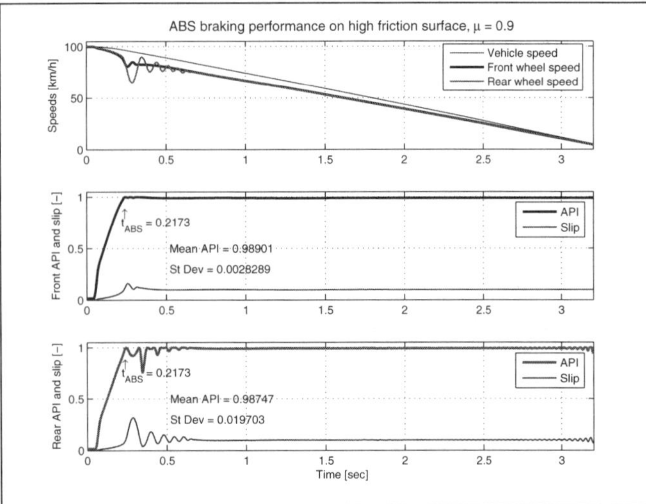

Figure 7: EMB controller performance on high friction surface, $\mu = 0.9$ (10% target slip)

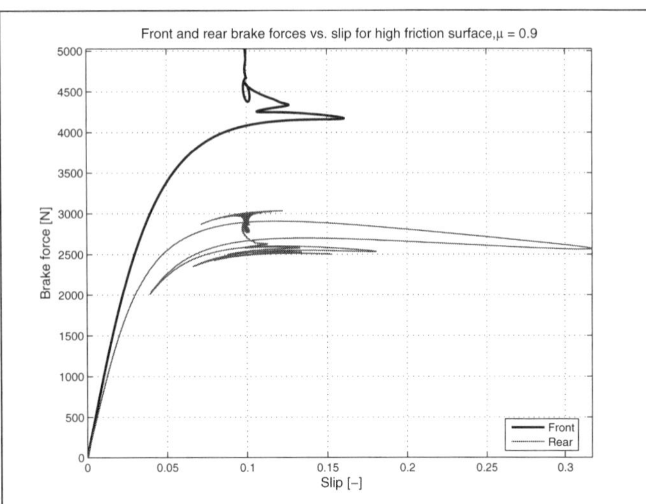

Figure 8: Force vs. slip plot for braking on high friction surface (10% target slip)

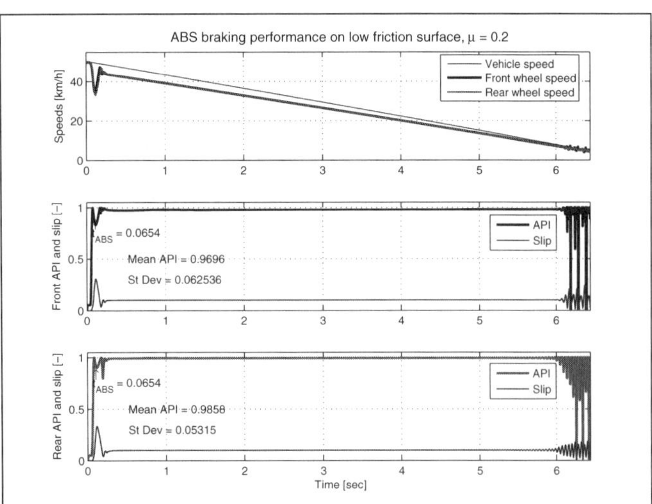

Figure 9: EMB controller performance on low friction surface, $\mu = 0.2$ (10% target slip)

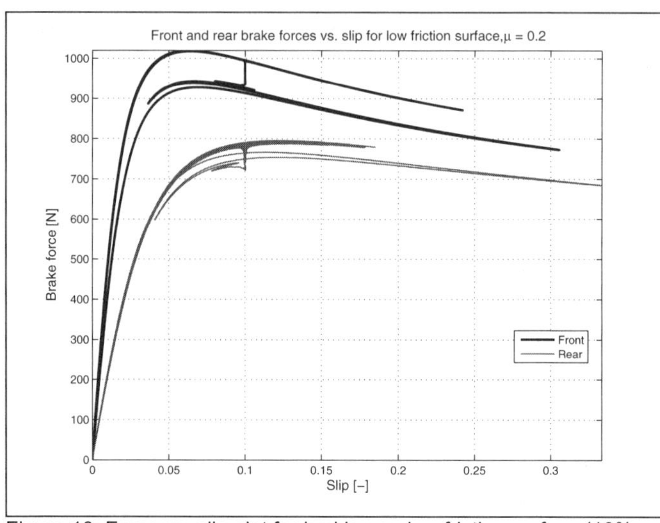

Figure 10: Force vs. slip plot for braking on low friction surface (10% target slip)

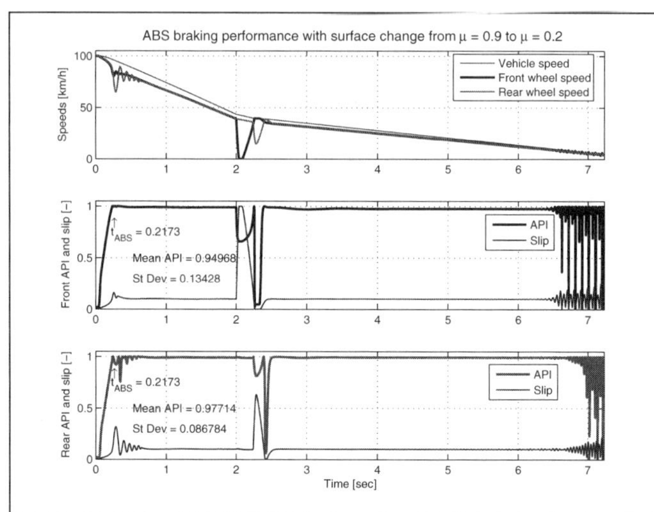

Figure 11: EMB controller performance with surface change from $\mu = 0.9$ to $\mu = 0.2$ (10% target slip)

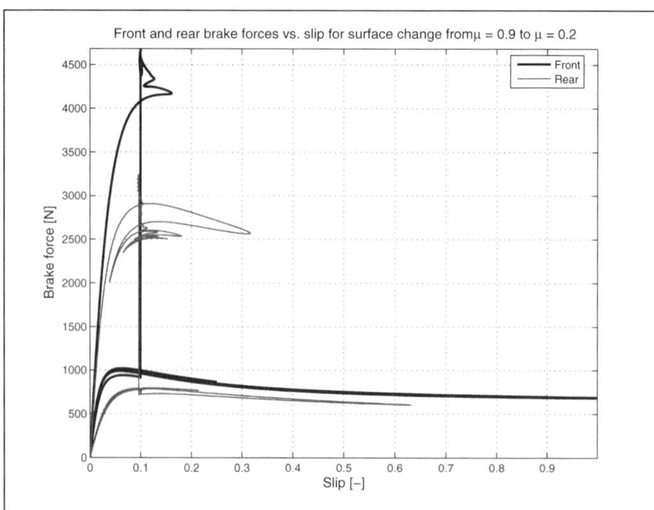

Figure 12: Force vs. slip plot for braking with high to low friction surface change (10% target slip)

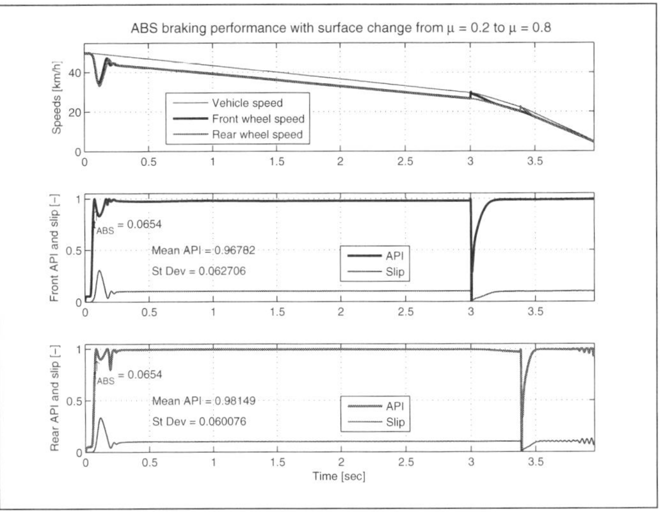

Figure 13: EMB controller performance with surface change from μ = 0.2 to μ = 0.8 (10% target slip)

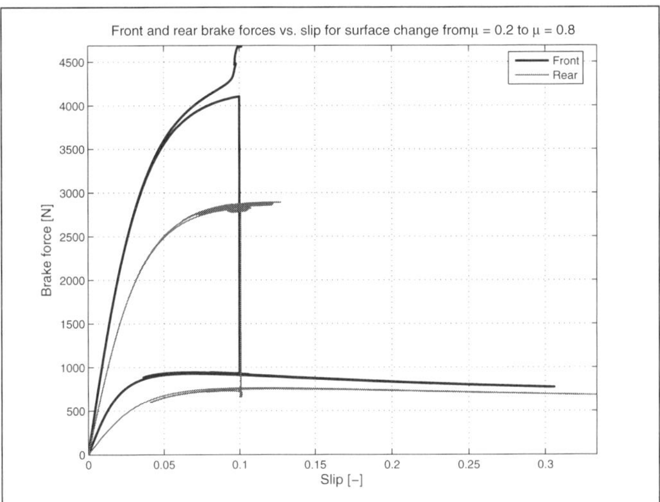

Figure 14: Force vs. slip plot for braking with low to high friction surface change (10% target slip)

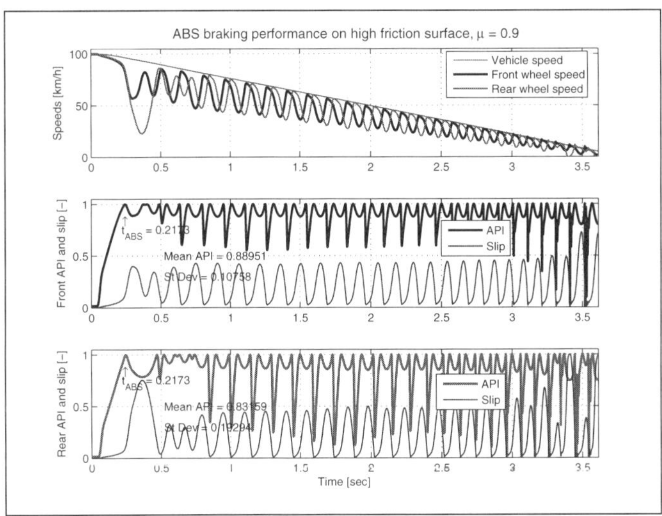

Figure 15: EMB controller performance on high friction surface, μ = 0.9 (20% target slip)

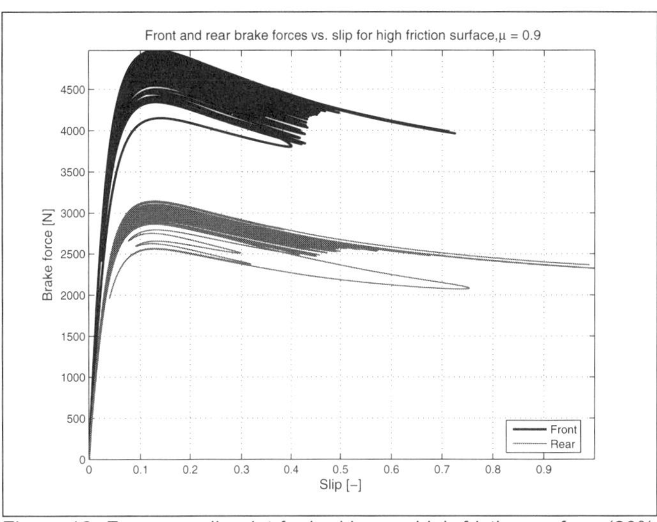

Figure 16: Force vs. slip plot for braking on high friction surface (20% target slip)

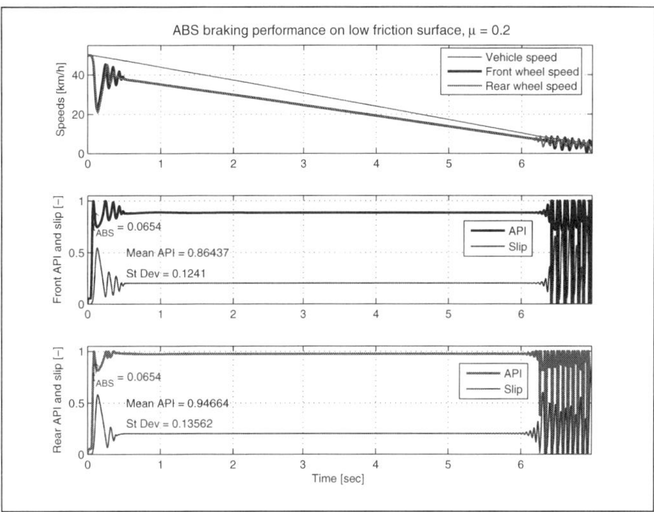

Figure 17: EMB controller performance on low friction surface, μ = 0.2 (20% target slip)

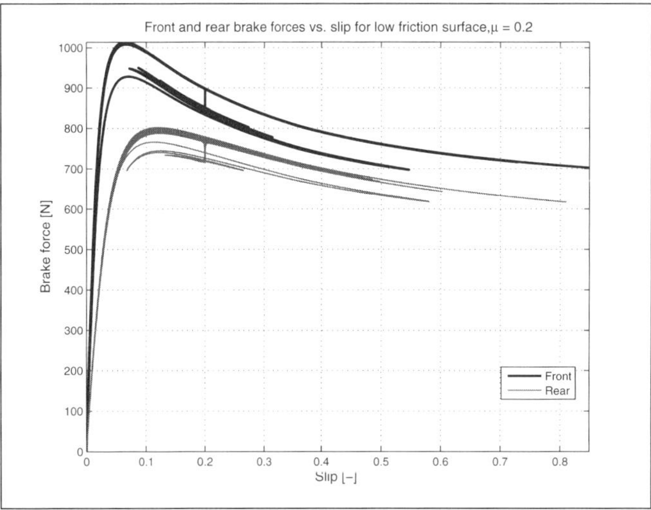

Figure 18: Force vs. slip plot for braking on low friction surface (20% target slip)

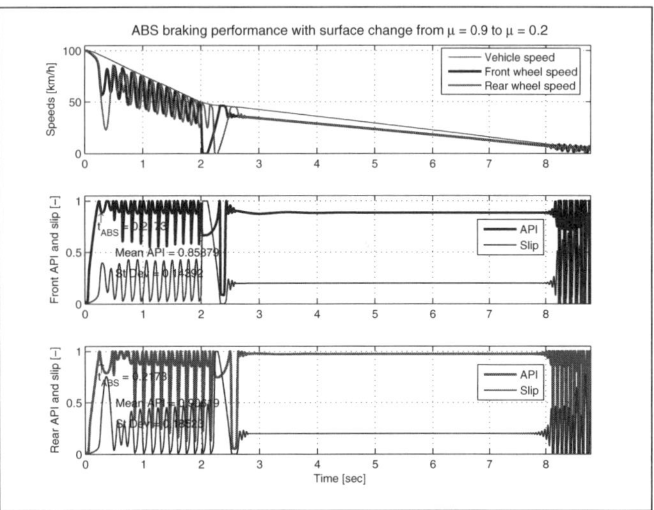

Figure 19: EMB controller performance with surface change from μ = 0.9 to μ = 0.2 (20% target slip)

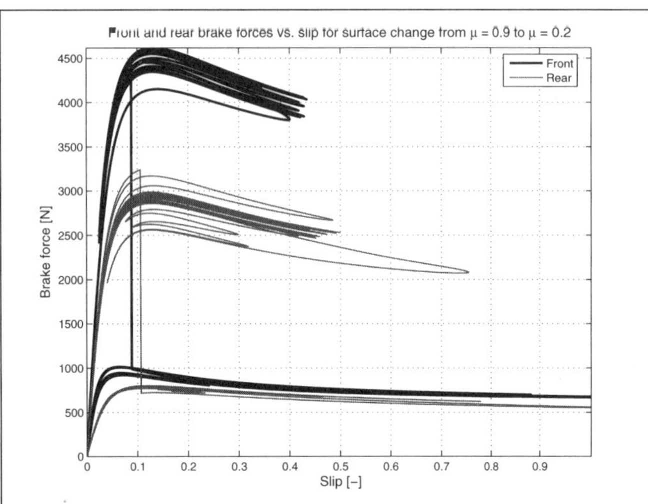

Figure 20: Force vs. slip plot for braking with high to low friction surface change (20% target slip)

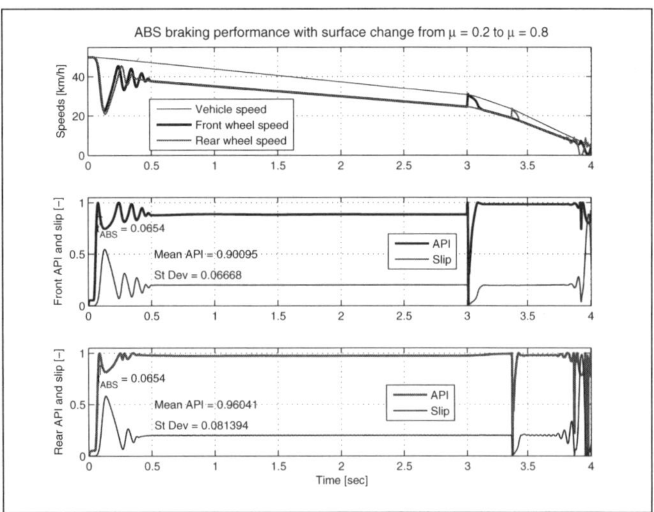

Figure 21: EMB controller performance with surface change from μ = 0.2 to μ = 0.8 (20% target slip)

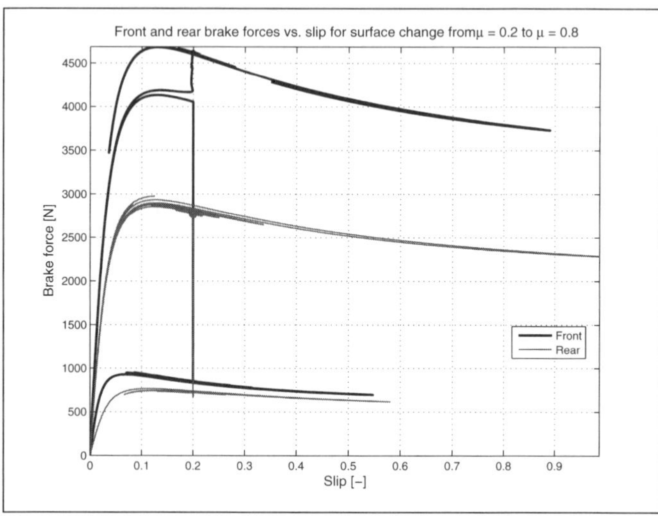

Figure 22: Force vs. slip plot for braking with low to high friction surface change (20% target slip)

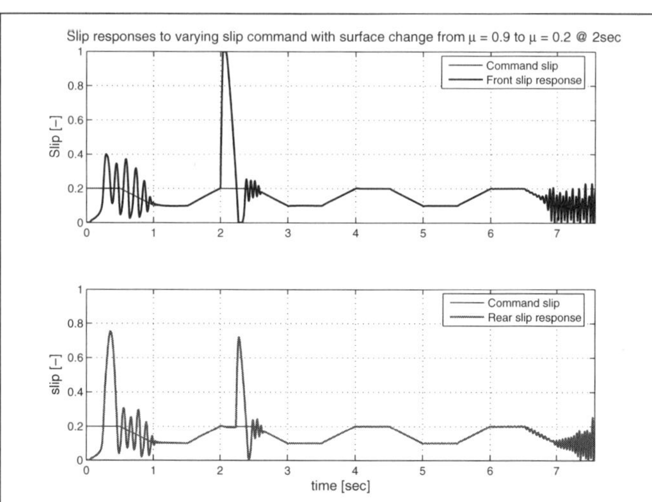

Figure 23: Response of the EMB controller to varying wheel slip command, with surface change from high to low friction surface after 2 sec

268

The Simulink half-car (bicycle) model used in the above analysis is shown in Figure 24 below.

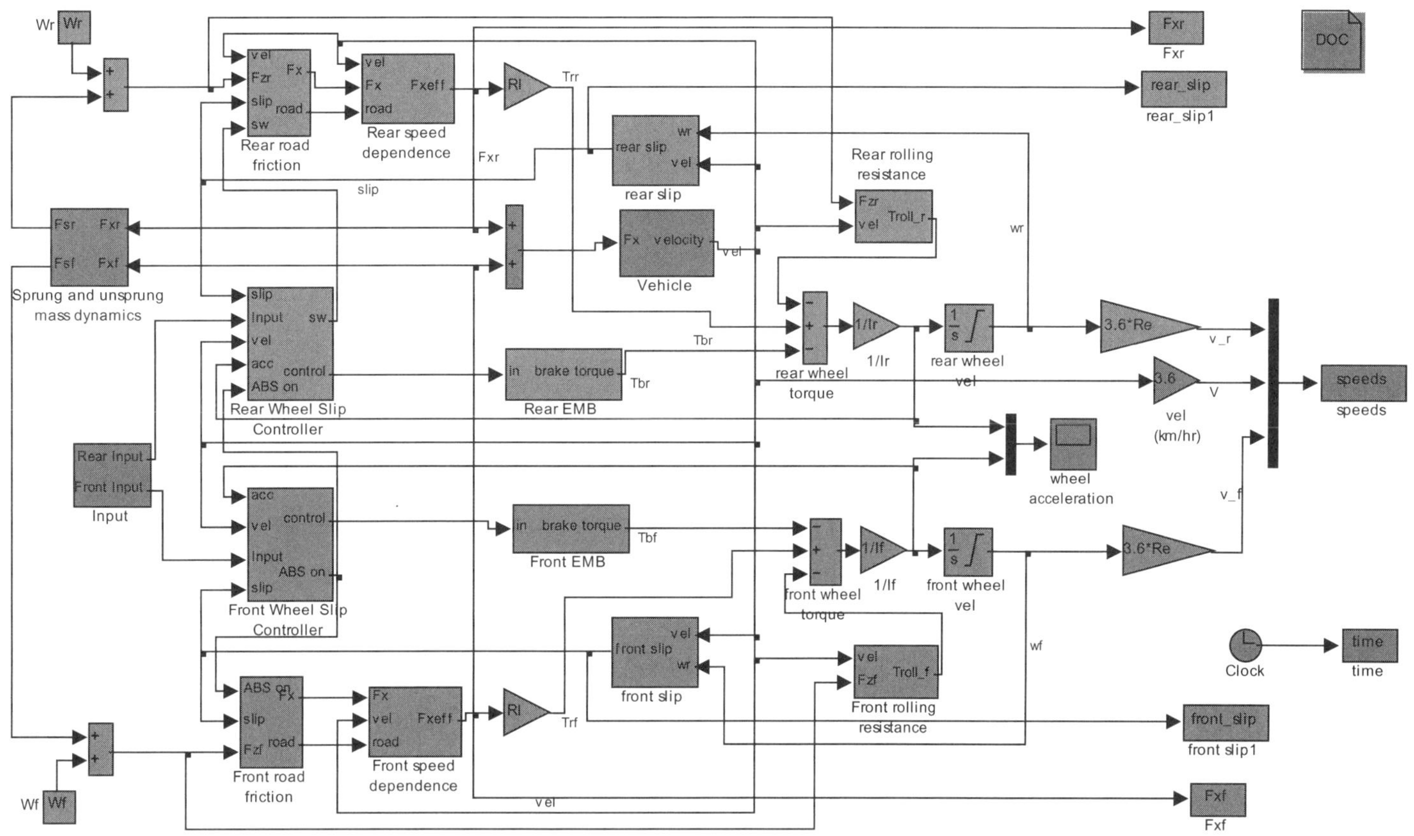

Figure 24: Simulink bicycle model representation of the vehicle

Hardware-in-the-loop Simulation for Traction Control and the Debugs of its Electric Control Unit

Huiyi Wang

Department of Automotive Engineering, Tsinghua University, China

ABSTRACT

A Traction Control System (TCS) is developed for the Anti-lock Braking system (ABS). A Hardware-in-the-loop simulation (HILS) method is used to help debug the control strategy of a TCS for a passenger vehicle.

The TCS HILS platform is extended from the ABS test rig by adding a throttle controller and two pressure control solenoid valves. First is established vehicle simulation model that includes a half vehicle dynamic submodel, an engine submodel, a manual transmission system submodel and the tyre submodel. Secondly, the control model of the TCS based on the logical threshold method is included in the same digital simulation software. The digital simulations of the passenger vehicle shows that the control model with different control thresholds as input is correct and it can be run in a real TCS's structure parameters and real electric control unit (ECU).

An ECU of TCS with an Infineon's 16-bit microcontroller C167CS as the main central processor unit is designed. The control parameters of the TCS's ECU are debugged in the HILS platform to fit the real passenger vehicle based on an experimental program.

INTRODUCTION

Traction Control System (TCS), is also called Acceleration Slip Regulation (ASR), is a functional extension of the Anti-lock Braking System(ABS). Its goal is to gain the maximum longitudinal force and lateral force during the accelerating period of a car. The relationship of adhesion coefficient and the slip rate of the TCS are similar to that of the ABS; the differences are one laying in the second quadrant and the other laying in the 1st quadrant once the relationships are shown in the chart from figure 1.

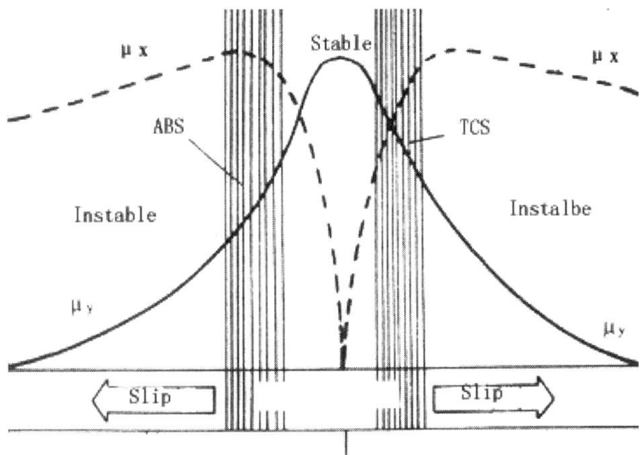

Figure 1 Relationships of the adhesion coefficient and the slip

The optimal values of the longitudinal and lateral coefficients are when the wheel slip rates are from 15% to 30%. In this case, the car has the ability of driving stability and the accelerative ability.

MODELLING of the VEHICLE for TCS

This paper deals with the case of the traction control when the car is running in a straightway. The 1/2 mathematic model of the vehicle is enough. The 1/2 car model is shown in Figure 2. There are 7 degrees of freedom: forward movement freedom X, vertical movement Z, pitch θ; fore wheel vertical movement Z11, wheel rotation ω 11, rear wheel vertical movement Z21 and wheel rotation ω 21.

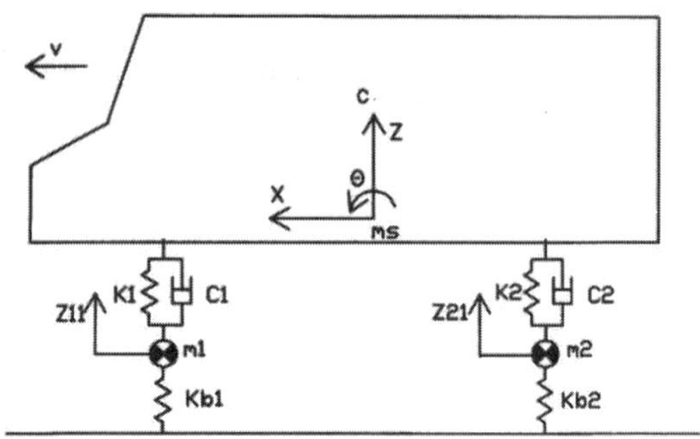

Figure 2 Half vehicle model

The dynamics of the half vehicle can be described as:

$$\frac{1}{2}m\ddot{x} + \frac{1}{2}m_s D_0 \ddot{\theta} = -F_{x11} - F_{x21} \tag{1}$$

$$\frac{1}{2}m_s \ddot{Z} + F_{s11} + F_{s21} = 0 \tag{2}$$

$$\frac{1}{2}J_y \ddot{\theta} + \frac{1}{2}m_s D_0 \ddot{x} - F_{s11}a + F_{s21}b = 0 \tag{3}$$

$$m_1 \ddot{Z}_{11} - F_{s11} + K_{b1}Z_{11} = 0 \tag{4}$$

$$m_2 \ddot{Z}_{21} - F_{s21} + K_{b2}Z_{21} = 0 \tag{5}$$

$$J_1 \ddot{\omega}_{11} = F_{x11}R_1 - P_{11} \tag{6}$$

$$J_2 \ddot{\omega}_{21} = F_{x21}R_2 - P_{21} \tag{7}$$

$$F_{x11} = \mu_{x11}(\frac{1}{2}m_s g \, b/_l + m_1 g - K_{b1}Z_{11}) \tag{8}$$

$$F_{x21} = \mu_{x21}(\frac{1}{2}m_s g \, a/_l + m_2 g - K_{b2}Z_{21}) \tag{9}$$

$$F_{s11} = K_1(Z - Z_{11} - a\theta) + C_1(\dot{Z} - \dot{Z}_{11} - a\dot{\theta}) \tag{10}$$

$$F_{s21} = K_2(Z - Z_{21} + b\theta) + C_2(\dot{Z} - \dot{Z}_{21} + b\dot{\theta}) \tag{11}$$

An engine model with two parameters was used in this paper. The two parameters are engine throttle and the rotation speed of the crank. The model can be described as:

$$T_i = C_e \cdot \zeta_{TH} \cdot \zeta_{AFI} \tag{12}$$

where,

$$C_e = a_1 \cdot \omega_e^2 + a_2 \cdot \omega_e + a_3 \tag{13}$$

$$J_e \cdot \dot{\omega}_e(t) = T_i(t) - T_f(t) - T_p(t) \tag{14}$$

where,

$$T_f = 0.0938\omega_e + 12.7 \tag{15}$$

Hardware-in-the-loop SIMULATION SYSTEM for TCS

Based on the original test system for ABS, the Hardware-in-the-loop Simulation (HILS) System was done by adding the engine model, the transmission model, and the actuator of the engine throttle.

The schematics of the HILS is shown in Figure 3.

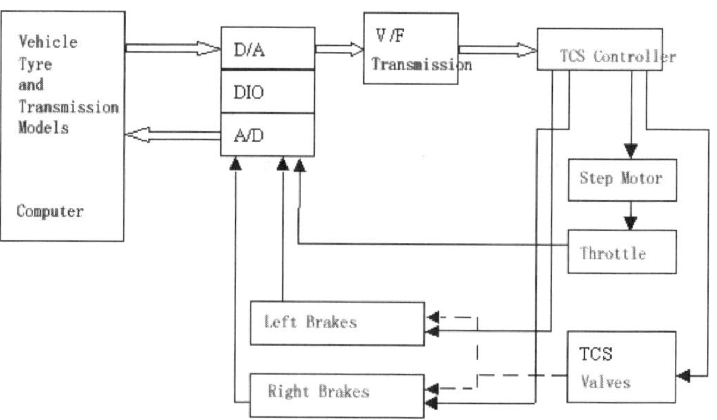

Figure 3 Schematics of the HILS for TCS

- Computer

It will perform the calculation of the vehicle model, the engine model, the transmission model, and the tyre model. The output will be the wheel rotation speed signal.

- I/O Interface

The high-speed multi-function acquisition card, performs Analog/Digital (A/D), Digital/Analog (D/A), Digital Input (DI), Digital Output (DO).

- Voltage/Frequency（V/F）Transformation Circuit

V/F transformation circuit will receive from the computer the wheel rotation speed signal, and transfers the voltage signal into a frequency signal, in a form of a pulse signal that emulates the real wheel rotation speed signal received from the wheel speed sensor.

- TCS Electric Control Unit（ECU）

ECU is the kernel of the TCS. It will work as in real vehicle.

- Sensors

Sensors in the HILS will check the pressu responses of the wheel cylinders and movement of the engine throttle.

- Brake System and Stepping Motor

Wheel cylinders and a stepping motor are the actuators of HILS, they are controlled by the TCS controller.

Hardware Structure of the ECU of TCS

The hardware of the ECU of TCS are composed from the input signal circuit, the valve drive circuit, the valve checking circuit, the power circuit, and the relay drive circuit. The main controller is C167CR 16-bit microcontroller. The relationship of different circuit is shown in figure 4. A prototype of ECU for TCS is shown in figure 5.

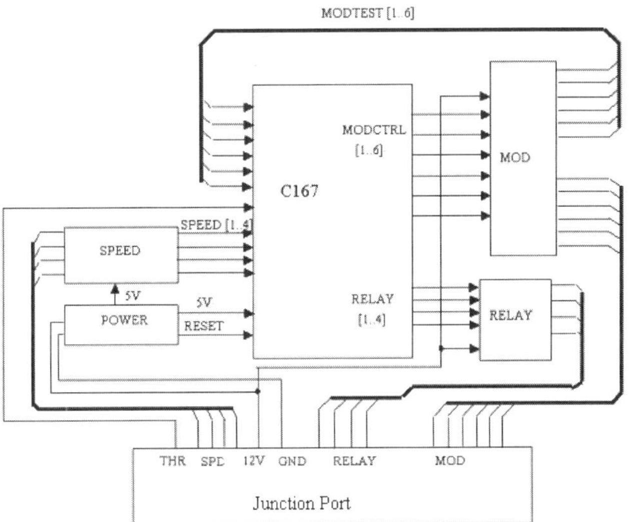

Figure 4 the structure of the ECU for TCS

Figure 5 Prototype of ECU for TCS

Software of the ECU of TCS

The software of ECU of TCS would acquire the wheel speed signals from the input signal circuit. The correlative control signals of the TCS would be sent out in time after calculating and analyzing the acquired wheel speed signals. The flowchart of the software of the ECU for TCS was shown in figure 6.

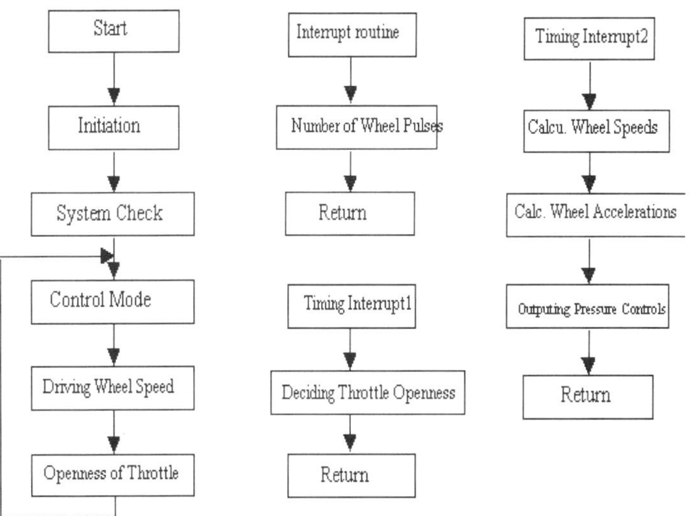

Figure 6 Flowchart of the software of ECU for TCS

The software of the ECU of TCS was compiled with the Keil C166 compiler. With the help of the DAVE 2.0 software of Infineon Company, the settings of registers of the C167CR and the C500 microcontroller were assigned easily. A primary program with Keil C code was obtained at the end by using of DAVE 2.0. The control law was added into the primary program. The input and output values of the software were debugged using an emulation device AX166 BONDOUT, produced by HITEX company.

RESULTS OF THE HILS

The digital simulation of the vehicle with TCS was done first. The digital simulation performance fasts that the HILS will perform.

The simulation started by a trigger signal from the pressure sensor. The wheel speeds were calculated, and then the values were sent to the V/F module that transforms the voltage values into the correlative frequencies of pulse signal that the real ECU would acquire. Then the ECU of TCS in the HILS worked as in the real vehicle acquiring the wheel speed signals, making decision of control, sending control instructions to the valve of the Hydraulic Control Unit (HCU), and the stepping motor of the engine throttle. At the same time, the responses of the pressure of the wheel cylinders and the throttle were fed back to the computer by the sensors and the Multi-function acquisition card. The computer will calculate the models of the vehicle, the engine, the transmission and the tyre again by setting the step and a new wheel speed signal.

Setting the maximum adhesion coefficient at 20%, the engine throttle opening is 100%, the start speed of the vehicle as 2m/s, and the acceleration time 10 second, the results of the HILS for the vehicle with TCS and without TCS were shown in figure 7~9.

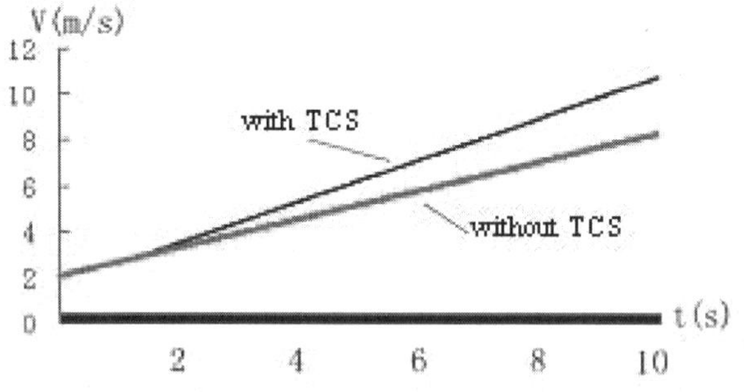

Figure 7 Comparison of wheel speed

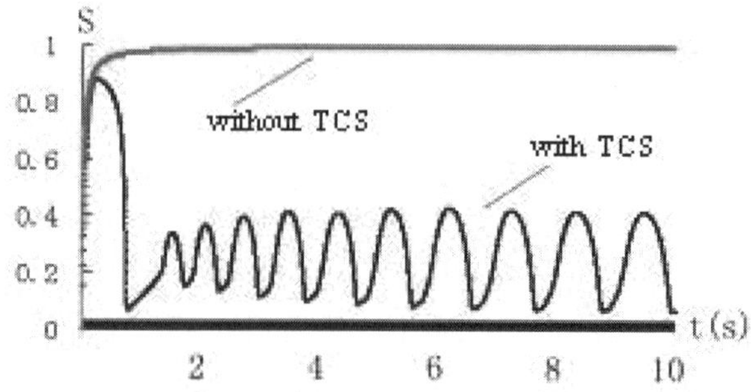

Figure 8 Comparison of slip ratio

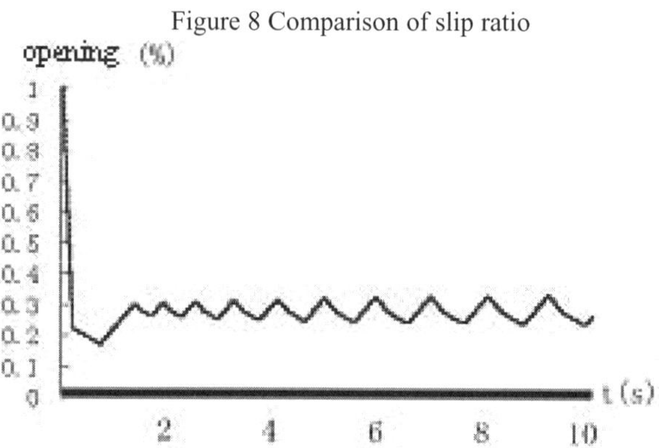

Figure 9 Change of engine throttle

From the above figures, we can find that:

The end speed (10.75m/s) of the vehicle with TCS after 10 second acceleration was larger than that (8.26m/s) without TCS. In the same time, the wheel slip ratio of the vehicle without TCS was in the instable range more than 50% that means the vehicle without TCS was dangerous due to the less lateral forces. On the contrary, the wheel slip ratio of the vehicle with TCS was in the stable range near to 20%; it ensured the stability of the vehicle.

CONCLUSION

The TCS, it is not only improving the acceleration performance, but also ensuring the steering stability of the vehicle. Based on the modeling of the vehicle with TCS, the HILS for vehicle with TCS was established. In the HILS, a real TCS controller was designed using an Infineon 16-bit microcontroller C167CR. The experimental results of the TCS HILS showed that the vehicle with TCS had higher speed and better steering stability than this one without TCS concerning the same acceleration time.

REFEERENCE

[1] Cheng Jun. Development with Infineon C166 Series 16-bit Microcontroller. Beijing: Publishing of Beijing University of Aviation, 2001(In Chinese)

[2] Dick D.Hoffman. The Corvette Acceleration Slip Regulation (ASR) Application with Preloaded Limited Slip Differential. SAE : 920642

[3] Heinz Leiber, Armin Czinczel. Antiskid System for Passenger Cars with a Digital Electronic Control Unit. SAE : 790458

[4] Ka C.Cheok. Fuzzy Logic Approach to Traction Control Design. SAE : 960957

[5] P.D.Rose. Automotive and Aerospace electronic Systems. Dependability Requirement. Microelectron. Reliab : Vol.36 , No.11/12 pp.1923-1929, 1996

[6] Jae-Bok Song, Byong-Cheol Kim and Dong-Chul Shin. Development of TCS Slip Control Logic Based on Engine Throttle Control. KSME International Journal : Vol. 13, No. 1, pp. 74-81, 1999

[7] Jae-Bok Song, Kyung-Seok Byun. Throttle Actuator Control System for Vehicle Traction Control. Mechatronics, Vol. 9, No. 5, pp. 477-495, 1999.

[8] Jae-Bok Song, Byung-Chul Kim, Hyo-Whan Chang. Slip Control Systems Based on Engine Throttle Control Approach. Proc. of 4th International Symposium on Advanced Vehicle Control, pp. 165-170, 1998.

[9] Yamada, Kiichi, Hashiguchi, Masayuki. Traction control system. Simulation analysis of the control system. International Journal of Vehicle Design v 12 n 1 1991 pp. 89-96 0143-3369.

[10] L-E.Berskiold,"Volvo's New System to Avoid Skidding", ISATA 84010

[11] H.-J Sch o pf and J.Paul, "ASR Acceleration Skid Control-A Further Contribution Towards Increasing The Active Safety of Daimler-Benz Vehicles", SAE Paper 885050

Dana Torque Vectoring Differential Dynamic Trak™

John Park and William J. Kroppe
Dana Corporation

ABSTRACT

This paper presents a novel torque-vectoring (biasing) differential system Dynamic Trak™ that can be applied to both the inter-axle and the inter-wheel differential systems. The Dynamic Trak™ has three multi-plate clutches. The center clutch located inside the differential case provides a limited-slip or complete lock-up capability. The two outboard clutches positioned at either sides of the differential case selectively adjust the torque flow to the left or right shafts/wheels. An electronic control unit and a hydraulic circuit unit control the three clutches, realizing the active management of the torque to the two output shafts/wheels. The Dynamic Trak™ can provide maximum of 100% torque add or subtraction to the wheels, while maintaining the open differential feature. A virtual model representing a passenger car equipped with the Dynamic Trak™ has been developed. The simulation results show the handling and stability enhancements by the Dynamic Trak™.

INTRODUCTION

The need for higher traction performance under slippery or difficult road conditions has been promoting the traction control system (TCS) and the all-wheel drive (AWD) system. The TCS generates a corrective yaw moment by selectively applying a brake to a certain wheel. On the other hand, the AWD avoids the saturation of tire traction force by distributing the traction torque to all wheels. Various configurations of the AWD has been developed.

One of the example is the on-demand AWD system. Some on-demand AWD systems are mainly a rear wheel drive system with added front wheel driveline that delivers a torque on an on-demand basis. The BMW's xDrive system [1] is one of the examples. Other AWD systems combine a front wheel drive system with an on-demand rear wheel drive system. Any on-demand AWD system employs a passive or active inter-axle coupling. The passive couplings are the limited-slip clutches such as the viscous couplings or the spring-biased multi-plate clutches. The active couplings are the controllable clutches based on hydraulic or electromagnetic power.

With the advent of the active yaw control (AYC) or vehicle stability control (VSC) systems, today's cars can adjust its yaw rate to avoid excessive over or under steering. However, they manipulate only the braking forces. Therefore, it is naturally the next evolutionary step to augment the AYC with the active torque vectoring driveline system.

BorgWarner's active torque management system [2] employs a bevel gear for propeller shaft and two electromagnetic clutches at either side of the output shafts. It does not have a differential gear. It can deliver a variable torque to the two rear wheels. It is basically an on-demand rear wheel drive system with inter-wheel torque vectoring capability. Ricardo's torque vectoring system [3] is a versatile device applicable to both the inter-axle differential and the inter-wheel differential. It employs an epicyclic gear train to reduce the clutch modulation torque.

The Dynamic Trak™ has been introduced to address these issues in a mechanically preferable manner. It is built upon a conventional open differential system so that it does not hinder the low speed cornering capability of the vehicle. In order to retain a torque-vectoring feature, three controllable clutches are added to the open differential. One is the center clutch engaging between the differential carrier and a half shaft. The others are the two outboard clutches, each of which are installed between the split half shafts. As such, the outboard clutches dictate the torque flow to the two output half shafts, while the center clutch ensures that the input power does not leak through the partially or fully disengaged outboard clutch. The clutches can be either hydraulic or electromagnetic. This paper focuses on the Dynamic Trak™ system with hydraulically modulated clutches.

In order to verify the practical value of the Dynamic Trak™, a hydro-mechanical model representing a single axle equipped with the Dynamic Trak™ and installed on a virtual dynamometer has been developed using the commercial multi-body dynamic analysis program ADAMS [4]. This subsystem level model is composed of the mechanical portion and the hydraulic portion. The model served as a design tool for the sizing of the clutch packs and the hydraulic components and also as a validation tool for the functionality of the Dynamic Trak™.

In addition to the subsystem level model, a full system model for a passenger car with Dynamic Trak™ system has been developed. This full-vehicle model served as a demonstration tool to prove the validity of the Dynamic Trak™ in terms of the vehicle dynamics enhancements. The simulations performed using the model are the cornering while accelerating, and the single lane change. The simulation results show the stabilizing and handling capabilities of the Dynamic Trak™, the torque vectoring differential system.

DYNAMIC TRAK™ SYSTEM

CONSTRUCTION – Figure 1 shows the overall view of the Dynamic Trak™ with its hydraulic circuit unit (HCU) mounted externally on the housing. Also shown in the figure are the input pinion shaft and the half shaft. The housing is cylinder shaped in general except the central cone-shaped protrusion for the pinion shaft assembly and the octagonal rear protrusion for the rear cover.

Figure 1. Overall view of Dynamic Trak™

Inside the housing are three divided chambers for the three clutches shown in Figure 2. The central chamber accommodates a conventional differential system and the center clutch. The two other chambers contain the outboard clutches. Each half shaft is split into the inner half shaft and the outer half shaft, which are coupled together by the outboard clutch.

The center clutch is a typical hydraulically engaged clutch. However, the two outboard clutches are the spring applied and hydraulically released (SAHR) clutches. Therefore, the outboard clutch is engaged by default by the spring forces, and is partially or fully disengaged by the hydraulic pressure.

The electronic control unit (ECU) controls the HCU and modulates the three clutches. Figure 3 shows the layout of the control system. The bicycle model takes the vehicle speed and steering wheel angle signals and then generates the reference-yaw rate that is an ideal quasi-static yaw rate. The control logic uses the yaw rate error (the difference between the actual yaw rate and the reference yaw rate) to determine the hydraulic pressures for the three clutches.

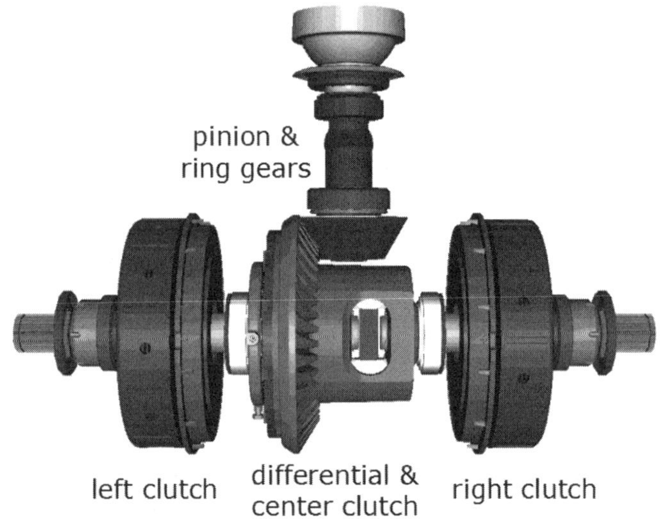

Figure 2. Mechanical structure of Dynamic Trak™

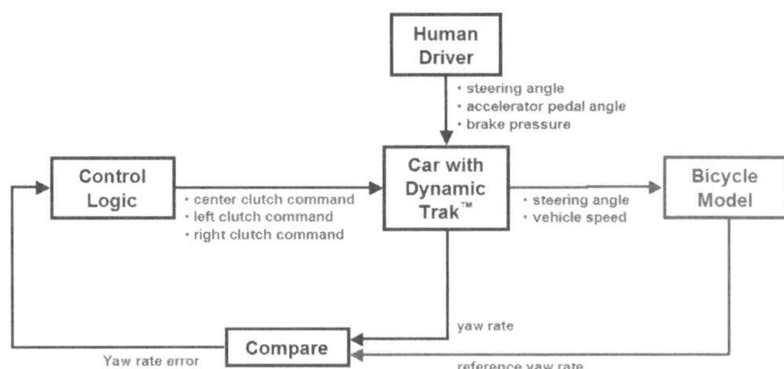

Figure 3. Control architecture

OPERATION – The Dynamic Trak™ can add or subtract the yaw moment to the car by vectoring (biasing) the torque flow to the left and right wheels. The positive engine power combined with the action of the Dynamic Trak™ generates the addition of the yaw moment. Whereas any engine braking and the Dynamic Trak™ action produce the subtraction of the yaw moment. Figure 4 is a schematic diagram for the torque vectoring through a controlled modulation of the three clutches.

The left most part of the curve in Figure 4 represents the 100% clock wise yaw add situation, realized by the full engagement of the center and left clutches and the full opening of the right clutch. In this 100% of the propeller shaft torque will be transmitted to the left wheel through the left clutch. On the other hand, the right most part of the curve represents the 100% counter clock wise yaw add situation by transferring 100% of the input torque to the right wheel.

The central portion of the curve in Figure 4 is the neutral or open differential situation in which the system behaves like a conventional open differential. In this case, both the left and right clutches are fully engaged by the clutch springs.

The intermediate sloped-portions of the curve represent the partial (larger than 0% but less than 100%) torque vectoring situations. In other words, any portions other than the above mentioned three extreme situations falls into this partial yaw add/subtract action.

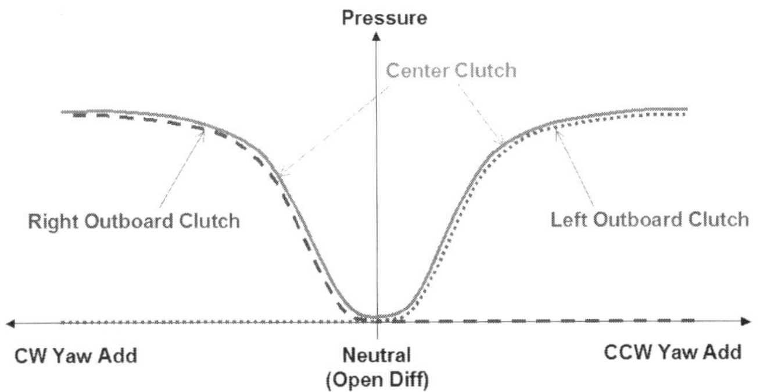

Figure 4. Modulation of three clutches

ADVANTAGES - Dynamic Trak™ torque vectoring system differentiates itself from the other systems in its use of the outboard clutches. While the other systems use the outboard clutches to brake the half shaft or to change the differential gear ratio, the Dynamic Trak™ use them to partially or fully cut off the torque flow to a specific half shaft. In other words, the torque-cut-off half shaft and the wheel are free to rotate on the road surface. This gives two distinct advantages to the Dynamic Trak™:

- While the fully locked up axle of the other systems disrupt the wheel speed sensor, the freely rotating torque-cut-off wheel of Dynamic Trak™ allows the wheel speed sensor to measure the correct road speed at the wheel.
- The freely rotating wheel allows the existing ABS, TCS, AYC or VSC to apply full braking force to the wheel, yielding a harmonious corporation of them with the Dynamic Trak™.

The fact that the Dynamic Trak™ retains the open differential and the built-in limited slip or full lock up capabilities produces additional advantages listed below:

- The Dynamic Trak™ provides a full range of torque vectoring from a conventional open differential to the 100% torque bias.
- The Dynamic Trak™ can provide not only the stability or handling enhancement but also the off-road capability.
- Dynamic Trak™ allows manual differential lock feature via a dash board mounted switch.

Finally, the Dynamic Trak™ may be implemented as a stand-alone system or be combined with any existing ABS or VDC (vehicle dynamics control) systems for improved stability and handling of passenger cars.

MODELLING AND SIMULATION

This section presents the virtual prototyping of the Dynamic Trak™ system in both the subsystem and the total system levels.

SUBSYSTEM-LEVEL MODEL – A virtual prototype for the Dynamic Trak™ installed to an axle dynamometer has been developed using the multi-body dynamic analysis program ADAMS [4]. Figure 5 shows the general configuration of the model. A drive motor applies a controlled motion or torque to the input (propeller) shaft of the Dynamic Trak™. Two absorbers are attached to the either ends of the output shafts. Also shown in the figure are the center clutch that is illustrated as a small cylinder geometry and the two sets of outboard clutches that are depicted as pairs of larger cylinders.

The subsystem model consists of two different physical portions: A mechanical portion and the hydraulic portion. The mechanical portion of the model is based on the multi-body dynamics, and the hydraulic portion is constructed as a connection of various one-dimensional modeling components. The connection between the mechanical and the hydraulic portions are through the hydraulic cylinder forces. In other words, any hydraulic force generated from the hydraulics portion would be transferred to the corresponding force element in the mechanical portion of the model.

Figure 5. Dynamic Trak™ and virtual dynamometer

Figure 6 shows the hydraulic circuit portion of the subsystem model. The pump generates the pressure, while the pressure relief valve maintains a predetermined system pressure. The accumulator reduces the required capacity of the hydraulic pump. The control valves distribute the hydraulic pressure to the three clutches. As was mentioned in the subsection CONSTRUCTION, any hydraulic pressure to the cylinder for the center clutch would engage the clutch, while the pressure to the outboard clutches would defeat the spring force to disengage the clutches.

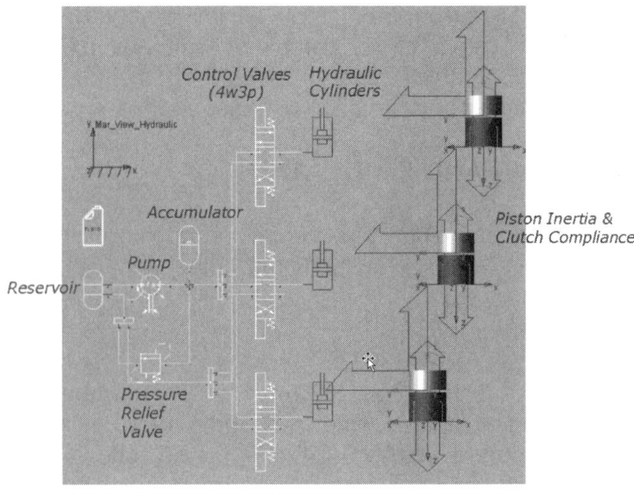

Figure 6. Hydraulic circuit

SUBSYSTEM-LEVEL SIMULATION – An open-loop simulation was performed using the hydro-mechanical subsystem model. Initially the system is at a standing still and then the motor spins up the system to a steady state. At 2 seconds, the center clutch is engaged and the left clutch is disengaged, while the right clutch remains engaged. Figure 7 shows the spool displacement of the left clutch valve, and Figure 8 shows the time-delayed flow rate to the left hydraulic cylinder.

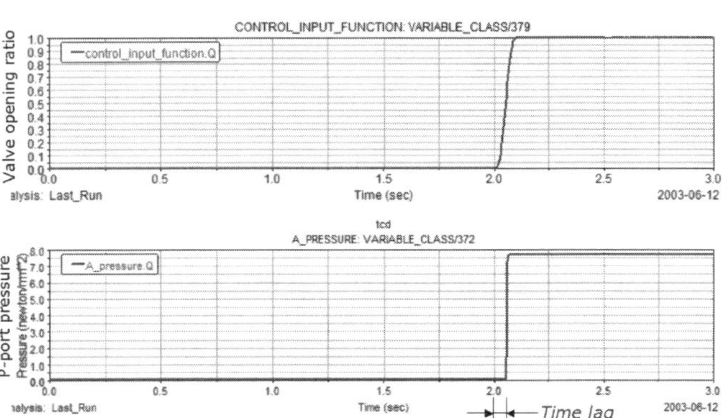

Figure 7. Left-clutch valve opening and the pressure build-up.

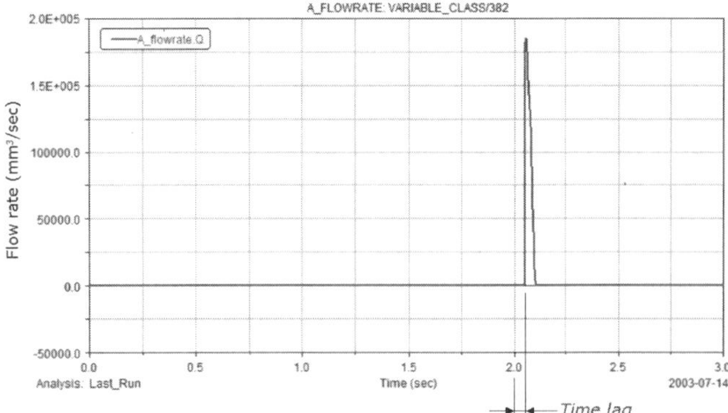

Figure 8. Flow rate to the left hydraulic cylinder.

Figure 9 shows the clamping forces of the center and the left clutches both of which receive a full hydraulic pressure at 2 seconds. The center clutch receives a full

clamping force locking or engaging the clutch. On the other hand, the SAHR type left clutch switches from a fully engaged to a fully disengaged status. The right outboard clutch that does not receive any hydraulic pressure remains to be locked. These changes in clutch states results in a 100% torque flow from the propeller shaft through the center and right clutches and finally to the right half shaft. Figure 10 shows this torque-vectoring event at around 2 second. This demonstrates the function of the Dynamic Trak™.

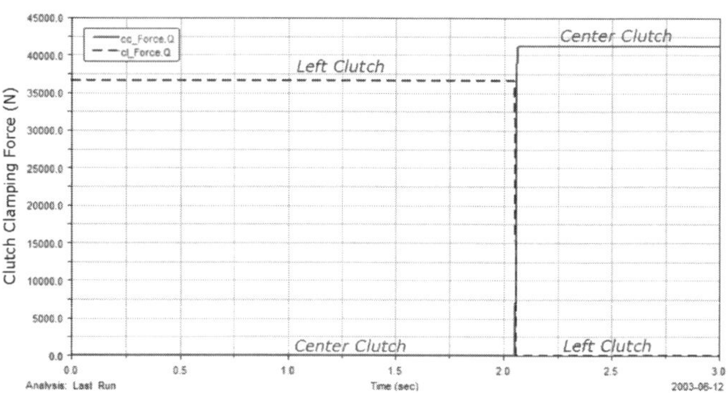

Figure 9. Clamping force of three clutches.

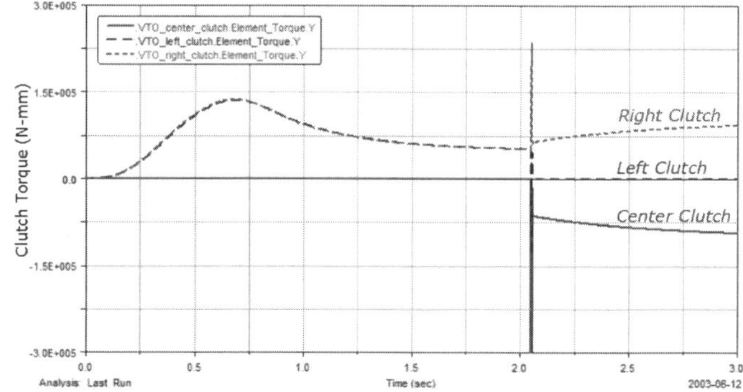

Figure 10. Transmitted torque across the three clutches.

FULL-VEHICLE-LEVEL MODEL – A virtual prototype for a rear wheel drive car with a Dynamic Trak™ has been developed. Figure 11 shows the full-vehicle model representing a typical full-size SUV. A torque is applied to the propeller shaft that drives the Dynamic Trak™. Two output shafts of the Dynamic Trak™ are connected to the rear wheels through the CV joint half shafts. An analytical tire model based on a combined slip generates the tire forces. A rolling resistance based on the empirical formula [5] was added to the tire forces. For simplicity's sake the suspension mechanism has been omitted in the model. The car body receives a speed-dependent aerodynamic drag [6].

FULL-VEHICLE-LEVEL SIMULATION – A simulation for a single lane change maneuver has been performed. The run conditions are a 113 km/hr (70 mile/hr) initial speed, a constant propeller shaft torque of 100 N-m, final gear ratio of 4.1, an open-loop steering command, a steering ratio of 17.5, and the feedback controlled Dynamic Trak™ action.

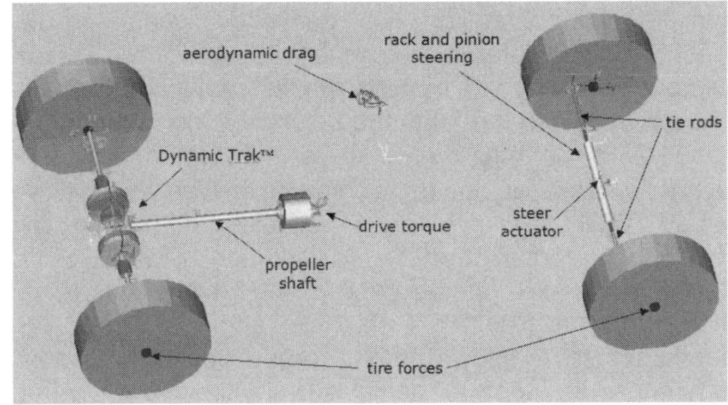

Figure 11. Full vehicle model.

Figure 12 shows the open-loop steering wheel angle command. Figure 13 shows the actual and the reference yaw rates. It is obvious that the reference yaw rate does not have any time lag and that the actual yaw rate has not only a time lag but also an overshoot. Figure 14 shows the yaw rate error defined as the difference between the actual yaw rate and the reference yaw rate. The controller determines the redistribution of the torque based on the yaw rate error signal.

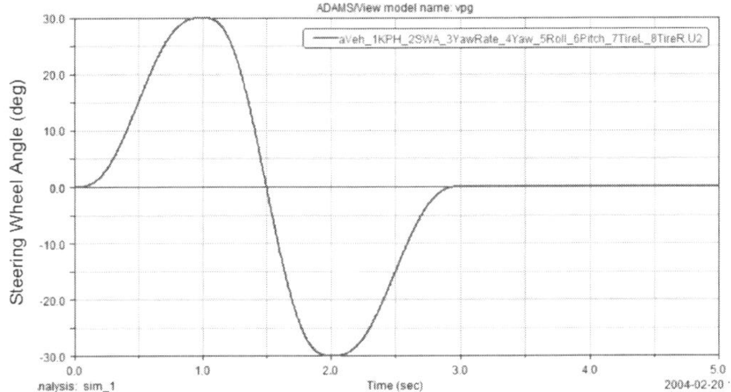

Figure 12. Steering wheel angle command.

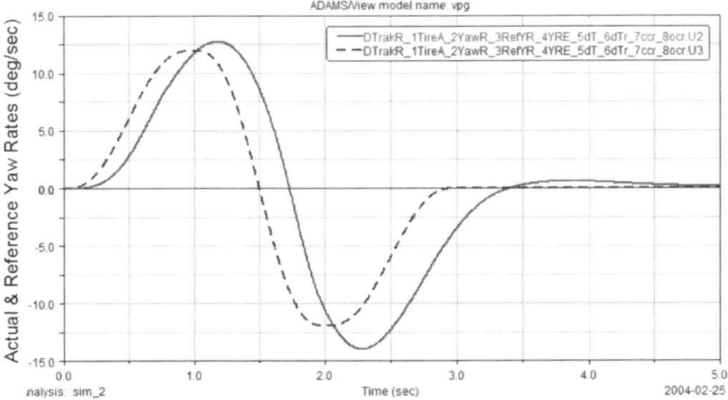

Figure 13. Actual yaw rate and reference yaw rate.

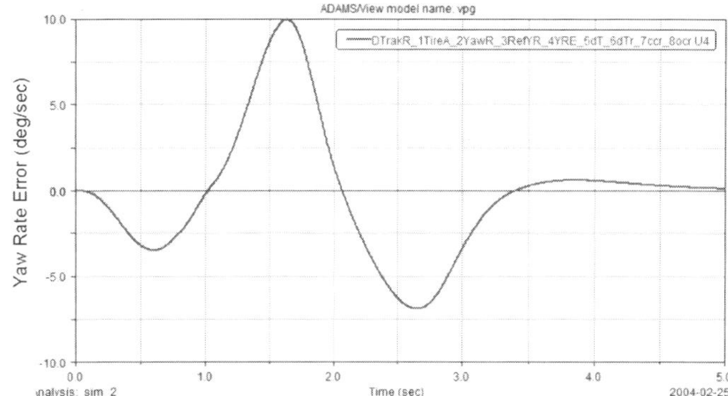

Figure 14. Yaw rate error.

Figure 15 shows the left and right shaft torques of the Dynamic Trak™. When the yaw rate error is positive (that is when the vehicle is yawing too much to the counter clock wise direction), the Dynamic Trak™ transfers more torque to the left tire. The resultant clock-wise yaw moment brings the vehicle back to the desired yaw condition. Figure 16 shows the tire longitudinal forces. It is evident that the rear tires have a fluctuation in their traction forces in accordance with the clutch torques in Figure 15. On the other hand, another simulation with a conventional open differential reveals a monotonous trend of the rear traction forces as shown in Figure 17.

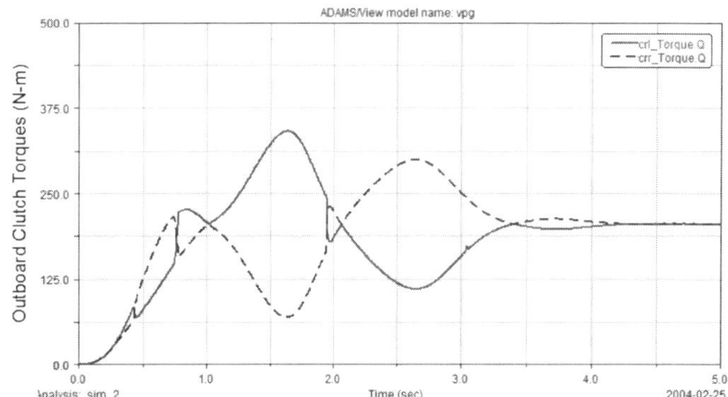

Figure 15. Left and right clutch torques.

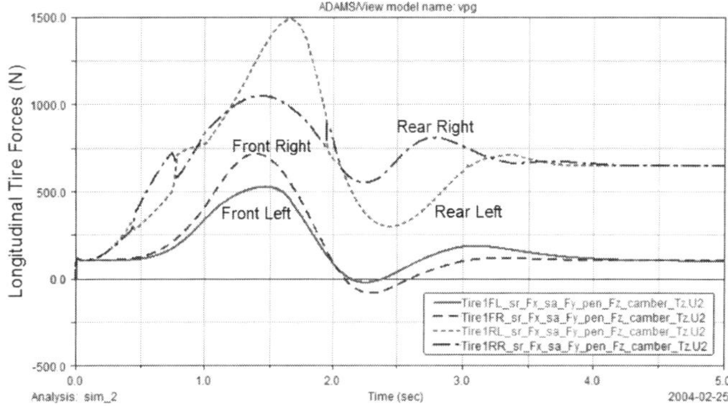

Figure 16. Tire longitudinal forces (Dynamic Trak™).

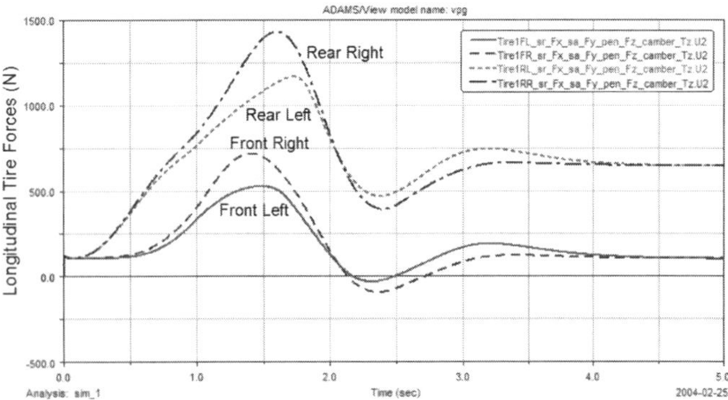

Figure 17. Tire longitudinal forces (open differential).

This redistribution of the traction forces by the torque vectoring capability of the Dynamic Trak™ allows the vehicle to have a different yaw rate when compared to a car equipped with a conventional open differential. Figure 18 shows the yaw rate comparison. Figure 19 is the comparison of the vehicle trajectories between a car with the Dynamic Trak™ and another car with the open differential. Since the time-integrated effect of the yaw rate is the vehicle position, the difference is more noticeable in the vehicle trajectory than in the yaw rate.

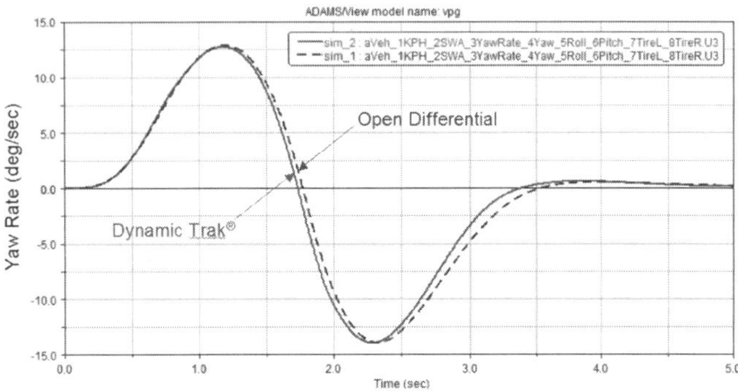

Figure 18. Yaw rate comparison.

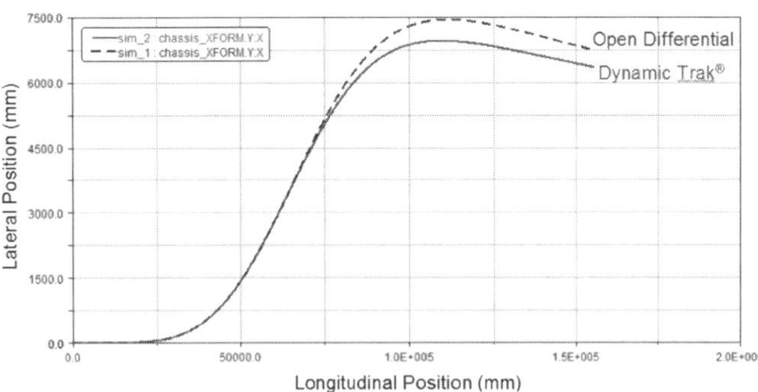

Figure 19. Vehicle trajectory comparison.

CONCLUSION

In this paper a novel torque vectoring differential system Dynamic Trak™ has been introduced which enhances not only the handling performance but also the off-road capability. The virtual prototypes for the sub-system and the full-car system have been presented to demonstrate the validity of the Dynamic Trak™ system. Simulation results indicate that the Dynamic Trak™ can enhance the stability and handling of passenger cars in a mechanically preferable manner. In addition, it can improve the off road mobility via its partial/full lock-up capability. Since it retains the differential gear, it can be used as a primary axle. It is anticipated that the Dynamic Trak™ is not only compatible with the brake-based TCS but also enhance its performance by reducing a torque transfer to a wheel that TCS is applying a brake. In the case of the limiting situations that involve a significant side slip angle, it would be necessary to employ another torque vectoring center differential in order to achieve an active yaw control capability.

ACKNOWLEDGMENTS

The authors would like to express their sincere appreciation to Mike Greene, Jun Yoshioka and Ralph Baxter of Dana Corporation who helped in the development of the torque-vectoring differential system, Dynamic Trak™.

REFERENCES

1. Gerhard Fischer, et al., *4x4 Drive Train with Vehicle Dynamic Control for the New BMW X3 and X5 MJ04*, 12th Aachen Colloquium, 2003.
2. John Peterson, et al., *Improving Vehicle Behavior Using a Dual Clutch Rear Axle*, Global Powertrain Congress, Ann Arbor, Michigan, August, 2003.
3. B Reynolds and J. Wheals, *Torque Vectoring Driveline: Design, Simulations, Capabilities and Control*, 12th Aachen Colloquium, 2003.
4. ADAMS program, MSC Software Corporation, 2003.
5. J. W. Wong, *Theory of Ground Vehicles*, Second Edition, John Wiley & Sons, Inc., 1993.
6. T. D. Gillespie, *Fundamentals of Vehicle Dynamics*, Warrendale, PA, SAE Publication, 1992.

CONTACT

John Park, Technical Specialist
  Phone: +1 419 887 3309   Fax: +1 419 887 5962
  E-mail: john.park@dana.com
William J. Kroppe, Director
  Phone: +1 419 887 3302   Fax: +1 419 887 5950
  E-mail: william.kroppe@dana.com
Surface mail address:
  Advanced Chassis and Suspension Systems, GETG Dana Corporation, 3939 Technology Drive, Maumee, Ohio 43537

An Empirical Model For Longitudinal Tire-Road Friction Estimation

Mingyuan Bian, Keqiang Li, Nenglian Feng and Xiaomin Lian
Tsinghua University

ABSTRACT

It's important to monitor the longitudinal friction at the tire/road interface for automotive dynamic control systems like ABS and ASR. Of all the tire friction models the empirical model provides a good illustration on longitudinal wheel forces. An improved exponential friction model based on vehicle driving states was proposed in this paper, the model can monitor the friction characteristics between the tire and road surface for longitudinal braking. Its validity was proven using experiments and comparison with the Pacejka Magic Formula (MF) model and others.

INTRODUCTION

It was well known that the vehicular driving and braking performances depend on the forces (torques) acting on wheels. The increasing of wheel forces (torques) can provide the vehicle a better driving or braking potential to some extent, but skidding will happen in the tire/road interface when wheel forces exceed the friction capacity. The study on longitudinal wheel friction characteristics on road surfaces has become an important part of some automotive dynamic control systems like ABS (Anti-lock Braking System) and ASR (Anti-Slip Regulation).

There are many factors affecting the friction between tire and road, such as road condition, tire, and vehicle driving states etc. Generally there are two methods to assess the longitudinal road friction, one is to evaluate the adhesive coefficient with experiments, the other one is to estimate it empirically. Special facilities for measurement of road surface adhesion were available in the UK, USA, Sweden and other countries since the 1970's[1]. The optical sensors were broadly used to assess the friction situation of the terrain and identify the road condition for automotive control (Holzwarth, Eichhorn, etc). Breuer

and Eichhnorn also used strain sensors outfitted on the tires to predict the friction characteristics of the terrain [2][3][4]. Coupled with these methods, the acoustic sensors are also applied to evaluate the longitudinal tire friction [5]. Measuring the friction coefficient with special facilities can provide a precise result, but it's always very expensive [6].

The relationship of some variables like vehicle speed, acceleration, wheel slip ratio, slip angle and longitudinal friction coefficient is well formulated with math functions. This is helpful for understanding the dynamic state between the wheel and road interface, and making a better control for longitudinal dynamics control systems. Two methods are commonly used to assess the road friction by math function, one is the theoretical model, and the other one is the empirical friction model.

The theoretical tire model is based on structure mechanics of the tire. The math formulations are always complex and require more fundamental understanding of the tire structure mechanics. So the application of theoretical tire model on describing longitudinal wheel friction characteristics is limited in practical applications.

OVERVIEW ON EMPIRICAL TIRE MODELS

The empirical tire models are based on recursive analysis of a large amount of experimental data that reflect the adhesive characteristics between the tire and the road surface. Because the empirical equations are established based on experimental results, they can estimate the friction coefficient very well for various road conditions and driving situations.

The Magic Formula proposed by Pacejka is the popular equation widely used in the research field of vehicle dynamics. It fits the wheel experimental data with a trigonometric formula, and totally expresses different driving situations, such as, longitudinal, lateral force, aligning torque or a combined condition. Because this formula is achieved by fitting experimental data, every parameter in the model has physical meanings. The commonly used MF model has the following form:

$$Y = y + S_v$$
$$y = D\sin(C\arctan(Bx - E(Bx - \arctan(Bx)))) \qquad (1)$$
$$x = X + S_h$$

In the study of longitudinal vehicle dynamics, Y can be assigned to longitudinal friction force between tire and road, and S is used to describe the wheel slip ratio. The factors B, C, D and E have the physical meanings as stiffness, shape, peak value and curvature factor, respectively.

The MF model can accurately compute the longitudinal and lateral wheel force, however, this equation always has high computational complexity because it is a non-linear polynomial composed of trigonometric functions with several parameters.

The research on mathematical models that estimate the friction coefficient with simpler equations has made some progress recently. A compact double linear model has been used in analyzing vehicular dynamics systems like ABS. The model is as follows:

$$\mu = \frac{\mu_h}{S_T}S \qquad S \le S_T$$
$$\mu = \frac{\mu_h - \mu_g S_T}{1 - S_T} - \frac{\mu_h - \mu_g}{1 - S_T}S \qquad S \ge S_T \qquad (2)$$

This model has a simpler shape and lower computational complexity than MF model. However, the friction coefficient and wheel slip ratio shows a complicated relation, and the vehicle dynamics parameters that affect such a relation change in real-time. So, it cannot precisely express the change of friction between the tire and the road surface with a pre-set μ_h and μ_g.

$$y = a_1 x^n + a_2 x^{n-1} + \cdots + a_{n-1}x + a_n \qquad (3)$$

The non-linear polynomial fit equation (equation 3) is also used sometimes to assess tire friction condition. This model can simulate the non-linear relationship of longitudinal friction coefficient change with wheel slip ratio in some driving condition, but it needs higher polynomial orders. The polynomial variables will be different in various driving situations for the difference of adhesive characteristics of road surface, which means that it needs a different equation for different driving situation. So, its application is limited.

$$\mu = A - BS - A \cdot \exp(-aS) \qquad (4)$$

Equation 4 is a polynomial tire model with an exponential function. It regard the tire model $\mu - S$ curve as the line

$$\mu_1 = A - BS$$ superimposed with the exponential curve

$$\mu_2 = -A \cdot \exp(-aS),$$ as Fig. 1 shows.

$$\mu = \mu_1 + \mu_2 \qquad (5)$$

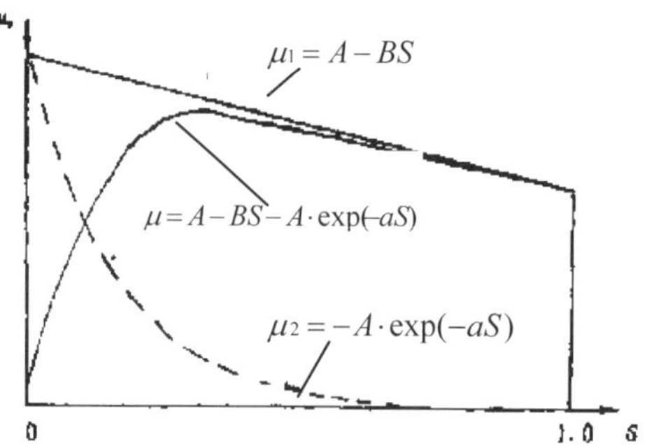

Fig. 1 Non-linear fit of friction

This model can precisely illustrate the non-linear changing trend of the longitudinal friction coefficient with wheel slip ratio in the stable area of the $\mu - S$ curve. But there will be a distinct linear trend of the curve because of the large value of a in the equation. So, an improved friction model needs to be put forward to describe the non-linear trend of the longitudinal friction between the tire and terrain.

THE IMPROVED EMPIRICAL FRICTION ESTIMATION MODEL

Road condition	Peak value of μ	Optimal slip ratio
Dry asphalt	0.85-0.98	0.12-0.2
Wet concrete	0.75-0.80	0.15-0.25
Wet asphalt	0.62-0.78	0.15-0.25
Wet gravel	0.78-0.85	0.15-0.25
Fresh snow	0.20-0.28	0.10-0.20
Ice	0.1	0.10-0.20

Table 1. parameters of longitudinal friction curve for general road conditions

Automotive anti-skid control systems (ABS/ASR) focus on controlling the longitudinal wheel forces to fully use the friction capacity of road surface. The $\mu - S$ curve shows different characteristics for various road conditions, as table 1 shows. So the new friction model must be able to reflect the difference of parameters for different road conditions.

A group of equations was achieved to express the characteristic parameters of $\mu - S$ curve on various driving situations with the following formulae by analyzing experimental data [1][7-8]:

$$\mu_{kp} = 0.92 \times 0.1304^{\sigma} + 0.002 \times \exp(\sigma) \cdot (64 - V) - 0.0426\sqrt{u}$$
$$\mu_{ks} = 0.677 \times 0.1553^{\sigma} + 0.002 \times \exp(\sigma) \cdot (64 - V) - 0.0426\sqrt{u} \qquad (6)$$
$$S_{xm} = 0.1532\sigma^3 - 0.4780\sigma^2 + 0.256\sigma + 0.1693 + 0.105\log(64/V)$$

Specially when $\sigma = 1.2$

$$S_{xm} = 0.12$$
$$\mu_{xp} = 0.62 - 0.09\sqrt{V} \qquad (7)$$
$$\mu_{xs} = 0.5 - 0.14\sqrt{V}$$

Where the parameters μ_{xp} and μ_{xs} respectively refer to the peak value and final value at 100% wheel slip ratio of friction coefficient of the road surface, the S_{xm} refers to the optimum of slip rate. V is the vehicle speed (Km/h), $u = \dfrac{Fz}{Fs}$ is the load coefficient of the wheel, F_z is the vertical load of the tire, and F_s is the rated load of the tire. σ is a factor to reflect different road conditions. Table 2 shows different values of σ.

Road condition	σ value
Dry asphalt	0
Wet asphalt	0.134
Wet earth	0.253
Fresh snow	0.6
Packed snow	0.75
Dry ice	1
Rainy road	1.2

Table 2.   Value of σ assigned to road condition

A new friction force model was proposed based on the tire model as the Equation 4 shows to enhance the non-linear characteristics of the $\mu - S$ curve when slip ratio has a large value. The first order linear function was replaced by an exponential function, so the new tire model is composed of two exponential functions and three parameters.

$$\mu = A(\exp(-bS) - \exp(-aS)) \qquad (8)$$

Some parameters of this formula were obtained by using the following equations:

$$A = \frac{\mu_{xs}}{1 - a\,e^{-aS_{xm}}} \qquad (9)$$

$$b = -\ln\left[1 - a\,e^{-aS_{xm}}\right] \qquad (10)$$

The resolution of parameter a_k was obtained by iteration of equation 11.

$$a_{k+1} = \frac{1}{S_{xm}} \ln\left[\frac{(\dfrac{\mu_{xp}}{\mu_{xs}} - S_{xm}) \cdot a_{k-1}}{\dfrac{\mu_{xp}}{\mu_{xs}} - 1} \right] \qquad (11)$$

Stop iteration and let $a = a_{k+1}$ when

$$\left| \frac{a_{k+1} - a_k}{a_k} \right| \leq \varepsilon = 10^{-3}$$

It can be inferred from the above equations that the wheel slip based tire-road friction estimation result does not only have a close relationship with the road condition, but also it is affected by some parameters that describe the vehicle running state, such as vehicle speed, acceleration and axle load shifting, etc. So, it's the premise to have clear knowledge on vehicle state in real-time to monitor the longitudinal friction situation of the wheel/terrain interface. In literature [8] the tire and vehicle dynamics model was established, and the vehicle driving states were estimated using the wheel information by means of a special algorithm. Since the estimating theory has been illustrated in literature [8] in detail, it's unnecessary to give a tautology in this paper.

COMPARISON RESULTS OF DIFFERENT FRICTION MODELS

A BJ130 light truck with the 7.50-16 tire was experimented to brake at the velocity of 64Km/h on a test bench. It was used to simulate the brake on dry asphalt road, vehicle speed was determined from the roller speed, wheel speed and cylinder pressure information were collected to estimate the slip condition between the tire/road interface. The collected data were disposed and fitted by several tire models illustrated in this paper. The results are shown in Fig. 2.

It can be seen from Fig. 2 that the Pacejka MF model compares well with the experimental data. The combined exponential polynomial model has a clear linear trend when the wheel slip ratio becomes large. Thus, it lacks practical application potential. The double exponential friction model proposed in this paper has a close fit to the results as does the Pacejka model, and it's error is small when compared with practical longitudinal friction coefficient.

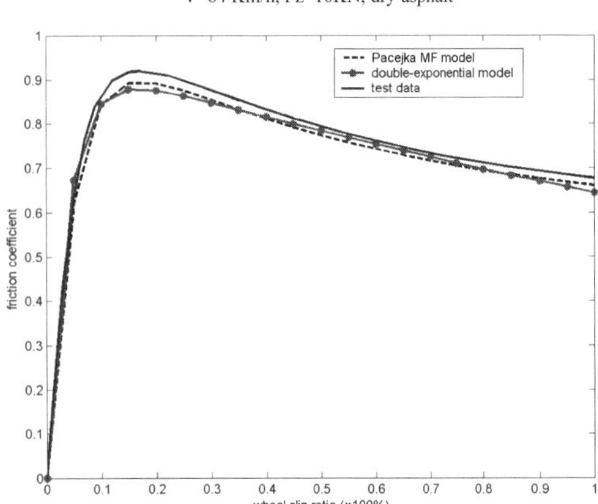

Fig. 2 comparison of different friction model

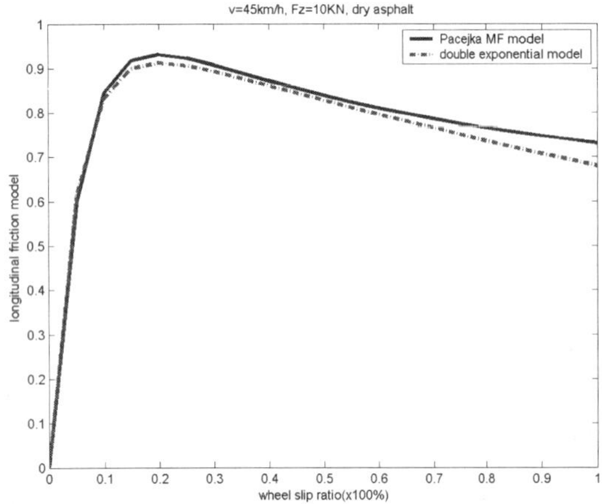

Fig. 3 comparison with MF model at low speed

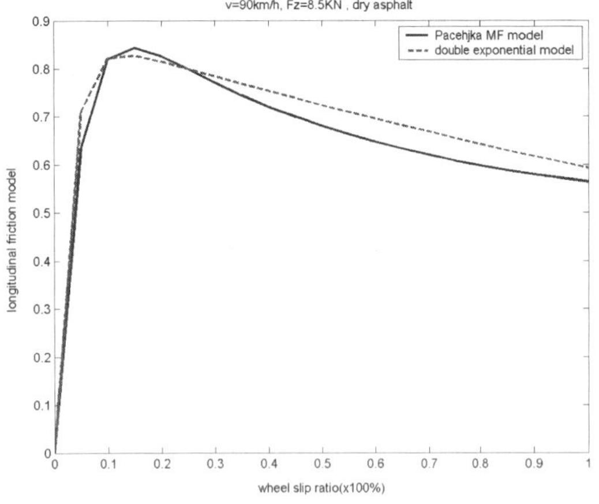

Fig. 4 comparison with MF model at high speed

Fig.3 and Fig.4 simulated comparison results of the double exponential friction model with the MF model at different driving conditions. It can be seen that the exponential model has results similar to the MF model. It has a clear non-linear changing trend in the non-stable area of the $\mu - S$ curve and can illustrate the friction characteristics between the tire and road surface. It has

fewer variables, a simpler shape and lower computational complexity than the Pacejka model, which makes it practical.

CONCLUSION

Of all the tire models that describe the longitudinal friction force of road surface, the empirical models can fit the practical data more closely than physical tire models, because they were established based on analyzing large amounts of experiment data.

The improved three-parameter-double exponential friction model proposed in this paper illustrates the longitudinal friction coefficient in a simpler way as a non-linear function of road condition, velocity, wheel load and slip ratio etc.

REFERENCES

1. Liu Zhaodu, "Mathematical Models of Tire-Longitudinal Road Adhesion and Their Use in the Study of Road Vehicle Dynamics", Journal of Beijing Institute of Technology, vol.5, no.2, pp.193-203, 1996.

2. Holzwarth F. and Eichhorn U., "Non-Contact Sensors for Road Conditions", Sensors and Actuators, June-August, pp.121-127, 1993.

3. Uno T., Sakai Y., Takagi J. and Yamashita T., "Road Surface Recognition Method using Optical Spatial Filtering", in Proceedings of AVEC'94, pp.509-515, 1994.

4. Breuer B., Eichhorn U. and Roth J., "Measurement of Tyre/Road-Friction Ahead of the Car and Inside the Tyre", in Proceedings of AVEC'92, pp.347-353, 1992.

5. Steffen Muller, Michael Uchanski and Karl Hedrick, "Slip-Based Tire-Road Friction Estimation During Braking", in Proceedings of IMECE'01, 2001 ASME International Mechanical Engineering Congress and Exposition, New York, USA, pp.1-11, 2001.

6. Ke Wei and Zhu Jun, "Testing Research on Road Surface Adhesion Coefficient," Inner Mongolian Highway & Transportation, No.3, pp31-33, 1996. (in Chinese)

7. F.C.Sun, M.Y.Bian and S.Z.Chen, "Monitoring of Longitudinal Friction Characteristics between the Tire and Road Surface", in Proceeding of the 6th International Symposium on Advanced Vehicle Control, AVEC'02, Sept 2002, Japan

8. Bian Mingyuan, "Research on Road Condition

Recognition and Vehicular Velocity Calculating Algorithm for Automotive Anti-Skid Control System (ABS/ASR)", Doctorate degree dissertation at Beijing Institute of Technology, March 2003.

Vehicle Dynamics and Torque Management Devices

Heinrich Huchtkoetter and Theodor Gassmann
GKN Driveline

ABSTRACT

Brake based traction control and electronic stability control continue to grow as an approach to enhance vehicle traction and stability. However, there is still a considerable contribution to the improvement of traction as well as to the longitudinal and lateral dynamics and stability of a vehicle provided by Torque Management Devices (TMD) such as limited-slip differentials and on-demand couplings. Specifically electronically controlled Torque Management Devices can offer significant improvements without compromising other systems like ABS (Anti-Lock Braking System) and ESP (Electronic Stability Program), a.k.a. DSC (Dynamic Stability Control), VDC (Vehicle Dynamics Control), IVD (Interactive Vehicle Dynamics) or TRACS (Traction and Stability Control), etc.

This paper explains various driveline concepts using electronically controlled Torque Management Devices ranging from on-demand (hang-on coupling) to full time AWD systems with a center differential. The influences of controlled TMD's on vehicle dynamics are shown for various driveline layouts and TMD arrangements. Results from vehicle starting and acceleration tests prove the traction improvements achievable with TMD's. Furthermore it is shown that TMD's also have a significant influence on vehicle handling, stability and safety in 4WD systems as well as in 2WD applications. Vehicle test results are presented from circle tests like throttle-off in a bend, which prove that the systems have a significant positive impact on stability and safety. The same applies to highly dynamic driving maneuvers like double lane changes. The results achieved with the Electronic Torque Manager (ETM) from GKN demonstrate the high potential of sophisticated controlled TMD's not only to enhance traction but also to offer significant vehicle dynamics improvements.

INTRODUCTION

In recent years TCS (Traction Control Systems) and vehicle stability control systems like ESP, VDC, IVD which use elements of the ABS systems have become more and more common in the market. As these systems have limitations, a tendency to use additional systems which help to overcome these limitations and to add additional features can be observed. Torque Management Devices like limited-slip differentials and on-demand couplings offer the required improvements. In particular, actively controllable TMD's complement brake based traction control and stability control systems and allow for further improvements. The following results from vehicle tests with various driveline configurations shall give an overview about the improvements TMD's can offer to vehicle traction, stability, safety and comfort.

DRIVELINE CONFIGURATIONS

Figure 1 shows the general driveline configurations on the market.

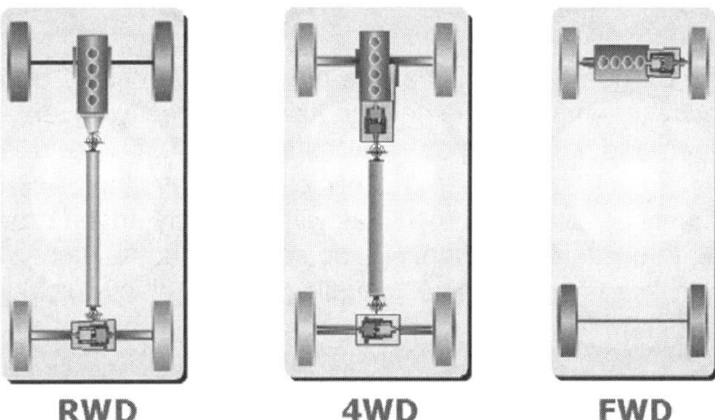

Figure 1: General Driveline Configurations

TMD's can be used in FWD, RWD as well as in AWD vehicles. In permanent 4WD vehicles TMD's are applicable as control units parallel to the differentials, i.e. as limited-slip differentials. In an on-demand AWD vehicle a TMD can also be used as "hang-on" coupling. Figure 2 shows principle sketches of these two 4WD configurations.

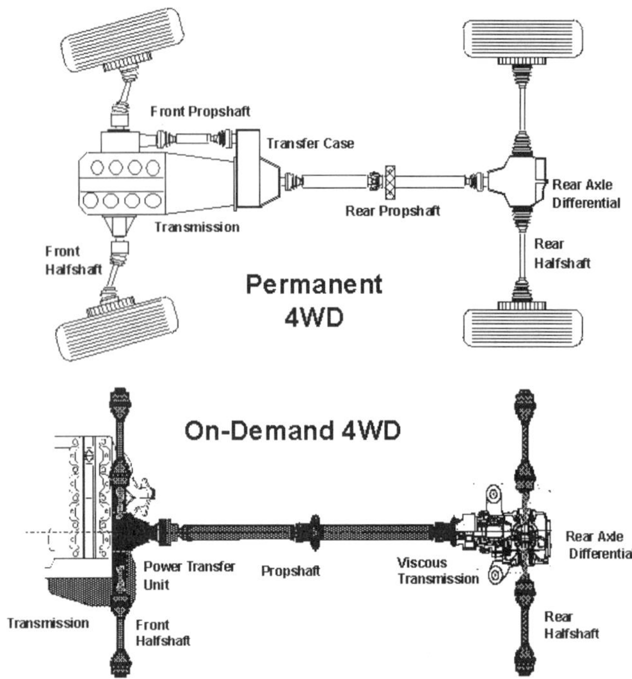

Figure 2: 4WD Driveline Configurations

In a permanent 4WD vehicle the gearbox output torque is fed into a transfer case with a center differential which distributes the torque with a static ratio (50:50 or unequal torque distribution) to the front and rear axles. In both axles the input torque then is distributed to the wheels by differentials with a static torque distribution of 50:50.

In an on-demand 4WD vehicle one axle is primarily and directly driven while the second axle is indirectly driven – on-demand – via a power transfer unit (PTU) and a coupling. On-demand 4WD vehicles are available with primarily driven front as well as with primarily driven rear axle. In on-demand couplings one can find dog-clutches, speed sensing passive couplings and electronically controlled units in the market.

All these different driveline configurations cause varying traction and vehicle dynamics. Therefore, this paper can only give some typical examples for the influence of driveline configurations on vehicle performance.

TORQUE MANAGEMENT DEVICES

Beside brake-based systems like Traction Control Systems (TCS), different types of Torque Management Devices are available in the market. The concepts are shown in Figure 3.

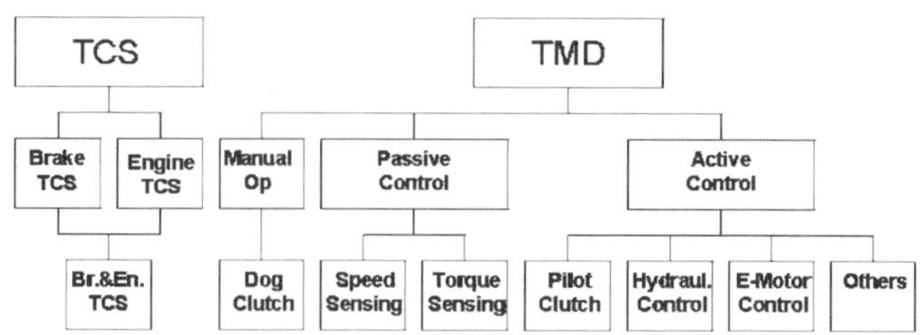

Figure 3: Concepts of Torque Redistribution Systems

Traction Control Systems react on spinning wheels either by reducing the engine torque output (Engine TCS) or by applying the brake of the spinning wheel (Brake TCS). However, it becomes more and more common to combine these systems to Engine & Brake TCS.

Torque Management Devices can be manually operated (dog clutch) or passively or actively controlled. Passively controlled Torque Management Systems are either torque sensing (helical gear type, multi-plate type etc) or speed sensing like the commonly known viscous coupling [1].

This paper mainly concentrates on actively controlled Torque Management Devices. Therefore these devices will be explained in more detail.

Several systems available are electromagnetically controlled by a pilot clutch. One example is shown in the Figure 4.

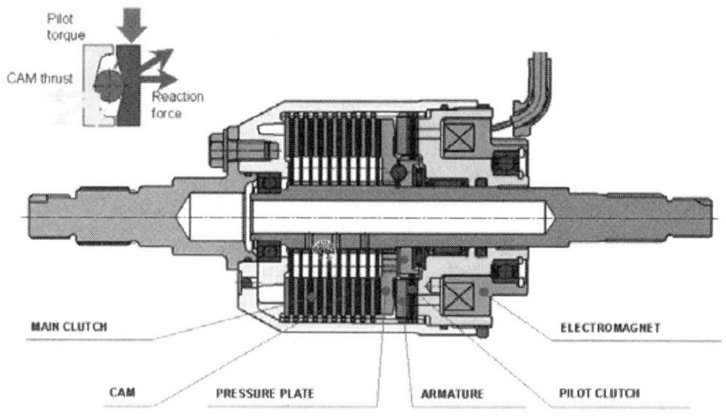

Figure 4: Electromagnetically Controlled Pilot Clutch Type TMD

In this system a small clutch pack - the pilot clutch - can be activated by an electromagnet. The magnetic force pushes the clutch plates in the pilot clutch pack together, which then activates a ball and ramp mechanism once a speed difference occurs between input and output. This ball and ramp mechanism acts on the main clutch pack thus providing the required locking torque.

Other systems in the market are activated hydraulically. Typically these systems contain a pump, which creates hydraulic pressure, under speed difference, that acts on a conventional wet clutch pack. By electronically altering this pressure via solenoids or other valves, the locking torque of the hydraulically controlled type TMD is controlled.

Another system is the electric motor controlled type TMD. An example for this is shown in Figure 5.

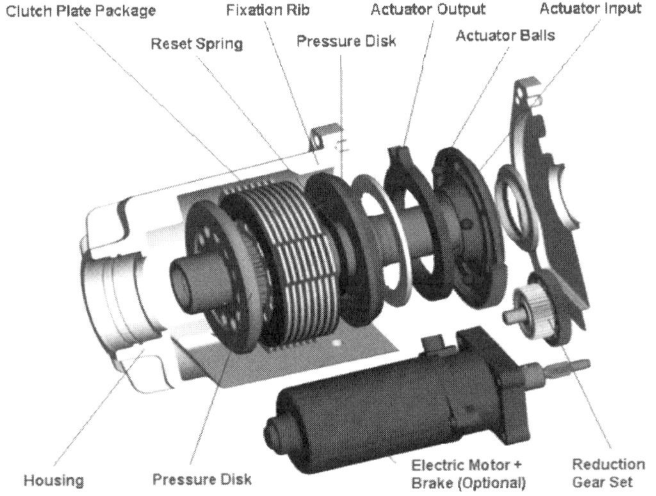

Figure 5: Electric Motor Controlled Type TMD

Here the torque supplied by an electric motor is transformed into a longitudinal force via a ball and ramp or cam mechanism, which activates a wet clutch pack. This means that the locking torque of the TMD is controlled by altering the electric motor current.

Additional information with respect to this type of electronically controlled device can be found in [1].

Other systems like torque-vectoring devices will not be explained in this paper.

RESULTS

INFLUENCE OF TMD ON TRACTION

On surfaces with different friction coefficients μ on the left and the right side of the vehicle (split-μ) open axle differentials have the disadvantage that they can only transmit the same (low) amount of torque to the high-μ side which is transmitted to the low-μ side. During acceleration on split-μ this typically leads to spinning wheel(s) on the low-μ side and insufficient acceleration.

Brake based traction control systems overcome this disadvantage by applying the brake of the spinning wheel. The brake torque at the spinning wheel results in an additional drive torque on the corresponding high-μ wheel, which gives the required traction improvement. But by this method a large portion of the available engine torque is wasted by braking the spinning wheel(s).

A similar situation is given when starting a 4WD vehicle on μ-transition. μ-transition is a road condition where the front wheels are positioned on low-μ and the rear wheels on high-μ or vice versa. With open center differential the wheels on low-μ tend to spin. Again brake based traction control systems brake the spinning wheels to produce extra driving torque at the high-μ wheels thus wasting engine torque.

A Torque Management Device like a limited-slip differential works in a different way: by partially or fully locking the corresponding differential, the available engine torque is redistributed to the wheel(s) with the better traction potential thus no energy is wasted.

The traction improvements, electronically-controlled TMD's are offering in comparison to brake based traction control systems respectively to a passive TMD, are demonstrated in the next chapters.

<u>Acceleration on Split-μ</u>

The first example is an acceleration test on a split-μ gradient with a permanent 4WD vehicle. The vehicle was equipped with TCS and two electronically controlled limited-slip differentials (e-motor controlled type TMD), one in the center differential and one in the rear axle. The results are shown in Figure 6.

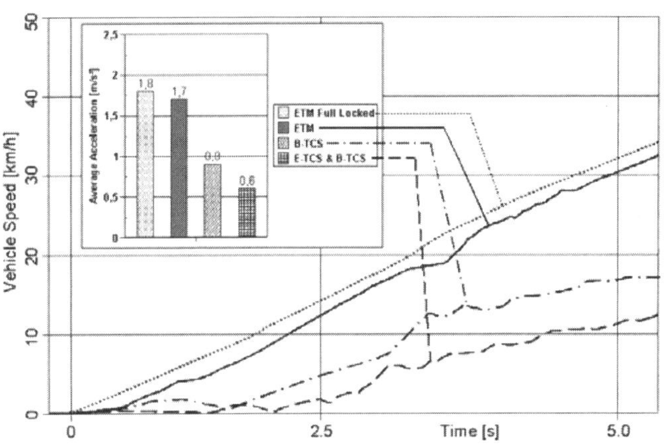

Figure 6: Acceleration Test Results on Split-μ with Permanent 4WD Vehicle

It is obvious that the slowest acceleration was produced with the vehicle with Engine & Brake TCS as torque is wasted by braking the spinning low-μ wheels and engine torque output is limited by the system. Switching off the engine control, leaving only Brake TCS active, lead to better acceleration. Further improvement was gained with the two active differentials (Electronic Torque Manager - ETM) activated and controlled automatically. Best acceleration was achieved by fully locking the active differentials.

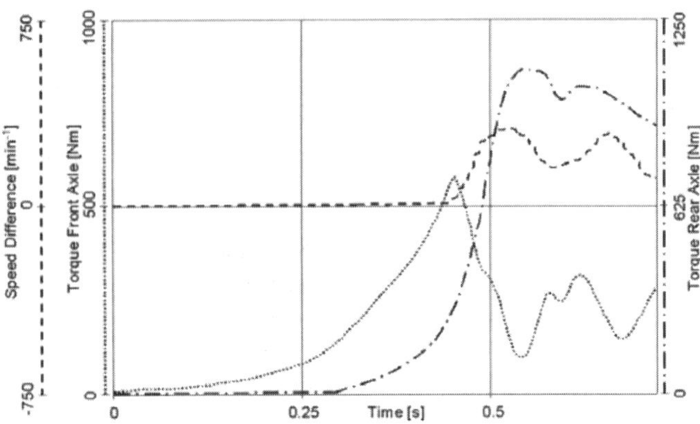

Figure 7: Acceleration Test Results on μ-Transition with On-Demand 4WD with Passive TMD

Figure 7 shows results from an acceleration test on μ-transition with an on-demand 4WD vehicle. This vehicle had a directly driven front axle and a viscous coupling on-demand driven rear axle. The front wheels were placed on ice and the rear wheels on dry pavement. Figure 8 shows the results for the same test maneuver with the same vehicle but with a controlled e-motor-type TMD as on-demand coupling.

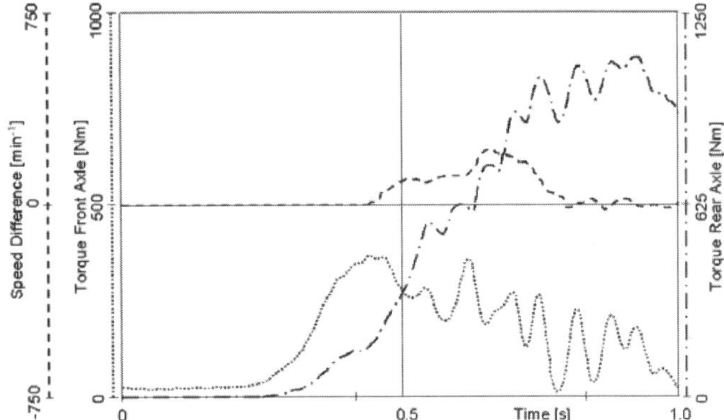

Figure 8: Acceleration Test Results on μ-Transition with On-Demand 4WD with Active TMD

Although the viscous coupling in this application is very stiff and reacts instantaneously, a speed difference between the coupling's input and output shaft of 300 min^{-1} (rpm) is required to achieve a locking torque of about 870 Nm (Figure 7). With an active on-demand system the torque increase shown in Figure 8 is faster than with the passive TMD and a higher torque of 1100 Nm is achieved at a lower speed difference of 200 min^{-1} (rpm). Figure 8 also shows that there was no delay in response of the TMD which was achieved by a pre-emptive feature in the control algorithm.

INFLUENCE OF TMD ON CORNERING

Steady-State Cornering

In general, passive TMD's increase, slightly, the understeering tendency which state-of-the-art-vehicles typically exhibit [2][3]. For actively controllable TMD's the vehicle's self-steering behavior during steady state cornering is therefore dependent upon the control strategy. Typically, active TMD's are not actuated during steady state cornering thus the self-steering behavior does not change.

Throttle-Off During Steady-State Cornering

When a vehicle is driven in a corner close to the maximum lateral acceleration, and the throttle is suddenly lifted, it typically exhibits an oversteering reaction. There are several physical reasons for this behavior but the most important is the weight re-distribution caused by the deceleration [4]. The front axle gets a higher load therefore the side slip angles at the front wheels are decreased while the rear axle load becomes lighter resulting in increased side slip angles at the rear wheels. This means the vehicle starts to oversteer.

Depending on the type of vehicle, weight and weight distribution, friction of the road, and other factors, this throttle-off oversteer can be so heavy that a normal driver is not capable to stabilize the vehicle again. ESP stabilizes the vehicle in an oversteering event by braking the outer front wheel. The braking force at this wheel creates an understeering yaw moment for the vehicle which counteracts the throttle-off oversteer.

TMD in general can improve throttle-off oversteer. Figure 9 shows the influence of various active TMD configurations on throttle-off oversteer.

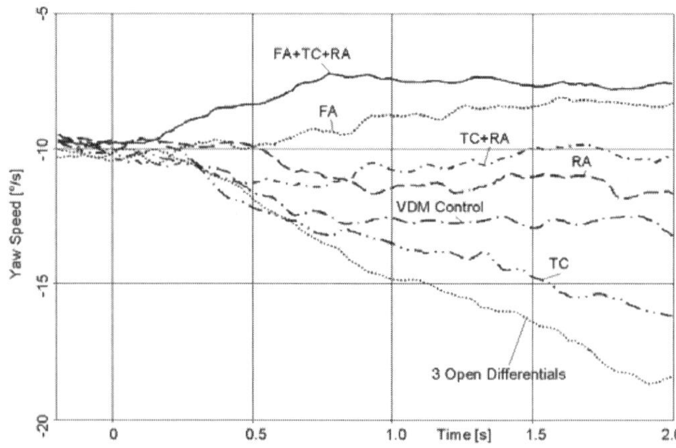

Figure 9: Results from Throttle-Off in a Bend Tests with Permanent 4WD Vehicle with Different Active TMD Configurations

The tests have been performed with a permanent 4WD vehicle with a 50/50 static torque distribution in the center differential. This vehicle was equipped with three active TMD's across the differentials. Each TMD could be switched on and off independently. Figure 9 shows the yaw speeds for various configurations during the throttle-off maneuvers performed on snow in "open-loop" mode, i.e. the steering wheel angle was kept constant throughout the test.

It is obvious that the vehicle with three open differentials exhibited the heaviest oversteering reaction as the yaw speed shows a significant increase ("3 Open Differentials" in Figure 9). Activating all three TMD's after throttle-off changes this oversteering even into an understeering reaction (yaw speed decrease after throttle-off) ("FA+TC+RA" in Figure 9). Activating only the TMD in the center differential leads to a small improvement of the oversteering reaction ("TC"), while activation of the TMD in the rear axle has a significant influence ("RA"). A further step is possible by activating the TMD's in the rear axle and the center differential simultaneously ("TC+RA"). But even this is exceeded by the use of a TMD in the front axle only ("FA"). The trace called "VDM Control" in Figure 9 shows the vehicle behavior tuned with the Vehicle Dynamics Module (VDM), which is part of GKN's control strategy for ETM. This performance was specifically tuned to the requirements of a car manufacturer. Generally it can be stated that this driveline configuration in combination with an intelligent control strategy allows for a tuning of the throttle-off behavior in the range between the trace "3 Open Differentials" and the trace "FA+TC+RA" shown in Figure 9.

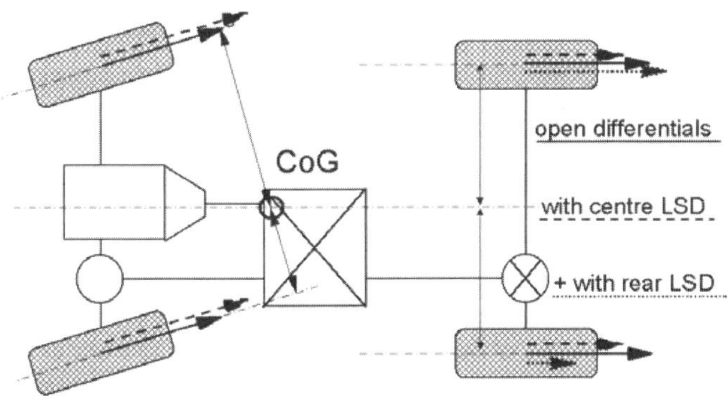

Figure 10: Influence of TMD on Throttle-Off in a Bend

Figure 10 shows a principle sketch of the vehicle's driveline. With open differentials, the braking forces at the wheels resulting from the engine's drag torque after throttle-off are equal. With a TMD added to the center differential the braking forces at the rear wheels become smaller while the forces at the front wheels increase as the rear wheels are rotating slower than the front wheels. The outer front wheel has a longer lever arm around the vehicle's CoG (Center of Gravity) than the inner wheel. Therefore, with a TMD the increased force at the front outer wheel causes – in direct comparison to the version with open differential – an understeering yaw moment. As the steering angle during the maneuver is

small also the influence of TMD is small as seen in Figure 9.

If the rear axle is equipped with a TMD the braking force at the outer rear wheel is greater than at the inner wheel due to the speed difference between outer and inner rear wheel. This unequal pair of forces creates an understeering yaw moment around the CoG which reduces throttle-off oversteer more significantly. On the front axle with TMD the outer braking force is larger than on the inner wheel, which creates a similar effect as on the rear axle. But, after throttle-off the front axle gets more vertical load. Thus the unequal forces can be "better transmitted to the ground" and the effect of a TMD on the front axle is more significant than with a rear TMD.

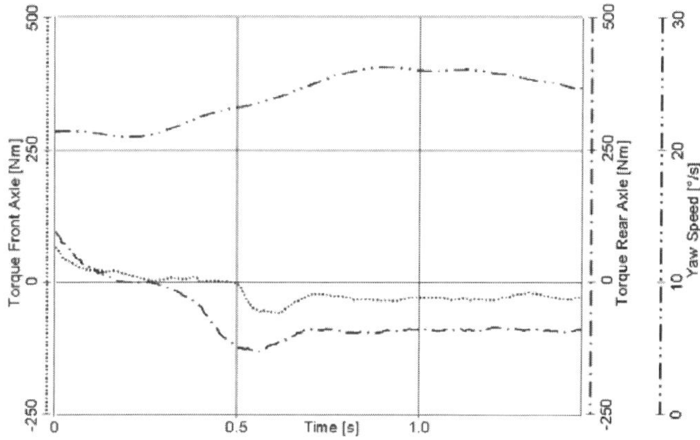

Figure 11: Results from Throttle-Off in a Bend Test with On-Demand 4WD Vehicle with Passive TMD

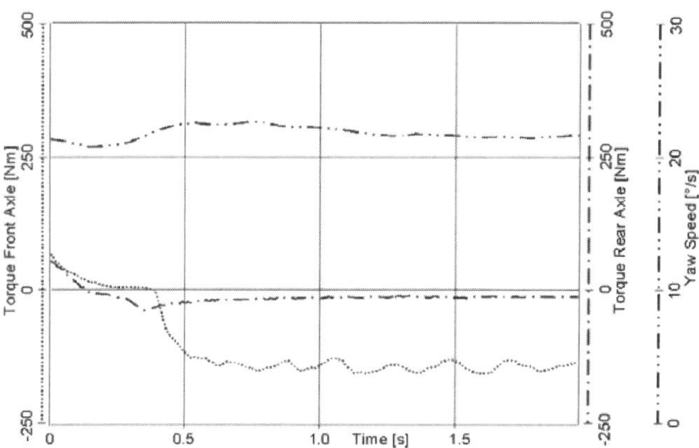

Figure 12: Results from Throttle-Off in a Bend Test with On-Demand 4WD Vehicle with Active TMD

Figures 11 and 12 show results from throttle-off tests on snow with an On-Demand AWD vehicle. Figure 11 shows the results for the vehicle with a passive TMD (viscous coupling) and Figure 12 for the same vehicle with an actively controllable on-demand coupling (E-Motor-Type).

The yaw-rate in Figure 11 shows an increase after initiation of throttle-off. Again, this indicates a significant oversteer reaction. In Figure 12 it is obvious that the increase of the yaw speed is much smaller, i.e. this vehicle variant oversteers less and is easier to control by the driver. It can also be seen from Figure 11 that the engine's drag torque is distributed to both axles. The resulting braking forces on the rear wheels reduce their lateral guidance which leads to increased sideslip angles and the observed oversteer.

The Vehicle Dynamics Module in the control algorithm of the active TMD detects this behavior and opens the on-demand coupling accordingly. As Figure 12 shows the engine's drag torque is transmitted to the front axle only while the rear axle torque is almost zero. So the rear wheels can transmit larger lateral forces, the sideslip angles are reduced and the throttle-off oversteer is significantly reduced.

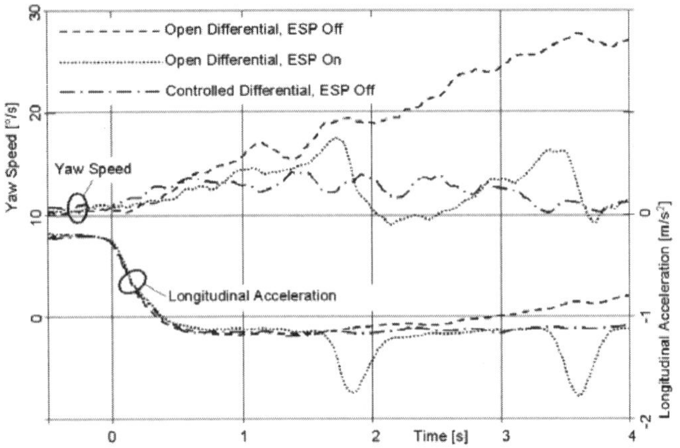

Figure 13: Results from Throttle-Off in a Bend Tests on Snow with FWD vehicle, Comparison of Active TMD and ESP Performance

Figure 13 shows an increase of the yaw speed after throttle-off for a FWD with open differential and deactivated ESP. As explained before a TMD improves this behavior by increasing the braking force on the outer driven wheel and decreasing it on the opposite side. The yaw speed trace for the ESP variant shows a similar behavior as for the TMD vehicle, i.e. the vehicle also oversteers less than without ESP. But an active TMD can refine the response of the vehicle.

At the bottom of Figure 13 the vehicle's longitudinal decelerations are shown for the three variants tested. It is obvious that with open differential and with active TMD the deceleration is quite steady while, with ESP, two deceleration peaks can be seen. These peaks result from the ESP activation when the outer front wheel is braked in pulses. These deceleration peaks may be annoying for the driver. In this situation, an active TMD offers a similar stability improvement as ESP, but it is smoother and therefore more comfortable.

<u>Acceleration During Steady-State Cornering</u>

Hard acceleration out of a bend especially with high-powered vehicles and/or on slippery surfaces, typically leads to significant changes of the steady-state self-steering behavior. A FWD vehicle understeers, RWD vehicles might exhibit a heavy oversteer and AWD vehicles show neutral, understeer or oversteer behavior dependent on driveline concepts, weight distribution and other parameters. TMD's are to improve this behavior as shall be demonstrated here on the example of an on-demand AWD vehicle only.

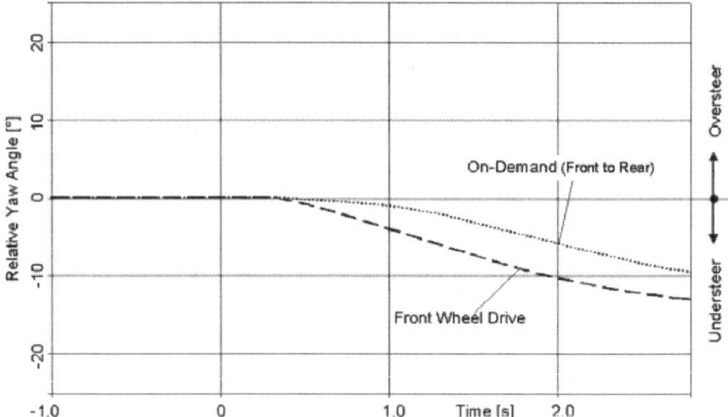

Figure 14: Results from Acceleration in a Bend Tests on Snow, Comparison of FWD and On-Demand 4WD Vehicle

In Figure 14, the so-called "Relative Yaw Angle" is plotted against time. The Relative Yaw Angle shows the actual yaw angle of the vehicle relatively to an absolute neutral reaction. A more detailed description can be found in [2]. It can be seen that the FWD version exhibits the expected understeering behavior. This is due to the fact that a tire cannot create the required lateral forces if a high longitudinal force is present. Therefore, when a FWD vehicle is accelerated hard in a corner on snow the traction forces minimize the lateral forces in a way that the vehicle understeers heavily.

In an on-demand AWD vehicle with primarily driven front axle the on-demand coupling distributes a portion of the drive torque to the rear axle. This creates lateral guidance at the front wheels, which reduces understeer. This is shown by the "on-demand" graph in Figure 14. Note that too much torque to the rear axle could result in oversteer. This is achievable with an active TMD and a sophisticated control strategy while a passive TMD has to be tuned to an acceptable compromise.

In other driveline concepts like RWD, permanent AWD etc. TMD's typically also have positive influences on the acceleration behavior in a bend. More details can be found in [3].

INFLUENCE OF TMD ON LANE CHANGE BEHAVIOUR

<u>Double Lane Change</u>

Double Lane Change maneuvers often create dangerous situations, as vehicles tend to exhibit instabilities, which cannot be handled by the majority of drivers. Figure 15 shows typical test results from a double lane change test with a permanent AWD vehicle.

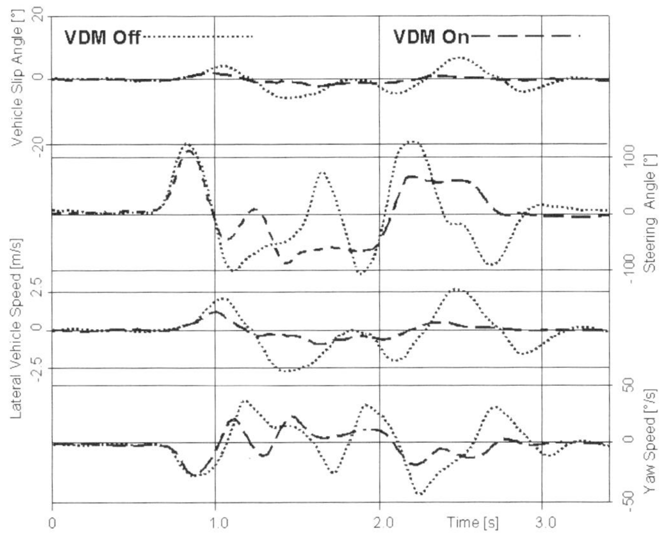

Figure 15: Results from Double Lane Change Tests on High-μ with Permanent 4WD Vehicle

In Figure 15 the vehicle with three open differentials is compared with the same vehicle with three active TMD's working as limited-slip devices across the differentials. It can be seen from the steering wheel angle traces that the required steering wheel inputs for this maneuver are smaller with the TMD's active than with three open differentials. Beyond it the version with three open differentials requires much more correction to recover the stability of the vehicle. This can be seen from the stronger oscillations of the steering wheel signal. The signals for the yaw speed, lateral vehicle acceleration and vehicle slip angle also show larger amplitudes and more oscillations with open differentials than with TMD's. The vehicle with TMD's active is more stable and much easier to control than the vehicle with three open differentials.

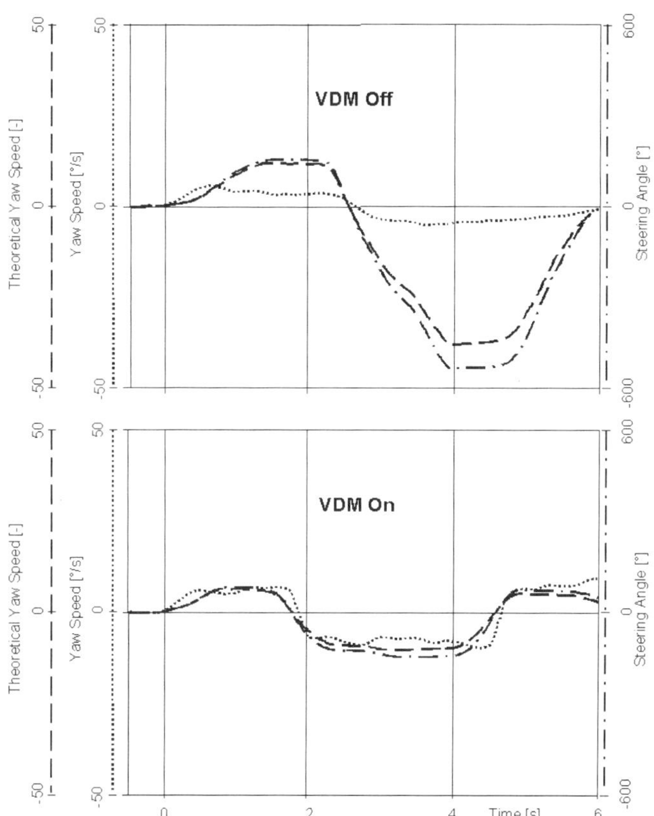

Figure 16: Results from Double Lane Change Tests on Wet Polished Granite with On-Demand 4WD Vehicle

Figure 16 substantiates the results from a double lane change maneuver under throttle-off condition performed on polished granite (friction similar to that of ice/snow). The test vehicle was equipped with an on-demand 4WD driveline with a directly driven front axle and the rear axle indirectly driven via an active on-demand E-motor type TMD. First the vehicle was driven with the Vehicle Dynamics Module switched off. It can be seen from the top graph in Figure 16 that although the driver feeds in very high steering angles the resulting yaw reactions are not sufficient to keep the vehicle on its intended path. There is a big difference between the theoretical yaw speed and the actual yaw speed. The theoretical yaw speed was calculated from the steering angle, vehicle speed, wheelbase etc. This behavior is the result of other tuning requirements, which do not allow the on-demand coupling to be activated in throttle-off situations. That means that the drag torque of the engine is transmitted to the surface by the front wheels only. The resulting large longitudinal forces at the front tires limit the potential for the lateral forces thus the vehicle cannot follow the steering input.

The VDM detects this behavior and activates the TMD accordingly. This results in a redistribution of the engine's drag torque from the front to the rear wheels, and the front wheels recover enough lateral force potential and the vehicle follows the driver's input more accurately. In the bottom graph of Figure 16 it can be seen that the steering input is smaller and the actual and the theoretical yaw speeds are closer to each other than without VDM.

ABS/ESP COMPATIBILITY

For ABS and even more for ESP it is important that the wheels can be controlled independently. Any partial or complete connection of an individual wheel with another wheel or axle via TMD deteriorates more or less the functionality of ESP and sometimes of ABS [5]. For actively controllable TMD's the requirements regarding ABS/ESP-compatibility therefore are: The switch-off times at an ESP or ABS event shall be as small as possible and the remaining or drag torque of the "open" TMD shall be as small as possible.

Drag Torque

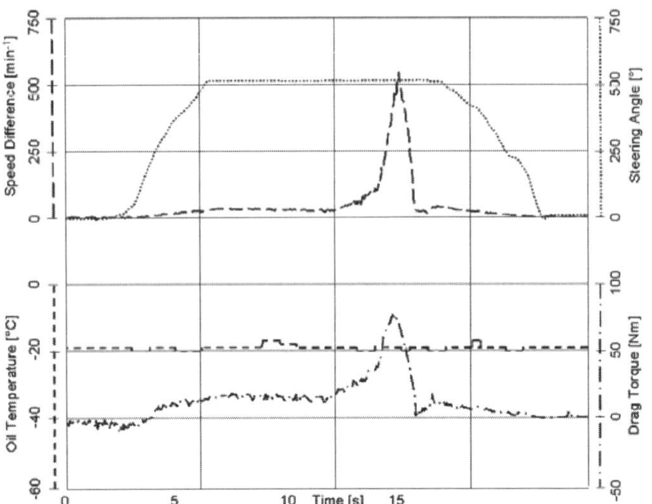

Figure 17: Drag Torque Measurement at -20°C in a Vehicle with E-Motor Type TMD (GKN ETM)

As drag torques of multi-plate clutch pack couplings are typically higher at lower temperatures, cold start tests with a vehicle equipped with an E-motor type TMD as on-demand coupling have been performed at around -20 °C. Figure 17 shows the results of these tests. It is obvious that the drag torque of the TMD tested does not exceed 20 Nm at small speed differences. Even at speed differences of more than 500 min^{-1} (rpm) the drag torque is smaller than 80 Nm.

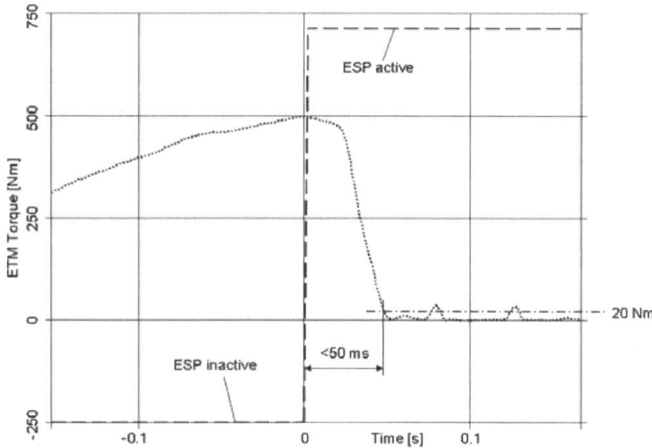

Figure 18: Switch-Off Time Measurement in a Vehicle with E-Motor Type TMD (GKN ETM)

Switch-off times were measured in the vehicle between ABS/ESP-active signal and the time when the torque falls below 20 Nm. That means it is the switch-off time of the complete system including electronic and mechanical response times. The switch-off time measured here was approximately 50 ms (see Figure 18). The time to build up 10 bar in the brake system at a temperature of -20 °C is 100 ms [6]. So the measured switch-off time is sufficient for ABS/ESP compatibility.

CONCLUSION

Torque management devices, especially when actively controlled, offer significant improvements to vehicle dynamics. With electronically controllable TMD's it is possible to improve traction and stability simultaneously without the necessity to make compromises as is typically required for passive devices. These active TMD's cannot replace ESP but extend the safety range in which ESP is not yet required to intervene. As ESP interventions are often rated as annoying, this is increasingly becoming an advantage, which should not be underestimated. The same applies for TCS with its pulsating action and waste of energy. A combination of these systems would enable the use of several synergy effects resulting in overall increased traction, vehicle stability, comfort and safety without affecting "fun to drive"-behavior.

A further step of enhancing the functionality of TMD's can be reached by the use of torque vectoring devices, which provide increased yaw control, but cost versus benefits of such a configuration needs to be understood.

REFERENCES

1. Heinrich Huchtkoetter, Heinz Klein: *The Effect of Various Limited-Slip Differentials in Front-Wheel Drive Vehicles on Handling and Traction*, SAE 960717, 1996

2. Theodor Gassmann, John A. Barlage: *Electronic Torque Manager (ETM): An Adaptive Driveline Torque Management System*, SAE 2004-01-0866, 2004

3. Michael Hoeck, Christian Gasch: *The Influence of Various 4WD Driveline Configurations on Handling and Traction on Low Friction Surfaces*, SAE 1999-01-0743, 1999

4. Adam Zomotor: *Fahrwerktechnik: Fahrverhalten*, Vogel Buchverlag Wuerzburg ISBN 3-8032-0774-7, 1987

5. Manfred Burckhardt: *Fahrwerktechnik: Radschlupf-Regelsysteme*, Vogel Buchverlag Wuerzburg ISBN 3-8023-0477-2, 1993

6. Thomas Straub: *Vehicle Stability Control (VSC) and It's Rapid Product Improvement Cycle to Meet a Wide Customer Base*, Active Safety in Vehicle Traction & Stability Systems TOPTEC Symposium, 27-28 September 1999 University of Technology Vienna, Austria

DEFINITIONS, ACRONYMS, ABBREVIATIONS

4WD: Four-Wheel Drive

ABS: Anti-Lock Braking System

B-TCS: Brake Control TCS

CoG: Center of Gravity

DSC: Dynamic Stability Control

ESP: Electronic Stability Program

E-TCS: Engine Torque Controlling TCS

ETM: Electronic Torque Manager

FA: Front Axle

FWD: Front-Wheel Drive

IVD: Interactive Vehicle Dynamics

LSD: Limited-Slip Differential

RA: Rear Axle

RWD: Rear Wheel Drive

TC: Transfer Case

TMD: Torque Management Device

TCS: Traction Control System

TRACS: Traction and Stability Control

VDC: Vehicle Dynamics Control

VDM: Vehicle Dynamics Module

Electronic Torque Manager (ETM®): An Adaptive Driveline Torque Management System

Theodor Gassmann and John A. Barlage

GKN Driveline

ABSTRACT

Today's automotive applications of brake based traction control (B-TCS) and dynamic stability control (DSC) continue to penetrate the market as an approach to enhance vehicle traction and stability. However, there is also a growing trend in the use of electronically controlled couplings and differentials.

The Electronic Torque Manager (ETM®), developed and manufactured by GKN Driveline, provides an innovative solution to driveline torque distribution as well as enhanced vehicle stability combined with adaptive control strategies.

The ETM®'s unique actuation system actively controls the torque transfer between axles or across an axle differential. The system consists of a ball ramp mechanism, in conjunction with a multi-plate clutch, directly actuated by an electric motor. An Electronic Control Unit (ECU) incorporating adaptive control strategies provides the required control.

This paper provides an overview of the ETM® working principle, functional characteristics, application scope, as well as key performance features.

INTRODUCTION

In recent years brake based traction control (B-TCS) and vehicle dynamic control systems (DSC) became more and more common in the market. This trend together with advanced engine, transmission and chassis control systems has forced the introduction of modern sensor and computer technology including CAN-Bus systems.

A wide range of sensor and vehicle information is now available on the CAN, which can be shared by the various control systems including active torque managing devices. This together with the reduced cost of electronic components makes actively controlled torque management devices more and more attractive.

Although DSC and B-TCS are effective ways to enhance traction and stability by using brake intervention, a significant improvement can still be achieved when combining them with limited-slip devices. Furthermore, On-Demand type AWD systems are becoming very popular which require a torque transfer coupling to control the torque flow between the axles.

Conventional passive limited slip devices are not fully compatible with brake based vehicle dynamic control systems. This requires compromises in the DSC performance or more likely the traction performance of the limited slip device.

The wide use of DSC together with the need for further traction and stability enhancement achieved by AWD systems with limited slip devices has triggered the move from passive LSD's to actively controlled torque management devices.

The overall performance of active torque management devices is not only defined by the performance of the mechanical system. The adaptive control strategy together with the available sensor signal on the CAN Bus are essential for their performance.

Today's active torque management devices (A-TMD) are expected to provide superior traction and to be fully compatible with DSC and B-TCS. This requires adaptive control as well as quick engagement and release time.

To achieve not only compatibility with current vehicle dynamic control systems, but act complementary with future DSC and support advanced integrated vehicle dynamic control strategies, the performance requirements with respect to control accuracy as well as activation and release time continue to rise.

WORKING PRINCIPLE

The actively controlled torque management device described in this paper uses an electric motor to engage a multi-plate clutch using a gear reduction and a ball ramp mechanism. Figure 1 gives an overview of the main mechanical parts of the described A-TMD.

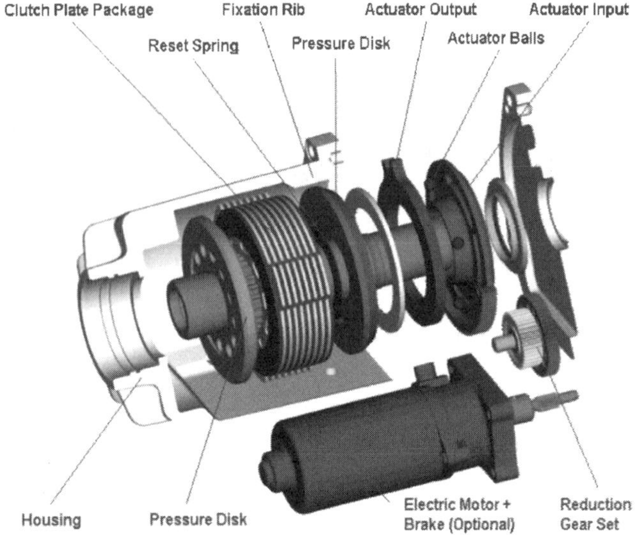

Figure 1: A-TMD Mechanical Components

The e-motor (electric motor) torque is multiplied by means of a reduction gear to drive the rotating input disc of a ball–ramp mechanism. The ball-ramp mechanism consists of two discs whereby one is rotationally fixed in the housing and the other one is driven by the e-motor. Balls are located in oppositely arranged unidirectional grooves with a defined slope (ramps). Relative rotation of the discs to each other drives the balls up the ramp forcing the discs apart, converting rotation into axial movement and the e-motor torque into axial force. The axial force is then applied to the multi-plate clutch through a pressure disc and a thrust bearing to generate the desired locking torque. The actuator itself is non-rotating, totally independent from the driveline rotation or speed difference across the clutch.

Unlike other actuation systems that use an electromagnetic actuated pilot clutch, the e-motor directly controls the actuation mechanism. A return spring forces the actuator into the open position when the e-motor is powered off and assures fail-safe behavior of the system under any circumstance.

The e-motor size as well as the gear ratio between motor and the ball-ramp actuator can be optimized for the specific application in terms of current consumption and engagement speed. The e-motor is optionally equipped with an electric brake feature, temperature and/or position sensors. The brake is mainly required in axle applications to allow full lock for extended periods of time.

The clutch pack is sized for the required locking torque. Various friction material/oil combinations for axle or transfer case applications have been validated but might have to be optimized for specific application needs.

The electric motor is controlled via an electronic control unit (ECU). By powering the e-motor, the ball-ramp mechanism directly engages the clutch. The design

specific relationship between e-motor torque and locking torque is used for the adaptive control.

When driven with DC current, the A-TMD exhibits an undesired torque hysteresis caused by mechanical friction in the actuator (solid line in Figure 2).

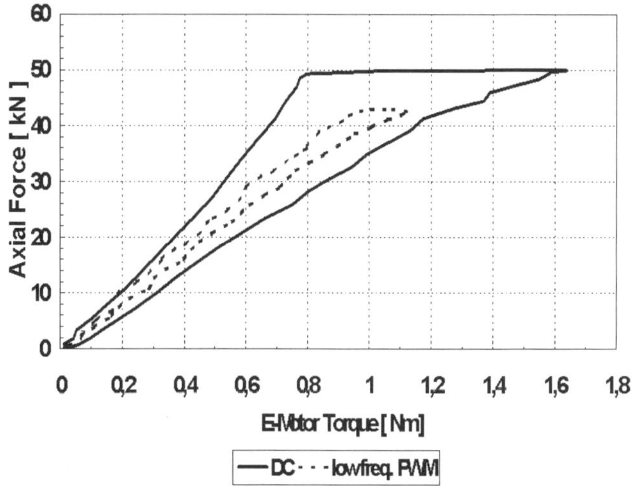

Figure 2: Effect of Low Frequency Modulation on Mechanical Hysteresis

By inducing a vibration in the actuator using low frequency Pulse Width Modulation (PWM) of the DC current the friction and hence the hysteresis can be significantly reduced (dotted line in Figure 2). The almost hysteresis-free actuator characteristic guarantees superior control behavior and control accuracy.

ENGAGEMENT AND RELEASE TIME

Quick engagement and rapid release are essential performance criteria to achieve superior traction as well as vehicle dynamic behavior and assure full DSC compatibility. Engagement and release time should not deteriorate at lower temperature to assure good performance and DSC compatibility, particularly in winter conditions.

Figure 3 shows a typical engagement event, demonstrating an engagement time of less than 150ms. Additional to the already quick engagement response, this direct actuation method is independent of speed difference direction (no indexing) and provides a truly pre-emptive engagement.

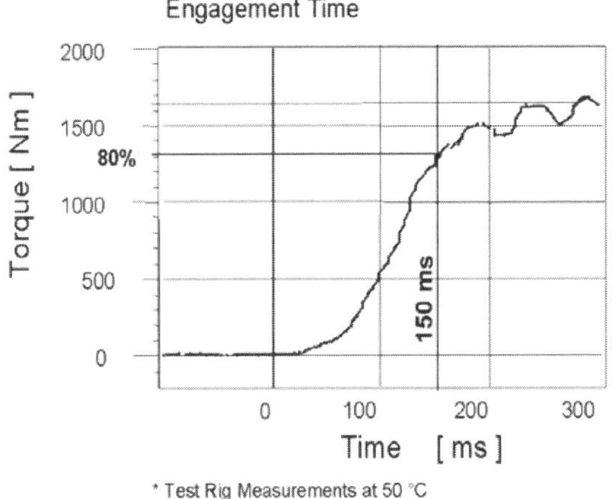

Figure 3: Engagement Time

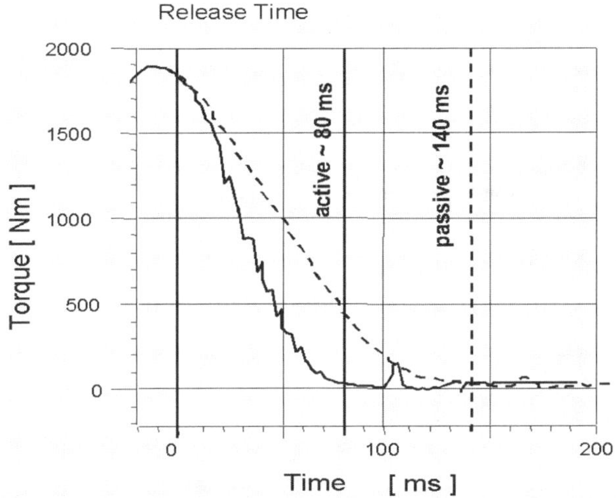

Figure 4: Release Time

Based on throttle acceleration and/or engine torque information the ETM can be locked before wheel or clutch slip even occurs. This allows pre-emptive control strategies to achieve superior traction with minimal wheel slip.

In continuous operation the engagement time is further enhanced by advanced control strategies using a combination of e-motor position information and Pulse Width Modulation (PWM).

Pump or electromagnetic coil driven active limited slip devices are released by switching off the coil or releasing the pressure. Similar release times are achieved with the e-motor driven A TMD when just switching off the e-motor. This is shown as "passive release" in Figure 4.

A significant reduction of the release time can be achieved when reversing the e-motor direction to actively back drive the actuator mechanism. Release times of less than 80ms for the "active release" (see Figure 4) demonstrate superior disengagement time. This assures full DSC compatibility under all conditions, particularly at low temperature.

DRAG TORQUE

The A-TMD drag torque characteristic is particularly important in modern On-Demand AWD systems in terms of ABS/DSC and mini spare wheel compatibility, two wheel towing and driveline losses during high speed driving.

The continued need to reduce fuel consumption of AWD vehicles will require the use of disconnectable driveline systems. In this case, low drag torque of the controlled coupling is essential to achieve the desired fuel saving effect.

The e-motor driven actuator is totally independent from the speed difference across the system and is not subject to any delta-speed induced self-locking effects. This is clearly apparent by the rig test results shown in Figure 5.

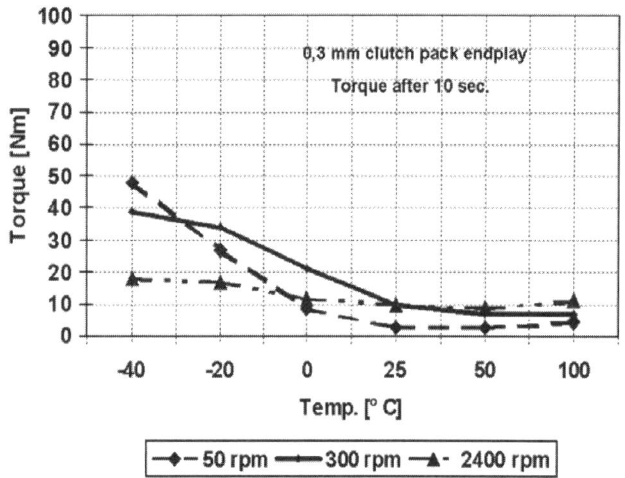

Figure 5: Drag Torque Rig Test Results

The low drag torque level even at low temperature demonstrates full mini spare wheel and two-wheel tow capability of the system. The slight increase of the drag torque over speed difference is related to the oil shear effect in the multi-plate clutch.

Further reduction of the remaining clutch drag can be achieved using measures known from automatic transmissions, like increased endplay or separation springs between the friction plates.

This low drag characteristic allows the presented A-TMD to be suitable for high efficiency part-time disconnectable AWD systems.

APPLICATIONS

The system described in this paper can be applied as an active limited slip device in conjunction with an open differential in axles and center differential or as an active coupling in so-called "Hang-On" drivelines (On-Demand type AWD).

Figure 6 shows an example using the system in an axle (currently in production). The compact layout of actuator and clutch pack as well as the flexible arrangement of the e-motor provides high locking torque even in restrictive packaging situations.

A cost and weight effective AWD layout is provided using the A-TMD to control the secondary driven axle in a Hang-On type AWD driveline.

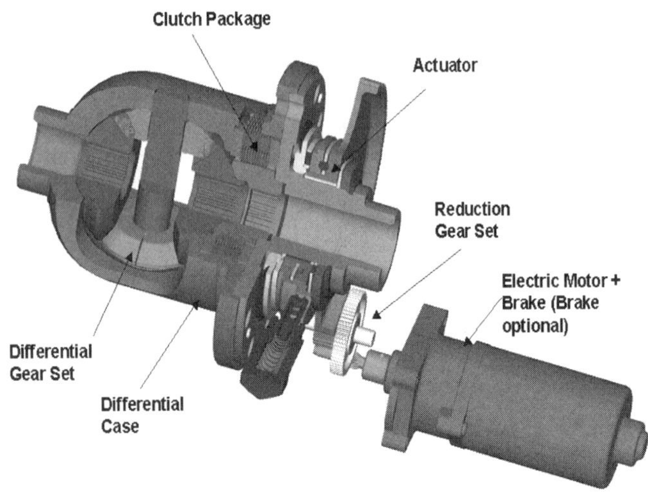

Figure 6: Axle Application

In FWD based AWD vehicles with transverse engines, the active coupling is typically located in front of the rear axle controlling the torque flow to the rear axle as shown in Figure 7.

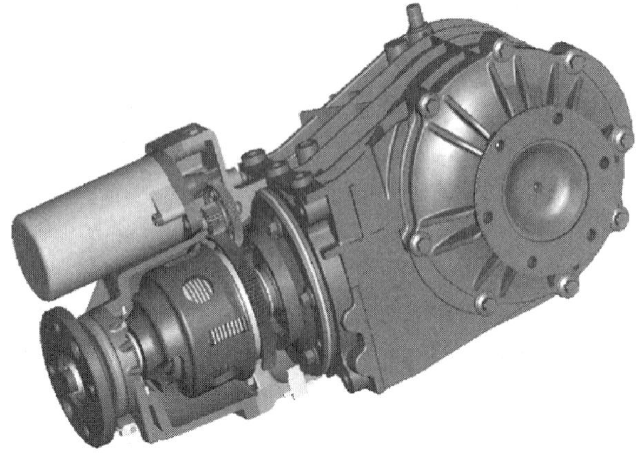

Figure 7: Hang-On Concept with A-TMD on Rear Axle

The high control accuracy of the A-TMD assures precise torque management below the maximum allowable driveline torque thereby allowing for downsizing of the secondary axle and driveline components (weight & cost benefits). Other mounting locations such as in a transfer case or propshaft are feasible as well.

CURRENT CONSUMPTION

The presented e-motor activated torque management device does not use kinetic energy from the driveline for engagement. This is key for the superior performance of the system described above, but requires more electric power compared to coil or pump driven systems.

The size of the e-motor and therefore the current consumption is defined by the required locking torque capacity and response time.

The result of a vehicle data acquisition from an active Hang-On AWD application with high winter road content is shown in Figure 8.

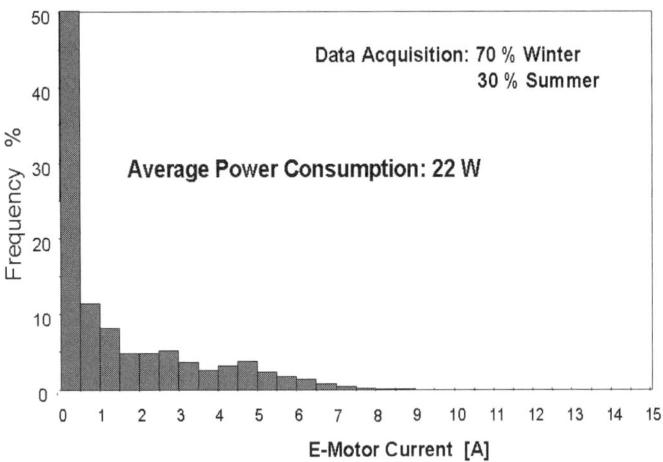

Figure 8: Electric Motor Current Consumption

Although the maximum current of the A-TMD peaks at approximately 15 A the average power consumption for the whole test was 22 W.

CONTROL

Modern CAN-Bus systems allow sharing of sensor signals and messages between the various electronic systems of a vehicle. The position of the A-TMD in the chassis/driveline control hierarchy as well as the available signals on the CAN defines the level of achievable control strategy.

Figure 9 shows two typical applications of an active TMD in terms of control hierarchy. As driveline master ECU, the system is configured to run vehicle level control strategies. The ECU reads all required information from the CAN bus and defines the level of locking torque to provide the desired traction and stability enhancement. In this case, the A-TMD acts autonomously during normal operation but sends status messages to other relevant control systems via the CAN.

Basic slip control is possible if only wheel speeds and throttle position are available on the CAN. More advanced control strategies can be applied, if signals like steering angle, YAW speed and longitudinal and transversal acceleration are available.

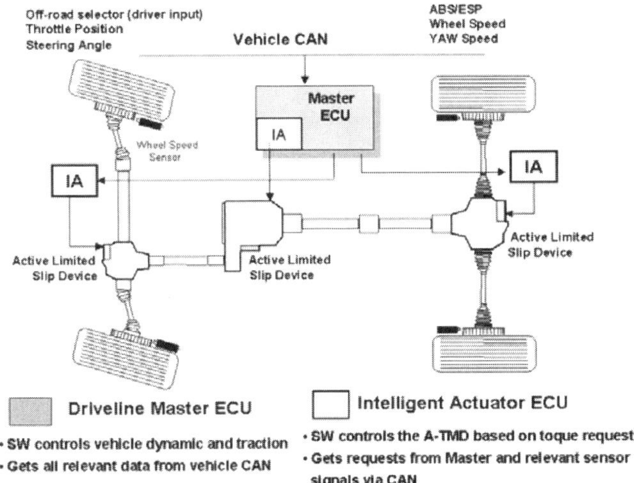

Figure 9: ECU Configurations

When the threshold of vehicle stability is exceeded the DSC or the chassis control system will then take command over the A-TMD. In this case, the A-TMD falls back into an intelligent actuator operation mode carrying out commands like active release or lock to a requested torque.

The A-TMD will also work in this operation mode when acting as intelligent actuator at the lower end of the hierarchy. In this case the integrated vehicle dynamic controller or any other appropriate controller in the powertrain will act as master and manages the overall driveline torque to achieve a maximum in traction and stability at all times.

The A-TMD will receive various requests from the master ECU such as locking torque, active release, brake apply etc. This requires a totally different set of Software than the vehicle level control strategy. In this case the ECU is configured to achieve high control accuracy and quick response, manage communication, diagnostic functions and fault actions, and to guarantee robust and reliable operation of the system.

Additional information with respect to vehicle level control strategy and vehicle dynamics can be found in [1].

CONCLUSION

The ETM® from GKN Driveline is based on an electric motor driven ball-ramp actuator mechanism to engage a multi-plate clutch for torque transfer.

Controlling the electric motor allows an accurate control of the generated locking torque. The low frequency modulation of the DC current minimizes mechanical hysteresis for optimum controllability.

The electric motor driven actuator works totally independent from the speed difference across the system, hence providing:
- Pre-emptive locking
- Low drag torque even at low temperature and high speed difference
- No indexing when changing speed difference direction
- Fail-safe behavior

The ability to drive the e-motor in reverse to actively back drive the system provides superior release time.

Quick engagement is achieved due to the rapid response of the actuator motor and advanced control strategies utilizing angular position of the motor.

Optimal packaging with high torque capacity is achieved by the compact actuator-clutch pack arrangement as well as the option to locate the electric motor in various positions.

The ETM® can be used for adaptive torque management between the axles in full time or Hang-On AWD systems or across an axle.

The superior performance of the ETM® in terms of response time, control accuracy and low drag torque makes the system suitable for today's and future driveline systems.

REFERENCES

1. Heinrich Huchtkoetter, Theodor Gassmann: *Vehicle Dynamics and Torque Management Devices*, SAE 2004-01-1058, 2004

DEFINITIONS, ACRONYMS, ABBREVIATIONS

AWD	- All Wheel Drive system
DSC	- Dynamic Stability Control
B-TCS	- Brake Traction Control System
LSD	- Limited Slip Device
A-TMD	- Actively controlled Torque Management Device
ETM®	- Electronic Torque Manager
DC	- Direct Current
ECU	- Electronic Control Unit
PWM	- Pulse Width Modulation
CAN	- Computer Array Network
IA	- Intelligent Actuator

The Advantages of an Electronically Controlled Limited Slip Differential

Jerry Kinsey
Dana Corporation, Torque-Traction Technologies, Inc.

ABSTRACT

Limited slip differentials offer significant performance benefits over standard differentials. Depending on the style and design, mobility and/or handling can be improved. Oftentimes, one must be improved at the expense of the other. For a speed sensitive limited slip differential, mobility can be improved by aggressive tuning. This would normally give adverse handling effects, due to the aggressiveness. If the differential is tuned for handling characteristics, mobility will be compromised, due to the moderate tuning. To optimize the performance, tuning must be adjustable according to the vehicle's input parameters. This paper will show the benefits of adjustable tuning in handling, mobility, and compatibility with other vehicle systems, such as ABS, Traction Control, and Stability Control.

INTRODUCTION

A limited slip differential can offer many benefits over a standard differential. A standard differential is limited by the traction available at the wheel with the lowest friction. A limited slip differential offers the ability to bias torque from the wheel with little traction to the wheel with better traction. This can offer improved mobility and handling characteristics. Traditional limited slip differentials have inherent limitations due to design. By electronically controlling the differential's output, mobility and handling can be optimized for any vehicle.

DIFFERENTIALS

Differentials were developed to allow two drive wheels to rotate at different speeds, e.g. when traversing a curve, while still transmitting power. A standard differential, because of the gearing, allows the drive wheels to rotate at different rates while still transmitting drive torque to each wheel. The drawback to this differential is that the torque at each wheel must be equal. Consequently, if one wheel encounters a slick surface, the wheel with

good traction is limited to the torque at the wheel on the slick surface. When one drive wheel slips, mobility is reduced or lost. Limited slip differentials were developed to address this problem. A limited slip differential has the ability to bias more torque to the wheel with better traction, when one wheel slips. Limited slip differentials come in three basic types. Locking differentials, which provide the ultimate in mobility because of the ability to bias all available torque to one wheel, will not be discussed. These differentials must be disabled when turning to prevent dangerous handling conditions. They provide no advantage to handling if disengaged, and detract from handling if allowed to engage.

Torque sensitive differentials bias torque as a function of the torque available at the slipping wheel. These differentials typically have fixed bias ratios for torque transfer, limiting their effectiveness. Figure 1 shows a typical graph of the torque at the slipping wheel vs the non-slipping wheel for a standard differential and a torque sensitive limited slip differential.

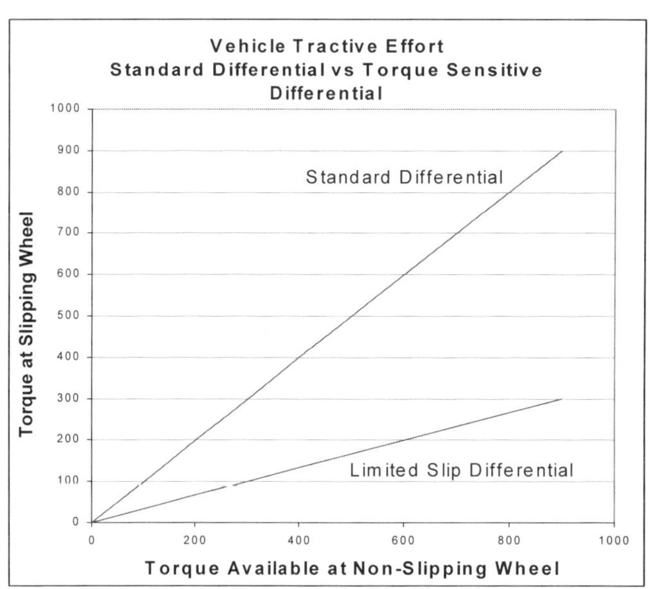

Figure 1: Tractive Effort for Standard And Torque Sensitive Limited Slip Differentials

Since these differentials bias torque as a function of the torque available at the slipping wheel, if a wheel is off the ground with no torque, the differential can bias nothing to the good wheel. For this reason, many torque sensitive differentials are designed with preload. Even with a wheel off the ground, there is some torque available to the wheel with good traction. Preload must be limited to prevent adverse handling effects in the vehicle, limiting its effectiveness. Figure 2 shows a comparison of a standard differential to a torque sensitive differential with preload.

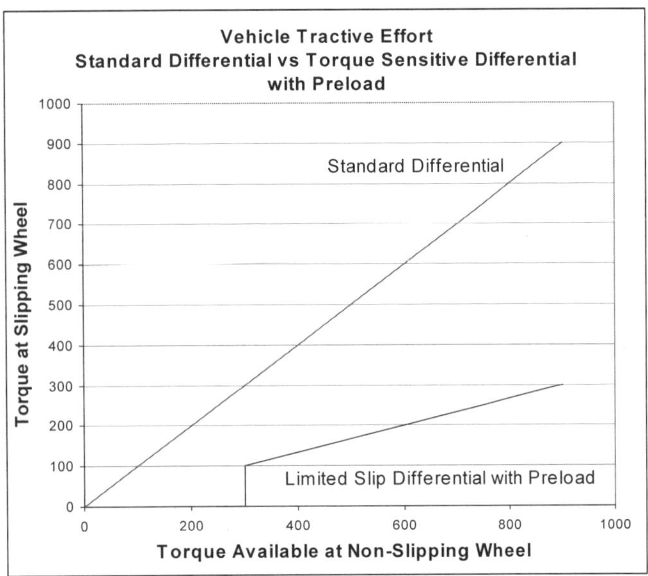

Figure 2: Tractive Effort for a Standard Differential and a Torque Sensitive Limited Slip Differential with Preload

Speed sensitive differentials bias torque based on the speed difference between the drive wheels. These differentials transfer a fixed amount of torque proportional to the speed difference. Depending on the design, they can be tuned for a specific vehicle. This allows the differential to provide improvements in mobility and possibly handling. Drawbacks to this style of differential are that, regardless of the external conditions, any time there is a speed difference, the differential will transfer torque. When a vehicle traverses a curve, this speed difference will provide some torque transfer, even though it is not necessary under normal conditions. More importantly, if Anti-Lock Braking, Traction Control, or Vehicle Stability systems are used, these systems artificially provide a speed difference by braking individual wheels. The differential could fight these systems by applying torque to the wheel that is trying to be slowed. This reduces the effectiveness of the differential and the control systems. To allow compatibility, the differential needs to be de-tuned, reducing the tractive effort of the vehicle. Figure 3 shows a graph of a typical hydraulic type speed sensitive

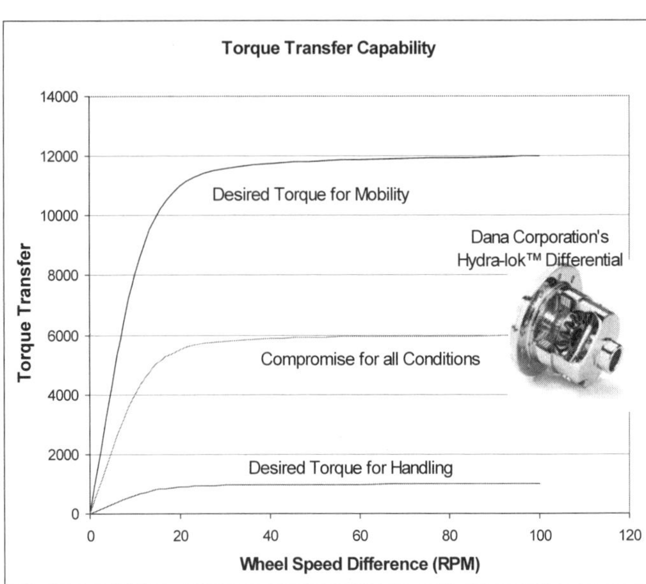

Figure 3: Torque Transfer Capability of a Speed Sensitive Differential

differential. Note that a viscous type speed sensitive differential would have a different shape curve, but still transfer torque as a function of the speed difference.

Optimal mobility and handling can only be achieved when the differential can be programmed to react differently to specific external conditions. Figure 4 shows the range of an electronically controllable differential. The values in Figures 1-4 are not meant to represent the capabilities or limitations of any specific differential, but are shown as typical curves for that style.

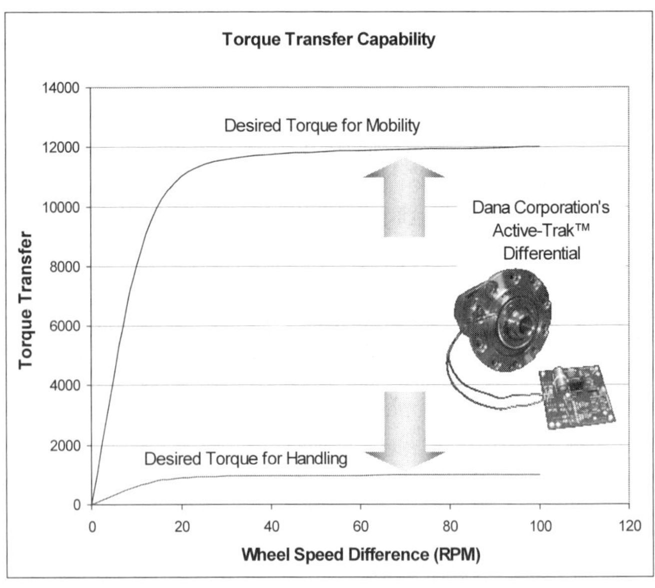

Figure 4: Torque Transfer Capability of an Electronically Controlled Speed Sensitive Differential

Dana Corporation's new Active-Trak™ differential is a gerotor style speed sensitive limited slip differential. This differential has the capability of modulating the pressure on the clutch pack to vary the tuning according to vehicle inputs and road conditions. This allows the differential to be aggressive for mobility and less aggressive for handling. It also offers the ability of disengaging for use with ABS and Traction Control systems. This paper will show the benefits of being able to modulate pressure for mobility, handling, and compatibility with other systems. It will show the advantages over a non-adjustable differential. Data will be provided for response time of a rear wheel drive system, along with braking data to show compatibility.

Design

The differential used for this testing is a hydraulic design. A gerotor pump spins when a speed difference is present. This pump draws fluid out of a sump and forces it into a reservoir. This reservoir pressurizes and applies a force to a clutch pack, providing the torque bias. For a passive design, a mechanical valve would be present to limit the torque for longevity of components, and tuning for the vehicle. In this design, for active control of the torque, an electronically controlled valve is employed. A small ball covers an orifice. An electro-magnetic coil is positioned above the ball. When the coil is energized, an armature, which is slightly offset, tries to align with the coil. This provides a force on the ball blocking the orifice. The amount of force on the ball controls the pressure that can be generated on the clutch pack. When the coil is disengaged, the ball can move away from the orifice allowing pressure to bleed off.

A Proportional-Integral-Differential (PID) controller is used to determine engagement. Several inputs are used to determine the condition of the vehicle. Inputs include individual wheel speeds, steering angle, throttle position, vehicle speed, brake status, transfer case mode, and temperature. Based on the inputs, theoretical wheel speeds are calculated. When the actual measured wheel speeds vary from the theoretical speeds by a pre-determined amount, the controller determines how much correction is needed. A Pulse Width Modulated (PWM) signal is sent to the coil. The duty cycle of this signal can be changed to provide a variable force on the ball. Additionally, inhibitors can be programmed into the software to prevent engagement under specified conditions. Conditions such as ABS, Traction Control, or Stability Control system engagement, as well as excessive temperature, mismatched tire sizes, and vehicle speeds can be used to prevent engagement. This allows aggressive tuning for mobility, but the torque transfer can be reduced for handling improvements or adjusted to specific conditions. When traveling around a

curve, theoretical wheel speeds will match actual wheel speeds, so the differential will not engage. If ABS, Traction Control, or Stability Control engages, the differential can be disabled to prevent interference, or can be adjusted to work simultaneously.

Current draw for the coil is low, less than three amps, allowing compatibility with present vehicle electrical systems. The compact design allows packaging in existing axle designs for easy retrofit. Control hardware and software have been partially developed for rear axle designs. Front axle and four wheel designs are being pursued.

Testing

All testing was done using a 2002 Toyota 4Runner. The front axle had a standard open differential. The rear axle was fitted with Dana Corporation's Active-Trak electronically controlled speed sensitive limited slip differential. Several terms are used in the following sections. For testing referring to the differential being off or disengaged, the power to the coil is off and the valve is in an open position where pressure will be bled off. This data is shown to represent an open or standard differential. While there is still some torque transfer, even when disengaged, it is low and is representative of an open differential. The data referring to automatic mode is conducted with the differential being controlled by the PID controller. The controller takes inputs from the sensors and determines whether engagement should occur. For the data referring to locked or manual mode, the controller is disabled, and a constant signal is sent to the coil to tell it to engage. This data is representative of an aggressively tuned passive speed sensitive limited slip differential. The controller sends a constant signal to the coil, which in turns acts like a mechanical valve, which is closed until the pressure exceeds the mechanical force. Testing was conducted at two facilities, the Keweenaw Research Center in Houghton, Michigan for winter testing and at Bosch Proving Grounds in New Carlisle, Indiana for warm weather testing. The coefficients of friction for the surfaces at Keweenaw Research Center were approximately 0.1 for the ice and 0.7 for the asphalt. For the Bosch Proving Grounds, the tile was approximately 0.3 and the asphalt was 0.7.

For mobility, as much torque transfer as possible is desired. This can have adverse effects on handling. A differential tuned aggressively for mobility can have ill effects during normal driving maneuvers. Excessive torque transfer during normal low speed maneuvers can cause unexpected results. Figure 5 shows a condition often referred to as "wheel hop." When excessive torque is transferred during a turn, the wheel can lock-up preventing normal rotation.

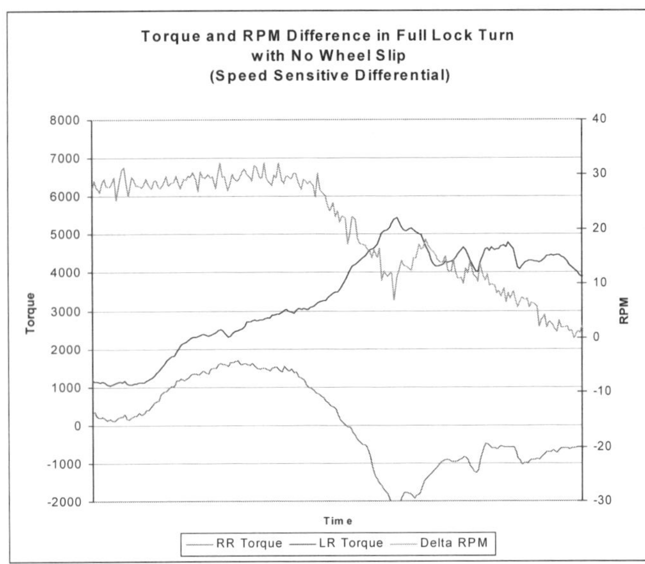

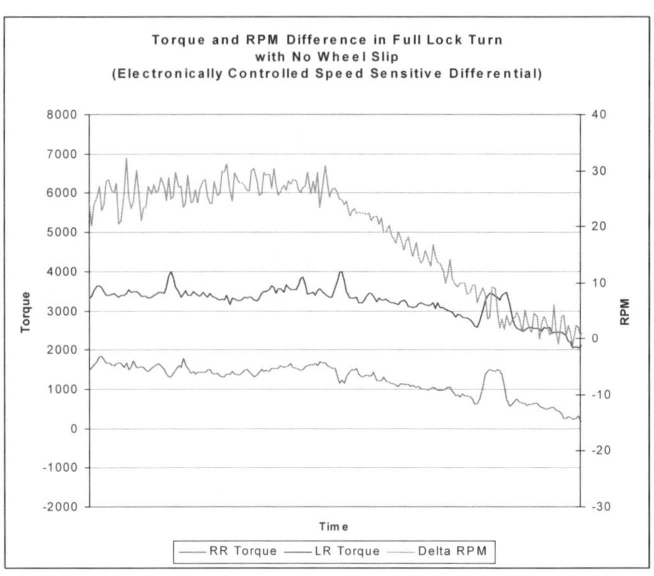

Figure 5: Wheel Torques in Full Lock Turn

Figure 6: Wheel Torques in Full Lock Turn

This plot shows that during a normal turn, when a speed difference is encountered, with no wheel slip, torque will be transferred. The torque for the right wheel becomes negative indicating the torque transfer is trying to equalize the wheel speeds. This shows up as a braking torque on the faster wheel. This can also be seen from the drop in speed difference between the wheels. This braking torque prevents normal wheel rotation, causing unpredictable handling.

An electronically controlled differential can be programmed to determine whether wheel slip is present and if not, prevent torque transfer from occurring. This will prevent these adverse reactions. Figure 6 shows the same plot with an electronically controlled differential.

In Figure 6, the same speed difference is present. Since the controller inputs show that no wheel slip is present, the differential does not engage, and the wheels continue to rotate smoothly showing only a small amount of torque transfer between the wheels.

Another example of the benefit of this tuning can be seen during hard acceleration on split μ surfaces. If a slick surface is encountered during driving, the wheel will spin sending a large amount of torque to the opposite wheel. An example of this is shown in figure 7.

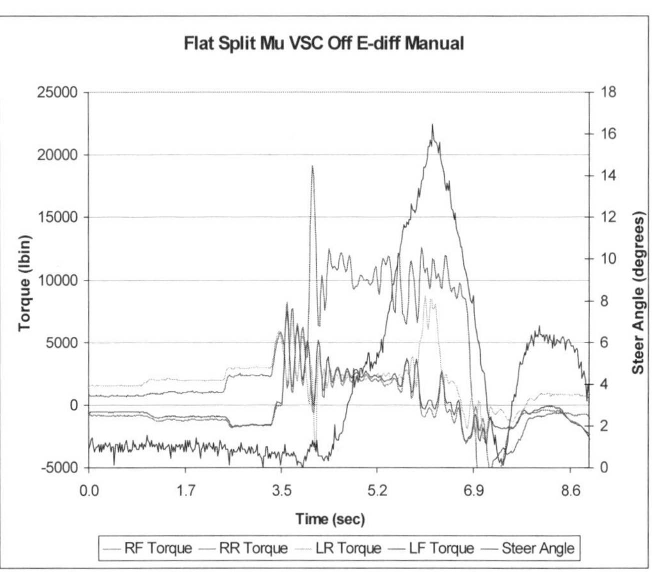

Figure 7: Split μ Acceleration - Speed Sensitive Differential

This plot shows the four wheel torques and steer angle. The steer angle is the angle of the front wheel and shows the amount of correction required due to the yaw moment of the vehicle. It can be seen that a very large torque spike occurs when the wheel slips and transfers torque to the wheel with good traction. This creates a large yaw moment, making the vehicle difficult to control. This data was collected with the differential actuator locked on to simulate a passive speed sensitive differential. Next, the actuator was set to automatic mode. Figure 8 shows a plot of the same vehicle and same test condition with an electronically controlled differential.

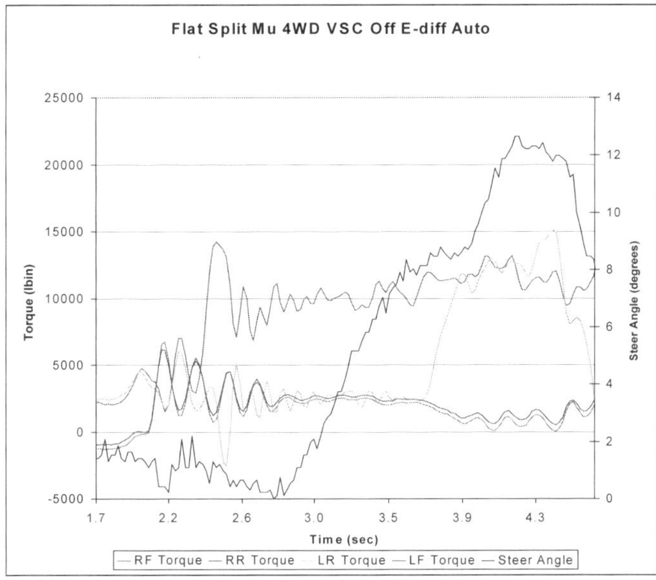

Figure 8: Split μ Acceleration - Electronically Controlled Differential

	Steering Correction (degrees)	Acceleration Rate (ft/s^2)
Traction Control On Active-Trak off	2.1	5.37
Traction Control Off Active-Trak Manual	17.0	7.81
Traction Control Off Active-Trak Auto	11.0	8.42
Traction Control On Active-Trak Auto	9.0	8.37

Table 1: Steering and Acceleration Comparison

This figure shows that the torque spike to the high traction wheel, due to proper PID tuning, has been significantly reduced. Consequently, the required steering correction is also reduced. These two figures show that with electronic controls, high torque can be obtained for mobility, but engagement can be controlled for handling.

It should be noted that with Traction Control engaged, the yaw moment was essentially eliminated, but this control comes at a price. With Traction Control engaged, the acceleration of the vehicle is reduced. The Traction Control operates by applying the brakes to independent wheels, while, in some cases, also decreasing engine power. This gives a very noticeable decrease in acceleration. Table 1 shows a comparison of yaw (based on steering correction) and acceleration rates for Traction Control, a speed sensitive differential (simulated by an electronically controlled differential in manual mode with a locked actuator), and an electronically controlled differential in automatic mode.

Note that these tests were performed at wide open throttle to evaluate worst case yaw rates. This table shows that with Traction Control alone, yaw is very small, but acceleration is low. With the differential manually engaged, high acceleration is achieved, but yaw is much higher. With the differential in automatic mode, excellent acceleration is still achieved, but yaw is significantly reduced. Last, with Traction Control Engaged and the differential in Automatic mode, steering correction is reduced even more, with almost no decrease in acceleration rate. With proper tuning, excellent tractive effort can be achieved while still maintaining acceptable handling.

ABS

Anti-Lock Braking has offered significant improvements in braking ability and safety. It is important that the differential be compatible with this system. The differential must advance the capability of the vehicle and not detract from performance or safety. In theory, there is the potential for compatibility issues. If a vehicle is braking on an uneven friction surface, one wheel can lock up. This creates a speed difference, with the wheel that is still rotating transferring torque to the wheel that has stopped. This torque transfer is trying to accelerate the wheel, while the brakes are trying to decelerate it. It is therefore important to ensure the systems will be compatible. It must be verified that the differential can be disabled before ABS is engaged, or ensure the differential action does not detract from the ABS function.

A typical ABS system has an effective response time of 200-300 ms. For compatibility, the differential must have a response faster than this. Figure 10 shows the Active-Trak differential disengagement response. From this graph, it can be seen that the response is 140 ms. This is from the time the signal is sent to the coil until the torque drops to the nominal level. The disengagement time for the differential is much quicker than the engagement time of the ABS, which ensures the systems will be compatible. Note that specific testing for ABS response was not done, but these numbers represent typical values given in manufacturer specifications. For reference, lines are shown in the ABS stop in Figure 11 indicating 200 and 300 ms. It can be seen that at 200 ms virtually no braking has occurred, and at 300 ms pressure has built and the wheel is just starting to slow.

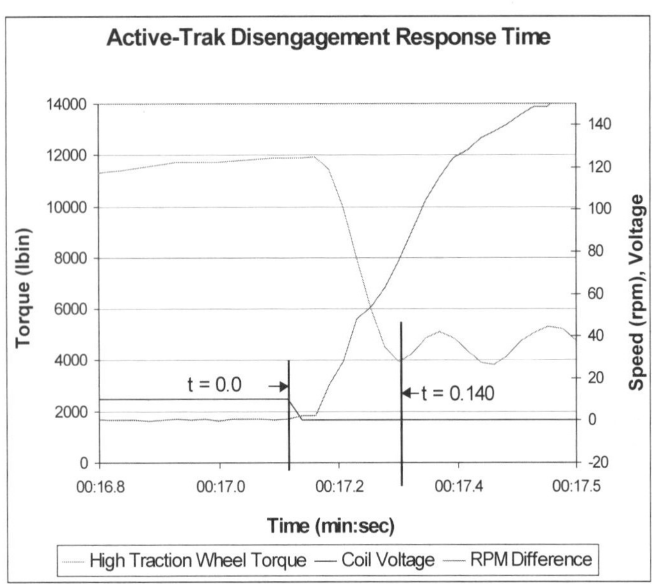

Figure 10: Active-Trak Disengagement Response Time

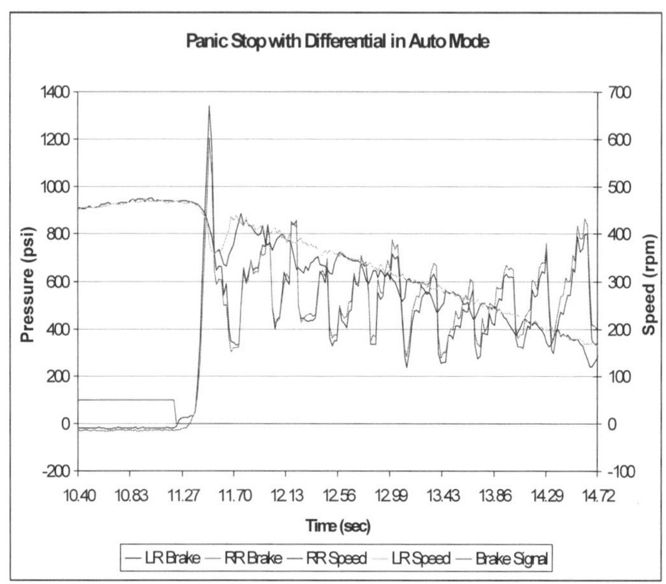

Figure 12: Panic Stop with Differential Engaged

The compatibility with the differential engaged was also investigated. As mentioned before, it is possible that the differential could reduce the braking effectiveness by sending torque to a wheel that is being braked. For this testing, panic stops were conducted from 40 mph on a split mu surface. The surfaces used were wet ceramic tile and asphalt. Figure 11 shows the stop with the differential disengaged. Figure 12 shows a similar stop with the Electronically controlled differential engaged.

In these graphs, the red curve is the brake signal, which is high until the brake is applied, then goes to zero. The purple and blue curves are the brake line pressures at each rear wheel, and the light blue and black curves are the rear wheel speeds. Both graphs look very similar. The deceleration rates of each test were calculated for comparison and are shown in Table 2.

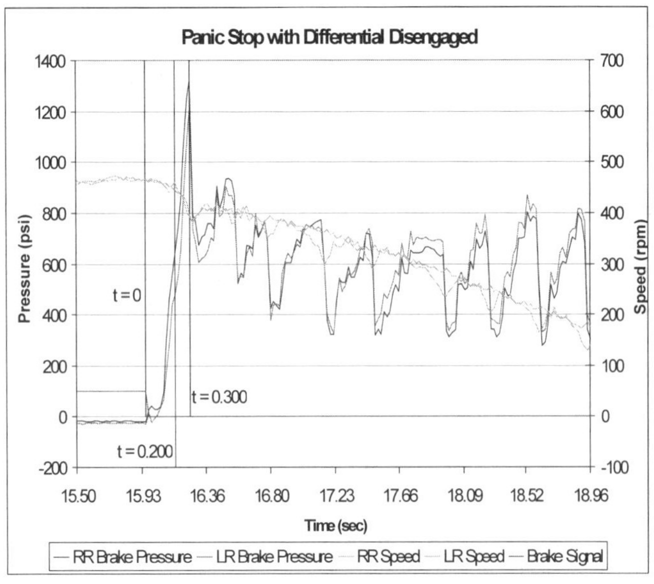

Figure 11: Panic Stop with Differential Disengaged

	Wheel Speed Drop (rpm)	Deceleration rate (ft/s^2)
Differential Disengaged	470 – 300	10.8
	470 - 95	11.6
Differential Engaged	470 – 300	11.1
	470 - 95	10.4

Table 2: ABS Deceleration Rates

It should be noted that these decelerations were obtained using the change in rpm and the time. These are not precise rates, but are meant to show a trend. This data indicates that the differential does not interfere with the ABS and provides comparable deceleration rates.

This data shows that there is potential to use both systems together without sacrificing the performance of either, but also that the differential disengagement is quick enough to be disabled if the manufacturer prefers to have the systems operate independent of each other.

TRACTION CONTROL

Although not as critical as ABS, compatibility with Traction Control offers benefits as well. As shown in the previous section, the disengagement time of the differential is much faster than the activation time of the brake system. As before, if the OEM prefers to have both systems, and keep them separate, they will be compatible since the differential can disengage quickly and allow normal operation of the Traction Control. Compatibility of Traction Control with an engaged differential was also tested.

The data in Figures 13 and 14 was recorded on a 40% split mu grade of ice and asphalt. Neither system, the traction control or limited slip differential, could climb the grade by itself. The differential was traction limited, and would transfer enough torque to slip the wheel on asphalt. The traction control system had to apply the brakes so much that forward momentum was stopped. When both systems were activated, the differential would engage before the traction control, but the traction control would prevent the high μ wheel from spinning excessively. This allowed the vehicle to climb the hill.

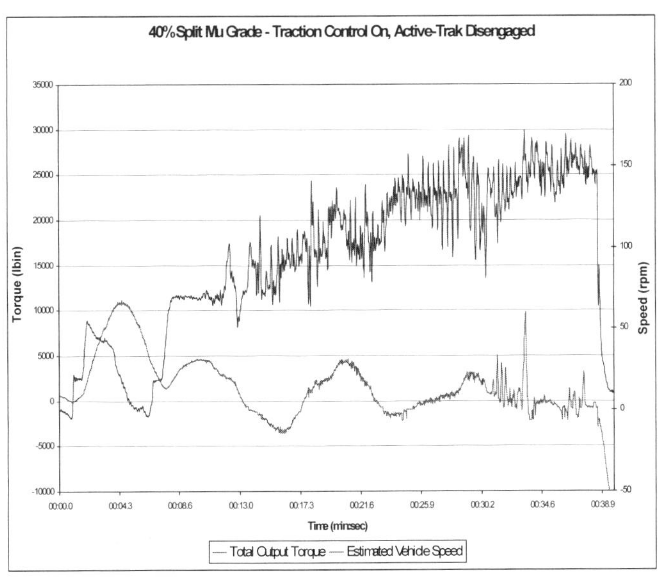

Figure 13: Traction Control On, Active-Trak Disengaged

It can be seen that in Figure 13, the torque output reaches a maximum and speed drops to zero, indicating a loss of momentum. In Figure 14, a higher total torque is generated, and forward momentum continues until the vehicle reaches the top of the hill. The combination of these two systems increases mobility of the vehicle significantly.

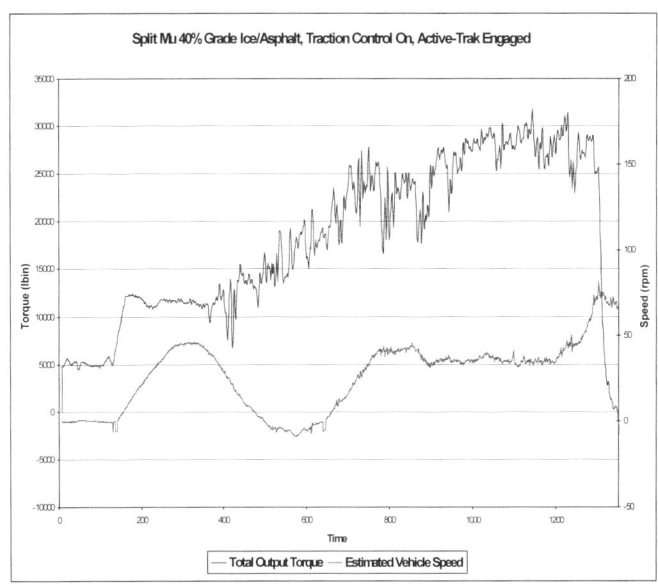

Figure 14: Traction Control On, Active-Trak Engaged

A second test was run on a flat split mu course. This is similar to the test shown earlier for yaw rates, but this is intended to show compatibility. The previous test had extremes for coefficients of frictions, to show worst case yaw rates. This test was run on the ceramic tile and asphalt course. The closer friction surfaces are more representative of normal driving conditions and are better suited to identifying compatibility issues. The first graph shown in Figure 15 is a split mu acceleration with the differential disengaged, similar to an open differential.

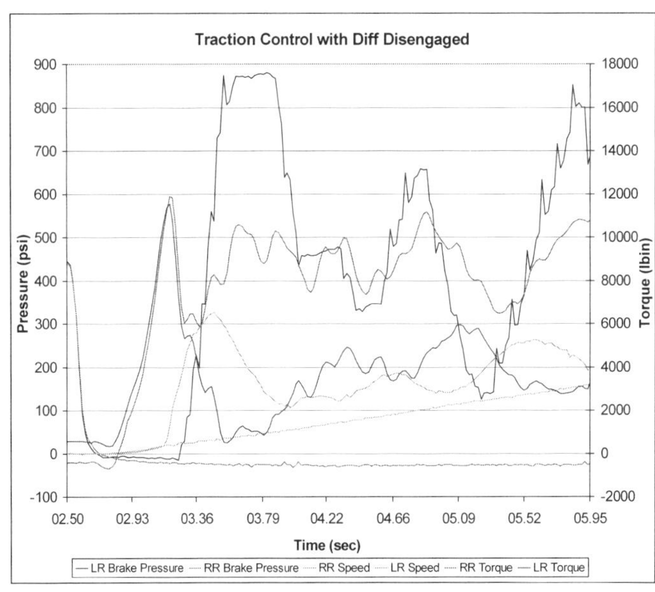

Figure 15: Traction Control with Active-Trak disengaged

This graph shows the left wheel (on the lower friction surface) begin to slip. At the same time, the torque at this wheel begins to drop off. The Traction Control causes the brake line pressure at this wheel to increase substantially. This pressure slows the spinning of the wheel and starts to generate torque.

Figure 16 shows the same test with an Electronically controlled differential in Automatic mode.

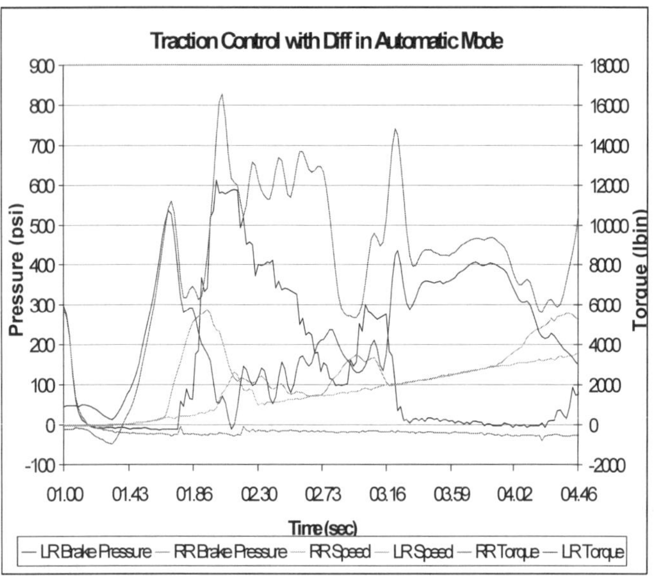

Figure 16: Traction Control with Active-Trak Engaged

This graph shows the same wheel slip. In this graph, it can be seen that brake pressure increases to a much lower value, while at the same time, the torque at the higher traction wheel increases to a much higher value. It can also be seen that the brake pressure drops off quicker, while the torque at the slipping wheel increases. This gives a higher total torque and a slightly faster acceleration rate.

For comparison, Figure 17 shows the differential with the valve manually locked on, similar to a passive limited slip differential.

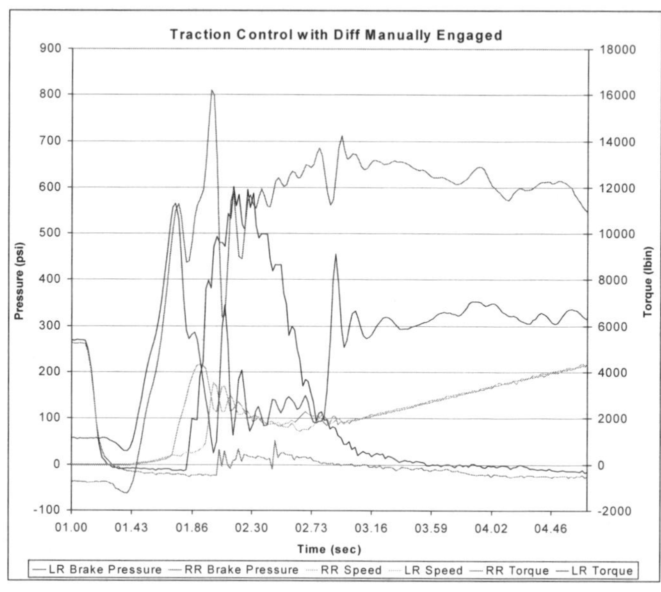

Figure 17: Traction Control with Active-Trak Manually Engaged

This graph shows similar curves to the previous graph, but has a slightly higher total torque, and faster acceleration. This has slightly higher values because of the tuning in the automatic mode. In Figure 16 at about 2.5 seconds, the wheel speeds are equal which tells the differential to disengage. This causes the torque to drop off momentarily. When the wheel begins to slip again, torque is applied and the vehicle continues to accelerate. With the valve locked on, there is still residual pressure in the system when the wheel speeds catch up to each other. At this point, when the wheels are at the limit of traction, pressure is retained in the differential until wheel slip is totally eliminated. With additional testing, it would be possible to fine tune the disengagement settings to be closer to the passive speed sensitive limited slip differential.

This data shows that the systems complement each other providing better mobility and acceleration than either system alone. When traction control is used alone, the braking of the wheels degrades performance. When a passive limited slip differential is used alone, large torque spikes can occur which may upset the stability of the vehicle. When an electronically controlled differential is engaged with Traction Control, both systems can work together to increase performance. The differential has a tendency to cause more wheel spin due to the large torque transfer. The Traction Control keeps the wheels from spinning wildly, but still allows some wheel slip, which causes the speed sensitive differential to transfer torque.

STABILITY CONTROL

Also very important is the compatibility with Stability Control. Although still fairly new, Stability Control offers safety improvements similar to ABS. Stability Control systems operate by braking individual wheels to counteract unintended vehicle yaw moments. No specific data was collected with regard to Stability Control systems. Based on the results of the Traction Control testing, it is reasonable to assume the systems could work together. Since the Stability Control system works on the same principle using the same components as Traction Control, the response time should be similar to the ABS and Traction Control. Further development would be required to test the compatibility. More importantly, it has been shown that the disengagement of the differential is faster than engagement of the braking system, so the differential could be disengaged to prevent interference.

CONCLUSIONS

This paper has shown tremendous benefits in having a controllable differential. Limited slip differentials can offer improvements over traditional standard differentials, but to reach the maximum potential of the differential and the vehicle, the differential must be able to react differently depending on the conditions. This paper shows that through proper design and tuning, a differential can be compatible with ABS, Traction Control, and Stability Control systems, while still offering improvements in mobility and handling. This combination would not be possible in a passive differential. The data provided shows that, if desired, the controllable differential can be disabled to prevent any interaction between the systems. The data also shows that with additional development, there is great potential for the controllable differential to work with the braking systems to provide better performance than either system alone. These improvements are not possible with a standard differential, or even a passive limited slip. The differential must be able to adjust to vehicle and driver inputs, as well as external conditions.

ACKNOWLEDGMENTS

Thanks to:

Tom Nahrwold for his help in Testing and Data Acquisition.

Jim Smith for his help in vehicle instrumentation.

Bryan McDonald for his help in software and controller design and tuning.

Anne Mitzel for her help in the coil design.

Wayne Borgen for his help in design.

John Grogg for his design work and initial development.

Ken Cooper and Jim Krisher for their help and support with the paper.

DEFINITIONS, ACRONYMS, ABBREVIATIONS

ABS: Anti-Lock Braking System

VSC: Vehicle Stability Control

μ: mu, the greek letter representing coefficient of friction

PID: Proportional Integral Differential

LF: Left Front Wheel

RF: Right Front Wheel

LR: Left Rear Wheel

RR: Right Rear Wheel

Performance, Robustness, and Durability of an Automatic Brake System for Vehicle Adaptive Cruise Control

Deron Littlejohn, Tom Fornari, George Kuo, Bryan Fulmer, Andrew Mooradian, Kevin Shipp, Joseph Elliott and Kwangjin Lee
Delphi Corporation

Margaret Richards
Michigan State University

ABSTRACT

Adaptive Cruise Control (ACC) technology is presently emerging in the automotive market as a convenience function intended to reduce driver workload. It allows the host vehicle to maintain a set speed and distance from preceding vehicles by a forward object detection sensor. The forward object detection sensor is the focal point of the ACC control system, which determines and regulates vehicle acceleration and deceleration through a powertrain torque control system and an automatic brake control system. This paper presents a design of an automatic braking system that utilizes a microprocessor-controlled brake hydraulic modulator. The alternatively qualified automatic braking means is reviewed first. The product level requirements of the performance, robustness, and durability for an automatic brake system are addressed. A brief overview of the presented system architecture is described. The control methodology of generating brake pressure via a hydraulic modulator to achieve the vehicle deceleration requested by ACC controller is then introduced. The paper includes a description of two Pulse Width Modulated (PWM) solenoid control designs and applications as an important technology to ensure the automatic braking performance. The implementation of moding the automatic brake system with ABS, Traction Control, and Vehicle Stability Control is revealed at the vehicle system level. Vehicle test data will be presented as insight to the braking performance and robustness. The control-related system durability will also be examined and discussed under vehicle testing profiles. Vehicle integration system test data summarizes and concludes the practice and value of the presented automatic brake system for vehicle adaptive cruise control.

INTRODUCTION

ACC is a system that uses a forward radar sensor to determine the distance between the host vehicle and a target vehicle. The system is intended to match the speed of the target vehicle by reducing the throttle and/or applying the brakes without requiring the driver to adjust the cruise control settings. Figure 1 shows the components and subsystems required to implement ACC.

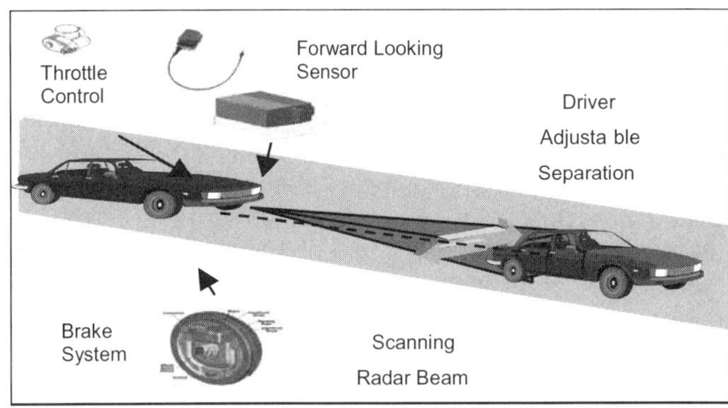

Figure 1 The Components of the ACC System

There are several ways to implement ACC on a vehicle and this paper will discuss some typical implementation schemes. The automatic braking function can be achieved by various means; some methods include an intelligent brake booster, electric calipers, stored hydraulic fluid pressure, etc. This paper details implementation based upon stored hydraulic fluid pressure; however, brief descriptions of other methods are given first. One method to regulate wheel brake pressure utilizes an intelligent/smart booster. Wheel pressure is regulated by a solenoid valve that controls air flow into the booster air chamber, which generates differential pressure across the booster's diaphram. This method generates little audible noise and is also very smooth; however, there is additional risk of mechanical failure plus the brake pedal drops during an autobraking event. One prior method of generating brake pressure with the hydraulic modulator is the traditional operation of traction control. When the pump, Prime Valve and Isolation Valve are energized together, brake fluid is delivered under pressure to the wheel brakes. The fluid flow from the master cylinder into the pump is regulated by the Prime Valves. The Isolation Valve is located on the high pressure side of the pump to prevent the fluid from returning to the master cylinder. The wheel brake pressure is regulated by the Apply Valves and Release Valves. The Apply valves are energized to block the in-flow of fluid to the brake to control the rate of the pressure increase, while the Release Valves are energized to control the rate of the pressure decrease. This method of controlling the wheel brake pressure is suitable for traction control and vehicle stability where noise and harshness are less of a concern. In the automatic braking mode of ACC, the noise and harshness would typically be unacceptable.

An alternate method to regulate the wheel brake pressure is through ON/OFF control of the Prime and Isolation Valves. Wheel pressure is increased by regulating the fluid supply to the pump through Prime Valve modulation and pressure is decreased by releasing fluid through the Isolation Valves. This method generates much less noise, but it is not as smooth as required for auto braking.

The chosen method to control the wheel brake pressure is through the use of a hydraulic pump (modulator), prime valve, and Variable Isolation Valve (VIV). The VIV operates as a pressure regulating valve where the blow-off pressure is proportional to the applied PWM duty cycle. To increase wheel brake pressure, the Prime Valve and pump are energized, and the PWM signal to control the VIV is set appropriately to hold/build the desired pressure. To release pressure, the VIV PWM duty cycle is reduced. This modulator control method is smoother, quieter, and more cost effective than the prior methods. The remainder of this paper gives some insight to the implementation, performance, robustness, and durability of this chosen method.

THE PRESENTED SYSTEM OVERVIEW

Hydraulic Mechanization

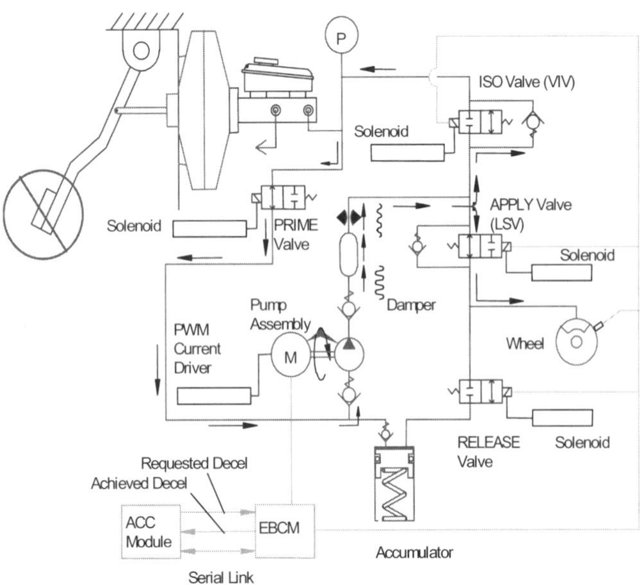

Figure 2 The Mechanization of Hydraulic Modulator

Figure 2 depicts the mechanization of the ABS/TCS/VSE controlled brake system capable of performing ACC Automatic Braking without driver input on the braking pedal. The ABS Controller commands the motor in the modulator to pump brake fluid from the master cylinder into the wheel brake lines through the energized and opened PRIME solenoid valves. ACC Automatic Brake utilizes PWM-driven variable isolation valves (VIV) to regulate the pressure level on the wheel brake hydraulic lines. Delphi VIV technology provides the characteristic of throttling the braking flow through the orifice of the solenoid valve with extremely low-pressure jumps. The result achieves smooth, uniform, and low-vibration vehicle deceleration. The VIV provides an attractive linear flow control of the solenoid valves where the blow-off pressure is proportional to the applied current. That overcomes the technologic limitations of conventional ON/OFF style solenoid valves, which are used in most of today's industrial market. Additional key features of the VIV hardware are the capabilities to match the deceleration control stability requirements of ACC automatic brake through synchronizing pressure control with ABS/TCS/VSE subsystems which basically utilize APPLY and RELEASE valves for wheel brake pressure regulation. To meet the low noise requirement, a PWM-driven motor control circuit is equipped to effectively minimize the noise level from pump piston impact.

Vehicle System Integration

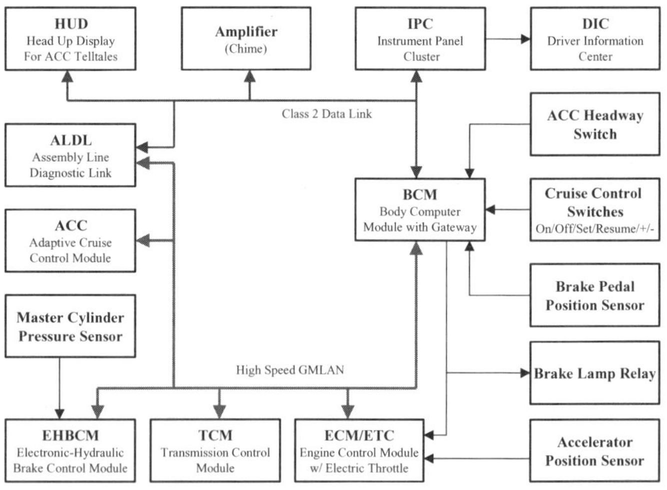

Figure 3 The Architecture of Vehicle System Integration

Figure 3 indicates the block diagram of implementing the ACC vehicle integration system. To maintain the vehicle headway and cruise speed, the ACC controller issues the requested deceleration or acceleration commands through a high-speed serial link to the EHBCM and ECM. The ACC Automatic Brake System is assigned to fulfill the task of achieving the desired deceleration level beyond the maximum decelerating capability available from the powertrain system. In ACC autobraking mode, when the driver applies the brake pedal, the automatic braking action is phased out. This transition is done and secured by monitoring the signals from brake pedal position sensor and master cylinder pressure sensor. Input to the electric throttle from the driver and the ACC module for speed maintenance is differentiated within the ECM. The ECM reports the driver's throttle input during ACC autobraking activation to the EHBCM via the high-speed serial link. When the ECM reports the driver's throttle input, ACC autobraking phases out. If the decel request from the ACC module persists once the driver's throttle input goes away, then ACC autobraking resumes. A brake lamp relay is included in the system for the purpose of signaling the tail brake warning light in the ACC automatic braking events.

AUTOBRAKING SYSTEM PERFORMANCE

Control Design

The key control design of the Automatic Brake system is to achieve the required performance through vehicle deceleration tracking control. The vehicle deceleration, which is derived from the fusion of wheel speed signals subject to possible wheel slips/spins or active brake control, is used as the feedback to guide the tracking schemes. To match the product requirements, a sequence of control sessions is adopted to take advantage of the dedicated hardware technologies, VIV and PWM Motor, to regulate the braking pressures at all 4 wheels. Figure 4 cascades the high-level flow of the wheel brake pressure control.

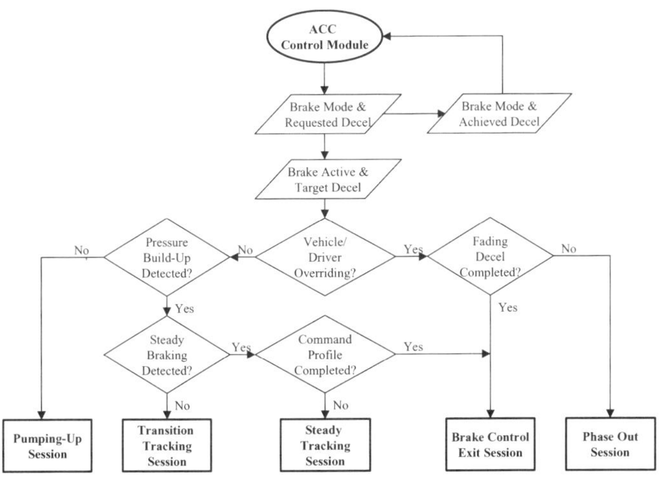

Figure 4 The Cascade of the Wheel Pressure Control

The pressure control scheme starts with a pumping up session that applies a temperature compensation based open-loop command to the VIVs and PWM Motor to maximize response time. After the detection of the pressure buildup, a closed-loop transient tracking session takes over to pilot the VIV blow-off pressure to achieve the vehicle deceleration target with minimum overshoot. Next, a steady tracking session with learning process follows to generate smooth, quiet, and low-vibration deceleration, which also helps to avoid misleading feedback due to rough road disturbances or emergency braking cycles. In the event of a driver's override or system failure, an open-loop phase out session is used to fade out the wheel brake pressures to satisfy the requirement for a smooth transition. In addition, a brake control exit session is executed to remove any residual upstream vacuum or downstream pressure relative to the modulator to ensure repeatable performance.

Vehicle Testing and Responses

The Automatic Brake control design is validated in the target production vehicle, which is equipped with Delphi's Adaptive Cruise Control Module with Forward Detection Sensor and Delphi's DBC7.2 brake system. The DBC7.2 brake system for the target production vehicle has capabilities for Antilock Braking (ABS), Traction Control (TCS), Electronic Stability Control (ESC) and Adaptive Cruise Control Automatic Braking. A variety of tests and evaluations were performed to achieve the production requirements on automatic braking deceleration – responsive, precise, smooth, low-noise, low-vibration, decent transition, uniform/stable, and robust. Fundamental tests, such as Rubber-Band Cruising, Lane Changes, Lane Cut-in, Tailing Stopping, Ride and Handling Cruise, Brake Overriding, Throttle Overriding, etc., were executed on proving ground tracks. Public Road Tests were done to verify the integrated system performance and to monitor the brake thermal reaction. The vehicles were exposed to low temperature and winter surfaces to confirm the response time and subsystem moding (ABS/ESC) with the Automatic Braking. Figure 5 shows data collected through a CAN Analyzer. The graph demonstrates the decel tracking response and precision during lane-cut-in tailing. It also displays the Automatic Braking capability to track the deceleration commands after transition from speed control mode (acceleration in this case) to braking mode.

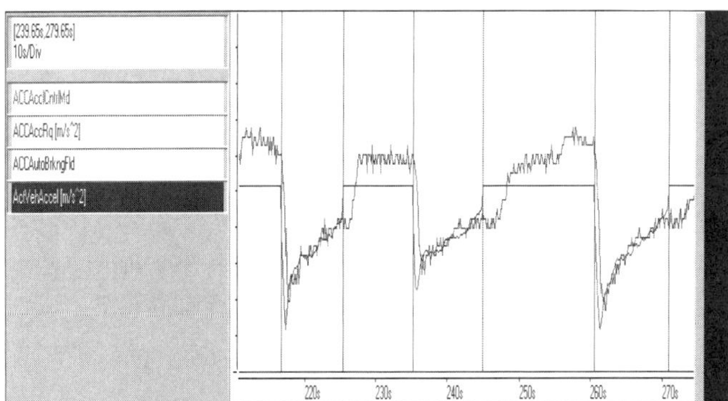

Figure 5 The Deceleration Tracking in Event of Lane-Cut-In Tailing.

AUTOBRAKING SYSTEM ROBUSTNESS

Modeling and Simulation Technology

The simulation of this system was run using a co-simulation between Matlab, Simulink, AMESim and CarSim. The brake control algorithm was generated using an s-function, in Simulink, which was created using Visual C++. This s-function derives the modulator commands based on the requested deceleration and the vehicle's behavior.

AMESim offers an environment for simulating mechanical, fluid and thermo-fluid systems. In this simulation, the pressure on each brake caliper was derived based on modulator commands, which are developed within the Simulink model. The AMESim model contains two pumps, one to run each side of the diagonally split braking system. The volumetric efficiency of each of these pumps was varied to understand the vehicle yaw and stability effects.

CarSim can be used for simulating and animating the behavior of four-wheel vehicles. The user may define a three-dimensional environment for the vehicle to travel in and various parameters of the vehicle. It can provide graphical outputs of specified parameters, a data file, or an animation of the vehicle in its environment. AMESim and CarSim generate s-functions when they are compiled; these s-functions are put into Simulink to interface with the brake control algorithm.

Simulation Results

When the model is run, CarSim allows the user to define the friction of the surface upon which the vehicle is traveling. Using this feature the simulated vehicle was driven on a straight path with a high coefficient of friction surface, to simulate a dry surface. It was run again with a split coefficient of friction surface, to simulate a vehicle that had its left wheels on a dry surface while its right wheels were on an icy surface.

CarSim also allows the user to vary the amount of steering. The tests were first run with a fixed steering wheel angle input. Then they were each run a second time with a simulated driver corrected steering. This simulated the driver having a one second preview. In each of the tests, the volumetric efficiency on the two pumps was varied to the most extreme case that is typically seen. After each condition had been tested, the yaw of the vehicle for each run was plotted. This shows the severity of changes in the pump efficiency based on yaw in a diagonally split braking application.

The yaw for a high coefficient of friction surface is shown in Figure 6.

The simulation was run for five seconds and the deceleration rate was requested when the time was equal to one second until the end of the simulation. The lower volumetric efficiency was on the pump that drives fluid to the left front and right rear brake corners, this explains the vehicle's movement to the right without the driver correcting the steering. The dark line depicts the case in which the simulated driver tried to correct steering. The driver over steers slightly, but is able to maintain the straight path with little effort. The lighter line is the case when the steering angle is fixed. This level of yaw is very minimal and could easily be corrected by a driver with very little effort.

This same test was run on a surface with a split coefficient of friction. The yaw in each case was again measured and plotted with respect to time. This plot is shown in Figure 7.

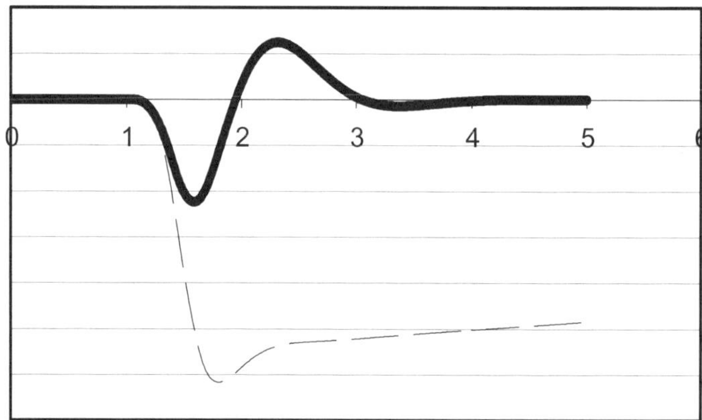

Figure 6 Yaw vs. Time on a Surface with High Coefficient of Friction.

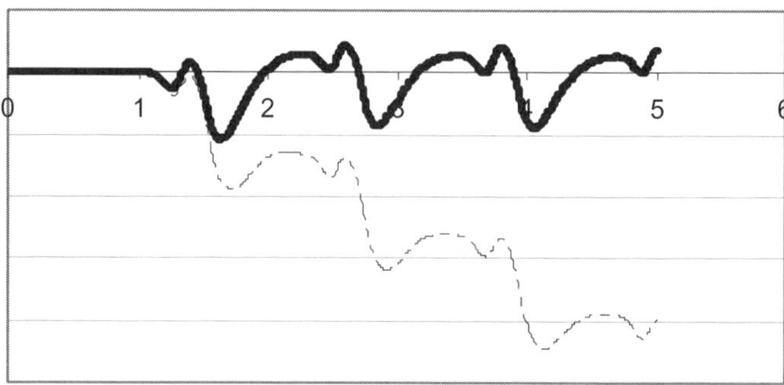

Figure 7 Yaw vs. Time on a Surface with Split Coefficient of Friction.

It is important to note that the right wheels of the vehicle are on a surface that is very slippery; this causes the vehicle to enter an antilock braking event when braking is requested. It is continually entering and exiting an antilock braking event for the remainder of the simulation. This causes the waves that are shown in Figure 7. Once again the right front and left rear brake corners are run by the greater pump efficiency. The dark line is the case in which the driver corrects the steering, while the lighter line is the instance in which the steering wheel is fixed. With the driver corrected steering the yaw is still very minimal, without the driver's correction the yaw is still within a safe range. Based on this simulation Delphi feels very confident in their brake control algorithm in this application.

AUTOBRAKING SYSTEM DURABILITY

Usage Profile Study

A question arose regarding what should be a reasonable hardware usage profile to be defined in the production specification for the presented ACC Automatic Brake System. This issue affects the system durability testing to meet the product platform requirements. Though a thought was to induct the profile by interpolating the conventional cruise control brake usage envelope based on SAE or an OEM data inventory, the deviations of driver's overriding projection and system turned-on acceptance ruled out using this method. There is also the complication of trying to estimate how many of the base brake applications that are required to disengage the conventional cruise control system will be inherited by the ACC system. Another plot was to refer the automatic braking profile to the existing market vehicles equipped with ACC systems, which use intelligent boosters as the automatic braking mechanisms. However, the discrepancies of the deceleration command limitation, speed range availability, powertrain braking capability, headway setting sensibility (sensitivity?), forwarding defense algorithm, and of course commercial secret obstacles this direct importing door.

Therefore a customer-supplier joint effort was initiated to study the usage profile in a development car equipped with the targeted product ACC system. The test vehicle was driven in a variety of situations and road conditions. There were 86 separate test drives completed, which resulted in 3,246 test miles of ACC activation being logged. A total of 847 Automatic Braking events were recorded during the 3,246 miles. The data was taken at different daily timing, at different speeds by different drivers in an attempt to simulate as many different driving situations as possible. The drive routes ranged from heavy urban street traffic in and around Detroit

Metro, Michigan and Dayton Metro, Ohio to high-speed rural interstate travels. Figure 8 shows the automatic event percentage and the deceleration request bands, which was from extrapolating the data across the standard 100,000 miles vehicle life cycle.

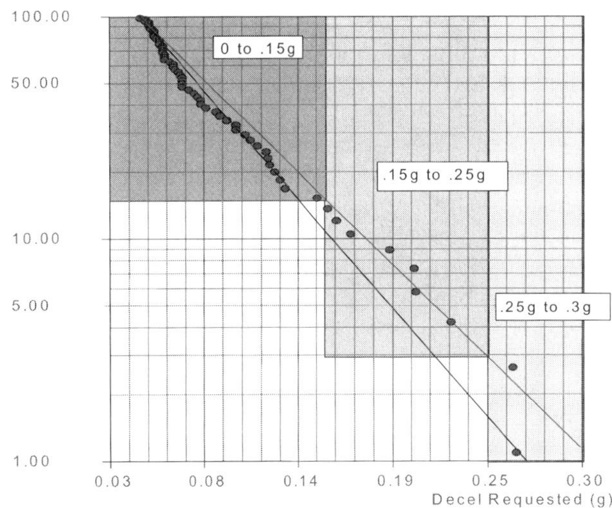

Figure 8 Autobraking Events vs Requested Decel

Component Durability Verification (DV)

In process of the presented Automatic Braking component Development Verification, the real-time vehicle control data corresponding to requested commands in the above study were broken down into 3 event-cycle-equivalent time domain profiles to fit practically for feasible chamber testing. Figure 9 displays the VIV valve input profiles which have linearly ramp-up and ramp-down signals to respectively represent the average energizing durations of pressure control. These 3 profiles were run in a chamber at four season temperatures evenly with the individually equivalent event cycle numbers and pre-set trigger waiting times. To minimize the hydraulic modulator noise, similar profiles with PWM commands proportional to these VIV signal levels were utilized to drive the hydraulic motor for the DV testing. The specification-met cycle numbers were accomplished in the chamber.

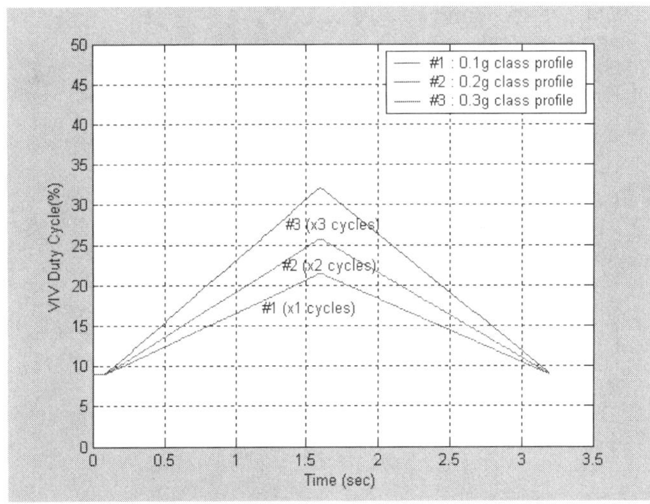

Figure 9 The VIV PWM Signals for the HCU DV Testing

CONCLUSION

There are several ways to implement an ACC automatic brake system for vehicle adaptive cruise control. The hydraulic-modulator-based implementation was examined here. The vehicle integration system testing verifies the presented automatic brake system performance. The component testing reveals its durability under brake life usage profiles. The simulation technology proves the automatic braking control robustness subject to driver, environment and manufacturer related disturbances.

ACKNOWLEGEMENTS

We wish to acknowledge the technician team in the Vehicle Test and Instrumentation departments who supported vehicle builds and testing.

CONTACT INFORMATION

Deron C. Littlejohn
Delphi Energy and Chassis Systems
M/S 483-3DB-210
12501 E. Grand River
Brighton, MI 48116-9326
Phone: 810-494-4428
Fax: 810-494-4458
Email: deron.littlejohn@delphi.com

REFERENCES

1. McLaughlin, S.: <u>Measurement of Driver Preferences and Intervention Responses as Influenced by Adaptive Cruise Control Deceleration Characteristics</u>, Thesis, Virginia Polytechnic Institute and State University, Blacksburg, Virginia; June 29, 1998
2. "Surface Transportation Act Boosts Intelligent Systems, Says I.T.S. America President," Top Tech Stories Washington, June 10, 1999 http://techmall.com.techdocs/TS980610-9.html
3. Hatipoglu, C.; Ozguner, U.; Sommerville, M.: "Longitudinal Headway Control of Autonomous Vehicles," Proceedings of the 1996 IEEE International Conference on Control Applications, New York, NY; 1996; p.721-6
4. Shein, E.; Mausner, E.: "Deployment and Commercialization of Cost and Safety-effective Autonomous Intelligent Cruise Control System," Microwaves and RF Conference Proceedings, Nexus Media, Swanley, UK; 1995; p. 124-31
5. Bryan Riley, George Kuo, Brian Schwartz, Jon Zumberge, Kevin Shipp; Development of a Controlled Braking Strategy for Vehicle Adaptive Cruise Control, SAE 2000 World Congress; March 2000
6.. Brakes for the Future, *Automotive Engineer*, September, 1998

DEFINITIONS, ACRONYMS, ABBREVIATIONS

ACC	Adaptive Cruise Control
ABS	Anti-lock Brake System
TCS	Traction Control System
ESC	Electronic Stability Control
EBCM	Electronic Brake Control Module
PWM	Pulse Width Module
VIV	Variable Isolation Valve
LSV	Linear Solenoid Valve

Adaptive Hydraulic Braking Traction Control for the 2003 Chevrolet Kodiak and GMC TopKick

Francis P. Tylenda
General Motors Corp.

Yuji Nakayasu
Continental Teves

ABSTRACT

The development and application of a traction control system using adaptive braking for the 2003 Chevrolet Kodiak and GMC TopKick are explained. Most traction systems use engine management to enable traction control, while the adaptive braking system can provide traction assist for either gas or Diesel powered vehicles from 14,000 lbs. to 33,000 lbs. GVW. The performance driven criteria that established the design requirements and the development of a new product to meet these objectives are discussed. Both the vehicle manufacturer and the traction controller supplier provided these criteria. The basic ABS and traction control hydraulic schematics will be described as they apply to the vehicles. The results of the development program will be compared to the criteria used to establish the goals, and the benefits of the traction control system will be discussed.

INTRODUCTION

General Motors (GM) began the redesign of its medium truck line, TopKick and Kodiak, by defining program "Wins". These program Wins are used to specify areas of performance, which a customer can perceive as Best in Class. One of the Wins is Unsurpassed Braking and Handling.

The goals of the Wins were established by using Voice of the Customer techniques and included both braking and handling responses. The characteristics included performance targets related to deceleration and items such as steering effort, turning radius and roll gain.

Figure 1. TopKick/Kodiak

In addition to the braking performance improvements, GM sought to offer an easily usable traction system for the lower range of GVW's in the truck range. The lower GVW's (16,000-19,500 lbs.) are always hydraulically braked, and hydraulic brakes are available up to 33,000 lbs. GVW. Electronic traction control was chosen for this application and is used on all GVW ratings of the TopKicks and Kodiaks with hydraulic brakes.

The Traction Control System (TCS) is similar to ABS in that it tries to maintain peak slip values at the rear wheels to optimize the vehicle dynamic performance in acceleration. The wheel slip is controlled by managing the brake torque alone to provide lateral stability and longitudinal acceleration.

GM partnered with its ABS supplier, Continental Teves, to develop a brake intervention only TCS used on hydraulic braked trucks. The functions provided by the system are:

1. Process wheel speeds
2. Derive dynamics
3. Determine TCS state
4. Determine optimum wheel slips
5. Control torques
6. Perform self-diagnostics, fail-safe operation, and interface with off-board manufacturing and service tools

BRAKE SYSTEMS

The TopKick and Kodiak use two different foundation brake designs and two actuation systems to cover the range of GVW's using hydraulic brakes.

The foundation brakes vary in size and basic design configuration. Pin slider and fixed, opposed piston designs are used.

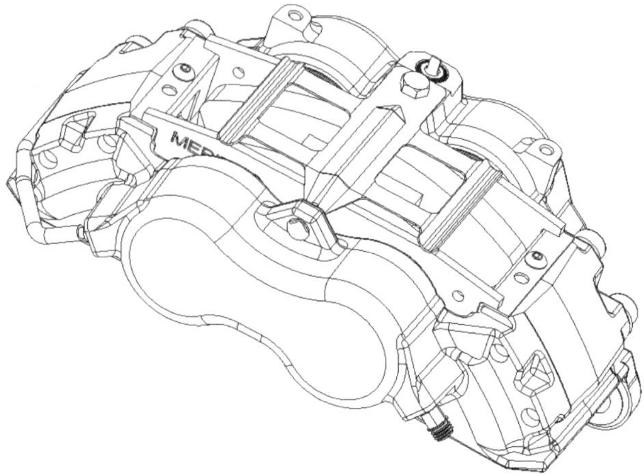

Figure 3. Foundation Brake Used with GVW's up to 33,000 lbs

The actuation systems differ based on the GVW range of the vehicle. The booster systems are hydraulically powered to provide the power assist for the brakes. Both boost systems operate on the same principle. The smaller capacity booster uses a pneumatic accumulator to provide emergency stopping in the event of loss of hydraulic pressure, while the larger system uses an integral electric motor and pump as a source of redundant pressure supply.

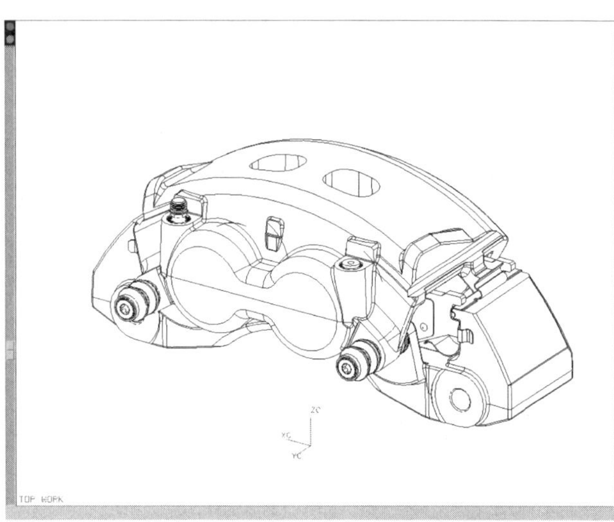

Figure 2. Foundation Brake Used with GVW's up to 19,500 lbs

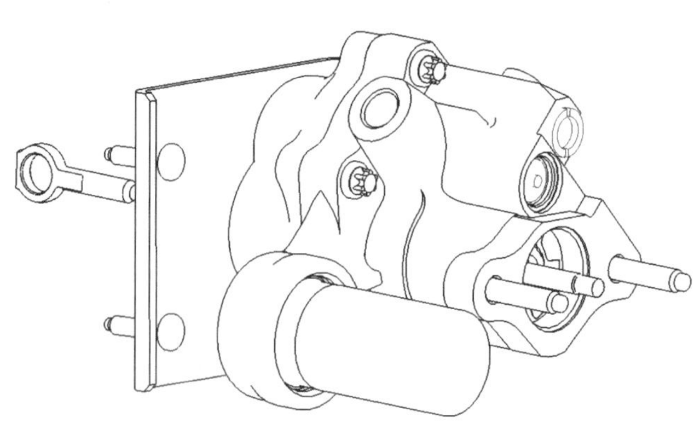

Figure 4. Brake Booster Used with GVW's up to 16,000 lbs.

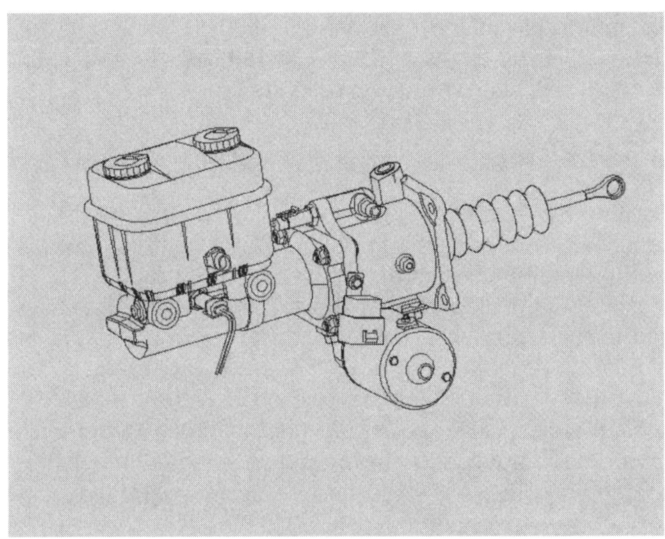

Figure 5. Brake Booster and Master Cylinder Used with GVW's from 16,000 lbs. – 33,000 lbs.

Differing brake sizes and master cylinders led to a challenging design problem faced by the two partners.

OPERATIONAL PERFORMANCE GOALS

GM approached the traction control design from a system perspective with vehicle level performance objectives. The performance goals were formulated to prioritize stability and acceleration in that order. The TCS is in operation at speeds up to 25 mph.

Vehicle level goals were established in the following areas:

1. Launch on uniform surfaces
2. Launch and acceleration on split surfaces
3. Gradability on split surfaces
4. Transition ability from traction to ABS
5. Transition capability on non-uniform surfaces

Each performance goal was superimposed with the additional requirement of maintaining a standard 12' lane and also limiting the driver's steering input. The maintenance of the lane with minimized steering input is in keeping with the Win of the program cited earlier, to enhance the ability of the driver to maintain vehicle stability in a wide variety of circumstances.

The ratios for acceleration compare the results of performance with TCS enabled versus the result obtained with TCS disabled. A brief definition of the vehicle level goals are as follows:

- Launch on uniform surface – the speed and time over a measured distance are compared with and without TCS enabled. The speed with traction will be greater than the speed without traction. Throttle is best effort, and steering inputs are also specified.

- Launch and acceleration on split surfaces – the acceleration on the split surface will be greater than the average of the acceleration on each individual surface with the traction active. In addition, steering input is specified. Speed will be greater with traction enabled than with the traction disabled.

- Gradability – the ability to launch on a 15% split coefficient grade fully laden. Launch can either be forward or reverse and will take place in 1.5 seconds or less.

- Transition traction to ABS – the time to maximum deceleration from TCS to ABS on a low coefficient surface when compared to the time to the same maximum deceleration on the same surface without TCS active.

- Traction capability on non-uniform surfaces – the measure of time to recognize and reduce tire overspin when transitioning from a high coefficient surface to a low coefficient surface, and the measure of the time to reach maximum acceleration when transitioning from a low coefficient surface to a high coefficient surface. The speed at which transition takes place is specified in both cases.

ABS & TCS FUNDAMENTALS

MK50 ABS/TCS HYDRAULIC SCHEMATICS

The chart in Appendix A shows the hydraulic schematics for the TopKick/Kodiak (front/rear brake split, rear wheel drive). For the front (ABS) circuit, there is a set of inlet valve (normally open) and outlet valve (normally closed) for each wheel brake. The inlet valve is energized, via electronic control unit (ECU) coil magnetic flux, to isolate the caliper pressure from the tandem master cylinder (TMC), and the outlet valve is energized to release the caliper pressure to the low pressure accumulator (LPA). The pump returns the fluid in the LPA to the TMC. For the rear (TCS) circuit, there are additional components in order to build up pressure at the calipers without brake pedal applications. The isolation valve (normally open) is energized to allow the pump to build pressure at the appropriate wheel(s). The valve contains an integrated pressure relief function to prevent system over-pressurization. The hydraulic TCS valve (normally open) provides an open path from the TMC reservoir through TMC to pump inlet for TCS operation. The valve closes during every brake pedal application through hydraulic pressure.

ABS (ANTI-LOCK BRAKE SYSTEM) OPERATION

The system constantly monitors all wheel speeds via wheel speed sensors and compares actual wheel speeds to the speed of the vehicle. ABS modulates the brake pressure so that each wheel receives the maximum possible brake force without any of the wheels locking. The brake force is adjusted in such way that a sufficiently large proportion of cornering force always remains. The goal of the ABS is to help maintain maneuverability, while enhancing stability and stopping distance.

When one or more wheels approach lock during braking, the corresponding inlet valve is closed and the outlet valve is opened. This closes off the fluid supply from the TMC and releases fluid to the LPA, which decreases the wheel brake pressure. The pump is also turned on to return fluid from the LPA to the TMC.

To obtain optimum braking, it is desirable to operate just on the verge of wheel lock. Therefore, pressure is built back up until the wheels approach lock again. To increase pressure, the outlet valve closes and inlet valve opens allowing fluid flow into the wheel brake. Since pressure is actually decreased and increased gradually in pulses, there is also a pressure hold phase where both inlet and outlet valves are closed. This isolates the wheel brake from the rest of the brake system. This cycle of pressure reduction/increase is repeated throughout the stop or until ABS control is no longer needed.

The pedal pulsation typically felt during ABS is the result of the fluid flowing in and out of the TMC. This is due to pump operation (rapid pulsation driving back the TMC pistons) and valve switching allowing fluid flow to the wheel brakes intermittently.

EBD (ELECTRONIC BRAKE DISTRIBUTION) OPERATION

EBD is a subsystem of the ABS. EBD replaces a traditional proportioning valve or load-sensing proportioning valve, which limits pressure to the rear axle. A load-sensing proportioning valve is typically used for truck applications in order to utilize rear brake force effectively at various loading conditions. However, the valve must be still set conservatively to account for all loading conditions. EBD limits rear wheel pressure only as needed, resulting in increased utilization of the rear brakes.

During EBD, the rear axle inlet valves close and outlet valves open. Like ABS, this lowers wheel brake pressure by moving some fluid to the LPA. Also like ABS, the pressure is modulated to achieve the most efficient braking.

The pump is not run during EBD. Fluid remains in the LPA until EBD is over; then it is returned to the TMC after the EBD control actuation ends.

TCS OPERATION

Positive slip (spinning) of one or more driven wheels causes the system to enter TCS control. TCS modulates the wheel brake pressure to control the wheel or wheels in a manner that will enhance vehicle stability and acceleration characteristics. The goal of the TCS is to control over-spinning drive wheels at a percentage of slip that provides stability and optimal vehicle acceleration. The system transfers drive torque via wheel brake application from the low μ (road/tire friction coefficient) drive wheel to the high μ drive wheel during split μ driving conditions.

Upon entering TCS control, the isolation valve is closed and the pump is turned on to build pressure at the appropriate wheel(s). The pump draws fluid from the TMC reservoir, through the TMC, brake tubes and the hydraulic TCS valve. Pressure in the HCU is built up to the opening pressure of the TCS valve pressure relief function, where excess fluid is bled off to the inlet side of the pump. This pressure is available to be applied to the wheel brakes as necessary to control the wheels through modulation of the inlet and outlet valves. Since the HCU pump draws fluid from the TMC reservoir by itself, the pump performance, or the brake pressure build up response time, is a function of flow resistance between the TMC reservoir and the pump. In the vehicle, the MK50 HCU installation position and the plumbing between the actuation system and the HCU were designed to minimize the suction flow resistance, which would increase significantly at low temperatures due to the increased viscosity of brake fluid. The system also continuously estimates the driven wheel brake lining temperature. When the lining temperature reaches a threshold, the TCS function shuts down to prevent the wheel brake from overheating until the temperature decreases enough to allow the TCS to be re-enabled.

BRAKE ONLY TCS VS. TCS WITH ENGINE MANAGEMENT

In addition to controlling the brake pressures at the driven wheels, the traction control system may interact with the engine controller and/or transmission controller extending the traction control function.

TCS with engine management:

  Advantages
  - Available for all vehicle speeds
  - "Smoother" system over BTCS Brake only Traction Control System

Disadvantages
- The sensation of engine reduction may be unpleasing to the aggressive driver

Brake only (BTCS):

Advantages
- No engine calibration necessary (reduces complexity of engine interface)
- Good vehicle accelerations on patchy/split μ surfaces versus a vehicle without BTCS

Disadvantages
- Speed threshold
- Comfort not as good as TCS

Due to the wide range of vehicle configurations for the TopKick/Kodiak project, including various engines (gas and Diesel with various displacements), a brake intervention only TCS was chosen in order to reduce the complexity.

THE DESIGN OF MK50 INTEGRATED CONTROL UNIT (ICU)

MK50 is a 4 hydraulic channel, 4 sensor ABS/TCS system developed for medium duty trucks and is based on an existing Teves ABS/TCS system, MK20E/MK25, for passenger cars and light trucks. The system is an add-on type integrated unit consisting of Hydraulic Control Unit (HCU) and Electronic Control Unit (ECU). See Figure 6.

The system was designed and developed in order to meet performance targets (vehicle performance, functional limits, reliability/durability) while minimizing the number of variations for vehicle configurations.

<u>Hydraulic Control Unit (HCU)</u>

The main HCU components are the aluminum housing with solenoid valves, low pressure accumulators (LPA's), and an integrated hydraulic pump driven by an electric motor. The HCU was sized to the caliper volume consumptions of medium duty trucks, which are four to ten times bigger compared to the consumptions of typical passenger cars and light trucks. Most of the components were newly developed for the TopKick/Kodiak range of vehicles. However, the interface to the ECU with solenoid valve coils and pump motor connector is kept the same as the one for the MK20E. It allowed Teves to use the same ECU chipset and housing from the MK20E, which reduced development and tooling cost. Besides the ECU interface, many of the interface dimensions on the aluminum housing to other key components were kept the same as MK20E/MK25 in order to simplify development and manufacturing. This allowed using existing production tooling and manufacturing processes for some components when carrying over existing components was not possible. Also, it enabled the Teves assembly plant to share the same HCU assembly line and some of the component assembly lines between MK50 and MK20E/MK25, despite the great size difference in the vehicle applications. The HCU is combined with the ECU to form a functional unit (ICU: Integrated Control Unit). See Figure 7.

Figure 6. MK50 View 1

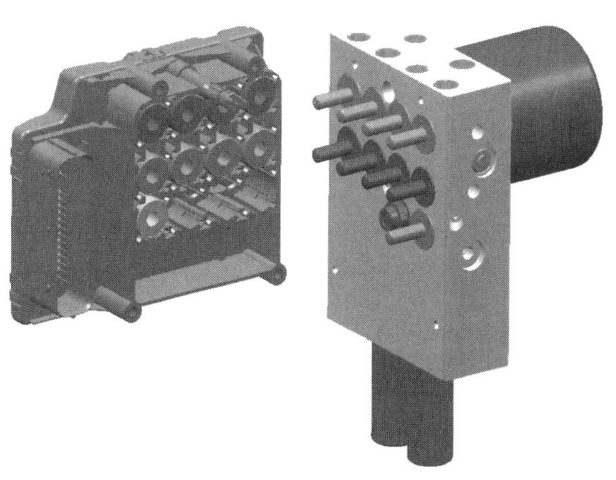

Figure 7. MK50 View 2

Despite the many variations of vehicle configurations; e.g. wheel bases, engines, actuation systems, calipers, tire sizes; the HCU shares many of the same components (pumps, valve orifice sizes, etc.) to cover all the vehicle configurations. The only component variation necessary was the sizes of LPA's corresponding to two types of actuation systems. For each actuation system, the LPA size needs to be large enough to hold brake fluid after a high μ to low μ transition, while small enough to insure generation of master cylinder line pressure if the fluid is lost in the LPA as a result of HCU internal leakage. Having the minimum number of HCU variations simplified the development effort at GM and Teves, as well as simplified the ICU assembly at Teves and vehicle installation at the GM plant.

<u>Electronic Control Unit (ECU)</u>

The ECU is comprised of a plastic housing with enclosed printed circuit board space, a connected coil box with 8 (for ABS system) or 9 (for ABS/TCS system) coils, a connector to the pump motor, and a connector to the vehicle. In its interior, the printed circuit board space accommodates a printed circuit board equipped with electronic components. This is connected to the coils, the pump motor connector, and the connector to the vehicle. As mentioned already, the MK50 HCU was designed in such a way that would require minimum changes to the existing MK20E ECU. The only main change made to the ECU was increased MOSFET size for the pump motor driver due to the higher motor current of the MK50 motor.

VEHICLE/SOFTWARE DEVELOPMENT

The MK50 software is developed for truck application, and is based on MK20E passenger car software. The application is unique compared to passenger cars and light trucks for the following reasons:

- Wide range of vehicle configurations for the same platform (GVW's, wheelbase lengths, center of gravity heights, tire sizes, brake caliper sizes, engines, etc.)

- Wide range of loading conditions (CVW vs. GVW, front vs. rear loading balance)

- Slow response time of tires to the brake pressure controls due to the larger tire rolling radius and increased wheel end inertia

The software developed for MK50 is highly adaptable to various vehicle configurations and loading conditions based upon wheel speed characteristics. As mentioned already, brake intervention only TCS, instead of TCS with engine management, was chosen for the project to reduce the complexity. Numbers of tests were performed with various vehicle configurations for various vehicle maneuvers throughout the development. The only difference in the software from vehicle to vehicle was a split for long or short wheelbase. The split for wheelbase only affected performance of the yaw torque control on split μ surfaces. It takes advantage of different response behavior of vehicles, depending on the wheelbase length, on split μ surfaces, and, as a result, it gives optimum stability and stopping distance on split μ surfaces. Other than that, no special programming or compensation was required for different caliper sizes, booster type, load conditions, etc. Performance was adaptable just based upon wheel speed characteristics.

Due to the slow response time of tire speeds to the brake pressure controls, the basic control algorithm was adjusted. Also, select-low ABS control for rear axle typically used for passenger cars was replaced with independent control. With select-low, in order to minimize brake force differences of the rear axle on split μ surfaces, the system controls the rear left and right brake pressures simultaneously in response to the wheel speed signal from the wheel on the lower friction coefficient side. The select-low strategy improves the stability on split μ surfaces while achieving relatively short stopping distance for passenger cars. However, for medium duty trucks, the select-low control did not optimize the stopping distance/stability relationship in many situations. Trucks have a much higher rear to front axle weight ratio when loaded and are much more likely to be actually loaded to GVW compared to passenger cars, which are not as likely to be loaded to GVW. When a truck has a load, even with weight transfer during braking, the rear axle still is carrying the majority of the weight. Therefore, independent rear pressure control strategy was used in order to utilize as much braking on the rear axle as possible. While reducing the stopping distances on split μ surfaces with independent rear pressure control, particular attention was paid to help maintain the stability of the vehicle. The algorithm for the front pressure control on split μ surfaces was adjusted in order to help maintain the stability of the vehicle.

TCS Performance Results

Performance Target = 100%

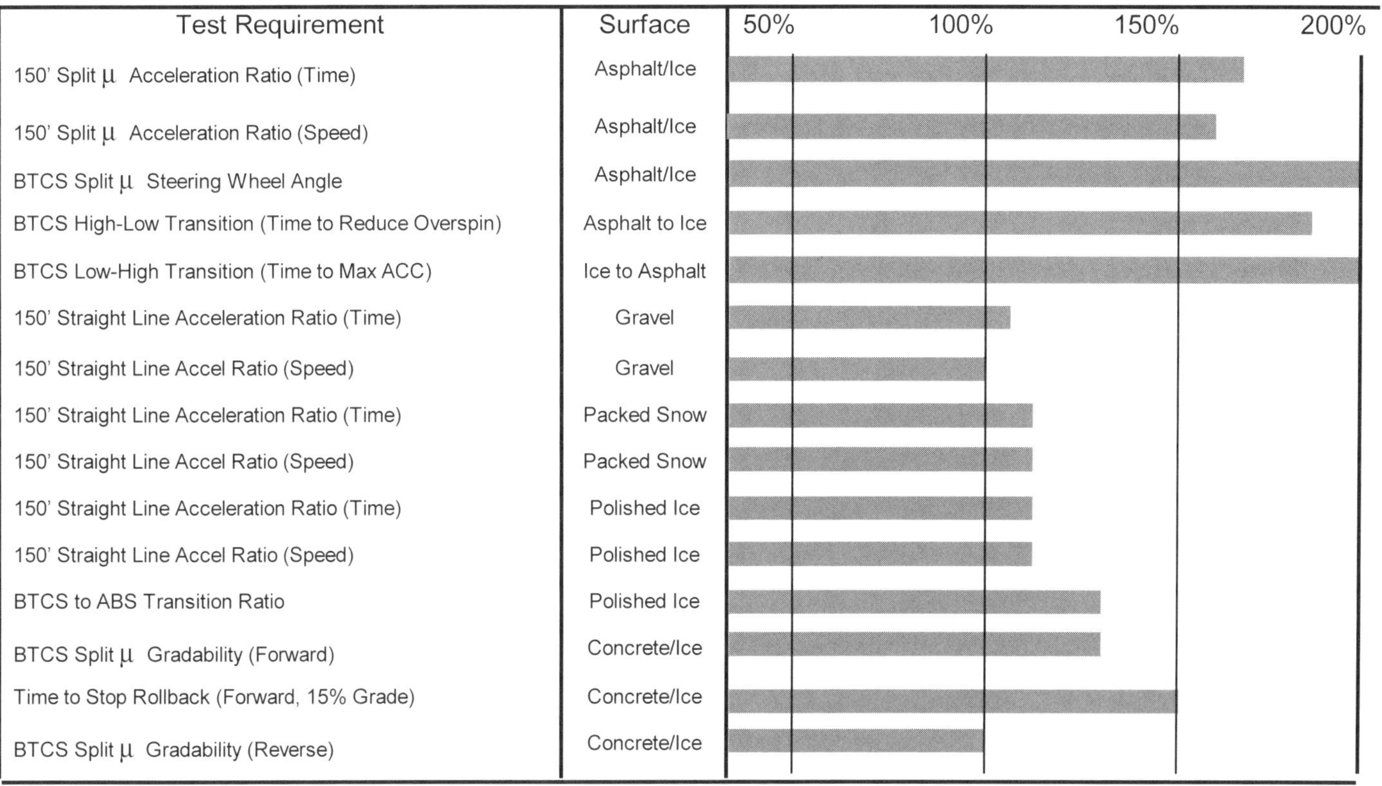

Figure 8

BENEFITS

On hydraulic braked commercial vehicles, the limited slip differentials are not driver controlled. Some mechanical systems affect the vehicle handling characteristics and may also cause mechanical damage to the axle drive components. Some installations of differential axle locks may not be warranted by the axle manufacturer. The choice of axles on which mechanical differential locks can be installed is limited to single speed applications. The brake only electronic traction control can be installed in any application without any unique components, other than the electronic control unit. The cost of installation of the adaptive system is a fraction of the mechanical system. The ease of use of the electronic system in a commercial truck eliminates the drawbacks of mechanical systems, and allows the driver to operate in all conditions. The availability of the electronically controlled traction assist is only one important feature that helps with the goal of making braking and handling a function of the truck's design.

CONCLUSION

The use of electronic traction control provides an economical means to assist the driver of commercial trucks in traction limited operating conditions. Its operation is automatic and provides flexibility in enabling the feature. It operates with gas or Diesel powered vehicles. The traction system has proven effective in launching the truck in adverse conditions and has responded to the "Win" requirement as requested by the Voice of the Customer.

ACKNOWLEDGEMENTS

Figure 2 – TRW

Figure 3 – Meritor

Figures 4 and 5 – Bosch

DEFINITIONS, ACRONYMS, ABBREVIATIONS

ABS: anti-lock brake system

BTCS – brake traction control system

CVW: curb vehicle weight

EBD: electronic brake force distribution

ECU: electronic control unit

GVW: gross vehicle weight

HCU: hydraulic control unit

LPA: low pressure accumulator

MOSFET: metal oxide semiconductor field effect Transistor

μ : coefficient of friction between tire and road

TCS: traction control system

TMC: tandem master cylinder

ICU: integrated control unit

APPENDIX A

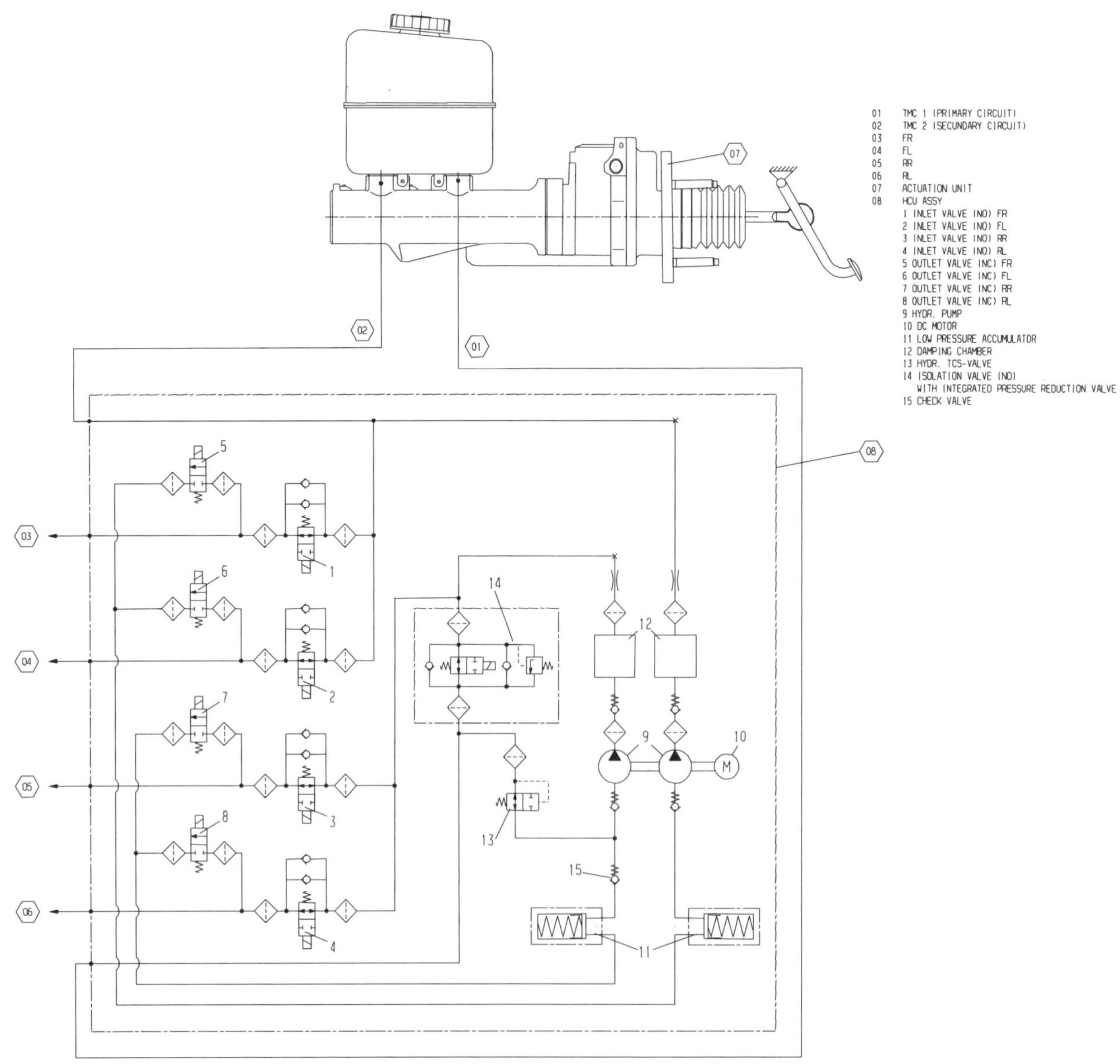

Improved Differential Function for Avoiding Slippage of Motor Vehicles in a Muddy Trench

S. S. Gill, N. S. Kalsi, Balraj Singh and Nirmal Singh
Department of Mechanical Eng., Beant College of Eng. and Tech

Neetu Singh
Department of Electronics Eng., Beant College of Eng. and Tech

ABSTRACT

Vehicles such as trucks, buses and cars often get struck up in muddy trenches or in snow with one of wheels in the trench and the other on level ground. With an open differential, driver's effort to pull the vehicle out normally ends up with one wheel, which is rotating still going deeper into the trench. Until now Limited Slip Differentials (LSD), Locked Differentials and Viscous Coupling etc. have been used to overcome this difficulty but with limited success.

Incorporating logic devices has developed a novel solution to this problem and the traction on the required wheel is achieved automatically. The system works independently of the conventional brakes.

Keywords: Logic devices, electro hydraulic servo valve, speed sensor and feed back devices

INTRODUCTION

Differential is a device, which allows the individual wheels to be driven at different speeds while transmitting torque to them at the same time. Suppose a car is taking left 90^0 turn as shown in figure 1.Then for an average diameter of 0.6 m of tyre and inside wheel turning radius of 4.3 m, there will be a difference of 9 revolutions between the left and right wheel.

For no skidding of right tyre, right side tyre must rotate faster than left tyre by 9 revolutions. This is easily done by using open type differential. But the things are different when one wheel, which is inside the mud rotating freely at high speed and the other wheel on level, ground being stationary .For this Limited slip differentials used earlier only limit or reduce wheel slippage but do not eliminate it. The degree of reduced wheel slippage is determined by the amount of friction applied to the side gears. Most of these limited slip differentials are designed to transmit only limited amount of torque and may not be working well in other conditions. In most of the LSD's, friction material is used in differential's clutch plate or cone clutch employed.

Although it is better than open differential but do not provide the maximum traction capabilities of locking differentials. So the design may not be ideal for every type of application. In TORSEN (stands for TORque SENsor) differential, it only locks the differential in a power application situation and has only torque sensing characteristics. Porsche also rejected the viscous coupling because they felt that the viscous coupling was too difficult to control and it had exponential rather than linear locking characteristics.

Locking differentials although had the advantage of both wheels moving at same speed in the mud, but skidding of tyre occurs when taking turns. According to Caroll Smith [6] the largest problem with locker is "Should a drive shaft or line joint fail under power, the car will turn in the direction away from failure at unbelievable speed and with inconceivable force". There is no recovery.

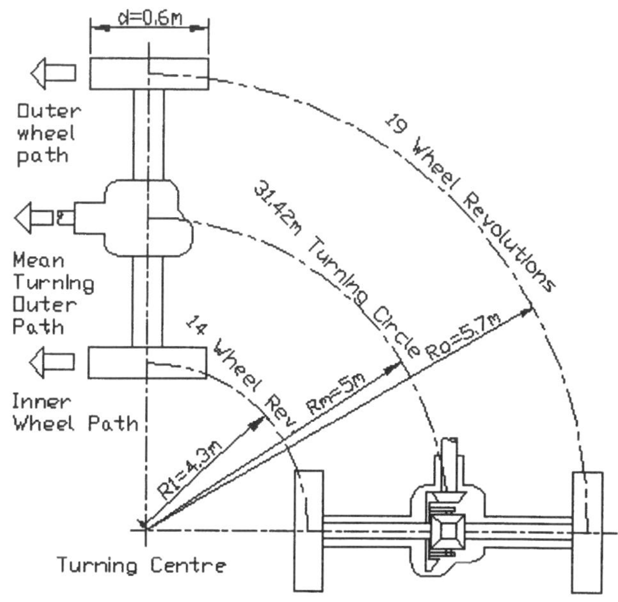

Figure 1: Illustration of Differential function

We have come out with a new design, which overcomes the difficulties of earlier differentials and incorporates logic devices. The detailed description and methodology follows hereunder.

WORKING PRINCIPAL

The operation of the system is based upon logic devices and control loop. Sensors are used to check the state of system or to measure certain quantity. Corrective action will be initiated by the system only if the difference of speed of two wheels has exceeded a particular set value. This set value is decided keeping in mind speed difference existing during normal or hard turning conditions. When slip on one tyre is more than the set value, an electrical signal is sent to the servo valve to apply brake to the appropriate side gear in the differential. The brakes are operated by using the pressured oil from either power steering or from automatic transmission. When the vehicle is moving straight ahead both wheels will have equal traction (indicated by the equal thickness of arrow line) and therefore brakes are not applied as shown in the *fig.2*.

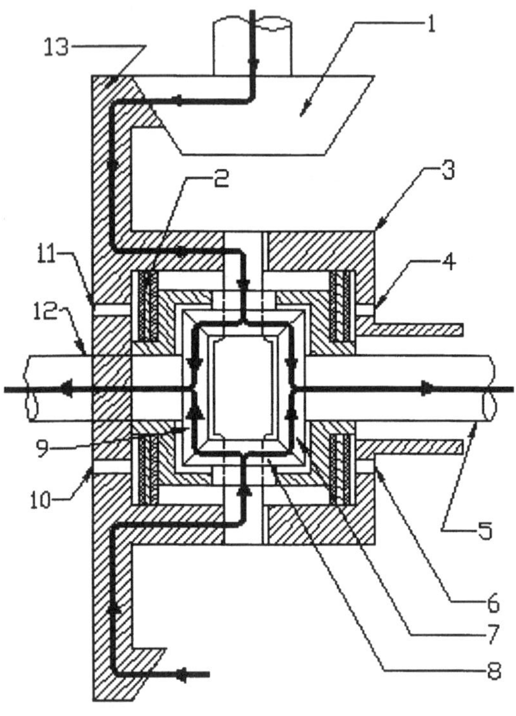

1.Axle Drive Pinion, 2.Brake, 3.Differential Case, 4,6,10 and 11.Passage for servo Hydraulic Pressure, 5.Right Axle Shaft, 7and 9.Differential Side Gears, 8.Differential Pinion, 12.Left Axle Shaft, 13.Axle Drive Gear.

Figure 2: Figure showing the torque being transmitted to the both wheels (indicated by the arrow line) when the vehicle is moving straight ahead and brakes not being applied to the side gears.

But when one wheel is in mud e.g. right wheel and because of differential action, the whole power from engine goes to this rotating wheel, brakes are needed to be applied to this wheel, so that power is transmitted to the other stationary wheel having traction. This way the vehicle comes out of the mud or snow. *Fig 3* shows an example where brakes have been applied to right wheel and most of the power therefore goes to the left wheel as can be seen from thickness of arrows.

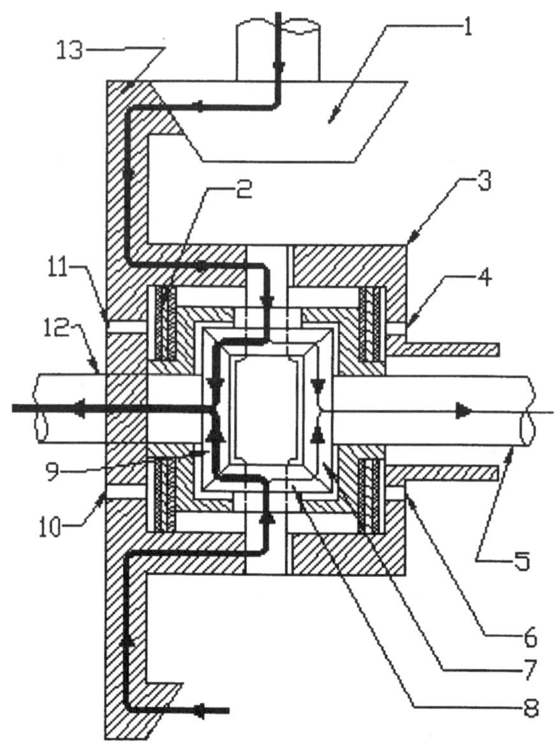

1.Axle Drive Pinion, 2.Brake, 3.Differential Case, 4,6,10 and 11.Passage for servo Hydraulic Pressure, 5.Right Axle Shaft, 7and 9.Differential Side Gears, 8.Differential Pinion, 12.Left Axle Shaft, 13.Axle Drive Gear.

Figure 3: Figure showing the torque being transmitted to the left wheel (thick arrow line) when brakes have been applied to the right side gear through servo valve.

SYSTEM FUNCTIONING - It uses the Body paragraph style and is identified with a header beginning the paragraph as shown here. For the new system to work effectively, various feed back devices and sensors have been used. They are
- Sensor to check if ignition is on (S_1).
- Clutch sensor (S_3)
- Gear sensor to check if gear is engaged or not. (S_2)
- Speed sensor at Right and Left wheel. (S_4)

The various conditions or situations to which this new electronic system has to operate are:
- Both Shafts rotating at same speed.
- Both shafts rotating but one rotating faster than the other.
- One shaft rotating and another is stationary.

- The electronic system will operate only if either condition three is satisfied or in case of condition two, if the speed difference of two wheels is greater than set value. The flow diagram for the present system is given in the figure 4.

The electronics system will keep on checking for the conditions stated above. This has been made possible by the use of control loop as shown in fig.5. The speed

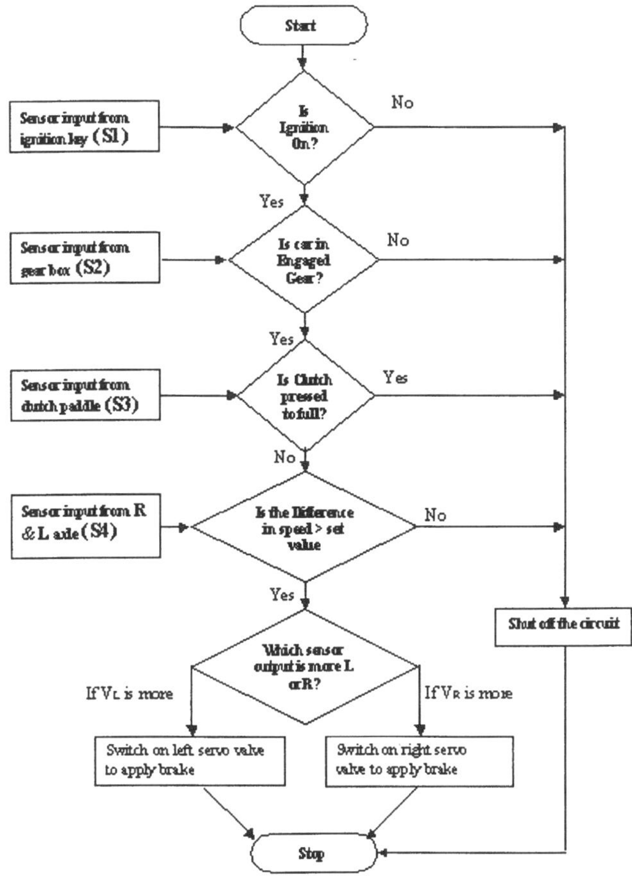

Figure 4: Flow diagram for the system

values from two speed sensors (Tachogenerator) will be fed to the comparator, which will compare the speed difference with the set value. So if this speed difference is greater than set value, then an appropriate signal

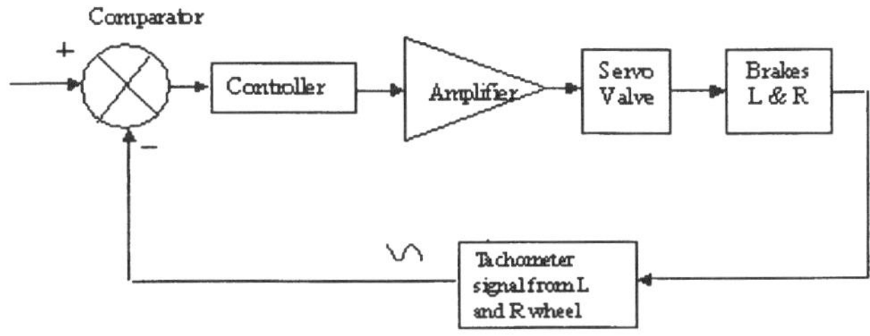

Figure 5: Control loop for the electronic system

through amplifier will be sent to the servo valve to apply brake on either left or right side gear in differential box

through the controller as shown in fig 6. We may need to use a 10bit controller or microprocessor for speed of up to 1200 rpm (wheel speed). For higher speeds, a 16bit or 32bit controller will be sufficient.

LOGIC CIRCUIT DIAGRAM: A logic circuit has been designed for the new system by combining logic devices such as AND, OR, NOR, XOR etc. and is shown diagrammatically in fig.7 below.

CONTROLLER : Controller sends the error signal to the appropriate servo valve (for right or left side gear) to apply brakes on the appropriate differential side gear. For example fig.3 shows servo brake has been applied to the right side gear and most of the power being driven to left wheel or left side gear in differential. The output from the controller is a signal to the electro hydraulic valve to apply the brakes on the side gear. This signal is passed through the amplifier before

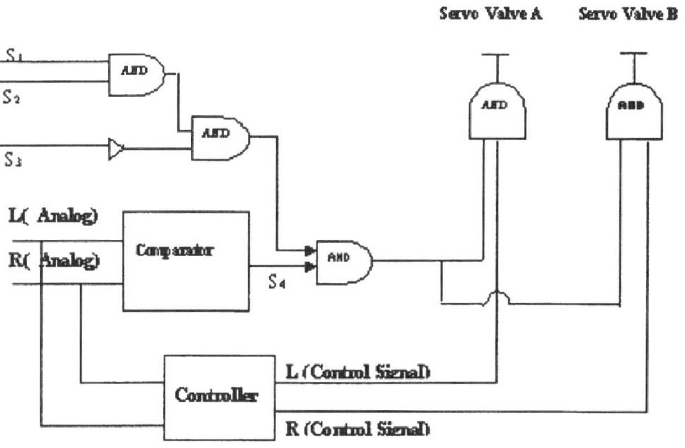

Figure 6: Figure showing arrangement of various devices in the present system

being fed to the servo valve.

WORKING OF CONTROLLER: Controller is meant to check whether the speed of right axle or left axle is more. The input given to the controller is the analog speeds of right and left axles. The output is digital signal, which is the '1' digital value according to the axle's speed. The various digital devices used in the controller are A/D converter, XOR gate, OR gate, AND gate and NOT gate. An A/D converter is used for this purpose, which converts the speed in rpm into digital signal. This is done both for left and right axle. Now the digital value of speed for both axles is compared bit by bit. Firstly the most significant bit is compared with the help of an XOR gate. The output for XOR gate is '1' only if the input is '0', '1' or '1', '0' respectively for left and right axles. This output of the XOR gate is feed into two AND gates. The other input for the AND gates are the input values of both left and right axel, until we find which one of the left or right axle speed is greater. Otherwise, the '0' bit is send for other XOR gates, which makes the output of other AND gates '0'. If the most significant bit for both left and right axles is same, then the next bit is compared. This process goes on, until we find which one of the left or right axle speed is greater. Otherwise, the '0' bit is send for other XOR gates, which makes the output of other AND gates '0'.

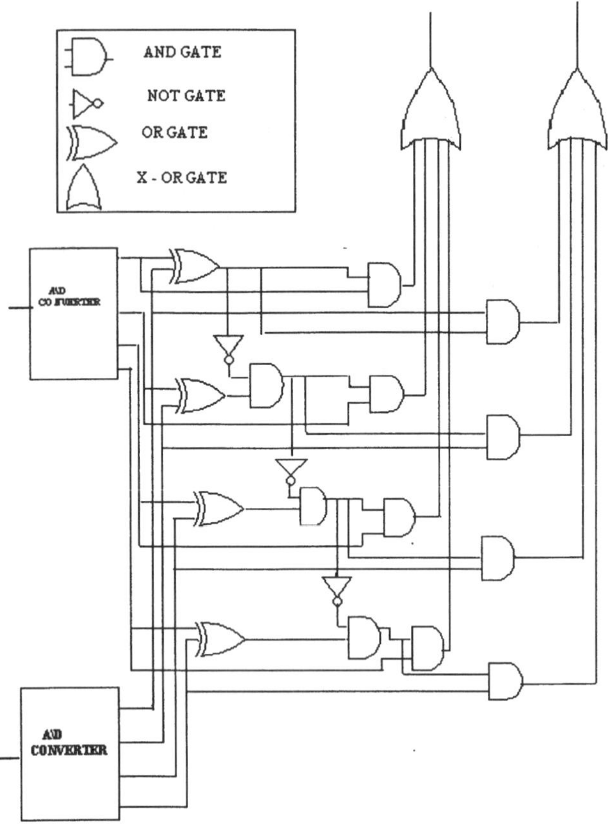

Figure 7: Logic circuit diagram for controller

CONCLUSION

This new system design based upon electronics is a novel idea and it is hoped that it is definitely better than limited slip differential and locking differential. Their disadvantages have been eliminated in this new methodology. The system is economical as it uses only logic devices and simple processor. The installation of this system is easy and no major design changes are required in the   differential.   No extra power sources are required for its operation as the pressurized oil from automatic transmission or power steering is used.   This system can later on be implemented on 4wd vehicles also.

REFERENCES

1.   Caroll Smith, Drive to Win, PP7 to 26
2.   Eliot Lim, Introduction to all wheel drive System, Feb 26 1999
3.   Heinz Heisler, Vehicle and Engine Technology, Vol-1, ELBS.
4.   Jacques Gordon, steering and suspension: Electronics help keep the black patch down. September, 1998
5.   Joseph Heitner, Second Edition, Automobile Mechanics, East-West Press Private Limited
6.   Kirpal Singh, Automobile Engg. , Volume1, 1997,Std. Publications & Distributors, New Delhi
7.   Off-Road.com Review: auburn gear limited slip diffrential.html (1999)
8.   Tom Morr, Traction for action: guide to lockers and limited Slip differentials (for off road trucks), April 1996
9.   Tom Tagiella, reviewers notebook, Sure Group Limited Slip differential, Jan 1999
10.   William H.Crouse, Donald L Anglin, Ninth Edition, Automobile Mechanics, Tata   Mc-Graw   Hill   Publishing Company Ltd, New Delhi.
11.   Wisconsin   Alumni   Research Foundation,   WARF: P991810S, Hybrid Brake System
12.   Automobile.com \ Troubleshooting
13.   www.enggineringtalk.com
14.   www.eskimo.com\reliot\awd.html
15.   www.drivetrain.com
16   www.howstuffwork.com\diffrential.html

CONTACT

ssgill@eudoramail.com

Improving Performance of a 6x6 Off Road Vehicle Through Individual Wheel Control

Andrew Jackson and David Crolla
University of Leeds

Adrian Woodhouse and Michael Parsons
QinetiQ

ABSTRACT

This paper presents a method of control for a 6x6 series-configured Hybrid Electric Off-road Vehicle (HEOV). The vehicle concerned is an eight-tonne logistics support vehicle which utilizes Hub Mounted Electric Drives (HMED) at each of its six wheel stations. This set-up allows Individual Wheel Control (IWC) to be implemented to improve vehicle handling and mobility. Direct Yaw-moment Control (DYC) is a method of regulating individual wheel torque to control vehicle yaw motion, providing greater stability in cornering. When combined with both a Traction Control System (TCS) and an Anti-lock Braking System (ABS) the tire/road interaction is fully controlled, leading to improved control over vehicle dynamics, whilst also improving vehicle safety.

INTRODUCTION

During the last ten years, an increased awareness of the automobile's negative effect on the environment has prompted the automotive industry to look at alternatives to the internal combustion engine powered vehicle. The Hybrid Electric Vehicle (HEV) is one particular solution to the problem. It relies on the combination of electric motors and an internal combustion engine in an integrated driveline. Although, perhaps not as ideal as the Electric Vehicle (EV), the HEV does however hold the potential for reduction in emissions whilst maintaining performance comparable to a conventional vehicle with none of the range restrictions present in the EV.

The hybridization of off-road, all-wheel drive vehicles is an area that has yet to have been exploited by the automotive industry. With the quick torque response that can be delivered by electric motors combined with the power and range of the internal combustion engine, the implementation of traction and power-train control systems to aid the mobility and performance of off-road vehicles is made even more viable.

The vehicle studied is an off-road, six wheel drive, four wheel steer, series configured, hybrid electric vehicle. The use of in-wheel motors gives rise to the utilization of individual wheel control. Whereas in a conventional vehicle this can only be realized through the use of torque-split devices such as limited-slip differentials or the control of individual brake actuators, the control of individual wheel torques in the HEV can be achieved almost exclusively in software. This in itself offers greater potential for control of vehicle performance. One particular control system that has been researched for both electrical [1-3] and conventional vehicles [4-6] is Direct Yaw-moment Control (DYC). DYC is a form of vehicle control that alters the torque applied to each wheel to improve vehicle handling and stability. By varying left and right wheel torques, yaw moments can be produced to help maintain a desired yaw rate. This gives the driver better control over the vehicle, reducing the need for driver intervention to maintain a desired course. The quick torque responses of the electric motor also offer good potential for traction control systems.

It is the combination of yaw-moment, traction and anti-lock braking control that is the subject of this paper. The developed mobility controller is implemented in MATLAB/SIMULINK^{tm} on a basic non-linear vehicle handling model utilizing a Dugoff tire model to determine longitudinal and lateral tire forces. A number of simulations are undertaken, testing the vehicle model on a number of road surfaces, under a variety of test maneuvers to assess the potential of the coordinated controller. Through the execution of this type of control, full Individual Wheel Control (IWC) is implemented. IWC allows the torque applied to each individual wheel station to be altered independently, depending on the vehicle's yaw-rate error and wheel-slip ratio. Independent Proportional Derivative (PD) traction and anti-lock braking controllers are implemented at each wheel while a fuzzy logic DYC controls the yawing motion of the vehicle body. The operation of the DYC alters depending on the state of the individual traction/anti-lock braking controllers.

The controller shows a marked improvement in vehicle handling over a conventional fixed-torque distribution system and also displays improved vehicle stability and safety.

VEHICLE MODEL

The vehicle model on which the controller is tested is a nine degree of freedom handling model shown in Fig.1. It includes longitudinal, lateral and yaw motion along with the rotation of the six wheels. As this model does not include the vehicle ride characteristics or model any load transfers due to cornering, acceleration or deceleration, the vehicle's roll, pitch and bounce motions are ignored. The model is implemented in the SIMULINK/MATLAB^{TM} environment, allowing easy manipulation of the vehicle parameters and inputs.

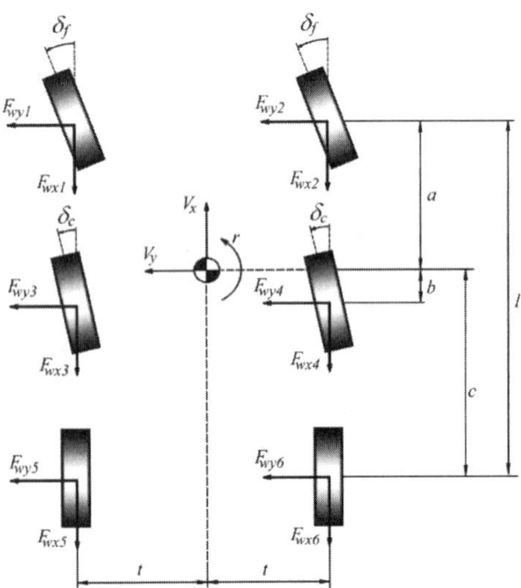

Fig. 1. Nine degree of freedom model

The equations for the vehicle motion are [7], [8] (see nomenclature):

$$M_b(\frac{dV_x}{dt} - V_y r) = F_{wx1} + F_{wx2} + F_{wx3} + F_{wx4} + F_{wx5} + F_{wx6} - F_d \quad (1)$$

$$M_b(\frac{dV_y}{dt} - V_x r) = F_{wy1} + F_{wy2} + F_{wy3} + F_{wy4} + F_{wy5} + F_{wy6} \quad (2)$$

$$I_z \frac{dr}{dt} = aF_{wy1} + aF_{wy2} - bF_{wy3} - bF_{wy4} - cF_{wy5} - cF_{wy6}$$
$$- tF_{wx1} + tF_{wx2} - tF_{wx3} + tF_{wx4} - tF_{wx5} + tF_{wx6} \quad (3)$$

The vehicles aerodynamic drag is given by:

$$F_d = \frac{1}{2} \cdot C_d \cdot \rho \cdot A_d \cdot V_x^2 \quad (4)$$

The wheel forces are derived using the Dugoff tire model [9], [8]. The equation for wheel rotation is:

$$\frac{d\omega_i}{dt} = \frac{1}{I_w}(T_i - r_w F_{wxi} - r_w F_{zi} RR) \quad (5)$$

MOTOR MODEL

The set-up of a series hybrid electric vehicle with HMEDs at each wheel allows the vehicle to be separated into two distinct systems, a power generation unit and an electric driveline. If the assumption is made that any power required at the motors is deliverable by the powertrain then for the purposes of controlling the vehicle dynamics, only the electric driveline need be modeled. The actual operation of the motors is ignored, and instead a torque-speed curve is used to determine the maximum torque that is available for a particular wheel speed. The HMED has a torque-speed curve as shown in Fig. 2 [10].

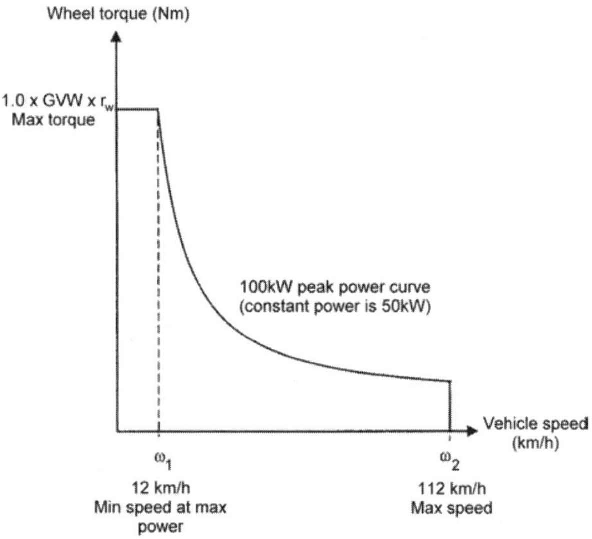

Fig. 2. Torque-speed curve for HMED

BRAKE MODEL

The method of braking utilized in the simulations is pure electrical braking. Although the HMED design incorporates a mechanical brake mechanism, the high torque capabilities of the electric motor means that electrical braking is sufficient for quick braking of the vehicle. Also, with the more accurate control of the applied torque, it offers good potential for an Anti-lock Braking System. The motor can provide regenerative braking up to the same power and torque limits as applied when driving. Any excess brake torque required in situations such as drive failure can be provided by the mechanical brake.

CONTROLLER DESIGN

It is the purpose of the controller to maintain a desired yaw-rate and to improve the vehicle's tractive performance. It utilizes six individual PD traction controllers. Depending on the on/off state of these controllers, the operation of a fuzzy logic DYC is altered. Driver torque demand calculations refer to the restriction of demand torque to within the boundaries of the torque-speed curve (fig. 2) before the control inputs are added. This prevents saturation of the torque demand, allowing DYC to still have an effect, even when the maximum torque is demanded by the driver's throttle input. Fig. 3. shows the operation of the torque demand calculations.

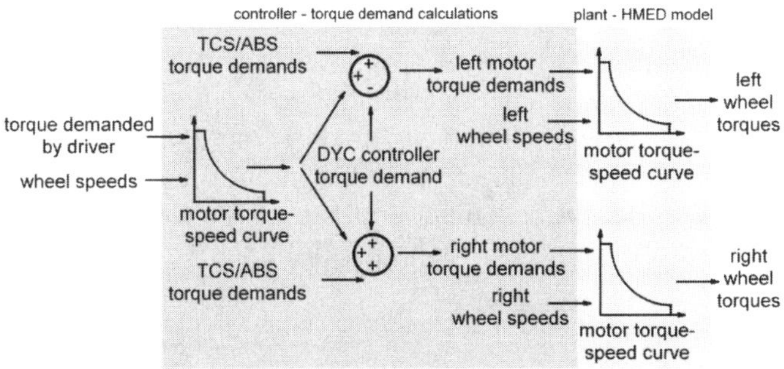

Fig. 3. Layout of torque demand calculations for control system and HMED's.

EVALUATION OF DESIRED YAW RATE AND WHEEL SLIP RATIO - In order to evaluate the effectiveness of both the control options presented below, it is necessary that a desired value for the yaw rate is derived. The desired yaw rate for these simulations is obtained from the steady-state behavior of the vehicle. A desired yaw rate is calculated, depending on steer input and vehicle speed by Eq. 6 [7].

$$r_{ss} = \frac{0.5(a-b)C_{\alpha f}C_{\alpha c} + 0.5(b+c)C_{\alpha c}C_{\alpha r} + (a+c)C_{\alpha f}C_{\alpha r}}{(a-b)^2 C_{\alpha f}C_{\alpha c} + (b+c)^2 C_{\alpha c}C_{\alpha r} + (a+c)^2 C_{\alpha f}C_{\alpha r} + M_b V_x^2 (cC_{\alpha r} - bC_{\alpha c} - aC_{\alpha f})} \cdot \delta_f$$

(6)

Using the parameters for the model concerned, this steady-state yaw rate will show a slight understeering characteristic. The desired handling characteristics of a vehicle is often a matter of driver preference. However, although no consideration to the drivers' preferred handling characteristics are presented in this paper, slight understeer is often desirable as the driver still maintains good control over vehicle handling which improves safety (oversteer would be undesirable from a safety perspective).

Wheel slip ratio is defined as:

$$s_i = \begin{cases} 1 - \dfrac{V_x}{r_w \omega_i} & \text{if } r_w \omega_i \geq V_x \quad \text{(in acceleration)} \\ 1 - \dfrac{r_w \omega_i}{V_x} & \text{if } r_w \omega_i < V_x \quad \text{(in braking)} \end{cases}$$

(7)

Although the maximum desired wheel slip alters depending on the road surface, prediction of this optimum value is a complicated matter. For the purpose of this paper a pre-defined value of $s_i = \pm 0.2$ is used,

this value promotes good responses on all road surfaces.

PD TRACTION CONTROLLERS - Simple PD controllers are implemented to control the wheel-slip ratio of each wheel. The traction controllers are only turned on when the wheel-slip ratio exceeds the desired value of 0.2. Two outputs are taken from the controller, its state and the torque reduction command. The state of the controller has four possible values: 0 = No wheel slipping, 1 = At least one right wheel slipping, 2 = At least one left wheel slipping and 3 = At least one left and one right wheel are slipping.

PD ANTI-LOCK BRAKING CONTROLLERS – Again, simple PD controllers are implemented to control the wheel-slip ratio of each wheel. During heavy braking (i.e. for an emergency stop), locking of the wheels is highly likely on a vehicle not utilizing an ABS. Wheel locking is undesirable and hence, by maintaining a particular wheel slip ratio, it is avoided. The Anti-lock braking controllers are only turned on when the wheel-slip ratio goes below the desired value of -0.2. Two outputs are taken from the controller, its state and the brake torque reduction command. The state of the controller has four possible values: 0 = No wheel slipping, -1 = At least one right wheel slipping, -2 = At least one left wheel slipping and -3 = At least one left and one right wheel are slipping.

FUZZY LOGIC DIRECT YAW-MOMENT CONTROL - The fuzzy logic controller developed has four input variables and two outputs. It is an extension of the one presented in [11]. The input variables are yaw-rate error, rate of change of yaw-rate error, the integral of yaw-rate error and the state of the traction controllers as presented in above. The outputs are the torque control values to the left and right wheels. These control torques are added to the torque demanded by the driver and to the torque reduction commands from the traction controllers. This results in a yaw-moment being created. The controller contains 49 rules as shown in table 1. Membership functions were tuned by trial and error. When all the traction and anti-lock braking controllers are inactive, the fuzzy controller implements DYC as presented in [11]. Operation of this controller is altered when any of the traction/anti-lock braking controllers become active. A similar control system for a 4WD electric vehicle is presented in [4], which also includes control of the vehicle wheel slips.

Key for Table 1:
Yaw Rate error: NB = Negative Big, NS = Negative Small, PS = Positive Small, PB = Positive Big
Rate of change of error: -ve = Negative, +ve = Positive
Integral of Error: -L = Negative Large, -S = Negative Small, +S = Positive Small, +L = Positive Large
Wheel-slip State: 0 = No wheel slipping, 1 – At least one right wheel slipping, 2 = At least one left wheel slipping and 3 = At least one left and one right wheel are slipping.
Right/Left Command Torques: GD = Greatly Decrease, D = Decrease, M = Maintain, I = Increase, GI = Greatly increase. The number in

brackets refers to a weighting of that particular rule. All other rule have unity weighting.

Inputs				Outputs	
Yaw rate error	Rate of change of error	Integral of Error	Wheel slip state	Right command torque	Left command torque
NB	-ve	-	0	D	I
NS	-ve	-	0	D	I
Zero	-ve	-	0	D	I
PS	-ve	-	0	I	D
PB	-ve	-	0	I	D
NB	Zero	-	0	GD	GI
NS	Zero	-	0	D	I
Zero	Zero	-	0	M	M
PS	Zero	-	0	I	D
PB	Zero	-	0	GI	GD
NB	+ve	-	0	GD	GI
NS	+ve	-	0	GD	GI
Zero	+ve	-	0	I	D
PS	+ve	-	0	GI	GD
PB	+ve	-	0	GI	GD
-	-	-L	0	GD (0.35)	GI (0.35)
-	-	-S	0	D (0.35)	I (0.35)
-	-	+S	0	I (0.35)	D (0.35)
-	-	+L	0	GI (0.35)	GD (0.35)
NB	-	-	3	GD	M
NS	-	-	3	D	M
Zero	-	-	3	M	M
PS	-	-	3	M	D
PB	-	-	3	M	GD
NB	-	-	2	M	GI
NS	-	-	2	M	I
Zero	-	-	2	M	M
PS	-	-	2	M	D
PB	-	-	2	M	GD
NB	-	-	1	GD	M
NS	-	-	1	D	M
Zero	-	-	1	M	M
PS	-	-	1	I	M
PB	-	-	1	GI	M
NB	-	-	-3	M	GI
NB	-	-	-3	M	I
Zero	-	-	-3	M	M
PS	-	-	-3	I	M
PB	-	-	-3	GI	M
NB	-	-	-2	GD	GI
NS	-	-	-2	D	I
Zero	-	-	-2	M	M
PS	-	-	-2	I	M
PB	-	-	-2	GI	M
NB	-	-	-1	M	GI
NS	-	-	-1	M	I
Zero	-	-	-1	M	M
PS	-	-	-1	I	D
PB	-	-	-1	GI	GD

Table 1. Fuzzy logic rules

SIMULATION RESULTS

The model and controller is simulated on two different road surfaces under a number of vehicle maneuvers. These maneuvers are simulated by inputting various steer inputs and vehicle demand speeds. The maneuvers are acceleration on a split μ road surface, hard braking on a wet road, hard braking during cornering and a lane change and J-turn during acceleration. The two road conditions on which the vehicle is simulated are dry asphalt ($\mu = 0.8$) and a wet road surface ($\mu = 0.5$). There is no modeling of the driver, hence controlled, open loop tests are performed (i.e. no corrective steering is applied by the driver).

ACCELERATION ON A SPLIT μ SURFACE - The vehicle is accelerated from 5km/h to full speed whilst the left set of tires are in contact with a dry road surface and the right tires are in contact with a wet road surface. The resultant yaw rates are shown in Fig. 4a, the vehicle path in Fig. 4b and the results for wheel slip of the six tires are shown in Figs. 4c and 4d for the non-controlled and controlled vehicle respectively. From Fig. 4b it can be seen that the implemented controller greatly reduces the instability produced from the split μ surface and the vehicle path remains almost unaffected. This is obvious from looking at Fig. 4a. This removes any need for driver intervention to maintain the desired course. The slip responses shown in 4d show that the controller manages to maintain wheel slip close to the desired value, hence increasing both longitudinal and lateral tire forces aiding vehicle stability and cornering performance.

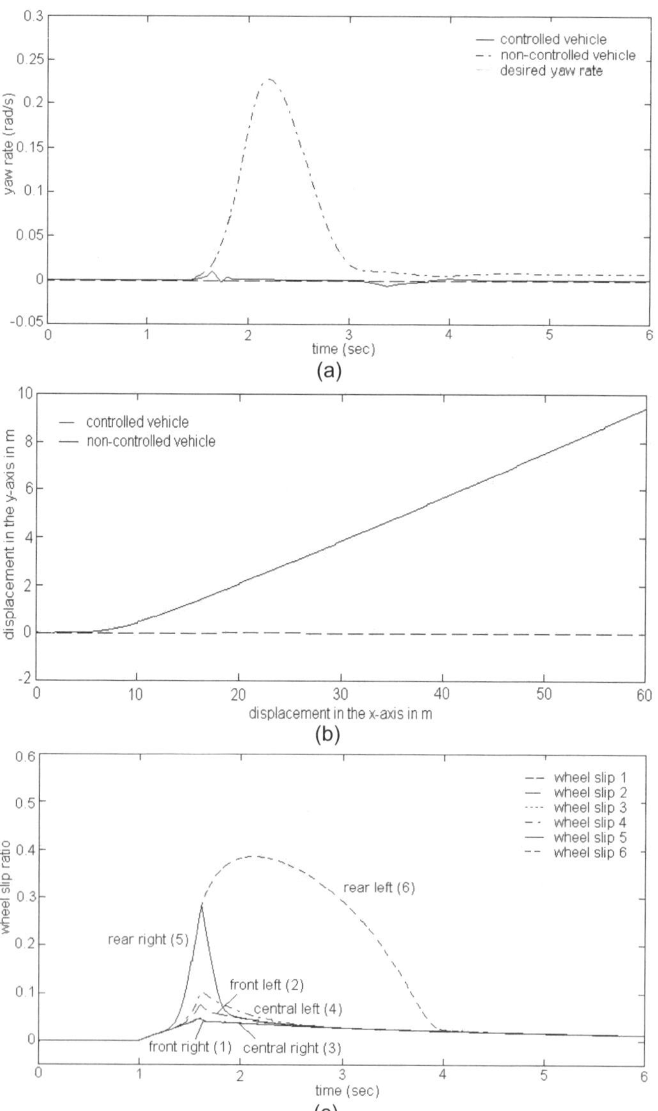

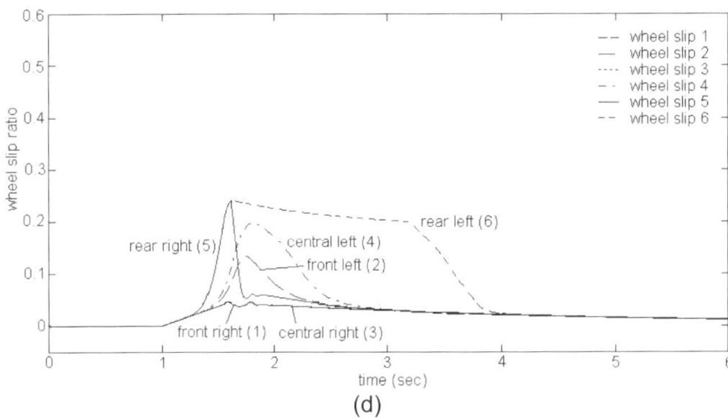

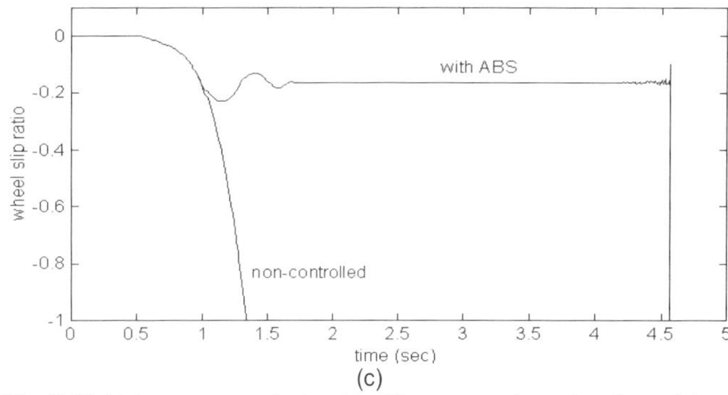

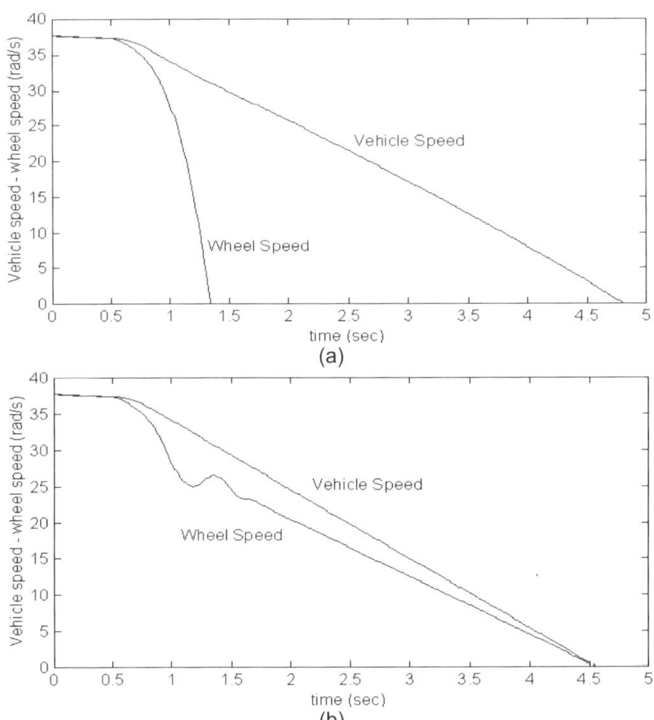

Fig. 4. (a) Yaw rate response to acceleration on a split μ surface with proposed control, (b) the vehicle path for the controlled and non-controlled vehicle, (c) the individual wheel slips on the controlled vehicle and (d) non-controlled vehicle.

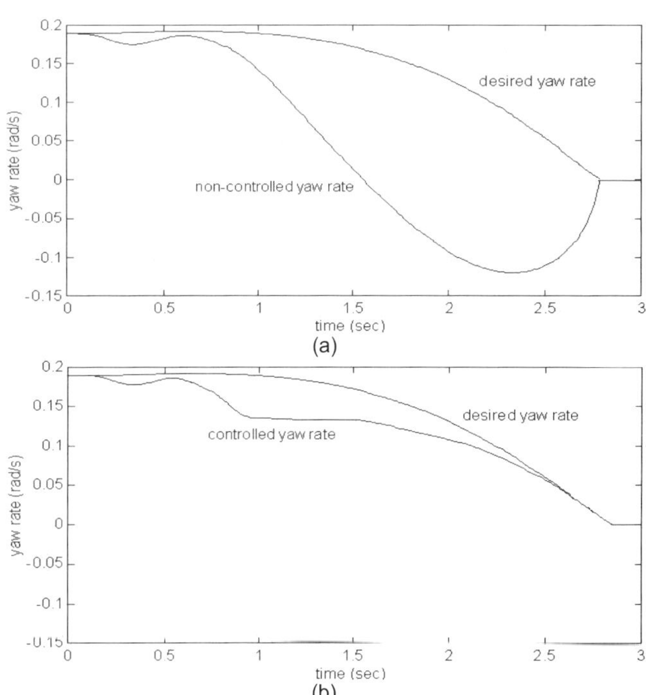

Fig. 5. Vehicle responses to hard braking on a wet road surface. (a) vehicle and wheel speed for non-controlled vehicle, (b) vehicle and wheel speed for controlled vehicle and (c) front right wheel slip for controlled and non-controlled vehicle.

HARD BRAKING ON A WET ROAD SURFACE - The vehicle is decelerated from 80km/h to 0km/h whilst the tires are in contact with a wet road surface. The resultant vehicle and wheel speeds are shown for controlled and non-controlled vehicle in are shown in Fig. 5a and 5b respectively and the results for wheel slip of the front right tire for the controlled and non-controlled vehicle are shown in figs. 5c. From fig. 5b and 5c it can be seen that the implemented controller stops the wheels from locking, maintaining the wheel slip at the desired value. This results in a reduction of the stopping time and hence stopping distance as a higher longitudinal tire force is available. However, stopping the wheels from locking has an even greater effect on the vehicles lateral stability as locked wheels cannot transmit nearly as much lateral tire force as a pure rolling tire. This can be seen in the next simulation.

HARD BRAKING DURING CORNERING - The vehicle is undertaking a steady state cornering maneuver on a dry road at 80km/h when at 0.1 seconds the vehicle brakes hard. The resultant yaw moments are shown in fig. 6a and 6b for non-controlled and controlled vehicle respectively, the vehicles side-slip velocity is shown in fig. 6c. When wheels lock, the lateral force generated by the tires tends to zero, this leads to instability and hence, is highly undesirable. As can be seen in fig. 6a, the vehicle actually starts yawing in the opposite direction to the desired yaw rate and from fig. 6c it an be seen that the vehicle has a significant lateral velocity, meaning that the vehicle is sliding out of the corner. When the controller is introduced this side-slip velocity is greatly reduced (the actual desired value is zero) and the vehicle yaw rate (fig. 6b) remains much closer to the desired value. It can be seen that by keeping the wheels from locking, lateral stability is improved and hence vehicle motion is improved.

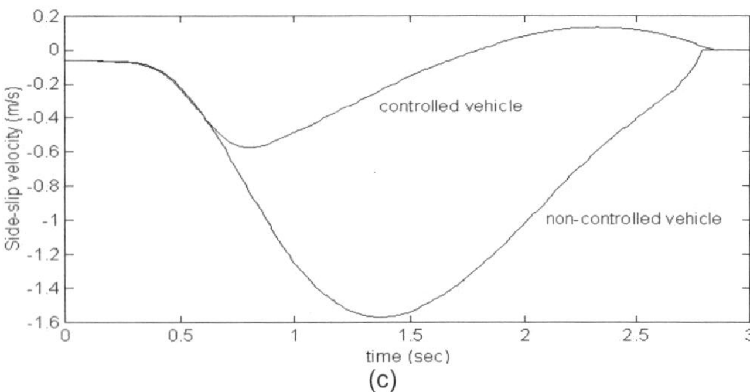

Fig. 6. Vehicle responses to hard braking on a dry road surface with a constant steer input. (a) yaw rate for non-controlled vehicle, (b) yaw rate for controlled vehicle and (c) side-slip velocity for both controlled and non-controlled vehicle.

J-TURN MANEUVER - The vehicle is simulated undertaking a J-turn, the steer input for this maneuver is a ramp input starting at 2s and reaching a steady value of 0.075rad at 3.5s. Fig. 7a and 7b show the yaw rate response when the J-turn is undertaken during quick acceleration starting at 1 second from a speed of 5km/h. It can be seen from these results that the controller offers improved yaw-rate tracking during acceleration or at steady speed. This is advantageous for both vehicle handling and stability.

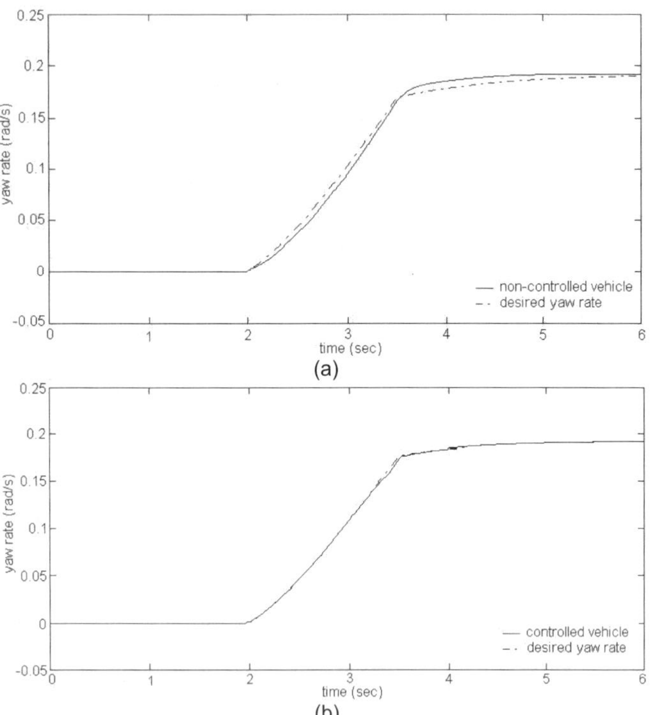

Fig. 7. Vehicle responses to J-turn steer input. (a) yaw rate response under acceleration for non-controlled vehicle and (b) controlled vehicle.

LANE CHANGE MANEUVER - The vehicle runs straight when at 2 seconds a sine wave input for the steer angle is applied, the sine wave has an amplitude of 0.1rad and a frequency of π. Fig. 8a and 8b show the yaw-rate response when the lane-change is undertaken during quick acceleration, starting at 1 second from a speed of 5km/h. Again it can be seen that the controller helps maintain the desired yaw rate close to the desired value under these conditions.

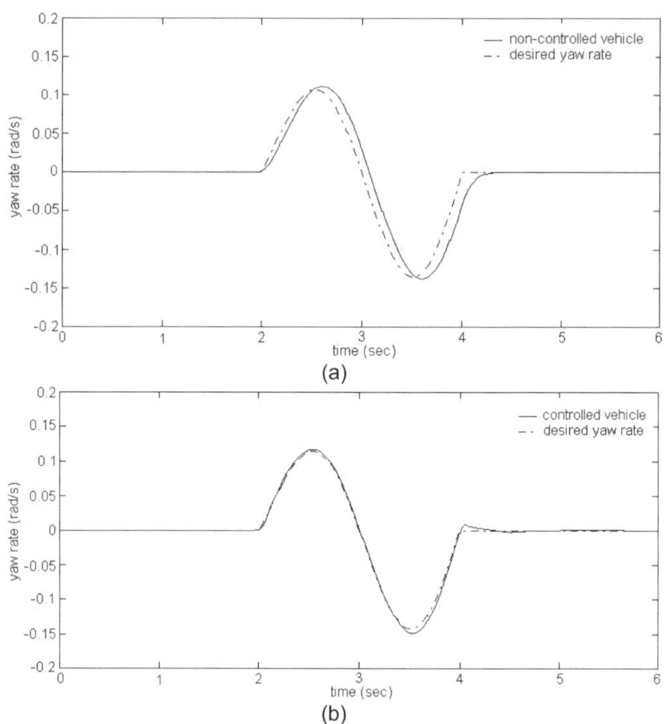

Fig. 8. Vehicle responses to lane change steer input. (a) yaw rate response under acceleration for non-controlled vehicle and (b) controlled vehicle.

Fig. 9. shows the vehicle yaw rate against the steer input for the steady velocity lane-change maneuver on the dry road. From this it can be seen that the controller offers less hysteresis than the vehicle without such control, which means shorter time delays between input steer angle and output yaw motion. This shorter response time will offer improved collision avoidance capability as driver demands are executed faster. This, along with the improved yaw rate tracking and reduced vehicle side-slip, helps improve vehicle handling, safety and stability.

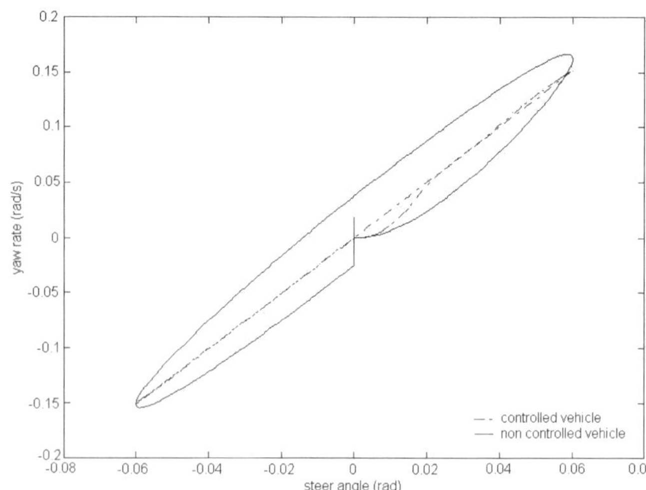

Fig. 9. Yaw rate against steer angle for the controlled and non-controlled vehicle.

CONCLUSION

This paper presents an intelligent controller for combined yaw-moment, traction and anti-lock braking control. Through coordinated use of these controllers, individual wheel control is implemented. Through computer simulation of a basic non-linear handling model, a controller has been developed which shows marked improvements in handling and stability over a conventionally controlled vehicle. The following conclusions have resulted:

- The controller accurately tracks the desired yaw rate on various road conditions and during heavy acceleration.

- Improved safety should be offered by the control system through rapid and accurate responses to steer inputs.

- The controller reduces the need for corrective steering, leading to more predictable vehicle handling. This is especially evident during acceleration on a split μ surface, where no driver intervention is required to maintain desired path.

- The use of anti-lock braking increases vehicle stability during heavy braking which improves stopping distances and more importantly the vehicles lateral stability.

- The potential of individual wheel control to improve vehicle handling, stability and tractive characteristics has been shown.

Future work will include extension of the vehicle model to include load transfer effects along with the suspension system model will have significant effects on the handling behavior of the vehicle and so is necessary to help further validate the potential of such a system. Also further fine tuning of the control system will be required to optimize the vehicle responses.

REFERENCES

1. Sakia, S., Sado, H. and Hori, Y., 1999, "Motion Control in an Electric Vehicle with Four Independently Driven In-Wheel Motors" IEE/ASME Transactions on mechatronics, Vol. 4, No. 1.
2. Shino, M., Wang, Y.Q. and Nagai, M., 2000, "Motion Control of Electric Vehicles Considering Vehicle Stability", Proceedings of AVEC 2000.
3. Yoshimura, T., Uchida, H., Nasu, H., Hino, J. and Ueno, R., 1997, "Traction force control of an electric vehicle in 2WS-4WD mode using fuzzy reasoning", Int. J of Vehicle Design, Vol. 18, No. 5
4. Brennan, S. and Alleyne, A., 1999, "Driver assisted yaw rate control", Proceedings of ACC 1999.
5. Yoshioka, T., Adachi, T., Butsuen, T., Okazaki, H. and Mochizuki, H., 1999, "Application of sliding-mode theory to direct yaw-moment control", JSAE review 20, pp. 523-529.
6. Jung, H., Kwak, B., Park, Y., 2000, "Improved directional stability in traction control system", Proceedings of AVEC 2000.
7. Crolla, D.A., Firth, G. and Horton, D., 1992, "An introduction to vehicle dynamics", University of Leeds, Department of mechanical engineering.
8. Huh, K., Jhang, K., Oh, J., Kim, J. and Hong, J., 1999, "Development of a simulation tool for the cornering performance analysis of 6WD/6WS vehicles", KSME International Journal, Vol. 13, No. 3, pp. 211-220.
9. Dugoff, H., Fancher, P.S. and Segel, L., 1970, "An analysis of tire traction properties and their influence on vehicle dynamic performance", SAE paper no. 700377.
10. Thompson, R.W., 1999, "Hub Mounted Electric Drive project at DERA and MST (UK)", All Electric Combat Vehicle Conference 1999.
11. Jackson, A.E., Brown, M.D., Crolla, D.A., Woodhouse A. and Parsons M., 2001, "Co-ordinated Mobility Control of a 6x6 Off-road Vehicle with Individual Wheel Control" Proceedings of EAEC 2001.

ACKNOWLEDGEMENTS

This work has been funded by the EPSRC and the UK Ministry Of Defence.

NOMENCLATURE

M_b is the sprung mass of the vehicle body in kg
I_z is the yaw moment of inertia in kgm^2
V_y is the vehicle lateral velocity in m/s
V_x is the vehicles longitudinal velocity in m/s
F_{wxi} are the longitudinal tyre forces in N
F_{wyi} are the lateral tyre forces in N
r is the vehicle yaw rate in rad/s
r_{ss} is the steady state yaw rate in rad/s.
F_d is the longitudinal aerodynamic drag in N
A_d is the frontal vehicle area in m^2
C_d is the aerodynamic drag coefficient
ρ is the density of air in kg
T_i is the individual wheel torque in Nm
ω_i are the rotational wheel speeds in rad/s
r_w is the effective wheel radius in m
I_ω is the wheels moment of inertia in kgm^2
F_{zi} is the vertical normal tyre force in N
RR is the rolling resistance factor
$C_{\alpha f,c,r}$ is 2x the tyres cornering stiffness in N/rad
s_i is the wheel slip ratio

Rotating Wheel Torque Meter with Telemetric Data Transmission

Thomas Berther and Hans Burkard
KISTLER Instrumente AG

ABSTRACT

A new system has been developed for measuring the driving- and braking-torques acting on passenger car wheels. The RWT (Rotating Wheel Torquemeter) is intended above all for vehicle dynamics applications, particularly the development of control systems for dynamic stability, traction control and ABS, as well as investigations of brake fade, brake judder, measurement of transmitted power and determination of friction values, coast down characteristics, and safety tests.

The modular system consists of three main components: The torque sensor, the transceiver head for telemetric signal transmission and the control unit located in the vehicle. The torque-sensing device can be fitted to the vehicle with the aid of hub and rim adapters. This allows rapid installation to different vehicles. The torque measurement is achieved via piezoelectric quartz force sensors. In addition to the torque signal, there is provision on the rotating element of the My sensor for installing four K-type thermocouple elements, e.g. for monitoring temperatures on the brake disk.

The measured data are sent from the transceiver head by telemetric signal transmission to the control unit in the vehicle. The transceiver head includes all components for the data acquisition, telemetry and telecommand. The power is supplied by batteries.

The control unit consists of a 32-bit processor and digital-, serial-, CAN- and analogue interfaces. Radio frequency links, based on a DECT transceiver module, as used in cordless phones, are set up between the control unit and the transceiver heads attached to the system. One control unit can serve up to four transceiver heads.

User access for commands and measured data is possible through CAN and RS232 interface. Analogue data output is also available.

The layout of the measuring system, technical data and examples of typical measurements are presented.

INTRODUCTION

The car industry's response to increasing demands for comfort and safety of cars are a series of new and innovative technical solutions, such as intelligent vehicle dynamics and traction control systems.

In order to develop and improve these new systems, the engineers require genuine road data. These will be gathered from rolling-road dynamometer tests or during actual road tests. This calls for having appropriate measuring tools to obtain original road data, either gathered during tests on a rolling road test stand or collected during test driving on public roads or proving grounds. Many engineers have been using the relatively complex 6 component wheel force measurement systems. Kistler has a long tradition in wheel force dynamometers and has developed two types of multi-component wheel dynamometers [1, 2, 3]:

Rotating Wheel Dynamometers (RWD) (Fig.1) are used mainly by vehicle manufacturers and tire makers, especially for collecting genuine road data. The newest generations embody the latest advances in multi-component force measuring techniques. Its mass has been greatly reduced. This brings the mass of a fully equipped test wheel close to that of the OE fitment road wheel [1].

Figure 1: The RWD, a 6 component measuring wheel for vehicles

Fixed Wheel Dynamometers (FWD), introduced more than 30 years ago and continuously improved, are ideal for measuring wheel loads on rolling roads. Worldwide, more than 80 units are being used by major tire makers.

However, in many applications driving and braking torque is requested only. The new **Rotating Wheel Torquemeter (RWT)** (Fig.2), is the ideal tool for the evaluation of driving- and braking-torque. Being based on quartz force sensors, the rugged RWT combines a wide measuring range with a high frequency response characteristic, has inherent low crosstalk and is very linear in its signal response. The RWT is the logical extension of wheel force measurement systems. into the domain of one-component instruments, suitable for measuring driving- and braking-torque.

Figure 2: The RWT on a VW Beetle

ROTATING WHEEL TORQUEMETER APPLICATIONS

The wheel torque meter is intended above all for vehicle development work, with special emphasis in the areas of
- Control systems for vehicle dynamics,
- ABS anti-lock braking and other braking effort distribution systems,
- Investigations of brake fade,
- Brake judder,
- Power transmission measurements,
- Determination of friction values,
- Coast down tests,
- Government safety tests (e.g. US Standard FMVSS 135).

All these different applications call for a universal sensor system, available at reasonable cost, easy to use, easy to adapt for specific vehicles, and which allows the measurement of very high as well as very low torque. Bearing all this in mind and in close co-operation with end users, KISTLER designed the RWT, taking full advantage of the benefits of having quartz element measuring devices.

DESIGN AND WORKING PRINCIPLE

The modular system consists the components:

1. The torque sensor
2. The transceiver head for telemetric signal transmission
3. The control unit located in the vehicle

Hub and rim adapters allow adaptations to different vehicle and rim combinations (Fig. 3).

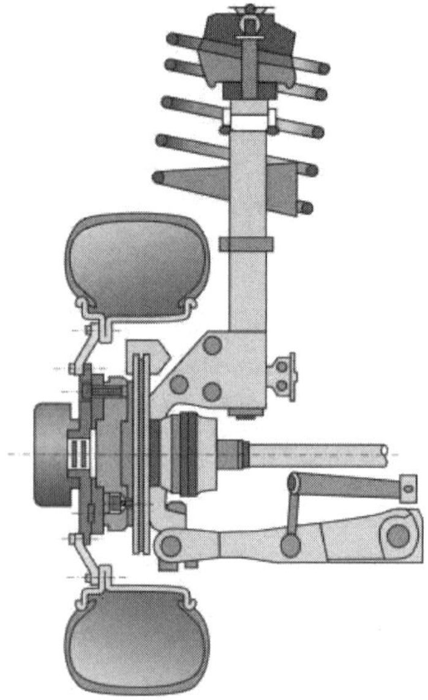

Figure 3: Cross section of a wheel equipped with a RWT

THE TORQUE SENSOR - The torque sensor is based on the proven design of the rotating wheel dynamometer RWD [1]. 8 quartz shear force sensors are mounted between two flanges. It is optimised for low mass, high rigidity and low effect of temperature variations. The modular design allows the original rim size and position (positive or negative offset) to be maintained by simply using car-specific rim and hub adapters and rim components from the BBS standard range (Fig. 4).

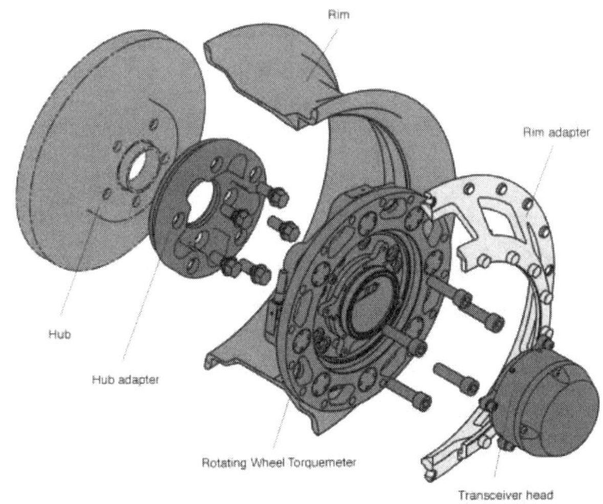

Figure 4: Modular design of the RWT with sensor, transceiver electronics and adaptation

In addition to the torque signal, there is provision on the rotating element of the My sensor for installing four K-type thermocouple elements. These can be used to suit individual applications, for instance for monitoring temperatures on the brake disk. The amplifiers for the charges and the thermocouple signals are integrated into the torque sensor. The thermocouple signal conditioning and correction includes a compensation circuit for the cold junction.

THE TRANSCEIVER HEAD - The measured data are sent from the transceiver head by telemetric signal transmission to the control unit. The transceiver head includes all components for the data acquisition, telemetry and telecommand, i.e. processor, signal conditioning circuits, a 16-bit-A/D converter with an 8-channel multiplexer and sender/receiver module with antenna. The power is supplied by batteries. The sampling rate is 1000 samples/sec for the torque, 8 samples/sec for the temperatures and 1 sample/sec for the battery status. To save battery power, the transceiver head is set to sleep mode - either by a command from the control unit, or, after a selected lapsed time interval, without any commands received from the control unit. Any movement of the head wakes it up again.

THE CONTROL UNIT - The control unit in the vehicle (Fig. 5) consists of a 32-bit processor and digital-, serial-, CAN- and analogue interfaces.

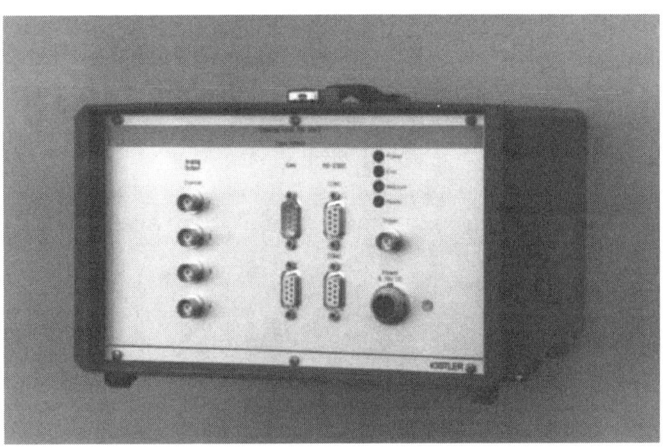

Figure 5: Control Unit with signal output connections

Radio links are set up automatically between the control unit and the transceiver heads. One control unit can serve up to four transceiver heads. The radio link is based on a DECT transceiver module as used in cordless phones. Several such systems can operate simultaneously without interfering with one another (Fig. 6).

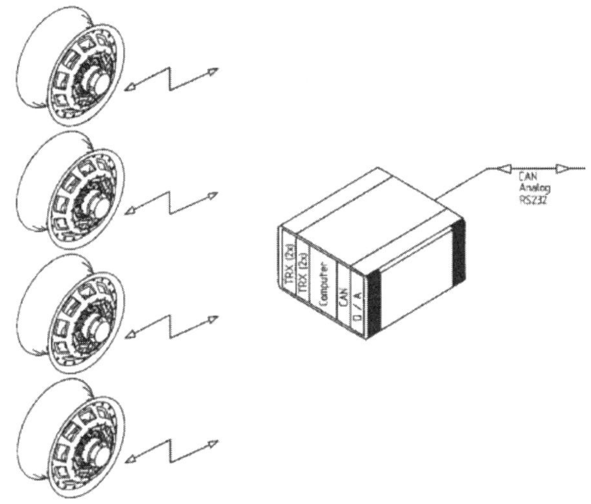

Figure 6: One control unit can handle up to four wheels simultaneously

After switching on, the transceiver heads automatically log on to the control unit. All sensor-specific data are downloaded onto the control unit. A few seconds later, the system is operational without any further user intervention. The synchronisation delay between the control unit and each transceiver head is less than 1 msec.

The control unit accepts commands - e. g. „start of measurement" through CAN-bus or RS232 interface. A digital input that serves as trigger input for e.g. the start of measurement sequence is also available. Measured data - converted into physical units - are output through the CAN or RS232 interface. An analogue output (1 per transceiver head) is also available.

OPTIONAL SLIP RING SIGNAL TRANSMISSION - For applications where angular information is required it is possible to mount a data transfer module, using slip rings and fixing arm, instead of the transceiver head (Fig. 7). This allows the use of a digital angle encoder (1024 pulses per revolution). Signal outputs from the torque sensor and the thermocouples are then available as analogue signals only.

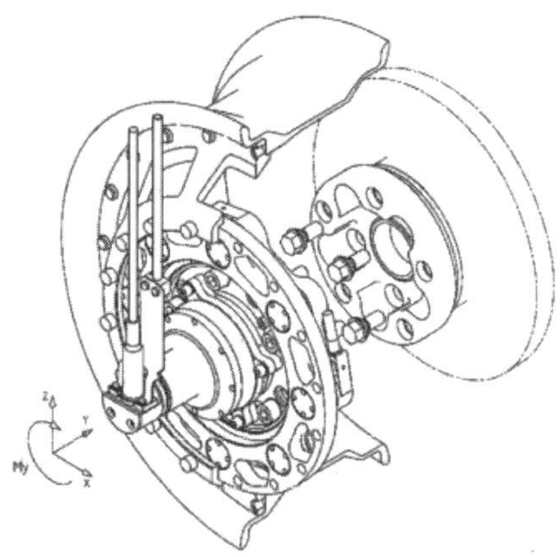

Figure 7: RWD with slip ring transmission module

CALIBRATION

The RWT is calibrated quasistatically on a 3-component calibration stand at Kistler (Fig. 8). This unique calibration stand is extremely rigid, allowing forces to be applied exactly in the desired direction. Consequentially, it is possible to determine with a high degree of accuracy sensitivity, linearity and hysteresis. Additionally also the true crosstalk values can be established very accurate. The calibration stand has a large operating space, making calibration possible not only along the coordinate axes through the centre of the RWT, but also at the geometric location of the centre of the tire footprint. This is achieved by using appropriate rigid lever arms for applying the calibration forces eccentrically.

Figure 8: Kistler's unique high precision three component calibration stand

Extensive dynamic tests on the rolling road test stand of Kistler complete the calibration procedures (Fig. 9).

Figure 9: Comparison measurements on the Kistler rolling test stand

On this test stand the behaviour of the RWT is compared with the fixed wheel dynamometer FWD used as reference sensor in the test stand. (Fig. 10).

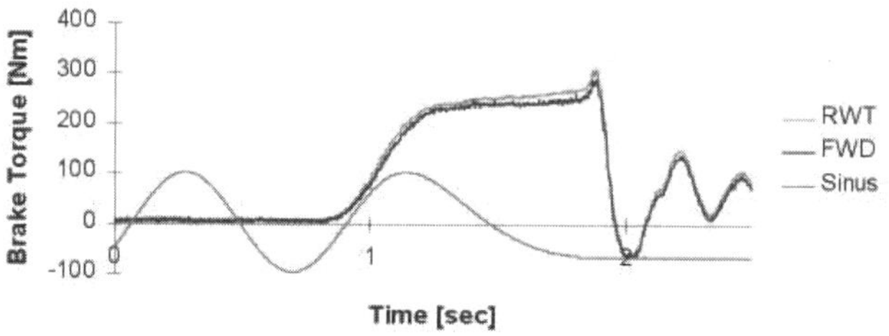

Figure 10: Dynamic comparison of RWT with FWD on rolling test stand

FEATURES AND TECHNICAL DATA

FEATURES – Additionally to the unique feature of the long distance wireless signal transmission system the RWT has several very unique features thanks to its piezoelectric design:

*   Two measuring ranges without compromising in ruggedness:
    The sensor can be used to measure very high brake torque of several kNm as well as very low torque of only a few Nm. This is possible thanks to quartz measuring technology with only one torque sensor and without compromises in reliability or safety of the test drivers.
*   The system allows automatic range switching.
*   Modular design with rim and hub adapters allow for easy adaptation to different cars. The wheel offset can be maintained. The installation on the vehicle is similar to a regular wheel change.

- After less than 15 minutes set-up time the system can deliver reliable test data (Easy to use).
- Low additional unsprung mass of approximately 2.5 kg ensures minimal influence on the dynamic properties of the vehicle under test.
- CAN-bus, RS-232C interface and analogue signal output.

SUMMARY OF THE MAIN TECHNICAL SPECIFICATIONS

Measuring Range

My, range I	±3000 Nm	(±26000 lb-in)
My, range II	±300 Nm	(±2600 lb-in)

Accuracy

Resolution, range I and II	≤ 1.5 Nm	(13 lb-in)
Sensitivity around 360°	≤ ±10 Nm	(±87 lb-in)
Linearity	≤ ±1.0 % range	
Hysteresis	≤ 1.0 % range	
Crosstalk, $F_y \rightarrow M_y$	≈ ±4 Nm/kN	(±0.16 lb-in/lb)
Crosstalk, $\Delta F_z \rightarrow M_y$	≈ ±5 Nm/kN	(±0.2 lb-in/lb)
Temperature offset	≤ 10 Nm/K	(87 lb-in/K)

Unsprung mass
Additional mass of fully equipped wheel
≈ 2.5 kg   (≈ 5 lb)

Wireless signal transmission
Sample rate, My   1000 samples/s
Data formats at control unit
Serial (RS-232C)   user selectable
CAN   user selectable
Analogue   ± 10 V

Slip ring signal transmission
My, analogue output signal   ± 3.5 V
4 x Temperature   0 ... 3.5 V

EXAMPLES OF MEASUREMENTS

Figure 11 shows data collected on a front wheel driven vehicle. During acceleration each gear shift can clearly be seen as a drop in driving torque. Additionally to the torque data the brake pressure has been recorded, too. This allows the correlation between brake pressure and braking torque.

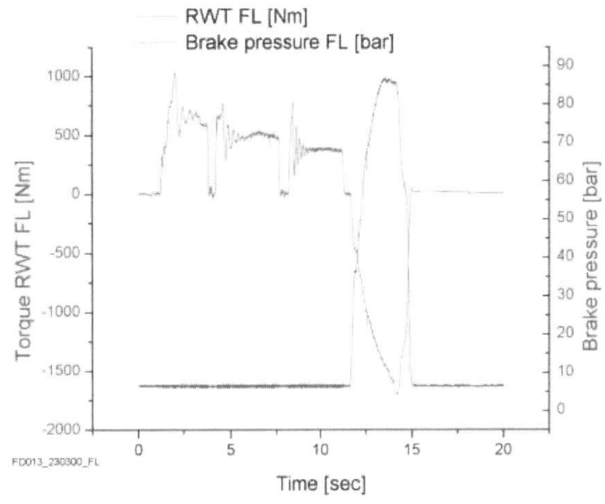

Figure 11:   Torque   and   brake   pressure measurement on a front wheel

Brake-torque measurements were taken at a RWD evaluation on the AT&G testing ground in Papenburg, Germany. Figure 12 shows a complete braking maneuver from 200 km/h to stop. During this maneuver the vehicles ABS systems takes control of the process. Figure 13 shows in more detail the effects of the ABS.

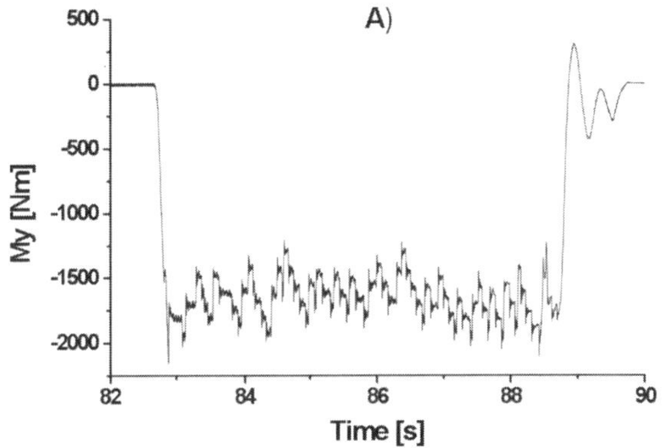

Figure 12: Brake torque measurements on a front wheel of a vehicle

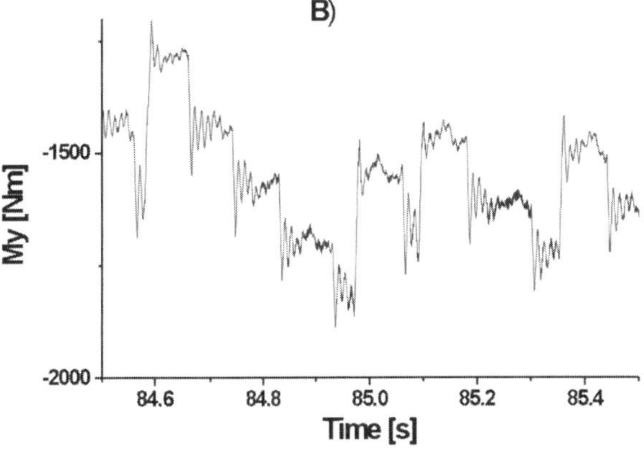

Figure 13: Detailed view of the brake torque measurement

CONCLUSION

The new RWT from Kistler combines over 30 years experience of wheel dynamometers and force measuring techniques with the latest possibilities of telemetric signal transmission. This allows use of the My torquemeter without any additional fixed arm equipment. The RWT is a powerful tool for torque measuring under real-life conditions. The quartz force sensors give the highest possible performance and the capability for accurately measuring highly dynamic events. The RWT is easy to install in place of a standard rim and ready for taking measurements within less than 15 minutes. Adapters for different hubs and rims make the unit very versatile. One single sensor covers all main application fields in development work such as driving- and braking-torque measurement, coast down performance measurement and government acceptance tests.

REFERENCES

1. Rotating Wheel Dynamometer with High Frequency Response; Hans Burkard, Christian Calame; Tire Technology International 1998, p. 154
2. Rotating Wheel Dynamometers with quartz sensors measuring force and moments; Dieter Barz, Reinhard Drew, Wolfgang Hübner; Automobiltechnische Zeitschrift ATZ 1/90
3. Piezoelectric Measuring Systems for Vehicle Development; Thomas Berther, Hans Burkard, Josef Stirnimann; Kistler Publication 20.203e, 6/99

CONTACT

Thomas Berther, KISTLER Instrumente AG, Winterthur, Eulachstrasse 22, CH-8408 Winterthur, Switzerland, Phone: +41-52-224 11 11, Fax: +41-52-224 11 11; E-mail: thomas.berther@kistler.ch

Traction Control Applications in Engine Control

Badih Jawad, Nabil Hachem, Sasa Cizmic, Janette Leese and William Bowerman

Lawrence Technological University

ABSTRACT

Traction control is an electronic means of reducing the wheel spin caused by the application of excessive power for the valuable grip. Wheel spin can result in loss control of the car, reduce acceleration and cause tire wear. In the front wheel tire the loss grip is experienced as underwater, where the front of the car 'pushes' forward, not turning as much as the driver has exposed by turning the tearing. In the rear wheels slip causing oversteer, where the rear of the car swings around, turning much sharper than the driver anticipated. The result of all these problems is that the driver starts loosing control of the vehicle, which is undesirable. With the new design of the Traction Control System, the amount of the wheel slippage is precisely controlled. In racing, this means corner can be taken constantly quicker, with system applying the maximum power possible while the driver remains in total control.

INTRODUCTION

The 2000 Formula SAE team has implemented the traction control system in 2000 formula car as one of the new designs on the vehicle. Implementation of Traction Control system is inexpensive method of eliminating one of the main barriers for maintaining constant speed at all times, having the acceleration without the slippage. In this situation, the driver is in total control of the vehicle. Also, the performance and overall capabilities of the vehicle are much more enhanced.

The Formula 2000 SAE team has decided to implement the Traction Control System on the vehicle by using the following parts:

- MoTec ECU

- Traction Control Multiplexer

- Two Sensors mounted on the front wheel

- Differential Speed Sensor mounted on the engine of the vehicle

Traction Control is one of the most important safety features on the vehicle in the new model cars, but since our interests are with racecars, the emphasis will be concentrated on this subject.

There are three types of traction control; the description of each system is as follow:

1) Limited slip differential: This system transfers engine torque to the wheel that has the best traction control in any given situation. It is not an electronic system, and generally does not perform as newer types of traction control. Modern limited–slip differentials are able to transfer the power to the good wheels before slippage occurs, however if both wheels are on the slippery surface, this system will not function properly.

2) Brake system Traction Control: It works just like ABS in reverse. This system uses the same sensors and hardware that ABS uses to apply the brakes and keep a wheel from spinning. Each wheel is individually controlled, but since our vehicle is not equipped with ABS, this system was automatically ruled out.

3) Drivetrain Traction Control: Our vehicle is equipped with this system which delays power delivery to the slipping wheel at any speed. Using the sensors mounted on the front wheels and the differential speed sensor will give the signal to the processor that might do one of the following actions:

   a)   close the throttle

   b)   cut the fuel supply

   c)   delay the spark timing or

   d)   shut the cylinders.

The most advanced system will perform all these actions, plus it pushes the accelerator pedals against the driver's foot to tell that it is working. Our Traction Control system will be using the retarding spark timing which will cut the power in all slippery situations, a switch will be provided to turn the system off for the situations where slippage is desired.

TRACTION CONTROL DESIGN

The traction control system is shown in the circuit diagram (Figure 1). It contains three sensors on the front two wheels of the vehicle. Differential speed sensor mounted on the engine, traction control mutliplexer and MoTec ECU unit. Sensors will transmit the signal to the Traction Control Multiplexer, whose function will be explained in the next paragraph, and finally the signal will be processed by the ECU. As the consequence of the spark will be retarded and power is cut to the engine.

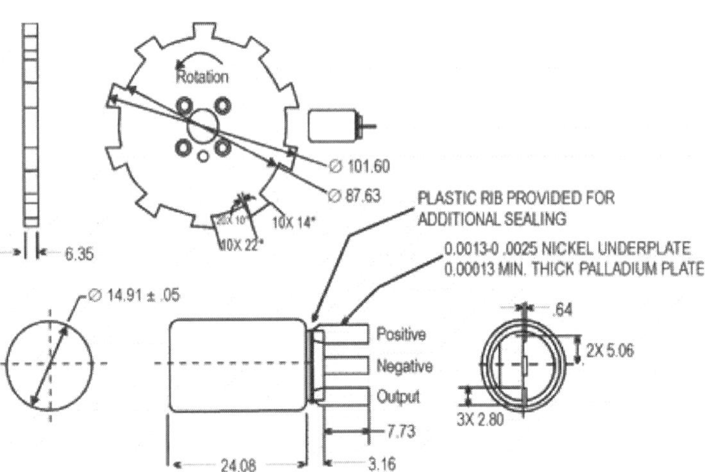

Figure 1.   Traction Control Diagram

Traction Control Multiplexer:

The MoTec Traction control is mounted between the ECU and the wheel speed sensors. The function of the MoTec unit is as follow:

-   It gathers the speed information from the four wheels, and

-   Calculates the ground speed from the fastest rolling wheel and speed difference from the driven wheels.

The MoTec Traction Control Multiplexer is very useful in allowing four separate channels to be brought to the input of the Multiplexer to the ECU (Figure 2).

From the schematic shown in Figure 3, we can see the connections of traction control Multiplexer to the left and right wheel as well as the connection to the back speed sensor providing the speed information about the rear wheels. Since our car has locked differential we will use just one sensor to measure its speed. On the diagram we can notice that we have provided the kill switch connected to pin 15, in case that we want to disable the Traction Control System. Also we can notice the connection with ECU (Figure 3) over the pins 26 and 27.

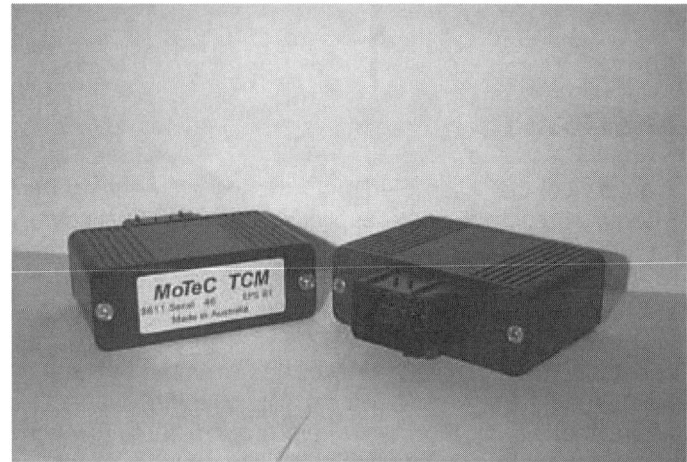

Figure 2.   M4 Pro Control Unit

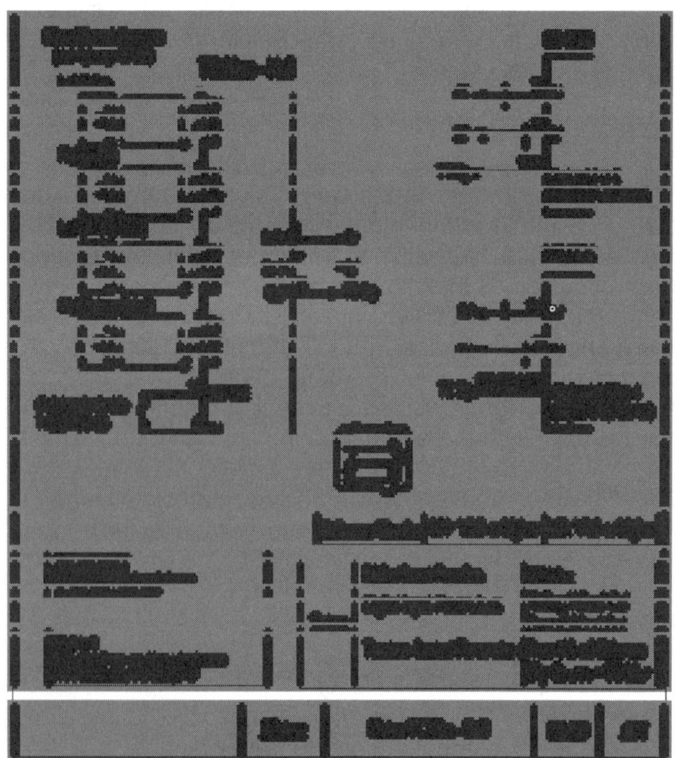

Figure 3.   MoTec Circuit Diagram

M4 Pro Control Unit

The M4 Pro is the top of our M4 range of Engine Management Systems, all of which use innovative, leading edge technology (Figure 4). The M4 Pro is intended for racing applications in all motor sports. The wide range of sensor inputs, control functions and extremely high speed makes it much more than just a fuel injection computer. Standard features include sequential injection for up to four cylinder piston engines and two rotor rotaries, and group fire injection for up to 12 cylinders. The Traction Control Multiplexer will be mounted under the drivers' seat by the M4 ECU unit, the ignition module and the fuel pump.

Figure 4. MoTec Unit

M4 Pro Features:

The M4 Pro can use nearly all original and after market ignition triggers, modules and coils, giving it unmatched flexibility. This avoids the cost and time needed to remanufacture the distributor in order to trigger ECU's that are not as advanced as the MoTeC system. The M4 Pro can be triggered by either a hall effect switch, a logic drive or a magnetic sensor, with various signal types and intelligent signal filtering insuring proper signal to noise ratio at any crank speed. Electromagnetic interference has a big effect on electronic devices. The MoTeC unit takes extensive measures to reject low impedance conducted interference and to shield from radiated interference. (Figure 5).

The heart of the M4 Pro is a Motorola 32 bit microcontroller operating at 33 MHz, and is built to ISO 9001 standards, which puts it on the leading edge of automotive control systems. The M4 Pro reads its sensors 2400 times a second and the entire control program are recalculated 200 times a second.

The MoTeC unique software is the programmable switch mode driver that can trigger any injector with up to 70%

less power usage than other products. This will draw less power from the electrical system and generates less heat inside the ECU. Also, it generates a comprehensive diagnostics report on injector current usage and any wiring open/short circuit after each firing of the injector.

There is a comprehensive group of computer tools available for the M4 Pro including: Engine setup, tuning diagnostics and utilities. Monitoring data logging and analysis. Utilities for loading new program code and enabling special features. Other M4 Pro features include the use of programmable flash memory chip to store the control program. This means that there is no program chip or EPROM to change or upgrade the program. When a new program feature is offered you simply send the M4 Pro the new code and the latest features are available to you. Flash memory has also been included for the internal data logging, hence, stored data will never be lost even if power to the ECU is removed.

Optional features on the M4 Pro include a high accuracy, fully temperature compensated wide-band lambda sensor facility and 512K of internal data logging memory. Both are provided for the first 6 hours running to assist with initial setup and tuning. For specific applications the M4 Pro is available with MIL-C-26482 circular connections.

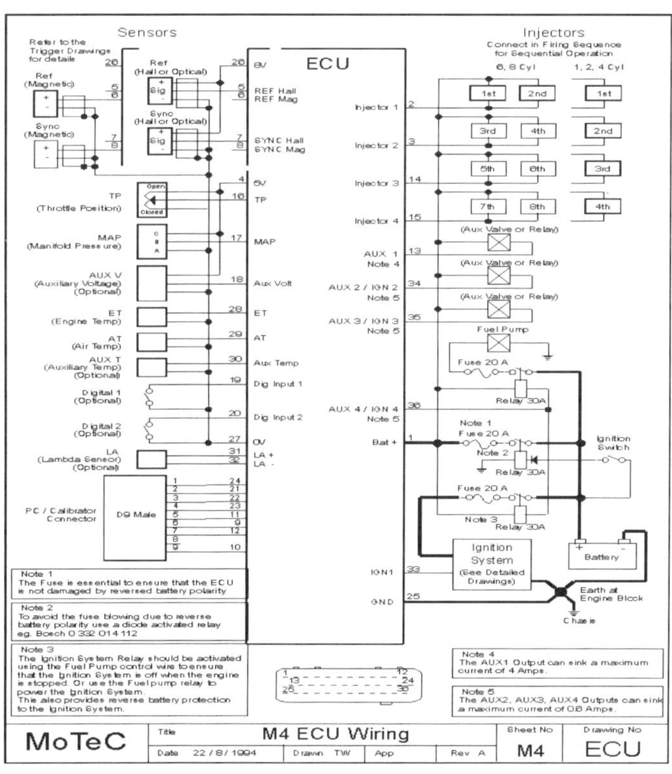

Figure 5.   M4 Pro Circuit Diagram

M4 Pro Characteristics:

The M4 Pro system has the following characteristics:

- Environmentally-protected building block

- Senses ferrous metal targets

- Digital current-sinking output

- Optional inverted output

- Better signal-to-noise ratio than variable reluctance sensors

- Excellent low speed performance - output amplitude is not dependent on target speed

- Fast operating speed (up to 100kHz)

- Reverse voltage and transient protected

- EMI resistant

-Wide operating temperature (-40°C to 150°C)

-Twist insensitive

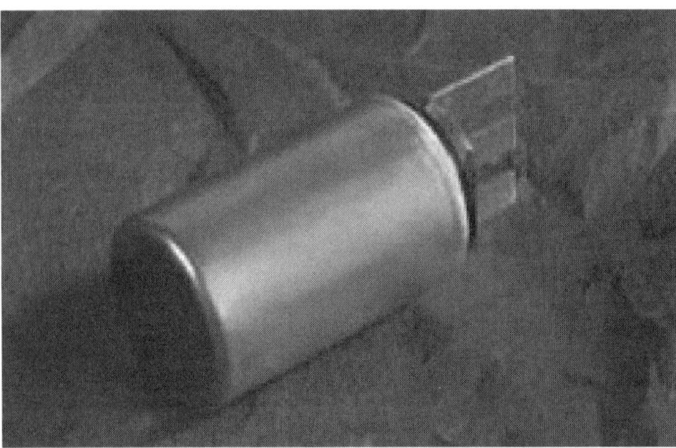

Figure 6.   15 Beta Tooth Gear Sensor

TYPICAL APPLICATIONS

The typical applications of the M4 Pro are: Camshaft and crankshaft speed, position sensing and transmission speed sensing. The Alpha 1.5P building block is a Hall effect gear tooth sensor that accurately senses the movement of ferrous metal targets. It employs a specially designed Integrated Circuit, discrete components and a magnet that are enclosed in an environmentally sealed stainless steel can. This package has been carefully designed for cost-effective 2.0 level packaging. (Figure 6)

The unit will function from a 4.5 to 24 VDC power supply. Its output is digital, current sinking or inverted is an option. Reverse polarity protection is standard. Built-in protection against pulsed transients (ISO 7637/1) is also included.

Optimum sensor performance is dependent upon the following variables that must be considered in combination:

- Target material, geometry and speed

- Sensor to target airspace

- Ambient temperature.

Sensor will be located on the front uprights using the bracket that will be holding the sensor in place. Bracket will be manufactured out of aluminum with a punched whole on the side that would be tightly fixed on the upright. From the backside we have used the Differential Speed Sensor already mounted on the Honda's engine. Since our vehicle is equipped with the locked differential, we are not using the other two sensors. Instead of that we are using the signal transmitted from the Differential Speed Sensor.

Spark Retarding

Cutting the power to the engine will be executed as we mentioned before by cutting the spark to the engine. This will stop any chances of a weak mixture occurring, but it carries its own potential problems due to a large quantity of unburned fuel traveling through the cylinder and out of the exhaust. This fuel can remove some of the oil lining inside the cylinder and pass it through the exhaust. Again this becomes a problem only if the fuel to one particular cylinder is cut for an extended time.

We have concluded that the best way to overcome this problem is to rotate the order in which the cylinders are cut. With retarding the ignition engine torque can retarded by 50 % in a very short period of time. The timing is adjusted by sending a set of ramp from the ignition map value. When wheel speed is detected the ignition timing is adjusted and power is reduced.

The best results would be achieved when the ABS system will be involved. This will not only prevent the spinning wheel but acts to prevent the limited slip differential action. We will get good results especially when the vehicle will be on the road with different and varying braking force coefficients. But, since our vehicle will not be equipped with ABS system, we had to rely only on reducing the power by using the spark retarding.

Conclusion

Using the Traction Control design that we introduced, we have shown that it maintains the same speed at all times, especially during the race, which was our goal. Using the Traction Control cornering the vehicle would be much more improved and above everything control would be accurate. This will improve performance of the vehicle. As shown on our 2000 vehicle the Traction Control design is compatible with other systems on the vehicle and as mentioned before it would be best used in parallel with ABS system.

Although, testing and final design verification will be implemented by the SAE 2001 Formula Team, here, we might add that Traction Control Systems are becoming increasingly present in today's automobiles. Therefore we should strive to improve the traction and stability of the vehicles by using Traction Control or ABS Systems on the vehicle.

ACKNOWLEDGMENTS

The authors would like to express their appreciation to the Lawrence Tech. 2000 Formula SAE team for their efforts in making this project a success. Also, the authors acknowledge Lawrence Tech. Administration and staff for their support and encouragement.

CONTACT

Dr. Badih Jawad of the mechanical engineering department at LTU is the faculty advisor in charge of the Formula SAE project. He may be reached at e-mail: **JAWAD@LTU.EDU**.

Dr. Nabil Hachem of the electrical engineering department at LTU is the faculty co-advisor of the Formula SAE project. He may be reached at e-mail: **HACHEM@LTU.EDU**.

Sasa Cizmic, Janette Leese, and William Bowerman are the electrical engineering students in the FSAE team, in charge of all electrical systems in the car.

References:

1. Kim Lyon, Matthias Phillip, and Erwin Grommes 'Traction Control for a Formula 1 racecar: Conceptual design, Algorithm development and Calibration Methodology (SAE 942475)

2. Bernard Booning, Reiner Folke, and Knut Francke 'Traction Control (ASR) Using Fuel-Injection Suppression–A Cost Effective Method of Engine-Torque Control' (SAE 920554)

3. Tom Denton, Colchester Institute, Colchester, Essex 'Automobile Electrical & Electronic Systems'

4. Ronald Jurgen 'Electronic, Braking, Traction, and Stability Controls'

STABILITY CONTROL SYSTEMS (VDC)

The Development of Active Geometry Control Suspension (AGCS) System

Sangho Lee, Hyun Sung and Unkoo Lee
Hyundai Motor Company

ABSTRACT

This paper presents the development of the active geometry control suspension(AGCS) system as the world-first, unique and patented chassis technology, which has more advantages than the conventional active chassis control systems in terms of basic concept.

The control approach of the conventional systems such as active suspensions(slow active, full active) and four wheel steering(4WS) system is directly to control the same direction with acting load to stabilize vehicle behavior resulting from external inputs, but AGCS controls the cause of vehicle behaviors occurring from vehicle and thus makes the system stable because it works as mechanical system after control action.

The effect of AGCS is the remarkable enhancement of avoidance performance in abrupt lane change driving by controlling the rear bump toe geometry.

INTRODUCTION

In recent years, the vehicle stability has been studied with strong interests together with the development of high-power vehicles. For example, it is well known that 4WS system enhances the high-speed cornering performance by adjusting a rear wheel steer angle. However, 4WS hardware is very complicated and expensive because it requires much power in controlling the system. Table 1 shows the comparison between the conventional active chassis control system and AGCS system.

In this paper, AGCS control concept has been introduced and its performances were analyzed with the ADAMS full vehicle model. The simulation results of AGCS system show the improved handling performance. In addition, the subjective tests were performed to evaluate the AGCS performance for lane change and steady-state cornering.

Table 1. Comparison between the conventional active chassis control system and AGCS

Item	Conventional active chassis system	AGCS
Control concept	phenomenon control	cause control of phenomenon
Control direction	same direction with acting load → excessive energy	perpendicular direction with acting load → small energy
Control feel	may be unnatural	is natural

AGCS SYSTEM CONCEPTS

The kinematics arrangement of the suspension link changes toe angle, which is largely affecting running stability, as the running condition changes.

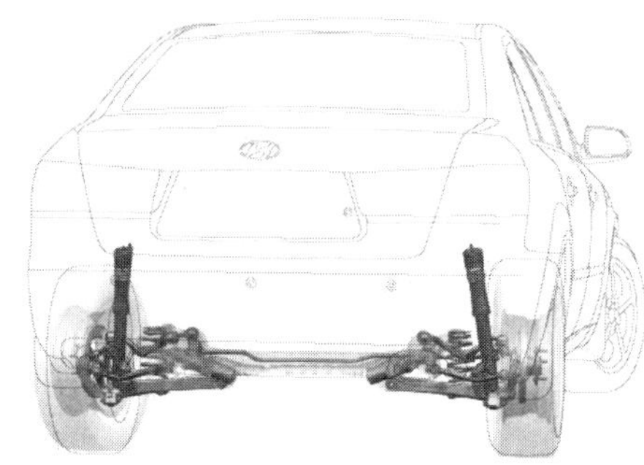

Figure 1. Hyundai AGCS vehicle

The AGCS system of Hyundai Motors is a device to optimize the toe angle of a rear wheel by controlling the position of a rear suspension link. It consists of actuators, control lever(these are mounted in rear subframe) and ECU as shown in Figure 2. ECU is adjusting actuator stroke based on the vehicle speed and steering angle. Then the control lever is rotating downward or upward around hinge. It moves the inboard mounting point of the rear wheel assist link to maintain optimal bump toe-in value.

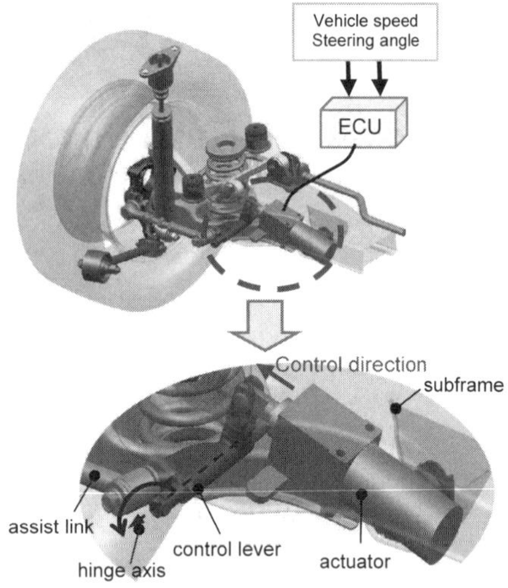

Figure 2. AGCS system layout

Sensor part has vehicle speed sensor and steering angle sensor. Vehicle speed sensor reads the vehicle's speed, and the steering angle sensor reads the driver's steering amount.

Control part commands the actuator by estimating lateral acceleration acting on the vehicle based on the data from the vehicle speed sensor and the steering angle sensor.

Actuator part is producing the optimal toe angle of the rear outside wheel by controlling the position of the link mounting point of rear wheel based on the command from the control part.

The concept of AGCS is intelligent, and it overcomes many negative points of conventional active suspension systems. AGCS has simple control logic and hardware equipment. Moreover, if energy supply goes off in conventional system the performance becomes failure mode but in AGCS, vehicle performance will be equal to passive suspension.

At situations accompanying rolling such as high speed cornering the AGCS system estimates lateral acceleration acting on the vehicle based on the data from vehicle speed sensor and steering angle sensor. Based on the estimated lateral acceleration, it moves the actuator to increase rear outside. (Figure 3)

Since abrupt steering at high speed produces large centrifugal force on the vehicle, the rear part of the vehicle slips outward.(AGCS OFF in Figure 4)

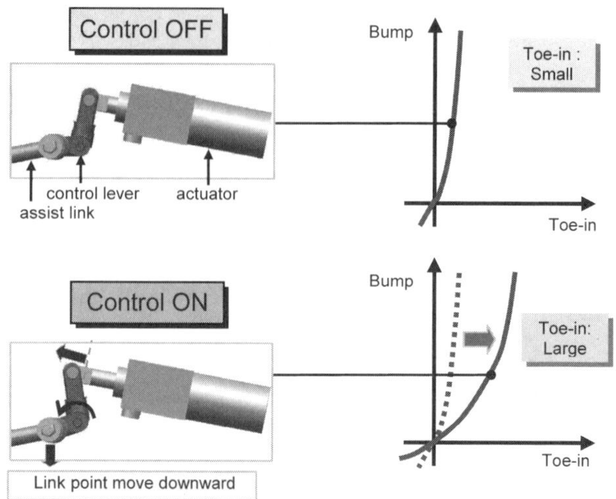

Figure 3. Bump toe characteristics

Thus, it is difficult for the vehicle to show stabilized steering performance. The AGCS system prevents the vehicle from leaning, which increases grip force of the rear wheels, thereby decreasing the rear-sway.(AGCS ON in Figure 4)

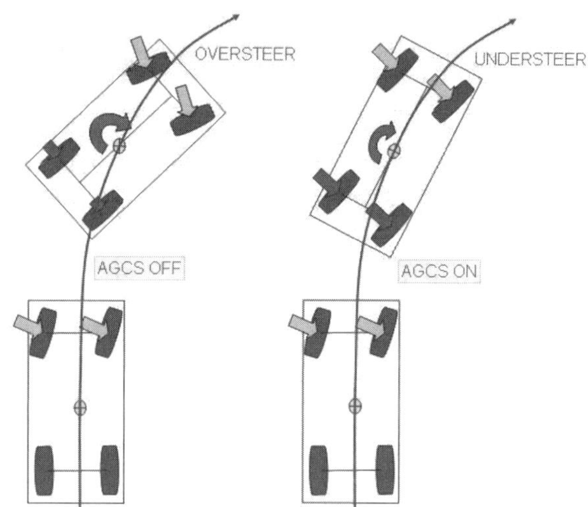

Figure 4. Vehicle trajectory in high-speed cornering

Because the control direction is perpendicular to the acting load such as in Figure 4, the system is inherently more efficient than conventional systems that control in the same direction as the acting load. The resultant consumption energy will be nearly zero, $W = F \bullet S \approx COS(90°) \approx 0$.

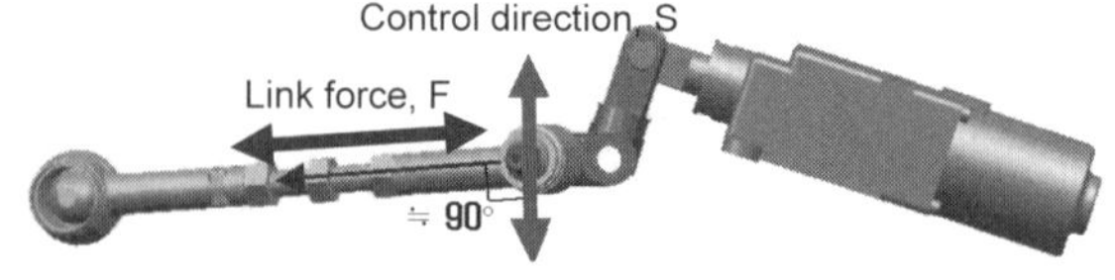

Figure 5. Control direction of AGCS

CONTROL LOGIC

The control logic of the AGCS system is investigated in this section. The objective of the logic philosophy was to maximize the positive effect of the AGCS on the precise situations. The most important aims to accomplish are :

• Actuation starting points at constant levels of lateral acceleration for all speeds in order to guarantee homogenous effect

• Definition of different control maps with increasing AGCS application strokes with increasing lateral acceleration levels for better progressivity and natural effect on the handling response of the vehicle

The definition of all the logic has been done bearing in mind the control parameters available for the set-up:

• Vehicle speed

• Steering wheel angle

• Steering wheel rate (actually this value is calculated)

Only an additional throttle position has been used for the consideration of the power-off in turn situation.

The idea to be applied in the control logic is to establish the starting point of the actuation of the AGCS at a certain level of lateral acceleration of the vehicle. This level will be determined based on the steady-state characteristics of the vehicle for different speeds, therefore with a given steering wheel angle and speed the system will establish the corresponding steady state acceleration and decide the actuation of the AGCS.

Additionally different mappings will be defined based on the steering wheel angle rate. This will inform about the level of non-steady state(or transient) situation of every driving condition of the vehicle. The more transient the situation will be the sooner and the more aggressive the system will act on the suspension. Figure 6 shows the control logic flow of the AGCS system.

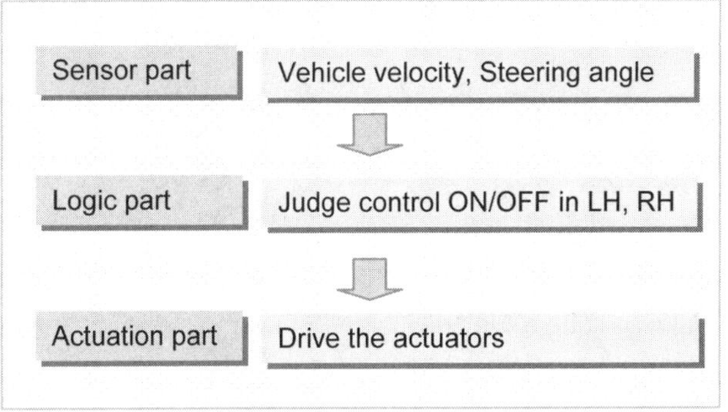

Figure 6. Control logic flow

SIMULATION AND TEST RESULTS

In this section, in order to verify the effectiveness of AGCS the step steer input simulation is performed using full vehicle ADAMS model. The vehicle model comprises double wishbone front suspension and multi-link rear suspension. (Figure 7)

The AGCS ADAMS model consists of assist link, control lever and actuator. Control lever is connected to the assist link with the bushing and also connected to the subframe by the revolute joint. The connecting part between the actuator and the control lever is modeled with a contact element of ADAMS. As the actuator generates translational motion, the control lever will rotate about the revolute joint, and assist link mounting point moves downward.

STEP STEER INPUT SIMULATION

Step steer input maneuver is simulated to investigate the transient response in AGCS OFF and ON condition. Vehicle speed is 140 KPH and steering wheel angle is 50 degrees. Figure 8 shows the results of vehicle response for the step steer input.

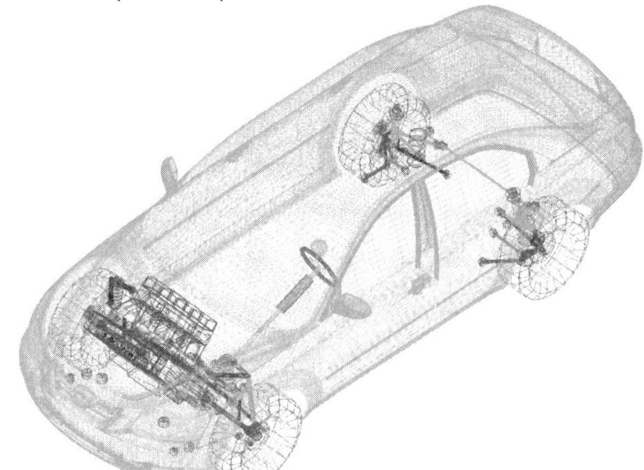

Figure 7. Full vehicle model (ADAMS)

It can be noted that the overall trends in AGCS ON condition have superior performance when compared to those in AGCS OFF condition. Especially, in case of maximum magnitudes of body side slip angle are considerably reduced in AGCS ON condition. (Table 2)

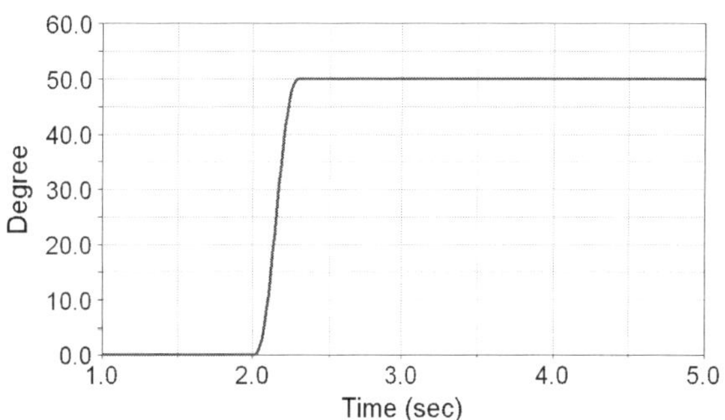

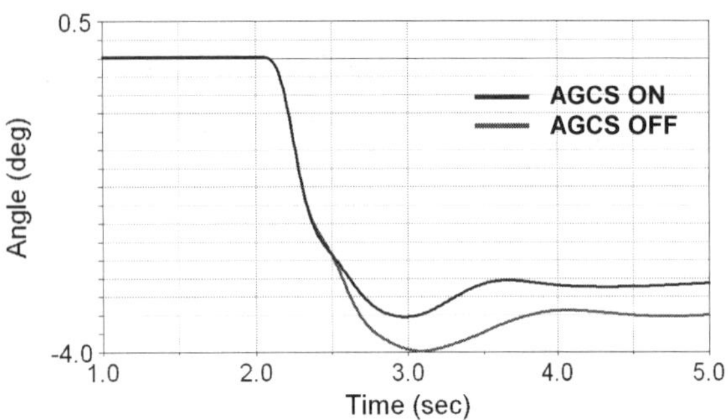

(a) Steering angle

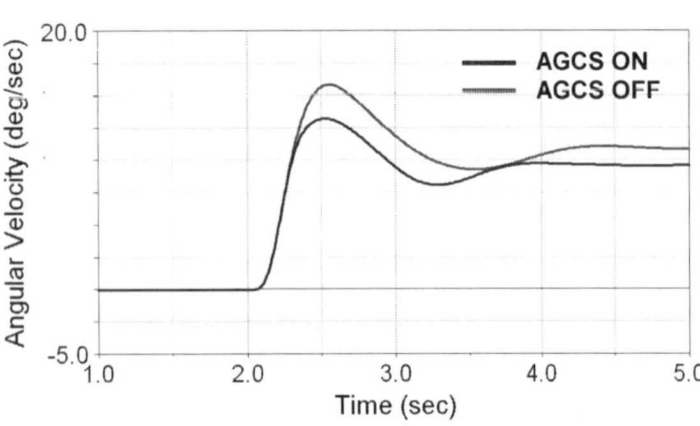

(b) Roll angle

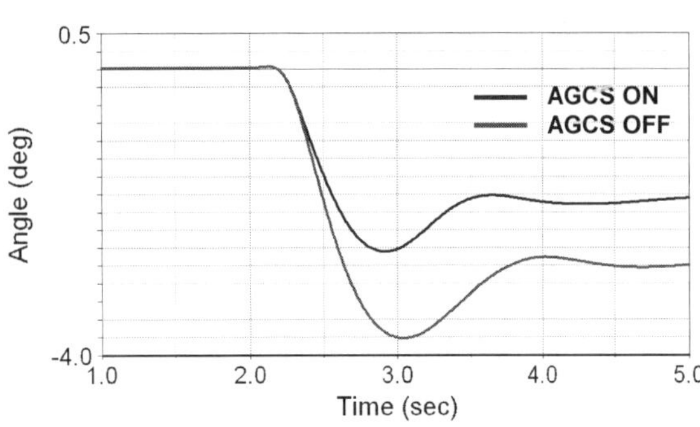

(c) Yaw rate

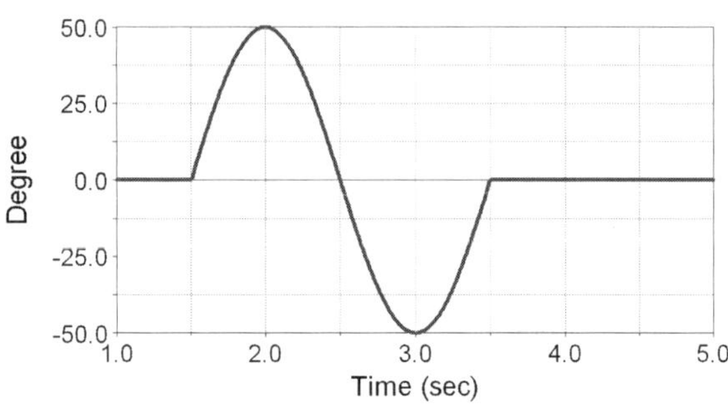

(d) Body side slip angle

Figure 8. Step steer input simulation results

Table 2. Maximum magnitudes comparison in step steer input simulation

Item	AGCS OFF	AGCS ON
Roll angle	-3.99 deg	-3.58 deg (10.3%↓)
Yaw rate	15.79 deg/s	14.52 deg/s (8.0%↓)
Side slip angle	-3.77 deg	-2.7 deg (28.43%↓)

SINUSOIDAL INPUT SIMULATION

Also, the sinusoidal input maneuver is performed to investigate the transient handling performance. Initial vehicle speed is 140 KPH and steering wheel angle is sinusoidal as in Figure 9 (a). Figure 9 (b)~(d) show the results of transient response for the sinusoidal input. Like the step steer input, it can be noticed that AGCS ON vehicle has superior handling performance than AGCS OFF vehicle. In particular, it can be found that magnitude reduction of side slip angle is large in case AGCS ON condition. (refer to Table 3)

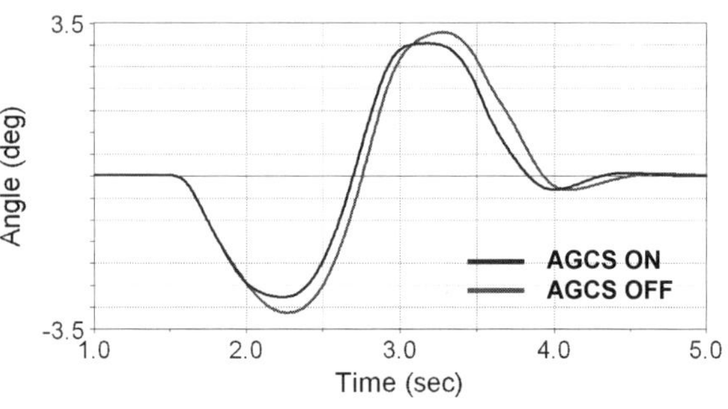

(a) Steering angle

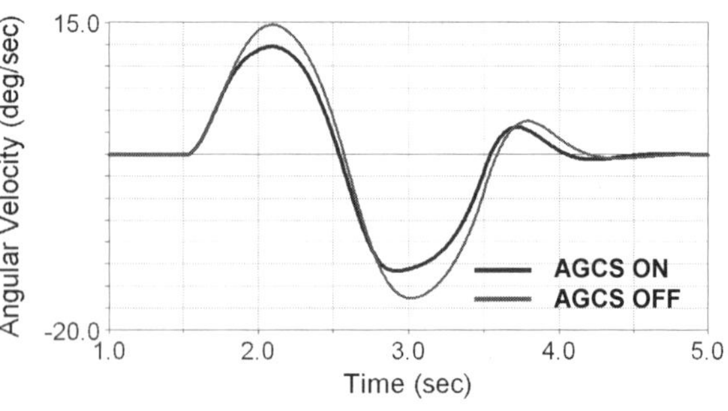

(b) Roll angle

(c) Yaw rate

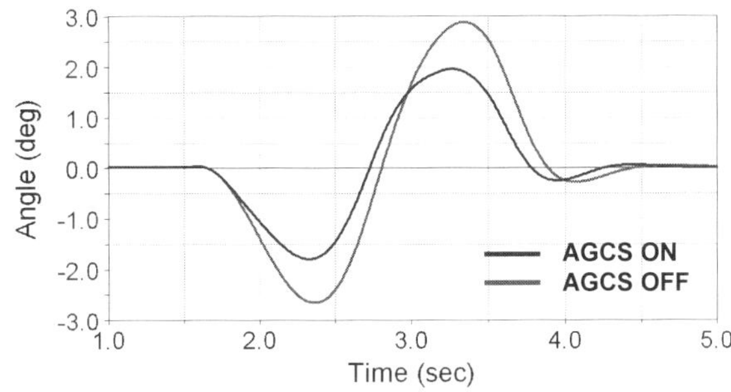

(d) Body side slip angle

Figure 9. Sinusoidal input simulation results

Table 3. Maximum magnitudes comparison in sinusoidal input simulation

Item	AGCS OFF	AGCS ON(reduction %)
Roll angle	3.29 deg	3.03 deg(7.9%↓)
Yaw rate	-16.38 deg/s	-13.29 deg/s(18.0%↓)
Side slip angle	2.88 deg	1.95 deg(32.29%↓)

SUBJECTIVE TESTS

In order to establish the level of influence of the AGCS effect, The rear suspension were installed with AGCS hardware such as actuator, control lever for road testing in Figure 10. (ECU is not seen in the Figure)

The subjective evaluations are performed to cover the principal aspects of the handling performance that are affected by AGCS system. Test driver tries two kinds of driving case, AGCS off and on condition for single lane change and quasi steady-state cornering.

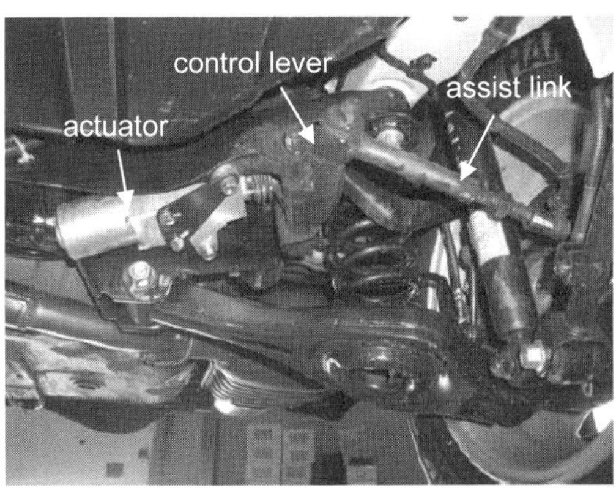

Figure 10. AGCS hardware mounted on the test car (rear suspension)

The vehicle characteristics are also evaluated when turning in steady-state or quasi steady-state situation. All the response properties and balance of the vehicle is marked.

The punctuation of the different points is from 1 – 10, and target as follow;

Rating	~3	4	5	6
Perception	Poor	Objection	Requires improvement	Barely acceptable

7	8	9	10
Fair	Good	Very good	Excellent

The summary table with all evaluations is shown in Table 4.

Table 4. Subjective ratings comparison (Reference 6)

		AGCS OFF	AGCS ON
Lane change	Response delay	7	7.2
	Response damping	6	7.5
	Rear slip generation	6.5	7.7
	Controllability at the limit	6.5	6.9
Quasi Steady-state cornering	Understeering tendency	6.5	7
	Front-rear balance	6.7	7.5
	Predictability of limit	7	7.2

Vehicle response during lane change is rather good at low-mid levels of lateral accelerations, delay, damping and response (roll a bit excessive, lack of damping). However, at higher lateral accelerations the poor yaw damping and excessive vehicle reaction makes the response too quick and difficult to control. Oversteer occurs at the limit and countersteer is necessary.

The vehicle with AGCS improves the transient reaction by increasing the yaw damping as the most important effect. Additionally some other improvements are also noticed like faster reaction of the car, and more progressive and understeer response.

Steady-state response is rather good with a understeering evolution. Predictability of limit is rather good, but increasing understeering would improve it.

The influence of AGCS is only reached when steering wheel inputs are higher than certain degrees. This is not a purely steady-state situation but could be considered as a slight transient and include it on this evaluation. The effect of AGCS is small, but in the right direction. The generation of rear slip is more controlled and the understeering tendency at the limit of adherence more pronounced. The results in a better predictability of grip limit, and improved balance of front and rear slips.

CONCLUSION

In this paper, AGCS to improve handling performance as the active chassis control system is suggested. The summaries of results are as follows:

1. The AGCS system is applied on the rear suspension to regulate bump toe-in using the electric actuator which varies the geometry of assist link, and enables to improve road grip and extend the slipping point.

2. Because the control direction is perpendicular to the acting load in regulating wheel toe-in, the system is inherently more efficient than conventional systems that control in the same direction as the acting load.

3. AGCS system has superior handling performance when compared to the conventional system in step input simulation results and subjective tests.

REFERENCES

1. U. K. Lee, U.S. PN 5,577,771
2. U. K. Lee, U.S. PN 5,700,025
3. U. K. Lee, U.S. PN 6,182,979
4. ADAMS/Car User's Guides, Mechanical Dynamics, Inc., 2002.
5. Namio, I. and Junsuke K. "4WS technology and the prospects for improvement of vehicle dynamics", SAE No. 901167.
6. IDIADA Technical Report, "Dry handling tuning of AGCS systems with Hyundai NF", July, 2004.

Hardware-in-the-Loop Testing of Vehicle Dynamics Controllers – A Technical Survey

Herbert Schuette and Peter Waeltermann
dSPACE GmbH

ABSTRACT

Hardware-in-the-loop (HIL) test benches are indispensable for the development of modern vehicle dynamics controllers (VDCs). They can be regarded as a standard methodology today, because of the extremely safety-critical nature of the multi-sensor and multi-actuator systems used in vehicle dynamics control. The required high quality standards can only be ensured by systematic testing within a virtual HIL environment before going into a real car.

This paper aims to provide a condensed technical overview of state-of-the-art HIL test systems for VDCs, which are currently widely used in passenger cars, in the form of ABS and TCS, as well as ESP, or integrated chassis control, which is just coming onto the market. First, a short introduction to the basic functionality of these types of ECUs is given, and the reasons why HIL testing is necessary and especially useful for VDCs are discussed. Since most of the dominant suppliers of VDC systems like Bosch, TRW, and Continental Teves, have the same principle system architecture, the closed loop HIL system (including I/O interfaces, signal conditioning, real-time models, user Interface, and test automation) can be generalized. However, there is still a possibility depending on the module being tested that one must include module specific resources to the setup.

A more detailed description of the currently used sensor signals for steering wheel position, lateral/longitudinal acceleration, yaw rate, wheel-speeds (active/passive/ intelligent) and the hydraulic pressure sensors (internal/external) is presented. For the synthetic simulation of these signals within a simulator setup, it is also necessary to take into account the diagnostic functions of the ECU, for example to properly simulate the self-adjusting procedures for some of these sensors.

Controlling brake forces at each wheel is the main task of such a system and it is normally performed by hydraulic valves actuated by solenoids. Since these devices are enclosed within the housing of the ECU itself, a special device, called a valve signal detection unit (VSD), is used to measure the valve actuation.

In order to close the loop, a model of the complete vehicle is required to provide consistent and precise sensor signals as a response to the current actuator and (simulated) driver input signals. Some parts of the model will be explained in more detail to show how it is possible to overcome the partly numerically stiff behavior of the governing equations and to emphasize their importance within test scenarios and standard test maneuvers.

In the last section of the paper, an example of the interactivity of systems like ESP, EPB (electrical parking brake), and ACC (automatic cruise control) will show how demanding today's chassis controls are and how HIL technology copes with these challenges.

INTRODUCTION

The history of hardware-in-the-loop (HIL) simulation in automotive engineering goes back to the 1980s. To begin with, HIL was widely used for testing individual components (e. g., new actuators for active suspensions), mainly in universities and research or advanced engineering departments. In the 1990s, an increasing number of electronic control units (ECUs) began to be installed in vehicles, and there was greater emphasis on testing these ECUs. The ECU itself was now the object under test. HIL test systems were increasingly used not only in advanced engineering and control design, but also in production development, where they gradually replaced the open-loop or stimulus simulators that were still prevalent at the time. The main focus of these black box tests was on function verification, and on testing self-diagnostics and hardware, to support production-level verification by automated regression tests. The number of ECUs in vehicles as well as their networks continued to rise throughout the 1990s. The breadboards that were used to test the ECUs proved to be inadequate to test these networks leading to an increase in the quality assurance problems. As a result, the breadboards were increasingly replaced by laboratory vehicles (= networked HIL systems connected to networked ECUs [1], [2]) allowing automated testing of the complete ECU network including ECU variants The result was a considerable improvement in quality with regard to distributed functions and inter-ECU communication.

The introduction of anti-lock braking systems (ABS) in production vehicles in 1979 by DaimlerChrysler made it possible to prevent the wheels from locking during braking maneuvers. This not only ensured that the vehicle remained steerable, it also allowed optimum utilization of the friction coefficient between the tires and the road surface, thereby reducing braking distances. The essential function of the ABS ECU is to use wheel speed to detect when the maximum static friction has been exceeded and if necessary reduce brake pressure at the wheel brake cylinders via magnetic valves to prevent the wheels from locking. The first HIL test systems for ABS ECUs were built at the end of the 1980s [3]. As these first ABS systems were optimized and HIL technology continued to develop, the mid-1990s saw a whole series of publications on the subject [4], [5], [6], [7]. Among other things, these illustrate how the spread of HIL was supported by increased processor performance, and also by special HIL hardware for I/O (e.g., fast wheel speed signal generation).

With a pure ABS system, the driver is largely dependent on steering to prevent under- or oversteering of the vehicle in a μ-split braking maneuver (in which the left and right vehicle sides have different friction coefficients). The obvious next step was to try to actively superimpose a stabilizing yaw torque on the vehicle by applying separate brake pressures to each wheel. In the mid-1990s, the first VDC/ESP (Vehicle Dynamics Control, Electronic Stability Program) systems were installed in production vehicles (DaimlerChrysler, March 1995). One of the early HIL systems for testing ESP ECUs at Audi [8] was based on a multiprocessor system with 5 DSP processors that computed the models for the powertrain, front/rear suspension, and front/rear wheels separately. The brake hydraulics were integrated as a real system with a pedal actuator, among other reasons because the processing power for computing brake hydraulics in real time was not available at that time.

Progress from the original ABS to ESP with integrated TCS (traction control system) and BAS (brake assistant system) presented HIL simulation with new challenges with regard to the sensors, actuators, and models.

ESP generally involves yaw rate and side slip angle controls that complemented one another. The VDC system achieves this by computing the reference values for the yaw rate and side slip angle via a single-track model with a simple tire model, whose inputs consist of the engine torque (from the engine management via CAN) or the vehicle speed, the brake pressure (pressure sensor in the hydraulics), and the steering angle (CAN node). A Kalman filter (estimator) can be used to determine the actual values for the yaw rate and the side-slip angle by means of a simple two-track model if the yaw rate and lateral acceleration are also additionally captured. A PID controller uses these values to compute a stabilizing yaw torque that is implemented by the underlaid wheel slip controller (Fig. 1).

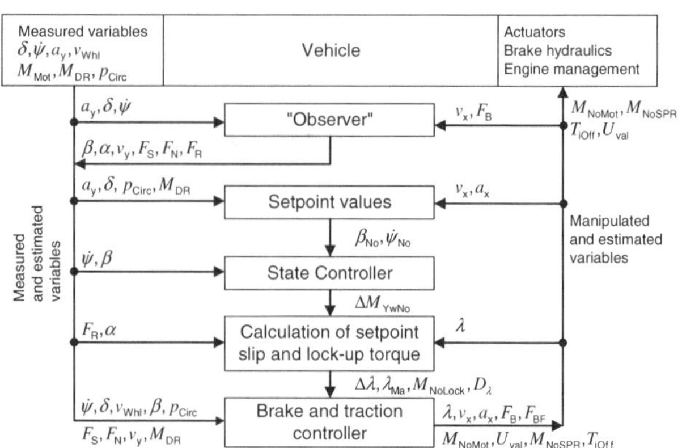

Figure 1: Functional architecture of VDC systems [9], [10].

The specific sensors and actuators commonly used for vehicle dynamics control, and the associated interfaces for HIL simulation, are described in greater detail in the section below.

The following section gives an outline of the necessary simulation models and appropriate modeling tools. Implementation aspects of real-time simulation are discussed, with some numerically critical submodels as examples.

The Applications chapter describes the design concept and hardware structure of current VDC HIL systems for a single ECU and for an ECU network. Additionally, HIL systems for EPAS and ACC systems are introduced in more detail. The final chapter discusses future challenges to HIL simulation by presenting current development trends in vehicle dynamics control (integrated chassis control, ICC).

I/O INTERFACES OF VDC SYSTEMS

The sensor and actuator signals that are typical of vehicle dynamics control system are shown in Fig. 2:

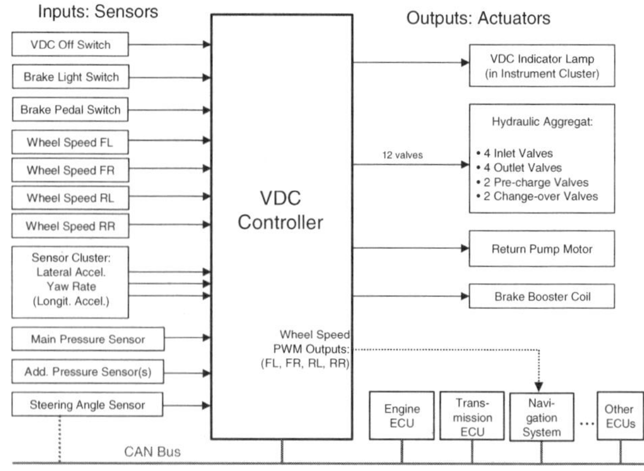

Figure 2: Overview of the sensors and actuators of a VDC system.

SENSORS

Vehicle dynamics control systems are extremely safety-critical components, so monitoring their sensors has high priority, from the point of view of both the electrical interface (current interface instead of voltage interface) and the signal interface (plausibility check using auxiliary variables). The most important sensor types are described below.

Steering Angle Sensor:

The steering angle is generally measured at the steering column and transmitted to the VDC system via the powertrain CAN bus. An alternative is to use encoder-type sensors, having two signals with a 90° phase shift (to detect the direction of rotation) directly connected to the VDC. To make the signal plausible, the steering angle sensor is given the necessary information during initialization. Zero position (= straight ahead) is when all 4 wheel speed sensors measure the same speed (within narrow confines). In addition, the steering angle sensor and the VDC module in each vehicle have to be made known to each other via a diagnostic device. This is done at the end of vehicle production, or later in a garage, e.g., when one of the two components has been replaced.

In HIL simulation, the steering wheel angle is determined in the driver model or in maneuver control. The sensor is simulated by generating either the appropriate CAN messages (CAN sensor) or a suitable encoder signal (e.g., by means of a fast signal processor, slave DSP).

Sensor Cluster:

To capture the lateral acceleration and yaw rate, sensor clusters are used. These are micro-mechanical devices that are usually located in the middle of the vehicle, close to the center of gravity. The variables that are captured, together with the wheel speeds, enable the VDC system to determine the actual movement of the vehicle (e.g., wheel-slip and side-slip angle). The vehicle's actual longitudinal speed is computed from the speed of the driveless wheels. As this is not possible with all-wheel-drive vehicles, the sensor cluster in these includes a sensor for longitudinal acceleration. The sensor cluster is connected to the VDC module either via a private CAN bus or via discrete lines. In either case, extensive initialization sequences are performed before the VDC system starts running. In some systems, these sequences can be repeatedly initiated by the VDC module, even while it is running. For plausibility checks, values such as the ratio of yaw rate and steering angle are used (when the vehicle moves slowly in a circle, a constant ratio emerges) as a reference. If the sensor cluster is a real component in the HIL system, correct positioning is essential, as otherwise gravitational acceleration is interpreted as constant lateral or longitudinal acceleration.

The variables longitudinal and lateral acceleration, as well as the yaw rate, are determined in the vehicle dynamics model, and must be fed to the VDC module in a suitable form. A sensor cluster with discrete signals can be simulated relatively simple via appropriate analog outputs. Far more work is needed for the CAN version. Via a private CAN bus, the VDC module starts an initialization process (initialization protocol, often proprietary for security reasons), to which the sensor cluster must react. A possible solution in HIL systems is to establish communication between the real VDC module and the real sensor cluster via the simulator, using a CAN gateway. The simulator reads in the initialization messages on a CAN channel, and outputs the data bytes directly and unchanged on a second CAN channel. In contrast, in normal operation the CAN messages from the sensor cluster are manipulated by insertion of the simulated variables longitudinal and lateral acceleration and yaw rate before being passed on to the VDC module. Fig. 3 shows the basic principle, and Fig. 4 an excerpt from the model.

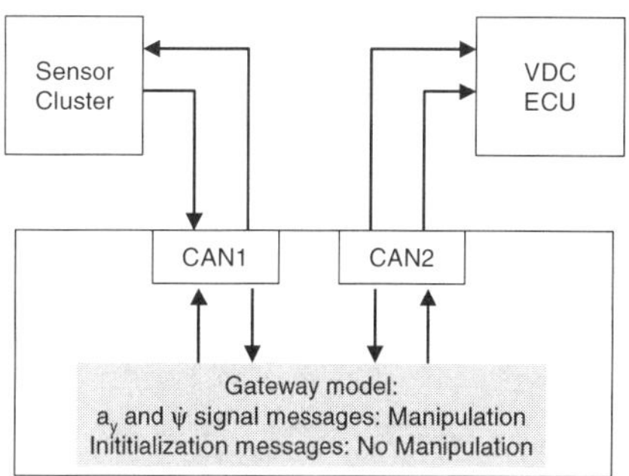

Figure 3: Sensor cluster gateway.

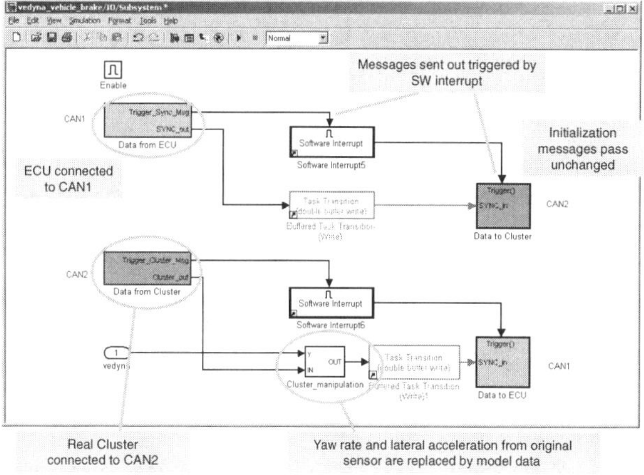

Figure 4: Sensor cluster gateway model for the manipulation of the cluster messages.

Brake Pressure Sensor:

The hydraulics aggregates have internal and/or external brake pressure sensors. However, both the pressure in the main brake cylinder and the brake pressures in the individual brake circuits are important measurement variables for the VDC system, so after the system is switched on, extensive initialization procedures are performed (again initiated by the VDC module), in which the sensor reports values that are intentionally outside the valid range several times in succession (out-of-range check) or that have a specified time behavior. Some VDC systems additionally have combined sensors that alternately transmit the brake pressure and the membrane temperature (multiplex signals). The brake pressures are computed in the brake hydraulics model and generated via analog outputs of the simulator. If the VDC module has internal pressure sensors, the valve flow capture has to be suitably prepared to stimulate them (see Fig. 7, Valve signal detection unit).

<u>Wheel Speed Sensors:</u>

In VDC applications passive, active and active intelligent wheel speed sensors can be distinguished. Passive sensor of variable reluctance type just generate a sinusoidal voltage signal and have been now replaced more and more by active sensors with a current interface (7/14 mA or 7/14/28 mA) which are less sensitive to signal noise and disturbances. Both sensor types generate a signal where the frequency is proportional to the wheel speed. One current development is active, intelligent wheel speed sensors, for example, based on the Infineon TLE4942 Hall IC (Bosch ESP). Sensors from Continental Teves function in a similar way (Fig. 5).

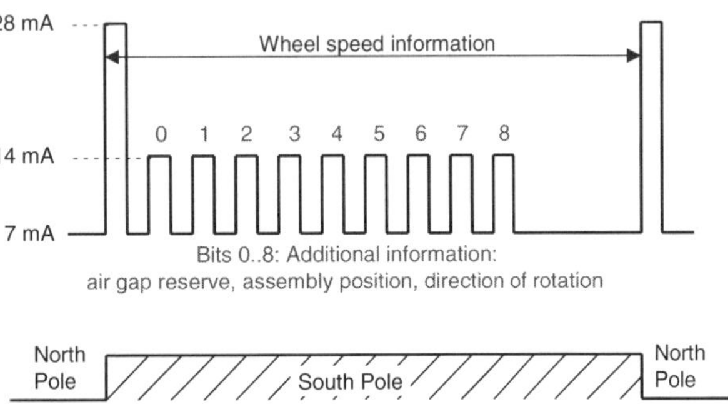

Figure 5: Protocol example of active intelligent wheel speed sensors.

These sensors have their own extensive diagnostics. The wheel speed data (frequency) is encoded in the 28-mA pulses, and the 7-mA and 14-mA pulses are used to transmit additional data by means of Manchester encoding. The additional data includes the air gap reserve, the assembly position, and the direction of rotation.

The wheel speed values for the individual signals for the sensor simulation are calculated in the wheel model. For the protocol for the active wheel speed sensors a slave processor running with a signal sample rate of approximately 4-6 µs is applied. This sample rate is required to achieve a good speed resolution even at high speeds of more than 300 km/h.

Additional inputs to the VDC module include the VDC off button, which can be used to switch off the ESP, brake light and brake pedal switches, and coding pins that can be used for hardware coding of the vehicle configuration (engine, transmission, etc.) in the wiring harness of the VDC module.

ACTUATORS

The actuators of a VDC system consist of the solenoids for controlling the valves of the two brake circuits and the return pump. Fig. 6 shows a hydraulics aggregate with the ECU and the solenoids (left), and the hydraulics block with the return pump. Another actuator is the brake booster coil, which is located in the brake booster.

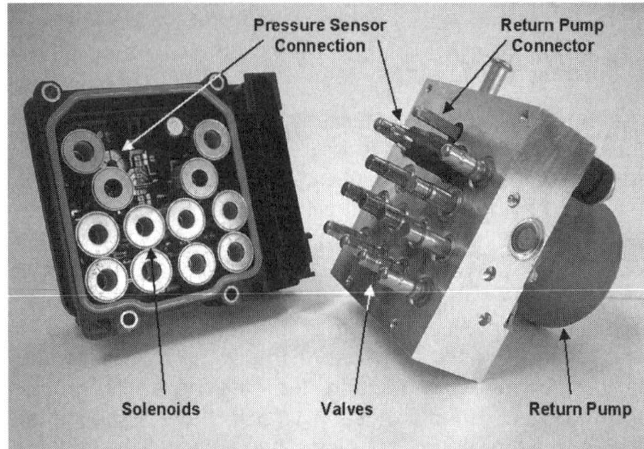

Figure 6: Hydraulic aggregate of an ESP system with solenoids, valves, and return pump.

<u>Valve System</u>

Today's VDC systems are implemented as hybrid technology in an integrated ECU, that is, the actual controller, the power stages and the valve coils are integrated in a plastic enclosure and mounted directly on the brake hydraulics block (Fig. 6). Similar configuration can be found also in transmission ECUs [11]. The problem for HIL simulation is that the electric control signal of the actuators (valves) cannot be captured directly. Instead, the solenoid current or the resultant magnetic field is determined by means of a valve signal detection unit (VSD) based on continuous Hall sensors.

Fig. 7 shows a VSD unit for two different ECUs. The metal pins shown contain the Hall sensors, whose output signals are amplified by signal conditioning inside the VSD enclosure. The VSD unit passes the captured solenoid signals on to the HIL simulator, where they are used as inputs for the brake hydraulics model. Fig. 7 also shows the spring pins used as contacts for the integrated brake pressure sensors, at top right.

Earlier valve systems basically used only switching valves, but today continuous valves are also used. This means that it is generally no longer sufficient to simply detect on and off states. Instead, the solenoid current (or the magnetic field) must be measured precisely by the Hall sensors, which requires good alignment of the magnetic field with the Hall sensor.

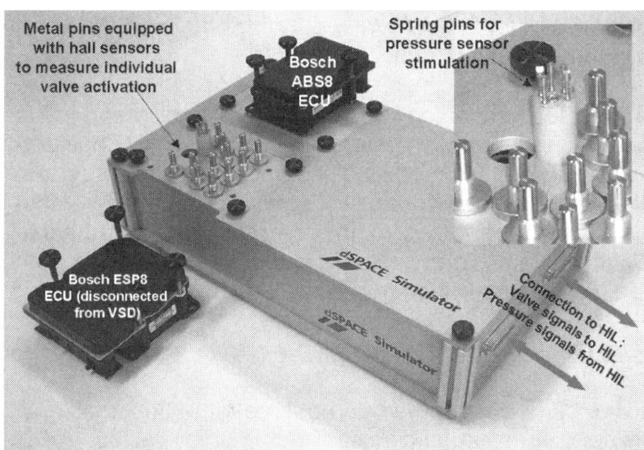

Figure 7: Valve signal detection (VSD) unit for two different VDCs (Bosch ESP8 and Bosch ABS8).

Return Pump

The return pump (see Fig. 6, right) is controlled directly by the VDC module. It ensures the necessary brake pressure in the event that the driver does not use the brake pedal (VDC intervention). The pump motor is specified for currents of up to 80 A, and the brief period of shutdown delay (in the ms range) for which the motor continues running when the ECU is switched off is used to diagnose defective pumps. Thus, for use in the HIL system, either a real pump motor can be installed, or alternatively, a substitute circuit (RC circuit) on the simulator's load boards, which has the same mechanical time constant as the real motor.

Brake Booster

Brake assist functionality requires high brake pressures to be reached fast, e.g., for panic braking, even if the driver does not apply much force on the pedal. Panic braking situations are detected via a fast pressure rise in the main brake cylinder or via a membrane position sensor in the brake booster. In this case, the VDC module actuates a booster coil to raise the brake pressure up to the limit of transferable brake torque. For HIL simulation, it is usually sufficient to "simulate" the solenoid by means of a resistance and to read in the booster control signal digitally as an input to the brake booster model.

Additional outputs of the VDC module include the VDC indicator, which tells the driver that active intervention is occurring or that the VDC is deactivated, and the four wheel speeds that are used, among other things, for interpolation of the GPS data by the navigation system (e.g., in tunnels).

CAN COMMUNICATION

In addition to actuators and sensors, the VDC system also needs to communicate with other ECUs in order to function properly. It mainly has to interact with the engine and transmission ECUs. Fig. 8 shows the most important data that is typically exchanged via the CAN bus.

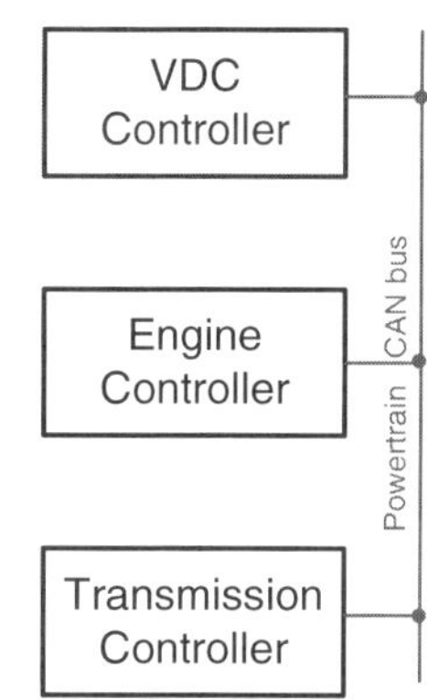

VDC receives:
- „real" engine torque
  (incl. all external commands)
- engine torque based on throttle position
- driver torque command
- pedal or throttle position
- selected gear
- gear shift information

VDC sends:
- „fast torque reduction" command
  (by ignition angle, throttle, and injection)
- „slow torque reduction" command
  (only by throttle)
- engine torque increase command
- gear shift command

Figure 8: CAN communication of VDC with other ECUs [12].

For HIL simulation, this means that the behavior of the engine and transmission ECUs on the CAN bus has to be simulated (restbus simulation). In one direction, there has to be a suitable reaction to messages from the VDC module (e.g., with regard to torque reduction, see the chapter on Simplified ECU Software Models), and in the other, the engine and transmission messages have to be returned correctly to the VDC module (with regard to torque and gear shifts).

VEHICLE MODELING

MODEL COMPONENTS, TOOLS, AND REAL-TIME SIMULATION

As already described above, the models for HIL simulation have a particularly important role to play, as they have to represent the dynamics of a vehicle precise enough to close the control loops to the real ECU. To put it more precisely, all the sensor signals detailed above have to be computed consistently and in the correct physical relation to the corresponding actuator signals.

The basic model components are outlined briefly below. Implementation variants, the importance of submodels for the testing of ECUs, and modeling tools are also described in individual cases.

Powertrain Model

To test VDC systems, a simple look-up table based combustion engine with downstream elastic powertrain, including clutch, Trilok torque converter, manual or automatic transmission, differential transmission, and elastic drive shafts is needed. This powertrain must be configurable for different variants, e.g., for rear, front, or all-wheel drive (Fig. 9).

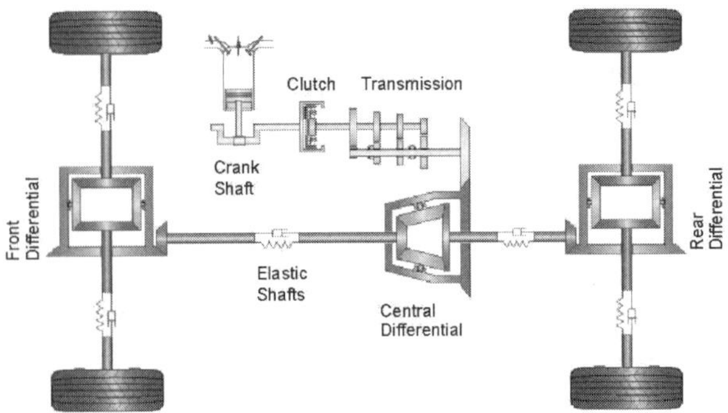

Figure 9: Powertrain model.

Vehicle Body Dynamics Model

Typically, the dynamics of a vehicle is described as a multi-body system which yields a set of differential equations of 2nd order. For a simplified approach the necessary physical abstraction with 10 degrees of freedom (DoFs), 6 for the body translation/rotation and 4 for the wheels, is shown in Fig. 10.

Although multi-body simulation tools with C code export are available (e. g., SimPack [13]), the multi-body simulation codes/equations that are currently used are still mainly derived manually, or only partly automatically generated and then optimized for real-time processing. One reason is that there is often insufficient support for selecting suitable coordinate systems for flexible integration of different axle kinematics into the simulation code of the vehicle dynamics. Another more important reason why these modeling tools are not more widely used in HIL is probably inadequate code efficiency, i.e., processing times are too high.

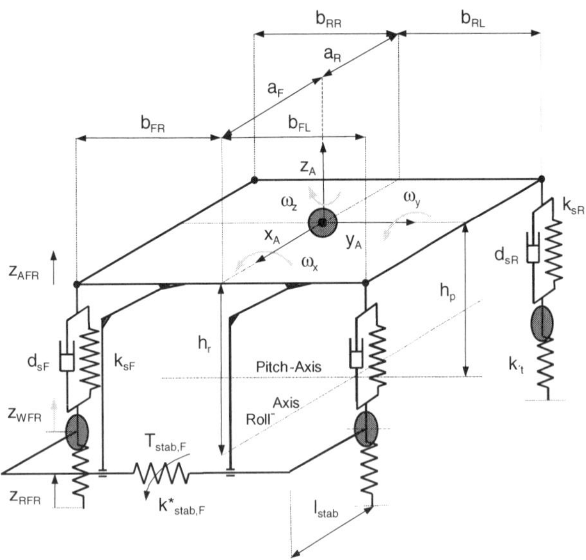

Figure 10: Body dynamics model.

End users of HIL systems are primarily concerned with testing an ECU. Thus, the user normally wants to apply a suitable real-time-capable model directly to achieve this

objective instead of the prior (and possibly time-consuming) use of a modeling tool.

As HIL test technology becomes more widespread in the field of model-based design, e. g., in advanced engineering departments, modeling tools for HIL simulation are likely to grow in importance. Especially since the components being modeled in these areas are new mechatronic systems, and they are generally not available on the market as off-the-shelf models.

Axle Dynamics/Kinematics Model

Axles can be computed in real time either kinematically, using their geometric design data, or by tables. Axles based on geometry (e.g., rear axle semi-trailing arm, McPherson strut, four-bar axle) are more complex to model, but have the advantage of being easier to parameterize once all the design data is available. As this is frequently not the case, axles are nowadays often mapped as tables, which represent the toe-in, camber, and caster angles, and the position of the wheel hub as a function of the wheel lift, and additionally for steered axles, as a function of steering rod displacement. The table data is either captured at an axle test bench or computed by kinematics simulation (e.g., with MSC.Adams [14]). This kinematics simulation is used more frequently, since process integration or the automation of axle test benches are often insufficient. For example, typically steering angle effects can not be taken into account.

Tire Models

The tire model is extremely important to VDC interaction, as it describes all the governing forces and torques between the vehicle and the road surface according to the current wheel load. Finally, exactly these forces and torques are the control objectives of the VDC system.

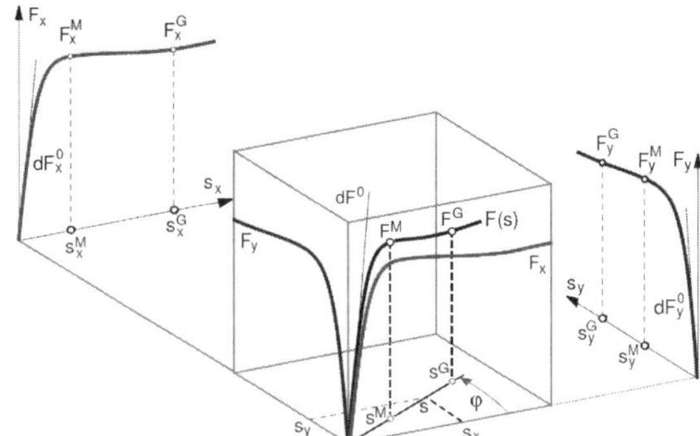

Figure 11: Easy to use tire model [15].

Some common tire models are the Easy-to-use Tire Model ([15], [16]) and the Magic Formula or Delft-Tire Model [17]. The latter is no doubt the most widely used, and all major tire manufacturers supply data sets for it. Both tire models include delayed, coupled longitudinal and lateral forces, self-aligning and bore torques. E. g. Fig. 11 shows the mutual dependencies of the longitudinal and the lateral forces within the Easy-to-use model.

Real-time implementation of these models is possible, but requires some effort, especially for starting from stand still, braking to stand still, and standing on an inclined road surface. Another aspect is the computation of the contact point or contact area, i.e., the interaction between the road surface and the tire. This topic will not be described in further detail here.

Steering Model

A model of the steering system is also indispensable for testing VDC systems. The model should take interventions by electric power-assisted steering (EPAS) into account. As the steering system is generally not the object under test, power-assisted steering can often be implemented by superimposing a speed-dependent steering force.

Brake Hydraulics Model

Unlike pure ABS systems, realistic testing of a VDC system requires a model of the brake hydraulics, as the system has to supply essential sensor variables (e.g., hydraulic pressure) in accordance with the control of the return pump, the pre-charge pump, and the solenoid valves (usually 12). It is mostly implemented by simple variable orifices (e.g., valves), which are coupled via hydraulic accumulators (e.g., wheel brake cylinders, damping chambers). An ESP 5.7 (see Fig. 12) from Bosch can be represented by five coupled, nonlinear differential equations of 1st-order, for each brake circuit. Effects such as hydraulic inductivity or cavitation are mostly ignored in HIL simulation.

If the structure of the hydraulic system changes frequently, or new types of hydraulic systems with as yet unknown modeling depths need to be examined, special

modeling tools can be useful. Tools such as AMESim [18] and DSHplus [19] allow hydraulic systems to be modeled at component level and C-code for real-time simulation, e.g., in the form of a Simulink S-function, to be exported.

The particular effects that have to be included are on-off-controlled and continuously controlled valves in current use and valves with orifices with variable flow areas (change-over or pre-charge valve). These valves are often not available in common hydraulics libraries. Submodels for the shutdown delay of the return pump, which is used for VDC self-diagnostics, must be included also.

The brake hydraulics and the vehicle are coupled via the brake force at the wheels. As it interacts with the rest of the vehicle, the hydraulic system must be regarded as a stiff subsystem, whose numeric integration will be described in greater detail below.

As already mentioned in the introduction, when more detailed tests are required (e.g., when HIL is used for function development), it makes sense, and is quite usual, to construct a complete real brake hydraulic system.

This can then be operated in an environmental chamber, for flexible studies on temperature effects, which cannot yet be simulated with a reasonable effort.

The core model parts described so far can be constructed reasonably easily from submodel libraries available on the market (e. g. [20]). However, this is usually not sufficient for HIL simulation. To make the models realistic enough for the required maneuvers, additional items are needed: a driver model, a road model, and a maneuver control.

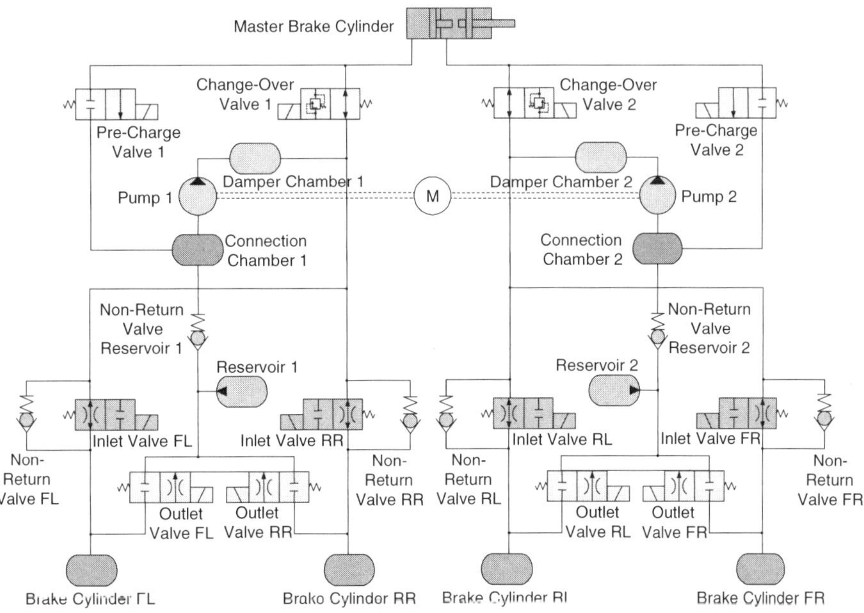

Figure 12: Brake hydraulics of ESP 5.7.

<u>Driver Model</u>

The driver model comprises a cycle driver (model-based controller) which is able to follow a specified speed profile, and a lateral controller, which guides the vehicle on the road via steering intervention. Optimum longitudinal speed adaptation also uses a road preview to ensure that lateral acceleration on the road stays within the permitted range. Some typical solution approaches to lateral control can be found in [21], [22], [23], [24], [25], [26], [27], [28], [29], and [30].

Cognitive driver models [31], [32] play only a minor role in actual HIL testing today. Driver models that optimize lap times in racing applications [33], or that map aggressive, sporty, or economical driver behavior in order to test adaptive transmission controls, could grow in importance in the future. New areas of applications will result in further challenges for driver models, e. g. for the test of Automatic Cruise Control (ACC) systems which require traffic flow simulation.

<u>Road Description</u>

Road models compute the longitudinal, lateral, and vertical profile of the road as inputs for the vehicle model and the driver. The interfaces between these submodels must be optimally adjusted to one another. The (horizontal) road profile is usually represented by straight, circular, clothoid, and spline segments. Each segment has friction coefficients, for example, and other surface properties (e.g. cross-groove profile and similar), which need to be entered conveniently via a graphical user interface [16]. Real time capable animation of the vehicle, road, and environment has also been a part of vehicle dynamics simulation in HIL for some years [34]. One important feature of good tool integration is therefore a road module that can generate a VRML representation of the road to avoid unnecessary manual work.

Data captured via GPS systems for specific test or racing tracks, or mountain passes, is also used for simulation, as an alternative to the road descriptions mentioned above.

<u>Maneuver Control</u>

Maneuver control basically comprises controls for the gas pedal, brakes, gear lever, and steering. Time histories captured in a vehicle are often fed in as stimuli to simulate test drives. Alternatively, it is usually possible to define these variables segment-wise over the time or over the traveled route. A distinction is usually made between longitudinal control (accelerate to 100 km/h at 70 % gas pedal, then brake with 60 % of the max. brake force, etc.) and lateral control (follow the road for 200 m or change the steering angle over time according to a specified steering angle table). Many experiments also require external or internal events to be included in the maneuver definition, e.g.: initiate braking if a specific speed or maximum lateral acceleration is reached, or if a specific CAN signal is sent by the ECU. Like the road description, consistency of these complex maneuver

controls can be ensured only by a suitable graphical user interface.

Fig. 13 depicts the overall structure of a vehicle dynamics model with the needed components and their essential dependencies.

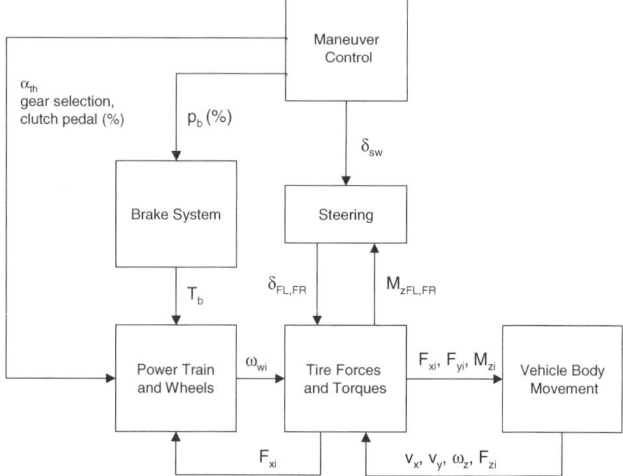

Figure 13: Vehicle dynamics model structure.

<u>Simplified ECU Software Models</u>

The last model components required are simplified models of ECUs or interfaces that are present in the real vehicle but typically not included in the simulator. To simulate the vehicle dynamics, for example, a starter model with appropriate starter control and a simplified control for the automatic transmission or Trilok torque converter is needed. In addition, the engine model needs an interface that carries out engine torque reduction as requested by the vehicle dynamics control via CAN (e. g., inverse engine map (LUT) with the engine speed (n_{mot}) and desired torque ($\tau_{desired}$) as input and the throttle valve position as the output).

NUMERIC ASPECTS

Because the drive shafts are relatively stiff, modeling the elastic powertrain provides a so-called stiff differential equation system, i. e., there are relatively low eigenvalues up to approx. 10 Hz and eigenvalues over 100 Hz that do not easily allow stable integration of the differential equations by an explicit Euler method at the currently typical sampling rates of 0.5-1 ms.

Using e. g. a Runge-Kutta integration method instead of an Euler is also not generally more suitable for this type of model, because although they promise greater precision at the expense of longer processing times, they achieve that precision only if the state variables behave smoothly in the sampling interval. However, due to dry friction and gear shifts in the vehicle models, this cannot be assumed to be the case.

Numeric stability problems also occur in brake hydraulics modeling, due to low hydraulic volumes and high bulk modulus ($E_{Oil}/V \approx 13.000$ bar/cm^3). This list of stiff submodels continues, e. g., combustion engines (mean value models) with small intake manifold volumes (< 2 liters) which yield to numeric problems due to high manifold pressure gradients.

In the past, a number of solution strategies have been used to meet the real-time requirements. For example, in the case of brake hydraulics and the powertrain, the Jacobian ($\partial f(x)/\partial x^T$) of the system ($\dot{x} = f(x,u)$) can be computed analytically in advance to stabilize the Euler integration (pseudo-implied integration [15], [35]). This method necessitates the additional computation of the Jacobian matrix and is therefore always tied to a fixed model structure (at least for the subsystem under observation). If the structure changes, these terms must be recomputed.

Another way is to use an implicit Euler method, though the computational power required for the solution (zero search) rises if the state variables have steep gradients and this generally violates the real-time condition. This can be counteracted at the expense of precision by limiting the number of zero search iterations to a fixed value. Modeling tools like Dymola [36] are able to distinguish between fast and slow subsystems and to choose an implicit Euler integrator for the fast system parts and an explicit one for the slow parts [37], which can considerably reduce computing time, especially, if the fast subsystem requires only a small amount of computation compared to the slow parts.

As regards hydraulics simulation, [38] and others have shown that small hydraulics accumulators for coupling hydraulic subsystems can be avoided. This results in a differential algebraic equation system (differential equations with algebraic boundary conditions) whose differential parts can be integrated far more stably and faster because the high eigenvalues are no longer present. The algebraic boundary conditions must also be solved iteratively. Limiting the number of iterations despite a loss in precision can be a very stable, real time capable solution for specific subsystems. The disadvantage is that in a block diagram -based tool such as Simulink, the algebraic equations cannot be inserted and solved efficiently except by hand-written code.

Simulink therefore offers another option to cope with numerically stiff systems: system parts with high eigenvalues can be grouped in subsystems for multiple evaluation within one sampling step (FOR iterator block). If a local block for integration within these systems is implemented (1/s-block), stiff systems can be solved elegantly (e.g., with tenfold oversampling) without having to derive additional equations and without imposing serious restrictions on downstream block-based modeling. The effectiveness of this method basically depends on selecting a suitable oversampled subsystem. Like all the other methods, it results in a correspondingly higher, but deterministic, processing time.

USER INTERFACE

Simulation/Modeling Environment, Parameterization and Variant Handling

It is important to implement the model in an open environment like MATLAB/Simulink. This allows the user to extend the model by more complex ones (e. g. for network simulators) or even replace parts of the model. Users can add proprietary tire models or new axle kinematics models, or vary the powertrain by Haldex coupling or a Torsen differential, if the HIL model is completely open and accessible and not in the form of C or Fortran code.

In addition to the GUIs for road parameterization and maneuver control described above, more suitable GUIs are needed for parameterizing the model components (powertrain, axles, tires, etc.). To ensure convenient variant handling, an integrated GUI component must bring together all the handling controls for offline simulation and for real-time simulation independently of the modeling tool. This final point is extremely important, as vehicles with greatly varying configurations e. g., with regard to the powertrain (transmission variants, different tires, etc.) have to be tested with the same ECU. For future systems, it will also be important to have an automation interface (COM, DCOM) for such a tool, so that higher-level automation tools can set all the necessary structure and parameter variants.

Typical vehicle models in current HIL applications contain all the model components mentioned above. As a rough estimate, this results in approx. 24 mechanical degrees of freedom for the powertrain, body motion, steering, and wheels. In addition, there are 8 states for tire dynamics and 10 states for the brake hydraulics. In other words, there is a minimum of 66 states without the driver or (software) ECU models. Models of similar or even higher complexity (see veDYNA [16] and CarSim [39]) can be executed on a real-time system (e.g., dSPACE DS1006 with 2.2 GHz AMD Opteron™ Processor [34]) at 65 -200 μs, so processing time today is no longer a limiting factor to modeling depth. A real HIL application including standard I/O and CAN bus restbus simulation for a common VDC module can be calculated in less than 200 μs execution time. The challenges to modeling for HIL applications therefore lie more in model parameterization and validation. The parameterization problem can be resolved by suitably integrating data acquisition into the electronics development process. In the future, model validation using measurements from real vehicles could be automated by means of optimization methods.

The final point to be made about modeling is that (as so often) the "whole" is more than the sum of the parts. One consequence of this is that domain-specific modeling tools and individual model libraries for hydraulics or powertrain components have so far been used mainly in niches of HIL simulation. The majority of systems today are based on complete vehicle dynamics models with optimally coordinated submodel components and additional visualization, parameterization and operating software.

TYPICAL TESTS ON VDC SYSTEMS

For testing the vehicle dynamics several standard test maneuvers are defined which are used in track testing on proving grounds with real vehicles as well as in HIL simulation. These tests can be distinguished in
- open-loop procedures with constant, fixed or predefined steering wheel angle, brake and accelerator pedal and
- closed-loop procedures where a driver (or driver model) takes the road information into account to keep the vehicle on a predefined course.

Most of the test procedures have been standardized by the ISO technical committee ISO/TC 22/SC 9: Vehicle dynamics and road-holding ability [40]. Typical test procedures for ABS are e. g.:
- part-brake and full-brake tests: brake performance test to evaluate the brake distance for 100 to 0 km/h (or 60 to 0 mph) under high-μ and low-μ conditions,
- μ-Split braking, i. e. straight-ahead braking on surfaces with split coefficient of friction (ISO 14512).

For TCS (traction control systems) mainly the acceleration on a flat or on an inclined road are studied under low-μ or μ-split conditions.

Finally for VDC systems the following tests are performed:
- double lane change maneuver (ISO 3888-1),
- steady-state circular driving behavior (ISO 4138),
- lateral transient response test (step steering wheel input, ISO 7401),
- braking in a turn (ISO 7975),
- braking in a turn with μ-split, or
- rollover protection test: NHTSA Road-Edge Recovery Test ("fish hook test").

These procedures are used to validate the vehicle model (without any control systems) as well as to validate vehicle dynamics control algorithms.

An important requirement to a vehicle dynamics simulation environment is therefore to support these types of tests procedures, or it should at least support the engineer to define complex maneuvers by means of a library of suitable maneuver steps (please refer to the chapter on maneuver control).

The following plots show a comparison between in-vehicle measurements and the corresponding simulation results for two driving maneuvers in order to validate the models for lateral (step steering wheel input, Fig. 14) as well as for longitudinal (straight ahead braking with ABS, Fig. 15) dynamics.

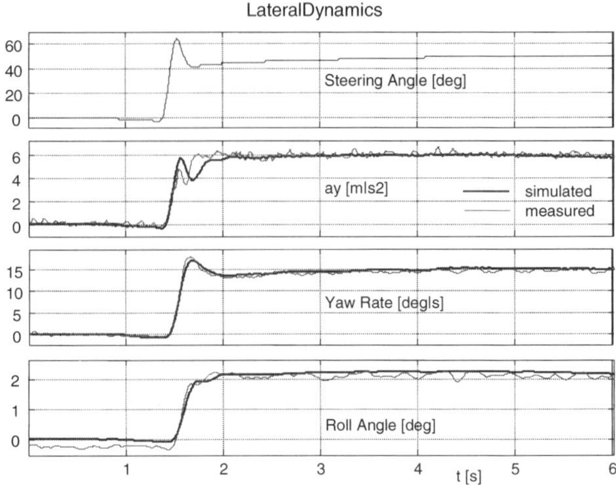

Fig. 14: Lateral step steering.

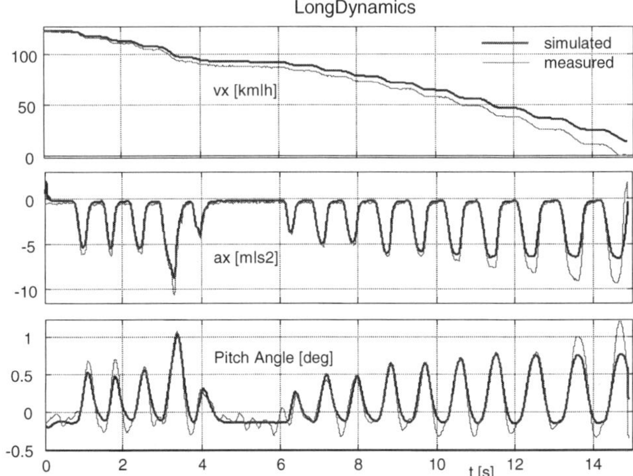

Figure 15: Longitudinal impulse brake.

In addition to testing the "normal behavior" of a vehicle, HIL simulation is also used to test the vehicle's behavior under faulty conditions, for example in cases where sensor or actuator failures occur. These tests can hardly be performed in real vehicles because it is difficult to set a specific error situation reproducible during driving. Additionally these test are dangerous for the test drivers and prototype vehicles can be destroyed as well.

APPLICATIONS

Application examples of HIL simulation for vehicle dynamics ECUs are presented below. The first is a typical simulator for testing *one* VDC module, the second for testing an ECU network. Special problems and possible solution approaches are also described.

SINGLE-ECU TEST SYSTEM FOR VDC MODULES

Fig. 16 shows a typical standard simulator of the kind frequently used in HIL tests on VDC systems. It consists of a simulator rack and valve current capture unit, on which the VDC module is fixed.

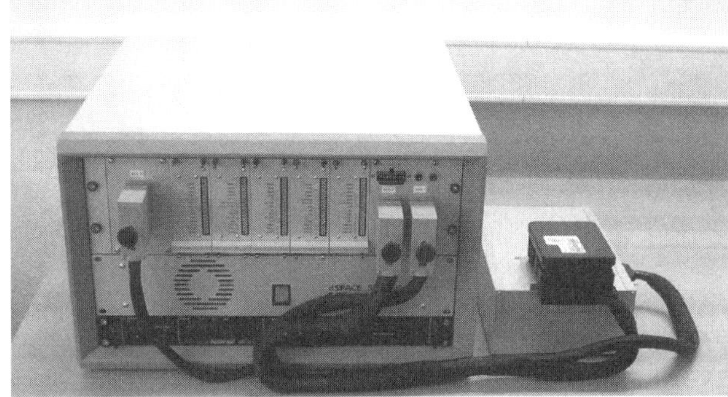

Figure 16: Typical mid size simulator for HIL testing of single ECUs. Device under test: ContiTeves MK60.

At the bottom of the rack is a remote-controlled power supply unit that simulates the battery and allows the battery voltage to be varied from within the real-time simulation (undervoltage and overvoltage tests, voltage drops during engine start, etc.)

The real-time processor hardware and the I/O hardware can be seen above that. The processor hardware in this system is a DS1006 Processor Board, equipped with an AMD Opteron® Processor at 2.2 GHz. This board computes the model components for vehicle dynamics simulation as described further above, and the sensor and actuator scaling. To handle the simulator's inputs and outputs, a DS2211 HIL I/O Board is used (more information can be found in [41]). This is the standard board for dSPACE HIL applications. It provides, for example, the entire I/O for the simulation of a typical 8-cylinder engine, e.g., crankshaft angle synchronous signals; CAN communication; and analog, digital, and PWM I/O, including the necessary signal conditioning. However, the board is not restricted to engine applications, but is used in a large variety of different areas, particularly in testing vehicle dynamics ECUs. For example, it has a fast digital signal processor that provides signal generation in a range of microseconds and is used for purposes such as simulating active, intelligent wheel speed sensors (see Fig. 5).

The top part of the simulator rack shown in Fig. 16 contains a load and failure simulation unit. The load boards can accommodate small resistance loads, e.g., the substitute load for the brake booster solenoid or the RC circuit for simulating the return pump. The failure simulation boards are used to generate electrical failures (open wire, cross short circuits, and short circuits to battery voltage (Vbatt) or ground), especially for testing diagnostic functions in the ECU (OBD II). The load unit can also contain special signal conditioning modules, such as a current source/sink for the active wheel speed sensors. The real-time hardware (signal processor) drives the module by an analog output signal, which the current source/sink module converts to a proportional current signal (maximum 30 mA).

On the software side, the entire dynamic model and the complex I/O (incl. CAN messages, etc.) are completely specified in MATLAB®/Simulink®, and the real-time-capable model code and I/O code are generated automatically (Real-Time Workshop®). Extensive Simulink block libraries are available for defining the I/O (dSPACE Real-Time Interface, RTI). The generated code is linked and downloaded to the real-time hardware.

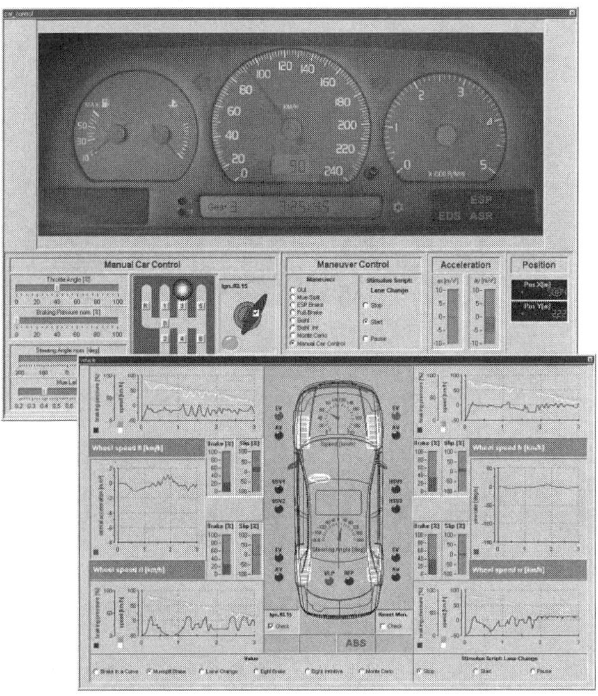

Figure 17: Layouts for instrumentation of the vehicle dynamics HIL experiment.

For experiment control, interactive operation, and instrumentation, ControlDesk [34] can be used. This provides online access to all real-time variables (parameters and outputs of all the blocks of the Simulink model). Fig. 17 shows two examples of application-specific layouts from the field of vehicle dynamics. If tests need to be performed automatically, AutomationDesk [42] is used. This provides project and test administration facilities, a graphical test editor, and test libraries for a great variety of different test scenarios.

To assess simulation results in vehicle dynamics, time histories need to be evaluated, and real-time, 3-D animation is also important. As with the model and I/O definitions, development engineers should not have to worry about the implementation details of the animation, but be able to concentrate on the actual task of optimizing vehicle dynamics. Programs such as MotionDesk (Fig. 18) make this possible by providing a Simulink library for 3D transformations, which allows fast, simple conversion of coordinate systems (e.g., of wheel-specific or body-specific coordinate systems to the world coordinate system required for the animation). Animated force and torque vectors and the ability to record video sequences (AVI format) also make it easier to evaluate the results. An important criterion for a real-time animation tool is that the animated bodies (geometric data, number, environment information) must be completely decoupled from the actual animation program. If this is not the case, it is not possible to properly integrate any desired geometries without recompiling the animation tool.

Figure 18: MotionDesk for the animation of the vehicle motion

Fig. 19 shows HIL simulation results obtained from an experiment (ABS functionality during a full brake maneuver) performed with a HIL simulator as it has been described above. The unit under test was a Bosch ESP.

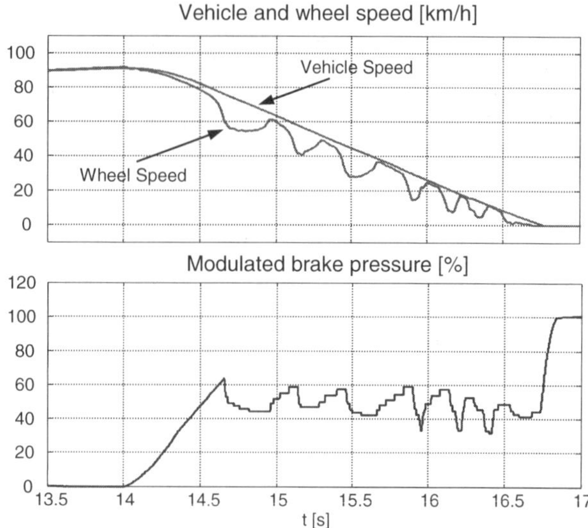

Figure 19: HIL simulation of a full brake maneuver with a real VDC system in-the-loop.

NETWORK TEST SYSTEMS (FULL-SIZE SIMULATOR FOR VDC MODULES)

As the number of ECUs installed rises, along with an increased level of networking and more sharing of sensors and actuators, test systems for vehicle dynamics control are also growing in complexity. While the tests run on individual ECUs are mainly hardware and function tests, network simulators are essentially used to test distributed functions and either the whole or parts of the CAN/LIN network.

The growing importance of testing ECU networks is also reflected in the growing number of publications on the subject [43]. Applications in powertrains and vehicle interiors at Opel [44], Audi ([1], [2]), DaimlerChrysler [45], Fiat/Elasis [46] and Ford [47] have been published, in addition to applications for racing vehicles [48], [49].

Fig. 20 shows an example of a network simulator for the powertrain and vehicle dynamics components of the Bugatti Veyron 16.4, a sports car with more than 1000 hp. Due to the low production numbers for this vehicle, only very few prototypes are available for ECU testing. Systematic use is therefore made of HIL technology [50]. The entire system is controlled interactively via ControlDesk, and 3-D animation of the driving behavior is visualized by MotionDesk.

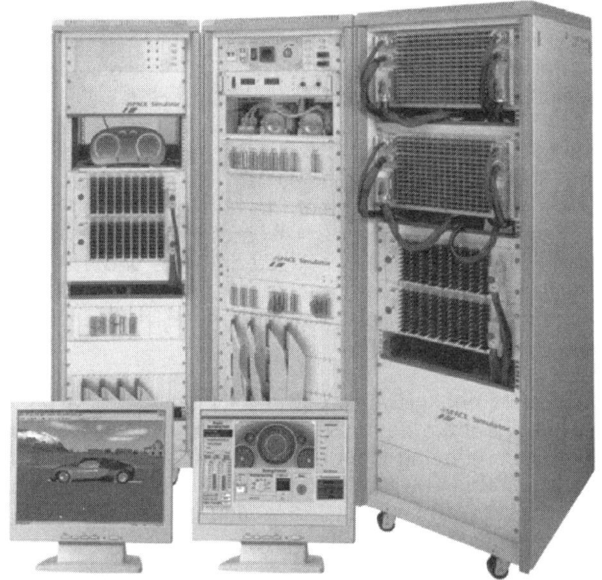

Figure 20: HIL simulator for testing the ECU network of the Bugatti Veyron 16.4.

In addition to large systems for testing an entire ECU network with the HIL simulator, smaller test systems are also frequently used to run only the vehicle dynamics ECUs, or two to three closely networked ECUs, on the simulator. Depending on the vehicle project platform and the OEM's or supplier's organizational structure, there exist different requirements and test system configurations. Some typical combinations are VDC systems combined with CDC, EPB, and also ACC and electronic steering systems.

TEST OF VDC-RELATED SYSTEMS

Finally, some typical problems with VDC-related ECUs and corresponding solution components are presented below.

Electronic Parking Brake (EPB)

In new vehicles, the electronic parking brake is replacing the classic mechanical parking brake. The stationary vehicle is braked either by means of a central actuator which actuates the brake calipers via linkage or by means of local electrical actuators on the wheels (e. g., TRW). EPB systems frequently have an inclination sensor integrated into the ECU, which prevents the vehicle from rolling backwards when started on a slope (with manual transmission).

Direct stimulation of the inclination sensor is particularly problematic for HIL simulation. One solution is to place the ECU on a turnable platform that is controlled by the simulator (see Fig. 21). The platform is given the current vehicle inclination by the real-time simulation and inclines the ECUs accordingly.

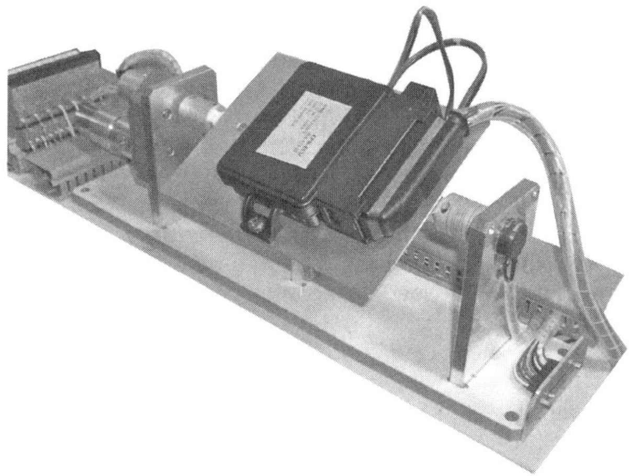

Figure 21: Turnable platform with mounted EPB ECU for stimulating an internal inclination sensor.

Due to the fact that a precise simulation, e. g. of the friction effects at the brake actuators, is very difficult, the brake forces are measured at the real brake calipers by means of force transducers. These forces are then used as inputs to the simulation model. Fig. 22 shows an EPB actuator that is mounted directly on the brake caliper, the force transducer replaces the real brake disk.

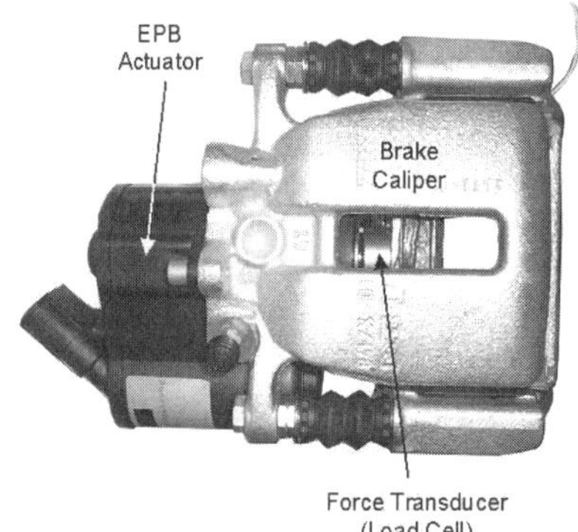

Figure 22: EPB actuator and measurement hardware for capturing the real forces.

Adaptive Cruise Control (ACC)

Common cruise control systems just keep the driver-selected set-speed. In addition to that, *adaptive* cruise control systems automatically adjust the vehicle speed to maintain a driver-selected distance from the vehicles ahead. Therefore, ACC systems are much more complex with respect to sensors and actuators. Typically, the systems use radar sensors (to detect vehicles ahead) and yaw and steering data (from the VDC system) to determine which other vehicles are in the vehicle's predicted path. A connection to the VDC system via CAN or via direct brake booster access is used to decelerate the vehicle.

For HIL simulation the radar sensor has to be simulated and a traffic flow simulation of a certain number of vehicles is required. This model has to generate the horizontal angles as well as the relative speeds and absolute distances of all vehicles in the range of the radar sensor. This data has to be transferred to the ACC controller. Typically radar sensor and ACC controller are located in the same enclosure so that a special version of the ACC system with open radar sensor interface is used for HIL testing purposes. One example of such a system is the Bosch Puma ACC emulator with CAN interface to the radar sensor. As a result, radar sensor simulation by means of CAN message generation can be performed.

The ACC system communicates with various controllers via CAN, for example:
- with the engine controller to accelerate or decelerate (active torque intervention),
- with the brake system (ESP) to actively brake the vehicle and get information on the driving direction (yaw rate, steering angle and vehicle speed),
- with the restraint system (airbags, seat belt tensioner) if a crash situation is very likely or even unavoidable (pre crash system), and
- with the instrument cluster to display the status of the ACC system.

This communication has to be handled by the HIL simulator by transmitting and receiving CAN messages and reacting accordingly with real-time models of the ECUs involved in the scenarios. In order to achieve a more realistic behavior of these ECUs and their CAN communication it is also possible to connect the ECUs as *real* components to a vehicle dynamics network simulator (please refer to the chapter "Network Test Systems").

The setup described above is usually sufficient for testing the ACC ECU as well as its CAN communication. But it is not possible to test the radar sensor itself, which is of course crucial for the overall system functionality. In order to test also the sensor in [51] a test facility called VEHIL is shown, in which the device under test can be a real vehicle with an ACC system. The car runs on a chassis dynamometer, while the vehicles of the surrounding traffic are simulated by special remote controlled autonomous vehicles. These so called moving bases get their input from a real-time traffic simulation.

Electrical Steering Systems

Basically two types of systems can be distinguished in this field:

The first ones are electrical power steering (EPS) systems. Instead of a hydraulic system the steering is supported by forces/torques that are generated electrically. The major advantages of this type of system are, that a complex hydraulics is not needed anymore and that it is very flexible with regard to the adaptation to different driving situations but also to different vehicles (packaging).

The second ones get their power still from a conventional hydraulics but a steering angle can be superimposed to the driver's steering input by an electrical system which is e. g. combined with a planetary gear. These systems (e. g. BMW AFS) can be used to stabilize the vehicle dynamics in critical driving situations.

Depending on the desired test scenarios three different interface levels between the EPS system and a HIL test bench can be identified:

- An interface on the signal level (Fig. 23, interface 1) is adequate, if the focus lies on software tests. It is often used by the suppliers of EPS systems, because for them it is easy to make the internal signals of the ECU accessible. Of course in such a setup a sophisticated actuator model (motor, transmission etc.) has to be provided to obtain the precise motor currents in dependency of the measured PWM signal, that is normally given to the power stages. Moreover, torque and position sensor signals have to be synthesized.

- If the power stages should also be tested, but the actuation system is not available as a real component, the test system must have a dynamic high current sink/source ($\geq$ 80 Amps), which is controlled by the real-time system according to the simulated motor (Fig. 23, interface 2). In this case the modeling effort is still high, but the mechanical demands are as low as for interface level 1.

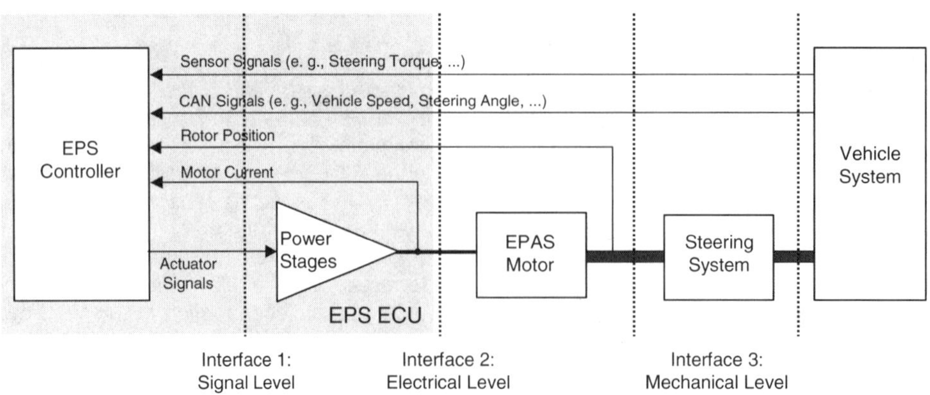

Figure 23: Typical architecture of EPAS systems and possible interfaces to HIL test benches.

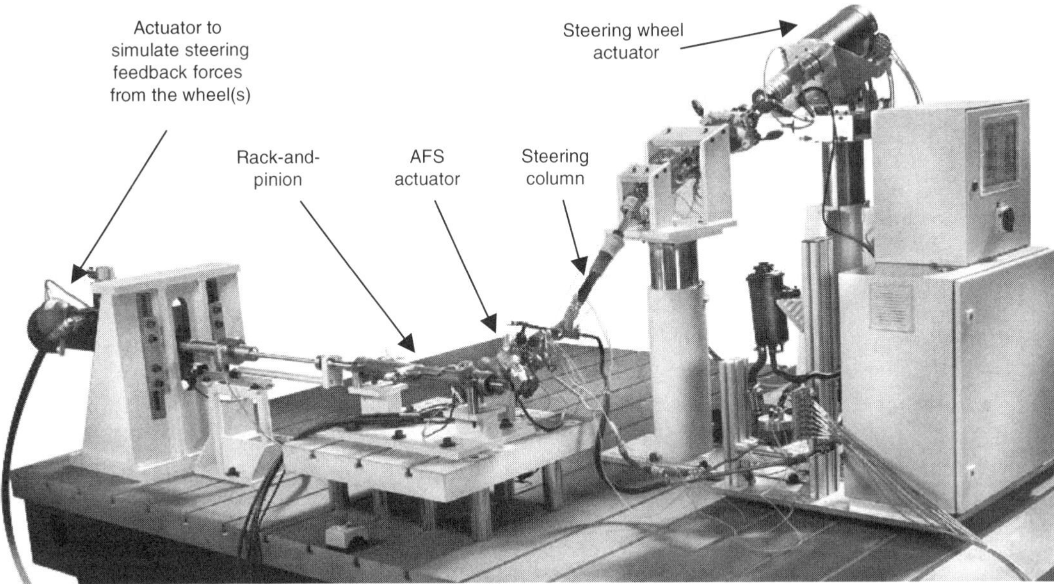

Figure 24: HIL test bench for Active Front Steering (BMW AFS [52], [53])

- On the interface level 3 (Fig. 23) the entire EPS system including the motor and maybe even the transmission are built up as real parts. In opposite to the scenarios above, no model of the steering system is needed here. But of course an electrical or hydraulic actuator system will be necessary to put an actual load on the steering wheel and on the rack or directly on the pinion. This means that the software efforts for modeling are replaced by a higher hardware effort (see Fig. 24). Such an approach makes sense especially if a suitable model can not be obtained or if the actuator system should undergo e. g. durability tests.

FURTHER DEVELOPMENTS AND OUTLOOK

The development departments of many OEMs and tier1 suppliers are currently working on further improvements to yaw rate control. They are combining the ESP/VDC with one or more other active steering or chassis systems, such as Active Front Steering (AFS), Electronic Power Steering (EPS), Continuous Damping Control (CDC), Active Body Control (ABC), Air Suspension System (ASS), Active Torque Control (X-Drive, BMW [54]) and Anti-Roll Control (ARC). These systems have names such as Integrated Chassis Control (ICC, OPEL [55]), Vehicle Dynamics Management (VDM, Bosch [56]), and Global Chassis Control (GCC, Continental Teves [57]), and consist of a super ordinate, coordinating software layer that generates reference values for the underlying control systems in such a way that positive control interferences are reinforced (e.g., yaw rate stabilization for shortening braking distances by >15 % in an ESP/AFS combination) and negative interference is suppressed by stricter separation of working areas. Global safety concepts and switch-off concepts can also be implemented.

It is in the nature of these systems that they mainly operate at critical vehicle dynamics limits and in safety-critical situations, which makes HIL simulation particularly useful for them. Network simulators similar to those used for powertrains and body electronics, made up of the HIL simulators available for lower-level subsystems, will be used more commonly in the future.

Current developments make it clear that the networkability of HIL simulators, their flexibility with regard to variant handling, their fast adaptability to new and additional ECUs, and their test automation abilities will become increasingly important for handling the complexity of the electronics in this field.

LITERATURE:

[1]   Gehring, J.; Schütte, H.: Automated Test of ECUs in a Hardware-in-the-Loop Test Bench for the Validation of Complex ECU Networks. Society of Automotive Engineers, SAE Paper No. 2002-01-0801.

[2]   James, A.; Rudolph, M.; Gehring, J.; Pöhlmann, T.: Hardware-in-the-Loop bei Audi-Antriebssträngen. ATZ 4/2002.

[3]   Bach, Th.: Real-time Simulation in Anti-lock Brake System Development based on a Personal Computer. VDI 6th International Congress "Measurement and Testing Techniques in the Automotive Industry", Berlin, April 28, 1992.

[4]   Hanselmann, H.: Advances in Desktop Hardware-in-the-Loop Simulation. SAE Paper 970932, 1997.

[5]   Bach, Th.; Schmitt, H; Schwanke, W; Tumbrink, H.-J.: ROADRUNNER Real-time simulation in anti-lock system development. SAE paper No. 950758.

[6]   Kiffmeier, U.; Otterbach R.; Schütte H.: Real-Time Simulation of a 3-D Vehicle Dynamics Model on the DEC Alpha Processor. 15th IMACS World Congress '97, Berlin, August 24-29, 1997.

[7]   Krohm, H.; Gheorghiu, V.: Hardware-in-the-Loop Simulation for an Electronic Clutch Management System. SAE paper No. 950420.

[8]   Grund, C.; Schineis, W.: A Hardware-in-the-Loop Test Bench for Wheel Slip Systems. ISATA `97, Florence, June 16-19, 1997, Paper No. 97VR022, June 1997.

[9]   Bosch: Automotive Handbook 5th Edition. Robert Bosch GmbH, Stuttgart, 2000.

[10]   Zanten, A. T. van; Erhardt, R; Pfaff, G; Kost, F; Hartmann, U; Ehret, T.: Control Aspects of the BOSCH-VDC. AVEC `96, Intern. Symposium on Advanced Vehicle Control, Aachen, June 24-28, 1996.

[11]   Michalsky, Th.; Büdenbender, M.: Testing Transmission ECUs With Integrated Sensors. Auto Technology, Volume No. 1, pp. 62-65, June 2001.

[12]   van Zanten, A.: Fahrdynamik & Fahrdynamik-regelsysteme (Vehicle Dynamics and Vehicle Dynamics Control Systems). IIR Training Seminar, Düsseldorf, Germany, 2004.

[13]   SimPack: See product information on http://www.simpack.com/

[14]   ADAMS: See product information on http://www.mscsoftware.com/products/products_detail.cfm?PI=413

[15]   Rill, G.: Simulation von Kraftfahrzeugen. Vieweg Verlag, Braunschweig, 1994.

[16]   veDYNA User's Guide, Tesis DYNAware, Munich, 2001, (http://www.tesis.de).

[17]   Pacejka, H.-B.: Tire and Vehicle Dynamics. SAE, Warrendale, PA, 2002.

[18]   AMESim: See product information on http://www.amesim.com/

[19]   DSHplus: See product information on http://www.fluidon.de/

[20]   TNO ADVANCE: See product information on http://www.automotive.tno.nl/VD/Docs/Brochure_advance.pdf

[21]   Sharp, R. S.; Casanova, D.; Symonds, P.: A Mathematical Model for Driver Steering Control, with Design, Tuning and Performance Results. Vehicle System Dynamics, 33 (2000), pp. 289-326, 2000.

[22]   MacAdam, Ch. C.: Application of an Optimal Preview Control for Simulation of Closed-Loop Automobile Driving. IEEE Transactions on Systems, Man and Cybernetics, Vol. SMC-11, No. 6, June 1981.

[23]   MacAdam, Ch.: The MSC Driver Model. Chapter 7, 1980.

[24]   Wolter, Th.-M: Ein hybrides Modell zur Kraftfahrzeugführung. ZMMS Spektrum Band 14, Fortschritt-Berichte VDI, Reihe 22, Nr. 9, VDI Verlag, Düsseldorf, 2002.

[25]   Vögel, M.: Fahrbahnmodellierung und Kursregelung für ein echtzeitfähiges Fahrdynamikprogramm. Diplomarbeit, TU München, 1997.

[26]   Majjad, R.: Hybride Modellierung und Identifikation eines Fahrer-Fahrzeug Systems. Dissertation Universität Karlsruhe, 1997.

[27]   Kiencke, U.; Nielsen, L.: Automotive Control Systems, For Engine, Driveline, and Vehicle. Springer Verlag, Berlin Heidelberg New York, 2002.

[28]   Butz, T.; Ehmann, M.; Stryk, O. v.; Wolter, Th.-M.: Realistische Straßenmodellierung für die Fahrdynamiksimulation in Echtzeit. ATZ, Heft 106 (2004), pp. 118–125, February 2004.

[29]   Riedel, A; Wurster, U: Fahrermodelle in der Praxis. Automotive Engineering Partners, March, 1998.

[30]   Prokop, G.: Modeling Human Vehicle Driving by Model Predictive Online Optimization. Vehicle System Dynamics 2001, Vol. 35, No. 1, pp. 19-53, 2001.

[31]   Jürgensohn, Th: Hybride Fahrermodelle. ZMMS Spektrum Band 4, Pro Universitate Verlag, Sinzheim, 1997.

[32]   Irmscher, M: Modellierung und Simulation von Motivationseinflüssen auf das Fahrerverhalten. ZMMS Spektrum, Band 12, Fortschritt-Berichte VDI, Reihe 22, Nr. 6, VDI Verlag, Düsseldorf, 2001.

[33]   Ehmann, M.; Butz, T.: Raceline Optimization and Driver Modeling for the Simulation of Race Car Dynamics in Real-Time. Race.Tech 2004 Conference, Munich, Oct. 14-15, 2004.

[34]   dSPACE MotionDesk, dSPACE ControlDesk, dSPACE Simulator. See product information on http://www.dspace.de/

[35]   Rill, G.: A modified implicit Euler Algorithm for Solving Vehicle Dynamic Equations. EUROMECH, Halle, March 1 – 4, 2004.

[36]   Dymola: See product information on http://www.dynasim.com/.

[37]   Schiela, A.; Olsson, H.: Mixed-mode Integration for Real-time Simulation. Modelica Workshop 2000 Proceedings, pp. 69-75, October 23-24, 2000.

[38]   Borchsenius, F.: Simulation ölhydraulischer Systeme. Fortschritt-Bericht VDI, Reihe 8, Nr. 1005, VDI Verlag, Düsseldorf, 2003.

[39]   CarSim Product Description, Mechanical Simulation Corporation (MSC), Ann Arbor, MI, 2001. (http://www.carsim.com/).

[40]   ISO technical committee ISO/TC 22/SC 9: Vehicle dynamics and road-holding ability http://www.iso.org/iso/en/CatalogueListPage.CatalogueList?COMMID=847

[41]   Schütte, H.; Plöger, M.; Diekstall, K.; Wältermann, P.; Michalsky, Th.: Testsystems in the ECU Development Process. ATZ/MTZ Special Issue, Automotive Electronics, pp. 16-21, March, 2001.

[42]   Lamberg, K.; Richert, J.; Rasche, R.: A New Environment for Integrated Development and Management of ECU Tests. Society of Automotive Engineers, SAE Paper No. 2003-01-1024.

[43] Wältermann, P.; Schütte, H.; Diekstall, K.: Hardware-in-the-Loop Test verteilter Kfz-Elektroniksysteme (Hardware-in-the-Loop Testing of Distributed Electronic Systems). ATZ 05/2004, pp. 416-425.

[44] Köhl, S.; Lemp, D.; Plöger, M.; Otterbach, R.: Steuergeräte-Verbundtests mittels Hardware-in-the-Loop-Simulation (ECU Network Testing by Means of HIL Simulation). ATZ 10/2003, S. 948.

[45] Honisch, A.; Hutter, A.; Schmid, H.: Vollautomatisierter Test von Steuergeräte-Netzwerken (Fully Automated Test of ECU Networks). ATZ/MTZ Special Issue, Mercedes-Benz A-Klasse, pp. 116-125, October, 2004.

[46] Plöger, M.; Ferrara, F. et al: Testing Networked ECUs in a Virtual Car Environment. SAE Paper No. 2004-01-1724.

[47] Heiming, B; Haupt, H.: Hardware-in-the-Loop Testing of Networked Electronics at Ford. SAE Paper No. 2005-01-1658.

[48] Waeltermann, P.; Michalsky, T. Held, J.: Hardware-in-the-Loop Testing in Racing Applications. SAE Motorsports Conference 2004, Dearborn, Mi, USA, SAE Paper No. 2004-01-3502.

[49] Urban, P.; Wältermann, P.; Henking, B.: Toyota Motorsport Racing Ahead with dSPACE (In German: Toyota Motorsport mit dSPACE auf der Überholspur. Automotive Engineering Partners 6/2001, pp. 40ff. English Version available via: http://www.dspace.de/ftp/papers/dspace_AutEP_0111_e_f22.pdf.

[50] Baeker, W.; Lichtenthäler, D.: Bugatti: Powerful Cars Need Powerful Tests. dSPACE News 01/2004, dSPACE GmbH, Paderborn (www.dspace.de).

[51] TNO VeHIL: Vehicle Hardware-in-the-Loop-Test-Facility. http://www.automotive.tno.nl/smartsite.dws?id=57.

[52] Abou-El-Ela, A.; Wachinger, M.; Krenn, M.: HIL Test Bench for BMW's Active Steering. dSPACE News 03/2004, pp.14-16.

[53] IABG: Intelligent Testing. Brochure. IABG Ottobrunn, Germany (www.iabg.de).

[54] Fischer, G.; Pfau, W.; Braun, H.-S.; Billig, C.: xDrive, Der neue Allradantrieb im BMW X3 und BMW X5. ATZ 2/2004, pp. 92-102.

[55] Holzmann, H.; Hahn, K.-M.: Application of HiL-Simulation in the Development Process of Modern Chassis Control Systems Considering Integrated Chassis Control (ICC). VDI-Bericht Nr. 1828, pp. 549-558, Steuerung und Regelung von Fahrzeugen und Motoren – AUTOREG 2004.

[56] Trächtler, A.: Integrated Vehicle Control Using Active Brake, Steering, and Suspension Systems. VDI-Bericht Nr. 1828, pp. 463-473, Steuerung und Regelung von Fahrzeugen und Motoren – AUTOREG 2004.

[57] Schwarz, R.; Rieth, P.: Global Chassis Control – Systemvernetzung im Fahrwerk (Global Chassis Control – Integration of Chassis Systems). at-Automatisierungstechnik, 7/2003, pp. 300-312.

ABBREVIATIONS

ABC:	Active body control
ABS:	Anti-lock braking system
ACC:	Adaptive cruise control
AFS:	Active front steering
BAS:	Brake assist system
CAN:	Controller area network
CDC:	Continuous damping control
DoF:	Degrees of freedom
DSP:	Digital signal processor
ECU:	Electronic control unit
EPAS:	Electric power-assisted steering
EPB:	Electronic parking brake
ESP:	Electronic stability program
EPS:	Electric power steering
GPS:	Global positioning system
HIL:	Hardware-in-the-Loop
ICC:	Integrated chassis control
ISO:	International Organization for Standardization
LIN:	Local interconnect network
LUT:	Look-up table
NHTSA:	National Highway Traffic Safety Administration
OBD II:	Onboard diagnostics 2nd generation
PWM:	Pulse width modulation
TCS:	Traction control system
VDC:	Vehicle dynamics control
VSD:	Valve signal detection unit

CONTACT

Dr. Herbert Schuette
Director Applications/Engineering
dSPACE GmbH, Paderborn, Germany.

E-mail: hschuette@dspace.de

Dr. Peter Waeltermann
Lead Engineer Applications/Engineering
dSPACE GmbH, Paderborn, Germany.

E-mail: pwaeltermann@dspace.de

Web: http://www.dspace.de

Enhanced Traction Stability Control System

Youssef A. Ghoneim, William C. Lin and Yuen-Kwok Chin
General Motors Corporation, Research and Development Center

David M. Sidlosky
General Motors Corporation, Product Development

ABSTRACT

This paper is directed to an Enhanced Traction Stability Control System (ETSC) that is based on the estimate of vehicle yaw rate and does not require yaw rate or lateral accelerometer sensors information. The validity of the yaw rate estimate is determined and used to select the appropriate control methodology. We estimate the vehicle yaw rate based on the measured speeds of the un-driven wheels of the vehicle, and we utilize various other conditions to determine if the estimated yaw rate is valid for control purposes. When it is determined that the yaw rate is valid, a combined closed-loop yaw rate feedback, and an open-loop feed-forward derivative control based on the driver input is employed. Whereas in conditions under which it is determined that the estimated yaw rate is not valid, an open-loop feed-forward control with a proportional, derivative and a diminishing integrator terms, is employed. In addition, we develop a bank angle compensation algorithm using the steering angle, vehicle speed, and the estimated yaw rate to compensate for the effect of banked road. Test results indicate marked enhancement of vehicle stability with ETSC when compared with ABS and TCS. Finally, we present test results to compare the performance of ETSC system to yaw rate feedback control only Electronic Stability Control System (ESC) using yaw rate and lateral accelerometer sensors information.

INTRODUCTION

Chassis control technology has achieved noteworthy progress, thanks to advancements in sensing and computing technologies as well as advances in estimation and control theories. One such enhancement is the control of the tire forces through the braking force-distribution control strategy [1-10]. Basic sensors used to supply information for the control algorithms in to supply information for the control algorithms.

The braking-force distribution control includes a steering wheel angle, wheel speeds, a lateral accelerometer, and a yaw-rate sensor. With the information obtained from these sensors, Electronic Stability Control Systems (ESC) are devised to enhance vehicle stability. Because of the price of these sensors, this ESC technology is limited to a smaller number of vehicles. In order to further reduce the cost of ESC suitable for less expensive vehicles three approaches emerged and warranted further investigation: (1) steering wheel sensor only, (2) yaw acceleration sensor, and (3) steering wheel sensor and lateral accelerometer. The first approach with steering wheel sensor only, the lowest cost version, is the main emphasis in this paper. The technology with steering-wheel angle sensor only can be obtained by eliminating the lateral accelerometer sensor and the yaw rate sensor with a bank-angle compensation algorithm and an estimate of the yaw rate obtained from the un-driven wheel speeds as follows:

$$\dot{\psi}_{est} = \frac{\omega_l - \omega_r}{2t} \tag{1}$$

where ω_l and ω_r are the left and right un-driven wheel speeds, and t is half the track of the vehicle.

Once the vehicle yaw rate is estimated, a yaw-rate feedback control can be devised.

However, when the vehicle is braking or when the vehicle exhibits an oversteer condition, the wheel speeds are corrupted. The yaw rate estimated from the wheel speeds deviates from the actual yaw rate of the vehicle and cannot be used. Therefore it is very important to determine when the estimated yaw rate obtained from the un-driven wheel is valid and when it is invalid.

When the estimated yaw rate obtained from the un-driven wheel is valid, a yaw-rate feedback control strategy similar to a two channel ESC can be used. Alternatively when the estimated yaw rate is invalid, a different control strategy using feed-forward control replaces the yaw-rate feedback control during the time when the estimated yaw rate is invalid.

In the following section, we discuss the ETSC control structure. Section 3 describes various ETSC algorithms. In Section 4, we compare the performance of ETSC to the performance of a two channel ESC under similar vehicle test scenarios.

2. ENHANCED TRACTION STABILITY CONTROL STRUCTURE

Figure 1 illustrates the ETSC implementation. Based on the driver steering input, a yaw-rate command interpreter determines the vehicle desired yaw rate. The lateral accelerometer is removed, and the yaw rate sensor is replaced with yaw rate estimation obtained from the un-driven rear wheels. When we determine that the yaw rate estimation is valid, a differential brake command to the wheel is computed through feedback yaw rate control is used to control the vehicle yaw rate. When the estimation is not valid especially during braking or when the vehicle oversteers, the differential brake control switches from a feedback yaw rate control to a feed-forward control based on the driver input.

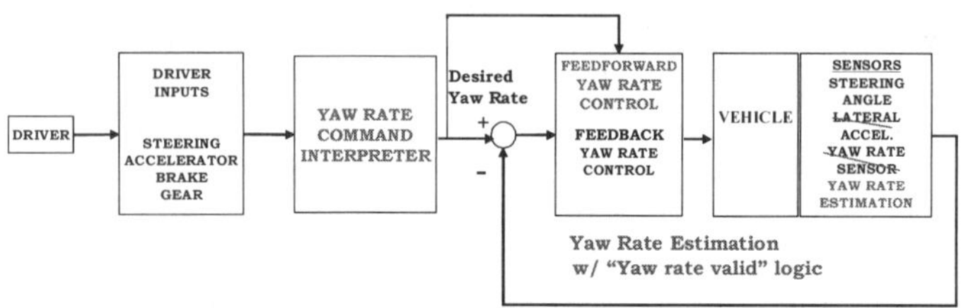

Figure 1   Enhanced Traction Stability Control (ETSC) structure

3. ENHANCED TRACTION STABILITY CONTROL ALGORITHMS

In this section, we present brief descriptions of the ETSC algorithms. Details can be found in the Appendices.

3.1 Yaw Rate Estimation

The estimation of the vehicle yaw rate can be accomplished using both wheel speeds and vehicle speed information as follows:

By referencing to Fig. 2, V_m is the instantaneous velocity of the center of gravity, and V_i, i=1,4, is the forward velocity measured at wheel i.

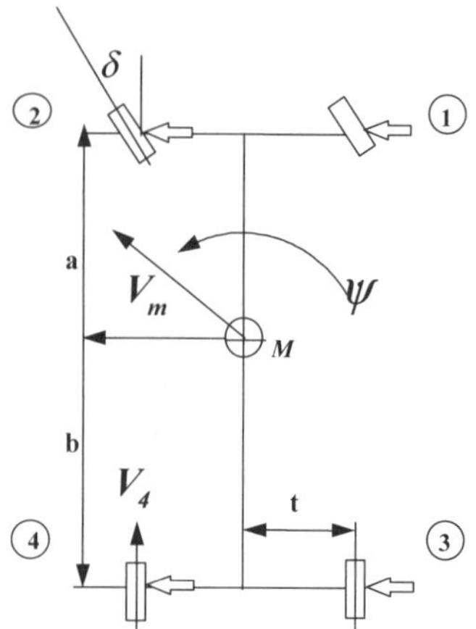

Figure 2   Yaw plane dynamical model

The forward velocity V_i is related to the instantaneous velocity of the center of gravity V_m by the following relationship:

$$V_i = V_m + (-1)^i t \dot{\psi}, i = 1,4 \qquad (2)$$

where t is half the track of the vehicle and $\dot{\psi}$ is the yaw rate of the vehicle.

The yaw rate can be determined from equation (2) using the two un-driven wheels 3 and 4,

$$V_4 = V_m + t \dot{\psi}$$
$$V_3 = V_m - t \dot{\psi}.$$

Therefore

$$\dot{\psi} = \frac{V_4 - V_3}{2t}. \qquad (3)$$

Assume that un-driven wheels are not braking then

$$V_i = \omega_i r_i, i = 3,4. \qquad (4)$$

Therefore, an estimate of the yaw rate is obtained from the un-driven wheel speeds 3, and 4 as follows:

$$\dot{\psi}_{est} = \frac{\omega_4 r_4 - \omega_3 r_3}{2t}. \qquad (5.a)$$

Equation (5.a) shows that the vehicle yaw rate can be computed as a function of the measured speeds of the un-driven wheels and the distance between the center of the wheel treads (i.e., the track). However, the estimate may fail to equal the actual vehicle yaw in a vehicle with a relatively high center of gravity due to the relatively high body roll of the vehicle. When a vehicle with large body roll and high center of gravity is driven under high lateral acceleration, the outside tire is compressed and the radius of the inside tire is increased. Therefore, the estimated yaw rate is higher than the actual yaw rate of the vehicle. It is desired to compensate for the yaw rate estimation, without a yaw sensor or lateral acceleration sensor. Therefore, we can control vehicle yaw rate even during conditions that have previously degraded the yaw rate estimation such as in vehicles with large body roll and high center of gravity. Therefore, a compensation term, based on vehicle speed and estimated yaw rate, is then used to adjust for the estimated yaw rate due to vehicle roll in high c.g. vehicles as follows:

$$\dot{\varphi}(k) = \frac{\dot{\psi}_{est}}{1 + KV_m \left| \dot{\varphi}(k-1)V_m \right|} \qquad (5.b)$$

where $\dot{\varphi}(k)$ is the adjusted estimated yaw rate at time k.

3.2 Yaw-Valid Logic

We have discussed in the last section, that the estimation of the vehicle yaw rate can be accomplished using information from the un-driven wheel speeds. However, when the vehicle is braking or when the vehicle exhibits an oversteer condition, the wheel speeds are corrupted. During these conditions, the yaw rate estimated from the wheel speeds deviates from the actual yaw rate of the vehicle, and the information obtained cannot be used any more to determine an estimate of the yaw rate of the vehicle. Therefore it is very important to determine when the estimated yaw rate obtained from the un-driven wheel is valid and when it is invalid. The validity of the estimated yaw rate is judged based on a logical analysis of the measured wheel speed information, braking information, and steering wheel angle. The measured speeds of the un-driven wheels are used to compute an average un-driven wheel speed and average un-driven wheel acceleration. The steering angle and the vehicle velocity may be used to determine a desired yaw rate, which is compared to the yaw estimate to find a yaw rate error. Based on these variables, the control reliably determines whether the estimated yaw rate is valid, and selects an appropriate control methodology in accordance with the determination.

To illustrate how the algorithm performs, Figure 3 illustrates a 50-mph random steering maneuver on snow using a vehicle equipped with both ESC and ETSC algorithms. Six seconds into the maneuver the vehicle starts sliding slightly, the rear wheel speeds information is corrupted, and the estimated yaw rate deviates from the actual yaw rate. At this point the algorithm detects

the deviation, and the yaw valid flag is set to 0 indicating invalid yaw estimate. At about 11.5 seconds the vehicle regains control and the yaw valid flag is set to 1, indicating that the estimated yaw rate is valid and can be used for control.

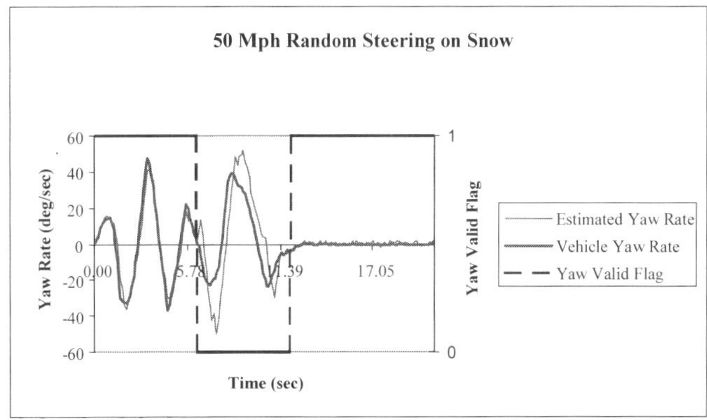

Figure 3 Yaw-valid logic

3.3 Yaw-Rate Feedback Control

Figure 4 illustrates the computation of the delta-velocity command for the feedback differential-brake control. The command is determined by summation of a proportional term based on yaw rate error and a derivative term based on derivative of yaw rate error. The delta-velocity is then computed after passing both terms through a Dead-Band as:

$$\Delta V_{\dot{\psi}} = K_{p\dot{\psi}}(\dot{\psi}_d - \dot{\varphi}) + K_{d\dot{\psi}}(\ddot{\psi}_d - \ddot{\varphi}) \qquad (6)$$

The proportional and derivative gains and the desired yaw rate are derived in [2].

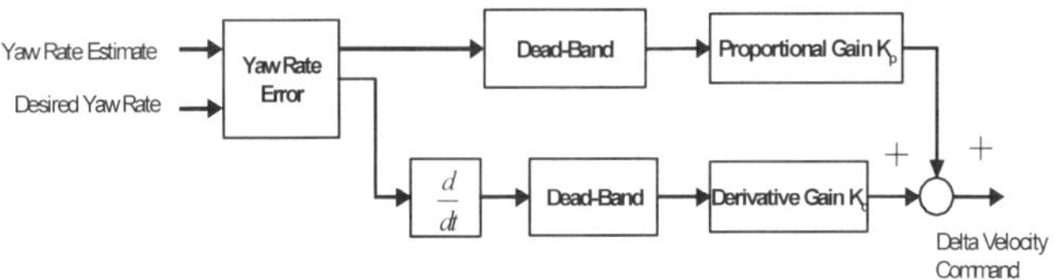

Figure 4 Block diagram of the ESC yaw-rate feedback control system

3.4 Open-Loop Feed-Forward Control

Figure 5 illustrates the computation of the delta-velocity command for the Feedforward differential-brake control. The control system uses vehicle input of steering wheel information and vehicle speed. The command is determined by summation of a proportional term based on desired yaw rate and a derivative term based on derivative of desired yaw rate. The proportional and derivative gains are determined as function of the vehicle speed. A Dead-band is determined where the control remains inactive until the Dead-band exceeds a predetermined threshold function of vehicle speed. The Dead-band is computed as:

$$DB_{ff} = \left| K_p \dot{\psi}_d + K_d \ddot{\psi}_d \right| \qquad (7)$$

A saturation level, function of the desired yaw rate is used to further reduce the aggressiveness of the control when the steering is excessive. Finally, the use of the diminishing integrator in the Feedforward control reduces the steady state error.   The delta-velocity is then computed as:

$$\Delta V_{\dot{\psi}ds} = K_p \dot{\psi}_{ds} + K_d \ddot{\psi}_{ds} - K_p K_i \left(\frac{1}{s - K_i} \right) \dot{\psi}_{ds} \qquad (8)$$

where

$\dot{\psi}_{ds}$ and $\ddot{\psi}_{ds}$ are the desired yaw rate, and the derivative of the desired yaw rate subjected to the dead-band and the saturation function. The derivation of the feedforward control is described in Appendix B.

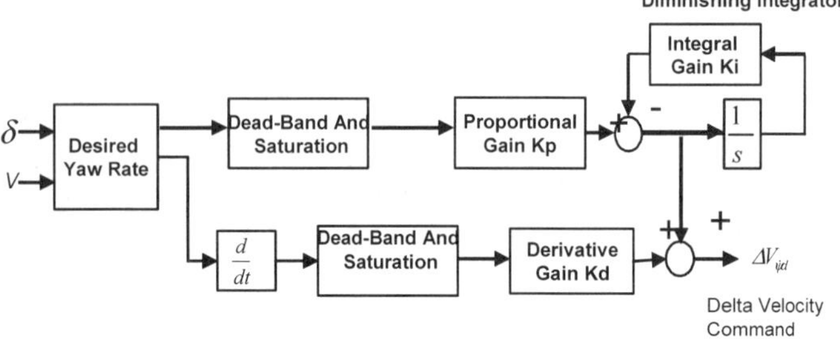

Figure 5   Block diagram of feed-forward control system

3.5 Bank Angle Compensation

When the vehicle operator is driving on a banked road, the operator introduces a correction to the steering angle to maintain the vehicle on the road. This correction by the operator may be interpreted by the command interpreter as a steering input by the vehicle operator and therefore the desired yaw command will indicate that the driver wishes to travel on the bank and not across it. In this case the active brake control might be triggered unnecessarily especially when the vehicle is in the linear range of operation. Therefore it is important to issue the yaw rate command allowing the vehicle operator to direct the vehicle to where he intended for it to go in a stable and consistent manner.

When the vehicle is equipped with lateral accelerometer and yaw rate sensor, such as in ESC equipped vehicles, bank angle compensation is accomplished by computing a bank compensation term function of yaw rate of the vehicle and the lateral acceleration. When the lateral acceleration information is not available, we compute for bank compensation without lateral acceleration information.

Define:

δ_b : the steering angle when the vehicle is driving on a banked surface.

δ_f : the steering angle with the bank effect being compensated

$$\delta_b = (\delta_f - K_u g \sin(\varphi_b)). \qquad (9)$$

When the vehicle is driving on bank surface, the desired yaw rate computed is given by

$$\dot{\psi}_{des} = \frac{V_m}{L + K_u V_m^2} \delta_b$$
$$= \frac{V_m}{L + K_u V_m^2} (\delta_f - K_u g \sin(\varphi_b)). \qquad (10)$$

To compensate for the bank effect we need to isolate the term

$$\frac{V_m}{L + K_u V_m^2} K_u g \sin(\varphi_b).$$

In the <u>linear region</u> of the vehicle operation the measured yaw rate when the vehicle is on a bank is approximately equal to:

$$\frac{V_m}{L + K_u V_m^2} \delta_f. \qquad (11)$$

The difference between the measured yaw rate, and the desired yaw rate, is the term needed to compensate for the effect of the bank.

Figures 6a and 6b respectively show the bank angle compensation for both the ESC and ETSC. Figure 6.a corresponds to vehicle operation on a road surface bank angle of 28 degrees, at a speed of 60 mph, over an interval of 25 seconds. The traces of Figure 6.b correspond to vehicle operation on a road surface where the bank angle changes from negative to positive, at a speed of 70 mph, over an interval of approximately 12 seconds. As demonstrated by the traces, the bank compensation of the ETSC provides a comparable performance, despite the absence of a lateral acceleration sensor.

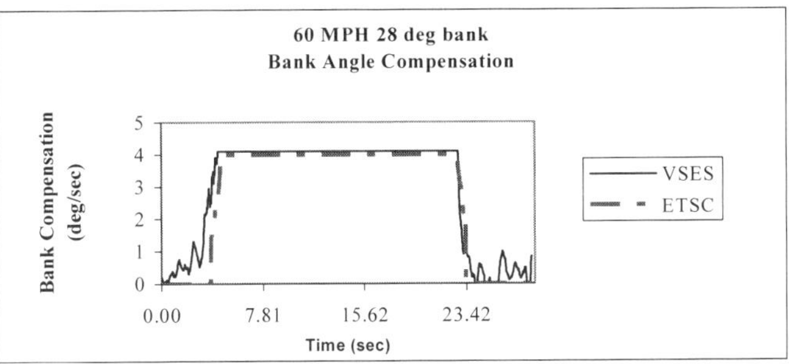

Figure 6.a Bank angle compensation using a vehicle equipped with ESC and ETSC on fixed road bank

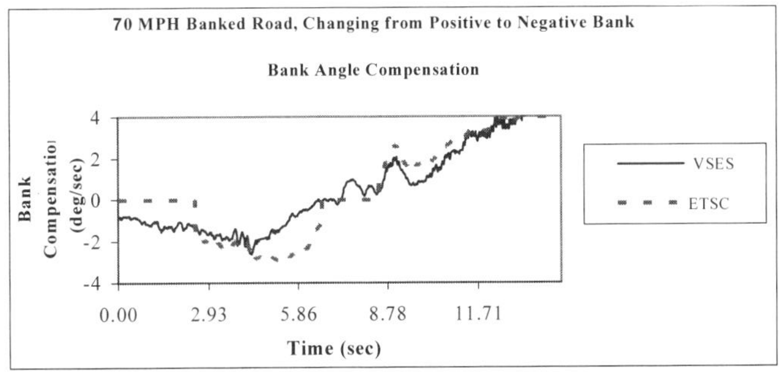

Figure 6.b Bank angle compensation using a vehicle equipped with ESC and ETSC on varying road bank

In summary, the ETSC bank angle algorithm compensates for yaw effects on banked road surface, without requiring a lateral acceleration sensor.

3.6 ENHANCED TRACTION STABILITY CONTROL
Entrance/Exit Criteria

The ESC is equipped with Entrance/Exit criteria for determining whether or not the control system is in the active brake control mode. The yaw rate error $\dot{\psi}_{error}$ and the rate of the yaw rate error $\ddot{\psi}_{error}$ are computed as follows:

$$\dot{\psi}_{error} = \dot{\psi}_d - \dot{\psi} \ , \quad \ddot{\psi}_{error} = (\dot{\psi}_{error}(t) - \dot{\psi}_{error}(t-T))/T$$

$$. \tag{12}$$

The following term $\left| G_p\dot{\psi}_{error} + G_d\ddot{\psi}_{error} \right|$ is compared to dead-band DB_{fb}.

If $\left| G_p\dot{\psi}_{error} + G_d\ddot{\psi}_{error} \right| > DB_{fb}$, and the vehicle speed is larger than certain threshold, then the entrance criteria sets the ABC flag, indicating that the active brake control is active.

In the case when the yaw rate sensor and lateral accelerometer sensors are removed, the ETSC

algorithm provides criteria to enter and exit the active brake control. The criteria have two operating schemes. The first operating scheme uses the ESC criteria to activate and terminate the active brake control. The second operating scheme uses the so-called desired yaw rate, and the rate of the desired yaw rate to determine the condition to activate and terminate the active brake control when the estimated yaw rate is determined to be valid. The ETSC uses the first scheme with the measured yaw rate replaced with the estimated yaw rate obtained from the un-driven wheel speeds. If the estimated yaw rate is determined to be invalid, the ETSC switches immediately to the second operating scheme.

In general for the Entrance condition: If the estimated yaw rate is valid, the criteria use the first scheme and the Entrance condition is determined by

$$\left| G_{p1}\dot{\psi}_{error} + G_{d1}\ddot{\psi}_{error} \right| > DB_{fb} \tag{13}$$

If the estimated yaw rate is invalid, the criteria switch to the second scheme, and the Entrance condition is determined by

$$| G_{p2}\dot{\psi}_{des} + G_{d2}\ddot{\psi}_{des} | > DB_{ff} \tag{14}$$

The dead-band, DB_{ff}, is vehicle speed dependent. Figure 7 illustrates the dead-band DB_{ff} as function of vehicle speed.

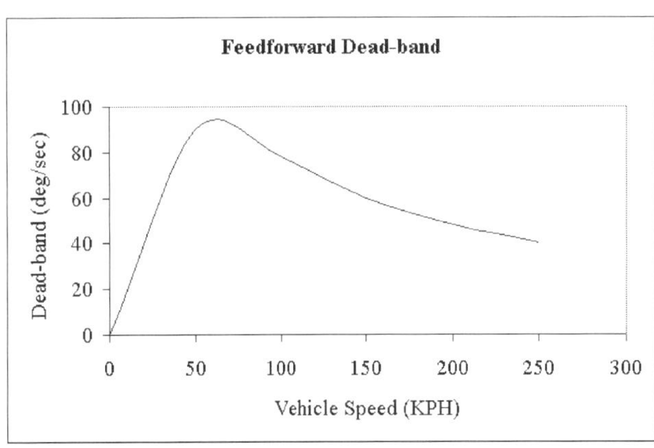

Figure 7 Dead-band for feed-forward control entrance criteria

In general for the Exit condition: If the estimated yaw rate is valid, the criteria use the first scheme and the Exit condition is determined by:

$$|\dot{\psi}_{error}| < ABC\_Exit1, \quad |\ddot{\psi}_{error}| < ABC\_Exit1\_dot \tag{15}$$

If the estimated yaw rate is invalid, the criteria switch to the second scheme, and the Entrance condition is determined by

$$|\dot{\psi}_{des}| < ABC\_Exit2, \quad |\ddot{\psi}_{des}| < ABC\_Exit2\_dot \tag{16}$$

Exit conditions for both scheme-1 and scheme-2 must be met for a predetermined length of time in order for the exit conditions to be satisfied.

4. VEHICLE TEST RESULTS

To evaluate the ETSC vehicle performance, we perform different maneuvers on various surfaces. First we compare a 45 mph double lane change maneuvers on snow with the ESC and the ETSC. Figures 8.a and 8.b illustrate the performance of the vehicle under both control strategies by comparing the vehicle yaw rate the steering angle and the side-slip in both cases

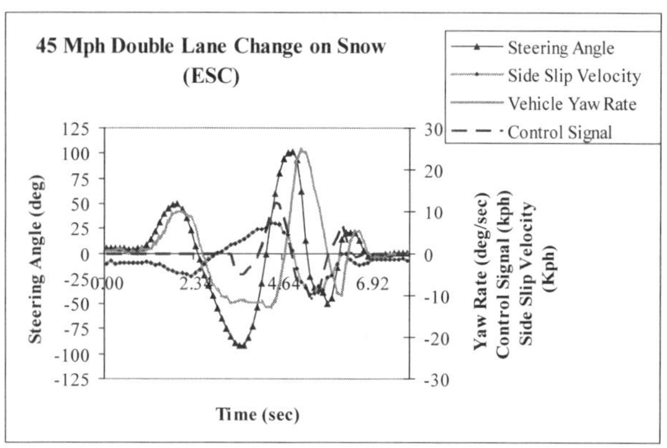

Figure 8.a Vehicle performance in a double lane change on snow with full-sensor ESC

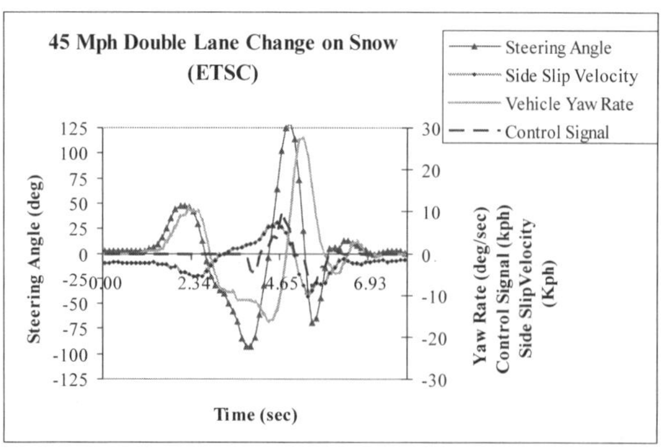

Figure 8.b Vehicle performance in a double lane change on snow with ETSC

Figures 9.a and 9.b illustrate another example of a 40 mph slalom steering on gravel road.

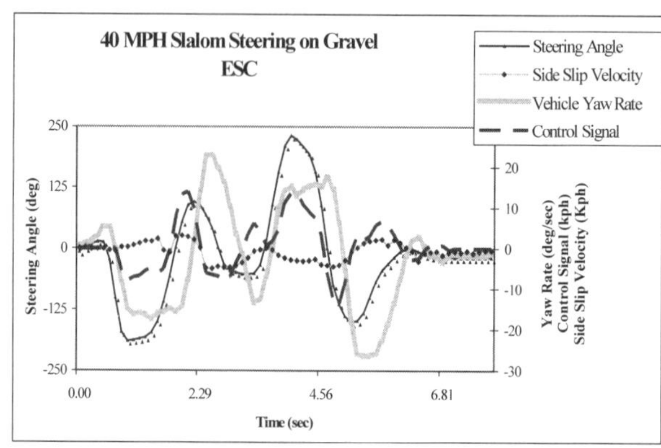

Figure 9.a Vehicle performance in a 40 mph slalom steering on gravel with ESC

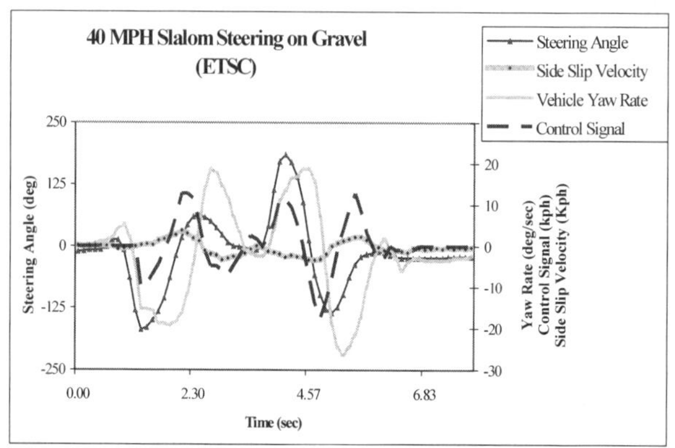

Figure 9.b Vehicle performance in a 40 mph slalom steering on gravel with ETSC

To investigate the performance of the ETSC during a braking maneuver, we perform a 45 mph single lane change on snow while braking with ESC and ETSC. The results of these two tests are shown in Figures 10.a and 10.b.

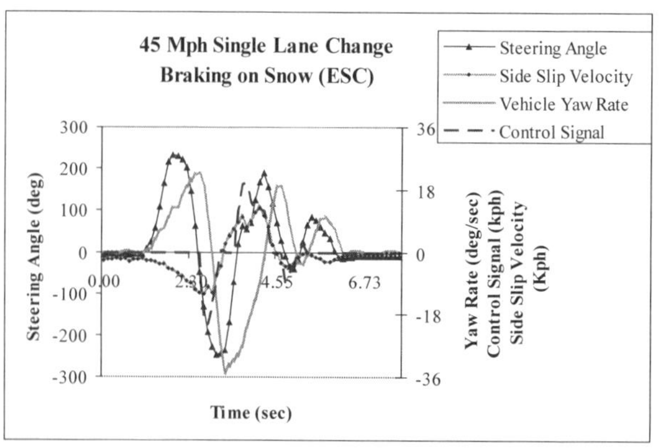

Figure 10.a Vehicle performance in a 45 mph single lane change during braking on snow with ESC

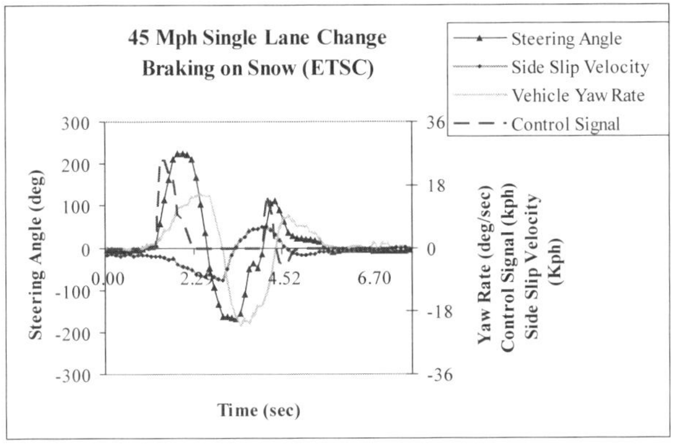

Figure 10.b Vehicle performance in a 45 mph single lane change during braking on snow with ETSC

Figure 10.b with ETSC shows a better vehicle behavior than the ESC vehicle. The reason is explained as follows. The feed-forward control is based on the driver steering angle. The steering angle leads the vehicle yaw rate in this case. Therefore, the control based on the steering angle is faster than the control based on the yaw rate. However, it is important to note that with the use of a feed-forward control, it is assumed that the driver is steering the vehicle in the right direction.

Finally, we compare the performance of both systems during a surface transition from ice to dry at 35 mph. Figures 11.a and 11.b show similar performance for both systems.

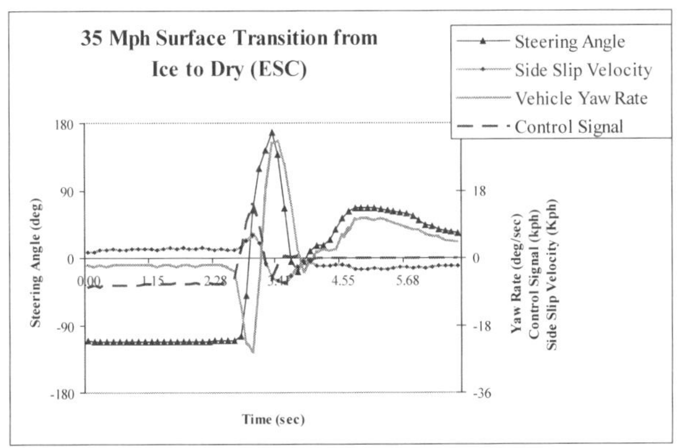

Figure 11.a Vehicle performance in a 35 mph surface transition from ice to dry with ESC

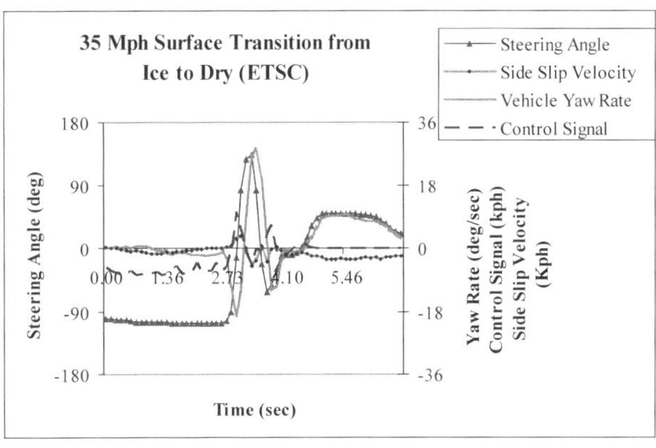

Figure 11.b Vehicle performance in a 35 mph surface transition from ice to dry with ETSC

CONCLUSION

Based on vehicle test results we conclude:

The proposed ETSC offers significant benefits to the driver in emergency maneuvers over Traction Control System. With a performance fairly close to a ESC during non-braking scenarios for two-wheel drive vehicle.

During vehicle braking, the feed-forward control of the ETSC offers the driver better vehicle stability as compared to the ABS control only.

The ETSC bank angle compensation algorithm eliminates unnecessary control activation especially when the vehicle is in the linear range of operation without lateral acceleration information while on a banked road.

REFERENCES

1. Hoffman, D., and Rizzo, M. "Chevrolet C5 Corvette Vehicle Dynamic Control System", SAE paper 980233, 1998.

2. Ghoneim, Y., Lin, W., Sidlosky, D., Chen, H., Chin, Y.-K, and Tedrake, M., "Integrated Chassis Control System to Enhance Vehicle Stability", Int. J. of Vehicle Design Vol.23, Nos.1/2, pp124-144, 2000.

3. Jost, K. (1996) 'Cadillac Stability Enhancement', Automotive Engineering, October 1996.

4. Leffler, H., Auffhammer, R., Heyken, R., and Roth, H. "New Driving Stability Control System with Reduced Technical Effort for Compact and Medium Class Passenger Cars", SAE paper 980234, 1998.

5. Nakazato, H., Iwata, K., and Yoshiyoka, Y. "A New System for Independently Controlling Braking Force Between Inner and Outer Rear Wheels", SAE paper 890835, 1989.

6. Inagaki, H., Akuzawa, K., and Sato, M. "Yaw Rate Feedback Braking Force Distribution Control With Control by Wire Braking System", Proceeding of 1992 AVEC, pp. 435-440, 1992.

7. Salman, M., Zang, Z., and Boustany, N., "Coordinated Control of Four Wheel Braking and Rear Steering", Proceeding of the 1992 American Control Conference, pp.6-10, 1992.

8. Pilutti, T., Ulsoy G., and Hrovat D. "Vehicle Steering Intervention Through Differential Braking", Proceedings of the American Control Conference, 1995.

9. Shibahata, Y., Abe, M., Shimada, K., and Furukawa, Y. "Improvement on Limit Performance of Vehicle Motion by Chassis Control", Proceedings of the 13th IAVSD Symposium on the Dynamics of Vehicles on Roads and Tracks, 1995.

10. Sawyer, C. A. "Controlling Vehicle Stability", Automotive Industries, 1995.

CONTACT

This is where main author information is typed, if desired, such as background, education, e-mail address, and web address. This is an optional section.

DEFINITIONS, ACRONYMS, ABBREVIATIONS

a_y: lateral acceleration of vehicle's center of gravity [m/s^2]

V_m: instantaneous velocity of vehicle's center of gravity [m/s]

V_i: forward velocity measured at wheel i [m/s]

δ : steering angle of the front wheels [rad]

t : half of track of the vehicle [m]

$\dot\psi$: yaw rate (yaw velocity) of vehicle [rad/s]

$\dot\psi_{est}$: yaw rate estimate of vehicle [rad/s]

$\dot\varphi$: adjusted yaw rate estimate of vehicle [rad/s]

$\ddot\psi_{est}$: rate of yaw rate estimate of vehicle [rad/s/s]

$\ddot\varphi$: adjusted rate of yaw rate estimate of vehicle [rad/s/s]

$\dot\psi_d$: desired yaw rate of vehicle [rad/s]

$\ddot\psi_d$: rate of desired yaw rate of vehicle [rad/s/s]

$K_{p\psi}$: proportional gain for yaw rate control [kph/(rad/s)]

$K_{d\ddot\psi}$: derivative gain for yaw rate control [kph/(rad/s/s)]

DB_{ff}: open-loop feed-forward control dead-band [deg/sec]

K_p: dead-band proportional gain

K_d: dead band derivative gain [1/s]

$K_{p\dot\psi d}$: proportional gain for open-loop feed-forward control [kph/(rad/s)]

$K_{d\ddot\psi d}$: derivative gain for open-loop feed-forward control [kph/(rad/s/s)]

K_u: vehicle understeer coefficient [deg/g/m/s^2]

φ_b: road bank angle [deg]

L: vehicle wheel base [m]

DB_{fb}: dead-band for yaw rate feedback control [deg/sec]

G_{p1}: proportional gain for yaw rate feedback entrance criteria

G_{d1}: derivative gain for yaw rate feedback entrance criteria [1/s]

G_{p2}: proportional gain for yaw rate Feed-forward entrance criteria

G_{d2}: derivative gain for yaw rate Feed-forward entrance criteria [1/s]

ω_i: rotational velocity measured at wheel i [rad/s]

r_i: tire radius at wheel i [m]

ω_l: rotational velocity of left un-driven wheel [rad/s]

ω_r: rotational velocity of right un-driven wheel [rad/s]

APPENDIX

(A) Adjustment of the Yaw Rate Estimation

Consider the following free body diagram

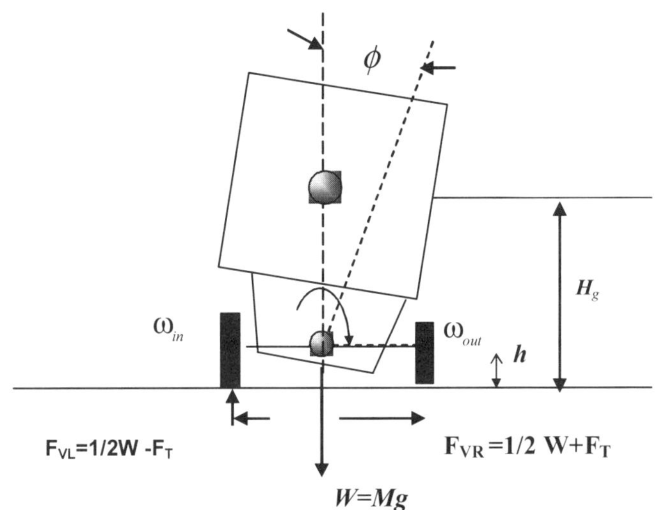

Where

M is the sprung mass

h is the height of the roll center

H_g is the height of the center of gravity

ω_{in},ω_{out} are the inside and outside tires rotational speeds

F_{VR}, and F_{VL} are the right and left wheel vertical forces

W is the weight force

T is the track of the vehicle

F_T is the lateral load transfer

F_S is the sprung mass side force=$M_S\,a_y$

Neglecting the effect of the unsprung mass, and taking the moments about the center of contact for the outside tire yields:

$$M_s a_y h - F_{VR} T + \frac{WT}{2} = M_s a_y h - F_T T = 0$$

(A.1)

where

R is the radius of the tire under normal condition (no vehicle roll)

a_y is the lateral acceleration of the vehicle

since the lateral acceleration in our case is not known it is approximated by

$$a_y \approx \dot{\phi} V_x$$

where $\dot{\phi}$ is the compensated yaw rate of the vehicle, and V_x is the vehicle speed calculated from the wheel speeds.

Therefore Equation (A.1) can be recast as:

$$F_T \approx \frac{M_s \dot{\phi} V_x h}{T}$$

(A.2)

Define the roll coefficient J_r

$$J_r = \frac{h}{H_g}$$

(A.3)

Therefore equation (A.2) can be written as

$$F_T \approx \frac{M_s \dot{\phi} V_x J_r H_g}{T}$$

(A.4)

Define

R as the unloaded radius of the tire, in practice it is measured through the unloaded tire circumference

R_l as the loaded radius of the tire, dependent due to load transfer, is the distant of the wheel axis from the ground.

The tire vertical deflection is quite closely proportional to the vertical force. We define the tire vertical stiffness k_t as

$$k_t = \frac{F_{VR} - F_{VL}}{\delta_t} = \frac{F_T}{2\delta_t}$$

(A.5)

k_t depends on the size, construction and inflation pressure. A typical value is 250N/mm. The tire vertical stiffness generally increases with load capacity, rim width.

Increasing speed increases the stiffness at about 0.004m/sec.

Hence

$$k_t = k_{to}(1+0.004V_x),$$

(A.6)

where V_x is in m/sec.

The inner and outer wheel radii are given by

$$R_i = R + \delta_t$$
$$R_{out} = R - \delta_t \tag{A.7}$$

The estimated yaw rate obtained from the non-driven wheel speeds at time k is given by

$$\dot{\varphi}_{est}(k) = \frac{R(\omega_{in} - \omega_{out})}{T} \tag{A.8}$$

The compensated yaw rate of the vehicle is given by

$$\dot{\varphi}(k) = \frac{(R - \delta_t)\omega_{in} - (R + \delta_t)\omega_{out}}{T}$$
$$= \dot{\varphi}_{est}(k) + \frac{\delta_t(\omega_{in} + \omega_{out})}{T}$$
$$\approx \dot{\varphi}_{est}(k) + \frac{\delta_t V_x}{2TR} \tag{A.9}$$

Where V_x is the vehicle speed and it is approximated by the average of the non-driven wheel speeds.

First substitute Equations (A.5 and A.6) into Equation (A.4) we get

$$\delta_t = \frac{M_s J_r H_g}{2T k_{to}(1 + 0.004 V_x)} \dot{\varphi}(k) V_x \tag{A.10}$$

Substitute Equation (A.10) into Equation (A.9)

$$\dot{\varphi}(k) = \dot{\varphi}_{est}(k) + \frac{M_s J_r H_g V_x}{(2T)^2 R k_{to}(1 + 0.004 V_x)} \dot{\varphi}(k) V_x \tag{A.11}$$

Multiply and divide the RHS of Equation (A.11) by $|\dot{\varphi}(k-1)|$

we get

$$\dot{\varphi}(k) = \dot{\varphi}_{est}(k) + K\dot{\varphi}(k)|\dot{\varphi}(k-1)|V_x \tag{A.12}$$

where

$$K = \frac{M_s J_r H_g}{(2T)^2 R k_{to}(1 + 0.004 V_x)|\dot{\varphi}(k-1)|} \tag{A.13}$$

Rearrange equation (A.7)

$$\dot{\varphi}(k) = \frac{\dot{\varphi}_{est}(k)}{1 + K V_x |\dot{\varphi}(k-1)V_x|} \tag{A.14}$$

The reason we defined the gain K as such in Equation (A.13), from our vehicle testing we found that the value of the gain K depends on the yaw rate and the vehicle velocity in a nonlinear fashion that is why we used a gain table rather than a direct computation of the gain.

Equation (14) is a rough approximation of the yaw compensation and we needed to adjust the gain K to improve the estimation.

(B) Vehicle Yaw-Rate Feedforward Control

Given the vehicle dynamic equation

$$\begin{bmatrix} \dot{v}_y \\ \ddot{\psi} \end{bmatrix} = \begin{bmatrix} a_{11} & a_{12} \\ a_{21} & a_{22} \end{bmatrix} \begin{bmatrix} v_y \\ \dot{\psi} \end{bmatrix} + \begin{bmatrix} b_{11} \\ b_{21} \end{bmatrix} \delta + \begin{bmatrix} b_{12} \\ b_{22} \end{bmatrix} \Delta f \tag{B.1}$$

the transfer functions between the steering input, δ, and yaw rate, r, can be derived:

$$G_1(s) = \frac{\dot{\psi}(s)}{\delta(s)} = \frac{b_{21}s + (a_{21}b_{11} - a_{11}b_{21})}{s^2 - (a_{11} + a_{22})s + (a_{11}a_{22} - a_{12}a_{21})} \tag{B.2}$$

The transfer function between the differential braking force, Δf, and yaw rate can be derived:

$$G_2(s) = \frac{\dot{\psi}(s)}{\Delta f(s)} = \frac{b_{22}(s - a_{11})}{s^2 - (a_{11} + a_{22})s + (a_{11}a_{22} - a_{12}a_{21})} \tag{B.3}$$

If the goal for the feedforward yaw-rate control is to have the desired yaw rate follow the steady-state command derived from the steering input, we have

$$\dot{\psi}_{ds}(t) = G_1(0)\delta(t) \tag{B.4}$$

and

$$\dot{\psi}_{ds}(s) = G_1(0)\delta(s) \tag{B.5}$$

Since the vehicle yaw rate is contributed by both the steering input and the differential braking force, we have

$$\dot{\psi}(s) = G_1(s)\delta(s) + G_2(s)\Delta f(s) \tag{B.6}$$

If the objective of the control is to achieve

$$\dot{\psi}(s) = \dot{\psi}_{ds},$$

we have

$$G_1(0)\delta(s) = G_1(s)\delta(s) + G_2(s)\Delta f(s) \tag{B.7}$$

Solving Equations (B.5), and (B.7) for $\Delta f(s)$ yields

$$\Delta f(s) = \frac{[G_1(0) - G_1(s)]\dot{\psi}_{ds}(s)}{G_2(S)G_1(0)} \tag{B.8}$$

$G_1(0)$ can be expressed by substituting the variable s in Eq. (B.2) with 0, that is:

$$G_1(0) = \frac{a_{21}b_{11} - a_{11}b_{21}}{a_{11}a_{22} - a_{12}a_{21}} \tag{B.9}$$

Substituting Eqs. (B.1), (B.2) and (B.9) into Eq. (B.8) yields

$$\Delta f(s) = \frac{s}{b_{22}}\left[1 + \frac{C}{s - a_{11}}\right] \tag{B.10}$$

where

$$C = -(a_{11} + a_{22} + \frac{b_{21}}{G_1(0)}) \tag{B.11}$$

The delta-velocity is then computed as:

$$\Delta V_{\dot{\psi}d} = \Delta f(s)\frac{1}{N_f K_u V} \tag{B.12}$$

Where N_f, K_u, and V are the normal force, the understeer coefficient, and the vehicle speed respectively.

The delta-velocity can be expressed in terms of the proportional, the derivative and the integral gain of the feed-forward control as follows:

$$\Delta V_{\dot{\psi}ds} = K_p\dot{\psi}_{ds} + K_d\ddot{\psi}_{ds} - K_p K_i(\frac{1}{s - K_i})\dot{\psi}_{ds} \tag{B.13}$$

where $K_p = \dfrac{C}{b_{22}N_f K_u V}$

$$K_d = \frac{1}{b_{22}N_f K_u V}$$

$$K_i = a_{11} \qquad\qquad\qquad\qquad \text{(B.14)}$$

These gains can be expressed in terms of the vehicle parameters as follows:

$$K_p = \frac{1}{2tV_m^2 N_f K_u}\left[\frac{I(c_{of}+c_{or})}{M}+b(bc_{or}-ac_{of})\right]-\frac{ac_{of}}{2tN_f}$$

$$K_d = \frac{I}{2tN_f K_u V_m}$$

$$K_i = -\frac{c_{of}+c_{or}}{MV_m}$$

$$\text{(B.15)}$$

Where I is the vehicle yaw moment of inertia, c_{of}, and c_{or} are the cornering stiffness of front and rear tires, M is the vehicle mass, a, and b are the distance from the vehicle center of gravity to the front and rear axles of the vehicle.

The form of the transfer function in Eq. (B.13) can be realized by a control structure composed of a derivative and proportional terms with a diminishing integrator as illustrated in Figure 5.

Vehicle System Modeling for Computer Aided Chassis Control Development

Judy Che, Mark Jennings and Alexander Zaremba
Ford Motor Company

ABSTRACT

As the complexity of automotive chassis control systems increases with the introduction of technologies such as yaw and roll stability systems, processes for model-based development of chassis control systems becomes an essential part of ensuring overall vehicle safety, quality, and reliability. To facilitate such a model-based development process, a vehicle modeling framework intended for chassis control development has been created. This paper presents a design methodology centered on this modeling framework which has been applied to real world driving events and has demonstrated its capability to capture vehicle dynamic behavior for chassis control development applications.

INTRODUCTION

The trend in the automotive industry is toward an increasing number of electronic features which are responsible for the overall dynamic behavior of a vehicle. Typically, a single vehicle program must integrate a number of electronic modules developed by a variety of sources either internally at the OEM or externally at suppliers. More and more these electronic features rely on a set of complex interactions between the powertrain, chassis, body, and other vehicle systems. Given the overwhelming complexity, there is a critical need to manage the vehicle system control development and integration process. As a result, model-based control development processes are being used more frequently to manage the complexity surrounding the integration of various electronic control features. The fundamental concepts of model-based control system design processes have been established through prior work [1], [2], [3]. Additionally, a description of the Vehicle Model Architecture (VMA) developed to support model-based Vehicle System Control design work and model-based vehicle system engineering activities in general have been discussed previously [4], [5].

This paper presents a design methodology involving a new modeling framework intended for model-based chassis control design processes. This new modeling framework has been applied to a real world driving test event to demonstrate its effectiveness for chassis control

development. The test event selected for this demonstration is a vehicle tip-in from rest on a snowy surface. The discussion begins with a description of the process for creating the modeling framework followed by details about the formulations of the individual models. Then, the procedures for setting up the vehicle simulations are presented. Finally, simulation results are compared against experimental vehicle data.

MODELING FRAMEWORK

VEHICLE MODEL ARCHITECTURE

The Vehicle Model Architecture (VMA) is the primary element of the model-based control systems design infrastructure developed at Ford Motor Company. This section contains a brief description of the Vehicle Model Architecture. A more detailed description of the VMA can be found in [4], [5].

The philosophy underlying the Vehicle Model Architecture is that a complete vehicle model including the driver and environment can be established using modular subsystems linked together with fixed I/O connections. The VMA provides a framework that enables model sharing between vehicle system engineering functions by facilitating the model integration process. The VMA is seen as a key enabler for model sharing not only between groups within a particular OEM organization, but also sharing across various OEM organizations and with supplier companies.

A key objective of the VMA is to maximize re-use of component models for system modeling activities and in particular, for vehicle system control design. This goal can be realized by defining a modeling architecture with a well defined set of interface requirements. In the context of the VMA, subsystem and component models from a variety of sources can be easily plugged into the architecture or framework provided the basic interface requirements are met. Models generated to fit those interfaces can automatically be re-used within the architecture for numerous system engineering purposes.

By adopting a generic structure aligned to physical vehicle subsystems it is easier to separate model development responsibilities and allocate them to appropriate domain experts. For example, a control engineer developing a traction control system may have extensive knowledge of braking systems but limited experience regarding engine control. In the past, the controls engineer might have been expected to produce an entire vehicle model for functional development work. By adopting a standard structure, combined with good subsystem model requirements, it is possible to create a separate work package for an engine model that can be assigned to a powertrain modeling expert. Taking an approach where subsystem models are produced and maintained by appropriate domain experts, development time will be reduced and state of the art models will be available for application to a broad spectrum of vehicle engineering activities.

The Vehicle Model Architecture consists of a high level breakdown of a vehicle into key subsystems and the definition of the interfaces between those subsystems. The architecture itself does not include any models representing system behavior. Rather, it provides a structure into which various component models can be inserted. The model architecture can be linked to libraries of subsystem and component models and corresponding parameter data. The subsystems that define the architecture would act as pointers to the underlying model libraries. Figure 1 illustrates this intended relationship between the architecture and model libraries.

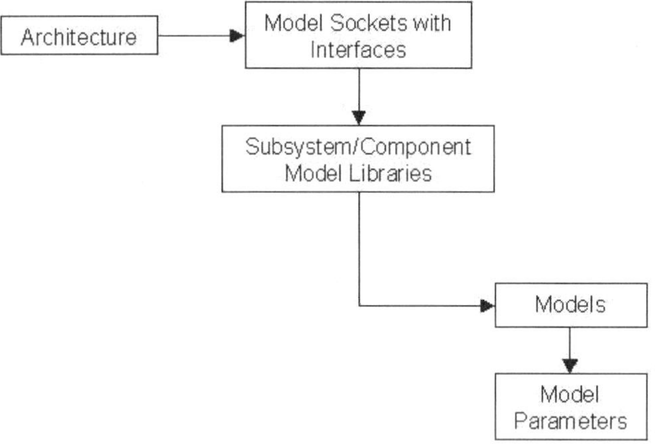

Figure 1: Relationships between model architecture, model libraries, and model parameter data

Top Level Subsystems

The VMA is defined by a highly modular, top-level subsystem breakdown of the vehicle including the signal connections between these subsystems. The top-level subsystems are:

- Driver
- Environment
- Electrical
- Auxiliaries
- Powerplant
- Transmission
- Driveline
- Chassis
- Braking
- Steering
- Vehicle System Controller
- Bus

The subsystem definitions do not prescribe any particular hardware implementation. Together the subsystems represent a conceptual breakdown of the vehicle according to key functions. The physical, hardware-based subsystems, *i.e.* Electrical, Auxiliaries, Powerplant, Transmission, Driveline, Chassis, Braking, and Steering, each have a local controller with the ability to accept and generate electrical control signals. Input control signals can modify internal subsystem states, while output signals can be used by other subsystems.

Subsystem Interfaces

In addition to specification of the top-level subsystems, the VMA specification defines the interfaces between these subsystems. The modular subsystems of the VMA are linked together with fixed I/O connections. Expansion of the model is possible by adding to, but never replacing, the fixed I/O with supplementary links.

An important feature of the VMA interfaces is a well-defined bus structure designed to distribute signals between subsystems. The bus structure enables efficient tracking of signals through the system, which simplifies model development, debugging, and browsing processes. Also, the signals can be added to the buses without changing subsystem interfaces. Signals are distributed to the top-level subsystems by the four large-scale signal buses defined in Table 1.

Global Plant Bus	The Plant Bus carries the physical inputs required by plant models located in each model top-level subsystem.
Global Controller Bus	The Controller or Communications Bus carries mixed control signals between the controller models located in each model top-level.
Global Driver Bus	The Driver Bus carries signals unique to the human driver-vehicle system interface.
Global Environment Bus	The Environment Bus carries signals unique to the environment-vehicle system interface.

Table 1: Signal buses for the Vehicle Model Architecture

Figure 2 shows the Matlab/Simulink®[1] implementation of the Vehicle Model Architecture with its breakdown into the key subsystems and the supporting bus structure.

---

[1] Matlab and Simulink are registered trademarks of the Mathworks Inc, Natick, MA.

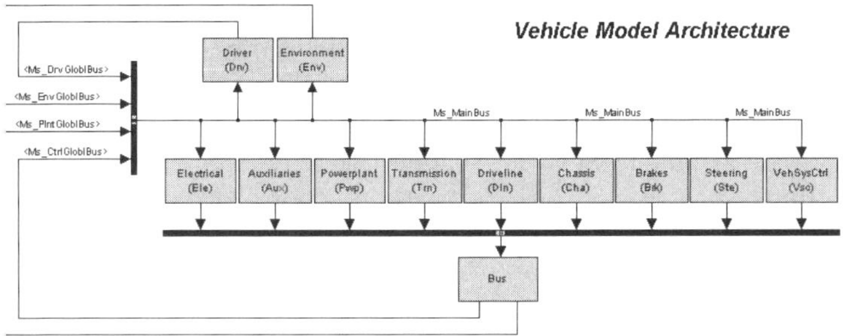

Figure 2: Top level implementation of the VMA in a Matlab/Simulink® environment

VEDYNA® MODELING FRAMEWORK

Another commonly used vehicle modeling framework for chassis control development is the veDYNA®[2] simulation tool set offered commercially by Tesis. The veDYNA® vehicle represents a multi-body system model that can accurately represent chassis dynamic response to a driver input on a variety of road surfaces. Figure 3 shows the system breakdown for the veDYNA® "splitted" vehicle model along with its interface and distributed bus structure in a MATLAB/Simulink® environment [6].

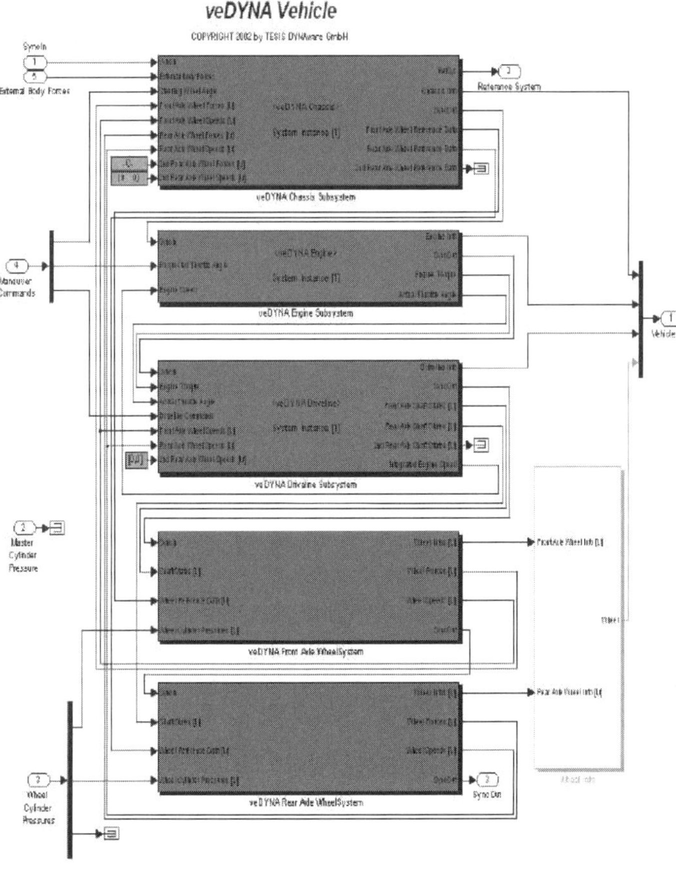

Figure 3: veDYNA® "splitted" vehicle model framework in a MATLAB/Simulink® environment

---

[2] veDYNA is the registered trademark of TESIS *DYNA*ware Technische Simulation Dynamischer Systeme GmbH, München, Germany.

The veDYNA® "splitted" vehicle model is broken down into the following key subsystems:

- Chassis
- Engine
- Driveline – including transmission
- Front Axle Wheel System
- Rear Axle Wheel System

INTEGRATED VMA - VEDYNA® FRAMEWORK

The key to the design methodology presented in this paper is the creation of a vehicle analysis framework tailored specifically to chassis control development applications. Creation of such a tool involves the integration of elements of the VMA modeling framework with those of the veDYNA® modeling framework. Within this combined framework, the veDYNA® chassis and front and rear axle wheel systems are combined into a superblock to represent the corresponding VMA chassis, brakes, and steering subsystems following guidelines set forth in the VMA Specification [2]. Additionally, the veDYNA® maneuver controller and road models have been implemented for the driver and environment subsystems, respectively. The signal interfaces for this combined framework comply with the VMA framework. Figure 4 shows the resulting combined VMA / veDYNA® vehicle modeling framework in a Matlab/Simulink® environment.

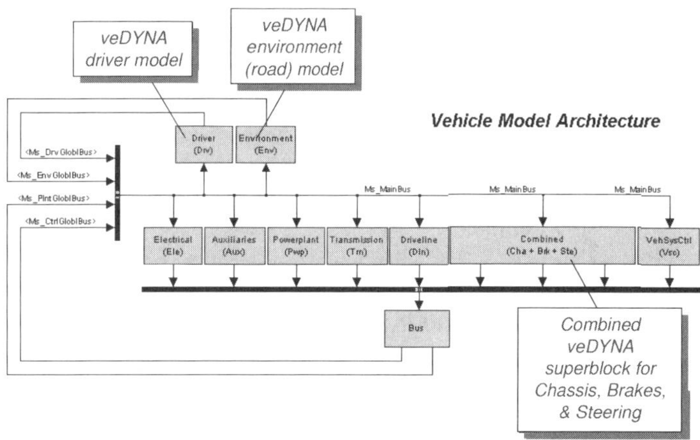

Figure 4: Combined VMA / veDYNA® vehicle modeling framework in a Matlab/Simulink® environment

MODEL FORMULATION

BASELINE VEHICLE FRAMEWORK

The creation of a modeling framework tailored to chassis control development begins with the assembly of a baseline vehicle framework under the Vehicle Model Architecture. For the methodology described in this paper, the Ford Powertrain model, referred to as the FPT model, is used to create the baseline framework. The FPT model is a Ford corporate standard tool used to predict fuel economy and performance for production vehicle programs. The fact that the FPT model supports vehicle programs makes it an excellent starting point for full vehicle behavioral modeling because of the extensive parameter database associated with it. The FPT model is compliant with the Vehicle Model Architecture and has been implemented in the Matlab/Simulink® environment, which is consistent with accepted tools and methods for control design. The individual FPT vehicle subsystem models shown in Figure 2 can be described as follows:

- Electrical - The primary function of the electrical system for vehicles powered by IC engines is to determine an alternator torque load for a given vehicle operating point.
- Auxiliary - The auxiliary system includes models of the engine driven accessories (*i.e.* mechanical fan, power steering pump, etc.) and models of the engine auxiliaries (*i.e.* oil pump, etc.).
- Engine - The engine model relies on calibration specific engine data files and ensures that the engine operates within its speed and torque constraints. For the conventional IC engine, the engine system combines the accessory torque from the auxiliary system model, the engine inertial torque, and the engine torque request to determine the required engine output torque.
- Transmission - The transmission system model involves the computation of transmission losses and the transformation of the engine output speed and torque into driveline input speed and torque. Computation of the transmission losses accounts for front pump load and spin and load based gear losses.
- Driveline - The driveline model transforms the transmission output speed and torque into driveline output speed and torque accounting for driveline inertial effects and driveline losses. Driveline models are available to simulate front-wheel, rear-wheel, four-wheel, and all-wheel drive vehicles.
- Chassis - The chassis system incorporates the component models necessary to represent the front, rear, and trailer wheels in the multipurpose wheel system.

Once the baseline VMA modeling framework has been created and populated with the FPT model components, individual subsystem components can be enhanced to meet the modeling requirements for a specific application. Typically higher fidelity plant and controller models capable of handling the complex powertrain and chassis interactions are required for chassis control development. The details of the enhanced models for the powerplant, driveline, chassis, brakes, steering, driver, and environment subsystems are presented in the following sections.

VEDYNA® CHASSIS, BRAKES, STEERING, DRIVER, AND ENVIROMENT MODELS

The enhanced chassis, brakes, steering, driver, and environment systems to be integrated within the VMA framework come from the veDYNA® simulation tool set. The veDYNA® vehicle is a multi-body system model that can accurately represent chassis dynamic response to driver inputs on a variety of road surfaces. These veDYNA® subsystems, also implemented in a MATLAB/Simulink® environment, have the appropriate level of fidelity and are widely used for chassis control development. The following section gives a brief description of the veDYNA® chassis, brakes, steering, driver and environment models. Refer to [6], [7], [8], [9] for a detailed description of the model formulations.

veDYNA® Chassis Model

In veDYNA®, the chassis is modeled as a multi-body system shown in Figure 5.

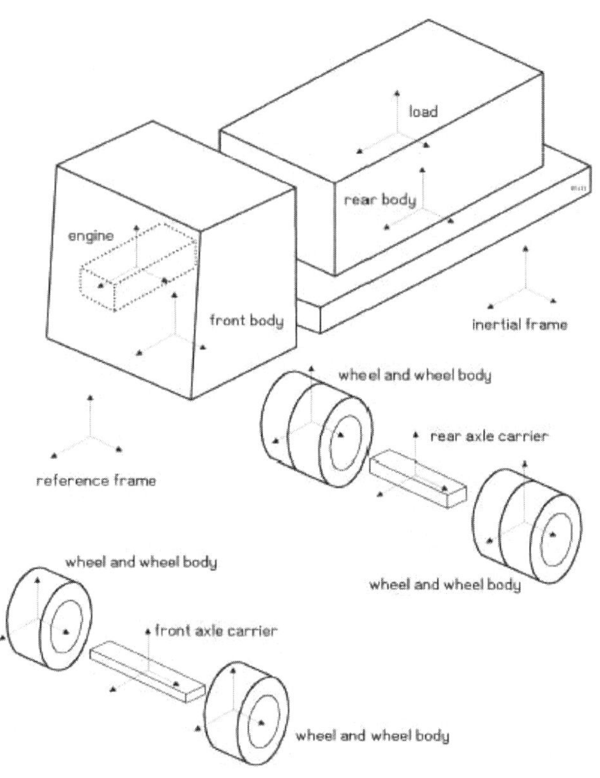

Figure 5: veDYNA® vehicle chassis multi-body system model configuration from [7]

Here, the bodies are connected by means of springs, dampers, bushings and rigid joints. The standard veDYNA® chassis elements are defined as follows [7]:

- Front body - Represents the main vehicle body which is assigned 6 degrees of freedom (DoF).
- Front axle - Contains three DoF including left and right wheel lift and steering angle. It consists of a front left and front right wheel body and a front axle carrier.
- Rear axle - Contains two DoF including left and right wheel lift. It consists of a rear left and rear right wheel body and a rear axle carrier.
- Wheel - Each wheel rotation is described as a one DoF system.
- Optional rear body - Contains one torsional DoF to model torsional stiffness of the vehicle body.

- Optional second rear axle - Contains two DoF including left and right wheel lift and consists of a second rear left and rear right wheel body.
- Optional free body - Accounts for a load, a driver body, or a cabin in the front of the vehicle and is assigned 6 DoF.
- Optional engine body - Contains 6 DoF.

veDYNA® Tire Model

The veDYNA® toolset contains a semi-empirical tire model called TM-Easy which represents the longitudinal and lateral tire force characteristics and the self-aligning torque as functions of a longitudinal slip and/or a lateral slip. Longitudinal and lateral tire properties are described by a few physical parameters such as tire radius, width, and stiffness and damping coefficients. Within TM-Easy, parameters describing the force characteristics can be calculated from empirical tire force curves.

veDYNA® Brakes Model

In the veDYNA® brakes model, the brake torque produced at each wheel, τ_{brk}, is calculated as follows:

$$\tau_{brk} = 2 * (P_{brk} * \mu * A_{brk} * R_{brk}) \quad (1)$$

where P_{brk} is the brake pressure, A_{brk} is the area of a single pad, R_{brk} is the brake radius and μ is the friction coefficient between brake pad and brake disk.

veDYNA® Steering Model

In the veDYNA® kinematic steering model, the steering attributes are defined in the form of kinematic tables. The steering ratio is a function of the steering input to the axle kinematics tables. The veDYNA® dynamic steering model also accounts for the stiffness and damping of the steering wheel column.

veDYNA® Road Model

The veDYNA® road model is comprised of both standard and advanced representations [7],[8]. The standard road characteristics include a road grade, a friction coefficient, and surface unevenness profile while the advanced road model includes a full definition of a horizontal layout and a vertical profile. The standard road is treated as an infinite plane with the following features:

- Grade - A grade is superimposed on the profile.
- Friction coefficient - The friction coefficient can be set on sections of the road by either applying a simple scaling factor or by modifying the tire data set to represent specific road conditions.
- Unevenness profile - A road surface unevenness profile can be applied to the entire road and can be defined using a Power Spectral Density.

An advanced road representation can be implemented by creating an entire three-dimensional road as a series of consecutive segments with specific characteristics. The definition of an advanced road contains both a horizontal path description as well as a vertical profile including road friction characteristics.

veDYNA® Driver Model

The veDYNA® driver model is defined with respect to the type of road model used. For a standard road profile, the veDYNA® maneuver controller can be used, while an advanced veDYNA® driver is required for an advanced road representation [7], [9].

The maneuver controller on a standard road can be used in either an open-loop or a closed-loop mode. In the open-loop mode, external data for the steering wheel angle, engine throttle, brake position and gear position can be input into the model. In the closed-loop mode, a combination of longitudinal and lateral controllers can be used to control vehicle speed, acceleration, steering, and overall trajectory. These closed-loop controls are implemented by means of PI and PID regulators.

The advanced driver model in veDYNA® is required to drive the vehicle along an advanced road profile. This driver attempts to keep the vehicle on a target point moving along a given target path.

DYNAMIC DRIVELINE MODEL

Full integration of the veDYNA® chassis model with the FPT transmission model requires the creation of a new dynamic driveline model shown in Figure 6.

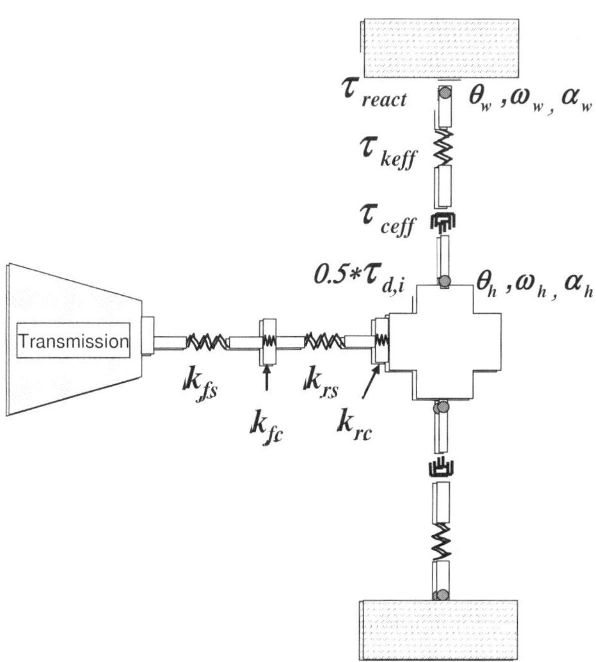

Figure 6: Dynamic drivetrain model configuration

This dynamic driveline model includes distributed and finite elasticity of the prop-shaft and distributed elasticity of the left and right half-shafts. The rear differential is modeled as a planetary gear set.

<u>Model Formulation</u>

The torque cascaded downstream from the transmission output shaft is divided evenly between the left and right half-shafts, so that:

$$\tau_{hf,l} = \tau_{hf,r} = \frac{1}{2}\tau_{d,i} \qquad (2)$$

where $\tau_{hf,l}$ and $\tau_{hf,r}$, are the torques at the left half-shaft and right half-shaft, respectively, and $\tau_{d,i}$ is the input torque to the differential. For the left and right sides, the reaction torque at each rear wheel, τ_{react}, can be computed by subtracting the elastic and damping torques from the half-shaft input torque τ_{hf}:

$$\tau_{react} = \frac{1}{2}\tau_{d,i} - \tau_k - \tau_c \qquad (3)$$

Here τ_k and τ_c, the torques due to the driveline stiffness and damping, respectively, can be computed as:

$$\tau_k = k_{eff}(\theta_w - \theta_h) \qquad (4)$$

and

$$\tau_c = c_{eff} \cdot \omega_h \qquad (5)$$

where k_{eff} is the effective driveline stiffness coefficient, c_{eff} is the effective driveline damping coefficient, ω_h is the rotational speed of the half-shaft, and θ_h and θ_w are the angular position of the half-shaft and the wheel, respectively. The effective stiffness coefficient for the left and right half-shafts can be calculated as follows:

$$k_{eff,l} = \frac{k_p R_{fd}^2 \cdot k_{hf,l}}{k_{hf,l} + k_p R_{fd}^2} \qquad (6a)$$

and

$$k_{eff,r} = \frac{k_p R_{fd}^2 \cdot k_{hf,r}}{k_{hf,r} + k_p R_{fd}^2} \qquad (6b)$$

where R_{fd} is the final drive ratio, $k_{hf,r}$ is the stiffness of the right half-shaft, $k_{hf,l}$ is the stiffness of the left half-shaft, and k_p is the combined stiffness of the prop-shaft which can be calculated from the stiffness of its components:

$$(k_p)^{-1} = \frac{1}{k_{fs}} + \frac{1}{k_{fc}} + \frac{1}{k_{rs}} + \frac{1}{k_{rc}} \qquad (7)$$

In Equation 7 k_{fs}, k_{fc}, k_{rs}, and k_{rc} are the stiffness coefficients for the front shaft, front coupling, rear shaft, and rear coupling, respectively.

The rotational accelerations at the left and right half-shafts, $\dot{\omega}_{hf,l}$ and $\dot{\omega}_{hf,r}$, respectively, can be calculated by applying LaGrange's Method [10] to define the dynamics of the planetary gear set as:

$$\dot{\omega}_{hf,l} = \frac{\tau_{hf,l}(J/4 + J_{eff,r}) - \tau_{hf,r}(J/4)}{(J/4)(J_{eff,r} + J_{eff,l}) + J_{eff,r} \cdot J_{eff,l}} \qquad (8a)$$

and

$$\dot{\omega}_{hf,r} = \frac{\tau_{hf,r}(J/4 + J_{eff,l}) - \tau_{hf,l}(J/4)}{(J/4)(J_{eff,r} + J_{eff,l}) + J_{eff,r} \cdot J_{eff,l}} \qquad (8b)$$

where J is the effective inertia of the input pinion;

$J_{eff,l}$ is the effective inertia of the planetary gear set connected to left half-shaft;

$J_{eff,r}$ is the effective inertia of the planetary gear set connected to the right half-haft.

The effective inertia values $J_{eff,l}$ and $J_{eff,r}$ are:

$$J_{eff,l} = J_l + J_r \qquad (9a)$$

and

$$J_{eff,r} = J_r + J_s \qquad (9b)$$

where J_r and J_s are the inertia values of the ring and sun of the planetary set, respectively, and J_l and J_r are the inertia values of the left and right half-shafts, respectively.

POWERPLANT MODEL

Three powerplant models were considered for this chassis modeling application. The first of these models comes from the veDYNA® toolset, while the second is the basic Ford Powertrain Model (FPT) engine model. The third model with the highest level of fidelity of those considered is a Ford internally developed model, referred to as a Control-Oriented Engine model (COE). In both the FPT and the veDYNA® engine models, the engine torque is interpolated from the steady-state engine map tables as a function of the throttle angle and the engine speed. The engine dynamics are represented by a 1st order time delay. The first two engine models considered are not adequate for chassis control development because they do not allow for fast engine torque reduction due to transmission and/or anti-lock braking (ABS) torque modulation requests. Fast torque control refers to the manipulation of either spark timing or fuel flow rates while base torque control refers to the control of throttle air flow. Fast torque control functionality in the engine model is required to represent accurately the dynamic interaction between subsystems for chassis control development.

The COE engine model was selected for this application primarily because it allows for both fast and base engine torque control. At its core, the COE model relies on steady-state engine mapping data with added dynamic elements representing the physics of manifold filling. However, a trade-off arises in implementing any higher fidelity model because of the increased parameter data requirement to support the model for any target vehicle application.

COE Engine Plant Model

The COE engine model is derived from the work of [12] and formulated as a spatially-lumped, cycle-averaged, quasi-steady representation of a spark-ignition engine. This control-oriented model captures continuous, nonlinear, low-frequency physics within the engine. The COE model primarily relies on regressions of engine dynamometer data to characterize throttle, engine, and exhaust gas recirculation (EGR) flows as well as torque generation phenomena. The formulation for manifold filling is based on the plenum air flow model described in [11]. The continuity equation for mass air flow into and out of the plenum is formulated as:

$$\frac{dm_{a,m}}{dt} = \dot{m}_{a,th} + \dot{m}_{a,egr} - \dot{m}_{a,eng} \quad (10)$$

where $m_{a,m}$ is the mass of air in the plenum, $\dot{m}_{a,th}$ is the mass air flow rate past the throttle body, $\dot{m}_{a,egr}$ is the EGR flow rate, and $\dot{m}_{a,eng}$ is the mass air flow rate into the engine.

The manifold pressure, P_m, and temperature T_m, are defined based on the gas energy conservation law as:

$$\frac{dP_m}{dt} = \frac{\gamma R T_a}{P_a V_m}(\dot{m}_{a,th} + \frac{T_e}{T_a}\dot{m}_{a,egr} - \frac{T_m}{T_a}\dot{m}_{a,eng}) \quad (11)$$

and

$$\frac{dT_m}{dt} = \frac{\gamma R T_a}{V_m} \cdot \frac{T_m}{P_m}(\dot{m}_{a,th} + \frac{T_e}{T_a}\dot{m}_{a,egr} \\ - \frac{T_m}{T_a}\dot{m}_{a,eng} - \dot{m}_{a,m}\frac{T_m}{\gamma T_a}) \quad (12)$$

where V_m is the manifold volume, R is the universal gas constant, T_e is the EGR gas temperature, T_a is the ambient temperature, P_a is the ambient pressure, and γ is the adiabatic gas constant.

Finally, the throttle, EGR, and engine flows are formulated based on empirical engine data. For example, engine flow for an IC engine with variable cam timing (VCT) can be characterized as follows:

$$\frac{dm_{a,eng}}{dt} = k_0(\theta_c) + k_P(\theta_c) \cdot P_m \\ + k_W(\theta_c) \cdot N + k_{PW}(\theta_c) \cdot P_m N \quad (13)$$

where $k_0(\theta_c)$, $k_P(\theta_c)$, $k_W(\theta_c)$, and $k_{PW}(\theta_c)$ are the regression coefficients as a function of the cam position, θ_c is the cam shaft angle, and N is the engine speed.

The engine torque generation model contains maps of the hot engine brake torque and the friction torque mapped against speed and load for nominal air fuel ratio (AFR), variable cam timing (VCT), and maximum brake torque (MBT) spark. Empirical correction factors are used to account for torque changes due to AFR and VCT variations and spark retard. The friction torque, Tq_{fric}, is derived from the measured motoring torque on a reference engine and can be characterized by 3rd order polynomial of the manifold pressure and engine speed as:

$$Tq_{fric} = k_0 + k_1 \cdot P_m + k_2 \cdot N + k_3 \cdot P_m N \\ + k_4 \cdot N^2 + k_5 \cdot P_m N^2 + k_6 \cdot N^3 \quad (14)$$

where $k_0,...,k_6$ are regression coefficients.

COE Engine Controller Model

The engine controller model represents the main engine control features which include:

- Idle speed control
- Fuel injector control
- Spark retard calculation
- Fuel cut-off strategy
- Target Lambda calculation
- EGR control
- VCT control
- Throttle angle control

The control modules listed above are designed to accept engine control calibration maps and constraints. The throttle control module is capable of accommodating an engine torque control strategy in which an accelerator pedal signal is interpreted as the engine torque request. After the torque arbitration which considers engine torque modulation requests from the transmission, ABS brakes, and an integrated vehicle control (IVC) system, the final base and fast torque commands are sent to the engine controller. Special throttle angle maps were developed to calculate the throttle command from the base torque request and the engine speed. The fast torque control, typically implemented by either spark retard or fuel cut-off, is used to satisfy the transmission, ABS, and IVC torque modulation commands.

SIMULATION SETUP AND RESULTS

In this section, simulation results for a vehicle test event are presented to demonstrate the benefits of the modeling approach described previously. The vehicle model based on the combined VMA/veDYNA® modeling framework will be referred to as the Enhanced Chassis Control (ECC) model. Simulation results for the ECC vehicle model are validated against experimental vehicle test data and compared against veDYNA® vehicle model simulation results. The comparison of the ECC model results against the veDYNA® simulation results are presented to highlight the benefit of implementing an enhanced COE engine model in the context of the combined VMA/veDYNA® modeling framework which provides a standard approach to accommodate complex system-to-system interactions. The test event selected for this exercise is a vehicle tip-in from rest on a snowy surface.

SIMULATION SETUP

The key to validating results for behavioral vehicle simulations is to 'drive' the vehicle model in the same way as the experimental test vehicle so that vehicle trace parameters can be compared. For these simulations, a number of vehicle trace parameters from the experimental test data are used as external inputs to drive the simulation models. The vehicle input parameters used for these simulations include:

- Engine throttle angle
- Transmission gear shifting time
- Brake cylinder pressures for each wheel
- Transmission fast-torque reduction request to the engine during shifting (ECC model only)

Figures 7a and 7b show some of the key vehicle test parameter time traces for the tip-in event from rest on a snowy surface. Initially, the vehicle is at rest with the brakes depressed. At one second into the event, the driver releases the brakes and tips into the accelerator.

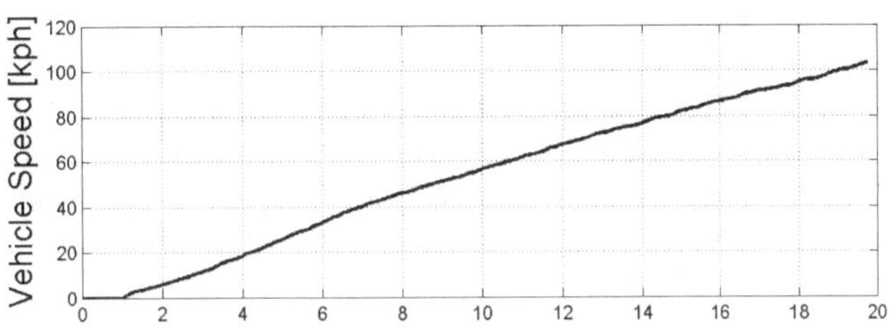

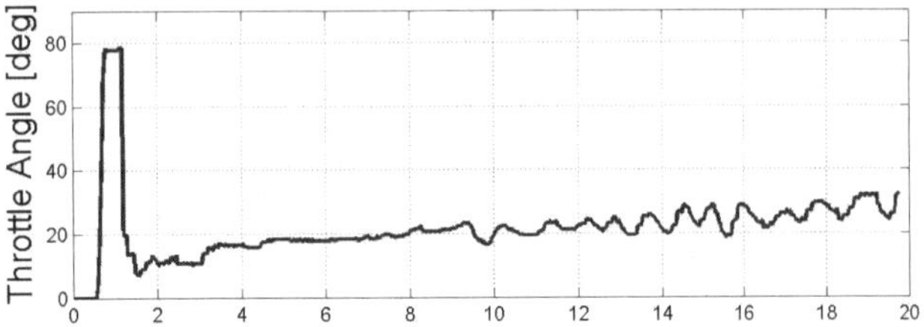

Figure 7a: Experimental vehicle speed and throttle angle time traces for the tip-in event from rest on a snowy surface

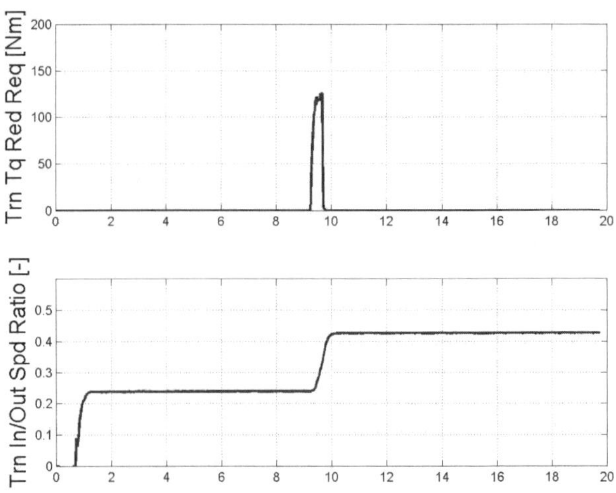

Figure 7b: Vehicle test parameters for a tip-in event from rest on a snowy surface

In some circumstances, the experimental driver steering wheel angle time trace data can also be used as an input for the vehicle simulation. Figure 8 shows the experimental driver steering wheel angle trace for the tip-in event from rest on a snowy surface.

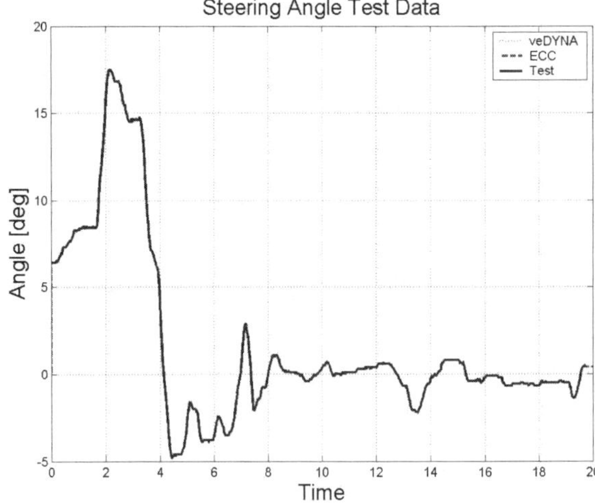

Figure 8: Experimental driver steering wheel angle trace from the tip-in event from rest on a snowy surface

For this particular test event, the experimental steering angle trace input caused the veDYNA® vehicle model to spin out, thus the experimental steering angle data is not suitable to drive these simulations. Figure 9 shows the longitudinal and lateral vehicle speed traces for the experiment test vehicle in green, for the ECC simulation results in red, and for the veDYNA® simulation results in blue. Here, the longitudinal velocity for the veDYNA® simulation goes negative indicating that the vehicle has spun 180° and is traveling in the negative X-direction. The negative lateral velocity means that the vehicle is traveling in the negative Y-direction as defined by the veDYNA® coordinate system. Notice that the ECC model was able to maintain a straight-line trajectory using the same experiment steering input.

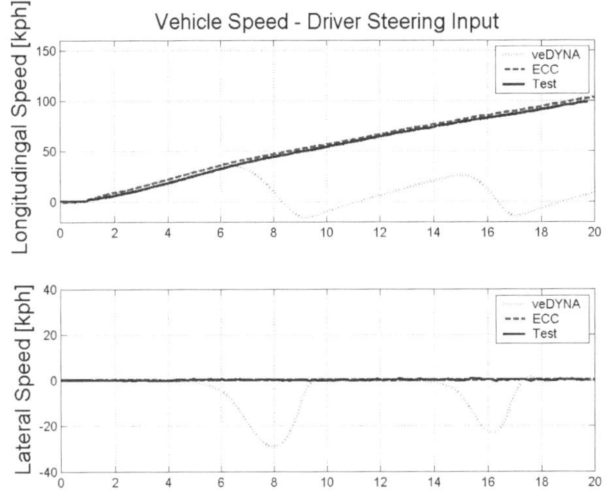

Figure 9: Longitudinal and lateral vehicle speed traces using experimental driver steering wheel angle as simulation input

For the simulation results presented in this paper, the veDYNA® straight-line maneuver controller was used to steer the vehicle models. The straight-line maneuver controller in veDYNA® uses a PI controller to minimize the lateral deviation of a vehicle reference point relative to a given trajectory.

In these simulations, the brake cylinder pressures for each wheel are used as inputs to mimic the effect of the vehicle ABS and traction control systems which are not included in the vehicle model. Figure 10 shows the brake pressure traces for each of the four wheels from the experimental vehicle test data.

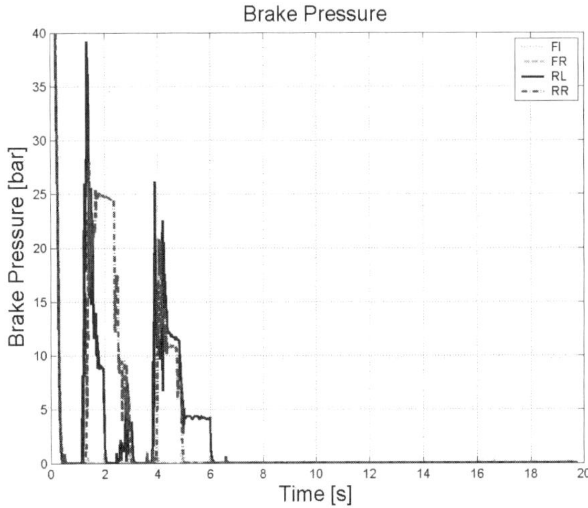

Figure 10: Experimental brake pressure traces for each for the four wheels

These vehicle simulations also use the veDYNA® standard road model which treats the road as an infinite plane with a characteristic grade, unevenness profile, and friction coefficient. For this test event, the road is characterized as a flat, smooth surface with a constant coefficient of friction of 0.45 to represent the snow covered road.

SIMULATION RESULTS

Vehicle simulation results for both the ECC vehicle model and the standard veDYNA® vehicle model for a tip-in event from rest on a snowy surface are presented in this section. Figure 11 shows the vehicle speed and the speeds of the four individual wheels for the experimental test vehicle. As expected for a rear-wheel drive vehicle, the left and right front wheel speeds line up well with the overall vehicle speed. The rear wheels, on the other hand, show rapid accelerations that indicate spinning especially early in the event as the vehicle begins to accelerate from rest.

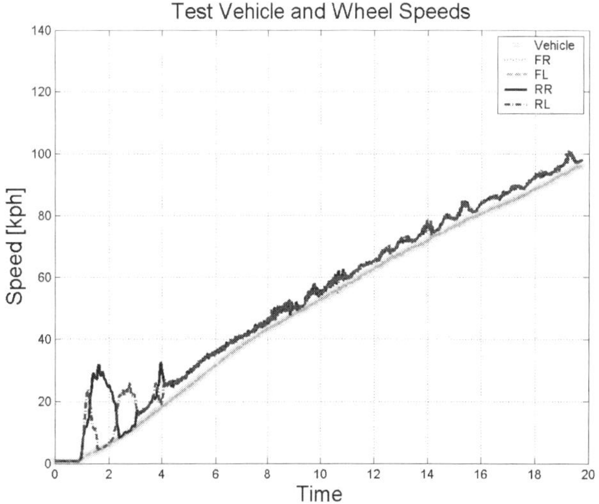

Figure 11: Experimental traces for vehicle speed and the speed for each of the four wheels

One of the key features of the enhanced COE engine model is the capability for fast engine torque control through spark timing adjustment. During the tip-in event from rest, the transmission sends the engine a request for fast torque reduction during a gear shift event. Figure 12 shows a number of torque signal traces from the experimental vehicle test including the driver demand torque, base and fast torque requests from the chassis controller, the transmission torque reduction request, the engine indicated torque, the friction torque, and the adjusted engine brake torque. The adjusted engine brake torque is the indicated torque minus the friction torque minus the transmission torque reduction request. The spike in the transmission torque reduction request at approximately nine seconds coincides with a transmission shift event.

The combined VMA/veDYNA® framework presented in this paper provides a standard approach to account for the system-to-systems interactions represented by the transmission torque reduction request to the engine. The basic veDYNA® modeling framework, on the other hand, does not provide a standard method for dealing with these interactions. In these simulations, the transmission torque reduction request to the engine is accounted for in the ECC model simulation results but not in the veDYNA® simulation results. The justification for such an assumption comes from the fact that torque modulation can be naturally achieved in the ECC model whereas it is not currently a feature available in the veDYNA® modeling tool.

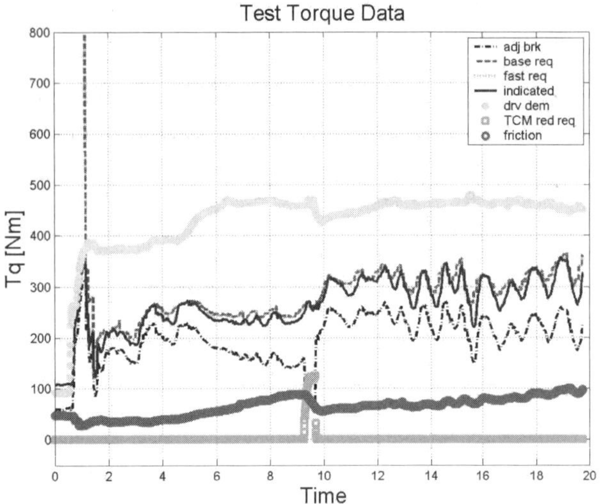

Figure 12: Experimental vehicle torque signal traces

Figure 13 shows the vehicle speed traces for the experimental test vehicle in green, for the ECC simulation results in red, and for the veDYNA® simulation results in blue. All three vehicles accelerate at a similar rate early in the vehicle test event. At approximately nine seconds, the vehicle speed of the veDYNA® simulation diverges significantly from both the ECC simulation results and the experimental vehicle test data. This deviation in the vehicle speed can primarily be attributed to the different treatment of the transmission torque reduction request as the transmission shifts from first to second gear during this time.

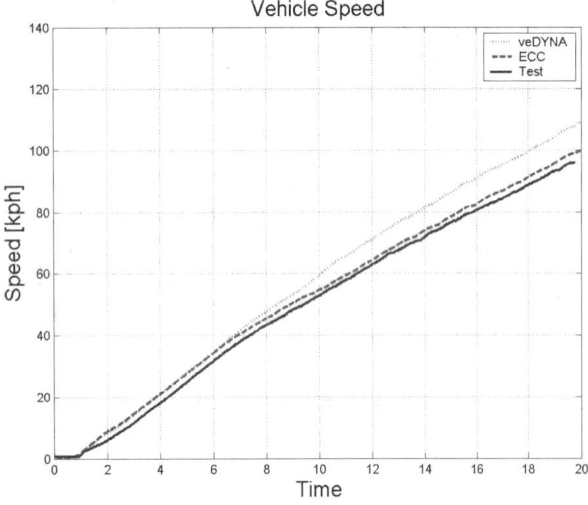

Figure 13: Vehicle speed trace for the tip-in event from rest

Figure 14 shows the correspondence in the ratios of the transmission output speed to the input speed for the test vehicle and both of the simulation models.

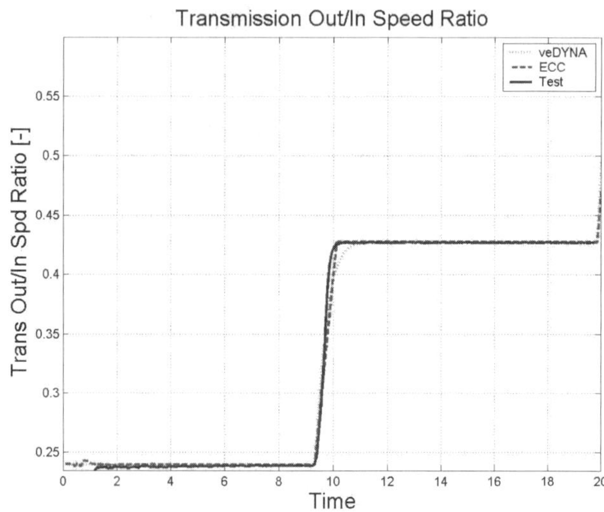

Figure 14: Ratio of the transmission output speed to the transmission input speed

Figure 15 shows a comparison of the engine speed for the test vehicle and the two simulation models. Both the VMA model and the vehicle test data show that the engine idles at approximately 600 rpm whereas the veDYNA® engine model begins the test with the engine initially at rest.

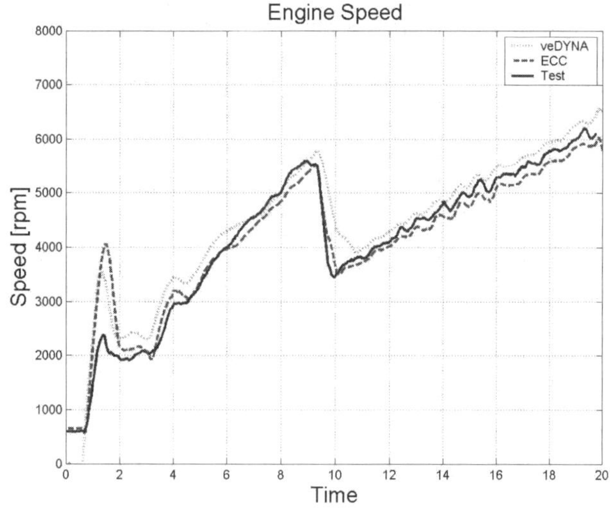

Figure 15: Engine speed trace for the tip-in event from rest

In these test cases, both the ECC model simulation results and the experimental vehicle test data account for the reduction in engine torque while the veDYNA® simulation results do not. Given these circumstances, the veDYNA® vehicle is expected to accelerate faster during the shift. Figure 16 shows the longitudinal vehicle accelerations for the three data sets, where the veDYNA® vehicle model clearly accelerates faster from approximately nine to eleven seconds as a result of the higher engine torque during the shift.

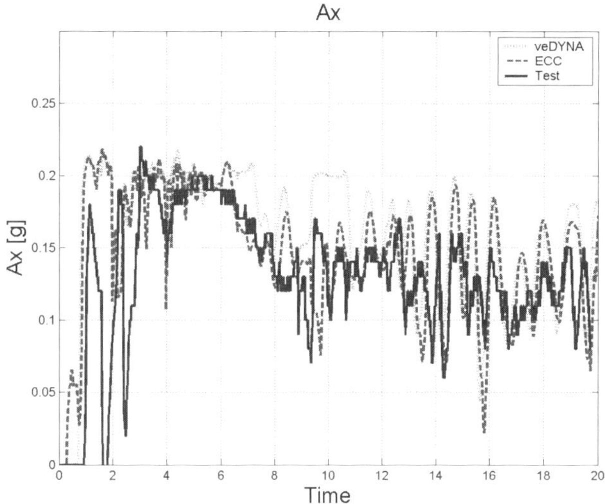

Figure 16: Vehicle acceleration time traces

The effect of the higher engine torque in the veDYNA® simulation results can also be seen in the individual wheel speed traces. Figure 17 shows a comparison of each individual wheel speed for the three cases.

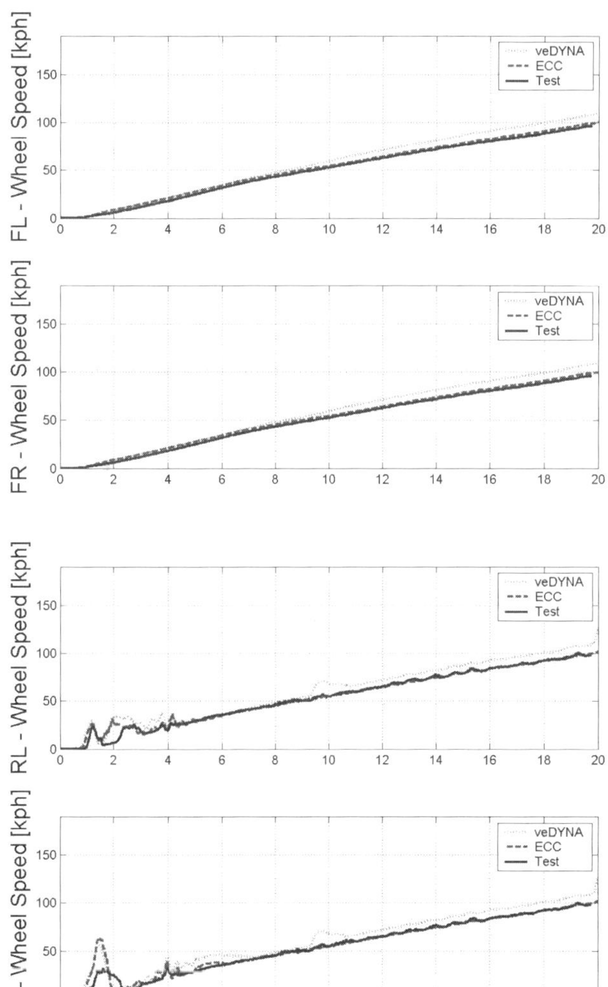

Figure 17: Wheel speed traces for the front left, front right, rear left, and rear right wheels, respectively

Both the front left and front right wheel speeds for all three cases show speed profiles similar to the overall vehicle speed. The rapid increases in the rear left and right wheel speeds during the first seven seconds indicate that the rear wheels are slipping. After seven seconds, the rear wheels have regained traction and have stop spinning. Then at approximately nine seconds the vehicle shifts from first to second gear. The higher engine torque in the veDYNA® simulation causes the rear wheels to spin again.

CONCLUSION

With the inclusion of increasingly complex control features in vehicles, application of model-based systems engineering processes for control development is crucial for delivering safe and reliable vehicles. Details of a combined VMA/veDYNA® modeling framework for chassis control development were presented. A vehicle model was created within this new framework through the integration of a number component models from a variety of sources. A tip-in event from rest was used to demonstrate the capability of the ECC model to capture vehicle dynamic behavior. The ECC vehicle model results were validated against experimental vehicle test data and the results from a commercial simulation tool. The simulation results showed the capability of the ECC vehicle model to accurately represent vehicle dynamic behavior for chassis control development applications.

ACKNOWLEDGMENTS

The authors gratefully acknowledge many of our colleagues who have contributed to this work especially Mark Crawford, Nick Darnton, Roger Graaf, Bradley Hieb, Carlo Miano, Glen Moore, Oliver Nehls, Charles Schepens, and John Shepherd.

REFERENCES

1. Butts, K., Cook, J., C. Davey, C., Friedman, J., Menter, P., Raman, S., Sivashankar, N., Smith, P., Toeppe, S., Automotive "Powertrain Controller Development Using CACSD," *Perspectives in Control: New Concepts and Applications*, Tariq Samad (ed.), IEEE Press, 2000.
2. Sivashankar, N., Butts, K., "A Modeling Environment for Production Powertrain Controller Development," *Proceedings of the 1999 IEEE International Symposium on Computer Aided Control System Design*, Hawaii, 1999.
3. Jennings, M., Tiller, M. and Butts, K., "Defining a Flexible Modeling Methodology for Design and Development of Automotive Powertrain Systems", *2001 ASME Computers In Engineering Conference*.
4. Chris Bellon, Peter Bennett, Peter Burchill, David Copp, Nick Darnton, Kenneth Butts, Judy Che, Brad Hieb, Mark Jennings, Timothy Mortimer, "A Vehicle Model Architecture for Vehicle System Control Design," *SAE World Congress Proceedings 2003*, Paper number 03AE-176.

5. Ken Butts, Bradley Hieb, Judy Che, Mark Jennings, Roger Graaf, Oliver Nehls, Chris Belton, Pete Burchill, David Copp, and Nick Darnton, *NPA Vehicle Model Architecture Specification Version 2.02*, Ford Motor Company internal document.

6. *ve-DYNA Standard 3.6 Simulink Manual*, TESIS *DYNA*ware Technische Simulation Dynamischer Systeme GmbH, München, Germany

7. *ve-DYNA Standard 3.6 Reference Manual*, TESIS *DYNA*ware Technische Simulation Dynamischer Systeme GmbH, München, Germany

8. *ve-DYNA Standard 3.6 Road Manual*, TESIS *DYNA*ware Technische Simulation Dynamischer Systeme GmbH, München, Germany

9. *ve-DYNA Standard 3.6 Driver Manual*, TESIS *DYNA*ware Technische Simulation Dynamischer Systeme GmbH, München, Germany

10. Arnold, V.I., *Mathematical Methods of Classical Mechanics, 2nd ed.* Translated by K. Vogtmann and A. Weinstein. New York: Springer-Verlag, 1989.

11. Heywood, J.B., 1988, *Internal Combustion Engine Fundamentals*, McGraw Hill, New York, p. 311-313.

12. Powell, B.K. and Cook, J.A., 1987, *Nonlinear, Low Frequency Phenomenological Engine Modeling and Analysis,* Proc. American Controls Conference, Vol. 1, pp 332-340.

A Vehicle State Detection Method Based on Estimated Aligning Torque Using EPS

Kenji Nakajima, Masahiko Kurishige, Masaya Endo and Takayuki Kifuku
Mitsubishi Electric Corp.

ABSTRACT

This paper proposes a vehicle state detection method for improving the stability of vehicles equipped with electric power steering (EPS) and electronic stability control (ESC) systems. ESC is an effective vehicle stability control system that operates within a vehicle's stability limitations. Generally ESC uses a vehicle state signal such as yaw rate. To enhance the ESC function so that it can alleviate understeer, a process that is capable of detecting understeer is required. This concept motivated us to develop a vehicle state detection algorithm based on estimated self-aligning torque using EPS. It is well known that maximum self-aligning torque occurs before maximum cornering force is reached. We have confirmed that the proposed algorithm can detect understeer earlier than conventional means based on vehicle yaw rate. We have also implemented this algorithm in an ESC electronic control unit (ECU) and conducted an actual vehicle test to confirm that a reduction in turning radius is realized compared with that using conventional ESC.

INTRODUCTION

Recently, vehicles equipped with EPS have come into widespread use. This is because EPS is better than hydraulic power steering (HPS) in that it helps improve fuel consumption and reduce emissions. However, most present power steering systems, including HPS, merely function to assist handling based on torque input from the driver. There are very few vehicles equipped with EPS systems that include the function of improving vehicle stability.[1] ESC is developed to avoid dangerous states that could lead to a vehicle spinning or drifting. ESC stabilizes the vehicle through the application of a braking system that independently controls braking at all four wheels (e.g., independent four-wheel brake system). In the case of understeer ESC controls braking to augment vehicle yaw moment and reduce vehicle speed. In case of oversteer, ESC controls braking to adjust vehicle yaw moment, which reduces the vehicle's sideslip angle. To enhance the ESC function so that it alleviates understeer a process for detecting understeer is required. Taking this situation into consideration, we developed a vehicle state detection algorithm based on estimated self-aligning torque using EPS. It is well known that maximum self-aligning torque occurs before maximum cornering force is reached.

This encouraged us to estimate a self-aligning torque that works to generate self-centering of the steering wheel, and develop a vehicle state detection algorithm based on the estimated self-aligning torque.

Then we applied this algorithm to an ESC system. Through actual vehicle testing, we confirmed that the system can detect understeer approximately one-half-second earlier than conventional ESC. Regarding the turning radius, a reduction of 1-2 meters is obtained compared to conventional ESC.

EPS MECHANISM AND MODELING

In EPS systems, a motor is applied to the steering mechanism via a reduction gear. This system has both frictional loss torque and spring characteristics or backlash in the steering mechanism.[2] During steering operation, self-aligning torque is generated by the tires and the road surface. Figure 1 shows an illustration of EPS operation. When the steering wheel is turned, a torque sensor located between the steering wheel and motor detects the steering torque. According to the signal from the torque sensor, the electric motor current is determined to reduce the steering torque. The torque of the electric motor is input to the steering shaft through a reduction gear and the product of the motor torque and gear ratio act as torque assistance for the driver. The sum of the driver's steering torque and this assisting torque rotates the steering shaft against the self-aligning torque from the tires and the frictional losses in the steering mechanism. To control the EPS system, the supply voltage and current to the motor are also detected in addition to the steering torque.

The steering mechanism operation is expressed by Eq. 1. Here, "Ttran" denotes the reaction torque at the steering shaft. This is the sum of the self-aligning torque from the tires "Talign" and the frictional loss torque of the rack-and-pinion, etc. "Tfric".

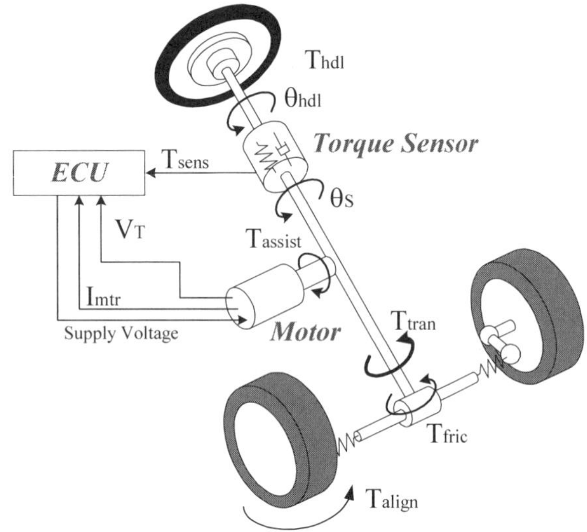

Figure 1 EPS mechanism

$$J\ddot{\theta}_S = T_{assist} + T_{hdl} - T_{tran} \qquad (1a)$$

where

$$T_{assist} = G_{gear} \bullet \left\{ K_T \bullet I_{mtr} - T_{f\_mtr} \bullet sgn(\dot{\theta}_S) \right\}$$
$$T_{tran} = T_{align} + T_{f\_rp} \bullet sgn(\dot{\theta}_S) \qquad (1b)$$

Equation (1a) and (1b) yield Eq. 2.

$$J\ddot{\theta}_S = T_{hdl} + G_{gear} \bullet K_T \bullet I_{mtr}$$
$$- T_{align} - \left(G_{gear} \bullet T_{f\_mtr} + T_{f\_rp} \right) \bullet sgn(\dot{\theta}_S) \qquad (2)$$

ESTIMATION ALGORITHM

OUTLINE OF ESTIMATION ALGORITHM

Our aim is to detect a vehicle's state using estimated self-aligning torque. A block diagram of this algorithm is shown in Fig. 2. As can be seen, first we estimate the self-aligning torque using the "Self-Aligning Torque Estimator". Next, according to the self-aligning torque estimated, the vehicle state is detected at the "Vehicle State Detection Block".

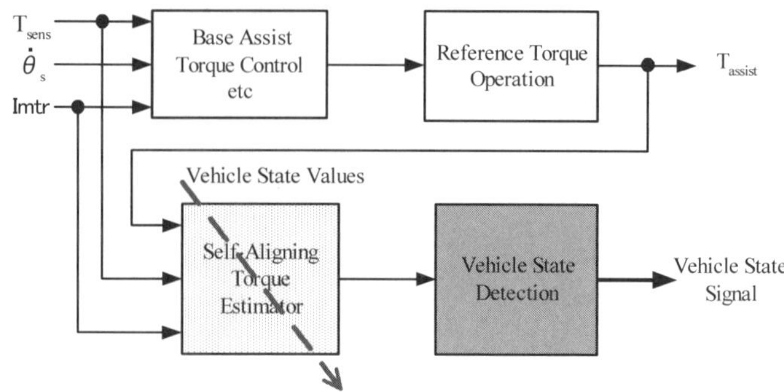

Figure 2 Estimation algorithm

SELF-ALIGNING TORQUE ESTIMATION

The newly developed Self-Aligning Torque Estimator estimates self-aligning torque based on the model of the steering mechanism shown in Eq. 2.[3] To eliminate the non-linear terms such as friction losses, etc., the proposed estimation method utilizes a filter that is adaptive to vehicle states . This allows the proposed estimation method to be used to estimate the self-aligning torque of various steering patterns.

ESTIMATION RESULTS

An example of the estimation results for self-aligning torque is shown in Fig. 3. In this example, the driver steers in a sinusoidal pattern at 0.2Hz and 1.0Hz. The lateral acceleration peak is about 0.2G and vehicle velocity is about 40km/h. Figure 3 shows the results, which is a comparison of measured self-aligning torque and that estimated using the proposed method. The upper figure shows the 0.2Hz pattern, and the lower the 1.0Hz pattern. As clarified in Fig. 3, the Self-Aligning Torque Estimator estimates the self-aligning torque well.[4] To measure the self-aligning torque, we used a sensor produced by Nissho Electric Works Co., Ltd.

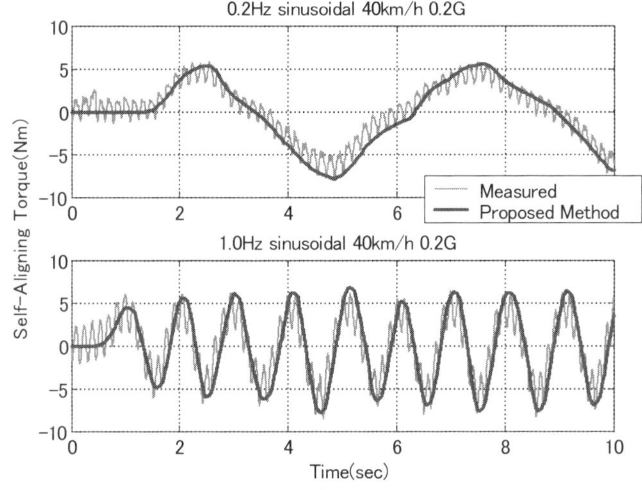

Figure3 Estimated self-aligning torque results

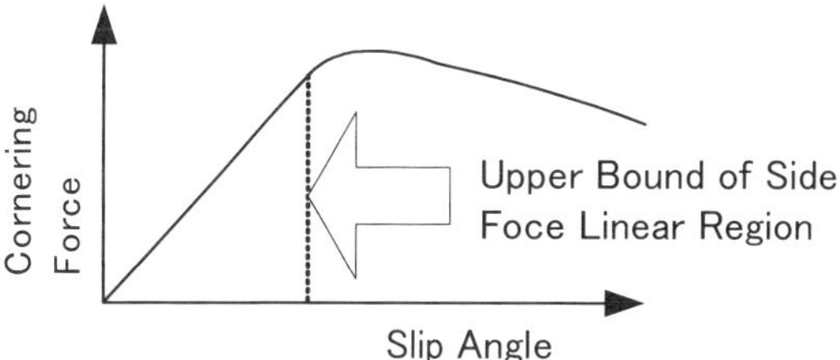

Figure 4a Slip angle vs. cornering force

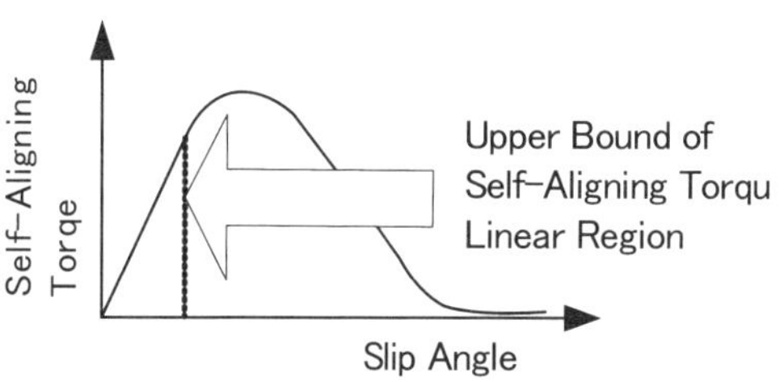

Figure 4b Slip angle vs. self-aligning torque

FEATURES OF SELF-ALIGNING TORQUE

CONCEPT OF VEHICLE STATE DETECTION

Self-aligning torque characteristics are very important. Figure 4 shows the characteristics of slip angle versus cornering force and slip angle versus self-aligning torque. It is well known that self-aligning torque reaches a maximum before the cornering force. When a vehicle is stable, self-aligning torque characteristics are linear, however during vehicle understeer, self-aligning torque characteristics become non-linear. The linear region of the self-aligning torque is smaller than that of the cornering force. Self-aligning torque is expressed by Eq. 3. It is possible to see the region where cornering force is increased while self-aligning torque is decreased. There is a reason why self-aligning torque is decreased prior to the reduction in cornering force. This is because the caster trail is reduced before cornering force begins to decrease. Self-aligning torque that is yielded by these products certainly begins to decrease before the cornering force does. It should be noted that the yaw rate is the integral of yawing moment with the same dimension of cornering force. Hence, the change in yaw rate signal occurs later than the change in cornering force. We utilized this characteristic to develop a vehicle state detection algorithm based on estimating the self-aligning torque using EPS.

$$T_s = 2\xi K_f \beta_f \qquad (3)$$

$$\xi = \xi_n + \xi_c \qquad (4)$$

OFFLINE SIMULATION USING VEHICLE TEST DATA

We examined the performance of the understeer detection method through simulation. The simulation used vehicle test data. The test conditions are shown in Table 1, and the test course is shown Fig. 5. The test vehicle is a small car equipped with ESC. We introduced an understeer situation by creating excessive steering at low μ load, with the vehicle speed approaching 45 km/h, and using a steering rate of 120 deg/s. We used this test data to evaluate the detection of understeer using the estimated self-aligning torque in our proposed algorithm.

Table 1 Test Conditions

Steering Patten	Turning left 120(deg/s)
Road Condition	μ =0.4 (Correspond to a packed snow road)
Velocity (km/h)	45
Understeer Detection	Conventional algorithm Proposed algorithm

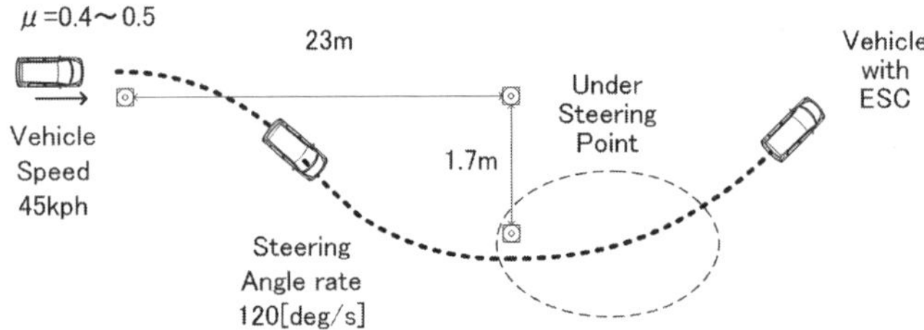

Figure 5 Test course

UNDERSTEER DETECTION RESULTS

The simulation results for detecting understeer are shown in Fig. 6. The conventional understeer detection algorithm is based on maximum yaw rate, and the proposed understeer detection algorithm is based on estimated self-aligning torque. We confirmed that the proposed algorithm could detect understeer as much as 0.5 second faster than the conventional algorithm through simulation.

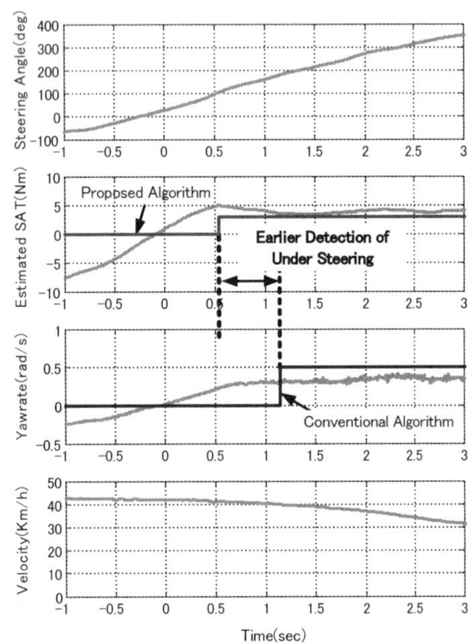

Figure 6 Simulation results

EXEPERIMENT

Understeer Detection

We tested understeer detection using estimated self-aligning torque and a test vehicle. We obtained results similar to the offline simulation results. Understeer detection using the estimated self-aligning torque method is faster than the conventional method using yaw rate. Figure 7 shows the test results. We confirmed that

understeer detection using estimated self-aligning torque is approximately one-half-second faster than using the yaw rate.

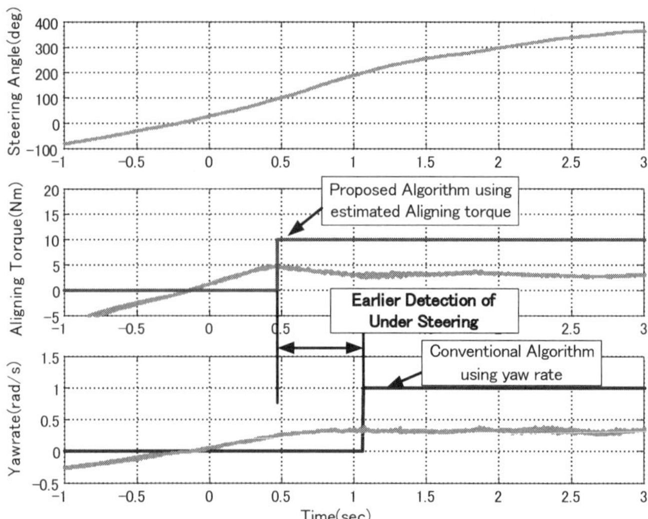

Figure 7 Vehicle test results

Example application of proposed understeer detection algorithm

System Structure

Figure 8 shows the general functions of an ESC system. ESC stabilizes vehicle dynamics by applying brake torque independently to each of the four wheels. In the case of understeer, the ESC system applies brake torque to prevent vehicle deviation from the lane. In the case of oversteer, the ESC system applies brake torque to prevent the vehicle from spinning. Figure 9 shows the system layout in the test vehicle. It supports the reinforcement function of understeer detection. The EPS ECU detects understeer using estimated self-aligning torque and transfers this signal to the ESC ECU. When the ESC ECU receives the understeer signal, it begins controlling brake torque independently to each of the four wheels in order to prevent understeer.

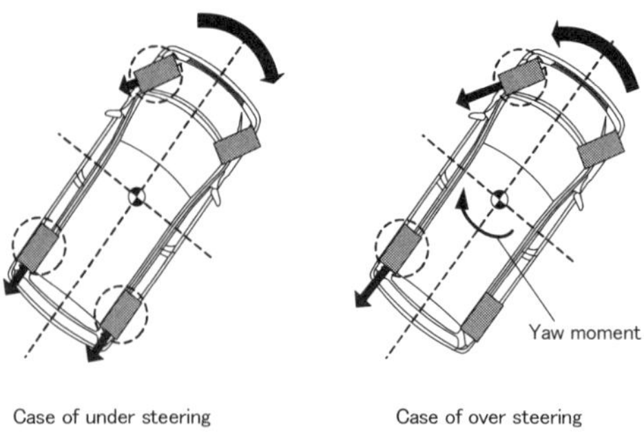

Figure 8 Functions of ESC[5]

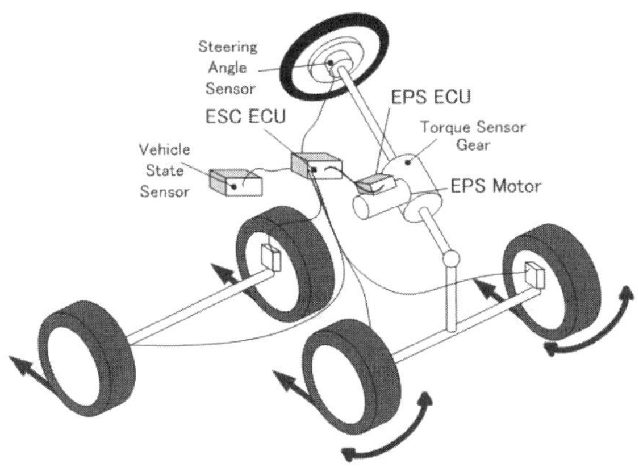

Figure 9 System layout

VEHICLE TEST CONDITIONS

We tested three types of vehicles (Vehicle without ESC, Vehicle with Conventional ESC, Vehicle with Proposed ESC using understeer detection signal). The test conditions are shown in Table 2, and the test course is shown in Fig. 10. Understeer was created by excessive steering input on a low μ road (μ =0.4), with the vehicle speed approaching 45 km/h, and using a steering rate of 120 deg/s.

Table 2 Test Conditions

Steering Patten	Turning left 120(deg/s)
Road Condition	μ =0.4 (Corresponding to a snow-covered road)
Speed (km/h)	45
Vehicle type	Without ESC Conventional ESC Proposed ESC

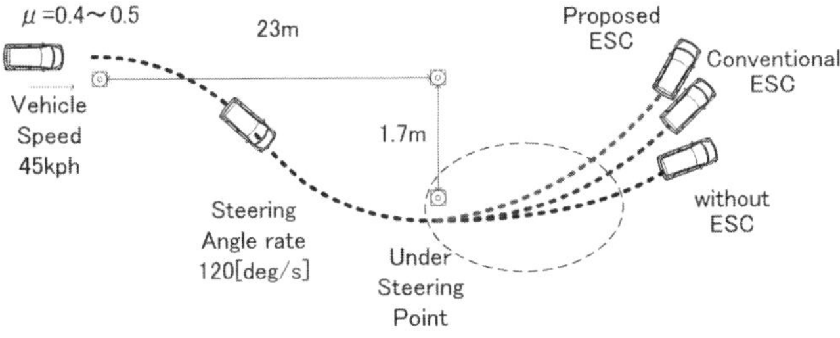

Figure 10 Test course

VEHICLE TEST RESULTS

The test results are shown in Fig. 11. Each test was performed under the same conditions. The vehicle with conventional ESC and proposed ESC were able to detect understeer using their respective detection algorithms. Both the vehicle with conventional ESC using yaw rate and the vehicle with the proposed ESC using estimated self-aligning torque detected understeer, and the ESC systems applies brake torque independently to each of the four wheels to prevent vehicle deviation from the lane. Hence, the speeds of these vehicles began to slow down sooner than the vehicle without ESC. Furthermore, the vehicle with the proposed ESC detected understeer and began controlling braking earlier than either of the other vehicles.

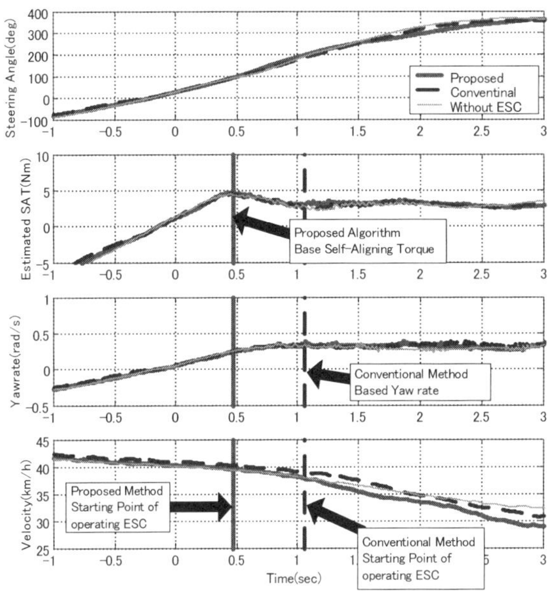

Figure 11 Test results

Figure 12 shows the test results for preventing understeer. The X axis denotes the time and the Y axis denotes the turning radius. The vehicle equipped with the proposed ESC had the smallest radius. About two seconds after the ESC began operating, the turning radius of the vehicle with the proposed ESC is approximately 5 meters smaller than that for the vehicle without ESC. As for the turning radius of the vehicle with conventional ESC, turning radius of the vehicle with the proposed ESC was approximately 2 meters smaller. These results show even further contributions to the proposed understeer detection algorithm based on self-aligning torque using EPS. The proposed ESC enables braking torque to be applied to each of the four wheels and vehicle deceleration faster than either of the other systems. Therefore the radius is the smallest compared with other test results.

407

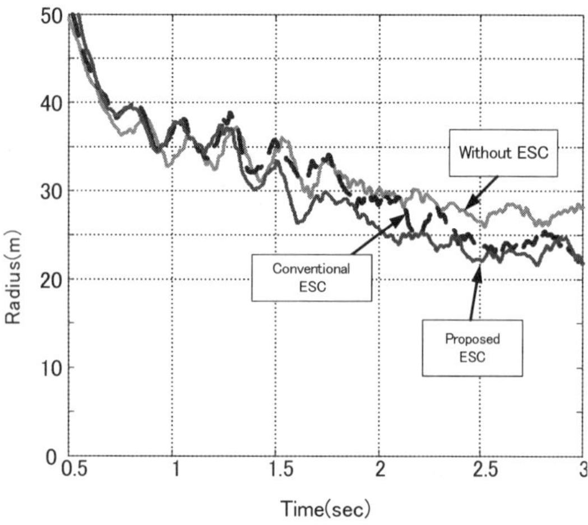

Figure 12 Vehicle turning radius results

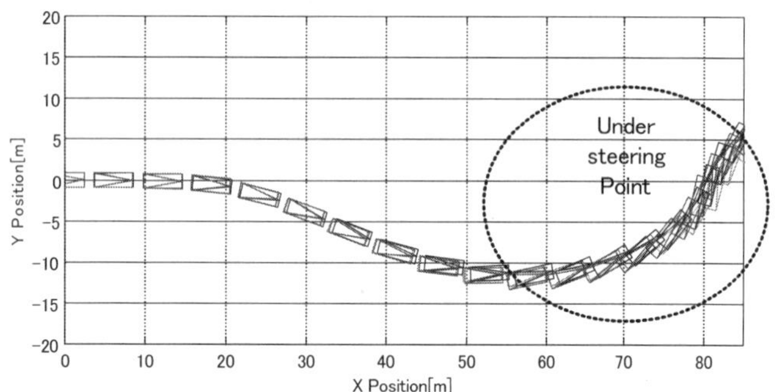

Figure 13 Vehicle track

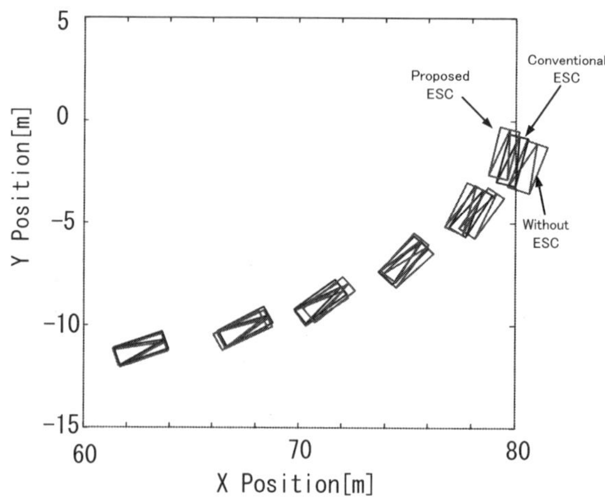

Figure14 Expansion of vehicle track at point of understeer

Figure 13 shows the vehicle tracks when understeer occurs. The region marked by the circle indicates the understeer point. The vehicles detect understeer in this area and applies brake torque each of the four wheels. Figure 14 shows an enlargement of the point at which understeer occurs. The vehicle without ESC show signs of understeer and deviated from the lane. However, the vehicles with the proposed ESC and conventional ESC showed improved turning radii. This is because the vehicles with the proposed ESC and conventional ESC adequately control braking torque. The conventional ESC applies braking torque to reduce understeer using the yaw rate. The proposed ESC applies braking torque to reduce understeer using estimated self-aligning torque when understeer is detected. It may be more natural than Conventinal ESC operation. Therefore vehicles with the proposed ESC run stably under these conditions, allowing the driver a more relaxed driving experience.

FUTURE DEVELOPMENT

As explained in the preceding sections, we have described an understeer detection system based on estimated self-aligning torque and the use of ESC to correct understeer. At the same time, detecting oversteer is as important as detecting understeer when controlling vehicle stability. Presently, we are working on the development of an oversteer detection algorithm that uses self-aligning torque, similar to that of understeer detection. We've just started to develop this algorithm and are refining it to enable further improvements in vehicle stability in the future.

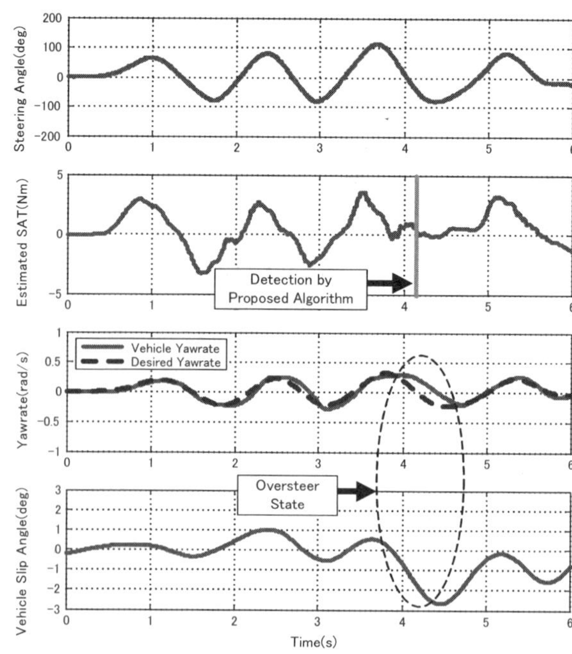

Figure15 Results of the proposed oversteer detection examination

Figure 15 shows the results of proposed oversteer detection conducted using a test vehicle. In this test, the driver steered in a sinusoidal pattern on a low μ road(μ=0.2). Vehicle speed was about 40km/h. In Fig. 15, the rapid increase in vehicle slip angle at approximately 4.1 second shows the oversteer state of the vehicle. We confirmed that proposed algorithm could detect oversteer.

The advantage of the proposed algorithm is that it does not require vehicle state signals such as yaw rate, etc. Therefore, the algorithm can be applied to vehicles equipped with EPS without additional cost. The EPS will then be able to provide assist torque to recover vehicle stability. In addition, we expect that the algorithm can be integrated into the ESC or active front steering (AFS) systems to improve vehicle stability.

CONCLUSION

A proposed algorithm to detect understeer was studied with the following results.

The proposed algorithm to detect understeer based on self-aligning torque can detect understeer faster than conventional detection using yaw rate.

We confirmed that a vehicle equipped with an ESC incorporating the proposed understeer detection algorithm that uses self-aligning torque, is capable of alleviating understeer faster than vehicles with conventional ESC or without ESC.

REFERENCES

[1] http://www.toyota.co.jp/Showroom/All_toyota_lineup/ crownmajesta/index.html   (In Japanese)
[2] Post J. W., "Modeling and Testing of Automobile Power Steering Systems for the Evaluation of On-Center Handling", Ph. D Dissertation, Clemson University, 1995
[3] Kurishige M. et al., "A New EPS Control Strategy To Improve Steering Wheel Returnability", SAE Paper 2000-01-0815, 2000
[4] Tanaka H. et al., "The Torque Controlled Active Steer For EPS". AVEC '04 Paper
[5] Toyota Crown Majesta Maintenance Manual

NOMENCLATURE

T_{sens}:   Mesured steering torque

I_{mtr}:   Motor Current

V_t: Terminal voltage of motor

θ_{hdl}:   Steering wheel angle

θ_s: Column axis angle

J: Moment of inertia of motor (at column axis)

T_{hdl}:   Driver's steering torque

T_{tran}:   Amount of reaction torque from below the

K_t: Torque constant of the motor

T_{f_mtr}:   Frictional loss of the motor

T_{f_rp}:   Frictional loss of the rack & pinion

G_{gear}:   Reduction gear ratio from motor to column axis

T_{assist}:   Torque assisted by motor (at column axis)

T_{align}:   Reduction torque of the tires (at column axis)

ξ_n: Pneumatic trail

ξ_c: Caster trail

Integration of Active Suspension and Active Driveline to Ensure Stability While Improving Vehicle Dynamics

Nicholas Cooper, David Crolla and Martin Levesley
University of Leeds

Warren Manning
Manchester Metropolitan University

ABSTRACT

Most active control systems developed for passenger vehicles are developed as safety systems. These control systems usually focus on improving vehicle stability and safety while ignoring the effects on the vehicle driveability. While stability is the primary concern of these control systems the driveability of the vehicle is also an important consideration. An example of compromised driveability in a stability control system is brake based active yaw control. Brake based systems are very effective at stability control but can have a negative impact on the longitudinal dynamics of a vehicle. The objective of the vehicle control systems developed for the future will be to preserve vehicle driveability while ensuring the stability of the vehicle. In this work, active suspension and active drivelines are developed as stability control systems that have a minimal impact on the driveability of the vehicle. The active control systems are developed as separate driveability and stability controls and then tested individually. Then the individual controls are integrated to create a multi-objective control system to improve driveability while maintaining stability. Enhancing driveability is a requirement but the primary goal of the controllers is to maintain stability. The controllers are tested with step steer and fishhook manoeuvres. Since many manoeuvres that require the intervention of a stability control system occur during evasive action, the controllers are tested at the limit handling conditions of the vehicle.

INTRODUCTION

The key characteristics of vehicle handling that concern the driver are sideslip stability, yaw rate tracking and maintaining vehicle speed. There are clear conflicts in these performance objectives and the objective of this work is to use vehicle control systems to enhance the driveability while ensuring vehicle stability to increase the performance envelope of the vehicle.

Vehicle lateral stability is monitored through the sideslip angle. Sideslip angle control is primarily used as a stability control, as shown by Abe for torque control and for roll control [1,2]. The problem with sideslip angle based controllers is that the sideslip angle is difficult to measure directly so an estimator must be used.

The yaw rate is easier to measure compared to sideslip angle and the controls can be used to direct the vehicle path more directly. Yaw rate controllers usually use a reference model control like the torque control in Motoyama [3]. An example of roll control is given by Williams [4]. Combined sideslip angle and yaw rate controllers combine the benefits of both individual control strategies. The sideslip angle control can be used to stabilize the vehicle while the yaw rate control can guide the vehicle on the desired path as demonstrated by Nagai [5].

Sideslip and yaw rate control can be achieved through a combination of brakes, steering, drivetrain and suspension. The braking system has seen a lot of development with active controls for vehicle safety [Abe, 1]. However, a result of using the braking system is a loss of velocity which is undesirable [Smakman, 6]. A more efficient control can be made by using a controlled drivetrain for variable torque distribution (VTD) [Esmailzadeh, 7]. A VTD system can consist of active differentials that are used to place the torque where it is most useful. By distributing the engine torque to the outside wheel of a vehicle in a corner, a pro-cornering moment can be created. Conversely, a contra-cornering moment can be created by driving the inside wheel of a vehicle.

The steering system has also seen a lot of development to improve handling. An active steering control can be very effective at low cornering forces but as the tyres saturate at high g's, steering controllers lose the ability to affect the handling [Nagai, 5]. Active suspension systems are useful at high cornering forces since they actively change the vertical tyre forces. However, full active suspension systems take a lot of energy to run and are very expensive [Everett, 8].

Roll moment distribution (RMD) is different type of suspension system control that can change the balance of a vehicle at lower power levels than the fully active system. It can be achieved by using active roll bars which can capture the relatively low frequency requirements of vehicle handling dynamics [Williams, 4]. The active roll bars can be stiffened or softened to move the roll stiffness from one axle to the other. By increasing the roll stiffness of an axle more load transfer across that axle is created. Since tyres are nonlinear and produce a relatively smaller amount of lateral force with increasing normal force, an axle with high load transfer will not be able to create as much lateral force as an equivalent axle with an even distribution of load transfer. In this way the potential force from an axle can be reduced or increased by changing the roll stiffness. When two axles are considered together the understeer/oversteer balance of a vehicle can be manipulated by changing the distribution of the total roll moment from the front to the rear of the vehicle or vice versa [Williams, 4].

The combined torque distribution and roll moment distribution controls follow the same type of control strategies as the individual controls. Sideslip angle control, as implemented by Smakman, is used to maintain vehicle stability through a phase plane approach [6]. The strategy is to use the roll moment control initially and only add the torque control when it is needed. This limits the intrusive nature of the torque control. Again, yaw rate controllers tend to be model following controllers as shown by Everett [8]. Here the rate of change of the yaw rate is used to determine which control system is dominant. If the rate of change is large, the manoeuvre would require the more powerful torque distribution control to stabilize the vehicle. A controller using both yaw rate and sideslip signals is developed by Kitajima [9]. The integration method used measures the potential effects of each control and then uses an algorithm to prioritize them to get the optimal solution.

This paper will investigate VTD driveability control and RMD stability control as individual systems and both together in an integrated system to create a multi-objective control to improve both stability and driveability. The controls will be evaluated in a steady state step steer manoeuvre as well as an accelerating fishhook manoeuvre.

VEHICLE MODEL

The vehicle modelled and parameters used are taken from the University of Leeds Formula SAE race car. This is an open wheel, single seat formula style race car with a 600cc 4 cylinder motorcycle engine and rear wheel drive.

The vehicle model is a nonlinear model. This model consists of eight degrees of freedom, longitudinal, lateral, yaw and roll motions as well as the four wheel spin degrees of freedoms. The equations describing the balance of forces on the vehicle body are similar to the equations for a bicycle model with the addition of the longitudinal and roll degrees of freedom as shown in equations (1) to (4).

$$M_{tot}(\dot{u} - vr) = \sum F_x \tag{1}$$

$$M_{tot}(\dot{v} + ur) + (am_f - bm_r)\dot{r} + m_b h_r \ddot{\phi} = \sum F_y \tag{2}$$

$$(am_f - bm_r)(\dot{v} + ur) + I_{zz}\dot{r} + I_{xz}\ddot{\phi} = aF_{yf} - bF_{yr} \tag{3}$$

$$m_b h_r(\dot{v} + ur) + (K_\phi - m_b g h_r)\phi + D_\phi \dot{\phi} + I_{xx}\ddot{\phi} = \\ d_f F_{yf} - d_r F_{yr} \tag{4}$$

Here M_{tot} is the total vehicle mass, m_f, m_r, and m_b are the front axle, rear axle and body masses, u is longitudinal velocity, v is the lateral velocity, r is the yaw rate, ϕ is the roll angle, F_{yf} and F_{yr} are the lateral tyre forces at the front and rear axles respectively, F_x is the longitudinal force, I_{zz} is the yaw moment of inertia, I_{xx} is the roll inertia, I_{xz} is the roll-yaw inertia product, K_ϕ is the roll stiffness, D_ϕ is the roll damping, a and b are the distances from the centre of gravity to the front and rear axles respectively, h_r is the height of the centre of gravity above the roll axis and d_f and d_r are the front and rear scrub derivatives.

The tyre model used is a nonlinear Pacejka "Magic Formula" tyre model that is dependent on vertical load, lateral slip, longitudinal slip and camber [10]. For simplification, the camber has been set to zero in the current vehicle model.

Although there are no pitch or heave degrees of freedom in the model, the vertical load on the tyres is dependent on load transfer due to roll and the lateral and longitudinal accelerations. The load on each tyre is defined as:

$$F_z = \frac{1}{2}\left(\frac{bM_{tot}g}{a+b} \pm \frac{M_{tot}h\dot{u}}{a+b}\right) \pm \\ \frac{M_{tot}(\dot{v} + ur)}{2t_f}\left(\frac{h_r K_{\phi f}}{K_\phi} + \frac{bh_{cf}}{a+b}\right) \tag{5}$$

where the two $\pm$ terms are determined by which tyre load is being calculated, t_f and t_r are the front and rear track widths, $K_{\phi f}$ and $K_{\phi r}$ are the front and rear roll stiffness and g is gravity. Aerodynamic drag and lift are not taken into account.

CONTROL SYSTEMS AND STRATEGIES

The VTD driveability controller is an internal model control (IMC) based on yaw rate. An IMC controller is used because it incorporates both feedforward and feedback control. IMC control was investigated for the stability controller but it was not robust enough so an RMD proportional-integral-derivative (PID) controller based on sideslip angle is used. There is also a longitudinal velocity controller to maintain the vehicle speed during the test manoeuvres. The speed controller is a PI controller based on u that controls the amount of torque available at the wheels from the engine.

The VTD driveability control is shown in Figure 1.

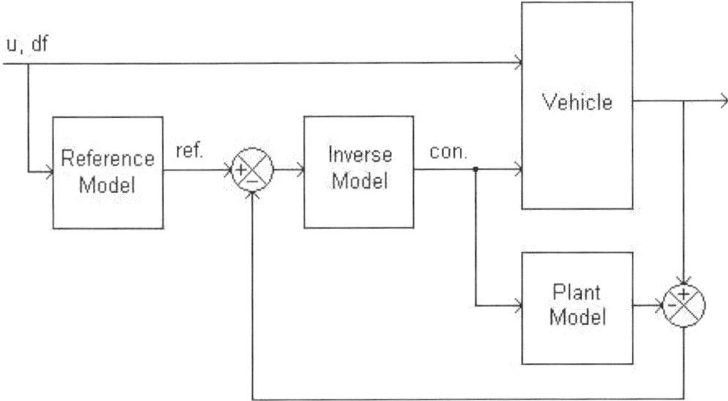

Fig. 1 Internal model control block diagram

The vehicle model is the full eight degree of freedom model described previously while the reference model is a linear two degree of freedom steady state model that determines the drivers desired reference yaw rate. The inverse model and plant model are linearized two degree of freedom models that include VTD. The inverse model determines the required value of C_{dr}, the coefficient of torque distribution at the rear axle, for a given desired yaw rate. C_{dr} is limited between a value of 0 and 1, where 1 represents all the torque distributed to the right side of the axle. The plant model has an input of zero for the steer angle so it calculates the yaw rate from the VTD. The purpose of the plant model is to compensate for any differences between the inverse model and vehicle model and to modify the reference yaw rate to remove the influence of the steer angle from the control signal. The reference yaw rate, r_{ref}, is given by a steady state model with linear tyres and two degrees of freedom, yaw rate and lateral acceleration, given by equation (6).

$$\begin{bmatrix} \dfrac{C_f + C_r}{U} & mU + \dfrac{aC_f - bC_r}{U} \\ \dfrac{aC_f - bC_r}{U} & \dfrac{a^2 C_f + b^2 C_r}{U} \end{bmatrix} \begin{bmatrix} v \\ r \end{bmatrix} = \begin{bmatrix} C_f \\ aC_f \end{bmatrix} [\delta_f] \qquad (6)$$

The inverse and plant models are given by equation (7). The inverse is found by inverting with respect to r and C_{dr}.

$$\begin{bmatrix} m & 0 \\ 0 & I_{zz} \end{bmatrix} \begin{bmatrix} \dot{v} \\ \dot{r} \end{bmatrix} + \begin{bmatrix} \dfrac{C_f + C_r}{U} & mU + \dfrac{aC_f - bC_r}{U} \\ \dfrac{aC_f - bC_r}{U} & \dfrac{a^2 C_f + b^2 C_r}{U} \end{bmatrix} \begin{bmatrix} v \\ r \end{bmatrix} = \begin{bmatrix} C_f & 0 \\ aC_f & tK \end{bmatrix} \begin{bmatrix} \delta_f \\ C_{dr} \end{bmatrix} \qquad (7)$$

An internal model control was also tried with roll moment distribution. It was capable of initially changing the vehicle behaviour as desired but did not prove robust enough since the tyre nonlinearities need to be included to properly model the roll moment distribution. However, the inverse model could not model the tyre nonlinearities sufficiently while still remaining simple enough to be easily invertable. Due to this, PID control was implemented for roll moment distribution.

The RMD stability PID control is given by:

$$C_{Kd} = c_d G_P |\beta| + c_d G_I \left| \int \beta dt \right| + c_d G_D \left| \frac{d\beta}{dt} \right| + C_{Kds} \qquad (8)$$

where β is the sideslip angle, G_D, G_P and G_I are the derivative, proportional and integral gains, C_{Kd} is the coefficient of roll moment distribution. This is limited to values between 0 and 1 with 1 representing all the roll moment at the front of the vehicle. C_{Kds} is the static coefficient of roll moment distribution and c_d, which is limited to values of -1 to 1, is used to determine the direction of the control input and is given by:

$$c_d = G_c \beta r \qquad (9)$$

Here G_c is a gain to determine the weighting on the sign change of the yaw rate. Due to the complex non-linear nature of the complete vehicle system and controllers, the gain selection process consisted of an extensive trial and error process where the optimum gain values were defined.

The integrated controller combines the VTD and RMD controllers and includes a PID sideslip angle based control to determine the priority of the controls as shown in Figure 2. There is also an adaptive offset gain that delays the transition from the initially dominant VTD driveability control to the RMD stability control as the vehicle sideslip angle increases. The adaptive offset gain allows the vehicle to run with the VTD driveability control until a certain sideslip angle is reached before the control dominance is transferred over to the RMD stability control.

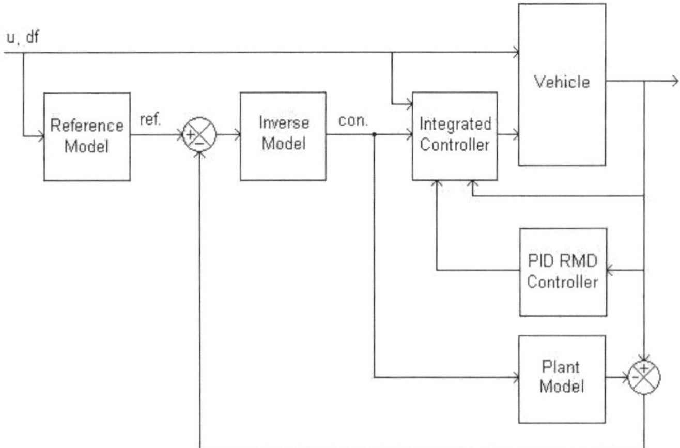

Fig. 2 Integrated Controller

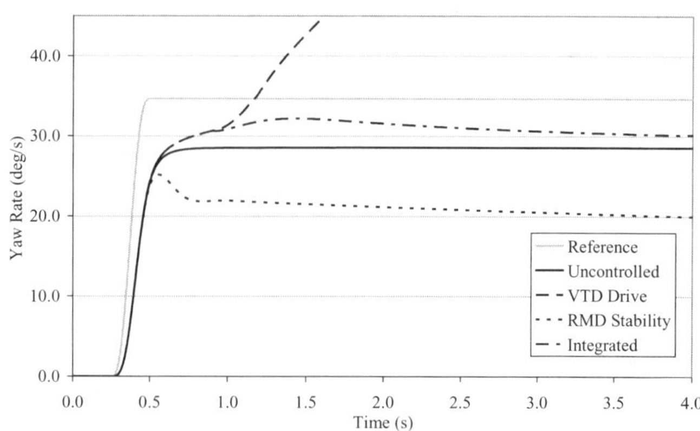

Fig. 3 Yaw rate results for the step steer manoeuvre

The critical sideslip angle and the speed of transfer from the driveability control to the stability control are adaptively determined based on the steer angle. The greater the steer angle the earlier the stability control takes precedence. The underlying transfer PID control is given by:

$$C_I = c_d G_P |\beta| + c_d G_I \left| \int \beta dt \right| + c_d G_D \left| \frac{d\beta}{dt} \right| \qquad (10)$$

Here C_I is the integration coefficient, with a value limited between 0 and 1 where 0 gives all precedence to the VTD driveability controller while a value of 1 gives the RMD stability controller dominance.

CONTROLLER EVALUATION AND RESULTS

The controllers are evaluated with two manoeuvres, a constant velocity step steer and an accelerating fishhook manoeuvre. The step steer manoeuvre shows the steady state response of the vehicle and is run at higher speed. The fishhook manoeuvre gives insight into the transient behaviour of the vehicle and the response from changing the steer direction and velocity and is run at a lower speed.

STEP STEER MANOEUVRE - The step steer is carried out at 30 m/s with 1.7 degree step input which results in a 1.4 g corner. Figure 3 shows the yaw rate results from the step steer manoeuvre.

The uncontrolled vehicle doesn't manage to match the reference yaw rate. It lags behind it and ends at a value of 28.6 deg/s. The VTD internal model driveability control tries to follows the yaw rate but goes unstable after around a second while the stability control shows the lowest yaw rate stabilising at 21.9 deg/s then decreasing. The Integrated control shows an increased yaw rate. It almost goes unstable and peaks at 32.2 deg/s before settling down to a stable yaw rate.

Figure 4 shows the sideslip angle phase plane results for the step steer manoeuvre. The uncontrolled vehicle stabilises at a sideslip angle of 2.2 deg with little overshoot. The VTD driveability control becomes unstable and the sideslip increase without bound. The smallest steady state sideslip angle, 0.9 deg, is achieved with the RMD control while the integrated control has a steady state value similar to the uncontrolled vehicle at 2.3 deg but shows a higher overshoot.

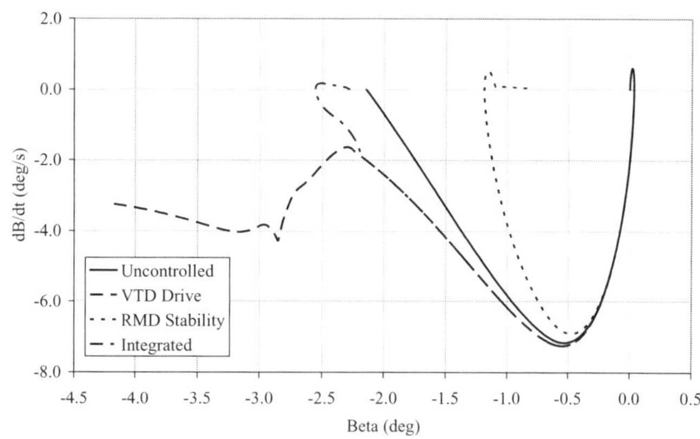

Fig. 4 Sideslip angle phase plane results for the step steer manoeuvre

FISHHOOK MANOEUVRE - The fishhook manoeuvre starts at 20 m/s and accelerates at 1 m/s^2. The steer input is a 3.0 degree sine step to the left then to the right. This manoeuvre also results in a lateral acceleration of about 1.3g. Figure 5 shows the yaw rate response to the fishhook manoeuvre.

The uncontrolled vehicle doesn't match the reference yaw rate in either part of the manoeuvre and initially stabilises at 34.5 deg/s before going unstable after 2.9 seconds. The VTD control does manage to push the yaw rate close to the reference yaw rate in both parts of the manoeuvre but drives the vehicle to an unstable state after 2.8 seconds. The RMD controller has the lowest yaw rate in both parts of the manoeuvre and settles at a yaw rate of 31.6 deg/s.

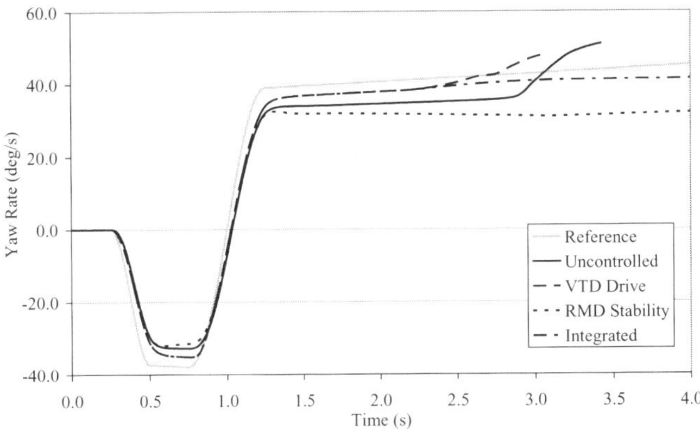

Fig. 5 Yaw rate results for the fishhook manoeuvre

The integrated controller follows the driveability controller and shows the same improved response time but stabilises at a yaw rate of 41.2 deg/s, without going unstable.

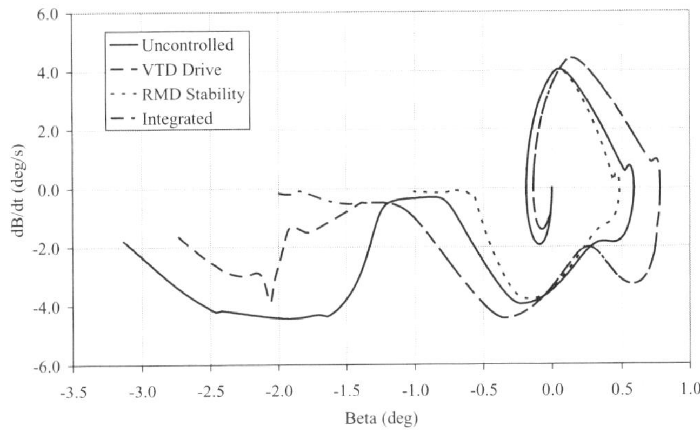

Fig. 6 Sideslip angle phase plane results for the fishhook manoeuvre

The sideslip angle phase plane results are shown in Figure 6. The uncontrolled vehicle shows an unstable response and never settles. As seen in Figure 4, the driveability controller also pushes the vehicle to an unstable state where the sideslip angle starts rapidly increasing. The stability controller shows a stable vehicle throughout the manoeuvre stabilising at 0.7 deg before slowly increasing as the vehicle accelerates. The integrated controller follows the driveability controller but breaks free before the VTD controller goes unstable and settles at 1.8 deg then also slowly begins to increase.

DISCUSSION OF RESULTS

The behaviour of the uncontrolled vehicle is characterised by the nonlinearities in the vehicle model resulting in a generally stable but slower response behaviour. The nonlinearities of the tyres and cross-coupling of the longitudinal and lateral motions prevent the uncontrolled vehicle from matching the performance of the reference model. This is seen in the yaw rate performance where the uncontrolled vehicle has a

slower response to the steer input and doesn't reach the same peak yaw rate.

The goal of the VTD driveability internal model control is to match the reference yaw rate through torque distribution across the rear axle. The step steer shows an interesting result. Although torque distribution is capable of inducing large yaw rates, the controller doesn't manage to enhance the response time of the vehicle. The controller only manages to approach the reference yaw rate once the vehicle is in the high g corner. The reason for this is that the step steer manoeuvre is carried out at a constant velocity. This means that there isn't a large torque input at the axle that can be distributed. Only in high g cornering, where a significant torque is used to maintain the constant velocity, is there enough torque to induce a large enough yaw moment to follow the reference yaw rate. The result is that during the constant velocity step steer manoeuvre the VTD driveability controller follows the same initial path as the uncontrolled vehicle and the controller cannot improve the response time.

Unlike the step steer, the fishhook manoeuvre shows that the response time of the vehicle can be improved if there is sufficient torque to induce a yaw moment. In the fishhook manoeuvre the VTD driveability control does improve the response time of the vehicle in the initial phases of the manoeuvre. This is because the manoeuvre is an accelerating manoeuvre with torque being input into the axle throughout the duration of the test. Not only is the response time improved, but like the step steer manoeuvre the vehicle does approach the reference yaw rate. Again, in the fishhook manoeuvre the reference yaw rate is more than the vehicle can actually handle and the controller pushes the vehicle into an unstable state. This is not a surprise since the driveability controller is only concerned with trying to match the reference yaw rate and does not take the stability of the vehicle into account at all.

The purpose of the RMD PID stability controller is to reduce the sideslip angle. The controller is successful at stabilising the vehicle. In both step steer and fishhook manoeuvres the stability controller shows the lowest sideslip angles and rate of change of the sideslip angle on the phase plane plot. However, the penalty is that the stability controller also reduces the yaw rate of the vehicle the most which is undesirable. The effects of the stability control are not dramatic when compared to the uncontrolled vehicle but when it is integrated with the VTD driveability control its' effects become more desirable.

The integrated controller aims to increase the performance envelope of the vehicle by tracking the yaw rate better than the uncontrolled vehicle but without losing stability. In the step steer test, the integrated controller shows improvement in following the yaw rate over uncontrolled vehicle. Before it matches the reference yaw rate the stability control takes precedence and stabilises the vehicle. Again, in the steady state manoeuvre the VTD control is not very powerful due to lack of available torque to distribute across the axle. Even though the integrated controller doesn't follow the reference yaw rate, there is an improvement in yaw rate following and the vehicle is also brought back to a stable state with a higher yaw rate than the uncontrolled vehicle.

The fishhook manoeuvre shows a better performance from the integrated controller. Since the manoeuvre is an accelerating manoeuvre there is torque available to be distributed to affect the yaw rate response from the beginning of the manoeuvre. The integrated controller matches the yaw rate tracking performance of the VTD driveability controller. The time response and yaw rate following are both improved compared to the uncontrolled vehicle. The difference between the driveability controller and the integrated controller comes when the driveability control pushes the vehicle into an unstable state while the integrated controller manages to bring the vehicle back to a stable state when the RMD stability control takes precedence. The integrated controller phases out the VTD and allows the RMD to induce some understeer to stabilise the vehicle and bring it back to a higher yaw rate than the uncontrolled vehicle.

Although the integrated controller performs well and increases the performance envelope of the vehicle while maintaining its stability, there is always room for improvement. The VTD internal model control is a very effective control, the only drawback is that it only becomes effective when there is torque given to the axle. The PID controls for both the RMD and the integration are not very sophisticated although they are very effective. It is difficult to determine the optimal gains in a scientific manner when a complex nonlinear system is being controlled and the result is a working control system but one with room for improvement. The adaptive offset gain for the integration control is also very effective but again, mapping the offset gain in a complex nonlinear system is difficult.

CONCLUSION

The uncontrolled vehicle shows good behaviour of a normal race car. As expected, it cannot match a linear reference model and shows slower response. The VTD driveability internal model control shows that the vehicle can be pushed to match the reference model, but only when there is sufficient torque input into the axle. The driveability control also shows that if vehicle stability is not taken into account then the control will send the vehicle into an unstable state. The RMD stability control is effective at reducing the sideslip angles and stabilising the vehicle but it also reduces the yaw rate of the vehicle at the same time. The integrated controller succeeded at matching the yaw rate while maintaining the stability of

the vehicle. Again, it is most effective when there is torque input into the axle to allow the VTD to induce large yaw moments but it does work to a lesser degree in steady state manoeuvres. An improvement to the control systems would be to use a more sophisticated control on the RMD stability control and better mapping of the offset gain in the integration or again, a more sophisticated control.

REFERENCES

1. Abe, M., et al.: "Improvement of Vehicle Handling Safety with Vehicle Side-Slip Control by Direct Yaw Moment", Vehicle System Dynamics, Vol. 33 (Suppl), pp. 665-679, (1999).
2. Abe, M.: "Roll Moment Distribution Control in Active Suspension for Improvement of Limit Performance of Vehicle Handling", Proceedings of AVEC 1992, pp. 378-383 (1992).
3. Motoyama, S., et al.: "Effect of Traction Force Distribution Control on Vehicle Dynamics", Vehicle System Dynamics, Vol. 22, pp. 455-464, (1993).
4. Williams, D.E., Haddad, W.M.: "Nonlinear Control of Roll Moment Distribution to Influence Vehicle Yaw Characteristics", IEEE Transactions on Control Systems Technology, Vol. 3(1), pp. 110-116, (1995).
5. Nagai, M., et al.: "Integrated Control of Active Rear Wheel Steering and Direct Yaw Moment Control", Vehicle System Dynamics, Vol. 27, pp. 357-370, (1997).
6. Smakman, H.: "Functional Integration of Slip Control with Active Suspension for Improved Lateral Vehicle Dynamics." Herbert Utz Verlag, Munich, (2000).
7. Esmailzadeh, E., G. R. Vossoughi, et al. (2001). "Dynamic Modeling and Analysis of a Four Motorized Wheels Electric Vehicle." Vehicle System Dynamics 35(3): pp163-194.
8. Everett, N.R., et al.: "Investigation of a Roll Control System for an Off-Road Vehicle", SAE Technical Paper Series 2000-01-1646, (2000).
9. Kitajima, K., Peng, H.: "H Infinity Control for Integrated Side-Slip, Roll, and Yaw Controls for Ground Vehicles", Proceedings of AVEC 2000, (2000).
10. Pacejka, H.B., Besselink, I.J.M.: "Magic Formula Tyre Model with Transient Properties", Supplement to Vehicle System Dynamics, Vol. 27, pp. 234-249, (1997).

CONTACT

Nicholas Cooper
University of Leeds
School of Mechanical Engineering
Leeds, LS2 9JT, UK
mennjc@leeds.ac.uk

Modelling and Simulation of Electric Stability Program for the Passenger Car

Huiyi Wang and Chunyu Xue

Department of Automotive Engineering, Tsinghua University, China

ABSTRACT

Electronic Stability Program (ESP), also named Vehicle Dynamic Control (VDC), extends the capabilities of the Anti-lock Braking System and the Traction Control System. Firstly, a vehicle dynamic model with 16 DOF, including vehicle body submodel, wheel submodel, force submodel, tire submodel, engine submodel, transmission submodel, also containing braking submodel, and steering submodel is built under SIMULINK software. After completing the simulation under normal mode, A Real-Time Windows Target model under the external mode is produced. The Real-Time Windows Target model for ESP is upgraded to run under Matlab/Simulink/RTW software, to complete the ESP's Hardware-in-the- Loop simulation (HILS) with the help of the I/O boards produced by the Advantech Co. and the I/O driver block library provided by the Real-Time Windows Target. In the end, an interface for controlling the simulation system is designed.

INTRODUCTION

Nowadays, the Anti-lock Braking System (ABS) is a standard device installed in a car from the product line. Recent years, people were paying more and more attention to the safety performance gradually with the increasing vehicle quantity and the increasing traffic number when they plan to buy a car. It is being a common knowledge that the ABS will ensure the safety of the driver and the passengers once the car is running on the wet or snow road. The new type car coming into the market with Traction Control System (TCS) is also being accepted by people, which is not only ensure the safety of the car during braking period, but also it can increase the traction performance and driving safety of the car during starting period. With the new car model GOLF IV came into the Chinese market by the First Auto Work (FAW), one of the biggest auto works in China in the summer 2003, the Electronic Stability Program (ESP) was known by the civilians gradually.

ESP is a fire-new safety system of car in a way. Hebb G. described an automotive dynamic system in 1988[1]. The system in his paper contained ABS, TCS, suspension system, and the steering control sub-system. Once the four sub-systems work together integrated, the car can be run well to keep the driver and the passenger to be comfort whenever it is in accelerating, braking, steering, or swerving.

In the 90s of last century, the technology of the ABS became mature and reliable. Contrarily, the cost of the ABS was decreasing. The ABS was supported to many vehicles and to be a standard device of the new car. More than 90% cars in the United States installed this device in later of 90s. Undoubtedly, the ABS saved many people in the bad weather or the wet road. The limit of the ABS was appearing slowly, that it couldn't support powerful performance during swerving or steering in high speed of the car. In the beginning of the 90s of last century, scientist of Robert Bosch GmbH began the study on the automotive dynamics, they established a system called Vehicle Dynamic Control (VDC)[2][3][4]. The configuration was to integrate the ABS, TCS, with the traction torque control system of engine to make the every sub-system worked well, it got the target of car running stability. Other company, such as Continent Teves GmbH, BMW Co. and HONDA Co., began the study of automotive dynamic also. These years, the scientists and the engineers like to call the automotive dynamic control system as Electronic Stability

Program (ESP).

MODELLING OF THE PASSENGER CAR WITH ESP AND Its OFFLINE SIMULATION

Before analyzing the ESP system, it is necessary to establish the mathematic model of the car with ESP.

When the ESP works, it will involve the longitudinal dynamic and lateral dynamic of the car. The mathematic model of the car will include more variables than the model of the car with ABS or TCS. The model contains not only the wheel speed variable, but also the lateral accelerator variable, yaw rate variable and the steering wheel angle variable. The mathematic model for all road situations is necessary to the Hardware-in-the-Loop Simulation (HILS) of the ESP. It will consider the load change of the four tyres during running, the steering running and the steering braking.

A schematic structure of a passenger car is shown in figure 1.

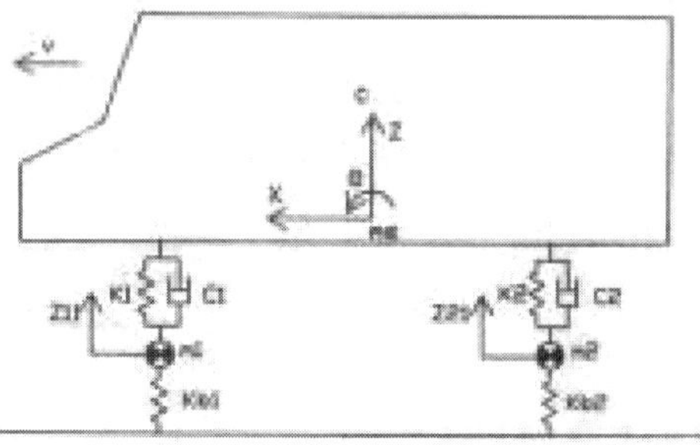

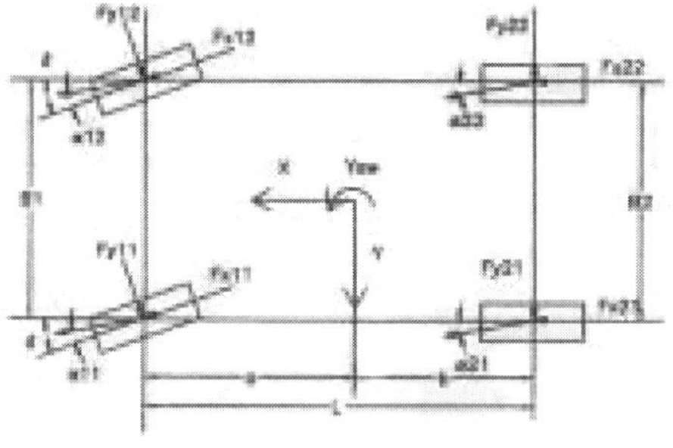

Figure 1 Schematic structure of a passenger car

There are 15 degrees of freedom, they are: longitudinal freedom X, lateral freedom Y, vertical freedom Z of the mass above the suspension, listing freedom φ, yawing freedom Yaw, pitching freedom θ, 4 vertical freedom of four wheels, 4 rolling freedom of four wheels, and the steering freedom of fore tyres.

The movement equation of the longitudinal dynamic of the car is:

$$m(\dot{u}-v\,\dot{Yaw})+m_s D_0\,\ddot{\theta}+m_s e_0\,\dot{Yaw}\,\dot{\varphi}$$
$$=(F_{x11}+F_{x12})\cos\delta-(F_{y11}+F_{y12})\sin\delta+F_{x21}+F_{x22} \tag{1}$$

The movement equation of the lateral dynamic of the car is:

$$m(\dot{v}+u\,\dot{Yaw})-m_s e_0\,\ddot{\varphi}+m_s D_0\,\dot{Yaw}\,\dot{\theta}$$
$$=(F_{x11}+F_{x12})\sin\delta+(F_{y11}+F_{y12})\cos\delta+F_{y21}+F_{y22} \tag{2}$$

The vertical movement equation of the mass above the suspension is:

$$m_s\,\ddot{Z}+F_{s11}+F_{s12}+F_{s21}+F_{s22}=0 \tag{3}$$

The yaw movement equation of the car is:

$$I_z\,\ddot{Yaw}=\frac{B1}{2}\left[(-F_{x11}+F_{x12})\cos\delta-(-F_{y11}+F_{y12})\sin\delta\right]+\frac{B2}{2}(-F_{x21}+F_{x22})$$
$$+a\left[(F_{x11}+F_{x12})\sin\delta+(F_{y11}+F_{y12})\cos\delta\right]-b(F_{y21}+F_{y22}) \tag{4}$$

The pitching movement of the mass above the suspension is described as:

$$J_y\,\ddot{\theta}+m_s D_0\left(\dot{u}-v\,\dot{Yaw}\right)-F_{s11}a-F_{s12}a+F_{s21}b+F_{s22}b=0 \tag{5}$$

The listing movement of the mass above the suspension is described as:

$$J_x\,\ddot{\varphi}-m_s e_0\left(\dot{v}+u\,\dot{Yaw}\right)+F_{s11}\frac{1}{2}B_{11}-F_{s12}\frac{1}{2}B_{11}+F_{s21}\frac{1}{2}B_{22}-F_{s22}\frac{1}{2}B_{22}=0 \tag{6}$$

According to the structure of the car shown in figure 1, the vertical movements of the four tyres are described as:

$$m_1\,\ddot{Z}_{11}-F_{s11}+K_{b1}Z_{11}=0 \tag{7}$$

$$m_1\,\ddot{Z}_{12}-F_{s12}+K_{b1}Z_{12}=0 \tag{8}$$

$$m_2\,\ddot{Z}_{21}-F_{s21}+K_{b2}Z_{21}=0 \tag{9}$$

$$m_2\,\ddot{Z}_{22}-F_{s22}+K_{b2}Z_{22}=0 \tag{10}$$

Similarly, according to the force schematic drawing in figure 2, the rolling movements of the four tyres are described as:

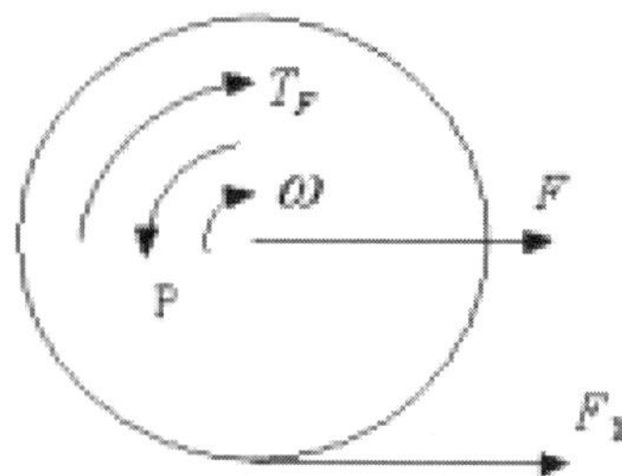

Figure 2 the forces of the rolling tyre

$$J_1 \dot{\omega}_{11} = -F_{x11}R_1 - P_{11} + T_{F11} \qquad (11)$$

$$J_1 \dot{\omega}_{12} = -F_{x12}R_1 - P_{12} + T_{F12} \qquad (12)$$

$$J_2 \dot{\omega}_{21} = -F_{x21}R_2 - P_{21} \qquad (13)$$

$$J_2 \dot{\omega}_{22} = -F_{x22}R_2 - P_{22} \qquad (14)$$

The forces on the suspensions are:

F_{s11}、 F_{s12}、 F_{s21}、 F_{s22} , they are decided by the following

four equations:

$$F_{s11} = K_1(Z - Z_{11} - a\theta + \frac{1}{2}B_{11}\varphi) + C_1(\dot{Z} - \dot{Z}_{11} - a\dot{\theta} + \frac{1}{2}B_{11}\dot{\varphi}) \qquad (15)$$

$$F_{s12} = K_1(Z - Z_{12} - a\theta - \frac{1}{2}B_{11}\varphi) + C_1(\dot{Z} - \dot{Z}_{11} - a\dot{\theta} - \frac{1}{2}B_{11}\dot{\varphi}) \qquad (16)$$

$$F_{s21} = K_2(Z - Z_{21} + b\theta + \frac{1}{2}B_{22}\varphi) + C_2(\dot{Z} - \dot{Z}_{21} + b\dot{\theta} + \frac{1}{2}B_{22}\dot{\varphi}) \qquad (17)$$

$$F_{s22} = K_2(Z - Z_{22} + b\theta - \frac{1}{2}B_{22}\varphi) + C_2(\dot{Z} - \dot{Z}_{22} + b\dot{\theta} - \frac{1}{2}B_{22}\dot{\varphi}) \qquad (18)$$

Selecting the tyre model

Tyre model is the important sub-model of the whole passenger car because the major force to the passenger car is come from the tyre. In this paper, a parameter-force_rate is defined to describe the road situation. Once the force_rate=0.8, it means the road is high-miu road, contrarily, force_rate=0.3 means the road is low-miu road.

After defining coefficient of the longitudinal force φ_x as

the ratio of the longitudinal force F_x and the vertical force

F_z, the relationship of the φ_x and the slip rate is shown in

figure 3. We can see that the longitudinal force coefficient has a maximum value during the slip rate closed to point 10%.

Similarly, after defining coefficient of the longitudinal force

φ_y as the ratio of the longitudinal force F_y and the vertical

force F_z, the relationship of the φ_y and the slip rate is shown

in figure 4.

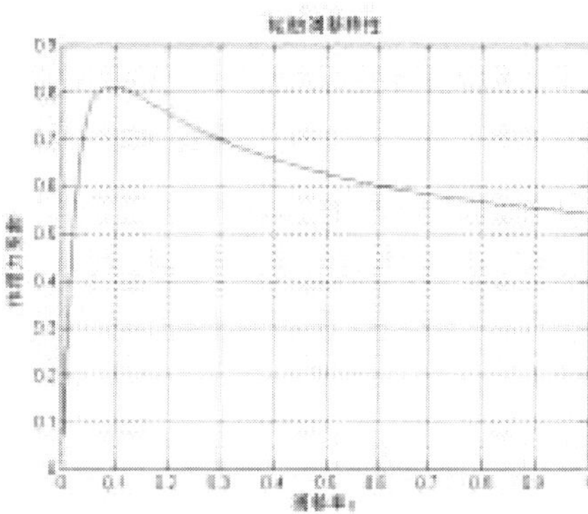

Figure 3 relationship of the φ_x and the slip rate

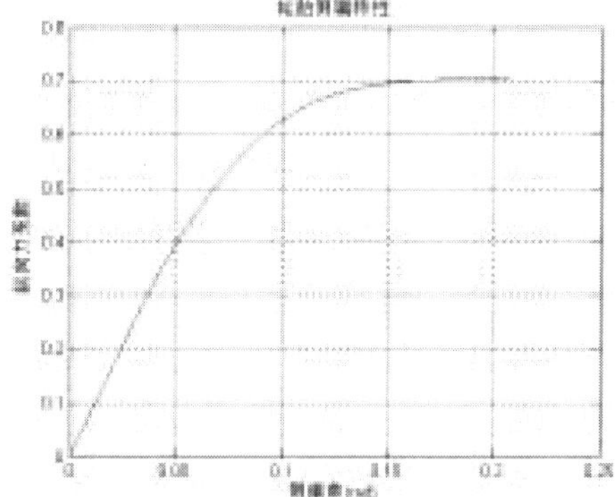

图 4 relationship of the φ_y and the slip rate

Adapting the SAE tyre reference system (figure 5), the side leaning angle s and longitudinal slip rates is described as:
1) Side leaning angle and longitudinal slip rates of the left front wheel are:

$$\alpha_{11} = \tan^{-1}\left(\frac{v + a\dot{Yaw}}{u - 0.5B_1\dot{Yaw}}\right) - \delta \qquad (19)$$

$$S_{x11} = \frac{\omega_{11}R_1 - u + 0.5B_1\dot{Yaw}}{u - 0.5B_1\dot{Yaw}} \qquad \text{when}$$

$$u - 0.5B_1\dot{Yaw} >= \omega_{11}R_1 \qquad (20)$$

$$S_{x11} = \frac{\omega_{11}R_1 - u + 0.5B_1\dot{Yaw}}{\omega_{11}R_1} \qquad \text{when}$$

$$u - 0.5B_1\dot{Yaw} < \omega_{11}R_1$$

2) Side leaning angle and longitudinal slip rates of the right front wheel are:

$$\alpha_{12} = \tan^{-1}\left(\frac{v + a\dot{Yaw}}{u + 0.5B_1 Yaw}\right) - \delta \qquad (21)$$

$$S_{x12} = \frac{\omega_{12}R_1 - u - 0.5B_1\dot{Yaw}}{u + 0.5B_1\dot{Yaw}} \qquad \text{when}$$

$$u + 0.5B_1\dot{Yaw} >= \omega_{12}R_1 \tag{22}$$

$$S_{x12} = \frac{\omega_{12}R_1 - u - 0.5B_1\dot{Yaw}}{\omega_{12}R_1} \quad \text{when}$$

$$u + 0.5B_1\dot{Yaw} < \omega_{12}R_1$$

3) Side leaning angle and longitudinal slip rates of the left rear wheel are:

$$\alpha_{21} = \tan^{-1}\left(\frac{v - b\dot{Yaw}}{u - 0.5B_2\dot{Yaw}}\right) \tag{23}$$

$$S_{x21} = \frac{\omega_{21}R_2 - u + 0.5B_2\dot{Yaw}}{u - 0.5B_2\dot{Yaw}} \quad \text{when}$$

$$u - 0.5B_2\dot{Yaw} >= \omega_{21}R_2 \tag{24}$$

$$S_{x21} = \frac{\omega_{21}R_2 - u + 0.5B_2\dot{Yaw}}{\omega_{21}R_2} \quad \text{when}$$

$$u - 0.5B_2\dot{Yaw} < \omega_{21}R_2$$

4) Side leaning angle and longitudinal slip rates of the right rear wheel are:

$$\alpha_{22} = \tan^{-1}\left(\frac{v - b\dot{Yaw}}{u + 0.5B_2\dot{Yaw}}\right) \tag{25}$$

$$S_{x22} = \frac{\omega_{22}R_2 - u - 0.5B_2\dot{Yaw}}{u + 0.5B_2\dot{Yaw}} \quad \text{when}$$

$$u + 0.5B_2\dot{Yaw} >= \omega_{22}R_2 \tag{26}$$

$$S_{x22} = \frac{\omega_{22}R_2 - u - 0.5B_2\dot{Yaw}}{\omega_{22}R_2} \quad \text{when}$$

$$u + 0.5B_2\dot{Yaw} < \omega_{22}R_2$$

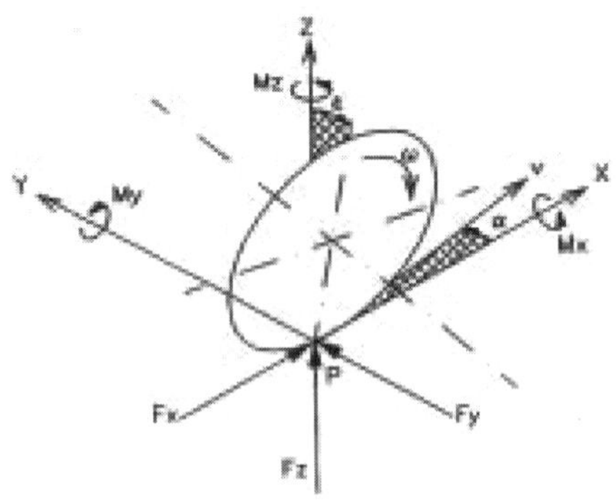

Figure 5 Tyre reference systems

The torque from the engine is transferred to the differential via the clutch, shift gears, and the transmission axle, the movement equation is:

$$(J_e \cdot i_g^2 \cdot i_o^2 + J_g \cdot i_o^2 + J_o) \cdot \dot{\omega}_d = \eta \cdot T_p \cdot i_g \cdot i_o - T_d \tag{27}$$

Considering the requirement of real-time running of the mathematic model of the passenger car, the above equation is revised as:

$$\eta \cdot T_p \cdot i_g \cdot i_o = T_d \tag{28}$$

$$T_d(1) = T_d(2) = 0.5T_d \tag{29}$$

Off-line Simulation

Before off-line simulation, the values of the parameters of the mathematic model were decided first. The tyre parameters are shown in the table 1.

Table 1 the parameters of the tyre

	A1	A2	A3	A4	A5	A6	A7	A8
Fy	-22.1	1011	1078	1.82	0.208	0	-0.354	0.707
Mz	-2.72	-2.28	-1.86	-2.73	0.110	-0.07	0.643	-4.04
Fx	-21.3	1144	49.6	226	0.069	-0.006	0.056	0.486

On the MATLAB/SIMULINK platform, the Runge-Kutta algorithm is selected to solve the mathematic model. The step of solving model is fixed as 0.0006ms.

The mathematic models of three cases of the passenger car were solved when the car is accelerating at first speed shift, braking at high speed (more than 80km/h), and accelerating with steering at fist speed shift.

Accelerating at first speed shift:

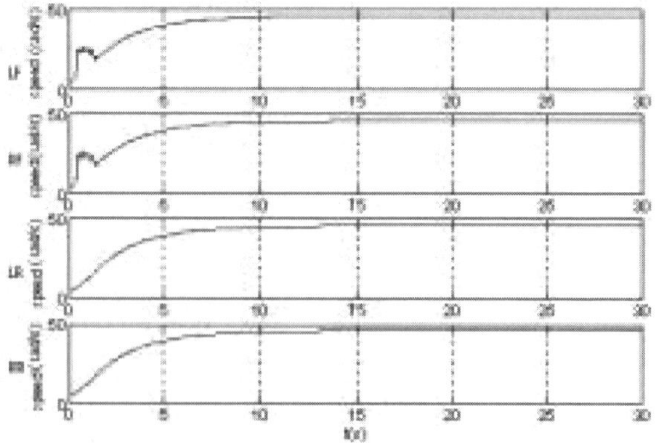

Figure 6 the wheel speeds of the passenger car during accelerating at I speed shift

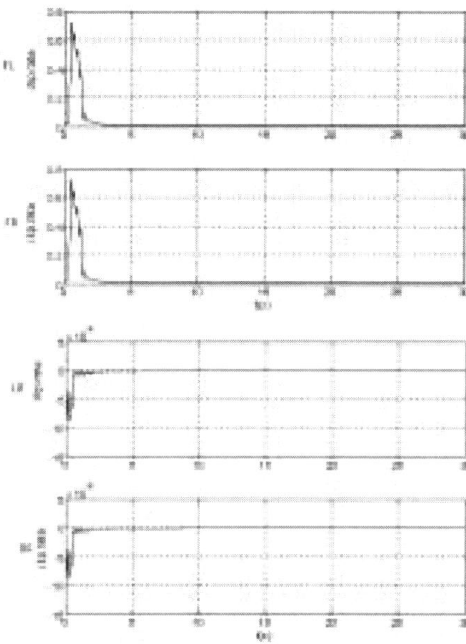

Figure 7 the slip rates of the passenger car during accelerating at
I speed shift

The front wheels had the obvious skidding during the accelerating in first speed shift without any control to the passenger car. It caused the accelerating distance longer.

REALTIME OF THE SYSTEM MODEL AND Its REALTIME SIMULATION

Basing the offline simulation, the model will be modified according to the requirement of the real time simulation. Before the real time simulation using the Real-Time Windows Target, the hardware of the HILS system were configured firstly.

The ESP HILS will perform the operation of the ESP in the real passenger car. The structure of ESP HILS system is shown in figure 8.

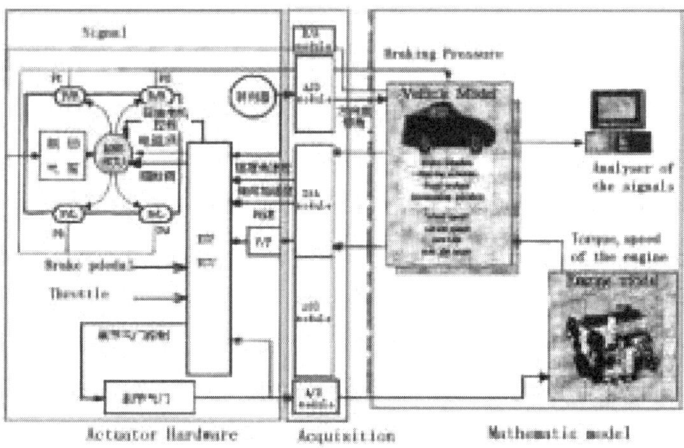

Figure 8 the structure of the ESP HILS system

The ESP HILS system contains three parts: outside hardware device, data acquisition device, and the model solving system. The outside hardware devices have the braking system, which is nonlinear and hard to be modeled, signal sensors, and correlated electronic circuits.

The process of the operation of the ESP HILS can be concluded as:

Step 1: Running the passenger car model, the signals of the driver's operating parameters, such as the openness signal of the throttle pedal, the braking signal, the steering wheel angle signal, etc. were transferred to the mathematic model through the Analog/Digital device.

Step 2: Sending the parameters of the passenger car to the ECU board through the Analog/Digital device again, the ECU will detect the related wheel speeds. After calculating the differential of the wheel speeds, the ECU will check the thresholds of the wheel speeds and the wheel acceleration. Once the measured wheel speed and/or the wheel rotation acceleration match the set threshold, the ECU will send the control signal for the pressure modulator.

Step 3: Controlling the desired control parameters, such as openness of the throttle, pressure of the braking system with the regulating of the actuators, the parameters of the passenger car will be changed, the situation of the passenger car will be a new values.

Step 4: Returning to the step 1, one time of the closed loop control is completed.

The ECU of the ESP has seven input signals normally; there are 4 wheel speeds, 1 lateral accelerator, 1 yaw rate, and 1 steering wheel angle. An acquisition device PCI 1710HG, produced by the ADVANTECH CO. was utilized to connect the mathematic model of MATLAB/SIMULINK environment and the hardware rig including the brake actuator, the pipe of the braking system, and the ECU of the ESP. In the same time, the operating situations of the pressure actuators of the ESP were shown in the signal analyzer checked by the digital input port of the PCI 1710HG. It will help the researchers to design the detailed ESP control strategy.

The pressures of the wheel cylinders of the passenger car were the main parameters to indicate the state of the passenger car. The braking torque of the wheel was the real parameter for the mathematic model of the passenger car. Considering the relationship of the braking torque of the wheel and the braking pressure is necessary. After a serial testing of a pair of disk brake and drum brake, the relationships of the braking torque and the braking pressure were concluded as:

$$Tb = -3.0488P2 - 95.4313P, \quad \text{for disk brake} \tag{30}$$

$$Tb = -0.4370P2 - 16.436P, \quad \text{for the drum brake} \tag{31}$$

INTERFACE OF THE SOFTWARE OF THE ESP HILS

The interface of the software is an important part of the ESP HILS system. Under the MATLAB environment, the interface of the ESP HILS software was designed using the GUI tools. The interface includes five parts: Initiation of the model, Control of the system, Data managing, Data plotting, and Connecting of hardware and the software. Figure 9 shows the designed interface of the ESP HILS.

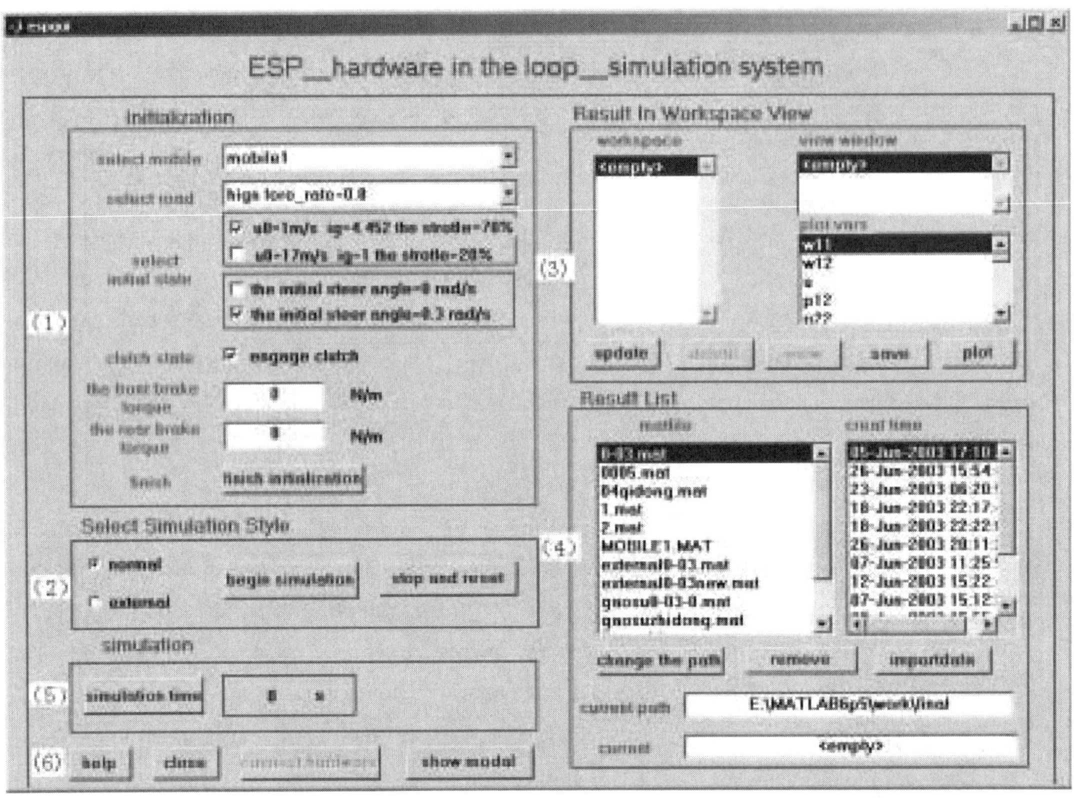

Figure9 the interface of the software of the ESP HILS

HILS OF THE PASSENGER CAR WITH ESP

The ESP HILS system is composed of three parts indeed: mathematic models for the PC computer, actuator hardware, and the ECU for controlling the ESP. The real ESP HILS system established in the State Lab of the Automotive Safety and Energy is shown in figure 10.

There are two PC in the ESP HILS system, the PC1 is for running the mathematic model in real time environment, and the PC2 is for running the ESP control program in the Infineon C167 simulator in PCI protocol interface. The signals of the engine throttle, the braking pressures were connected to the PC1 through the acquisition device. The real time state of the passenger car was calculated, the real time wheel speed values were transferred to the ESP

control board-the C167 simulator in digital mode by the V/F module. The ESP control board in PC2 would perform the control strategy according to the wheel speed signals in digital mode.

The operating process of the ESP HILS system can be concluded as:

Step 1: Power on the PC1 and PC2.

Step 2: Run the MATLAB software, and then load the GUI program.

Step 3: Select the running mode as "external", and then initialize the correlative parameters: the type of the vehicle, the road adhesion coefficient, the initial speed of the vehicle, etc.

Step 4: Press the button to begin the simulation.

Step 5: Select the desired variables to be shown in PC1 once the "simulation finish" prompt appearing. In the same time, you can select the desired variables to be saved in the hard disk.

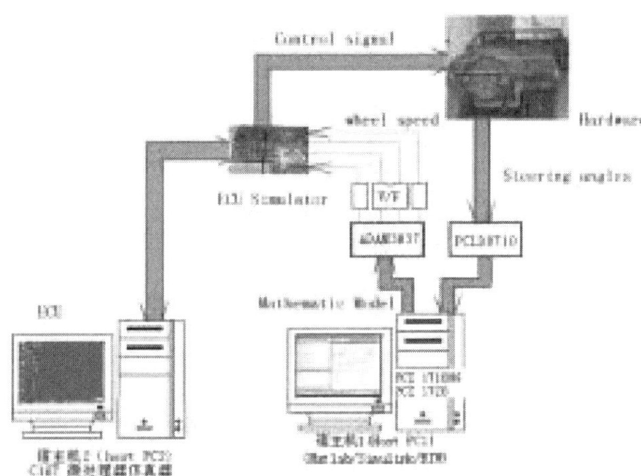

Figure 10 the real ESP HILS system

CONCLUSION

The ESP extends the capabilities of the Anti-lock Braking System and the Traction Control System; it was fixed on the car in Chinese market just recent days. In this paper, the primary researches on the ESP fixed on a passenger car were accomplished.

The mathematic model of a passenger car with ESP was established, it included more than 15 DOF.

The off-line simulation of the ESP was performed, that examined the mathematic model based on the MATLAB/SIMULINK platform.

The HILS system was established considering the nonlinear characteristics of the hydraulic actuator, the real time requirement of the system.

An interface designed for the HILS helped the user to operate the HILS.

REFERENCES

[1].Heeb G, Van Zanten. System Approach to Vehhicle Dynamics Control. FISITA 1998, No. 885107, Deptroit.

[2].Anton T.van Zanten, Rainer Erhardt and Georg Pfaff. VDC,The Vehicle Dynamic Control System of Bosch. Robert Bosch GmbH. SAE950759

[3].Anton T.van Zanten, Rainer Erhardt, Albert Lutz, Wilfried Neuwald and Hartmut Bartels.Simulation for the Development of the Bosch-VDC. Robert Bosch GmbH. SAE960486

[4].Anton T.van Zanten, Rainer Erhardt, Klaus landesfeind and Georg Pfaff. VDC Systems Development and Perspective. Robert Bosch GmbH. SAE 980235

[5].Y.Shibahata, K.Shimada and T.Tomari. Improvement of Vehicle Maneuverability by Direct Yaw Moment Control. Vehicle System Dynamics, 22(1993), pp.465-481

[6].Anton T.van Zanten. Evolution of Electronic Control Systems for improving the Vehicle Dynamic Behavior. Robert Bosch GmbH.   AVEC02

[7].Anton T.van Zanten. Bosch ESP Systems: 5 Years of Experience. SAE 2000-01-1633

[8].Michael Fodor, John Yester and Davor Hrovat. Active Control of Vehicle Dynamics. Ford Motor Company, 0-7803-5086-3/98 IEEE

[9].H.E.Tseng, D.Madau, B.Ashhrafi, T.Brown and D.Recker. Technical Challenges in The Development of Vehicle Stability Control System. Ford Motor Company, Visteon Automotive Systems, Proceeding s of the 1999 IEEE, International Conference on Control Applications.

[10].H.E.Tseng, D.Madau, B.Ashhrafi, T.Brown and D.Recker.   The Development of Vehicle Stability Control at Ford. IEEE/ASME Transactions on Mechatronics, Vol.4, No.3, September 1999.

[11].Egbert Bakker, Lars Nyborg and Hans B. Pacejka. Tyre Modelling for Use in Vehicle Dynamics Studies. SAE870421.

[12].Egbert Bakker, Hans B. Pacejka and Lars Nyborg.   A New Tire Model with an Application in Vehicle Dynamics Studies.   SAE890087

[13].Ken Koibuchi, Masaki Yamamoto, Yoshiki Fukada and Shoji Inagaki. Vehicle Stability Control in Limit Cornering by Active Brake.   SAE 960487, Toyota Motor Corp.

[14].Yoshiyuki Yasul, Kenji Tozu, Noriaki Hattori and Masakazu Sugisawa. Improvement of Vehicle Directional Stability for Transient Steering Maneuvers Using Active Brake Control. SAE 960485, Aisin Selld Co., Ltd

[15].Chen Jiong, Wang Huiyi, and Song Jian. Research of the ABS's Control Strategy under the Virtual Environment Road and Traffic Technology. Vol.19, No.6, 2002 (In Chinese)

Evaluation Criteria for AWD Vehicles System Analysis

P. Borio, U. Delcaro, S. Frediani and C. Ricci
Centro Ricerche Fiat

G.Caviasso
Fiat Auto Innovation & Component Development

ABSTRACT

Handling behaviour is a very important aspect of modern cars, which need to reach high levels of lateral acceleration maintaining good stability and driveability. The presence of an AWD powertrain changes the response of the vehicle and then it is very important to qualify the car behavior in terms of perceived performances through objective indexes. The paper presents the development of evaluation criteria for AWD vehicles objective assessment; simulation tools and SW standard procedure has been optimised for AWD performances virtual evaluation and different architectures (locking, self-locking, visco-coupling, active differentials) are compared in terms of longitudinal, lateral and cross-coupling dynamics performances. Through the proposed methods it is possible to make vehicle system analysis using simulation model and standard procedures to classify different AWD architectures. It is also a powerful tool for target setting-deployment-achieving in order to traduce AWD vehicle performance in terms of architecture choice and component specifications. The same criteria could be used to support control strategies development and on road AWD systems experimental tuning through simulation analysis. In the paper the application of the method to a virtual upgrade of the Alfa Romeo car is also described: the AWD layouts of interest have been simulated on the virtual car, as a support to the choices of a new Fiat platform.

AWD VEHICLES PERFORMANCES ANALYSIS

Speaking about the handling performance of a car, it is necessary to distingue between longitudinal, lateral performances and cross coupling (see figure 1).

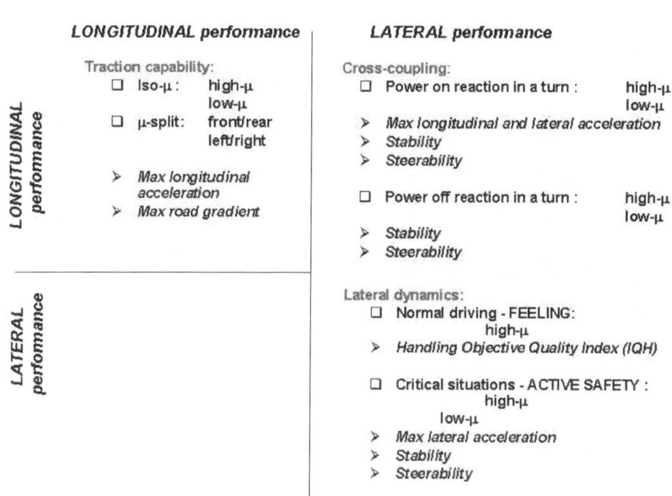

Figure 1: Handling performances of a car

LONGITUDINAL PERFORMANCE

The longitudinal performance regards traction capability in straight line (maximum acceleration) and the maximum road gradient which can be passed in different cases:

- Iso μ: the adhesion coefficient is the same under the four wheels: all the adhesion value are considered, from low (i.e. icy or snowy ground) to high (dry asphalt, good grip).

- μ-Split: the adhesion coefficient is different under the wheels. Depending on the archetype in study, it might be interesting to consider left/right split (it is the case of on axle differential), or front/rear split (central differential).

Taking in account of the pitch transfer load and of the geometry of the car (wheelbase, CG height, mass), the maximum performance on iso-μ conditions of the car is described by $a_x = \mu$.

In order to quantify the performance loose with respect to the maximum one, the TCI (Traction capability Index) has been created, as difference between the area held by the ideal performance and the obtained performance. The upper limit is defined by the engine: considering the vehicle in 1st gear, knowing the gearboxes transmission rates and taking into account of the powertrain efficiency, it is possible to convert the maximum engine power into a maximum possible acceleration. In figure 2 the sketched blue area represents the TCI: the higher TCI is, the best the performance is.

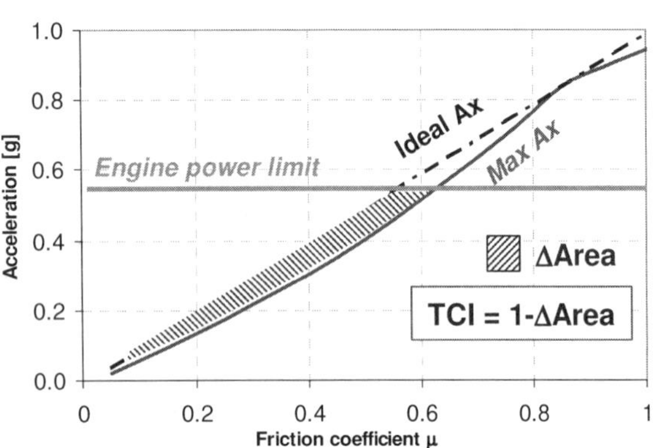

Figure 2: TCI definition

LATERAL DYNAMICS

The lateral dynamics performance must be seen from two different corners: normal driving (feeling area) and critical situations (car at limit, active safety).

- Feeling area: it is considered to be in the low-middle lateral acceleration values, on high μ. In this conditions the handling feedback to the driver is studied by IQH (Handling objective Quality Index) [1].

- Critical situations: in this case the car is near the limit, and maximum lateral acceleration reachable becomes very important, since they may really improve the active safety of the car.

The lateral vehicle dynamics can be detected with the collaboration of professional drivers, who could describe vehicle behavior through their sensations. A different engineering approach could be defined, finding a correlation between objective measurements during testing and subjective evaluations: this is the approach which has led to the development of IQH index. The application of the related methodology (collecting data in standardized ISO tests and analysing them in standard procedures) permits to evaluate the perceived handling quality of a car, in terms of:

- Steering wheel activity (IAV)

- Quickness of car response (IRV)

- Car feedback graduality (IPI)

- Roll motion (ICO)

- Roll motion velocity (IVC)

Usually the IQH ranking of a car is represented by a spider graph, like shown in figure 3.

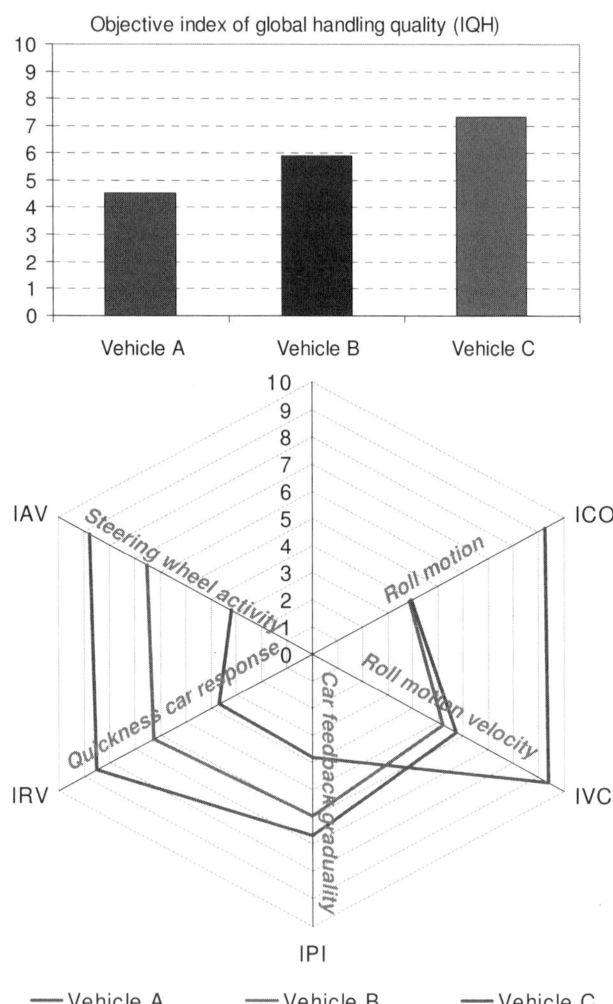

Figure 3: IQH ranking of three different cars

The IQH methodology is a powerful way to compare the different 4WD archetypes, in term of perceived handling quality of the car. The IQH methodology is based on a set of standard maneuvers: from the collected (or simulated) time histories a set of synthetic parameters is calculated, obtaining through statistical models and correlations the indexes of perceived quality. The set of maneuvers used in this approach is composed by:

- Steady state circular test, according to ISO 4138 [2]

- Step steering wheel input, according to ISO 7401 [3]

- Double lane change according to ISO/DIS 3888 [4]

To complete the analysis, frequency sweep (according to ISO 7401) are processed.

LATERAL DYNAMICS IN CRITICAL SITUATIONS

At high lateral acceleration the IQH methodology is not applicable, and the point is not the good feeling from the car, but the active safety. The car should reach great values of lateral acceleration without strong oscillations, permitting its control by steer wheel. In general, two aspects become really important: stability and steerability. The car must be stable, without spin phenomena (stability), and must respond to driver's steering input (steerability). These conditions guarantee a good safety near the limit of the car, i.e. in a critical situation like an obstacle avoiding at high velocity. The tests usually used to evaluate the limit are:

- Step steer input at high steering angle values (ISO 7401)

- Slow ramps (according to ISO 4138)

In figure 4 the time histories of two cars are superimposed, in a step steer input at limit. The vehicle "A" has stronger oscillation than vehicle "B", with high amplitudes and low damping.

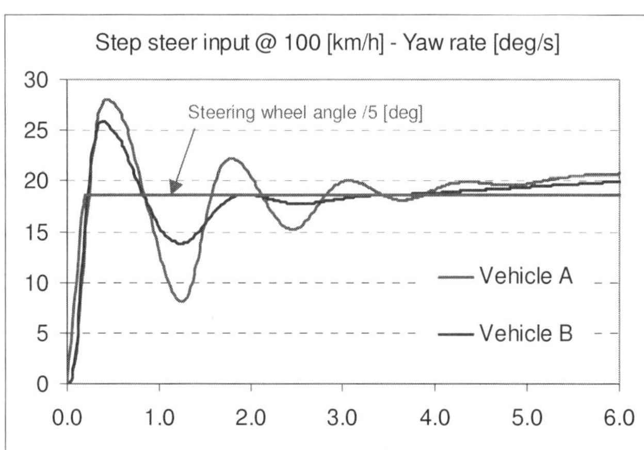

Figure 4: Step steer input at limit

The results coming from the analysis of slow ramps are understeer and side slip curves. The vehicles are compared in terms of maximum lateral acceleration taking into account of the sideslip angles values and of the understeer curve. An all wheel drive powertrain could strongly change the limit of the car, especially if the same components are controlled (i.e. C.R.F.

CYMENTO© [5] active differential or electromechanical coupling). In particular, the rear active differential of C.R.F. is able to split a certain amount of torque regardless of the torque coming from the engine and the relative speed of the output shafts.

CROSS COUPLING

When the cross coupling is studied, both longitudinal and lateral performance have to warrant a good reaction of the car: this is the case of power in or power off in a turn. The car could follow the path increasing or decreasing the radius, with yaw and sideslip oscillations. At the same time if the car has to be powerful during a power in, the maximum performance is required not only in X-direction, but also in Y-direction. To investigate this behaviors, the test usually done are:

- Power on/off in steady state cornering (according to ISO 9816 [6])

- Steering pad on constant radius (R=10-40-100 [m]) with constant acceleration/deceleration

To qualify the reaction of the car during the power off, the ISR index will be used. This index takes into account the yaw and sideslip response after the throttle transient, and the higher is the index value, the better is the response of the car in terms of deviation from its original path.

SIMULATION TOOLS

The all wheel drive layouts in study are evaluated as described below (see figure1) by the use of softwares developed at Centro Ricerche Fiat, each of them dedicated to particular performances (figure 5).

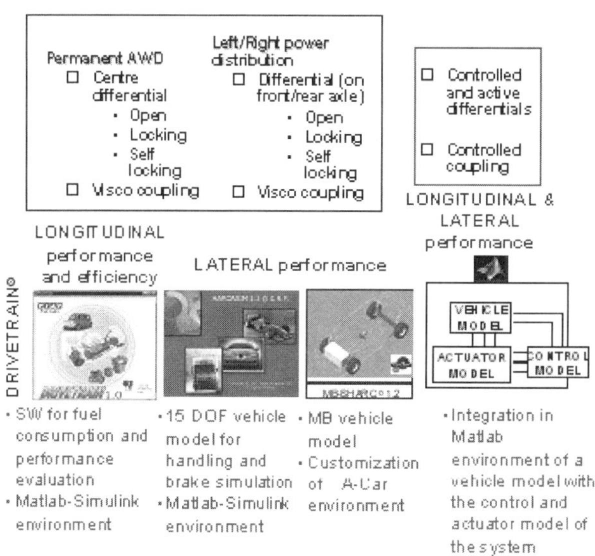

Figure 5: Simulation tools

DRIVETRAIN© is a software of Centro Ricerche Fiat, written in Matlab environment. Its goal is to estimate longitudinal performances and fuel consumption of a car.

Harcasim© code is a 15 DOF model to simulate the lateral dynamics in Matlab/Simulink environment. It is under constant development at Centro Ricerche Fiat. All maneuvers necessary to evaluate IQH index are available in Harcasim© code, so starting from the vehicle general data (dimensions, weight, suspensions characteristics, tyres characteristics), a complete simulation can be done. MB SHARC© is an integrated handling and ride comfort environment based on multibody vehicle modelling, born like a customization of Adams/Car, containing all the IQH maneuvers. These tools are very useful to examine the AWD archetypes to provide a ranking: with DRIVETRAIN© it is possible to optimize the torque distribution (front/rear or left/right) and the locking factors to obtain the best longitudinal performance. Then, by HARCASIM© code the impact of this choice on the lateral dynamics is evaluated. If a zoom on a vehicle subsystem is needed, MB SHARC© multibody vehicle models are usually used. Nowadays, the application of active systems in all wheel drive archetypes is more and more diffuse, so the integration between actuators/controls and the vehicle model is necessary. Matlab has been chosen to melt together these aspects, providing an integrated environment to evaluate the active subsystem contribution to the handling behaviour of the car, once the actuators models and controls algorithms are known. This task in particular is described in this paper: two layouts with active rear differential are compared to the same passive ones.

ALL WHEEL DRIVE SYSTEM ANALYSIS: APPLICATION

This example regards an application to a virtual upgrade of an Alfa Romeo car, the characteristics of which are represented in figure 6.

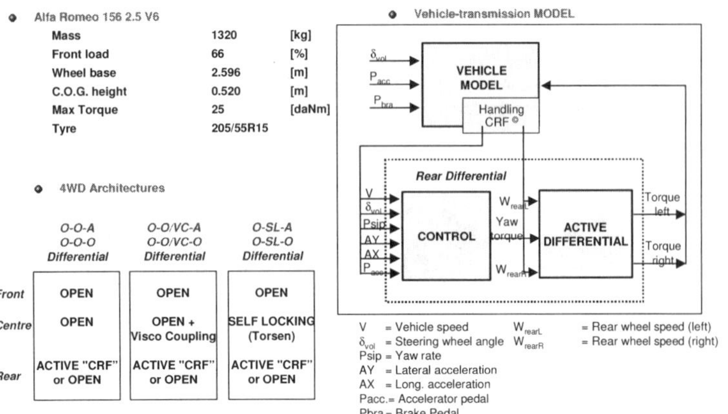

Figure 6: Application of system analysis - layouts

The reference vehicle is front axle drive and both suspensions and tyres characteristics are known. The

first layout upgrade is to consider the vehicle as a 4x4 one, with three open differentials or the active differential on the rear axle. The second layout presents a visco coupling at center and open or active on the rear, then the last has a self locking (Torsen) on the center and an open or the active on the rear. Since the active differential is composed by actuators and controls, the integrated Matlab environment has been necessary to do the analysis (as shown in figure 8 and explained below). The first task examined has been the traction capability by DRIVETRAIN © software, as shown in figure 7.

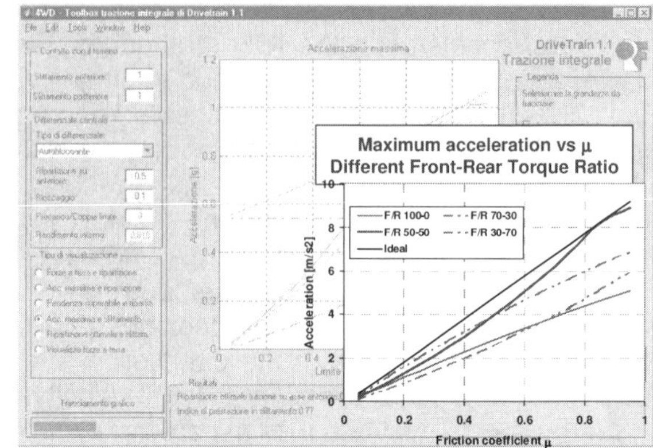

Figure 7: Traction capability: front/rear torque transfer

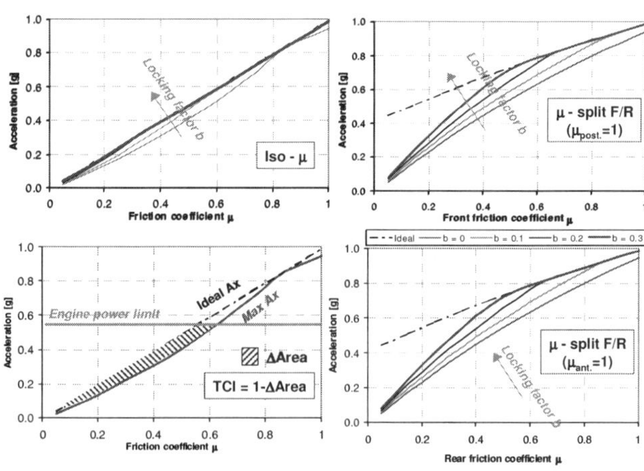

Figure 8: Traction capability – Central differential locking factor (torque rate = 50%)

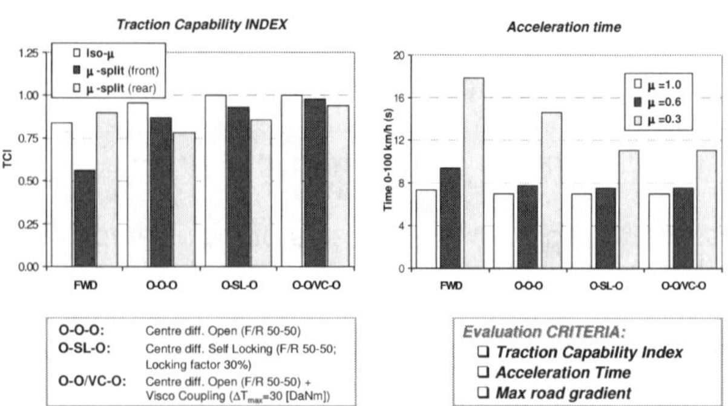

Figure 9: Traction capability – Ranking

Considering the (μ,ax) charts and the TCI index as indicators of the performances, the different layouts have been compared. When the central diff is a self locking,

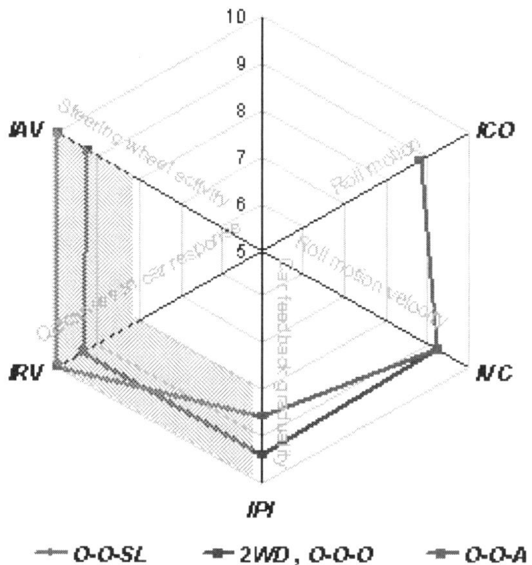

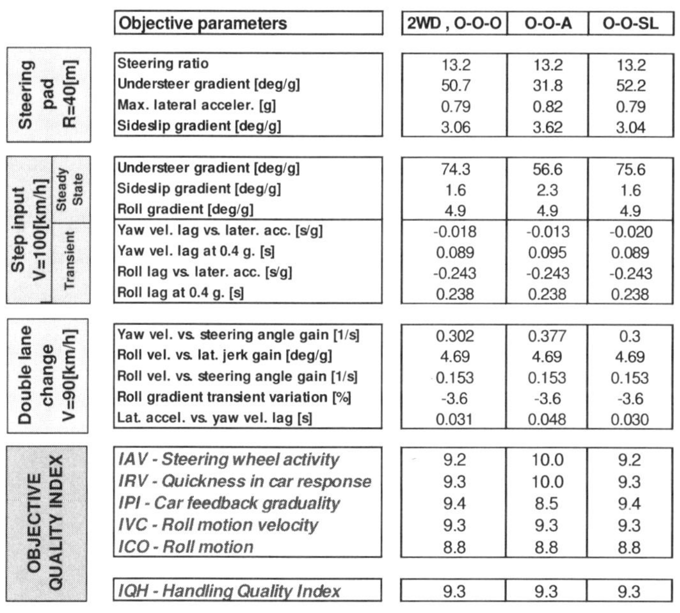

	Objective parameters	2WD , O-O-O	O-O-A	O-O-SL
Steering pad R=40[m]	Steering ratio	13.2	13.2	13.2
	Understeer gradient [deg/g]	50.7	31.8	52.2
	Max. lateral acceler. [g]	0.79	0.82	0.79
	Sideslip gradient [deg/g]	3.06	3.62	3.04
Step input V=100[km/h] — Steady State	Understeer gradient [deg/g]	74.3	56.6	75.6
	Sideslip gradient [deg/g]	1.6	2.3	1.6
	Roll gradient [deg/g]	4.9	4.9	4.9
Transient	Yaw vel. lag vs. later. acc. [s/g]	-0.018	-0.013	-0.020
	Yaw vel. lag at 0.4 g. [s]	0.089	0.095	0.089
	Roll lag vs. later. acc. [s/g]	-0.243	-0.243	-0.243
	Roll lag at 0.4 g. [s]	0.238	0.238	0.238
Double lane change V=90[km/h]	Yaw vel. vs. steering angle gain [1/s]	0.302	0.377	0.3
	Roll vel. vs. lat. jerk gain [deg/g]	4.69	4.69	4.69
	Roll vel. vs. steering angle gain [1/s]	0.153	0.153	0.153
	Roll gradient transient variation [%]	-3.6	-3.6	-3.6
	Lat. accel. vs. yaw vel. lag [s]	0.031	0.048	0.030
OBJECTIVE QUALITY INDEX	IAV - Steering wheel activity	9.2	10.0	9.2
	IRV - Quickness in car response	9.3	10.0	9.3
	IPI - Car feedback graduality	9.4	8.5	9.4
	IVC - Roll motion velocity	9.3	9.3	9.3
	ICO - Roll motion	8.8	8.8	8.8
	IQH - Handling Quality Index	9.3	9.3	9.3

Figure 10: Lateral dynamics - IQH approach

the 50% torque rate is decided, based on the weight balance and the geometry of the car, and the locking factor can be optimized for this mission: in figure 10 a comparison with different locking factors is shown. In the (μ,ax) graph, it is shown that increasing the locking factor is useful to reach higher performances, both on iso-μ conditions and μ-split ones. Also the TCI index (see figure 9) and the acceleration time (0-100 km/h) show a strong increasing as the locking factor increase.

When the visco coupling is present at the center, TCI index is the best in all situations, while the layouts with active rear differential are expected to be more powerful in lateral and cross coupling situations. Thus, basing the performance evaluation on the TCI index and acceleration time, once known the future mission of the car (sport car, sport utility vehicle etc..) the first tuning of

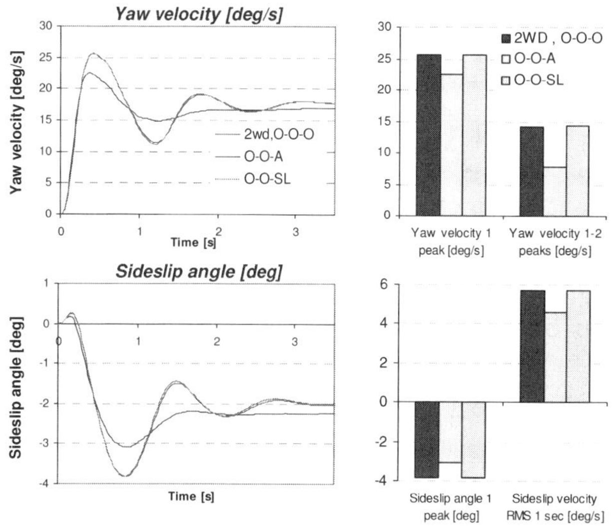

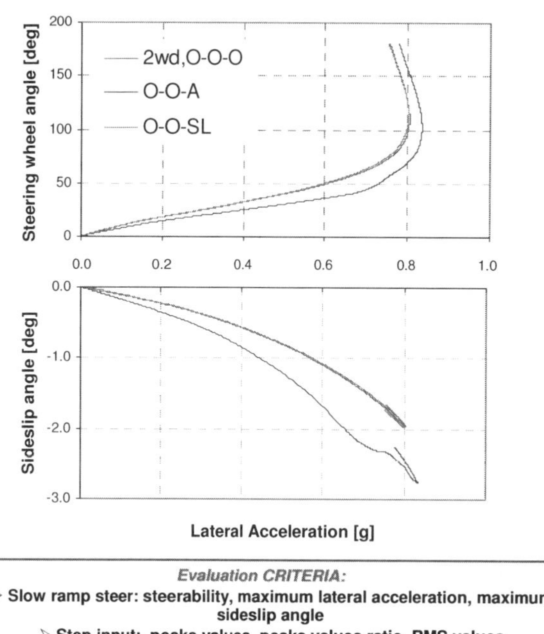

Figure 11: Lateral dynamics – active safety (limit)

the powertrain can be done. Next step is to evaluate the influence of the chosen layouts on the lateral dynamics. As explained below, this can be done by IQH index (see figure 10). In the spider graph, it is shown that the passive layouts cannot change the feeling area: in fact, the 2WD vehicle has the same handling perceived quality of the passive 4WD car with three open differentials or with a self locking on the rear axle. This is because the engine torque required to complete the IQH maneuvers is just a little amount of the maximum one, so the differentials cannot transfer torque enough to create a significant yaw momentum. This results in a similar behavior between 2WD car and passive 4WD. The presence of the active differential on the rear allows to obtain different perceived handling and, depending on the control algorithm, all the filled area of the spider graph can be modified. This is because the device is able to transfer torque from one to the other rear wheel, regardless the engine torque, and this results in a significant yaw moment which can heavily modify the perceived behavior of the car. The same effects are shown at the limit, where the 4WD active vehicle has different response with respect to 2WD or passive 4WD (see figure 11). In the step steer at limit, it is clear that the active vehicle has the lowest amplitudes and frequencies of yaw oscillations; sideslip peak value is lower than the 2WD's one. This is because the passive layout cannot create a yaw momentum, as the active layout can do. The 4WD active vehicle has the highest values of lateral acceleration, but the higher values of sideslip angle too. In conclusion, at the limit the passive 4WD vehicle is really similar to 2WD one, in term of oscillations, maximum lateral acceleration and maximum sideslip angle values. If the active differential is mounted on the rear axle, the vehicle behavior could heavily change, and the changes may be optimized by the tuning of the active system control. When the cross coupling behavior is in study, two are the interesting tasks to study: the traction capability and the reaction of the car in a tip in/out in a turn. In this application, the first

Steering pad on constant radius

10 [m] Ax = 0.3 [g]

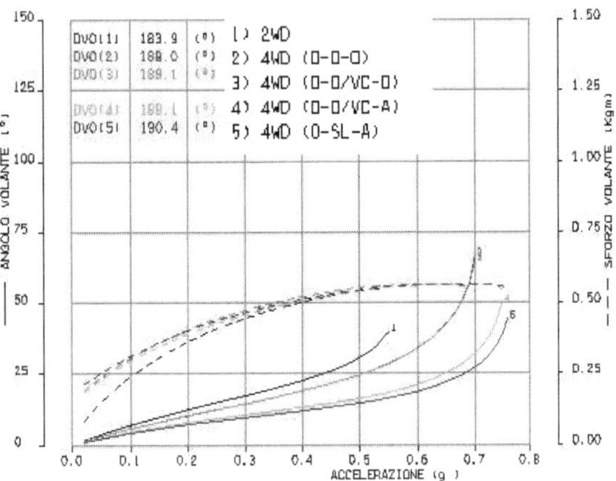

Figure 12: traction capability in a turn

task is studied through steering pad on constant radius, with constant acceleration. This particular procedure consists in performing a quasi static analysis at constant radius with a constant (positive or negative) longitudinal acceleration. This approach is very useful to simulate the cross coupling, even if it is not suitable during track tests. Figure 12 shows four AWD layouts compared to the 2WD car in a 10 meters steering. The dotted lines represent the steer wheel torque required, while the solid ones represent the steer angle vs lateral acceleration. The 2WD vehicle has the highest understeer and the lowest lateral performance (maximum value of Ay is less

than 0.6 [g]). If the car were 4WD with three simple open differentials, the performance would increase strongly: understeer became lower and limit lateral acceleration higher than in the first case.

The same performance were obtained if a central visco coupling were used. Passing to an AWD active layout, the understeer still decrease and lateral acceleration reaches the highest values when the central differential is a self locking one. This analysis shows that a great improvement of performance could be obtained passing from 2WD to a simple AWD layout. If the AWD architecture became more complicate, then the performance could be improved. To quantify the traction capability, the lateral acceleration at maximum longitudinal acceleration is usually calculated. Speaking about power off in a turn, the layouts in study have been compared as shown in figure 13. In the upper part of it, the time histories of yaw rates are shown, consequently to the power off. The 2WD car time history has a peak and then a drift , both due to the power off. The passive layouts shows the typical reaction to the throttle release: the car tends to reduce the bend radius. This behavior becomes more and more critical when the car is four wheel drive, in particular if a self locking differential is mounted at the center, as shown by the ISR index.

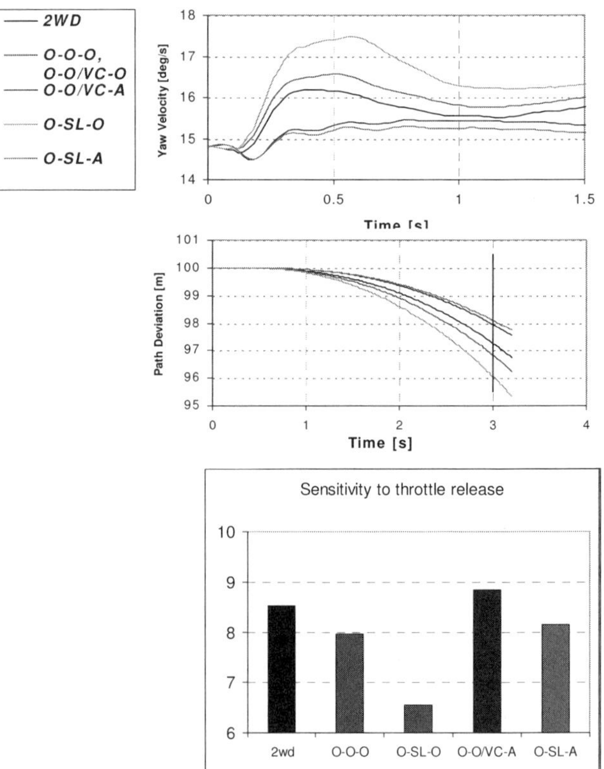

Figure 13: Power off in a turn

If the car has three open differentials, the locking is impossible and the handling of the car is similar to the 2WD's one. In fact if a visco coupling is present, since no slip is detected the car behaves like a 4WD with three open differentials, or, in other words, like the 2WD car. If an active differential is present on the rear axle, the car behavior is heavily changed, as shown in figure 15, by the tuning of the active system parameters. In the lower figure 15, the radial change in radius due to the throttle release is shown: the plot helps to quantify how the passive 4x4 cars follows a different path with a smaller radius than the 2WD, while the active archetypes show particular reactions, due to the control setup.

APPLICATION CONCLUSIONS

The analyzed archetypes has been classified in terms of longitudinal, lateral and cross coupling behavior, through the described performance indexes and the final ranking is shown in figure 17. With respect to the 4WD-three open differential, the analysis has shown:

- Longitudinal performance:

The presence of a self locking or visco coupling on the center allows better performances in all conditions (except in left right split, because of the layout). If the central device is a free differential or a self locking, the performances in the left right split are increased because of the device torque transfer in such conditions.

- Lateral performance on high grip:

Only the presence of the active device on the rear allows to change both the normal driving feeling and the active safety. In particular, this device permits a tuning of the vehicle response, based on the control strategies. If a central self locking is present, the performance at high lateral acceleration becomes worst, because of a loosing in stability, due to locking phenomena.

- Cross coupling:

The central self locking increases the power on performance, but the reaction in power off becomes worst. The active device increases a lot steerability (power on), stability (power off) and permits to personalyze the car response.

CONCLUSIONS

A methodology for the comparison of longitudinal and lateral performances of two AWD vehicles, in terms of maneuvers and significant parameters, has been developed. Software standard procedures for AWD performances virtual evaluation has been created. These results help to start up and optimize:

- Vehicle system analysis using simulation model and standard procedures to classify different AWD architectures;

- Target setting-deployment-achieving in order to traduce AWD vehicle performance in terms of architecture choice and component specifications.

REFERENCES

1. Data, S. and Frigerio, F. Objective evaluation of handling quality, Proc. Instn Mech. Engrs, Part D: Journal of Automobile Engineering, 2002, vol.216, issue (D4), pp.297-306.
2. ISO 4138 Road Vehicles – Steady state circular test procedure
3. ISO 7401 Road Vehicles – Lateral transient response test method.
4. ISO 3888-1 Road Vehicles – Test procedure for a severe lane change manoeuvre
5. R.Avenatti, S.Campo, L. Ippolito: "A rear active differential: Theory and practice of a new type of controlled splitting differential and its impact on vehicle behaviour", GPC '98, Detroit, Ottobre 1998.
6. ISO 9816 Passangers cars - Power-off reactions of a vehicle in a turn - Open-loop test method

AWD Configurations positioning vs O-O-O

		O - SL - O	O - O/VC - O	O - O - SL	O - O - A
LONGITUDINAL Performance	Iso - μ	+	+	=	=
	μ-split (F/R)	+	+ +	=	=
	μ-split (L/R)	=	=	+	+
LATERAL Performance	Normal driving FEELING	=	=	=	Feeling personalization + +
	Critical situations ACTIVE SAFETY	=	=	-	Higher lateral acceleration + + Increased stability & steerability
CROSS COUPLING Performance	Power on in a turn	+	=	+	Higher lateral & long. performance + + Increased steerability Power on vehicle response person.
	Power off in a turn	-	=	-	Increased stability + + Power off vehicle response person.

Figure 14: Performance comparison

Integration Chassis Control (ICC) Systems of Mando

Youseok Kou, Wanil Kim, SangHo Yoon, JeongWoo Lee and Dongshin Kim

Mando Corp., South Korea

ABSTRACT

This paper presents the integrated chassis control (ICC) system under development at MANDO. By sharing the sensor and control information through the communication link among the existing 2 or more chassis subsystems, the integrated chassis control system improves vehicle performance and reduces cost for the sensors and related wiring. ICC consists of continuously variable damping control system(CDC), rear toe angle control system(AGCS) and electronic stability program (ESP). In ICC, the steering angle and yaw control information of ESP are delivered to the other systems through the CAN interface, and co-operative control strategy overrides each subsystem to improve the vehicle handling performance and stability. The effectiveness of both integrated chassis control systems are illustrated by a computer simulation and vehicle test on dry asphalt, snow road surface and so on.

INTRODUCTION

Studies on the integrated chassis control system are recently active in analyzing each system and optimizing the performance. In this paper the characteristics of the respective systems (CDC, AGCS, and ESP) are verified through a simulation and vehicle test, and various approaches of the integrated chassis control methods are proposed. ICC focuses on the improvement of the braking performance and the handling performance, while giving the driver comfort as well as stability.

	CDC	AGCS	ESP
Control Target	Damping Force	Rear Toe Angle	Braking Force
Function	Comfort Handling	Handling Stability	Stability Safety

Table.1. Chassis control system list for ICC

BODY

1. Basic principle

***Braking performance**

Given that the road condition and tire characteristics are the same, the vehicle's stopping distance would be affected by adjusting the damping force since the load on the axle is dependent on the vehicle load transfer based on the damping force.

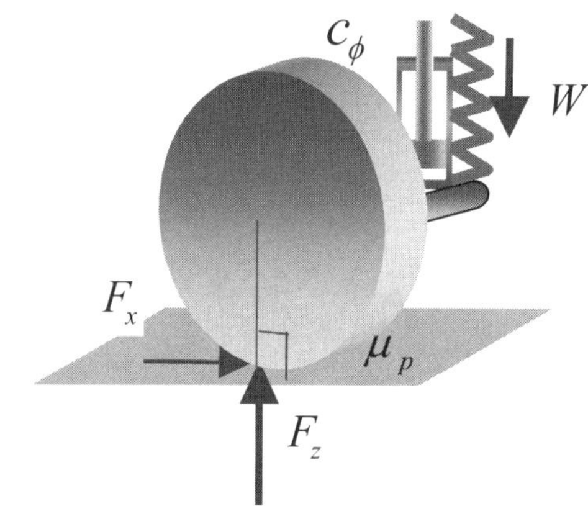

Fig 1.   Braking force vs. Normal force on an axle.

$$F_x = \mu_p W = \mu_p (W + W_d)$$

$$F_x = \mu_p \cdot F_z , \quad \Delta F_z = G(c_\phi)$$

F_x : Braking force, μ_p : Coefficient of friction

F_z : Normal force, C_ϕ : Damping coefficient

W : Weight of the vehicle

***Handling stability**

Adequate cornering force on the tire should be applied for a stable cornering performance in order to provide the driver with comfort and safety.

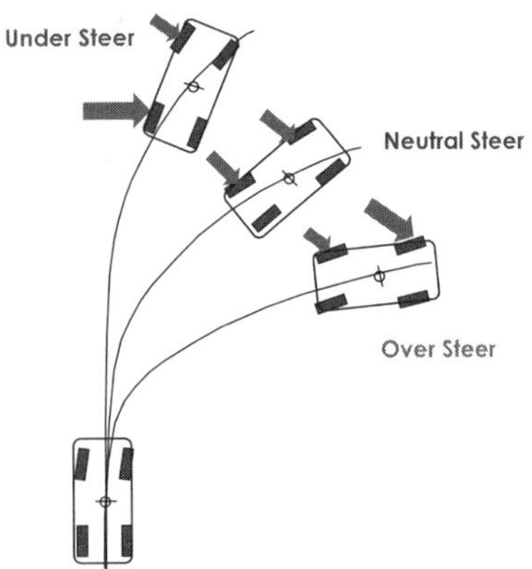

Fig 2. Vehicle dynamics

Three major cornering behaviors of the vehicle are described in Fig. 2. It is well known that neutral steer or a little under-steer enables the vehicle to embrace the stable handling characteristics.

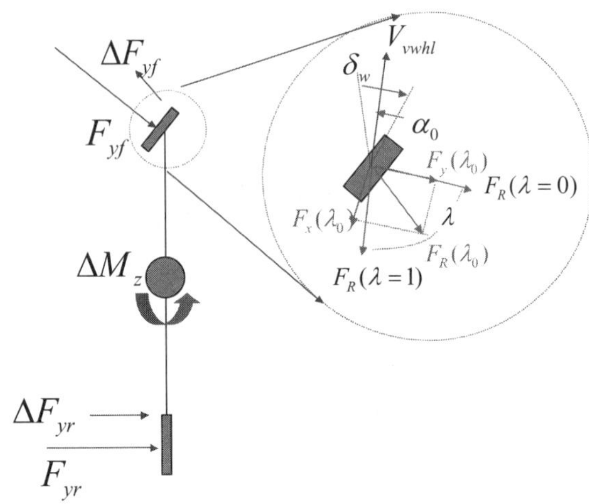

Fig 3. Rotation of tire force

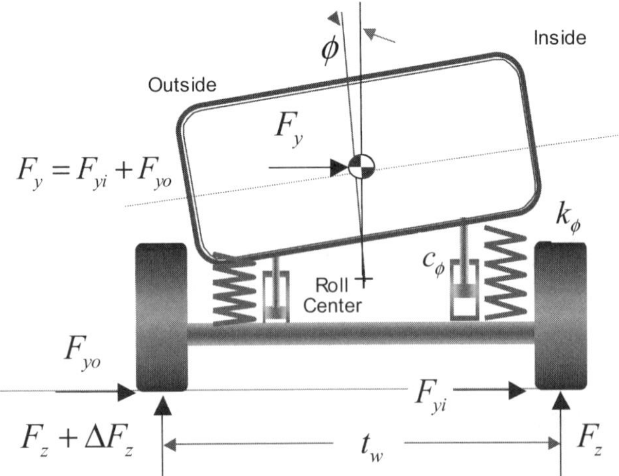

Fig 4. Force analysis of a simple vehicle in cornering

As shown in Fig 3. and Fig 4., the longitudinal slip of the tire caused by braking effects the corrective yaw moment and the balance of roll moments distributed on the axles that depend on the roll stiffness and damping altering handling performance. Corrective yaw moment is described as a function of lateral force as follows,

$$\Delta M_z = \Delta F_{yf} \cdot a - \Delta F_{yr} \cdot b$$

Correlation between lateral force F_y and chassis parameters such as damping coefficient, slip angle and longitudinal slip are induced as follows,

$$\Delta F_{z\_F} = \left[k_\phi \cdot \phi + c_\phi \cdot \rho \cdot \left(\frac{d\phi}{dt} \right) \right] / t_w$$

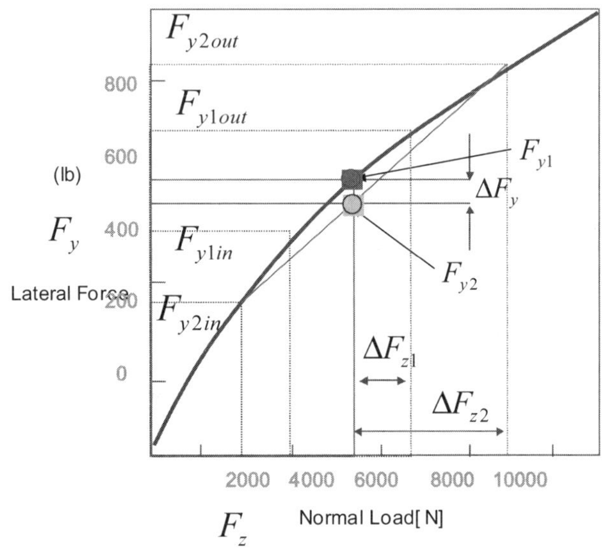

Fig 5. Lateral force variation

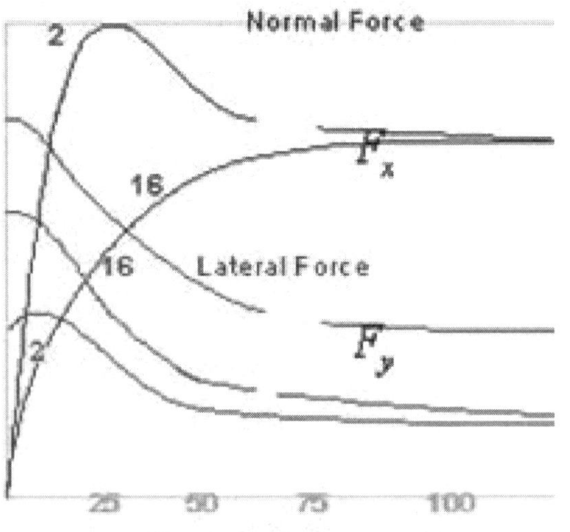

Fig 6. Braking and lateral force as a function of longitudinal slip

Given that the other conditions have been eliminated, lateral force is inversely proportional to the longitudinal slip, λ , and the damping coefficient, c_ϕ , and is proportional to the slip angle, α . The mechanism is given by :

$$\Delta F_y = f(c_\phi, \alpha, \lambda)$$

$$\Delta F_y \propto c_\phi^{-1}, \Delta F_y \propto \alpha, \Delta F_y \propto \lambda^{-1}$$

The longitudinal slip, λ , depends on the braking force and the damping coefficient, c_ϕ , depends on the damping force and the slip angle, α , depends on the toe angle.

2. System Configuration

ICC system is composed of ESP, CDC and AGCS that control parameters λ, c_ϕ, α , adjusting the characteristics of the chassis. ECU with the sensors and the continuous variable damper with solenoid valve constitute the CDC system. AGCS is composed of ECU, the DC motor and the reducer. ESP system, which is composed of an ECU and a hydraulic unit, automatically activates the braking force on the respective axle.

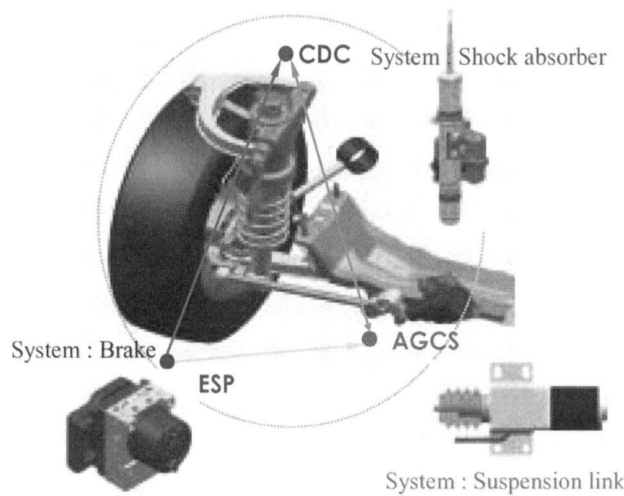

Fig 7.   System configuration

ESP system calculates the dynamic characteristics of the vehicle from the sensors that are installed in the vehicle and adjusts the magnitude of the wheel slip using braking force. CDC system judges the drivers' intentions through the steering and speed sensors and control damping force to adjust the weight transfer. AGCS system alters the toe angle by adjusting the geometric characteristics of the suspension link.

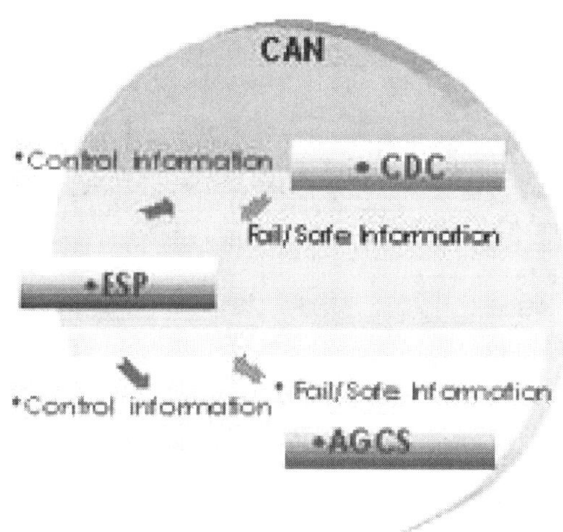

Fig 8.   ICC system networking

ICC enables each system to share the sensor information, the control information and fail-safe information via a network based on CAN(Control Area Network). ICC system helps the vehicle to operate stably by the driver as if supporting the ball to maintain its position on the semi-column. ESP system helps the vehicle to maintain stability in extreme cases such as a spin and drift while CDC system and AGCS system make a dominant role of stabilizing the vehicle at the area where the relation between cornering force and slip angle is linear.

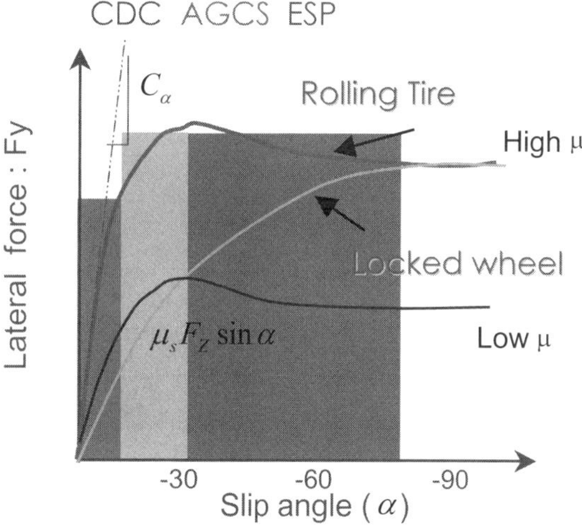

Fig 9.   Tire cornering force property

The schematic diagram about the operating area of the respective system is illustrated in Fig. 10. at the base of the dynamic characteristics of the tire.

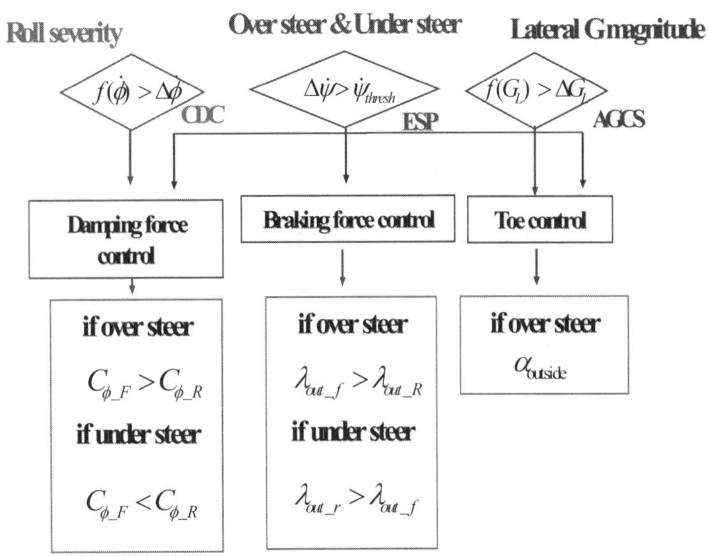

Fig 10. ICC control strategy

The flow chart of ICC control is illustrated in Fig. 11. In a CDC system, the severity of roll is judged from the steering input and vehicle speed sensor information and the magnitude of the damping force is adjusted. AGCS system controls the geometry of the rear suspension based on the estimated value of the lateral acceleration. ESP system judges over-steer and under-steer of the vehicle using steering input, lateral acceleration and yaw rate and then controls the braking force of the respective wheel to prevent spin-out and drift. CDC and AGCS can cooperate with ESP using the information transmitted by the ESP system.

3. Computer Simulation Analysis

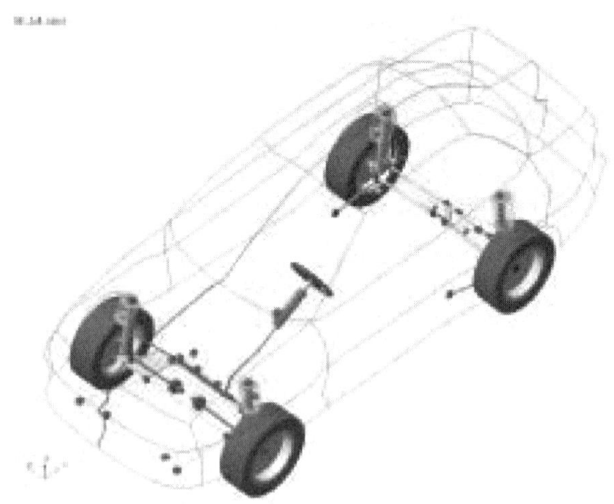

Fig 11. ICC Vehicle Model using ADAMS program

As shown on the vehicle model of Fig. 11, the dynamic simulation is performed to verify the feasibility of the system and evaluate the effects of the systems. As shown in Fig. 12, the analysis result using the Adams dynamic simulation on the braking situation shows that the stopping distance gets shorter as the damping force is adjusted to minimize the variation of normal force.

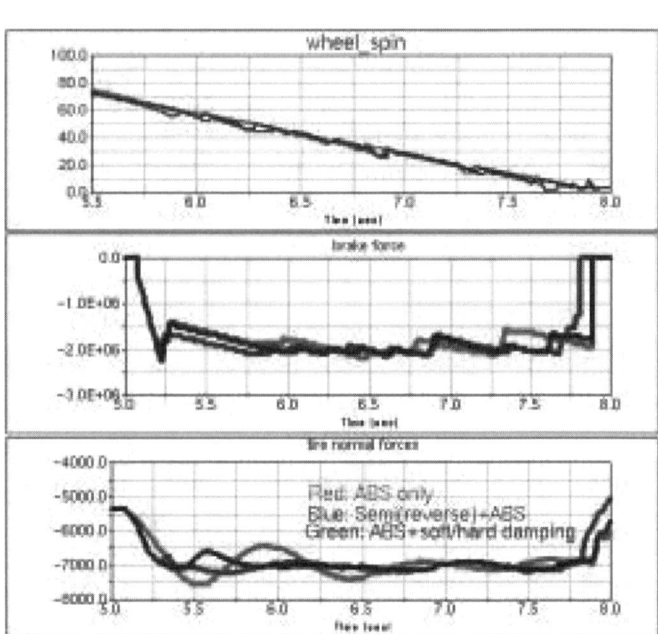

Fig 12. Stopping distance according to the damping force.

Model	Stopping Distance	Comparison
ABS+ Soft damping	39.7 (m)	-
ABS+ Rebound hard damping	39.3 (m)	-36(cm)
ABS+ Compression hard damping	39.1 (m)	-58(cm)

Table.2. Simulation result of stopping distances according to damping set

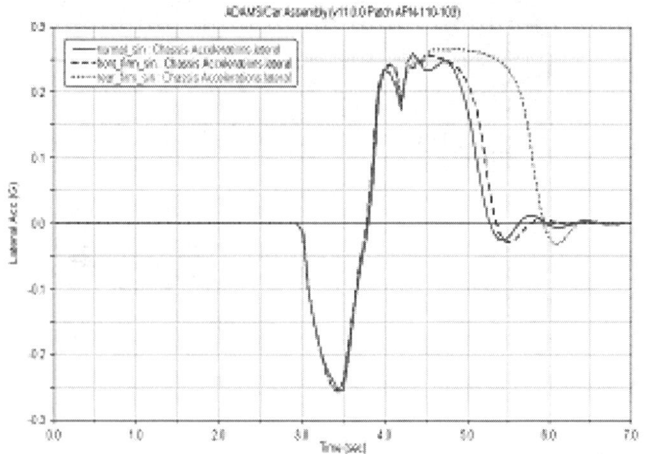

Fig 13. Handling characteristics according to the damping force

The analysis result shows that difference between the front and the rear damping affects the cornering force of the vehicle. As shown in Fig. 13. The bigger damping force on the rear side compared to the front, the bigger lateral acceleration. Especially the condition is low μ road condition and therefore the vehicle can be easily affected by the damping variation under the above lateral acceleration that is relatively small at the high μ road condition

4. Vehicle test

Vehicle test is performed on a road with a coefficient of friction, μ, of 0.3~0.4. The characteristics of the system are compared according to the control of the respective system,. When the same steering input and speed condition are applied, the lateral acceleration and yaw are measured in order to analyze the behavior of the vehicle.

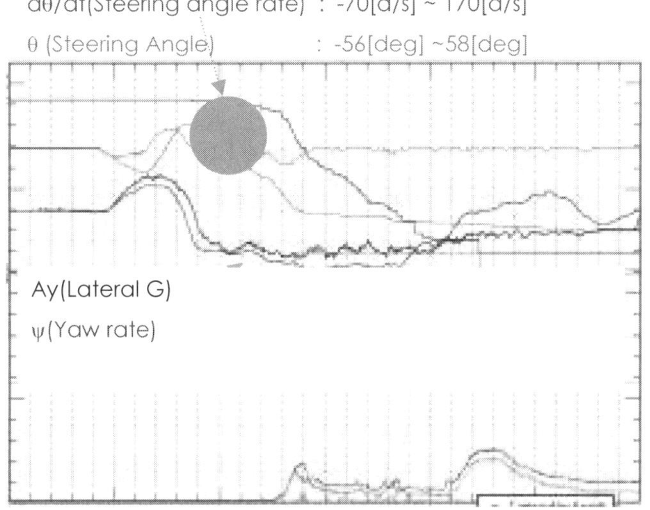

Fig 14. No system in snow road (@60[kph])

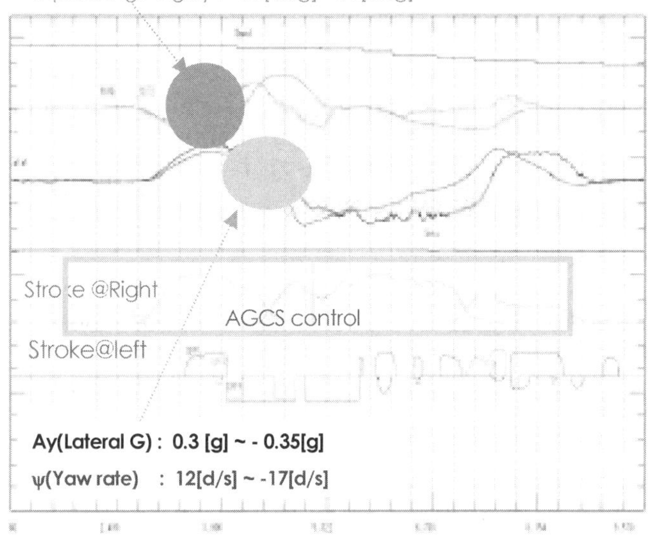

Fig 15. AGCS control in snow road (@70[kph])

The average drivers do not prevent spin-out of the vehicle in the condition of the general lane change on the snow road as shown in Fig. 14. However, AGCS provides the vehicle more stability in spite of its limits in the performance on low μ road condition. As shown in Fig 15, the lateral acceleration of the vehicle reaches the limit of the road friction where the vehicle is about to slip and in that case AGCS system changes the stroke of the actuator to adjust the suspension link which can give rise to under steer or over steer of the vehicle. However this AGCS system or CDC system can't make the vehicle stable in case of the severe maneuver on the snow or ice road   (μ < 0.4) where the lateral acceleration of the vehicle can easily go over the limit of the road friction.

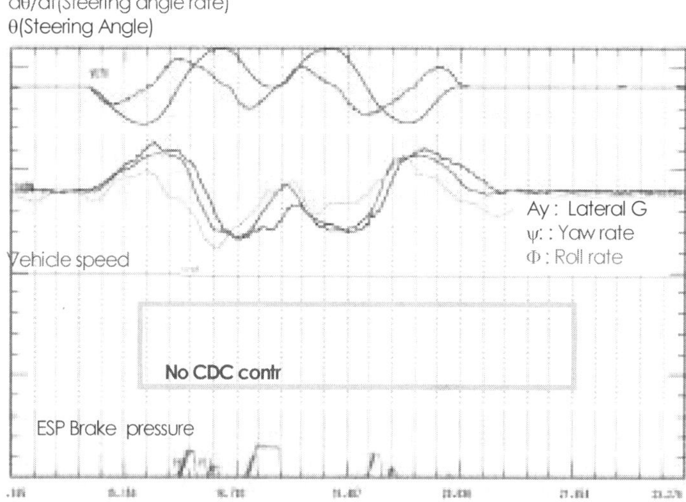

Fig 16. Only ESP system

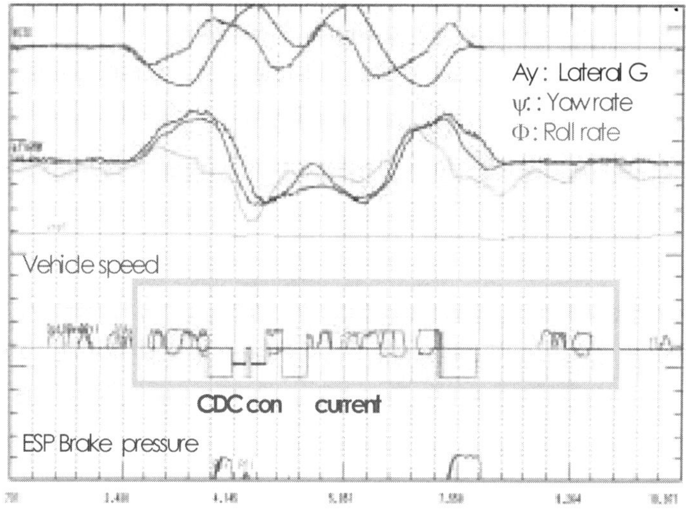

Fig 17. CDC system contribution to ESP system

While ESP system without CDC system can make the vehicle stable through 5 times braking intervention, ICC system can not only provide the driver more comfort but also make the vehicle stable by only 2 times braking intervention as shown in Fig. 16 and Fig 17.CDC system adjusts the damping force to provide the vehicle under steer based on the vehicle dynamic information transferred from the ESP system. In this case, we can avoid the intervention between CDC or AGCS and ESP system.

The control process of the ICC is as follows. At the first step CDC system makes the vehicle maintain the flat body with a hard damping force and AGCS adjusts the stroke of the actuator in order to change the toe angle at the rear side and finally ESP system stabilizes the vehicle compensating the balance of yaw moment through the braking force. 'A' is the current output of CDC and 'B' is the stroke of the AGCS 'actuator and 'C' is the brake pressure of the wheel controlled by ESP.

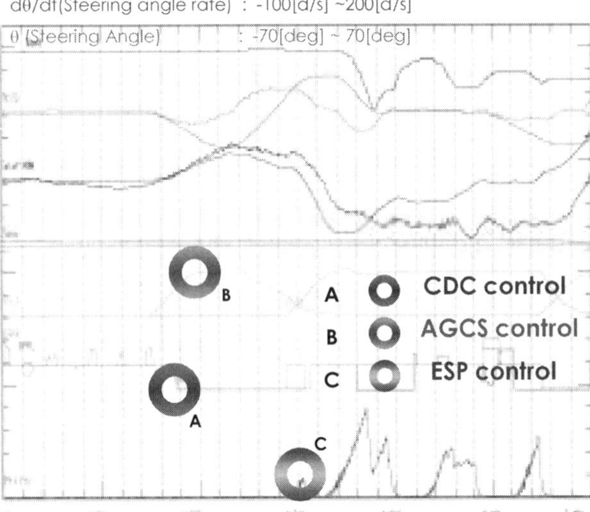

Fig 18. ICC system control process.

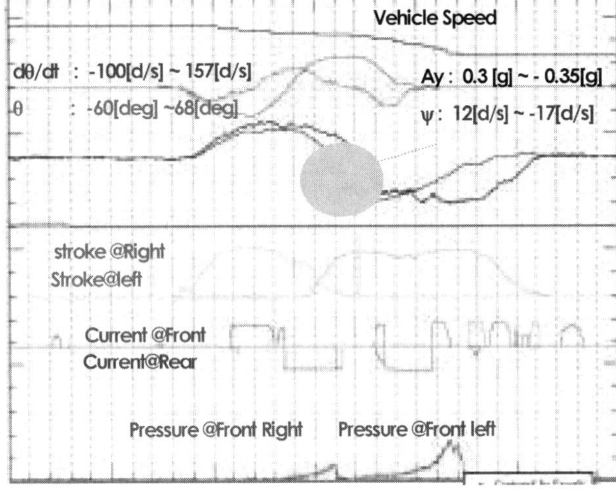

Fig 19. Lane change in snow road (@70[kph])

The result of ICC vehicles test shows stable cornering without spin-out and drift as shown in Fig. 19. CDC and AGCS system help the vehicle to maintain stability at the initial stage of the maneuver and then ESP system contributes to stability of vehicle in severe conditions.

CONCLUSION

In this paper, we analyzed the characteristics of CDC, AGCS, and ESP that composes the Mando's ICC and discussed their contributions to the vehicle performance through a dynamic simulation. Mando's ICC system, which includes CDC, AGCS and ESP is proposed to enhance the stability and safety of the vehicle and an optimized control algorithm and networking method has been developed. At the road holding area where the relation between the tire cornering force and the slip angle is linear, the suspension control system is effective in providing the driver with comfort and stability, but suspension system has limits in stabilizing the vehicle on the low μ condition. In such conditions, the ESP system comes into control and takes a dominant role in providing stability and safety. Sensitivity analysis between the suspension control system and the braking control system will be discussed and the optimal parameters that improve vehicle performance will be discussed on high μ condition based on the present research results. Research about synergy through the various combinations of the developing chassis systems will be practiced to minimize braking distance and enhance stability and safety. It is imperative to analyze the problems that may occur when the respective system activates independently and the vehicles tests are also followed to get the optimal tuning parameter of the integrated control.

REFERENCES

SAE TECHNICAL PAPER SERIES "VDC, The Vehicle Dynamics Control System of Bosch" .Robert Bosch GmbH.950759.

Journal : Mark O. Bodie and Aleksander, "Close Loop Yaw Control of Vehicle Using Magneto-Rheological Dampers", 2000 SAE, 2000-01-0107, 2000.

Thomas D.Gillespie, "Fundamentals of Vehicle Dynamics", SAE, 1992

Katsuhiko Ogata, "Discrete-time Control System", Prentice Hall,1994

CONTACT

Senior Research Engineer of Central R&D Center in Mando Corporation

Precision Mechanical Engineering Hanyang University

Email: yskou@mando.com

Telephone:(82)313005284

Accelerometer Design for Vehicle Control Safety System

Mark Harrison
Denso International America, Inc.

Yuzuru Otsuka and Minekazu Sakai
Denso Corporation

ABSTRACT

In order to reduce traffic accident casualties, sophisticated safety systems have been developed and are continuously being upgraded in today's passenger vehicles. One system showing growth in the global automotive industry is a feature currently available on high-end passenger cars, Vehicle Stability Control (VSC). VSC can control side slipping, an unstable phenomenon which can lead to critical accidents. VSC systems are multi-functional systems that include an acceleration sensor to detect forces applied to the vehicle. Acceleration sensors sometimes referred to as G sensors are indispensable and are one of the key sensors for vehicle safety systems. New safety systems require acceleration sensors with high sensitivity and accuracy. We have achieved these increased requirements by adopting a unique stacked IC structure. This paper discusses the small-sized simple acceleration sensor that uses a stacking structure technology and switched capacitor circuit for VSC systems that we have developed.

INTRODUCTION

Automotive safety is broadly classified as preventive safety and crash safety. Preventive safety employs VSC and Antilock Brake System (ABS). Airbag technology is typically used for crash safety. It is expected that a growing number of VSC systems will be installed. A typical VSC system diagram is shown in Figure 1 below.

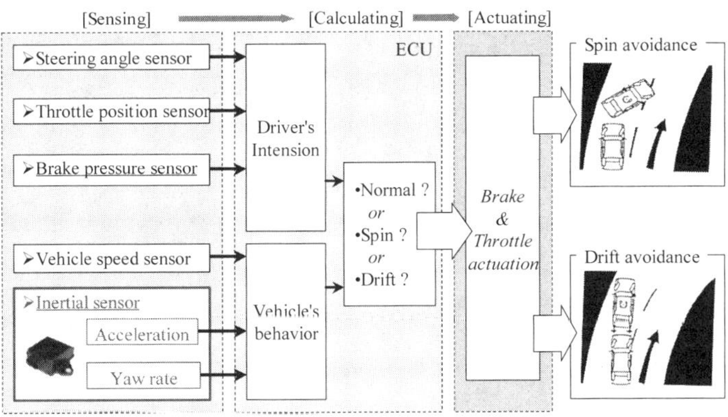

Figure 1: VSC system concept

VSC systems detect the condition most likely to cause sideslip based on data from the steering angle sensor, yaw rate sensor and acceleration sensor. The VSC system uses the data to drive the brake actuator and throttle actuator, which automatically control brake oil pressure for each wheel and engine output.

In this system, the inertial sensor is the key to detecting vehicle sideslip. Figure 2 below shows the inertial sensor, which is composed of a yaw rate sensor that detects angular velocity of a car and an acceleration sensor that detects vehicle acceleration.

Figure 2: Inertial sensor

Research and development of capacitive acceleration [1][2] and yaw rate sensors [3][4] has intensified due to Micro Electrical Mechanical Systems (MEMS) technology advances. The MEMS capacitive sensor is highly sensitive, superior in DC detection, has better temperature characteristic and self-testing is relatively easy to conduct, when compared to the piezoresistive or piezoelectric sensors. The detection range of a VDC acceleration sensor is at about the +/- 1.5G level, so a small-sized, highly sensitive and accurate MEMS sensor is highly desirable.

Therefore, we investigated the following: film thickening of structural body using SOI [5], anodic bonding technology for silicon and glass [6], and the use of a trench refill process [7]. For improved accuracy [8][9] the inventive use of an anchor structure was in order to reduce the effect from heat and packaging stress.

For the MEMS trend, we started a new type of acceleration sensor development. Figure 3 shows the concept. For our downsizing goals a stack structure is used. The circuit chip and sensor chip are stacked on the ceramic package. To super-sensitize the sensor chip a SOI wafer is used and for improved accuracy we targeted the circuit chip and package technology.

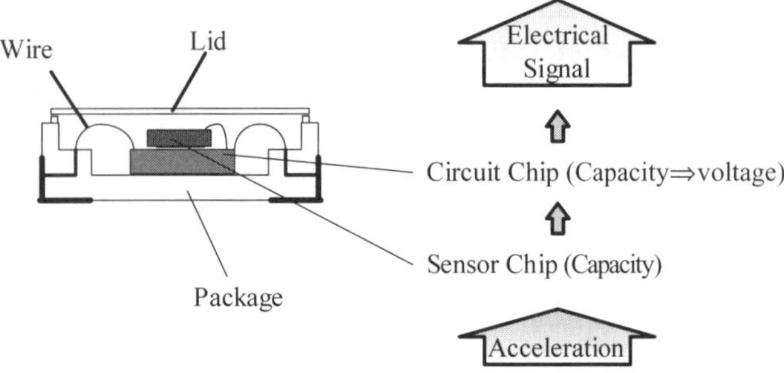

Figure 3: Concept for acceleration sensor

Since the yaw rate sensor and acceleration sensor are constructed with comb-tooth electrodes, our first step was to develop the acceleration sensor.

The keys for development are sensing element design to detect acceleration, circuit design to convert the capacitance variance into voltage, and package design.

SENSING ELEMENT DESIGN

Two key points for developing the sensing element are high sensitivity and high reliability. Therefore, we selected SOI structure made from monocrystal Si to thicken the capacitive electrodes for high sensitivity and to avoid the material fatigue for high reliability. To reduce cost, the cavity under the electrode is formed by backside etching process adopted from common SOI wafers. Figure 4 shows the birds-eye view of a developed sensing element. The movable sensing element is composed of a proof mass, a movable electrode that mutually opposes the capacitive electrode, and a beam spring on a movable element. Multiple movable electrodes are formed from side to side of the mass, and deflect with the mass according to the applied acceleration.

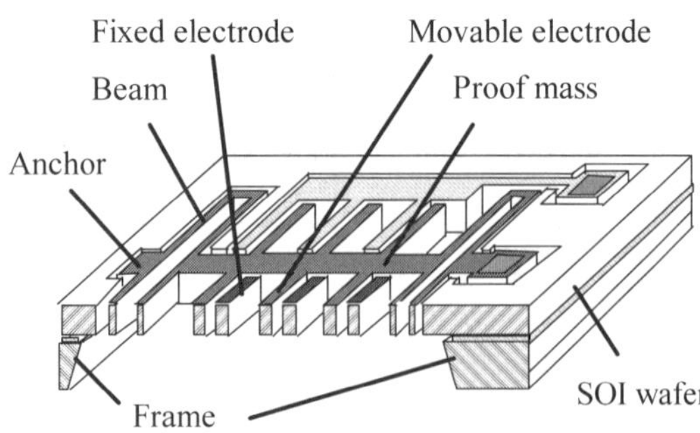

Figure 4: Sensing element

Figure 5 shows a Scanning Electron Microscope (SEM) photo of electrodes used to detect capacitance variation. 3μm film thickening was achieved on a 15-μm monocrystal electrode. With the use of a monocrystal Si structure, however, the film thickening keeps the electrode from deforming.

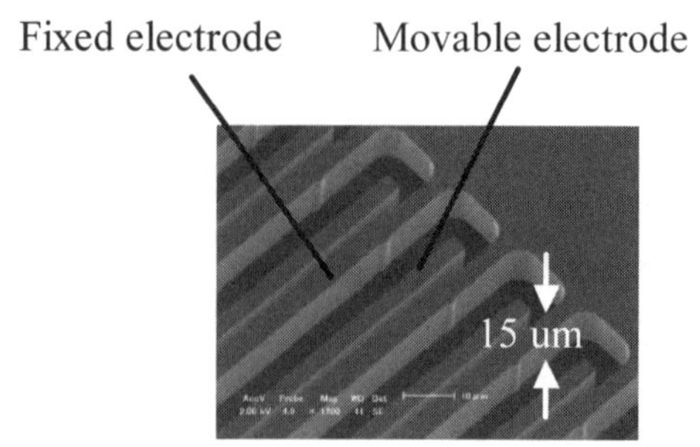

Figure 5: SEM electrode picture

The principle of acceleration detection is illustrated in Figure 6. Capacitance (C_0) is expressed as $C_0 = \varepsilon \cdot S/d$, where d is the distance between the movable electrode and the fixed electrode before the acceleration is applied. S is the area of the opposed electrode of a movable element and fixed electrode. When acceleration is applied, inertial force displaces the mass and movable electrode of the movable element, and the beam is deformed. The deformed beam changes the static distance between the movable electrode and fixed electrode. Δd and capacitance value between a movable electrode and a fixed electrode can be mathematically represented as $\Delta C = 2C_0 \cdot \Delta d/(d^2 - \Delta d^2)$. The acceleration is detected by converting capacitance variation into a voltage value with a C/V conversion circuit. The increase of ΔC provides for an increase in super-sensitivity.

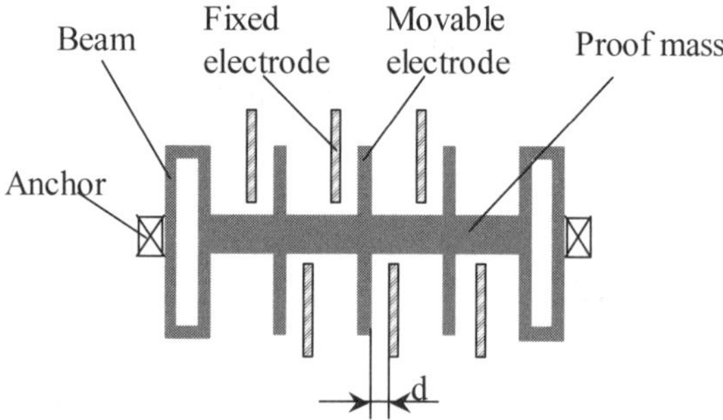

Capacitance: $C_0 = \varepsilon \cdot S/d$ (S: Electrode opposed area)

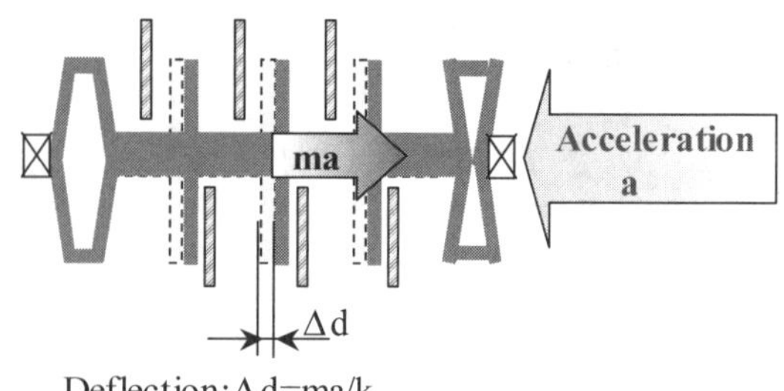

Deflection: $\Delta d = ma/k$
Capacitance variation: $\Delta C = 2C_0 \Delta d/(d^2 - \Delta d^2)$

Figure 6: Operation principle of acceleration sensor

For super-sensitization, however, the following issues are generated.

To heighten sensitivity, the spring constant needs to be reduced or the capacitance (Co) increased. Yet, this weakens the spring restoring force (k) and strengthens the electrostatic gravity (Fe) that works between electrodes. The weaker restoring force of spring than that of electrostatic gravity causes malfunction due to sticking of electrodes. Also, an electrode too long for the capacitance increase can cause sticking, as the electrode is easily deformed. Therefore, we designed an optimal structure using simulation. Figure 7 shows the analysis flow chart. The dimensions of the element structure are set. Then mechanical analyses of the mass, spring constant and electrical analysis of the capacitance between the electrodes is conducted. The sensitivity was evaluated based on the acquired parameter using tools such as MATLAB and Saber.

In order to accurately reproduce sensitivity and frequency characteristics, a comb-tooth structure model is segmented, and a spring-mass model is added to the movable electrode, mass and fixed electrode respectively. Figure 8 shows the diagram, with a flow chart to reproduce sticking.

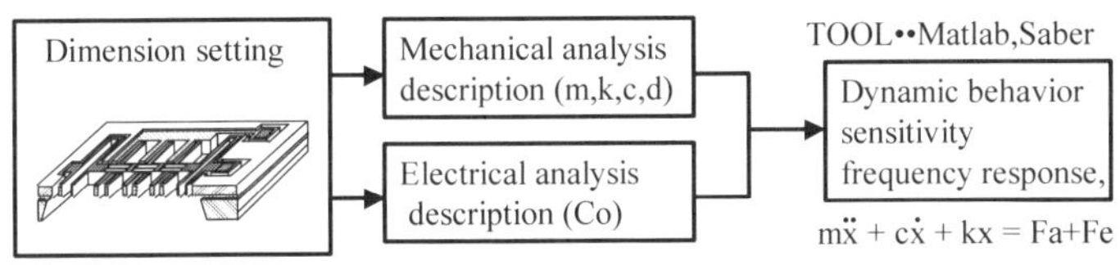

Figure 7: Analysis flow chart

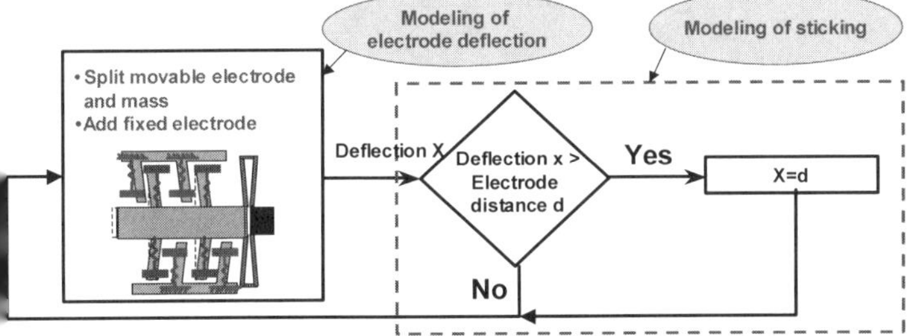

Figure 8: MATLAB model

The electrode structure analysis result is shown in Figure 9 using the above diagram. Electrode width and electrode length are determined to be 4μm and 350μm respectively, with consideration of the restriction of chip size and processing technology as well as sticking. As a result, capacitance value is determined as 0.74pF.

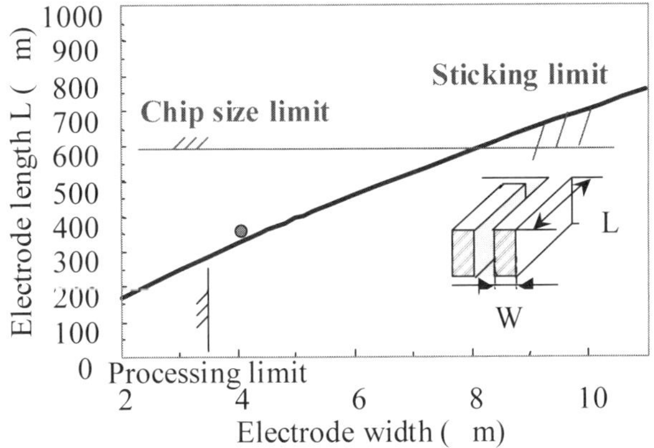

Figure 9: Electrode structure analysis result

CIRCUIT DESIGN

The circuit designing point is to select a C/V conversion circuit suitable for higher accuracy and to add an adjustment circuit. The circuit block shown below in Figure 10 is composed of a C/V conversion circuit, sample hold and various adjusting circuits.

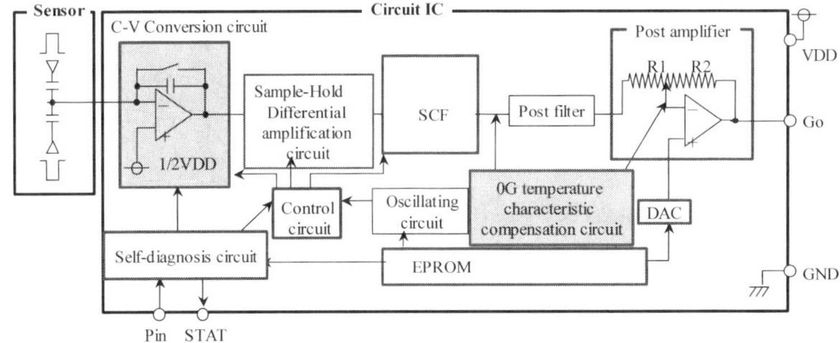

Figure 10: Circuit block diagram

Parasitic capacitance is inherent in the C/V conversion circuit due to the wiring from the pad to the movable/fixed electrode, between support substrates, on the sensor chip and on the signal processing circuit chip. The output of the sensing circuit must be free of parasitic capacitance. Also, the initial stage of the C/V conversion is desirable to control noise. For these reasons, we selected a switched-capacitor circuit for the micro-capacitance sensing circuit. Figure 11 shows the operation principle of a switched-capacitor.

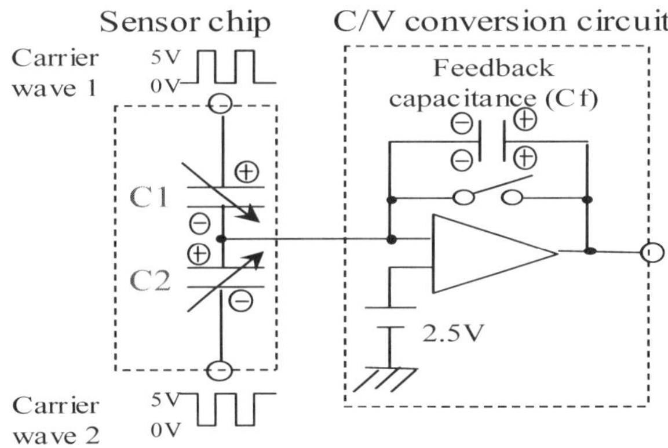

Figure 11: Micro-capacitance detective circuit

The (C1-C2) differential is dependant on the deflection of the movable electrode and is proportional to acceleration. Therefore capacitance variation, equal to acceleration variation, can be detected by the variation of this electric charge volume. Output voltage is described as follows:

$$\text{Output} = (C1-C2) \cdot 5v/Cf,$$

Where Cf is feedback capacitance.

For improved accuracy, a 0G temperature compensation circuit is embedded. The temperature compensation circuit was adjusted with an EPROM. Higher accuracy was achieved by compensating the 0G temperature characteristic before the gain circuit in the later stages of the conditioning circuits. Figures 12 and 13 show 0G output comparison of before and after temperature compensation.

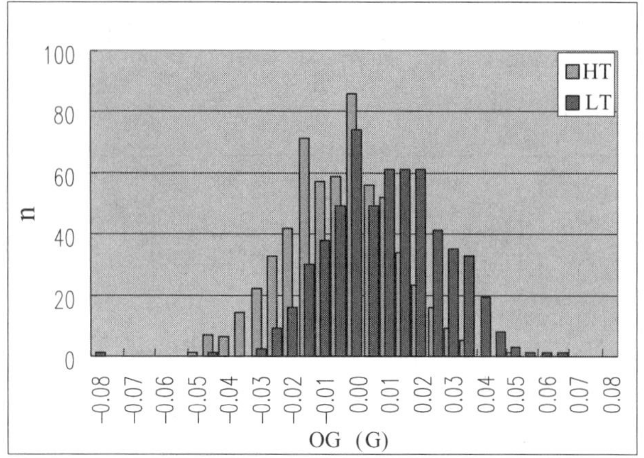

Figure 12: 0G output distribution at −30 and +85 degrees before temperature compensation

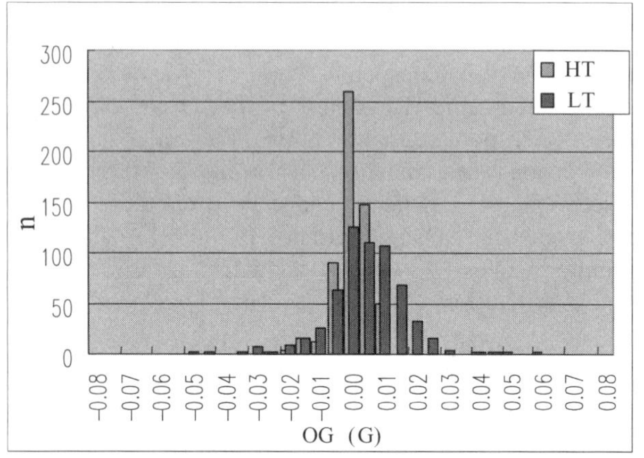

Figure 13: 0G output distribution at −30 and +85 degrees after temperature compensation

Finally a 20% plus operation check was added using a self-test circuit to check element operation.

PACKAGE DESIGN

The point of the package is to design a structure for higher accuracy. An acceleration sensor is normally mounted onto an ECU substrate with solder. Consequently, heat deflection impact is transferred to the sensor chip and the sensor element is deformed and output varies. In order to avoid these effects, we adopted a stack structure where the circuit chip is deployed under the sensor chip. Figure 14 below shows the G sensor's internal configuration. The sensing chip and conditioning circuit chip use a stack structure to significantly downsize the overall package. The completely sealed ceramic package with metal lid is a remarkable 6.0 x 6.7 mm.

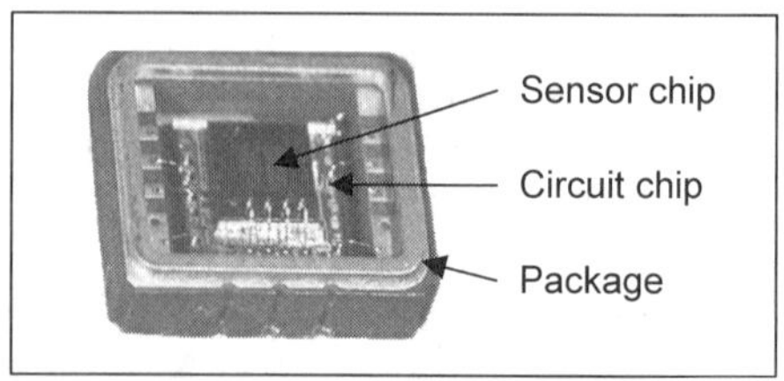

Figure 14: Stack structure

For even higher accuracy, as shown in Figure 15, structural deflection of the sensing chip due to temperature (i.e. variance of electrode distance due to thermal stress) can be minimized with the adoption of stack structure shown in Figure 14. The thermal stress to the sensor chip is a function of the mounting adhesive. In order to reduce the effect of thermal stress, we reduced the elastic modulus of the adhesive under the sensor chip. Using FEM analysis, we determined the relationship between acceleration output and the adhesive Young's modulus.

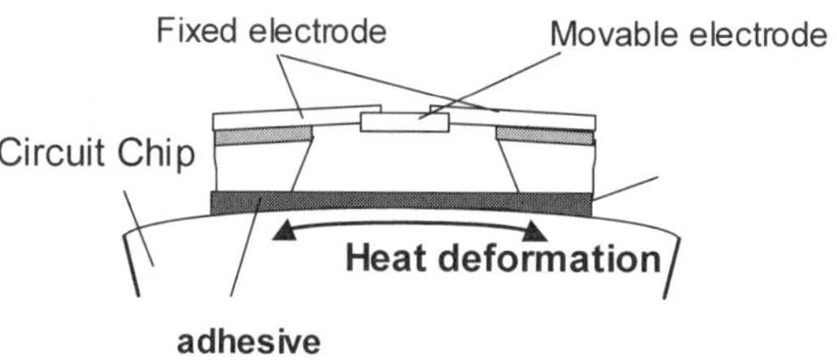

Figure 15: Heat deflection effect

We found the curve of the sensing element when varying temperature by use of the stack structure model shown in Figure 16. Also we found the capacitance variation by analyzing capacitance of the same circuit. With this, the adhesive was decided.

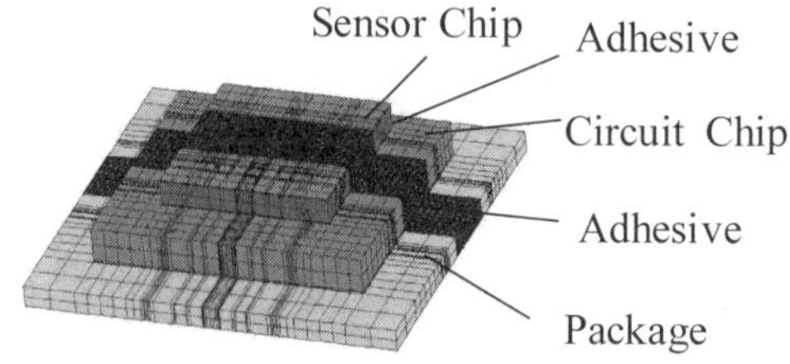

Figure 16: FEM analysis model

Figure 17 shows the results of the 0G output characteristic for Young's modulus of the adhesive and adhesive thickness. We found that by reducing the adhesive's Young's modulus, and using thinner adhesive thickness, the 0G output is less susceptible. As a result, adhesive of 100 MPa or less Young's modules shall be used.

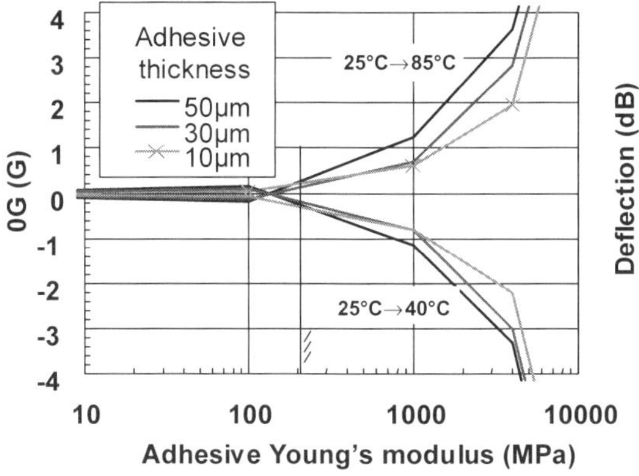

Figure 17: Adhesive Young's modulus and 0G analysis result

The adhesive selection was confirmed after analysis of the transfer function for acceleration of stack structure.

EVALUATION RESULT

Evaluation results for green and aged sensors have been examined for the developed acceleration sensor. The developed acceleration sensor has achieved sufficient accuracy and quality for the VSC systems.

Some of the test results are introduced as follows.

Figure 18 shows the output characteristics of the developed acceleration sensor with applied acceleration from –2.0G to +2.0G. It shows quite linear characteristics within approximately 1% in the full-scale. The characteristics indicate that the acceleration sensor is applicable for VSC systems.

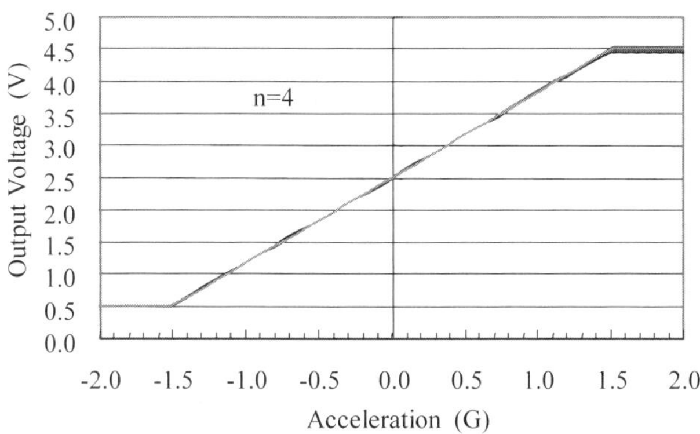

Figure 18: Output characteristics for the acceleration

Figures 19 and 20 show the temperature dependency of the accelerometer, which is one of the most significant characteristics. This data indicates that both 0G offset and sensitivity of the acceleration sensor are stable with temperature drift owing to the innovative new technologies as mentioned above.

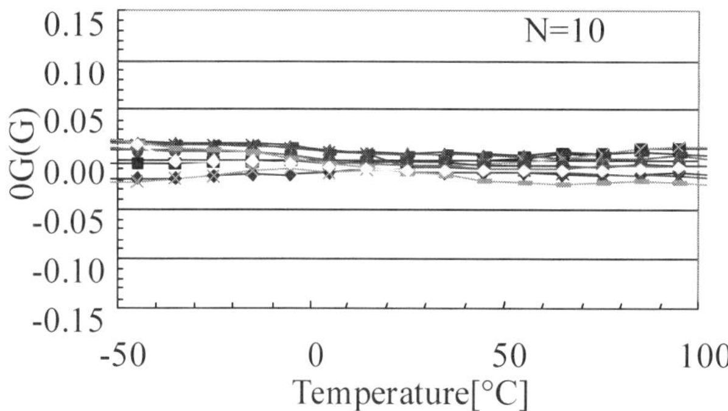

Figure 19: 0G temperature characteristics

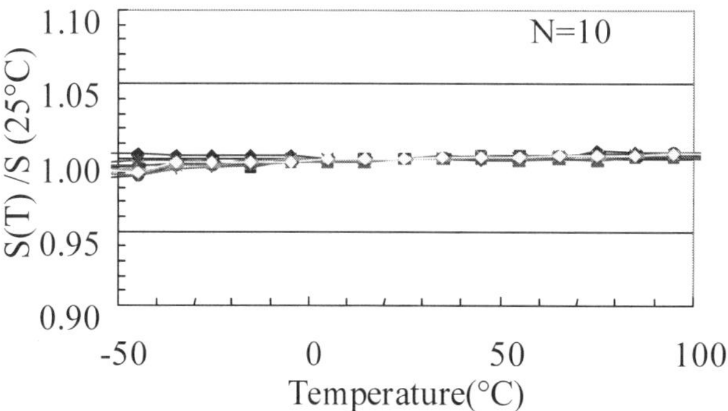

Figure 20: Sensitivity temperature characteristics

Figure 21 shows high temperature operation durability test results, which is one of the most severe tests because of the simultaneous application of both high temperature (+85 degrees) and vibrating acceleration (+/- 1G at 10Hz). As you can see below, there is almost no variation after more than 4 million cycles. It consequently indicates that the sensor is of sufficient quality for commercialization.

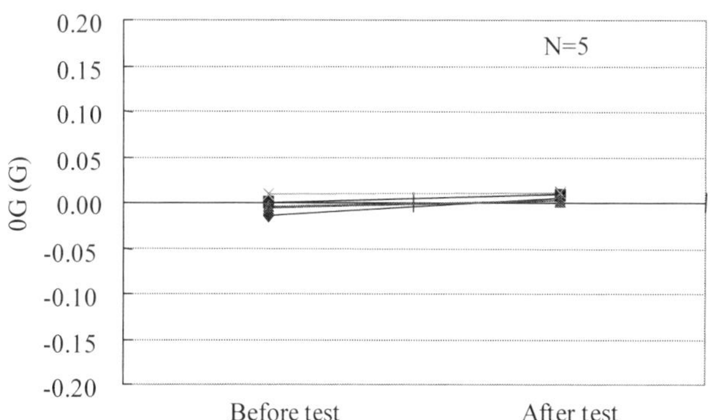

Figure 21: Durability test result

In addition, Table 1 is a sample of the acceleration sensor specifications for the VSC currently developed.

<u>Absolute Maximum Ratings</u>

Power supply voltage	-0.3 to +6V
Storage temperature	-40 to +85°C

<u>Operating Conditions</u>

Operating voltage	5 ± 0.25V
Measuring range	14.7 m/s² [±1.5G]
Operating temperature	-30 to +85°C

<u>Electrical Characteristics</u>

Sensitivity	$\frac{4}{147}$ VGS V/(m/s²) ± 5% (1.333V/G ± 5% @5V)
Offset	$\frac{1}{2}$ VGS V ± 0.98(m/s²) (2.5V ± 0.1G @5V)
Cross axis sensitivity	within 5%
Current consumption	within 10 mA
Output random noise	within 20mVp-p
Output impedance	within 20 ohm

Table 1: VSC acceleration sensor specifications

Finally, figure 22 shows an inertial sensor printed circuit board with the developed capacitive acceleration sensor mounted.

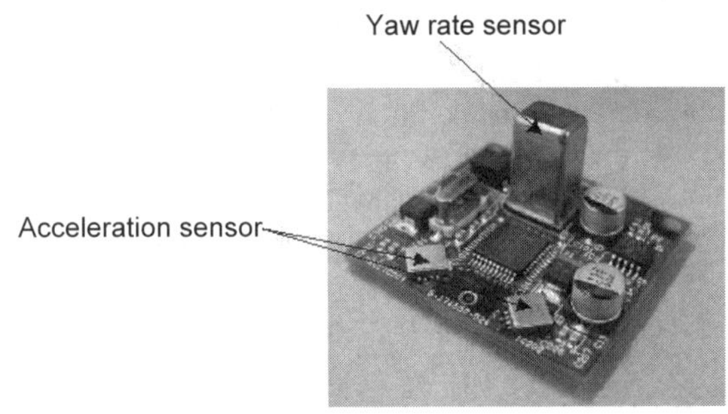

Figure 22: Inertial sensor printed circuit board

CONCLUSION

We have worked on the development of a super-sensitized, highly accurate capacitive acceleration sensor for VSC systems and gained the following results.

1. Using an SOI wafer, we have achieved a capacitive acceleration sensor having electrodes of monocrystal Si with 15 μm electrode thickness.

2. To improve the sensitivity of the sensing element, we increased the capacitance of the sensing element and optimized the spring-mass system, taking into account the electrode deflection and the electrostatic effect between the facing movable and fixed electrodes.

3. For higher accuracy, we developed better sensor temperature characteristics by adding a 0G temperature adjusting circuit and lowering the Young's modulus of the adhesive.

4. The super small packaging size of 6.0 x 6.7 mm is achieved by adopting a stack structure for the sensing chip and the conditioning circuit chip.

Through the above approaches, we developed a super-sensitized, highly accurate capacitive acceleration sensor for VSC systems. In the future, we are going to promote the development of a wide variety of sensors based on this technology.

REFERENCES

1. M. Offenberg, H. Munzel, D. Schubert, O. Schatz, F. Larmer, E. Muller, B. Maihofer, and J. Marek, "Acceleration Sensor in Surface Micromachining for Airbag Applications with High Signal/Noise Ratio" SAE1996, 960758.

2. Analog Dialogue, Vol. 33, No. 1, January, 1999 "Dual Axis, Low *g*,Fully Integrated Accelerometers".

3. Seokyu Kim, Byeungleul Lee, Joonyeop Lee, and Kukjin Chun, "Gyroscope Array with Linked-Beam Structure", pp30-33, MEMS 2001.

4. Moorthi Palaniapan, Roger T. Howe and John Yasaitis, "Performance Comparison of Integrated Z-Axis Frame Microgyroscopes", pp482-485, MEMS 2003.

5. Jack D. Johnson, Seyed R. Zarabadi , John C. Christenson and Tracy A. Noll, "Single Crystal Silicon Low-G Acceleration Sensor", SAE2002-01-1080

6. Junseok Chae, Haluk Kulah, and Khalil Najafi, "A Hybrid Silicon-on-Glass Lateral Micro-Accelerometer with CMOS Readout Circuitry, pp 623-626, MEMS2002.

7. Junseok Chae, Haluk Kulah and Khalil Najafi, " An In-Plane High-Sensitivity, Low-Noise Micro-G Silicon Accelerometer", O_6_1, pp466-469, MEMS2003.

8. Daniel Lapadatu, Soheil Habibi, Bjørg Reppen, Guttorm Salomonsen, Terje Kvisterøy, "Dual-Axes Capacitive Inclinometer/Low-G Accelerometer for Automotive Applications", pp34-37, MEMS 2001.

9. Jonathan Hammond, Andrew McNeil, Rick August and Dan Koury, "Inertial Transducer Design For Manufacturability and Performance At MOTOROLA" 1D3.3, TRANSDUCERS '03.

Four-wheel-steering Control Strategy and its Integration with Vehicle Dynamics Control and Active Roll Control

Mauro Velardocchia, Andrea Morgando and Aldo Sorniotti
Politecnico di Torino, Department of Mechanics

ABSTRACT

The paper presents a 4-wheel-steering (4WS) control strategy devoted to reduce the turn diameter for small longitudinal speed values and to obtain a yaw rate damping effect in dynamic manoeuvres. Moreover, the 4WS active system conceived produces compensation both for lateral wind and road irregularities. The main results obtained through a functional vehicle model are presented. 4WS was integrated with a Vehicle Dynamics Control (VDC), which was improved for turn while braking manoeuvres. The results due to integration were very good, with a reduction of both systems interventions. Finally, a VDC-4WS-Active Roll Control (ARC) integration was tried, based on only one reference body yaw rate for all the active systems. The main results obtained are presented and discussed.

BASIC 4WS CONTROL STRATEGY

The 4ws control strategy was based on simulations results determined through a vehicle model implemented in Matlab-Simulink®. The model has ten degrees of freedom, six due to body, one rotational for each wheel. The vehicle considered is a middle-size two front wheel drive sedan. First target consisted in to reduce the steering diameter, quite consistent (12.5 m) because of the mechanical scheme. The target was to obtain the same values (about 10.5 m) of a similar car rear wheel drive. The desired performance was obtained by steering in counter-phase rear wheels of about 4.5°, as scheme of Figure 1 shows. Figure 2 is a comparison, in terms of centre of gravity trajectories, between 2WS and 4WS vehicle models, in correspondence of the maximum steering wheel angle.

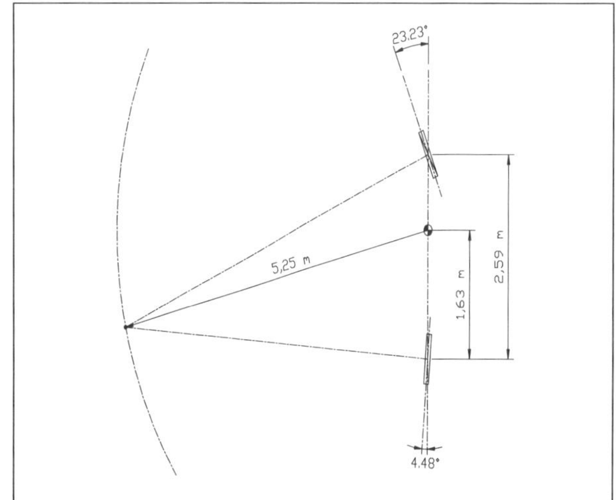

Figure 1: All Wheels Steering vehicle sketch.

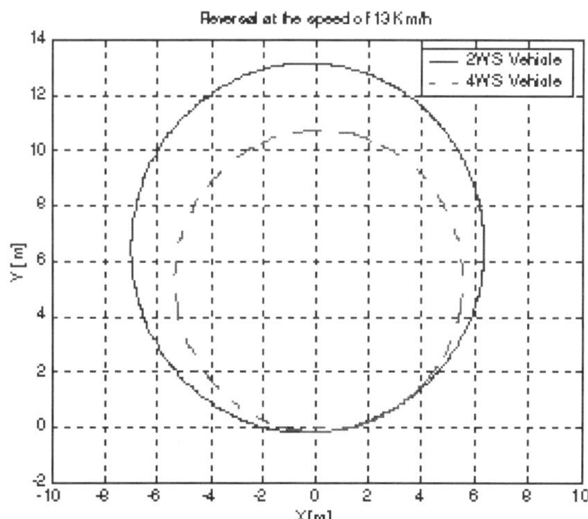

Figure 2: Comparison of Centre of Gravity trajectories.

The second and main task was to investigate the actual 4WS effectiveness in dynamic manoeuvres, characterized by tyres working near their physical adherence limit. As various authors [2, 3] stated, from a theoretical point of view, 4WS would not seem very useful in the typical dynamic manoeuvres used to test, e.g., a VDC. However, it is very efficient from the point of view of sweeping steer manoeuvres, to improve frequency response, and considering small lateral acceleration values. First test was on a feed forward control strategy: for low vehicle speeds rear wheels are counter steered, the opposite at high speeds, to increase vehicle stability. Figures 3, 4 and 5 are about a step steer manoeuvre in high adherence conditions, with a linear ratio of the front-to-rear steering angles.

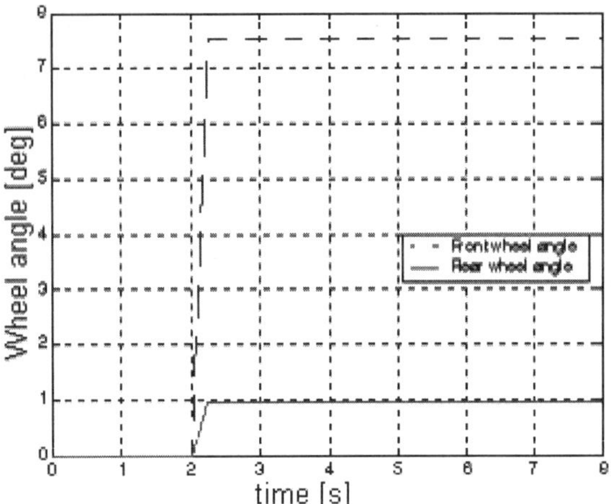

Figure 3: Step steer manoeuvre, high adherence.

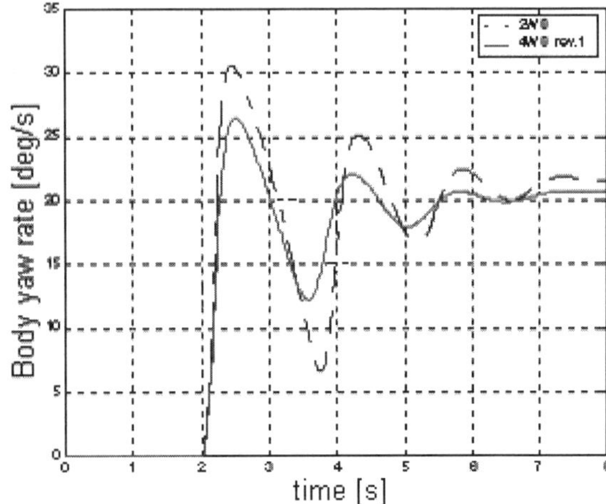

Figure 4: Step steer manoeuvre, high adherence.

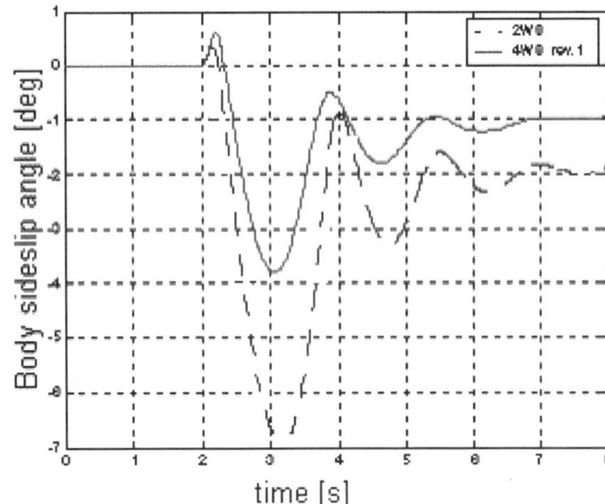

Figure 5: Step steer manoeuvre, high adherence.

Particularly consistent is the improvement in the peak value of body sideslip angle. On the other side, such a control strategy does not guarantee a significant improvement in yaw rate oscillations and adds a consistent delay in vehicle response to the step steering input. When considering a dynamic manoeuvre, two parameters are fundamental:

 A. the rapidity of yaw rate variation (the biggest, the better is the vehicle);

 B. the oscillations minimization

4WS basic-approach-strategy decreases the first (A) and improves a little the second (B). The vehicle so equipped behaves worse than a VDC-equipped 2WS vehicle. It is necessary, in dynamic manoeuvres, to introduce a closed-loop strategy. A first solution can be based on a yaw rate controller. Supposing the vehicle is provided with the typical sensors of a VDC-equipped car:

$$\dot{\psi}_{nom} = \left[\frac{vel_x \dfrac{\delta}{R_s}}{L + K_u vel_x^2} \right] \qquad (1)$$

$$\dot{\psi}_{max} = \frac{\mu g}{V} \qquad (2)$$

$$\dot{\psi}_{reference} = \min(\dot{\psi}_{nom}, \dot{\psi}_{max}) \qquad (3)$$

$$\Delta\dot{\Psi} = \dot{\Psi}_{measured} - \dot{\Psi}_{reference} \qquad (4),$$

where understeer coefficient K_u depends on body lateral acceleration and estimated adherence conditions (see list of symbols for reference).

When (4) overreaches an offset, VDC would intervene. Considering a stand-alone 4WS, rear steering angle is directly proportional to (4). Instead of a heavy intervention of brakes, like that typical of a braking active control, the 4WS here gives origin to a smooth rotation of rear wheels.

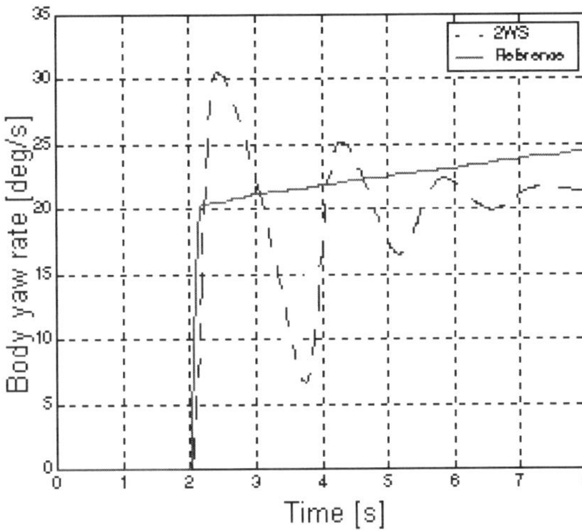

Figure 6: Step steer manoeuvre, high adherence.

Figure 6 is the comparison between 2WS and reference 4WS body yaw rate values in a step steer; Figure 7 plots the two contributions of (1) and (2).

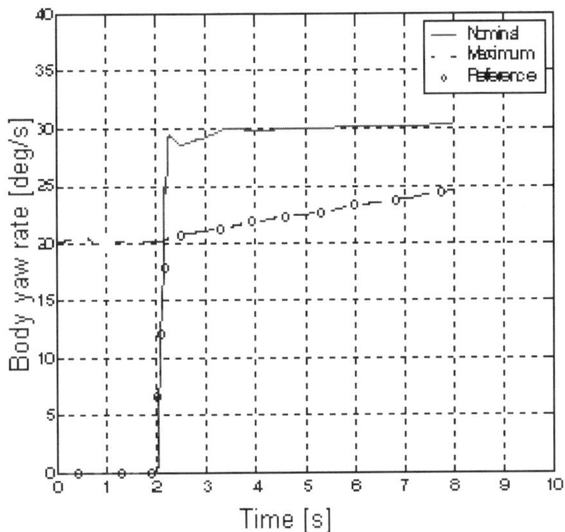

Figure 7: Comparison of reference body yaw rate values.

In the first part of the manoeuvre, the vehicle shows a delay: rear wheels are counter steered (Figure 8), then are heavily steered in phase. A damped oscillation characterize the manoeuvre; the stationary value of rear steering angle is negative since the reference body yaw rate is less understeering than the real vehicle one.

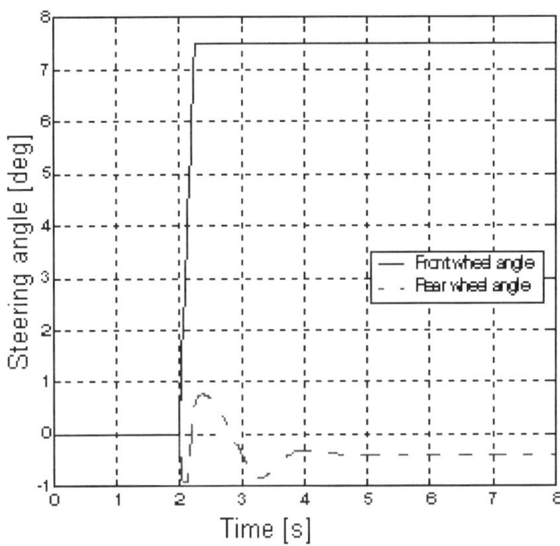

Figure 8: Step steer manoeuvre, high adherence.

Figure 9, 10 and 11 show that both targets A and B are successfully reached, by using 4WS potentiality also in this kind of manoeuvre. Figure 11 puts in evidence that also vehicle final speed has increased compared to that of the 2WS car, because of the reduced lateral slips.

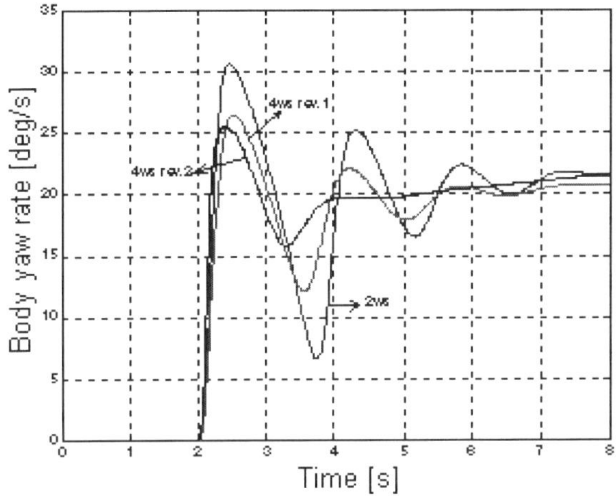

Figure 9: Step steer manoeuvre, high adherence.

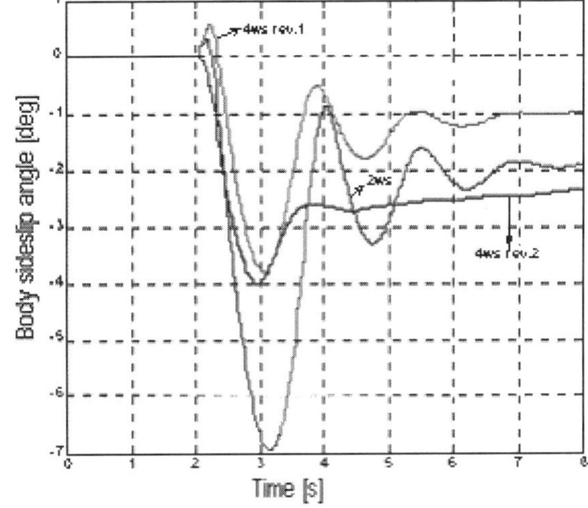

Figure 10: Step steer manoeuvre, high adherence.

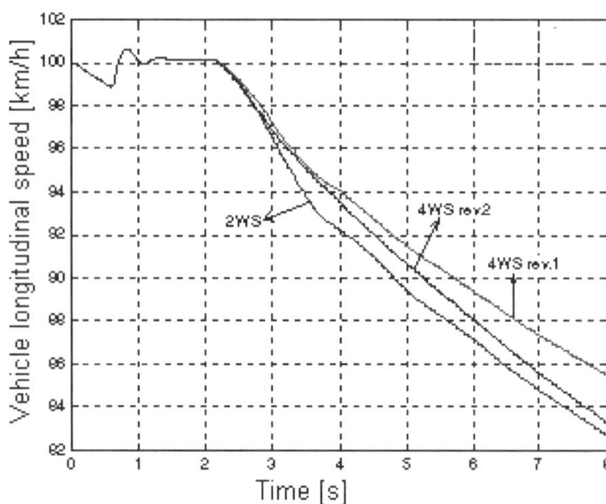

Figure 11: Step steer manoeuvre, high adherence.

The same manoeuvre in low adherence conditions introduces some more problems, because of a delay in adherence estimation. Such delay influences body yaw rate reference values, as Figure 12 presents, at the manoeuvre beginning. This delay was much more tolerable for VDC since it had also a body sideslip angle based control strategy, giving origin to consistent wheel pressures in the phase on the incorrect adherence estimation. Such a kind of emergency control is not suitable for 4WS, because it requires a progressive variation of rear wheels steering angles.

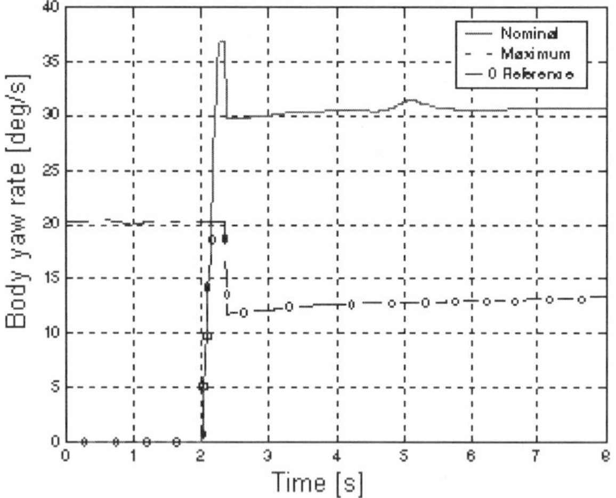

Figure 12: Step steer manoeuvre, low adherence.

There is absolutely no improvement from the body yaw rate point of view (Figure 13) in dynamic conditions on a low adherence ground.

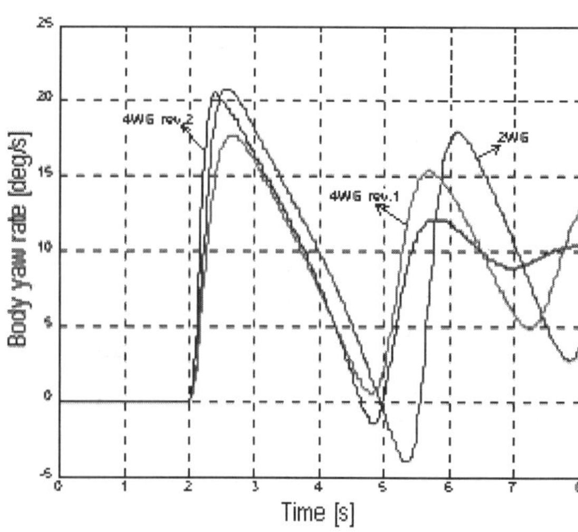

Figure 13: Step steer manoeuvre, low adherence.

NEW 4WS CONTROL STRATEGY

A new 4WS control strategy was ideated, evaluating the adherence as a function of the only lateral acceleration. Formulas (1)-(4) were still used to actuate rear steering. In such a way, (2) gets the most important term to estimate body yaw rate, instead of (1). The estimated adherence can be given in the form of a look-up-table, having the best shape in relation with the pre-fixed targets of vehicle dynamics. Figure 14 shows three different example of ways to set this table: rev.3 is about a strategy to increase mainly vehicle stability (as a consequence, the estimated adherence is lower than the measured lateral acceleration), rev.4.1 is about a strategy to reduce vehicle response delay for low levels of lateral acceleration and increase stability for high values (for low lateral accelerations, the estimated adherence is bigger than the measured lateral acceleration, whereas for high lateral accelerations, the estimated adherence is lower than the measured lateral acceleration), rev.4.2 is about a strategy to reduce understeering phenomena in all the conditions (the estimated adherence is always bigger than the measured lateral acceleration) . Figure 15 shows a comparison, in terms of vehicle dynamics, between the settings of Figure 14.

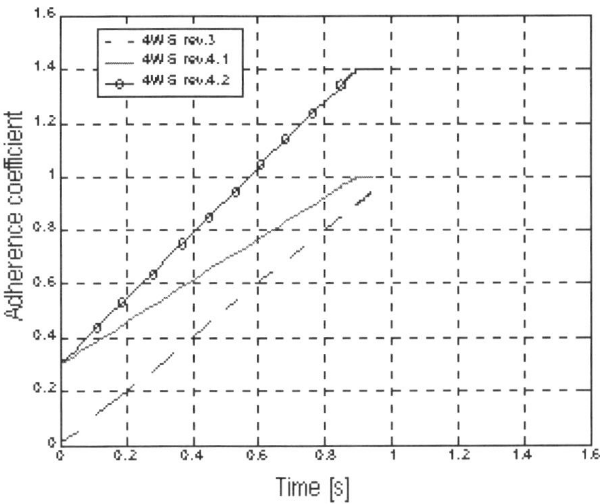

Figure 14: Adherence estimated through different look-up table setting.

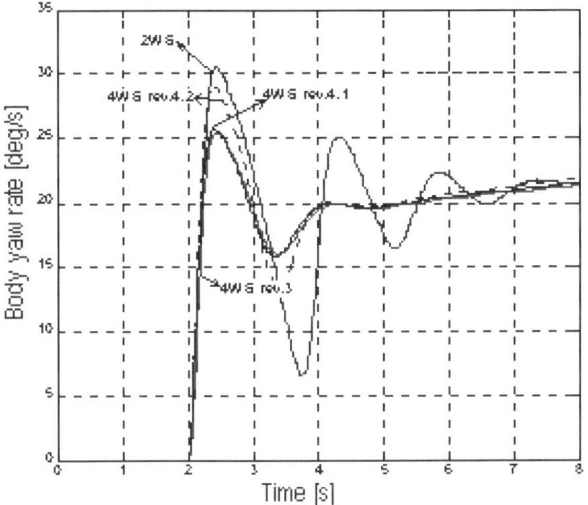

Figure 15: Step steer manoeuvre, yaw rate response depending on look-up table setting.

This strategies gave excellent results in all the manoeuvres tested, at whatever adherence, gear ratio or throttle angle value. The benefits on vehicle dynamics get more evident in low adherence conditions. Figures 16 and 17 are about a double step steer manoeuvre in low adherence conditions. The yaw rate ideal response would have the same shape of the steering wheel input. Passive car is abundantly out of the driver control, as it is clear by looking at the final positive peak of body yaw rate, corresponding to a zero-value of steering wheel angle. In dynamic conditions, quite low values (no more than 2°) of rear steering angle are sufficient to stabilize the vehicle. They are smaller than those computed to reduce the steering diameter, presented at the beginning of the paper. In ramp steer manoeuvres, 4WS has not a big influence (Figure 18), especially with a 1:1 correspondence between lateral acceleration and adherence.

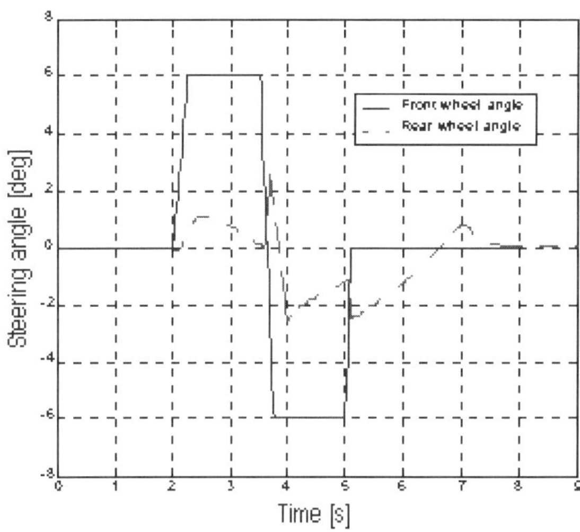

Figure 16: Double step steer manoeuvre, low adherence conditions.

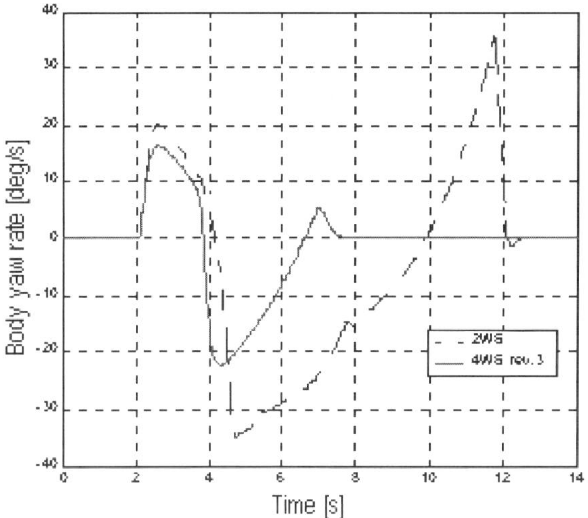

Figure 17: Double step steer manoeuvre, low adherence conditions.

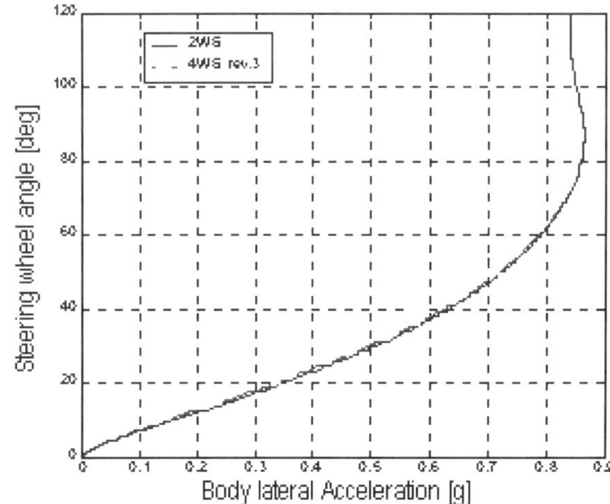

Figure 18: Ramp steer manoeuvre, high adherence.

The same control strategy was tested in straight ahead (Figures 19 and 20) and in turn manoeuvres (Figure 21) with disturbances phenomena due to lateral wind.

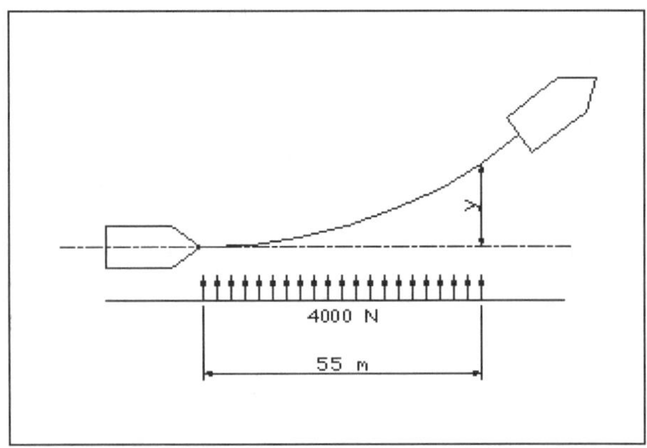

Figure 19: Straight ahead manoeuvre standardization.

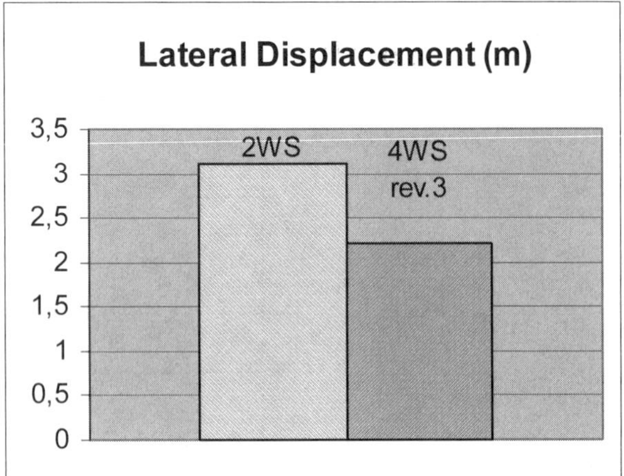

Figure 20: Straight ahead manoeuvre results.

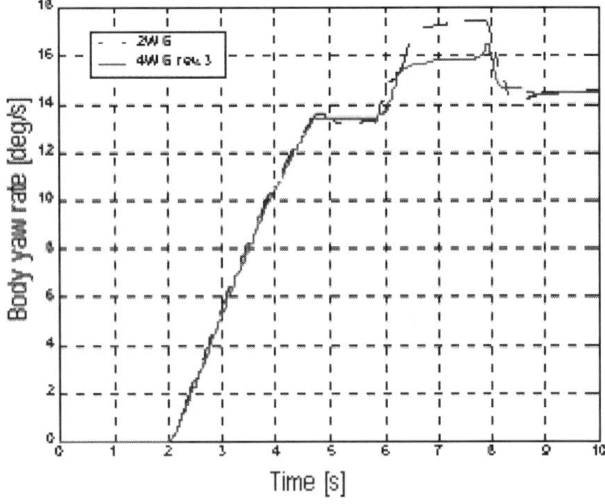

Figure 21: Turn manoeuvre, high adherence.

Also vehicle oversteering trend, due to longitudinal load transfer during braking while turning, was significantly reduced (Figure 22). Reference body yaw rate computed by this 4WS control strategy corresponds to the stationary value it would have for the same lateral accelerations and steering wheel angles, without considering the dynamic effect related to body sideslip rate.

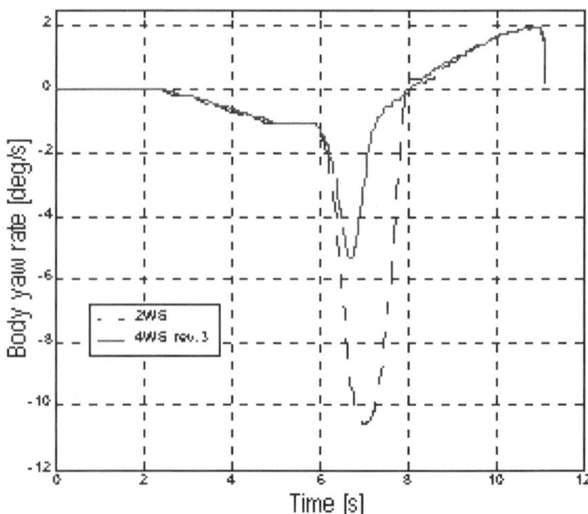

Figure 22: Turn with braking action manoeuvre.

ELASTOKINEMATICS REAR SUSPENSION CHARACTERISTICS INFLUENCE

Since rear steering wheel angles values are small, it was necessary to investigate about the rear suspension elastokinematics characteristic influence on 4WS. It was supposed to have a rear double-wishbone suspension. The attachment points were fixed and kinematics characteristics were designed. The main targets were to reduce half track variations (Figure 23), have a strong camber compensation while rolling (Figures 24 and 25), have a near to zero toe angle change during wheels vertical motions. The latter task was obtained through the method suggested by [8].

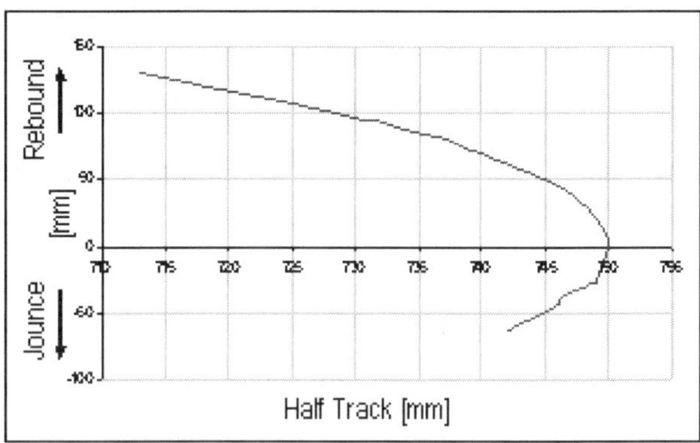

Figure 23: Half track variation due to suspension travel.

The half-track variation is important because it gives origin, during jounce and rebound, to a wheel lateral slip rate, which modifies the value of tyre slip angle. Figure 26 shows toe variation, computed by the 2WS vehicle model having the designed suspension, during a transient manoeuvre. Figure 27 puts in evidence the very toe variation small influence on 4WS behaviour. The various elastokinematics main characteristics resulted negligible changed.

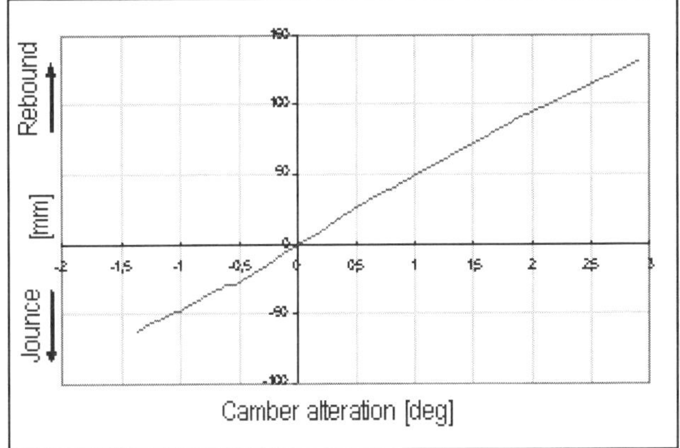

Figure 24: Camber angle variation due to suspension travel.

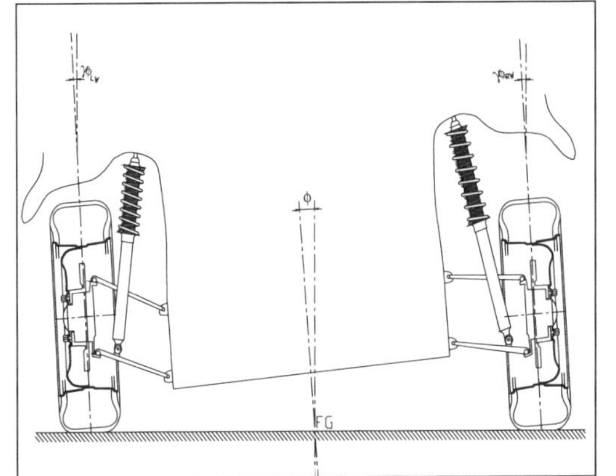

Figure 25: Camber compensation, suspension scheme.

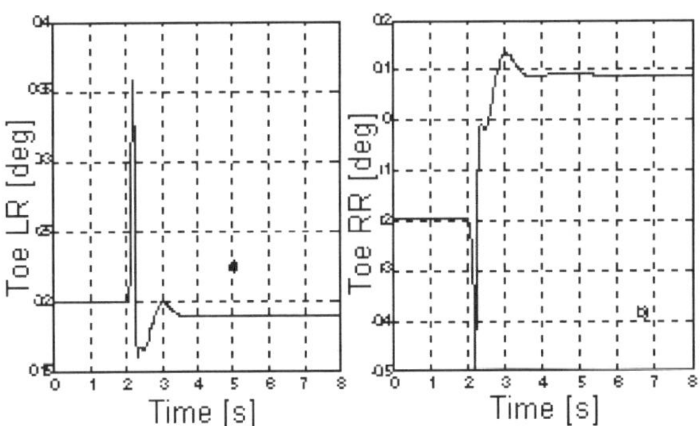

Figure 26: Toe variation during transient manoeuvre.

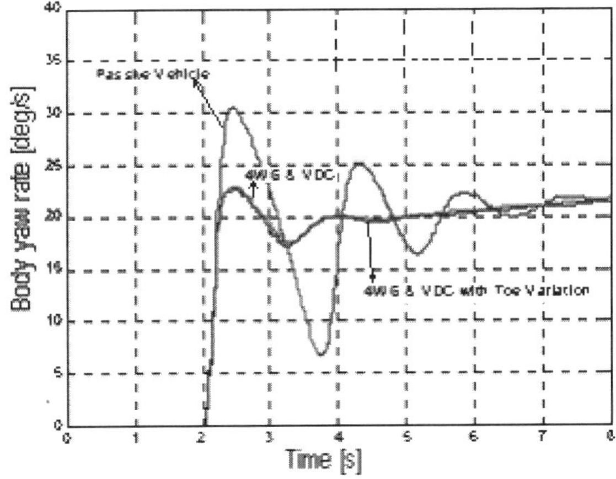

Figure 27: Toe variation influence on vehicle dynamics.

INTEGRATION WITH VEHICLE DYNAMICS CONTROL (VDC)

First part demonstrated that a VDC strategy based on a high/low adherence estimation is not suitable for 4WS, which must have a smooth adherence estimation. VDC strategy was therefore modified using 4WS reference body yaw rate evaluation. Integration between VDC and 4WS consisted in having only one high level control strategy, computing a common reference body yaw rate, according to the available sensors, and two different actuation strategies, one for each control system. All the tested manoeuvres demonstrated the better performance of the integrated system than the stand-alone device.

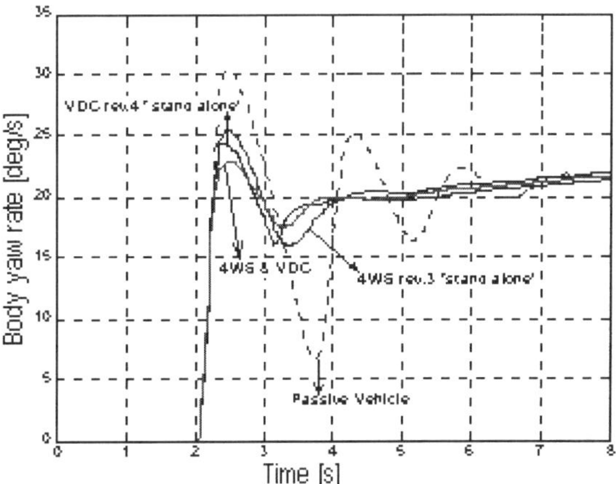

Figure 28: Step steer manoeuvre, active control integration influence.

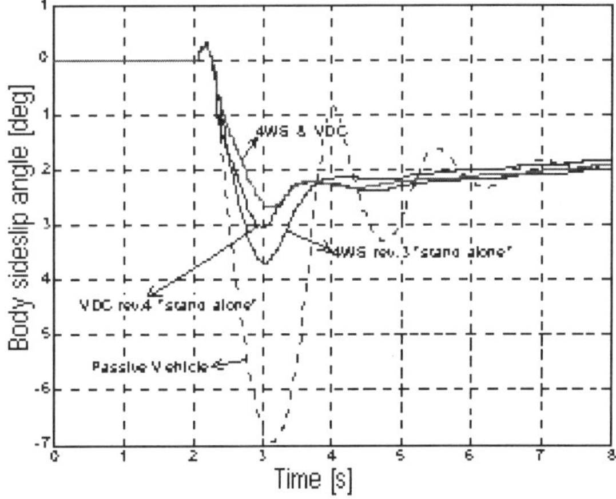

Figure 29: Step steer manoeuvre, active control integration influence.

When designing a control strategy, it is fundamental to verify its effectiveness for all possible conditions, including load changes due to passengers or luggage. The Table reports the variation of the most important parameters when changing vehicle weight. Figure 30 shows that active vehicle dynamic behaviour variation due to different loads, from the point of view of the peak values of body yaw rate, is minor that considering passive vehicle variation.

Vehicle Parameters	Standard Load	Full Load
M [kg]	1310	1610
J_z [kg * m^2]	2201	2876
a [m]	0.96	1.235
b [m]	1.63	1.355

Table 1: Vehicle main parameters depending on different load conditions

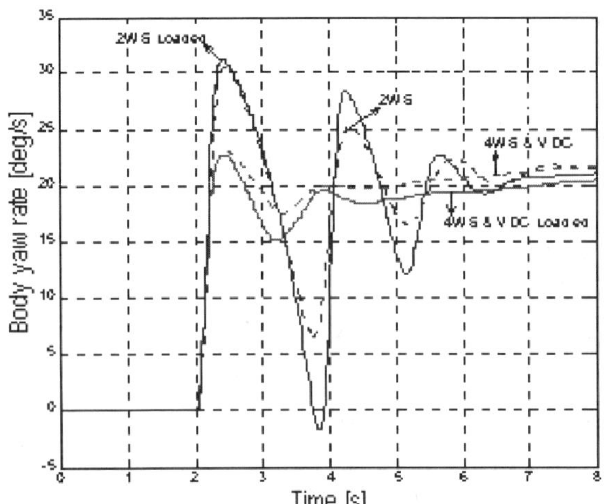

Figure 30: Active vehicle dynamic behaviour variation due to different loads

INTEGRATION WITH ACTIVE ROLL CONTROL (ARC)

4WS and VDC were also integrated with an Active Roll Control system, characterized by a rear active bar [13].

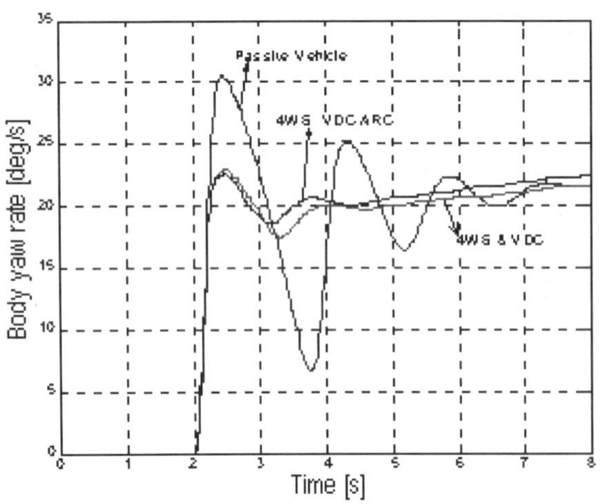

Figure 31: Step steer manoeuvre, ARC integration influence

The integration technique is the same used to manage the 4WS co-operation with VDC. The active bar permits a reduction of understeering phenomena in ramp steer manoeuvres (Figure 32).

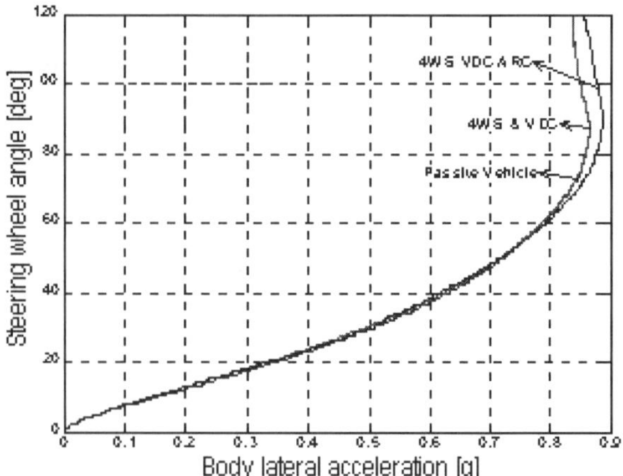

Figure 32: Ramp steer manoeuvre, ARC integration influence

CONCLUSIONS

A 4-wheel-steering (4WS) control strategy was presented, capable of turn diameter reduction and yaw rate damping effect in dynamic manoeuvres. Both lateral wind and road irregularities were positively compensated by the developed strategy. 4WS integration with a Vehicle Dynamics Control (VDC) and Active Roll Control (ARC) based on only one reference body yaw rate for all the active systems was described. The main results obtained justify the increasing interest about 4WS due especially for safety and performance.

ACNOWLEDGEMENTS

Politecnico di Torino Research Team wish to thank FIAT Auto to have sustained the present work both through funds and with technical suggestions.

LIST OF SYMBOLS

J_z: Vehicle inertia around vertical (z) axis.
M: Vehicle mass.
R_s: Front wheels steering ratio.
δ: Steering wheel angle.
ψ: Vehicle body yaw angle.
vel_x: Vehicle longitudinal speed.
V: Vehicle absolute speed.
K_u: Understeer coefficient.
L: Vehicle wheel base.
μμ: Adherence coefficient.

BIBLIOGRAPHY

[1] T. MIMURO, M. OHSAKI, H. YASUNAGA and K. SATOH, *Four Parameter Evaluation Method of Lateral Transient Response*, Research Department Passenger Car Engineering Centre, Mitsubishi Motors Corp. SAE Technical Paper 901734

[2] K. SHIMADA, Y. SHIBAHATA, Comparison of three Active Chassis Control Methods for Stabilizing Yaw Moments, Honda R&D Co., Ltd., SAE Technical Paper 940870

[3] M. YAHAMOTO, *Active Control Strategy for Improved Handling and Stability*, Toyota Motor Corp. SAE Technical Paper 911902

[4] H. SATO, H. KAWAI, M.ISIKAWA, H. IWATA and S. KOIKE, *Development of Four Wheel Steering System Using Yaw Rate Feedback Control*, Toyota Motor Corp. SAE Technical Paper 911922

[4] Y. LIN, *Improving Vehicle Handling Performance by a Closed-Loop 4WS Driving Controller*, UC Berkeley - Path. SAE Technical Papers 921604

[5] M.B. GERRARD, *Roll Centres and Jacking Forces in Independent Suspension - A First Principles Explanation and a Designer's Toolkit,* Randle Engineering Solution Ltd. SAE Technical Paper 1999-01-0046

[6] H. KAWAKAMI, H. SATO, M. TABATA, H INOUE, and H. ITIMARU, *Development of Integrated System Between Active Control Suspension, Active4WS, TRC and ABS,* Toyota Motor Corp. SAE Technical Paper 920271

[7] C. H. SUH *Suspension Analysis with Instant Screw Axis Theory, Department of Mechanical Engineering*, University of Colorado, Boulder CO, SAE Technical Paper 910017

[8] J. REIMPELL, H.STOLL, J.W.BETSLER, *The Automotive Chassis*, Second edition, SAE International 2001

[9] N. IRIE and J. KUROKI, *4WS Technology and the Prospects for Improvement of Vehicle Dynamics*, Nissan Motor Co., Ltd. Yokosuka, Japan. SAE Technical Papers 901167

[10] A. TANEDA and T. Yamanaka, *Design of Actuator for Active-Rear-Steer System,* Aisin Seiki Co., Ltd. SAE Technical Paper 981114

[11] K. FUJITA, K. OHASHI, K. FUKATAMI, S. KAMEI, Y. KAGAWA and H.MORI, *Development of Active Rear Steer System Applying H∞-µ Synthesis,* Toyota Motor Corporation, SAE Technical Paper 981115

[12] M. MITSCHKE, E. AHRING, *Control Loop for Driver-Vehicle with Four Wheel Steering,* Technische Universität Braunschweig, Institut für Fahrzeugtechnik, Hans-Sommer-Str. 4, 38106 Braunschweig, GERMANY

[13] D. DANESIN, P. KRIEFF, M. VELARDOCCHIA, A. SORNIOTTI, *Active Roll Control to Increase Handling and Comfort,* SAE Technical Paper 2003-01-962

[14] Y. YAMAMURA, K. ITO, T. EDUCHI, O. SATO, *Robust Control System for Electric Rear Wheel Steering Actuator,* Nissan Motor Company Ltd., Kanagawa, Japan.

CONTACT

Prof. Mauro Velardocchia, Ph. Dr. St. Andrea Morgando, Ph. Dr. St. Aldo Sorniotti
Department of Mechanics Politecnico di Torino
C.so Duca degli Abruzzi 24 -10129 Torino ITALY
Tel: +39 011 564 6931 ; +39 011 564 6915
Fax: +39 011 564 6999
Email:mauro.velardocchia@polito.it
andrea.morgando@polito.it
aldo.sorniotti@polito.it

A Supervisory Control to Manage Brakes and Four Wheel Steer Systems

Edward J. Bedner, Jr. and Hsien H. Chen
Delphi Corporation

ABSTRACT

This paper presents the development of coordinated control of vehicle systems, specifically for controlled brakes and controlled steering systems. By utilizing a control structure to oversee a four-wheel-steer (4WS) system and a brake-based vehicle stability enhancement (VSE) system, it is possible to achieve improvements in vehicle stability and driver workload/comfort, and to reduce compromises in vehicle handling. The coordinated control is designed to leverage the unique strengths of 4WS and VSE, and to prevent conflicts between them. Vehicle test results prove the viability of the concept.

INTRODUCTION

Motivated by the demand for safety and performance, control systems for trucks and SUV's have made remarkable progress in recent years to influence the overall motion of the vehicle. Such systems include controlled brakes (ABS, TCS, VSE), controlled steering (4WS, EPS), controlled suspensions (dampers, springs, roll bars), and controlled drivetrains (engines, transmissions). The introduction of 4WS systems for trucks and SUV's is the most recent example of such technology evolution. With so many systems gaining in authority and in sophistication, it is now recognized that there is a need to better manage multiple systems to achieve optimal performance and to eliminate conflicts in regions of overlap. A control structure is proposed to provide such a means to mediate between multiple systems. Figure 1 shows a functional block diagram representation of the control concept.

Both controlled brake systems with ABS/TCS/VSE and controlled steering systems with 4WS have large authority over the yaw-plane motion of the vehicle, directly influencing longitudinal, lateral, and yaw motions. As first priority, the coordinated control ensures there is no conflict between the two, i.e. that both are working to achieve the same longitudinal, lateral, and yaw motions for the vehicle.

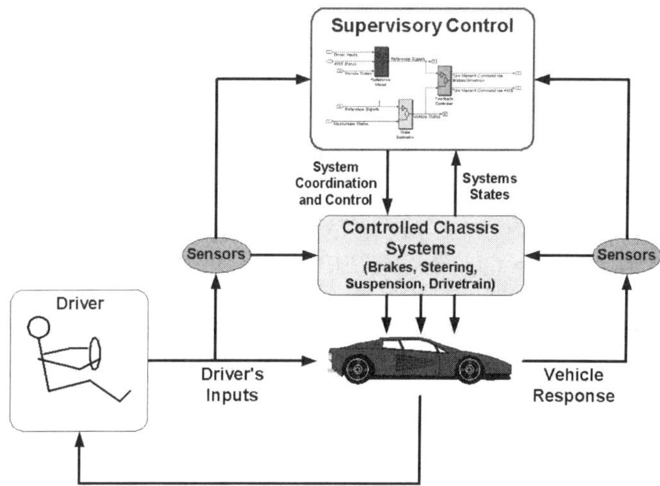

Figure 1: Functional diagram of the control concept

Secondly the coordinated control provides a means to optimize overall performance by appropriately distributing workload to each system as needed, taking advantage of the unique strengths of each system. In doing so we see the reduction of some compromises that typically hinder each stand-alone system.

This paper is organized in the following way. First an overview is given for controlled steering and brake systems, which includes a description of tire force generation and its impact on vehicle motion. The second section provides motivation for control coordination by discussing the dissimilar capabilities and limitations of each system, and by showing examples of potential conflicts. In the third section, the coordinated control design is presented, showing the structure and discussing the considerations for workload distribution. Lastly, vehicle test results are shown for 2 maneuvers that demonstrate the benefits of control coordination, specifically an emergency lane-change on snow and a split-coefficient braking event.

OVERVIEW OF SYSTEMS AND PHYSICAL PRINCIPLES

The following section presents a review of active steering and braking systems, as well as how they generate forces at the wheels and how the forces contribute to the overall motion of the vehicle.

STEERING

Steering systems are beginning to incorporate various actuation elements, such as 4WS that is available today, and also Active Front Steer (AFS) and Steer By Wire (SBW) in the near future. These systems are able to directly control the steer angle of one or more wheels, and they provide new types of functionality.

Focusing on 4WS for this paper, the active element of 4WS provides a way to steer the rear wheels of a vehicle, while the front wheels are steered with a traditional passive system under control of the driver. Using Delphi's QUADRASTEER™ 4WS as an example, the rear wheels are steered together by a rack with tie rods, and the QUADRASTEER™ system can steer the rear wheels up to 12 degrees for certain applications. A brushless motor drives the rear steer rack through gears, and there is a return-to-center spring mechanism that returns the wheels to center in the event of a system failure. The system includes an electronic controller with power electronics and a microcontroller, along with sensors for handwheel angle and for motor position, and a communication bus provides vehicle speed and other data.

Looking next at how steering systems generate lateral forces at the wheel, refer to Figure 2, which shows a plan view of a single wheel and vectors representing force and velocity. When the tire's direction of travel is at an angle α to the wheel plane, a lateral force F_y develops at the tire-road contact patch. The lateral force, or cornering force, is known to be a function of the tire slip angle α as depicted in Figure 2. The relationship of force and tire slip angle is also affected by other factors such as road surface friction, normal load, and inflation pressure. Furthermore, any camber angle of the tire will also contribute to lateral force, but it will be considered negligible for this discussion.

Next, the impact on vehicle motion is considered. Figure 3 shows a bicycle model diagram of a vehicle with 4WS. The figure shows lateral forces on the rear wheels due to 4WS motion. Specifically, the rear wheels are steered to an angle δ_R with respect to the vehicle coordinate frame. The vehicle is operating with a side-slip angle β_R at the rear axle. The rear tire slip angle α_R is then:

$$\alpha_R = \beta_R - \delta_R$$

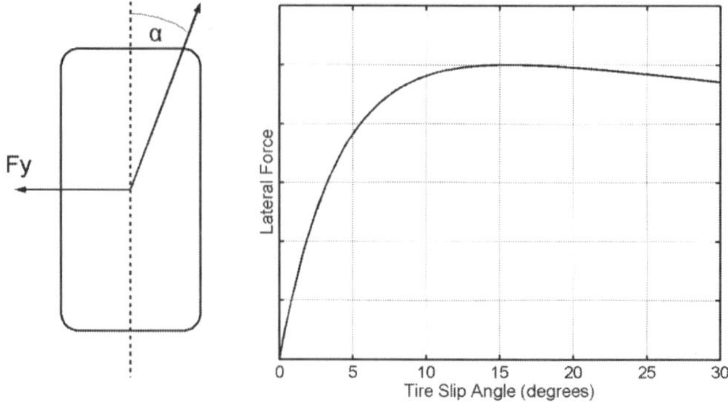

Figure 2: Tire slip angle and lateral force

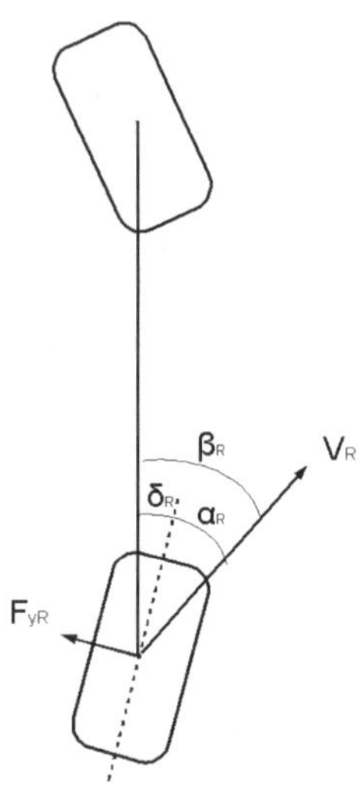

Figure 3: Bicycle model

The magnitude of rear lateral force is thus determined by the tire slip angle α_R. The rear lateral force F_{yR} has 3 impacts on the vehicle's yaw-plane motion:

1. On yaw moment:
    $$I_{zz}\, r' = a\, F_{yF}\, \cos(\delta_F) - b\, F_{yR}\, \cos(\delta_R)$$

2. On lateral acceleration:
    $$M\, a_y = F_{yF}\, \cos(\delta_F) + F_{yR}\, \cos(\delta_R)$$

3. On longitudinal acceleration:
    $$M\, a_x = F_{yF}\, \sin(\delta_F) + F_{yR}\, \sin(\delta_R)$$

Symbol definition:

I_{zz}	Yaw Moment of Inertia
r'	Yaw Acceleration
a	Distance from CG to Front Axle
b	Distance from CG to Rear Axle
M	Vehicle Mass
a_y	Lateral Acceleration
a_x	Longitudinal Acceleration

Therefore, 4WS affects overall vehicle motion through the generation of rear tire slip angles α_R and rear lateral forces. It is important to recognize that the lateral forces and subsequent vehicle motion are limited by several operating conditions including the side-slip angle β_R of the vehicle and the surface coefficient of friction, and by the actuation range of motion.

BRAKES

For some time now, controlled brake systems have offered the ability to regulate the longitudinal slip/spin of the tire, to help maintain directional control of the vehicle while also optimizing deceleration and acceleration performance of the vehicle. More recently, brake system features were extended to improve vehicle stability through what is referred to as Vehicle Stability Enhancement (VSE). The purpose of VSE systems is to provide corrective yaw moment to the vehicle to help the driver maintain control during severe oversteer and understeer conditions. The corrective yaw moment is generated by manipulating tire forces through the control of longitudinal slip/spin of the wheels.

Delphi's Traxxar™ brake system is an example of a controlled brake system that includes VSE. The features of the system are the hydraulic modulator, the sensors for wheel speeds, yaw rate, lateral acceleration, handwheel angle, and master cylinder pressure, and an electronic control unit. The electronic controller has power electronics and a microcontroller, along with a communication bus that links to the powertrain system and other vehicle systems.

Looking next at how braking affects forces at the wheel, refer to Figure 4 which shows plan views of 2 wheels. For both cases, the tires are operating at the limit of adhesion as represented by the dashed ellipse. In the left figure, the tire is experiencing maximum cornering force F. In the right figure, braking force is also applied in addition to cornering force. Compared to the left figure, the force vector F has been rotated due to the application of brake force. The lateral component F_y is reduced and the longitudinal component F_x is increased.

The longitudinal force is a function of the longitudinal slip/spin as well as the tire slip angle. The relationship of force and slip/spin is also affected by other factors such as the road surface friction, the normal load, and the tire inflation pressure.

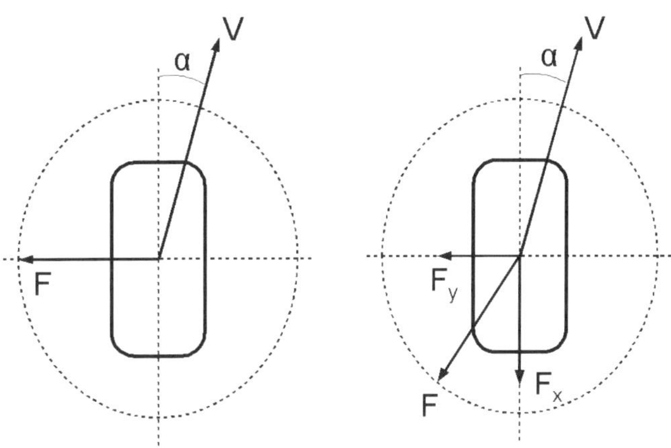

Figure 4: Development of tire forces during cornering without braking (left) and with braking (right)

Next, the impact on vehicle motion is considered. Figure 5 shows a plan view diagram of a vehicle that is in an impending oversteer condition. There are cornering forces at each wheel, and the VSE system has also generated a braking force at the outside front wheel. The brake force from VSE activation has 3 impacts on the yaw-plane motion of the vehicle:

1. On yaw moment
2. On lateral acceleration
3. On longitudinal acceleration

Thus, VSE influences overall vehicle motion by manipulating tire forces through the control of longitudinal slip/spin of the wheels.

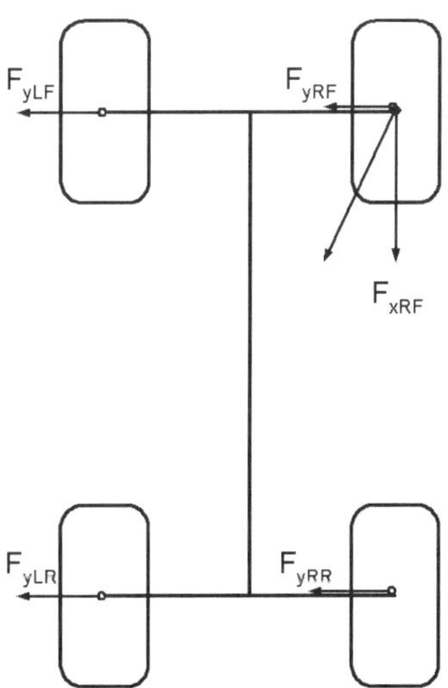

Figure 5: Forces acting on the vehicle during cornering and VSE activation

COMPARISON OF SYSTEM CAPABILITIES

The following section discusses the capabilities and limitations of each control system, specifically for yaw moment generation, intrusiveness, and bandwidth.

The first item for comparison is the capability to generate yaw moment. It has been shown [1] that large yaw moments can be generated by longitudinal tire slip/spin control via brakes and driveline actuations and by lateral tire slip control via 4WS. However the yaw moment magnitude depends on several factors. For example, on a dry asphalt surface where the peak lateral force occurs at large tire slip angles (e.g. α_{PEAK} = 15 degrees), and with rear steer capable of +/- 12 degrees actuation, it is possible to maintain maximum lateral force on the rear tires for side-slip angles up to β_R = 15+12 = 27 degrees.

Likewise on a slippery surface such as ice, the peak lateral force occurs at a much smaller tire slip angle (e.g. α_{PEAK} = 4 degrees), thus the rear steer is capable of holding maximum lateral force for side-slip angles up to β_R = 4+12 = 16 degrees. For β_R greater than 16 degrees, the rear lateral force typically decreases rapidly according to the shape of the tire's lateral force curve [2]. By these examples, we note yaw moment generation via rear steer control depends on the operating slip angle of the vehicle, the surface friction, and the maximum amount of actuation travel.

Another item for comparison is the level of "intrusion" or disturbance to the driver. In some cases, brake-based yaw moment generation may be perceived as intrusive to the driver. That is, brake actuations cause vehicle deceleration, modulator noise, and pedal pulsation that may be objectionable to the driver. To reduce the possibility of unnecessary brake activation and unnecessary disturbance to the driver, the brake-based VSE systems are designed with a control deadband, which prevents activation during all nominal driving conditions.

In contrast, steer-based systems are much less perceptible to the driver since there is little or no audible or tactile feedback and since they cause insignificant deceleration of the vehicle. For these reasons, little or no deadband is needed for steering control, and thus it is possible for the system to react sooner to small perturbations. In this way, a steering system provides a more preventative action for perturbations that might have otherwise grown larger.

For a final comparison item, actuation bandwidth should be considered, especially for certain operating conditions. Both brake and steer system bandwidths are largely functions of actuator designs, and the bandwidths can be comparable. For systems containing hydraulic fluid, responses at low temperatures become critical as fluid viscosity decreases. For systems containing electric motor actuation, operating voltage may be a critical factor for system response.

Given that steering and braking control systems have different strengths and weaknesses, it is natural to consider their combination to take advantage of the strengths of each. The following sections discuss other motivations and benefits of control coordination.

MOTIVATION FOR COORDINATION

As a first step in realizing the benefits of systems integration, it is necessary to ensure that individual systems are not in conflict. Since both brake systems and steering systems have authority over the generation of yaw moment, it is important to provide coordination so that both systems are generating yaw moment in the same direction with the appropriate magnitude and phase relationship. Without coordination, conflicts can arise that lead to undesirable vehicle response and sub-optimal performance. The following section examines 2 cases that lack coordination.

First, consider 2 systems, each with a reference model as part of their control structure. However, in this case neither reference model comprehends the operation of the other system. The reference models generate target state values that are frequently not achieved due to the lack of systems coordination, which leads to excessive activations as systems try unsuccessfully to track the improper target states.

Taking the case of QUADRASTEER™ 4WS steer as an example, the basic mode of control is to steer the rear wheels as a speed-dependent function of front wheel angle. Furthermore, this basic operation is selected by the driver for one of three possible modes: Normal 4WS Mode; Trailer Tow 4WS Mode; and Off Mode. In each mode the relationship of rear wheel angle and front wheel angle is different, so the vehicle's handling response will also depend on the selected mode. Additionally there are other conditions that will affect the ultimate rear wheel angle such as low-speed swing-out reduction, mode transitions, and fail actions.

If the brake-based system's reference model does not comprehend all of the possibilities noted above, then there will be instances when its reference states do not match the actual states of the vehicle due to the rear steer activity. In these instances the brake-based system will activate and apply brake forces to create a yaw moment change that is actually unnecessary. The driver would perceive this as a nuisance or a problem with the brake system.

A solution to this problem is to establish a common reference model that comprehends all possible activities of the steer and brake systems, thereby ensuring the validity of the reference states. The need for a common comprehensive reference model is an element of the proposed control structure, which is presented in the next section.

In a second case involving the lack of coordination, there is sub-optimal vehicle response due to mismatch of feedback control. In this case, first assume that there is no conflict in reference model matching. That is, a common reference model exists that comprehends all activity of the constituent systems. Next assume separate and unique feedback control mechanisms that are driven by state errors for yaw rate and side-slip. The designs of these separate controls may be vastly different [3], possibly yielding dissimilar yaw moment commands, even though each one alone may be an acceptable design that provides acceptable vehicle performance.

In Figure 6, the plots on the left are an example of vehicle test data in which the 4WS and the brake-based systems were implemented with separate and different feedback control methods that are uncoordinated. The maneuver was an emergency lane change on groomed snow. In this data we see instances where the individual yaw moment commands are dissimilar, to the extent of being in opposition at times.

For the purpose of comparison, the plots on the right side of Figure 6 show the case of control coordination where the yaw moment command is identically the same for the 2 systems. With coordination of feedback control, the overall vehicle response is notably better (less side-slip angle of the vehicle, less driver steering workload), and there is less actuation of brakes and rear steer.

The above demonstrates the need for a common feedback control. The shared feedback control is another element of the proposed control structure, presented in the following section.

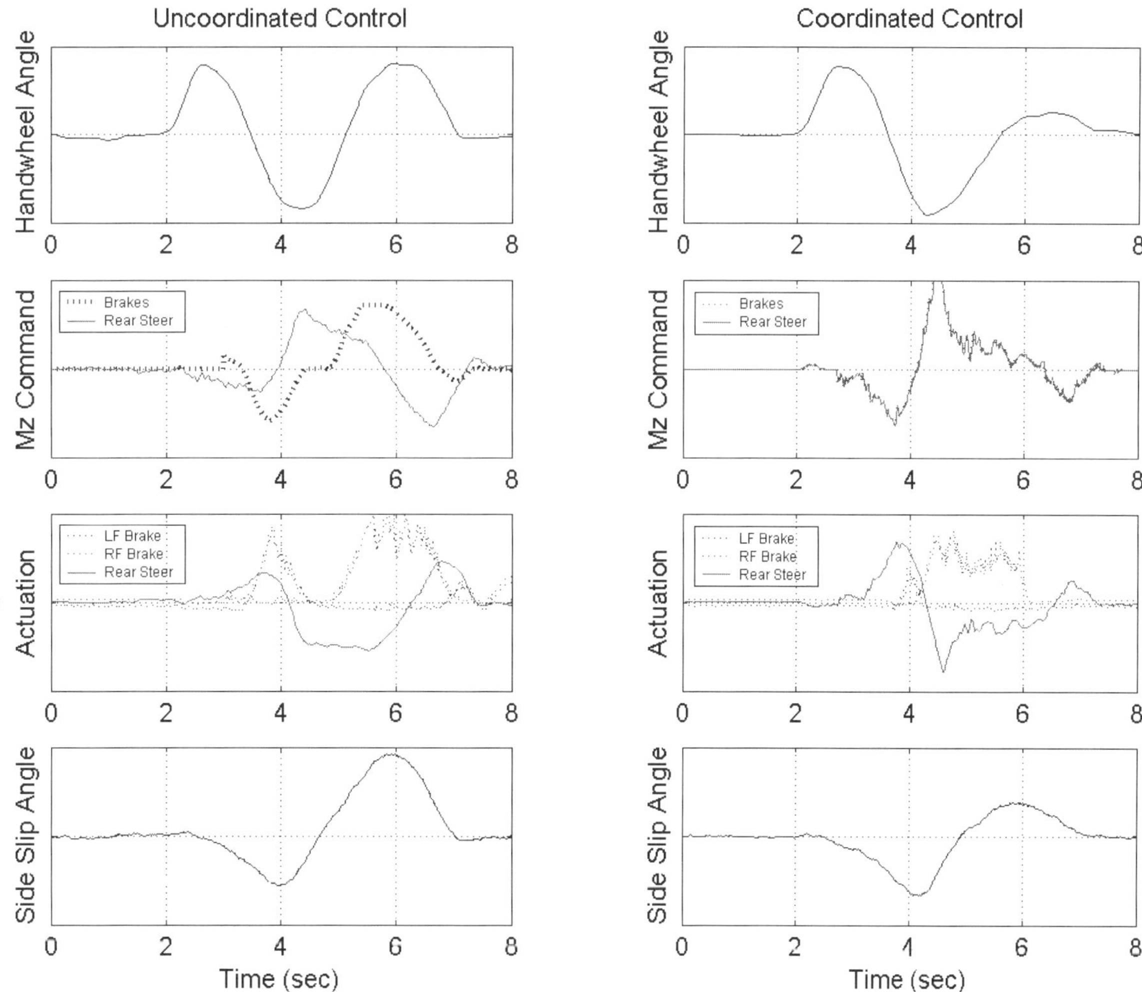

Figure 6: Vehicle data for uncoordinated and coordinated control, for an emergency lane change maneuver on snow

CONTROL DESIGN

In this section a control structure is proposed to coordinate multiple chassis systems. There are several possible ways to achieve system coordination, and for this evaluation a supervisory control structure is employed. The structure is shown in Figure 7, which includes a reference model, a state estimator, and a feedback control.

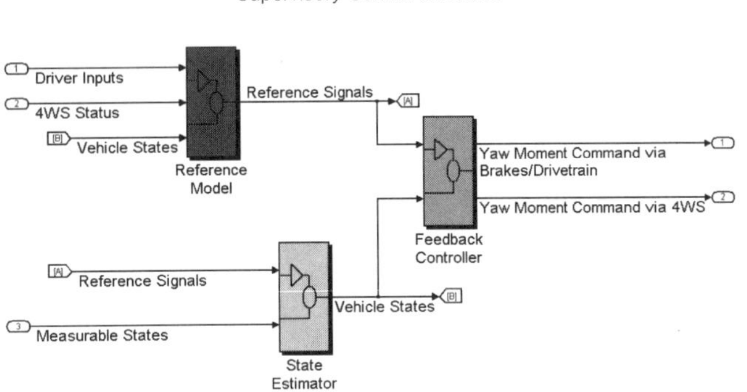

Figure 7: Supervisory control block diagram

The control is designed to achieve 2 goals:

1. Elimination of conflicts among constituent systems.
2. Optimization of vehicle performance by leveraging the strengths of each system.

An advantage of this type of structure is its scalability for a given set of sensors and actuators. For the purposes of this paper, the actuation systems are brakes and steering, thus the control will focus on the coordination of yaw-plane motion. Alternatively, if other systems were available, such as controlled suspension systems, then the coordinated control could be expanded to also include ride modes (heave, pitch, roll). For this discussion, we will concentrate solely on the management of motion in the yaw-plane.

The first element of the control design is the reference model. The purpose of the reference model is to determine desired states of the vehicle based mainly on driver's inputs of handwheel, brake, and throttle. In the case of yaw-plane control, the outputs of the reference model are the desired states of yaw rate and side-slip.

Unique to this application is the inclusion of the present status of the rear steer system, which is an additional input to the reference model. As noted in a previous section, a common reference model must account for the basic operation of individual systems, such as the

rear steer mode (Normal, Trailer Tow, Off) and other conditions (Fail Action, Low Speed Swing-Out Reduction, etc.). By including the present status of the rear steer system as an additional input to the reference model, the output desired states are then valid for all nominal operating conditions.

A second element of the control design is the state estimator. The purpose of the state estimator is to determine the actual states of the vehicle, based on sensor signal inputs. For yaw-plane motion, the primary sensor inputs are yaw rate, lateral acceleration, wheel speeds, handwheel angle, engine torque, master cylinder pressure, brake pedal force/position, and rear wheel angle. The state estimator performs diagnostic checks of the sensor signals, removes biases as necessary, and also calculates other states and conditions such as side-slip angle, vehicle velocity, longitudinal tire slip/spin, road bank angle, road friction, and reverse motion.

An advantage of a common state estimator is the availability of more measured states than that of the individual systems. This gives improvement in quality of the estimated states, since more information is available.

The third element of the control design is the feedback control (FBC). In the case of yaw-plane motion control, it is the duty of the FBC to compare the desired and actual states of the vehicle, and to then generate the command for change in yaw moment. Several control strategies are known such as those cited in [3].

As noted in a previous section, the use of a common feedback control in the supervisory control implementation optimizes the integrated control of multiple systems. The proposed FBC provides this capability. The common yaw moment command from the FBC provides the necessary coordination and eliminates potential conflicts between individual systems as seen in an earlier example.

It is also the duty of the FBC to distribute the yaw moment command to the constituent systems in an optimal way. To determine the criteria for yaw moment distribution, the design of the FBC must consider the capabilities and limitations of each system, specifically:

- The ability to generate yaw moment
- Intrusiveness to the driver
- Bandwidth or responsiveness

These items were discussed earlier, and their relation to control distribution is presented next.

The FBC must first determine the existing operating conditions, and then decide if one system or the other or both must be activated. It is understood that if the present condition is controllable by either steer alone or by brake alone, then priority is given to steer due to its non-intrusive nature. Brake activation has an undesirably intrusive aspect, so it is given second priority.

In making the actuation decision, the FBC uses information on road surface friction and on the degree of actuation saturation. Saturation means that the actuation of a system has reached a maximum in term of yaw moment generation, and can't generate anything more. In the case of steering, saturation may be reached due to road friction or due to actuator end-of-travel.

For example, if steer had been acting alone and had subsequently reached a saturation level in yaw moment generation, the brakes are activated to provide additional yaw moment as needed. The steer saturation point is known to vary based on surface friction. Specifically, steer saturation occurs at small tire slip angles for icy road surfaces, but at much larger tire slip angles on gravel and on dry surfaces. Thus the FBC will require relatively less activity from brakes when operating on gravel and on dry surfaces than on ice.

Furthermore, the responsiveness of the steering and brake systems is affected by temperature and voltage conditions. Knowledge of these conditions can be applied to influence the distribution of control.

As a side note, the complementary aspects of the systems' response characteristics can also influence the selection of mechanical components. For example, brake precharge mechanisms are sometimes used to improve hydraulic response for low temperature conditions. With steering also available to help generate yaw moment, the hydraulic response requirement could be relaxed, perhaps to the point of not needing the precharge mechanism.

This section has described the design of the coordinated control structure and presented considerations for the distribution of the yaw moment command. The coordinated control has been evaluated in simulation and in vehicle tests, and the following section shows some typical results.

RESULTS

This section presents some results of the evaluation of the coordinated control. The control structure is implemented as a supervisory controller in a rapid prototyping environment and tested on a full size pick-up truck. The truck is equipped with a QUADRASTEER™ 4WS system and a Traxxar™ brake control system, and it also is fitted with additional safety equipment such as a roll cage, outriggers, and 5-point harnesses.

The following results are from 2 common maneuvers, specifically an emergency lane change maneuver on a snow surface and a panic brake maneuver on a split-coefficient surface (ice on one side and dry concrete on the other). Both maneuvers require significant management of the vehicle's yaw moment, thus highlighting the capability of the control systems.

For each maneuver, two control configurations are evaluated and compared. One configuration is a baseline control configuration and the other is the coordinated control configuration for VSE and 4WS actuation.

EMERGENCY LANE CHANGE ON SNOW

Figure 8 shows the vehicle test data for the emergency lane change maneuver on groomed snow. For this maneuver, the truck was driven at constant speed, and the driver applied a large and quick handwheel input to initiate the lane change followed by a second handwheel input to try to straighten out in the adjacent lane.

The 3 upper plots in Figure 8 show data for the baseline control configuration, i.e. VSE is acting alone to stabilize the vehicle and there is no control coordination. It is noted that the 4WS system was operating in its basic mode of control wherein rear wheel angle is strictly a function of handwheel angle and vehicle speed.

For this case, the data shows that the driver had to apply a third handwheel input to counteract the side-slip angle of the vehicle. This third handwheel input is referred to as countersteer during the recovery phase. It should be noted that the VSE system provided some benefit by allowing time for the driver to react to recover the vehicle. Without VSE, the vehicle would have been unmanageable and the driver would have likely lost control, resulting in a spin-out condition.

The 3 lower plots in Figure 8 show data for the coordinated control configuration, i.e. both VSE and 4WS are working in a coordinated way to stabilize the vehicle. Comparing the results for the two control configurations, we can make the following observations:

- For the coordinated control case, the vehicle has less side-slip angle during the recovery phase, implying an improvement in overall stability.
- The driver's handwheel input is much less during the recovery phase for the coordinated control configuration, implying an improvement in driver workload.
- For the coordinated control case, there is less brake actuation. The actuation workload is shared between 4WS and brakes, with 4WS given first priority. This implies an improvement in overall comfort (reduction in intrusion) since there is less brake activity.
- The 4WS and brake activation are well coordinated. Both impart a change in yaw moment that is in the same direction (no opposition).

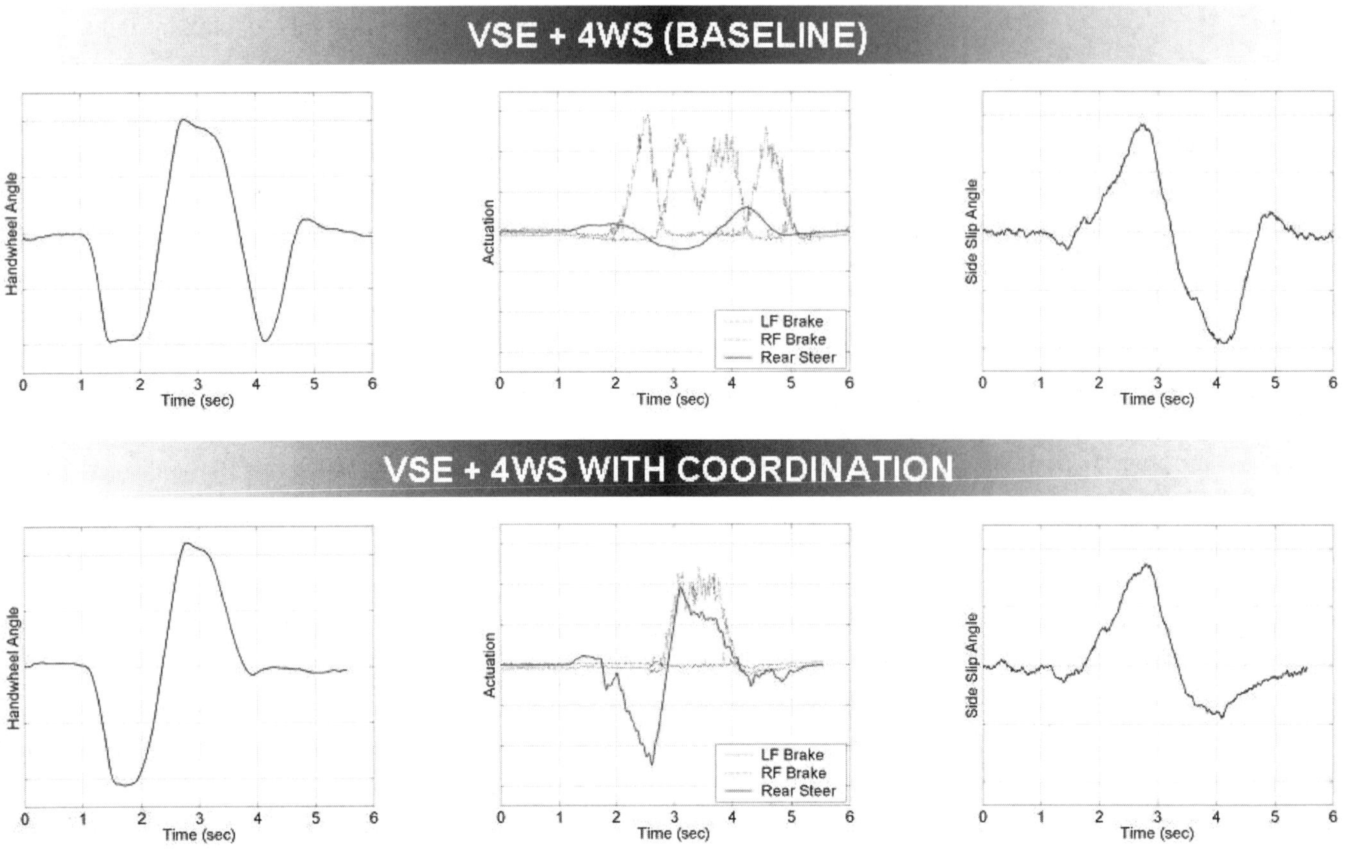

Figure 8: Vehicle test data for the lane change maneuver on snow

These results demonstrate a reduction in the typical compromises of stability control systems, which is represented by the diagram of Figure 9. VSE systems are tuned with a trade-off between stability and driver comfort as represented by the lower line in the figure. For example a highly stable system is less comfortable (more intrusive) as depicted by point A. Conversely a more comfortable system provides less stability as depicted by point B. The tuning engineer is constrained to points along the line, trading comfort for stability. By adding 4WS and using a control structure to coordinate both 4WS and VSE, the vehicle performance can go beyond the line, providing both higher comfort and higher stability, depicted by point C.

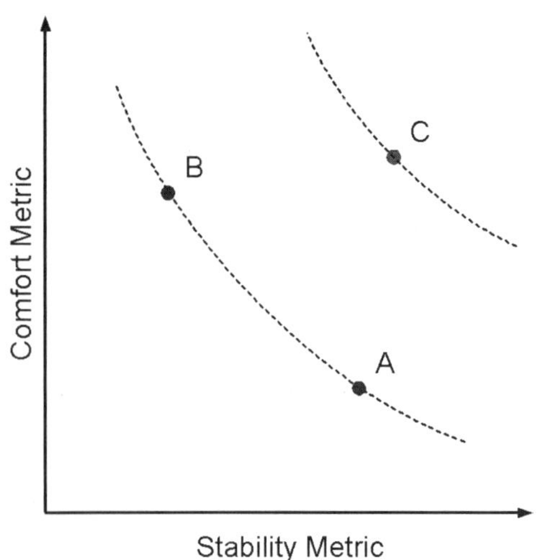

Figure 9: Trade-off of stability and comfort

PANIC BRAKING ON SPLIT-COEFFICIENT SURFACE

The control of vehicle motion on split-coefficient road surfaces is a challenging problem due to the large difference in side-to-side longitudinal forces. During braking and acceleration, the force difference causes large rotation (yaw) of the vehicle that can lead to spin-out, even for skilled drivers. Controlled brake systems with ABS and TCS have the ability to automatically manage the longitudinal forces and help maintain directional stability. With most ABS today, vehicle stability is mainly managed by limiting the apply rate of brake pressure to the front wheel on the high-coefficient side, but at the expense of deceleration performance.

Thus, brake control systems are tuned for a compromise between deceleration and lateral stability for split-coefficient events, as depicted by Figure 10. In one extreme, ABS can be tuned to give maximum stability at the expense of deceleration, represented by point A in the figure. In another extreme, ABS can be tuned to give high deceleration at the sacrifice of directional stability, represented by point B in the figure.

The addition of 4WS systems now offers another means to manage directional motion. By carefully coordinating the control of the brake and steering systems, vehicle performance is optimized for maneuvers on split-coefficient surfaces by allowing the brake system to generate maximum deceleration and using the steering system to maintain directional stability. This is shown as point C in the figure, which represents another example of a reduction in the typical compromises of stability controls.

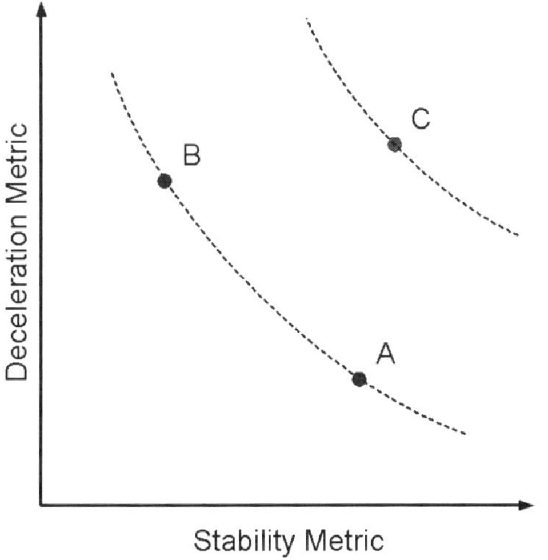

Figure 10: Trade-off of stability and deceleration

Figure 11 shows the vehicle test data for the panic braking maneuver on a split-coefficient surface, where the road friction coefficient is significantly different between the left and right side wheels. In this case, ice was on the right side and dry concrete was on the left side. The vehicle's initial velocity was 70 kph, and the driver initiated braking at time = 1 second.

The 3 upper plots in Figure 11 show data for the baseline control configuration, i.e. ABS is acting alone to stabilize the vehicle and there is no 4WS activation and no coordinated control. For this case, the driver had to apply a large handwheel input to counteract the brake-induced yaw moment, to keep the vehicle on a straight course. It should be noted that the ABS system provided some benefit by allowing time for the driver to react to stabilize the vehicle. Without ABS, the vehicle would have been unmanageable and the driver would have likely lost control, resulting in a spin-out condition.

The 3 lower plots in Figure 11 show data for the coordinated control configuration, i.e. both ABS and 4WS are working together in a coordinated way to stabilize the vehicle. Comparing the results for the two configurations, we can make the following observations:

- For the coordinated control case, the vehicle has better deceleration and is able to stop in a shorter distance, due to the fact that ABS is tuned to achieve optimum deceleration.
- The driver's handwheel input is much less for the coordinated control configuration, implying an improvement in driver workload. The reduction in driver countersteer is due to the fact 4WS provides most of the lateral force needed to balance the brake-induced yaw moment.

CONCLUSION

This paper described a method to coordinate a 4WS and a brake system to achieve improvements in vehicle performance. Since both systems have authority over yaw-plane motion, it is essential to coordinate them to eliminate conflicts and to leverage their strengths.

Each system influences yaw-plane motion differently. Brakes directly affect longitudinal forces at the tires, while 4WS affects lateral forces at the rear tires. It was shown that the ability to generate yaw moment depends on several factors, including vehicle side-slip angle and surface friction. The 4WS system is less intrusive than brakes, and system bandwidths are affected differently by temperature and by operating voltage.

A supervisory control structure implementation was presented which met the requirements, namely the need for coordinated control. In the design, consideration was given to the noted strengths of each system, which drove the strategy for control distribution.

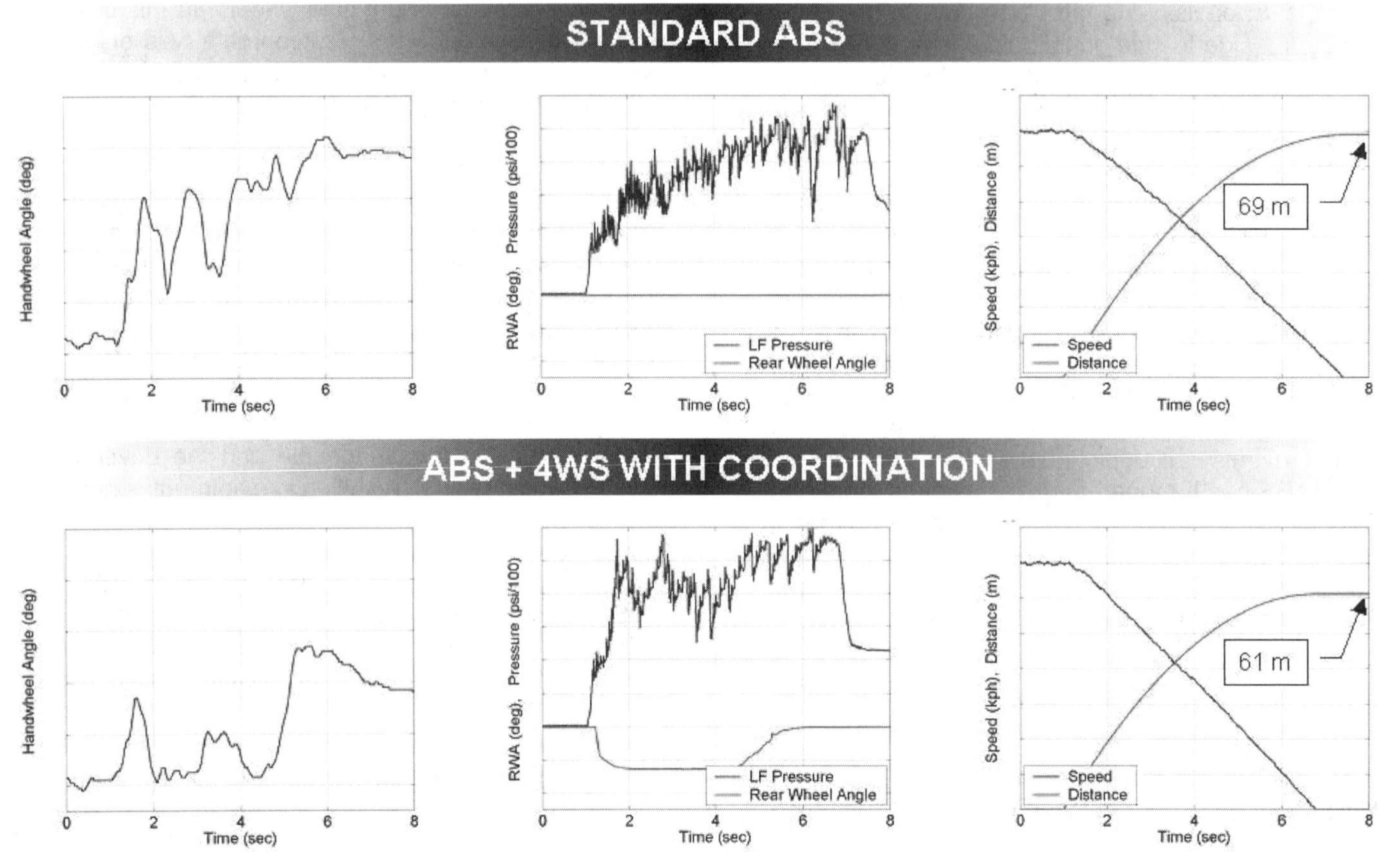

Figure 11: Vehicle test data for the split-coefficient braking maneuver

Test results were presented for 2 maneuvers that require yaw moment management. In both cases, the results for the coordinated control of brakes and 4WS were compared to typical brake-only operation. The coordination method achieved significant improvements in terms of stability, driver workload, driver comfort, and deceleration. The coordinated control overcomes the typical compromises in stability vs. comfort and in stability vs. deceleration.

Lastly, the ideas presented here are applicable to trucks as well as passenger cars. The ideas can be extended to other actuation systems such as Active Front Steer, Active Dampers, and Active Roll Control, given the scalability of the coordinated control structure.

ACKNOWLEDGMENTS

The authors gratefully acknowledge the contributions and support from members of the Innovation Center, the QUADRASTEER™ Product Group, and the Chassis Systems Group of Delphi.

REFERENCES

1. Shimada, K. and Y. Shibahata, "Comparison of Three Active Chassis Control Methods for Stabilizing Yaw Moments," SAE 940870.
2. Takahashi, T., et al, "The Modeling of Tire Force Characteristics of Passenger and Commercial Vehicles on Various Road Surfaces," Proceedings of AVEC 2000, pages 785-792.
3. Hac, A., "Evaluation of Two Concepts in Vehicle Stability Enhancement Systems," 31st ISATA, 1998.

Closed-Loop Evaluation of Vehicle Stability Control (VSC) Systems using a Combined Vehicle and Human Driving Model

Taeyoung Chung and Kyongsu Yi
Hanyang University

Jeongtae Kim
Hyundai Mobis

Jangmoo Lee
Seoul National University

ABSTRACT

This paper presents a closed-loop evaluation of the Vehicle Stability Control (VSC) systems using a vehicle simulator. Human driver-VSC interactions have been investigated under realistic operating conditions in the laboratory. Braking control inputs for vehicle stability enhancement have been directly derived from the sliding control law based on vehicle planar motion equations with differential braking. A driving simulator which consists of a three-dimensional vehicle dynamic model, interface between human driver and vehicle simulator, three-dimensional animation program and a visual display has been validated using actual vehicle driving test data. Real-time human-in-the loop simulation results in realistic driving situations have shown that the proposed controller reduces driving effort and enhances vehicle stability.

INTRODUCTION

Human-in-the-loop evaluation method plays an essential role in the design of Vehicle Stability Control (VSC) systems. Because of interactions with the vehicle, VSC system and driver, it is more effective to assess VSC system by human-in-the-loop evaluation than using general open-loop evaluations based on prescribed steering and velocity profile [1]. A driving simulator has been used to evaluate the performance of VSC systems and to investigate interactions between the VSC and a human driver.

In this paper, sliding control law for VSC systems based on 3-DOF vehicle planar motions has been evaluated using the proposed human-in-the-loop method. Most of previous studies have used a two-step design approach for the design of VSC [2][3][4]. First, the direct yaw moment control input is designed using various methods such as linear quadratic optimal control, sliding control

theory and fuzzy logic without considerations of vehicle longitudinal motion. Then, brake inputs for the production of the direct yaw moment are computed. Since lateral and longitudinal motions of a vehicle are governed by longitudinal and lateral tire forces, a 3-DOF vehicle yaw plane model has been used in the design of the VSC [8]. Compared to the 2-DOF vehicle model used before, 3-DOF model includes the effect of differential braking for stability control on vehicle longitudinal dynamics. The brake control inputs have been directly derived from the sliding control law based on yaw plane vehicle model with differential braking.

The driving simulator based on full vehicle model is validated via comparing actual vehicle test results. The controller is tested under circular track driving and emergency lane changing situations.

DRIVING SIMULATOR VALIDATION

This section describes the vehicle dynamic model used in real-time simulations and hardware configurations of the driving simulator. Although this method does not exactly predict the response of the physical and mental systems of the human driver, the model does correlate well with trends observed from vehicle test results of driver responses under certain driving conditions.

VEHICLE DYNAMIC MODEL

The proposed vehicle dynamic model consists of a vehicle body, suspension, nonlinear tire model and powertrain models [5]. The full vehicle dynamic model performs real-time simulations in the driving simulator. A schematic diagram of vehicle model is shown in Figure 1. The vehicle body with 6 degrees of freedom is connected to 4-quarter car suspensions with roll bar. The powertrain model includes an engine, a torque converter model validated by actual test data, an automatic transmission with gearshift map and driving shafts. Pacejka's tire model has been used to compute the longitudinal and lateral tire forces of each wheel.

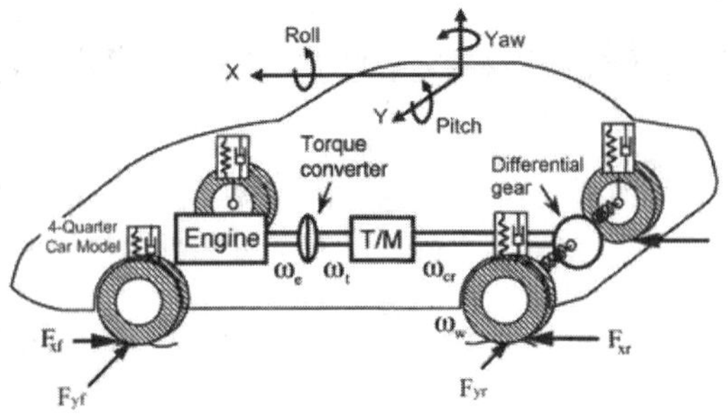

Figure 1. Schematic diagram of vehicle model

CONFIGURATIONS OF DRIVING SIMULATOR

A real-time simulation environment of the driving simulator is implemented by using Matlab xPC target that provides real-time kernel on the target PC [6]. As can be seen from Figure 2, we can modify the vehicle model and VSC algorithm using the Host PC and upload simulation code to the Target PC. During real-time simulations, Target PC calculates the variables of the vehicle model controlled by VSC algorithm with sensed driver reactions.

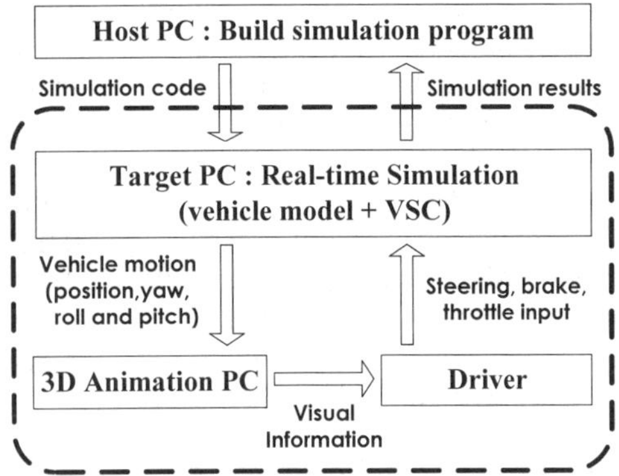

Figure 2. Real-time Hardware configuration

Figure 3. Interface of the driving simulator

Then, it provides vehicle motions such as position, yaw, roll and pitch angle to 3D Animation PC, which displays visual information of a driving situation. Driver responses are acquired through brake pressure sensor, steering wheel angle sensor and throttle angle sensor.

Figure 3 shows the vehicle-driver interface of the driving simulator. Wide-view monitor shows a 3D animation of a driving situation. Actual vehicle parts such as dashboard, steering wheel, brake pedal, accelerator pedal and driver's seat are included.

TEST RESULTS OF DRIVING SIMULATOR

Vehicle test data on a 30m-radius circular road have been used to validate the driving simulator. Figure 4 shows comparisons between vehicle test data and simulator driving data. Vehicle speed was increased until the vehicle drifted out from the circular road of 30m-radius as shown in Figure 4 (B). Figures 4 (A) and (B) show that the driver steers abruptly at a vehicle speed of 55kph in order to follow the track. Comparisons of lateral accelerations and yaw rates of the vehicle are shown in Figures 4 (C) and (D).

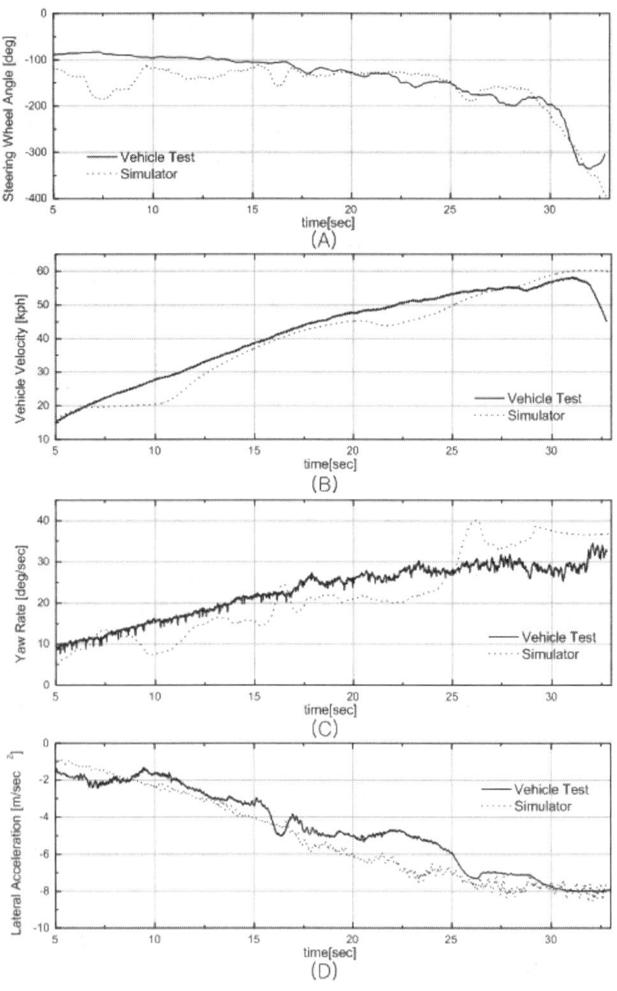

Figure 4. Comparison between simulator and test data

When vehicle speed is increased similar to the actual vehicle test results, driver steering responses are quite similar at drift out speed. The comparison between the driving simulator and vehicle test results shows that the proposed driving simulator is very feasible for describing actual vehicle dynamic behaviors. It shows that the driving simulator reproduces actual driving conditions appropriately.

DIFFERENTIAL BRAKING STRATEGY

A differential braking control law for the VSC based on three-degree of freedom vehicle planar motions has been evaluated in this study.

VEHICLE YAW PLANE MODEL

The vehicle model used in this study is a 3-DOF vehicle yaw plane representation with differential braking and is shown in Figure 5.

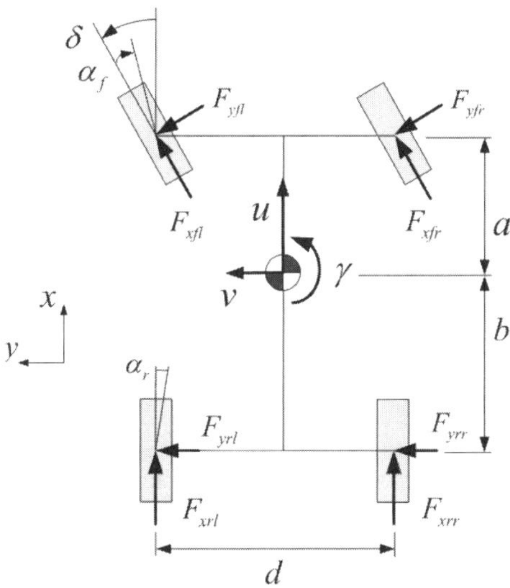

Figure 5. Vehicle yaw plane model

Equations of motions describing vehicle yaw plane model are as follows:

$$m\dot{u} = F_{xr} + F_{xf}\cos\delta - F_{yf}\sin\delta + mrv$$

$$m\dot{v} = F_{yr} + F_{xf}\sin\delta + F_{yf}\cos\delta - mru \qquad (1)$$

$$I_z\dot{r} = aF_{xf}\sin\delta + aF_{yf}\cos\delta - bF_{yr}$$
$$+ \frac{d}{2}(F_{xfr} - F_{xfl})\cos\delta + \frac{d}{2}(F_{xrr} - F_{xrl})$$

where u is longitudinal vehicle velocity, v is lateral vehicle velocity and γ is yaw rate. F_x and F_y are longitudinal and lateral tire forces. The front and rear tire

slip angles, α_f and α_r are average slip angles of the left and right wheels. δ is steering angle input. The subscript fr, fl, rr and rl represent front right, front left, rear right and rear left, respectively. Other vehicle parameters are listed in the appendix.

The tire longitudinal forces can be controlled by wheel braking control. Since the wheel dynamics is much faster than vehicle body dynamics, assuming front driving, the tire longitudinal forces can be approximated as

$$F_{xf} = F_{xfl} + F_{xfr} \cong \frac{T_s}{r_{wf}} - \left(F_{Bfl} + F_{Bfr}\right)$$
$$F_{xr} = -\left(F_{Brl} + F_{Brr}\right) \qquad (2)$$

Where T_s represents driving shaft torque and F_B's are braking forces of each wheel. For brake proportioning, the brake force of the rear wheels needs to be controlled as follows.

$$F_{Br} = \frac{a\,g + a_x h}{b\,g - a_x h} F_{Bf} = f_x(a_x) F_{Bf} \qquad (3)$$

Where a_x is vehicle longitudinal acceleration

SLIDING CONTROL LAW

The desired value of the yaw rate is computed based on the driver's steering input and vehicle longitudinal speed using a linear tire model as follows [2].

$$r_{des} = \frac{u}{(a+b)\left(1+u^2/u_{ch}^2\right)}\delta \qquad (4)$$

Where u_{ch} is characteristic velocity of the vehicle. Since the lateral acceleration is limited by the friction coefficient between the tires and the road, the desired yaw rate is also limited under the value of $\mu g/u$. The sliding surface is defined as follows.

$$s = r_{des} - r + \rho \cdot \beta \qquad (5)$$

Where the scalar, s, is a combination of sideslip angle, β, and the difference between the desired yaw rate and actual yaw rate. ρ is a positive constant. The controller derived from (5) achieves the basic objectives and parametric robustness of VSC and robustness more successfully than a controller built on either side slip angle or yaw rate alone [7].

The control objective is to keep the scalar, s, at zero and this can be achieved by choosing the control law such that

$$\dot{s} = -K \cdot s \tag{6}$$

Where K is a positive constant.

The derivative of S is defined using (1), (2) and (3) as

$$\dot{s} = \dot{r}_{des} - \dot{r} + \rho \cdot \dot{\beta} = \frac{\partial r_{des}}{\partial u}\frac{du}{dt} - \dot{r} + \rho \cdot \dot{\beta} \tag{7}$$

Derivative of scalar variable S can be formulated using equations (1) and (2) [8].

BRAKE INPUT COMMAND

Braking control input of each wheel have been directly derived from (7) as in [8]

$$P_{Bfl,com} = \frac{r_{wf}}{K_{Bf}} F_{Bfl} = \frac{r_{wf}}{K_{Bf}} \frac{N_{l,b}}{D_{l,b}}$$
$$P_{Bfr,com} = \frac{r_{wr}}{K_{Br}} F_{Bfr} = \frac{r_{wr}}{K_{Br}} \frac{N_{r,b}}{D_{r,b}} \tag{8}$$

where $P_{Bfl,com}$ and $P_{Bfr,com}$ are brake pressure input commands for front left wheel and front right wheel, respectively. The brake pressure commands for the rear wheels are computed based on the front wheel brake commands for brake proportioning [8]. $N_{lr,b}$ and $D_{lr,b}$ which represent the numerator and denominator of F_{Bfl} and F_{Bfr}, are derived from (7) directly. An examples of the equations to calculate $N_{l,b}$, $D_{l,b}$ are described in the Appendix. When a brake input is calculated as a negative value for a wheel, no control input is applied to the wheel.

CLOSED-LOOP EVALUATIONS OF VSC SYSTEMS

The proposed vehicle stability controller has been evaluated via closed-loop method using the driving simulator. The driver conducted circular track driving and emergency lane change cases with or without the proposed VSC system.

CIRCULAR TRACK

Figure 6 shows the test results. The driver increases vehicle speed and corrects the steering angle to make the vehicle follow the track with a 30 m radius of curvature in a clockwise direction. Figure 6 (A) shows the vehicle velocity. Without VSC, the driver's steering input is diverged at 60kph as shown in figure 6 (B). With

VSC, the longitudinal velocity is limited by the brake input and the yaw rate follows the desired value. Figure 6 (D) shows the brake torque applied to front wheels. The brake torques of the right wheels has been controlled in order to stabilize the vehicle.

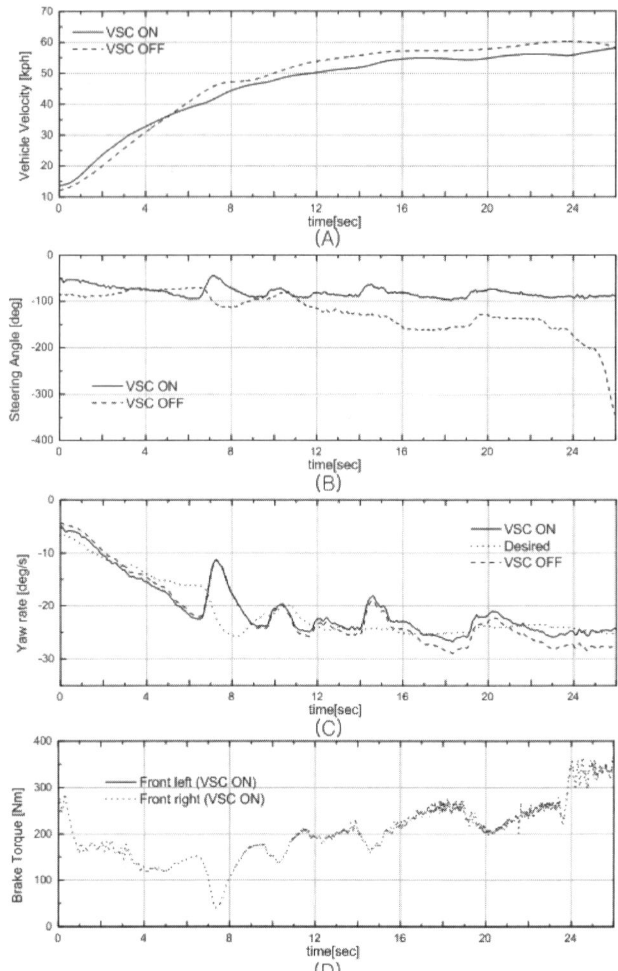

Figure 6. Vehicle responses (circular track)

EMERGENCY LANE CHANGE MANEUVERS

A single lane change test was conducted on high-μ road at 120kph and 60kph. Under an assumption of sudden cut-in situation of another vehicle or other obstacles, the driver performs the single lane change maneuver within 2 seconds.

Lane change without braking

The driver changes lanes promptly when the vehicle speeds are about 120kph. Without VSC, a driver steers more to change the lane, and has difficulty in stabilizing after the lane change (Figure 7(A)). With VSC, the brake inputs are generated on both wheels in turns during the lane change maneuver as shown in Figure 7(B). To prevent wheel locking, front brake torques are limited to about $110\ Nm$, based on the friction coefficient of the road. The desired yaw rate is limited to about the peak steering angle, but the yaw rate follows the desired value with VSC. The behavior of the controlled vehicle is more stable with respect to yaw rate and body slip angle as shown in Figure 7 (C) and 7 (D). It can be seen that the controller exhibits superior handling performance. The driver feels better as his steering effort is reduced to stabilize the vehicle with the VSC.

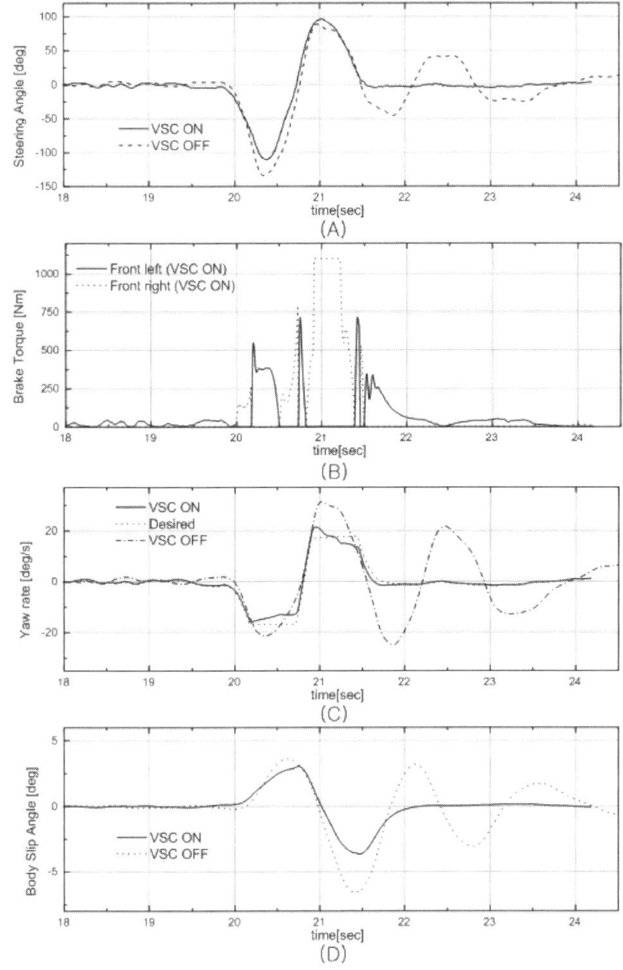

Figure 7. Vehicle responses (lane change)

Lane change, full braking

In this case, the driver changes a lane at 60kph with full braking. Because the brake inputs of left and right wheels are equal in the uncontrolled case (Figure 8 (B)), the vehicle spins out at about 7sec. In case of the VSC ON, brake inputs have been applied alternately to prevent vehicle spinning. The yaw rate and body slip angle diverges at that time (Figure 8 (C) and 8 (D)). In the case of VSC ON, the yaw rate follows desired value, and body slip angle is regulated under 10 degrees.

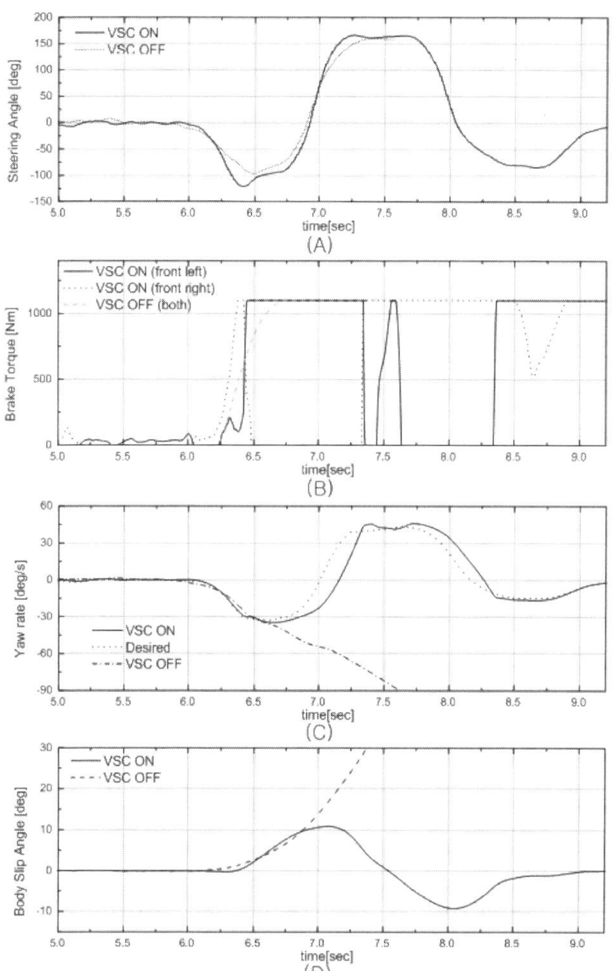

Figure 8. Vehicle responses (lane change)

Figure 9 shows the trajectories of each case. Without VSC, the vehicle goes unstable and spins out during full braking. The controlled vehicle trajectory shows transient response but does not lose controllability.

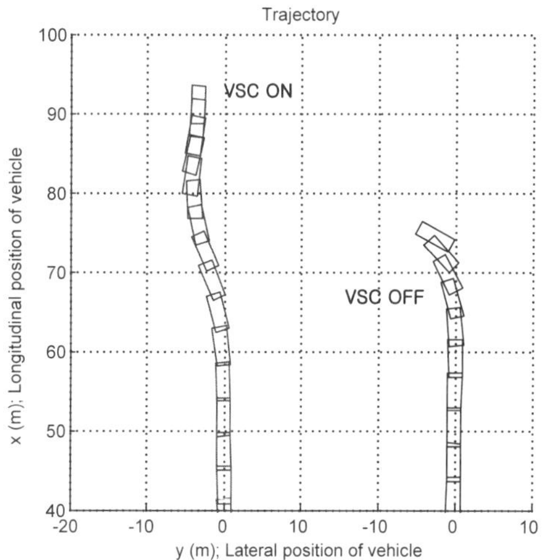

Figure 9. Vehicle trajectories (lane change)

CONCLUSION

A closed-loop method has been used to evaluate a VSC system. Human driver-VSC interactions have been investigated under realistic operating conditions in the laboratory. A vehicle simulator which consists of a three-dimensional vehicle dynamic model, human driver-vehicle interface, three-dimensional animation program and a visual display has been used.

Real-time human-in-the-loop simulation results for realistic driving situations have shown that the proposed controller reduces driving effort and enhances stability of the vehicle. Since the human driver–vehicle-VSC interactions can be evaluated under the realistic driving conditions in the laboratory, the closed-loop evaluation method proposed in this paper can be a powerful tool for the development of the VSC systems.

ACKNOWLEDGMENTS

This research is jointly supported by the Hyundai Mobis and the Brain Korea 21 Project (BK21).

REFERENCES

1. Chen, B. and Peng, H., "Differential-Braking-Based Rollover Prevention for Sport Utility Vehicles with Human-in-the-loop Evaluations," Vehicle System Dynamics, Vol. 36, No. 4-5, pp. 359-389, 2001.
2. Van Zanten, A.T., Erhardt, R., Pfaff, G., Kost, F., Hartmann, U. and Ehret, T., "Control Aspects of the Bosch-VDC," AVEC'96, pp. 573-607, 1996.
3. Tseng, H. E., Ashrafi, B., Madau, D., Brown, T. A. and Recker, D., "The Development of Vehicle Stability Control at Ford," IEEE/ASME Transactions of Mechatronics, Vol. 4, No. 3, September, 1999.
4. Pilutti, T., Ulsoy, G and Hrovat, D., "Vehicle Steering Intervention Through Differential Braking," Transactions of the ASME, J. of Dynamic Systems, Measurement, and Control, Vol. 120, No. 3, pp. 314-321, 1998.
5. Ha, J., Chung, T., Kim, J., Yi, K. and Lee, J. "Validation of 3D Vehicle Model and Driver Steering Model with Vehicle Test," Spring Conference Proceeding of KSAE, Vol. II, pp. 676-681, 2003.
6. Lee. S., Kim. Y., Park, K and Kim, D., "Development of Hardware-in-the-Loop Simulator and Vehicle Dynamic Model for Testing ABS", SAE Paper No. 2003-01-0858, 2003.
7. Uematsu, K. and Gerdes, J.C., "A Comparison of Several Sliding Surfaces for Stability Control," AVEC'02, Paper No.20024578, 2002.
8. Yi, K., Chung, T., Kim, J. and Lee, J., "An Investigation into Differential Braking Strategies for Vehicle Stability Control," ImechE. (to be published), 2003.

APPENDIX

A : Vehicle Parameters

a **Distance from c.g. to front axle,** m

b **Distance from c.g. to rear axle,** m

d **Track width,** m

μ **Friction coefficient between road and tire**

g **Gravity acceleration,** $m/\sec^2$

m **Vehicle mass,** kg

C **Tire cornering stiffness,** N/rad

I_z **Moment of inertia about yaw axis,** kgm^2

h **Height of the mass enter**

r_w **Effective radius of wheel, m**

K_B **Brake gain,** m^3

P_B **Brake pressure,** N/m^2

B : Subscripts

f, r : **Front, Rear**

fr, fl : **Front right , Front left**

rr, rl : **Rear right, Rear left**

com : **Control command**

C : Brake Input Command of left front wheel (While full braking)

$$N_{l,b} = -K \cdot s - \Gamma(\delta,u,v,r) - \left(A \cdot \frac{1}{m} - \frac{a}{I_z} \right) \delta \frac{T_s}{r_{wf}}$$

$$+ f_x(a_x) \frac{T_s}{2r_{wf}} \left(-A \frac{\delta}{m} - \frac{d}{2I_z} \right)$$

$$- F_{Bfr} \left(\left(-A \cdot \delta \cdot \frac{1}{m} \pm \frac{d}{2I_z} \right) + \frac{a}{I_z} \delta - \rho \cdot \frac{1}{mu} \delta \right)$$

$$- F_{Brr} \left(-A \cdot \delta \cdot \frac{1}{m} \pm \frac{d}{2I_z} \right)$$

$$- \rho \cdot \left[\frac{1}{mu} \left\{ C_r \alpha_r + C_f \alpha_f + \frac{T_s}{r_{wf}} \delta \right\} - r \right]$$

$$D_{l,b} = \left(-A \frac{\delta}{m} \pm \frac{d}{2I_z} \right) (1 + f_x(a_x)) + \frac{a}{I_z} \delta - \rho \frac{1}{mu} \delta$$

where,

$$A = \frac{\left(1 - u^2/u_c^2\right)}{(a+b)\left(1 + u^2/u_c^2\right)^2}$$

$$\Gamma(\delta,u,v,r) = Arv\delta - A\delta^2 \frac{1}{m} F_{yf} - \frac{a}{I_z} F_{yf} + \frac{b}{I_z} F_{yr}$$

If a driver does not applied brake input, the braking force of right wheels, F_{Bfr}, F_{Brr}, must be neglected.

ESC II – ESC With Active Steering Intervention

Peter E. Rieth and Ralf Schwarz
Continental Teves AG & Co. oHG

ABSTRACT

Through the decline in prices of semiconductor technology, automotive applications have boosted and can be characterized in technological leaps, the first example was the introduction of ABS in 1978 leading on to the equipment with ESC in 1995.

In the next technological leap forward, the cross-linking of today´s many and varied, largely stand-alone chassis control units, is envisaged. This technological leap forward in networking, the functional integration of brakes and steering, is described by the ESC II system.

INTRODUCTION

As prosperity and the standard of living in our society grows, the desire for more comfort and enjoyment on the one hand and more protection and security on the other grow too – in all areas of daily life.

With reference to the automobile as an indispensable element of our personal mobility, these needs have placed their mark in all areas of vehicle design. The prerequisite for advanced functions is, apart from permanent improvement (evolution), the introduction of newly available technologies for large-scale production applications.

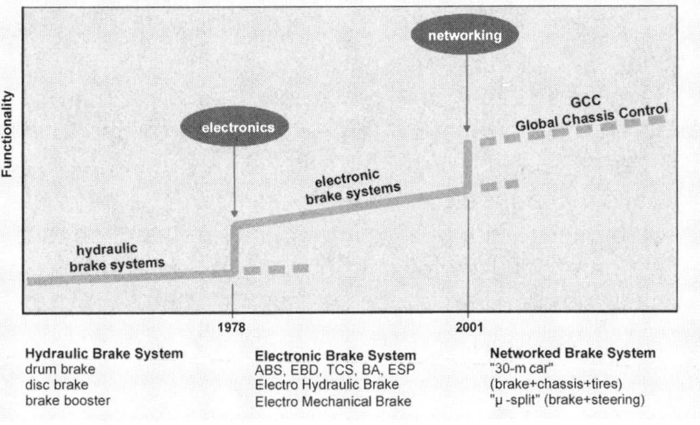

Fig. 1:   Technology change as a basis for functional leaps

This makes the next leap (revolution) in development first possible. A typical example of such a technological leap forward is the introduction of electronics in the chassis with the installation of the anti-lock brake system ABS in 1978 (Fig. 1).

This trend to electronification has been further boosted by the decline in prices of semiconductor technology in recent years. Today, electronics already account for 22 percent of the value of each automobile (18 percent hardware, 4 percent software). An increase to 35 percent (22 percent hardware, 13 percent software) is anticipated for the year 2010. Luxury class vehicles today have as many as 70 electronic control units serving safety, comfort and increasingly, motoring enjoyment.

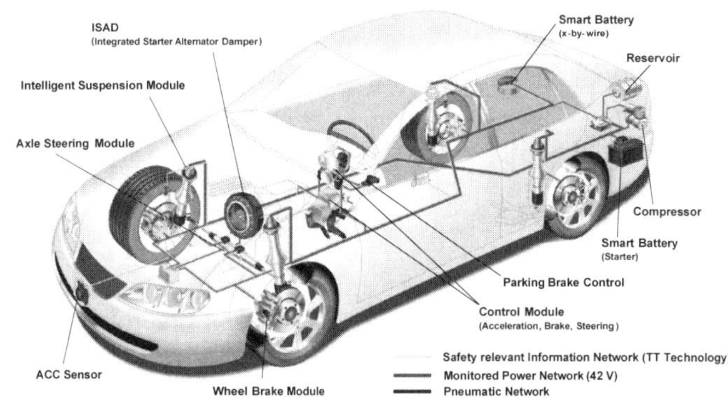

Fig. 2: Global Chassis Control with "Dry Chassis" Components

As a rule they are currently operated as stand-alone units. With regard to the chassis of the future, this global networking approach is termed Global Chassis Control (GCC, see Fig. 2).

A significant step forward in the direction of chassis networking is described in this paper and titled ESC II, the functional integration of brakes and steering.

ELECTRONIC STABILITY CONTROL ESC

SYSTEM AND COMPONENTS

Today´s standard, state of the art ESC system consists of the brake subsystem namely hydraulic wheel brakes, brake actuation (vacuum booster, tandem master cylinder and fluid reservoir), wheel speed sensors and ESC modulator with integrated electronic controller. To monitor driver behavior a steering angle sensor is integrated into the steering column, a pressure sensor measures brake circuit pressure. To monitor the reaction of the vehicle, a yaw-rate and lateral acceleration sensor, mounted into one housing - the so-called sensor cluster, is installed.

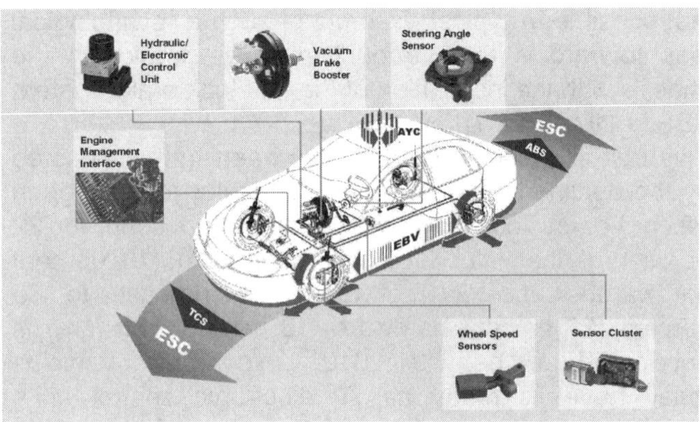

Fig. 3: Electronic Stability Control ESC – System Layout

BASIC PRINCIPLE

Inventors sought to develop an affordable, effective protection against swerving motions of the automobile. ESC was introduced with this function in 1995 and was offered in several luxury sedans.

ESC development was spurred by the realization that accidents involving skidding cars are almost always serious, as it is not the nose or front end but the soft side of the car, which crashes into an obstacle. Apart from the great effectiveness, economy was a point in favour of this system, because it is based on the hydraulics of a four-wheel traction control system TCS, which can actively access each of the four wheel brakes without the driver actuating the pedal.

Driver, vehicle and environment constitute a control loop in which the driver intervenes by means of steering wheel, accelerator and brake pedal. With the Electronic Stability Control ESC, a particularly highly developed assist system now enters this control loop [1].

Aided by sensors, the ESC software gets a picture of the driver's wishes every 10 milliseconds and monitors vehicle response to his commands or to disturbances coming from outside. If the car behaves stably, that is to say if its behavior complies with the driver's wishes, ESC does not intervene to make any corrections. It does this only when it detects a difference between driver intention and vehicle behavior, which the driver alone cannot make up for, or only with the greatest effort.

If the vehicle condition needs correcting, ESC acts within fractions of a second on the wheel brakes, selecting just one wheel, or several wheels, depending on the requirements. By braking the wheels individually, ESC influences tire slip and - indirectly - longitudinal and transverse forces at each wheel, thereby producing a yawing torque which optimizes the side-slip angles of the vehicle and the slip angle of the wheels in the desired direction.

ESC is in permanent readiness, whether one is driving at a constant speed, braking or accelerating, on smooth or rough surfaces, on dry asphalt or on polished ice, where, of course, its effectiveness is strongly limited by the low coefficients of friction. Even μ-split and rapid changes of direction in succession cannot outwit ESC.

In addition to intervening in the braking of individual wheels, ESC can reduce engine torque, effectively diminishing a speed, which is too high in a particular situation. Whatever control strategy ESC opts for depends entirely on the driving conditions. In one case, briefly reducing engine output may suffice to maintain driving stability. In another case, ESC must throttle down the engine and, for several seconds brake one, two, or even three wheels, with variable force and for different lengths of time. Through suitable programming of the software, engineers are also able to impress a characteristic behavior of ESC to create the handling impression desired by the vehicle manufacturer.

Despite its capabilities, ESC cannot prevent a vehicle going off a bend taken at too high a speed. But in contrast to a vehicle without ESC, the car with ESC does not go into an uncontrolled skid as long as ESC can utilize the potential for suitable longitudinal or transverse forces on one or more wheels. What remains is a lateral drift out of the line of travel, which the driver can handle more easily than a spin.

Thus, ESC is more than just the sum of ABS, TCS, electronic brake-force distribution EBD and engine drag control (a Brake Assist BAS can be integrated, but does not have to be), because these systems exclusively control the longitudinal slip of the wheel and concentrate primarily on vehicle dynamics in the direction of travel. ESC, on the other hand, also takes yaw motion into account.

As described earlier, the Electronic Stability Control is based on the hydraulics of an automobile equipped with ABS and fourwheel TCS (Fig. 3). But additional hydraulic components are needed to actively access the wheel brakes. The perception of driver intention and vehicle behavior likewise calls for further components, in this case sensors.

FURTHER FUNCTIONS

Active Rollover Protection

ARP mainly targets sport utility vehicles and vans, which enjoy growing popularity. These are vehicles with a high center of gravity in relation to their track width and accordingly, have a greater tendency to tip over. ARP recognizes critical rollover situations (arising, for example, from a rapid steering movement, typical of an evasive maneuver) and reduces the lateral force by braking the front outer wheel, thereby reducing side forces and simultaneously cutting back the vehicle speed. The result is considerably less danger of tipping. The involvement of the steering as described in the following chapter, with its potential for rapid reduction of lateral force, can effectively support this brake intervention.

<u>Trailer Stabilization</u>

The trailer stabilization (TSP) is able to conclude from the behavior of the towing (tractor) vehicle that the trailer is snaking. Before directional stability and road adhesion reach critical limits due to this snaking, the current TSP applies the brakes of the vehicle combination in a forceful but controlled manner
.

ESC II – DRIVING DYNAMICS, NEXT GENERATION

Similar to ABS, which was further developed into the Traction Control System (TCS) in 1986, the Electronic Stability Program (ESC) in 1995, networking leading to Global Chassis Control will proceed step by step.

As an important step in this direction, ESC II, the next generation of the described driving dynamics control system ESC, utilizes active steering intervention for the first time and thereby not only provides greater handling safety but also creates new dimensions of driving pleasure and ride comfort. The system makes the vehicle easier to control at critical limits and offers more agility for handling.

SYSTEM AND COMPONENTS

The ESC II system consists of the subsystems brake & engine intervention and now, in addition, steering intervention, with its internal and external sensors (Fig. 4).

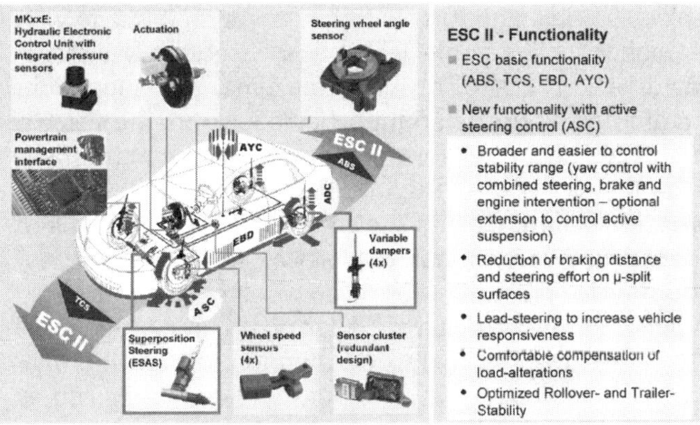

Fig. 4: ESC II - based on MKxxE Technology

As an electronic brake system, the Mk60E5/Mk25E5 (advancement of today's ESC systems with controlled analogue inlet and cut-valves and 5 pressure sensors – Mk60E5 for light and the Mk25E5 for heavy vehicles) is utilized. These systems possess internal pressure sensors for redundant acquisition of the four wheel brake pressures and the driver's intention, required for several ESC II functions.

SUPERPOSITION STEERING

The superposition steering is the most innovative component of ESC II. It permits superposing a further steering angle, in addition to the angle which the driver sets using the steering wheel, by means of a two-stage planetary gear integrated in the steering column and driven by an electric motor. This angle, which influences the steering effort and driving dynamics, is thus made up of the driver-set angle and the superposed angle set by the planetary gear. Depending on the direction of rotation of the planet carrier, which is adjusted by an electric motor, more steering angle (steering becomes more direct) or less steering angle (steering becomes more indirect) is applied to the wheels.
If the system or the energy supply fails, the planet carrier is locked by means of a locking device, which closes without current. The driver has then a conventional hydraulic power steering.

Figure 5 shows a prototype of a superposition steering with its components. The acronym ESAS stands for Electric Steer Assisted Steering [2].

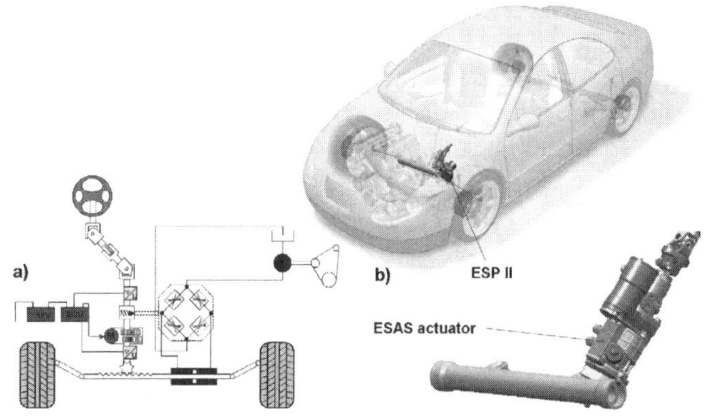

Fig. 5: Steering enhancement with ESAS (steering angle superposition)
System layout (a) and actuator (b)

A conventional hydraulic open-center power steering unit (HPS) serves as a basic steering system for steering torque assistance. If this HPS is designed as a parameter steering, i.e., if it permits influencing the steering-assist torque, it will even be able to compensate for the additional torque – minimal, but present – introduced into the system by the adjustment of the superposition gear.
The original motivation for developing superposition steering systems was the desire to resolve the compromise, which current steering systems represent

with their fixed designed ratio showing slightly progressive steering angle over increasing steering lock. Due to this fixed steering ratio, a vehicle can either be designed for sportiness (direct steering ratio less than 1:12) or comfort (indirect steering ratio of as much as 1:20 for off-road vehicles). A sporty design entailing little steering effort with fewer turns of the wheel for handling and parking can feel too sensitive at high speeds (which is exactly what sportier cars attain) – dangerous wrenching of the steering wheel may be the result. A comfortable design with an indirect ratio makes parking and maneuvering of all things, an uncomfortable issue requiring large steering effort. With the superposition steering, which enables influencing the steering ratio electronically, the ratio can be set dependent upon the vehicle speed.

But the interface between the steering and other systems provides further potential for functions going beyond the basic functions of a variable, speed-sensitive steering ratio. This applies particularly to cross-linking with ESC. At the beginning of the ESC introduction, vehicle stabilization was the main concern. Increasingly though, carmakers are taking an interest in the "how" of vehicle stabilization as an expression of their brand philosophy. Integration of the superposition steering into driving dynamics control opens up new dimensions for providing the vehicle with certain characteristics by means of software. This is due to the particular fact that the conventional ESC brake intervention is perceived by the driver as decelerating the vehicle and consequently can only be applied in the critical vehicle dynamics range. However, intervention by way of the steering, if harmonically balanced, remains unnoticed by the driver and can be executed also in the non-critical range, e.g., to influence vehicle self-steering behavior and thus the vehicle characteristics. Finally, at critical limits, the steering can be combined with brake intervention as a very effective means of influencing the horizontal dynamics of the vehicle – even postponing noticeable brake intervention until later, or making it completely unnecessary.

To enable easy integration of superposition steering units of various steering manufacturers, the interface of ESC II electronics with the steering has been designed with respect to "Cartronic" conformity [3].

ESC II – STEERING CONTROL FUNCTIONS

The basic function of the superposition steering, namely the variable steering ratio, has been described. This function merely requires transmitting the vehicle speed information from the ESC controller to the steering control unit. The value-added functions additionally gained from integrating the superposition steering into the driving dynamics control system ESC II are explained in detail in the following sections.

<u>Handling support (lead steering)</u>

The handling support function addresses the vehicle characteristics described at the beginning, driving pleasure and driving safety. Basically, it takes into account that a vehicle responds to rapid steering commands with a delay owing to the elastic suspension elements such as tires and rubber mounts and owing to vehicle inertia about the vertical and longitudinal axis. With respect to agility, this is sensed by the driver as indirect reaction and an impairment of driving pleasure. But the phase delay of vehicle response can also lead to safety critical situations: If, for instance, a driver has to evade an obstacle quickly, as a rule he will tend to "oversteer" due to the delay in vehicle response. When the vehicle then reacts, the reaction will be more severe than expected by the driver – the vehicle becomes unstable.

From the vehicle speed, the steering angle and the lateral acceleration, ESC II computes a desired vehicle response in the form of a nominal yaw acceleration. This desired vehicle behavior is defined by the vehicle manufacturer to conform with it´s handling philosophy and is stored as software parameters in the ESC II control unit. While the vehicle is on the move, it`s actual responsive behavior is constantly monitored by the yaw rate sensor. Dependent upon the phase delay between desired and actual reaction, the ESC II control unit then briefly requests a lead angle from the superposition steering.

The vehicle reaction is accelerated; at the same time via the yaw acceleration the closed control loop also compensates for changed vehicle loads, tire types and chassis wear.

YAW TORQUE COMPENSATION

The braking action on surfaces with different coefficients of friction on each side of the vehicle (µ-split braking) provides a particularly vivid example of the effectivity of ESC II.
In vehicles without electronic brake systems, unequal braking forces build up on the two sides of the vehicle during severe braking on the described surface. The vehicle is subjected to yaw torque favoring the side with the higher friction coefficient, and consequently turns in that direction. Since, at the same time, the supporting lateral force is lost when the wheels lock up, the vehicle becomes uncontrollable. Today's electronic brake systems recognize the µ-split situation and build up the "pressure difference" on the front axle only to a limited extent and on top of that, only slowly. This is carried out in order to prevent overstress of the driver for his necessary countersteering. Additionally, for reasons of stability, at the rear axle (for lateral control of the yaw torque) the pressure level of both brakes is adjusted to the lower pressure level ("select low"). Thanks to these two measures, despite the extreme situation, the vehicle remains controllable even for the average driver. Simultaneously however, an extended braking distance

must be accepted because of the slower pressure build-up.

ESC II provides the additional option of electronic steering intervention. The yaw torque originating from the different braking forces can be compensated for by automatic rapid application of opposite lock according to the situation. The driver can continue steering in the desired direction and is not required to countersteer. The steering wheel position agrees with the desired course of the vehicle – unlike the present ABS and ESC systems, which require the driver himself to apply opposite lock. The rapid yaw torque compensation by the steering simultaneously permits nearly undelayed brake pressure buildup at the front wheels and a modified rear axle brake pressure control. Braking distance is appreciably reduced – by as much as 15 percent, depending on the difference in friction coefficients and conventional ABS/ESC tuning.

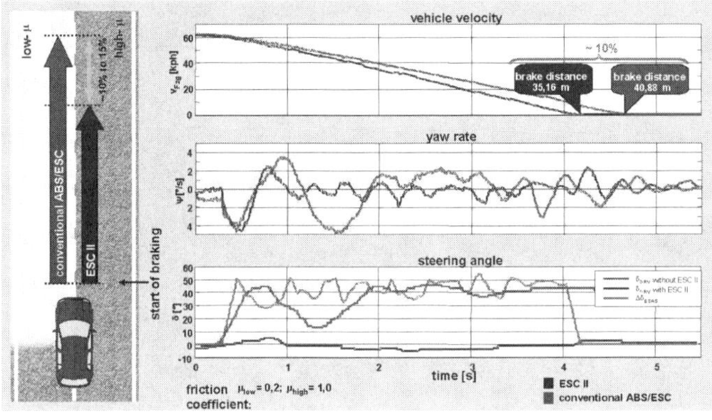

Fig. 6: μ-split braking with/without ESC II

Figure 6 shows the results of μ-split braking at a vehicle speed of 60 km/h with conventional ABS/ESC and ESC II. The smaller steering effort for the driver, the reduced yaw motion and the shorter braking distance for the vehicle with ESC II (blue curves) is clear to see compared with the production vehicle (red curves). One can also recognize, how the ESC II vehicle's superposed steering angle, plotted in green (scaling shown in reference to the steering wheel), is quickly built up for yaw torque compensation and throughout the braking, constantly adjusted depending on the yaw reaction of the vehicle.

The subsystems brake and steering then can devote themselves to their original functions: the brake serves to minimize braking distance, the steering countersteers.

ESC II can also intervene in the steering with a stabilizing effect when moving off or driving on surfaces with different friction coefficients, to compensate for the disturbing yaw torque caused by the differing tractive forces at the wheels.

Handling safety and ride comfort are consequently improved by yaw torque compensation.

Yaw rate control

With the integration of the steering in the yaw rate control, ESC II in contrast to the standard ESC, has a further effective means, apart from the brake, of intervening in the vehicle horizontal dynamics. The critical driving dynamics range is extended and becomes more easily controllable (ESC II automatically applies opposite steering lock in the event of vehicle oversteer). Since the steering intervention remains unnoticed by the driver, it can be applied earlier, thereby avoiding critical situations with respect to stability. Should dynamic instability nevertheless occur, ESC II can intervene more effectively with the steering because of its greater effective leverage compared to the brake (wheelbase larger than track width). However, brake intervention to reduce vehicle speed remains indispensable in these situations.

The basic principle of the yaw rate control of ESC II relies on the familiar yaw rate control of a standard ESC. From the vehicle speed, the steering angle and the lateral acceleration, by means of the so-called reference model, a desired yaw rate is computed as main reference input. Through parameterization of this reference model the vehicle manufacture again has the opportunity to lend the vehicle a specific character. In the case of ESC II this has more far-reaching significance for the manufacturer since ESC II is already capable of intervening in the handling in situations, which are not critical in terms of driving dynamics.

The actual vehicle behavior is measured by a yaw-rate sensor. This is designed as a redundant unit in ESC II, as is the lateral acceleration sensor, since the yaw rate control operates at much smaller yaw rate deviations than the standard ESC; sensor errors cannot be detected in time by the plausibility checks familiar from the standard ESC, for example cross-checking against wheel speed information.

Based on the deviation between measured yaw rate and desired yaw rate (from the reference model), ESC II computes a correction torque designed to force the vehicle back onto its intended course.

Based on a typical ESC test maneuver, the double lane change shown in figure 7a, the different steering angles in the ESC II system are shown in Figure 7c. Clearly recognizable is the almost continuous steering control action (red). The driver's steering angle (green), compared with the cumulative steering angle (blue), i.e., the steering angle, which the driver would have had to apply to perform the maneuver on the same path without ESC II, is appreciably more harmonic in its curve. The steering effort is considerably smaller.

Figure 7b shows the principle of this distribution and the changed control entry thresholds compared with a standard ESC.

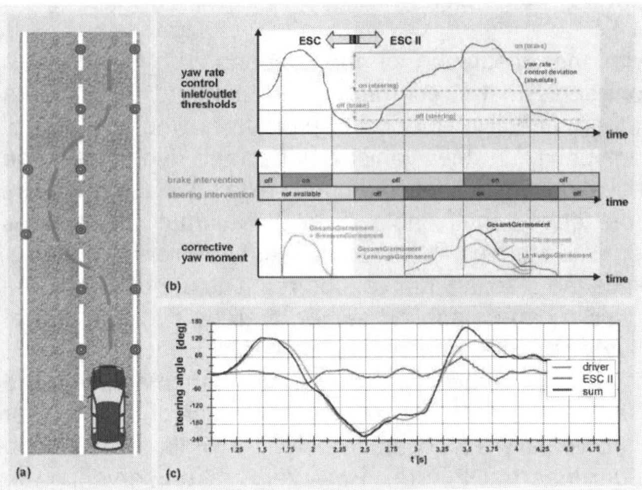

Fig. 7: Yaw Rate Control, double lane change

In summary, the following advantages from the integration of the steering in the yaw rate control can be established: more effective vehicle stabilization, extended critical limits, reduced steering effort, and fewer abrupt, vehicle-decelerating brake interventions. The yaw rate control of ESC II thus enhances driving safety, comfort and pleasure.

FUNCTION INTEGRATION

A major goal during the development of ESC II was to distribute the functions in the direction of Global Chassis Control [4].

Today, the chassis subsystems, as described in the introduction, for reasons of historically evolved relations are handled by development departments of the vehicle manufacturers and suppliers organized along component lines, and are built as stand-alone systems also for strategic purchasing reasons. Thus, in a vehicle with superposition steering, active roll bars, ESC and electronic differential, as many as four independent horizontal dynamics controllers can be installed, each with its own driving state estimator, its own reference behavior calculator and its own driving state controller. On top of that, several sensors are often installed to determine the same driving state variable (e.g. lateral acceleration) – this is because the responsibility for the signals lies with the system supplier.

The functional structure designated the "coexistence approach" outlines this distribution. Since the single controllers, depending on the system they control, pursue different aspects of the control strategy (comfort, handling and safety), their radius of action must be coordinated so that they cannot influence each other negatively. In a way they must be separated a safe distance from another. It is impossible to attain an optimal overall result from control efforts. A further decisive drawback of this approach is the exponential increase in application effort as the number of stand-alone systems grows. If, for example, at a certain point in development the parameterization of one of the controllers changes, all other systems then must be re-released in order to ensure that the change has not influenced the vehicle characteristics, thereby resulting in undesirable behavior in the interaction of the systems.

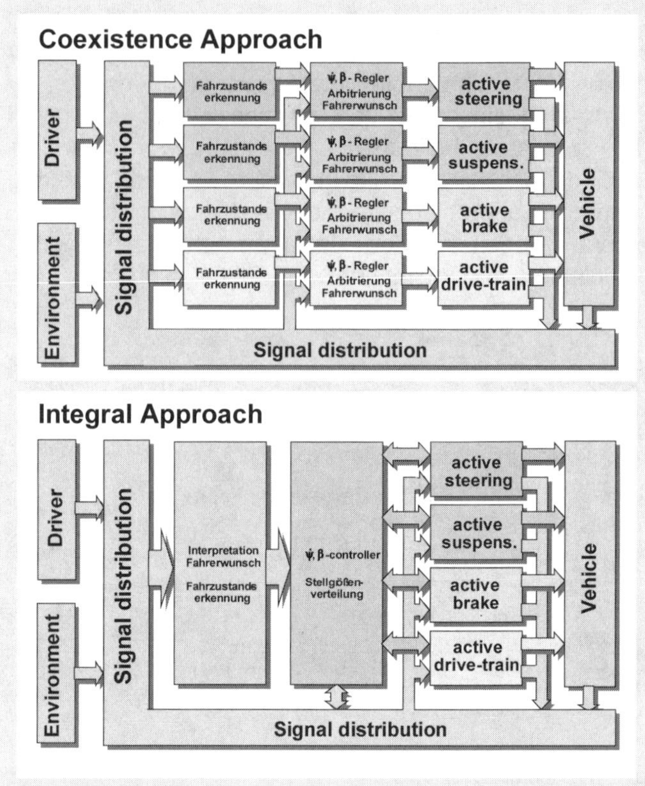

Fig. 8: Methods for function distribution: Coexistence approach and integrated approach

Considering the effort and expense that goes into operational tests for the approval of horizontal dynamics control systems, this procedure is no longer acceptable in the medium term, with the ever shorter development times and increasingly late availability of prototypes.

With ESC II, a function distribution designated as the "integral approach" (see Fig. 8) is realized. This approach assigns one basic function to each of the individual systems, steering, brakes, suspension and drivetrain. In reference to horizontal dynamics, this basic function remains limited to feed forward control, e.g., speed-dependent steering ratio or lateral-acceleration-dependent brake force distribution right/left. At the same time, these functions are involved in a constant interchange with the horizontal dynamics controller in ESC II and report their current actuating limits and dynamics to it. From the driver commands and the vehicle dynamics variables, the central horizontal dynamics controller computes in parallel a desired vehicle behavior and compares this with the actual vehicle behavior, ascertained by means of standardized sensors. If this comparison shows a need for corrective yaw torque, keeping the functions described before in mind, the central controller distributes this torque among the individual actuators based on its knowledge of the driving state, the driver's intention and the actuating limits and dynamics of said actuators.

The advantage of this approach is that the complexity of coordination remains manageable and optimal results in terms of horizontal dynamics control can nonetheless be obtained. Thus, the basic tuning of the subsystems can continue to be handled in small teams between the manufacturers and suppliers. The overall tuning likewise involves only one team from the manufacturer and one from the supplier specifically responsible for the central horizontal dynamics control.

CONCLUSION

The benefits of an ESC system are without doubt decisive to significantly reduce the danger of traffic accidents and eventual fatalities. The effects of a high ESC installation rate in a number of car models in Europe, especially in Germany, have shown that these fatatilities can be reduced by at least 15%.
With ESC II the first step is taken on the way to the completely networked, holistically controlled chassis. ESC II not only opens up new dimensions of handling safety, ride comfort and driving pleasure by integrating

the steering, and optionally the suspension, but also offers considerable potential for savings in terms of application effort compared with stand-alone systems. Other cost advantages result from opportunities for savings with respect to the number of sensors and to control unit performance compared with stand-alone systems. The interfaces between the subsystems are "Cartronic"-conforming so that the vehicle manufacturer remains free in his choice of subsystem suppliers for steering and suspension.

REFERENCES

[1] Rieth, P.; Drumm, S.; Harnischfeger, M.: Electronic Stability Program, The brake that steers, Verlag Moderne Industrie 930575
[2] Nell, J.; Rieth, P.; Bayer, R.; Böhm, J.; Linkenbach, S.; Hoffmann, O.: Erlebbarer Kundennutzen durch Erweiterung heutiger hydraulischer Lenksysteme und deren systemtechnische Umsetzung. PKW Lenksysteme, Haus der Technik, Essen 2-3 April 2003
[3] Kallenbach, R.; Hermsen, W.: CARTRONIC Systemarchitektur zur Vernetzung elektronischer Fahrzeugsysteme, VDA Technischer Kongress, Stuttgart, 20-21 March 2002
[4] Schwarz, R.; Rieth, P.: Global Chassis Control – Systemvernetzung im Fahrwerk, Autoreg, Mannheim 15-16 April 2002

A Failsafe Strategy for a Vehicle Dynamics Control (VDC) System

Mauro Velardocchia and Aldo Sorniotti
Politecnico di Torino, Department of Mechanics

ABSTRACT

The paper presents a failsafe strategy conceived for a Vehicle Dynamics Control (VDC) system developed by the Vehicle Dynamics Research Team of Politecnico di Torino. The main equations used by the failsafe algorithm are presented, especially those devoted to estimate steering wheel angle, body yaw rate and lateral acceleration, each of them fundamental to correctly actuate the VDC. The estimation is based on redundancy; each formula is considered according to a weight depending on the kind of maneuver. A new recovery algorithm is presented, which does not deactivate VDC after a sensor fault, but substitutes the sensor signal with the virtually estimated value. The results obtained through simulation are satisfactory. First experimental tests carried out on a ABS/VDC test bench of the Vehicle Dynamics Research Team of Politecnico di Torino confirmed the simulation results.

VDC CONTROL STRATEGY

VDC control strategy described in this paper was conceived through simulation, by using ten degrees of freedom vehicle model (six for the car body, one rotational for each wheel), implemented in Simulink®. Car body dynamics was determined by using Lagrange's method. Suspensions consist of four springs, two torsion springs and taking into account the antidive phenomena. Powertrain, driveline and braking systems were modeled according to functional equations. Tires were modelled by using Pacejka's formulas. The model was experimentally validated through road tests.

Politecnico di Torino VDC control strategy [1] considers a standard control based on body yaw rate and body sideslip angle. Its main purpose consists of making vehicle behavior in transient maneuvers as much close to vehicle behavior in stationary maneuvers. The control strategy, tested on a vehicle model having an understeering behavior, computes a reference body yaw rate value given by:

$$\dot{\psi}_{reference} = \left[\frac{v_x \dfrac{\delta_l}{i_l}}{l + K_u v_x^2} \right] \quad (1)$$

where K_u is defined through a table and depends both on vehicle body lateral acceleration and the estimated adherence condition [1]. The table to define K_u was conceived considering a number of maneuvers simulation results. Depending on the difference (and also on its rate) between the reference body yaw rate and the measured body yaw rate, the control strategy computes the desired pressure levels of the calipers. Generally, body yaw rate can be reduced with the application of brakes on external wheels to the bend. The opposite is used to determine a body yaw rate increase. Figures 1, 2, and 3 summarize the vehicle dynamics improvement in comparison to the passive vehicle, in an abrupt step steer maneuver.

This VDC also consists of a second control strategy based on body sideslip angle estimation [2], which significantly acts when body sideslip angle or rate overreach well defined values [1].

VDC HYDRAULIC UNIT MODEL

A detailed braking system hydraulic model, with a VDC hydraulic unit, was demonstrated out by using the commercial software AMESim®. Figure 4 is an example of a model developed for a standard ABS/VDC pump group, characterized by a cam, two pistons equipped with unidirectional valves, a low pressure accumulator and a high pressure accumulator. The model permits investigating different phenomena like, e.g., pedal feeling. During a panic brake maneuver, as presented in Figure 5, typical pedal oscillations are evident due to tandem master cylinder pressure oscillations.

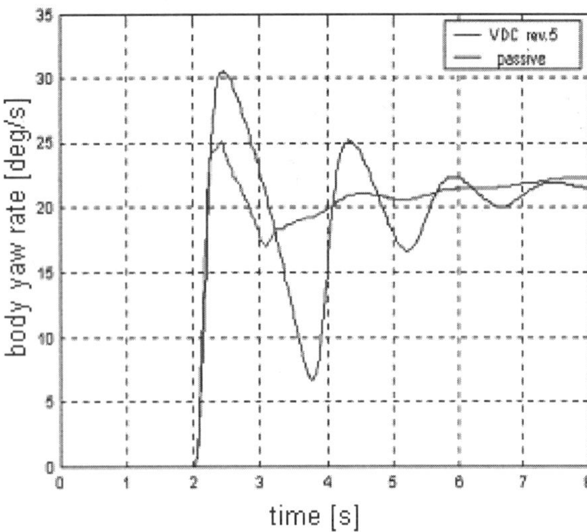

Figure 1: Step steer maneuver

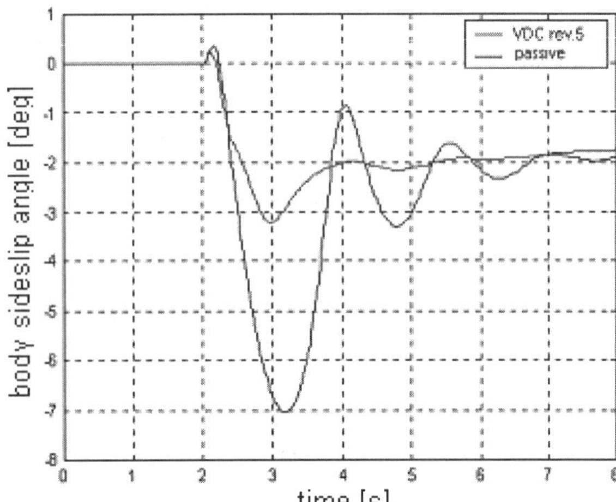

Figure 2: Step steer maneuver

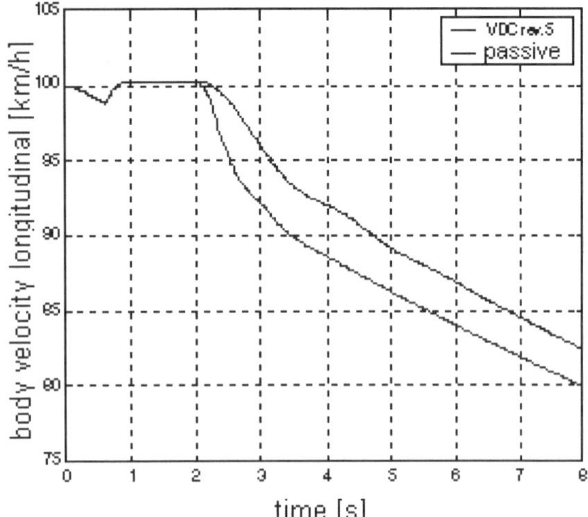

Figure 3: Step steer maneuver

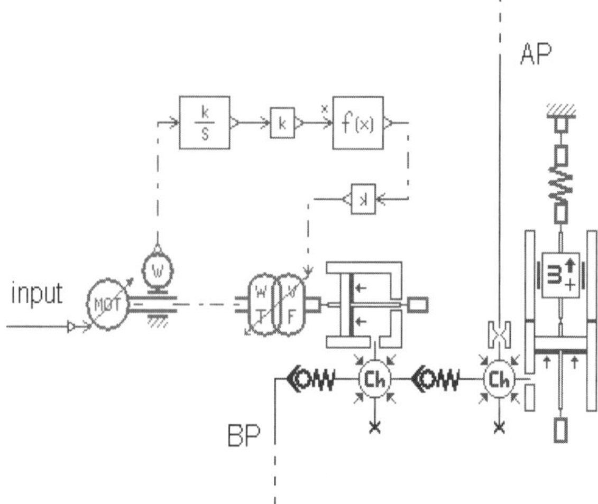

Figure 4: Hydraulic pump AMESim® simple model

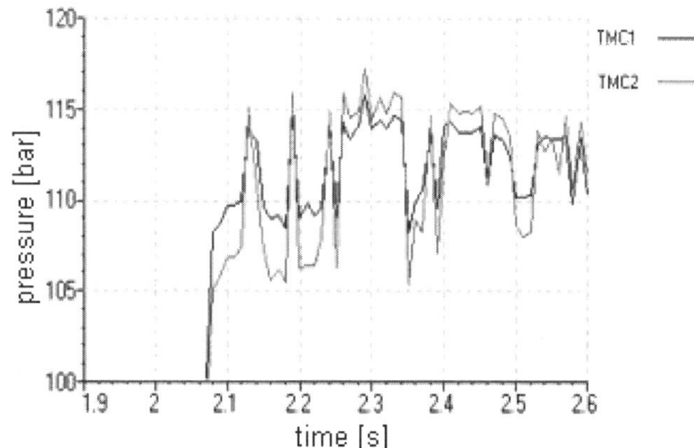

Figure 5: Tandem master cylinder pressures vs. time

Through the detailed model it was possible to investigate the actuation strategy of the valves and the motor pump. To experimentally validate the model, a test bench for braking systems was built at Department of Mechanics, Politecnico di Torino. It consisted of the complete hardware of a braking system, including ABS/VDC units, and a vehicle running in real time on a PC. This bench can be used for the following tasks:

1. To characterize a passive braking system, in particular to validate boosters and calipers models;
2. To study the behavior of commercial standard ABS/VDC systems and validate the models of their hydraulic units;
3. To develop new control strategies to be tested on a physical hardware.

Figure 6 is a picture of a booster tested on the bench. The figure presents the main components: the booster (1) fed by the tank (6), the pressure sensors at master cylinder (3, 4) and calipers. The sensor signals are digitally recorded and they are input for the vehicle model, determining its behavior.

Figure 6: Braking Systems test bench

FAULT INJECTION THROUGH SIMULATION AND EXPERIMENTATION

Extensive fault injection tests were performed to determine the hazard of each fault, by simulating faults on all VDC components, including valves and sensors.
Figure 7 and 8 are about a simulated step steer maneuver, with a significant VDC intervention. Figure 7 plots wheel-braking torques, due to VDC intervention (the driver does not push brake pedal): left wheels (FL and RL) are internal to the bend and, right wheels (FR and RR) are external to the bend. At the beginning of the maneuver, the target is to have a fast increase in body yaw rate; as a consequence, brakes are applied on the internal wheels. When real body yaw rate gets bigger than reference body yaw rate, application of brakes is on the external wheels. Since the front right pressure reduction valve does not open due to a solenoid fault, which is not detected by the failsafe mechanism, there is no front right caliper pressure reduction until rear right wheel is also not subjected to VDC intervention anymore. Since VDC hydraulic model works together with a vehicle model, it is possible to evaluate the influence of the various faults on vehicle dynamics. In the Figure 8, a high reduction of body yaw rate is observed, due to the fault. The overall vehicle dynamic behavior nevertheless is not dangerous.

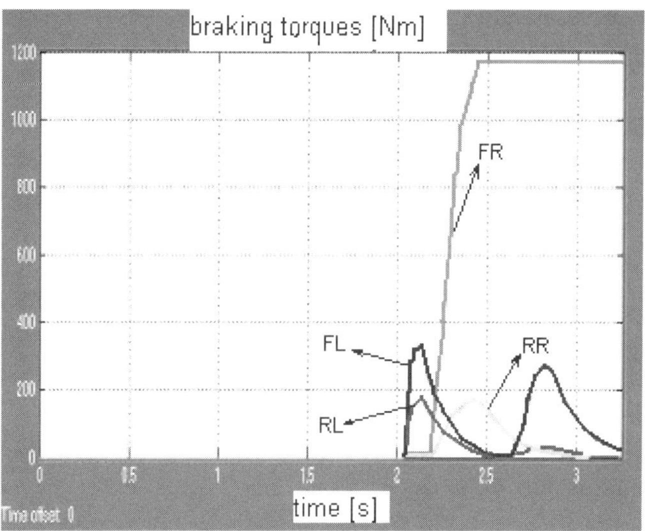

Figure 7: Front right pressure reduction valve always open

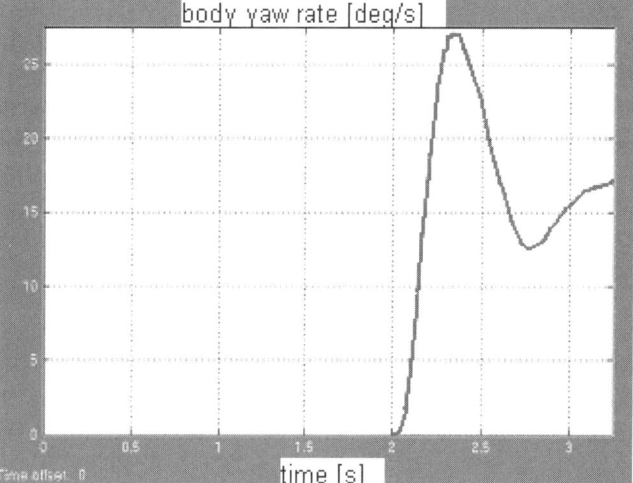

Figure 8: Front right pressure reduction valve always open

Fault injection campaigns were also performed on the test bench, employing a commercial VDC. Figure 9 is the result of a test performed on Politecnico di Torino braking systems bench. The procedure involves a panic brake maneuver on a commercial ABS equipped vehicle, driven at an initial speed of 100 km/h. During the test, the right front wheel speed signal was deactivated; as a consequence, ABS was deactivated by the failsafe strategy. The Electronic Brake Distribution (EBD) continued working, as indicated by the low pressure level of the rear wheels. Figure 9 also shows a consistent offset, in dynamic conditions, between tandem master cylinder pressure levels and caliper pressure levels, due to pipes and ABS hydraulic unit orifices.

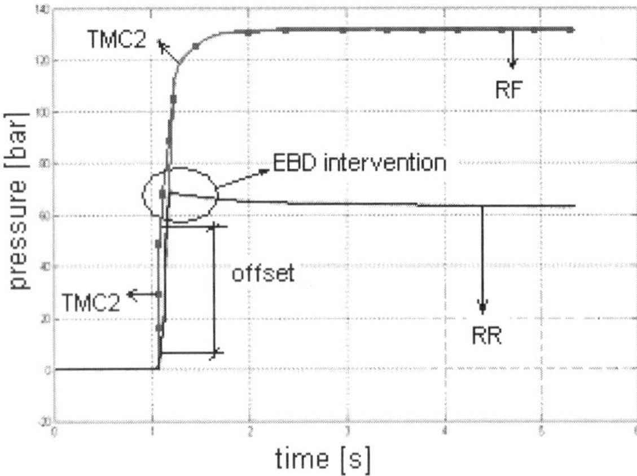

Figure 9: Panic Brake Maneuver on Politecnico braking systems' test bench

A National Highway and Traffic Safety Administration (NHTSA) research [3] showed the following breakdown of ABS-systems problems: 41% sensors, 16% valves, 8% computers, 3% incorrect installation, 1% electrical connections, 30% electromagnetic wave interference. Through fault injection, it is possible to associate the related hazard, in the form of an index, to each fault. By using statistical data, like those of the NHTSA, the frequencies of each of the faults can be estimated. As a result, the product of fault danger index and fault frequency, determines the numerical index for the faults. A risk matrix, associating the indices to their respective faults, can be prepared. The index with the largest value is given most attention to prevent the related fault.

Secondly, fault injection campaigns permit to understand the maximum time between the fault and the signal of the warning lamp in the vehicle cockpit. In the current production vehicles, it is quite common to have warning lamp's lighting phenomena due to short duration problems. They absolutely do not provoke any problem related to safety, but have a consistent impact on customers' satisfaction. Due to complexity of the experimental tests, it is nearly impossible to calibrate the warning lamp indicator with the best possible strategy by road tests. Hence, simulation is not only useful to define VDC stand-alone failsafe strategy but also the connection of the VDC with the whole vehicle (considering also Controller Area Network - CAN).

BASIC VDC FAILSAFE STRATEGY

Every failsafe strategy is based on redundancy and compatibility of different signals. In this paper, faults of the main sensors typical of VDC (body yaw rate, body lateral acceleration, steering wheel angle) are dealt with. The equations used to estimate body yaw rate are [4]:

$$\dot{\psi}_1 = \frac{v_{fr} - v_{fl}}{S} \tag{2}$$

$$\dot{\psi}_2 = \frac{v_{rr} - v_{rl}}{S} \tag{3}$$

$$\dot{\psi}_3 = \frac{a_{lat}}{v_{ref}} \tag{4}$$

$$\dot{\psi}_4 = \frac{\delta_L}{i_L} \frac{v_{ref}}{l + K_u (v_{ref})^2} \tag{5}$$

Similar formulas are also necessary to determine lateral acceleration of the body and steering wheel angle. Each formula has different fields of validity. They were tested in a variety of maneuvers, both dynamic and stationary, while braking and, driving in all possible adherence conditions. For example, the first formula (2) should work for non-slipping front wheels, the second formula (3) for non-slipping rear wheels, the third (4) with a positive value of vehicle speed, the fourth (5) mainly for step steer maneuvers at high speed values. A weighted average value of these formulas is a good approximation of the actual measure. To obtain reliable results, it was necessary to change the weights depending on the kind of maneuver. Everything becomes much more complex when this method is used on several correlated magnitudes, as it happens in this case. A fault is detected when the difference between the body yaw rate value measured by the sensor and the estimated value exceeds an offset depending on the kind of maneuver. Figures 10, 11, 12 compare the yaw rate of:

- a car with a perfectly working VDC (no fault);
- the same car without VDC (passive);
- the same car with a VDC which does not detect the faults;
- the same car having a diagnostic strategy deactivating VDC ('On Off') as soon as the fault is detected (as it is generally done).

The maneuver is a step steer with a fault of body yaw rate sensor at the beginning of steering wheel rotation for the last three vehicles. It is evident that the third vehicle is particularly dangerous: VDC detects a near-to-zero value of body yaw rate, whereas reference body yaw rate is quite big, since steering wheel angle is consistent (2). The pressures of the wheels internal to the bend are consistently increased, giving origin to instability phenomena, not controllable by the driver.

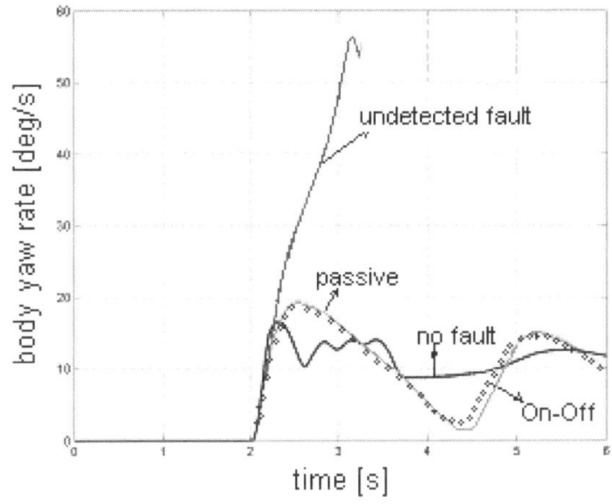

Figure 10: Step steer, 100 km/h, 60° steering wheel angle, low adherence

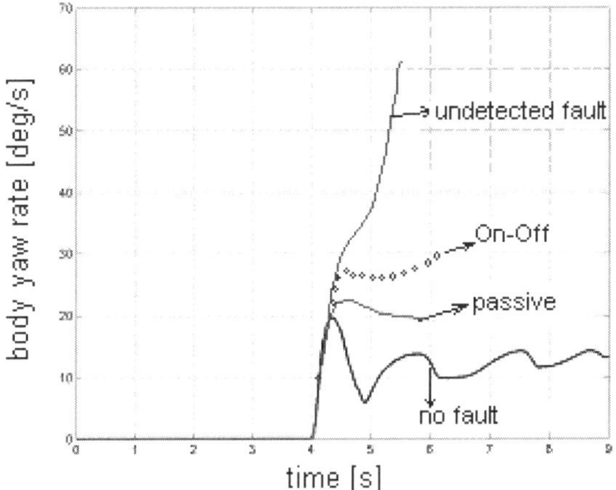

Figure 11: Step steer, 100 km/h, 60° steering wheel angle, low adherence, gas released

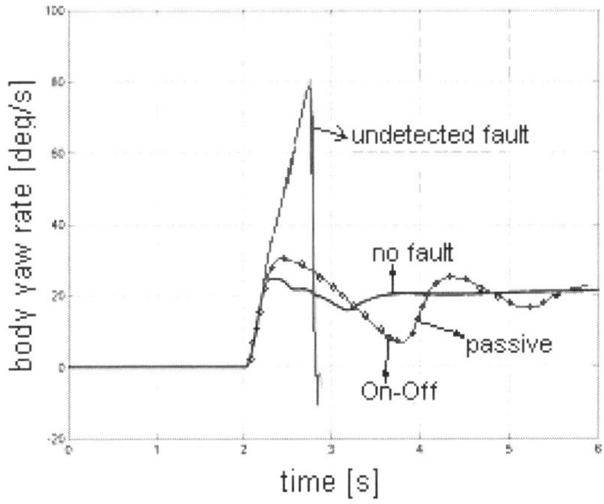

Figure 12: Step steer, 100 km/h, 100° steering wheel angle, high adherence

INNOVATIVE VDC FAILSAFE STRATEGY

When a sensor of a commercial VDC system has a fault, the failsafe algorithm should recognize it. As a consequence, VDC recovery strategy should prevent dangerous events. For example, commercial VDC deactivate body yaw rate control when a body yaw rate sensor fault is detected. Figures 13 and 14 are tests carried out considering the same VDC equipped with different failsafe and recovery strategies. The purpose of this paper is to present a recovery algorithm based on virtual sensors, which uses, when a fault of a sensor is detected, instead of the measured body yaw rate, the signal from the body yaw rate estimator (formulas (2), (3), (4), (5)).

Body yaw rate control is 'On' before and after the fault: this new strategy will be presented as 'On On', in comparison to the traditional 'On Off' (body yaw rate control is 'On' before the fault and 'Off' after the fault). Figures 13 presents a step steer maneuver in high adherence conditions. The fault is introduced into the body yaw rate signal at the beginning of the rotation of the steering wheel. The innovative strategy ('On-On1', the other lines are the same of Figure 12) gives origin to the poor results than the classical 'On Off'. body yaw rate estimation described by equations (2)-(5) deviated by a large amount from the true value in extreme dynamic maneuvers. The conclusion is that equations (2)-(5) are reliable to detect a fault of the actual sensor but they are ineffective if they are substituted with the sensor itself in the control strategy.

Consequently, to carry out an efficient recovery based on virtual sensors when an actual sensor is definitively in fault, new formulas were necessary to estimate, e.g., body yaw rate in dynamic conditions. A first example is presented in Figures 14 and 15. During an abrupt transient state, the vehicle is considered as a mass-spring-damper system (through a properly defined transfer function) relative to body yaw rate behavior, with the typical oscillations, not perceived by eq. (2)-(5).

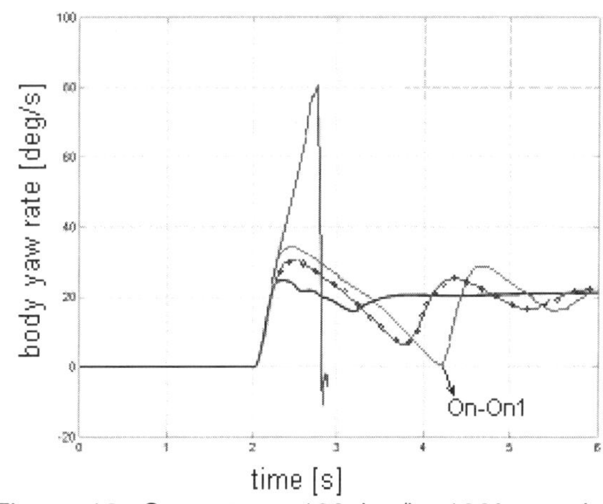

Figure 13: Step steer, 100 km/h, 100° steering wheel angle, high adherence

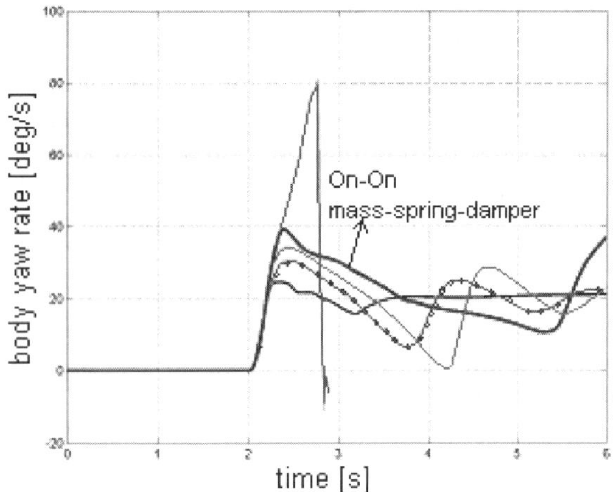

Figure 14: Step steer, 100 km/h, 100° steering wheel angle, high adherence

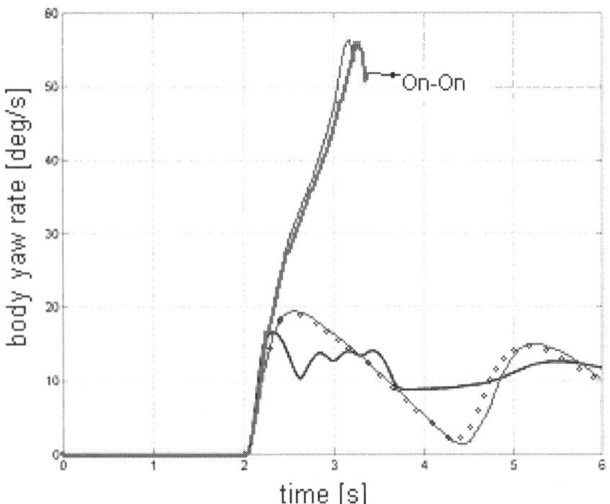

Figure 15: Step steer, 100 km/h, 60° steering wheel angle, low adherence

Figure 16 uses a more sophisticated body yaw rate estimation algorithm, creating artificial oscillations (generated by gains and transfer functions) of vehicle body yaw rate according to steering wheel and throttle valve inputs. The results are very good in all the maneuvers tested, with body yaw rate failures occurring at all instants. Body yaw rate was the most difficult to be estimated, since its behaviour is strictly connected to body sideslip angle by the simplified equation:

$$a_y = v(\dot{\psi} + \dot{\beta}) \qquad (6)$$

Following fast input on the steering wheel, lateral acceleration reaches the target value with a delay, while the body yaw rate and sideslip rate oscillate. Figure 16 shows the results obtained with the 'On On2' strategy applied to both body yaw rate and lateral acceleration, by simulating a fault at lateral acceleration sensor. The 'On On2' strategy revealed the same results of a perfectly working VDC, as confirmed by Figures 17 and 18, about step steer at different throttle opening levels. In some cases (Figure 17), there was also an improvement of VDC behaviour with the virtual sensor instead of the well working sensor. It put forth in evidence that new improvements were possible to the defined VDC control strategy.

All the figures, also those related to lateral acceleration sensor faults, plot body yaw rate versus time. In fact, for the same maneuver, lateral acceleration changes very little between a good and a bad VDC (or between a passive car and a car equipped with VDC). Large deviations of the estimated lateral acceleration from the real value give origin to consistent problems from the point of view of vehicle body yaw rate, not of lateral acceleration. Body yaw rate is the main measurable parameter which permits evaluating VDC performance. For example, active and passive vehicle of Figure 17 have very different values of body yaw rate and very similar values of body lateral acceleration.

Figure 19 is about a double step steer manoeuvre, with a fault of lateral acceleration sensor immediately after the first steering wheel rotation. It was adopted from the 'On On2' strategy for steering wheel angle sensor. In a failsafe strategy, the evaluation time and the maximum admissible offset are fundamental. The evaluation time can be defined as the time delay between fault occurrence and its identification. The maximum possible offset can be defined as the maximum difference between the estimated and the measured value of the magnitude without failsafe intervention. Table 1 reports examples of the evaluation times and the maximum possible offsets, in the case of the steering wheel sensor.

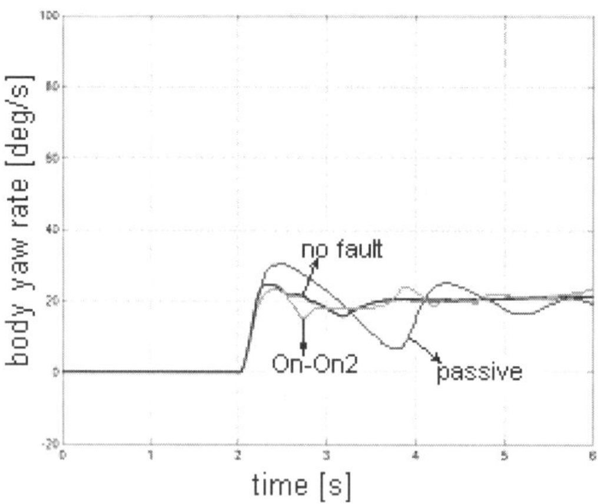

Figure 16: Step steer, 100 km/h, 100° steering wheel angle, high adherence

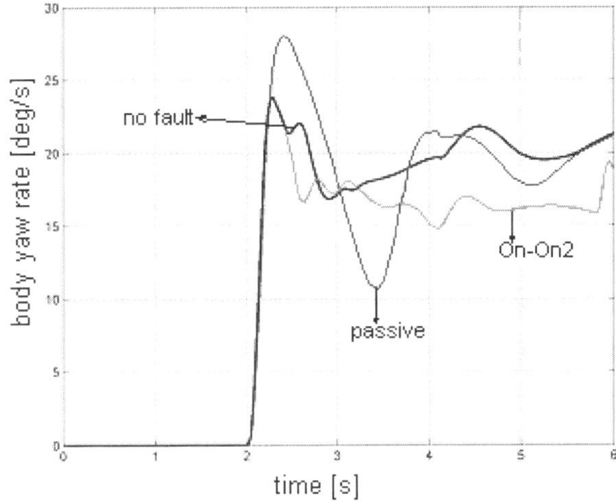

Figure 17: Step steer, 100 km/h, 100° steering wheel angle, high adherence, throttle valve fully open

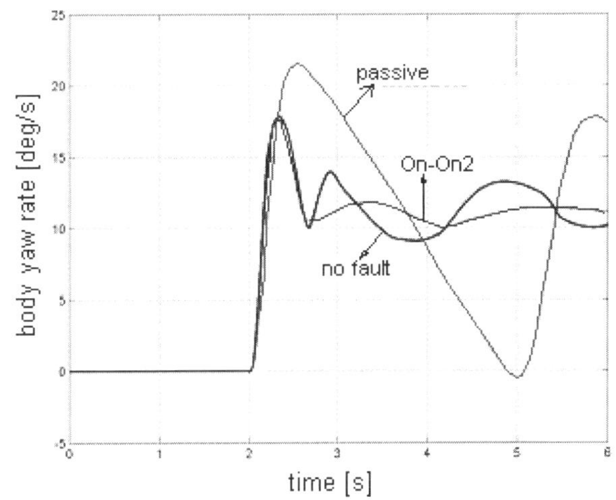

Figure 18: Step steer, 100 km/h, low adherence, throttle valve fully open

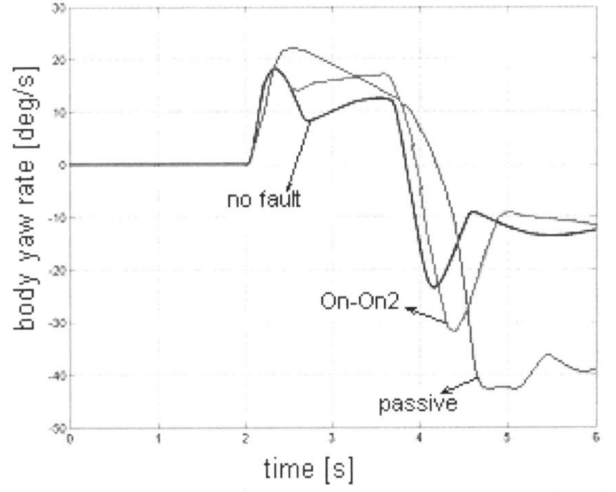

Figure 19: Double step steer maneuver, 100km/h, low adherence

Table 1: Tuning parameters of the failsafe algorithm

Maneuver	evaluation time	maximum offset
Semi stationary	20 ms	0.37 rad
Dynamic	80 ms	0.54 rad

Figures 20, 21 and 22 summarize the final results obtained through fault injection on the different sensors with the complete 'On On' strategy, extended to body yaw rate, lateral acceleration and steering wheel angle. The algorithm firstly recognizes the sensor affected by a fault, and then activates the recovery, continuing to perform yaw rate control. Of course, these are only a few examples of all the simulations performed to verify the 'On On' algorithm. It was also necessary to vary the time and the duration of the faults, in addition to driver's inputs and adherence conditions.

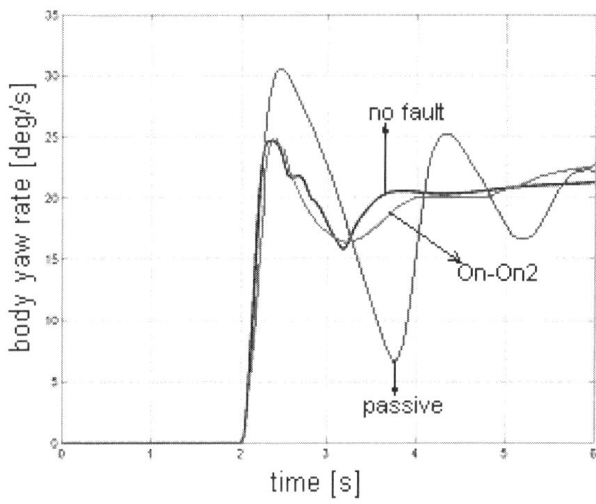

Figure 20: Steering wheel angle sensor fault during a step steer, 100 km/h, 100° steering wheel angle, high adherence

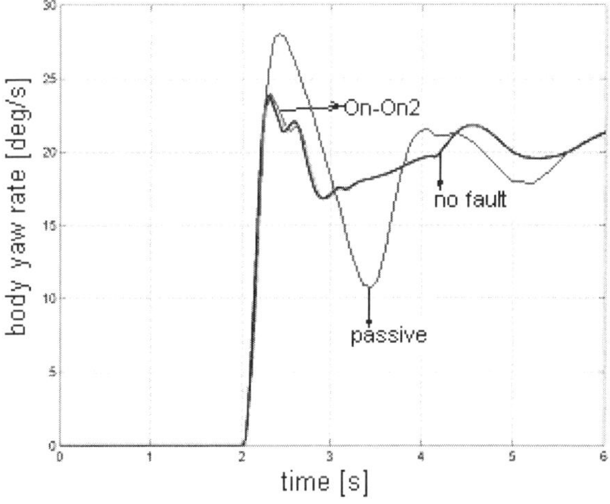

Figure 21: Lateral acceleration sensor fault during a step steer, 100 km/h, high adherence, throttle valve fully open

485

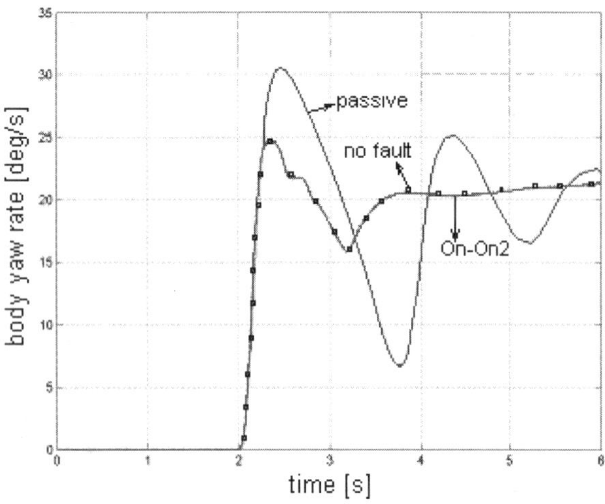

Figure 22: Step steer, 100 km/h, high adherence; steering wheel angle invalid data at t = 2.150 s (due to a simulated temporary fault of CAN)

CONCLUSIONS

This paper demonstrates that it is possible to conceive a chassis control system with failsafe and recovery strategy, even when a fault at one of the sensors is detected, by substituting the sensor signal in fault with the estimated values of the signal. The methodology adopted was applied considering a VDC as test case. The experimental validation on Politecnico di Torino test bench confirmed the very promising results obtained through simulations.

LIST OF SYMBOLS

a_{lat}: vehicle body lateral acceleration
i_L: steering ratio
K_u: corrective factor
S: half track
v: vehicle body speed
v_{ref}: vehicle speed estimated by VDC
v_x: vehicle longitudinal speed
β: vehicle body sideslip angle
δ_L: steering wheel angle
ψ: vehicle body yaw angle
FR: front right
FL: front left
RR: rear right
RL: rear left

BIBLIOGRAPHY

1. Morgando, Sorniotti, Velardocchia, 'Vehicle Dynamics Control based on Yaw Rate and Sideslip Angle', EAEC 2003, Paris
2. Hac, Simpson, 'Estimation of vehicle slip angle and yaw rate', SAE 2000-01-696
3. Limpert, 'Brake Design and Safety', 2nd Edition, SAE 2001
4. Fennel, Ding, 'A Model-Based Failsafe System for the Continental TEVES Electronic Stability Program (ESP)', SAE 2000-01-1635
5. Faye, Spelsberg, 'Improving the Dynamic Driving Behavior of Lightweight Vehicles', SAE 2000-01-0108
6. Kawagoe, Suma, Watanabe, 'Evaluation and Improvement of Vehicle Roll Behavior', SAE 97, Nr. 970093
7. Sasaki, Nishimaki, 'A Side-Slip Estimation Using Neural Network for a Wheeled Vehicle', SAE 2000-01-0695
8. Kubota, Ushijima, Brown, 'Correlation of Driver Confidence and Dynamic Measurements and the Effect of 4WD', SAE 95, Nr. 950972

ACKNOWLEDGEMENTS

The authors of Politecnico di Torino Research Team on Vehicle Dynamics wish to thank FIAT Auto to have sustained the presented work both through funds and technical suggestions.

CONTACT

Prof. Mauro Velardocchia
Politecnico di Torino
Department of Mechanics
C.so Duca degli Abruzzi 24
10100 Torino
ITALY
Tel: +39 011 564 6931
Fax: +39 011 564 6999
Mail: mauro.velardocchia@polito.it

Ph. Dr. Student Aldo Sorniotti
Politecnico di Torino
Department of Mechanics
C.so Duca degli Abruzzi 24
10100 Torino
ITALY
Tel: +39 011 564 6915
Fax: +39 011 564 6999
Mail: aldo.sorniotti@polito.it

JSAE 20037164

SAE 2003-01-2883

Vehicle Stability Improvement Using Fuzzy Controller and Neural-Network Slip Angle Observer

Mohammad Durali and Yusef Bahramzadeh
Center of Excellence for Design, Robotics and Automation
Mechanical Engineering Department
Sharif University of Technology
P.O. Box 11365-9567, Tehran, Iran
durali@sharif.edu

ABSTRACT

This article describes the design and implementation of a fuzzy controller developed for improving car stability by controlling car side-slip angle. The strategy has been to estimate the slip angle by a trained neural network and to determine an appropriate force arrangement on the wheels to produce the necessary yaw moment to limit car side slip control. A seven degrees of freedom car model including nonlinear tire behavior is used in design stage. The results were then validated on a full car model in ADAMS having 156 DOF and including elements nonlinearities and flexibilities. The simulations show the capability of the designed controller in improving stability of the car in sever maneuvers.

INTRODUCTION

In recent years many new subsystems are developed for improving stability of passenger cars. These subsystems employ the car basic capabilities such as; brake, ABS, power train, and steering system to bring the car into a relatively more stable state in the outbreak of conditions. These systems apply a direct yaw moment to the car by producing uneven brake or traction forces on the wheels towards avoiding the car further instability. The usual states used for these control actions are lateral or side slip angle and yaw angle. From stability point of view, the side slip angle is a direct indication of car tendency towards instability, because at large side slip angles, the car will face over steering condition which can further lead to spinout and rollover. The car side slip angle is linearly related to yaw moment for small slip angles, but its relation becomes nonlinear as side slip angle increases.

To sustain car stability in maneuvers, the car side slip angle should be kept in within a certain limit. Shibahana [1] used slip angle methods (β Methode) to study car instability. The important result of his research was that in large slip angles the sensitivity of yaw rate to steering input decreases very sharply. Large side slip angles are the angles for which the tires' friction-slip curve reaches its maximum saturated value. Recent advancements in control and estimation of slip angle has limited this parameter to three degrees. For slip angles larger than this the steering angle input does not produce yaw moment with the same rate as before. Shibahana showed that by appropriate arrangement of brake forces on the tires and hence controlling each wheel's slip, the car yaw and side slip can be controlled.

Vanzanten [2] presented a dynamic controller that could control slip angle and yaw rate at the same time. He showed that this controller can considerably improve car behavior in critical driving conditions. Inagaki [3] studied the stability of the car in critical conditions by using the phase curve between slip angle and its derivative. He showed the car side slip angle can be controlled by controlling car yaw moment using brakes.

Abe et al. [4] presented a method by which side slip angle was controlled using a sliding mode controller. In addition an active estimator was used in their control loop that could estimate the car slip angle. Yoshika et al. [5] did a similar work in which a variable structure system (vss) sliding mode controller was used and the slip angle was estimated using least square method. Buckholtz [6] presented a fuzzy controller to control yaw rate in which necessary slip angle on each wheel to control the car yaw angle was controlled.

The capabilities of fuzzy controllers in dealing with complex nonlinear systems have made these controllers a useful tool in different vehicle control systems. Most estimation methods involve solution of the system's differential equation that are normally timely and impractical in control purposes. However,

Neural network method presents many possibilities in modeling, identification and estimation of dynamic systems, and can offer reasonable accuracy and calculation speed at the same time. This makes them ideal as an observer in high speed control loops.

In this article a fuzzy controller is designed for controlling a vehicle lateral slip in a safe range. This is accomplished by producing direct yaw moment by on the vehicle by tire forces. The output of the controller is the tires' side slip angle. The vehicle's slip angle should also be estimated by a reliable method.

CAR DYNAMIC MODEL

The preliminary model used here is a 7 degrees of freedom model representing car longitudinal and lateral velocities, car yaw rate, and rotational speeds of the four wheels. The governing equations

The car longitudinal Motion:

$$M(\dot{V}_x + V_y r) = F_{x1} \cos \delta_1 + F_{x2} \cos \delta_2 - F_{y1} \sin \delta_1 - F_{y2} \sin \delta_2 + F_{x3} + F_{x4} \tag{1}$$

The car lateral Motion:

$$M(\dot{V}_y - V_x r) = F_{x1} \sin \delta_1 + F_{x2} \sin \delta_2 + F_{y1} \cos \delta_1 + F_{y2} \cos \delta_2 + F_{y3} + F_{y4} \tag{2}$$

The car rotational motion:

$$\begin{aligned} I_z \dot{r} = tf\,(&F_{x1} \cos \delta_1 - F_{x2} \cos \delta_2 \\ &- F_{y1} \sin \delta_1 + F_{y2} \sin \delta_2 \\ &+ t_r(F_{x3} - F_{x4}) \\ &+ l_f(F_{x1} \sin \delta_1 + F_{x2} \sin \delta_2 \\ &+ F_{y1} \cos \delta_1 + F_{y2} \cos \delta_2\,) \\ &- l_r(F_{y3} + F_{y4}) \end{aligned} \tag{3}$$

The value of the longitudinal force acting on each wheel depends the normal wheel reaction as well as its longitudinal and lateral slip. The normal ground reaction results from static weight distribution and the load transfer due to lateral and longitudinal accelerations. Therefore:

$$F_{zfr} = \frac{M}{2L}[gl_r - a_x h + \frac{2a_y l_r}{T_f}] \tag{4}$$

$$F_{zfl} = \frac{M}{2l}[gl_f - a_x h - \frac{2a_y l_y}{T_r}] \tag{5}$$

$$F_{zrr} = \frac{M}{2l}[gl_r + a_x h + \frac{2a_y l_y}{T_f}] \tag{6}$$

$$F_{zrl} = \frac{M}{2l}[gl_f + a_x h - \frac{2a_y l_f}{T_r}] \tag{7}$$

The longitudinal slip for each wheel can be determined from Eq. 8:

$$\lambda_i = \frac{V - r\omega_i}{V} \quad i = 1..4 \tag{8}$$

Therefore the slip angle for front and rear wheels can be formulated as in Equ.s 9 to 12

$$\alpha_1 = \delta_1 - \tan^{-1}\left(\frac{V_y + l_f r}{V_x + t_f r}\right) \tag{9}$$

$$\alpha_2 = \delta_2 - \tan^{-1}\left(\frac{V_y + l_f r}{V_x - t_f r}\right) \tag{10}$$

$$\alpha_3 = -\tan^{-1}\left(\frac{V_y - l_f r}{V_x + t_r r}\right) \tag{11}$$

$$\alpha_4 = -\tan^{-1}\left(\frac{V_y - l_r r}{V_x - t_r r}\right) \tag{12}$$

THE WHEEL MODEL:

Eliminating the wheel rolling resistance, the equations of motion of the wheels can be written as:

$$I_{wi}\dot{\omega}_i = F_{xi} r - T_i \quad i = 1..4 \tag{13}$$

It is useful to remind that the slip angle is also defined as:

$$\beta = \tan^{-1}\frac{V_y}{V_x} \tag{14}$$

TIRE MODEL:

The slip forces on the wheels are the result of combined longitudinal and lateral slip. It was therefore decided to use modified Dogoff slip model to determine the forces [8,9]. The inputs to this model are:

a) normal tire force
b) tire side slip angle
c) tire longitudinal slip
d) and car longitudinal speed.

The equations determining the tire slip forces were entered in Matlab/Simulink and solved during the simulation process.

CONTROL STRATEGY

The car stability in sever maneuvers is heavily affected by lateral slip angle. Increased sideslip angle will reduce the yaw moment from one hand and decreases the driver comfort from other hand, resulting inappropriate driving reactions. If this angle can be kept within a permissible limit, the yaw rate gain and lateral acceleration rate will also be limited. Therefore it may be advisable to control the slip angle instead of the yaw rate. The controller determines the slip angle and its deviation from the ideal value. Based on this the force required by each wheel to contribute to the necessary yaw moment on the car is determined. The value of wheel slip is measured by the ABS subsystem.

In this article a Sliding Mode controller is used to control each wheel slip. A Fuzzy controller is used to control the vehicle slip angle.

FUZZY CONTROLLER DESIGN

The controller is normally designed so that the slip angle does not exceed 5°. The optimum slip angle can be estimated using a two degrees of freedom model. To simplify the problem, the base slip angle will be taken equal to zero, however any nonzero value for base slip angle could be considered. If the car slip angle changes suddenly, the most effective stabilization action is the application of an outward yaw moment. This can be achieved by applying brake on the out-front wheel (Ingaki [3]). This is because this brake force will cause an outward yaw moment around car center of mass on one hand , and on the other hand it will reduce the tire's side force due to the action of frictional circle. Therefore it is advised to apply the control brake force on front wheels.

The fuzzy control system described in figure 2has two inputs that are the slip angle and its rate. The slip angle rate helps in better performance of the controller and faster approach to a stable state. The fuzzy membership functions are shown in Figure 3. The controller determines the essential value of the front wheels slip that are to produce the correcting moment. The non-fuzzy membership functions for longitudinal wheel slip is shown in Figure 4. The large number of non-fuzzy ownership functions used is because of the sensitivity of tire force calculations compared to longitudinal slip. One should notice the increased calculation time due to added ownership functions.

THE FUZZY CONTROLLER RULES

Fifteen rules are presented for control of slip angle (Fig. 5). These rules are designed on the basis of expert human reactions to control slip angle. In general, the slip applied to the tires is depends on the vehicle deviation from ideal conditions, in way that sufficient correcting moment can be produced by tire forces.

NEURAL NETWORK OBSERVER FOR SIDE SLIP

The presented car model is a nonlinear system, therefore in estimating the states such as side slip angle, methods capable of estimating nonlinear systems must be used. A neural network is a computational tool that can perform nonlinear mapping between input and output. This performance is optimized by adjustment of neural functions and weights towards minimizing the error between the out put data from the real model and the estimated output by the network.

The two main steps of a neural network design are choice of; appropriate network structure, and neural functions in different layers. This will include making decision on the number of layers, the number of neurons on each layer, the neuron functions, and in most cases output data feedback to improve the network performance. For dynamic models of mechanical systems, output data feedback with one and two time step delays has proven to yield very good results. Also feeding delayed input signals along with the input signal itself has proven to be effective in reducing the estimation error by the network.

The neural network designed for estimation of slip angle is shown in Figure 6. This network has two hidden layers with three and five neurons having nonlinear sigmoid neuron function. This structure was obtained by trading between estimation error and computation time required for the network. The output of the network is the estimated slip angle. The inputs are; longitudinal and lateral accelerations along with their one and two time step delays, the one and the two time step delays of the out put signal, and the steering wheel angle.

The network is trained using 'back propagation of error' method and to adjust weights and bias in the neurons Levenberg-Marquadt optimization scheme was used.

SIMULATIONS:

To check the controller performance, a 7-degrees of freedom model for the car was primarily used. At a later stage the controller was checked on a full car model on ADAMS. A lane change maneuver at a speed of 72 Km/h, with amplitude of 6° and velocity of 2 rad/s was applied to the model (Fig. 7). Figure 8 shows the car side-slip angle for uncontrolled and controlled situations and figure 9 shows the car slip angle versus yaw angle for the two modes.

It can be seen that the slip angle is limited by the action of the produced yaw moment and the system has remained stable for the given maneuver. It can also be noted that the car yaw angle is controlled indirectly by the controller action. Figures 11 and 12 show the comparison between slip angle reference value and its estimated value by neural network for lane change and J-turn maneuvers. The estimation error is less than 5% that has enough accuracy for being used in the control loop. This will eliminate the need for a yaw angle sensor to determine the slip.

SIMULATIONS USING ADAMS

To validate the results, the designed controller was combined with a full car model in ADAMS. This model includes 104 DOF of rigid bodies and 52 DOF for moving parts. The rear axle of the car under study is a compound crank axel that acts like a torsion spring in the system. Therefore the ADAMS model will also include flexible element dynamics. This part of the model done using ADAMS/Flex and was combined with the original model. This will considerably increase the computation and simulation time. The springs, shock absorbers, their joints, and even the impact bumpers are all considered with their nonlinear behavior in the model. The steering system, of rack and pinion type, is also fully modeled and incorporated in the car model.

Figures 13, 14 and 15 show the ADAMS simulation results which very well validates the results obtained from the simplified model. It also proves that the controller has been well capable of reducing the car slip angle and limiting the way angle.

CONCLUSIONS

The design details of a controller for controlling side slip of the car by direct application of yaw moment was presented in this paper. A seven DOF car model including modified Dogoff nonlinear tire model was used for simulation of the car behavior. To correct the car side slip angle, a fuzzy controller is used. The objective of this fuzzy controller is to determine the slip on each wheel hence prescribing an appropriate force distribution on the tires to produce the necessary yaw moment for slip angle control. The simulation results show that the controller will perform very well to stabilize the car in critical driving conditions. This will also improve the car yaw angle in rapid maneuvers. This will eliminate the need for separately controlling the yaw angle. To estimate the slip angle and hence avoiding measurements by expensive sensors a special neural network estimator has been designed and trained in this project. The simulations show that training the network by the data obtained from a seven DOF model can result in an accurate enough estimator for incorporation in the real control loop.

REFERENCES

1- Shibahata Y.,Shimida K., Tomari.,
"Improvement of Vehicle Maneuverability by Direct Yaw Moment Control" , J. of Vehicle System Dynamics, 22, pp. 465-481(1993)
2- Anton Van Zanten , "VDC The Vehicle Dynamic Control System of Bosch", SAE,950759
3- Inagaki, S. et al, "Vehicle Stability Control in Limit Cornering by Active Brake", SAE960487
4- Abe, M. et al., "Improvement of Vehicle Handling Safety with Vehicle Side-slip Control by Direct Yaw Moment", J. of Vehicle System Dynamics, Vol. , (1999) pp. 665-679
5- Yoshioka, T, et al , "Application of sliding-mode theory to direct yaw-moment control", JSAE Review 20, (1999), pp 523-52
6- Kenneth R. Buckholtz, Delphi Automotive System, "Use of Fuzzy Logic in Wheel Slip Assignment-part 1: Yaw Rate Control", SAE, 2002-01-1221
7- Durali, M. and Kasaiezadeh, A.R., "A Neural Network approximation of nonlinear car model using ADAMS simulation results", SAE2001-01-3324
8- Wong, J.Y., "Theory of Ground Vehicles", John Whily and Sons, Inc. (1993).
9- Hac, A. and Bode, M., "Improvements in vehicle handling through integrated control of chassis systems", J. of Vehicle Design, Vol.29, (2002), pp23-48
10- Neural Network Toolbox-3.0 for use with MATLAB, Mathworks, 1998.

NOMENCLUTURE

F_{xi} : tire longitudinal force

F_{yi} : tire lateral force

F_{zi} : tire normal force

g : gravity acceleration
h : height of vehicle center of gravity
I_z : yaw moment of inertia
L : wheel base
l_f : front axle distance to c.g.
l_r : rear axle distance to c.g.
M : total mass of vehicle
r : yaw rate
T : braking torque
t_f : front track
t_r : rear track
V_x : longitudinal velocity
V_y : lateral velocity
α_i : tire slip angle
β : vehicle body side slip angle
δ : steering angle
ω_i : rotational velocity of wheel

FIGURES

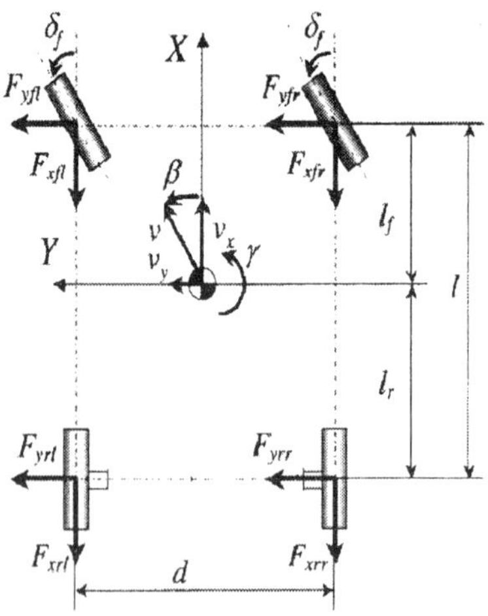

Fig. 1: Vehicle preliminary model

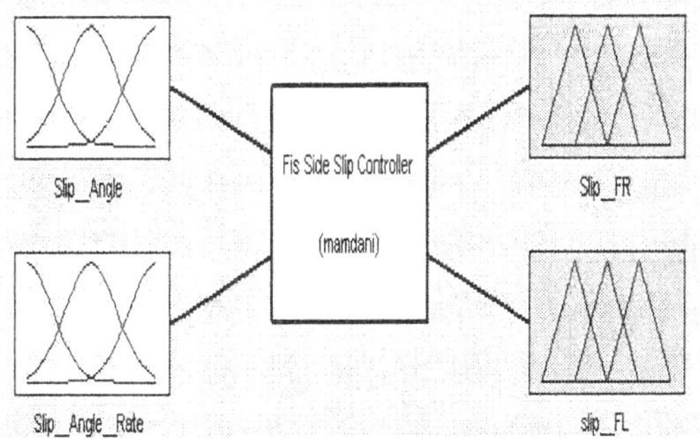

Fig.2: Fuzzy Side Slip Control

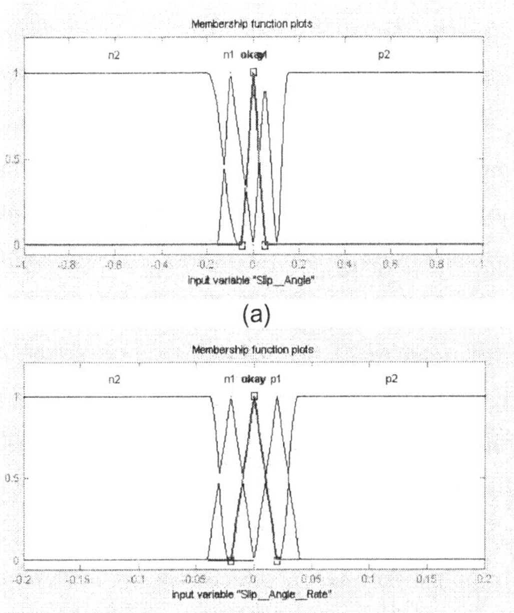

(a)

(b)

Fig.3: Input membership function, a) Slip Angle, b) Slip Angle Rate

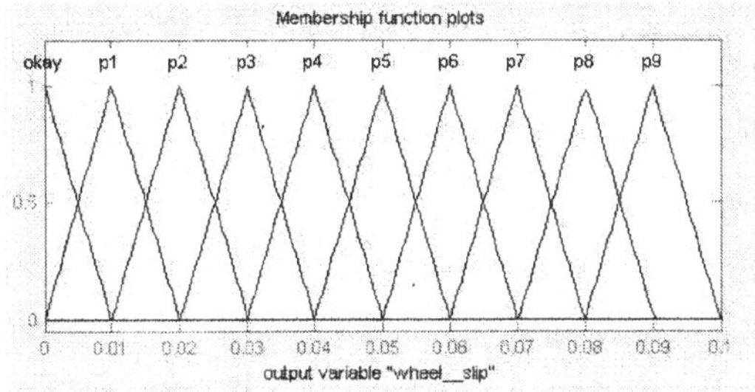

Fig.4: Output membership function "Wheel Slip

1. If (Slip_Angle is okay) and (Slip_Angle_Error is okay) then (Slip_FR is okay)(slip_FL is okay) (1)
2. If (Slip_Angle is okay) and (Slip_Angle_Error is p1) then (Slip_FR is p1)(slip_FL is okay) (1)
3. If (Slip_Angle is okay) and (Slip_Angle_Error is p2) then (Slip_FR is p2)(slip_FL is okay) (1)
4. If (Slip_Angle is okay) and (Slip_Angle_Error is n1) then (Slip_FR is okay)(slip_FL is p1) (1)
5. If (Slip_Angle is okay) and (Slip_Angle_Error is n2) then (Slip_FR is okay)(slip_FL is p2) (1)
6. If (Slip_Angle is n1) and (Slip_Angle_Error is n1) then (Slip_FR is okay)(slip_FL is p4) (1)
7. If (Slip_Angle is n1) and (Slip_Angle_Error is n2) then (Slip_FR is okay)(slip_FL is p5) (1)
8. If (Slip_Angle is n2) and (Slip_Angle_Error is n1) then (Slip_FR is okay)(slip_FL is p6) (1)
9. If (Slip_Angle is n2) and (Slip_Angle_Error is n2) then (Slip_FR is okay)(slip_FL is p7) (1)
10. If (Slip_Angle is p1) and (Slip_Angle_Error is p1) then (Slip_FR is p4)(slip_FL is okay) (1)
11. If (Slip_Angle is p1) and (Slip_Angle_Error is p2) then (Slip_FR is p5)(slip_FL is okay) (1)
12. If (Slip_Angle is p2) and (Slip_Angle_Error is p1) then (Slip_FR is p6)(slip_FL is okay) (1)
13. If (Slip_Angle is p2) and (Slip_Angle_Error is p2) then (Slip_FR is P7)(slip_FL is okay) (1)
14. If (Slip_Angle is p2) then (Slip_FR is p4)(slip_FL is okay) (1)
15. If (Slip_Angle is n2) then (Slip_FR is okay)(slip_FL is p4) (1)

Fig.5 : Fuzzy Controller Rule Table

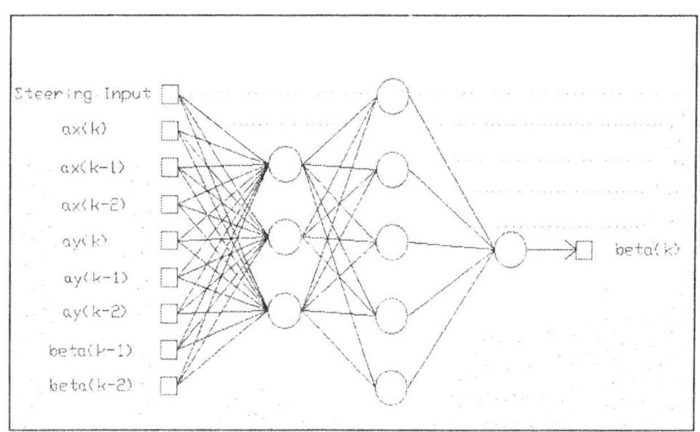

Fig. 6: Neural network structure

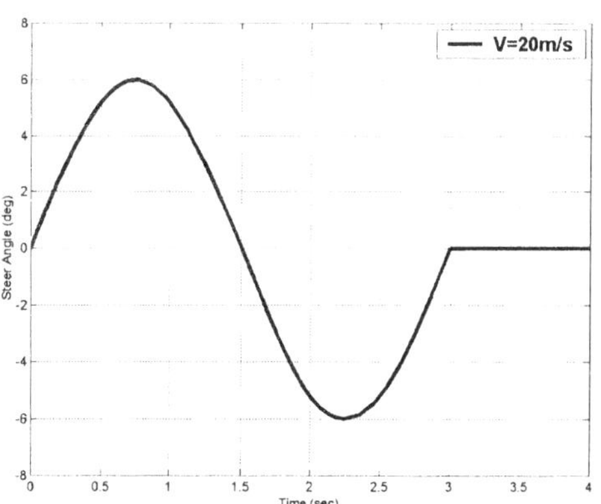

Fig. 7:Steering Input

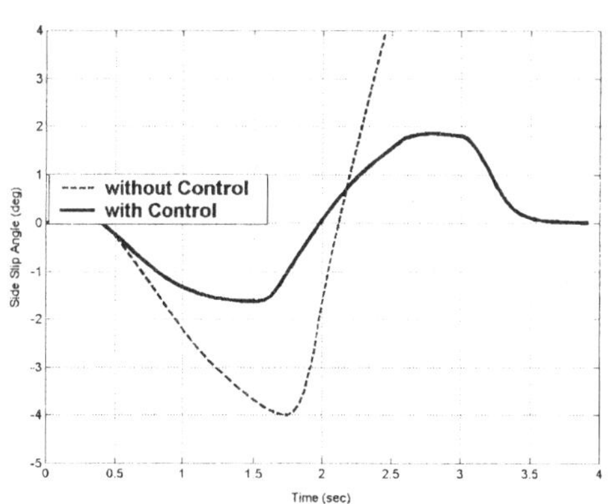

Fig.8: Side Slip vs. Time

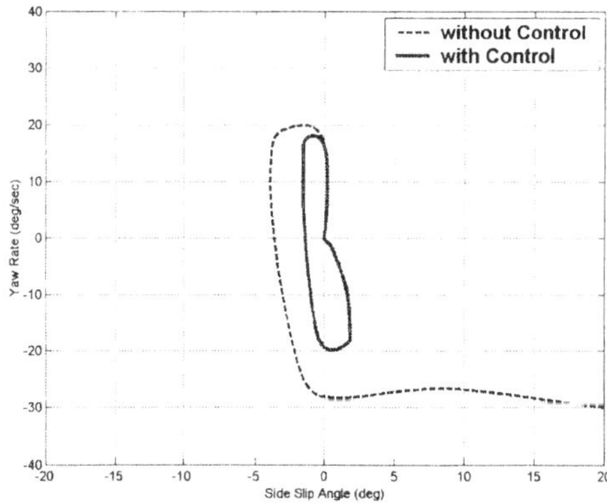

Fig.9:Yaw rate vs. Side slip angle

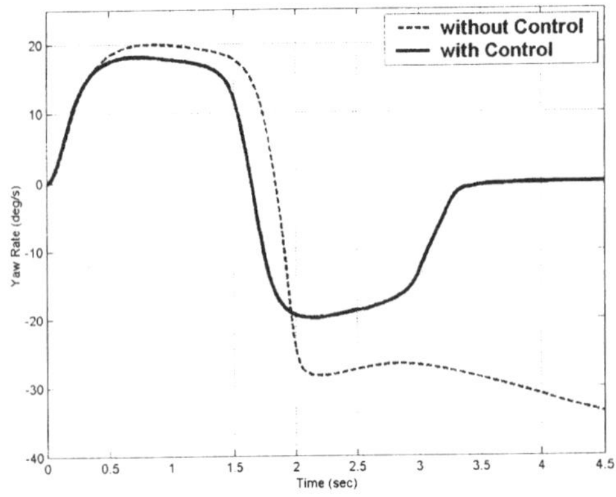

Fig.10: Yaw rate vs Time

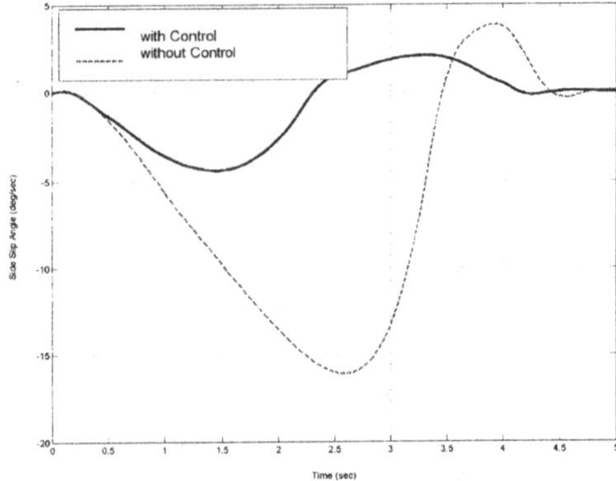

Fig.13 : Side slip vs Time (ADAMS CAR)

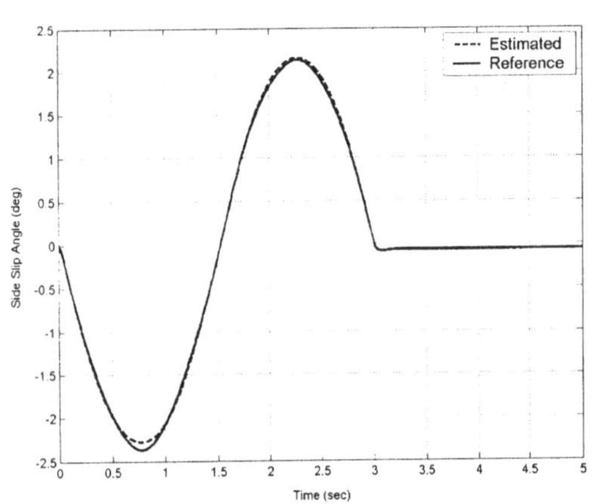

Fig.11: Side slip angle estimation in lane change

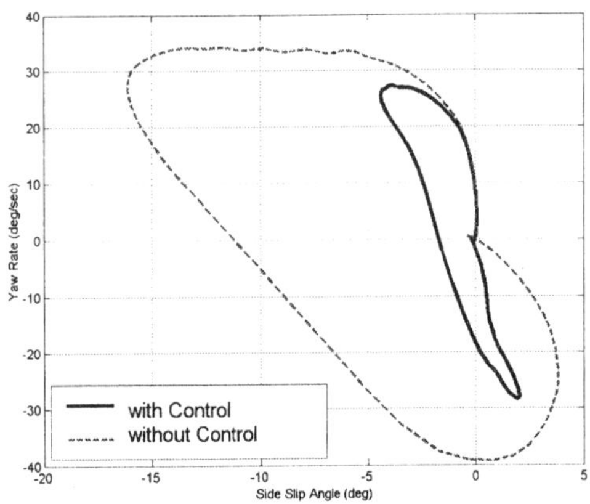

Fig 14 :Yaw rate vs side slip angle(ADAMS CAR)

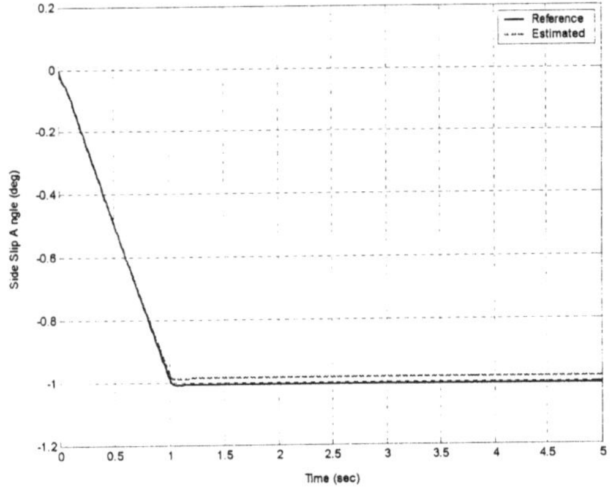

Fig.12: Side slip angle estimation in J turn

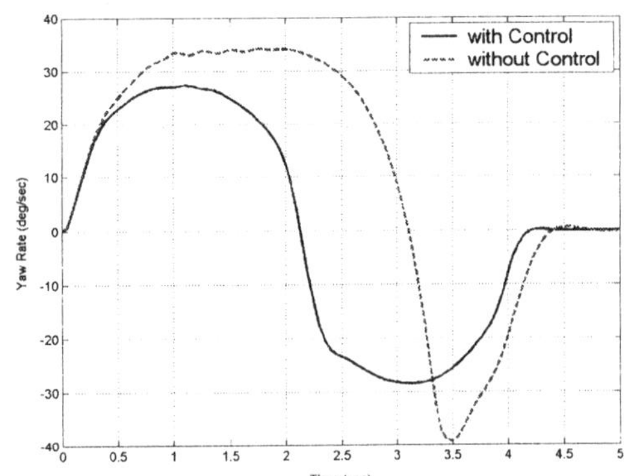

Fig.15 :Yaw rate vs Time(ADAMS CAR)

Stability Assist System for a Two– Motor–Drive Electric Vehicle using Fuzzy Logic

Farzad Tahami
Jovain Electrical Machines Co., IRAN

Shahrokh Farhanghi
University of Tehran, IRAN

Reza Kazemi
IKCO Automobiles., IRAN

ABSTRACT

This paper presents a novel method for motion control and driver stability assist system of an electric vehicle with independently driven wheels. A combination of two controllers for yaw rate and wheel slip is used to improve yaw stability of a two-wheel-drive electric vehicle. Vehicle speed is estimated using a multi-sensor data fusion method. To overcome the uncertainties in tire-road friction, a Fuzzy logic approach is employed for both controllers. The effectiveness of the proposed control method is evaluated by simulation.

INTRODUCTION

Nowadays a great deal of effort is underway to develop better electric vehicles. It seems as if the problem of battery or fuel cell will be overcome in the near future, and pure electric vehicles will be commercially accepted.

An innovative configuration for pure electric and series hybrid vehicles is a multi-wheel-drive vehicle, in which each wheel is driven individually by motor-wheels. This type of EV has the benefit of eliminating driveline components in the vehicle including the transmission, the differential, the universal joints and the drive shaft. Apart from many advantages of this configuration, unequaled independent wheel control allows vehicle dynamic control to assist the driver with path correction. That results in enhancing cornering and straight-line stability and providing enhanced drive safety. In fact, the yaw rate of a car is influenced by disturbance torques resulting from crosswind, breaking and acceleration on a μ-split road, and so on. A conventional front-wheel steering system cannot guarantee the vehicle stability on slippery roads [1]. An electric vehicle with individually driven wheels at the left and the right side has the advantage of another

steering control input, i.e. torque steering. Stability improvement, using torque steering is usually addressed as Direct Yaw-moment Control (DYC). It has been proved that DYC provides good yaw control ability even compared to four-wheel steering [2]. In fact the yaw moment resulting from differential longitudinal tire force in the left and the right wheels, is insignificantly affected by lateral acceleration [3]. On the contrary, the yaw moment generated by four-wheel steering (4WS) decreases as the lateral acceleration is increased [3].

A lot of papers dealing with Vehicle Stability Assist (VSA) systems and automatic steering assume a single-track model for the vehicle [4-16]. With this simple model, vehicle behavior on a μ-split road cannot be evaluated. Some papers perform even more simplification by assuming linear characteristics for the tire-road friction, either in longitudinal or lateral directions or both [5, 6, 11, 14, 16, and 17]. Hence, the tire saturation is ignored.

In this paper, a more accurate vehicle dynamic model, as well as a tire dynamic model, is considered. To overcome the uncertainties in tire-road characteristics and vehicle parameters, a fuzzy logic approach is employed. A fuzzy logic controller is used to control the yaw rate of an electric vehicle equipped with two individual electric motors in the rear and a conventional front wheel steering. The disturbances in yaw moment are counter-balanced by a differential torque applied to the driven wheels. Another fuzzy controller is employed in order to prevent the driven wheels entering unstable region due to the additional torque applied by the yaw controller.

VEHICLE MODEL

A four-wheel rigid chassis is considered for the vehicle, where two electric motors are mounted on the rear wheels. The suspension system is neglected for the sake of simplicity.

Considering the forces acting on the vehicle body as depicted in Figure 1, the motion equations can be written as follow:

$$M\dot{V}_x = \left(F_{xfr} - F_{Rfr}\right)\cos\delta + \left(F_{xfl} - F_{Rfl}\right)\cos\delta + F_{xrr} + F_{xrl} - F_{yfr}\sin\delta - F_{yfl}\sin\delta - F_{Rrl} - F_{Rrr} + F_c\sin\beta - C_d V_x^2 \tag{1}$$

$$M\dot{V}_y = F_{yfr}\cos\delta + F_{yfl}\cos\delta + F_{yrr} + F_{yrl} + \left(F_{xfr} - F_{Rfr}\right)\sin\delta + \left(F_{xfl} - F_{Rfl}\right)\sin\delta - F_c\cos\beta \tag{2}$$

$$J\dot{r} = l_f\left(F_{yfr}\cos\delta + F_{yfl}\cos\delta + F_{xfr}\sin\delta + F_{xfl}\sin\delta\right) - l_r\left(F_{yrr} + F_{yrl}\right) + d\left(F_{xfr}\cos\delta + F_{xrr} + F_{yfl}\sin\delta\right) - d\left(F_{xfl}\cos\delta + F_{xrl} + F_{yfr}\sin\delta\right) + M_{Zfr} + M_{Zfl} + M_{Zrr} + M_{Zrl} \tag{3}$$

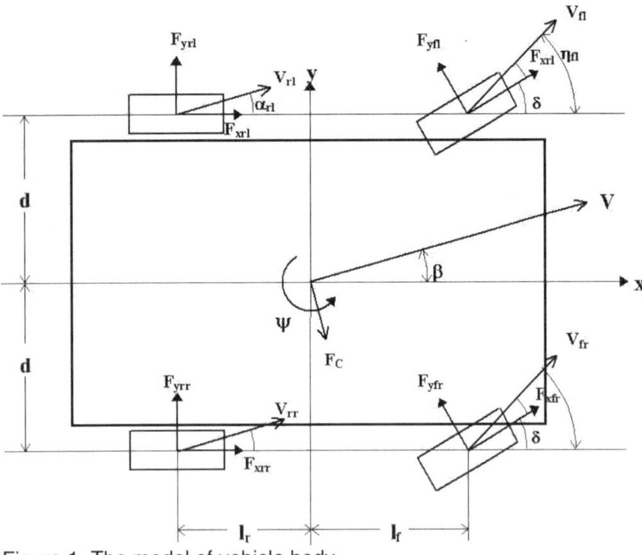

Figure 1. The model of vehicle body

Where M and J imply the vehicle mass and the moment of inertia about vertical axis, and l_f, l_r and d are dimensional parameters as indicated in Figure 1. Indices fr, fl, rr, rl indicate front-right, front-left, rear-right and rear-left wheels, respectively. C_d is the longitudinal aerodynamic drag coefficient and M_z refers to the wheel self-aligning torque. F_R is the rolling resistance of wheels.

Although the roll and pitch effects are not included in the model, their influence on the normal tire force is considered as weight-shift:

$$m_{fr} = \frac{l_r}{2(l_f + l_r)}\left(M + \frac{hF_c}{dg} - \frac{hMa_x}{l_r g}\right) \tag{4}$$

$$m_{fl} = \frac{l_r}{2(l_f + l_r)}\left(M - \frac{hF_c}{dg} - \frac{hMa_x}{l_r g}\right) \tag{5}$$

$$m_{rr} = \frac{l_f}{2(l_f + l_r)}\left(M + \frac{hF_c}{dg} + \frac{hMa_x}{l_f g}\right) \tag{6}$$

$$m_{rl} = \frac{l_f}{2(l_f + l_r)}\left(M - \frac{hF_c}{dg} + \frac{hMa_x}{l_f g}\right) \tag{7}$$

In which h is the height of the center of gravity and F_c is the centrifugal force.

TIRE MODEL

A précised tire model like the one described in [18] is too complicated for simulation purposes. Hence we use an analytical formula to express longitudinal tire friction as follows [19]:

$$\mu_x = \frac{F_x}{F_z} = \mu_{max} K_s \left\{ 1 + \left(\frac{1.582}{K_s} - 1\right) e^{(1-\frac{\lambda}{\lambda_T})} \right\}(1 - e^{-\frac{\lambda}{\lambda_T}}) \tag{8}$$

Where F_z is the vertical force and parameters μ_{max}, K_s and λ_T (the slip corresponding to maximum friction coefficient) can be obtained from experiments. Typical parameters for various slip angles are shown in Table 1.

Slip angle	λ_T	K_s	μ_{max}
0	.15	.66	1
2	.17	.754	.88
4	.19	.86	.78
8	.23	.91	.75

Table 1. Typical parameters for tire model

Another simple formula is used for lateral tire force with respect to slip ratio and slip angle:

$$\mu_y = \frac{F_y}{F_z} = A(\lambda)(1 - e^{-\alpha/\alpha_0}) \tag{9}$$

Where $A(\lambda)$ and α_0 are selected to best fit the curves of the measured values. Table 2 shows typical values for A with respect to λ.

Slip (λ)	0	.05	.09	.13	.16	.22	.30	.37	.55	.70	.90	1.0
A		.95	.90	.74	.61	.50	.38	.26	.20	.14	.125	.10	.075

Table 2. Typical values for A

Figure 2 depicts the longitudinal and lateral friction coefficients derived from the above equations, with parameters corresponding to Tables 1 and 2 and taking α_0=.08.

Note that these formulations are used only for simulation purposes. In experiment a tire model is not required, due to model free controllers that are used.

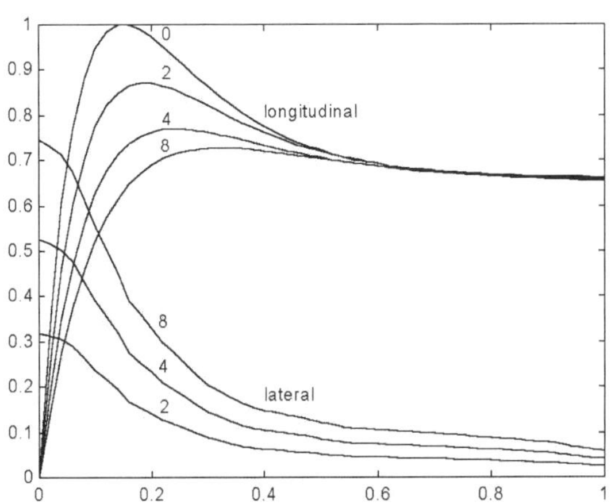

Figure 2. Typical friction coefficients obtained from equations 8 & 9

ELECTRIC MOTOR MODEL

The dynamic response of modern motor drives is much faster than vehicle dynamics. Hence an electric motor and its inverter can be simply modeled as:

$$G(s) = \frac{T}{T^*} = \frac{Ke^{-sT_t}}{(1+sT_e)} \approx \frac{K}{(1+sT_e)(1+sT_t)} \tag{10}$$

Where T_t is the delay due to inverter and T_e is the electrical time constant of electric motor.

CONTROLLER DESIGN

YAW RATE CONTROLER - An uneven longitudinal tire force may have a significant undesired effect on yaw dynamics. In this paper a differential torque is applied to the driven wheels in order to compensate the disturbance yaw moment.
Rearranging equation 3 for a rear-wheel driven EV and neglecting the longitudinal force of non-driven wheels one can write:

$$J\dot{r} = l_f (F_{yfr} + F_{yfl})\cos\delta - l_r (F_{yrr} + F_{yrl})$$
$$+ d(F_{yfl} - F_{yfr})\sin\delta + d\Delta F \tag{11}$$
$$+ M_{Zfr} + M_{Zfl} + M_{Zrr} + M_{Zrl}$$

Where ΔF is the difference in longitudinal forces of driven wheels:

$$\Delta F = F_{xrr} - F_{xrl} \tag{12}$$

Hence, the yaw rate can be directly controlled by applying a differential input torque (in tire stable region).

From the steady state cornering theory of a bicycle model we know that the yaw velocity of a vehicle satisfies the following equation [20]:

$$\frac{r}{\delta} = \frac{V/L}{1 + \frac{KV^2}{L}} \tag{13}$$

Where K is the understeer gradient and L is the vehicle wheelbase.

A neutral steer (K=0) will be obtained if the yaw rate is kept up with the following value:

$$r_d = \frac{\delta V}{L} \tag{14}$$

A fuzzy logic controller is used to keep the yaw rate in its desired value, r_d.

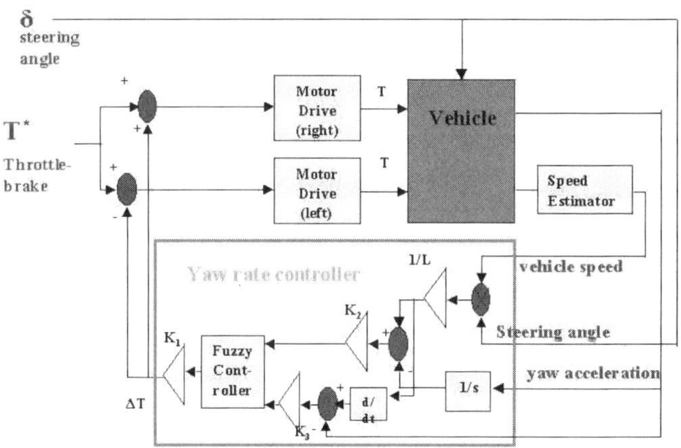

Figure 3. Yaw rate controller

The error e=r-r_d and the change in error are applied to a fuzzy controller. The output of the controller is the deviation in the applied torque to the motors. Figure 3 shows the block diagram of the vehicle model and the yaw rate controller.

The normalized membership functions for fuzzification of the controller inputs and defuzzification of the controller output are depicted in Figure 4.

Table 3 shows the rule base of the Fuzzy controller.

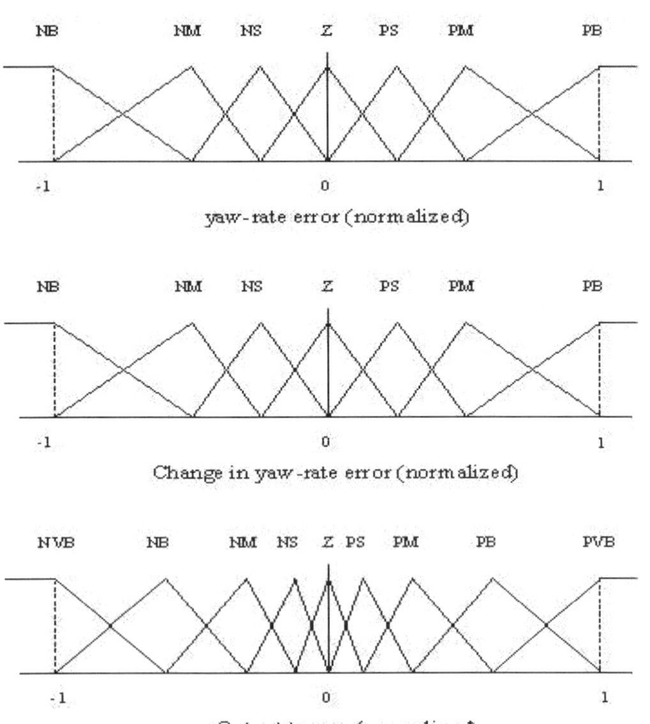

Figure 4. Membership functions for the yaw rate controller

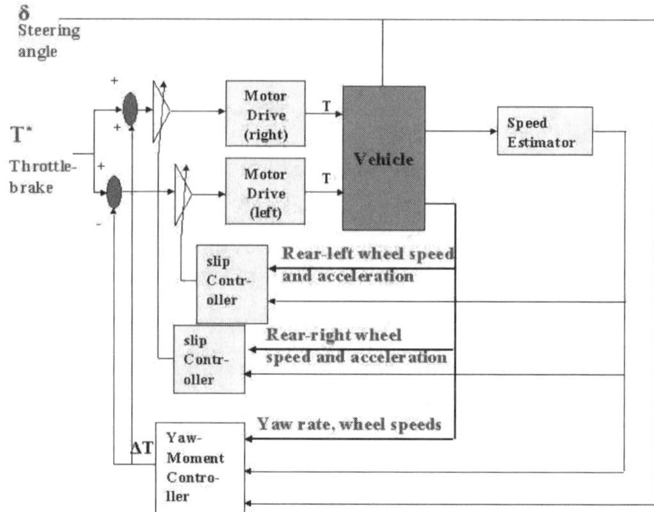

Figure 5. Overall block diagram of control system

Figure 6 shows the membership functions. The rule base of the Fuzzy slip controller is tabulated in Table 4.

Wheel acceleration times the sign of slip	Slip (unsigned)						
	Z	S	C	M	RB	B	VB
NB	U	U	U	U	U	U	NU
NM	U	U	U	U	U	NU	M
Z	U	U	U	U	NU	M	Z
PM	U	U	M	S	S	Z	Z
PB	U	NU	S	Z	Z	Z	Z

Table 4. Fuzzy rule-base for the slip controller

Change in yaw-rate error	yaw-rate error						
	NB	NM	NS	Z	PS	PM	PB
NB	NVB	NVB	NVB	NB	NM	NS	Z
NM	NVB	NVB	NB	NM	NS	Z	PS
NS	NVB	NB	NM	NS	Z	PS	PM
Z	NB	NM	NS	Z	PS	PM	PB
PS	NM	NS	Z	PS	PM	PB	PVB
PM	NS	Z	PS	PM	PB	PVB	PVB
PB	Z	PS	PM	PB	PVB	PVB	PVB

Table 3. Fuzzy rule-base for the yaw rate controller

SLIP CONTROLER - The additional torque applied by the yaw controller may cause the tire to enter into its unstable region, hence a slip ratio controller is included in the yaw control loop. A Fuzzy controller for each driven wheel is used to keep the slip ratio within its stable region. The inputs to the controller are the wheel slip and the wheel angular acceleration. The Fuzzy rules regard the latter input as a virtual criterion for the direction of slip variations. The output of the controller is the amount of torque weakening that should be devoted to the torque command of the motors, in order to prevent the tire entering into its saturation region. A block diagram of the overall control system and the vehicle model is depicted in Figure 5.

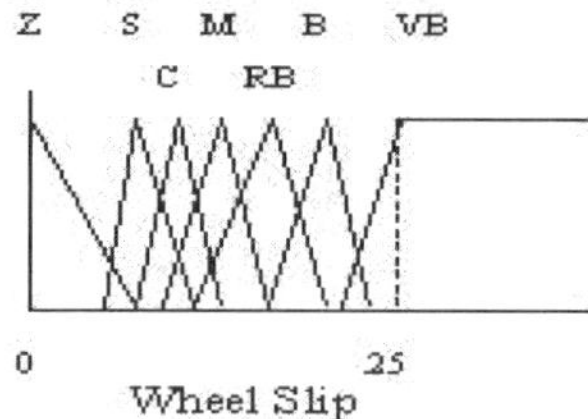

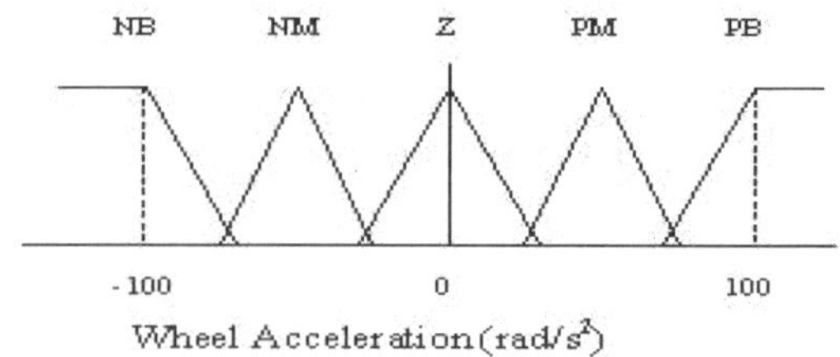

Figure 6. Membership functions for slip controller

SPEED ESTIMATOR

A crucial point in the development of wheel slip control systems is the determination of the vehicle speed. To obtain very accurate results, optical or microwave sensors can be used. However, these sensors are very expensive and can not be used for this application. In this paper the vehicle speed is estimated. A data fusion method is used for this purpose. The wheel linear speeds are used as pseudo-sources of vehicle speed. An additional accelerometer is embedded in the vehicle in order to measure the vehicle longitudinal acceleration. The integral of the acceleration is regarded as another input. The inputs are all fed into an estimator, where a fuzzy logic determines which input is more reliable. Inputs are weighted and averaged to give the estimated speed. Figure 7 shows the block diagram of the estimator. The estimator consists of two stages. In the first stage, which is addressed as preprocessing, the wheel slips are calculated using the previous estimated vehicle speed and the measured wheel slips. To avoid drifting of integrator, the offset of the accelerometer is rejected by subtracting the derivation of the estimated speed from the measured acceleration and then low-pass filtering.

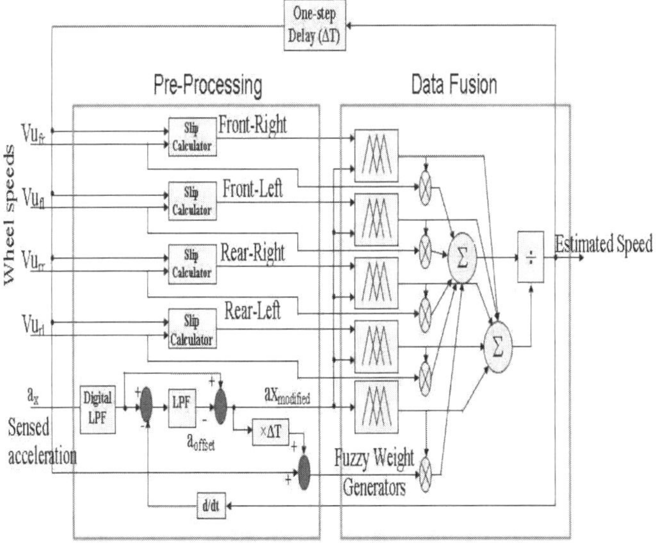

Figure 7. Block diagram of speed estimator

In the next stage, the wheel linear speed and the vehicle speed, which has been carried out by integration, are weighted using fuzzy rules. The weighted values are averaged to give the vehicle speed.

Figure 8 depicts the membership functions used in fuzzy weight generators. In Table 5 the rule sets, which are used for the weight generators, have been depicted. These rules have been established on basis of linguistic terms such as:

- In cruise driving, the integral of the vehicle acceleration is not reliable, since the accelerometer signal is comparable to offset and noise of the measuring circuit.
- In braking situation all wheels are weighted low, because of large wheel slips.
- In acceleration situation, the rear wheels are less weighted than the front wheels, because the vehicle is rear-wheel drive.

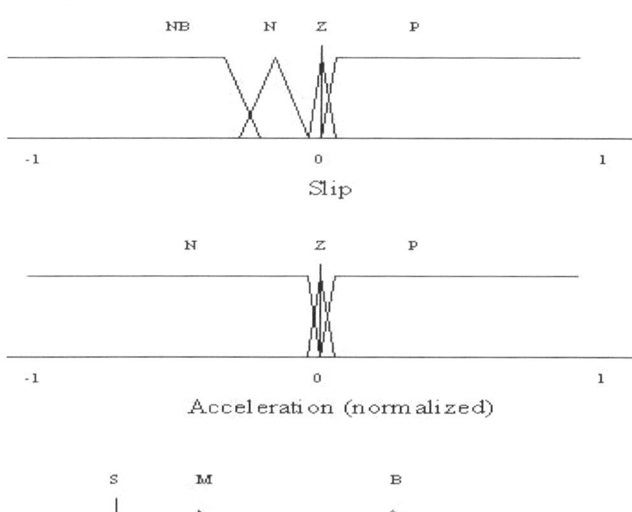

Figure 8. Membership functions for fuzzification of inputs and the output of speed estimator

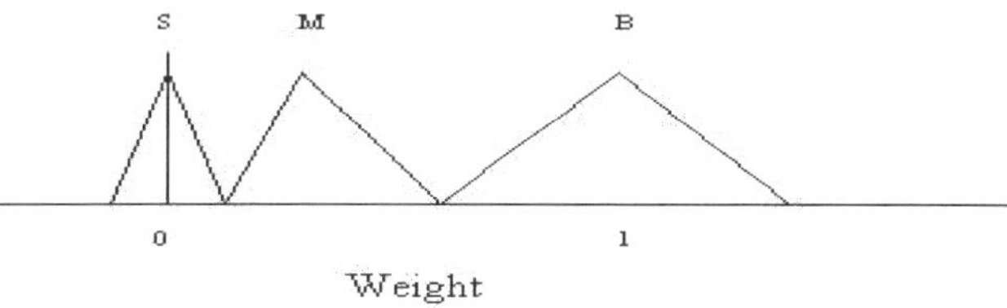

	slip			
Vehicle acceleration	NB	N	Z	P
N	S	H	M	M
Z	S	M	H	H
P	S	S	M	H

(a)

	slip			
Vehicle acceleration	NB	N	Z	P
N	S	H	M	S
Z	S	M	H	M
P	S	S	M	H

(b)

acceleration		
N	Z	P
H	S	H

(c)

Table 5. Weighting rule sets for: a) front wheels, b) rear wheels, and c) vehicle acceleration

SIMULATION RESULTS

A series of computer simulations was carried out to evaluate the performance of the proposed control system. The parameters of the vehicle model are tabulated in Table 6.

Parameter	Symbol	Unit	Value
Vehicle mass	M	Kg	1704
Moment of inertia of sprung mass about z axis	I_{zz}	Kgm^2	2619
Track width	2d	m	1.54
Distance from front axle to CG	l_f	m	1.01
Distance from rear axle to CG	l_r	m	1.68
Height of CG	h	m	.32
Effective radius of wheels	R_w	m	.305
Aerodynamic drag coefficient	C_d	$N / (m/s)^2$	.3

Table 6. Parameters of the vehicle model

LANE CHANGE ON A SLIPPERY ROAD - The result of a lane change maneuver for a 2-meter lateral displacement at 60 km/h on unpacked snow is illustrated in Figure 9. This figure compares lane change with and without controller.

As depicted in Figure 9a, with the controller on board, the desired lateral displacement achieved after 140 meters of longitudinal distance. Without the controller system, the vehicle passes the target lateral displacement and additional efforts are required to return the car into the path. Obviously the controller successfully helps the driver to handle the lane change.

Figure 9b shows that the controller keeps the vehicle's yaw rate thoroughly near the yaw rate reference. The control system applies additional torques to rear wheels as depicted in Figures 9c and 9d. The torques are rather differential.

BRAKING ON μ-SPLIT ROAD - Next simulation performs braking at 60 km/h on a μ-split road (dry pavement in the right side and unpacked snow in the left). A regenerative braking is assumed; hence a constant torque is applied to the driven wheels by the electric motors.

Simulation results are depicted in Figure 10. One can see that the control system successfully prevents the car to deviate from its path. The controller response is so fast that the lateral displacement is negligible when the car enters the split road. Even though the deviation is acceptable without slip controller, but as shown in Figure 10d the rear left tire is saturated and blocked even earlier than in case of no controller at all, which results in lacking of lateral stability.

Figure 10b shows that the rate of speed reduction is maximum, in the case of both controllers involved and

minimum, when slip controller is removed. Figures 10e and 10f show the applied torques to the wheels.

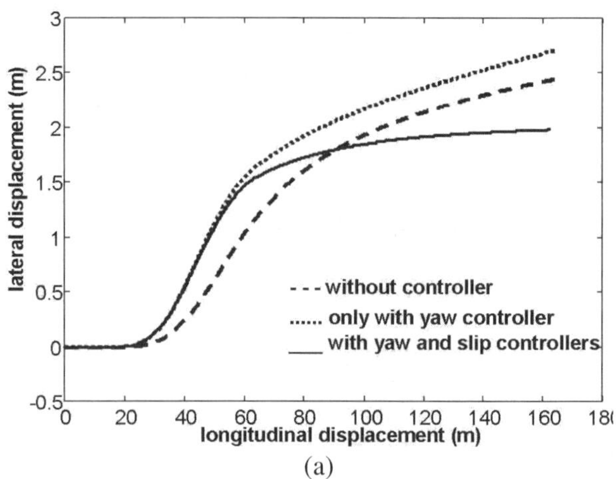

(a)

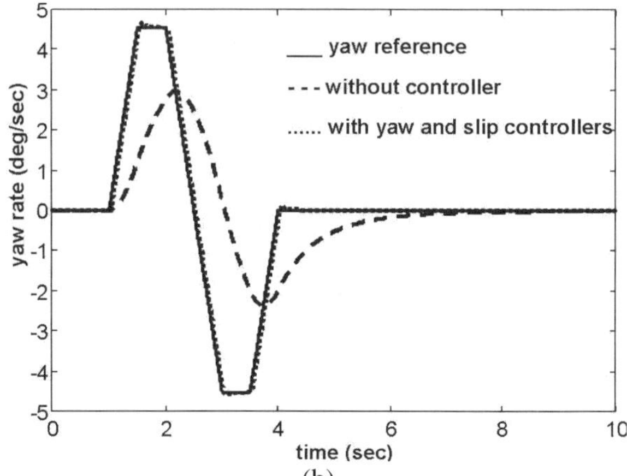

(b)

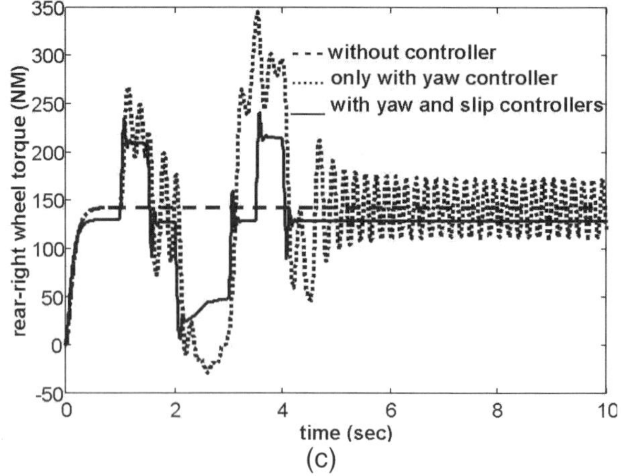

(c)

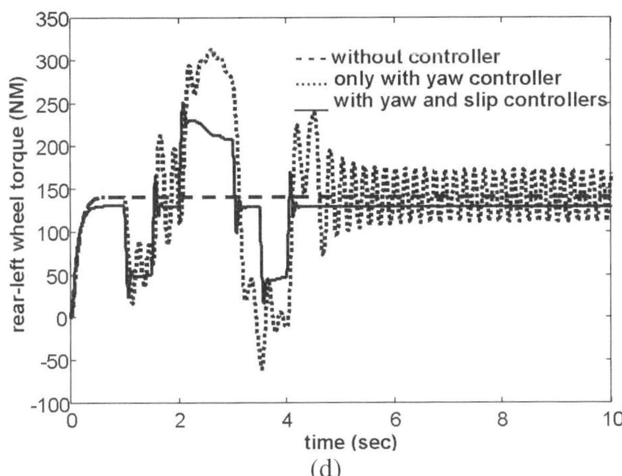

Figure 9. Lane change on a slippery road: a) Vehicle trajectory, b) Yaw rate and yaw reference, c) Applied Torque to the Rear-right wheel, d) Applied Torque to the Rear-left wheel

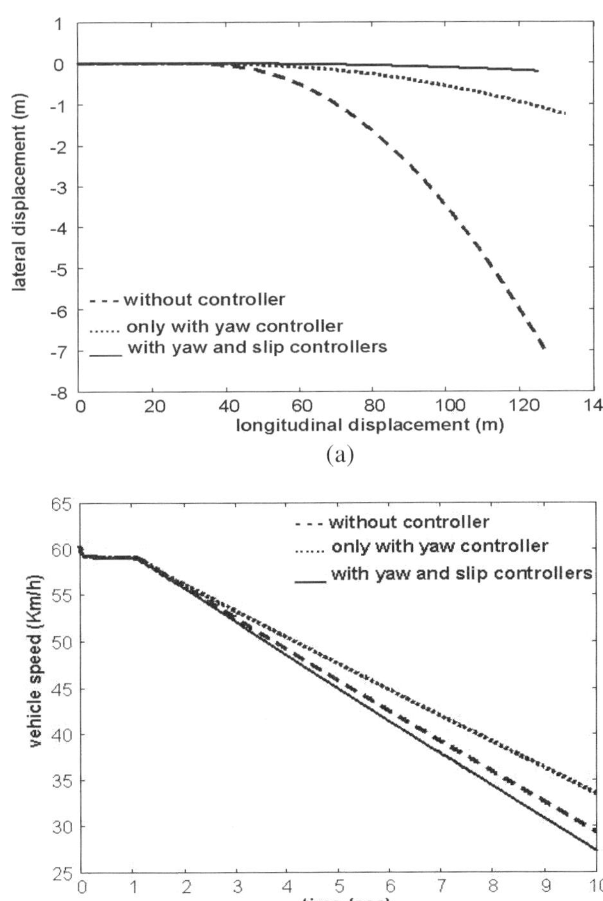

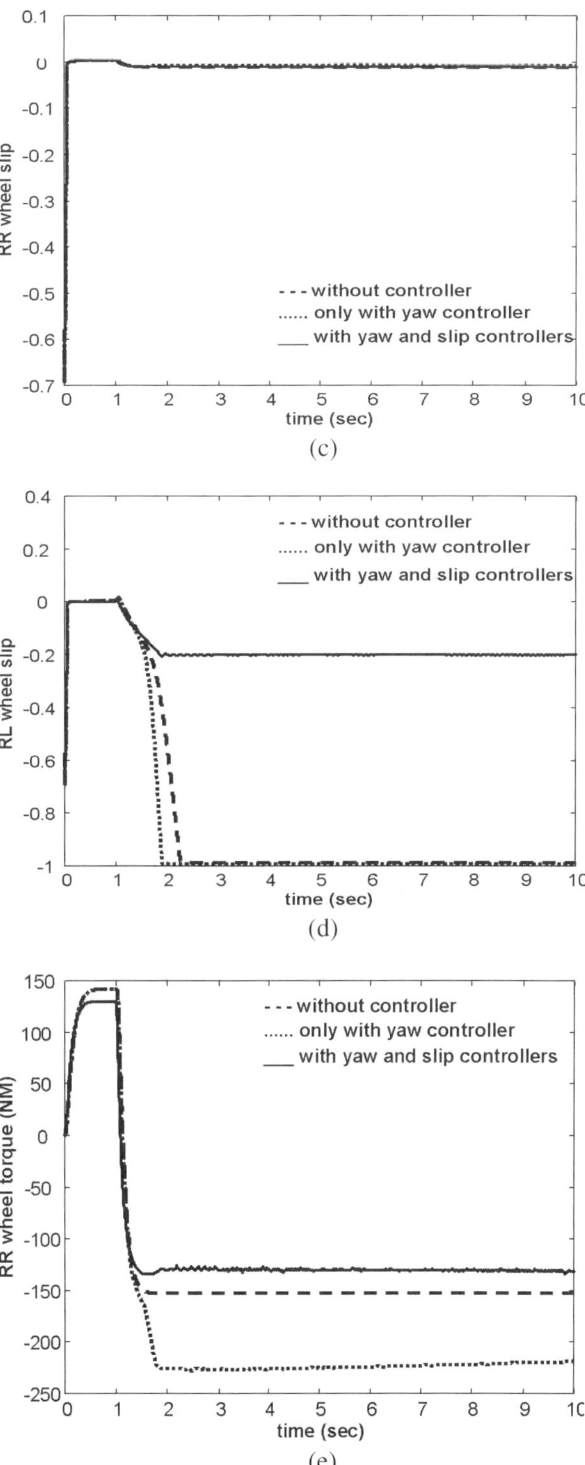

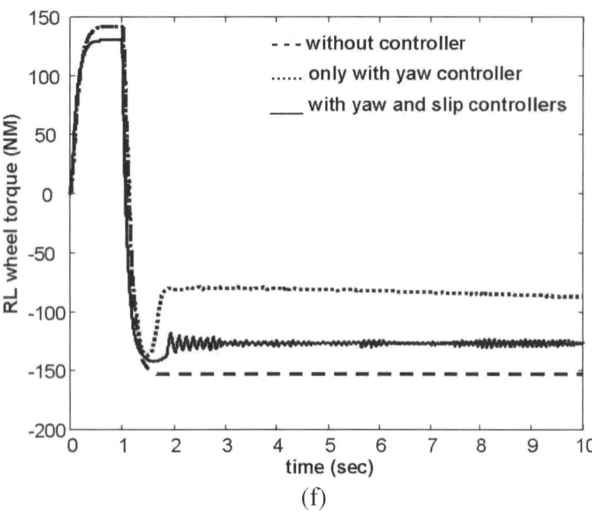

Figure 10. Braking on a µ-split road: a) Vehicle trajectory, b) Vehicle speed, c) Rear-Right wheel slip, d) Rear-Left wheel slip, e) Applied Torque to the Rear-Right wheel, f) Applied Torque to the Rear-Left wheel

CONCLUSION

A novel driver stability assist system for two-motor-drive electric vehicles was introduced. The system is based on Direct Yaw-moment Control and employs a Fuzzy yaw rate controller and individual Fuzzy slip controllers for each driven wheel, within the yaw control loop. Employing model-free Fuzzy controllers avoids the uncertain tire-road characteristics to affect the system performance. A multi-sensor fuzzy-based data fusion was introduced in order to estimate the vehicle speed. Simulation results show excellent performance of the proposed control system on slippery roads. The control strategy can be easily extended to four-wheel-drive electric vehicles.

REFERENCES

1. N. Matsumoto and M. Tomizuka, "Vehicle Lateral Velocity and Yaw Rate Control with two Independent Control Inputs," *in proc. American Control Conference*, pp.1863- 1875, 1990.
2. T. Yoshioka, et al, "Application of Sliding-mode Control to Control Vehicle Stability," *in proc AVEC'98*, 1998.
3. S. Motoyama, et al, "Quantitive Evaluation for Yaw Control Ability," *in proc. AVEC'98*, 1998.
4. J. Ackermann, J. Guldner, W. Sienel, R. Steinhauser, and V. Utkin, "Linear and nonlinear controller design for robust automatic steering," *IEEE. Trans. Control Systems Technology*, vol. 3, pp. 132-143, March 1995.
5. J. Langheim, J. Fetz, "Driving Behavior of a vehicle with two Induction Motors for the rear wheels," *in Proc. Conf. EVS-11*, pp1-11, 1994.
6. M. Depoorter, "Driver Assisted Control Strategies: Theory and Experiment," *in proc. ASME*, 1998.
7. D. Wang, et al, "A Lateral Control Scheme for a 4 wheel Steering AGU," *in proc. IEEE international conference on Intelligent Vehicles*, 1998.
8. M. Tomizuka and P. Hingwe, "Vehicle Lateral Control for Automated Highway Systems," *in proc AVEC'98*, 1998.
9. W. Sienel, "Estimation of the Tire Cornering Stiffness and its Application to Active Car Steering," *in Proc. Conference on Deceision & Control*, Dec. 1997.
10. E. Lim, K. Hedrick, "Lateral and Longituidal Control Coupling for Automated Vehicle Operation," in *Proc ACC*, June 1999.
11. M. Nagai, et al, "Integrated Robust Control of Active Rear Wheel Steering and Direct Yaw Moment Control," *in Proc. Vehicle System Dynamics Supplement 28*, pp. 416-421, 1998.
12. J. Guldner, V. Utkin, and J. Ackermann, "A Sliding Control Approach to Automatic Car Steering," in *Proc. American Control Conference*, Baltimore, June 1994.
13. H. Leffler, "Consideration of Lateral and Longituidal Vehicle Stability by Function Enhanced Brake and Stability Control System," *SAE paper 940832*, 1994.
14. U. Chong, et al., "Torque Steering Control of 4-Wheel Drive Electric Vehicle," in *Proc. Power Electronics in Transportation*, Oct. 1996.
15. S. Sakai, H. Sado, Y. Hori, "Motion Control in an Electric Vehicle with Four Independently Driven In-Wheel Motors," *IEEE/ASME Trans. on Mechatronics*, vol. 4, No. 1, March 1999.
16. J. Ackermann, "Robust car steering by yaw rate control," in *Proc. Conf. on Deceision and Control*, Honolulu, Hawaii, 1990, pp. 2033-2034.
17. S. Sakai, Y. Hori, "Robustified Model Matching Control for Motion Control of Electric Vehicle," in *Proc. 5th AMC Conf.*, 1998.
18. E. Bakker. H. B. Pacejka and L. Lidne, "A New Tire Model with an Application in Vehicle Dynamics Studies," *SAE transactions: Journal of passenger cars*, no.98, pp. 451-439, 1989.
19. Z. Fan, Y. Koren and D. Weke, "Practical Rule-Based Vehicle Traction control," *SAE paper 942402*, 1994.
20. T. Gillespie, "Fundamentals of Vehicle Dynamics," *Society of Automotive Engineers*, 1992.
21. A. Halvaei, R. Kazemi & S. Farhanghi, "Yaw Moment Control of a Two-Wheel-Drive EV, Part I: Vehicle yaw moment control using Fuzzy Logic (in Persian)," 8th Iranian Conference on Electrical and Electronics Engineering, Isfahan, Iran 2000.

A Predictive Control Algorithm for a Yaw Stability Management System

Sohel Anwar
Visteon Corporation

ABSTRACT

Generalized predictive control (GPC) is a discrete time control strategy proposed by Clark et al [1]. The controller tries to predict the future output of a system or plant and then takes control action at present time based on future output error. Such a predictive control algorithm is presented in this paper for yaw stability management of an automobile. Most of the existing literature on the yaw stability management systems lacks the insight into the yaw rate error growth when the automobile is in a understeer or oversteer condition on a low friction coefficient surface in a handling maneuver. Simulation results show that the predictive feature of the proposed controller provides an effective way to control the yaw stability of a vehicle.

INTRODUCTION

Yaw Stability Control (YSC) systems have been established in the automotive industry as a safety/performance feature. YSC generally prevents the vehicle from under-steering and over-steering in a handling maneuver (e.g. lane change, slalom, etc.), particularly on a low coefficient surface. It also helps the driver maintain yaw stability of the vehicle in a high G handling maneuver.

A predictive controller is presented in this paper. The predictive nature of the control algorithm would provide an insight into the incipient yaw instability that can be controlled with appropriate actuation system. As a result, vehicle yaw stability performance (i.e. response, etc.) would significantly improve, particularly on low coefficient surfaces. The control algorithm compares the vehicle yaw rate from a production grade yaw rate sensor with a desired value (which is computed based on vehicle speed and steering wheel angle). If the yaw rate error (the difference between the desired and measured yaw rate) exceeds a certain threshold, a controlling yaw moment is calculated based on a predictive control strategy. This yaw torque command is then translated into actuator command(s).

Sato et al [2] investigated a four wheel steering system with the use of yaw rate feedback and steering angle feedforward control. When the vehicle deflects due to a sudden side wind, road surface disturbance, or abrupt braking, steering is automatically corrected through the rear wheel to significantly improve forward stability. Shibahata et al [3] discussed a chassis control strategy for improving the limit performance of vehicle motion. They studied the effects of braking force distribution on a vehicle's lateral and longitudinal directions. It was claimed that controlling the lateral distribution of the braking force on the front wheels was effective for the improvement of the vehicle stability while that on the rear wheels was effective for extending the limit of vehicle motion.

Wang et al [4] presented a method to improve the handling and stability of vehicles by controlling yaw moment generated by driving / braking forces. Yaw moment was controlled by the feedforward compensation of steering angle and velocity to minimize the side slip angle at the vehicle center of gravity. They provided simulation results and scaled experimental results to verify their claims. Wang and Nagai [5] discussed an integrated control system providing high performance within tires' strong nonlinear areas with an adaptability to the changing road and other conditions, by optimally controlling the front and rear steering angles and the yaw moment, based on the information of system parameters. Simulation results were provided to prove the claims. Savkoor and Chou [6] investigated the application of active aerodynamic devices for suppressing parasitic motion and for improving the response of vehicles to steering, within the scope of the linear dynamic behavior. The improvements in the performance of the base-line vehicle that were achievable by the application of direct yaw and roll moments by applying either an open loop control pre-filter or a state feedback control law based on LQR design. They observed that the control strategy yielded a superior performance but demanded unreasonably large moments from the actuators in the context of available aerodynamic forces. They also observed that the demand on direct yaw and roll moment of actuators is modest when the actuators are controlled using the LQR feedback only and if the control design was used to track a desired yaw rate trajectory and simultaneously to reduce the parasitic rolling motion.

Nagai et al [7] presented an integrated control system of active rear wheel steering and yaw moment control using braking forces. Considering the tire friction circle, the control system was designed using model matching control theory to make the vehicle performance follow a desired dynamics model even during large decelerations or lateral accelerations. They provided simulation results to verify the claims made. Park and Ahn [8] described an H_∞ yaw moment control scheme using brake torque for improving vehicle performance and stability specially in high speed driving. The controller was designed to minimize the difference between the performance of the actual vehicle behavior and that of its model behavior under disturbance input. An eight DOF vehicle model was used to verify the enhancement claims on vehicle performance and stability.

Drakunov et al [9] investigated the application of sliding mode control on the yaw stability control for an automobile. The control law was based on optimum search for minimum yaw rate via sliding mode control. The developed algorithm determined the level of vehicle stability through the used of measured vehicle states and then intervened if necessary through individual wheel braking to provide added stability and handling predictability. Hac and Bodie [10] discussed a method of improving vehicle stability and emergency handling using electronically control chassis systems. They analyzed a simple nonlinear vehicle model in the yaw plane to show that the vehicles can become unstable during portions of handling maneuvers performed at or close to the limit of adhesion. They also showed that small changes in the balance of tire forces between front and rear axles may affect vehicle yaw moment and stability. They presented preliminary test results for a vehicle with integrated closed loop control of brakes and suspension, performing typical handling maneuvers.

The present work utilizes the predictive characteristics of the GPC to derive a yaw stability control algorithm. The control algorithm is based on a linearized vehicle model. This model is then discretized via a bilinear transformation. The control algorithm is then simulated using SIMULINK/TRUCKSIM model. Simulation results show that the predictive controller is effective in minimizing the understeer and oversteer conditions.

LINEARIZED VEHICLE MODEL

Yaw Stability Control (YSC) System has been around on the high-end cars for a number of years. The effectiveness of these systems varies widely depending on the system design, road conditions and driver's response. Most of these systems are based on empirical

data and heavily dependent on testing. In the present investigation, a more systematic approach is taken to develop an YSC system based on a linearized vehicle model and a predictive control algorithm.

Figure 1 illustrates the forces acting on the tire contact patches for a vehicle during a handling maneuver. The yaw dynamics for the vehicle in such a maneuver can be described with following equation [12]:

$$I_{zz}\frac{dr}{dt} = a(F_{yFL}\cos\delta_1 - F_{xFL}\sin\delta_1 + F_{yFR}\cos\delta_2 -$$

$$F_{xFR}\sin\delta_2) + b(F_{yRR} + F_{yRL}) + c(F_{yFL}\sin\delta_1 +$$

$$F_{xFL}\cos\delta_1 + F_{xRL}) - d(F_{yFR}\sin\delta_2 + F_{xFR}\cos\delta_2$$

$$+ F_{xRR}) + M_z$$

$$(1)$$

where
I_{zz} = Vehicle yaw inertia
M_z = Control yaw moment
F_{xFL} , F_{yFL} , F_{xFR} , F_{yFR} , F_{xRL} , F_{yRR}, F_{yRR}, = Tire contact patch forces in x- and y-directions as illustrated in Figure 1.
δ_1, δ_2 = Road wheel angle for the front wheels
a, b, c, d = Contact patch locations from the vehicle C.G.

Now the following assumptions are made to simplify equation (1):

- Road wheel angle for the front left tire is equal to the road wheel angle for the front right tire
- The force in x-direction is very small in a non-braking situation

With the above assumptions, equation (1) can be re-written as:

$$I_{zz}\dot{r} = a(F_{yFL} + F_{yFR})\cos\delta + b(F_{yRR} + F_{yRL}) +$$

$$cF_{yFL}\sin\delta - dF_{yFR}\sin\delta + M_z$$

$$(2)$$

It is further assumed that the normal force on the left and right side of the vehicle is same, i.e. normal force on the front left contact patch is the same as that on front right contact patch, etc.. However, the friction coefficient is assumed to the different for each contact patch. Also, the lateral friction forces are assumed to linearly vary with the slip angle [12].

$$\alpha_{FL} = \alpha_{FR} = \alpha_F ; \alpha_{RL} = \alpha_{RR} = \alpha_R$$

$$F_{yFL} = c_{FL}\alpha_F ; F_{yFR} = c_{FR}\alpha_F ; F_{yRL} = c_{RL}\alpha_R ; \quad (3)$$

$$F_{yRR} = c_{RR}\alpha_R$$

where c_{FL}, c_{FR}, c_{RL}, & c_{RR} are the cornering coefficients from a two track vehicle model. α_{FL}, α_{FR}, α_{RL}, & α_{RR} are slip angles associated with each wheel. μ_{FL}, μ_{FR}, μ_{RL}, & μ_{RR} are the friction coefficients associated with each road-tire contact patch.

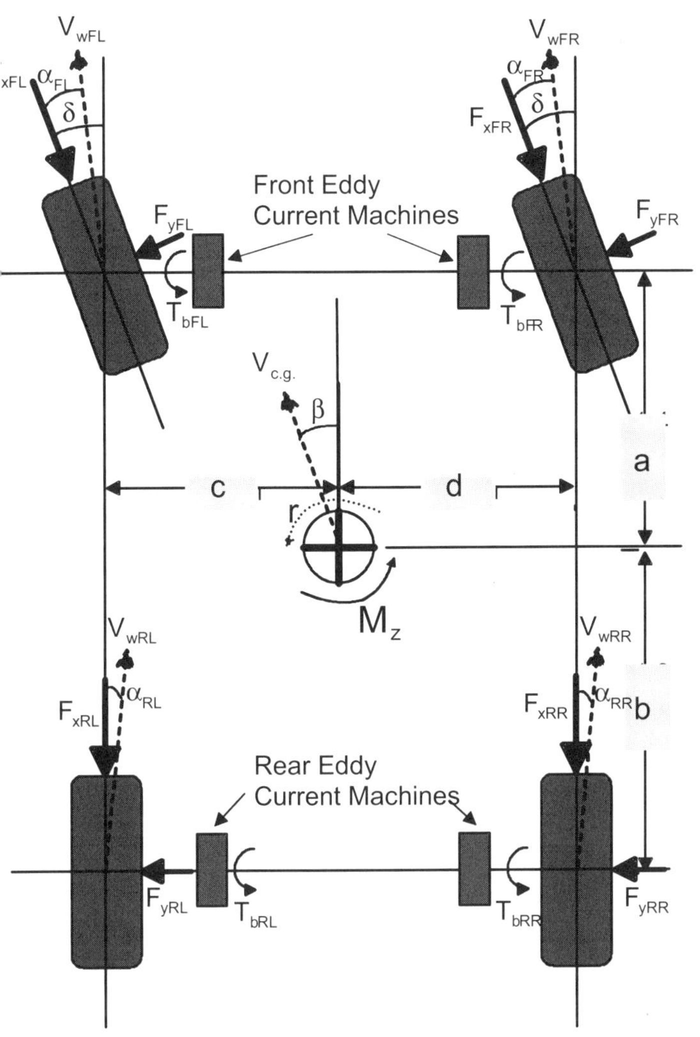

Figure 1: Schematic representing vehicle yaw

With above simplification, the following yaw dynamics equation is obtained:

$$\dot{r} = \frac{1}{I_{zz}}\left[\begin{array}{l} a(c_{FL} + c_{FR})\alpha_F \cos\delta + b(c_{RL} + c_{RR})\alpha_R + \\ (c*c_{FL} - d*c_{FR})\alpha_F \sin\delta + M_z \end{array} \right] \quad (4)$$

Now, the slip angles can be related to the body side slip angle, road wheel angle, and the yaw angle by the following relationship (figure 1).

$$\alpha_F = (\delta - \beta - \frac{r}{V_{cg}}a)$$

$$\alpha_R = (-\beta + \frac{r}{V_{cg}}b) \quad (5)$$

Substituting the above relationship (4) in equation (5), the following equation is obtained:

$$\dot{r} = \frac{1}{I_{zz}}\left[\begin{array}{l} a(c_{FL} + c_{FR})(\delta - \beta - \frac{r}{V_{cg}}a)\cos\delta + \\[6pt] b(c_{RL} + c_{RR})(-\beta + \frac{r}{V_{cg}}b) + \\[6pt] (c*c_{FL} - d*c_{FR})(\delta - \beta - \frac{r}{V_{cg}}a)\sin\delta + M_z \end{array} \right] \quad (6)$$

Assuming the $c_{FL} = c_{FR} = c_F$ and $c_{RL} = c_{RR} = c_R$, the following equation is obtained:

$$\dot{r} = \frac{1}{I_{zz}}\left[\begin{array}{l} \{2ac_F \cos\delta + (c-d)c_F \sin\delta\}\delta - \{ \\[6pt] \{2ac_F \cos\delta + (c-d)c_F \sin\delta\}\frac{a}{V_{cg}} - \frac{2b^2 c_R}{V_{cg}}\}r - \\[6pt] \{2ac_F \cos\delta + (c-d)c_F \sin\delta + 2bc_R\}\beta + M_z \end{array} \right] \quad (7)$$

Now, the side slip and state equation is obtained as follows [12].

$$V_{cg}(\dot{\beta} + r) = \frac{1}{m_{cg}}\left[\begin{array}{l} (F_{xFL} + F_{xFR})\sin\delta - (F_{yFL} \\ + F_{yFR})\cos\delta - (F_{yRL} + F_{yRR}) \end{array} \right]\cos\beta +$$

$$\frac{1}{m_{cg}}\left[\begin{array}{l} (F_{xFL} + F_{xFR})\cos\delta + (F_{yFL} + F_{yFR})\sin\delta + \\ (F_{xRL} + F_{xRR}) \end{array} \right]\sin\beta \quad (8)$$

Again, assuming that the forces in x-direction in a non-braking situation are very small, we obtain:

$$\dot{\beta} = -\frac{1}{V_{cg}m_{cg}}\left[\begin{array}{l} (F_{yFL} + F_{yFR})\cos\delta + \\ (F_{yRL} + F_{yRR}) \end{array} \right]\cos\beta$$

$$+ \frac{1}{V_{cg}m_{cg}}\left[(F_{yFL} + F_{yFR})\sin\delta\right]\sin\beta - r \quad (9)$$

Further simplifying and substituting the relationship between slip angle and lateral forces, the following equation is obtained:

$$\dot{\beta} = -\frac{1}{V_{cg}m_{cg}}\left[\begin{array}{l} (C_{FL} + C_{FR})(\cos\beta\cos\delta - \sin\beta * \\ \sin\delta)\alpha_F + (C_{RL} + C_{RR})\alpha_R \cos\beta \end{array} \right] - r \quad (10)$$

Now substituting the slip angle equation,

$$\dot{\beta} = -\frac{1}{V_{cg}m_{cg}}\left[\begin{array}{c}(C_{FL}+C_{FR})(\delta-\beta-\dfrac{r}{V_{cg}}a)*\\[2mm]\cos(\beta+\delta)+\\[2mm](C_{RL}+C_{RR})(-\beta+\dfrac{r}{V_{cg}}b)\cos\beta\end{array}\right]-r \qquad (11)$$

Since the above equation is nonlinear in β, it is assumed that variation of β is very small about the operating value. With this assumption, the above equation can further be simplified as follows:

$$\dot{\beta} = -\frac{1}{V_{cg}m_{cg}}\left[\begin{array}{c}(C_{FL}+C_{FR})(\delta-\beta)\cos\delta-\\[2mm](C_{RL}+C_{RR})\beta-\\[2mm](C_{FL}+C_{FR})\dfrac{r}{V_{cg}}a\cos\delta+\\[2mm](C_{RL}+C_{RR})\dfrac{r}{V_{cg}}b\end{array}\right]-r \qquad (12)$$

Using the assumption that $C_{FL}=C_{FR}=C_F$ and $C_{RL}=C_{RR}=C_R$, the following equation is obtained:

$$\dot{\beta} = -\frac{1}{V_{cg}m_{cg}}\left[\begin{array}{c}2C_F(\delta-\beta)\cos\delta-2C_R\beta-\\[2mm]2C_F\dfrac{r}{V_{cg}}a\cos\delta+2C_R\dfrac{r}{V_{cg}}b\end{array}\right]-r \qquad (13)$$

Combining the equations for r and β, the following state equations are obtained:

$$\begin{bmatrix}\dot{\beta}\\\dot{r}\end{bmatrix}=\begin{bmatrix}-\dfrac{2C_F\cos\delta+2C_R}{V_{cg}m_{cg}} & -\dfrac{2aC_F\cos\delta-2bC_R}{V_{cg}^2m_{cg}}-1\\[4mm]-\dfrac{\{2ac_F\cos\delta+(c-d)*\\c_F\sin\delta-2bc_R\}}{I_{zz}} & -\dfrac{\{\{2ac_F\cos\delta+(c-d)*\\c_F\sin\delta\}a-2b^2c_R\}}{V_{cg}I_{zz}}\end{bmatrix}\begin{bmatrix}\beta\\r\end{bmatrix}+$$

$$\begin{bmatrix}\dfrac{2C_F\cos\delta}{V_{cg}m_{cg}}\\[4mm]\dfrac{2ac_F\cos\delta+(c-d)c_F\sin\delta}{I_{zz}}\end{bmatrix}\delta+\begin{bmatrix}0\\[2mm]\dfrac{1}{I_{zz}}\end{bmatrix}M_z \qquad (14)$$

Now for the sake of simplicity, let us linearize the above equation about $\delta=0$. The following equation of obtained:

$$\begin{bmatrix}\dot{\beta}\\\dot{r}\end{bmatrix}=\begin{bmatrix}-\dfrac{2(C_F+C_R)}{V_{cg}m_{cg}} & -\dfrac{2(aC_F-bC_R)}{V_{cg}^2m_{cg}}-1\\[4mm]-\dfrac{2(ac_F-2bc_R)}{I_{zz}} & -\dfrac{2(a^2c_F-b^2c_R)}{V_{cg}I_{zz}}\end{bmatrix}\begin{bmatrix}\beta\\r\end{bmatrix}+$$

$$\begin{bmatrix}0\\[2mm]\dfrac{1}{I_{zz}}\end{bmatrix}M_z \qquad (15)$$

Therefore the plant dynamics (vehicle yaw dynamics) can be represented by the following set of equations:

$$\begin{bmatrix}\dot{x}_1\\\dot{x}_2\end{bmatrix}=\begin{bmatrix}a_{11} & a_{12}\\a_{21} & a_{22}\end{bmatrix}\begin{bmatrix}x_1\\x_2\end{bmatrix}+\begin{bmatrix}b_1\\b_2\end{bmatrix}u \qquad (16)$$

$$y=\begin{bmatrix}c_1 & c_2\end{bmatrix}\begin{bmatrix}x_1\\x_2\end{bmatrix}$$

where

$$x_1=\beta$$

$$x_2=r$$

$$a_{11}=-\frac{2(C_F+C_R)}{V_{cg}m_{cg}}$$

$$a_{12}=-\frac{2(aC_F-bC_R)}{V_{cg}^2m_{cg}}-1$$

$$a_{21}=-\frac{2(ac_F-2bc_R)}{I_{zz}}$$

$$a_{22}=-\frac{2(a^2c_F-b^2c_R)}{V_{cg}I_{zz}}$$

$$b_1=0$$

$$b_2=\frac{1}{I_{zz}}$$

$$c_1=0$$

$$c_2=1$$

A transfer function representation of the above state-space system is given by the following equation:

$$\frac{R(s)}{M_z(s)}=\frac{\dfrac{1}{I_{zz}}(s-a_{11})}{s^2-(a_{11}+a_{22})s+(a_{11}a_{22}-a_{12}a_{21})} \qquad (17)$$

The above transfer function can be discretized in order to obtain a discrete time transfer function. A bilinear transformation is utilized for this purpose.

$$\frac{R(z)}{M_z(z)}=\frac{(n_0+n_1z^{-1}+n_2z^{-2})}{(d_0+d_1z^{-1}+d_2z^{-2})} \qquad (18)$$

Where

$$n_0 = \frac{1}{I_{zz}}(2T - a_{11}T^2)$$

$$n_1 = -\frac{2a_{11}T^2}{I_{zz}}$$

$$n_2 = -\frac{1}{I_{zz}}(2T + a_{11}T^2)$$

$$d_0 = T^2(a_{11}a_{22} - a_{12}a_{21}) - 2T(a_{11} + a_{22}) + 4$$

$$d_1 = 2T^2(a_{11}a_{22} - a_{12}a_{21}) - 8$$

$$d_2 = T^2(a_{11}a_{22} - a_{12}a_{21}) + 2T(a_{11} + a_{22}) + 4 \tag{19}$$

In the above set of equations, T represents the sample time.

PREDICTIVE CONTROL LAW

Like most of the YSC algorithm, the proposed control algorithm also requires the knowledge of desired vehicle yaw rate, given the steering angle and vehicle speed. The objective of the controller is to track the desired yaw rate by minimizing the a sum of future yaw rate errors.

$$J = \sum_{j=0}^{N}[r_{des}(t+j) - r(t+j)]^2 \tag{20}$$

where
J = Yaw rate performance index for the vehicle
N = Prediction horizon
$r_{des}(t+j)$ = Desired yaw rate at time (t+j)

$r(t+j)$ = Predicted yaw rate at time (t+j)

Generalized predictive control (GPC) utilizes Diophantine type discrete mathematical identities to obtain predicted plant output in the future. In addition to its predictive capabilities, GPC has been shown to be robust against modeling errors and external disturbances [1].

In the following section, a discrete version of the GPC (Generalized Predictive Control) is derived.

The transfer function in equation (18) can be rewritten as:

$$(d_0 + d_1 z^{-1} + d_2 z^{-2})R(z) = (n_0 + n_1 z^{-1} + n_2 z^{-2})M_z(z) \tag{21}$$

Now the Diophantine prediction equation (j-step ahead predictor) is given by,

$$E_j(z^{-1})(d_0 + d_1 z^{-1} + d_2 z^{-2})\Delta + z^{-j}F_j(z^{-1}) = 1 \tag{22}$$

where,
$E_j(z^{-1})$ = A polynomial in z^{-1} with order (j-1)
$F_j(z^{-1})$ = A polynomial in z^{-1} of degree 1.

Multiplying both sides of equation (22) by $r(t+j)$ and rearranging,

$$r(t+j) = F_j r(t) + E_j(n_0 + n_1 z^{-1} + n_2 z^{-2})\Delta M_z(t+j-1) \tag{23}$$

The objective function can now be rewritten in matrix format as,

$$J = [R_{Des} - R]^T[R_{Des} - R] \tag{24}$$

where,

$$R_{Des} = [r_{Des}(t+1) r_{Des}(t+2)........r_{Des}(t+N)]$$

$$R = [R(t+1)R(t+2)........R(t+N)]$$

where

$$R(t+1) = F_1 r(t) + G_1 \Delta M_z(t)$$

$$R(t+2) = F_2 r(t) + G_2 \Delta M_z(t+1)$$

$$. \tag{25}$$

$$.$$

$$R(t+N) = F_N r(t) + G_N \Delta M_z(t+N-1)$$

where

$$G_j(z^{-1}) = E_j(z^{-1})(n_0 + n_1 z^{-1} + n_2 z^{-2})$$

The predicted slip equations can be re-written in a matrix format as follows :

$$R = G*U + f$$

$$where$$

$$G = \begin{bmatrix} g_0 & 0 & . & . & 0 \\ g_1 & g_0 & . & . & 0 \\ . & . & . & . & . \\ . & . & . & . & . \\ g_{N-1} & g_{N-2} & . & . & g_0 \end{bmatrix}$$

$$U = [\Delta M_z(t)\Delta M_z(t+1)....\Delta M_z(t+N-1)]^T$$

$$f = [f(t+1)f(t+2)....f(t+N)]^T$$

$$f(t+1) = [G_1(z^{-1}) - g_{10}]\Delta M_z(t) + F_1 r(t)$$

$$f(t+2) = z[G_2(z^{-1}) - z^{-1}g_{21} - g_{20}]\Delta M_z(t) + F_2 r(t)$$

$$..$$

$$..$$

$$G_i(z^{-1}) = g_{i0} + g_{i1}z^{-1} + ... \tag{26}$$

The objective function can now be rewritten as follows:

$$J = [R_{Des} - f - GU]^T[R_{Des} - f - GU] \tag{27}$$

Minimization of the objective function yield the following predictive control law:

$$U = [G^T G]^{-1} G^T (R_{Des} - f) \qquad (28)$$

In the above equation, U is a vector. To obtain the control law at present time, only the first element of U is used. Therefore the control law is given by,

$$\Delta M_z(t) = \Delta M_z(t-1) + g^T (R_{Des} - f) \qquad (29)$$

where g^T is the first row of $[G^T G]^{-1} G^T$

Equation (29) is the predictive control law for the anti-lock braking system.

Now, the control moment can be generated via a number of actuation systems. In this particular invention we present an electromagnetic brake based yaw control system. The yaw moment is generated by selectively energizing these EM brakes, which are located at the four corners of the vehicle. Let's first develop the control law in terms of M_z then we will present the electromagnetic means to deliver the yaw moment. There are two situations that accompany yaw instability: a) Understeer condition and b) Oversteer condition. In an understeer condition the absolute value of the vehicle yaw rate r is always smaller than the absolute value of desired vehicle yaw rate r_{des}. In an oversteer condition, the absolute value of the vehicle yaw rate r is always larger than the absolute value of desired vehicle yaw rate r_{des}. In an understeer condition, the control moment is generated by applying braking torque on the inner wheels whereas in an oversteer condition the control yaw moment is generated by applying braking torque on the outer wheels. Now the amount of braking torque on the wheels is dictated by the control yaw torque M_z. In any of these two vehicle dynamic conditions, either both wheels or one wheel (on one side) can be braked to generate M_z. In case of braking only one wheel, however, it has been observed that braking the front wheel is more effective in an oversteer condition whereas braking the rear wheel has been found to be more effective in an understeer condition. From an optimal control point of view, it is recommended to use only one wheel to generate the control moment.

Based on the above analysis, the control yaw moment can be related to brake torques as follows (applied braking forces act only in the direction of tire longitudinal axes).

Assuming counterclockwise positive,

$$M_z = cF_{xFL} \cos\delta - aF_{xFL} \sin\delta - dF_{xFR} \cos\delta -$$

$$aF_{xFR} \sin\delta + cF_{RL} - dF_{RR} \qquad (30)$$

Understeer Condition:
Vehicle turning counterclockwise: Brake rear left wheel

$$M_z = cF_{xRL} = c\frac{T_{bRL}}{R}$$

$$T_{bRL} = \frac{R}{c} M_z \qquad (31)$$

Vehicle turning clockwise: Brake rear right wheel

$$M_z = dF_{xRR} = d\frac{T_{bRR}}{R}$$

$$T_{bRR} = \frac{R}{d} M_z \qquad (32)$$

Oversteer Condition:
Vehicle turning counterclockwise: Brake the front right wheel

$$M_z = (d\cos\delta - a\sin\delta)F_{xFR}$$

$$= (d\cos\delta - a\sin\delta)\frac{T_{bFR}}{R}$$

$$T_{bFR} = \frac{R}{(d\cos\delta - a\sin\delta)} M_z \qquad (33)$$

Vehicle turning clockwise: Brake the front left wheel

$$M_z = (c\cos\delta - a\sin\delta)F_{xFL}$$

$$= (c\cos\delta - a\sin\delta)\frac{T_{bFL}}{R}$$

$$T_{bFL} = \frac{R}{(c\cos\delta - a\sin\delta)} M_z \qquad (34)$$

Since the electromagnetic brakes torque is a function of rotor speed, in certain situations these actuators may saturate. In case of actuator saturation, one wheel actuator may not be able to deliver the requested yaw moment. In this condition, both front and rear wheel actuators can be used to generate the requested yaw moment.

First we need a torque estimation algorithm for the eddy current machines. Reference [11] describes a torque estimation algorithm for an eddy current machine given the rotor speed and excitation current as follows:

$$T_{est} = f_0(\omega) + f_1(\omega) * i + f_2(\omega) * i^2$$

$$(35)$$

where
T = retarding torque
i = retarder feedback current

$$f_i(\omega) = a_{i0} + a_{i1}\omega + a_{i2}\omega^2 \qquad (36)$$

a_{i0} , a_{i1} , a_{i2} = identified parameters
ω = rotor speed

If the estimated EM torque for the desired wheel is less than requested torque in equations (31) through (34), then the EM actuator on the other wheel on the same side of the vehicle is energized as well. The amount of required energy for the actuator will depend on the requested torque.

Understeer Condition:
Vehicle turning counterclockwise: Brake both front and rear left wheels

$$M_z = (c\cos\delta - a\sin\delta)F_{xFL} + cF_{RL}$$

$$= (c\cos\delta - a\sin\delta)\frac{T_{bFL}}{R} + c\frac{T_{bRL}}{R}$$

$$If\, T_{bRL} > T_{estRL}, then\, T_{bRL} = T_{estRL} \qquad (37)$$

$$T_{bFL} = \frac{RM_z - cT_{estRL}}{(c\cos\delta - a\sin\delta)}$$

Vehicle turning clockwise: Brake both front and rear right wheels

$$M_z = -(d\cos\delta + a\sin\delta)F_{xFR} - dF_{RR}$$

$$= -(d\cos\delta + a\sin\delta)\frac{T_{bFR}}{R} - d\frac{T_{bRR}}{R}$$

$$If\, T_{bRR} > T_{estRR}, then\, T_{bRR} = T_{estRR} \qquad (38)$$

$$T_{bFR} = -\frac{RM_z + dT_{estRR}}{(d\cos\delta + a\sin\delta)}$$

Oversteer Condition:
Vehicle turning counterclockwise: Brake both front and rear right wheels

$$M_z = -(d\cos\delta + a\sin\delta)F_{xFR} - dF_{RR}$$

$$= -(d\cos\delta + a\sin\delta)\frac{T_{bFR}}{R} - d\frac{T_{bRR}}{R}$$

$$If\, T_{bFR} > T_{estFR}, then\, T_{bFR} = T_{estFR} \qquad (39)$$

$$T_{bRR} = -\frac{RM_z + (d\cos\delta + a\sin\delta)T_{estFR}}{d}$$

Vehicle turning clockwise: Brake both front and rear left wheels

$$M_z = (c\cos\delta - a\sin\delta)F_{xFL} + cF_{RL}$$

$$= (c\cos\delta - a\sin\delta)\frac{T_{bFL}}{R} + c\frac{T_{bRL}}{R}$$

$$If\, T_{bFL} > T_{estFL}, then\, T_{bFL} = T_{estFL} \qquad (40)$$

$$T_{bRL} = \frac{RM_z - (c\cos\delta - a\sin\delta)T_{estFL}}{c}$$

Once the torque command has been calculated for each EM retarder based on the above equations, the current command to the respective eddy current machine can be generated for a given wheel speed (assuming that actuator is not saturated) can be obtained as follows [11].

$$I_{XY} = \frac{(T_{bXY} - f_0^{xy}(\omega))}{f_1^{xy}(\omega)} \qquad (41)$$

where
I_{XY} = Current command to FL, FR, RL, or RR eddy current machine
T_{bXY} = Desired torque for the FL, FR, RL, or RR eddy current machine
f_0^{xy}, f_1^{xy} = Speed dependent retarder parameters

In case of actuator saturation, however, the performance of the yaw control system will be somewhat limited.

Equations (31) through (41) represent the control law for the yaw management system proposed in this paper. It is assumed that means of estimating the tire-road friction coefficient and normal force on the each tire is available. Also, the desired yaw rate is also assumed to be known a priori either via experimental data or data from previous developments.

SIMULATION RESULTS

The above control law (31)-(41) has been implemented on full vehicle model in a simulation environment. The results for simulation runs are discussed in this section. Since these equations will provide yaw stability control functionally based on a predetermined desired yaw rate, it is necessary to have this data a priori. Also, for a smooth vehicle ride and handling it is desired to activate the yaw moment controller based on a certain threshold yaw rate error. In the following simulation, this yaw rate

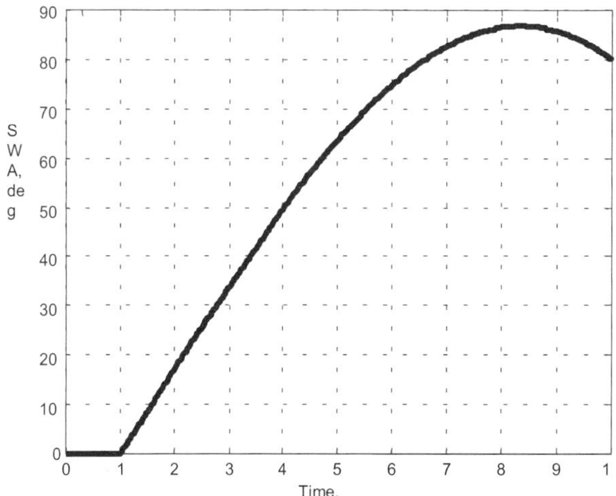

Figure 2 Steering wheel angle in a U turn maneuver on snow without yaw stability control.

error threshold is set at 5 degrees per second.

The vehicle model has 14 degrees of freedom (DOF) and this is implemented using TRUCKSIM software with a SIMULINK interface for controller implementation. A vehicle speed estimator, which is not the subject of this paper, is utilized to obtain the vehicle speed. Wheel speed is directly obtained for the vehicle model. In reality, the wheel speed will be obtained from the sensors.

Figures 2 though 5 show the braking simulation results for a U turn maneuver on snow without any yaw stability control. These figures show the baseline performance of the vehicle.

It is evident from figure 5 that the vehicle oversteers heavily before spinning out on snow. The vehicle yaw rate far exceeds the desired yaw rate for the given steering angle and vehicle speed.

Now simulation results with the predictive yaw stability controller are presented in this section. The following

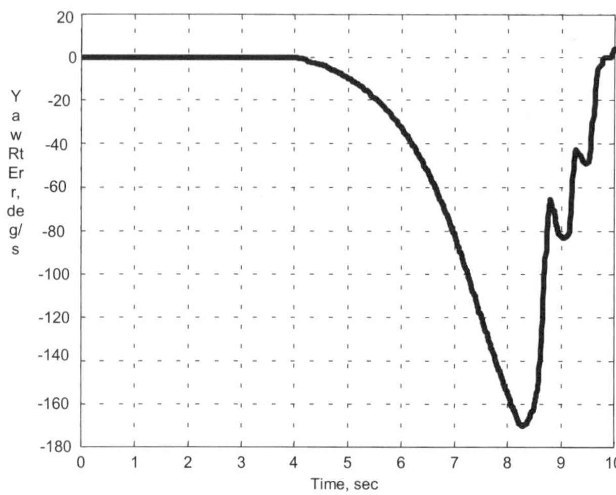

Figure 5 Yaw rate error in a U turn maneuver on snow without yaw stability control.

Control Horizon = 1
Speed threshold for YTC activation = 4 kph
Yaw rate error threshold for YTC activation = +/- 5 degrees

Figures 6 and 7 show the steering wheel angle and vehicle speed for simulations with stability control. The steering wheel input is the same at the previous case. However, the vehicle speed is a function of the controller performance. Hence, the vehicle speed profile differs from the previous case.

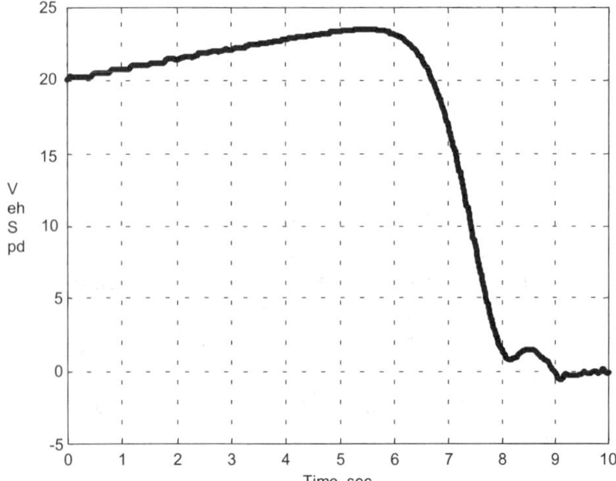

Figure 3 Vehicle speed in a U turn maneuver on snow without yaw stability control.

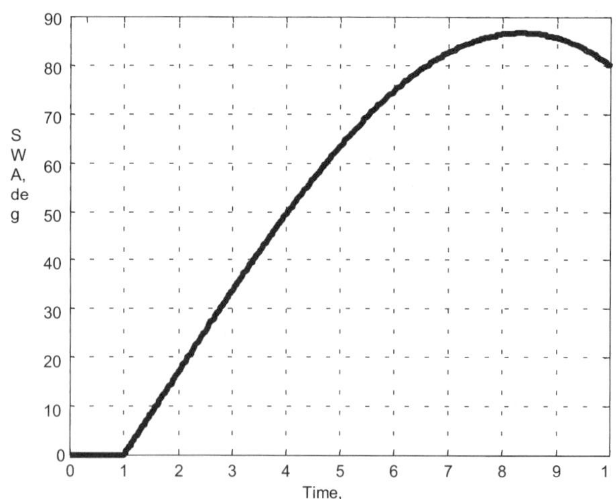

Figure 6 Steering wheel angle in a U turn maneuver on snow with yaw stability control.

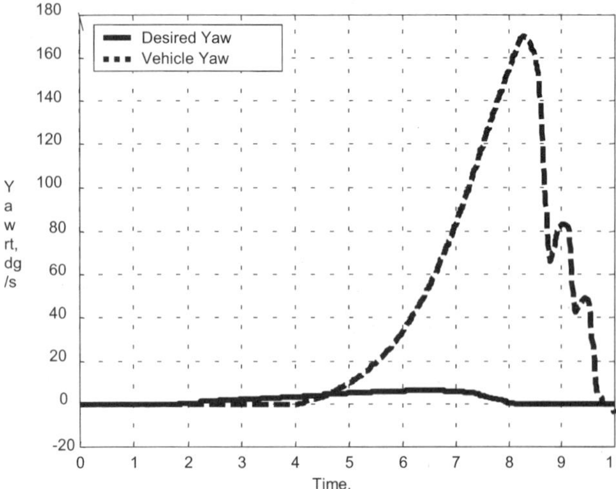

Figure 4 Desired and simulated yaw rate in a U turn maneuver on snow without yaw stability

controller parameters has been used in simulation:

Prediction Horizon = 3

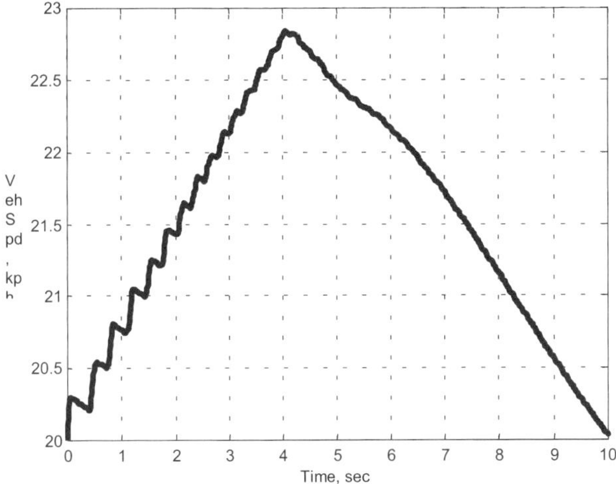

Figure 7 Vehicle speed in a U turn maneuver on snow with yaw stability control.

The normal force on each wheel is estimated from the static weight distribution and dynamic weight transfer in an acceleration/deceleration event.

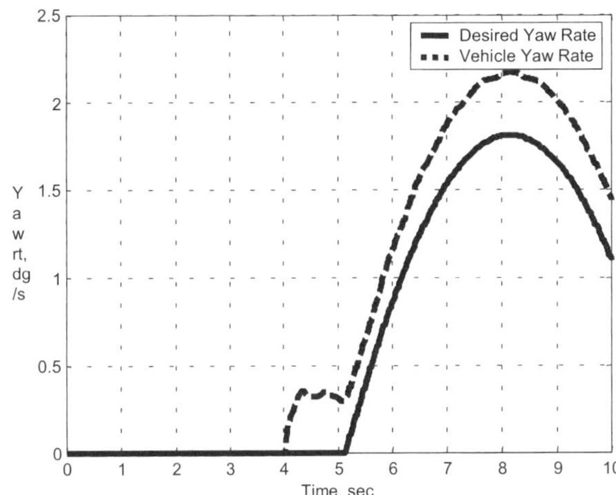

Figure 8 Desired and simulated yaw rate in a U turn maneuver on snow with yaw stability control.

Figure 8 and 9 show desired and simulated vehicle yaw rate and the yaw rate error respectively in a U turn maneuver on snow. It is clear that the yaw rate error is very small compare to the previous case. The vehicle yaw rate tracks the desired yaw rate well and as a results the vehicle remains stable throughout the maneuver. This is the result of the sub-optimal performance due to Figure 10 and 11 show the control yaw torque for the vehicle and the wheel braking torque command from the controller respectively.

CONCLUSIONS

A generalized predictive control law has been derived for a simplified linear vehicle model for a brake based yaw stability control system. The predictive nature of the controller has been utilized to predict the yaw rate error

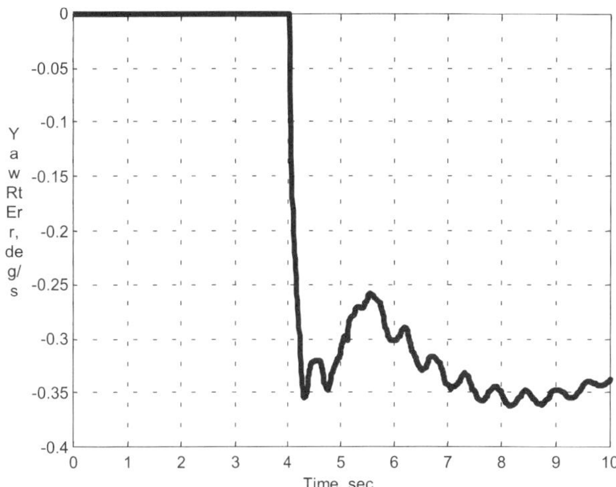

Figure 9 Yaw rate error in a U turn maneuver on snow with yaw stability control.

growth, which is then utilized to derive the control law. Simulation results show that the vehicle can be effectively stabilized in an oversteer/understeer condition on snow surface using the predictive controller. The desired yaw rate tracking performance is very good as evident from the simulation results. Further investigation of the controller is necessary to compare the performance enhancements for other yaw stability controllers with the proposed predictive controller presented in this paper.

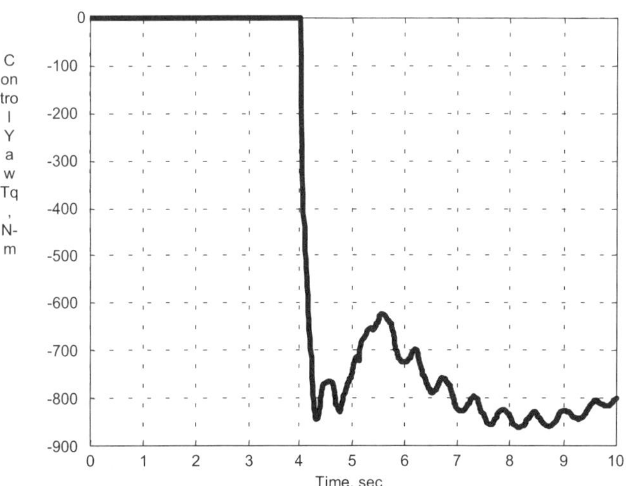

Figure 10 Control yaw torque to stabilize the vehicle in a U turn maneuver on snow with YTC

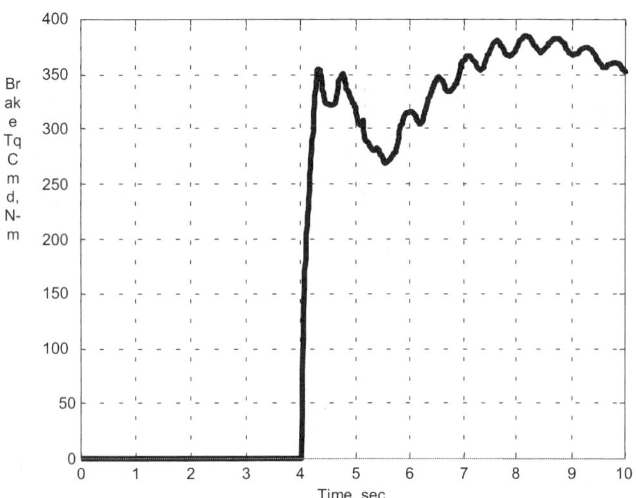

Figure 11 Front right brake command in a U turn maneuver on snow with YTC control (Oversteer).

REFERENCES

1. Clarke, D.W, Mohtadi, C., Tuffs, P.S., "Generalized Predictive Control – Part I. The Basic Algorithm", Automatica, Vol. 23, No. 2, pp. 137-148, 1987.
2. Sato, Hiroki, Kawai, Hiroyukilsikawa, and Hitisikoike, Masarulwata, "Development for four wheel steering system using yaw rte feedback control", 1991, SAE technical paper series – Passenger Car Meeting and Exposition, September 16-19, 1991, Nashville, TN, USA.
3. Shibahata, Y., Shimada, Abe, M., Shimada, K., Furukawa, Y., "Improvement on limit performance of vehicle motion by chassis control", Vehicle System Dynamics, v 23, n SUPPL, 1994.
4. Wang, Yu-qing, Morimoto, Tadashi, and Nagai, Masao, "Motion Control of front-wheel-steering vehicles by yaw moment compensation (Comparison with 4WS Performance)", Transactions of the Japan Society of Mechanical Engineers, Part C, v 60, n571, Mar, 1994, pp.912-917.
5. Wang, Yuqing and Nagai, Masao, "Integrated control of four-wheel-steer and yaw moment to improve dynamic stability margin", Proceedings of the IEEE Conference on Decision and Control, v 2, December 11-13, 1996, Kobe, Japan, pp 1783-1784.
6. Savkoor, Arvin, R. and Chou, C.T., "Application of aerodynamic actuators to improve vehicle handling", Vehicle System Dynamics, v 32, n4, 1999, pp. 345-374.
7. Nagai, Masao, Yamanaka, Sachiko, and Hirano, Yutaka, "Integrated control of active rear wheel steering and yaw moment control using braking forces", JSME International Journal, Series C, v 42, n 2, 1999, pp 301-308.
8. Park, Jong Hyeon and Ahn, Woo Sung, "H-Infinity yaw moment control with brakes for improving driving performance and stability", Proceedings of the 1999 IEEE/ASME International Conference on Advanced Intelligent Mechatronics, AIM, September 19-23, 1999, Atlanta, GA, USA, pp 747-752.
9. Drakunov, Sergey V., Ashrafi, Behrouz, and Rosiglioni, Alessandro, "Yaw Control Algorithm via Sliding Mode Control", Proceedings of the American Control Conference, v 1, Jun 28–30, 2000, Chicago, IL, USA, pp 580-583.
10. Hac, Aleksander and Bodie, Mark, O., "Improvements in vehicle handling though integrated control of chassis systems", International Journal of Vehicle Design, v 29, n 1-2, 2002, pp 23-50.
11. . S. Anwar, "An Enhanced Torque Model of an Eddy Current Electric Machine for Automotive Braking Applications", Technical Report B600-011, Vehicle Dynamics Systems Engineering Department, Visteon Corporation, April, 2002.
12. Kiencke, U. and Nielsen, L., "Automotive Control System for Engine, Driveline, and Vehicle", SAE International, 2000.

Sohel Anwar received his Ph.D. in Mechanical Engineering from University of Arizona in 1995. Dr. Anwar worked as a consultant to Caterpillar, Inc. from 1995 to 1999 where he was responsible for the development of prototype control systems for earthmoving machinery. Since 1999 he has been with the Chassis Advanced Technology Department in Visteon Corporation as a Chassis Controls Engineer where his responsibilities include control system design & development, system modeling & simulation, and software tool development. Dr. Anwar received his B.S.M.E. and M.Sc.E. from Bangladesh University of Engineering & Technology (B.U.E.T.) in 1986 and 1988 respectively. He taught at B.U.E.T. as a Lecturer in Mechanical Engineering from 1986 to 1988. He received his M.S.M.E. from Florida State University in 1990.

Direct Yaw Control of an All-Wheel-Drive EV Based on Fuzzy Logic and Neural Networks

Farzad Tahami
Jovain Electrical Machines Co.

Reza Kazemi
IKCO Automobiles

Shahrokh Farhanghi
University of Tehran

ABSTRACT

A novel driver-assist stability system for all-wheel-drive Electric Vehicles is introduced. The system helps drivers maintain control in the event of a driving emergency, including heavy braking or obstacle avoidance. The system comprises a Fuzzy logic that independently controls wheel torques to prevent vehicle spin. A neural network is trained to generate the required yaw rate reference. Another Fuzzy system for each wheel controls the slip to ensure vehicle stability and safety. Furthermore a new vehicle speed estimator is employed for slip estimation. The intrinsic robustness of fuzzy controllers allows the system to operate in different road conditions successfully. Moreover, the ease to implement fuzzy controllers gives a potential for vehicle stability enhancement.

INTRODUCTION

In-wheel-motors are revolutionary new electric drive systems that can be housed in vehicle wheel assemblies. The design eliminates traditional drive-train components, including the transmission, the differential, universal joints and driveshaft as well as a central motor. By placing the motors in the wheels, there is more space for batteries, auxiliary power unit or fuel cells, along with more cargo and passenger space. It also provides space to meet an ever increasing demand for onboard electrical systems, fuel cell equipment, and components. The electric drive system can be used in battery powered electric vehicles, both autonomous and grid connected hybrid electric vehicles and fuel cell vehicles. The system also reduces overall vehicle weight, lowers costs, and provides highly efficient and effective regenerative braking. It is ideally suited for drive-by-wire vehicle of the future and provides important new safety features through significantly increased vehicle control. Indeed, the distributed control system and the direct drive features provide independent wheel control in both acceleration and braking, results in easy integration of ABS, traction and stability control systems. Torque of each wheel can be adjusted according to road grip and the amount of power or braking needed to maintain constant traction. Independent driven wheels provides another steering control input, i.e. the torque steering. The vehicle yaw rate can be controlled by this steering method, which is usually addressed as Direct Yaw-moment Control (DYC), in order to assist the driver with path correction and to enhance straight-line stability. It has been proved that DYC is more effective in enhancing vehicle stability than four wheel steering [1]. Actually, the yaw moment resulting from difference in longitudinal tire force of left and right wheels is insignificantly influenced by lateral acceleration. On the contrary, the yaw moment generated by four-wheel steering decreases as the lateral acceleration increases [2].

In this paper, a vehicle dynamic enhancement system for all-wheel drive electric vehicles with a conventional front wheel steering is presented. This vehicle dynamic control system assists the driver with path correction, thus enhancing cornering and straight-line stability and providing enhanced safety. The system comprises of a Fuzzy logic controller in order to control the yaw rate. Another Fuzzy controller is employed in order to prevent the wheels being saturated due to the additional torque applied by the yaw controller.

The yaw controller impels the vehicle to pursue a reference yaw rate. In previous papers the authors employed a simplified vehicle model to generate a reference yaw rate [3, 4]. In this paper the yaw reference is generated by a well trained neural network. The neural net is trained on a high adhesion road condition.

CONTROL SYSTEM DESIGN

YAW RATE CONTROLLER - In an oversteer condition, the vehicle rotates more than the driver commands and this could result in the rear of the vehicle becoming "loose" or sliding outward. Understeer conditions, or "pushing," results when the front tires of the vehicle are not cornering as fast as the rear of the vehicle. Apart from these sources of instability, an uneven longitudinal tire force may have a significant undesired effect on vehicle yaw motion.

In above conditions the yaw control system intervenes either by increasing or decreasing the torque at each individual wheel. This creates a counter yaw movement that will rotate the vehicle in the opposite direction of the spin and maintaining the driver's control.

Considering the forces acting on the vehicle, as depicted in Figure 1, one can formulate the vehicle yaw motion as below:

$$
\begin{aligned}
I_z \dot{r} = {} & (I_x - I_y)pq + L_f [F_{x1}\sin(\delta_{f1}) \\
& + F_{y1}\cos(\delta_{f1}) + F_{x2}\sin(\delta_{f2}) \\
& + F_{y2}\cos(\delta_{f2})] - L_r (F_{y3} + F_{y4}) \\
& + \frac{T_f}{2}[F_{x1}\cos(\delta_{f1}) - F_{y1}\sin(\delta_{f1}) \\
& - F_{x2}\cos(\delta_{f2}) + F_{y2}\sin(\delta_{f2})] \\
& + \frac{T_r}{2}(F_{x3} - F_{x4}) + \sum_{i=1}^{4} M_{zi}
\end{aligned}
\tag{1}
$$

Where r, p and q denote the angular velocities corresponding to yaw, roll and pitch angles. I_x, I_y, and I_z are the vehicle moment of inertia about the x, y, and z axes and M_z is the tire self-aligning torque.

Rearranging equation 1 and assuming small and equal steering angles for the front wheels and no steering for the rear wheels one can write:

$$
\begin{aligned}
I_z \dot{r} = {} & (I_x - I_y)pq + L_f F_{yf} - L_r F_{yr} \\
& + \frac{T_f}{2}\Delta F_{xf} + \frac{T_r}{2}\Delta F_{xr} + \sum_{i=1}^{4} M_{zi}
\end{aligned}
\tag{2}
$$

Where ΔF_{xf} and ΔF_{xr} are the differences in the longitudinal force of left and right wheels in the front and the rear:

$$\Delta F_{xf} = F_{x1} - F_{x2}$$
$$\Delta F_{xr} = F_{x3} - F_{x4}$$

And F_{yf} and F_{yr} are the total lateral forces in the front and rear wheels.

$$F_{yf} = F_{y1} + F_{y2}$$
$$F_{yr} = F_{y3} + F_{y4}$$

Hence, applying a differential input torque to the right and left side wheels can directly control the yaw rate.

A fuzzy logic controller is used to keep the yaw rate in a desired value, r_d. The error ($e=r-r_d$) and the change in error are applied to a fuzzy controller. The output of the controller is the deviation in the applied torque to the motors. Figure 2 shows the block diagram of the vehicle stability assist system.

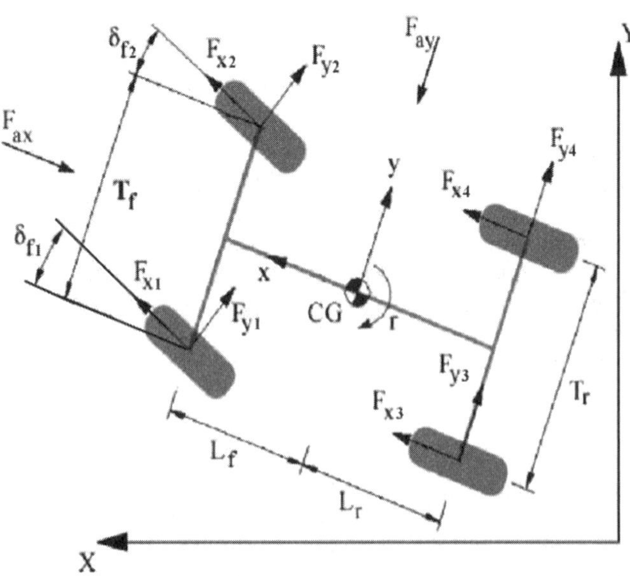

Figure 1. Forces acting on vehicle in horizontal plane

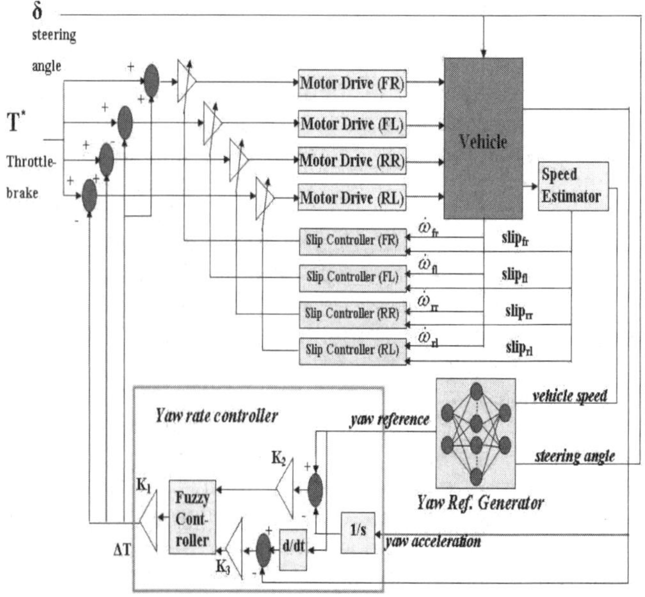

Figure 2. Block diagram of control system

YAW REFERENCE GENERATOR – From the steady state cornering theory of a single track model, it is known that the yaw velocity of a vehicle satisfies the following equation [5]:

$$r_d = \frac{V}{L + KV^2}\delta \qquad (3)$$

Where K is the under-steer gradient, L is the vehicle wheelbase, V is the vehicle speed and δ is the wheel turn angle. A desired yaw rate is assumed to be attainable with neutral steer $(K=0)$, while a correction term is included to modify the result.

$$r_d = \frac{V}{L + KV^2}\delta + r_{correction}(\delta, V) \qquad (3)$$

The correction term is produced by a feed-forward neural network, which is trained to approximate the desired yaw rate. Certain types of neural networks are universal approximators. A properly trained neural network with adequate neurons can simulate any continues function on a compact domain to any given degree of accuracy. It has to be obtained by iteratively processing of training data with the learning algorithm. Figure 3 shows the neural network used for yaw rate reference generation. The net consists of two hidden layers having 30 and 20 neurons. The neural net maps the vehicle speed and the steering angle to give the yaw rate reference. The training process is performed by a back-propagation learning algorithm on a high adhesion road under a sine waveform maneuver in different velocities. Hence, the yaw rate generator learns to demand the same yaw behavior from the vehicle on slippery roads as it is on adhesive roads.

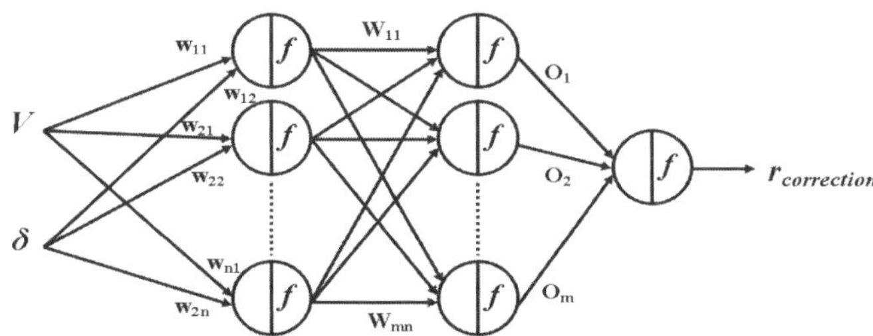

Figure 3. Yaw reference generator

SLIP CONTROLLER – The additional torque applied by the yaw controller may saturate the tire force, hence a slip ratio controller is included in the yaw control loop. A Fuzzy controller for each wheel is used to keep the slip ratio within tire stable region. The inputs to the controller are the wheel slip and the wheel angular acceleration. The Fuzzy rules regard the latter input as a virtual criterion for the direction of slip variations. The output of the controller, as shown in Figure 2, is the amount of torque weakening that should be devoted to the torque command of the motors, in order to prevent the tire to enter into its saturation region.

SPEED ESTIMATOR – All the required control signals, but the vehicle speed, can be obtained from sensors, at a reasonable price. These sensors measure steering wheel angle, wheel speeds and yaw rate. Vehicle speed however, should be estimated anyhow. In this paper a data fusion method is used for this purpose, the wheel linear speeds are used as pseudo-sources of vehicle speed. An additional accelerometer is embedded in the vehicle in order to measure the vehicle longitudinal acceleration. The integral of the acceleration is regarded as another input. The inputs are all fed into an estimator, where a Fuzzy logic determines which input is more reliable. Inputs are weighted and averaged to give the estimated speed. Figure 4 shows the block diagram of the estimator. The estimator consists of two stages. In the first stage, which is addressed as preprocessing, the wheel slips are calculated using the previous estimated vehicle speed, wheel angular velocities and yaw rate. The wheel angular velocities are transferred to the vehicle center of gravity in order to calculate the wheel slips.

$$V_{RFR} = V_{UFR} - \dot{\psi}\frac{T_f}{2}$$

$$V_{RFL} = V_{UFL} + \dot{\psi}\frac{T_f}{2}$$

$$V_{RRR} = V_{URR} - \dot{\psi}\frac{T_r}{2} \qquad (3)$$

$$V_{RRL} = V_{URL} + \dot{\psi}\frac{T_r}{2}$$

To avoid drifting of integrator, the offset of the accelerometer is rejected by subtracting the derivation of the estimated speed from the measured acceleration and then low-pass filtering.

In the next stage, the wheel linear speeds and the vehicle speed obtained by integration are weighted using Fuzzy rules. The weighted values are averaged to give the vehicle speed.

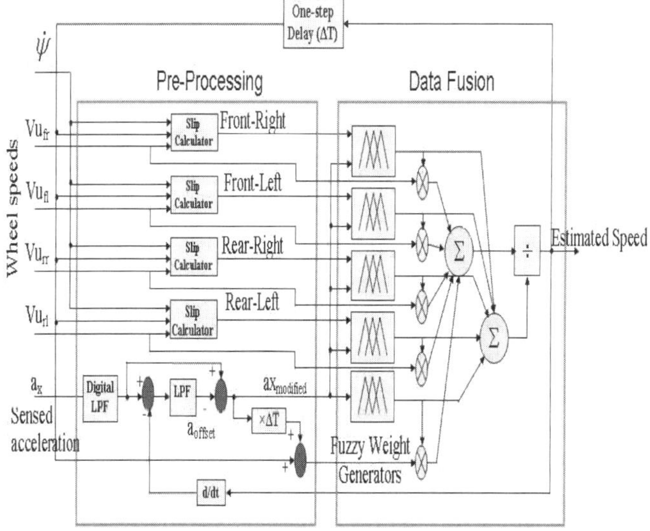

Figure 4. Block diagram of the speed estimator

Figure 5 depicts the membership functions used in Fuzzy weight generators. In Table 1 the rule sets, which are used for the weight generators, are tabulated. These rules have been established on basis of linguistic terms such as:

- In cruise driving, the integral of the vehicle acceleration is not reliable and minimally weighted, since the accelerometer signal is comparable to offset and noise of the measuring circuit. The wheels are weighted according to their slip.

- In braking situation if a wheel has a small negative slip, then it is slightly weighted, otherwise it is minimally weighted.

- In acceleration, when a wheel has a positive or zero slip it will be slightly weighted, otherwise it will be minimally weighted.

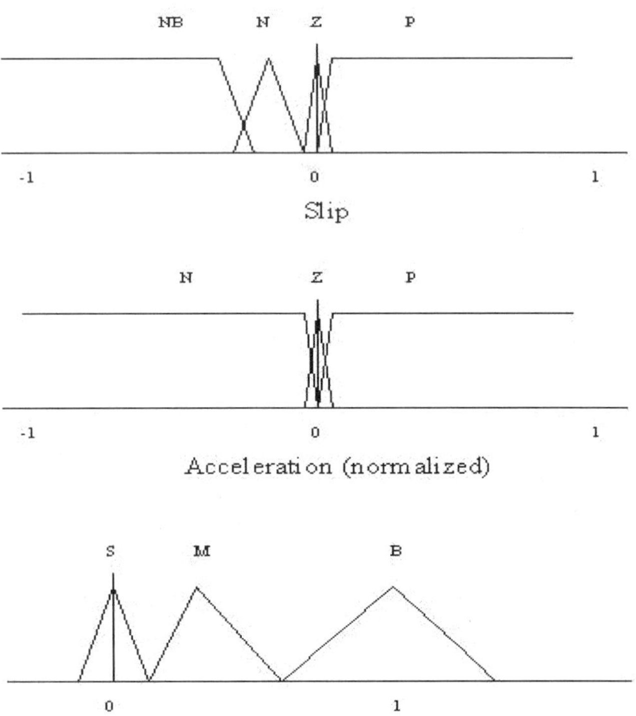

Figure 5. Membership functions for fuzzification of inputs and the output of speed estimator

slip

vehicle acceleration		NB	N	Z	P
	N	S	B	M	S
	Z	S	M	B	M
	P	S	S	M	B

(a)

acceleration

N	Z	P
B	S	B

(b)

Table 1. Weighting rule sets for: a) wheels speed, b) Integration

SIMULATION

VEHICLE MODEL – A fourteen-degree of freedom model is used for the simulation purposes, where 6 degrees are devoted to the chassis motion and eight degrees are assigned to the wheels rotational and vertical movement. Fig. 6 shows different coordinates that are used for the modeling purpose. All movements of the vehicle body are given with the centre of gravity coordinate system, while there is a coordinate system for each individual wheel. The parameters of the vehicle model are tabulated in Table 2.

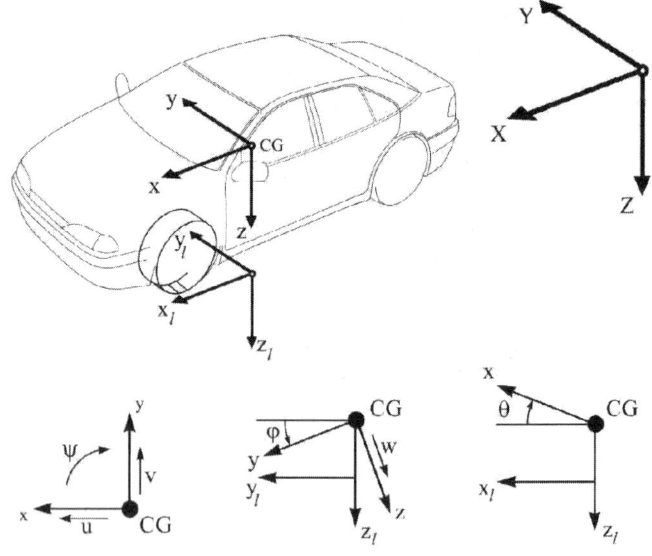

Figure 6. Coordinate systems for vehicle modeling

Parameter	Symbol	Unit	Value
Vehicle total mass	M_t	Kg	1482.7
Front Unsprung mass	M_{uf}	Kg	95.5
Rear Unsprung mass	M_{ur}	Kg	108.8
Moments of inertia of total car about X axis	I_x	Kgm^2	346.73
Moments of inertia of total car about Y axis	I_y	Kgm^2	1675.8
Moments of inertia of total car about Z axis	I_z	Kgm^2	1808.8
Moment of inertia of wheels	I_w	Kgm^2	2.11
Track width	T	m	1.4375
Distance from front axle to CG	L_f	m	1.2247
Distance from rear axle to CG	L_r	m	1.4373
Height of CG	h_{cg}	m	0.5253
Effective radius of wheels	R_w	m	0.285
Spring constant of front springs	K_{sf}	N/m	15400
Spring constant of rear springs	K_{sr}	N/m	19000
Spring constant of tires	K_u	N/m	175000
Damping coefficient of front dampers	C_{sf}	Ns/m	1150
Damping coefficient of rear dampers	C_{sr}	Ns/m	6000
Damping coefficient of tires	C_{uf}	Ns/m	50
Aerodynamic drag coefficient	C_d	$N/(m/s)^2$	.32

Table 2. Vehicle Parameters used in simulation

NEURAL NETWORK TRAINING – Prior to apply the yaw rate control, we should train the neural network. The goal is achieved by a simulation run. The vehicle accelerates from standstill up to a maximum speed then brakes down to standstill, while a sine-wave form driving is performed in order to excite different dynamics.

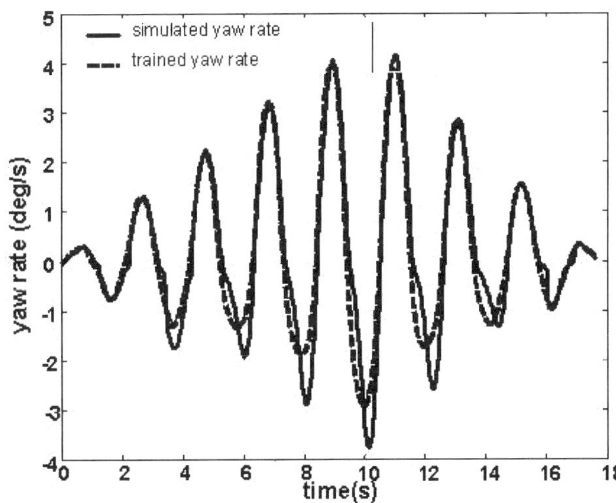

Figure 7. Results of training process

Figure 7 shows the result of training process. During the training, the yaw reference generator learns to become close to the yaw rate for different vehicle speeds and steering angles.

SIMULATION RESULTS – A series of computer simulations are carried out to evaluate the performance of the proposed control system. The results of simulation are explained in the following sections.

Speed Estimation – To evaluate the performance of the speed estimator, a braking action is simulated. The vehicle is initially moving with a speed of 60 km/h on a slippery road, then suddenly brakes. In the simulation run the rear wheels are blocked and stop turning during the vehicle deceleration and the front wheels experience considerable slip values. Figures 8a to 8c show the simulation results. The estimation is rather desirable.

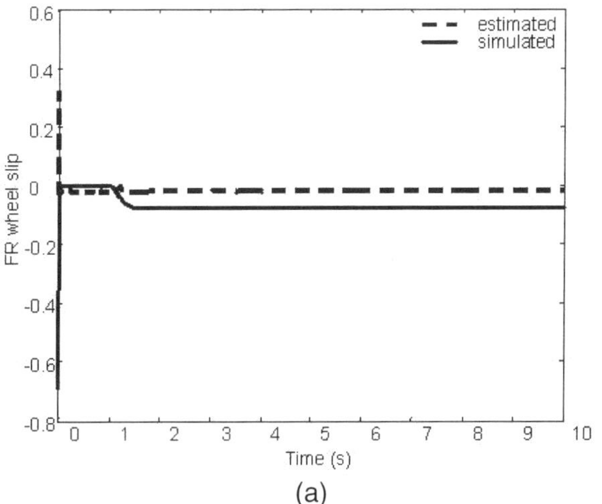

(a)

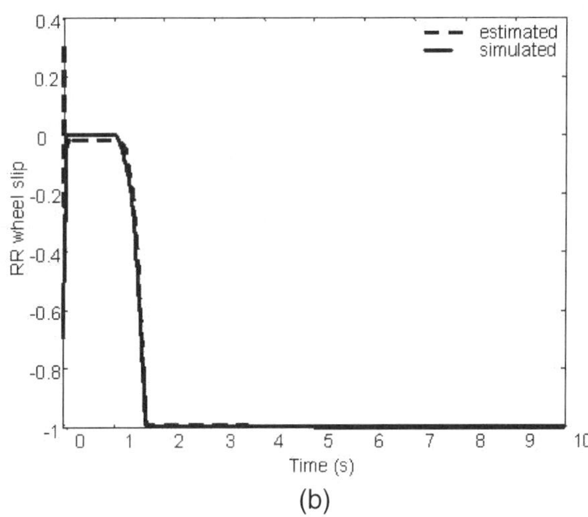

(b)

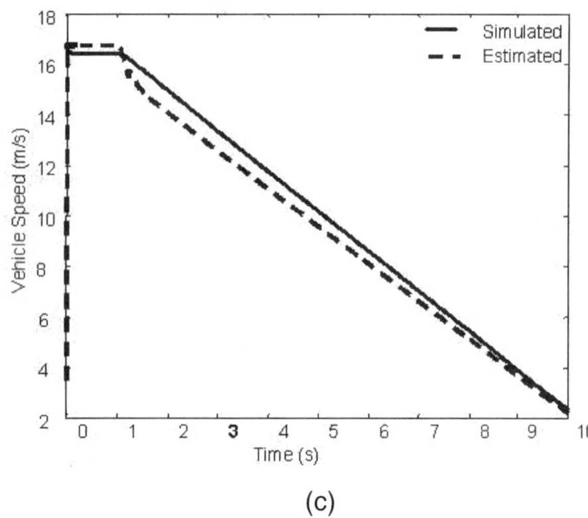

(c)

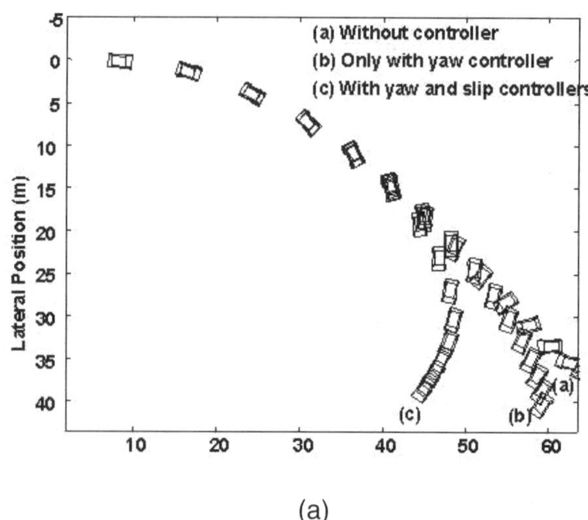

(a)

Figure 8. Estimated speed and wheel slips; a) front-right wheel slip, b) rear-right wheel slip, c) vehicle speed

<u>Braking While Turning</u> – As an example, vehicle turning while decelerating on a slippery road is examined. The vehicle initially moves with a velocity of 60 km/h. A sudden change in steering angle amounts to a wheel turn angle of 0.1 radians. At the same time the driver induces a heavy braking to the vehicle. Figure 9 shows the simulation results. One can see in Figure 9a that the control system prevents the vehicle to spin. Without the controllers the vehicle turns around. Figure 9b compares the yaw rate with the reference value. It is clearly found that the yaw rate is nearly closed to that of desired dynamic model. With only the yaw rate controller and without the slip controllers some offset remains. On the other hand, using the proposed control system, the side slip angle in the vehicle body stays definite as depicted in figure 9c. Without the controllers the vehicle spins, hence the side slip indomitably increases. The front wheel slips are shown in figures 9d and 9e. Without the slip controller, no matter whether the yaw controller is used or not, the wheels are blocked. However, with both the yaw and slip controllers, the wheel slips remain in the tire stable region, retaining steerability during turning.

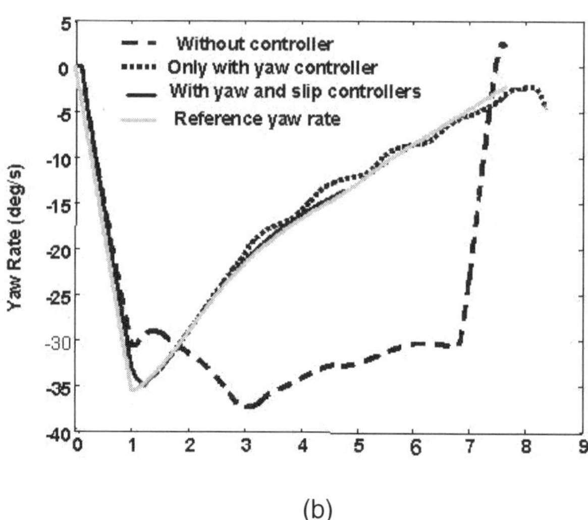

(b)

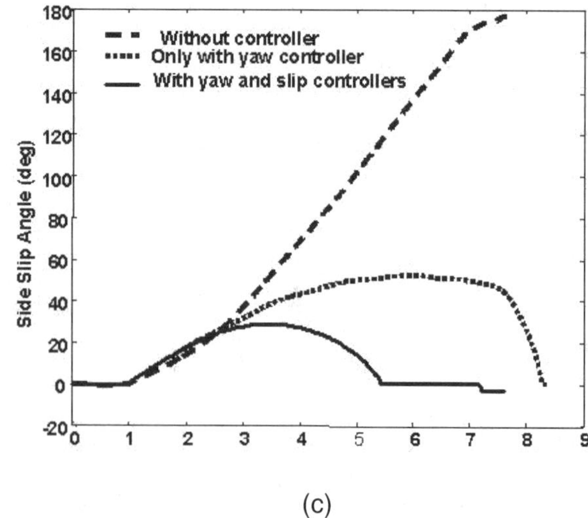

(c)

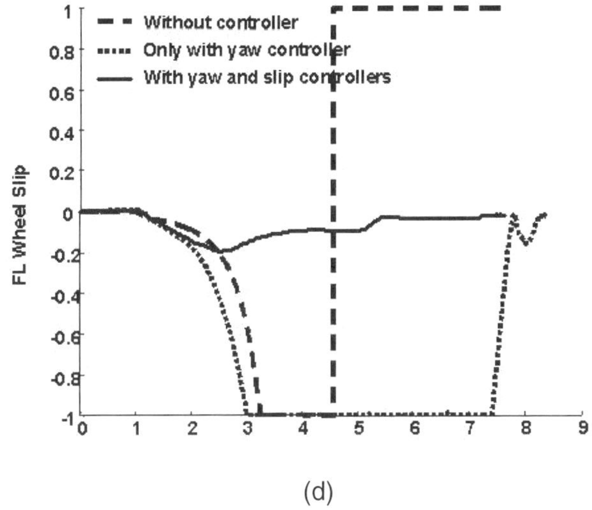

(d)

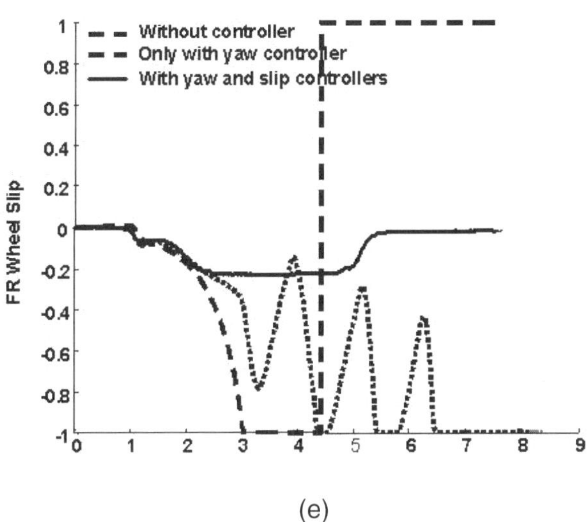

(e)

Figure 9. J-turn maneuver; a) Vehicle trajectory, b) yaw rate, c) side-slip angle, d) Front-left wheel slip, e) Front-right wheel slip

CONCLUSION

A novel driver-assist stability system for a four in-wheel drive electric vehicle was introduced. The system is based on Fuzzy logic Direct Yaw-moment Controller and individual slip controllers for each wheel using fuzzy logic. A neural network was introduced to generate the required yaw reference. The effectiveness of the proposed controller was evaluated by simulation. A 14-degree of freedom vehicle model and the well-known magic tire formula [6] was employed in the simulation program. Simulation results show excellent performance of the proposed control system on slippery roads.

REFERENCES

1. T. Yoshioka, et al, "Application of Sliding-mode Control to Control Vehicle Stability," *in proc AVEC'98*, 1998.
2. S. Motoyama, et al, "Quantitive Evaluation for Yaw Control Ability," *in proc. AVEC'98*, 1998.
3. F. Tahami, et al, "Fuzzy Based Stability Enhancement System for a Four-Motor-Wheel Electric Vehicle," SAE Dynamics & Stability Conference, Detroit, May 2002.
4. F. Tahami, S. Farhanghi and R. Kazemi, "Stability Enhancement of a Two-Motor-Drive Electric Vehicle Using Fuzzy Logic (in Persian)," 8st *Iranian Conference on Electrical Engineering*, Tabriz, Iran, May 2002.
5. T. Gillespie, "Fundamentals of Vehicle Dynamics," *Society of Automotive Engineers*, 1992.
6. E. Bakker. H. B. Pacejka and L. Lidne, "A New Tire Model with an Application in Vehicle Dynamics Studies," *SAE transactions: Journal of passenger cars*, no.98, pp. 451-439, 1989.

The Development of an Advanced Control Method for the Steer-by-Wire System to Improve the Vehicle Maneuverability and Stability

Se-Wook Oh and Seok-Chan Yun
Department of Chassis Module R & D, Hyundai Mobis

Ho-Chol Chae, Seok-Hwan Jang, Jae-Ho Jang and Chang-Soo Han
Hanyang Univ.

ABSTRACT

This paper presents the advanced control method for the SBW (Steer-by-Wire) system to improve the vehicle maneuverability and stability. In this paper, the steering wheel motor control for improvement of the driver's steering-feel is designed to change the steering-feel according to vehicle velocity and steering angle. The steering wheel motor controller creates approximately same steering-feel of the commercial power steering system and also improves driver's steering-feel as the vehicle state by using steering wheel reactive torque map. In the front wheel motor control algorithm, understeer and oversteer propensity are used as effective factors for controlling the vehicle's maneuverability and stability. As a result, vehicle yaw rate and lateral acceleration could be controlled.

INTRODUCTION

An SBW system is one part of the DBW system that the automobile industry will research in future. The new DBW system will replace mechanical and hydraulic systems for steering, braking, suspension, and throttle functions with electronic actuators, controllers, and sensors. The DBW system will improve overall vehicle safety, driving convenience and functionality significantly. An SBW system is one in which the conventional mechanical linkages between the steering wheel and the front wheel are removed and the system is operated by electronic actuators. The SBW system has many merits compared with a conventional steering system with a mechanical linkage. To begin with, the SBW system can reduce a vehicle's weight by reducing the number of necessary parts which can lead to energy reduction effectiveness. In addition, the danger of a driver being crushed when there is a front-end collision is eliminated as there is no steering column. Finally, most valuable merit is that it permits automatic steering and vehicle stability control to be free. For this reason, the study of the SBW system has proceeded. Hayama and Nishiazki (2000) proposed the SBW system using direct yaw moment control based on vehicle stability. Czerny et al. (2000) investigated fail-safe logic using 2 micro controllers to compensate for the Electronic Control Unit (ECU) of an SBW system. Segawa et al. (2001) proposed yaw rate and lateral acceleration control for improving vehicle stability. This paper discusses a steering wheel motor control algorithm to improve driver's steering feel and then base control and high performance control algorithms are proposed to improve vehicle maneuverability and stability. In here, the base control means feed forward control which does not give feedback on vehicle states of a front wheel for low cost control, namely no sensor used. Furthermore, high performance control is controlling the front wheel using accurate vehicle states by feedback vehicle states.

SBW SYSTEM

An SBW system modeling used to develop an SBW system controller and a vehicle model applied to the SBW modeling are explained in this chapter. In an SBW system, conventional mechanical linkages between a steering wheel and front wheel are eliminated and the SBW system is controlled by an ECU through electric wires. Fig. 1 shows the overall modeling of the SBW system.

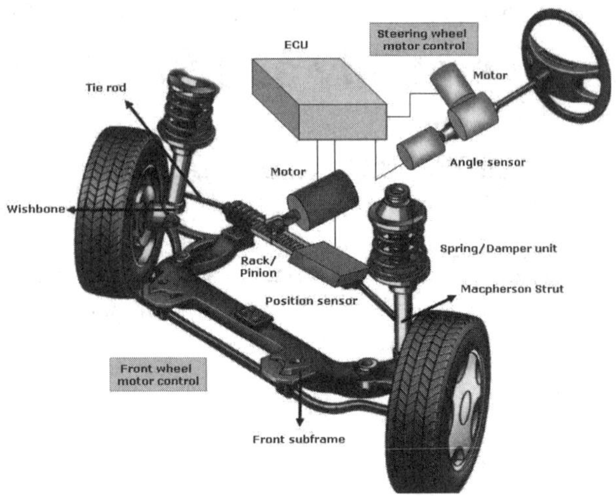

Fig. 1 Steer-by-Wire system

The SBW system is divided into steering wheel and front wheel and consists of two electronic actuators assisting the operation of two parts. These two electronic actuators receive input signals from an ECU and then one actuator generates reactive torque to the steering wheel and the other actuator steers the front wheel following the driver's will.

STEERING WHEEL MODELING

A steering angle sensor and torque sensor are located in the steering wheel. The data obtained from these two sensors are transmitted to the ECU and generate output signals to control the steering wheel's reactive torque. The steering wheel, motor generating reactive torque to the driver and steering wheel column are modeled using the bond graph method. The reason for using this method is that the steering wheel is composed of mechanical and electric systems. Therefore, the bond graph method easily expresses both the mechanical and the electric systems' energy flow together. Fig. 2 shows the bond graph modeling of the steering wheel.

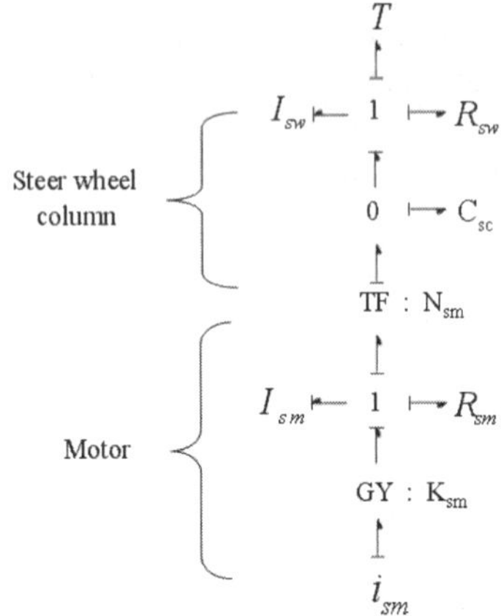

Fig. 2 Bond graph modeling of the steering wheel

The modeling element of the bond graph consists of driver's steering torque, steering reactive torque motor, and a steering column connecting the steering wheel and the motor. The stiffness of the motor shaft is ignored because it is much smaller than that of the steering column. Following Eq. (1) and (2) show the equations of the steering wheel modeling.

$$T = I_{sw}\ddot{\theta}_{sw} + R_{sw} \cdot \dot{\theta}_{sw} + C_{sc}(\theta_{sw} - N_{sm} \cdot \theta_{sm}) \tag{1}$$

$$\begin{aligned} K_{sm} \cdot i_{sm} &= I_{sm}\ddot{\theta}_{sm} \\ &+ R_{sm} \cdot \dot{\theta}_{sm} + N_{sm}\{C_{sc}(N_{sm} \cdot \theta_{sm} - \theta_{sw})\} \end{aligned} \tag{2}$$

Eq. (1) indicates the driver's torque input and Eq. (2) indicates the steering wheel motor torque input

FRONT WHEEL MODELING

The torque from the front wheel motor is transmitted to the front tires through the front wheel system consisting of a front wheel steering motor, rack & pinion gear, and a tie rod. In addition, there is a rack bar displacement sensor and the data obtained from it are transmitted to the ECU. Therefore, the desired steering wheel and front wheel angle can be calculated. Fig. 3 shows the bond graph modeling of the front wheel.

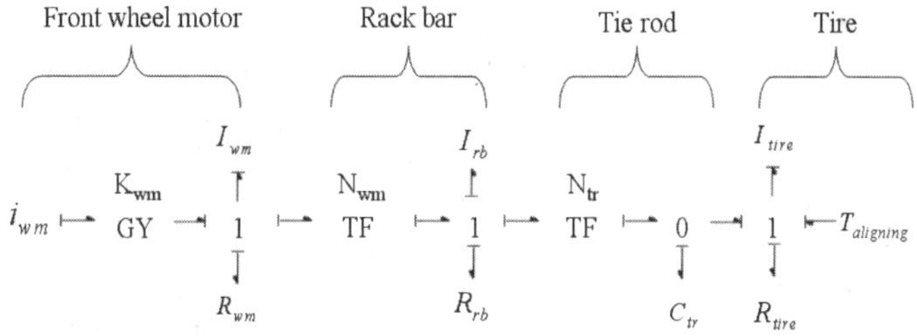

Fig. 3 Bond graph modeling of the steering wheel

In front wheel modeling, only the tie rod stiffness is considered as it has the greatest effect a vehicle state. Eq. (3) and (4) show the equations of the front wheel modeling.

$$\begin{aligned} i_{wm} \cdot K_{wm} &= \left(I_{wn} + I_{rb}N_{wm}^{2}\right)\ddot{\theta}_{wm} + \left(R_{wm} + R_{rb}N_{wm}^{2}\right)\dot{\theta}_{wm} \\ &+ N_{wm}N_{tr}C_{tr}\left(N_{wm}N_{tr} \cdot \theta_{wm} - \theta_{tire}\right) \end{aligned} \tag{3}$$

$$T_{aligning} = I_{tire}\ddot{\theta}_{tire} + R_{tire} \cdot \dot{\theta}_{tire} + C_{tr}\left(\theta_{tire} - N_{wm}N_{tr}\theta_{wm}\right) \tag{4}$$

Eq. (3) indicates the front wheel motor torque input and Eq. (4) indicates the front wheel aligning torque from the front tire.

SBW SYSTEM CONTROL

The SBW system controller is divided into the steering wheel motor control and the front wheel motor control. The purpose of the steering wheel motor control is to improve driver's steering feel by generating reactive torque. The purpose of the front wheel motor control is to steer the front wheel angle appropriately for improving the vehicle's maneuverability and stability. Fig. 4 shows the base control algorithm discussed in this chapter. In the base control, the least number of sensors are used and the front wheel is controlled using feed forward control. In addition, advanced control method using many sensors for accurate vehicle state feedback is proposed in this chapter.

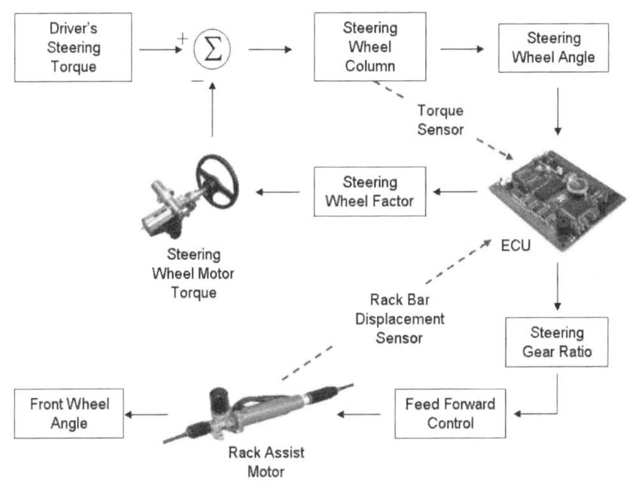

Fig. 4 Base control algorithm

STEERING WHEEL MOTOR CONTROL

The basic purpose of the steering wheel motor control is to generate reactive torque like a real commercial vehicle when the driver steers. Furthermore, it makes the steering wheel easy to steer at low speeds or when parking the vehicle and to make steering wheel tight at high speeds for improving the driver's steering feel by adjusting reactive torque. To determine the steering feel, vehicle speed and steering wheel angle are used for steering wheel motor torque. According to the steering wheel angle, steering reactive torque is determined as a function and it is multiplied by steering wheel angle parameter K_α. Therefore, steering reactive torque is changed according to K_α as shown in Fig. 5

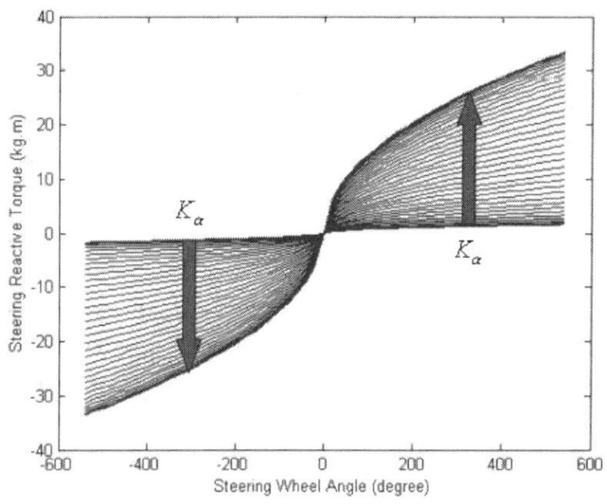

Fig. 5 Steering reactive torque according to steering wheel angle when changing K_α

As same with Fig. 5, according to the vehicle velocity, steering reactive torque is also determined as a function and it is multiplied by vehicle velocity parameter K_β. Therefore, steering reactive torque is also changed according to K_β as shown in Fig. 6

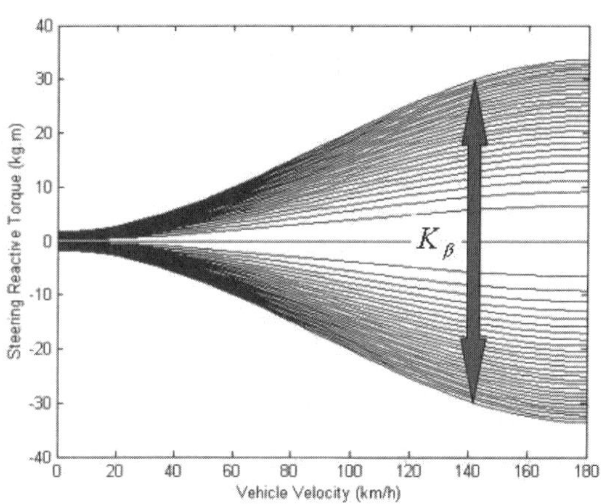

Fig. 6 Steering reactive torque according to vehicle velocity when changing K_β

As shown in Fig. 6, steering reactive torque is very small at low speeds to make driver easy to steer when parking the vehicle. In addition, steering reactive torque is converged at high speeds to prevent excessive torque. In conclusion, steering reactive torque can be determined by changing only two control parameters, such as K_α and K_β. By adjusting those two control parameters, overall steering reactive torque map is designed as shown in Fig. 7

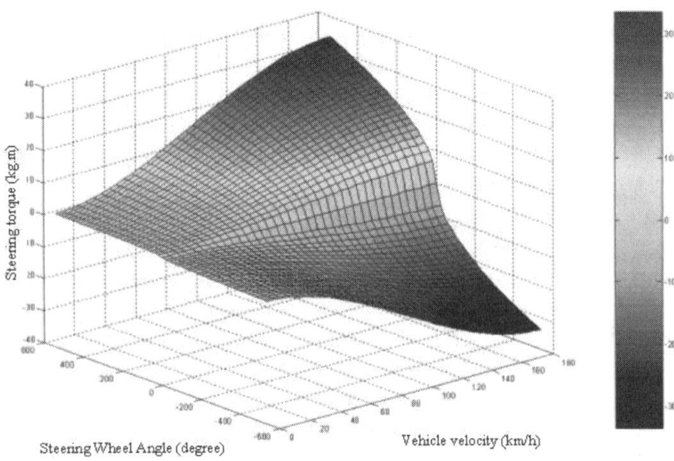

Fig. 7 Torque map for the steering wheel motor control

As shown in Fig. 7, at low speeds & parking, the steering wheel motor torque is kept small and almost same respect to steering wheel angle. However, at high speeds, the steering wheel motor torque is increased according to the vehicle velocity & steering wheel angle. As a result, the vehicle designer can control the steering wheel motor torque of the SBW system by using this kind of torque map.

FRONT WHEEL MOTOR CONTROL

To improve vehicle stability, understeer propensity is controlled in this study. The control method can also improve vehicle maneuverability at the same time. Understeer is the front wheels' sideway slip, resulting in a wider line, when the vehicle cornering forces exceed tire performance. Understeer propensity control methods to improve vehicle performance are as follows: Fig. 8 shows the control methods as a graphic illustration.

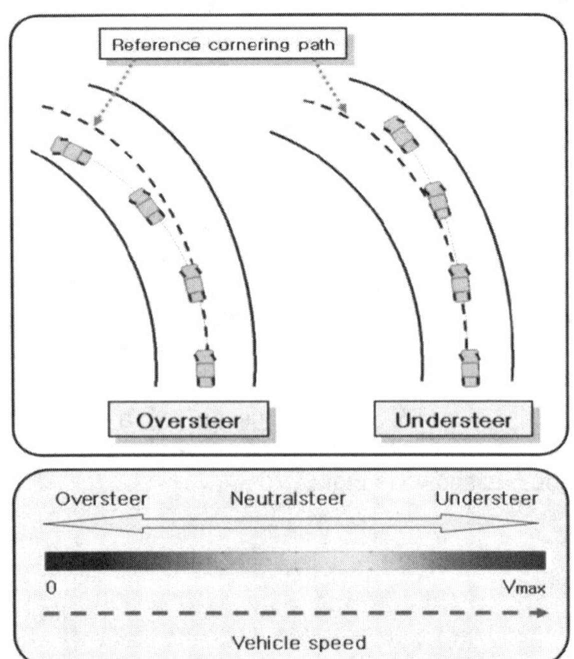

Fig. 8 Understeer propensity control methods

The first method is to improve vehicle maneuverability by driving a vehicle with oversteer. The reason is that vehicle's yaw rate and lateral acceleration at a low speed is not important for vehicle stability. On the other hand, with front wheel steering, the vehicle's quick response with respect to the driver's steering is more important for improving the vehicle performance. The second method is to prohibit rapid steering which causes vehicle instability at high speeds by increasing understeer propensity according to increasing vehicle speed. In other words, variable gear effects are realized in the SBW system without mechanical linkage. Fig. 9 shows a 2 D.O.F bicycle model used for the control methods in this paper.

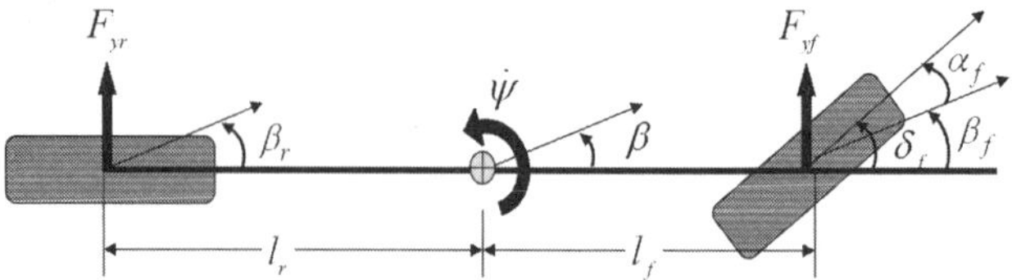

Fig. 9 2 Degree of freedom bicycle model

Eq. (5) represents the understeer gradient equation with the bicycle's steady state cornering.

$$K = \frac{1}{a_y}\left(\delta_f - 57.3\frac{L}{R}\right)$$

(5)

$K\,(\deg/g) : Understeer\ gradient$

where

$K = 0$; *Neutral Steer* $(\alpha_f = \alpha_r)$

$K > 0$; *Understeer* $(\alpha_f > \alpha_r)$

$K < 0$; *Oversteer* $(\alpha_f < \alpha_r)$

In Eq. (5) R is the radius of vehicle cornering and a_y is the vehicle's lateral acceleration. If R is much bigger than the length $(l_f + l_r) = L$ between wheel bases of the vehicle, R becomes $R \cong L/\delta_f$. In addition, vehicle lateral acceleration $a_y = V^2/R$ then Eq. (5), the understeer gradient equation becomes:

$$\delta_c = \left(\frac{L}{L + KV^2}\right)\delta_f$$

(6)

where,

$$\delta_f = 57.3\frac{L}{R} + K\frac{V^2}{R}, \quad \delta_c = 57.3\frac{L}{R}, \quad R = \frac{L}{\delta_c}$$

From Eq. (6), understeer propensity is increases as vehicle velocity increases because the cornering front wheel angle is reduced as vehicle velocity increases. By using this fact, vehicle maneuverability and stability can be improved at the same time by setting appropriate K according to vehicle velocity, for instance, setting K as a negative number at low speeds or parking for oversteer propensity and setting K as a positive number at high speeds. Therefore, feed forward control would be possible if an appropriate K is selected as a gain parameter of the control algorithm according to the vehicle designer's needs. Fig. 11 shows the feed forward control algorithm.

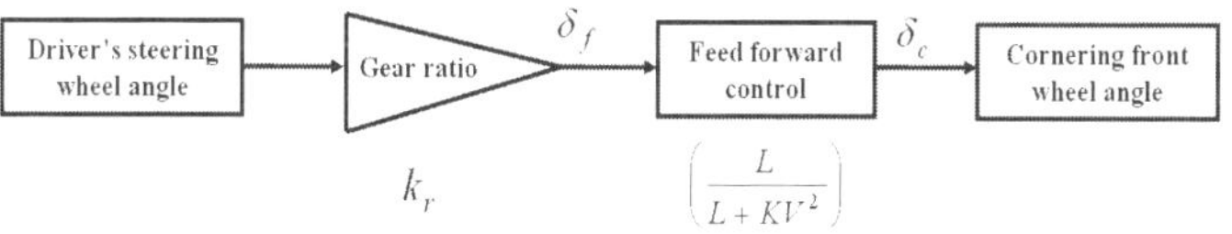

Fig. 9 Feed forward control algorithm

As shown in Fig. 9, the feed forward control does not have to give feedback regarding vehicle states, lowering costs by not using sensors will be possible. Fig. 10 shows the results of lane change simulation with feed forward control describing the yaw rate variation according to changing understeer gradient K from 0 to 0.01 for understeer propensity when vehicle velocity is 80km/h.

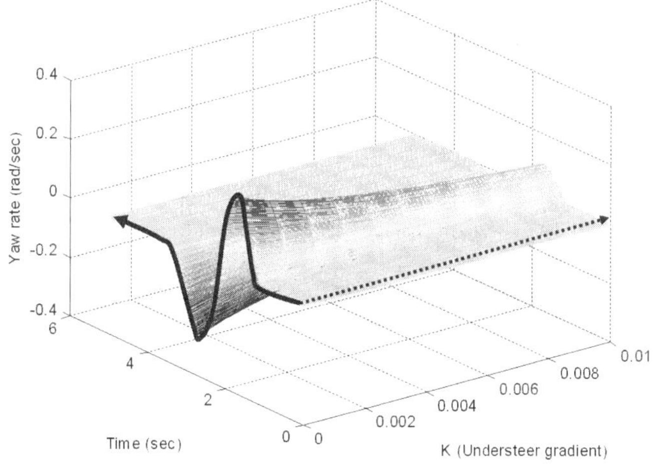

Fig. 10 Lane change simulation with feed forward control for understeer.

As shown in Fig. 10, vehicle stability can be improved when the understeer gradient is controlled at high speeds because the yaw rate is reduced gradually according to increasing the understeer gradient. Fig. 11 shows the results of the lane change simulation with feed forward control describing the yaw rate variation according to changing the understeer gradient K from 0 to -0.01 for oversteer propensity when the vehicle's velocity is 40km/h.

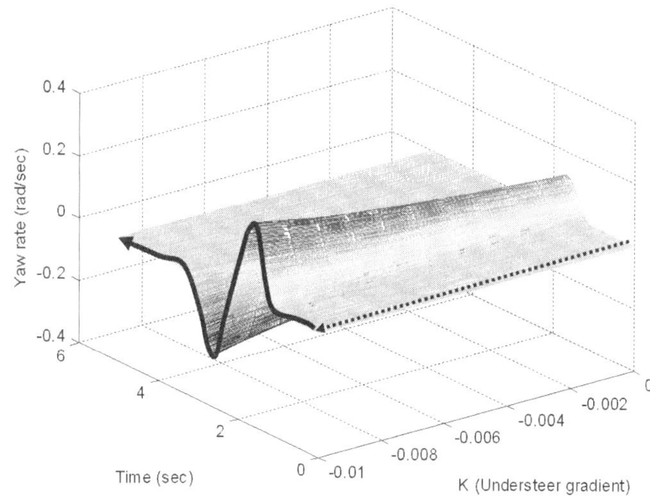

Fig. 11 Lane change simulation with feed forward control for oversteer.

As shown in Fig. 11, vehicle maneuverability can be improved when the negative understeer gradient is controlled at low speeds because the yaw rate is increasing gradually as decreasing the understeer gradient. Because the vehicle's quick response for front wheel steering with respect to the driver's steering caused by oversteer propensity improves vehicle maneuverability. By using the understeer gradient control method, vehicle maneuverability and stability control be improved if the vehicle designer optimizes the understeer gradient as a control parameter according to the type of vehicles and drivers.

PROPOSED ADVANCED CONTROL METHOD

Advanced control method is a more accurate and advanced control method controlling the front wheel by adding high quality sensors to the base control system and gives exact state feedback of the vehicle. The steering wheel reactive torque motor for improving the driver's steering feel is controlled by the desired output through vehicle dynamics after obtaining the vehicle state from a torsion bar torque and vehicle velocity sensor. The front wheel steering motor for improving vehicle maneuverability and stability is also controlled by the desired output through vehicle dynamics after obtaining the vehicle states from a yaw rate and lateral acceleration sensor. The control concept of the advanced control method is as follows.

(1) The limit of vehicle instability factors like roll over, yaw moment, and rear sway needs to be set.

(2) When the vehicle exceeds instability limit, the vehicle dynamics control system operates to decrease vehicle instability factors by controlling motor torque using sensor feedback data.

Therefore, as a part of the advanced control, Active Roll Stability Control (ARSC) method is constructed in this paper. To define the rollover threshold, vehicle rolling dynamics is used and the rollover threshold can be derived.

$$\frac{a_y}{g} = \frac{t}{2h} \tag{7}$$

By using the rollover threshold, additional torque gain K_T can be determined as follows

$$K_T = K_P \left(a_y - \frac{t}{2h} g \right) \tag{8}$$

where,

K_T = Additional torque gain

K_P = Proportional gain to control K_T

a_y = Vehicle lateral acceleration

By using the additional torque gain K_T, when vehicle lateral acceleration exceeds roll over threshold, additional torque can be added to prevent rollover. Therefore, what the vehicle designer has to do is only to adjust proportional gain K_P for controlling K_T. Following Fig. 12 shows the overall SBW control algorithm with ARSC method.

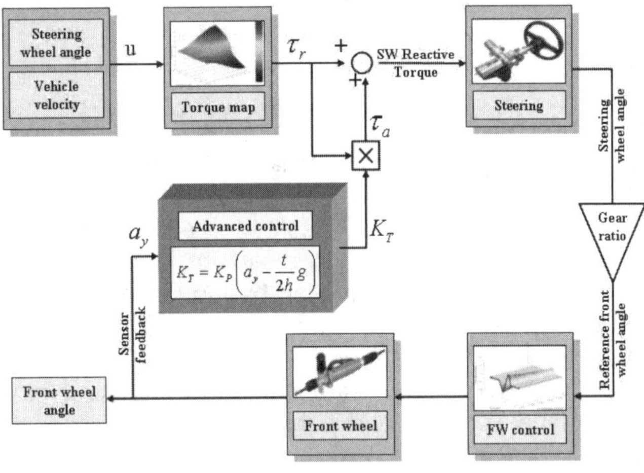

Fig. 12 SBW control algorithm with ARSC method.

Fig. 13 shows the slalom test results to verify the ARSC method. For the input steering wheel angle, 60 degree sinusoidal angle is inputted in the test and vehicle speed is 80km/h. Therefore, the rollover occurrence and how it is compensated using ARSC method can be seen. The control concept is that when the vehicle's lateral acceleration exceeds the vehicle's instability limit, rollover threshold, additional reactive torque is added to the steering. Therefore, the driver experiences difficulties in steering the vehicle and exceeding the rollover threshold is prevented.

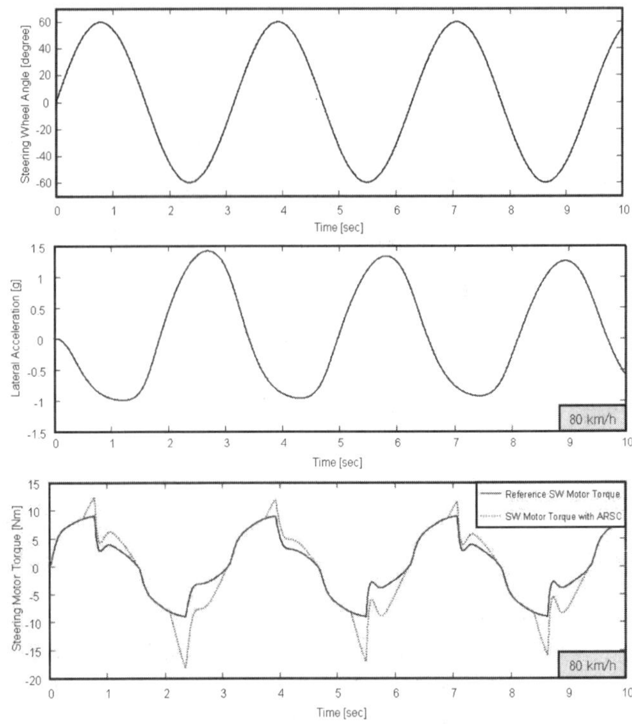

Fig. 13 ARSC method simulation at 80 km/h

As shown in the simulation result, when the lateral acceleration exceeds the instability limit, the additional torque is added to the steering wheel motor. Actually the additional torque proportional gain, K_p, is set to 2 in this simulation. Therefore, the rollover prevention effect can be adjusted by changing K_p. In conclusion, vehicle's rollover could be prevented by using ARSC method.

CONCLUSIONS

(1) The SBW system was modeled by using the bond graph method.

(2) Steering wheel motor torque map is established to improve driver's steering feel.

(3) Understeer propensity control method was used to improve vehicle's maneuverability and stability.

(4) Advanced control method is proposed and ARSC algorithm is designed to prevent the vehicle's rollover.

REFERENCES

1. Se-Wook Oh, Tong-Jin Park and Chang-Soo Han, "The Design of a Controller for the Steer-by-Wire System," FISITA 2002 World Automotive Congress,2002.
2. Se-Wook Oh, Tong-Jin Park, Jae-Ho Jang, Seok-Hwan Jang, and Chang-Soo Han, "Electronic Control Unit for the Stter-by-Wire System Using a Hardware-In-the-Loop-Simulation System," FISITA 2002 World Automotive Congress, 2002.
3. Tong-Jin Park, Se-Wook Oh, Jae-Ho Jang and Chang-Soo Han, "The Design of a Controller for the Steer-by-Wire System Using the Hardware-In-the -Loop -Simulation System," Proceedings of the 2002 SAE automotive Dynamics Stability conference, pp. 305-310
4. Ji-Hoon Kim, Jae-Bok Song, 2002, "Control logic for an electric power steering system using assist motor," Mechatronics.
5. Masaya Segawa, Shiro Nakano, Osamu Nishihara, Hiromitsu Kumamoto, 2001, "Vehicle stability control strategy for steer by wire system," JSAE Review 22 (383-388).
6. Ryouhei Hayama, Katsutoshi Nishizaki, "The vehicle stability Control Responsibility Improvement Using Steer-by-Wire," Proceedings of the IEEE Intelligent Vehicles Symposium, 2000.
7. Sanket Amberkar, Joseph G. D'Ambrosio and Brian T. Murray, "A system-Safety Process For By-Wire Automotive Systems," SAE international Congress, paper 2000-07-1056, 2000.
8. Leonard Segel, 1998, "On the Lateral Stability and Control of the Automobile as Influenced by the Dynamic of the Steering System," ASME 65WA/MD-2.

New Generation of Inertial Sensor Cluster for ESP- and Future Vehicle Stabilizing Systems in Automotive Applications

R. Willig and M. Mörbe

Robert Bosch GmbH, Automotive Equipment, Division Chassis Systems, Product Group Sensors

ABSTRACT:

In 1995 Robert Bosch GmbH (RB) started the mass production of the first VDC-System (Vehicle Dynamics Control system) for vehicles, today called ESP (Electronic Stability Program). This ESP-System went beyond ABS and Traction Control Systems and offered consumers unsurpassed driving confidence and safety.

The key part of this system was a first generation Yaw Rate Sensor DRS 50/100, based on a metal vibrating cylinder. The second generation DRS MM1, introduced in 1998, based on silicon micromachining and included an integrated linear acceleration sensor element.

For new additional functions of ESP and of future high dynamic and high performance vehicle stabilizing systems, like Hill Hold Control (HHC) or Steer by Wire (SbW) BOSCH develops the third generation, a flexible and cost-effective Inertial Sensor Cluster with a modular concept for hard- and software, called DRS MM 3.x.

For all inertial sensor elements silicon surface micromachining is used. A fully digital signal processing with delta-sigma ($\Delta\Sigma$-) modulation in closed or open loop control is implemented into ASICs, the communication between the sensor elements and the internal microcontroller takes place via a SPI-interface. Internal logical links and self monitoring, CAN-interface between the Sensor Cluster and the ECUs and the possibility of internal redundancy for high safety-relevant, high dynamic and performance systems meet the future requirements.

Design, basic functions, modular concept, safety features and system requirements of the new Inertial Sensor Cluster DRS MM 3.x are presented.

Keywords: Inertial Sensor Cluster, Yaw Rate Sensor, Acceleration Sensor, Silicon Surface Micromachining, Modular Concept, ESP, HHC, ACC, 4w, ROM, EAS, ROSE, ASC, SbW.

INTRODUCTION

In 1995 Robert Bosch GmbH (RB) started the mass production of the first VDC-System (Vehicle Dynamics Control system) for vehicles, today called ESP (Electronic Stability Program). The key part of this system was a first generation Yaw Rate Sensor DRS 50/100, based on a metal vibrating cylinder with piezo-electric transducers. [12], (Figure 1)
The second generation DRS MM1, introduced in 1998, based on a combination of silicon bulk and surface micromachining with a electromagnetic drive and capacitive detection and included an integrated linear acceleration sensor element for measuring the lateral acceleration of the vehicle. [13], (Figure 1)

ESP is a safety system for road vehicles which controls the dynamic vehicle motion in emergency situations.
Vehicle handling at the physical limit of adhesion between the tires and the road is extremely difficult. In such situations the driver may be supported by controlling the longitudinal and lateral forces on the tires. The ESP- system of Bosch does that by controlled braking of individual wheels which makes the vehicle motion approach the nominal motion intended by the driver. It uses signals to determine the driver's intent, like steering wheel angle, brake pressure, engine torque and other signals to derive the actual motion of the vehicle, like angular velocity of the car around it's vertical axis (Yaw Rate) and lateral acceleration.
The support of the driver is not limited to coasting conditions. Also during full braking, partial braking, engine drag, free rolling and acceleration of the vehicle the system supports the driver in all safety critical situations, within the physical limits.

For new additional functions of ESP and of future high dynamic and high performance vehicle stabilizing systems or comfort systems, like Hill Hold Control (HHC), Navigation (Navi, Travel Pilot), Adaptive Cruise Control (ACC), Four Wheel Drive (4w), Roll over Mitigation

(ROM), Electronic Active Steering (EAS), Roll Over Sensing (ROSE), Active Suspension Control (ASC) and Steer by Wire (SbW) BOSCH develops the third generation, a flexible and cost-effective Inertial Sensor Cluster called DRS MM 3.x.

These new system-functions require additional inertial measuring data of the dynamic behaviour of the vehicle like angular acceleration, longitudinal acceleration, tilt angle, acceleration of the z–axis, angular velocity of the x-axis and redundancy of the inertial data in the case of high safety relevant systems.

The demands for this new Sensor Cluster are higher robustness, more compactness, easier applicability, higher accuracy and better signal to noise ratio (SNR), insensitivity against external interference, high signal refresh rate, extendability for cross-system applications, higher system availibility and reliability and an integral safety concept. This leads to a modular concept for hard- and software of the Sensor Cluster.

Design, basic functions, modular concept, safety and monitoring concept and system requirements of the new Inertial Sensor Cluster DRS MM 3.x are presented.

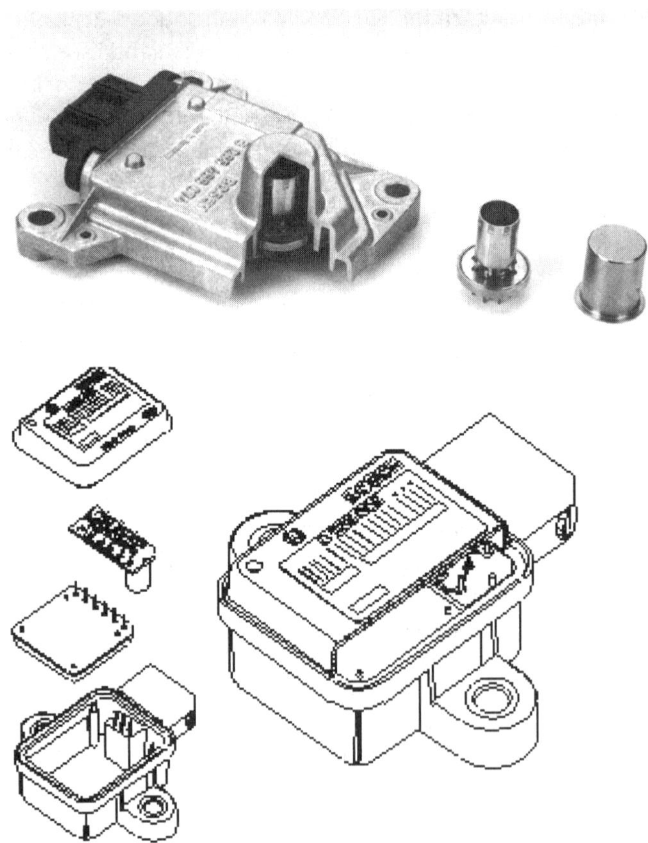

Figure 1: Housing of the sensor, DRS 50/100 (1995) and DRS MM 1.x (1998)

BASIC FUNCTION AND DESIGN OF THE INERTIAL SENSOR ELEMENTS AND READOUT ELECTRONICS

Angular velocity sensor measuring element and readout electronic module

The new designed sensor measuring element for measuring angular velocity or rate of rotation belongs to the type of the well known CVGs, Coriolis Vibratory Gyroscope ([1] – [19]).

It works like an inverse tuning fork configuration with two linear in-plane orthogonal oscillation modes, a so called primary or drive mode and a secondary or detection mode, where the electrostatically excited motion (v) of the primary mode (Xdrive) is a linear antiphase oscillation of two coupled spring-mass structures. The secondary in-plane mode (Ydetection) of the two coupled spring-mass structures is excited due to Coriolis inertial forces produced as a result of applying a rate of rotation (Ω) perpendicular to the plane. The oscillation of the drive mode is provided by electrostatic forces at comb drive electrodes as well as the sensing of the motion of this mode while the detecting of the Coriolis acceleration (a_c) in the secondary mode is accomplished capacitively via interdigitated electrodes. (Figure 2)

$$\text{Coriolis acceleration:} \quad -\vec{a}_c = 2\,\vec{\Omega} \times \vec{v}$$

with a_c: Coriolis acceleration (m/s²)

       Ω: Angular velocity (Yaw rate) (s⁻¹)

       v: Vibration velocity of drive mode (m/s).

To fullfil various demands the sensor measuring element is designed as two spring-mass structures, which are mechanically coupled. Both modes are oscillating in plane at one natural resonant frequency of the structures, nominal 15 kHz with constant amplitude.

Silicon surface micromachining is used for a more compact size and cost reduction according to the improved Bosch foundry process. (Figure 2)

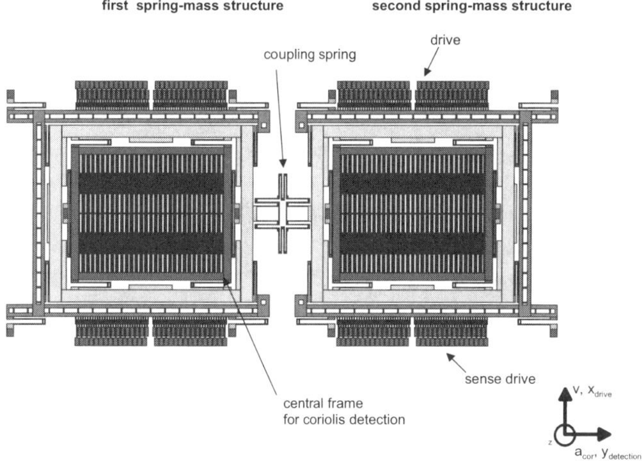

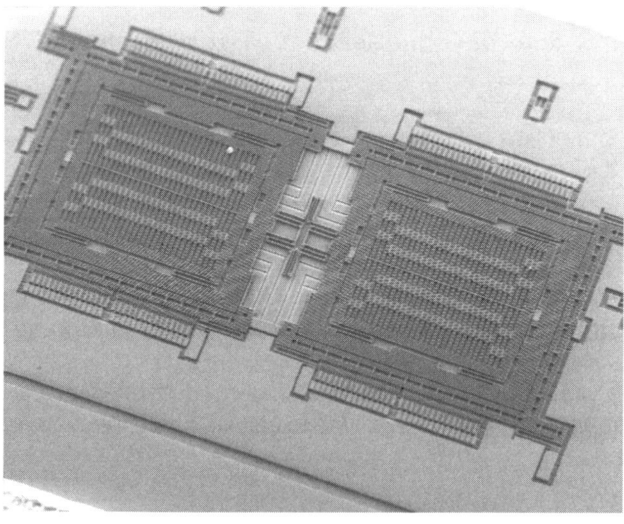

Figure 2: Sensor measuring element, design and SEM foto

To be insensitive against external mechanical interferences and overloads in the vehicle like linear and angular vibrations a high natural resonant frequency of the micromachined structure is used, which is out of the range of power spectrum of the interference in the vehicles. To increase resolution, signal to noise ratio, linearity and bandwidth a high Q-factor, a double resonant oscillation (only one frequency of drive and detection mode) and a closed loop principle are neccessary and are implemented into the sensor readout electronic module (ASIC). A so-called multibit force rebalance control loop including an electromechanical $\Delta\Sigma$-modulator (high sampling rate and quantisation, noise shaping) will electrostatically drive the Coriolis force to zero. For a high Q-factor residual dynamic gas damping must be minimized due to an oscillation at low internal pressure. Mass and stiffness imperfections during manufacturing processes require an electrostatic stiffness tuning for the double resonant oscillation, a mechanical decoupling of the drive and detection mode of both spring-mass structures to suppress mechanical crosstalks and parasitic effects and an electronic out-of-phase (quadrature) rejection.

The new sensor readout electronic module (ASIC) contains a drive and detection mode loop with the following functional blocks :
analog frontends, digital backend, programmable read only memory (PROM) and serial Peripheral Interface (SPI). (Figure 3)

The analog frontends are for fully differential C/V converting and common mode controlling with SC-technique (switch capacitor technique) and mainly for early digitizing the signals with a differential multibit flash-A/D converter.

The drive loop consists of an analog frontend with the ADC, a digital bandpass filter, an AGC (automatic gain control), a lowpass filtering, a PI controlling and a DAC, a charge pump, a PLL (phase locked loop) with 8 and 40 MHz clocks and a start up circuit.
The functions of the drive loop guarantee a very fast open control loop for starting up the oscillation of the micro-mechanical sensor measuring element in x-direction and then the controlling of the oscillation at one resonant frequency with constant amplitude.

In order to measure the Coriolis force the electronics of the detection mode loop and the micro-mechanical sensor measuring element are combined to form the electromechanical $\Delta\Sigma$-modulator. The detection loop contains the same analog frontends as in the drive loop and a forced feed back loop. A synchronous demodulation and filtering, a frequency adjustment done by electrostatic stiffness tuning of the detection mode, quadrature control loop, output filtering, compensation of parameter shifts with temperature and calibration of offset and scale factor, a 40 bit wide PROM to program the coefficients for adjustments, compensations and calibrations, and a temperature sensor belonging to the digital backend.
A SPI (Serial Peripheral Interface = synchronous serial data link with 2 MBaud) ensure the communication and data transfer between the ASIC and the sensor internal microcontroller with 8 or 40 bit frames.
Self-testing during power on and self- monitoring on ASIC level are realized for 5V supply voltage, bond wires, range overflow of control parameters and a digital backend test.

The sensor measuring element and the ASIC are cased in a plastic surface mount package. This packaged device is fully calibrated, temperature compensated and tested. (Figure 4)

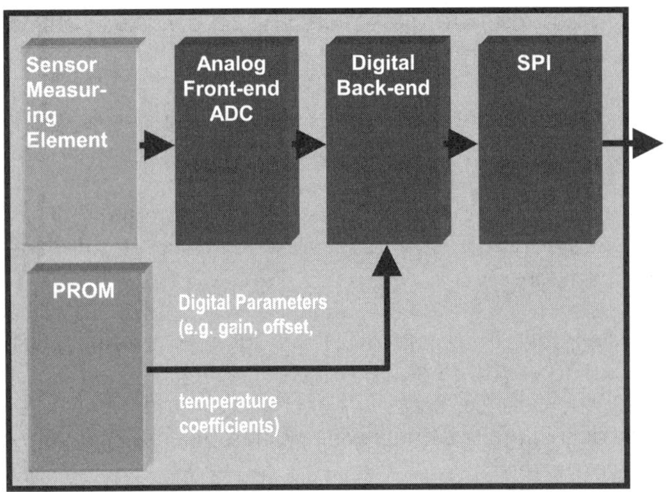

Figure 3: Functional blocks of the sensor readouelectronic module

Figure 4: Sensor measuring element and ASIC in plastic surface mount package

Linear acceleration sensor measuring element for lateral and longitudinal acceleration and readout electronic module

For the sensor measuring element of linear lateral and longitudinal acceleration an in plane spring-mass structure is used according to the improved Bosch silicon surface microfabrication technology. (Figure 5)

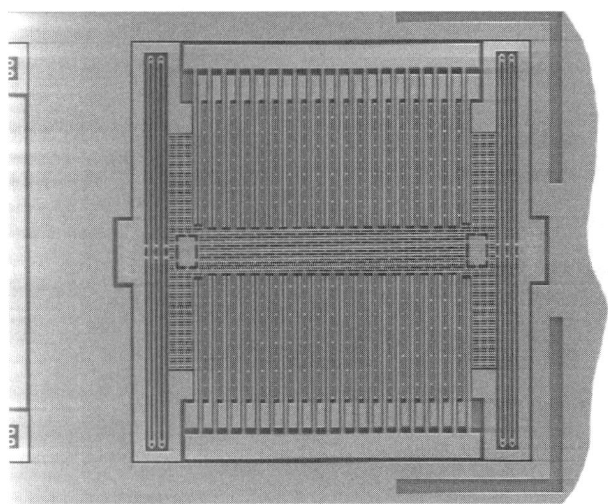

Figure 5: SEM foto of the sensor measuring element

The deflection of the spring-mass structure in the sensitive axis due to external acceleration forces is detected with a differential capacitive comb structure. The spring-mass structure is designed for a high mechanical g-range (70 g) and together with the new sensor readout electronic module (ASIC) a variable low g measuring range (2 g) with low noise (10 mg rms) and high accuracy is possible.

In this way external high mechanical interferences cause no clipping of the mechanical structure or of the signal processing.

The ASIC contains the same analog frontend like the ASIC of the sensor measuring element for angular velocity, with a fully differential charge to voltage conversion and a correlated double sampling (CDS) to achieve high dc stability.

The fully digital signal processing includes a mechanical open loop electrical closed loop $\Delta\Sigma$-modulation, filtering, offset adjust with temperature compensation, output filter and finally a scale factor calibration. A temperature sensor, 8 MHz oscillator and a 24 bit PROM are realized just as self- testing and -monitoring and data transfer via SPI.

The packaging type is the same as the package of the angular velocity sensor element.

Also a two channel version is developed with two separate spring-mass structures on one chip with sensitive axes 90 degree to each other and one sensor readout electronic module with two output signals.

DESIGN, MODULAR CONCEPT AND SAFETY FEATURE OF INERTIAL SENSOR CLUSTER DRS-MM 3.X

The packaged sensor measuring elements and ASICs, see figure 4, now called sensor elements are mounted on a PCB together with a microcontroller, EEPROM, voltage regulator, watch dog, quartz, etc.
The PCB is fixed in a plastic Sensor Cluster housing which is sealed by a protective cover. A four pin connector is used for external power supply (12 V) of the Sensor Cluster and for data transfer via CAN. The Sensor Cluster can be installed in passenger compartment or trunk and the housing can be fixed with two holes for bolts. (Figure 6)

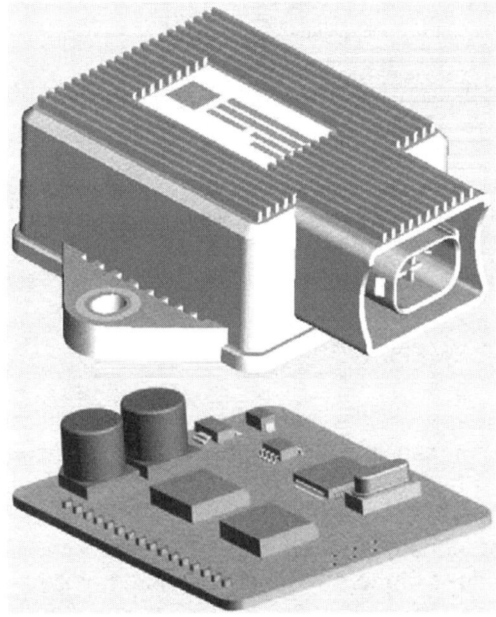

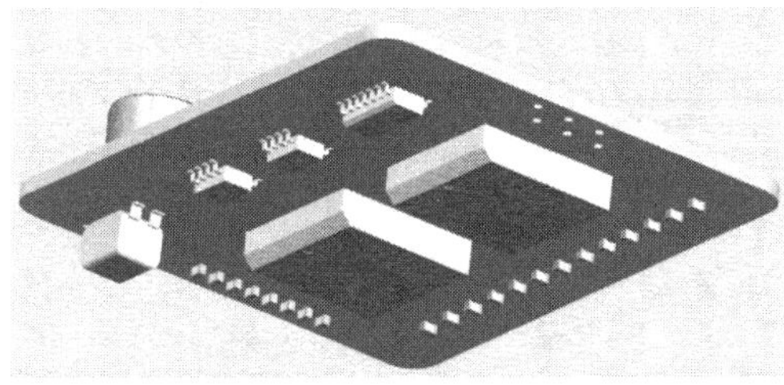

Figure 6: Housing of the Sensor Cluster DRS MM 3.x and PCB with components

Microcontroller, sensor elements, etc. are supplied with controlled and monitored 5 V power supplies. The internal data transfer and communication between the various and extendable sensor elements and the microcontroller are done by SPI, a synchronous bidirectional serial data link with 2 Mbaud, with four wires. Various data and status informations, like the measured inertial signals, control loop parameters and temperature signals, are transmitted via SPI to the microcontroller, checked by the microcontroller and transformed in a CAN-matrix and transmitted via CAN to the System-ECU after a request. The content of the CAN-identifier-matrix is also extendable and the synchronous CAN communication time is variable between 5 and 20 ms. (Figure 7)

Depending on system requirements various sensor elements can be mounted accordingly to the PCB, all SPI participants are addressed and the software (SW) of the microcontroller can check and handle all these different configurations. Due to this a modularity in HW and SW is cost-effectively possible and offers many different extendable versions of the Sensor Cluster DRS-MM 3.x with distinctive features in different inertial measuring data.

The Sensor Cluster internal safety and monitoring concept base on
- sensor element internal self-testing and -monitoring of control parameters at the drive mode (AGC, PLL), the detection mode, the digital backend and SPI-communication on ASIC level (sensor element and SPI failures) during power-on and normal mode,
- furthermore a sensor element initialisation and signal monitoring on microcontroller level (sensor element and SPI failures) with sensor element ordinary inertial signals (Yaw rate or Lateral acceleration), additional signals and control loop parameters and temperature signals.

Microcontroller itself is testing and monitoring (µC and CAN failures) periodically CPU, CAN, SPI, Oscillator, ADC, watchdog, undervoltage and overvoltage (12V and 5V), RAM, ROM and EEPROM.
The Sensor Cluster internal various status informations of the results of internal testing and monitoring are transmitted via CAN together with each inertial measuring data and the internal failure detection time is less 25 ms.

An additional step for high safety and availibility of the system is a high CAN communcation time and a signal monitoring, filtering and plausibility checks on System-ECU level and a redundancy of sensor elements in the Sensor Cluster DRS-MM 3.x.

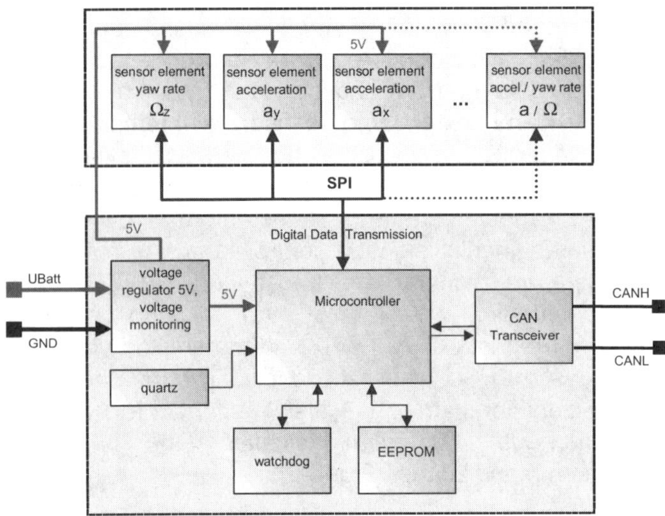

Figure 7: Schematic circuit diagram of Sensor Cluster DRS MM 3.x

SYSTEM REQUIREMENTS AND ASPECTS

Different system applications with new additional functions require the simultaneous measurement of inertial signals, rates of rotation and linear accelerations, of the dynamical behaviour of the vehicle about the three orthogonal axes with different measuring ranges, bandwidth or accuracy.

Comfort systems like car navigation (Travel Pilot) and vehicle guidance (ACC), where Yaw Rate and linear longitudinal and lateral acceleration with a small signal bandwidth and high accuracy is required,
or passive safety systems like restraint systems (Airbag) and Roll over sensing (ROSE) with Roll and Yaw Rate and low and high g-acceleration in all three axes
or active safety systems like ESP (incl. ABS and TCS), HHC, ROM, EHB, ASC and EAS
or future high safety relevant, high dynamic and high performance stabilizing systems like four wheel steering and steer by wire (SbW) with Yaw, Roll Rate and angular acceleration and low g-acceleration in redundancy and with high signal bandwidth, high accuracy and very low noise, needs often the same inertial signals but with different measuring ranges, signal quality, safety and dynamic demands.

The Inertial Sensor Cluster DRS MM 3.x is fit for these requirements of a growing diversity. The modular concept of HW and SW has furnished information for different systems and different equipment rate for vehicle platforms or models and is extendable in sensor measuring element equipment rate to enable add-on functionality and adaptation to new system architecture. (Figure 8)

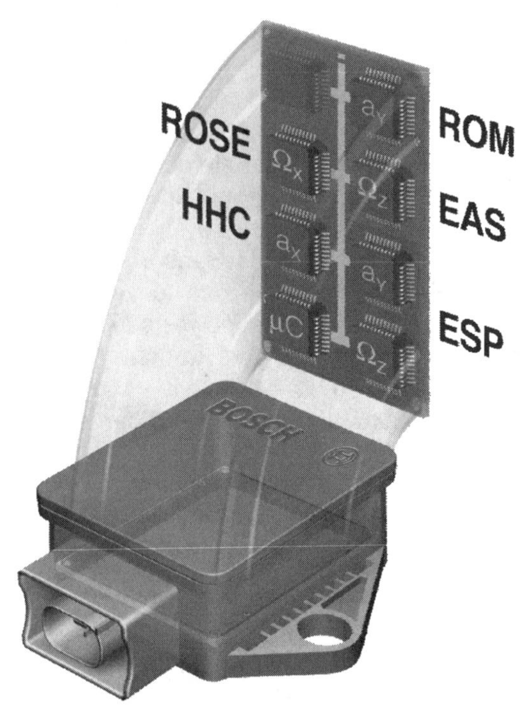

Figure 8: Modular Concept of Sensor Cluster DRS MM 3.x

FUTURE TRENDS

The first step in the direction of a kinematic sensor platform for sensing the complete vehicle dynamics is presented with the Inertial Sensor Cluster DRS-MM 3.x.
Future complex cross-system applications will require an intelligent sensor platform (ISP), which is sensing all inertial values of the three main axes internally, receiving external sensor signals like wheel speeds and forces, steering angle, engine torque, braking pressure, tire pressure, actuator states and informations of the driver assistance systems (radar, video, GPS) and then calculating or estimating the dynamic values like side slip angle, speed over ground in longitudinal and lateral direction, yaw rate, road inclination, road uphill gradient, etc. and controller values.
These applications will need high data volume, high update rates and safe data transfer, why time triggered data networks like TT-CAN, TTP or FlexRay will be required. [20]
For applications using only ESP without additional system functions the sensor elements of Yaw Rate and Acceleration can be integrated into the Electronic Control Unit of the ESP Hydraulic Unit.

SUMMARY

For new additional functions of ESP and for future high dynamic and high performance vehicle stabilizing systems, like Hill Hold Control (HHC) or Steer by Wire (SbW) BOSCH develops the third generation, a flexible and cost-effective Inertial Sensor Cluster called DRS MM 3.x.
These new system-functions require additional inertial measuring data about various axes of the dynamic behaviour of the vehicle.

The requirements for this new Sensor Cluster are higher robustness, more compactness, easier applicability, higher accuracy and better SNR, insensitivity against external interference, high signal refresh rate, extendability for cross-system applications, higher system availibility and reliability and an integral safety concept. This leads to a modular concept for hard- and software of the Sensor Cluster.

For all inertial sensor measuring elements silicon surface micromachining is used. To be insensitive against external mechanical interferences and overloads in the vehicle like linear and angular vibrations the micromachined structures are carefully designed. To increase resolution, signal to noise ratio, linearity and to suppress mechanical crosstalks and parasitic effects a fully digital signal processing with SC-, CDS-, and $\Delta\Sigma$-technique in closed or open loop control is implemented into ASICs.

The communication between the sensor elements and the internal microcontroller takes place via a SPI-interface.

Internal logical links, self testing and self monitoring, CAN-interface between the Sensor Cluster and the System-ECUs and the possibility of internal redundancy for high safety-relevant, high dynamic and performance systems meet the future safety relevant requirements.
Design, basic functions, modular concept, safety and monitoring concept and system requirements of the new Inertial Sensor Cluster DRS MM 3.x are presented.

ACKNOWLEDGMENTS

The authors would like to thank all their colleagues of Corporate Research and Advanced Engineering and of Automotive Equipment, Division Automotive Electronics and Division Chassis Systems for the design of the complete sensor system, design and layout of the sensor measuring elements and sensor readout electronic modules.

REFERENCES

[1] M.W.Putty, K.Najfi, "A Micromachined Vibrating Ring Gyroscope", Solid-State Sensor and Actuator Workshop, June 13-16,1994.

[2] J. D. Johnson, S. Z. Zarabadi, D. R. Sparks, "Surface Micromachined Angular Rate Sensor", SAE Technical Paper Series, 950538.

[3] J. Bernstein S.Cho, A. T. King, A. Kourepenis, P. Maciel, M. Weinberg, "A Micromachined Comb-Drive Tuning Fork Rate Gyroscope", 0-7803-0957-2/93, 1993 IEEE.

[4] K. Funk, A.Schilp, M. Offenberg, „Surface-micromachining of Resonant Silicon Structures", Transducers `95, 519-News, page 50.

[5] M. Hashimoto, C. Cabuz, K. Minami, M. Esashi, "Silicon Resonant Angular Rate Sensor Using Electromagnetic Excitation and Capacitive Detection", Technical Digest of the 12th Sensor Symposium, 1994.

[6] K. Tanaka, Y. Mochida, M. Sugimoto, K. Moriya, T. Hasegawa, K. Atsuchi, K. Ohwada, " A Micromachined Vibrating Gyroscope", Sensors and Actuators A 50 (1995).

[7] Y. Cho, B. M. Kwak, A. P. Pisano, R. Howe, "Slide film damping in laterally driven microstructures", Sensors and Actuators A, 40 (1994).

[8] M. Offenberg, F.Lärmer, B.Elsner, H.Münzel, W. Riethmüller, "Novel Process for a Monolithic Integrated Accelerometer", Transducers 95, 148 - C4.

[9] K. H.-L. Chau, S.R. Lewis, Y. Zhao, R. T. Howe, S. F. Bart, R. G. Marcheselli, "An integrated Force-Balanced Capacitive Accelerometer for Low-G Applications", Transducers 95, 149 - C4.

[10] M. Offenberg, B. Elsner, F. Lärmer, Electrochem. Soc. Fall-Meeting 1994, Ext. Abstr. No 671.

[11] M. Offenberg, H. Münzel, D. Schubert, " Acceleration Sensor in Surface Micromachining for Airbag Applications with High Signal/Noise Ratio, SAE Technical Paper, 960758.

[12] A. Reppich, R. Willig, " Yaw Rate Sensor for Vehicle Dynamics Control Systems", SAE Technical Paper 950537 (1995)

[13] M. Lutz, W. Golderer, J. Gerstenmeier, J. Marek, B. Maihöfer, D. Schubert, " A Precision Yaw Rate Sensor in Silicon Micromachining ", SAE Technical Paper 980267 (1998)

[14] A.W. Leissa : Vibration of shells, NASA SP-288 (1973)

[15] G.B. Warburton : Vibration of thin cylindrical shells, Journal Mechanical Engineering Science, Vol 7 No 4 (1965)

[16] C.H.J. Fox and D.J.W. Hardie : Harmonic response of rotating cylindrical shells, Journal of Sound and Vibration, 101 (4) (1985)

[17] J.S. Burdess : The dynamics of a thin piezoelectric cylinder gyroscope, Proc. I. Mech. E., Vol. 200 No c4 (1986)

[18] P.W. Loveday, " Analysis and Compensation of Imperfection Effects in Piezoelectric Vibratory Gyroscopes ", Diss. Virginia Poytechnic Institute and State University, Blacksburg, Virginia, 1999.

[19] J. A. Geen, "A Path to Low Cost Gyroscopy ", Solid-State Sensor and Actuator Workshop,June 8-11, 1998 .

[20] D. Sparks et al., "Multi-sensor modules with data bus communication capability", SAE paper 1999-01-1277, 1999.

Development of Mando ESP (Electronic Stability Program)

Dongshin Kim, Kwangil Kim, Woogab Lee and Inyong Hwang
Mando Corporation

ABSTRACT

This paper describes the MANDO MGH ESP (Electronic Stability Program) and consists of the control philosophy, hydraulic actuator and the simulation and test results. The ESP system controls the dynamic vehicle motion in the emergency situation such as the final oversteer and understeer and allows the vehicle to follow the course as desired by the driver. The MANDO MGH ESP is integrated with the existing MANDO MGH ABS/TCS, which is improved with the more information and controls both brake pressure and engine torque for the optimal performance. The look-up tables are emphasized to have the accurate target yaw rate of the vehicle and obtained from vehicle test for the whole operation range of the steering wheel angle and vehicle speed. The wheel slip control is applied for the yaw compensation and the target wheel slip is determined by error between the target yaw rate and actual yaw rate. The hydraulic actuator doesn't use the pre-charge element and its body block is same as that of MANDO MGH TCS. Since the ESP has a high severity level and robust control is required, thousands of failsafe and robustness tests are carried out in the laboratory using hardware-in-the-loop simulation and on the various kinds of test tracks. Both computer simulation and test results are showed in this paper.

INTRODUCTION

Audi [1] has shown that the number of spin out accidents increases rapidly with increases in vehicle speed. Between 80km/h and 100km/h more than 40% of all accidents with personal injures are related to spin out. Above 100km/h nearly every accident occurs because of spin out. Once the vehicle begins to spin out, the driver is not able to control the vehicle and finally severe accidents take place. This kind of spin-out can't be solved by ABS/TCS as well as suspension or steering controller.

The ESP system makes use of the brake and engine intervention to keep the vehicle on its intended track under most road conditions.

Although ABS and TCS control the wheels of the vehicle at the limit of adhesion between the tire and road surface in the longitudinal direction of wheel, the ESP system deals with the longitudinal and lateral tire dynamics beyond the nonlinear motion of the vehicle. In Europe, 5 million ESP are expected to produce annually by 2004, and in the U.S ESP is just catching on.[2] On the other hand, in South Korea the demand of ESP is slowly rising.

Now the six suppliers offer ESP worldwide and MANDO is going to produce the ESP system for the front wheel drive car and 4 wheel drive sport utility vehicle in the year 2004. Mando has researched the ESP system since 1995 and carried out the winter test in Sweden and New Zealand every year.

CONTROL PHILOSOPHY

VEHICLE REFERENCE MODEL

MANDO ESP system controls the dynamic vehicle motion in the emergency situation such as the final oversteer/understeer and allows the vehicle to follow the course as desired by the driver. Therefore it is very important to determine the desired course, which is decided by the vehicle reference model. The first reference model is based on the fundamental physics. Formula (1) and (2) show the 2 DOF linear equations of motion according to the yaw rate and side slip angle used for the first reference model.[3]

$$I_z \dot{r} = N_\beta \beta + N_r r + N_\delta \delta_{sw} \tag{1}$$

$$mV(r + \dot{\beta}) = Y_\beta \beta + Y_r r + Y_\delta \delta_{sw} \tag{2}$$

However, since the vehicle motion is highly nonlinear, the simple linear equation is not enough to describe the whole vehicle motion in the various kinds of situation, especially unstable motion in the tire adhesion limit. For the nonlinear system design, it is well known that the look-up table is very effective and the table of the vehicle reference model should be determined by the actual vehicle tests on high and low friction road surface. The look-up table for the first vehicle reference is composed of 20 by 20 matrix according to the steering wheel angle and vehicle speed. The vehicle tests for each applied vehicle should be carried out to complete the look-up table.

On the other hand when the steering wheel is turned excessively in relation to the existing vehicle speed on the low-μ surface, the desired yaw rate which is predetermined by the first reference model is not adequate to consider as the criterion for the vehicle control because if the vehicle follows the 1st model, its side slip angle increases and the vehicle pushes out in a strong understeering manner. The control of the vehicle motion makes sense only if the adhesion of the wheels on the road surface permits the required yaw moment to act on the vehicle. Therefore, the reference model should be prevented from being selected as the preset value under all circumstances, according to the 1st reference model.

And the maximum yaw rate should be limited to the value which is available on the road surface. The limitation value is determined by the coefficient of friction between the tire and road surface and vehicle speed.

In case of stationary motion which has the small or zero steering angle rate on the low-μ surface, the 1st reference model is still effective as long as the difference between the measured lateral acceleration and the target limited acceleration is below a threshold. Therefore, a look-up table is required to define the enter condition for application of the yaw limitation and should be determined by the actual vehicle tests. Figure 1 shows the look-up table.

There exists a special condition, U-turn, in which the above 1st reference model and yaw limitation don't have enough information to control the understeering of the vehicle.

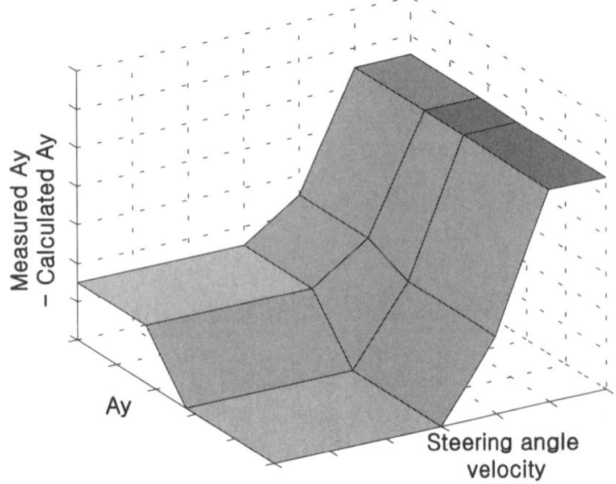

Figure 1. A look-up table for the enter condition of the yaw limitation.

Therefore an additional reference model is needed using the side slip angle rate.

Consequently, The three reference models are used to minimize the following performance index J under all driving condition.

$$J = \int \left(\begin{array}{c} K_P (\dot{\psi}_{measured} - \dot{\psi}_{desired}) \\ + K_D \dfrac{d(\dot{\psi}_{measured} - \dot{\psi}_{desired})}{dt} \end{array} \right) dt \qquad (3)$$

WHEEL SLIP CONTROL

The PD gains K_P, K_D of equation (3) depend on the vehicle speed and the driving situations (understeering and oversteering). When reaching the maximum lateral acceleration the tire adhesion limit will be exceeded first at the front axle for an understeering vehicle and first at the rear axle for an oversteering vehicle which means the slip angle increases uncontrolled at the respective axle. For understeering vehicle this results in a decrease of the side slip angle and because of that a relative decrease of the rear side slip angle. The corresponding decrease of the rear cornering force leads to a stabilization of the vehicle on a large path radius with a smaller lateral acceleration. For oversteering vehicle in contrast this results in an increase of the side slip angle and therefore a relative rise of the front slip angle. The corresponding increase of the front leads to the dangerous spin-out. So, the PD gains in the performance index J are less weighted for the understeering case since the compensation of understeering can make the vehicle the excessive oversteering. In order to minimize the performance index J, the brake and engine intervention are allowed. In case of the brake intervention, if the brake pressure is increased too much, the wheel goes to lock-up and the driving comfort is reduced and controllability doesn't become good, especially on ice road surface.

Therefore we have to monitor and limit the wheel slip value for the vehicle body control. Figure 2 shows the relationship between the yaw moment (or yaw acceleration) compensated by one wheel brake intervention and the wheel slip ratio. These are obtained by step steering input and say the limitation values of the controlled wheel slip. The figure says that even though the wheel slip increase, the yaw compensation moment doesn't increase after the front outside wheel slip reaches approximately 30% on the dry asphalt, and rear inside wheel slip does approximately 10%, respectively. On the ice road surface the saturation points are 10% for respectively. On the ice road surface 10% for rear, 60% for front, rear wheel and 60% for front wheel.

To achieve the target wheel slip which is determined from the performance index J, the wheel pressure is modulated by using solenoid valve which is controlled with Linear Flow Control (LFC) concept instead of on-off (digital) method to reduce the actuation noise and improve the brake pedal feeling when the driver applies the braking.

Since the amount of yaw moment compensation should be changed according to the coefficient of friction of the road surface, the estimation process for determining the coefficient of friction of the road surface is inevitable. The coefficient of friction can be determined only after the incipient instability of the vehicle begins. Since no estimated coefficient of friction is not available before the side slip of the vehicle is built up, the coefficient of friction $\mu = 1$ is set at first. The incipient instability is detected by the comparison between the first reference vehicle model and measured yaw rate.

Since the amount of yaw moment compensation should be changed according to the coefficient of friction of the road surface, the estimation process for determining the coefficient of friction of the road surface is inevitable. The coefficient of friction can be determined only after the incipient instability of the vehicle begins. Since no estimated coefficient of friction is not available before the

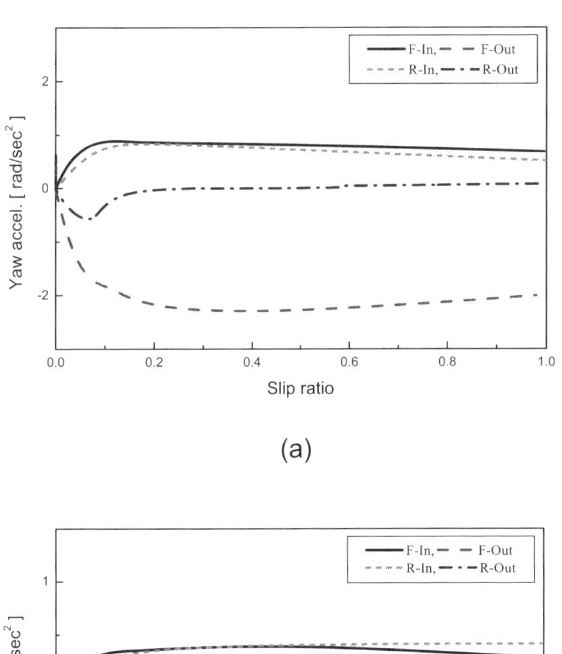

(a)

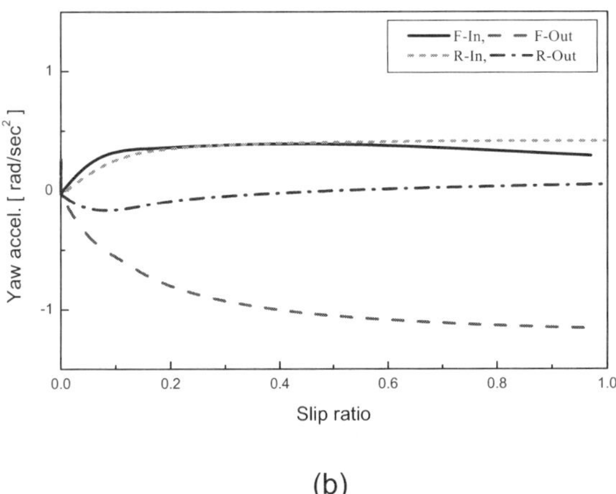

(b)

Figure 2. Relationship between the compensated yaw acceleration (moment) and wheel slip ratio. (a) Dry asphalt (b) Ice road

side slip of the vehicle is built up, the coefficient of friction $\mu = 1$ is set at first. The incipient instability is

detected by the comparison between the first reference vehicle model and measured yaw rate.

VEHICLE REFERENCE SPEED

The vehicle reference speed calculated only with the wheel speed sensors is not accurate enough to control the vehicle body in case of cornering maneuver because ESP makes use of the yaw rate of the vehicle body as a reference, although it is quite good for the individual wheel control of ABS, TCS. So, the actual yaw rate value should be included in the calculation of vehicle reference speed.

Especially, 4WD vehicle makes the determination of the vehicle reference speed more difficult because a free rolling wheel doesn't exist any more. To solve this problem, a longitudinal accelerate sensor is used and on the hill road, a special strategy is needed.

ENGINE INTERVENTION

For the enhancement of vehicle stability, the brake intervention is the most effective method, but the engine intervention also helps reduce the instability of the vehicle although its response is slow. Especially, in the stationary motion such as driving on the circle, the engine torque reduction prevents the vehicle from push-out and spin-out in advance, which makes the brake intervention minimized or not to be needed any more. In the continuous slalom maneuvering, the engine torque reduction increases the yaw rate of the vehicle, but its rate is so slow that the final oversteering is diminished and the vehicle becomes stabilized by the reduced vehicle speed.

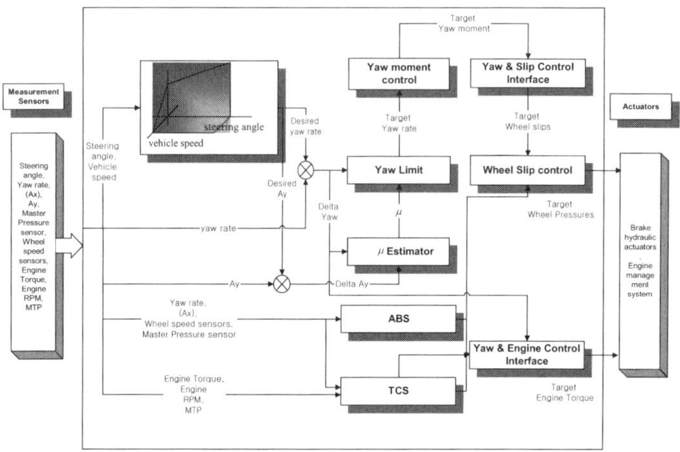

Figure 3 shows the control flow diagram of MANDO ESP system.

ABS AND ESP INTEGRATED CONTROL

If the vehicle is in an anti-lock braking mode and ESP mode simultaneously, the control concept of the front

537

outside wheel for oversteering and the rear inside wheel for understeering is same as ESP only case, but the control strategy for the their diagonal wheels should be changed. That is, the side force at the diagonal wheel to the ESP control wheel should be increased to reduce the understeering & oversteering of the vehicle and it is achieved by reduction of the target slip ratio at the diagonal wheel. While the brake is applied during lane change, the understeering occurs. At this situation ESP helps the anti-lock brake control by increasing the pressure at the rear inside wheel and decreasing the pressure at the front outside wheel.

HYDRAULIC UNIT (HU)

ESP hydraulic unit was introduced with the pressure rise rate assisting component such as the pre-charge pump, active booster or pre-charge accumulator, which were effective especially in the low temperature environment. MANDO also started the ESP project with the pre-charge accumulator. The wheel pressure rise rate was fast enough to compensate the excessive yaw moment of the vehicle body, even in the temperature $-30\,℃$. The volume and the number of the elements of HU, however, needed to be reduced and the pre-charge accumulator could be eliminated by the optimization of HU.

MANDO MGH25 ESP without the pre-charge accumulator makes the higher wheel pressure rise rate than that of the previous models with pre-charge accumulator in the normal & low temperature ($-30\,℃$). Figure 4 shows MANDO MGH25 hydraulic unit and Figure 5 the sensors used for ESP.

STRATEGY AT FAILURE

The failsafe logic is composed of three parts, that is, basic monitoring, self testing, and model based sensor monitoring. The basic monitoring includes most of already proven monitoring logic of MANDO ABS/TCS. It monitors the physically admissible function of the wheel speed sensors, the electrical functions of solenoid valves and pump motor, the power supply for the additional sensors and ECU, the connector pins and CAN communication. For these failsafe, hundreds kinds of tests should be carried out. The second part, self testing logic is related to the sensor's own specification. The third part, model based sensor monitoring is very effective in this kind of system which makes use of many sensors. The steering wheel angle, yaw rate, lateral acceleration and wheel speeds have the physical relationship with each other. The model based sensor monitoring is accomplished by the residual concept – the difference between sensor output signal and analytical redundancy (reference).

Figure 4. Mando MGH25 ESP HECU

Figure 5. Mando MGH25 ESP Sensors

The residual depends on both the faults and model uncertainties and should be determined so as to be sensitive to faults and robust against the model uncertainties. The sensor offsets are corrected based on the residual value and if the residual is beyond a threshold, the ESP warning lamp is switched on.

The yaw control system fault tolerance for a single sensor is one of the important things and is defined as follows. In case of steering angle sensor failure, yaw control function is not available and does a graceful exit. For yaw rate sensor failure, yaw control function is available while switching to feed forward control based only on steering and vehicle reference speed for 4WD vehicle and uses the estimated yaw rate for 2WD. For lateral acceleration sensor failure, yaw control is available only with the estimated value.

SIMULATION AND TEST RESULTS

COMPUTER SIMULATION

For off-line simulation, ADAMS is used and for hardware in the loop simulation (HILS), AC-104 with MATRIXx and ETAS LabCar system are used. Figure 6 shows the simulation results of the double lane change on the snow road surface, where the linear predictive driver model is used and the driver tries to avoid the collision. Without ESP, both yaw rate and side slip angle become so large

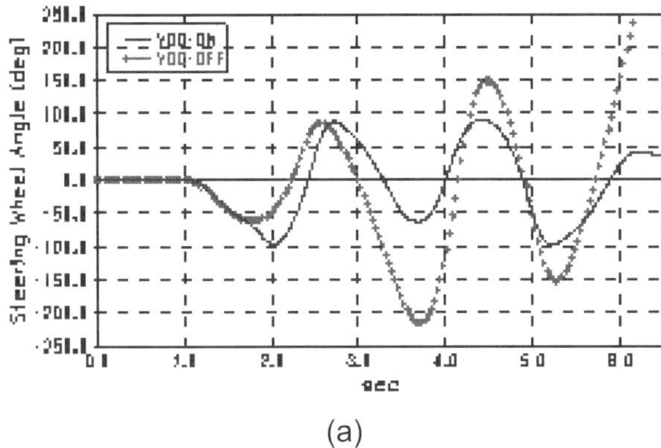

(a)

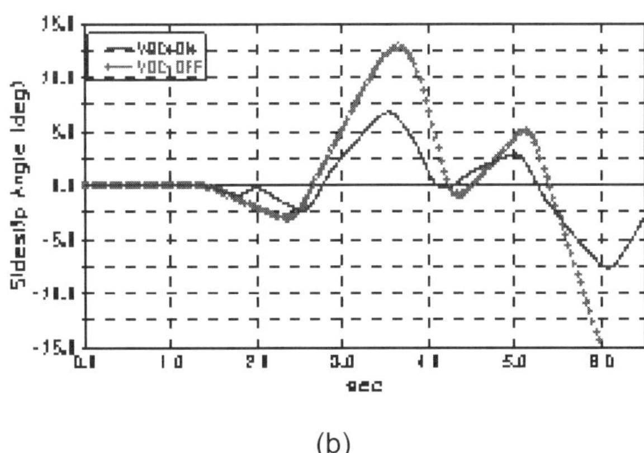

(b)

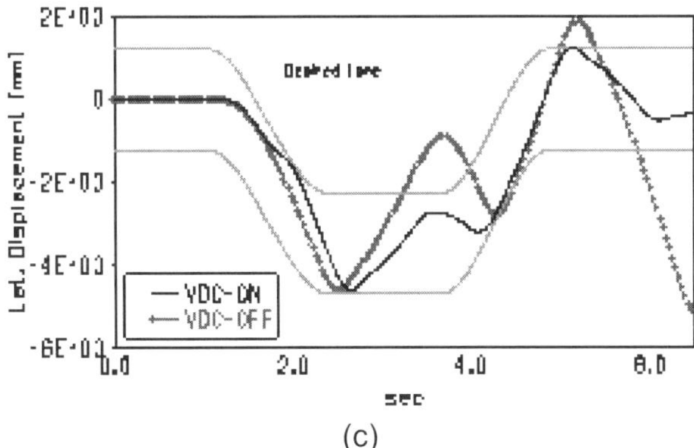

(c)

Figure 6. Double lane change simulation results (a) steering wheel angle, (b) side slip angle, (c) Trajectories

that the driver has to steer heavily to compensate the excessive vehicle motion. The large countersteering causes another countersteering and the steering wheel angle becomes larger and larger and finally the vehicle goes to spin out. Figure 6 (c) illustrates the trajectories of the vehicle for without ESP and with ESP, which can be used as the performance assessment of ESP.

VEHICLE TESTS

Figure 8 shows the slalom on the snow road surface. The vehicle without ESP loses its stability after 1 cycles of steering input, but the vehicle with ESP follows the desired steering input well.

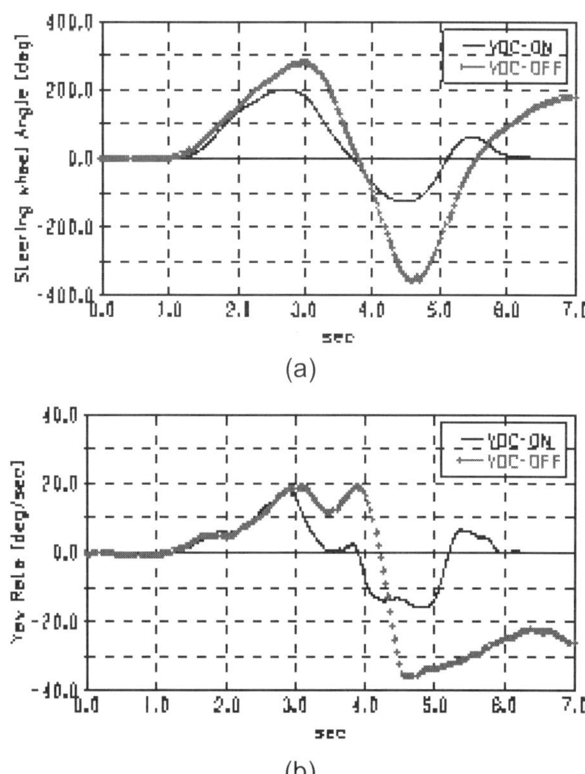

(a)

(b)

Figure 7. Lane change test results on ice road surface (a) steering wheel angle, (b) yaw rate

Figure 9 shows braking in a turn test results on the snow road surface. The longitudinal acceleration for the vehicle with ESP is bigger than that for the vehicle without ESP and the stopping distance is reduced with the help of ESP. Figure 7 shows the lane change test results on the ice road surface and the vehicle with ESP can follow the desired course although the vehicle without ESP goes out of control.

CONCLUSION

The performance of ESP can be different from car to car and from sedan to sport utility vehicle even though the control logic is same because the vehicle reference model is included in the control logic. Therefore, it takes a lot of efforts to set the vehicle reference model as accurate as possible and the control logic structure should be made so as to adjust the appropriate parameters

In an assessment of ESP, it must be shown to which extend the active safety of the driver is improved in reality. A comparison with ABS using braking in a turn test shows that the stability of the vehicle is enhanced and the stopping distance is reduced.

For the application project the objective evaluation method should be determined and the cooperation between the supplier and customer would be helpful to settle down the criteria for the assessment of ESP system. For fine tuning the sensitivity of the ESP controller can be adjusted to each driving condition. A higher sensitivity is limited so as to prevent unnecessary activation on the abnormal conditions such as rough road, banked road and the vehicle parameter variations on tire and loading conditions. Actually the unnecessary intervention gives the driver trouble. So the robustness tests on the various kinds of road conditions, vehicle conditions and driving situations are very important and actually hundreds of test are carried out.

REFERENCES

1. van Zanten, A.T., et al, "VDC System Development and Perspective", SAE980235, 1998
2. Automotive News, Nov.20, 2000
3. Milliken, W, et al, Race car vehicle dynamics, 1995

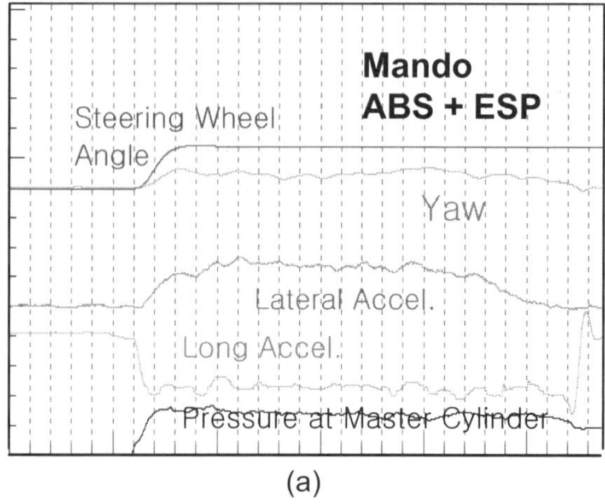

(a)

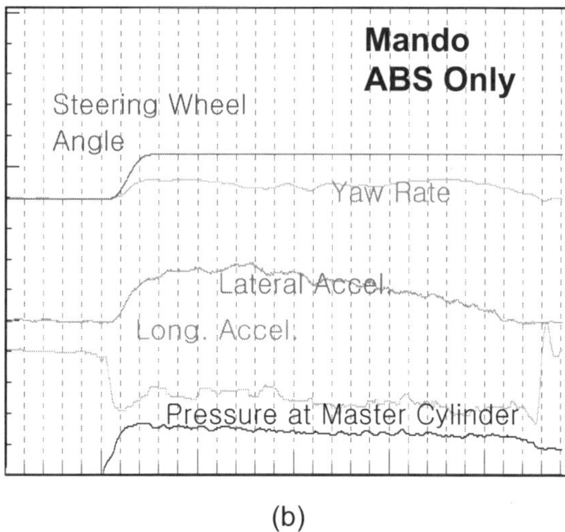

(b)

Figure 9. Braking in a turn test results on snow road surface
(a) MANDO ABS only, (b) MANDO ABS+ESP

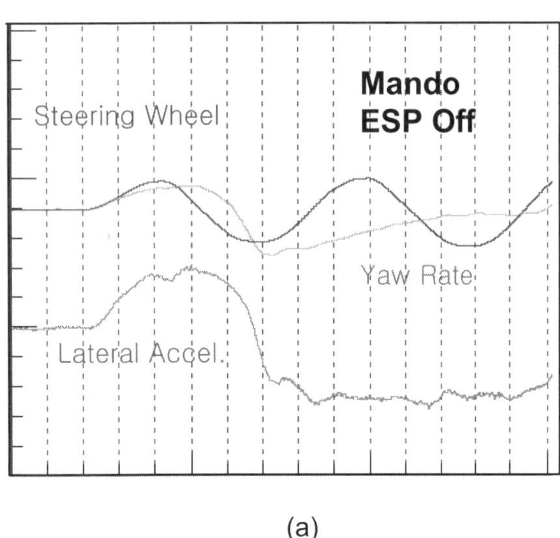

(a)

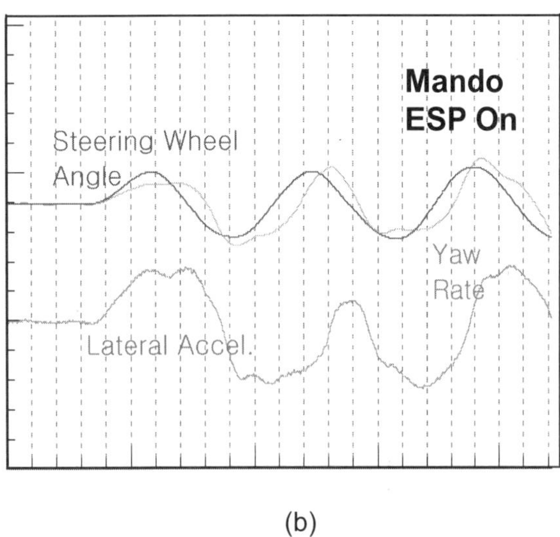

(b)

Figure 8. Slalom test results on snow road surface
(a) ESP Off, (b) ESP ON

Fuzzy Based Stability Enhancement System for a Four-Motor-Wheel Electric Vehicle

Farzad Tahami
Jovain Electrical Machines Co.

Reza Kazemi
IKCO Automobiles

Shahrokh Farhanghi
University of Tehran

Behzad Samadi
IKCO Automobiles

ABSTRACT

The stability of a four motor-wheel drive electric vehicle is improved by independent control of wheel torques. An innovative Fuzzy Direct Yaw Control method together with a novel wheel slip controller is used to enhance the vehicle stability and safety. Also a new speed estimator is presented in this paper, which is used for slip estimation. The intrinsic robustness of fuzzy controllers allows the system to operate in different road conditions successfully. Moreover, the ease to implement fuzzy controllers gives a practical solution for vehicle stability enhancement.

INTRODUCTION

Motor-wheels are the revolutionary new electric drive systems that can be housed in vehicle wheel assemblies. All-Wheel-Drive systems have been recognized as a break-through concept that will have a major impact on future electric and hybrid vehicle design. The drive system is versatile and can be configured for all types of electric vehicles including battery only, plug-in hybrid, autonomous hybrid and fuel cells. Motor-wheels permit packaging flexibility by eliminating the central drive motor and the associated transmission and driveline components, including the transmission, the differential, the universal joints and the drive shaft. In-wheel motors provide many advantages in manufacturing electric vehicles such as:

- More space for the installation of batteries, or auxiliary power unit for fuel cells.
- Better weight distribution.
- Low floor possibility for urban buses due to elimination of the axle and differential.
- Built-in 4WD capability.
- More tractive force and braking regenerative power due to all-wheel-drive capabilities.

The unequaled independent torques applied to the wheels provides another important aspect of such a vehicle. That is the ability of vehicle dynamic control to assist the driver with path correction, thus enhancing cornering and straight-line stability and providing enhanced safety. In fact, an electric vehicle with independent driven wheels provides another steering control input, i.e. the torque steering. Controlling of the yaw rate of a vehicle by utilizing this steering method is usually addressed as Direct Yaw-moment Control (DYC). It has been proved that DYC is more effective in enhancing vehicle stability than four wheel steering [1]. Actually, the yaw moment resulting from differential longitudinal tire force in the left and the right wheels is insignificantly influenced by lateral acceleration. On the contrary, the yaw moment generated by four-wheel steering decreases as the lateral acceleration increases [2].

In this paper, a control strategy for a driver assist stability system for a four in-wheel drive electric vehicle with a conventional front wheel steering is presented. The system comprises of a fuzzy logic controller in order to control the yaw rate. The disturbances in yaw moment are counter-balanced by a differential torque applied to the driven wheels of both sides. Another fuzzy controller is employed in order to prevent the wheels to enter saturation region due to the additional torque applied by the yaw controller.

Fuzzy controllers have been already proved to have good performance in two-motor-wheel drive vehicles [3, 4].

VEHICLE MODELING

VEHICLE MODEL - A fourteen-degree of freedom model is used for the simulation purposes. Where, 6 degrees are devoted to the chassis motion and eight degrees are assigned to the wheels rotational and vertical movement. Figure 1 shows different coordinates that are used for the modeling purpose.

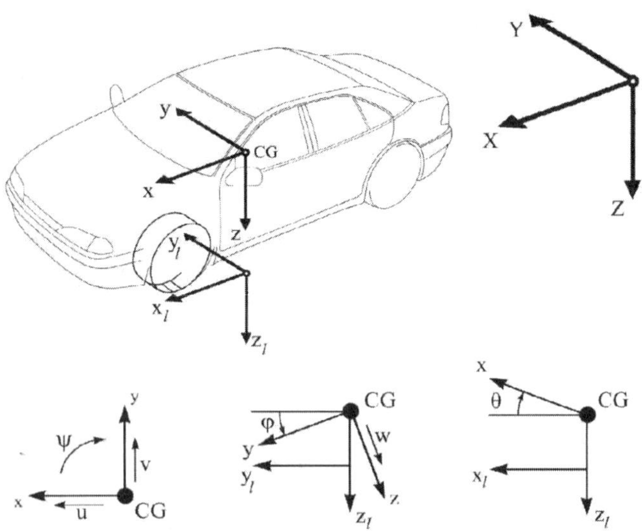

Figure 1. Coordinate systems for vehicle modeling

Where u, v and w are associated with longitudinal, lateral and vertical velocities and ψ, φ and θ denote the yaw, roll and pitch angles.

Suspension Model – A model of the suspension system is shown in Figure 2. It is assumed that the suspension is independent for each wheel. The suspension model consists of a spring and damper. Parameters K_{si} and C_{si} correspond to the spring stiffness and the damping factor. A virtual spring and damper assembly (corresponding to K_{ui} and C_{ui}) is considered for the tire elasticity. Z_s is the height of the center of gravity of the sprung mass. Z_{ui} is assigned for the height of unsprung mass and Z_{ri} is due to road roughness.

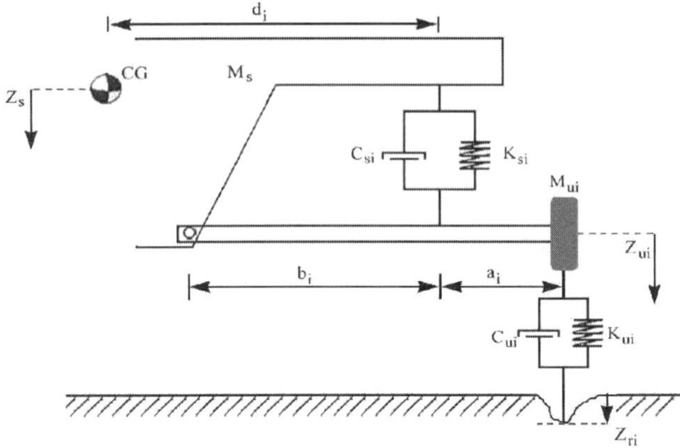

Figure 2. The model of suspension system

Tire Model – The accuracy of the simulation results is predominantly determined by the accuracy of the tire model used. The desired simulation for the application herein concerned can be carried out by a steady state tire characteristics. In this paper the simulation is performed using the well known magic formula [5]. The longitudinal force is carried out, in this formulation, from the following formula:

$$F_x = D\sin(C\arctan(B\Phi)) + S_v \qquad (1)$$

Where,

$$\Phi = (1 - E)(\lambda + S_h) + (E/B)\arctan(B(\lambda + S_h))$$

In which λ is the wheel slip and other parameters depend on the normal load and factors, which are determined from experience.

There are similar formulas for the lateral force and the self-aligning torque. Figure 3 shows typical longitudinal and lateral forces obtained from the magic formula.

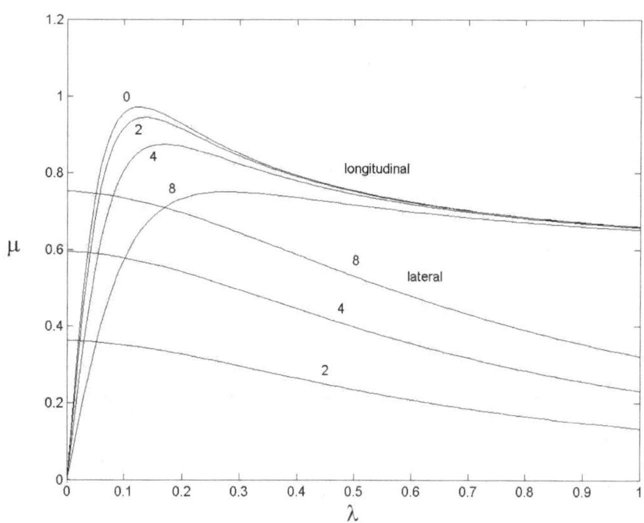

Figure 3. Typical tire characteristics at different slip angles

DYNAMIC EQUATIONS – Considering the forces acting on the vehicle, as depicted in Figure 4, one can write the vehicle motion equations in horizontal plane as below:

$$M_t(\dot{u} + qw - rv) = M_s h(\dot{q} + pr) + F_{x1}\cos(\delta_{f1})$$
$$- F_{y1}\sin(\delta_{f1}) + F_{x2}\cos(\delta_{f2}) \qquad (2)$$
$$- F_{y2}\sin(\delta_{f2}) + F_{x3} + F_{x4} - F_{ax}$$

$$M_t(\dot{v} + ru - pw) = -M_s h(\dot{p} + qr) + F_{x1}\sin(\delta_{f1})$$
$$+ F_{y1}\cos(\delta_{f1}) + F_{x2}\sin(\delta_{f2}) \qquad (3)$$
$$+ F_{y2}\cos(\delta_{f2}) + F_{y3} + F_{y4} - F_{ay}$$

$$I_z \dot{r} = (I_x - I_y)pq + L_f[F_{x1}\sin(\delta_{f1})$$
$$+ F_{y1}\cos(\delta_{f1}) + F_{x2}\sin(\delta_{f2})$$
$$+ F_{y2}\cos(\delta_{f2})] - L_r(F_{y3} + F_{y4})$$
$$+ \frac{T_f}{2}[F_{x1}\cos(\delta_{f1}) - F_{y1}\sin(\delta_{f1}) \qquad (4)$$
$$- F_{x2}\cos(\delta_{f2}) + F_{y2}\sin(\delta_{f2})]$$
$$+ \frac{T_r}{2}(F_{x3} - F_{x4}) + \sum_{i=1}^{4} M_{zi}$$

Where r, p and q denote the angular velocities corresponding to yaw, roll and pitch angles, M_t is vehicle total mass, M_s is the sprung mass and I_x, I_y, and I_z are the vehicle moment of inertia about the x, y, and z axes. The parameter h is the distance from roll axis to sprung mass centre of gravity and F_{ax} and F_{ay} are the aerodynamic drag coefficients.

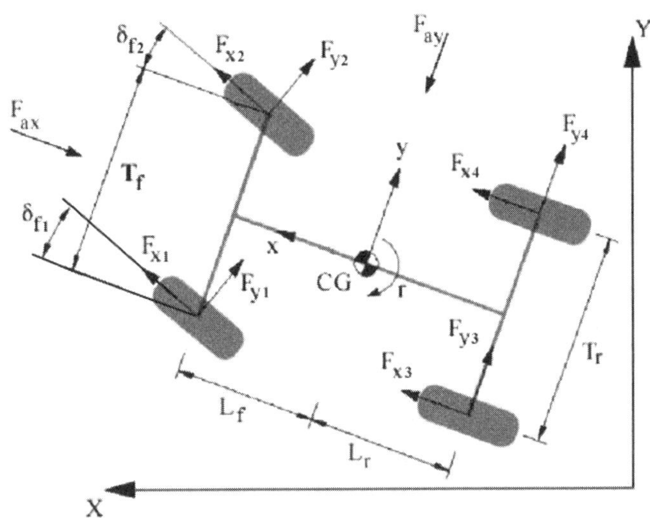

Figure 4. Acting forces in horizontal plane

Now, considering the acting torques on the vehicle body, the sprung mass motion can be carried out as follows.

The vehicle roll motion satisfies the following equations:

$$I_{xs}\dot{p} + (I_{zs} - I_{ys})qr = \sum_{i=1}^{4}\overline{R}_{ri}F_{si} + M_s gh\varphi$$
$$- M_s h(\dot{v} + ru - pw) \qquad (5)$$

Where I_{xs}, I_{ys} and I_{zs} are the moment of inertia of the vehicle sprung mass about different axes and $\overline{R}_{ri}$ is a coefficient corresponding to suspension geometry of the i_{th} wheel as follows:

$$\overline{R}_{r1} = -d_1 + (d_1 - b_1)\frac{a_1}{a_1 + b_1}$$

$$\overline{R}_{r2} = d_2 - (d_2 - b_2)\frac{a_2}{a_2 + b_2}$$

$$\overline{R}_{r3} = -d_3 + (d_3 - b_3)\frac{a_3}{a_3 + b_3}$$

$$\overline{R}_{r4} = d_4 - (d_4 - b_4)\frac{a_4}{a_4 + b_4}$$

Dimensional parameters a_i, b_i and d_i have been depicted in Figure 2. F_{si} is the force acting on suspension system of the wheel number i:

$$F_{s1} = K_{s1}(Z_{u1} - Z_s + L_f\sin(\theta) + d_1\sin(\varphi))$$
$$+ C_{s1}(w_{u1} - w + L_f q\cos(\theta) + d_1 p\cos(\varphi))$$
$$F_{s2} = K_{s2}(Z_{u2} - Z_s + L_f\sin(\theta) - d_2\sin(\varphi))$$
$$+ C_{s2}(w_{u2} - w + L_f q\cos(\theta) - d_2 p\cos(\varphi))$$
$$F_{s3} = K_{s3}(Z_{u3} - Z_s - L_r\sin(\theta) + d_3\sin(\varphi))$$
$$+ C_{s3}(w_{u3} - w - L_r q\cos(\theta) + d_3 p\cos(\varphi))$$
$$F_{s4} = K_{s4}(Z_{u4} - Z_s - L_r\sin(\theta) - d_4\sin(\varphi))$$
$$+ C_{s4}(w_{u4} - w - L_r q\cos(\theta) - d_4 p\cos(\varphi))$$

Where w_{ui} is the vertical speed of the i_{th} wheel.

Pitch rate (q) is carried out as in below:

$$I_{ys}\dot{q} + (I_{xs} - I_{zs})pr = -L_f(\frac{F_{s1}}{R_{r1}} + \frac{F_{s2}}{R_{r2}})$$
$$+ L_r(\frac{F_{s3}}{R_{r3}} + \frac{F_{s4}}{R_{r4}}) + h_{cg}\sum_{i=1}^{4} X_i \qquad (6)$$

Where h_{cg} is the height of the center of gravity and the parameter R_{ri} is defined with respect to the suspension geometry of the i_{th} wheel:

$$R_{ri} = \frac{a_i + b_i}{b_i}$$

The motion equation for the sprung mass in the vertical direction can be written as follows:

$$M_s(\dot{w} + pv - qu) = \sum_{i=1}^{4}\frac{F_{si}}{R_{ri}} \qquad (7)$$

The vertical movement of each unsprung mass is expressed by the following equation:

$$M_{ui}\dot{w}_{ui} = K_{ui}(Z_{ri} - Z_{ui}) + C_{ui}(w_{ri} - w_{ui}) - \frac{F_{si}}{R_{ri}} \quad (8)$$

ELECTRIC MOTORS AND DRIVES MODELING

Permanent Magnet Synchronous (PMS) motors are the most popular motors for in-wheel applications. The developed torque of a salient pole PMS motor in d-q coordinates is:

$$T = P\varphi i_q + P(L_d - L_q)i_d i_q \quad (9)$$

Where P is the number of poles and φ denotes the magnetic flux linkage. L_d, L_q and i_d, i_q are the inductances and currents in d and q directions.

Figure 5 shows the block diagram of a typical electric drive system for PMS motors. Note that the flux control loop (i_{sd} control loop in figure 5) is slower than the torque control loop and the flux is a slow varying variable. Furthermore, since the dynamic response of modern motor drives are much faster than wheel dynamics, and considering the dominant poles of the closed loop system, an electric motor and its drive can be simply modeled as:

$$G(s) = \frac{T}{T^*} = \frac{1}{\left(1 + 2\xi s + 2\xi^2 s^2\right)} \quad (10)$$

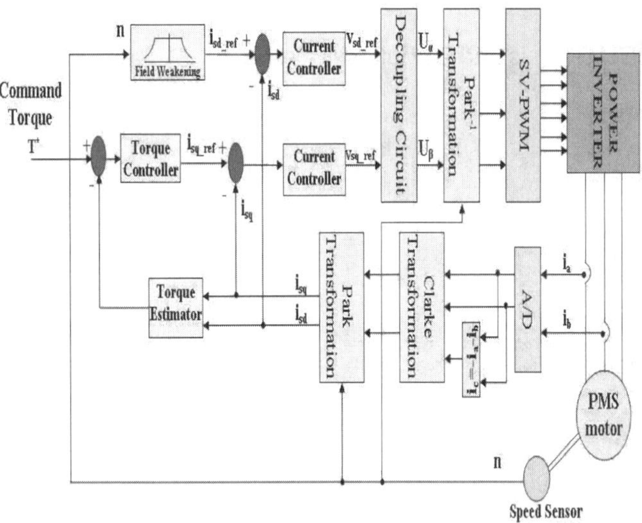

Figure 5. Block-diagram of an electric drive system for PMS motors

CONTROLLER DESIGN

YAW RATE CONTROLLER - An uneven longitudinal tire force may have a significant undesired effect on yaw motion. In this paper a differential torque is applied to the driven wheels of the right and left sides in order to compensate the disturbance yaw moment.

Rearranging equation 4 and assuming small and equal steering angles for the front wheels and no steering for the rear wheels one can write:

$$I_z \dot{r} = (I_x - I_y)pq + L_f F_{yf} - L_r F_{yr}$$
$$+ \frac{T_f}{2}\Delta F_{xf} + \frac{T_r}{2}\Delta F_{xr} + \sum_{i=1}^{4} M_{zi} \quad (11)$$

Where ΔF_{xf} and ΔF_{xr} are the differences in the longitudinal force of left and right wheels in the front and the rear:

$$\Delta F_{xf} = (F_{x1} - F_{x2})$$
$$\Delta F_{xr} = F_{x3} - F_{x4}$$

And,

$$F_{yf} = F_{y1} + F_{y2}$$
$$F_{yr} = F_{y3} + F_{y4}$$

Hence, the yaw rate can be directly controlled by applying a differential input torque to the right and left wheels (in tire stable region).

From the steady state cornering theory of a bicycle model, it is known that the desired yaw velocity of a vehicle satisfies the following equation [6]:

$$r_d = \frac{V/L}{1 + \frac{KV^2}{L}}\delta \quad (12)$$

Where K is the under-steer gradient and L is the vehicle wheelbase:

$$L = L_f + L_r$$

A fuzzy logic controller is used to keep the yaw rate in its desired value, r_d.

The error $e = r - r_d$ and the change in error are applied to a fuzzy controller. The output of the controller is the deviation in the applied torque to the motors. Figure 6 shows the block diagram of the vehicle model and the yaw rate controller.

The normalized membership functions for fuzzification of the controller inputs and defuzzification of the controller output are depicted in Figure 7.

Table 1 shows the rule base of the Fuzzy controller.

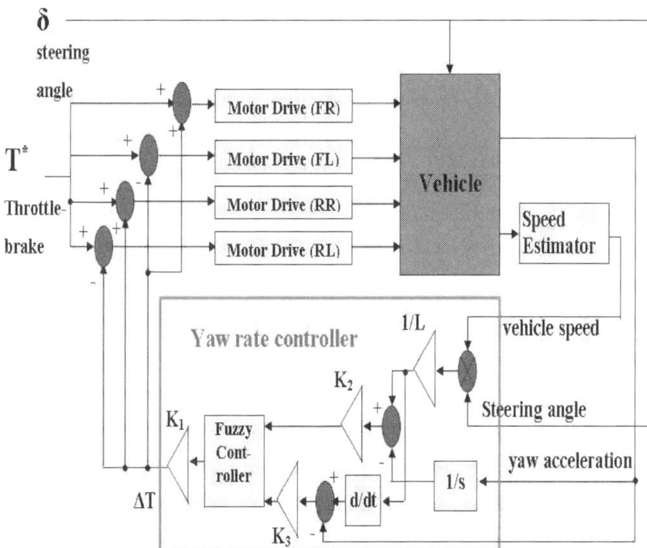

Figure 6. Yaw rate controller

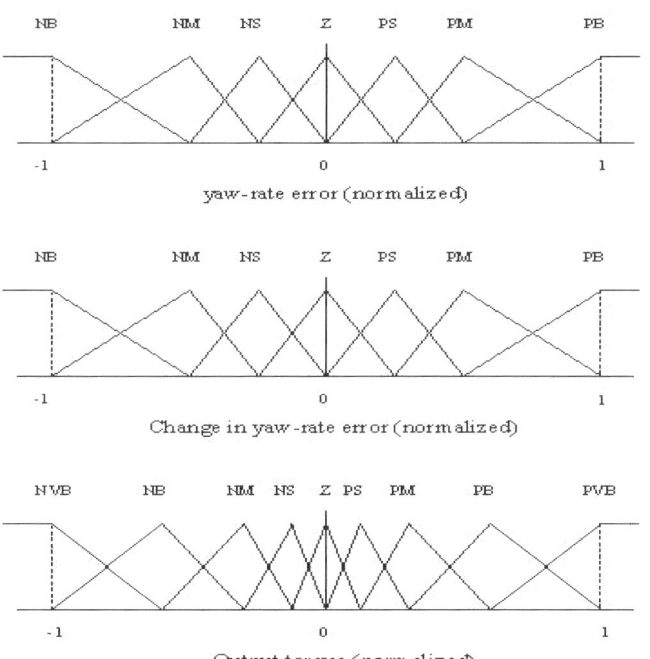

Figure 7. Membership functions for the yaw rate controller

	yaw-rate error						
Change in yaw-rate error	NB	NM	NS	Z	PS	PM	PB
NB	NVB	NVB	NVB	NB	NM	NS	Z
NM	NVB	NVB	NB	NM	NS	Z	PS
NS	NVB	NB	NM	NS	Z	PS	PM
Z	NB	NM	NS	Z	PS	PM	PB
PS	NM	NS	Z	PS	PM	PB	PVB
PM	NS	Z	PS	PM	PB	PVB	PVB
PB	Z	PS	PM	PB	PVB	PVB	PVB

Table 1. Fuzzy rule-base for the yaw rate controller

SLIP CONTROLLER - The additional torque applied by the yaw controller may saturate the tire force, hence a slip ratio controller is required in the yaw control loop.

A Fuzzy controller for each wheel is used to keep the slip ratio within its stable region. The inputs to the controller are the wheel slip and the wheel angular acceleration. The Fuzzy rules regard the latter input as a virtual criterion for the direction of slip variations. The output of the controller is the amount of torque weakening that should be devoted to the torque command of the motors, in order to prevent the tire to enter into its saturation region. A block diagram of the overall control system is depicted in Figure 8.

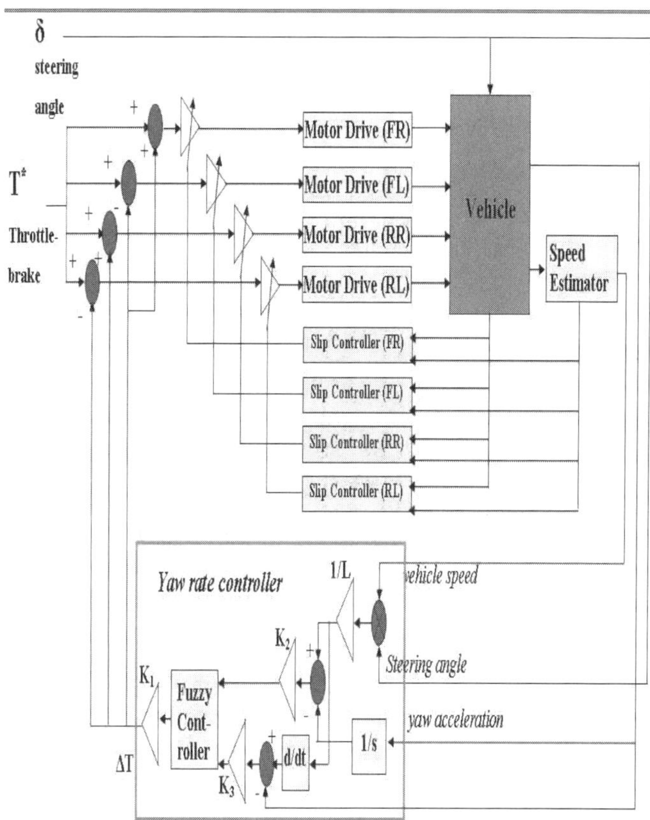

Figure 8. Overall block diagram of control system

Figure 9 shows the membership functions. The rule base of the Fuzzy slip controller is tabulated in Table 2.

	Slip (unsigned)						
Wheel acceleration times the sign of slip	Z	S	C	M	RB	B	VB
NB	U	U	NVB	NB	NM	NS	Z
NM	U	U	U	U	NU	M	Z
Z	U	U	U	U	NU	S	Z
PM	U	U	NU	S	Z	Z	Z
PB	U	NU	S	Z	Z	Z	Z

Table 2. Fuzzy rule-base for the slip controller

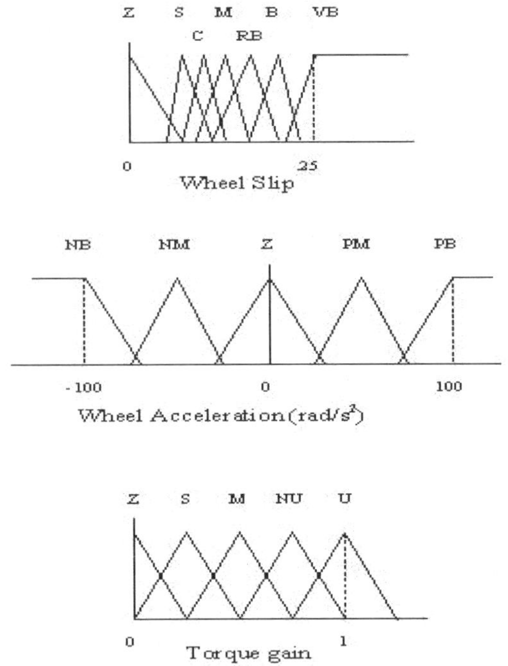

Figure 9. Membership functions for slip controller

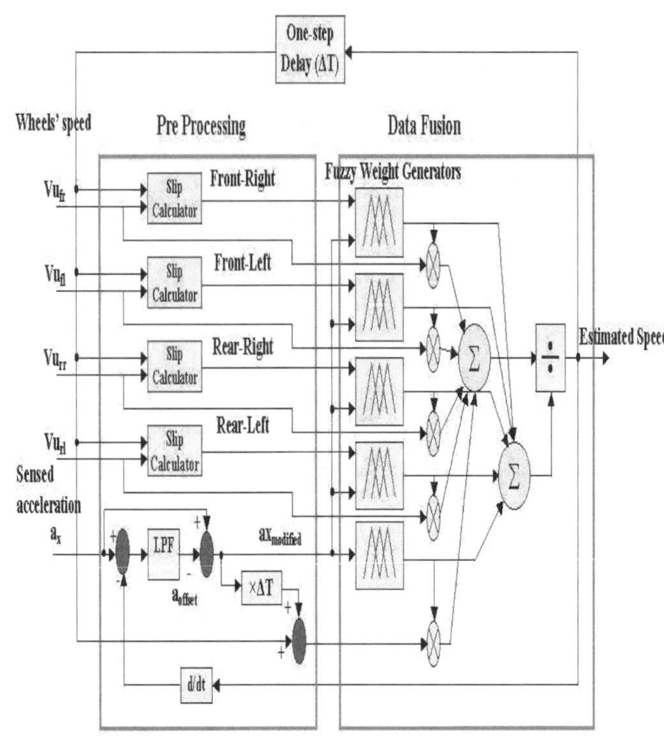

Figure 10. Block diagram of the speed estimator

SPEED ESTIMATOR

All the required control signals, but the vehicle speed, can be obtained from sensors, at a reasonable price. These sensors measure steering wheel angle, wheel speeds and yaw rate. Vehicle speed however, should be estimated anyhow. In this paper a data fusion method is used for this purpose, the wheel linear speeds are used as pseudo-sources of vehicle speed. An additional accelerometer is embedded in the vehicle in order to measure the vehicle longitudinal acceleration. The integral of the acceleration is regarded as another input. The inputs are all fed into an estimator, where a fuzzy logic determines which input is more reliable. Inputs are weighted and averaged to give the estimated speed. Figure 10 shows the block diagram of the estimator. The estimator consists of two stages. In the first stage, which is addressed as preprocessing, the wheel slips are calculated using the previous estimated vehicle speed and the measured wheel speeds. Meanwhile, the offset of the accelerometer is rejected by subtracting the derivation of the estimated speed from the measured acceleration and then low-pass filtering.

In the next stage, the wheel linear speed and the vehicle speed, which has been carried out by integration, are weighted using fuzzy rules. The weighted values are averaged to give the vehicle speed.

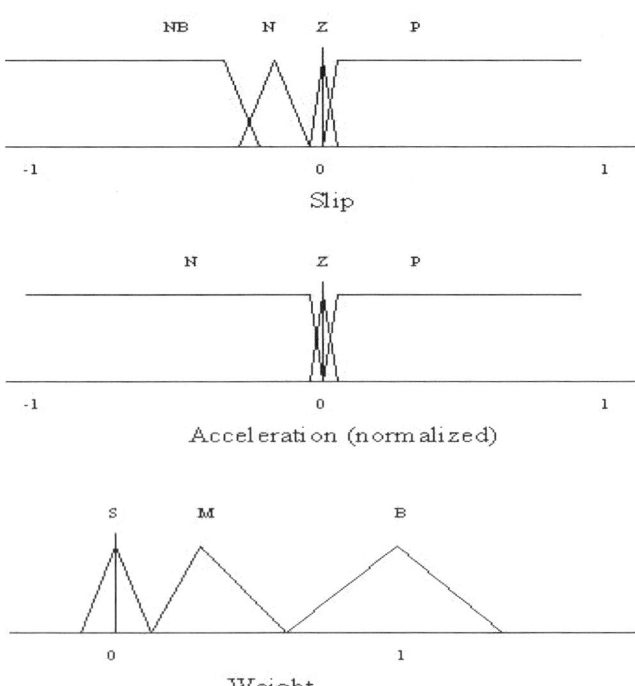

Figure 11. Membership functions for fuzzification of inputs and the output of speed estimator

Figure 11 depicts the membership functions used in fuzzy weight generators. In Table 3 the rule sets, which are used for the weight generators, are tabulated. These rules have been established on basis of linguistic terms such as:

- In cruise driving, the integral of the vehicle acceleration is not reliable, since the accelerometer signal is comparable to offset and noise of the measuring circuit.
- In braking situation all wheel speeds are weighted low, because of large wheel slips.

<table>
<tr><td rowspan="5" style="writing-mode: vertical-lr">Vehicle acceleration</td><td></td><td colspan="4">slip</td></tr>
<tr><td></td><td>NB</td><td>N</td><td>Z</td><td>P</td></tr>
<tr><td>N</td><td>S</td><td>H</td><td>M</td><td>S</td></tr>
<tr><td>Z</td><td>S</td><td>M</td><td>H</td><td>M</td></tr>
<tr><td>P</td><td>S</td><td>S</td><td>M</td><td>H</td></tr>
</table>

(a)

	acceleration	
N	Z	P
H	S	H

(b)

Table 3. Weighting rule sets for: a) wheels speed, b) Integration

SIMULATION RESULTS

A series of computer simulations was carried out to evaluate the performance of the proposed control system. The parameters of the vehicle model are tabulated in Table 4.

Parameter	Symbol	Unit	Value
Vehicle total mass	M_t	Kg	1482.7
Front Unsprung mass	M_{uf}	Kg	95.5
Rear Unsprung mass	M_{ur}	Kg	108.8
Moments of inertia of total car about X axis	I_x	Kgm^2	346.73
Moments of inertia of total car about Y axis	I_y	Kgm^2	1675.8
Moments of inertia of total car about Z axis	I_z	Kgm^2	1808.8
Moment of inertia of wheels	I_w	Kgm^2	2.11
Track width	T	m	1.4375
Distance from front axle to CG	L_f	m	1.2247
Distance from rear axle to CG	L_r	m	1.4373
Height of CG	h_{cg}	m	0.5253
Effective radius of wheels	R_w	m	0.285
Spring constant of front springs	K_{sf}	N/m	15400
Spring constant of rear springs	K_{sr}	N/m	19000
Spring constant of tires	K_u	N/m	175000
Damping coefficient of front dampers	C_{sf}	Ns/m	1150
Damping coefficient of rear dampers	C_{sr}	Ns/m	6000
Damping coefficient of tires	C_{uf}	Ns/m	50
Aerodynamic drag coefficient	C_d	$N/(m/s)^2$	.32

Table 4. The specifications of the vehicle used in simulation

LANE CHANGE ON SLIPPERY ROAD - The results of a lane change maneuver for a 3-meter lateral displacement at 60 km/h on unpacked snow are illustrated in Figure 12. It is assumed that the driver applies the same steering effort as on an adhesive road. This figure compares lane change with and without controllers.

As depicted in Figure 12a, with the controller on board, the desired lateral displacement achieved after 60 meters of longitudinal distance. Without the control system, the vehicle cannot achieve the target lateral displacement and additional efforts are required to return the car into the path. Obviously the controller successfully helps the driver to handle the lane change.

Figure 12b shows how the controller keeps the vehicle's yaw rate thoroughly near the yaw rate reference. The control system applies additional torques to wheels as depicted in Figures 12d to 12g. The torques in opposite sides are rather differential.

BRAKING ON μ-SPLIT ROAD - Next simulation performs braking at 60 km/h on a μ-split road (dry pavement on the right side and unpacked snow on the left). A regenerative braking is assumed; hence a constant torque is applied to the driven wheels by the electric motors. For better illustration of results, no driver model is considered and the steering wheel is assumed to be fixed.

Simulation results are depicted in Figure 13. One can see that the control system successfully prevents the car to deviate from its path. The lateral displacement and lateral acceleration is negligible when the car brakes on the split road. Even though, in the absence of slip controller the deviation is acceptable, but as shown in Figures 13d and 13e the left side tires are saturated and blocked even earlier than in case of no controller at all, which results in lacking of lateral stability.

Also, while the rate of speed reduction with no controller is slightly better than the controlled vehicle, but the car starts turning around after braking, as can be seen in Figure 13a.

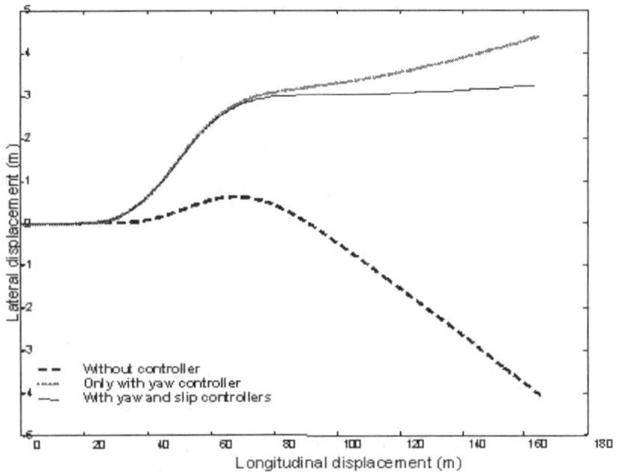

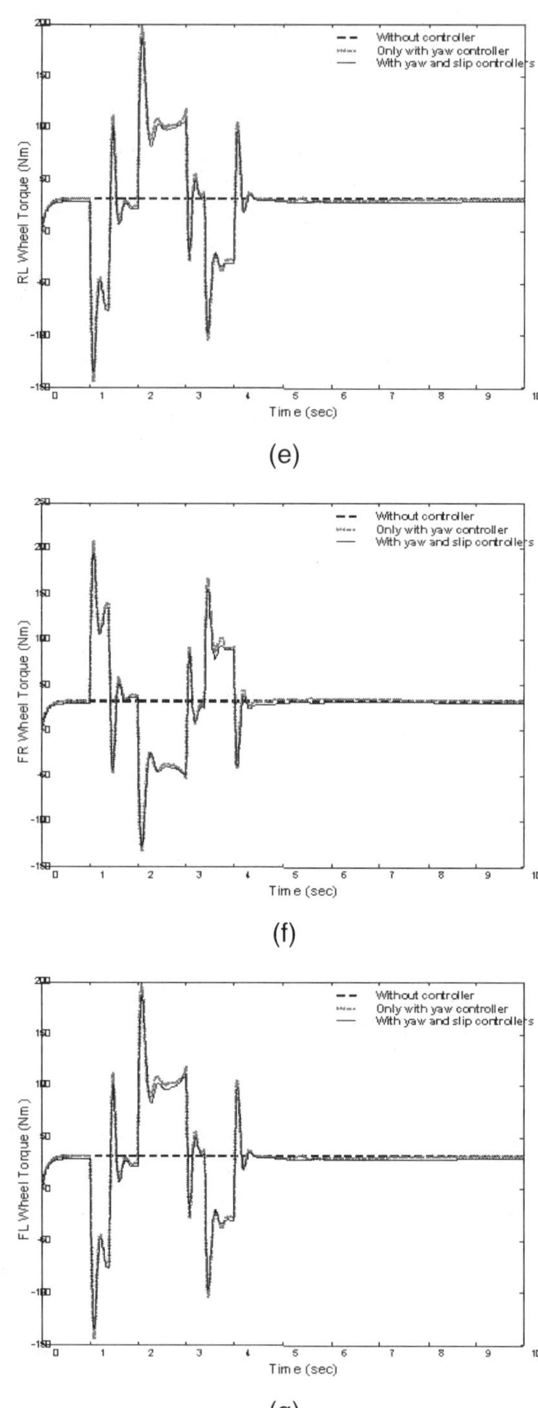

Figure 12. Lane change on a slippery road: a) Vehicle trajectory, b) Yaw rate and yaw reference, c) Vehicle side slip angle, d) Applied Torque to the Rear-Right wheel, e) Applied Torque to the Rear-Left wheel, f) Applied Torque to the Front-Right wheel, g) Applied Torque to the Front-Left wheel.

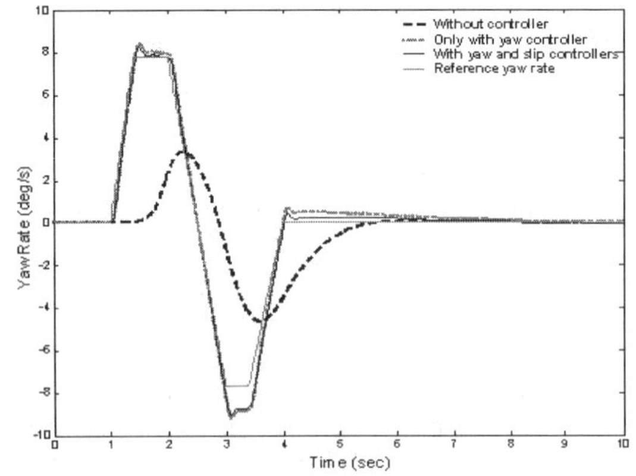

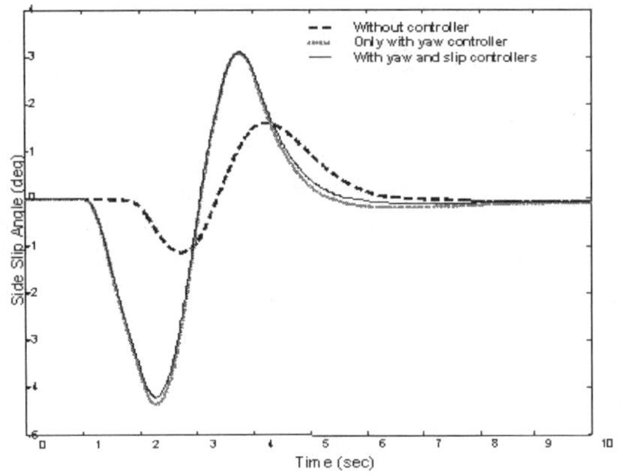

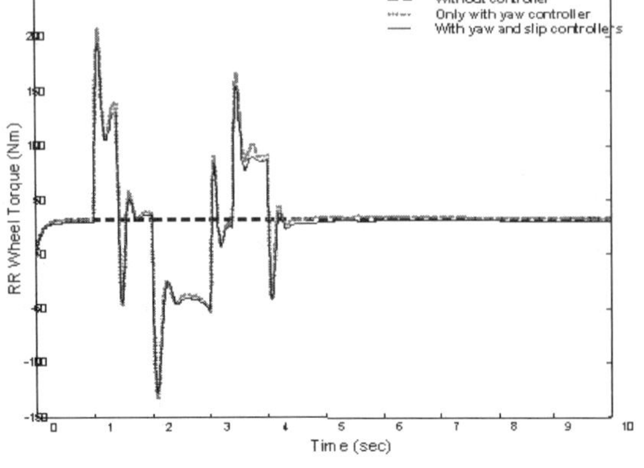

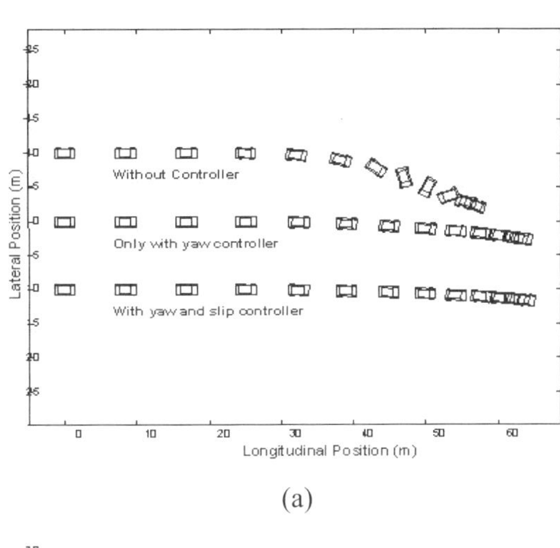

(a)

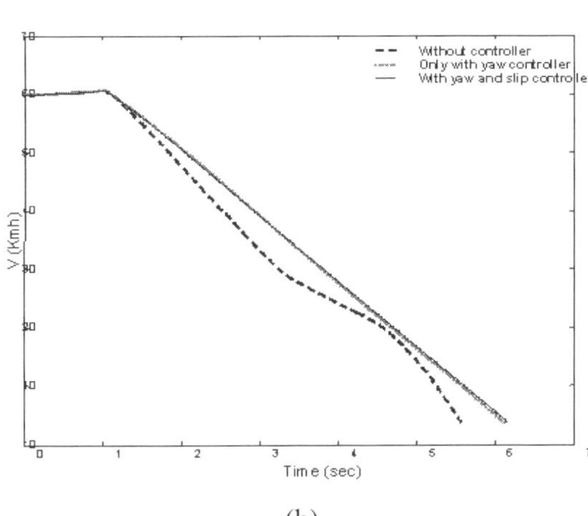

(b)

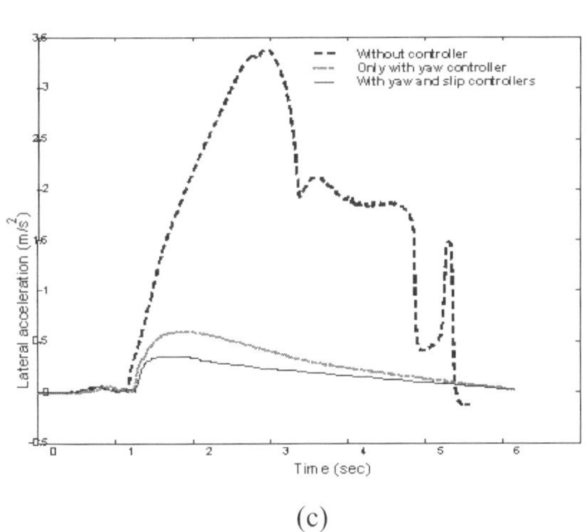

(c)

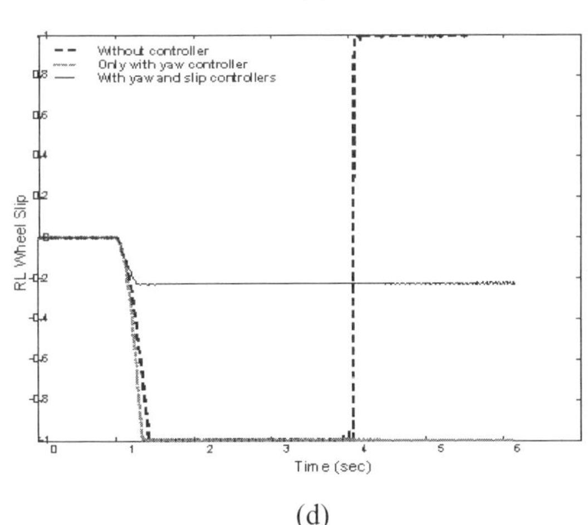

(d)

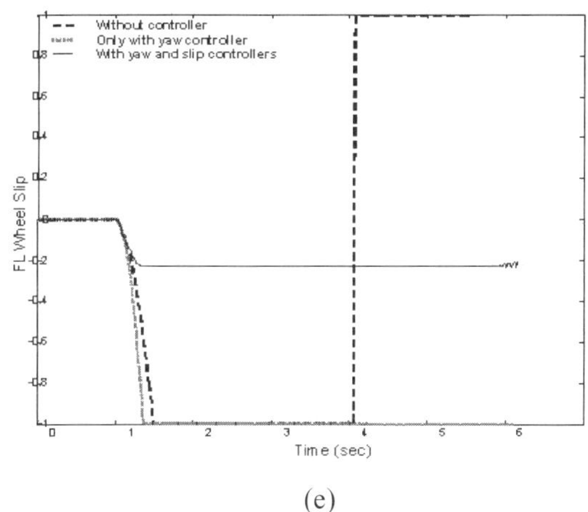

(e)

Figure 13. Braking on a μ-split road: a) Vehicle trajectory, b) Vehicle speed, c) Lateral acceleration, d) Rear-left wheel slip, e) Front-Left wheel slip

CONCLUSION

A novel driver-assist stability system for a four in-wheel drive electric vehicle was introduced. The system is based on Fuzzy logic Direct Yaw-moment Controller and individual slip controllers for each wheel using fuzzy logic. A multi-sensor data fusion method was introduced in order to estimate the vehicle real speed. The effectiveness of the proposed controller was evaluated by simulation. A 14-degree of freedom vehicle model and the well-known Pacejka tire formula was employed in the simulation program. Simulation results show excellent performance of the proposed control system on slippery roads.

REFERENCES

1. T. Yoshioka, et al, "Application of Sliding-mode Control to Control Vehicle Stability," *in proc AVEC'98*, 1998.
2. S. Motoyama, et al, "Quantitive Evaluation for Yaw Control Ability," *in proc. AVEC'98*, 1998.
3. A. Halvaei, R. Kazemi & S. Farhanghi, "Yaw Moment Control of a Two-Wheel-Drive EV, Part I: Vehicle yaw moment control using Fuzzy Logic (in Persian)," *8th Iranian Conference on Electrical and Electronics Engineering*, Isfahan, Iran 2000.
4. F. Tahami, S. Farhanghi and R. Kazemi, "Stability Enhancement of a Two-Motor-Drive Electric Vehicle Using Fuzzy Logic (in Persian)," *1st Iranian Conference on hybrid and electric vehicles*, Tehran, July 2001.
5. E. Bakker. H. B. Pacejka and L. Lidne, "A New Tire Model with an Application in Vehicle Dynamics Studies," *SAE transactions: Journal of passenger cars*, no.98, pp. 451-439, 1989.
6. T. Gillesple, "Fundamentals of Vehicle Dynamics," *Society of Automotive Engineers*, 1992.

Trade-offs for Vehicle Stability Control Sensor Sets

Mike Babala, Gary Kempen and Paul Zatyko
TRW Automotive

ABSTRACT

Customers of new vehicles expect their vehicle to provide reliable operation. One path vehicle manufacturers have chosen to meet this expectation is to offer their customers advanced braking systems. Antilock Brakes (ABS) and Traction Control (TC) are two advanced braking systems that have evolved to a point at which many OEM's offer them as standard equipment. Size, weight, and performance have also improved to the point of near transparent operation in many cases. The current direction of braking system evolution is in making Vehicle Stability Control (VSC) widely available as well. VSC adds the ability to assist the driver in negotiating understeer and oversteer, by adding corrective braking and engine torque to the vehicle as appropriate.

A large percentage of VSC system modeling is related to the sensors chosen to provide driver and vehicle dynamic information to the system's electronic control unit (ECU). This paper will introduce the reader to the trade-offs of Sensor Sets affecting vehicle dynamics and driver modeling.

INTRODUCTION

ABS, TC, and VSC all offer the driver enhanced vehicle controllability under adverse driving conditions. This is accomplished by adding a layer of electrohydraulic control between the driver's response and the actual response of the vehicle. Whenever this type of control is introduced between a driver and the vehicle, reliable real-time modeling of the vehicle dynamics is required to assist the driver appropriately in all driving conditions. With a VSC system, the intervention offered makes this modeling reliability even more essential.

Five primary sensors are required by the VSC system to appropriately assist the driver in controlling the vehicle through understeer, oversteer, and all road conditions. Figure 1 depicts understeer and oversteer, along with corrective forces that must be applied to generate a corrective torque in the vehicle. Steering wheel angle is essential to model which way the driver would like to turn. Yaw rate and lateral acceleration are required to calculate sideslip angle, the measure of whether the vehicle is in oversteer or understeer. Also required is a pressure transducer for determining how much braking

the driver is requesting, and wheel speed sensors for calculation of individual tire slip. The combination of these sensors, working cooperatively in a VSC system, is defined here as a Sensor Set.

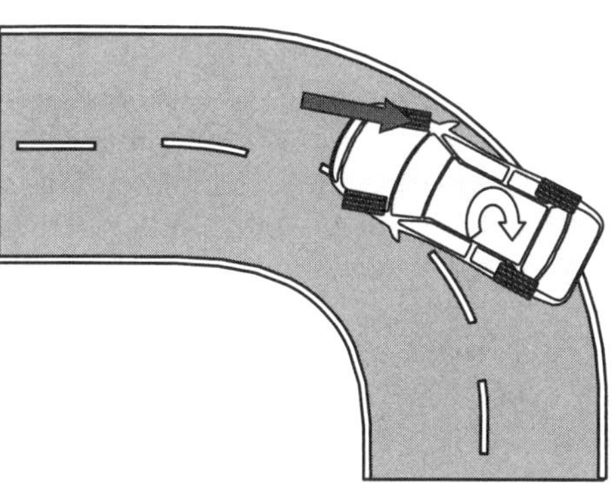

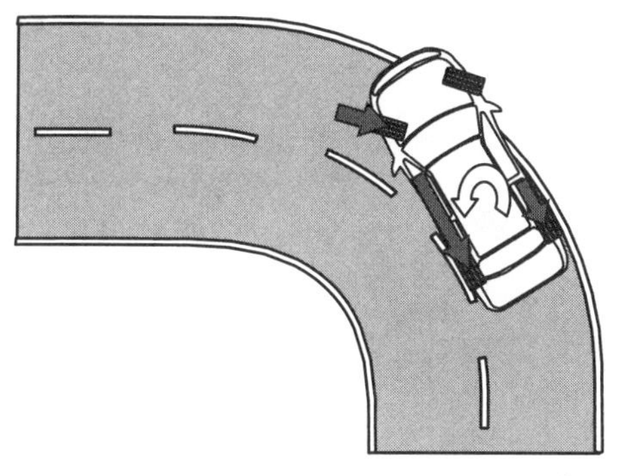

Figure 1. Vehicle oversteer and understeer.

The goal of this discussion is to offer the reader insights to the trade-offs of these Sensor Sets and how they affect performance of a VSC system. Requirements for accurate modeling of vehicle dynamics will be discussed first, including types of error, delays, and environmental concerns. Technology types for each of the sensors will next be examined, with an evaluation of strengths and weaknesses of the technologies. Finally, potential combinations of sensors for a complete and accurate physical estimation of vehicle dynamics will be evaluated, from the perspective of value to the buyer of the vehicle.

VEHICLE MODELING REQUIREMENTS

Component size and weight have been concerns in the automotive field since its beginnings. These, and other key variables related to sensing and their impact on system design, must be carefully balanced to achieve the optimal performance characteristics desired by a driver of the vehicle. With a VSC system, the performance of the vehicle is directly related to the accuracy of the VSC model. Therefore, various inaccuracies in the sensors must be accounted for in the design stage. Manufacturing processes, sensor material, software filtering algorithms, and environmental extremes can all contribute to the many different types of error.

Factors such as fixed sensor offset and noise can typically be compensated for with software algorithms; however, non-linearities and offset drift due to thermal properties of a sensor can be problematic and require a much more in-depth analysis. For example, a sensor that has a higher gain at cold temperatures will cause modeling inaccuracies when the system is operating on ice. A stability control based on that model would need to have wider activation thresholds to avoid inappropriate intervention. Further, hardware or software filters used to reduce noise may also introduce a phase lag in the sensor signal that must be considered.

Non-linearities and offsets are the major types of error typically challenging an engineer developing an accurate model of dynamics; however, just as important are the more subtle issues of initialization and diagnostics. Consistent performance is an important factor when designing any control system – especially vehicle controls, where the driver must be able to predict what the vehicle will do when negotiating traffic and road conditions. If a VSC system requires a certain amount of driving in order to initialize its model of the vehicle dynamics, it cannot perform control during that initialization time. From the driver's perspective, the partial system availability means the vehicle's performance is not consistent between typical driving and the initialization time following key-on; therefore, initialization time must be kept to a minimum.

Diagnostics must also be quick in order to maintain consistent vehicle performance. When a sensor doesn't operate according to specification, the signal sent to the control system will be corrupt. If the controller is unable to detect this condition, the vehicle model will also be corrupt; the system response to this undetected error of the vehicle model could cause a destabilizing activation of the VSC system. Therefore, customer value of diagnostics may not be immediately apparent, but to the experienced designer, is of the same importance as operational performance.

As with all automotive applications, environment is a key concern for sensors. Different sensor types may respond to temperature changes and vibration in varying ways, from various forms of additional error to outright failure of the part. These types of variables must also be considered in order to achieve the best customer value.

This leads us to the question, how accurate is accurate enough? Modeling calculations require enough information accuracy to deliver the expected level of performance to the customer. Additional information may provide no added value. Performance expected by the driver of a minivan is not the same as that of a sports car. Consequently, selecting the optimal Sensor Set for a specific level of accuracy is a function of customer expectations for the platform in question. Net accuracy, initialization, diagnostics, and environment are all forces requiring balance.

SENSING TECHNOLOGY AVAILABLE

The breakthrough sensing technologies associated with a basic performance level of VSC were developed recently. In the late 90's, MEMS (micro electro mechanical structures) technology had matured to the point where yaw rate and pressure sensors became feasible for automotive applications. This was the final milestone that allowed VSC to be integrated on a production passenger vehicle.

In the few short years between then and now, a further maturation has allowed a proliferation of options. A brief description of the technologies now available for VSC systems is given below, along with the strengths and weaknesses of each. The reader may find this information useful in understanding design opportunities when choosing Sensor Sets for VSC systems.

YAW RATE SENSORS

Yaw rate sensors measure the rate of angular displacement through the use of Coriolis force. Coriolis force is experienced when a linearly moving object is displaced through a rotating reference frame. This functional principle is the same regardless of the yaw rate sensor's application. In automotive applications, achieving a sustained linear motion in a sense element is accomplished by driving a small resonant structure into motion electronically. Perturbations of the element caused by Coriolis force are then electronically conditioned into an output signal.

MEMS commercialization has made yaw rate sensors widely available, encouraging high application rates of systems that would benefit them. Quartz and silicon are the materials commonly used for sense elements today. Air, liquid, metal alloys, and ceramic elements have also been proposed; however, none have been able to meet the requirements of the automotive marketplace. While silicon has a greater potential for miniaturization than quartz, its output signal is smaller and requires more amplification. The increased amplification can lead to a decrease in signal-to-noise ratio.

Vibration and shock are also key technological constraints of the current generation of yaw rate sensors. Since the principle of operation is based on an electromechanical vibration of the sense element, external vibrations and shocks can cause sensor disturbances. As such, more severe limitations must be considered for the sensor design, package isolation, and location in the vehicle. Controlling thermal drift with predictability is another consideration affecting both silicon and quartz sense elements. These are just a sampling of the important considerations leading to size and performance trade-offs of the yaw rate sensor.

LOW G ACCELERATION SENSORS

Most low G acceleration sensors produce ratiometric outputs based on the deflection of a seismic mass. The mass can take several forms, using both piezoelectric and capacitive sensing technologies. Early designs of low G accelerometers employ macro structures, which deflect under acceleration causing a change in capacitance. The capacitance change is processed into a signal proportional to acceleration. This approach is still used in new accelerometer designs.

The earlier devices use ceramics and metals for the sense element. These materials have stable and predictable thermal behaviors. With the wide commercialization of MEMS, microstructures of bulk and poly silicon are now available which have significantly reduced size. Along with these size advances have come other challenges. The signal produced by a MEMS element is very small and therefore can be influenced by factors such as packaging and manufacturing techniques. These effects can manifest themselves as non-linearities and unpredictable thermal behavior. The benefits in size and packaging must be balanced with increased non-linearities and thermal drift.

STEERING ANGLE SENSORS

Steering angle sensors exist in many forms and have been developed from several technologies to measure steering wheel position. This permits a high degree of design freedom, and currently sensors are available with a wide range of output types and signal quality. Unlike an accelerometer or yaw rate sensor, which are both dynamic sensors and are independently mounted to the vehicle, a steer angle sensor determines driver intent and is usually directly mounted to the steering column. Resolution, accuracy and signal range are important characteristics of any steering angle sensor, and considerations for each should be based upon what is required for a specific vehicle platform.

Steering angle sensors can be designed to output data either as the absolute position of the steering wheel or as the relative position based upon a reference point. The most desirable sensor will output data as an absolute position from lock to lock with sufficient resolution to discern between mechanical dither and rotational movement. However, due to the need for internal gearing or a microprocessor to count turns, this solution may be prohibitive in many applications. In the latter case, the sensor electronics may require a quiescent current at key-off, which must also b

consideration in overall system design. Advantages of using an absolute position sensor include, decreased initialization time, distinct output signal with respect to position, and improved diagnostic capabilities.

The opposite end of the performance spectrum is a simple encoder. This type of sensor is designed to output two separate digital signals in quadrature. Direction of travel is determined at the system ECU by monitoring signal phases. These sensors are widely available, however the system ECU must count pulses and perform correlation operations to other sensors to initialize and determine the position of the hand wheel at key-on. After initialization, the ECU must also continue to count pulses and use software to monitor the steering wheel angle. These operations all require overhead within the software operating system.

PRESSURE SENSORS

Pressure sensors used in hydraulic braking systems have recently undergone significant improvements in the area of size reduction. Both silicon and thin film sensing technologies are used in these high-pressure applications. The sensor's operating principle is based on a deflecting diaphragm, which produces a proportional change in capacitance, or resistance, of a sense element bonded to the diaphragm. This signal change is conditioned and amplified with electronics for use in system modeling. Size reductions have allowed higher degrees of system integration, which results in the sensor being exposed to potentially more severe environments. The new environments may have higher temperature gradients or greater levels of electromagnetic interference.

While thin film elements have excellent linearity and thermal predictability, their output signals are small and require significant amplification. This amplification leads to decreased signal-to-noise ratios. In comparison, the silicon elements require less amplification, but poor linearity leads to a higher level of electronic conditioning to produce useful pressure data.

WHEEL SPEED SENSORS

Wheel speed sensors measure the rotational speed of the vehicle wheels and provide a frequency based output signal. The sensors are either a passive or active magnetic based technologies. They can be designed to sense a ferrous based target or a magnetically charged ring. Depending on the package size and the mounting location, the sensor can be discretely mounted or integrated with the wheel bearing pack. Temperature, airgap, and cable routing are all important characteristics that must be considered for an application.

Passive sensors employ a copper wound coil, a magnet, and flux concentrator, sensing wheel speed through the use of Faraday's Law. The sensor produces an analog waveform based upon the time rate of change of magnetic flux flowing through the copper coil. Therefore, as the speed of the vehicle wheel increases both the frequency and the amplitude of the output signal will also increase. Two main drawbacks of passive sensors are their inability to sense zero vehicle speed, and the necessity of a large magnet, which requires a relatively large package size. This type of sensor has seen wide use in many vehicle platforms, however the signal quality is very susceptible to mechanical vibration and any imperfections with the target.

Active sensors use Hall effect or magnetoresistive devices to sense the rotational speed of the vehicle wheel, and usually require a bypass capacitor for ESD protection and noise suppression. They can be designed to sense either a ferrous target or a magnetically charged ring, and they contain signal conditioning circuitry – allowing them to provide a current-based output signal that is much more stable over airgap and temperature changes. A load resistor is needed to convert the current output to a voltage signal for the system ECU. Benefits of active sensors include the ability to work down to zero speed, a discrete output that is insensitive to airgap, and in some designs, provide reverse direction detection and even airgap diagnostics.

SENSOR SETS FOR VEHICLE DYNAMICS MODELING

THE SENSOR MATRIX

MATRIX

The need to develop a more accessible understanding of the sensing tradeoffs was driven by the requirements of multiple customer platforms. Because very few of the sensors involved have matured to a simple ideal technology, our team developed a matrix to better describe how the sensor interactions were related to our ability to model vehicle dynamics.

From this matrix, consensus building for selection of appropriate Sensor Sets is a much simpler task. If the desired performance characteristics are known, potential combinations shown on the matrix can quickly guide the reader to an ideal Sensor Set for a particular dynamics modeling application. Expanding the matrix enough could ultimately define system requirements for many applications.

Key: ✓ performance desired; * sensor selection required

VSC Sensor Interaction Matrix

	Distinguishing Characteristics	Value	1	2	3	4	5	6	7	8	9	10	11	12	13	14	15	16
Modeling Accuracy	precision, phase lag, thermal stability	good	✓	✓														
							✓	✓										
					✓	✓					✓	✓						
		better							✓	✓					✓	✓		
													✓	✓				
																	✓	✓
		best																
Initialization Time		good	✓		✓						✓		✓					
		better		✓		✓						✓		✓				
		best					✓	✓	✓	✓					✓	✓	✓	✓
Diagnostics[1]	complexity (low complexity is referred to as best)	good	✓		✓													
				✓		✓					✓		✓					
		better					✓		✓			✓		✓				
								✓		✓					✓		✓	
		best														✓		✓
Inertial Sensor (Yaw + Lat. Accel)	offset accuracy, gain accuracy	Good	*	*	*	*	*	*	*	*								
		Better									*	*	*	*	*	*	*	*
Steering Angle Sensor — Relative	resolution, linearity	Good	*	*							*	*						
		Better			*	*							*	*				
Steering Angle Sensor — Absolute	resolution, linearity	Good					*	*							*	*		
		Better							*	*							*	*
Pressure Sensor	offset accuracy, gain accuracy	Good / Better	Minimal impact on inertial modeling accuracy, initialization time & diagnostics. Primary impact is on driver braking modeling.															
Wheel Speed Sensor	reverse detection, low speed	Good	*		*		*		*		*		*		*		*	
		Better		*		*		*		*		*		*		*		*

Row groups: "Desired Inertial Modeling Performance" spans Modeling Accuracy, Initialization Time and Diagnostics; "Sensor Choice" spans Inertial Sensor, Steering Angle Sensor, Pressure Sensor and Wheel Speed Sensor.

[1] actual diagnostic performance is virtually identical for all apps; "complexity" refers to additional software, support circuitry, etc. necessary to meet required diagnostics.

Rather than try to quantify levels of accuracy or performance in the matrix, we have left it to the reader to determine what level of detail is important to their application. An additional benefit to this approach is that changing technology will not invalidate the basic matrix; after an update it will retain all its original effectiveness. Inertial sensing technology in particular is still in a relatively high state of flux, and different applications have different requirements and constraints. We therefore felt it would be more useful to the reader to explicitly state that Sensor Sets must be evaluated to gain a complete understanding of the tradeoffs involved. In the matrix, "good" means the low end of acceptable performance. "Best" means that any further performance gains would either not add significant value, or the cutting edge of technology has been reached.

It is important to note that software algorithms, electronic interfaces, and other factors can in some cases significantly affect modeling performance of a sensor system. As such, the current form of the Matrix primarily assists the reader in two ways: quickly zeroing in on target technologies for further investigation, and communicating tradeoffs of the sensors in a more accessible form than typical requirements documents.

To use the matrix:

- From the *Value* column, choose *Desired Inertial Modeling Performance* ('good', 'better', and 'best' in this matrix) for each of the *Distinguishing Characteristics* listed in the top section.
- Find the *Sensor Set* column that is closest in performance.
- If there's a perfect match, refer to the sensor selections listed below in the *Sensor Choice* section; if not, choose an acceptable compromise.

MATRIX USAGE EXAMPLES

We have listed examples below reflecting a few desired performance characteristics, and evaluations of Sensor Sets that are well suited to the application.

Example 1: High accuracy modeling

Sensor Set 11 of Matrix

An engineer wishes to design a VSC system for an application targeted towards an experienced driver, on a performance-driving vehicle that is expected to be high volume. Translated to characteristics of *Desired Inertial Modeling Performance,* high model accuracy for the driver's experience is the primary focus. Initialization time is deemed to be flexible, because the target driver is the type who will generally be aware of the state of the vehicle. Resources required for diagnostic complexity

can afford to be high, because of high volume production.

Reading the Matrix, Sensor Sets 15 and 16 are the peak performance choices for this platform, with 11 and 12 as close runners up. The engineer chooses to avoid Set 12 and 16, deciding that the extra bells and whistles associated with reverse detection and low speed operation were not enough benefit over 11 and 15 for the driver of the vehicle. Allowing a short initialization delay between key-on and system availability are deemed an acceptable tradeoff here, which the engineer interprets as favoring the simplicity of a relative steering angle sensor. The sensor's simplicity can then be compensated for by a well-developed diagnostic design – the resources required to develop the diagnostics are acceptable on this high-volume application.

The engineer decides to go with Sensor Set #11. With the current level of technology, a high accuracy gyro, with a high resolution relative encoder steering angle sensor, variable reluctance wheelspeed sensors, and a basic pressure transducer should be able to detect vehicle inputs on a smooth, continuous basis for this application. This would significantly reduce the digital effects of a more basic system, while requiring additional resources to develop diagnostics for the simplicity of the steering sensor.

Example 2: Tradeoffs

Sensor Set 6 of Matrix

Another engineer designs a VSC system for a less-experienced driver, with a system focus of immediate VSC availability, and accessible technology for a high-volume program where engineering time is somewhat limited. The engineer interprets the key target characteristic as "best" initialization. Modeling accuracy is negotiable, but because of the desire for accessible technology and limited engineering time, extra features should be avoided. Diagnostic complexity is also negotiable because of the competing restrictions of high volume and limited resources. However, the program is seen to slightly favor a more simple diagnostic because of the limited engineering time; the designer does not wish to take risk with diagnostics.

From a quick read of the Matrix, Sensor Sets 5-8 and 13-16 stand out as meeting the initialization requirement. Sets 13-16 are quickly ruled out as having unnecessary features, with only a slight difference in return on diagnostic development. Choosing between these proves very difficult, as they are a fine balance of where to put the system complexity, in sensor hardware or elsewhere in electronics or software. The engineer favors a low-complexity diagnostic solution, leaving Sets 6 and 8. At that point, there is room for flexibility; tests are ordered to determine which of the two is more feasible for the platform under design.

Driveline Torque-Bias-Management Modeling for Vehicle Stability Control

Chia-Shang Liu, Vincent Monkaba, Hualin Tan, Clive McKenzie, Hyeongcheol Lee and Sophia Suo

Driveline Systems Department Visteon Corporation

ABSTRACT

This paper describes the modeling of driveline systems with electronically controllable torque biasing devices. It focuses on the application of torque distribution for vehicle yaw stability control. Three types of driveline torque bias arrangements as well as the concepts of the yaw moment control with respect to these devices are introduced. To systematically derive the equations of motion of driveline systems, the techniques of Bond Graphs are applied in this paper. The derived driveline model is further integrated into a full vehicle model developed in ADAMS to monitor the vehicle responses with different driveline torque bias strategies. Since the controller is developed in MATLAB and the vehicle model is built in ADAMS, the methodology of co-simulation is applied to communicate the control algorithms with the vehicle model in synchronize time. Simulation of the developed model to emulate winter test scenarios is presented and is used to validate the torque control algorithms as well.

INTRODUCTION

Vehicle safety systems have evolved significantly in the auto industry in the past two decades. These include passive systems (for example air bags) and active systems (for example antilock brake systems). One of the main trends in the recent development of vehicle safety systems is the vehicle dynamics control to maintain vehicle stability.

It is known that the tire slip angles, and consequently, the vehicle slip angle might increase rapidly without a corresponding increase or even decrease in lateral forces if a vehicle reaches its physical limit of adhesion between tires and the road [1]. Since most drivers have few experiences in operating the car under this situation, they might have difficulty in controlling the vehicle as they expected [2]. The goal of vehicle the stability enhancement system is to bring the vehicle into predictable vehicle behaviors so that drivers could maintain better control of the vehicle.

The control strategy behind the current vehicle stability control system is referred to as yaw moment control. Yaw moment can be generated either by applying the brake forces at the wheels or by varying the torque split through the powertrain to create an offsetting yaw moment. Most vehicle stability control systems in the market are brake based [3-7]. Since this method is essentially a brake strategy, the vehicle speed is compromised. By comparison, the torque split based yaw moment control has a better vehicle speed performance and is gaining more visibility. It was noted that the emphasis of Four-Wheel Drive (4WD) or All-Wheel Drive (AWD) systems has shifted from traction performance enhancement to on-road stability and handling performance improvement [8-10].

At Visteon the applications of torque biasing devices to maintain vehicle directional stability as well as to improve traction performance have been developed for several years. To reduce critical development time and cost, modeling has been essential to provide the upfront analysis and evaluation of the influence of control logics and torque biasing devices on vehicle handling responses. This paper describes the modeling of virtual prototyping driveline systems with electronically controllable torque biasing devices. The main focus is to determine the effect of torque split controls on vehicle yaw stability. Three different types of driveline torque bias configurations are studied. The yaw moment controls with respect to these devices are introduced as well. The techniques of Bond Graphs [11] are applied in this paper to formulate the equations of motion of driveline systems.

The derived driveline model is further integrated into a full vehicle model developed in ADAMS. The methodology of co-simulation [12-13] is applied to communicate the control algorithms with the vehicle model in synchronize time, since the control module is developed in MATLAB and the vehicle model is built in ADAMS. Finally the verification of the control strategies through simulation will be presented.

TORQUE BIASING DEVICE OVERVIEW

Three types of torque biasing configurations are studied in this paper. They include Electro-Magnetic Coupler (EMC), twin coupler and Electro-Magnetic Limit Slip Differential (EMLSD).

EMC –

A conceptual cross section of the electro-magnetic coupler is shown in Figure 1. The major components of the device include the main clutch, the pilot cutch, ball and ramp assemblies and solenoids.

The external Electronic Control Unit (ECU) controls the current to the magnetic coils of the solenoid to generate electromagnetic forces. The magnetic force attracts the pilot to create friction force at the pilot clutch, which will activate the mechanism of the ball and ramp to engage the main clutch assembly. The torque can then be transferred from the front prop-shaft to the rear prop-shaft or vice versa depending on the driveline architecture.

TWIN COUPLER –

The schematic diagram of the twin coupler is shown in Figure 2. It is an electro hydraulic controllable device, which contains clutch assemblies, solenoid valves, gerotor pumps, and valves.

The ECU controls two solenoid valves to regulate the pressure generated by the gerotor pumps. The pressure built on the piston will generate a thrust force that engages the clutch assembly. The torque can then be transferred from the ring gear to the half shaft that is engaged.

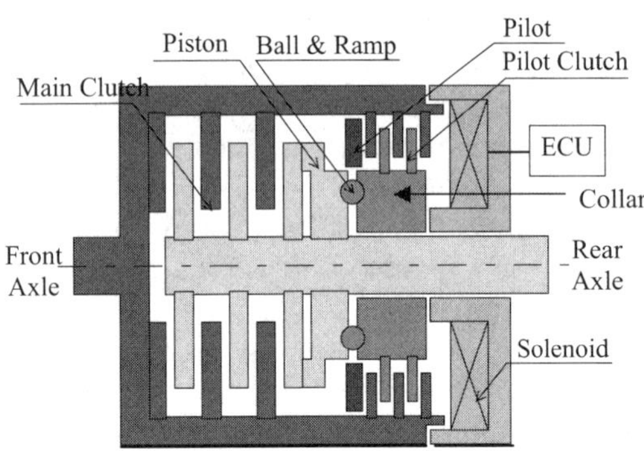

Figure 1. Electro-Magnetic Coupler

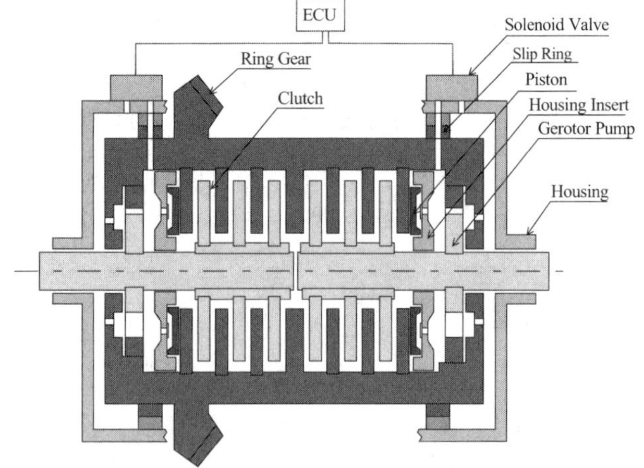

Figure 2. Twin Coupler

EMLSD –

The EMLSD system has the same mechanisms as the EMC except that a differential gear set is included as shown in Figure 3.

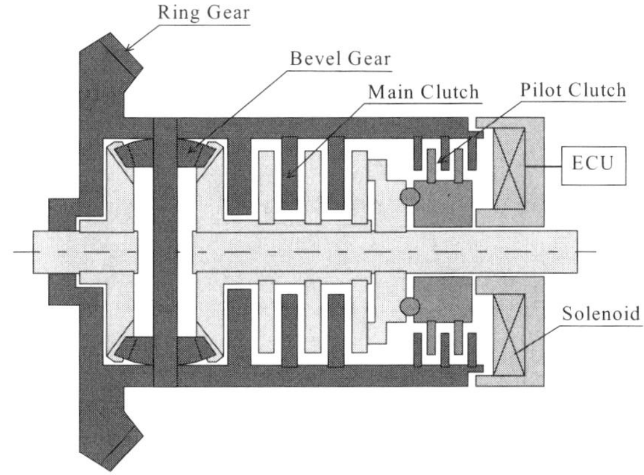

Figure 3. EMLSD

The ECU regulates the current to the solenoid to engage the main clutch that will decrease the rotational speed of the fast half shaft.

APPLICATIONS FOR YAW CONTROL

The external forces affecting vehicle dynamics mainly come from the forces generated at the tire-road contact patch. By controlling the distributions of the applied torques at the wheels, the tire forces change, consequently the dynamic responses of the vehicle change.

The frequently used criterion for vehicle stability control is the vehicle yaw rate. Alternating, the torques applied at the wheels will create a correcting moment to offset the yaw rate. The generated yaw moments due to the applications of the aforementioned torque biasing devices will be addressed as follows.

APPLICATION OF EMC DEVICE –

The EMC replaces the center differential mechanism in an All Wheel Drive (AWD) vehicle. The devices will transfer torques from the front differential to the rear prop shaft for a Front-Wheel-Drive (FWD) based vehicle.

In reference to Figure 4, a FWD-based AWD vehicle with the EMC is shown. Assuming that the vehicle is turning right. It is known in the tire literatures that tire steering/lateral forces will reduce when accelerating or braking during cornering [1]. If more driving torque is transferred to the rear wheels and no loss of traction of each tire, the cornering forces of front wheels will increase. This would generate a net yaw moment in the clockwise direction that the vehicle under-steer tendency will be decreased as shown in Figure 4.

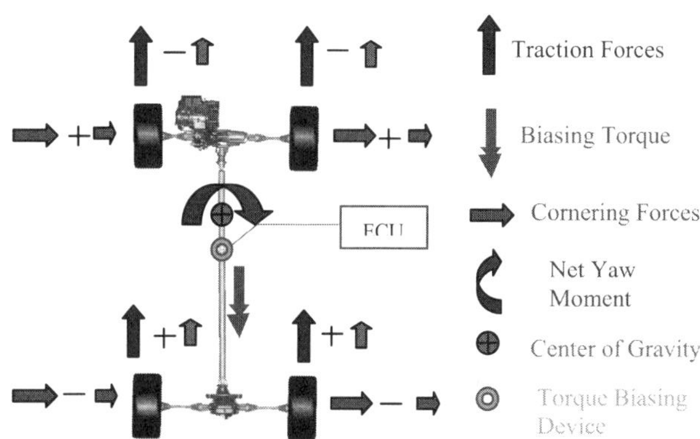

Figure 4. Yaw Moment Control of Electro-Magnetic Coupler

APPLICATION OF TWIN COUPLER –

The twin coupler replaces the differential gear set at the rear axle. It is able to impart different torque to right and left rear wheels from the prop shaft as desired.

Figure 5 shows a FWD based AWD vehicle with the twin coupler installed at the rear axle.

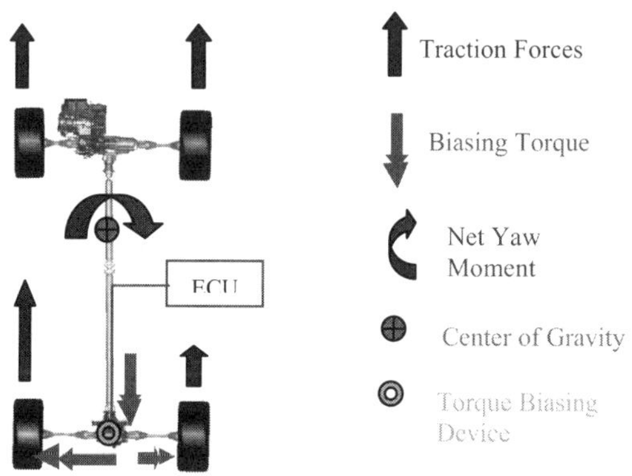

Figure 5. Yaw Moment Control of Twin Coupler

To simplify the analysis, the effect of lateral steering forces is neglected. If there is more torque transmitted to the left rear wheel, generating larger traction forces at the left tire, a net yaw moment in the clockwise direction will be created. Assuming that the vehicle is making a right turn, this will decrease the under-steer tendency of the vehicle.

APPLICATION OF EMLSD –

The EMLSD augments the function of the open differential set with active limited slip differential effect. A RWD vehicle with the EMLSD differential installed at the rear axle is shown in Figure 6.

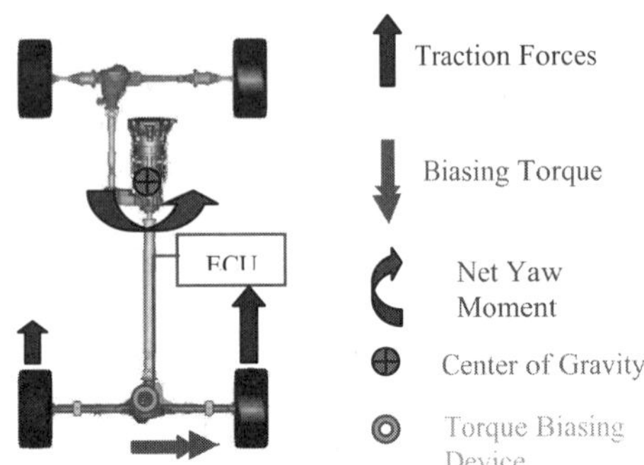

Figure 6. Yaw Moment Control of EMLSD

It is assumed that the vehicle is making a right turn. The effect of lateral steering forces is neglected. Since the vehicle makes a right turn, the left (outside) wheels will have faster speed than the right (inside) ones. The device only transfers torque from the outside wheel to the inside wheel, thus a net yaw moment is generated in counterclockwise direction. This would increase the under-steer tendency of the vehicle. If less under-steer is desired, less torque should be transferred to the inside wheel.

DRIVELINE SYSTEM MODELING

This section describes the mathematical modeling of the driveline system with the three different torque-bias configurations. The model development is based on a pictorial approach in different energy domain systems, which is referred to as bond graphs [11]. Bond graphs derive the state equations with the measurement of the energy state of a system that always has physical meaning. It could provide the physical insight. Among the modeled torque biasing devices, two of them are FWD-base AWD vehicles. The EMLSD is assumed to be installed in a Rear Wheel Drive (RWD) vehicle.

DRIVELINE WITH EMC –

A schematic driveline system with the electro-magnetic coupler (EMC) is shown in Figure 7.

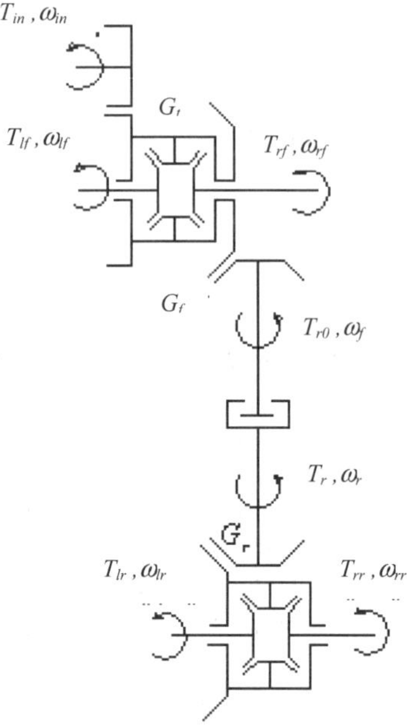

Figure 7. Driveline Model with EMC

The bond graphs of the above model are depicted in Figure 8. It is a FWD-based AWD vehicle in which the solenoid regulates the amount of torque transmitted to the rear prop shaft.

The tire dynamics and details of the torque biasing devices are not modeled. From the bond graphs shown in Figure 8, the equations of motion are derived as follows:

Equations (1) to (4) describe the torque transmitted from the transmission to front wheels

$$T_{rf} = \frac{T_{in}}{2G_t} - \frac{T_{r0}G_f}{2} \tag{1}$$

$$T_{lf} = T_{rf} \tag{2}$$

$$\omega_{in} = \frac{\omega_{rf} + \omega_{lf}}{2G_t} \tag{3}$$

$$\omega_f = G_f\left(\frac{\omega_{rf} + \omega_{lf}}{2}\right) \tag{4}$$

Equations (5) to (8) describe the torque transmitted to the rear half shafts.

$$T_{r0} = J_f \alpha_f + T_r \tag{5}$$

$$T_{rr} = G_r\left(\frac{T_r - J_r \alpha_r}{2}\right) \tag{6}$$

$$T_{lr} = T_{rr} \tag{7}$$

$$\omega_r = G_r\left(\frac{\omega_{rr} + \omega_{lr}}{2}\right) \tag{8}$$

where

T_{in} is the input driving torque from the transmission, ω_{in} denotes the transmission speed, T_{r0} is the torque exerted on the front prop shaft to be transferred to rear axles, T_r is the driving torque exerted on the rear prop shaft which is regulated by the electro-magnetic coupler, $T_{lf}, T_{rf}, T_{lr},$ and T_{rr} are the driving torques at left front, right front, left rear and right rear half shafts, respectively,

$\omega_{fl}, \omega_{fr}, \omega_{rl}$ and ω_{rr} are the angular speeds, of front left, front right, rear left and rear right half shafts, respectively, ω_f and ω_r are the angular velocity of the front and rear prop shafts, respectively, α_f and α_r are the angular acceleration of the front and rear prop shafts, respectively, J_f and J_r are moment inertia of the front and rear prop shafts, respectively, G_t is the transmission final gear ratio, G_f is the transfer box gear ratio, and G_r denotes the final gear ratio of the rear axle.

The bias torque T_r regulated by EMC depends on the command signals from ECU and the speed difference between the front and the rear prop shaft. Its system characteristics could be provided by experiments as a look up table. Or a simplified assumption could be made on the torque biasing devices, for example, if assuming

$$T_r = \frac{f(x)}{1 + f(x)} \times \frac{T_{in}}{G_f G_t} \tag{9}$$

then it will have 50:50 torque split when the torque biasing device receives full ECU command.

where $0 \leq x \leq 1$, $0 \leq f(x) \leq 1$, $f(1) = 1$, $f(0) = 0$, $x = 1$ when ECU gives full command, and $x = 0$ without any command.

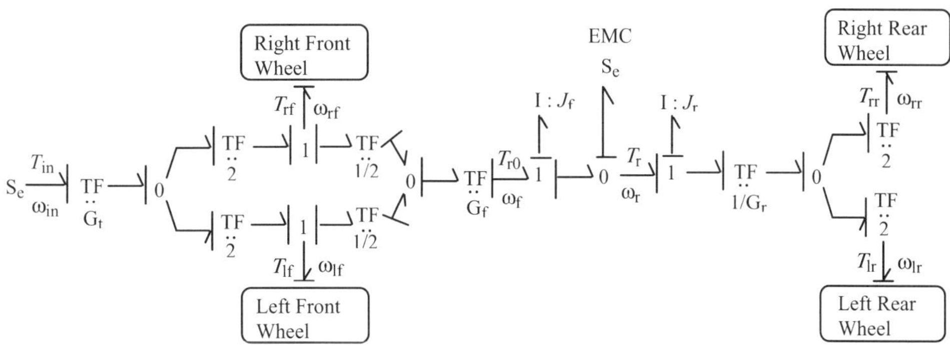

Figure 8. Bond Graphs of Driveline with EMC

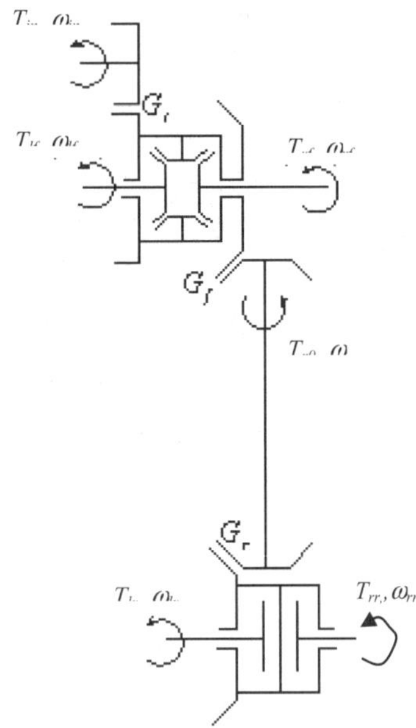

Figure 9. Driveline Model with Twin Coupler

DRIVELINE WITH TWIN COUPLER –

The schematic driveline model with the twin coupler is shown in Figure 9. It is a FWD based AWD vehicle in which the twin device regulates the torques transferred to the rear left or right half shafts, independently.

The bond graphs for the driveline model in Figure 9 are shown in Figure 10. On the basis of Figure 10, the equations of motion are derived as follows.

The equations for the torque transmitted from engine to the front half shafts are the same to equations (1) to (4).   The equation to describe the torque transferred to the rear wheels is

$$T_{r0} = J\alpha + \frac{T_{rr} + T_{lr}}{G_r} \tag{10}$$

The biasing torques at the rear half shafts T_{rr} and T_{lr} are determined by the system characteristics of the twin device.   A simplified assumption is applied to T_{rr} and T_{lr}, referring equations (11) and (12), so that the torque split will be 50:50 when both solenoids receive full command from ECU or only one receives full command and the other one is deactivated.

$$T_{rr} = \frac{T_{in}f^2(x_r)}{G_f G_t\left(f(x_l) + f(x_r) + f^2(x_l) + f^2(x_r)\right)} \tag{11}$$

$$T_{lr} = \frac{T_{in}f^2(x_l)}{G_f G_t\left(f(x_l) + f(x_r) + f^2(x_l) + f^2(x_r)\right)} \tag{12}$$

where $0 \le x_l, x_r \le 1$, $0 \le f(x_l), f(x_r) \le 1$, $f(1) = 1$, $f(0) = 0$, $x_l = 1$ when ECU gives full command to the left solenoid valve, $x_r = 1$ when ECU gives full command to the right solenoid valve, and $x_l = 0$, $x_r = 0$ without command signals.

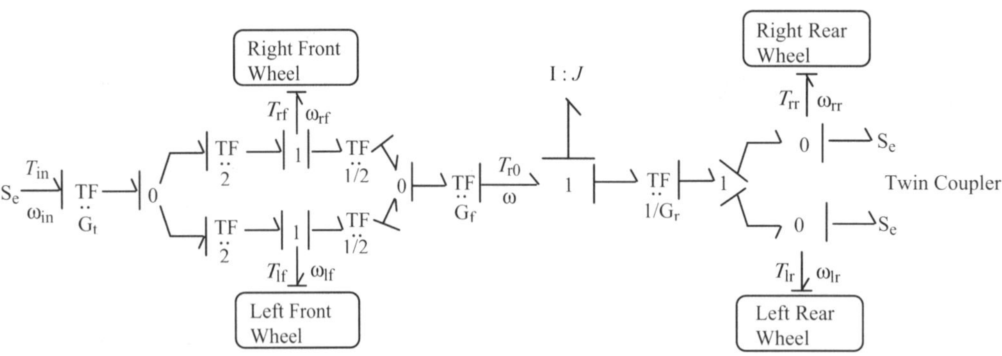

Figure 10. Bond Graphs of Driveline with Twin Coupler

DRIVELINE WITH EMLSD –

The driveline model with the EMLSD is shown in Figure 11. Different from the previous model it is a RWD vehicle.

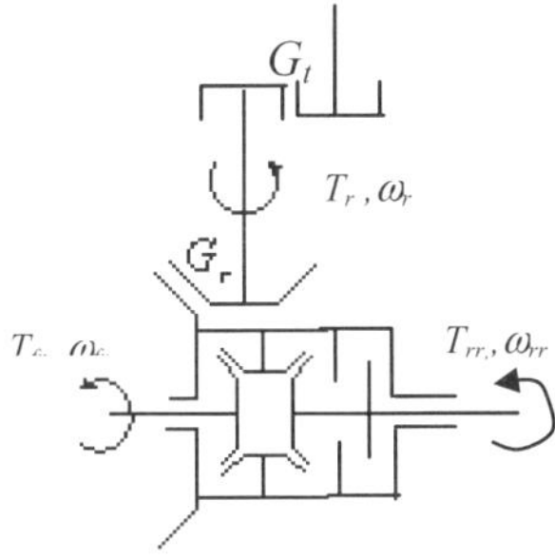

Figure 11. Driveline Model with EMLSD

The bond graphs of this model are presented in Figure 12. The tire dynamics are expressed in the bond graphs as well.

The equations of motion can be derived on the basis of bond graphs. The torque transmitted from the transmission to the rear shafts is described by

$$T_r = \frac{T_{in}}{G_t} \tag{13}$$

$$\omega_{in} = \frac{\omega_r}{G_t} \tag{14}$$

$$T_{rr} = \frac{(T_r - J_r \alpha_r)G_r + T_d}{2} \tag{15}$$

$$T_{lr} = \frac{(T_r - J_r \alpha_r)G_r - T_d}{2} \tag{16}$$

$$\omega_r = G_r \omega_{rin} \tag{17}$$

$$\omega_{rin} = \frac{\omega_{rr} + \omega_{lr}}{2} \tag{18}$$

where T_d is the biasing torque regulated by the EMLSD, and ω_{rin} is the ring gear angular speed.

The tire dynamics are described by

$$J_{rr} \alpha_{rr} = T_{rr} - F_{xrr} R_{rr} \tag{19}$$

$$J_{lr} \alpha_{lr} = T_{lr} - F_{xlr} R_{lr} \tag{20}$$

where J_{rr}, J_{lr} are moment of inertia of right and left tires, R_{rr}, R_{lr} are rolling radius of right and left tires, α_{rr}, α_{lr} are angular acceleration of right and left tires, and F_{xrr}, F_{xlr} are longitudinal forces of right and left tires.

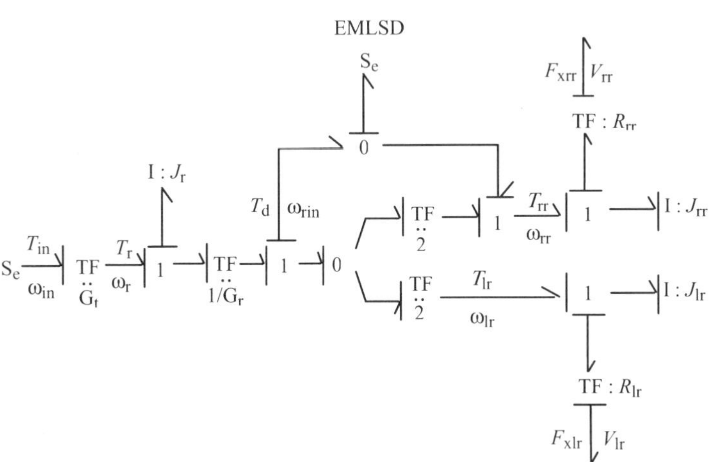

Figure 12. Bond Graphs of Driveline with EMLSD

Equation (18) shows that if it can force $\omega_{rr} = \omega_{rin}$, then it will have $\omega_{rr} = \omega_{lr} = \omega_{rin}$, which means that the differential system is locked. The biasing torque T_d will serve the function of locking the differential as described above.

A simplified procedure is then applied to derive T_d. First, considering the processes of locking the differential system. Substitute equation (13) into equations (15) and (16) and from equations (19) and (20), which will have

$$J_{rr}\alpha_{rr} = \frac{T_0 + T_d}{2} - F_{xrr}R_{rr} \tag{21}$$

$$J_{lr}\alpha_{lr} = \frac{T_0 - T_d}{2} - F_{xlr}R_{lr} \tag{22}$$

where

$$T_0 = (\frac{T_{in}}{G_t} - J_r\alpha_r)G_r \tag{23}$$

Assuming that the EMLSD will lock the wheel within Δt second and $J_{rr} = J_{lr}$.

Integrating equations (21) and (22) of both sides and rearranging the equations, which will have

$$\frac{1}{J_{rr}} \int_t^{t+\Delta t} \left(\frac{T_0 + T_d}{2} - F_{xrr}R_{rr} \right) d\tau$$
$$= \omega_{rr}(t + \Delta t) - \omega_{rr}(t) \tag{24}$$

$$\frac{1}{J_{rr}} \int_t^{t+\Delta t} \left(\frac{T_0 + T_d}{2} - F_{xlr}R_{lr} \right) d\tau$$
$$= \omega_{lr}(t + \Delta t) - \omega_{lr}(t) \tag{25}$$

where t denotes the time that the system is activated.

Applying the concept of mean value of T_d, T_0, and tire forces, it will lead to equations (24) and (25) as

$$\left(\frac{\overline{T}_0 + \overline{T}_d}{2} - \overline{F}_{xrr}\overline{R}_{rr} \right)\Delta t = J_{rr}\left(\omega_{rr}(t + \Delta t) - \omega_{rr}(t)\right) \tag{26}$$

$$\left(\frac{\overline{T}_0 - \overline{T}_d}{2} - \overline{F}_{xlr}\overline{R}_{lr} \right)\Delta t = J_{rr}\left(\omega_{lr}(t + \Delta t) - \omega_{lr}(t)\right) \tag{27}$$

where the head bar denotes the mean value that satisfies $\int_t^{t+\Delta t} g(\tau)d\tau = \overline{g}\Delta t$.

When the differential system is locked, it will have $\omega_{rr}(t + \Delta t) = \omega_{lr}(t + \Delta t)$. Then $\overline{T}_d$ can be derived by subtracting equation (27) from equation (26) as

$$\overline{T}_d = \frac{J_{rr}\left(\omega_{lr}(t) - \omega_{rr}(t)\right)}{\Delta t} + \left(\overline{F}_{xrr}\overline{R}_{rr} - \overline{F}_{xlr}\overline{R}_{lr}\right) \tag{28}$$

From equation (28), it shows that input torque T_0 disappears and the biasing torque consists of two parts on the right side of the equation when the differential locks.

The first part of equation (28) is related to the speed difference between the left and right wheels and the second part is related to the difference of tire/road contact resistant torques between the two tires.

Since the system will act like an open differential set if no command signals come from the ECU, a weighting function $f(x)$ can be added at equation (28) to represent the conditions between no command and full command as

$$\overline{T}_d = f(x)(\frac{J_{rr}\left(\omega_{lr}(t) - \omega_{rr}(t)\right)}{\Delta t} + \tag{29}$$
$$\overline{F}_{xlr}\overline{R}_{lr} - \overline{F}_{xrr}\overline{R}_{rr})$$

where $0 \le x \le 1$, $0 \le f(x) \le 1$, $f(1) = 1$, $f(0) = 0$, $x = 1$ when the ECU gives full command, and $x = 0$ without any command.

Since it is known that the friction torques transmitted at the clutch assembly have the phenomenon of stick/slip, the same concept can be applied to equation (29), such that

If $\Delta\omega < \Delta\omega_{sat}$ then

$$\overline{T}_d = f(x)(J_{rr}K\frac{\Delta\omega}{\Delta\omega_{sat}} + \overline{F}_{xlr}\overline{R}_{lr} - \overline{F}_{xrr}\overline{R}_{rr})$$

(30)

if $\Delta\omega \ge \Delta\omega_{sat}$ then

$$\overline{T}_d = f(x)(J_{rr}K + \overline{F}_{xlr}\overline{R}_{lr} - \overline{F}_{xrr}\overline{R}_{rr})$$

(31)

where $K = \dfrac{\Delta\omega}{\Delta t}$ is assumed to be a constant,

$\Delta\omega(t) = \omega_{lr}(t) - \omega_{rr}(t)$ is the speed difference between left rear and right rear wheels, $\Delta\omega_{sat}$ denotes the saturation point of the speed difference. Equations (30) and (31) could then be used to approximate the biasing torque T_d.

It should be noted that equations (9), (11), (12), (30) and (31) are only theoretical based biasing torques. The true biasing torque characteristics should resort to experiment results. The physical limit of the transferable biasing torques should be considered in the model too.

ADAMS FULL VEHICLE MODEL AND CONTROL STRATEGY

FULL VEHICLE MODEL –

The full vehicle model was developed in ADAMS. The model was verified via kinematics and compliance correlation as well as dynamic-event correlation [14].

Several subsystems were used to build the full vehicle model, which include vehicle body, road and tire model, front suspension, rear suspension, steering system, and power train system. The power train system is modified to

integrate the torque biasing devices as we studied before. The schematic of the ADAMS full vehicle model is shown in Figure 13.

Figure 13. ADAMS Full Vehicle Model

CONTROL STRATEGY –

The objective of vehicle stability control is to bring the vehicle back to a more controllable driving condition if undesired vehicle behavior is detected. A reference yaw rate γ_d is used as the criterion to determine control actions, which is shown in equation (32)

$$\gamma_d = \frac{v_x\delta_f}{(l + K_h v_x^2 / g)}$$

(32)

where δ_f is the front wheel steering angle, K_h is a stability factor, l is the wheelbase length, and v_x is the vehicle longitudinal speed.

The yaw rate sensor or a state observer is applied to measure the real yaw rate γ of vehicles. By comparing the actual yaw rate with desired yaw rate $\gamma - \gamma_d$, which is referred to as error signals, the controller will instruct torque-biasing devices to apply the appropriate yaw moment for yaw motion offset [10] [13].

SIMULATION RESULTS

This section applies the virtual vehicle model with the torque biasing devices to verify the developed control algorithms.

The simulation assumes a double lane change event on a packed snow road. The tire/road friction coefficient μ is equal to 0.3. The model presented here is a RWD vehicle with electromagnetic limited slip differential. The initial vehicle speed is 46 km/hour. The vehicle is accelerated at 0.018g. A virtual driver is built in ADAMS to mimic the response of the driver to maintain the vehicle on the desired path.

The yaw responses of vehicles with and without yaw stability controls are shown in Figure 14. The reference yaw rate is based on equation (32) for the control algorithms.

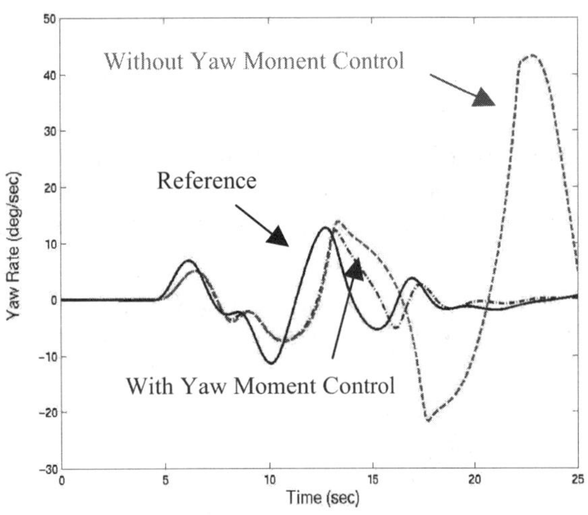

Figure 14. Yaw Responses of Vehicle with/without Control

The vehicle with yaw moment control maintains its stability to follow the desired yaw rate. The vehicle without yaw moment control loses its stability and spins.

A comparison of the vehicle longitudinal speed is shown in Figure 15.

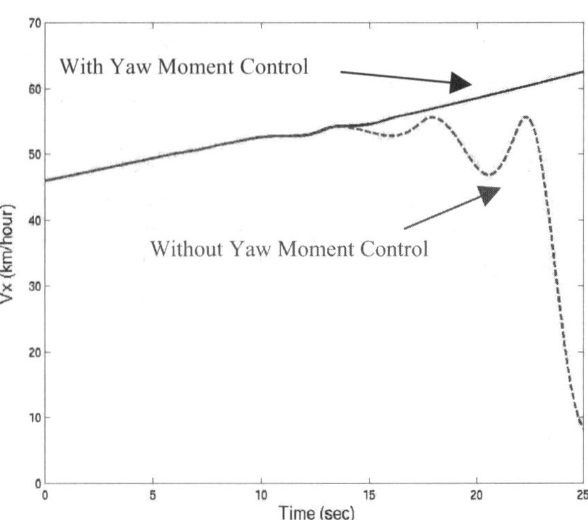

Figure 15. Vehicle Longitudinal Velocity with/without Control

As shown in Figure 15, the speed of the vehicle with yaw moment control is maintained. The vehicle without yaw control shows an adverse speed performance due to the spin of the vehicle. More simulation results could be referred to [13].

CONCLUSIONS

The functions as well as the concepts of yaw moment controls of three different torque-biasing devices have been introduced. The mathematical models of the driveline system with different biasing torque control strategies were presented as well.

On the basis of developed mathematical models, a virtual prototyping vehicle to simulate winter test events was used to validate the applicability of driveline torque controls to maintain vehicle stability. These results could serve as the upfront analysis of the control strategies and the influence of torque biasing devices on vehicle responses before building the real hardware.

In this paper, quasi-static biasing torque models were integrated into the full vehicle system. To accurately measure the dynamic effect of these devices on vehicle responses, detailed models of the torque biasing devices should be developed for evaluation.

REFERENCES

1. T. D. Gillespie, *Fundamentals of Vehicle Dynamics*, Warrendale, PA, SAE Publication, 1992.
2. Y. Shibahata, K. Shimada, and T. Tomari, "Improvement of Vehicle Maneuverability by Direct Yaw Moment Control," Vehicle System Dynamics, no. 22, pp. 465-481, 1993.
3. A. T. van Zanten, "Bosch ESP Systems: 5 Years of Experience," SAE paper 2000-01-1633, 2000.
4. J. H. Park, and W. S. Ahn, "H_∞ Yaw-Moment Control with Brakes for Improving Driving Performance and Stability," Proceeings of the 1999 IEEE/ASME Int. Conf. on Advanced Intelligent Mechatronics, pp. 747-752, Atlanta, USA, September 1999.
5. K. Koibuchi, M. Yaamoto, Y. Fukuda, and S. Inagaki, "Vehicle Stability Control in Limit Cornering by Active Brakes," SAE paper 960487, 1996.
6. A. T. van Zanten, R. Erhardt, and G. Pfaff, "VDC, The Vehicle Dynamics Control System of Bosch," SAE paper 950759, 1995.
7. S. Matsumoto, H. Yamaguchi, H. Inoue, and Y. Yasuno, "Improvement of Vehicle Dynamics Through Braking Force Distribution Control," SAE paper 920645, 1992.
8. K. Matsuno, R. Nitta, K. Inoue, K. Ichikawa, and Y. Hiwatashi, "Development of a New All-Wheel Drive Control System," Seoul 2000 FISITA World Automotive Congress, Seoul, Korea.
9. S. Mohan, "All-Wheel Drive/Four-Wheel Drive Systems and Strategies," Seoul 2000 FISITA World Automotive Congress, Seoul, Korea.
10. M. Ohba, H. Suzuki, and T. Yamamoto, "Development of a New Electronically Controlled 4WD System: Toyota Active Torque Control 4WD," SAE paper 1999-01-0744, 1999.
11. D. C. Karnopp, D. L. Margolis, and R. C. Rosenberg, *System Dynamics: a Unified Approach*, John Wiley & Sons Inc., 1990.
12. A. S. Elliott, "A Highly Efficient, General Purpose Approach for Co-Simulation with ADAMS," 2000 MDI North American User Conference, Novi, Michigan, USA.
13. C. S. Liu, V. Monkaba, H. Lee, T. Alexander, and V. Subramanyam, "Co-simulation of Driveline Torque Bias Controls," SAE paper 2001-01-2782, 2001.
14. V. Subramanyam, V. Monkaba and T. Alexander, "A Unique Approach to All-Wheel Drive Vehicle Dynamics Model Simulation and Correlation", SAE paper 2000-01-3526, 2000.

Understanding the Interaction Between Passive Four Wheel Drive and Stability Control Systems

Syun K. Lee and Nancy M. Atkinson
Ford Motor Co.

ABSTRACT

The purpose of this paper is to describe and define the interaction between a brake based stability control system and a passive coupler (viscous coupling unit) inside the transfer case of a Four-Wheel Drive (4WD) vehicle. This paper will focus on the driveline system and the impact that a stability control system can have on it. It will provide understanding of torque transfer on 4WD vehicles that are equipped with a brake based stability control system and use this knowledge to recommend ways to reduce the undesirable torque transfer interaction between the two systems. These recommendations can be readily applied to future 4WD/AWD vehicles to improve compatibility between the two systems.

INTRODUCTION

The future of most vehicles will include some type of brake based stability control system. This feature is becoming a very common option on vehicles in the same way that anti-lock brake systems are today. Applications of this technology have been predominately in 2WD vehicles or 4WD vehicles with an open center differential inside the transfer case up to now [1]. However, in recent months brake based stability control systems have been successfully integrated in 4WD vehicles equipped with torque transfer technologies other than open center differential.

Passive couplers in general provide an attractive means of torque transfer in 4WD vehicles. Passive couplers provide excellent reliability with proven past performance at a relatively low cost as compared to electronically activated clutch systems. Since passive couplers are not electronically controlled, they cannot be disconnected during a stability control activation event. Inability to disconnect the two outputs (primary and secondary) at the transfer case with a passive coupler was thought to potentially reduce the effectiveness of the stability control system. The central issue of this paper is the interaction between a passive coupler inside the transfer case and a brake based stability control system on a 4WD vehicle. This paper will describe and quantify using equations the interaction between the two systems and recommend a means by which to limit the negative interaction between the two systems.

BACKGROUND

A brake based stability control system is an ABS based control system that works by applying one or more of the vehicle's brakes independent of the driver to create a corrective yaw moment on the vehicle during an oversteer or understeer condition. Currently most automotive manufacturers have implemented stability control systems using open center differential transfer cases on 4WD vehicles. Open center differentials allow stability control systems to function with minimal torque transfer side effects that can potentially reduce the system's ability to generate corrective yaw moment on the vehicle during a stability control activation event.

When a passive coupling device such as a viscous coupling is the chosen method to transfer torque between primary and secondary axles, it is believed that it could potentially reduce the effectiveness of the stability control system due to inability to completely disconnect the secondary axle from the primary axle. Typically during a stability control activation event, one wheel will be driven to deep slip (50% reduction in wheel speed) to generate the maximum corrective yaw moment on the vehicle. If a passive coupling transfers a significant amount of braking or negative value torque to the opposite axle (i.e. negative torque to the rear axle when front wheel is being braked) during a stability control activation event, it can act to potentially reduce the effectiveness of the stability control system. This study will compare and provide understanding of torque transfer during a stability control activation event with both passive coupler and open center differential inside the transfer case.

UNDERSTANDING THE TRANSFER CASES USED IN THIS STUDY - Two transfer cases will be examined in this paper. The first transfer case utilizes an open center planetary differential with 35% of the torque being output to the secondary axle and 65% of the torque being output to the primary axle during all driving conditions. In this study the primary axle is the rear axle, and the secondary axle is the front axle. The torque input from the transmission drives the planetary carrier, while the sun gear drives the front driveline output, and the ring gear drives the rear driveline output. The second transfer case in this study utilizes the same center planetary differential; however, a viscous coupling unit (passive coupling used in this study) directly couples the input to the front output (see Figure 1) [2][3].

Definitions:

$\tau_{differential}$ - Torque going into the 35/65 planetary differential

τ_{input} - Torque going into the transfer case

τ_{vcu} - Viscous torque generated by the passive coupling due to speed difference.

$\tau_{front-d/s}$ - Front driveshaft torque

$\tau_{rear-d/s}$ - Rear driveshaft torque

τ_{loss} - Internal losses in the transfer case (We will assume this value is zero in this study)

ω_{input} - Transfer case input shaft speed

ω_{front} - Transfer case front output speed

ω_{rear} - Transfer case rear output speed

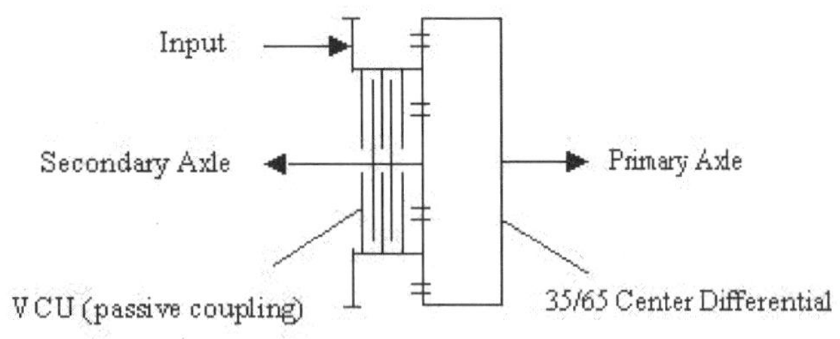

Figure 1 Schematic of the AWD w/VCU Transfer Case

<u>35/65 Center Differential with Passive Viscous Coupling (VCU) AWD Transfer Case:</u>

Speed Equation

1) $\omega_{input} = 0.65\ \omega_{rear} + 0.35\ \omega_{front}$

Torque Equation

2) $\tau_{input} = \tau_{front-d/s} + \tau_{rear-d/s} + \tau_{loss}\ (0)$

3) $\tau_{input} = \tau_{differential} + \tau_{vcu}$ or $\tau_{differential} = \tau_{input} - \tau_{vcu}$

4) $\tau_{front-d/s} = 0.35 * \tau_{differential} + \tau_{vcu}$

5) $\tau_{rear-d/s} = 0.65 * \tau_{differential} = 0.65 * (\tau_{input} - \tau_{vcu})$

<u>35/65 Open center differential AWD T-case:</u>

Speed Equation:

1) $\omega_{input} = 0.65\ \omega_{rear} + 0.35\ \omega_{front}$

Torque Equation:

2) $\tau_{input} = \tau_{front-d/s} + \tau_{rear-d/s} + \tau_{loss}\ (0)$

6) $\tau_{input} = \tau_{differential}$

7) $\tau_{front-d/s} = 0.35 * \tau_{differential}$

8) $\tau_{rear-d/s} = 0.65 * \tau_{differential}$

DISCUSSION

DEFINING AND CALCULATING $\tau_{\%REARD/S-SKID}$ [4] - To generate the maximum corrective yaw moment on a vehicle during a stability control activation event, all four wheels ideally need to work independently of each other. When braking of one front wheel causes the other three wheels to experience deep slip, it can potentially reduce the effectiveness of the stability control function. In 4WD terminology, potential degradation in performance of the stability control system will be a function of the amount of braking torque transferred to the opposite axle (rear axle in this study) during a stability control activation event. Variable $\tau_{\%reard/s-skid}$, is defined as the rear driveshaft torque as percentage of $\tau_{reard/s-skid}$, will be used throughout this paper to quantify the level of potential degradation in the stability control function induced by driveline.

Potential degradation in performance of the stability control system increases as $\tau_{\%reard/s-skid}$ increases. A 4WD system that transfers 0% of $\tau_{reard/s-skid}$ ($\tau_{\%reard/s-skid} = 0\%$) to the rear axle during a stability control activation event will experience no degradation in the stability control function induced by driveline. A 4WD system that transfers 100% of $\tau_{reard/s-skid}$ ($\tau_{\%reard/s-skid} = 100\%$) to the rear axle during a stability activation event can potentially experience significant degradation in the stability control function induced by driveline.

This paper will quantify the $\tau_{\%reard/s\text{-}skid}$ variable on the two transfer cases studied during a stability control activation event, and identify means to reduce $\tau_{\%reard/s\text{-}skid}$ to improve the performance of the stability control system on 4WD vehicles with a passive coupler.

Definitions:

$\tau_{reard/s\text{-}skid}$ - Rear driveshaft torque required to skid both rear wheels

$\tau_{\%reard/s\text{-}skid}$ - Rear driveshaft braking torque as percent of τ reard/s-skid

M_R - Mass over rear axle (Kg)

R - Tire static loaded radius (meter)

μ - Coefficient of friction between tire and ground surface

A - Axle Ratio (A/R)

g - Gravity (9.8 m/sec^2)

Note: Assume 100% Axle Efficiency

Equations:

9) $\tau_{reard/s\text{-}skid}$ = Skid Torque @ Rear Driveshaft = $(M_R * R * \mu * g)/A$

10) $\tau_{\%reard/s\text{-}skid}$ = $|(\tau_{rear\text{-}d/s})$ / (Skid Torque @ Rear Driveshaft)| * 100 = $|[(\tau_{rear\text{-}d/s}) / (M_R * R * \mu * g)/A]| * 100$

TWO SCENARIOS - In an oversteer condition, the stability control system will usually brake the outer front wheel (right/front wheel for this study) into deep slip (see Figure 2). During this stability control activation event, the transfer case front output speed can react to braking of right/front wheel in two likely ways. Assuming conventional open differentials in the primary and secondary axles, the two scenarios are:

1) No change in the transfer case front output speed in respect to the transfer case input and transfer case rear output speeds. Theory: Left/Front (L/F) wheel will speed up to compensate for the drop in the Right/Front (R/F) wheel speed. Net outcome is that transfer case front output shaft speed does not change and it rotates at the same speed as the transfer case input and rear output speeds.

2) Decrease in the transfer case front output speed in respect to the transfer case input and rear output speeds. Theory: Left /Front (L/F) wheel does not speed up to compensate for the drop in Right/Front

(R/F) wheel speed. Net outcome is that transfer case front output speed will drop with respect to transfer case rear output speed .

Vehicle testing was conducted to validate the two scenarios above and to better understand the torque transfer interactions on two transfer cases studied.

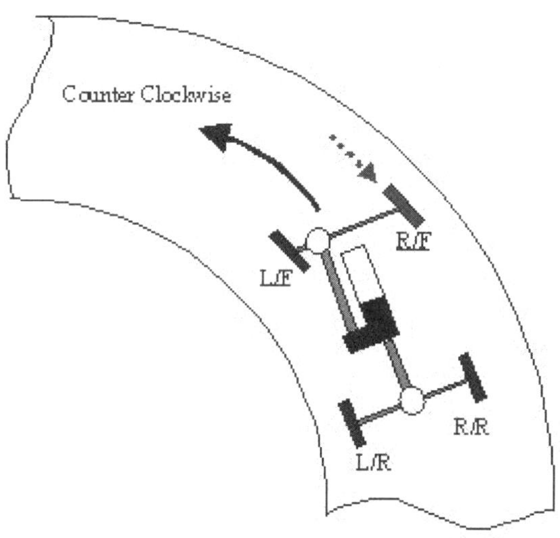

Figure 2 Throttle Release in a Turn

VEHICLE TEST - A 35/65 (front/rear) open center differential transfer case and a 35/65 (front/rear) center differential with a viscous coupling transfer case were evaluated back to back on the same vehicle on both low and high-mu surfaces. The test vehicle was equipped with conventional open differential in the primary and secondary axles. The maneuvers that were run were double lane change, constant radius turns on throttle, throttle release in a turn, and braking in a turn. During these events the following data was measured: individual wheel speeds, brake pressure at each wheel, brake pressure from each master cylinder, front and rear driveshaft torque, lateral and longitudinal vehicle velocity, lateral and longitudinal vehicle acceleration, yaw rate, slip angle, and steering wheel angle.

Note: All Driveshaft speeds reported are an average of the corresponding axles' average wheel speeds.

Graphs 1 & 2 show all four wheel speeds versus time for both open center differential and center differential with a viscous coupling transfer cases. Graphs 3 & 4 show driveshaft speeds and torques versus time for both open center differential and center differential with a viscous coupling transfer cases. The data was collected on a packed snow surface while releasing the accelerator throttle in a counter clockwise constant radius turn at a speed of approximately 48 km/hr.

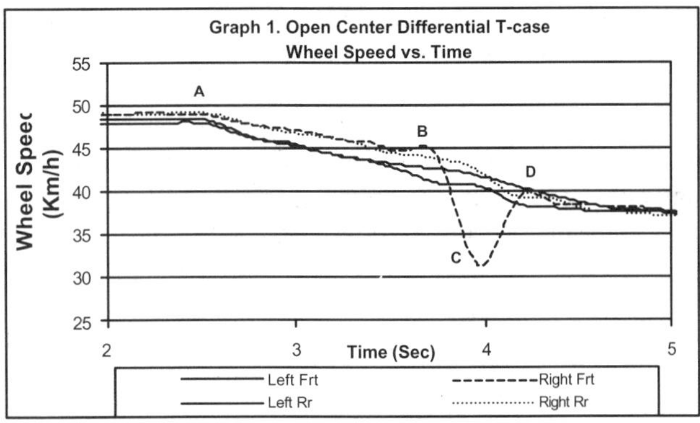

Graph 1. Open Center Differential T-case
Wheel Speed vs. Time

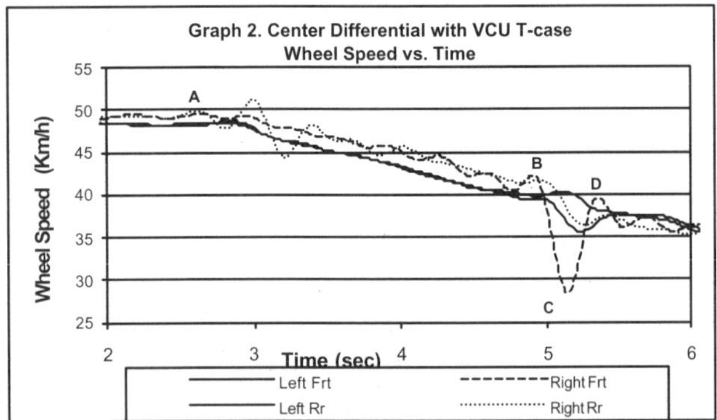

Graph 2. Center Differential with VCU T-case
Wheel Speed vs. Time

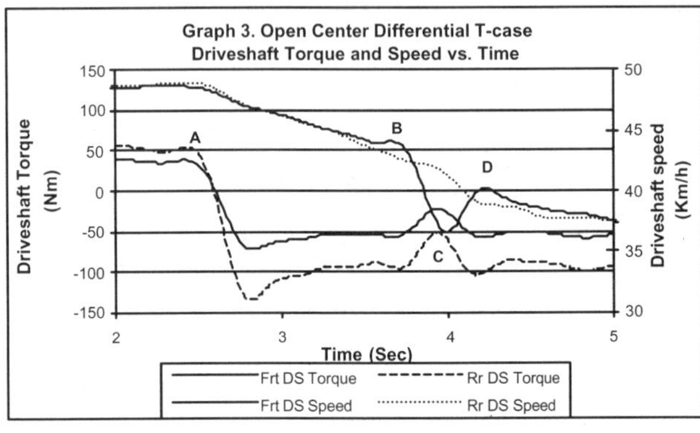

Graph 3. Open Center Differential T-case
Driveshaft Torque and Speed vs. Time

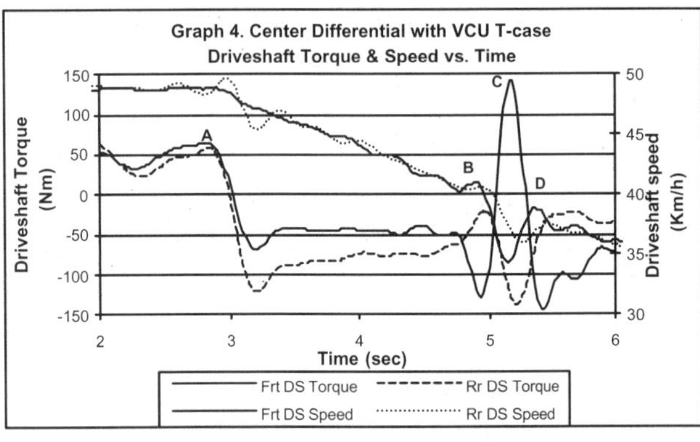

Graph 4. Center Differential with VCU T-case
Driveshaft Torque & Speed vs. Time

The sequence of events for the open center differential transfer case (Graphs 1 &3) follows:

A – Throttle release occurs and all four wheel speeds decrease at a linear rate. Also, the front and rear driveshaft torques change to a negative value from engine braking.

B – Start of a stability control activation event. R/F wheel is being driven to deep slip.

C – Maximum slip on R/F wheel is achieved. Three remaining wheel speeds continue to decrease at a linear rate. $\tau_{\text{rear-d/s}}$ reaches –50 Nm at point C.

D – End of a stability control activation event. All four wheels are rotating at the same speed.

The sequence of events for the center differential with VCU transfer case (Graphs 2 &4) follows:

A – Throttle release occurs and all four wheel speeds decrease at a linear rate. Also the front and rear driveshaft torques change to a negative value from engine braking.

B – Start of a stability control activation event. R/F wheel is being driven to deep slip.

C – Maximum slip on R/F wheel is achieved. Unlike the open center differential transfer case example, the speed difference between the transfer case front and rear outputs causes a positive torque application on the front driveshaft and a negative torque application (larger braking torque) on the rear driveshaft (Graph 4). A positive torque application on the front axle causes the L/F wheel to spin-up and a negative torque application on the rear axle causes the rear wheels to slow down at non-linear rate (Graph 2). $\tau_{\text{rear-d/s}}$ reaches -140 Nm at point C.

D – End of a stability control activation event. All four wheels are rotating at the same speed.

UNDERSTANDING OF SPEED AND TORQUE TRACES - The open center differential transfer case clearly exhibited scenario #2 (decrease in transfer case front output speed in respect to transfer case rear output speed) as shown on Graph 3. At point C, when maximum slip was achieved on the R/F wheel, the transfer case rear output speed was relatively unaffected by the drop in the transfer case front output speed (Graph 3). The open center differential transfer case allowed front and rear transfer case outputs to rotate at different speeds without significantly altering the transfer case torque outputs (Graph 3).

The center differential with a viscous coupling transfer case also exhibited scenario #2, but showed signs of scenario #1 as well. Graph 4 clearly shows decrease in transfer case front output speed in support of scenario #2; however, some signs of scenario #1 (increase in L/F wheel speed) can be also seen in Graph 2. The L/F wheel speed, however, does not increase enough to compensate for the full drop in R/F wheel speed to classify it as true scenario #1. Since the transfer case front output speed is an average of both left and right front wheel speeds multiplied by the axle ratio, the net reduction in the transfer case front output speed is reduced with the viscous coupling transfer case as compared to the open center differential transfer case with identical 50% slip on the R/F wheel.

Also, unlike the open center differential transfer case, the speed difference between the transfer case front and rear outputs generated viscous torque transfer which affected both the transfer case front and rear output speeds. A positive viscous torque transfer to the front axle caused the L/F wheel to speed up and negative viscous torque transfer to the rear axle caused both rear wheels to decrease at a non-linear rate. $\tau_{\%reard/s\text{-}skid}$ was approximately a factor of three higher with the viscous coupling transfer case than with the open center differential transfer case. One can expect more driveline-induced degradation in the stability control function with the viscous coupling transfer case than with the open center differential transfer case.

One potential benefit of a viscous coupling over an open differential by itself was uncovered by this study. During stability control activation events, the viscous coupling would send positive torque to the front driveline causing the inside front (L/F) wheel to speed up while the outer front (R/F) wheel was being braked by the stability control system. This effect can result in an incremental counter yaw moment which does not exist in an open center differential transfer case.

In summary, the vehicle test data above provided real-time understanding of torque transfer on the two transfer cases studied and confirmed that braking of one front wheel would result in overall drop in transfer case front output speed. The remainder of the analysis below will follow scenario #2.

ANALYTICAL ANALYSIS OF SCENARIO #2 WITH CENTER DIFFERENTIAL WITH A VISCOUS COUPLING TRANSFER CASE - Braking of one front wheel during a stability control activation event on a 4WD vehicle will result in decrease of the front driveshaft speed as seen above in the vehicle test section.   Since the center differential with a viscous coupling transfer case used in this study must meet the 35/65 differential speed equation, the following relationship must hold at all times:

Speed Equation:

1) $\omega_{input} = 0.65\ \omega_{rear} + 0.35\ \omega_{front}$

and

$\omega_{front} < \omega_{input} < \omega_{rear}$ (for a rear overrun condition)

Torque Equation:

Substituting Equation (5) for $\tau_{rear\text{-}d/s}$ in Equation (10):

11) $\tau_{\%reard/s\text{-}skid} = |[0.65 * (\tau_{input} - \tau_{vcu})] / (\text{Skid Torque @ Rear Driveshaft})| * 100$

Since Skid Torque @ Rear Driveshaft is a function of parameters not directly controlled by the 4WD and stability control sytems, Equation (11) can be further simplified by only leaving in variables that the two systems have direct control over (τ_{input} & τ_{vcu}).

12) $\tau_{\%reard/s\text{-}skid} \approx |(\tau_{input} - \tau_{vcu})|$ or $f(\tau_{input}, \tau_{vcu})$

In order to reduce $\tau_{\%reard/s\text{-}skid}$ and minimize interaction between the two systems, Equation (12) states that variable $(\tau_{input} - \tau_{vcu})$ must be driven as close to zero as possible during a stability control activation event. This can be accomplished by increasing τ_{input} independently or decreasing τ_{vcu} independently; however, a systems engineering approach where both τ_{input} and τ_{vcu} are simultaneously optimized will provide the most effective solution with least amount of compromises. Details on how τ_{vcu} and τ_{input} can be modified to minimize the interaction between the two systems are below.

Minimizing τ_{vcu}: τ_{vcu} is a function of two variables:

1) Speed difference that stability control system induces between the primary and secondary axle during a stability control activation event. Employing a different stability control strategy/calibration, which reduces the speed difference between the primary and secondary axle during a stability control activation event will minimize τ_{vcu}. For example, an optimal wheel slip value during a stability control activation event should be investigated that balances both stability control system's counter yaw generation and viscous torque generation. The goal would be to identify an optimal wheel slip value (other than the standard 50% slip value typically used today) specifically for the passive coupling 4WD vehicle applications that balances both the counter yaw generation by the stability control system and degradation in stability control function induced by viscous torque generation.

2) Tuning of the viscous coupling unit. Detuning the viscous coupling will also reduce viscous torque generated during a stability control activation event.

Controlling τ_{input}:

τ_{input} in a majority of the vehicles today is directly controlled by driver's throttle input. However with advent of electronic throttle control system, it can also be varied depending on conditions independent of the driver's input. With development, an electronic throttle control system can be used to vary τ_{input} and offset τ_{vcu} generated during some stability control activation events to minimize negative torque transfer to the rear axle.

A system engineering approach, where two systems are jointly developed to provide best vehicle performance, is required to maximize synergy and reduce undesirable interaction between the Brake Based Stability Control and the 4WD Systems.

CONCLUSION

1) The question of what actually happens to the front driveshaft speed during a stability control activation event was answered in this study. This study confirmed through vehicle test data that scenario #2 (decrease in front transfer case output speed) occurs during a stability control activation event with open center differential and center differential with viscous coupling transfer cases. However, the center differential with a viscous coupling transfer case showed less reduction in transfer case front output speed due to viscous torque transfer than open center differential transfer case.

2) $\tau_{\%reard/s\text{-}skid}$ during the stability control activation event was approximately a factor of three higher with the center differential with a viscous coupling transfer case as compared to an open center differential transfer case. One can expect more driveline-induced degradation in the stability control function with the viscous coupling transfer case than open center differential transfer case due to the viscous coupling resisting the speed difference across the transfer case outputs by biasing torque front to rear.

3)This study did uncover potential benefits of a viscous coupling over an open center differential. During stability control activation events, the viscous coupling would send positive torque to the front driveline. This positive torque application on the front driveline caused the inside front (L/F) wheel to speed up while the outer front (R/F) wheel was being braked by the stability control system. Simultaneously applying drive (positive) torque to the ground on the L/F wheel and brake (negative) torque to the ground on the R/F wheel during a stability control activation event can generate incremental counter yaw moment which does not exist in an open center differential transfer case.

4) Means to minimize $\tau_{\%reard/s\text{-}skid}$ on the center differential with a viscous coupling transfer case during stability control activation event were discussed. Based on the analysis and vehicle test data, the following steps could be taken to limit the amount of interaction between the stability control system and a passive coupling transfer case:

- Modify the brake based stability control strategy/calibration to reduce speed difference generated across the passive coupling during a stability control activation event.

- A less aggressive viscous coupling unit (detuned viscous coupling) could be implemented to lower the amount of viscous torque transferred during a stability assist activation event.

- Incorporate Electronic Throttle Control System within the stability control strategy to reduce $\tau_{\%reard/s\text{-}skid}$ by varying the τ_{input} for optimal stability control performance.

This paper provided fundamental understanding of torque transfer on 4WD vehicles equipped with a brake based stability control system. The recommendations provided in this paper should be considered and developed in future applications of these technologies for improved compatibility.

REFERENCES

1. E. Herb, H. Krusche, E. Schwartz, H. Wallentowitz, *Stability-Control And Traction-Control At Four-Wheel-Drive Cars,* SAE Technical Paper, Series 885007 (1988)
2. Wolfgang Peschke, *A Viscous Coupling in the Drive Train of an All-Wheel-Drive Vehicle,* SAE Technical Paper, Series 860386 (1986)
3. Satoshi Ashida, Fumio Ueda, Akihiko Ichikawa, Yoshiyuki Furuta, Toshiaki Kuribayashi, *The Development of Fluid for Small-Sized and Light Weight Viscous Coupling,* SAE Technical Paper, Series 981446 (1996)
4. Thomas D. Gillespie, *Fundamentals of Vehicle Dynamics,* SAE Publications (1992)

Influence of Active Chassis Systems on Vehicle Propensity to Maneuver-Induced Rollovers

Aleksander Hac
Delphi Automotive Systems

ABSTRACT

The purpose of this paper is to evaluate through simulations the effects of active chassis systems on vehicle propensity to rollover caused by aggressive handling maneuvers. A 16 degree-of-freedom computer model of a full vehicle is used for this purpose. It includes models of active chassis systems and the associated control algorithms, and allows for simulation of vehicle dynamic behavior under large roll angles. The controllable chassis systems considered in this investigation are active rear steer, brake based vehicle stability enhancement system and active anti-roll bar. The maneuvers used in simulation are the double lane change and the fishhook maneuvers with increasing steering amplitudes. The vehicle represents a midsize SUV with a marginal static stability factor of 1.09 and aggressive tires. The results of simulations demonstrate that the uncontrolled vehicle rolls over in both maneuvers when the steering angle is sufficiently large. Each active control system significantly increases rollover stability – either the vehicle cannot be rolled over regardless of the magnitude of the steering angle, or the amplitude of the steering angle necessary to rollover the vehicle is markedly increased.

INTRODUCTION

With growing popularity among consumers of vehicles with high centers of gravity, evaluation of rollover propensity of these vehicles becomes an issue of increasing importance. Fundamentally, there exists two classes of tests designed to predict or evaluate vehicle tendency to rollover: static tests involving measurements of vehicle parameters (usually distances), which are related to vehicle rollover behavior, and dynamic tests in which vehicle is put through a set of severe handling maneuvers, which may induce two wheel lift off. Static tests usually provide a simple indicator, such as a static stability factor, tilt table ratio, side pull ratio or critical sliding velocity. In most cases these measures do not include the effects of suspension and tire compliance. Worst still, they do not incorporate any effects of electronically controlled chassis systems, such as stability enhancement systems, active steer systems, or controllable suspensions. This is a very serious limitation as such systems become quite common.

There are good reasons to believe, as well as significant body of evidence indicating that active chassis control systems markedly increase vehicle resistance to rollover. For example, Marine et al. (1999) have shown by analyzing vehicle behavior in a lane change maneuvers that a likelihood of two wheel lift off is increasing with increasing steering angle and vehicle yaw rate. Thus vehicle oversteer is an important factor contributing to rollover. Since brake based stability enhancement systems reduce vehicle tendency to oversteer and reduce lateral slip velocity of vehicle, they can be expected to reduce probability of rollover. This conjuncture was confirmed by a simulation study performed by Ungoren et al. (2001), in which SUVs were subjected to aggressive handling maneuvers. While the vehicle without the control system experienced two wheel lift off in several maneuvers, the vehicle with the brake based stability enhancement system turned on did not exhibit tipping tendencies in the same maneuvers and generally experienced lower roll angles.

Meanwhile other systems are investigated or are under development, which specifically target the rollover prevention. For example, Palkovics et al. (1998) describe a system for commercial trucks, which detects when outside wheels are about to lift off during cornering by judicious application of brakes and throttle and observing wheel slip. The system then applies front brakes to reduce the lateral acceleration. A similar system was proposed by Wielenga (Wielenga, 1999; Wielenga and Chace, 2000), in which front brakes are applied when lateral acceleration of vehicle exceeds a threshold. Since lateral acceleration alone is a poor predictor of impending rollover (Marine et al., 1999), Eisele and Peng (2000) considered a control algorithm, in which brakes were applied when a linear combination of roll angle, roll rate and lateral acceleration exceeded a threshold. This anti-rollover feature was added to the existing vehicle stability control algorithm, which primarily controls vehicle motion in the yaw plane.

Ackerman at al. (1999) proposed a system in which the height of vehicle center of mass was estimated on line, and active steering and braking were applied when the lateral acceleration approached a threshold value derived from a static stability factor. With advent of these and other active rollover prevention systems, the static measures of rollover stability will become even less useful.

Dynamic rollover testing typically consists of aggressive handling maneuvers, involving rapid steering and sometimes braking, performed on dry smooth surface (Garrot et al., 1999). In this type of tests, vehicle behavior is affected by many design variables not considered in the static tests, such as the effect of suspension and tires, and the influence of active chassis systems, which significantly affect vehicle behavior at the limit. However, dynamic rollover tests are dangerous and expensive. Thus only limited number of tests can realistically be performed and there is always a possibility that a vehicle, which performed well in a few tests, could roll over under slightly modified maneuver. In addition, it is difficult to achieve consistent results. The main reason is that the vehicle is on the verge of loosing stability, which by the very definition is a condition when small changes in the inputs, disturbances, vehicle parameters, or environment can result large changes in the output, that is the outcome of the test. Therefore there is a need to supplement the tests results with the results of simulation. Simulation can also be a useful tool in predicting rollover propensity of vehicle at a design stage.

In this paper, the influence of three presently available controlled chassis systems on vehicle rollover resistance is investigated. The chassis systems considered in this study are the active rear steer (ARS), the brake based vehicle stability enhancement (VSE) system and the active roll bar, referred to as dynamic body control (DBC) system. The main purpose of the first two systems is to improve vehicle yaw response, that is the balance between responsiveness and stability in the yaw plane. The DBC system is primarily used to improve the balance of ride and handling, in particular to reduce vehicle body roll during cornering maneuvers. The control algorithms considered here do not include any measures specifically targeting rollover prevention. The vehicle responses to a double lane change and Fishhook steering inputs are simulated for a passive vehicle and vehicles equipped with each one of the active chassis systems. The maneuvers are repeated with increasing amplitude of steering angle until either the vehicle rolls over or the maximum steering angle is reached. The vehicle parameters represent a midsize SUV with all independent suspension and a static stability factor of 1.09.

The paper is organized as follows. In the next section, the vehicle model used in simulations is briefly described. The fundamental control objectives are then explained for each system. Subsequently, the results of simulations are presented followed by conclusions.

VEHICLE MODEL

The vehicle model used in this study was developed as a tool for testing of active chassis control systems at the development stage, as well as for hardware in the loop simulations. The model has a total of 16 degrees of freedom, excluding the dynamics of subsystems. Vehicle body is modeled as a rigid body, which can perform three translations and three rotations. In the case of all independent suspensions each wheel has two degrees of freedom: rotation about the lateral axis and vertical translation. In the case of a rigid axle, the axle has two degrees of freedom, corresponding to axis roll and heave. In addition, both front and rear wheels can be steered. There are two types of inputs to the model: driver inputs, such as steering wheel angle, brake pedal force and throttle position, and environmental inputs, such as road displacement under each wheel, road inclination angles, and surface coefficient of adhesion between each tire and the road.

The model features simplified models of brake, powertrain and steering systems, which relate driver brake, throttle and steering inputs to wheel torques and front steering angle. The brake system model includes master cylinder, modulator, calipers and hydraulic circuit. It determines brake torque applied to each wheel from the brake pedal force. The brake model includes ABS and TCS functions in simplified forms. The powertrain model includes a model of an automatic transmission. It uses a map of engine torque and gear shift pattern and takes into account inertia of the engine and the drivertrain to determine the driving torque on the driven wheels. The model of the front steer system relates the steering wheel angle to the front wheel angle. A simple model of the actuator for active rear steer is also included. The model also includes simplified versions of vehicle stability enhancement, active rear steer and active roll bars algorithms. The tire model used here is a parametric model, which can be considered a modification of Duggoff's model (Bernard et al., 1977, Wong, 1993). It includes the effects of tire normal load on tire longitudinal and lateral stiffness coefficients and on the surface coefficient of adhesion. In addition, the surface coefficient of adhesion is a function of the velocity of sliding of the wheel with respect to the road. The effect of dynamic delay in building the lateral tire force is modeled by a first order filter, whose time constant depends on vehicle speed and relaxation length of the tire. The model has been validated against vehicle test data.

The model has a number of features, which make it suitable for simulating vehicle rollover maneuvers up to about 45 degrees of roll angle measured with respect to road surface. Specifically,

1. The model does not use a small roll angle assumption. This requires additional transformation of variables between the vehicle body fixed reference frame and the frame attached to the road surface, as well as modification of equations of motion when two wheels lift off the ground.

2. The effects of suspension jacking forces are included. These are vertical components of the forces in lateral links of suspension. These links are usually not parallel to the ground, so that during cornering the forces acting along these links have vertical components, which do not cancel out. As a result, they have a tendency to lift vehicle center of gravity during hard cornering.

3. Nonlinerities in suspension stiffness characteristics, including the bushings, and in damping characteristics are modeled using look up tables. The nonlinearities in suspension stiffness characteristics are important in modeling rollover events, since suspension stiffness characteristics are usually progressive. Consequently during cornering maneuvers compression of outside suspension is less than the extension of the inside suspension, which causes upward shift of vehicle center of gravity. The effect of suspension nonlinearities is particularly significant when vehicle is fully loaded, since in this case the outside suspension can be fully compressed during heavy cornering. This generates high jacking forces, which lift the body up. Concurrently, the normal tire forces increase, which may increase lateral forces and lateral acceleration.

4. The model includes the effect of change in vehicle half-track width during cornering resulting from lateral deformation of tires, suspension kinematics and lateral compliance of suspension. The reduction in vehicle half-track width under dynamic conditions, combined with increase of vehicle center of gravity height due to jacking effects of suspension affects rollover stability.

5. The model permits simulation of vehicle response with payloads, as long as the mass, location and inertial properties of payload are known. The model determines the static suspension deflections and calculates the variables of interest with respect to the new operating point, while maintaining proper limits on suspension deflections. This is important for proper modeling of bottoming out of suspension as discussed above.

6. The tire model includes the dependency of tire longitudinal and lateral stiffness coefficients and the surface coefficient of adhesion on the normal load. The character of these relationships has a dramatic influence on the rollover dynamics, because at the rollover threshold the normal forces of the outside tires are about double their static values. In addition, the effect of tire camber angle on tire forces is described through extrapolation of existing models, since the data for very large camber angles is not readily available for automotive tires.

ACTIVE CHASSIS CONTROL SYSTEMS

The active chassis systems considered in this paper are the brake based vehicle stability enhancement system, active rear steer system, and active roll bar. In what follows we provide a brief description of their basic functions. More detailed descriptions can be found in the references cited below.

VEHICLE STABILITY ENHANCEMENT (VSE) systems are active safety systems that control the response of vehicle to steering inputs at or near the limit of adhesion by selectively applying brakes to individual wheels independently of the driver. The main objective of control is to make the vehicle more predictable and easier to control by the driver in emergency maneuvers by reducing the difference between the vehicle behavior at the limit and in the linear range of handling behavior. This is accomplished by correcting excessive oversteer or understeer through selective application of brakes to one or more wheels. In order to perform these tasks the system must have means to determine the desired response of vehicle, compare it with measured or estimated response and, if a sufficient discrepancy between the two is detected, apply corrective action.

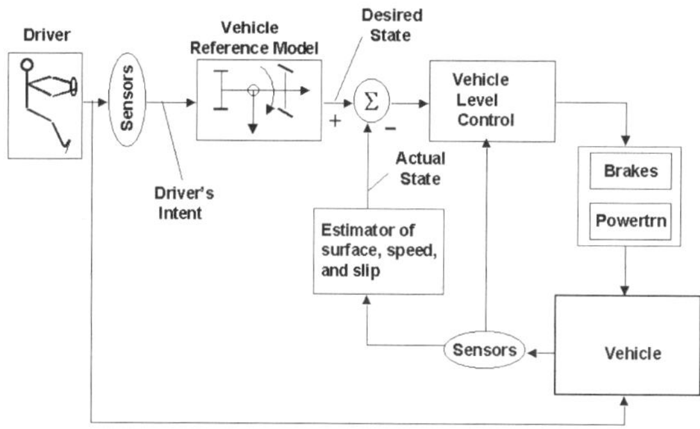

Figure1. Functional Diagram of Vehicle Stability Enhancement System

A typical control system consists of the following building blocks, illustrated in Figure 1.

1) Sensors, which measure driver inputs and vehicle response. The measured driver inputs typically include steering angle, brake pedal force and throttle position. The vehicle response is measured in terms of lateral acceleration, yaw rate and wheel speeds, from which vehicle reference speed is derived. There are usually additional sensors within brake and powertrain subsystems.

2) Vehicle reference model, which generates the desired vehicle response in terms of the desired yaw rate and the desired side slip angle or side slip rate, using primarily driver inputs and vehicle reference speed.

3) Estimation block, which provides estimates of vehicle reference speed, the surface coefficient of adhesion and usually vehicle side slip angle and side slip rate. Other estimation functions may be performed by subsystems, for example estimation of driving torque at the driven wheels.

4) Vehicle level control block, which compares the desired values of yaw rate and usually side slip angle with the measured or estimated values and calculates the necessary correction. The correction is typically expressed in terms of corrective yaw moment applied to the vehicle, or desired wheel slip correction. To generate these signals, closed loop control of yaw rate and usually side slip angle or side slip rate is used.

5) System level controllers, which include powertrain controller and brake modulator. They employ local control loops to achieve the target values of wheel slips or wheel torques as determined by the vehicle level controller.

More detailed descriptions of the vehicle stability enhancement systems can be found in van Zanten et al. (1995) or Hac (1998). An excellent coverage of issues involved in estimation of sides slip angle is provided by Nishio et al. (2001).

What is important from the viewpoint of rollover resistance, is that the stability enhancement algorithms control vehicle response in the yaw plane and specifically vehicle yaw rate and slip angle or slip rate. They do not include any explicit considerations of vehicle roll angle. Nevertheless, by limiting the vehicle side slip angle and therefore lateral velocity, they make it less likely for a vehicle to roll over either with or without tripping mechanism. In a tripped rollover a minimum sliding velocity (about 6 m/s for a typical SUV) is required to trip the vehicle and tip it over. During untripped rollover, maximum lateral forces on both outside tires are needed to generate peak lateral acceleration necessary to initiate rollover. The maximum lateral forces are achieved on dry surface at large tire side slip angles, typically in the range of 10-20 degrees. Such large sideslip angles cannot be achieved at the rear axle without a large vehicle side slip angle (oversteer). This can be confirmed by analyzing the rollover data provided by Marine et al. (1999), which shows that in maneuvers with two wheel lift off the magnitudes of yaw rate were significantly larger than in the maneuvers in which vehicle remained stable, while the peak lateral accelerations remained roughly the same. It must be concluded that in the maneuvers with rollover stability problem, vehicle typically experienced heavy oversteer with large side slip angles, even though the side slip angles were not recorded.

ACTIVE REAR STEER (ARS) is another type of active chassis control system, which limits vehicle oversteer and improves handling. The main objectives of this system are to enhance vehicle maneuverability at low speeds, improve stability at high speeds and improve vehicle transient response to steering inputs. The first objective is achieved by steering the rear wheels out of phase with the front at low speeds, since this reduces the radius of turns. Improving stability requires limiting vehicle tendency to oversteer. Vehicle oversteer, followed by a loss of control, is often caused by a rapid change in sign of the steering angle at high speeds. This causes a quick change in the direction of the lateral force of the front axle, while the rear axle force, which lags the front, still acts in the opposite direction. During transient, both lateral forces, which are opposite in signs, generate a large yaw moment, which begins to rotate the vehicle more rapidly than driver intends, eventually leading to oversteer and possibly spin out. In order to advance the phase of rear lateral force to match closer that of the front axle, the rear wheels must be steered in phase with the front ones. This makes the vehicle more stable in quick evasive maneuvers, since vehicle yaw rate and its rate of change are reduced.

It is known that in emergency lane change maneuvers both objective task performance measures and driver's subjective ratings of handling quality improve when the phase lags between the steering angle input and lateral acceleration and yaw rate responses are kept small. In addition, it is desirable to keep the lags in lateral acceleration and yaw rate approximately equal throughout the entire range of speeds. The lateral acceleration and yaw rate are related via the following kinematic expression

$$a_y = \dot{v}_y + v_x \Omega \qquad (1)$$

where a_y is lateral acceleration, Ω is yaw rate, v_x and v_y are longitudinal and lateral velocities, respectively. Thus both a_y and Ω are in phase when the derivative of lateral velocity is kept low, which can be approximately achieved by keeping the lateral velocity small. It is easy to show (Furukawa et al., 1989) that for a simplified bicycle handling model, the steady state value of lateral velocity can be brought to zero over the entire speed range if the rear wheels are steered in proportion to the front wheels, that is

$$\delta_r = K_{ff}(v_x)\delta_f \qquad (2)$$

where δ_f and δ_r are and the front and rear steering angles, respectively, and the feedforward gain, K_{ff}, is the following function of velocity:

$$K_{ff0} = \frac{-b + \dfrac{Ma}{C_r l}v_x^2}{a + \dfrac{Mb}{C_f l}v_x^2} \qquad (3)$$

Here a and b denote the distances of vehicle center of mass from front and rear axles, respectively, $l=a+b$ is vehicle wheelbase, M is total vehicle mass, C_f and C_r are the cornering stiffness coefficients of front and rear axles, respectively. The above gain as a function of vehicle speed is illustrated in Figure 2 for a midsize SUV.

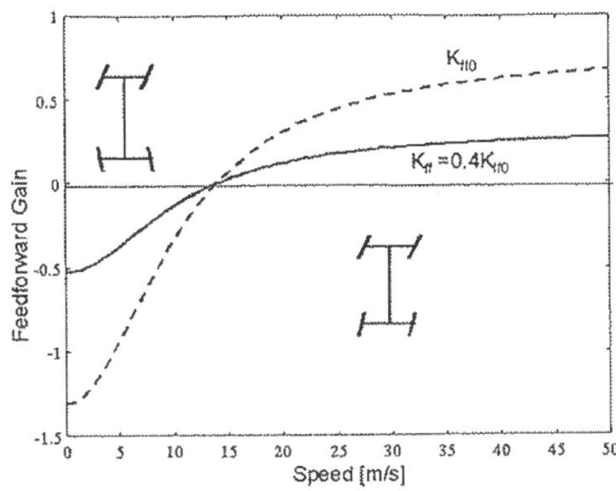

Figure 2.Feedforward Gain for Active Rear Steer System

The gain is negative at low speeds and positive at high speeds, which is consistent with the requirements of vehicle maneuverability at low speeds and stability at high speeds The gain computed from equation (3) calls for a large steering angle at the rear wheels at low speeds when the front wheels are steered sharply. It also yields very low, often subjectively objectionable, yaw rates in quick transients maneuvers performed at high speeds. In practice therefore the gain is scaled down, with the precise shape determined by vehicle tuning. Feedforward control of the rear steering angle according to equation (2) appears to be the most commonly used. In addition, a feedback control loop can be employed, as described by Fujita et al. (1998). In this study only a feedforward control of rear steering angle is assumed, with the gain being a scaled down version of that given by equation (3) with the scale factor of 0.4. Similarly to the VSE system, the active rear steer system does not specifically target vehicle rollover, but an improvement in rollover resistance is expected as a byproduct of limiting vehicle tendency to oversteer, which is a precondition of most rollovers.

DYNAMIC BODY CONTROL. The third type of active chassis system considered in this paper is an active body roll control system, which is here referred to as dynamic body control (DBC). This type of system is particularly beneficial for SUVs. Design of passive suspensions for this type of vehicle poses significant challenges because of wide spectrum of conditions under which vehicle operates. During off road use vehicle may be subjected to large ground inputs, usually of low frequency. This calls for large ground clearance and large axle articulation to maintain traction in these conditions, thus requiring also large suspension travel. At the same time, the vehicle must achieve acceptable handling and ride characteristics during operation on paved roads. This leads to design compromises, often resulting in large body roll angles during cornering maneuvers. The purpose of dynamic body control is to improve these trade-offs. The main design goals are to maintain or improve axle articulation over uneven terrain,

to improve ride quality during road use and to reduce body roll angle during cornering.

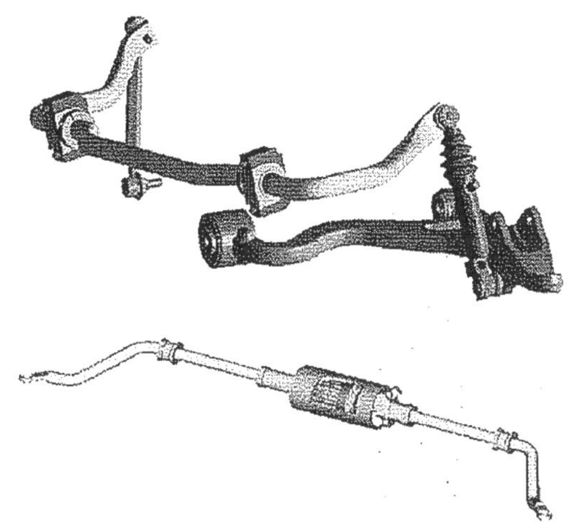

Figure 3. Active Roll Bars with Linear and Rotary Actuators

In the dynamic body control system, linear or rotary hydraulic actuators act on conventional roll bars to provide forces that resist vehicle roll, as shown in Figure 3. The actuators are controlled by two-way valves, which determine the chamber of the actuator to which the hydraulic pressure generated by a pump is supplied. During normal, straight line driving the actuators are not pressurized, so that left and right roll bar assemblies can rotate relatively freely with respect to each other, without generating a significant roll resisting moment. This improves vehicle response to road inputs on one side of vehicle. During cornering, pressurized fluid is supplied to properly selected chambers of the actuator, creating a torsional moment in the roll bar that opposes the roll motion of vehicle body. The control algorithm uses measured lateral acceleration of vehicle to determine the desired pressure in the chambers of front and rear actuator (Everett et al., 2000). Depending on the particular system hardware, sensors, and control algorithm, the roll resisting moment may be divided between front and rear axle in constant or varied proportions. In more advanced systems, the proportions of the roll moment can be varied in real time as a function of vehicle yaw response (measured by sensor) relative to the desired yaw rate (Everett et al., 2000). In this study a simpler system responsive to lateral acceleration and with fixed roll moment distribution is considered.

Practical DBC systems operate under a number of constraints of which the limits on the power of the system (limited by the power of the hydraulic pump) and the maximum torque (limited by the size of actuators and hydraulic pressure) are the most important from the viewpoint of this study. Due to the limit on the magnitude of torque, the vehicle body experiences some roll during steady-state cornering with large lateral acceleration, which is considered desirable by virtue of providing additional feedback to the driver. The limit of the power supplied influences vehicle response in quick transient maneuvers, when sudden changes in the magnitude and/or direction of the lateral acceleration result in fast roll rates and large power demand to reduce roll angle. Both of these limitations influence vehicle propensity to maneuver-induced rollovers.

RESULTS OF SIMULATIONS

In order to evaluate the effects of active chassis systems on vehicle tendency to rollover during emergency handling maneuvers, a series of simulations were conducted. The vehicle parameter data used in this study represents a midsize SUV with all independent suspension and a static stability factor of 1.09. The selected maneuvers discussed here are the double lane change maneuver and the fishhook maneuver with the initial speed of 30 m/s (67 mph). The steering patterns for both maneuvers are illustrated in Figure 4. Each of these maneuvers can lead to heavy oversteer, which is a contributing factor in rollovers. All simulated maneuvers were performed as open loop steering control maneuvers, without any driver model representing driver's reaction to vehicle response. Each of the maneuvers was repeated a number of times with the same speed of entry, and the same steering pattern, but with increasing amplitude A of the steering angle. As the amplitude increased, the vehicle eventually either rolled over, or the maximum steering angle of 540 degrees was reached. The rate of change of the steering angle was limited to 1000 deg/s, which approximately corresponds to the maximum rate that can be generated by human drivers.

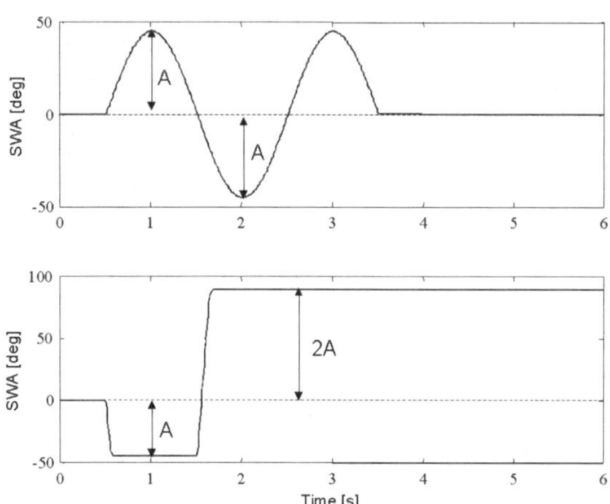

Figure 4. Steering Patterns for a Double Lane Change and Fishhook Maneuver

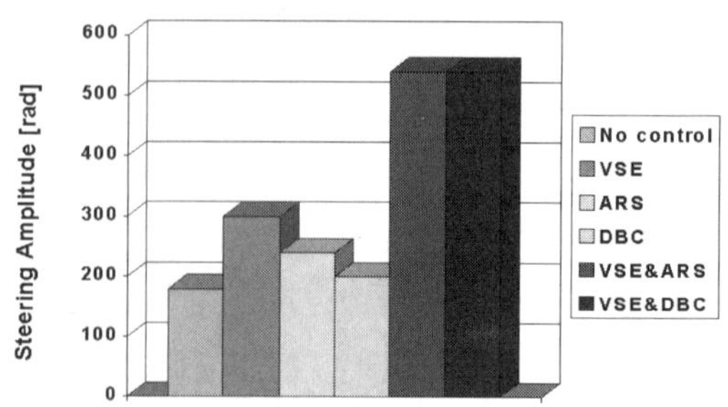

Figure 5. Maximum Steering Angle Amplitudes without Rollover for Double Lane Change Maneuver

The results of simulations performed for the double lane change maneuver are summarized in Figure 5. It shows the maximum amplitudes of the steering angle for each vehicle, for which the maneuver could be performed without rolling over. The vehicles considered are a passive vehicle with no active control systems, the same vehicle with vehicle stability enhancement system (VSE), with active rear steer system (ARS), with dynamic body control (DBC) system, with both VSE and ARS, and finally with both VSE and DBC systems. The maximum steering angle amplitude that can be reached without inducing rollover for the passive vehicle is 180 degrees, but at that steering angle vehicle developed a very large side slip angle, just over 50 degrees. For the vehicle with VSE, the maximum steering amplitude of 300 degrees could be achieved without rollover, for the vehicle with ARS the steering amplitude at the rollover threshold was 240 degrees, and for vehicle with DBC it was 200 degrees. The vehicles with two active systems, either VSE and ARS or VSE and DBC, did not rollover regardless of the amplitude of the steering angle. The vehicle with VSE and ARS also exhibited the smallest side slip angle.

As an example, the results obtained for the vehicle without control and with VSE system are illustrated in Figure 6 for the amplitude of the steering angle of 270 degrees. The passive vehicle experiences extreme oversteer in the second turn and rolls over; the simulation is terminated when the roll angle reaches 1 radian (57.3 degrees). The vehicle with VSE system develops smaller peak lateral acceleration and remains stable, but the peak roll angle is quite large (11.3 degrees) and the vehicle experiences a very brief two wheel lift off. Note also that the vehicle side slip angle is rather large, reaching a maximum of 11 degrees. It is possible that a more aggressive control of side slip angle in this vehicle would also reduce vehicle tendency to tip off. None of the control algorithms used here were fine-tuned for the particular vehicle used in simulations, so it is possible that the performance could be further improved.

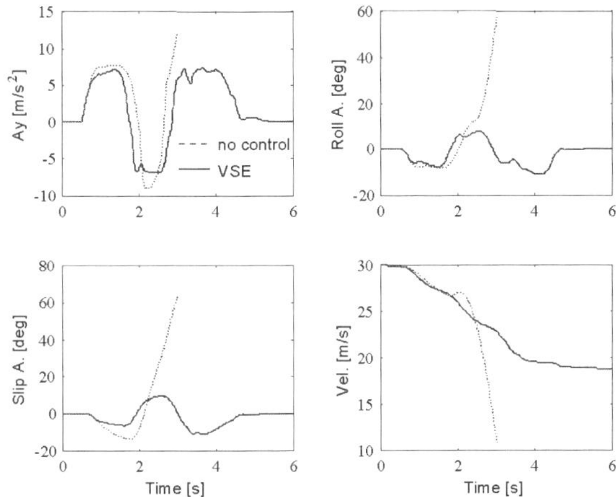

Figure 6. Vehicle Response in Double Lane Change Maneuver at Steering Amplitude of 270 Degrees without Control and with VSE

The results obtained for the fishhook maneuver are illustrated in Figure 7 in terms of the maximum amplitude of steering angle, which can be applied without causing rollover. Note that according to the definition of the amplitude A for the fishhook maneuver (Figure 4), the maximum steering angle is actually 2A. Thus the angle of 270 degrees indicates that the vehicle cannot be rolled over in this maneuver. The maximum steering angle amplitude for the passive vehicle was 40 degrees and for the vehicle with DBC it was 55 degrees. All the remaining vehicles were able to negotiate this maneuver without rolling over regardless of the magnitude of the steering angle. The vehicle with VSE system, however, was more stable than the one with ARS. It exhibited consistently smaller vehicle side slip angles and no wheel lift off, while the vehicle with ARS occasionally experienced one or two wheel lift off.

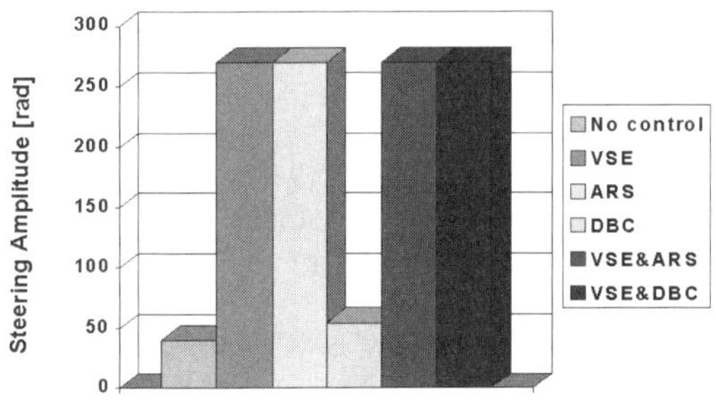

Figure 7. Maximum Steering Angle Amplitudes without Rollover for Fishhook Maneuver

The results in the case of the steering angle amplitude of 120 degrees for both systems are illustrated in Figure 8.

The vehicle with ARS displays much higher side slip angle than the vehicle equipped with VSE; it also has significantly higher peak lateral acceleration and roll angle.

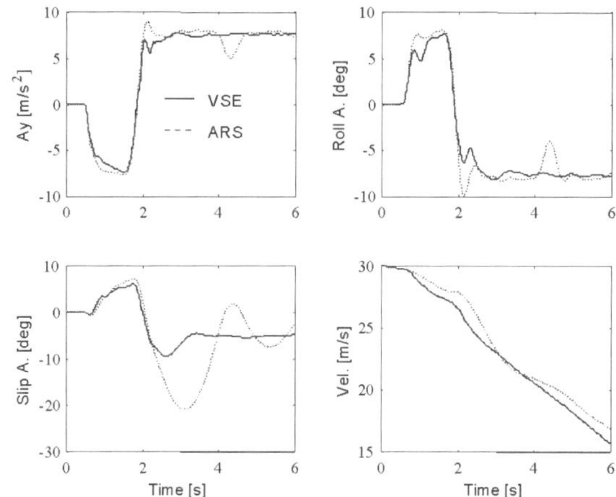

Figure 8. Vehicle Response in Fishhook Maneuver at Steering Amplitude of 120 degrees with VSE and with ARS

In Figure 9 the results obtained for vehicles with VSE and DBC systems are compared for the steering angle amplitude of 55 degrees. At steady state, vehicle with DBC system experiences much smaller roll angle, about half of that for vehicle with VSE. However, during rapid reversal of steering angle and lateral acceleration, the hydraulic DBC system cannot keep up with the quick change in roll angle and the roll angle exhibits a significant overshoot. For larger steering angle this yields to rollover of vehicle. The quickness of response of DBC system is limited primarily by the power of the pump, which in this study was set to 2 kW. Resistance to maneuver induced rollovers of vehicle with DBC system can be significantly increased if this design constraint is relaxed, e.g. by increasing the size of the pump or by including an accumulator or another energy storage device.

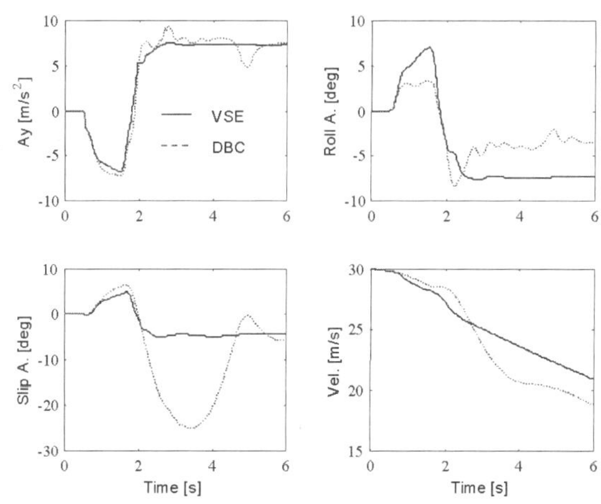

Figure 9. Vehicle Response in a Fishhook Maneuver at the Steering Amplitude of 55 Degrees with VSE and DBC Systems

In the case of fishhook maneuver, further simulations were conducted, in which the height of the center of gravity of the controlled vehicle was progressively raised, until the vehicle started to roll over at the same steering angle amplitude as the passive vehicle. It was found that presence of active brake control brings about improvement in vehicle resistance to maneuver-induced rollovers that is equivalent to increase in the static stability factor by 12%.

CONCLUSION

In this paper effectiveness of active chassis systems in preventing maneuver induced rollovers on smooth roads was evaluated through numerical simulations. The systems considered were brake based VSE system ARS system and DBC system. All three systems improved vehicle resistance to rollovers, but DBC system was the least effective primarily due to insufficient speed of response limited by the power of hydraulic pump. In all cases the steering angle necessary to roll over the vehicle equipped with one or two of the active chassis systems had to be significantly increased as compared to the passive vehicle, or the rollover could be avoided altogether regardless of the steering angle. The VSE system improved vehicle stability more than the ARS system did, and the vehicle with both of these control systems was the most stable of all, showing no tendency to rollover and very small side slip angles. It was found that presence of active brake control brings about improvement in vehicle resistance to maneuver-induced rollovers that is equivalent to increase in static stability factor by about 12%.

REFERENCES

1. Ackerman, J., Bunte, T., and Odenthal, D., 1999, "Advantages of Active Steering for Vehicle Dynamics Control", Proceedings of 32nd ISATA, Automotive Mechatronics Design and Engineering, Vienna, pp. 263-270.
2. Bernard, J. L. , Segel, L., and Wild, R. E., 1977, "The Shear Force Generation During Combined Steering and Braking Maneuvers", SAE paper No. 770852.
3. Eisele, D. D. and Peng, H., 2000, "Vehicle Dynamics Control with Rollover Prevention for Articulated Heavy Tracks", Proceedings of AVEC, Ann Arbor, MI, 2000.
4. Everett, N. R., Brown, M. D., Crolla D. A., 2000, "Investigation of a Roll Control System for an Off-road Vehicle", SAE paper No. 2000-01-1646.
5. Fujita, K., Ohashi, K., Fukatani, Kamai, S., Kagawa, Y., and Mori, H., 1998, "Devolepment of Active Rear Steer System Applying H$_\infty$ - μ Synthesis", SAE paper No. 981115.
6. Furukava, Y., Yuhara, N., Sano, S., Takeda, H., Matsushita, Y., 1989, "A Review of Four Wheel Steering from Viewpoint of Vehicle Dynamics Control", Vehicle System Dynamics, Vol. 18, pp. 151-186.
7. Garrott, W. R., Howe, J. G. and Forkenbrock, G., 1999, "An Experimental Examination of Selected Maneuvers that May Induce On-Road Untripped, Light Vehicle Rollover – Phase II of NHTSA's 1997-1998 Vehicle Rollover Research Program"
8. Hac, A., "Evaluation of Two Concepts in Vehicle Stability Enhancement Systems", 1998, Proceedings of 31st ISATA, Automotive Mechatronics Design and Engineering, Vienna, pp. 205-212.
9. Marine, M. C., Wirth, J. L. and Thomas, T. M., 1999, "Characteristics of On-Road Rollovers", SAE paper No. 1999-01-0122.
10. Nishio, A., Tozu, K., Yamaguchi, H., Asano, K., Amano, Y., 2001, "Development of Vehicle Stability Control System Based on Vehicle Sideslip Angle Estimation", SAE paper No. 2001-01-0137.
11. Palkovics, L., Semsey, A., Gerum, E., 1998, "Roll-Over Prevention System for Commercial Vehicles – Additional Sensorless Function of the Electronic Brake System", Proceedings of AVEC, 1998.
12. Ungoren, A. Y., Peng, H. and Milot, D. R., 2001, "Rollover Propensity Evaluation of an SUV Equipped with a TRW VSC System", SAE paper No. 2001-01-0128.
13. Van Zanten, A. T., Erhardt, R., and Pfaff, G., 1995, "VDC, The Vehicle Dynamics Control System by Bosch", SAE paper 950759.
14. Wielenga, T. J., 1999, "A Method for Reducing On-Road Rollovers – Anti-Rollover Braking", SAE paper No. 1999-01-0123.
15. Wielenga, T. J. and Chase, M. A., 2000, "A Study in Rollover Prevention Using Anti-Rollover Braking", SAE paper No. 2000-01-1642.
16. Wong, J., Y., 1993, "Theory of Ground Vehicles", John Wiley Inc., New York.

A New Control Strategy for Vehicle Active Suspension System Using PID and Fuzzy Logic Control

Yu Fan, Li Jun, Feng Jinzhi and Zhang Jianwu
Institute of Automotive Engineering, Shanghai Jiao Tong Univ

ABSTRACT

Since the nonlinearity which inherently exists in vehicle system need to be considered in active suspension control law design, a new control strategy is proposed for active vehicle suspension systems by using a combined control scheme, i.e., respectively using a PID controller and a fuzzy logic controller in two loops. In this paper, the investigation is mainly focused on vehicle ride comfort performance and simulations in straight running operating condition are presented. The control goal is to minimize vehicle body vertical and pitch accelerations for passenger comfort. The control system consists of two parallel control loops. One loop, using PID control, is to minimize vehicle body vertical acceleration; and the fuzzy logic controller is to minimize pitch acceleration and meanwhile to attenuate vehicle body vertical acceleration further by tuning weighting factors. Based on a four degree-of-freedom nonlinear vehicle model, the algorithm is implemented and simulations are carried out in different road disturbance input conditions. Simulation results show that the control strategy is very effective in reducing peak values of vehicle body accelerations, especially within the most sensitive frequency range of human response and also with good stability even if the system is subject to a discrete event input, i.e., a sudden change of road conditions, such as a pothole, an obstacle or a step input. Compared with conventional passive suspensions and an active vehicle suspension by using a linear and fuzzy logic controls, the new designed control system can improve vehicle ride comfort performance significantly and offer better system robustness.

INTRODUCTION

Active suspension is an advanced vehicle control system developed in recent years. Compared with conventional passive suspensions, it is very effective in improving vehicle ride comfort and handling stability. A key task for active suspension design is to determine a control law, which is capable of giving good system performance, and various approaches to derive the control scheme have been proposed by many researchers [1~10].

The optimal control, i.e., LQG control, and robust control have been introduced for active suspension application and some good performances been achieved on the assumption of a linear vehicle model [1~3]. In fact, vehicle is a complicated, nonlinear system with uncertainties of itself. And also operating conditions are changeable, e.g., the changes of road irregular excitation input with the variation of road surface roughness and of vehicle speed. So the control approaches for active suspensions based on the linear assumption of vehicle model have difficulties in practical application for good performance and robustness. However, for some practical systems including nonlinear elements, which cannot be expressed accurately in mathematics, the fuzzy logic control, being recently developed and needless of mathematical model and also with ability of emulating human logic, has been proved to be a good choice. Some researchers have successfully applied fuzzy logic control in active suspension control law design and satisfactory performances have been achieved [4~8]. Yoshimura used a combined controller, i.e., a linear and fuzzy logic controller in active suspension system and good performances have been obtained [5]. But in his controller, the linear feedback, in essence, is a single proportional control and the overshoot of response might be resulted by system inputs in some severe disturbance cases. Since the integral and differential of the errors between system actual output and reference output can be useful in reducing the steady-state error and overshoot of system, these information should be utilized in order to ensure the control precision of system.

Therefore, in the present paper, a new control scheme for vehicle active suspension system by using PID and fuzzy logic control is proposed. The goal is to minimize vertical and pitch accelerations of vehicle body for passenger comfort. The designed controller consists of two parallel control loops. One is PID control which takes the difference between calculated vehicle body vertical acceleration and reference acceleration value as input, the other loop is fuzzy logic control, whose main function is to reduce the pitch acceleration and also to attenuate vertical acceleration further by tuning weighting factors. Based on a four degree-of-freedom nonlinear vehicle model, the algorithm is implemented. Simulations are carried out in many different road excitation input conditions and the comparing tests are also made. Results in both time and frequency domains show that present control strategy is more effective in reducing peak values of vehicle body vibration caused by irregular road excitation, especially within the most sensitive frequency range of human response.

Furthermore, simulations are also carried out in some discrete event input conditions in order to examine system robustness. Compared with the system by using a linear and fuzzy logic controller, i.e., the scheme proposed by Yoshimura, the new designed active suspension can offer better system robustness.

1 VEHICLE MODEL

A half vehicle model is shown in Fig.1, in which the nonlinearity of spring and dumper are considered based on experimental data for a typical, practical car suspension. The motion equations for vehicle body, front and rear wheels are respectively given by,

$$m_c \ddot{x}_c + c_{2f}(\dot{x}_{2f} - \dot{w}_f) + k_{2f}(x_{2f} - w_f) + c_{1r}(\dot{x}_{1r} - \dot{x}_{2r}) +$$
$$k_{1r}(x_{1r} - x_{2r}) - f_f - f_r = 0 \qquad (1)$$

$$I_c \ddot{\theta} + c_{1r}(\dot{x}_{1r} - \dot{x}_{2r}) + k_{1r}(x_{1r} - x_{2r}) - c_{1f}(\dot{x}_{1f} - \dot{x}_{2f}) -$$
$$k_{1f}(x_{1f} - x_{2f}) - f_r + f_f = 0 \qquad (2)$$

$$m_f \ddot{x}_{2f} + c_{2f}(\dot{x}_{2f} - \dot{w}_f) + k_{2f}(x_{2f} - w_f) - c_{1f}(\dot{x}_{1f} - \dot{x}_{2f}) -$$
$$k_{1f}(x_{1f} - x_{2f}) + f_f = 0 \qquad (3)$$

$$m_r \ddot{x}_{2r} + c_{2r}(\dot{x}_{2r} - \dot{w}_r) + k_{2r}(x_{2r} - w_r) - c_{1r}(\dot{x}_{1r} - \dot{x}_{2r}) -$$
$$k_{1r}(x_{1r} - x_{2r}) + f_r = 0 \qquad (4)$$

in which,

$$x_c = \frac{x_{1f} b + x_{1r} a}{l}$$

$$\theta_c = \frac{x_{1r} - x_{1f}}{l}$$

$$l = a + b$$

where

m_c, m_f, m_r — masses respectively for vehicle body, front wheel and real wheel

x_c, x_{2f}, x_{2r} — vertical displacements respectively for vehicle body c.g., front wheel and rear wheel

x_{1f}, x_{1r} — vertical displacements of the front and the rear of vehicle body

a, b — distances respectively from front axle and rear axle to vehicle body c.g.

l — wheelbase

θ_c — vehicle body pitch angle

f_f, f_r — front and rear suspension control forces

w_f, w_r — road input displacement respectively at front and rear wheels

f, r — subscript, respectively represents front and rear

In simulations, the nonlinear damping coefficients

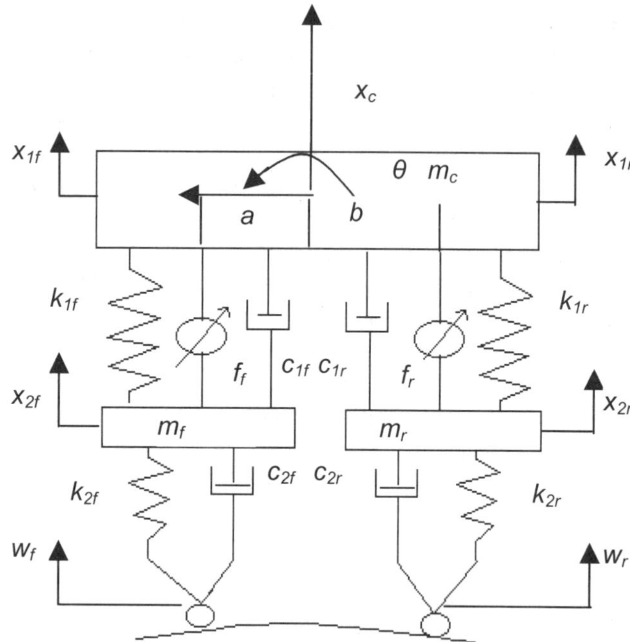

Fig.1 Half Vehicle model

c_{1f}, c_{1r} and nonlinear spring stiffness k_{1f}, k_{1r} are chosen based on practical experiment data. The relationships between damping forces F_{ci} and relative velocity between body and wheels, i.e., $\dot{x}_{1i} - \dot{x}_{2i}$ ($i = f, r$), are shown in Fig.2. And the relationships between suspension spring restoring forces F_{ki} and relative displacement between body and wheels, i.e., $x_{1i} - x_{2i}$ ($i = f, r$), are shown in Fig.3.

The road excitation input w_i, in form of a filtered white noise process, is used in the paper, with spectral density as,

$$S_w(\omega) = (\sigma^2 / \pi) \alpha v / (\omega^2 + \alpha^2 v^2) \qquad (5)$$

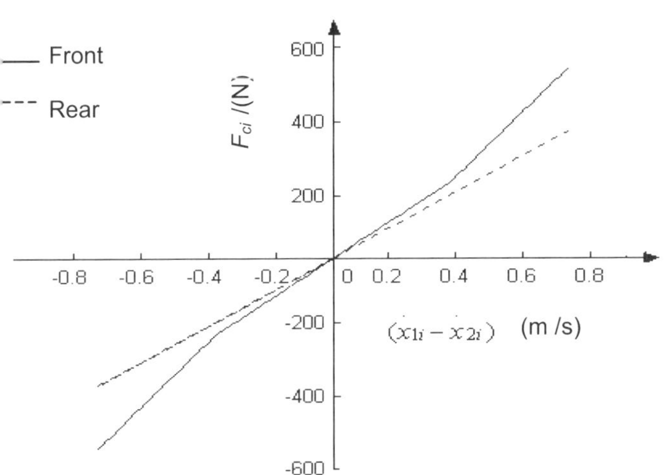

Fig.2 Damping forces of the front and rear suspension

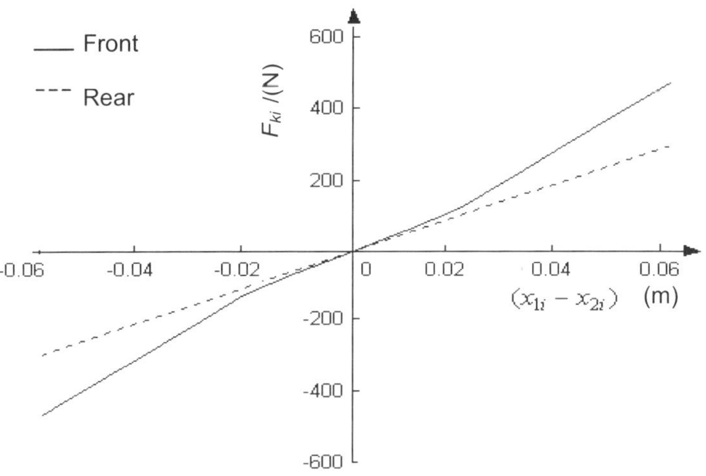

Fig.3 Restoring forces of the front and rear suspension

Where

S_w — power spectral density

ω — wave number

σ^2 — variance of the road input

v — vehicle speed

α — positive constant

Hence, the road input, w_i , can be given by,

$$\dot{w}_i + \alpha v w_i = \xi_i \quad (i = f, r) \tag{6}$$

in which ξ_i is a white noise process with zero mean value. The covariance function can be described by,

$$E[\xi_i(t)\xi_j(t-\tau)] = 2\pi A \delta(\tau) \quad \text{(when } i = j)$$
$$= 2\pi A \delta(\tau - t_l) \quad \text{(when } i \neq j) \tag{7}$$

in which τ is time delay of rear wheel road input, equal to wheelbase/vehicle speed, A denotes the intensity of road input and $\delta(\cdot)$ is the Dirac's delta function.

Combing the road input model, i.e., equation (6) and vehicle model, i.e., equation (1) ~ (4), the simulation model for active suspension system can be rewritten in a vector matrix form as below,

$$M\ddot{x} + C(t)\dot{x} + K(t)x = Df + Ew \tag{8}$$

where the state vector, active control and road excitation are respectively given by,

$$x = [x_{1f} \quad x_{2f} \quad x_{1r} \quad x_{2r}]^{\mathrm{T}}$$
$$f = [f_f \quad f_r]^{\mathrm{T}}$$
$$w = [w_f \quad \dot{w}_f \quad w_r \quad \dot{w}_r]^{\mathrm{T}}$$

and

$$M = \begin{bmatrix} m_c b/l & 0 & m_c a/l & 0 \\ I_c/l & 0 & -I_c/l & 0 \\ 0 & m_{2f} & 0 & 0 \\ 0 & 0 & 0 & m_{2r} \end{bmatrix}$$

$$C(t) = \begin{bmatrix} c_{1f} & -c_{1f} & c_{1r} & -c_{1r} \\ ac_{1f} & -ac_{1f} & bc_{1r} & -bc_{1r} \\ -c_{1f} & c_{1f}+c_{2f} & 0 & 0 \\ 0 & 0 & -c_{1r} & c_{1r}+c_{2r} \end{bmatrix}$$

$$K(t) = \begin{bmatrix} k_{1f} & -k_{1f} & k_{1r} & -k_{1r} \\ ak_{1f} & -ak_{1f} & bk_{1r} & -bk_{1r} \\ -k_{1f} & k_{1f}+k_{2f} & 0 & 0 \\ 0 & 0 & -k_{1r} & k_{1r}+k_{2r} \end{bmatrix}$$

$$D = \begin{bmatrix} 1 & 1 \\ a & b \\ -1 & 0 \\ 0 & -1 \end{bmatrix} . \quad E = \begin{bmatrix} 0 & 0 & 0 & 0 \\ 0 & 0 & 0 & 0 \\ k_{2f} & c_{2f} & 0 & 0 \\ 0 & 0 & k_{2r} & c_{2r} \end{bmatrix}$$

2 CONTROLLER DESIGN

In suspension design, body acceleration and body attitude are obviously important for ride comfort and stability performances. In present paper, assuming a straight-running condition, the control strategy is to minimize vertical acceleration and pitch movement of vehicle body. Because of the nonlinearity and uncertainty in vehicle system and also of the difficulty in modeling, the authors proposed a control scheme, expressed as a sum of PID controller and fuzzy controller as below,

$$f_i = u_{PIDi} + u_{Fi} \quad (i = f, \ r) \tag{9}$$

where u_{PIDi} is PID controller output variable for the reduction of vertical body acceleration and u_{Fi} is fuzzy controller output mainly accounting for minimizing pitch acceleration, but also along with the reduction of vertical body acceleration to some extend. The PID control is calculated as,

$$u_{PIDi} = K_{Pi}(\ddot{X}_c - \ddot{x}_c) + K_{Ii}\int(\ddot{X}_c - \ddot{x}_c)dt + K_{Di}\frac{d}{dt}(\ddot{X}_c - \ddot{x}_c)$$

$$(i = f, \ r) \tag{10}$$

in which

$\ddot{X}_c$ —the output value of reference acceleration, which is taken as zero

K_{Pi} , K_{Ii} and K_{Di} — respectively denote proportional, integral and differential coefficients

The main function of the first term in equation (10), i.e., the proportional part of the PID controller, is to speed up system response, implying to follow the reference body acceleration value rapidly. While the tasks of the second term, i.e., integral part, and third term, i.e., differential part, is to reduce system steady-state error and overshoot respectively. Fuzzy control u_{Fi} is mainly for the reduction of vehicle body pitch and it can be obtained by fuzzy control algorithm described bellow,

$$u_{Fi} = G_{Fi} \cdot \gamma_i \qquad (i = f, \ r) \quad (11)$$

in which

γ_i — fuzzy controller output variable of front or rear suspension

G_{Fi} — control gain

For the derivation of fuzzy logic control output γ_i, the input variables β_{1i} and β_{2i} are respectively assumed as,

$$\beta_{1i} = (x_{1i} + \kappa_{1i}\theta_{ci})/\eta_{1i}$$

$$\beta_{2i} = (\dot{x}_{1i} + \kappa_{2i}\dot{\theta}_{ci})/\eta_{2i} \quad (i = f, \ r)$$

in which

κ_{1i} , κ_{2i} — weighting factors

η_{1i} , η_{2i} — scaling factors

From the above equation, it can be seen that the fuzzy controller input variable also includes the terms of body displacement and body velocity. Correspondingly, by tuning the weighting factors, the vertical and pitch movements of vehicle body can be further controlled. By tuning the scaling factors, the input values of fuzzy controller are limited in the range of [-1, 1] in the implementation of the simulation algorithm.

Since the front and rear active suspension controllers are designed in same procedure, only one (front or rear suspension) fuzzy controller design is described. Therefore, the fuzzy controller input variables can be simply described as

$$\beta_1 = (x_1 + \kappa_1\theta_c)/\eta_1 , \ \beta_2 = (\dot{x}_1 + \kappa_2\dot{\theta}_c)/\eta_2 .$$

For a typical fuzzy controller with two inputs and one output, the control rules R^j can be expressed as,

R^j: if β_1 is A_j and β_2 is B_j, then γ is C_j

where β_1 and β_2 are input variables, γ is output variable, A_j, B_j, C_j are corresponding input and output fuzzy sets whose membership functions are assumed to be identical and shown in Fig.5.

The design procedure of the fuzzy logic controller include three main parts, i.e., fuzzification, fuzzy reasoning, defuzzification. Each part is respectively presented in detail as below.

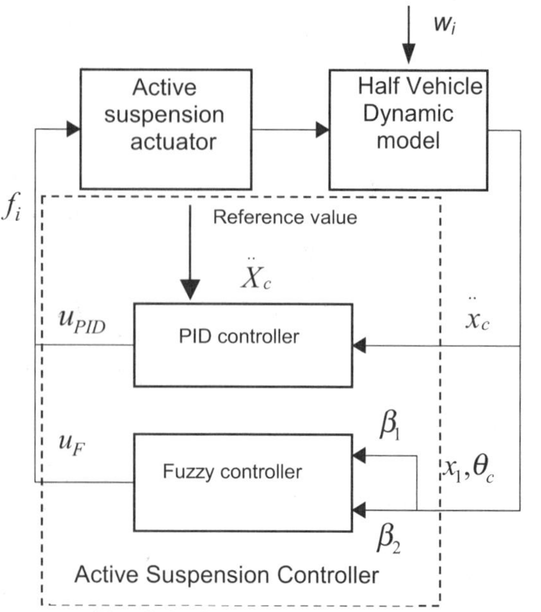

Fig.4 Active suspension control system

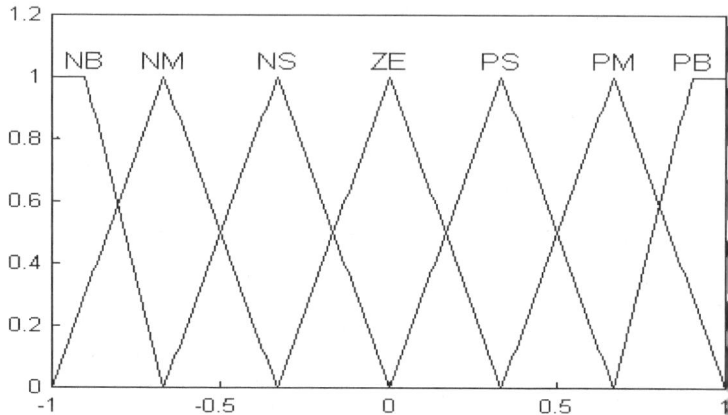

Fig.5 Membership functions of input &output fuzzy sets

(1) Fuzzification
The fuzzification process of input variable can be expressed by following equations,

$$\lambda_{1j} = \mu_{Aj}(\beta_1^0), \ \lambda_{2j} = \mu_{Bj}(\beta_2^0)$$

where

$\mu_{Aj}(\cdot)$, $\mu_{Bj}(\cdot)$ — respectively represent the membership functions of A_j and B_j

$\lambda_{1j}, \lambda_{2j}$ — degree of fitness of input value β_1^0, β_2^0, i.e., measurements of input variable β_1 and β_2

(2) Fuzzy reasoning
In present study, *Mandain* method is used in fuzzy reasoning process, given as,

$$\lambda_j = \lambda_{1j} \wedge \lambda_{2j}$$

where λ_j denotes the degree of fitness of fuzzy reasoning results, and symbol $\wedge$ conducts minimum operation. In each control rule R^j, the membership function for γ can be expressed as $\lambda_j \wedge \mu_{Cj}(\gamma)$, in which $\mu_{Cj}(\cdot)$ denotes the membership function of output C_j.

(3) Defuzzification

For the improvement of reasoning and computation speed, the product-sum gravity method [11] is used to obtain the defuzzified value γ. The product-sum-gravity method gives the defuzzified value γ^0 is,

$$\gamma^0 = \sum \gamma_j \lambda_j S_{Cj} / \sum \lambda_j S_{Cj}$$

where

S_{Cj} — area of $\mu_{Cj}(\gamma)$

γ_j — distance from center of gravity of S_{Cj} to point of zero

The empirical knowledge is used to construct the fuzzy control rules for suspension control system. And the fuzzy rules are described by language values as below,

(a) If β_1 and β_2 are both positive (negative) big, then γ negative (positive) big;

(b) If the absolute value of β_2 and β_1 is both small, then the absolute value of γ is relatively small;

(c) If β_1 is positive big (negative big) and β_2 is negative big (positive big), then the absolute value of γ is small;

(d) If β_1 is positive big (negative big), then γ is negative (positive) medium;

(e) If the absolute of value β_1 is small, then γ is small.

To summarize, the above fuzzy rules expressed by language values are shown in Table 1.

3 SIMULATION RESULTS AND ANALYSIS

Except for suspension sprung stiffness and damping coefficient, which have been given in Figs 2 and 3, the vehicle parameters chosen in the simulations are presented in Table.2 and the selected controller parameters are presented in Table.3,

In order to examine the effectiveness of the proposed control scheme, simulations are carried out in many different cases, e.g. in different road input conditions and with different control schemes. In the present paper, the results of the two control schemes, i.e., the linear combined with fuzzy control scheme and PID combined with fuzzy control scheme, are compared with the results of a conventional passive suspension system. The three different cases are respectively denoted as,

Case.A Active suspension by using the new proposed control scheme;

Case.B Active suspension using the linear and fuzzy logic control scheme proposed by Yoshimura;

Case.C Conventional passive suspension.

Simulation results in time domain, including vehicle body vertical acceleration and pitch acceleration, are compared for the three cases, respectively presented in Fig 6 and Fig 7. Compared with the results of the passive suspension system, significant improvements can be obtained by active suspension systems. The frequency domain results in Figs 8 and 9 show the improvements more clearly. From the figures, it can be found that the new proposed scheme offer the best performance. In the most sensitive frequency range of human response, the power spectral densities both for vertical and pitch accelerations are reduced compared with results of Case B. Particularly, in low frequency range, the new proposed scheme shows very significant improvement.

Not only for ride comfort, the vehicle overall performance for suspension design are also compared in Table.4, including the root mean square values of vertical body c.g. acceleration ($\ddot{x}_c$), pitch acceleration ($\ddot{\theta}_c$), front and rare tire deflections ($x_{2i} - w_i$), front and rear suspension working space ($x_{1i} - x_{2i}$). From the Table.4, it can be seen that, compared with the results of passive one, the vertical body accelerations are reduced to 49.04% and 63.34% and pitch accretions are reduced up to 20.51% and 30.78% respectively by using Case A and Case B. Simulations show the very similar results for r.m.s. values of tire deflection in the three cases. Although the r.m.s. values of suspension working space in Case A are relatively large, but no larger than 50% compared with the results in Cases B and C. This result is consistent with the previous study conclusion, i.e., the benefits of active suspension will increase with the increase of available suspension working space. Compared with the system by using a linear and fuzzy logic controller, i.e., the scheme proposed by Yoshimura, the new designed active suspension could offer slightly better performances.

Furthermore, simulations are also carried out in some discrete event input conditions in order to examine system robustness. Compared with passive suspension system, much better performance can be found even if the system is subject to a sudden change of road conditions, such as a pothole, an obstacle or a step input. Fig.11 and Fig.12 show the system response by using the proposed control scheme for active suspension and a passive suspension while the vehicle running at the speed of 20 k/m on the road with a discrete obstacle. The road input is illustrated as Fig.10 with a bump with amplitude of 0.085 meters. Because of the sudden bump disturbance, the amplitudes of acceleration, $\ddot{x}_c$ and $\ddot{\theta}_c$, change very sharply. Compared with the results of passive suspension, the proposed active suspension can reduce the peak values of accelerations considerably with the reduction of 39% for vertical acceleration and 40% for pitch acceleration. Even though the results of Case B are not presented in the figures, the simulations still show the slight better performances of Case A than those of Case B.

4 CONCLUSIONS

A new control strategy is proposed for active vehicle suspension systems by using a combined control scheme, i.e., respectively using a PID controller and a fuzzy logic controller in two loops. Based on a four degree-of-freedom nonlinear vehicle model, the algorithm is implemented and simulations are carried out in different road disturbance input conditions. Simulation results in both time and frequency domains show that present control strategy offer the best performance. Compared with the linear and fuzzy controller scheme, the proposed scheme is more effective in reducing peak values of vehicle body vibration caused by irregular road excitation, especially within low frequency range, i.e., the most sensitive frequency range of human response. The system robustness is also examined in some severe road input conditions, such as, a sudden change of road condition. Simulation in a road with bump input show that the peak values of vertical acceleration and pitch acceleration can be respectively reduced to 39% and 40% compared with the results of a passive system.

Hence, the new proposed control strategy is proved to be more effective in improving ride comfort and more robust even when the system is subjected to some severe input conditions.

REFERENCES

1   Ray, L. R. Robust linear optimal control laws for active suspension systems. ASME J. Dyn. Systems, Measurement Control. 1992, 114(6): 592-598

2   Hrovat, D. Optimal active suspension structures for quarter car vehicle models. Automatica.1990, 26(5): 845-860

3   Mohamed M.Elmadany and Zuhair S.Abduljabbar. Linear Quadratic Gaussian control of a quarter-car suspension. Vehicle System Dynamic, 1999, 32: 479-497

4   Li, Q., Yoshimura, T. and Hino, J. Active suspension with preview of large-sized buses using fuzzy reasoning. Int. J. of vehicle Design, 1998, 19(2): 187-198

5   Yoshimura, T. Nakaminami, K. *et al.* Active suspension of passenger cars using linear and fuzzy logic controls. Control Engineering Practice, 1999, 7: 41-47

6   Chou, J. H., Chen, S. H. and Lee, F.Z. Grey-fuzzy control for active suspension design. Int. J. of vehicle Design, 1998, 19(1): 65-77

7   Ghazi Zadeh, A., Fahim, A. and E1-Gindy, M. Neural network and fuzzy logic applications to vehicle systems: Literature survey. Int. J. Vehicle Design, 1997, 18(2): 132-193

8   Yeh, E. C and Tsao, Y. J. A fuzzy preview control scheme of active suspension for rough road. Int. J. of Vehicle design, 1994, 15(1/2): 166-180

9   Yu Fan, Crolla, D. A. Wheelbase preview optimal control for active vehicle suspensions. Chinese Journal of Mechanical Engineering, 11(2): 122-129

10 Li Jun, Yu Fan and Zhang Jianwu. Fuzzy Neural Networks Control of a Semi-active Suspension System with Dynamic Absorber. SAE Technical Paper Series，SP-1558，Paper No.2000-01-3077.

11 Kandel, A. and Langholz, G. (Eds). Fuzzy control systems. 1993, London: CRC Press.

CONTACT

YU FAN

Inst. of Automotive Engineering
School of Mechanical Engineering
Shanghai Jiao Tong University
1954 Huashan Rd, Shanghai 200030,
P.R.China

LI JUN

Inst. of Automotive Engineering
School of Mechanical Engineering
Shanghai Jiao Tong University
1954 Huashan Rd, Shanghai 200030,
P.R.China
Email: jli927@mail1.sjtu.edu.cn

APPENDIX:

Table.1 Fuzzy rules

γ		β_1						
		NB	NM	NS	ZE	PS	PM	PB
	NB	PB	PB	PM	PM	PS	ZE	ZE
	NM	PB	PM	PM	PS	ZE	ZE	NS
	NS	PM	PM	PS	PS	ZE	NS	NS
β_2	ZE	PM	PS	ZE	ZE	NS	NS	NM
	PS	PS	ZE	ZE	NS	NS	NM	NM
	PM	PS	ZE	NS	NS	NM	NM	NB
	PB	ZE	NS	NS	NM	NM	NB	NM

Table.2 Vehicle parameters for simulation

Notation	Value	Unit
m_c	480	Kg
I_c	620	kgm^2
K_{2f}	160	kN/m
K_{2r}	160	kN/m
c_{2f}	2	kN/(m/s)
c_{2r}	2	kN/(m/s)
m_{2f}	25	Kg
m_{2r}	25	Kg
a	0.871	M
b	1.469	M
v	25	m/s
α	0.04	1/m
A	10^{-6}	M

Table.3 Controller parameters for simulation

Notation	Value	Notation	Value
K_{Pf}	210	η_{2f}	0.03
K_{Pr}	160	η_{1r}	0.02
K_{If}	10	η_{2r}	0.03
K_{Ir}	8	κ_{1f}	24.2
K_{Df}	10	κ_{2f}	20.6
K_{Dr}	10	κ_{1r}	-26.2
G_{Ff}	195	κ_{2r}	-19.6
G_{Fr}	155	η_{1f}	0.02

Table.4 rms value of simulation results

	Case A	Case B	Case C	Units
$\ddot{x}_c$	0.1181	0.1524	0.2406	m/s^2
$\ddot{\theta}_c$	0.0032	0.0058	0.0156	rad/s^2
$X_{2f}\text{-}w_f$	0.0013	0.0013	0.0012	m
$X_{2r}\text{-}w_r$	0.0013	0.0012	0.0011	m
$X_{1f}\text{-}X_{2f}$	0.0054	0.0051	0.0037	m
$X_{1r}\text{-}X_{2r}$	0.0053	0.0051	0.0035	m

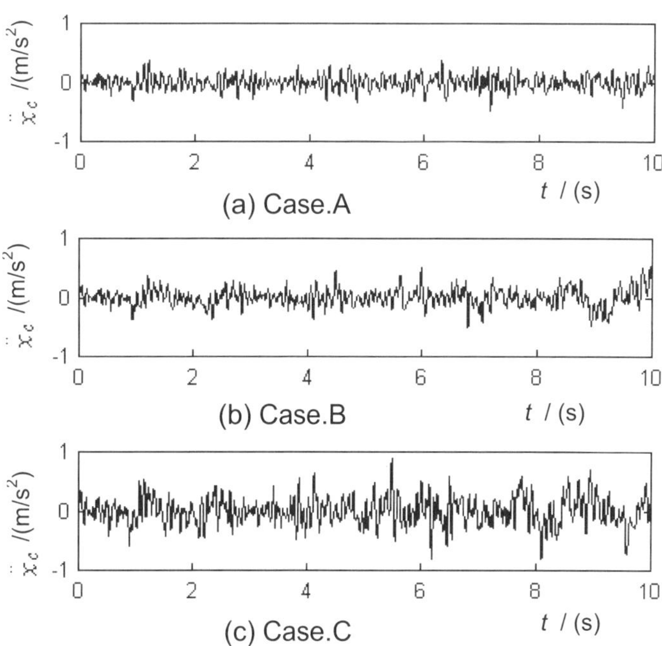

(a) Case.A

(b) Case.B

(c) Case.C

Fig.6 Vertical acceleration of vehicle body

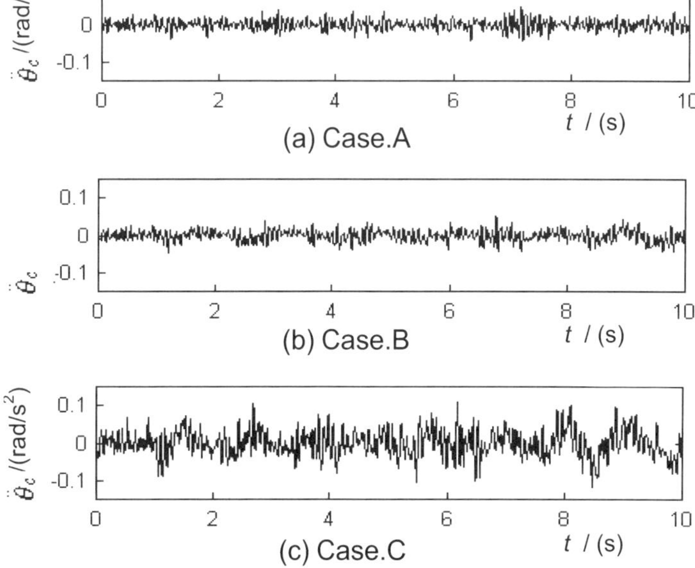

(a) Case.A

(b) Case.B

(c) Case.C

Fig.7 Pitch acceleration of vehicle body

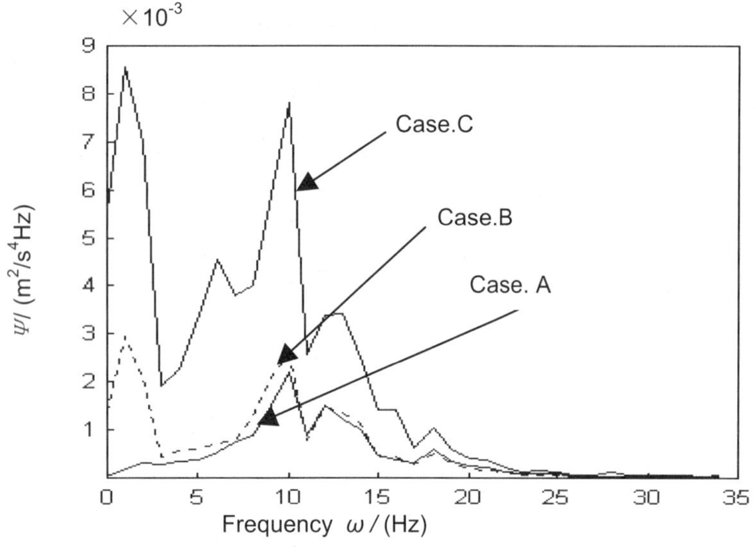

Fig.8 Spectral density of vertical acceleration of vehicle body

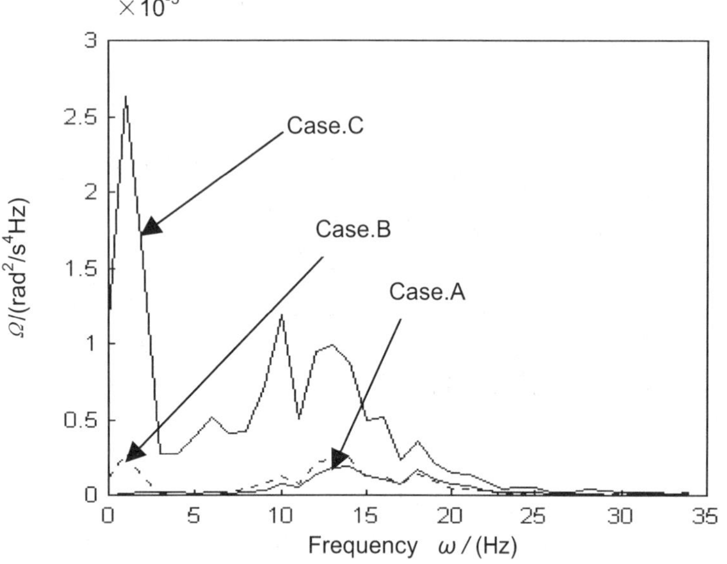

Fig.9 Spectral density of pitch acceleration of vehicle body

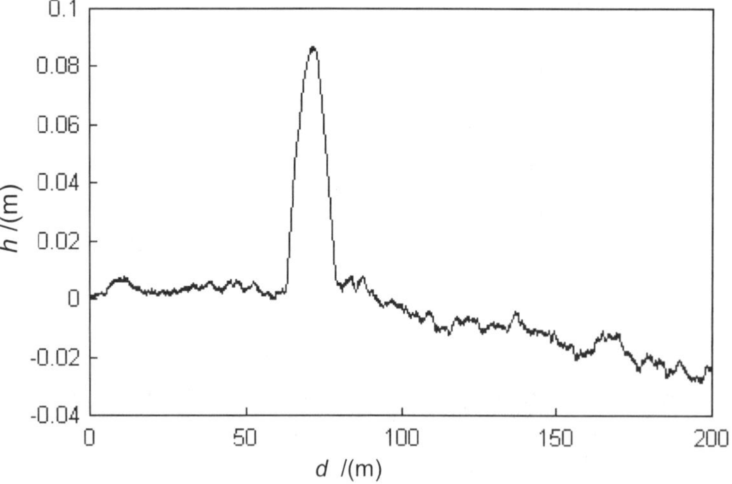

Fig.10 Typical road disturbance with a bump

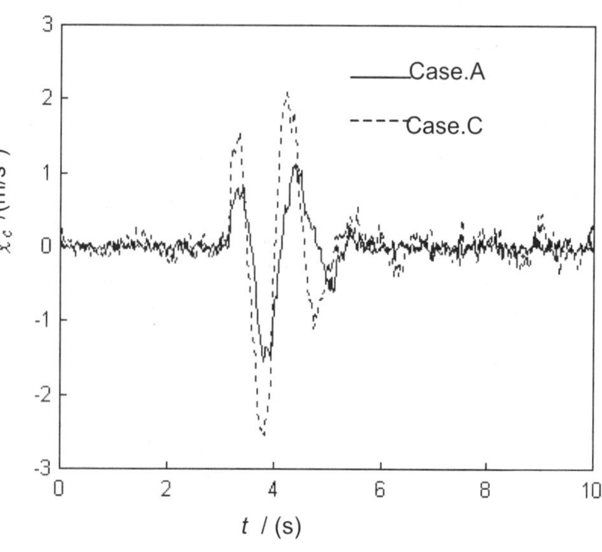

Fig.11 Vertical acceleration of vehicle body

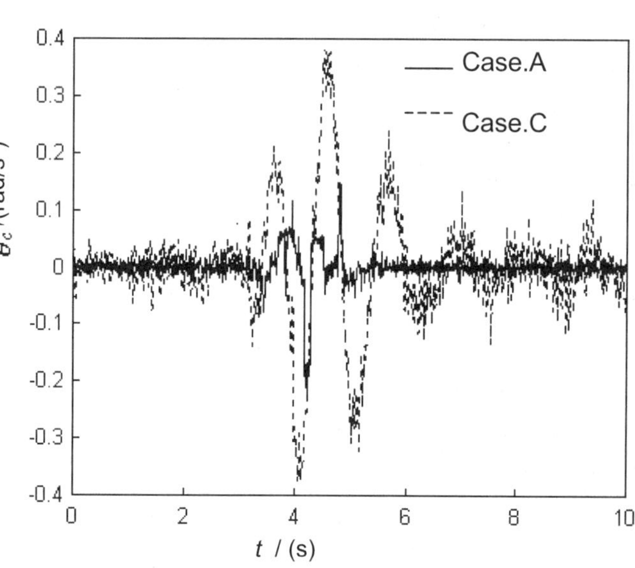

Fig.12 Pitch acceleration of vehicle body

Research of Driver Assistance System for Recovering Vehicle Stability from Unstable States

Xinpeng Tang
Faculty of Traffic Science and Engineering
Huazhong Univ. of Science and Technology

Kenichi Yoshimoto
Department of Mechano-Infomatics, Univ. of Tokyo

ABSTRACT

Recently, the direct yaw-moment control system, in which braking or driving torque is distributed appropriately to the tires on either side, has been put to practical use. Such systems are called VDC or VSC, and are available on the market. These systems aim to prevent the vehicle from falling into an unstable state, but cannot bring a destabilized vehicle back to a stable state. On the other hand, it is a well-known fact that highly skilled drivers are capable of steering a destabilized vehicle back into control. In this paper, we propose a driver-assistance system that automatically steers the front wheels in order to recover a vehicle that is spinning out. When a vehicle enters a tailspin, the system takes over the steering. Once the vehicle recovers, control is given back to the driver. The issue of how to return control from automatic to manual is of particular interest.

1. INTRODUCTION

It is a well-known fact that highly skilled drivers are capable of steering a destabilized vehicle back into control. In this report, we propose a driver-assistance system that automatically steers the front wheels in order to recover a vehicle that is spinning out. When vehicle enter a tailspin, the system takes over the steering. Once the vehicle recovers, control is given back to the driver. An important issue is the timing at which control is given back to the driver. In order to study this, the proposed assistance system is implemented on a driving simulator. Subjects were asked to drive in a number of different scenarios while we collected data. The subjects answered a questionnaire after the trails.

2. VEHICLE MODEL

The two-wheel model of vehicle dynamics is used. The dynamic equation is given by:

$$mu_a(\dot{\beta}+\omega) = F_{y1} + F_{y2} \qquad (1)$$

$$I\dot{\omega} = l_1 F_{y1} - l_2 F_{y2} \qquad (2)$$

For the front and rear side slip angles, we use the following approximations:

$$\beta_1 = \delta - (\beta + \frac{l_1\omega}{u_a}) \qquad (3)$$

$$\beta_2 = \frac{l_2\omega}{u_a} - \beta \qquad (4)$$

Here,
ω Yaw rate
u_a Vehicle speed
β Side slip angle of the center of mass
β_1 Side slip angle of the front wheels
β_2 Side slip angle of the rear wheels
δ Front steering angle
F_{y1} Lateral force acting on front wheels
F_{y2} Lateral force acting on rear wheels
m Vehicle mass
I Vehicle moment of inertia around yaw axis
l_1 Distance between center of mass and front axis
l_2 Distance between center of mass and rear axis

When the steering angle and the front and rear side slip angles are small, we can assume that the lateral force is equal to the cornering force, and that the cornering force is proportional to the side slip angle. The dynamic equation becomes[2].

$$mu_a(\dot{\beta}+\omega)=2(K_1\beta_1+K_2\beta_2) \qquad (5)$$

$$I\dot{\omega}=2(l_1K_1\beta_1-l_2K_2\beta_2) \qquad (6)$$

Since this paper deals with extraordinary conditions, we must model non-linear effects. For this reason, we obtained the lateral force on each tire using Fiala's formula, taking into consideration the changes in each tire's vertical load and its friction circle. The lateral forces obtained in this manner were used in (I) and (2) to obtain the vehicle's motion.

We also modeled the steering actuator using the following first order system:

$$\frac{d\delta_1}{dt}=-\frac{1}{T_1}\delta_1+\frac{1}{T_1}\delta_c \qquad (7)$$

Here δ_1 is the front steering angle and δ_c is steering angle command given to the actuator. T_1 is a time Constance.

3. SYSTEM DESIGN

A linear model is used to design the assistance system. The state equation for (5), (6), and (7) is,

$$\dot{x}=Ax+Bu \qquad (8)$$

here, A, B, x and u are given by,

$$A=\begin{bmatrix} a_{11} & a_{12} & a_{13} \\ a_{21} & a_{22} & a_{23} \\ 0 & 0 & -\dfrac{1}{T_1} \end{bmatrix}, B=\begin{Bmatrix} 0 \\ 0 \\ \dfrac{1}{T_1} \end{Bmatrix} \qquad (9)$$

$$x=\begin{Bmatrix} \beta \\ \omega \\ \delta_1 \end{Bmatrix}, u=\delta_c \qquad (10)$$

Where,

$$a_{11}=-\frac{2(K_1+K_2)}{mu_a}$$

$$a_{12}=-\frac{2(K_1l_1-K_2l_2)}{mu_a^2}-1$$

$$a_{13}=\frac{2K_1}{mu_a}$$

$$a_{21}=-\frac{2(K_1l_1-K_2l_2)}{I}$$

$$a_{22}=-\frac{2(K_1l_1^2+K_2l_2^2)}{Iu_a}$$

$$a_{23}=\frac{2K_1l_1}{I}$$

Here K_1 is side power of front wheel and K_2 is side power of rear wheel.

If we use a weighted quadratic norm for the performance index as given below, an optimal regulator can be designed for the linear system.

$$J=\int_0^\infty (x^TPx+u^TQu)dt \qquad (11)$$

$$P=\begin{bmatrix} \rho_1 & 0 & 0 \\ 0 & \rho_2 & 0 \\ 0 & 0 & \rho_3 \end{bmatrix}, Q=\rho_4 \qquad (12)$$

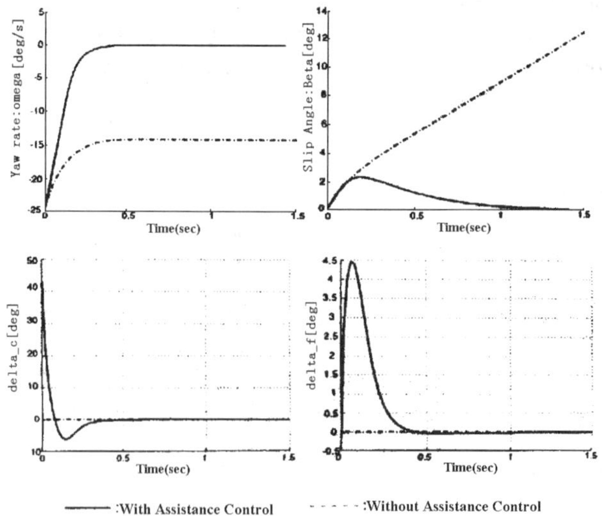

Fig.1 Result of Computer Simulation-Response to Disturbance

We selected a stabilizing feedback controller by performing computer simulations of the non-linear vehicle model under control by optimal regulators. The controller identified through these simulations was used in the assistance system, Fig.1. Shows the result of computer simulations using the selected controller. In the simulations, we assume that the Coulomb friction coefficient between the road and tire is 0.3. The vehicle is travelling along a straight line at a constant speed of 25[m/s] and that a yaw rate of -25[deg/s] is given instantaneously to the vehicle. The dash-dotted line represents the case with no control. And the solid line shows the motion of the vehicle under control of the assistance system. This assistance system was implemented on a driving simulator as show in Fig.2. The steering reaction torque is calculated and applied to the steering wheel at all times regardless of whether the assistance system is on or off. When the assistance system is on, the driver's steering actions are ignored; i.e., even if the driver turns the steering wheel, the tires do not respond to that command. Also, when the assistance system switches off, the driver's steering input is immediately re-engaged.

Fig.2 Assistance System

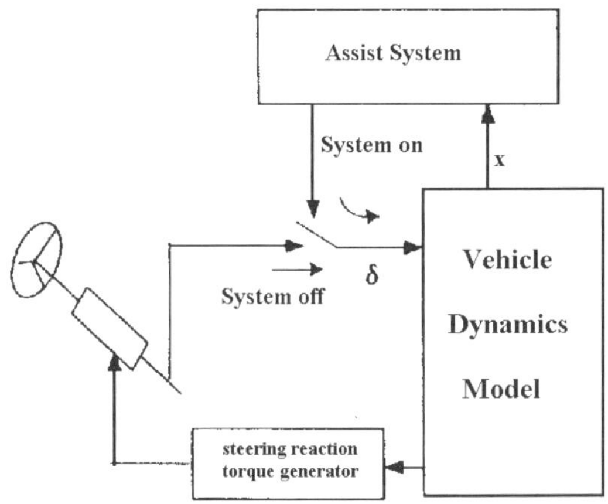

4. EXPERIMENTATION USING A DRIVING SIMULATOR

4.1. OBJECTIVE OF THE EXPERIMENTS

We have designed a driver assistance system using an optimal regulator, and have verified its effectiveness through computer simulation using a non-linear dynamic model, Nevertheless, automobiles are driven by human drivers, and, if we were to use the assistance system at all times, it would interfere with the driver and become a hindrance to smooth driving. For this reason, we perform experiments in this section to determine the entry condition for the assistance system to take over the system from the driver, and the exit condition for the system to relinquish control. The goal in this section is to find out through experimentation what the entry and exit conditions should be in order for the vehicle to smoothly recover from an unstable state without causing interference between the driver and the system. The experiments were done using a driving simulator developed in-house.

4.2. INVESTIGATION ON STRAIGHT COURSE

4.2.1. Method of Experiment

We selected a realistic scenario in which a vehicle can easily go out of control. A linear horizontal course 5 [m] wide was selected. In order to simulate a snow-covered road. The friction coefficient of 0.3 was selected. While the subject is driving along the road, a yaw rate of -25[deg/s] was applied instantaneously. The aim here is to simulate a scenario in which a tailspin occurs as a result of some sudden disturbance such as abrupt steerage. First, a preliminary experiment was done to determine the basic entry and exit condition. The basic entry condition was determined to be $\omega \geq 28$[deg/s] or $\beta \geq 5$[deg]. The basic exit condition was determined to be $\omega \leq 1$[deg/s] and $\dot{\omega} \leq 1$[deg/s^2]. During the preliminary experiments, we found that interference between the driver and the system hardly occurred when the system engaged, but was prominent when the system disengaged and control was returned to the driver. So, We used these entry and exit conditions and varied the timing of the exit. Seven cases with different exit timings were tested, and the drivers were asked to evaluate the effectiveness of each variation. In order to eliminate the effect of the driver becoming familiar with the test scenario, we varied the timing of the disturbance for each run.

Case1: Exit immediately upon satisfying the condition.

Case2: Exit 0.2 second after satisfying the condition.

Case3: Exit 0.4 second after satisfying the condition.

Case4: Exit 0.5 second after satisfying the condition.

Case5: Exit 0.6 second after satisfying the condition.

Case6: Exit 0.8 second after satisfying the condition.

Case7: Exit 1.0 second after satisfying the condition.

At the beginning of the experiment, the subjects were given the test scenario with the assistance system disabled so that they can experience the tailspin. The subjects were 10 male drivers, all in their twenties, with 2 to 10 years of driving experience after obtaining their driver's licenses. The driving frequency of the subjects varied from those who hardly drove at all to those who drove approximately once a week. After experiencing all the cases, the subjects were asked in a questionnaire to select the one that they felt best enabled them to drive smoothly with a good feel.

4.2.2. Results of the Experiment

Fig.3 through Fig.6 shows the driving trajectory for cases 1,4, and 7 of subject A, who is deemed to exhibit behavior that is representative of all the subjects. The horizontal and vertical axes of the figures are in meters. The horizontal axis represents distance along the course, and the vertical axis represents the lateral position along the width of the road. The "X" symbol represents that point at which the disturbance was introduced. (The approximate locations of the disturbance points are 240[m], 400[m], 370[m], and 300[m], in the order of the figure numbers.) This subject reported Case 4 to be the best one. From the experimental data, we can calculate the following performance index that was used in the controller design:

$$ J = \frac{1}{N} \sum_{k=1}^{N} (x_k^T P x_k + u_k^T P u_k) \qquad (13) $$

Where N is the number of data points. Fig.7 shows the calculated performance index for all seven cases for subject A. This, too, shows that Case 4 is the best.

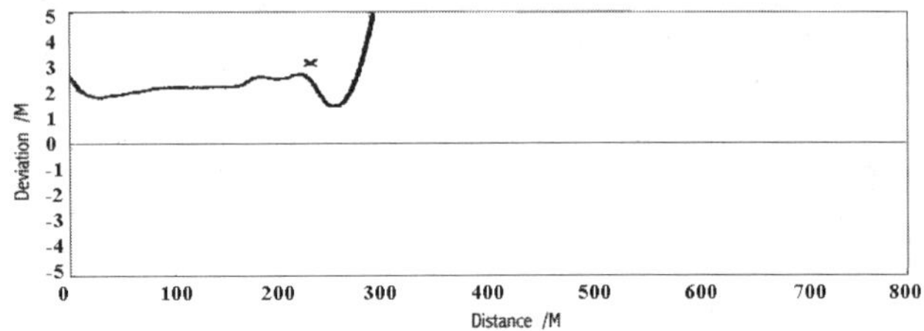

Fig.3 Vehicle Trajectory (Without Assistance System)

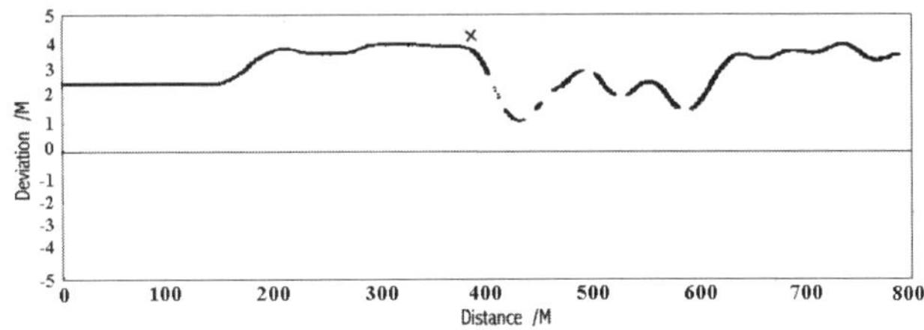

Fig.4 Vehicle Trajectory (Case 1)

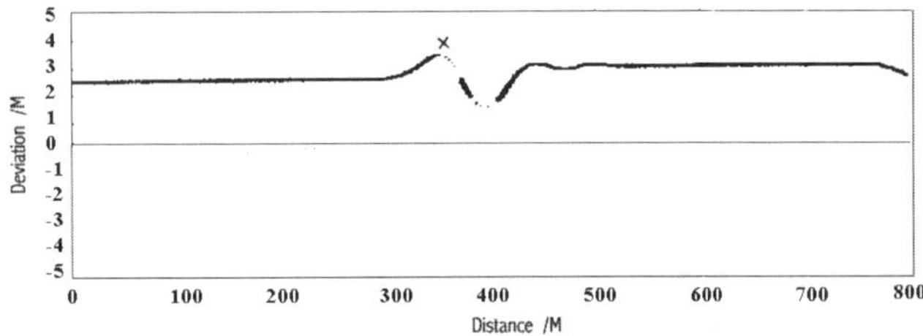

Fig.5 Vehicle Trajectory (Case 4)

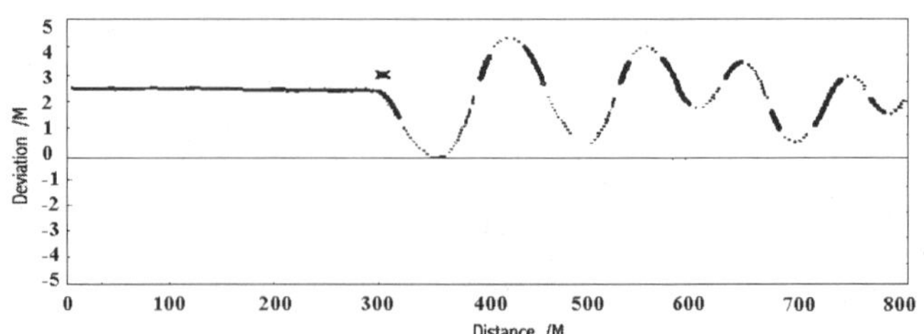

Fig.6 Vehicle Trajectory (Case 7)

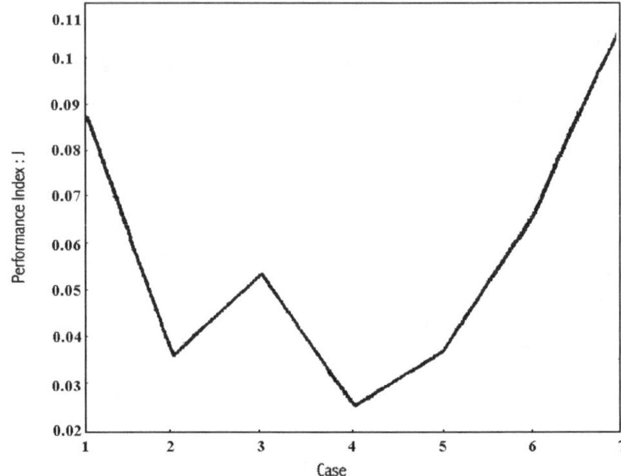

Fig.7 Performance Index Calculated from Test Data

4.2.3. Evaluation/Analysis of the Experimental Results

A tendency common to all the subjects was that cases 1,2,6 and 7 received poor ratings. This can be explained in the following manner. For cases 1 and 2, the system is able to stabilize the vehicle shortly after it enters the tailspin. The driver, not being as quick as the assistance system, will continue with a recovery maneuver even after the assistance system has restored the vehicle to a stable state. For this reason, the vehicle exhibits an oscillatory behavior, and dose not stabilize in a smooth manner. There is also an explanation for cases 6 and 7. The assistance system is not designed to steer the vehicle in a particular direction, so the user must control the direction of the vehicle. When the system's latency time is much longer than the driver's response time, the driver is not able to steer the vehicle to the desired direction during the latent period. For case 3,4, and 5, the evaluation differed from subject to subject. The fact that the infrequent driver chose 5 is indication that the latency time depends on the driver's skill level, and that skilled drivers tend to prefer a short latency time.

4.3. INVESTIGATION ON CURVED COURSE

In the previous section, we performed the preliminary experiment and determined the entry condition to be ω≥28[deg/s] or β≥5[deg]. In the case of a curved course, we subtract the yaw rate due to the curve in the entry condition.

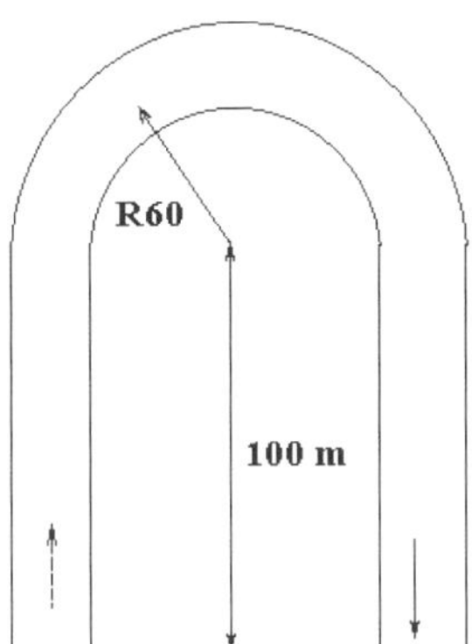

Fig. 8 Curved Test Course

4.3.1. Method of Experiment

The test course is horizontal and is 5 meters wide. The course layout is given in Fig.8. As a result of preliminary experiment, we determined the entry condition to be $\omega-\omega_r \geq 30$[deg/s] or $\beta \geq 5$[deg], and the exit condition to be $\omega \leq 1$[deg/s] and $\dot{\omega} \leq 1$[deg/s^2]. Here, ω_r is the steady-state yaw rate for the curve. We know from the previous section that a latency time is needed, but, in the case of a curved course, any latency will result in the vehicle deviating form the course very quickly. For this reason, we used automatic steering to change the vehicle's direction during the latent time after the assistance system has completed its task. In order to investigate how to smoothly recover the vehicle under such a scheme, we used three different cases for the automatic steering and two different latency times. We had drivers try all six combinations of steering types and latency times in order to try to find a combination that gives smooth transition to driver control with little interference. For automatic steering, we used the following three schemes after the exit condition was satisfied: (a) continue to apply the assistance system, (b) steer the vehicle with steering wheel angle δ_0. and, (c) steer the vehicle with steering wheel angle $1.5*\delta_0$, where δ_0 is the angle of the steering wheel immediately before engaging the assistance system. For each of these cases, latency times of 0.5 second and 1.0 second were tested. The steering angles δ_0 and $1.5*\delta_0$ respectively represent the cases when the vehicle is command to recover the deviation from the coursed by the assistance system. These test conditions are summarized in Table 1 and are illustrated in Fig.9. We used 10 subjects varying in driving frequency from almost none to once a week. The subjects were asked to try all of the test cases and to evaluate how smoothly control was given back to them from the assistance system. We used a five-point evaluation system with 5 being very smooth and 1 being very difficult to drive due to interference.

Table 1: Summary of Test Cases

Latency time	Continue to use assistance system	δ_0	$1.5*\delta_0$
0.5 [s]	Case 1	Case 3	Case 5
1.0 [s]	Case 2	Case 4	Case 6

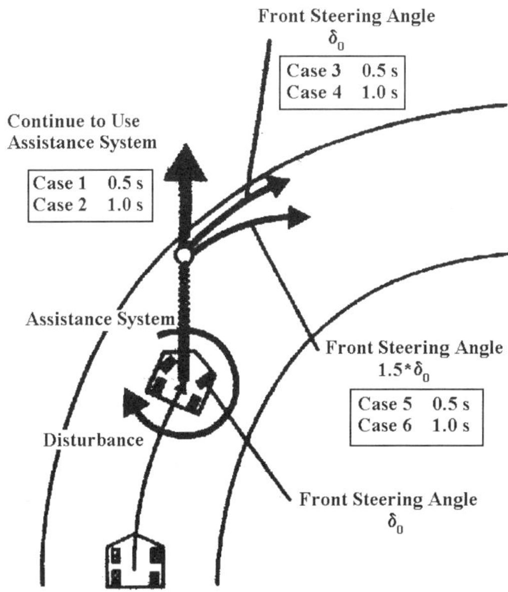

Fig.9 Illustration of Experiment 2

4.3.2. Result of Experiment

Table 2 shows the result of the questionnaire from the experiment. The point total over the 10 subjects is given for each system configuration. The maximum possible total is 50, and the minimum possible total is 10.

Table 2: Result of Questionnaire

Latency time	Case	Continue to use assistance system	δ_0	$1.5*\delta_0$
0.5[s]	Case1	30		
0.5[s]	Case3		48	
0.5[s]	Case5			36
1.0[s]	Case2	18		
1.0[s]	Case4		39	
1.0[s]	Case6			29

4.3.3. Evaluation/Analysis of the Experimental Results

We will first compare the results of the two different latency times. The results show that, for all steering angles, 0.5 second received a higher score than 1.0 second. This indicates that many drivers felt that 1.0 second was too long. Since we know that interference occurs if we use a latency time of 0, we expect that the optimal latency time to be between 0 second and 1 second. So, we can see that the latency time is necessary, but if it is too long, it hinders the ability to smoothly return control to the driver. Next we compare the steering angles. For both latency times, using δ_0 (i.e. case 3 and 4) showed the highest rating. The fact that δ_0 showed a higher rating than $1.5*\delta_0$ suggests that the driver does not feel comfortable with the assistance system automatically changing the direction of the vehicle. In fact, we can see from the test data for $1.5*\delta_0$ that many of the subjects applied corrective steerage in the opposite direction after the assistance system was disengaged. This can be explained as follows. The driver generally tries to apply corrective steerage to redirect the vehicle after the tailspin, however, in cases 5 and 6, the assistance system has already corrected the direction, so the vehicle's direction is overcompensated and heads towards the inside of the curve. As a result, the opposite corrective action becomes necessary. From this observation, it is clear that cases 5 and 6 result in interference and thus are not desirable. The cases where the assistance system was continued showed poor ratings. This is because the counter-steering of the assistance system is applied for a long time, resulting in increased deviation from the course, which results in the need for a longer time to return to the course. In particular, case 2 results in a very large course deviation, and some drivers had the assistance system engage multiple times before finally settling on the course. Seven out of ten subjects evaluated case 2 with the worst possible 1 point. From these analyses, we can conclude that case 3, which uses a latency time of 0.5 seconds and a steerage of δ_0, is the best system which offers offering the least interference between the driver and the assistance system. This result offers a guideline for when to return control from an automatic system to manual steering.

5. CONCLUSION

In this study, we designed a driver assistance system in which the front wheels of a vehicle are automatically steered. We verified through simulation that the system could be used to recover a vehicle from a tailspin.

We installed this driver assistance system on a driving simulator and performed experiments to study the timing to switch between automatic and manual steering. Analysis of the results showed that:

(a) A latency time is needed when the assistance system returns control to the driver.

(b) The latency time depends on the driver's skill.

(c) The latency time cannot be made too long.

REFERENCES

1. S.Doi, E. Ono, S. Hosoe, "Anti-spin Control by H_∞ Control-Part1: Application to Active Front Wheel Steering Control System" in proceedings of the Japanese Society of Automotive Engineers Conference, No.954 (1995) pp141-144.
2. M. Abe, "Dynamic and Control of Automobiles" Sankaido, 1992.

A Coordination Approach for DYC and Active Front Steering

M. Selby, **W. J. Manning, M. D. Brown and D. A. Crolla**
School of Mechanical Engineering, University of Leeds, United Kingdom

ABSTRACT

Integrating chassis control systems can lead to improvements in the safety, efficiency of action and overall production of a modern car. The sharing of information between chassis sub-systems allows the controller to take the optimum course of action since it has more than one option to affect the dynamics of a vehicle. This paper investigates the principle of coordination of chassis subsystems by selecting active steering and yaw stability control. A controller that coordinates the action of active front steering(AFS) and direct yaw moment control(DYC) is proposed.

Preliminary results for the coordinated controller using limit handling tests suggest that such an integrated approach can lead to overall improvements in vehicle dynamic response.

INTRODUCTION

In the past, controllers designed to affect vehicle handling have been given a specific area of the dynamics to control, e.g. ABS controls the longitudinal deceleration of the vehicle. Coordination of sub systems, intelligent data fusion, and integrated control of the vehicles motion can lead the way to improved control of the dynamics of the vehicle [1...4]. Such a controller, knowing the state of the vehicle, can select its action based on current operating conditions.

Significant research has been reported on the integration of two or more controllers [5...9], e.g. DYC and four wheel steer (4WS). These integrated schemes all aim to improve the handling dynamics of the vehicle.

A number of approaches are possible for the coordinated control of vehicle handling dynamics. One possibility is to look at the vehicle as a single dynamic system with a set of inputs and outputs and design a global controller for all aspects that single system. However, the mathematical complexity of a vehicle makes this a very difficult task. An alternative, more pragmatic, approach is to use and extend existing chassis control systems

VEHICLE MODELLING

A non-linear vehicle handling model was developed for this study. This consists of a 4 degree of freedom(d.o.f) lumped-parameter model (Figure 1) that includes longitudinal, lateral, yaw and roll motions with quasi-static longitudinal load transfer effects included. A roll bar was also included. Equations of motion for this model are presented with parameter values in the appendix.

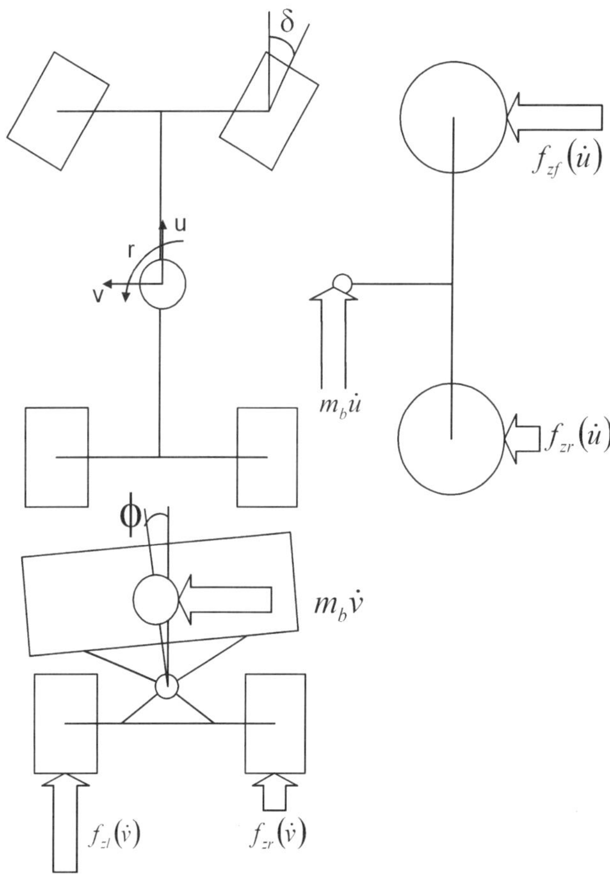

Figure 1 4-d.o.f model of a car showing tire forces and velocities

---

*menmase@leeds.ac.uk

TYRE MODEL

The tire forces are a property of the longitudinal and lateral wheel slip. Longitudinal slip ratio is the deviation from a state of pure rolling. Lateral slip angle is the angle between the plane of the tire and the tires direction of travel. If the wheel is steered, then lateral slip is a function of the wheels lateral and longitudinal wheel velocities and steer angle. The tires are the dominant force-generating component of a vehicle. As such, an accurate model of the non-linear behavior of the tires is essential.

The characteristic relating tire force, slip angle, slip rate and tire vertical load has been modeled using the "Magic Formula" [10].

DRIVER MODEL -

Whilst a linear driver model was used to assess the relative merits of AFS and active rear steering (ARS), one was not used in this work. Closed loop tests play an important part in the analysis of limit handling, however few driver models have been developed to account for the non-linearities of a driver or a vehicle in the limit handling regime. For this reason, it was not used to investigate the interactions between AFS and DYC.

As an alternative, open loop tests that excite similar behavior in the vehicle model have been identified. The test used in this work is a sine steer input of increasing amplitude. Steady state results are also examined.

STAND-ALONE CONTROLLER DESIGN

Initially, the two controllers were developed and examined individually.

AFS Controller

The active steering system must aim to improve the handling qualities of the vehicle under normal driving compared to the passive vehicle, and at the same time reduce the need for the DYC system to become active. In stabilizing the vehicle, DYC slows the vehicle. This effect must be kept to a minimum and the driver must feel that one is being supported rather than overruled. Further to this, excessive use of DYC to control the lateral dynamics of the vehicle will lead to shorter tire life.

In order to choose a steering controller, two situations were selected. The first was the lane change: a transient maneuver that drivers experience regularly that is challenging both from drivers' point of view and from a vehicle dynamics point view. It is not however a limit handling situation. Secondly, the ability to reject external disturbances is a highly desirable property of any vehicle and for this reason, a side wind force applied laterally to the vehicle was simulated.

In order to select one steering system for coordination with DYC, AFS or ARS, controllers where selected from the literature [12...15] and then compared on a 2 d.o.f vehicle model and linear driver model. Only the conclusions from this work are presented here. Six control systems were selected for comparison, two active front wheel steering systems, and four active rear wheel steering systems. The controllers tested where limited in complexity by the fact that a linear model was being used. The controllers selected are listed in appendix 1.

In deciding which scheme to use a number of standard metrics where used. These include lateral tracking error, heading error, hand wheel rate, lateral acceleration, and side slip.

The metrics all took the form a sum squared value and are then normalized by the passive vehicle with no control, 1. Figure 2 shows the results for the tracking and heading metrics.

$$ \text{metric} = \frac{1}{\text{normalising constant}} \int_0^{t_{stop}} \text{property } dt \quad (1) $$

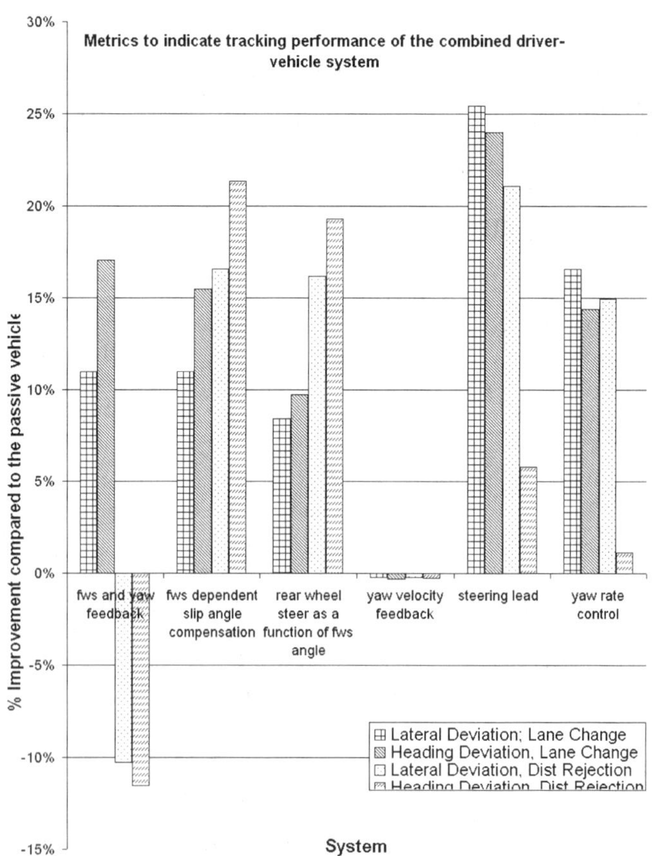

Figure 2 Evaluation of steering controllers

It found that AFS provided significant control over the lateral dynamics of the vehicle, Figure 2. It was decided that AFS is an appropriate system to coordinate with the limit-handling controller, DYC, because the improvements are significant compared with the passive vehicle. The AFS system showed considerable improvement over the passive vehicle, Figure 2.

This paper proposes a scheme for integrating AFS and DYC in a way that exploits the benefits of both controllers in different parts of the vehicle's operating regime.

The task of the AFS controller is to extend the region over which the driver can maneuver the vehicle safely and assist vehicle designers in defining the handling qualities of a vehicle. When stability is in question, the stability controller can apply a stabilizing yaw moment through the steering system.

To achieve this, a reference model controller is proposed as shown in Figure 3.

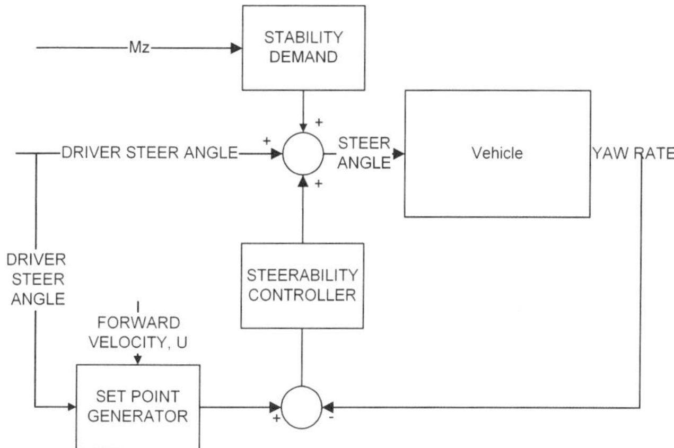

Figure 3 AFS Schematic

This control approach allows the tailoring of the vehicle handling properties up to 0.5g. The approach used is commonly demonstrated in the open literature [3,16…18]. The reference chosen here is yaw velocity. It possible to use a reference model of other properties of the vehicles dynamics in this way, i.e. sideslip angle. The benefits achieved will vary depending on which property is chosen.

Figure 4 shows that the approach is an effective one, Here the linear region of handling has been extended by approximately 50% in terms of vehicle speed. This leads to a vehicle that will handle in a more predictable manner at higher speeds and higher lateral accelerations.

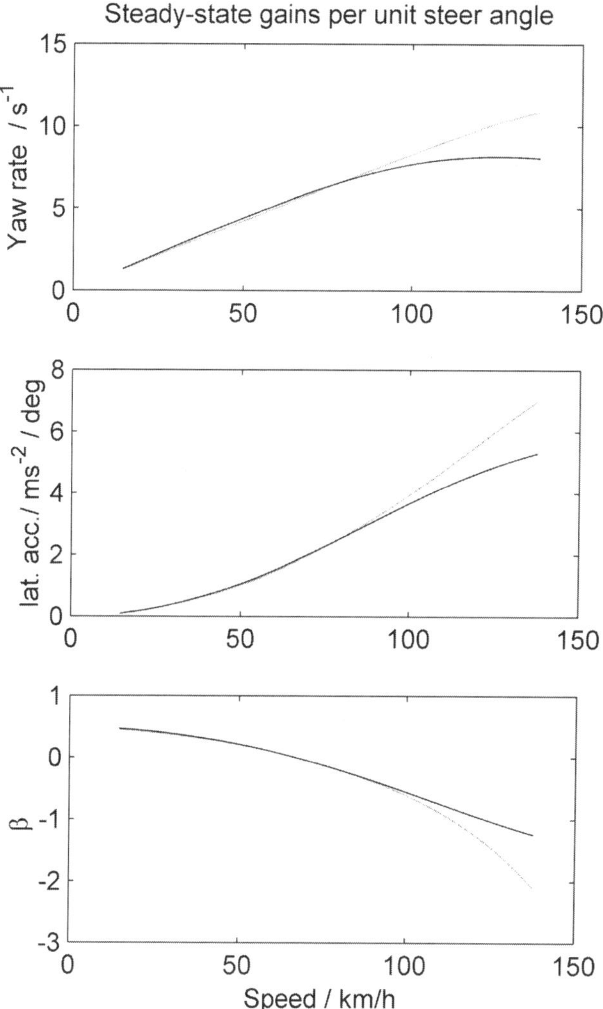

Figure 4 Steady state gains. PASSIVE(--), AFS(..)

<u>DYC Controller</u>

Extensive literature [5…9,16…18] exists describing the combination of braking based DYC systems and active steering systems. DYC is a limit handling stability control system; it does not affect vehicle handling under normal driving situations. When a vehicle is about to become unstable DYC applies a corrective yaw moment to the vehicle through the brakes. In order to simplify the situation, brake intervention will be considered at one wheel only. The amount of braking applied is defined by a simple proportional controller. This is in conjunction with a rule-base to determine which wheel should be braked. The approach is described in Smakman[18].

Pro-cornering moment	Rear inner wheel braked
Contra cornering moment	Front outer wheel.

It is possible to state whether or not a vehicle is stable by defining a stability region on the phase plane for vehicle sideslip angle β. It is also possible to calculate the appropriate restoring moment by determining the distance of vehicle from this stabile region. The stability controller looks at the vehicles states, β and $d\beta/dt$ and determines whether or not the vehicle is operating in a stable region, the approach is used by Smakman[18]. Stability is said to be in question if the vehicle lies either above or below a stable region. If the vehicle is not stable, DYC applies a restoring moment. The distance from the stable region determines the control effort.

<u>Results from uncoordinated controllers</u>

The two controllers presented here were both found to produce desirable effects on the dynamics of the vehicle, i.e. AFS extended the linear region of the vehicle dynamics, and DYC stabilized the vehicle in extreme maneuvers. Once both are implemented in parallel, it becomes clear some significant interaction takes place. Figure 5 shows the results for the two standalone controllers and then both systems on the same vehicle simultaneously. The test performed is a single sine input, a maneuver similar to a lane change. All three vehicle complete the test. When used in isolation AFS has the smallest effect on the longitudinal dynamics and DYC, as expected slows the vehicle. When the two systems are combined, the interaction is such that the interference with the longitudinal dynamics is increased, the phase lead effect created by the AFS system is reduced, and the required stabilizing yaw moment is of increased magnitude. These are all undesirable effects. It would be expected that in combining a steering system.

When combining these two systems, AFS and DYC, a reduction in the interaction with the longitudinal dynamics of the vehicle for a given maneuver. It is clear that some coordination is required to combine these complementary systems successfully.

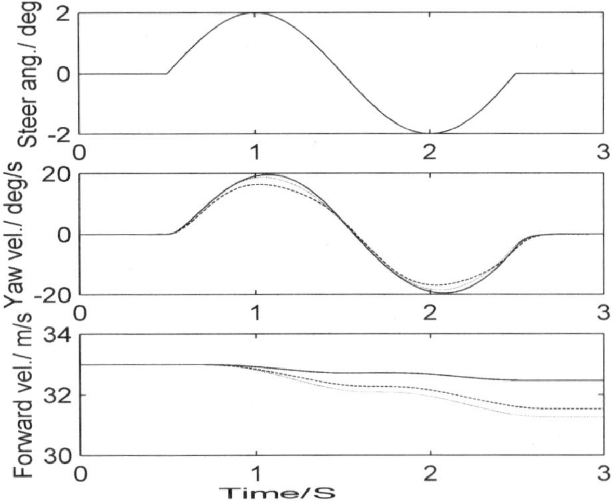

Figure 5 Single sine steer input. Stand alone controllers, AFS (-), DYC(- -), BOTH(...).

COORDINATED CONTROL

In order to coordinate controllers in this case, AFS and DYC, it must be stated what the primary objectives for each system are. In this case, it is clear that DYC is a stability controller. Once the vehicle begins to enter an unstable region, the DYC system provides a stabilizing yaw moment.

The proposed approach in coordinating the two controllers is that AFS is used to improve the vehicle handling properties at low to mid lateral accelerations and the DYC is only used when the stability limit of the vehicle is approached. The available pro and contra cornering moments available from each system are compared in Figure 6.

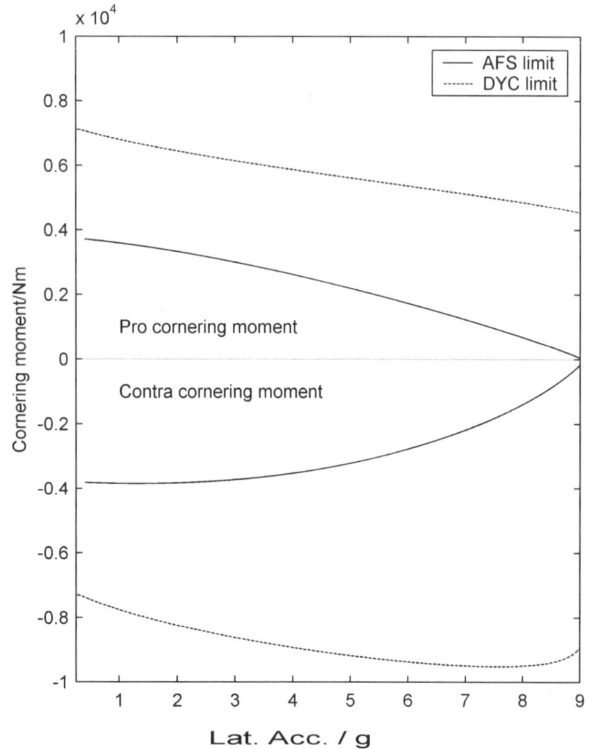

Figure 6 Actuator limits as a function of lateral acceleration

It can be seen that that the AFS system has a comparatively small ability to affect the lateral dynamics of the vehicle as the cornering limit is approached.

It has been decided that the AFS system should aim to extend the linear region of handling of the vehicle in question. In this way, it contributes to the safety of the vehicle by assisting the driver in steering the car. Its aim is to help the driver not to get into critical handling situations.

Smakman[18] comes to this conclusion in combining active suspension control for steerability and DYC for stability control as a similar trend is seen, the active suspension system has very little ability to affect the lateral dynamics of the vehicle as the cornering limit is approached.

The proposed approach to coordinating AFS and DYC is shown in the block diagram in Figure 7.

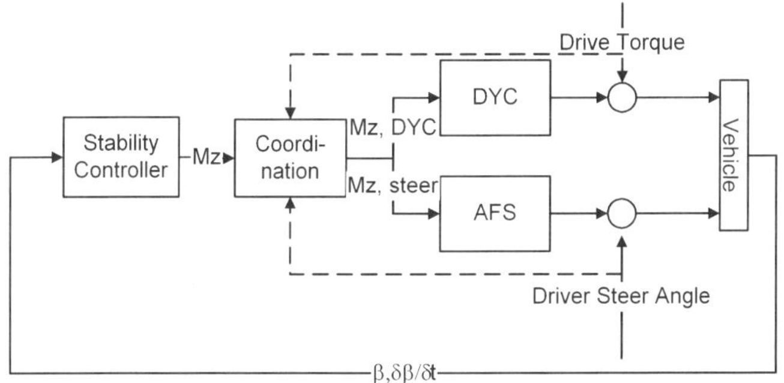

Figure 7 Proposed approach to coordination

<u>Coordination Controller</u>

After studying the combined stand-alone case, it was found that a suitable coordination scheme is a simple rule base that distributes the control demands of the two individual systems.

These rules are:

- Any large demands from the stability controller should be given entirely to the DYC system.

- Smaller demands should be shared between the AFS and DYC, except when the front wheel steer angle is already close to the maximum force generating slip angle of the tire.

- When switching between controllers, the rate at change between the two systems, must not be too large, otherwise instability occurs.

- Yaw moment demands to the AFS system must take priority over steerability considerations.

These rules are the product of investigations carried out in simulation.

Figure 8 shows yaw rate response and lateral acceleration response to an increasing amplitude sine input. The main feature of interest is that for a virtually identical response in the lateral dynamics, the interaction with the longitudinal dynamics is reduced considerably. This is shown by the higher final forward velocity in the coordinated system.

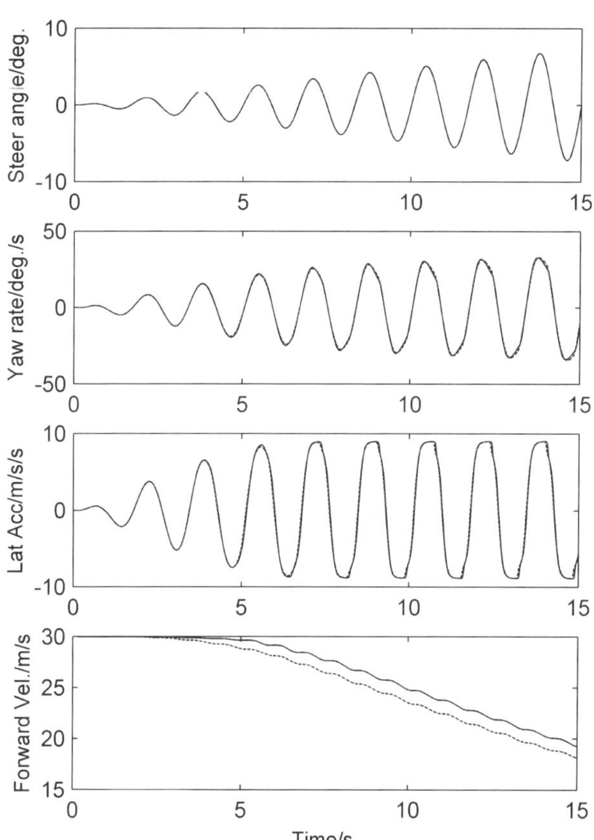

Figure 8 Increasing amplitude sine steer input, Multi-standalone(...), Coordinated approach(--)

CONCLUSION

It has been shown that by coordinating the operation of two controllers, AFS and DYC, additional benefits in overall vehicle handling behavior can be achieved. The AFS can be used effectively in the low to mid range lateral acceleration regime to influence the handling balance of the vehicle. Specifically, it can be used tune the understeer/oversteer parameter, for example, over this lateral acceleration range in a way that is obviously not possible with a passive vehicle. Once the stability limit of the vehicle is approached, the DYC assumes authority, with its brake intervention strategy then also leading to a reduction in vehicle speed.

Further work is now planned to extend this coordination approach to explore whether further integration is possible between vehicle chassis control systems

REFERENCES

1. S H Yu, J J Moskwa, "A global approach to vehicle control: coordination of four wheel steering and wheel torques", J. of Dynamic systems and measurement, 1994, Vol. 1116, pp659-667

2. S E Shladover, "Review of the state of development of advanced vehicle control systems", Vehicle System Dynamics 1995 Vol. 24 pp551-595

3. H Wallentowitz, "Longitudinal control and interaction with other systems", Smart Vehicles,pp252-275

4. G Roppenecker, H Wallentowitz, "Integration of chassis control systems, What is possible, What makes sense, what is under development", Vehicle System Dynamics 1993 Vol. 22 pp283-298

5. M Abe, "Vehicle Dynamics and Control for improving handling and active safety: From 4WS to DYC", Dept. of Mechanical Systems Engineering, Kanagawa Institute of Technology.

6. M Abe, N Ohkubo, Y Kano, "A Direct Yaw Moment Control For Improving Limit Performance Of Vehicle Handling- Comparison And Cooperation With 4WS", Vehicle System Dynamics Supplement 1996 Vol. 25 pp3-23

7. Y Wang, M Nagai, "Integrated control of four wheel steer and yaw moment to improve dynamic stability margin", Proceedings of the 35th Conference on decision and control, 1996, pp1783-1784

8. Y Furukawa, M Abe, "Advanced Chassis Control Systems for Vehicle Handling and Active safety", Vehicle System Dynamics 1997, Vol. 28 No.2-3,pp59-86

9. M Salmon,1990, "Coordinated control of braking and steering", ASME symposium on Advanced Automotive Technologies, Vol.108,pp69-79

10. H B Pacejka, I J M Besselink, "Magic Formula tyre model with Transient properties", Vehicle System Dynamics, 1997, Vol. 27, pp234-249

11. K Senger, "The influence of a four wheel steering vehicle", Vehicle System Dynamics

12. K Fujita, "Development of an active rear steer system applying H∞ synthesis", SAE 98-1115

13. A Taneda, "Design of actuator for active rear steering", SAE 98-1114

14. J Sridhar, "A comparitive study of four wheel steering models using the inverse solution", Vehicle System Dynamics

15. W Kramer, M Hackl, "Potential functions and benefits of Electronic active steering",XXVI Fisita Congress 1996, paper B0304

16. M Abe, Y Kano, K Suziki, Y Shibahata, Y Furkawa, "An experimental validation of side-slip control to compensate vehicle lateral dynamics for a loss of stability due to non-linear tyre characteristics", Proc. AVEC 2000

17. A T van Zanten, "Bosch ESP Systems: 5 Years of experience", SAE 2000-01-1633

18. H T Smakman, "Functional integration of slip control with active suspension for improved lateral vehicle dynamics", PhD thesis, Delft University of technology, The Netherlands. 2000

CONTACT

Mark Selby
Vehicle Dynamics and Control
School of Mechanical Engineering
University of Leeds
Leeds
LS2 9JT
UK
menmase@leeds.ac.uk
Tel: +44 (0)1132 332170
Fax: +44 (0)1132 424611

APPENDIX

EQUATIONS OF MOTION

$$Q_x = F_{xL} + F_{xR} = m_t\left(\dot{u} - rv\right)$$

$$Q_y = F_{yF} + F_{yR} = m_t\left(\dot{v} + ru\right) + \left(am_f - bm_r\right)\dot{r} + m_s h\ddot{\phi}$$

$$Q_\psi = aF_{yF} - bF_{yR} = \left(am_f - bm_r\right)\left(\dot{v} + ru\right) + I_{zz}\dot{r} + I_{xz}\ddot{\phi}$$

$$Q_\phi = 0 = I_{xx}\ddot{\phi} + C_\phi\dot{\phi} + \left(K_\phi - m_b gh\right)\phi + m_s h(\dot{v} + ru) + I_{xz}\dot{r}$$

VEHICLE PARAMETERS: -

u : longitudinal velocity

v : lateral velocity

ψ, r : Yaw angle, velocity

ϕ : Roll angle

$Q_{x,y,\psi,\phi}$: Generalized forces in x,y,ψ,ϕ

$F_{x(L,R)}$: Longitudinal tyre forces, left and right

$F_{y(F,R)}$: Lateral tyre forces, front and rear

$m_{t,b,f,r}$: Mass total, body, front unsprung, rear unsprung

a,b : Front/Rear distance from CoG to axle

h : Roll axis to body CoG

I_{zz}, I_{xz} : Vehicle inertias

C_ϕ, K_ϕ : Roll damping, stiffness

g : acceleration due to gravity

VALUES

$m_{t,b,f,r}$: 1800,100,100 kg

a,b,h : 1.55,1.51,0.36m

I_{zz}, I_{xx} : 1240,3670kgm^2

C_ϕ, K_ϕ : 50kNmsrad^{-1},2500Nmrad^{-1}

g : 9.81ms^{-2}

SYSTEM NUMBER FOR COMPARISON OF AFS AND ARS CONTROLLERS: -

1. Purely passive system.

2. Rear wheel steer angle as a function of front wheel steer angle.

$$\delta_r = \frac{C_f amU^2 - 2C_f C_r b(a+b)}{C_r bmU^2 + 2C_f C_r a(a+b)} \delta_f$$

3. Rear wheel steer angle as a function front wheel steer angle and yaw velocity feedback.

$$\delta_r = C_1 Ur - \delta_f \quad \text{with} \quad C_1 = \frac{m}{2(a+b)}\left(\frac{b}{C_f} + \frac{a}{C_r}\right)$$

4. Rear wheel steer angle as a function of the yaw velocity.

$$\delta_r = C_2 Ur \quad \text{with} \quad C_2 = \frac{m}{2(a+b)}\left(\frac{a}{C_r} - \frac{b}{C_f}\right)$$

5. Rear wheel steer angle from front wheel steer angle dependent slip angle compensation.

In this case, the rear wheel steer angle is a Laplace function of the front wheel steer angle.

$$\frac{\delta_r}{\delta_f} = \frac{-C_f UI \cdot s + C_f amU^2 - 2C_f C_r b(a+b)}{C_r UI \cdot s + C_r bmU^2 + 2C_f C_r a(a+b)}$$

6. Front wheel steering angle, Steering Lead

$$\delta_f = \delta_s + G_{sl}\frac{d\delta_s}{dt}$$

7. Yaw rate controlled by active front wheel steering

$$r_d(s) = \frac{1}{1+T_{ref}.s} r_{dss}(\delta_s)$$

This first term, r_{dss}, is the desired steady yaw rate based on current steering angle. This is then passed through a first order lag with T_{ref} is chosen such that the r_d just leads the actual vehicle yaw rate. An additional steer angle is calculated, based on this yaw rate error.

$$\delta_f = \delta_s + K_c(r_d - r)$$

Stability Control of Combination Vehicle

Pahngroc Oh, Hao Zhou and Kevin Pavlov
Visteon Corporation

ABSTRACT

This paper discusses the development of combination vehicle stability program (CVSP) at Visteon. It will describe why stability control is needed for combination vehicles and how the vehicle stability can be improved.

We propose and evaluate controller structures and design methods for CVSP. These include driver's intent identification, combination vehicle status estimation and control, and fault detection / tolerance. In this paper, the braking and steering dynamics of car-trailer and tractor-semitrailer combinations, and the brake systems which should be used extensively to increase the stability of combination vehicles are presented. Also our development platform is introduced and the combination vehicle simulation results are presented. The definition of combination vehicles in this paper includes car-trailer and commercial tractor-semitrailer combinations since their vehicle dynamics are based on the same equations of motion.

INTRODUCTION

After having increased the safety of combination vehicle braking by the means of ABS, the next step is to enhance its dynamic stability. Although the dynamic stability control of combination vehicles is a key factor in making highways safe, relatively little attention has been paid to this area due to the complexity of the vehicle dynamics.

The modeling of combination vehicles is essential in determining the dynamic behavior of combination vehicles and in designing stability control systems that increase safety and improve steering characteristics. Chen and Tomizuka developed two types of dynamic model for tractor-semitrailer [2]. Their complex simulation model depicts the lateral, yaw and roll motions in detail and the simplified control design model considers the lateral and yaw motions only. They used the simplified nonlinear model to design a controller based on input-output linearization [3]. Chen proposed a backstepping based approach to design a controller which gives freedom of design specification in both yaw and lateral

dynamics [4]. Wang designed the robust H_∞ lateral controller based on the linearized vehicle model of tractor-semitrailer [8]. Bosch introduced the structure of Electronic Stability Program (ESP) for commercial vehicles in 1998 and an independent control strategy of yaw control and roll-over prevention was proposed [9]. Klein, Tepper and Fait studied lateral directional stability of tow dolly type combination vehicles [6]. They observed that tow dolly stability is much higher than conventional hitch supported trailers.

In this paper, we propose the general architecture of CVSP with context diagram. It includes vehicle status estimation and main control algorithms for braking torque balance, yaw torque control, and roll moment damping. The simulation results demonstrate examples of combination vehicle maneuver behaviors with and without CVSP.

PROBLEM STATEMENTS AND STRATEGIES

One dangerous behavior of a combination vehicle is jackknifing. When the trailer wheels or drive wheels lock up during braking, the rear of tractor-trailer tends to come forward, which results in a jackknife. Some disturbances such as side wind gusts cause the lateral oscillation of trailer. This type of instability can be observed when the trailer reaches the critical speed at which it begins to oscillate. Another typical reason for highway accidents involving combination vehicles is roll-over. Since the driver becomes unable to control the motion of vehicle by steering when the articulation angle exceeds the critical limit, most of these accidents are not preventable just by driver's skill or a warning device. While ABS may reduce the chance of jackknifing and help the driver maintain steering control during emergency braking, some questions still remain for improving the combination vehicle's directional behavior and preventing its loss of stability. From these needs, the following three strategic concepts are applied to our CVSP controller design.

BRAKING FORCE BALANCE – The balance of braking torques between two vehicle bodies is required for stable behavior of combination vehicles. The trailer control module shall decide the braking force distributions based on the vehicle status information. It can be obtained from hitch force sensor, articulation angle sensor, and trailer wheel built-in wheel speed sensors. Hence, the combination vehicle compatibility can be improved by the torque balance control module.

YAW TORQUE CONTROL (YTC) – When the CVSP detects understeer or oversteer, it applies the corresponding brakes to help the combination vehicle turns to be more stable. Defining the desired yaw rate of the car ($\dot{\psi}_d$) and the desired articulation angle (η_d) as following

$$\dot{\psi}_d = f(\delta, \vec{v}), \quad \eta_d = g(\delta, \vec{v}) \qquad (1)$$

Note that the targets of yaw rate and articulation angle are determined based on a dynamic combination vehicle model. Braking torque differentiation shall be decided by the following algorithm,

$$IF \quad \left(c_1|\dot{\psi}_d - \dot{\psi}| + c_2|\eta_d - \eta|\right) > \gamma_{yaw}$$

$$Then \quad \tau_{yaw} = \begin{bmatrix} K_{\dot{\psi}} & K_{\eta} \end{bmatrix} \begin{bmatrix} \dot{\psi}_d - \dot{\psi} \\ \eta_d - \eta \end{bmatrix} \qquad (2)$$

Figure 1 shows two representative examples of YTC.

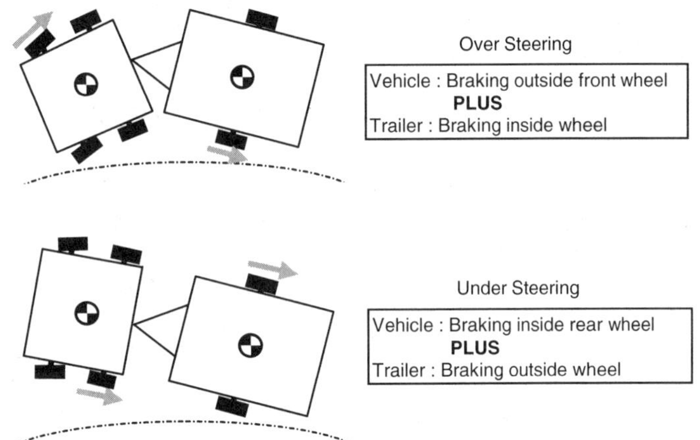

Figure 1. CVSP examples for oversteer and understeer

ROLL MOMENT DAMPING – The CVSP has a body roll damping algorithm, where ϕ is the roll angle, $\dot{\phi}$ is the roll rate, and a_y is the lateral acceleration.

$$IF \quad \left|c_3\phi + c_4\dot{\phi} + c_5a_y\right| > \gamma_{roll}$$

$$Then \quad \tau_{roll} = \begin{bmatrix} K_{\phi} & K_{\dot{\phi}} & K_{a_y} \end{bmatrix} \begin{bmatrix} \phi \\ \dot{\phi} \\ a_y \end{bmatrix} \qquad (3)$$

MATHEMATICAL MODEL

In this section the equations of motion of tractor-trailer are presented. Since ADAMS is used for combination vehicle behavior simulation, this mathematical model will be used for linear and nonlinear controller design.

KINEMATICS – The coordinate systems for tractor-trailer are shown below. {XYZ} represents the globally fixed inertial reference coordinate; $\{x_{u1}y_{u1}z_{u1}\}$ is the tractor's unsprung mass coordinate and z_{u1} axis passes through C.G. of the tractor; $\{x_{s1}y_{s1}z_{s1}\}$ is body-fixed at the tractor's center of gravity and $\{x_{s1}y_{s1}z_{s1}\}$ has roll motion relative to $\{x_{u1}y_{u1}z_{u1}\}$; $\{x_2y_2z_2\}$ is fixed on C.G. of the trailer.

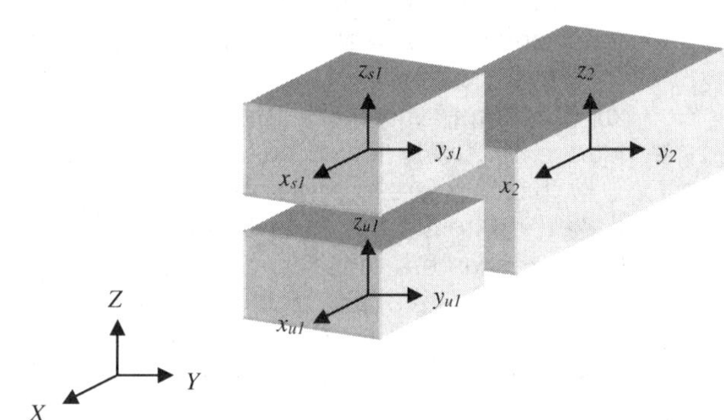

Figure 2. Coordinate system

The coordinate transformation matrices are the following where ψ is the tractor yaw angle w.r.t. {XYZ}, ϕ is the tractor roll angle w.r.t. $\{x_{u1}y_{u1}z_{u1}\}$, and η is the articulation angle between tractor and trailer.

$$[R]_{XYZ}^{xyz_{u1}} = \begin{bmatrix} \cos\psi & \sin\psi & 0 \\ -\sin\psi & \cos\psi & 0 \\ 0 & 0 & 1 \end{bmatrix} \qquad (4.1)$$

$$[R]_{xyz_{u1}}^{xyz_{s1}} = \begin{bmatrix} 1 & 0 & 0 \\ 0 & \cos\phi & \sin\phi \\ 0 & -\sin\phi & \cos\phi \end{bmatrix} \qquad (4.2)$$

$$[R]_{xyz_{s1}}^{xyz_2} = \begin{bmatrix} \cos\eta & \sin\eta & 0 \\ -\sin\eta & \cos\eta & 0 \\ 0 & 0 & 1 \end{bmatrix} \qquad (4.3)$$

GOVERNING EQUATIONS – The equations of motion are derived from Lagrange's equation,

$$\frac{d}{dt}\frac{\partial L}{\partial \dot{q}} - \frac{\partial L}{\partial q} = Q \qquad (5)$$

where L is Lagrangian, q is the generalized coordinate and Q is the generalized force. The Lagrangian is defined as $L = T_1 + T_2 - V$, where T_1 is the kinetic energy of tractor, T_2 is the kinetic energy of trailer and V is the potential energy. The change of potential energy of the tractor is mainly affected by roll motion and that of the trailer is due to the roll and pitch motions. The equations below are the kinetic and potential energies of the tractor and trailer, where h_{1z} and h_{2z} are the distance of the C.G. of tractor and trailer from the roll center, respectively.

$$T_1 = \frac{1}{2}m_1\vec{v}_{CG1}\cdot\vec{v}_{CG1} + \frac{1}{2}\vec{\omega}_{s1}\cdot\vec{I}_1\cdot\vec{\omega}_{s1} \qquad (6.1)$$

$$T_2 = \frac{1}{2}m_2\vec{v}_{CG2}\cdot\vec{v}_{CG2} + \frac{1}{2}\vec{\omega}_{s2}\cdot\vec{I}_2\cdot\vec{\omega}_{s2} \qquad (6.2)$$

$$V = m_1 g h_{1z}(\cos\phi - 1) + m_2 g h_{2z}(\cos\phi - 1) \qquad (6.3)$$

Also the linearized model of a combination vehicle can be obtained from Equation (5) by performing a small perturbation on all vehicle variables. It will be used for designing robust linear controllers.

CONTROLLER

This section consists of two parts. In the first we discuss the structure of the trailer controller which has been developed. It can be either a stand-alone system or an add-on system to the existing ABS. It has three main control algorithms for braking torque balance, yaw torque control, and roll moment damping, which are described in the previous section. The second focuses on the combination vehicle state estimator. Since sometimes it may not be technologically possible or economically feasible to measure all requested vehicle states, its estimation is needed. A schematic diagram, with the specific input / output ordering for CVSP, is shown in Figure 3.

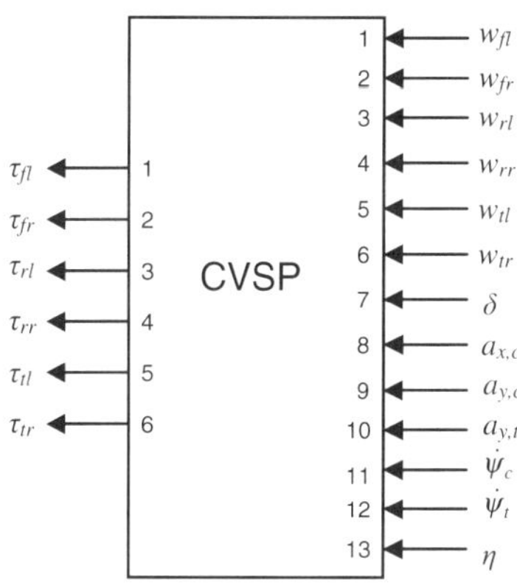

Figure 3. Schematic diagram of CVSP

CONTEXT DIAGRAM – A top level of trailer controller context diagram is shown in Figure 4. Its child diagram (not shown) contains driver's intent determination, combination vehicle status estimation, wheel force / torque distribution, and fault detection and tolerance algorithms.

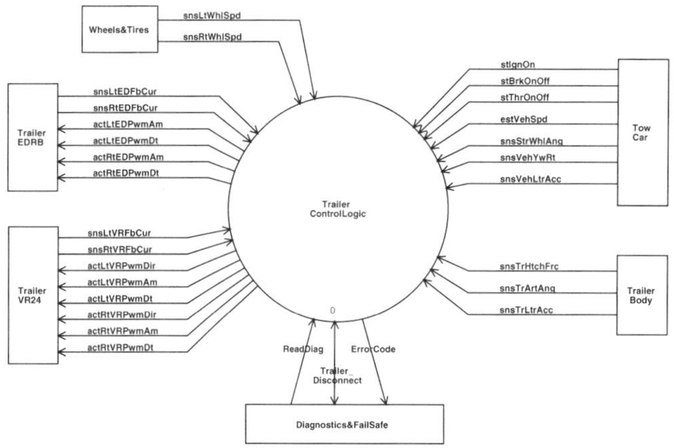

Figure 4. Context diagram for trailer controller

ESTIMATOR STRUCTURE – The principal structure of combination vehicle status observer can be seen in Figure 5. To remove the noises (high frequency and impulsive noises) of real time sensor data, digital low pass filter is applied.

$$X_{fk} = \sum_{i=1}^{k-1}a_i X_{fi} + \sum_{i=1}^{k-1}b_i X_{oi} \qquad (7)$$

where, X_o is the measured value before filtering, X_f is the filtered value and a_i, b_i are constants.

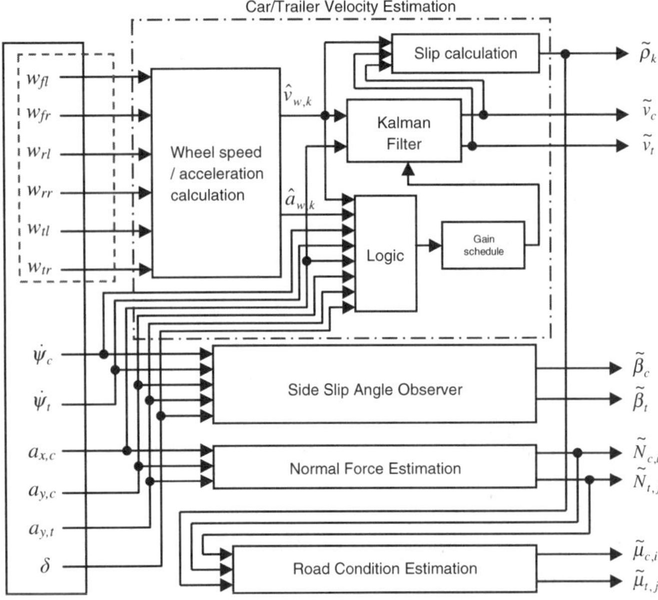

Figure 5. Combination vehicle status estimator

DEVELOPMENT PLATFORM ARCHITECTURE

A prototype trailer is fabricated as an initial investigation development platform to test CVSP. Figure 6 shows the prototype trailer architecture and its potential applications. The built-in capability of various types of electric brake systems allows it to evaluate the individual performance and blending characteristics. An electric drum brake is shown in the left column in Figure 6. The expected advantages and outcomes are the verification of dynamic modeling of tractor-semitrailer in Advanced Highway Systems (AHS), intelligent electric drum brakes and trailer stability enhancement. Since it is built based on the independent rear suspension module of Ford Mustang, its system can be easily applied to the independent rear wheel control for passenger car.

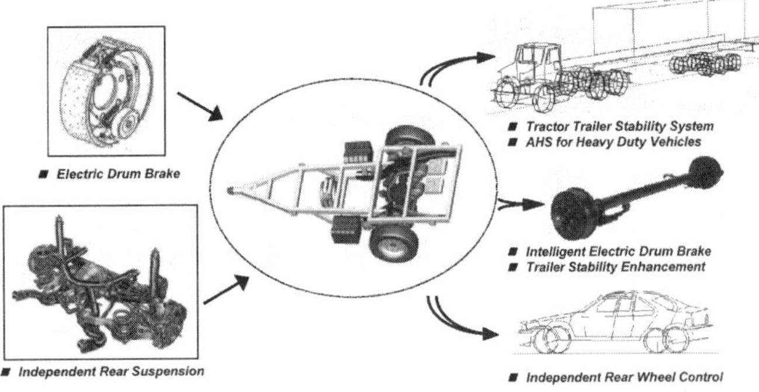

Figure 6. Prototype trailer and its potential applications

SIMULATIONS

In this section we describe how the Combination Vehicle Stability Program (CVSP) is folded into the ADAMS car and trailer simulation package. The combination vehicle simulations were run to evaluate the truck-trailer performance during steady-state Straight Line Drive, Lane Change, and Slalom maneuvers. For the realistic emulation, the centerline of the trailer is intentionally located at an offset to the centerline of the truck to create an asymmetric setup and unbalanced loading condition.

- Straight Line Braking Maneuver – The vehicle-trailer is moving at a speed of 55 mph when a 0.5g deceleration is applied. The hitch point load disturbance induced by braking during the straight line drive maneuver would cause a jack-knife type of response from the truck-trailer system. Figure 7 shows the simulation results of a straight line braking behavior in which the performance of a combination vehicle without CVSP is compared with that of the same vehicle with CVSP. The corresponding braking torques and tire normal forces of each wheel of truck and trailer are shown in Figure 8. The change of tire normal force reflects the longitudinal and lateral weight transfer during the simulation. From Figure 8a it can be observed that the truck-trailer system drifted to the side and became unstable during the braking. Figure 9 shows the trajectories of the hitch point load projected along the three axle of the vehicle coordinate system. For the combination vehicle without CVSP (Figure 9a), the longitudinal force acting on the trailer from the hitch point during braking generates the yaw moment due to the asymmetry and the trailer starts to rotate CW. The reaction force on the back of the truck would tend to force the truck to drift towards CCW direction. These plots illustrate that the truck-trailer system can be stabilized during emergency braking situations with CVSP.

- Lane Change Maneuver – Figure 10 shows the lane change behaviors of the combination vehicle equipped with CVSP system as compared with those of the combination vehicle without CVSP. The initial speed of combination vehicle is 80 mph and the steering wheel input is given in Figure 11a. Figure 11 shows the lateral acceleration, side slip angle, yaw rate, and roll angle of the truck, respectively. Note that the solid line represents the case of without CVSP and the dotted line represents that of with CVSP. The instability of the combination vehicle without CVSP is shown by the fast increase of yaw rate and side slip angle. We also observed that an incipient instability can be corrected by CVSP during lane change maneuver.

- Slalom – The simulation results of a slalom, which includes lateral acceleration, yaw rate, and side slip angle, are shown in Figure 12. The given steering wheel angle input is described in this figure.

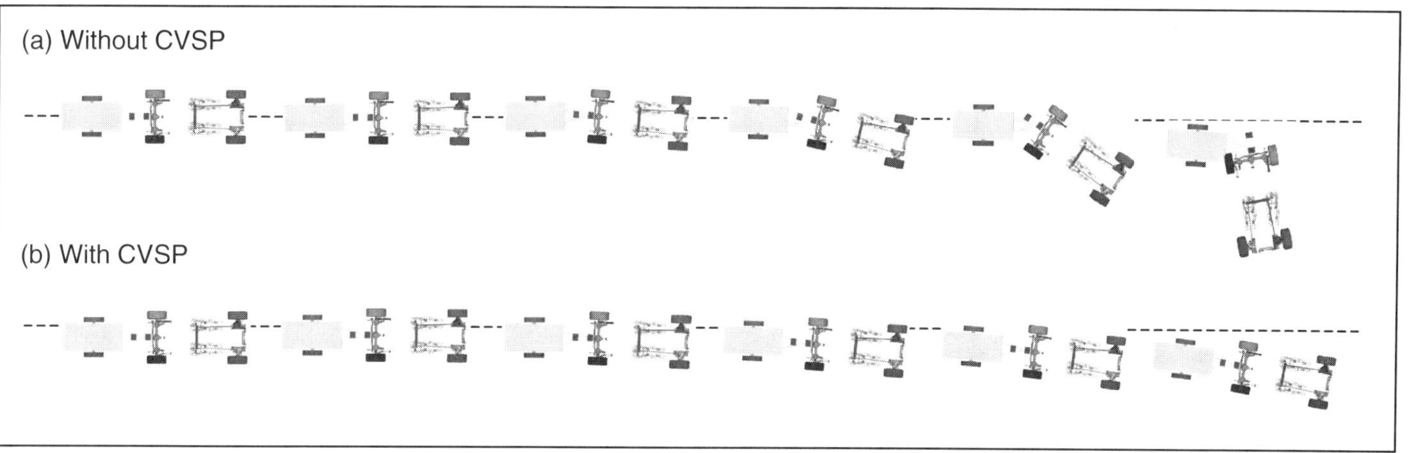

Figure 7. Combination Vehicle Behavior - Straight Line Braking

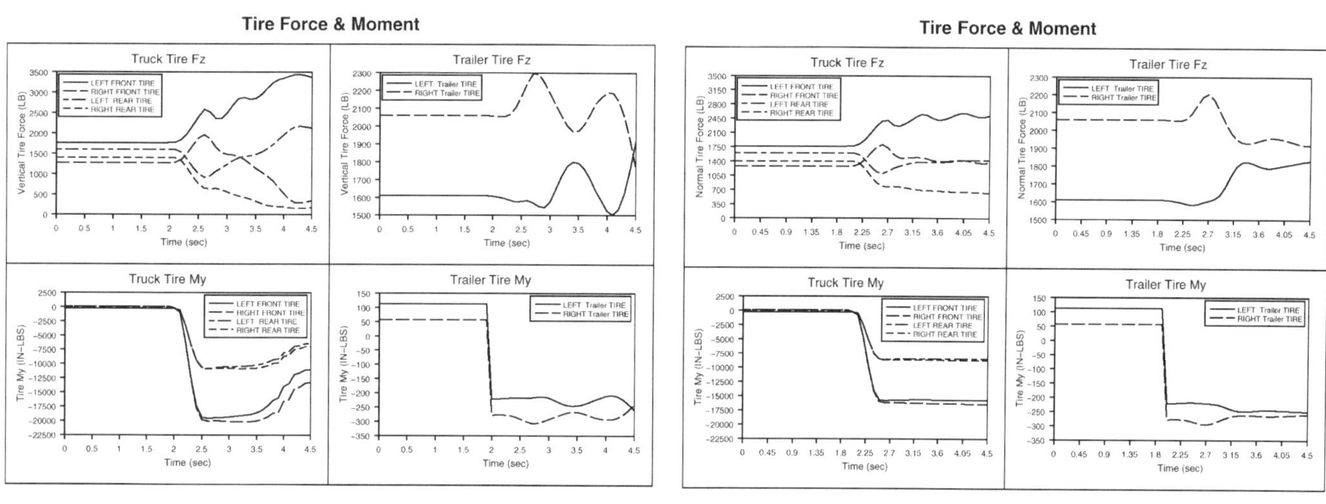

(a) Without CVSP　　　　　　　　　(b) With CVSP

Figure 8. Truck-Trailer tire normal forces (F_z) and braking torques (M_y) – Straight Line Braking

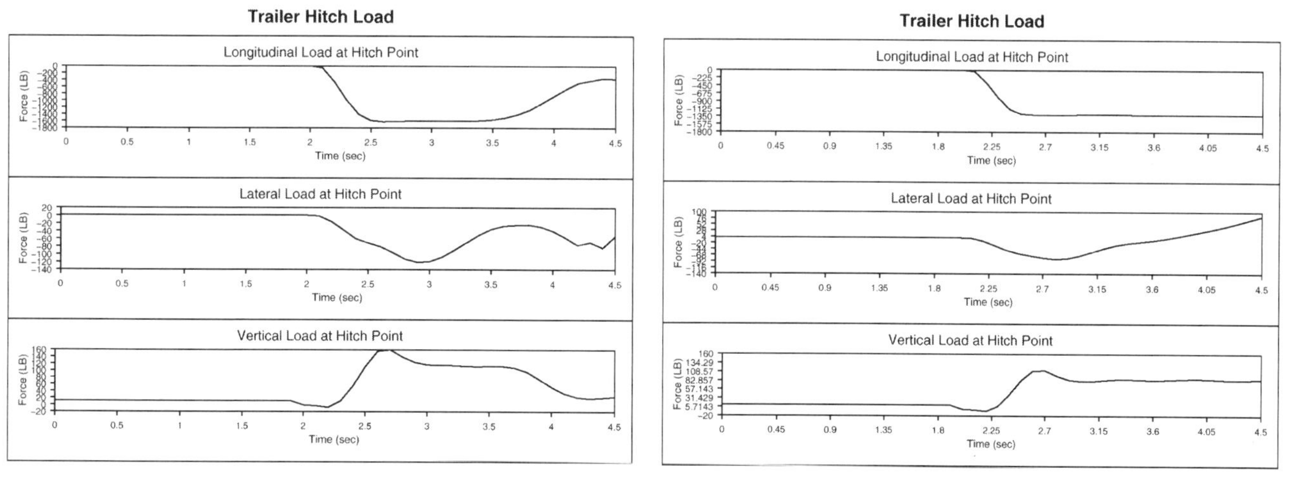

(a) Without CVSP　　　　　　　　　(b) With CVSP

Figure 9. Trailer hitch loads – Straight Line Braking

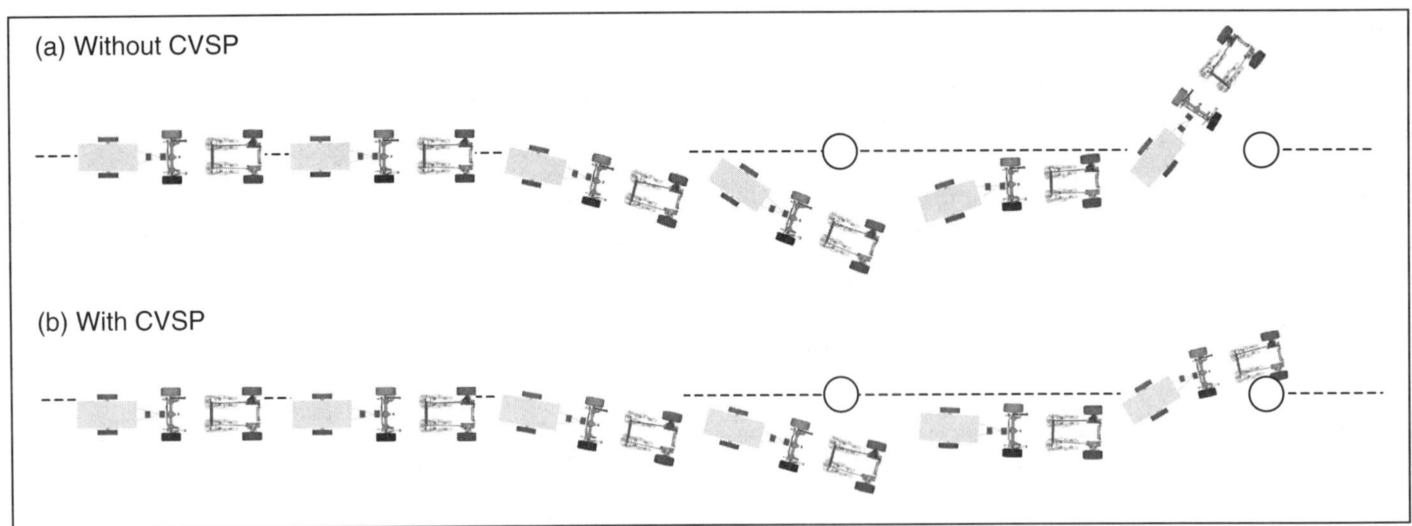

Figure 10. Combination Vehicle Behavior – Lane Change

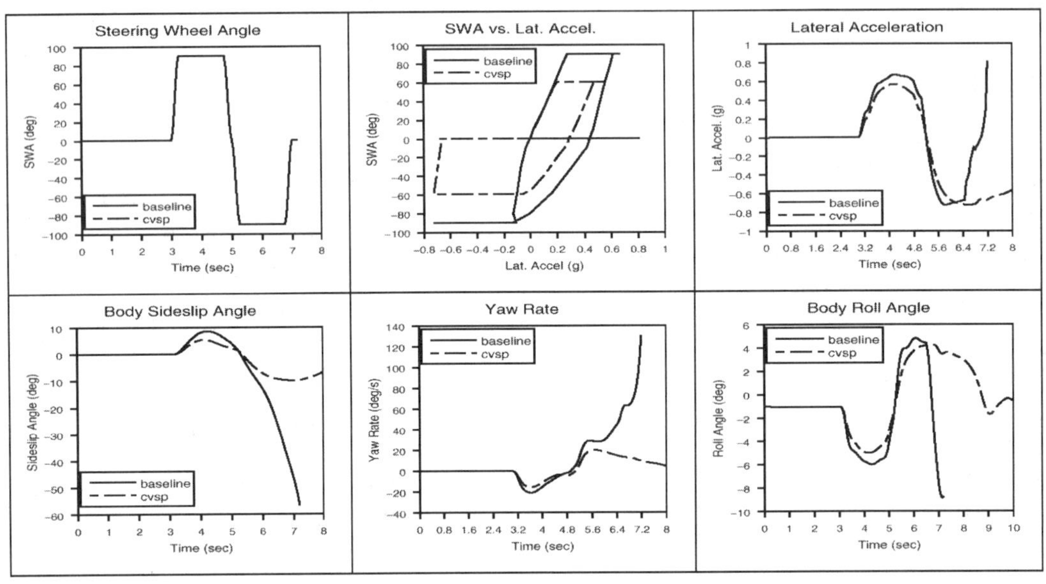

Figure 11. Lane Change Maneuver

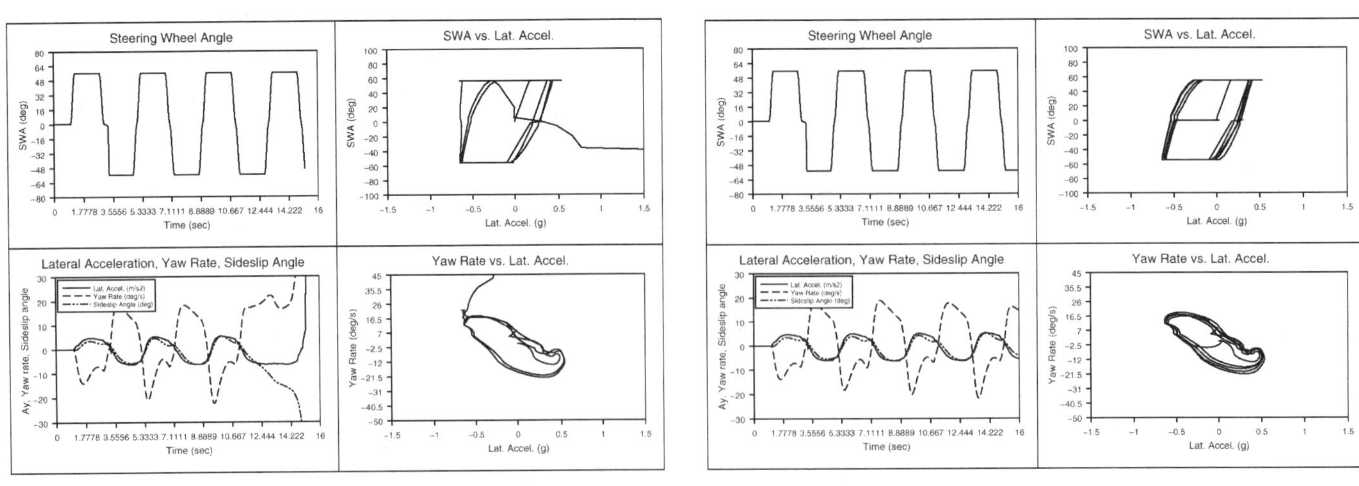

Figure 12. Slalom at 55 mph

CONCLUSION

This paper introduced the needs for a combination vehicle stability program (CVSP) and its proposed structure. The stabilizing properties of CVSP controller are investigated after implementing it into the combination vehicle simulation package.

How do observations like these help the control engineer of the combination vehicle stability program? If combination vehicles are to be formed from the non-uniformities such as the offset of hitch connection and varying C.G. location, as would seem likely, it becomes the task of the stability controller designer to optimize the combination vehicle in the face of these non-uniformities and uncertainties. While the concept of combination vehicle braking control using CVSP approach is promising, more research work is needed to realize this area. The effects on combination vehicle dynamics are usually highly coupled and the systems tend to be very nonlinear. Therefore, much research is required such areas as combination vehicle control structure, redundancy and fault-tolerance requirements, diagnostics and maintainability of the combination vehicle stability systems.

ACKNOWLEDGMENTS

The authors would like to acknowledge the contributions of Scott Funke and Mark Plansinis to this work.

NOMENCLATURE

$a_{x,c}$ Longitudinal acceleration of car (m/s^2)

$a_{x,t}$ Longitudinal acceleration of trailer (m/s^2)

$a_{y,c}$ Lateral acceleration of car (m/s^2)

$a_{y,t}$ Lateral acceleration of trailer (m/s^2)

δ Steering wheel angle (deg)

$\tilde{\beta}_c$ Estimated side slip angle of car

$\tilde{\beta}_t$ Estimated side slip angle of trailer

$\tilde{v}_c$ Estimated velocity of car (m/s)

$\tilde{v}_t$ Estimated velocity of trailer (m/s)

$\dot{\psi}_d$ Desired yaw rate of car (deg/s)

$\dot{\psi}_c$ Yaw rate of car (deg/s)

$\dot{\psi}_t$ Yaw rate of trailer (deg/s)

η Articulation angle (deg)

$\tilde{N}_{c,i}$ Estimated normal forces of car

$(i = fl, fr, rl, rr)$

$\tilde{N}_{t,j}$ Estimated normal forces of trailer ($j = tl, tr$)

$\tilde{\mu}_{c,i}$ Estimated road condition of car

$(i = fl, fr, rl, rr)$

$\tilde{\mu}_{t,j}$ Estimated road condition of trailer ($j = tl, tr$)

$\tilde{\rho}_k$ Estimated tire slip ($k = fl, fr, rl, rr, tl, tr$)

w_k Measured wheel pluses ($k = fl, fr, rl, rr, tl, tr$)

τ_k Wheel torque (Nm, $k = fl, fr, rl, rr, tl, tr$)

$\hat{v}_{w,k}$ Calculated wheel velocity (rad/s)

$\hat{a}_{w,k}$ Calculated wheel acceleration (rad/s^2)

fl Index of front left wheel of car

fr Index of front right wheel of car

rl Index of rear left wheel of car

rr Index of rear right wheel of car

tl Index of left wheel of trailer

tr Index of right wheel of trailer

REFERENCES

1. Beyer, C., Schramm, H., Wrede, J., "Electronic Braking System EBS – Status and Advanced Functions", *SAE 982781*
2. Chen, C. and Tomizuka, M., "Dynamic Modeling of Articulated Vehicles for AHS", *Proc. Of American Control Conference, 1995*
3. Chen, C., "Steering and Independent Braking Control of Tractor-Semitrailer Vehicles in AHS", *Proc. Of IEEE Conference on Decision and Control, 1995*
4. Chen, C., "Back Stepping Design of Nonlinear Control Systems and It's Application to Vehicle Lateral Control in AHS", *Ph.D dissertation UC Berkeley, 1996*
5. Eisele, D., Peng, H., "Vehicle Dynamics Control with Rollover Prevention for Articulated Heavy Trucks", *Proc. Of AVEC 2000, 5th Int'l Symposium on Advanced Vehicle Control*
6. Hyun, D., Langari, R., Ochoa, J., "Vehicle Modeling and Prediction of Rollover Stability Threshold for Tractor-Semitrailers", *Proc. of AVEC 2000, 5th Int'l Symposium on Advanced Vehicle Control*
7. Klein, R., Teper, G., Fait, J., "Lateral / Directional Stability of Tow Dolly Type Combination Vehicles", *SAE 960184*
8. Wang, J., M., Tomizuka, "Analysis and Controller Design Based on Linear Model for Heavy-Duty Vehicles",
9. Wang, J., M., Tomizuka, "Robust H_∞ Lateral Control of Heavy-Duty Vehicles in Automated Highway System",
10. Znten, A., Erhardt, R., Pfaff, G., "VDC, The Vehicle Dynamics Control System of Bosch", *SAE 950759*

Development of Vehicle Stability Control System Based on Vehicle Sideslip Angle Estimation

Akitaka Nishio and Kenji Tozu
Aisin Seiki Co., Ltd

Hiroyuki Yamaguchi, Katsuhiro Asano and Yasushi Amano
Toyota Central R&D Laboratory, Inc.

ABSTRACT

For the vehicle stability control system, which improves vehicle yaw stability for limit driving condition, accurate vehicle sideslip angle detection is one of key issues to confirm the system performance. In this paper, an estimation method of vehicle sideslip angle is described. It is necessary that the vehicle sideslip angle estimation must be robust against road surface condition change, road bank, sensor error, brake force effect, driver's operation and so on. We have developed the sideslip angle estimation method for compensating the driving condition changes which disturb the sideslip estimation. The vehicle sideslip angle estimation using a combination of vehicle model observer and pseudo integral is proposed. The vehicle model observer includes tire non-linear characteristics for fitting actual tire characteristics. Moreover, road-tire friction, road bank and vehicle spinout judgments are developed for increasing the estimation accuracy. The developed sideslip estimation method is evaluated through full-scale vehicle tests. It was verified that this sideslip estimation experimentally confirms the system performance and robustness.

INTRODUCTION

Recently, the vehicle stability control system, which maintains vehicle stability for active safety, has been getting popular in the automotive market [1][2]. The control system detects the vehicle sideslip information, which shows the vehicle yaw attitude, and then manages yaw moment by controlling brake forces independently and actively. Generally, the vehicle sideslip information is derived from onboard sensors [3]-[6]. The estimated sideslip information should be robust against condition changes. We have developed the vehicle sideslip angle estimation using a combination of vehicle model observer

and pseudo integral to ensure robustness of the control system [7][8].

SYSTEM CONFIGURATION

Figure 1 shows a configuration of the vehicle stability control system. The objective of the control system is to restrict extreme oversteer and understeer behaviors. Yaw rate, lateral acceleration, steering wheel angle and vehicle speed are detected using onboard sensors to evaluate vehicle behavior. The vehicle sideslip angle estimated from these sensor signals is used for vehicle stability judgment. If the vehicle behavior becomes unstable or unexpected, the control system intervenes a driver's operation through actuators to keep the vehicle stable.

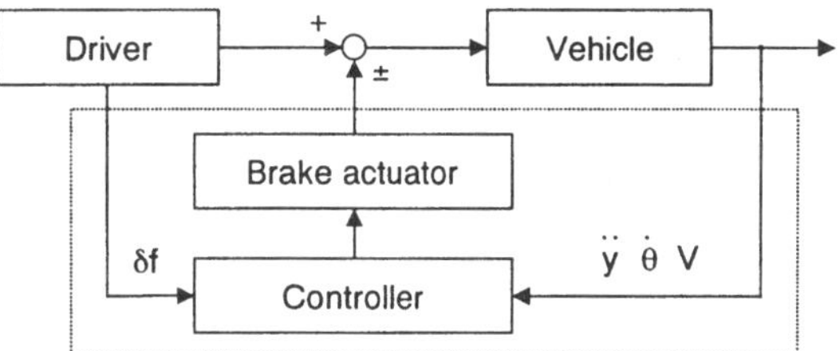

Figure 1. Configuration of vehicle stability control system

VEHICLE SIDESLIP ANGLE ESTIMATION

There are two methods to estimate the vehicle sideslip angle. One is to use a vehicle model observer and another is a pseudo integral. The vehicle model method is relatively robust against sensor error including zero drift and gain change. However, robustness against road condition change and operation disturbance while the system is activated is a problem for the vehicle model observer. On the other hand, the pseudo integral method is robust against road friction change and operation disturbance. However, it should be necessary to minimize a piling-up integral error. We have developed an estimation method for vehicle sideslip angle using a combination of the vehicle model observer and pseudo integral shown in Figure 2. This estimation algorithm includes road-tire friction, road bank and vehicle spinout judgments to estimate sideslip angel accurately and robustly. In the following section, the sideslip angle estimation based on vehicle model observer and pseudo integral is described.

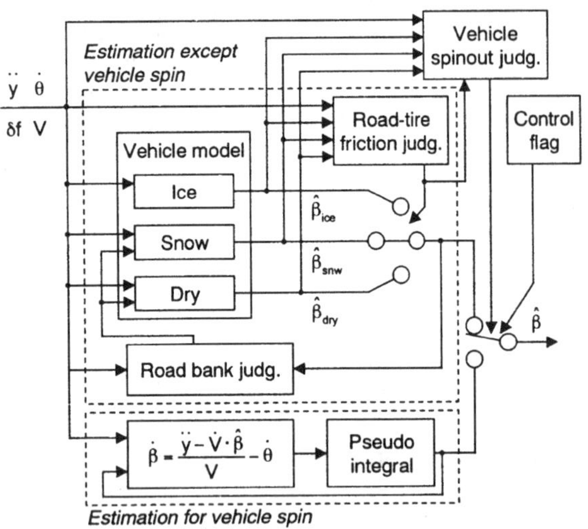

Figure 2. Sideslip angle estimation system combined vehicle model observer with pseudo integral

ESTIMATION BASED ON VEHICLE MODEL OBSERVER

There are two requirements for the sideslip angle estimation based on vehicle model observer. One is to easily express non-linear characteristics of the vehicle due to tire characteristics. Another is that calculation volume has to be small to program within a processor on the onboard ECU. A discrete system is necessary for a typical observer. The calculation volume for the discrete system may eventually become large when considering the system non-linearity. Therefore, we have developed the observer without the discrete system to easily describe the vehicle non-linear characteristics, which is shown in Figure 3.

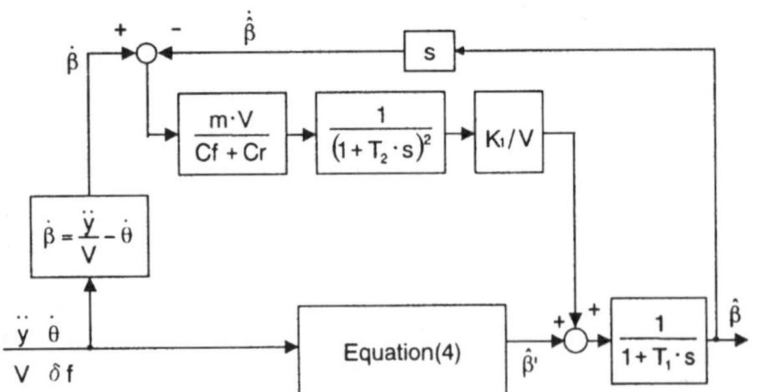

Figure 3. Sideslip estimation based on vehicle model observer

Vehicle model non-linearity

A method which models the vehicle non-linearity due to tire characteristics is described. Figures 4 and 5 show a vehicle model and its front tire characteristics. Based on the bicycle vehicle model in Figure 4, state equations for vehicle motion are:

$$m \cdot V \cdot (\dot{\beta} + \dot{\theta}) = -Cf \cdot \left(\beta + \frac{Lf \cdot \dot{\theta}}{V} - \delta f \right) - Cr \cdot \left(\beta - \frac{Lr \cdot \dot{\theta}}{V} \right) \quad (1)$$

$$I \cdot \ddot{\theta} = -Cf \cdot Lf \cdot \left(\beta + \frac{Lf \cdot \dot{\theta}}{V} - \delta f \right) + Cr \cdot Lr \cdot \left(\beta - \frac{Lr \cdot \dot{\theta}}{V} \right) \quad (2)$$

where Cf and Cr are cornering powers of front and rear tires. Non-linearity of the cornering power Cf is approximated using a fold line shown in Figure 5. The front cornering power Cf is given as slopes Cf_P and Cf_Q in the areas P and Q, respectively. The area P or Q is judged based on the estimated vehicle sideslip angle $\hat{\beta}$. Non-linearity of the rear cornering power Cr is similarly approximated.

A relationship among the vehicle speed V, sideslip angular velocity $\dot{\beta}$, yaw rate $\dot{\theta}$ and lateral acceleration $\ddot{y}$ is:

$$\ddot{y} = V \cdot (\dot{\beta} + \dot{\theta}) \quad (3)$$

This relationship is substituted for equation (1). Considering the tire characteristics approximation in Figure 5, the vehicle sideslip angle is:

$$\hat{\beta}' = -\frac{\left(m \cdot \ddot{y}[i] + \dfrac{(Cf \cdot Lf - Cr \cdot Lr) \cdot \dot{\theta}[i]}{V[i]} - Cf \cdot \delta f[i] - Cf \cdot \beta f - Cr \cdot \beta r + Ff + Fr \right)}{(Cf + Cr)} \quad (4)$$

where i represents sampling timing, βf and Ff are sideslip angle and cornering force for the front tire at the knee point in Figure 5. Similarly, βr and Fr are for the rear tire characteristics. Here, βf, βr, Ff and Fr are constant values. The estimated sideslip angle is obtained by substituting each sensor value at present sampling timing for equation (4). Therefore, non-linear tire characteristics can be easily described in equation (4) without the discrete system.

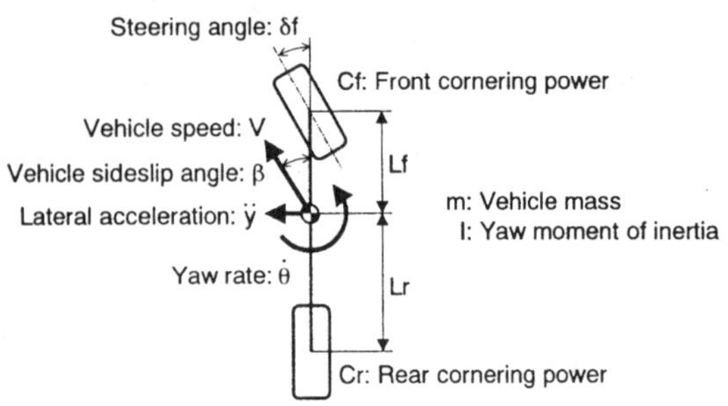

Figure 4. Vehicle model

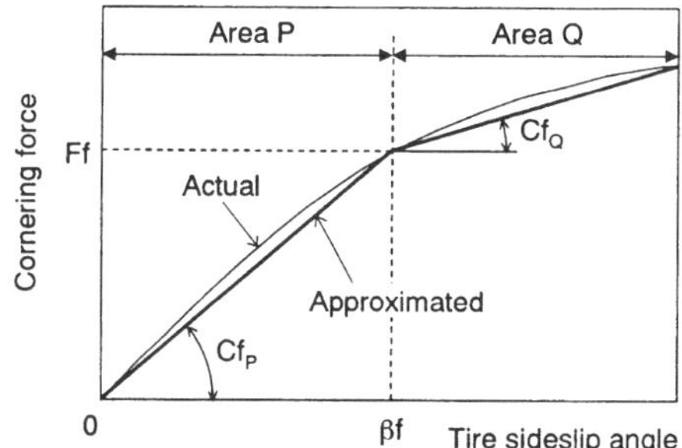

Figure 5. Approximated tire characteristics for vehicle model (Front tire)

<u>Vehicle observer configuration</u>

The estimation based on the vehicle model may include errors since vehicle constants are varied due to vehicle mass change (loaded or unloaded), tire characteristics change (tire exchange) and so on. The observer is constructed for equation (4) to reduce the estimation errors. Assuming $\mathbf{x} = [\beta \ \dot\theta]^T$, $u = \delta f$ and $y_p = \dot\beta$, the same-dimensional observer based on equations (1) and (2) is given by:

$$\dot{\hat{\mathbf{x}}} = \mathbf{A}\cdot\hat{\mathbf{x}} + \mathbf{B}\cdot u + \mathbf{Kp}\cdot(y_p - \hat{y}_p) \tag{5}$$

$$\hat{y}_p = \mathbf{C}\cdot\hat{\mathbf{x}} + \mathbf{D}\cdot u \tag{6}$$

where $\mathbf{A}$, $\mathbf{B}$, $\mathbf{C}$ and $\mathbf{D}$ indicate matrixes comprising vehicle specification (bold letters indicate vector or matrix), $\mathbf{Kp}$ is an observer gain, $\hat{\mathbf{x}}$ is an estimated state amount and $\hat{y}_p$ is an estimated output value. y_p is given by the measured sideslip angular velocity $\dot\beta \ (= \ddot{y}/V - \dot\theta)$. Integrating equation (5) with $\hat{\mathbf{x}}(0) = \mathbf{0}$, $\hat{\mathbf{x}}$ is obtained as:

$$\hat{\mathbf{x}} = \int \mathbf{A}\cdot\hat{\mathbf{x}} + \mathbf{B}\cdot u \, dt + \mathbf{Kp}\cdot\int(y_p - \hat{y}_p) \, dt \tag{7}$$

Multiplying both sides of equation (7) with $\mathbf{C'}=[1, \ 0]$, equation (7) becomes:

$$\hat{\beta} = \mathbf{C'}\cdot\left(\int \mathbf{A}\cdot\hat{\mathbf{x}} + \mathbf{B}\cdot u \, dt + \mathbf{Kp}\cdot\int(y_p - \hat{y}_p) \, dt \right) \tag{8}$$

where $\hat{y}_p$ is given as $\dot{\hat{\beta}}$, which is a differentiation of the estimated sideslip angle $\hat{\beta}$ at previous sampling timing. The first term at right side of equation (8) indicates the estimated sideslip angle based on equations (1) and (2). Substituting $\hat{\beta}'$ for equation (4), equation (8) is given by:

$$\hat{\beta} = \hat{\beta}' + K_1\cdot\int(\dot\beta - \dot{\hat{\beta}}) \, dt \tag{9}$$

where, $K_1 = \mathbf{C'}\cdot\mathbf{Kp}$ (10)

Since the sideslip angular velocity $\dot\beta$ includes sensor noise and drift components, the integration of equation (9) may not be converged. The estimated sideslip angle error $\Delta\beta$ is defined by:

$$\Delta\beta = \beta - \hat{\beta} \tag{11}$$

Omitting non-linearity, equations (1) and (4) yield:

$$m\cdot V\cdot(\dot\beta + \dot\theta) = -(Cf + Cr)\cdot\hat{\beta}' + \xi \tag{12}$$

where,
$$\xi = -\frac{(Cf\cdot Lf - Cr\cdot Lr)\cdot\dot\theta}{V} + Cf\cdot\delta f \tag{13}$$

Based on a relationship between equations (9) and (11), equation (12) is rearranged as:

$$m\cdot V\cdot\left(\dot{\hat{\beta}} + \Delta\dot\beta + \dot\theta\right) = -(Cf + Cr)\cdot\left(\hat{\beta} - K_1\cdot\int\Delta\dot\beta \, dt\right) + \xi \tag{14}$$

Thus, equation (14) becomes:

$$m\cdot V\cdot\Delta\dot\beta = (Cf + Cr)\cdot K_1\cdot\int\Delta\dot\beta \, dt + \varepsilon \tag{15}$$

where,
$$\varepsilon = -(Cf + Cr)\cdot\hat{\beta} + \xi - m\cdot V\cdot\left(\dot{\hat{\beta}} + \dot\theta\right) \tag{16}$$

Assuming the value ε is sufficiently small, the following relationship is derived from equation (15):

$$\int\Delta\dot\beta \, dt \propto \frac{m\cdot V\cdot\Delta\dot\beta}{Cf + Cr} \tag{17}$$

Using equation (17), equation (9) is rearranged as:

$$\hat{\beta} = \hat{\beta}' + \frac{K\cdot\Delta\dot\beta\cdot m\cdot V}{Cf + Cr} \tag{18}$$

where,
$$K = \frac{K_1}{V} \tag{19}$$

For calculating $\Delta\dot\beta$ in equation (18), it is necessary to differentiate the estimated sideslip angle value $\hat{\beta}$. To assure system stability against differentiation noise, the filters are added: a first-order low pass filter G_1 (time constant $T_1=0.159$ sec) for the estimated sideslip angle $\hat{\beta}$ and a second-order low pass filter G_2 (time constant $T_2=0.318$ sec) for $\Delta\dot\beta$. The estimated sideslip angle becomes:

$$\hat{\beta} = G_1\cdot\left(\hat{\beta}' + \frac{G_2\cdot K_1\cdot\Delta\dot\beta\cdot m}{Cf + Cr}\right) \tag{20}$$

In equation (20), Cf and Cr including the tire non-linearity in Figure 5 are used. Here, the cornering powers Cf and Cr, which indicate tire characteristics, are predetermined values. The value ε becomes large when the road-tire friction changes. Thus, the road-tire friction judgment is necessary.

Road-tire friction judgment

For compensating road-tire friction change, the online estimation method to identify tire force characteristics based on an adaptive observer has been developed. To minimize a time lag due to the noise filter for tire characteristics identification, we propose to prepare several tire models which can be offline-identified.

Figure 6 shows a block diagram of road-tire friction judgment with three tire models including dry, snow and ice surfaces. The road-tire friction is judged based on time history of lateral acceleration due to steering operation since a saturation of lateral acceleration is directly corresponding to the road-tire friction. Lateral acceleration waveforms based on each road-tire friction (dry, snow and ice surfaces) are estimated online and compared with a signal from the lateral accelerometer. The most similar road-tire friction is selected as a current road surface.

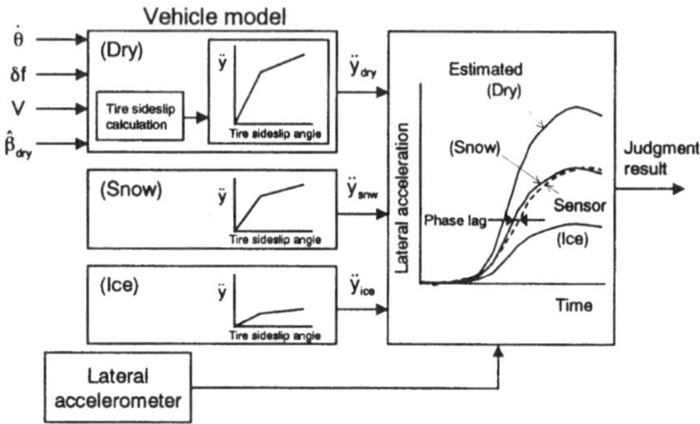

Figure 6. Road-tire friction judgment

A phase lag in the lateral acceleration waveform should be restricted for the accurate road-tire friction judgment. Therefore, the lateral acceleration based on the bicycle vehicle model is used for estimating the lateral acceleration waveform. The estimated sideslip angle is prepared for each road-tire friction since each estimated sideslip angle is needed for the lateral acceleration calculation based on equation (4). When a deviation between lateral acceleration and a product of yaw rate and vehicle speed is large, the road-tire friction is set as dry asphalt to avoid the judgment mistake.

Road bank judgment

When the vehicle runs on a bank, the lateral accelerometer detects a component of the gravity due to the road bank. Therefore, the lateral acceleration offset due to the road bank is compensated based on a deviation between the estimated and actual lateral acceleration. The gravity component of lateral acceleration on the road bank is derived from the deviation between actual lateral

acceleration detected by the lateral accelerometer and a product of yaw rate and vehicle speed. However, this deviation is also generated under driving on low μ surface. The road bank judgment is necessary so that sensor error compensation is only applied for driving on the road bank. Figure 7 shows a block diagram of the road bank judgment. The deviation between a differential of estimated sideslip angle and sideslip angular velocity calculated from sensor signals is calculated. The estimated sideslip angular velocity excludes the offset error and one from sensor signals includes the offset error. The road bank is judged when this sideslip angular velocity deviation is large.

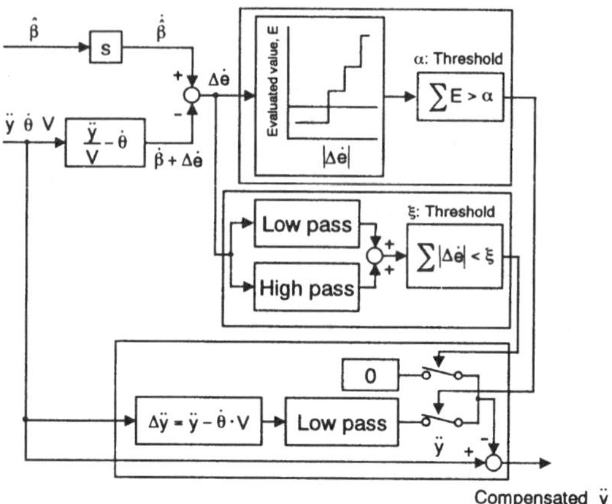

Figure 7. Road bank judgment

Vehicle spinout judgment

The vehicle model observer cannot estimate the sideslip angle accurately when the vehicle spins out. The integral of sideslip angular velocity from sensor signals is used for estimating the sideslip angle. Therefore, it is necessary to judge whether the vehicle spins out or not. The judgment of vehicle spinout is based on a step-change of the sideslip angular velocity. This vehicle spinout judgment should be done immediately to minimize the integral error. Figure 8 shows the vehicle spinout judgment. The estimated lateral acceleration shown in Figure 6 is used for the vehicle spinout judgment since the actual lateral acceleration saturates at limit cornering even when the vehicle does not spin out. Tire characteristics used for the estimated lateral acceleration are selected without saturation to set a judgment threshold. The judgment threshold is adjusted based on the road-tire friction.

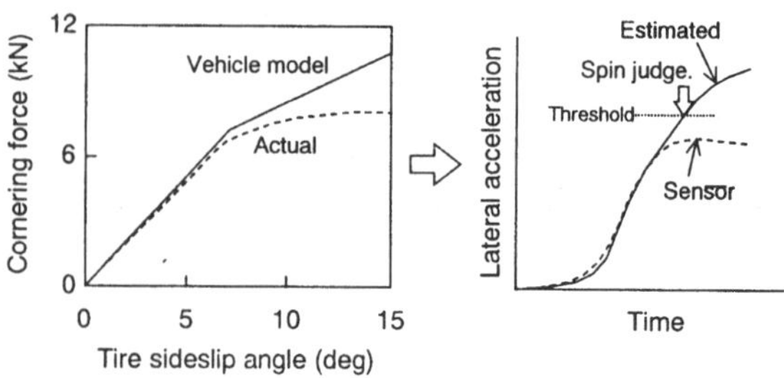

Figure 8. Vehicle spinout judgment

ESTIMATION BASED ON PSEUDO INTEGRAL

When the vehicle spinout is judged, the vehicle sideslip angle is calculated based on a pseudo integral using the following equation:

$$\hat{\beta} = \frac{T}{1+T \cdot s} \cdot \dot{\beta} \qquad (21)$$

where s is Laplacian. Equation (21) is discretized for processing through a digital computer.

$$\hat{\beta}[i] = K_2 \cdot \hat{\beta}[i-1] + K_3 \cdot \dot{\beta}[i] \qquad (22)$$

Here, K_2 and K_3 are constant values. Considering the vehicle deceleration at the vehicle spinout, the vehicle sideslip angular velocity is calculated using the sensor signals.

$$\dot{\beta}[i] = \frac{\ddot{y}[i] - \hat{\beta}[i-1] \cdot \dot{V}[i]}{V[i]} - \dot{\theta}[i] \qquad (23)$$

Sensor errors are compensated to improve accuracy for the pseudo integral. For the sensor error compensation, an offset error due to the road bank and sensor drift error are corrected. The lateral accelerometer signal is compensated using a low pass filter based on a deviation between lateral acceleration and a product of yaw rate and vehicle speed for the offset error. When the vehicle sideslip angular velocity is extremely small, the sensor drift is compensated to converge toward zero value using the following equation:

$$\hat{\beta}[i] = MED\left(\hat{\beta}[i-1] + K_4, 0, \hat{\beta}[i-1] - K_4\right) \qquad (24)$$

where MED( ) is a function to select medium value and K_4 is a constant. The estimation is reset to prevent increasing the integral error. The estimation reset is executed when the control system is not activated and the vehicle behavior is small.

ESTIMATION PERFORMANCE

Vehicle tests were conducted for verifying the sideslip angle estimation system. Lane change, slalom, steady-state turn and J-turn tests on the dry asphalt, packed snow and ice surfaces were performed. The estimated sideslip angle was compared with the sideslip angle detected from the optical velocity sensor as a reference. For evaluating the system robustness, vehicle tests were conducted on a bumpy mountain road covered with snow and ice and a split μ road. Tire change and load change tests were also carried out.

Example results of the above-mentioned vehicle test are shown in Figures 9 to 12. In these results, the estimation system is only activated to evaluate performance of the estimation system. Figure 9 shows results for a slalom steering input on a icy road. The road-tire judgment was adequately executed to estimate sideslip angle accurately. Figure 10 shows results for a steady-state turn with vehicle spinout on packed snow surface. The road-tire friction and vehicle spinout judgments were surely done. Figures 11 and 12 shows for the effects of road bank and tire and load changes, respectively. It was confirmed that the estimation system has enough robustness through these test results.

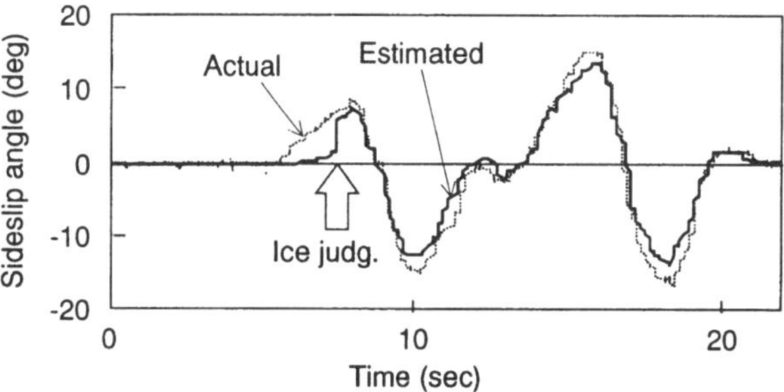

Figure 9. Effect of road-tire friction judgment (Slalom on ice road, Experimental)

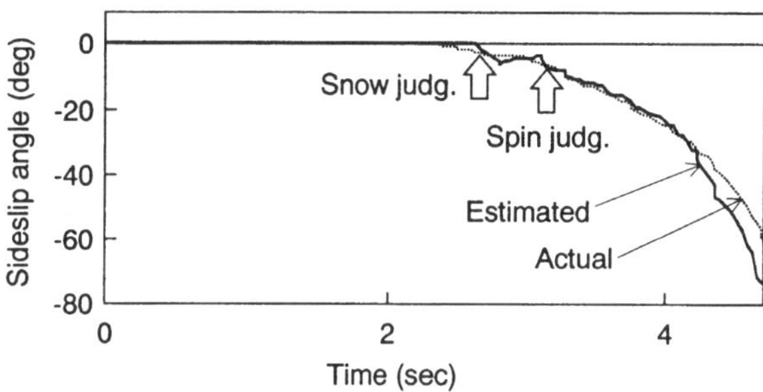

Figure 10. Limit steady-state turn with vehicle spinout on packed snow road (Experimental)

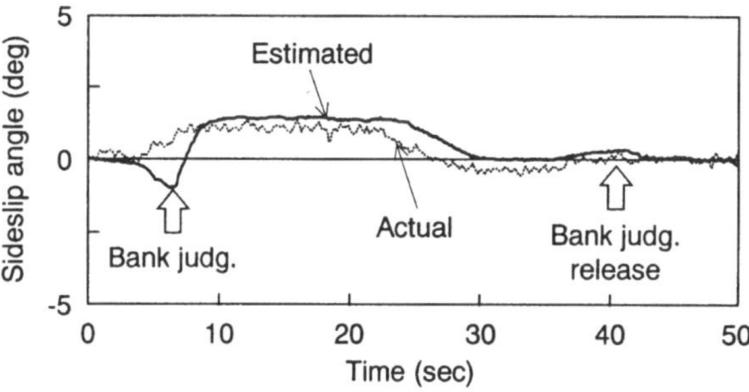

Figure 11. Effect of road bank judgment (Experimental)

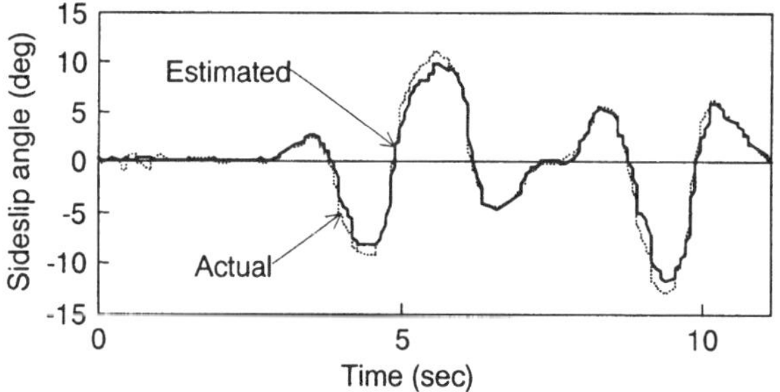

Figure 12. Effect of tire characteristics and load changes (Studless tire and additional 140kg weight, Experimental)

LIMIT OVERSTEER CONTROL

The vehicle stability control system includes the limit oversteer and limit understeer controls. The limit oversteer control is to restrict extreme oversteer behavior. The limit understeer control is for restricting extreme understeer. Figure 13 shows a configuration of the limit oversteer control. The limit oversteer controller is compromises of the vehicle sideslip angle estimation and control mode and control wheel selections. For the sideslip angle estimation, the method shown in Figure 2 is used. The control mode selection judges whether the intervention control is necessary or not. This judgment is conducted based on a relationship between the vehicle sideslip angle and angular velocity shown in Figure 13. To prevent unnecessary control operation due to the sensor noise and road disturbance, the control is not activated when $|\beta|<K_5$. Also, the control is not operated in the areas R_1 and R_2 in Figure 13 to prevent the driver's understeer feeling since the control within the areas R_1 and R_2 restricts vehicle behavior changing to zero β.

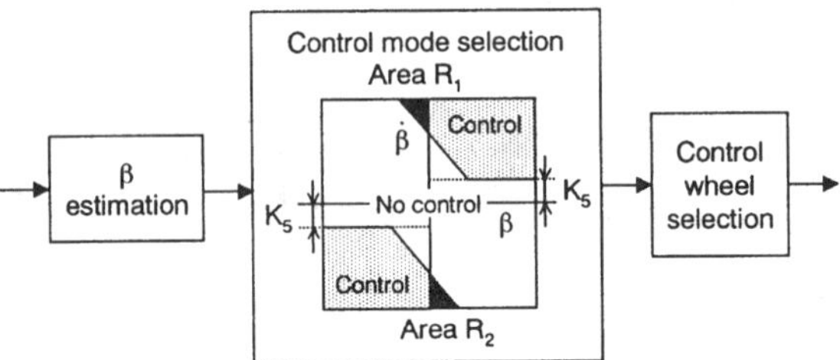

Figure 13. Configuration of limit oversteer control

For the control wheel selection in Figure 13, it is decided which wheel should be controlled to generate the outward yaw moment for improving vehicle stability. Figure 14 shows the effect of yaw moment generation for limit cornering condition [9]. Figure 14 indicates the yaw moment against slip ratio for each wheel and components due to brake and cornering force changes. The front-outer wheel brake is very effective in stabilizing the vehicle motion since both yaw moment changes due to the brake and cornering force changes work in the outward direction. For the rear-outer wheel, the brake force component is outward and the cornering force component is inward. The total yaw moment change acts up to only 10 % of the slip ratio. Therefore, the front-outer wheel is selected for brake control since this wheel is most effective. Moreover, additional control of the rear-outer wheel generates more outward yaw moment to recover the vehicle into stable area quickly for high μ surface since slip control is relatively easy on high μ surface.

Figure 15 shows a hydraulic circuit of brake actuator designed for a FF vehicle to manage each wheel cylinder pressure independently. A pump driven by an electric motor can apply active pressure for each wheel cylinder. Control solenoid valves modulate each wheel cylinder pressure independently to control the vehicle yaw moment. A solenoid valve equipped with a vacuum booster introduces ambient pressure to improve active pressure response.

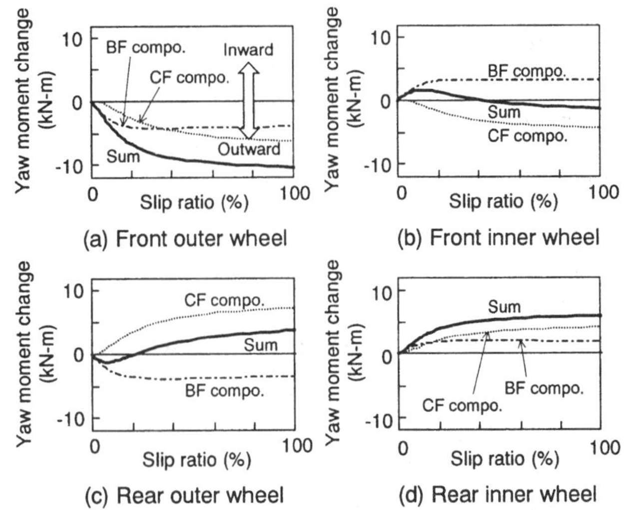

Figure 14. Yaw moment generation due to brake control for each wheel (Simulation)

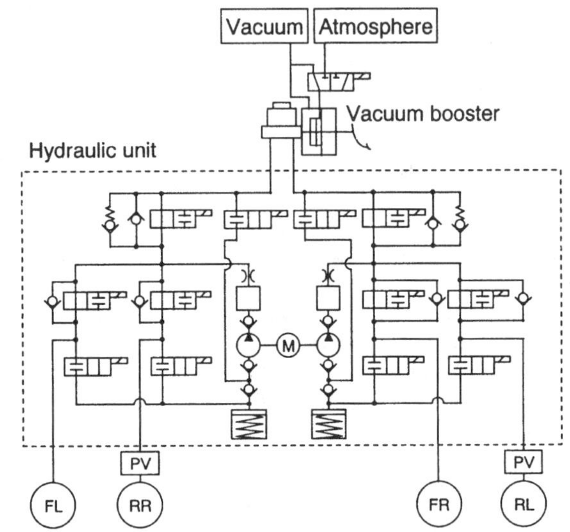

Figure 15. Hydraulic circuit of brake actuator

SYSTEM EVALUATION USING TEST VEHICLE

The vehicle tests were conducted to verify the control system on actual road surfaces. Figures 16 and 17 show for limit cornering test results with and without driver's counter steer operation. The vehicle sideslip angle is accurately estimated without the effect of the driver's counter steer operation. The control system surely restricts the vehicle oversteer behavior to maintain vehicle stability for the both cases.

An interaction between the driver operation and control intervention was surveyed. Figure 18 shows slalom test results. Figure 19 is a frequency analysis of the steering wheel operation shown in Figure 18. Focusing on a power spectrum density of the driver's steering wheel operation, the driver turned the steering wheel with a constant rhythm and no corrective steering operation to pass the slalom task course. Thus, it was verified that the brake control intervention did not affect the driver steering operation.

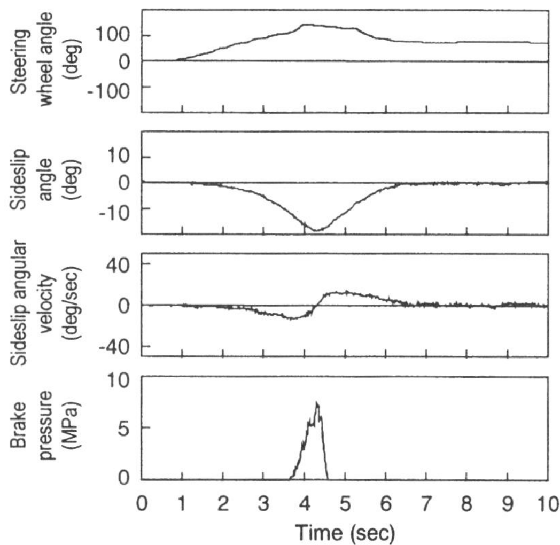

Figure 16. Limit steady-state turn without driver's counter-steer operation on dry asphalt road (Experimental)

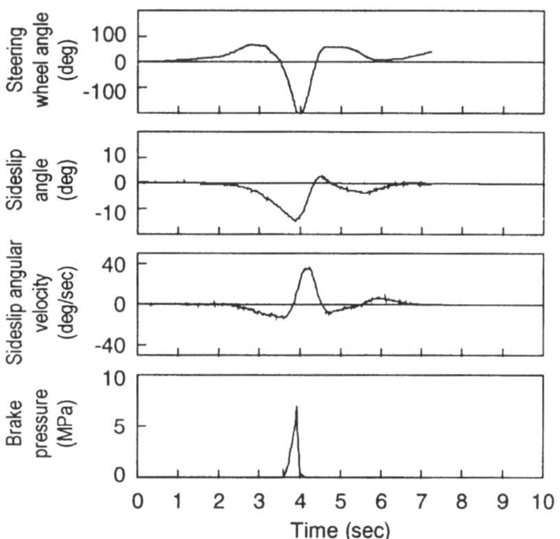

Figure 17. Limit steady-state turn test with driver's counter-steer operation on dry asphalt road (Experimental)

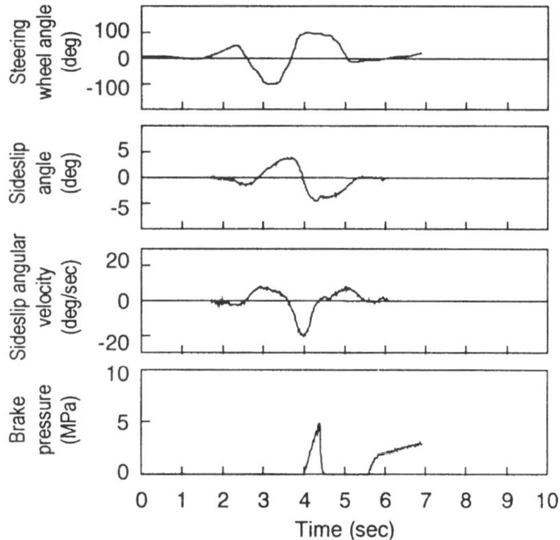

Figure 18. Pylon slalom test on packed snow road (Experimental)

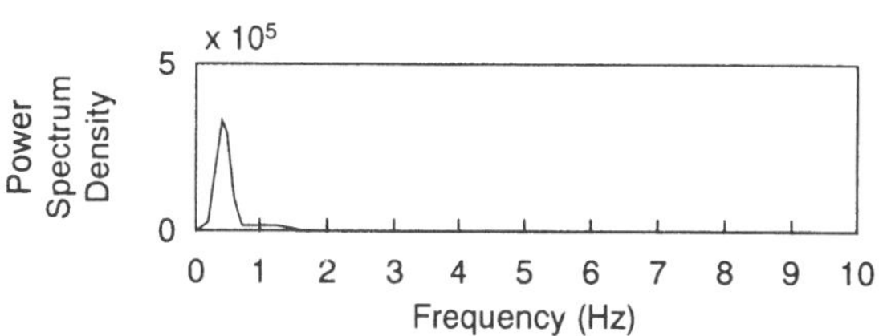

Figure 19. Power spectrum density of steering wheel operation for pylon slalom on packed snow road (Experimental)

SUMMARY

The estimation system for the vehicle sideslip angle based on a combination of the vehicle model observer and pseudo integral has been developed. The estimation system is robust against disturbances including road-tire friction change, road bank and vehicle spinout and ensures estimation accuracy of vehicle sideslip angle.

The limit oversteer control based on the developed sideslip angle estimation is evaluated through vehicle tests. It is verified that the control system surely restricts oversteer behavior without affecting the driver's steering wheel operation.

NOMENCLATURE

m	Vehicle mass
I	Yaw moment of inertia
Lf, Lr	Distance between front/rear axle and center of gravity
Cf, Cr	Front/rear cornering power
δf	Front wheel angle
Ff, Fr	Front/rear cornering force at knee point of approximated tire force characteristics shown in Figure 5
$\beta f, \beta r$	Front/rear tire sideslip angle at knee point of approximated tire force characteristics shown in Figure 5
β	Vehicle sideslip angle
$\hat{\beta}$	Estimated vehicle sideslip angle
$\hat{\beta}'$	Vehicle sideslip angle estimated from vehicle model
$\dot{\beta}$	Vehicle sideslip angular velocity
$\dot{\hat{\beta}}$	Vehicle sideslip angular velocity calculated from $\hat{\beta}$
$\Delta\beta$	Deviation between β and $\hat{\beta}$
$\Delta\dot{e}$	Deviation between $\dot{\beta}$ and $\dot{\hat{\beta}}$
$\ddot{y}$	Lateral acceleration
$\dot{\theta}$	Yaw rate
V	Vehicle speed
$\dot{V}$	Vehicle deceleration
$MED(\ )$	Function to select medium value

REFERENCES

[1] K. Koibuchi, M. Yamamoto, Y. Fukada and S. Inagaki, "Vehicle stability control in limit cornering by active brake", SAE, No. 960487, 1996

[2] A. T. Van Zanten, R. Erhardt and G. Pfaff, "VDC, The vehicle dynamics control system of Bosch", SAE, No. 950759, 1995

[3] H. Yamaguchi, K. Asano, Y. Yasui and T. Ito, "The estimation method of side slip angle", JSAE, No. 9637078, 1996 (in Japanese)

[4] Y. Fukada, "Estimation of vehicle slip-angle with combination method of model observer and direct integration", AVEC, 9836626, 1998

[5] M. Kaminaga and G. Naito, "Vehicle body slip angle estimation using adaptive observer", JSAE, No. 9832693, 1998 (in Japanese)

[6] Y. Furukawa, Y. Shibahata, M. Abe, A. Kato, K. Suzuki and Y. Kano, "Estimation of vehicle side-slip angle for DYC by using on-board-tire-model", AVEC, No. 9837049, 1998

[7] H. Yamaguchi, K. Asano, Y. Amano, K. Tozu and A. Nishio, "Development of vehicle spin control (First report) - The estimation method of sideslip angle", JSAE No. 20005242, 2000 (in Japanese)

[8] A. Nishio, T. Tozu and H. Yamaguchi, "Development of vehicle spin control system (Second report) - Vehicle spin restriction control based on β estimation", JSAE No. 20005249, 2000 (in Japanese)

[9] Y. Yasui, K. Tozu, N. Hattori and M. Sugisawa, "Improvement of vehicle directional stability for transient steering maneuvers using active brake control", SAE, No. 960485, 1996

Rollover Propensity Evaluation of an SUV Equipped with a TRW VSC System

Ali Y. Ungoren and Huei Peng
University of Michigan

Danny R. Milot
TRW Inc.

ABSTRACT

In this paper, a simulation-based dynamic rollover evaluation procedure is described. This work is based on the worst-case methodology developed at the University of Michigan, and is the result of a collaborated research project between the University of Michigan and TRW Inc. The target vehicle studied in this paper is a large production volume SUV. This vehicle is equipped with a production-intent TRW Vehicle Stability Control (VSC) system. The main goals of this paper are to (i) study the rollover propensity of this SUV, as influenced by vehicle and environment parameters such as vehicle speed, road condition, etc.; and (ii) investigate whether, and by how much, does the VSC system influence the rollover propensity of this SUV. The modeling, evaluation procedure, and preliminary evaluation results are reported.

INTRODUCTION

The growing popularity of Sports Utility Vehicles (SUV) in the last decade together with their higher rollover probability has necessitated a closer look at regulations to reduce fatalities in their rollover crashes. NHTSA reported that there were 10,857 rollover deaths in cars, vans and trucks last year, up from 9,771 in 1998, which seems to be closely related to the increased population of SUVs and light trucks. Recently (May 2000), the National Highway Traffic Safety Administration (NHTSA) announced its plan to include a vehicle measure of rollover resistance as an addition to the 2001 New Car Assessment Program (NCAP). NHTSA determined that a vehicle's "Static Stability Factor" (SSF) is the most reliable indicator of rollover risk of un-tripped single-vehicle crashes. NHTSA expects that its new rollover information program will motivate manufacturers to produce safer, more stable vehicles. Currently, the U.S. Congress is temporarily holding the publication of this new "star rating" system. Obviously, the rollover propensity of SUV is an important issue that has caught the attention of consumers, government agencies, car companies, and major suppliers.

For assessing the on-road, untripped rollover propensity of a vehicle, there were predominantly two types of metrics proposed in the literature. The NHTSA dynamic test program uses a set of rollover evaluation maneuvers and examines the severity and frequency of vehicle Two-Wheel-Lift (TWL) under these maneuvers. The Consumer Union's double lane change driving test also falls into this "dynamic testing" category. The "static" type of rollover metrics, on the other hand, is usually based on the simple measurement of vehicle parameters related to its roll behavior. For example, Static Stability Factor, Tilt Table Angle, Tilt Table Ratio, Critical Sliding Velocity and Side Pull Ratio all fall into this category. In general, static testing results may not completely incorporate the effects of controllers (e.g., vehicle stability control systems and ABS), transient tire behavior, and so forth.

In literature it has been shown that a properly designed active control system could reduce the risk of rollover. Four-wheel steering (4WS), active suspensions, passive suspensions accompanied by anti-roll bars, suspensions with variable stiffness and differential braking have all been studied for their effect on roll control. For heavy articulated trucks with multiple axles, the steering of tractor wheels was shown to improve lateral stability (Furleigh et al. 1988). Steering the rear axles of the tractor proportional to the front axles, with a proportionality constant dependent on the forward speed, reduced the lateral acceleration of the trailer at high-speed obstacle maneuvers. Cech et al. (2000) proposed two configurations--anti-roll and active roll suspensions for better cornering performance. The Anti-roll design uses hydraulically activated anti-roll bars implemented with passive suspensions. The latter offers anti-roll control as well as independent control of each suspension. Another approach for reducing the roll angle is to vary the suspension stiffness without adding any energy to the system (Constantine et al. 1994). Palkovics et al. (1999) proposed a Roll-Over Prevention (ROP®) system for commercial vehicles. Wheel speed and lateral acceleration are used to estimate the wheel lift-off. Full (left and right hand sides) braking of the vehicle is activated if a wheel lift-off is detected. In another study, Wielenga (1999) proposed an Anti-Rollover Braking (ARB™) system, which utilizes differential braking (braking on one side of the vehicle tires). If an impending rollover is predicted, either due to the existence of high lateral acceleration or the direct sensor detection of wheel lift-off, differential braking is applied to slow down the vehicle as well as to reduce the lateral acceleration.

The development of these advanced roll control systems are likely to accelerate in the coming decade. A forward-looking government-defined regulation/standard should incorporate a well-designed dynamic test procedure to obtain a more accurate assessment of vehicle rollover propensity, especially for those vehicles equipped with advanced control systems.

VEHICLE MODELS

Upon the request of the OEM, we are not going to mention the model name of the target SUV studied in this paper. Rather, it will be referred to as "Vehicle A" in the remainder of this paper. For comparison purposes, we will provide un-compensated (no VSC control) vehicle simulation results of a benchmark vehicle—a 1998 Jeep Cherokee. The benchmark vehicle will be referred to as "Vehicle B" from now on. The mathematical models for these two vehicles are described in the following:

Vehicle A:

- The vehicle dynamics model was created using the AutoSim software by Mechanical Simulation Corporation (MSC). The model was created for TRW as an extension to MSC's CarSim product.

- The model has 14 DOF (6 DOF for the sprung mass, 2DOF for each of the axles, and 1 DOF for each of the wheels).

- The model also includes a production-intent VSC control algorithm incorporated as a C-code module. In addition to the control algorithm, a production intent hydraulic control unit model is used to provide the active brake torque control. A lumped parameter powertrain model is also used for interaction with the VSC system for understeer control.

Vehicle B:

- The mathematical model is created using the TruckSim software (also the product of MSC), and is based on vehicle parameters published by the Vehicle Research and Testing Center of the NHTSA [Salaani et al. 1999]. This TruckSim model was verified against VRTC published test results (Chen and Peng 1999) at 2 different vehicle speeds with steering and braking inputs that generate lateral acceleration as high as 0.6g.

- The model also has 14 DOF (identical to those of Vehicle A model).

Both models include nonlinear models for tires, suspension springs, shocks, brakes, etc. The models are capable of predicting 3D forces and motions of vehicles in response to throttle (Vehicle A only), braking and steering inputs. The Vehicle A model also includes a powertrain model developed by TRW.

WORST-CASE EVALUATION PROCEDURE

Over the last two decades, computer-aided engineering has made tremendous progress in the automotive industry. One of the most noticeable trends is the growing popularity of computer-based simulation models to replace prototypes or early production vehicles. These models are now used extensively in crashworthiness, NVH, dynamic testing, etc. The worst-case evaluation method developed at the University of Michigan (Ma and Peng 1999) aims to use optimization techniques to identify weaknesses of vehicles and the worst-case scenarios (steering input, braking input, operational parameters, environmental parameters) that can be iteratively figured out by trial-and-error or by intuitive reasoning. This simulation-based approach provides a cheaper alternative to the existing test-matrix based approach. This is because test engineers' experience may be quite valuable for simple single-input-single-output vehicle response. However, as the (vehicle) dynamic model becomes very complex and/or nonlinear, finding the weaknesses of a vehicle design becomes strenuous and more trial-based even for experienced engineers. Therefore, a standard practice is to define a richer and richer test matrix, in hope that by testing enough samples, the relative performance of a vehicle design can be reasonably understood. Yet, the essence of a comprehensive evaluation of the "safety" aspect of a new vehicle before its production enforces the use of more effective means to identify its "worst-case" behavior, which may or may not be captured in the test matrix. Therefore, the simulation-based approach seems to be more viable compared with the test-matrix based approach.

To this end, the worst-case evaluation methodology, introduced in this paper, proves to be an efficient method. This method provides a systematic analysis of all the available data about the dynamic system and displays the conditions under which the system fails to meet the desired requirements. In the following section, we give the mathematical model of the dynamic system, which is an SUV in our case. By using the worst-case methodology, it is possible to determine the worst inputs (or more precisely, disturbance inputs, w(t)) and the optimal control inputs (u(t)) opposing the disturbance. In this study, we will be focusing on the identification of the worst disturbances (w(t)) to the vehicle, which are assumed to include hand-wheel steering angle and brake pedal force. The worst-case disturbance inputs aim to minimize, while the control inputs try to maximize "safety". In this paper, we will focus only on the vehicle rollover problem. In other words, w(t) tries to maximize vehicle roll angle. The control input u(t), when it exists, is assumed to be the brake pressure command provided by the VSC system.

In the mathematical form, the worst-case problem is defined as follows. The dynamic system of interest is assumed to be described by the nonlinear equation $\dot{x} = f(x,w,u,t)$ (Figure 1), where w(t) and u(t) are the "bad" and "good" (disturbance and control) inputs. The state vector x includes all vehicle dynamic variables, including, for example, sprung mass roll angle, vehicle side slip angle, etc.

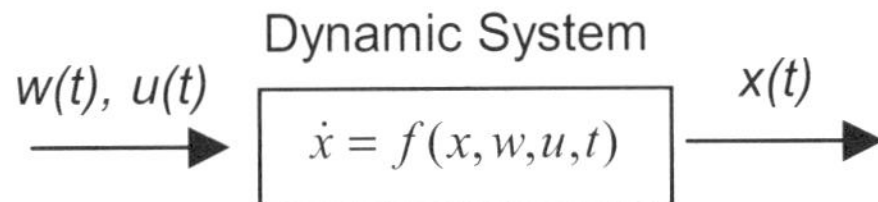

Figure 1 The abstract nonlinear system

The objective of the disturbance input is to maximize a performance index in the following general form:

$$J = \phi[x(t_f)] + \int_{t_0}^{t_f} L(x,u,v,w,t)\,dt$$

where the functions $\phi(x)$ and $L(x,u,v,w,t)$ are defined for different "safety" purposes. For example, for this paper, since the focus is on vehicle rollovers, these two functions will include heavy penalty weighting on vehicle roll angle and/or roll angle rates. Studies on the solutions of the above optimization problem date back to as early as the 17th century. If the dynamic system model is nonlinear, finding an analytic solution to the above problem is not trivial. Therefore, numerical methods are usually adopted. The worst-case methodology reduces the above problem into a two-point boundary value problem (TPBVP), and solves the resulting equations iteratively to identify the optimal value for the performance index, and the associated w(t) signal. Through this process, in theory we can also solve the optimal value for u(t), which could be viewed as the "model" for a good control signal. However, in this paper, we will focus on the worst-case disturbance generation only, in other words, u(t) will not be optimized. Instead, it is assumed to be either non-existent (for a vehicle with no-control) or generated by a pre-determined control algorithm.

As in all iterative numerical methods, achieving the global optimal solution is usually not guaranteed, and the quality of the final solution highly depends on the initial guesses (**IG**, of w(t)). A successful numerical scheme usually includes two kinds of IG generation mechanisms: Initial guesses that are believed to be close to a very good local optimal point, based on which, a search in a small neighborhood usually yields fruitful results (the so-called shot-gun IG), and a systematic, grid-search type of IG which may or may not be followed by a neighborhood search (the so called machine-gun IG).

In this paper, initial guesses are obtained by three different ways: the standard maneuvers from literature

(e.g., rollover evaluation maneuvers used by VRTC), intuition ("common-sense" maneuvers suggested by experienced engineers) or by iterative dynamic programming (IDP, Luus 2000). The IDP approach is a little more systematic compared to the other methods because the initial guesses generated by this method are more likely to converge to the global optimal solution, or at least a very good local optimal point. Therefore, we will be utilizing IDP as a major tool for the initial guess selection and TPBVP approach for the worst-case solution. The algorithm for the overall numerical method is outlined below.

<u>Outline of the Worst-case Algorithm:</u>

1. Define cost function with proper weighting factors.

2. Select vehicle operational parameters (initial speed, weight, etc.) and environmental parameters (e.g. road friction) and input ranges (e.g., steering wheel angle limited to ±100 degrees, brake pedal force limited between 0 and 150lbf, etc.)

3. Divide the problem horizon of [0, t_f] into P time stages.

4. Pick N grid points for each of the inputs (steering and braking) in the range specified in step 2.

5. Start from (P-1)th time point and evaluate the performance index at the time interval from (P-1) to t_f for each of the N grid points picked for that input. Save the input for the optimal cost function J at this time interval and step backward to the preceding time point (P-2)th to evaluate the optimal J of the input from (P-2)th to t_f. Repeat the evaluation of the optimal J at each of the time intervals in the same manner, by stepping backward each time to the previous time point until the overall optimal J of the input is determined for the entire time interval from 0 to t_f.

6. Reduce the allowed input range.

7. Refine IG (go to step 4), or go to step 8 below.

The seven steps described above help to find a good initial guess. Meanwhile, standard testing maneuvers such as J-turn, brake in a J-turn, fishhook maneuvers and random inputs satisfying the input constraints can also be used as initial guesses. Finally, worst-case solutions identified for a vehicle under different operating conditions (speed, road condition) or from different but similar vehicles can also serve as candidates of initial guesses.

Searching Process:

8. Use the initial guess obtained above to simulate the dynamic model forward in time from t=0 to t=t_f.

9. Save the state histories, evaluate necessary gradient information.

10. Update the identified disturbance signal according to the gradient information.

11. Continue iteration (go to step 6) until the desired solution is obtained.

Based on the algorithm defined, a worst-case evaluation program is created by using the MATLAB software package from the Mathworks Inc. A schematic diagram showing how this evaluation method works is given in Figure 2 with its user interface shown in Figure 3, and the animation example shown in Figure 4. The core of this simulation package is the dynamic model of the vehicle with proper input signals defined and all its states and state derivatives available as outputs. The optimization routine then searches for the worst-case inputs in an iterative manner. Equations of the vehicle dynamics need not be known exactly and/or be shown in state equation form. However, if they are given, the speed of convergence to the solutions will improve considerably. Depending on the definition of the performance index (J), worst-case scenarios can be determined for different aspects of the vehicle performance (roll, yaw stability).

SIMULATION RESULTS

Worst Case Maneuvers: The worst-case methodology described in the previous section is used to determine hand-wheel steering and brake pedal force combinations for SUV rollover propensity assessment. With appropriate saturation bounds such as steering angle and steering rate imposed, these input signals then simulate the worst-possible handling/braking inputs from human drivers. (in terms of rollover).

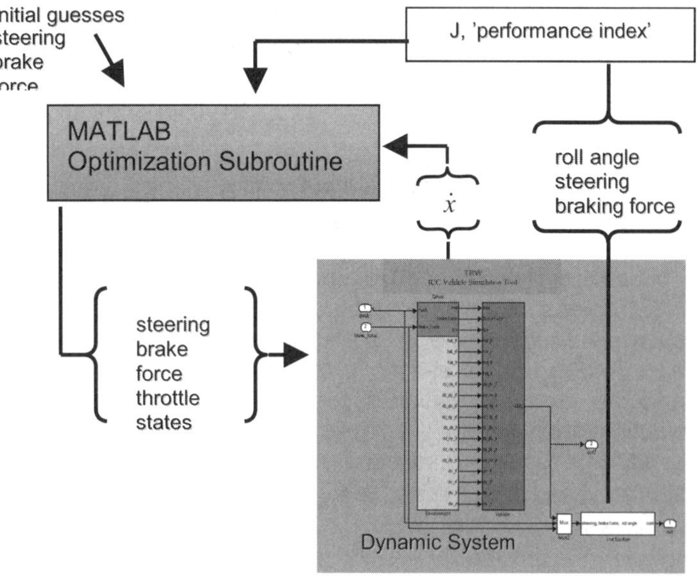

Figure 2: Architecture of the worst-case evaluation program

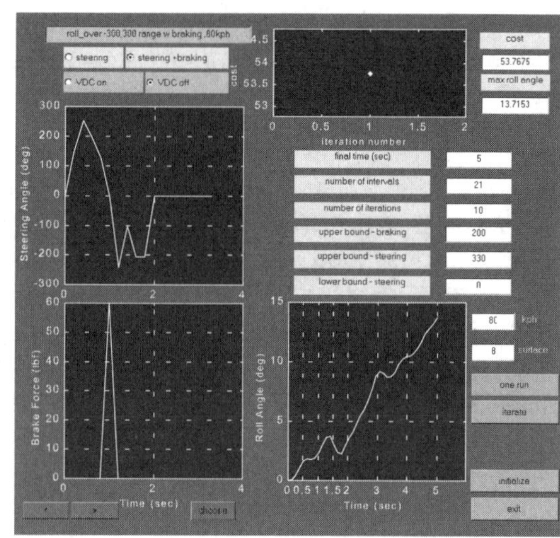

Figure 3: User interface of the worst-case evaluation program.

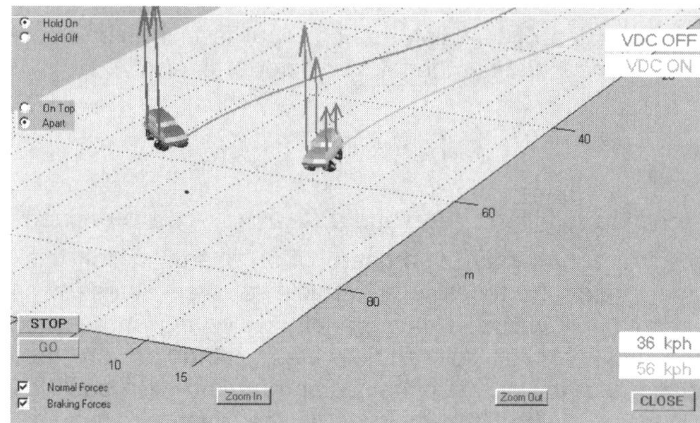

Figure 4: A screen shot from the VSC-on & VSC-off comparison animation, where arrows indicate tire normal forces.

Standard Maneuvers: The dynamic testing procedure used in the NHTSA Phase II study (Howe et al. 1999) included several maneuvers that were intuitively selected to represent maneuvers preceding rollover accidents. We will adopt several of these maneuvers, including the J-Turn, fishhook #1 and #2 in our simulations (see appendix for details of these maneuvers). The steering amplitudes of these maneuvers are sometimes scaled-down to make the comparison meaningful.

SIMULATION MATRIX

Case Studies:

1. Under the same steering angle and brake pedal force limit (200 lbf for pedal force, and [-45,45] degrees for steering), the relationship between maximum roll angle and vehicle speed is studied for Vehicle A under different maneuvers.

2. The relationship between allowable hand-wheel steering angle and maximum roll angle is studied for Vehicle A.

3. Under the same steering angle and brake pedal force limit (200 lbf for pedal force, and for steering, see Table 2), the relationship between surface conditions and maximum roll angle is studied for Vehicle A under different maneuvers.

4. Rollover stability metric comparisons of Vehicle A and Vehicle B are studied.

5. Comparison of maximum hand-wheel steering angles before rollover for Vehicle A and Vehicle B, at different vehicle speeds.

PRELIMINARY SIMULATION RESULTS

Case 1: Our results (Table 1) show that increasing the vehicle speed leads to an increase in maximum roll angle. We see that the worst-case maneuvers found by our method (targeting VSC-off conditions) generate much higher roll angle than the standard maneuvers. It is also clear that the VSC system is effective in reducing the risk of rollover for these worst-case maneuvers since the maximum rollover angle is dramatically decreased in its presence.

Table1: Vehicle speed vs. maximum roll angle for Vehicle A

speed (kph)			80	100	120
WC[1]	VSC-off	max roll angle (deg)	4.22	5.53	rollover
	VSC-on		1.43	1.28	3.04
Standard[2] (VSC-off)			1.57	2.18	3.03

Simulation conditions: High friction surface for tires; worst-case with respect to steering and braking (WC: Worst-case maneuvers, VSC-on: Worst-case maneuvers with the control system on).

Case 2: Table 2 shows that increasing the hand-wheel steering angle range leads to an increase in maximum roll angle. At each of the steering angle ranges, the maximum roll angle for the worst-case maneuvers is greater than for the standard maneuvers (with steering magnitudes scaled to the same saturation limit). It should be emphasized again that these "worst case" maneuvers are trained to target the VSC-off vehicles. The results found in the row labeled "VSC-on" thus only serve to show that under these three extreme maneuvers, the vehicle equipped with VSC will have much reduced roll angle.

Table2: Allowable hand-wheel steering angle range vs. maximum roll angle for Vehicle A.

hand-wheel steering angle range (deg)	[-30,30]	[-45,45]	[-60,60]

WC[1]	VSC-off	max roll	2.95	4.22	5.35
	VSC-on		1.92	1.43	2.04
Standard[2] (VSC-off)			1.03	1.57	2.17

Simulation conditions: High friction surface; vehicle speed is 80 kph; worst-case with respect to steering and braking.

Case 3: Increasing surface friction leads to an increase in maximum roll angle (Table 3), and thus elevates the risks for rollover. Similar to the previous cases, worst-case maneuvers are more effective in showing the vehicle's lateral stability under these conditions. On a high friction surface, having the VSC system in Vehicle A protects the vehicle from rollover even when the worst-case maneuvers are applied to the vehicle.

Table3: Surface condition vs. max. roll angle for Vehicle A.

surface condition			low friction	high friction
WC[1]	VSC-off	max roll angle (deg)	3.24	rollover[3]
	VSC-on		1.38	3.04[4]
Standard[2] (VSC-off)			2.58	3.03[5]

Simulation conditions: Vehicle speed is 120 kph; hand-wheel steering angle range is [-45,45], worst-case with respect to steering and braking.

Case 4: In this case study, vehicle parameters and steering limits are deliberately selected not to generate rollover, but nevertheless exhibit significant roll motions. The maximum roll angle is subsequently used as an indication of relative vulnerability to rollover. Under a predefined range of test conditions, multiple "worst-case" results were identified (each of which is a local minimum point in the overall optimization problem). When the uncontrolled Vehicle A and Vehicle B, under their respective worst-case maneuvers are compared, the maximum roll angles of these two SUVs are consistent with what would be obtained using "static" type of metrics. The Static Stability Factors (SSF) for Vehicle A and Vehicle B are 1.12 and 1.08 respectively. These numbers indicate that Vehicle A is expected to be less prone to rollover than Vehicle B. Consistent with this result, our worst-case methodology yields greater maximum roll angles for Vehicle B than for Vehicle A (Fig. 5). This finding is not surprising. However, it should be mentioned that there are plenty of examples in which a "worst-case" maneuver generated by our algorithm that targets the weakness of Vehicle A actually generates higher roll motion on Vehicle A than Vehicle B. An example maneuver is shown in Figure 6. It can be seen that even though the SSF of Vehicle A is 4%

---

[1] WC maneuver for VSC-off.

[2] Five standard maneuvers: J-Turn with and without pulse braking, fishhook #1 and #2, resonant steer maneuver. These maneuvers were used in dynamic testing procedure for the NHTSA Phase II study.

[3] See appendix.
[4] See appendix.
[5] See appendix.

higher, it rolls over at this maneuver while (the supposedly less safe) Vehicle B does not even experience wheel lift-off. In other words, the susceptibility of SUVs to rollovers cannot be concluded from one or a handful of dynamic maneuvers.

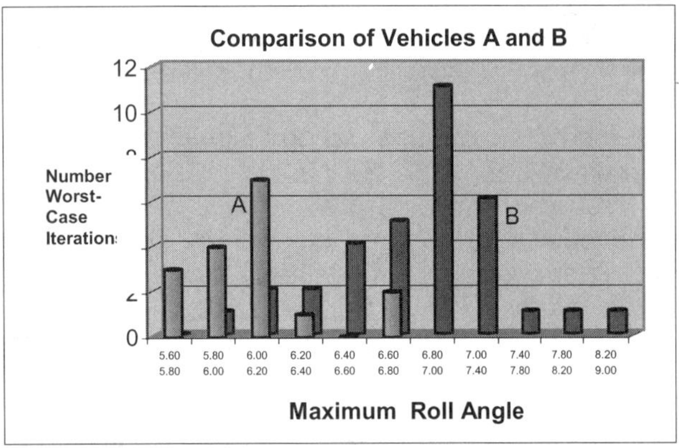

Figure 5: Comparison of maximum roll angles obtained through worst-case evaluation for Vehicle A and Vehicle B. Simulation conditions: Vehicle speed is 100 kph; hand-wheel steering angle range is [-100,100] degrees; high friction surface for tires; worst-case with respect to steering and braking; VSC off.

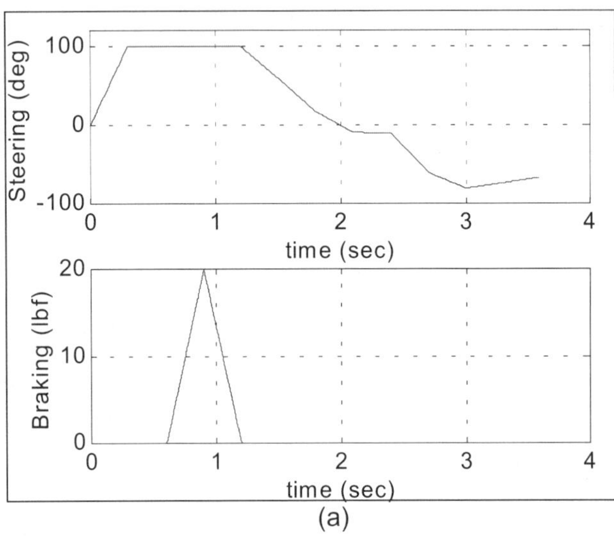

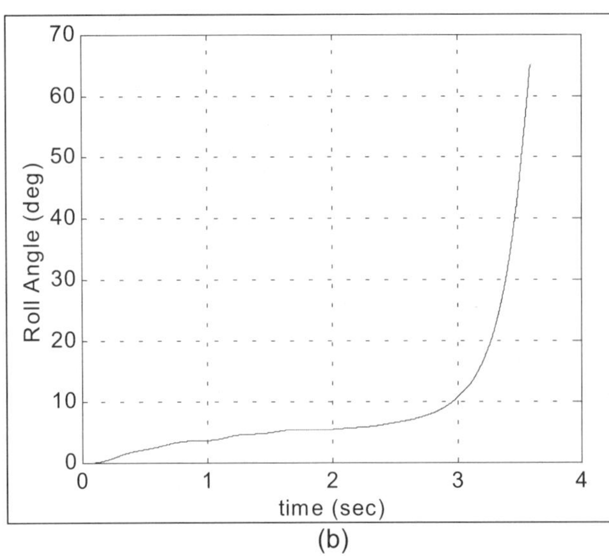

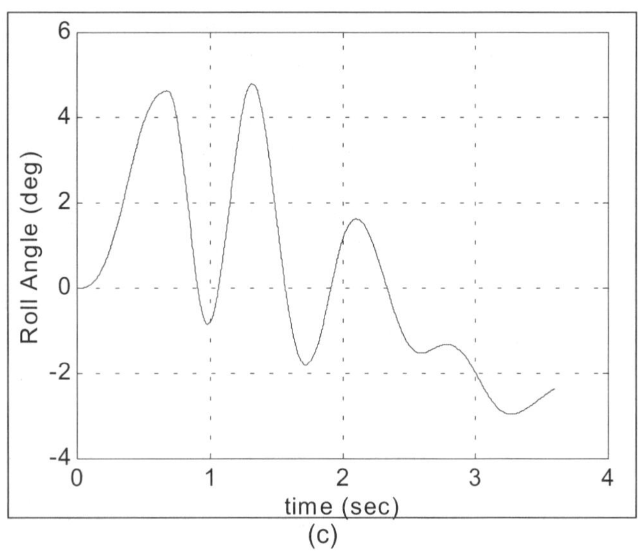

Figure 6: Example worst case input (a) and the roll response from vehicle A (b) and Vehicle B (c). Initial vehicle speed=100kph, on high friction surface.

Case 5: In this case study, maximum hand-wheel steering angle values before rollover are compared for the Vehicle A and Vehicle B. Worst-case evaluation method is used to search for the minimum hand-wheel steering angle range that leads to rollover. After achieving rollover for the initial range of the hand-wheel steering angle, range is reduced until rollover was unachievable. This procedure is repeated for the four different values of vehicle speed. It can be seen from figure 7 that up to around 105 kph Vehicle A seems to be harder to rollover than Vehicle B, whereas at higher speeds relative propensity of rollover for the vehicles reversed, showing that rollover propensity is speed dependent.

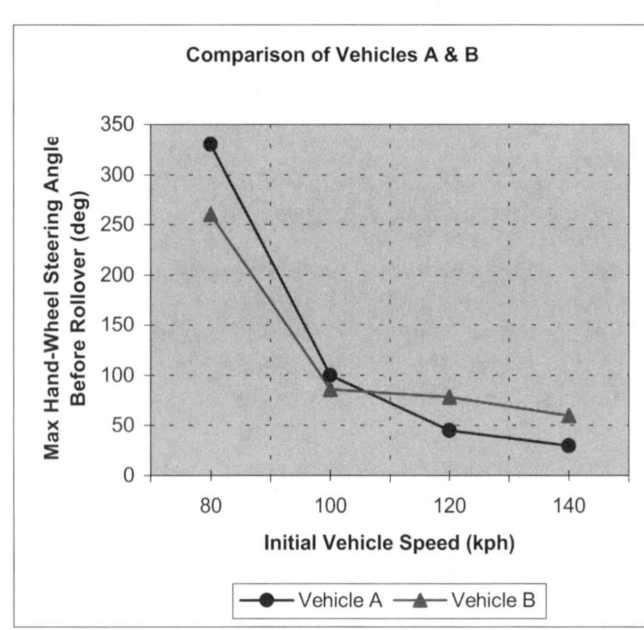

Figure 7: Comparison of Maximum Hand-wheel Steering Angle Values before rollover for the Vehicles A & B. Simulation conditions: High friction surface; worst-case with respect to steering and braking; VSC off.

DISCUSSION

The goals of this paper are to study the effects of vehicle and environment parameters and the influence of VSC systems on the rollover propensity of Sport Utility Vehicles. Five cases were studied in this paper, in the first three cases, worst-case evaluations are performed for different simulation conditions (speed, hand-wheel steering angle and surface friction). Our results show that for all three parameters tested in this study, the maximum roll angle increases along with the magnitude of the parameter. Thus, an increase in the road friction, in the speed or steering range of the vehicle elevates the risks of rollover. For all these cases, VSC was found to reduce the maximum roll angles significantly to about half of the corresponding VSC-off cases. In the fourth case, VEHICLE A and VEHICLE B were compared in terms of their propensity to rollover. The results show that VEHICLE B is more likely to rollover than VEHICLE A as would be expected from the SSF values of the two vehicles. In the last simulation case, maximum hand-wheel steering angle values before rollover for Vehicles A and B are compared under various forward speeds. Results show that rollover propensity is speed dependent.

The results presented in this paper are important since it shows that in many cases a VSC system reduces the maximum roll angle by improving the vehicle's lateral stability. Besides, the dynamic rollover evaluation procedure used to determine the worst-case maneuvers for SUVs brings a new dimension to be considered by the government in the establishment of the "star rating" system. Based on previous studies, NHTSA has decided that from the two types of metrics (static and dynamic) available to test a vehicle's tendency to untripped rollover, the dynamic testing method was less advantageous. The dynamic testing procedure used in the NHTSA Phase II study included five maneuvers (J-Turn with and without pulse braking, fishhook #1 and #2, and resonant steer maneuver) that were intuitively selected to represent maneuvers preceding rollover accidents. These maneuvers were tested on 12 vehicles including 4 SUVs and the results showed that the dynamic testing system did not predict the rollover tendency of an SUV better than the static testing method, which is based on a very simple physical principle. Moreover, a specific model of an SUV could become prone to rollover in response to a maneuver whose characteristics are not included among the chosen five. Therefore, dynamic test procedure was determined to be too complex and does not add enough benefit beyond simpler approaches such as SSF.

A better dynamic testing approach to determine the vehicle rollover propensity is to first find maneuvers that destabilize the vehicle most effectively. Assessing the risk of the vehicle's rollover under the conditions of these worst-case maneuvers will inform us about the safety of this particular SUV. However, such an approach requires an accurate dynamic model and is computationally intensive. As already stated, speed, surface friction and steering range, the operating parameters studied in this paper, are shown to influence vehicle motions. Thus, SSF is an insufficient indicator of a vehicle's lateral stability. Moreover, this index also fails to reflect the potential performance of vehicles equipped with a VSC system. As indicated before, for vehicles with a VSC system, the maximum roll angles for worst-case maneuvers could be significantly reduced. With a VSC system, it is possible for an SUV with a low SSF to become more stable than one with a higher SSF but no VSC. The final case study, and in particular, Figure 6, clearly indicates that it is not easy to assess a vehicle's rollover stability. Therefore, even though SSF is simple and captures the average behavior of vehicles without control systems, it is not the best way to measure the untripped rollover risks of a vehicle.

The simulation-based dynamic rollover evaluation procedure presented in this paper introduces an alternative method to measure the rollover risks of SUVs. Our method does not have the disadvantages of other dynamic testing procedures because it finds the characteristics of the worst-case maneuvers for a vehicle through iterative simulations. Moreover, this method takes the operating parameters and any control systems the vehicle might have into account while finding the worst-case maneuvers. Therefore, the maximum roll angles found by applying these maneuvers to the vehicles are a more reliable indicator of a vehicle's resistance to rollover than SSF.

CONCLUSIONS

The results obtained in our study have two main messages. First, Vehicle Stability Control system is a product that potentially could improve vehicle stability against rollovers, especially for SUVs. Because of its implications with respect to consumer safety, this control system constitutes a new issue to be considered. Before the U.S. Congress (and National Academy of Sciences) approves the new "star rating" system and endorses SSF as the indicator of a vehicle's rollover risk, the difference a control system makes in reducing the rollover risk of the vehicle should be considered. Having VSC has a dramatic impact on the roll behavior of Vehicle A. It seems to elevate the vehicle's resistance to rollover. The failure to recognize the benefits of these innovative control systems might inadvertently discourage vehicle manufacturers from equipping their vehicles with VSC or other innovative devices. This, in turn, could negatively impact consumer safety. Our worst-case methodology suggests an alternative to existing "dynamic" and "static" types of rollover metrics. Whereas using SSF can be considered to be a better approach than dynamic testing in evaluating the rollover risks, our method introduces a new simulation-based procedure that finds the worst case maneuvers for rollovers. This method has its advantages compared with SSF mainly because the effect of vehicle parameters, environment parameters and on-board control systems such as VSC are included in its calculations. More importantly, the worst-case method identifies the weak links of vehicles and/or the active control systems, and thus could be a valuable tool in the vehicle design process.

ACKNOWLEDGMENTS

This research project is supported by TRW Inc. and a grant from the National Science Foundation. The authors also wish to thank M. Sayers and S. Riley of MSC for making TruckSim available for this research.

REFERENCES

Čech, I., "Anti-Roll and Active Roll Suspensions," *Vehicle System Dynamics*, no 33, pp. 91-106, 2000.

Chen, B. and Peng, H., "A Rollover Warning Algorithm for Sports Utility Vehicles," Proceedings of the 1999 American Control Conference, San Diego, CA.

Constantine, C.J. and Law, A.H., "The Effects of Roll Control for Passenger Cars during Emergency Maneuvers ", SAE Paper No. 940224, 1994.

Furleigh, D.D, Vanderploeg, M.J. and Oh, C.Y., "Multiple Steered Axles for Reducing the Rolloever Risk of Heavy Articulated Trucks ", SAE Paper No. 881866, 1988.

Salaani, M.K., Guenther, D.A., Heydinger, G.J., "Vehicle Dynamics Modeling for the National Advanced Simulator of a 1997 Jeep Cherokee," SAE paper # 990121.

Howe, J. G., Garrott, W. R., Forkenbrock, G., "An Experimental Examination of Selected Maneuvers That May Induce On -Road, Untripped, Light Vehicle Rollover- Phase II of NHTSA's 1997-1998 Vehicle Rollover Research Program," NHTSA Technical Report, July 1999.

Luus, R., "Iterative Dynamic Programming, " Chapman & Hall/Crc, Monographs and Surveys in Pure and Applied Mathematics, 2000.

Ma, W. and Peng, H., "A Worst-Case Evaluation Method for Dynamic Systems," *ASME J. of Dynamic Systems, Measurement and Control*, Vol.121, No.2, June 1999, pp.191-199.

Palkovics,L., Semsey, A. and Gerum, E., "Roll-Over Prevention Systems for Commercial Vehicles – Additional Sensorless Function of the Electronic Brake System," 4th International Symposium on Advanced Vehicle Control, Nagoya, Japan, 1998.

Wielenga, T.J., "A Method for Reducing On-Road Rollovers – Anti-Rollover Braking ", SAE Congress, SAE Paper No. 1999-01-0123, 1999.

APPENDIX

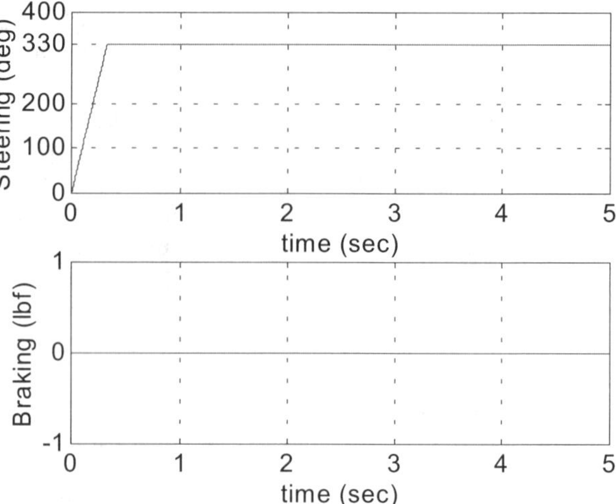

Figure A1: J-Turn (NHTSA Tech. Report 1999).

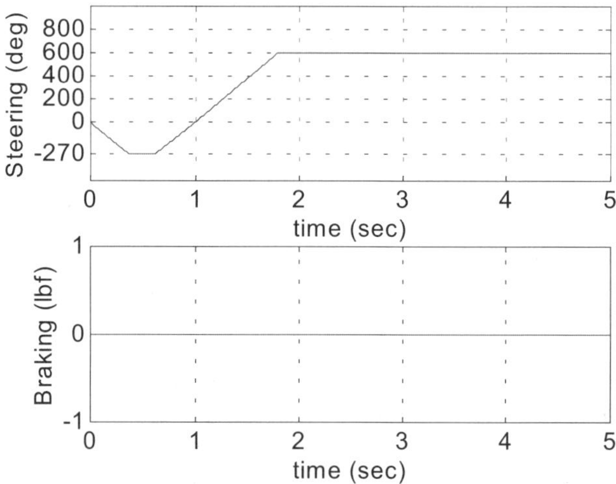

Figure A2: Fishhook #1(NHTSA Tech. Report 1999).

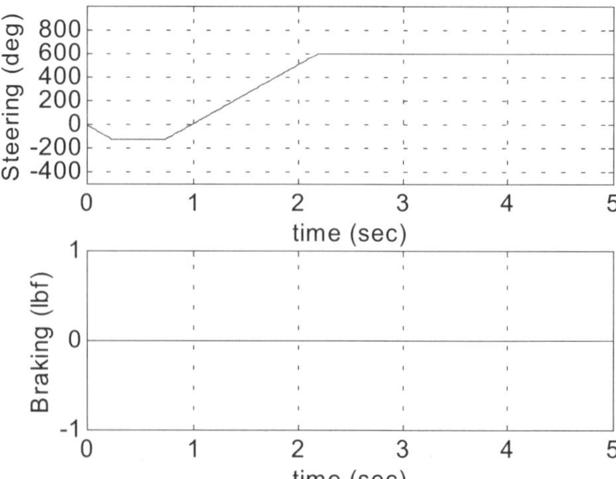

Figure A3: Fishhook #2(NHTSA Tech. Report 1999).

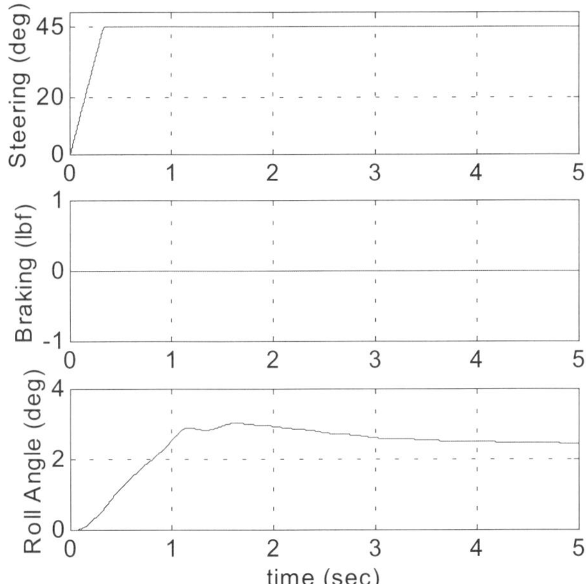

Figure A4: J-Turn scaled to [-45 ,45] deg. steering angle range. (steering as defined in figure A1 * 45/330). Example roll angle obtained with vehicle speed=120 kph, high friction surface, VSC-off.

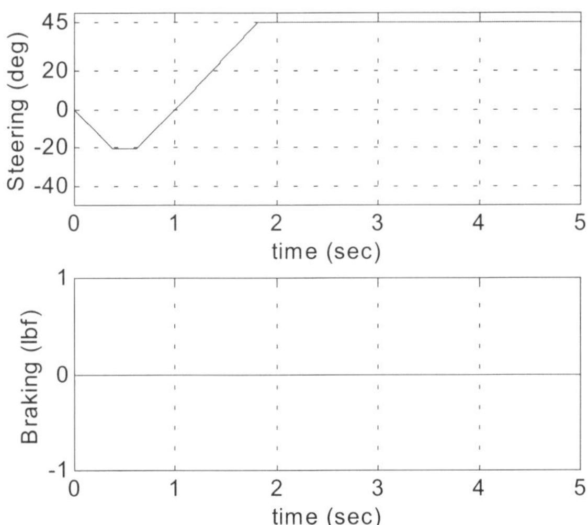

Figure A5: Fishhook #1 scaled to [-45 ,45] deg. steering angle range.

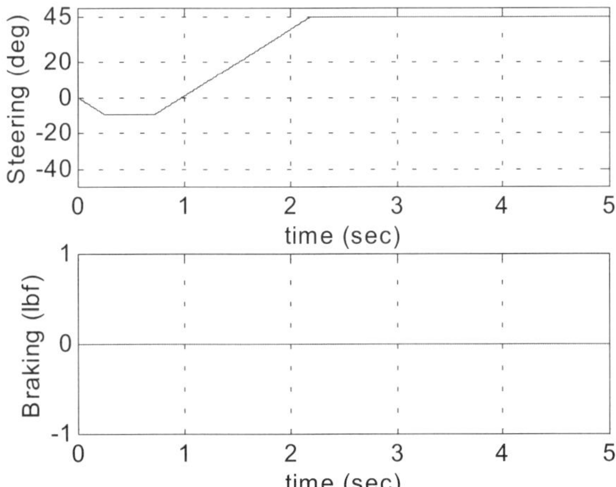

Figure A6: Fishhook #2 scaled to [-45 ,45] deg. steering angle range.

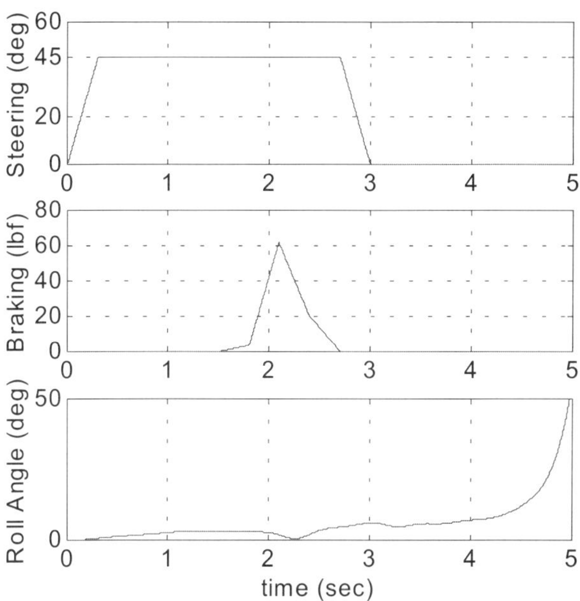

Figure A7: Vehicle speed is 120 kph; hand-wheel steering angle range is [-45,45], worst-case with respect to steering and

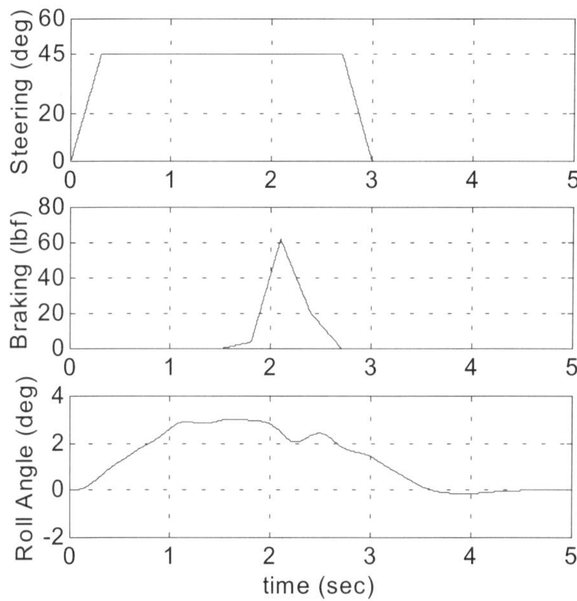

Figure A8: Vehicle speed is 120 kph; hand-wheel steering angle range is [-45,45], worst-case with respect to steering and braking, high friction surface, VSC-on.

WHAT THE FUTURE HOLDS

What the Future Holds

Active Development Underway in Many Areas

Future trends in the areas of electronics braking, traction, and stability controls are discussed in many of the papers in this book. The following are a few examples:

"The development departments of many OEMs and tier 1 suppliers are currently working on further improvements to yaw rate control. They are combining the ESP/VDC with one or more other active steering or chassis systems, such as Active Front Steering (AFS), Electronic Power Steering (EPS), Continuous Damping Control (CDC), Active Body Control (ABC), Air Suspension System (ASS), Active Torque Control (X-Drive, BMW) and Anti-Roll Control (ARC). These systems have names such as Integrated Chassis Control (ICC Opel), Vehicle Dynamics Management (VDM, Bosch), and Global Chassis Control (GCC, Continental Teves), and consist of a super ordinate, coordinating software layer that generates reference values for the underlying control systems in such a way that positive control interferences are reinforced (e.g., yaw rate stabilization for shortening braking distances by ›15% in an ESP/AFS combination) and negative interference is suppressed by stricter separation of working areas. Global safety concepts and switch-off concepts can also be implemented. It is in the nature of these systems that they mainly operate at critical vehicle dynamic limits and in safety-critical situations, which makes HIL simulation particularly useful for them. Network simulators similar to those used for powertrains and body electronics, made up of the HIL simulators available for lower-level subsystems, will be used more commonly in the future." (2005-01-1660)

"Many different concepts ensure the high safety within a braking system. There is no one feature that by its own can meet all the requirements of the future applications. However, by combining different functions in a chipset of microcontrollers with extended safety features, smart power system IC and advanced smart power switches/current regulators combine the individual advantages to a cost efficient system. Presupposition for an optimized solution is the availability of microcontrollers, sensors and a variety of semiconductor technologies and packaging options.

The combination in a system and the partitioning has to be defined in close cooperation between the semiconductor vendor and the braking system supplier." (2004-01-0251)

"Future complex cross-system applications will require an intelligent sensor platform (SP), which is sensing all inertial values of the three main axes internally, receiving external sensor signals like wheel speeds and forces, steering angle, engine torque, braking pressure, tire pressure, actuator states and informations of the driver assistance systems (radar, video, GPS) and then calculating or estimated the dynamic values like side slip angle, speed over ground in longitudinal and lateral directions, yaw rate, road inclination, road uphill gradient, etc. and controller values.

"These applications will need high data volume, high update rates and safe data transfer, why time triggered data networks like TT-CAN, TTP or FlexRay will be required." (2003-01-0199)

"The next generation of cars will consist of a high number of networked electronic controls units (ECUs) and significantly more complex software modules and control applications than today's models. Beside applications like engine control, air condition control, and anti-theft systems, which are already available in today's cars, the first steps towards the introduction of safety-relevant steer-by-wire and brake-by-wire systems will be undertaken…Since all these systems have conflicting requirements to the underlying network protocol (latency, predictability, throughput..), the straight-forward way would be to use autonomous busses and networks for every kind of distributed system within the car body (ultra-available safety-relevant systems, non-safety-relevant control systems, entertainment and media systems).

"This adds unwanted complexity and cabling overhead (and therefore cost) to the overall car electronc. So it is legitimate to think about alternative architectures, e.g. a common backbone network, which can be used by all ECUs. Nevertheless, the integration to a single in-car network must not impair nor compromise the service of any safety-relevant system." (2004-01-1734)

"The ability to link and extend the anti-lock braking system (ABS) and traction control system (TCS) to an electronic stability program (ESP) will greatly improve the safety and durability of the vehicle. The ESP is designed to detect a difference between the driver's control inputs and the actual response of the vehicle. When differences are detected, the system intervenes by providing braking forces to the appropriate wheels to correct the path of the vehicle. This automatic reaction is engineered for improved vehicle stability by reducing over-steering and under-steering skidding. The TCS will dramatically improve the safety during severe cornering and on slick or low-friction road surfaces." (2004-01-2926)

"Many embedded processors with the modern vehicle are moving beyond the 16-bit microcontroller to very complex 32-bit microcontrollers. The 32-bit microcontrollers of the next generation will require an increasing amount of S/W written in high level languages. The H/W design is relying on integration and very complex S/W to solve the new problems in engine management. The communications between each of these systems will increase the demand for data to be shared between ECUs around the car." (2004-01-2926)

"Automotive scientists and engineers also predict the integration of the E-M ride-by-wire (RBW) or x-by-wire (XBW) automotive mechatronic control system's steering (diversion), driving (propulsion), braking (dispulsion) and absorbing (suspension) technologies into integrated safety systems (ISS)," (2001-01-3321)

The following quotations are from the December, 2005 issue of *Automotive Engineering International*:

"The Flexray network took another step forward when Freescale Semiconductor rolled out a computer chip. As the infrastructure for Flexray grows, plans for the expected application of the bus are firming up.

"Freescale's MFR4200 provides 10 Mbits per second on each of its two channels, while also housing memory in a 64-pin package. This is the latest in a slow ramp up for the network, which provides higher bandwidth and determinism not offered by CAN (Controller Area Network) devices. Fujitsu Microelectronics unveiled a Flexray controller chip that employs Bosch

technology shortly after Freescale's announcement, and Philips Semiconductors recently teamed

with Freescale to create a common protocol for controller chips.

"Though Flexray was designed with drive-by-wire applications in mind, its first usage is

likely to be in less-critical areas.

'The first cars with FlexRay will go in the direction of replacing CAN where it's

inappropriate because CAN isn't able to do a time-bounded application,' said Rainer Makowitz,

Systems Engineer Manager Automotive at Freescale. (Terry Costlow, p. 32)

"Drive-by-wire is making some impact in throttle control and parking brakes, but there is not a

strong push to convert braking or steering to electronic technologies. Problems with an early

braking rollout may further slow progress for the technology.

"Once a hot topic, by-wire technology is not currently set for any major steering or braking

programs in the near future. The German automakers, which often lead the push to new

technologies, are not racing to replace hydraulic and mechanical linkages with electronic controls.

'BMW has been keen on brake- and steer-by-wire, but they've decided that nothing is

happening until 2015,' said Paul Hansen, President of Paul Hansen Associates.

"Mercedes-Benz produced a brake system but had problems with it. 'Mercedes had to

recall its brake-by-wire system,' said Mark Fitzgerald, Senior Analyst at Strategy Analytics. 'It

doesn't help when something touted as a benefit to drivers ends up being the cause of a recall.'

' There's pretty high penetration for throttle-by-wire, and there's a big push to convert

parking brakes to electronic controls,' Fitzgerald said. 'You don't need a giant brake lever, and

when you start to drive, the brake will automatically shut off so you're not driving with the brakes

on.' (p. 54)

"Siemens VDO Automotive says its electronic wedge brake (EWB) will enable the launch of

brake-by-wire production for vehicles with a 12-V vehicle electrical system by the end of this

decade. The EWB, which is said to offer considerable safety and comfort advantages over a hydraulic brake, is based on innovative technology developed by eStop, a recent acquisition. During the braking operation, a brake pad attched to a wedge is pressed between the brake caliper and the brake disk. As the wheel turns, the wedge effect is automatically intensified. This allows any level of braking power with a minimum of intricacy. Vehicles employing the electronic wedge brake solution will have an intelligent wheel-braking module fitted on each wheel. Based on the principle of self-energization, the braking effect builds up very rapidly and the intelligent control prevents any danger of the wedge blocking. This principle of 'unstable' control structures was taken from high safety-critical systems for aviation and aerospace applications. The EWB eliminates the need for components such as hydraulic pipes, brake cylinders, brake boosters, or antilock braking control units. (p. 42)

Ronald K. Jurgen, Editor